AA002446

International Conference on Synthesis, Modeling, Analysis and Simulation Methods and Applications to Circuit Design and 16th Conference PhD Research in Microelectronics and Electronics (SMACD/PRIME 2021)

Online
19-22 July 2021

ISBN: 978-1-7138-3572-1

Printed from e-media with permission by:

Curran Associates, Inc.
57 Morehouse Lane
Red Hook, NY 12571

Some format issues inherent in the e-media version may also appear in this print version.

Copyright© (2021) by VDE VERLAG GMBH
All rights reserved.

Printed with permission by Curran Associates, Inc. (2026)

For permission requests, please contact VDE VERLAG GMBH
at the address below.

VDE VERLAG GMBH
Bismarckstr. 33
P.O.B. 12 01 43
10625 Berlin, Germany

Phone: +49 30 34 80 01 - 0
Fax: +49 30 34 80 01 - 9088

kundenservice@vde-verlag.de

Additional copies of this publication are available from:

Curran Associates, Inc.
57 Morehouse Lane
Red Hook, NY 12571 USA
Phone: 845-758-0400
Fax: 845-758-2634
Email: curran@proceedings.com
Web: www.proceedings.com

SMACD / PRIME 2021

International Conference on Synthesis, Modeling, Analysis and Simulation Methods and Applications to Circuit Design

and

16th Conference on PhD Research in Microelectronics and Electronics

19 – 22 July 2021, Online Event

SMACD·PRIME 2021
ERFURT·GERMANY

Sponsors

cādence®

Technical Sponsors

Institutional Sponsors and Organizers

SMACD / PRIME 2021

International Conference on Synthesis, Modeling, Analysis and Simulation Methods and Applications to Circuit Design

and

16th Conference on PhD Research in Microelectronics and Electronics

19 – 22 July 2021 Online Event

General Chairs:
Ralf Sommer, IMMS GmbH & TU Ilmenau, Germany
Stefan Heinen, RWTH Aachen University, Germany

Organizer:
Information Technology Society (VDE ITG)

VDE VERLAG GMBH

SMACD / PRIME 2021 | 19 – 22 July 2021, Online Event

Bibliographic Information of the German National Library
The German National Library lists this publication in the National Bibliography;
detailed bibliographic data are available on the Internet at **http://dnb.dnb.de.**

© 2021   VDE VERLAG GMBH · Berlin · Offenbach, Bismarckstraße 33, 10625 Berlin
www.vde-verlag.de

All rights reserved.

Any utilization in breach of the strict limits of copyright law, without the prior approval of the publisher, is prohibited. Reproductions of common names, brand names, trademarks etc. in this publication are not subject to the acceptance that these names could be regarded as free or could be used by anyone, even without particular marking, in the sense of the trademark and brand protection legislation. Publication does not imply that the solutions described are not protected by intellectual property rights (e.g. patents and utility models). The publisher assumes no liability for the correctness and practicability of the programs, circuits, and any other arrangements and instructions published, nor for the correctness of the technical content of this publication. The up-to-date valid versions of the relevant statutory and official regulations and technical regulations (e.g. VDE body of regulations) have to be respected.

Produced in Germany

SMACD 2021

International Conference on Synthesis, Modeling, Analysis and Simulation Methods and Applications to Circuit Design

The 2021 edition of the International Conference on Synthesis, Modeling, Analysis and Simulation Methods and Applications to Circuit Design (SMACD) was originally planned in Erfurt, Germany, but will be held virtually, as a forum devoted to modeling, simulation and synthesis for Analog, Mixed-signal, RF (AMS/RF) and multi-domain (nanoelectronics, biological, MEMS, optoelectronics, etc.) integrated circuits and systems. Experiences with modeling, simulation and synthesis techniques including machine-learning and artificial intelligence in diverse application areas are also welcomed. Objective technologies include CMOS, beyond CMOS, and, Morethan-Moore such as MEMS, power devices, sensors, passives, etc.

SMACD 2021 is Technically Co-sponsored by IEEE, IEEE CEDA and IEEE CAS. The conference proceedings will be submitted for inclusion in IEEExplore.

PRIME 2021

16th Conference on PhD Research in Microelectronics and Electronics

PRIME has been established over the recent years as an important conference where PhD students and post-docs with less than one year post-PhD experience can present their research results and network with experts from industry, academia and research. PRIME 2021 will feature conference program reflecting the wide spectrum of research topics in Microelectronics and Electronics, building bridges between various research fields. In addition to the technical sessions, opportunities for the conference attendees will be the keynote talks, workshops, social events and conference proceedings will be submitted for IEEE Xplorer.

PRIME 2021 is seeking original papers in the areas given on the Call for Papers.

Committees

■ SMACD 2021

General Chair
Ralf Sommer, IMMS GmbH & TU Ilmenau, DE

Technical Program Chairs:
Giulia Di Capua, University of Cassino and Southern Lazio, IT
Günhan Dündar, Boğaziçi University, TR
Francisco V. Fernandez, University of Seville, ES
Ricardo Martins, IT/IST-University of Lisbon, PT
Jürgen Scheible, Reutlingen University, DE

Publication Chairs:
Nuno Lourenço, Instituto de Telecomunicações, PT
Piero Malcovati, University of Pavia, IT

Publicity Chair:
Fernando García Redondo, Arm Ltd., UK

Competition Chairs:
Engin Afacan, Kocaeli University, TR
Gürkan Sönmez, Dialog Semiconductors, DE

Industrial Liaison Chair:
Rafael Castro López, CSIC, ES
Eric Schäfer, IMMS GmbH, DE
Georg Gläser, IMMS GmbH, DE

Finance Chair:
Volker Schanz, VDE ITG, DE
Local Arrangements Chairs:
Hatice Altintas, VDE Conference Services, DE
Anne Arlt, IMMS GmbH, DE

Technical Committee
Engin Afacan, Gebze Technical University
Manuel Barragán, TIMA Laboratory
Francesco Brandosinio, Infineon Technologies
Rafael Castro-Lopez, CSIC
Pier Cavallini, Dialog Semiconductor
Hung-Ming Chen, National Chiao Tung Universit"
Catherine Dehollain, Ecole Polytechnique Federale de
    Lausanne
Mohammed Dessouky, Mentor Graphics Egypt
Giulia Di Capua, University of Cassino and Southern Lazio
Alex Doboli, State University of New York at Stony Brook,
    Department of ECE
Günhan Dündar, Bogazici University
Mourad Fakhfakh, University of Sfax, Tunisia
Nicola Femia, University of Salerno
Helena Fino, Universidade NOVA de Lisboa
Fernando García-Redondo, Arm Ltd. Research Department
Georg Gläser, IMMS GmbH
Helmut Gräb, Technical University of Munich
Francesco Grasso, Università di Firenze – DINFO
Lars Hedrich, University of Frankfurt
Eckhard Hennig, Reutlingen University
Nuno Horta, Universidade de Lisboa
Anton Klotz, Cadence Design Systems
Adam Kubacak, Allegro Microsystems
Gildas Leger, Instituto de Microelectronica de Sevilla
    (IMSECNM-CSIC)

Po-Hung (Mark) Lin, National Chiao Tung University
Jose María Lopez-Villegas, University of Barcelona
Marie-Minerve Louerat, UPMC
Nuno Lourenço, Instituto de Telecomunicações
Patrick Mäder, Technische Universität Ilmenau
Stefano Manetti, University of Firenze
Ricardo Martins, Instituto de Telecomunicações - Instituto
Superior Técnico, Ulisboa
Montserrat Nafría, Universitat Autonoma Barcelona
Vojin Oklobdzija, University of California
Kemal Ozanoglu, Bogazici University
Fábio Passos, Instituto de Telecomunicações
Shared Reza, Sandia National Laboratories
Elisenda Roca, Instituto de Microelectronica de Sevilla, CNM,
    CSIC and Universidad de Sevilla
Ben Rodanski, Southern Cross Drones
Carlos Sánchez-López, Autonomous University of Tlaxcala
Jürgen Scheible, Reutlingen University
Guoyong Shi, Shanghai Jiao Tong University
Ralf Sommer, Technical Universtiy of Ilmenau
Haralampos Stratigopoulos, Sorbonne Universités, UPMC,
    Univ. Paris 6, CNRS, LIP6
Sheldon Tan, UC Riverside
Esteban Tlelo-Cuautle, INAOE
Francisco V. Fernandez, IMSE
Gerd Vandersteen, Vrije Universiteit Brussel – Dept. ELEC
Mustafa Berke Yelten, Istanbul Technical University,
    Electronics and Communications
Lihong Zhang, Memorial University of Newfoundland

■ PRIME 2021

General Chair
Stefan Heinen, RWTH Aachen University

Technical Program Chairs
Gianluca Giustolisi, University of Catania

Publication Chairs
Piero Malcovati, University of Pavia

Industrial Liaison Chair
Rafael Castro López, CSIC, Spain

Local Arrangements Chairs
Anne Arlt, IMMS GmbH
Hatice Altintas, VDE-Conference Services

Steering Committee
Franco Maloberti, University of Pavia, IT
Catherine Dehollain, EPFL, CH
Alberto Gola, Power Integrations Switzerland GmbH, CH
Bernd Deutschmann, TU Graz, AT
Elena Blokhina, Univ. College Dublin, IE
Gunhan Dundar, Bogazici University, TR
Anton Klotz, Cadence, DE
Ravinder Dahiya, University of Glasgow, UK
Nuno Horta, University of Lisbon, PT
Salvatore Pennisi, University of Catania, I
Sandro Carrara, EPFL, CH

Conference Topics

■ SMACD 2021

Modeling
- Compact modeling
- Performance and behavioral modeling
- Power and thermal modeling
- Reliability and variability modeling
- Multi-domain modeling (fluidic, bio, chemical, optical, mechanical)
- EMC & analog signal integrity
- RF/microwave/mm-wave modeling
- Linear and non-linear model order reduction
- Circuit & system theory and application
- Automated model generation Simulation, verification and test

Simulation, verification and test
- Behavioral simulation
- Numerical and symbolic simulation methods
- Simulation and estimation methods of variability and yield
- Analysis of reliability effects (aging, electromigration, stress...)
- High-frequency simulation techniques
- Multilevel simulation techniques
- Multi-domain simulation
- Formal and functional verification
- Functional safety
- ESD

Synthesis
- Multilevel design methodologies
- Synthesis methods of multi-domain circuit and systems
- Physical synthesis
- High-frequency circuit and system design
- Low-power and energy-aware design techniques
- Parasitic-aware synthesis
- Variability-aware & reliability-aware design
- Performance monitors & self-healing techniques
- Test and design-for-test methods
- Optimization methods applied to circuit and system design
- Procedural design automation

Applications of modeling, simulation and synthesis in:
- Automotive
- Bio-electronics
- Security
- Internet-of-things
- Communications
- Renewable systems
- Energy management
- Engineering education

■ PRIME 2021

- Micro/Nanoelectronics
- Semiconductors
- Analog/Digital Signal Processing
- Computer Aided Design
- Analog/Digital/Mixed/RF IC Design
- Integrated Power ICs
- Sensors/Systems and MEMS
- Semiconductor Memories
- RF, Microwave and mm-wave Circuits
- VLSI and SoC Applications
- Visual Signal Processing
- Energy Harvesting
- Automotive Electronics
- Flexible Electronics
- Technical Trends & Challenges
- Biomedical Electronics

Contents

Tutorials

T1 **Design Meets EDA: Gaps and Countermeasures in Analog/Mixed-Signal IC Design**
Benjamin Prautsch[1], Reimund Wittmann[2], Frank Schenkel[3], Johannes Koelsch, Christoph Grimm[4], Gunter Strube[5]
[1]Fraunhofer IIS/EAS, Dresden, Germany; [2]Imst Gmbh, Kamp-Lintfort, Germany; [3]Muneda Gmbh, Unterhaching, Germany; [4]TU Kaiserslautern, Kaiserslautern, Germany; [5]Blu Business Development, Gröbenzell, Germany

T2 **Qubit Gate Quality Loss Through Circuit Parasitics and Noise** 19
Lotte Geck[1], Stefan van Waasen[1,2]
[1]Central Institute of Engineering, Electronics and Analytics, Electronic Systems, Forschungszentrum Jülich GmbH, Jülich, Germany; [2]Faculty of Engineering, Communication Systems, University of Duisburg-Essen, Germany

T3 **Modeling Power Supply Noise in RF SoCs** 24
Jonas Meier, Florian Menke, Lantao Wang, Tim Lauber, Ralf Wunderlich and Stefan Heinen
Chair of Integrated Analog Circuits and RF Systems, RWTH Aachen University, Aachen, Germany

T4 **High-Performance Flexible and Printed Electronics Based on Inorganic Semiconducting Structures** 30
Abhishek Singh Dahiya, Dhayalan Shakthivel, Yogeenth Kumaresan and Ravinder Dahiya
Bendable Electronics and Sensing Technologies (BEST) Group, University of Glasgow, Glasgow, UK

T5 **On-Time RFIC Development with Fast EM Simulation and Integrated Design Flow**
Fadoua Gacim; Anton Klotz, Cadence Design Systems, Germany

T6 **Metastable states in ECL- and CML designs and their consideration by design of TDCs with picoseconds resolutions**
Gerald Kell, Daniel Schulz, Technische Hochschule Brandenburg; FB Informatik und Medien, Germany

T7 **Optimizing Neural Networks for Embedded Hardware** 36
Domenik Helms[2], Karl Amende[3], Saqib Bukhari[4], Thies de Graaff[2], Alexander Frickenstein[1], Frank Hafner[4], Tobias Hirscher[3], Sven Mantowsky[4], Georg Schneider[4], and Manoj-Rohit Vemparala[1]
[1]BMW AG, Munich, Germany; [2]OFFIS e.V., Oldenburg, Germany; [3]Valeo, Kronach, Germany; [4]ZF Friedrichshafen AG, Saarbr¨ucken/Friedrichshafen, Germany

T8 **The Essential Role of Procedural Approaches in Electronic Design Automation** 42
Daniel Marolt[1], Jürgen Scheible[1], Göran Jerke[2], Vinko Marolt[2]
[1]Electronics & Drives, Reutlingen University Reutlingen, Germany; [2]Automotive Electronics, Robert Bosch GmbH Reutlingen, Germany

SMACD 2021

A1 | Competition (1)
Chair: Engin Afacan, Gebze Technical University, Turkey

A1.1 Trash or Treasure? Machine-learning based PCB layout anomaly detection with AnoPCB ... 48
Henning Franke[1], Paul Kucera[1], Julian Kuners[1], Tom Reinhold[2], Martin Grabmann[2], Patrick Mäder[1], Marco Seeland[2] and Georg Gläser[1]
[1]Technische Universität Ilmenau, Germany; [2]IMMS Institut für Mikroelektronik- und Mechatronik-Systeme gemeinnützige GmbH (IMMS GmbH), Ilmenau, Germany

A1.2 A Deep Learning Toolbox for Analog Integrated Circuit Placement 52
António Gusmão[1,2], António Canelas[1], Nuno Horta[1,2], Nuno Lourenço[1] and Ricardo Martins[1]
[1]Instituto de Telecomunicações, Lisboa, Portugal; [2]Instituto Superior Técnico – Universidade de Lisboa, Lisboa, Portugal

A1.3 A Differential Evolution based Methodology for Parameter Extraction of Behavioral Models of Electronic Components. 56
Gazmend Alia[1,2], Andi Buzo[1], Daniel Ludwig[1], Linus Maurer[2] and Georg Pelz[1]
[1]Infineon Technologies AG, Bundeswehr University Munich, Munich, Germany; [2]Bundeswehr University Munich, Munich, Germany

A2 | Machine Learning SS
Chair: Ralf Sommer, TU Ilmenau & IMMS GmbH, Germany

A2.1 Machine Learning in the Analog Circuit Simulation Loop 60
Petar Tzenov and Ahmed Sokar, Infineon Technologies AG, Neubiberg, Germany

A2.2 Bringing Structure into Analog IC Placement with Relational Graph Convolutional Networks. 64
António Gusmão[1,2], Nuno Horta[1,2], Nuno Lourenço[1] and Ricardo Martins[1]
[1]Instituto de Telecomunicações, Lisboa, Portugal; [2]Instituto Superior Técnico – Universidade de Lisboa, Lisboa, Portugal

A2.3 Machine Learning in Charge: Automated Behavioral Modeling of Charge Pump Circuits ... 68
Martin Grabmann[1], Christian Landrock[2] and Georg Gläser[1]
[1]IMMS Institut für Mikroelektronik- und Mechatronik-Systeme gemeinnützige GmbH (IMMS GmbH), Ilmenau, Germany; [2]X-FAB Global Services GmbH, Erfurt, Germany

A3 | RF Systems (1)
Chair: Nuno Lourenço, Instituto de Telecomunicações, Portugal

A3.1 Frequency-Limited Reduction of RLCK Circuits via Second-Order Balanced Truncation ... 72
Olympia Axelou, Dimitrios Garyfallou and George Floros, Department of Electrical and Computer Engineering, University of Thessaly, Volos, Greece

A3.2 A Mixed Time-Frequency RF Simulation Technique Based on Numerical Time-Slot Partitioning 76
Jorge Oliveira, School of Technology and Management, Polytechnic of Leiria, Leiria, Portugal and Instituto de Telecomunicações, University of Aveiro, Aveiro, Portugal

A3.3 Application of Asymmetric Crosstalk Harnessed Signaling on 3D Hexagonal Interconnect Arrays 80

Daniel Iparraguirre[1] and José Delgado-Frías[2]
[1]Intel Corporation, Hillsboro, OR, USA; [2]Washington State University, Pullman, WA, USA

A4 | Competition (2)
Chair: Engin Afacan, Gebze Technical University, Turkey

A4.1 Adaptive Test Bench Generation, Simulation and Parameter Extraction for AMS Circuitry 84

Alexander Meyer, Leon Weihs, Ralf Wunderlich and Stefan Heinen, Integrated Analog Circuits and RF Systems Laboratory, RWTH Aachen University, Aachen, Germany

A4.2 Monitoring Analog Circuit Performance using Adaptive Filters and RSM-based Behavioral Models 88

Maike Taddiken, Steffen Paul and Dagmar Peters-Drolshagen, Institute of Electrodynamics and Microelectronics (ITEM.me), University of Bremen, Germany

A5 | Device Modelling (1)
Chair: Nicola Femia, University of Salerno, Italy

A5.1 A Compact Model for Scalable MTJ Simulation 92

Fernando García-Redondo[1], Pranay Prabhat[1], Mudit Bhargava[2] and Cyrille Dray[3]
[1]Arm Ltd, Cambridge, UK; [2]Arm Inc, Austin, USA; [3]Arm Ltd, La Paros, France

A5.2 A Quantitive Analysis of the Recovery Effect in Batteries from Datasheets 96

Alberto Bocca, Yukai Chen, Wenlong Wang, Alberto Macii, Enrico Macii and Massimo Poncino Department of Control and Computer Engineering, Politecnico di Torino, Torino, Italy

A6 | RF Systems (2)
Chair: Günhan Dündar, Bogazici University, Turkey

A6.1 A Phase Error Correction Algorithm for RF Energy Harvesters Using Two Antennas 100

Ali Dogus Güngördü, Didem Erol, Alican Çaglar and Mustafa Berke Yelten Istanbul Technical University, Electronics and Communications Engineering, Istanbul, Turkey

A6.2 Robust Design Methodology for RF LNA including Corner Analysis 104

Antonio Dionisio Martínez-Pérez[1], Francisco Aznar[2], Guillermo Royo[1], Pedro Martinez[1] and Santiago Celma[1]
[1]Group of Electronic Design (GDE), Universidad de Zaragoza, Zaragoza, Spain; [2]Group of Electronic Design (GDE), Centro Universitario de la Defensa, Zaragoza, Spain

A6.3 Event-Driven Modeling and Simulation of 5G NR-Band RF Transceiver in SystemVerilog .. 108

Chan Young Park and Jaeh Kim, Elelctrical and Computer Engineering Department, Seoul National University, Seoul, Korea

A7 | Wireless Power SS
Chair: Giulia Di Capua, University of Cassino and Southern Lazio, Italy

A7.1 Sensitivity analysis in dynamic WPT systems based on non-intrusive stochastic methods ... 112
Paul Lagouanelle[1,3], Giulia Di Capua[4], Nicola Femia[5], Fabio Freschi[3], Antonio Maffucci[4], Lionel Pichon[1] and Salvatore Ventre[4]
[1]Group of Electrical Eng., Paris, CNRS, CentraleSupélec, Université Paris-Saclay, Gif-sur-Yvette, France; [3]Department of Energy, Politecnico of Torino, Turin, TO, Italy; [4]Department of Electrical and Information Eng., University of Cassino and Southern Lazio, Cassino, FR, Italy; [5]Department of Information and Electrical Eng. and Applied Math., University of Salerno, Fisciano, SA, Italy

A7.2 Performance Analysis of IPT Systems for Electric Vehicles Dynamic Battery Charging 116
Giulia Di Capua[1], Luca De Guglielmo[2] and Nicola Femia[2]
[1]Department of Electrical and Information Eng., University of Cassino and Southern Lazio, Cassino, FR, Italy; [2]Department of Information and Electrical Eng. and Applied Math., University of Salerno, Fisciano, SA, Italy

A7.3 Coil Geometry Modeling and Optimization for a Bidirectional Wireless Power Transfer System 120
Simon Nigsch, Falk Kyburz and Kurt Schenk, Institute for Energy Systems, Eastern Switzerland University OST, Buchs SG, Switzerland

A7.4 Impact of the Pad Geometry on System-Level Performance Indicators in WPT Systems for Electrical Vehicles 124
Antonio Maffucci[1], Salvatore Ventre[1] and Alberto Delgado Exposito[2]
[1]Dep. of Electrical and Information Engineering, Univ. of Cassino and Southern Lazio, Cassino, Italy; [2]Centre of Industrial Electronics (CEI), Universidad Politécnica de Madrid, Madrid, Spain

A8 | Complex System Analysis II
Chair: Jürgen Scheible, Reutlingen University, Germany

A8.1 A Probe Placement Method for Efficient Electromagnetic Attacks 128
Minmin Jiang and Vasilis Pavlidis, Advanced Processor Technologies group, Department of Computer Science, University of Manchester, Great Britain

A8.2 Dealing with hierarchical partitioning in bottom-up design methodologies 132
F. Passos[1,2], Pablo Saraza-Canflanca[1], Rafael Castro Lopez[1], Elisenda Roca1 and Francisco V. Fernandez[1]
[1]Instituto de Microelectrónica de Sevilla, IMSE-CNM (CSIC/Universidad de Sevilla), Sevilla, Spain; [2]Instituto de Telecomunicações, Lisboa, Portugal

B1 | Analog Circuit and System Engineering
Chair: Helmut Graeb, Technical University of Munich, Germany

B1.1 Modeling and Optimization of Supply Sensitivity for a Time-Domain Temperature Sensor 136
Jun Tan, IMMS Institut für Mikroelektronik- und Mechatronik-Systeme gemeinnützige GmbH (IMMS GmbH), Ilmenau, Germany

B1.2 Noise behavior in current mirror circuit based on CNTFET and MOS Devices 140
Roberto Marani[1] and Anna Perri[2]
[1]National Research Council of Italy (CNR), Institute of Intelligent, Industrial Technologies and Systems for Advanced, Manufacturing (STIIMA), Bari, Italy; [2]Department of Electrical and Information Engineering, Electronic Devices Laboratory, Polytechnic University of Bari, Bari, Italy

B1.3 A g_m/I_D Sizing Method for High-speed Multi-stage Operational Amplifiers with Feedforward-only Compensation 144
Qixu Xie, Guoyong Shi and Yaoyao Ye, Dept of Micro/Nano Electronics, Shanghai Jiao Tong University, Shanghai, China

B1.4 Hybrid Capacitor-less LDO with Switched-Mode Dead-Zone Control. 148
Nellie Laleni, Andreas Tsiougkos and Vasilis Pavlidis, Department of Electrical and Computer Engineering, Aristotle University of Thessaloniki, Greece

B2 | Behavioral Analysis
Chair: Georg Gläser, IMMS GmbH, Germany

B2.1 Verilog-A model development of a DC–DC boost controller with autonomous optimization . . 152
Davide Severin[1], Giovanni Capodivacca[1], Bernard Blaise Tchodjie Tchamabe[1], Andi Buzo[2] and Cristian-Vasile Diaconu[3]
[1]Infineon Technologies, Italy; [2]Infineon Technologies, Germany; [3]Infineon Technologies, Romania

B2.2 Analog Circuit Abstraction to SystemC-AMS Secured by Affine Forms 156
Ahmad Tarraf and Lars Hedrich, Institute for Computer Science, Goethe University Frankfurt, Germany

B2.3 Simulating the impact of Random Telegraph Noise on integrated circuits 160
Pablo Saraza-Canflanca[1], Eros Camacho-Ruiz[1], Rafael Castro-Lopez[1], Elisenda Roca[1], Javier Martin-Martinez[2], Rosana Rodriguez[2], Montserrat Nafria[2] and Francisco Fernandez[1]
[1]Instituto de Microelectrónica de Sevilla, IMSE-CNM (CSIC/Universidad de Sevilla), Sevilla, Spain; [2]Electronic Engineering Department (REDEC) group, Universitat Autònoma de Barcelona (UAB) Barcelona, Spain

B3 | Simulation Methods
Chair: Günhan Dündar, Bogazici University, Turkey

B3.1 Connecting Energy Storages from Tool Independent, Signal-flow Oriented FMUs 164
Meik Ehlert[1], Jan Michael[1], Christian Henke[1], Ansgar Trächtler[2], Matthias Kalla[3], Bakr Bagaber[3], Bernd Ponick[3] and Axel Mertens[3]
[1]Scientific Automation, Fraunhofer Institute for Mechatronic Systems, Design IEM, Paderborn, Germany; [2]Heinz-Nixdorf-Institute, University of Paderborn, Paderborn, Germany; [3]Institute for Drive Systems and Power Electronics, Leibniz Universität, Hannover

B3.2 Adaptive Simulation with HDL Control Module for Frequency Converting Circuits 168
Zoltan Tibenszky, Martin Kreißig, Corrado Carta and Frank Ellinger, Technische Universität Dresden, Dresden, Germany

B3.3 Step Size Determination Approach for Aging Simulations in Analog Ics. 172
Engin Afacan, Department of Electronics Engineering, Gebze Technical University

B4 | Procedural Design Automation
Chair: Daniel Marolt, Reutlingen University, Germany

B4.1 Schematic Generation of Programmable Analog Neural Networks for Signal Proccessing . . . 176
Florian Aul, Nikoletta Katsaouni, Lukas Krischker, Sascha Schmalhofer, Marcel H. Schulz and Lars Hedrich
[1]Institute for Computer Science, Goethe University Frankfurt, Germany; [2]Institute for Cardiovascular Regeneration, Goethe University Frankfurt, Germany

B4.2 Generators, Templates, and Code Generation for Flexible Automation of Array-Style Layouts 180
Benjamin Prautsch[1], Reimund Wittmann[2], Uwe Eichler[1], Uwe Hatnik[1] and Jens Lienig[3]
[1]Fraunhofer IIS/EAS, Institute for Integrated Circuits, Division Engineering of Adaptive Systems, Dresden, Germany; [2]IMST GmbH, Kamp-Lintfort, Germany; [3]Dresden University of Technology, Dresden, Germany

B4.3 Improvement of Simulation-Based Analog Circuit Sizing using Design-Space Transformation 184
Matthias Schweikardt and Jürgen Scheible, Electronics & Drives, Reutlingen University, Reutlingen, Germany

B4.4 Machine Learning Based Procedural Circuit Sizing and DC Operating Point Prediction 188
Yannick Uhlmann[1], Michael Essich[2], Matthias Schweikardt[1], Jürgen Scheible[1] and Cristóbal Curio[2]
[1]Electronics & Drives, Reutlingen University, Reutlingen, Germany; [2]Cognitive Systems, Reutlingen University, Reutlingen, Germany

B5 | Optimization Methods
Chair: Francisco Fernandez, Instituto de Microelectrónica de Sevilla, IMSE-CNM (CSIC and Universidad de Sevilla), Spain

B5.1 Surrogate-Assisted Multi-objective Differential Evolution based on Gaussian Process for Analog Circuit Synthesis. 192
Sen Yin, Wenfei Hu, Ruitao Wang, Zhikai Wang, Jian Zhang and Yan Wang, Institute of Microelectronics, Tsinghua University, China

B5.2 A fast Structural Synthesis Algorithm for Op-Amps based on Multi-Threading Strategies . . 196
Inga Abel, Clara Kowalsky and Helmut Graeb, Technical University of Munich, Munich, Germany

B5.3 An Essay on the Next Generation of Performance-driven Analog/RF IC EDA Tools: The Role of Simulation-based Layout Optimization 200
Ricardo Martins[1], António Gusmão[1,2], António Canelas[1], Fábio Passos[1,3], Nuno Lourenço[1] and Nuno Horta[1,2]
[1]Instituto de Telecomunicações, Lisboa, Portugal; [2]Instituto Superior Técnico, Universidade de Lisboa, Portugal; [3]Dialog Semiconductors, Lisboa, Portugal

B6 | Complex System Analysis I
Chair: Jürgen Scheible, Reutlingen University, Germany

B6.1 An Efficient Modeling Approach for Large Ring Oscillator Based Ising Machines 204
Markus Graber, Nico Angeli and Klaus Hofmann, Integrated Electronic Systems Lab, Technical University of Darmstadt, Darmstadt, Germany

B6.2 The Merging Technique to Simulate Synchronization Mode of Coupled Oscillators 208
Sergey Rusakov and Mark Gourary, IPPM, Russian Academy of Sciences, Moscow, Russia

B7 | Digital Circuit and System Engineering
Chair: Rafael Castro Lopez, Instituto de Microelectrónica de Sevilla, IMSE-CNM (CSIC/Universidad de Sevilla), Spain

B7.1 RTL Implementation of MCMC-based Constraints Solver 212
Moemen Ahmed[1], Youssef Ahmed[1], Younan Nagy[1], Manar Adbel-Rahman[1], Khaled Salah[2],
M. Watheq El-Kharashi[1], Ayub Khan[2]
[1]Department of Computer and Systems, Faculty of Engineering, Ain Shams University, Cairo, Egypt;
[2]Siemens Digital Industries Software, Fremont, USA

B7.2 A study of SRAM PUFs reliability using the Static Noise Margin 216
Eros Camacho-Ruiz, Pablo Saraza-Canflanca, Rafael Castro-Lopez, Elisenda Roca, Piedad Brox
and Francisco Fernandez, Instituto de Microelectrónica de Sevilla, IMSE-CNM (CSIC and
Universidad de Sevilla), Sevilla, Spain

**B7.3 Design and Optimization of a Control Algorithm for a Digital Low-Dropout Regulator in
System-on-Chip Applications** 220
Benedikt Ohse[1] and Jun Tan[2]
[1]Ernst-Abbe-Hochschule Jena, Jena, Germany; [2]IMMS Institut für Mikroelektronik- und
Mechatronik-Systeme gemeinnützige GmbH (IMMS GmbH), Ilmenau, Germany

B7.4 A Differential Public PUF Design for Lightweight Authentication 224
Shengyu Duan[1,2] and Gaole Sai[3,4]
[1]School of Computer Engineering and Science, Shanghai University, China; [2]State Key Laboratory
of Computer Architecture, Institute of Computing Technology, Chinese Academy of Sciences, China;
[3]Guangdong Provincial Key Lab of Robotics and Intelligent System, Shenzhen Institutes of Advanced
Technology, Chinese Academy of Sciences, China; [4]CAS Key Laboratory of Human-Machine
Intelligence-Synergy Systems, Shenzhen Institutes of Advanced Technology, China

B8 | Device Modelling (2)
Chair: Nuno Lourenço, Instituto de Telecomunicações, Portugal

B8.1 Organic Transistor Parameter Estimation and Accurate Modeling for Process Optimization 228
Rosalba Liguori, Gian Domenico Licciardo and Luigi Di Benedetto, Department of Industrial
Engineering University of Salerno, Fisciano (SA), Italy

**B8.2 Bias Temperature Instability Characterization and Modelling for 0.18 um CMOS under
Extreme Thermal Stress Conditions** 232
Yen Tran[1,2], Toshihiro Nomura[1], Mohamed Salim Cherchali[1], Claire Tassin[1], Yann Deval[2] and
Cristell Maneux[2]
[1]Etudes et Production Schlumberger, Clamart, France; [2]Laboratoire IMS, Universite de Bordeaux,
Talence, France

PRIME 2021

C1 | Digital Circuits and Subsystems
Chair: Miguel Garcia-Bosque, University of Zaragoza, Spain

C1.1 Run-Time Adaptive Hardware Accelerator for Convolutional Neural Networks 236
Cristian Sestito[1], Fanny Spagnolo[1], Pasquale Corsonello[1] and Stefania Perri[2]
[1]Department of Informatics, Modeling, Electronics and System Engineering – University of Calabria, Italy; [2]Department of Mechanical, Energy and Management Engineering – University of Calabria, Italy

C1.2 Design and Analysis of a Leading One Detector-based Approximate Multiplier on FPGA ... 240
Salvatore Scarfone[1], Fabio Frustaci[1] and Stefania Perri[2]
[1]Department of Informatics, Modeling, Electronics and System Engineering – University of Calabria; [2]Department of Mechanical, Energy and Management Engineering – University of Calabria

C1.3 Extending a RISC-V core with an AES hardware accelerator to meet IOT constraints 244
Anthony Zgheib, Olivier Potin, Jean-Baptiste Rigaud and Jean-Max Dutertre, Mines Saint-Etienne, CEA-Tech, Centre CMP, Gardanne, France

C1.4 Memristive Logic-In-Memory Implementations: A Comparison 248
Pietro Inglese, Elena Ioana Vatajelu and Giorgio Di Natale, TIMA Laboratory, France

C2 | Data Converters (1)
Chair: Alexander Meyer, RWTH Aachen University, Germany

C2.1 A 12-bit 100 MHz SAR ADC in 110-nm CMOS for MAPSs 252
Silvia Tedesco, INFN of Turin, Italy

C2.2 A Timing Skew Correction Technique in Time-Interleaved ADCs Based on a $\Delta\Sigma$ Digital-to-Time Converter 256
Gabriele Bè, Mario Mercandelli and Luca Bertulessi, Politecnico di Milano, Italy

C2.3 A low-noise high-speed comparator for a 12-bit 200-MSps SAR ADC in a 28-nm CMOS process 260
Luca Ricci, Luca Bertulessi and Andrea Bonfanti, Politecnico di Milano, Italy

C3 | Data Converters (2)
Chair: Alexander Meyer, RWTH Aachen University, Germany

C3.1 A 2GS/s 10-bit Time-Interleaved Capacitive DAC for Self-Interference-Cancellation Application 264
Mazyar Abedinkhan Eslami, Danilo Manstretta and Rinaldo Castello, Univerity of Pavia, Italy

C3.2 Implementation of a Low Power Decimation Filter in a 180nm HV-CMOS Technology for a Neural Recording Front-End 268
Markus Sporer, Nicolas Graber, Steffen Moll, Stefan Reich and Maurits Ortmanns, University of Ulm, Germany

C4 | Automotive
Chair: Michael Hanhart, RWTH Aachen University, Germany

C4.1 **Analog Baseband Filter and Variable-gain Amplifier for Automotive Radars in 22 nm FD-SOI CMOS.** 272
Andres Seidel[1], Songhui Li[1], Laszlo Szilagyi[1], Corrado Carta[1], Jens Wagner[1,2] and Frank Ellinger[1,2]
[1]Chair for Circuit Design and Network Theory, Technische Universität Dresden, Germany;
[2]CeTi, Center for Tactile Internet, Technische Universität Dresden, Germany

C4.2 **A Highly Linear High-Voltage Compliant Current Output Stage for Arbitrary Waveform Generation** 276
Felix Schwarze, Florian Protze, Frank Ellinger and Christian Matthus, Technische Universität Dresden, Germany

C4.3 **A RISC-V-based System on Chip for High-Speed Control in Safety-Critical 650 V GaN-Applications** 280
Mike Richter[1], André Lüdecke[1], Yoon-Cue Lee[1], Alexander Stanitzki[1], Alexander Utz[1], Günter Grau[2], Holger Kappert[1] and Rainer Kokozinski[1]
[1]Fraunhofer Institute for Microelectronic Circuits and Systems (IMS), Duisburg, Germany;
[2]advICo microelectronics GmbH, Recklinghausen, Germany

C4.4 **An Approach to Online Wear Out Monitoring of PCB Interconnects in Safety-Critical Systems** 284
Saeid Yazdani[1], Werner Wolz[1], Rainer Engelhardt[2], Christian Schott[1], Ulrich Heinkel[1] and Daniel Kriesten[3]
[1]TU Chemnitz, Germany; [2]Steinbeis GmbH, Chemnitz, Germany; [3]Hochschule Mittweida, Germany

C5 | Sensing Circuits (1)
Chair: Markus Sporer, University of Ulm, Germany

C5.1 **Experimental Investigation of Dielectric Loss Induced Noise in Charge Detection Systems for Cosmic Dust** 288
Sebastian Kelz, Markus Groezing and Manfred Berroth, Institut für Elektrische und Optische Nachrichtentechnik, University of Stuttgart, Germany

C5.2 **Generalized comparison of the accessible emission limits of flash- and scanning LiDAR-systems** 292
Roman Burkard[1], Reinhard Viga[1], Jennifer Ruskowski[2] and Anton Grabmaier[1]
[1]University of Duisburg-Essen, Germany; [2]Fraunhofer IMS, Duisburg, Germany

C5.3 **A Mixed-Precision Binary Neural Network Architecture for Touch Modality Classification** . . 296
Hamoud Younes[1,2], Ali Ibrahim[1,2], Mostafa Rizk[2] and Maurizio Valle[1]
[1]University of Genova, Italy; [2]Lebanese International University, Bekaa, Lebanon

C6 | Sensing Circuits (2)
Chair: Markus Sporer, University of Ulm, Germany

C6.1 **A CMOS SPAD pixel with an integrated mixed-signal rotatory TDC** 300
Sergio Moreno, Victor Moro and Angel Dieguez, University of Barcelona, Spain

C6.2 Germanium – InGaZnO heterostructured thin-film phototransistor with high IR photo-response 304
Hichem Ferhati, Fayçal Djeffal and A Bendjerad, University of Batna, Algeria

C7 | Power Circuits and Harvesting
Chair: Sebastian Kelz, University of Stuttgart, Germany

C7.1 Integrated Hysteretic Controlled Regulating Buck Converter with Capacitively Coupled Bootstrapping 308
Francarl Galea, Owen Casha, Ivan Grech, Edward Gatt and Joseph Micallef, University of Malta

C7.2 Single-Inductor Dual-Output Buck Converter with Charge Recycling 312
Kemal Ozanoglu and Gunhan Dundar, Bogazici University, Istanbul, Turkey

C7.3 Design of an integrated Maximum Power Point Boost Converter for PV Submodules 316
Léon Weihs, Michael Hanhart, Leo Rolff, Ralf Wunderlich and Stefan Heinen, RWTH Aachen University, Germany

C7.4 Design of a High PSRR Multistage LDO with On-Chip Output Capacitor 320
Jonas Zoche, Michael Hanhart, Jan Grobe, Léon Weihs, Leo Rolff, Ralf Wunderlich and Stefan Heinen, RWTH Aachen University, Germany

C8 | Circuits for Clock Generation and Optimization
Chair: Jonas Meier, RWTH Aachen University, Germany

C8.1 Skew and Jitter Performance in CMOS Clock Phase Splitter Circuits 324
Lorenzo Scaletti, Angelo Parisi and Luca Bertulessi, Politecnico di Milano, Italy

C8.2 Entropy Analysis of RO-based Physically Unclonable Functions 328
Guillermo Diez-Senorans, Miguel Garcia-Bosque, Carlos Sanchez-Azqueta and Santiago Celma University of Zaragoza, Spain

C8.3 On the Behavior of a Wide Set of Oscillators: PUFs or TRNGs? 332
Miguel Garcia-Bosque, Abel Naya, Guillermo Díez-Señorans, Carlos Sánchez-Azqueta and Santiago Celma, University of Zaragoza, Spain

C8.4 A 55 MHz Integrated Crystal Oscillator with Chirp Injection Using a 28-nm Technology ... 336
Lantao Wang, Adrian Arnold, Jonas Meier, Markus Scholl, Ralf Wunderlich and Stefan Heinen RWTH Aachen University, Germany

D1 | RF Circuits and Systems (1)
Chair: Christopher Nardi, RWTH Aachen University, Germany

D1.1 A low-power 26.56-GHz LC-based DCO for multi-gigabit communication systems 340
Pablo Jiménez-Fernández[1], Óscar Guerra[1], Rocío Del Río[1], Alberto Rodríguez-Pérez[2] and Enrique Prefasi[2]
[1]Instituto de Microelectrónica de Sevilla, Spain; [2]KDPOF, Spain

D1.2 A Wide-Tuning-Range 55 GHz CMOS VCO on 22 nm FD-SOI Technology 344
Zoltán Tibenszky, Corrado Carta and Frank Ellinger, Chair of Circuit Design and Network Theory,
Technische Universität Dresden, Germany

D1.3 A Fully Integrated 28 GHz Class-J Doherty Power Amplifier in 130 nm BiCMOS 348
Simone Veni[1], Michele Caruso[2], David Seebacher[2], Andrea Neviani[1] and Andrea Bevilacqua[1]
[1]University of Padova, Italy; [2]Infineon Technologies, Villach, Austria

**D1.4 A Scalable CPW Circuit Model in Advanced CMOS Technologies for mm-Wave
Frequencies.** 352
Carla Moran Guizan[1], Peter Baumgartner[1] and Stefan Heinen[2]
[1]Intel Deutschland, Germany; [2]RWTH Aachen University, Germany

D2 | Biomedical Circuits (1)
Chair: Catherine Dehollain, Ecole Polytechnique Federale de Lausanne, Switzerland

**D2.1 A Sub-1μA Low-Power Low-Noise Amplifier with Tunable Gain and Bandwidth for EMG
and EOG Biopotential Signals** 356
Rafael Vieira[1], Ricardo Martins[1], Nuno Horta[1], Nuno Lourenço[1] and Ricardo Póvoa[1,2]
[1]Instituto de Telecomunicações, Lisboa, Portugal; [2]Escola Superior Náutica Infante D. Henrique,
Paço de Arcos, Portugal

D2.2 Transistor Downscaling toward Ultra-Low-Power, sub-100 μm^2 and sub-Hz Oscillators 360
Gian Luca Barbruni[1], Chiara Bielli[2], Danilo Demarchi[2] and Sandro Carrara[1]
[1]Ecole Polytechnique Federale de Lausanne, Switzerland; [2]Politecnico di Torino, Italy

**D2.3 Electronic solution to compensate the effects of the temperature and the humidity on the
measurements of a capacitive sensor dedicated to an injection insulin pen** 364
Sylvain Joly[1], Albrecht Lepple-Wienhues[1] and Catherine Dehollain[2]
[1]Valtronic Technologies, Switzerland; [2]Ecole Polytechnique Federale de Lausanne, Switzerland

D3 | Biomedical Circuits (2)
Chair: Catherine Dehollain, Ecole Polytechnique Federale de Lausanne, Switzerland

D3.1 A scalable spike detection method for implantable high-density multielectrode array 368
Mattia Tambaro[1], Elia Arturo Vallicelli[2], Gerardo Saggese[3], Andrea La Gala[4], Marta Maschietto[1],
Alessandro Leparulo[1], Antonio Strollo[3], Marcello De Matteis[4], Andrea Baschirotto[4] and Stefano
Vassanelli[1]
[1]University of Padova, Italy; [2]INFN, Milan, Italy; [3] University of Naples „Federico II", Italy;
[4]University of Milano, Bicocca, Italy

**D3.2 Current-reuse Low-Power Single-Ended toDifferential LNA for Medical Ultrasound
Imaging.** 372
Olivia Mirea, Carsten Wulff and Trond Ytterdal, NTNU: Norwegian University of Science and
Technology, Trondheim, Norwa

D4 | RF Circuits and Systems (2)
Chair: Lantao Wang, RWTH Aachen University, Germany

D4.1 Low Power High Linearity 14-23 GHz SiGe HBT Downconversion Mixer 376
Syed Sharfuddin Ahmed and Hermann Schumacher, University of Ulm, Germany

D4.2 A Mixer-Embedded Low Noise Amplifier for Mixer-First Direct-Conversion Wake-Up Receivers. 380
Christopher Nardi, Alexander Kronig, Ralf Wunderlich and Stefan Heinen, RWTH Aachen University, Germany

D4.3 Make Some Noise: Energy-Efficient 38 Gbit/s Wide-Range Fully-Configurable Linear Feedback Shift Register 384
Christoph Wagner[1], Georg Gläser[2], Thomas Sasse[1], Gerald Kell[3] and Giovanni Del Galdo[1,4]
[1]Technische Universität Ilmenau, Germany; [2]IMMS GmbH, Ilmenau, Germany; [3]Technische Hochschule Brandenburg, Germany; [4]Fraunhofer IIS, Ilmenau, Germany

D4.4 Every Clock Counts – 41 GHz Wide-Range Integer-N Clock Divider 388
Christoph Wagner[1], Georg Gläser[2], Gerald Kell[3] and Giovanni Del Galdo[1,4]
[1]Technische Universität Ilmenau, Germany; [2]IMMS GmbH, Ilmenau, Germany; [3]Technische Hochschule Brandenburg, Germany; [4]Fraunhofer IIS, Ilmenau, Germany

D5 | Devices and Reliability
Chair: Stefan Heinen, RWTH Aachen University, Germany

D5.1 Modeling Ni/β-Ga$_2$O$_3$ SBD interface properties 392
Madani Labed[1], Nouredine Sengouga[1], Afak Meftah[1], Jun Hui Park[2], Sinsu Kyoung[3], Hojoong Kim[2] and You Seung Rim[2]
[1]Laboratory of Semiconducting and Mettalic Materials, Biskra university; [2]Department of Intelligent Mechatronics Engineering, and Convergence Engineering for Intelligent Drone; [3]Research and Development, Powercubesemi Inc.

D5.2 Performance assessment of a new low-cost RF sputtered Schottky diode based on a-Si/Ti structure 396
Hichem Ferhati, Fayçal Djeffal, A Bendjerad and A Benhaya, University of Batna, Algeria

D5.3 Digitally Programmable Potentiometer Multistage Architecture with Switch Independent Linearity 400
Giorgiana Catalina Ilie[1,2], Cristian Tudoran[2], Otilia Neagoe[2], Gheorghe Pristavu[1] and Gheorghe Brezeanu[1]
[1]On Semiconductor, Romania; [2]University "Politehnica" of Bucharest, Romania

D5.4 Reliability Investigation of 0.18μm CMOS for OilField Applications 404
Yen Tran[1,2], Toshihiro Nomura[1], Mohamed Salim Cherchali[1], Claire Tassin[1], Yann Deval[2] and Cristell Maneux[2]
[1]Etudes et Production Schlumberger, Clamart, France; [2]Laboratoire IMS, Universite de Bordeaux, France

D6 | Analog Circuits and Qubit Interfaces
Chair: Lotte Geck, Forschungszentrum Jülich, Germany

D6.1 A Cryogenic High-Voltage Amplifier for Ion Traps 408
Michael Sieberer[1], Christoph Sandner[1] and Peter Hadley[2]
[1]Infineon Technologies, Villach, Austria; [2]Graz University of Technology, Austria

D6.2 Cryogenic RF Transimpedance Amplifier in 22 nm SOI-CMOS for Control of a Qubit 412
Ricardo Heinen[1,2], Dennis Nielinger[1], Christian Grewing[1], Ralf Wunderlich[2] and Stefan Heinen[2]
[1]Forschungszentrum Jülich GmbH, Jülich, Germany; [2]RWTH Aachen University, Germany

D6.3 A First Order-Curvature Compensation 5ppm/°C Low-Voltage & High PSR 65nm-CMOS
Bandgap Reference with one-point 4-bits Trimming Resistor 416
Edoardo Barteselli[1], Luca Sant[2], Richard Gaggl[2] and Andrea Baschirotto[1]
[1]University of Milano – Bicocca, Italy; [2]Infineon Technologies, Villach, Austria

D6.4 Resource Efficient Sub-V_T Level Shifter Circuit Design Using a Hybrid Topology in
28 nm 420
Saikat Chatterjee and Ulrich Rueckert, CITEC, Bielefeld University, Germany

Qubit Gate Quality Loss Through Circuit Parasitics and Noise

Lotte Geck*, Stefan van Waasen*,†

*Central Institute of Engineering, Electronics and Analytics, Electronic Systems,
Forschungszentrum Jülich GmbH, Jülich, Germany
†Faculty of Engineering, Communication Systems, University of Duisburg-Essen, Germany
l.geck@fz-juelich.de

Abstract—This work aims to find signal sequences that implement high-quality logic gates for single GaAs qubits in a quantum computer. To achieve that an optimization of the gate quality with a scalable electronic architecture in mind is done. In addition, a behavioral model of the qubit and the control electronics in the Q-Interface is implemented. The electronic model includes typical integrated circuit impairments such as process variations and parasitic capacitors. While the initial quality of the optimized gates is high, simulations with the implemented model show partly large quality reductions for some impairments. A major quality loss is for example caused by parasitic capacitors, but some non-ideal effects can be mitigated through pre-distortion, among others.

Index Terms—quantum computing, parasitics, noise

I. INTRODUCTION

Quantum computing is a topic gaining a lot of attention due to its potential to solve currently intractable computational tasks in relevant industry and research fields. The basic processing unit of such a computer is quantum mechanical in nature and called a qubit. Nevertheless, a large portion of the quantum computer hardware is classical, as seen in Fig. 1. A lot of the depicted electronics are digital and can be built similar to their classical computer counterparts, but new electronics are necessary in the Q-Interface. In this part the electronics to operate and interface the qubits are included. For the control of most qubits, precise analog signals are necessary among other things. How exactly these electronics look depends on the qubit implementation. Examples for qubits types are trapped ions, superconducting qubits, but also different types of semiconductor qubits. Electrical control signals for these types can be for example quadrature amplitude modulated pulses, but also other high-frequency pulses and adaptable DC voltages or currents. Next to the control, also the readout circuitry for the qubits is included in the Q-Interface. The readout transports information from the quantum mechanical domain back to the classical one. How that is done is again dependent on the qubit type but can for example include impedance measurement.

To realize a universal programmable quantum computer supposedly millions to tens of millions of qubits are necessary. In today's few qubit experiments high quality lab equipment produces the precision control signals for the qubits. These settings are good for research purposes but a complete quantum computer implementation needs solutions that are more

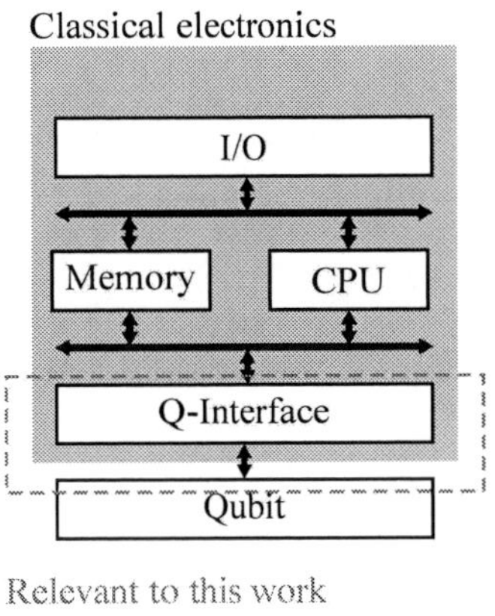

Fig. 1. Hardware of a Quantum Computer.

efficient in terms of form factor and power consumption. A promising approach is to use application-specific integrated circuits (ASICs) with conventional standard industry CMOS processes [1], [2]. This would transform current experimental electrical setups to a scalable part of the Q-Interface.

In a typical ASIC design flow with a clear use case, the requirements that assure performance quality of the circuit are defined in detail. This is not the case for electronics to control qubits, as the research on qubits is still ongoing. A comparison with the current experimental setup is also only partly helpful as a well-constructed arbitrary waveform generator (AWG) naturally has much higher capabilities than an ASIC with a very small form factor. To enable high-quality circuit design at this stage of the qubit development, an approach to co-simulating circuits and qubits has been made [3] for one qubit type.

This approach is very helpful to check how a specific circuit design operates together with a qubit, but a systematic study of different architecture options and potential impairments on the performance quality is difficult. On top of that, the control signals and circuits can be interdependent and thus specifications can be signal dependent. This work uses a detailed behavioral modeling approach to explore how non-ideal and parasitic effects in control electronics affect the quality of qubit logic gates (see Fig. 2). The first part of the model describes how the qubit behaves in the presence of its main control

© VDE VERLAG GMBH · Berlin · Offenbach

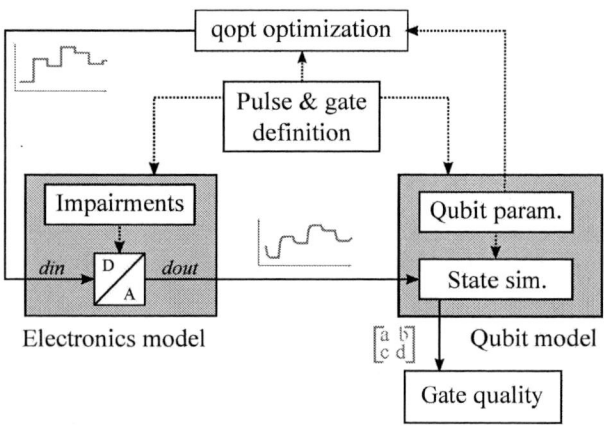

Fig. 2. Model and optimization concept in this work.

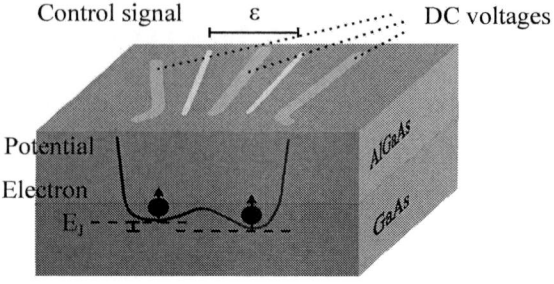

Fig. 3. GaAs qubit cross section.

signals ('Qubit model'). As a qubit realization, this work refers to the well-established gallium-arsenide (GaAs) S-T_0 qubit [4], but the principles of this work can be applied to other qubits as well. The basics of a GaAs qubit and the qubit state simulation are explained in Sec. II.

For the control electronics only the most significant part of the Q-Interface is modeled here, which is the high-frequency digital-to-analog converter (DAC) in the output stage. The DAC architecture and included non-ideal effects are described in Sec. III. Next to gate simulation, also signal optimization specific to the electronics is explored here (Fig. 2). Both topics are treated in Sec. IV and afterwards a conclusion is drawn.

II. QUBITS AND THEIR GATES

A. GaAs qubit

The GaAs qubit implementation is based on localizing two electrons and using the combined spin to encode information. In a first step, the electron localization is achieved through the utilization of a semiconductor heterostructure of aluminum-gallium-arsenide (AlGaAs) and GaAs. The conduction band structure of the two materials leads to a two dimensional electron gas (2DEG) at the material interface. In a second step the 2DEG is locally depleted with DC voltages applied to electrodes on top of the structure (Fig. 3). For one qubit about six DC electrodes are necessary (not all visible in Fig. 3). The resulting potential distribution forms two potential wells next to each other connected by a tunnel barrier. This is also called a double quantum dot.

For qubit operation one electron is loaded into each dot. With the addition of an external magnetic field and a magnetic difference field, a spin-dependent system of two energy levels is created in the double dot. A detailed derivation is not in the scope of this work, but can for example be found in [5]. The two distinct energy levels correspond to the singlet S and triplet T_0 spin state of the electrons. Whether the qubit is in a purely S or T_0 state or in a superposition of both states, can be controlled with a voltage control signal (Fig. 3). The control signal manipulates the energy difference between the dots E_J

and the relative tunnel barrier height, which changes the spin state. In this work the focus is on the control signal and how exactly it has to look like in order to achieve a specific spin state change. The DC voltages are not considered here.

B. Qubit gates

For logic operation, the qubit spin state is encoded into information. Here, the state S is encoded as a $\vec{0}$ and the T_0 state is $\vec{1}$. With this, the qubit state is defined as the state vector $\vec{\Psi}$:

$$\vec{\Psi} = \alpha\vec{0} + \beta\vec{1} = \alpha \begin{pmatrix} 1 \\ 0 \end{pmatrix} + \beta \begin{pmatrix} 0 \\ 1 \end{pmatrix} \quad \text{with} \quad \alpha, \beta \in \mathbb{C}. \quad (1)$$

The description highlights the superposition of the two basis states $\vec{0}$ and $\vec{1}$ through the amplitudes α and β. This is in contrast to classical logic where the information is discrete and either 0 or 1. Therefore, one qubit can contain more information than a classical bit. However, the deciding advantage of a quantum computer lies in the so called entanglement of several qubits. The states of entangled qubits are not independent of each other such that a manipulation of one qubit leads to a change in all entangled qubits. This ability in combination with dedicated quantum algorithms leads to the great potential speed-up of a quantum computer in comparison to a classical computer. To describe n entangled qubits, a state vector with the length 2^n is necessary. For bigger qubit numbers this leads to huge vectors and highlights why the simulation of a larger quantum processors is not possible, even with a supercomputer.

A logic quantum gate is reversible, deterministic and mathematically described through the matrix multiplication

$$\vec{\Psi} = U \ \vec{\Psi}_0, \quad (2)$$

with a starting state $\vec{\Psi}_0$. The values in U are influenced through the control signals applied to the qubit. A qubit gate is therefore not an implemented hardware block like in classical logic, but a specific signal sequence. This also means that a qubit is a processing and memory element at the same time. While the continuous state does have the advantage of containing more information, the disadvantage is that it is very perceptible to errors in the gate and distortions in the corresponding signal sequence. On top of that these errors are also more difficult to detect and correct compared to classical logic. A high quality gate is therefore essential for successful quantum computation.

A typical way to derive U is to assume that the input signals are piecewise constant. The qubit state at the time t_M is then given by

$$\vec{\Psi}(t_M) = \prod_{k=1}^{M} dU_k \cdot \vec{\Psi}_0. \tag{3}$$

The dU pieces of U are defined with the so-called Hamiltonian H which describes the energy levels of a qubit. The Hamiltonian is specific to the qubit implementation and is here given for the spin states of the GaAs S-T_0 qubit [6]:

$$dU_k = \exp\left(-i\hbar H/2\Delta t_k\right) \tag{4}$$

$$= \exp\left(-i\frac{\hbar}{2}\begin{pmatrix} \omega_J(\epsilon_k) & \Delta\omega_z \\ \Delta\omega_z & -\omega_J(\epsilon_k) \end{pmatrix}\Delta t_k\right). \tag{5}$$

The Hamiltonian includes the constant $\Delta\omega_z$ that is coming from a magnetic difference field ΔB_z. The term ω_J is calculated from the exchange energy E_J, which is dependent on the applied voltage of the control signal ϵ_k. The relation between ω_J and ϵ_k is exponential and given through

$$\omega_J(\epsilon_k) = \omega_s \exp\left(\epsilon_k/\epsilon_0\right). \tag{6}$$

The constants ω_s and ϵ_0 were determined experimentally through fitting in [4], [7].

With the dependency of the gate matrix U from the voltage ϵ described, the next step is to define the target gate. In classical logic two different universal gates exist, the NAND and the NOR gate. Such an easily implementable universal gate does not exist in quantum computing. However, there are different definitions of low complexity universal gate sets. Such a set for example includes different single qubit gates and one two qubit gate [8]. As it is not clear yet what set will be used in a quantum computer implementation, in a first step basic functionality for one qubit will be studied in this work.

All qubit gate sets will need the ability to rotate the state vector $\vec{\Psi}$ around any of the Cartesian axes. Thus, the target gates U_{target} here are rotations around the x,y and z-axis by an angle of $\phi = \pi/2$. The corresponding U matrices are given in Eq. 7,8.

$$U_{x2} = \frac{1}{\sqrt{2}}\begin{pmatrix} 1 & -i \\ -i & 1 \end{pmatrix} \quad U_{y2} = \frac{1}{\sqrt{2}}\begin{pmatrix} 1 & -1 \\ 1 & 1 \end{pmatrix} \tag{7}$$

$$U_{z2} = \frac{1}{\sqrt{2}}\begin{pmatrix} 1-i & 0 \\ 0 & 1+i \end{pmatrix} \tag{8}$$

The implementation accuracy of U is limited in reality and a cause for errors in the quantum computation. The better the gate the fewer errors have to be corrected, which is highly desirable. The qubit gate quality benchmark is given through the so-called fidelity F. The fidelity compares the real U with the target U_{target} and is calculated with (from [4], derived with [9]):

$$F = \frac{1}{4}|\text{Tr}[U_{target}^T U]|^2 \quad \text{with } F \in [0, 1]. \tag{9}$$

In practice, mostly the infidelity $Inf = 1 - F$ is stated, which is ideally 0. In general, a maximum infidelity for qubit gates in a quantum computer is determined by the capabilities of the employed quantum error correction (QEC). For the popular surface code for example, an estimated maximum infidelity of 10^{-2} is acceptable [10]. Best achieved fidelity values for single and two qubit gates for GaAs qubits are $5 \cdot 10^{-3}$ [4] for single gates and 10^{-1} [11] for two qubit gates with the potential to go into the range of 10^{-3} [12]. These values were reached with sophisticated experimental setups, but for the implementation of quantum computer it is crucial to achieve high fidelity gates with more efficient circuits.

III. DAC MODEL AND PARASITICS

The DAC that generates the qubit control signal ϵ in this work is a Capacitive Divider DAC (Cap DAC), as shown in Fig. 4. The typical operational amplifier at the output is not necessary as the qubit input impedance is very high ohmic. This specific DAC topology has been chosen because it showed a superior power and area efficiency for qubit control applications in earlier works [13] and can be made scalable to large numbers of qubit.

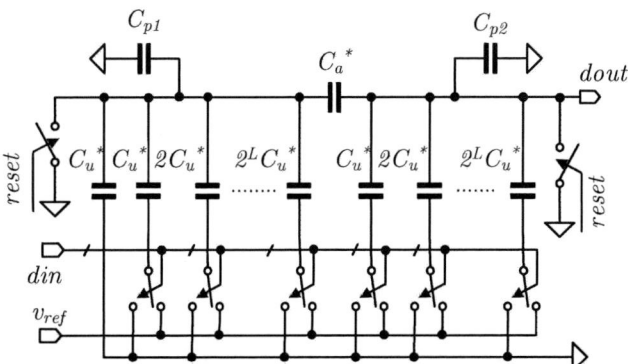

Fig. 4. Capacitive DAC with process variations C^* and parasitic circuit elements C_{p1} and C_{p2}.

The DAC is built with unit capacitances C_u and an attenuation capacitance C_a. These capacitances are subject to variations in comparison to their nominal value. This is indicated in Fig. 4 by a *. The reason for the variation is a non-perfect manufacturing process. The resulting capacitance values are described through a technology-dependent normal distribution. For this model a standard deviation of $\Delta\sigma_C^* = 3\%$ is used which is taken from a 65 nm CMOS technology manual. This translates to a deviation of $\Delta C_{3\sigma} = 9\%$ for the so-called process corner (Tab. I).

TABLE I
PARAMETERS

σ_J	$v_{n,RMS}$	C_p	$C*$	f_{BW}
300 ps	10 μV	50 fF	9 %	600 MHz

Another unavoidable impairment in the DAC implementation is a parasitic capacitive coupling to ground, which is included with C_{p1} and C_{p2} (Fig. 4). The value for both capacitors is set to $C_{p1} = C_{p2} = 50\,\text{fF}$ (Tab. I). This is a somewhat optimistic value in the range of a possible C_u value [13].

The impact of the process variations and the parasitic capacitors can be seen in the transfer characteristic of the DAC. It is depicted in Fig. 5 ('Cap') for a resolution $res = 11\,\text{bit}$ and a reference voltage of $v_{ref} = 8\,\text{mV}$. These specifications were loosely derived from experimental setups. In case of only process variations ('C^*') the nonlinearity of the DAC is increased, while the parasitic capacitors introduce a gain error ('C_p'). The curve ('all') includes both.

Even without any addition of parasitics or process variations, the transfer characteristic of a Cap DAC is not an ideal line. This is due to its architecture and the fact that an odd resolution is chosen.

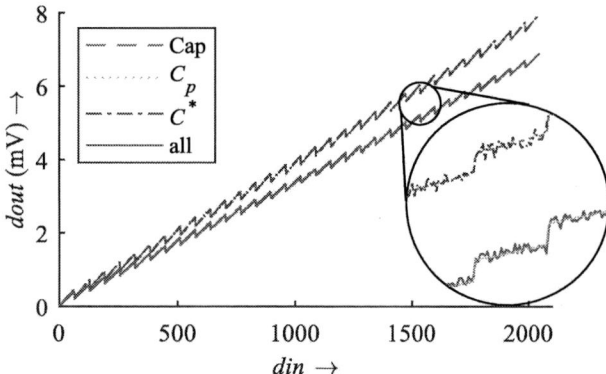

Fig. 5.   DAC transfer characteristic with and without impairments.

Other interferences for a DAC operation are jitter, noise on the reference voltage $v_{n,ref}$ and low-pass behavior. They are included in the electronics behavioral model used in this work as depicted in Fig. 6.

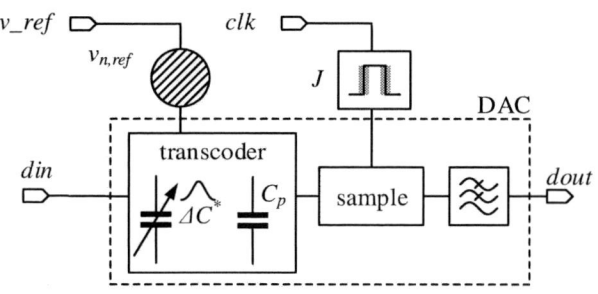

Fig. 6.   Behavioral model of the Cap DAC. At the output the qubit is connected, the input is a 11 bit binary signal.

Jitter is a typical issue in clocked systems and as such included here with random jitter J. For this model, it is assumed that the input is driven through the same clock signal as the DAC itself and thus no wrong input codes can occur. The jitter is here modeled with a normal distribution for the time difference at which the clock edge happens. Here the standard deviation of the distribution is set to $\sigma_J = 300\,\text{ps}$ which is approximately in the range of laboratory equipment.

Unavoidable in any system with a reference source is noise on the reference. In this model white gaussian voltage noise $v_{n,ref}$ is included. The amount of noise is set to $V_{n,RMS} = 10\,\mu\text{V}$, which should be acceptable from experimental experience. The bandwidth f_{BW} of the low-pass included in the model is set to $f_{BW} = 600\,\text{MHz}$ in order to minimize noise and not overly distort the signal.

IV.   Gate optimization and simulation

The pulse shapes that are used to control a GaAs qubit are based on sequences of rectangular samples of varying amplitude. In this work, one pulse has 16 samples with a sample frequency of 300 MHz.

The goal is to find a set of amplitudes for each target gate that implements the wanted gate operation best. That means minimizing the infidelity under certain boundary conditions, which is a classical optimization problem. In this work, the qopt package [14] was used to do the pulse optimization.

Next to the pulse length and the sample frequency, one boundary condition concerns the signal range. The voltage should conform to $\epsilon \in [-1.4, 1.4]\,\text{mV}$ due to the limited range of the fit used for Eq. 6. On top of that a quantization of 11 bit on a total voltage range of $v_{range} = 8\,\text{mV}$ is included in the optimization.

The qopt package also makes it possible to include certain transfer functions in the optimization. In this case, the DAC low-pass behavior with a bandwidth of 600 MHz is included. In order to include this, an oversampling factor is defined such that a pulse is made out of $16 * oversampling = 160$ samples. A section of an optimized pulse with these settings is shown in Fig. 7.

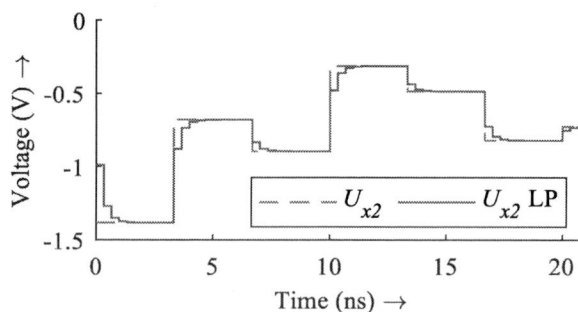

Fig. 7.   Section of optimized pulse for $Ux2$ and corresponding pulse with low-pass behavior.

The results of the optimization for all selected gates are shown in Fig. 8. Depicted are the signals including the transfer function. The resulting fidelities for these signals applied to the qubits directly are $Inf_{U_{x2}} = 2.3 \cdot 10^{-4}$, $Inf_{U_{y2}} = 3.6 \cdot 10^{-4}$, and $Inf_{U_{z2}} = 1.8 \cdot 10^{-3}$ (first row in Tab. II). These fidelities are not going to be reached in practice because the input signal in reality is not piecewise constant, but they are a good starting point. The infidelity results here are comfortably below the threshold associated with the surface code.

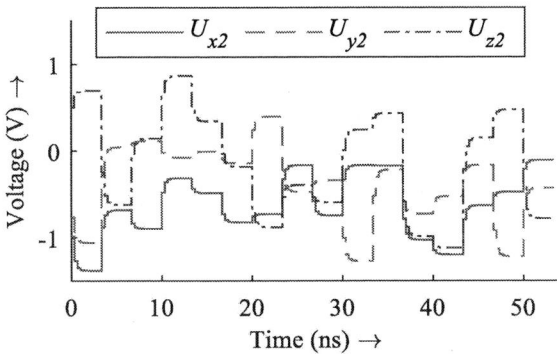

Fig. 8. Optimization result signals including low-pass behavior.

When the optimized 16 sample long digital signals are applied at the *din* input of the behavioral model (Fig. 6), the resulting infidelity increases as expected (second row in Tab. II). The increase is especially large for U_{y2} and U_{z2}, and they are then above the QEC threshold. However, the increase could be mitigated if the non-ideal transfer function of the Cap DAC architecture is taken into account. This happens through a pre-distortion of the digital signal such that the analog output voltage matches the wanted one as well as possible (third row in Tab. II). Unexpectedly, the infidelity decrease does not apply for U_{z2}. The reason for that could be that some distorting effects through the non-piecewise constant signal get compensated with the non-ideal transfer characteristic. More detailed studies are necessary, for example with varying oversampling factors.

The rows four to eight in Tab. II show the simulated infidelity results for different impairments alone and a combination of all of them. Overall, the inclusion of the parasitic capacitances produces the biggest infidelity increase compared to jitter, process variations and reference noise. This is due to the significant gain error the parasitics introduce in the DAC and the resulting signal distortion. A strategy to alleviate the distortion impact can be to include the parasitic capacitances in the pre-distortion.

V. CONCLUSION

On one hand, the results show that non-ideal effects can significantly reduce the qubit gate quality to the point that is above the values set by QEC. On the other hand, it is also clear that a systematic study of the electronics-qubit system and co-optimization of the signals is useful and can improve the gate quality. Especially the parasitic capacitors were identified as a source of infidelity, which will now be included in future optimization processes. Concerning the different gates, varying sensitivities are observed with a high sensitivity to non-ideal effects displayed by the U_{z2} gate. This is in accordance with other work [15] and is subject to further improvements.

TABLE II
INFIDELITIES OF DIFFERENT CONFIGURATIONS

Infidelity	Gate		
Config	U_{x2}	U_{y2}	U_{z2}
direct	$2.3 \cdot 10^{-4}$	$3.6 \cdot 10^{-4}$	$1.8 \cdot 10^{-3}$
no predist.	$6.4 \cdot 10^{-3}$	$1.0 \cdot 10^{-1}$	$2.0 \cdot 10^{-1}$
no imp.	$4.3 \cdot 10^{-3}$	$4.1 \cdot 10^{-2}$	$4.8 \cdot 10^{-1}$
J	$2.5 \cdot 10^{-2}$	$8.5 \cdot 10^{-2}$	$9.1 \cdot 10^{-1}$
Cp	$2.8 \cdot 10^{-1}$	$5.2 \cdot 10^{-1}$	$4.3 \cdot 10^{-1}$
C^*	$1.1 \cdot 10^{-2}$	$4.3 \cdot 10^{-1}$	$9.8 \cdot 10^{-1}$
v_n	$4.1 \cdot 10^{-3}$	$3.3 \cdot 10^{-2}$	$8.4 \cdot 10^{-1}$
all	$2.9 \cdot 10^{-1}$	$5.1 \cdot 10^{-1}$	$3.6 \cdot 10^{-1}$

ACKNOWLEDGMENT

L. Geck thanks Julian D. Teske for helpful support in using the qopt package.

REFERENCES

[1] B. Patra et al., "19.1 a scalable cryo-CMOS 2-to-20GHz digitally intensive controller for 4×32 frequency multiplexed spin qubits/transmons in 22nm FinFET technology for quantum computers," in *2020 IEEE International Solid- State Circuits Conference - (ISSCC)*, Feb. 2020.

[2] L. L. Guevel et al., "19.2 a 110mK 295μW 28nm FDSOI CMOS quantum integrated circuit with a 2.8GHz excitation and nA current sensing of an on-chip double quantum dot," in *2020 IEEE International Solid- State Circuits Conference - (ISSCC)*, Feb. 2020.

[3] J. van Dijk, A. Vladimirescu, M. Babaie, E. Charbon, and F. Sebastiano, "SPINE (SPIN emulator) - a quantum-electronics interface simulator," in *2019 IEEE 8th International Workshop on Advances in Sensors and Interfaces (IWASI)*, Jun. 2019.

[4] P. Cerfontaine, T. Botzem, D. P. DiVincenzo, and H. Bluhm, "High-fidelity single-qubit gates for two-electron spin qubits in GaAss," *Physical review letters*, vol. 113, no. 15, p. 150501, Jul. 2014.

[5] M. Shulman, "Entanglement and metrology with singlet-triplet qubits˝, Ph.D. dissertation, Grad. School Arts & Sciences, Harvard University, Boston, USA, 2015.

[6] P. Cerfontaine et al., "Closed-loop control of a GaAs-based singlet-triplet spin qubit," *Nature Communications*, vol. 11, no. 1, Aug. 2020.

[7] O. E. Dial et al., "Charge noise spectroscopy using coherent exchange oscillations in a singlet-triplet qubit," *Phys. Rev. Lett.*, vol. 110, no. 14, Apr. 2013.

[8] M. Nielsen and I. Chuang, "Quantum computation and quantum information˝, 8. print., Cambridge [u.a.], Cambridge Univ. Press, 2005, pp.191.

[9] M. A. Nielsen, "A simple formula for the average gate fidelity of a quantum dynamical operation," *Physics Letters A*, vol. 303, no. 4, pp. 249–252, Oct. 2002.

[10] A. G. Fowler, A. M. Stephens, and P. Groszkowski, "High-threshold universal quantum computation on the surface code," *Physical Review A*, vol. 80, no. 5, Nov. 2009.

[11] J. Nichol et al., "High-fidelity entangling gate for double-quantum-dot spin qubits˝, *npj Quantum Information*, vol. 3, 2017

[12] P. Cerfontaine, R. Otten, M. Wolfe, P. Bethke and H. Bluhm, "A high-fidelity gateset for exchange-coupled singlet-triplet qubits˝, *Phys. Rev. B*, vol. 101, no. 15, Apr. 2020.

[13] L. Geck, A. Kruth, H. Bluhm, S. van Waasen, and S. Heinen, "Control electronics for semiconductor spin qubits," *Quantum Science and Technology*, vol. 5, no. 1, p. 015004, Dec. 2019.

[14] J. D. Teske, P. Cerfontaine, and H. Bluhm, "qopt: An experiment-oriented qubit simulation and quantum optimal control package." [Online]. Available: https://github.com/qutech/qopt

[15] L. Geck, J. P. van Dijk, F. Sebastiano, S. van Waasen, and S. Heinen, "Influence of electrical noise on GaAs qubit gate quality", unpublished

Modeling Power Supply Noise in RF SoCs

Jonas Meier, Florian Menke, Lantao Wang,
Tim Lauber, Ralf Wunderlich and Stefan Heinen
Chair of Integrated Analog Circuits and RF Systems
RWTH Aachen University
Kopernikusstrasse 16, D-52074 Aachen
Email: mailbox@ias.rwth-aachen.de

Abstract—With rising integration densities and design complexities, the verification effort of modern Systems-on-Chip is rising even faster. Additionally, with the shift towards digital-centric circuits in smaller process nodes, supply noise has become a critical factor as it can degrade the performance of the remaining analog and RF components drastically. This paper provides an overview on recent advances in modeling power supply noise on block level. Considerations to transfer these approaches for simulating noise on system level are highlighted. An All-Digital PLL SoC is used as an example to showcase the insights that can be derived even when using simple models. The position and causes for spurs arising due to coupling on the power supply lines are investigated, giving designers a guideline on where to focus their optimization efforts.

Index Terms—power supply noise, system modeling, mixed-signal, RF SoCs, SystemVerilog, ADPLL, LDO

I. Introduction

The demand for low-power and low cost communication systems is driving the development of integrated circuits to ever rising integration densities, growing design complexities and shrinking feature sizes. At the same time, the requirements of modern communication standards are also growing, making the reliability of a modern System-on-Chip (SoC) more critical than ever. By integrating analog, digital and RF components, the verification effort to ensure a 'first time right' design grows even faster. Additionally, with the shift towards digital-centric circuits in smaller process nodes, supply noise has become a critical factor as it can degrade the performance of the remaining analog and RF components drastically. The analytical determination of the effects of power supply noise on large systems is very complex and spice-like simulations are often not feasible to determine the effects of power supply noise (PSN) on chip level. Event-driven simulations have become important for the verification of large systems but need manually build models for the analog components, which now also need to consider supply modulation.

II. Causes and Effects of PSN in RF Circuits

Various sources of PSN exist in intergated circuits. E.g. electromagnetic interference from external sources, ground bounces on-chip or from the PCB can cause fluctuations of the supply voltage. The most common source in mixed-signal environments is simultaneous switching noise (SSN). In the digital logic cells, the current consumption depends on the switching activity, transition time and output load. Charging and discharging the gates of successive logic cells leads to an IR-drop on the supply voltage. Each activity triggers further switching in successive cells. Since the digital signal processing is in general clocked by a reference clock, the current drawn by the circuit is periodic, causing a regular disturbance of the supply voltage that can be described as PSN. Due to the close proximity with the other subcircuits, the noise may also effect analog and RF blocks on the design through direct cross-coupling. Additionally, a similar affect might be caused by the analog circuits themselves. Due to periodically changing input signals, the power consumption of e.g. divider, power amplifiers etc. is not constant but varies with the periodicity of the input signals.

A common criteria to measure a RF circuit's performance is the necessary signal to noise ratio (SNR) to function correctly.

$$SNR = \frac{P_{Signal}}{P_{Noise}} \quad (1)$$

Power supply noise coupling through supply lines or parasitics in the signal path needs to be considered as an additional contribution to the circuits noise power P_{Noise} and degrades the SNR.

Another important performance criteria, especially in digital-centric and clocking circuits, is an accurate timing. A varying supply voltage directly affects the speed of a circuit. The propagation delay is affected by a variation of charging or discharging of e.g. successive gate capacitors. One metric to measure the timing variation is the time interval error (TIE). It is the deviation of a signal's crossing point from its ideal value. Especially for periodic signals, jitter is a common parameter to define a circuits performance. It is given by the peak TIE over time as shown in Fig. 1.

According to [2], there are different jitter metrics, among which two metrics of interest are presented here. First, there is the absolute jitter, which is used for oscillators, where it describes the deviation $\Delta\tau_n$ from the ideal signals crossing point τ_n and is proportional to the oscillators phase error ϕ, as shown in equations 2 and 3.

$$\tau_n = nT_0 + \Delta\tau_n \quad (2)$$

$$\Delta\tau_n = -\frac{\phi(\tau_n)}{\omega_0} \quad (3)$$

© VDE VERLAG GMBH · Berlin · Offenbach

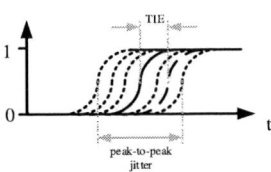

Fig. 1: Graphical explanation of TIE and jitter. [1]

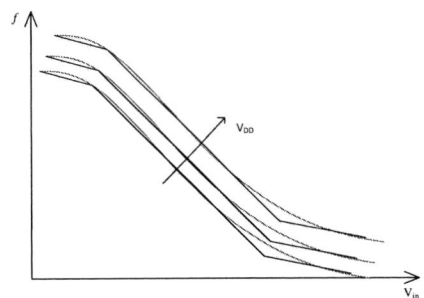

Fig. 3: Fitted frequency voltage relation for linear interpolation [4].

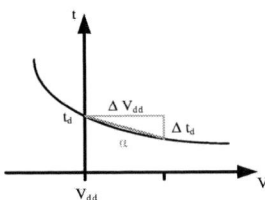

Fig. 2: Delay calculation of alpha factor.

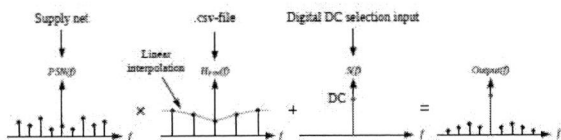

Fig. 4: Implementation of transfer function for supply noise in [7].

A second metric is the period jitter, which is the variation from a nominal clock signals period, i.e. $\tau_n - \tau_{n-1}$, in terms of clocked systems.

III. MODELING APPROACHES ON BLOCK LEVEL

Modeling approaches considering the effects of PSN on block level can be categorized in two groups. One solution models the signal transfer function in a parameterized way. The influence of a varying supply voltage then affects these parameters. A second group of approaches treats the supply as additional input variable of the block's transfer function.

A. Parameter Variation

Modeling power supply induced jitter by varying the propagation delay according to the changes of the supply voltage is a common approach for inverter-like structures, such as delay chains and oscillators. Assuming the voltage variation is small, using a linear relation is often sufficient. The α-model does exactly this. The factor α describes the linear relation between supply and delay variation:

$$\alpha = \frac{\Delta t_d / t_d}{\Delta V_{dd}/V_{dd}} \quad (4)$$

To extract α, a DC voltage offset is applied, and the delay variation measured as shown in Fig. 2:

$$\Delta t_d = \alpha \frac{\Delta V_{dd}}{V_{dd}} t_d \quad (5)$$

In [3] this technique is used to model the inverter delay variations to analyze the jitter in a ring-oscillator.

A similar approach is described in [4], where the influence of supply variation on the oscillator's frequency instead of internal delays is considered. Since the relation is not linear over the whole operating range, a piece-wise linear approach is chosen instead. This is done for different DC offsets on the supply voltage. The actual influence of the supply voltage on the frequency is then obtained, by linear interpolating between those characterized curves, as shown in Fig. 3. An alternative approach that does not assume a quasi-static supply voltage is the use of sensitivity functions as in [5], [6]. The sensitivity of the modeled parameter is characterized by applying an noise impulse at different times relative to the input event. By convoluting this transfer function with the arbitrary noise source used during simulation, the supply noise effects are represented.

B. Transfer Functions

A second approach to handle the influence of a variable supply voltage on a circuit is to treat the supply as a regular input and model its transfer function. An example where this parameter is of special importance is e.g. in linear drop-out regulators where the power supply rejection ratio is used to describe this behavior. The same approach can also be applied to other analog circuits. In [7] the supply transfer function of a complex filter is considered to analyze the effects of PSN on a $\Delta\Sigma$-ADC. The supply noise is convoluted with the supply transfer function and superpositioned with the regular signal path as shown in Fig. 4. Here a linear system relation was assumed, but the idea can also be extended to cover nonlinear circuit behavior. Cascading the LTI transfer functions with nonlinear transfer functions allows to describe

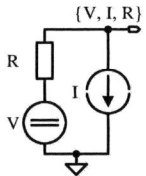 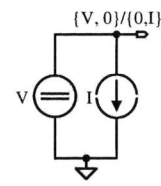

(a) The structure of EEnet drivers.

(b) The structure of IVnet drivers.

Fig. 5: Driver of supply net types.

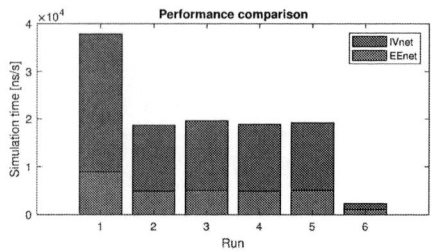

Fig. 6: Simulation speed of the different nettypes compared.

a wide range of nonlinear system behavior. These Wiener-Hammerstein models are well suited to describe analog circuit behavior.

An alternative to describe nonlinear circuits is using Volterra series, which can be thought of as Taylor series with memory [8]. In [9] it is modified in a way that accounts for dynamic supply deviations and used to model RF power amplifiers. A drawback of this approach is that a large number of parameters are needed to describe the system.

In addition to the the block oriented LTI or Wiener-Hammerstein models and the Volterra series, there are several black-box approaches. One example is the useage of neural networks [10]–[12]. The main disadvantage of these methods is that they only rely on the measured data and do not allow any insight into the behavior. Therefore, special care has to be taken to ensure the training data covers all possible usecases.

IV. CONSIDERATIONS FOR SYSTEM LEVEL MODELING

Since all blocks will be connected to a supply net, the supply will be one of the most active nets in the circuit. Any change in a block, where the power consumption is considered will cause a re-evaluation of the supply net and possibly trigger new events in other parts of the design. Thus, special care has to be taken to consider if subcircuits need to be immediately sensitive to a variation of the supply voltage or if an evaluation at the next data signal change is sufficient.

Furthermore, a simple, less computationally heavy model for the supply speeds up the simulation. A lumped model for the supply net will result in faster simulation at the possible cost of some accuracy. Also, a tree-shaped hierarchical net is to be preferred to a meshed modeling approach, which, due to its feedback structure, will need several iterations to settle for a final value.

A suitable net representation is the electrical equivalent net (EEnet) provided by Cadence. It models a voltage source, series resistance and parallel current source for each port as seen in Fig. 5a. The resolution function works according to Kirchhoff's law.

This is however not an optimal solution. In general the supply only has one voltage driving circuit attached to it, e.g. a LDO. All other blocks act as loads. Thus a simplified nettype, called IVnet, has been implemented. It provides

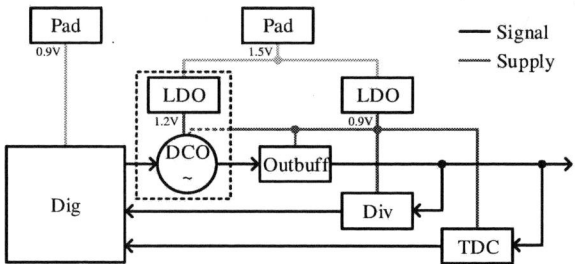

Fig. 7: The ADPLL from a supply perspective.

the blocks with the one voltage, driven by the LDO. All current values from sub-circuits attached to the net are simply summed up. The total current is passed to the driving circuit for evaluation. Using this simplified resolution function can increase simulation speed by approximately a factor of 4 as can be seen in Fig. 6.

V. SAMPLE APPLICATION: ADPLL SoC

As an example, an ADPLL SoC is modeled and the effects of supply variation are investigated. The PLL is a fractional PLL featuring a class-B DCO, tunable from 9 GHz to 11 GHz, with a $\Sigma\Delta$-modulator to increase the resolution to 15 kHz steps. A 2D Vernier-line based TDC is used for fractional phase measurements.

In Fig. 7 the major circuit blocks are shown with focus on the supply domains. Aside from the digital supply, which is directly provided from external, two analog supply domains are present. One supplies the DCO core with 1.2 V, the other provides the TDC, buffers and dividers with a 0.9 V supply. Both supply domains are fed by a low noise LDO as presented in [13].

A. LDO Model

The model recreates the toplevel hierarchy of the LDO by replacing each of the sub-blocks, which are shown in Fig. 8, with a macromodel. These are implemented in a simplified way. The transistor is implemented like the standard model in literature, with triode-, saturation- and off-region. The error amplifier is modeled with its gain (G), as well as maximum and minimum current saturation (I_{max}, I_{min}) (see Fig. 9). For

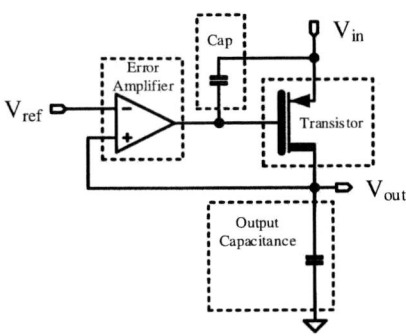

Fig. 8: The LDO modeling schematic.

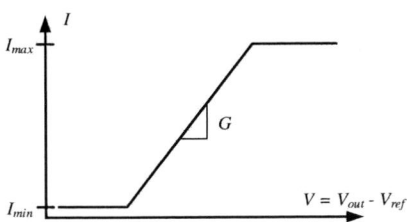

Fig. 9: The error amplifier's voltage current relation.

the capacitors, an efficient event driven model was developed, which integrates the current into the device to a voltage with $V_{cap} = \frac{1}{C} \int I_{cap} dt$. The algorithm integrates the current in discrete steps (6), where t_n represents the time at the n-th step. The time until the next step (7) is determined either by the next event or by a fixed time step, which is derived by the maximum voltage change tolerance V_{tol} with (8).

$$V_{cap,n} = \frac{1}{C} I_{cap,n-1} \Delta t + V_{cap,n-1} \quad (6)$$

$$\Delta t = t_n - t_{n-1} \quad (7)$$

$$t_{step} = \frac{V_{tol} C}{I_n} \quad (8)$$

With this implementation, the time steps are dynamically adopted, depending on the input signal and the number of events is kept small. On the other hand, the model reacts delayed on changes, because $I_{cap,n-1}$ is used.

B. Delay Modeling and Current Consumption

Apart from their functionality, the effects of PSN on buffer, divider and TDC are similar. Thus the modeling approaches for these blocks share the same principles. The functional behavior is separated from the propagation delay and power consumption. The aggregated delay and current is then modeled in a consecutive buffer block after the functional description.

To keep the computational effort on the supply calculations low, simple models are chosen to approximate the circuits behavior. To model the delays, the α-model presented in sec. III-A is implemented.

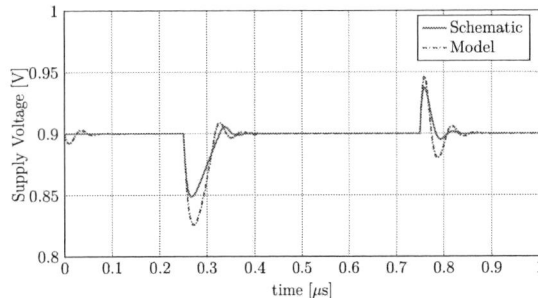

Fig. 10: Comparison of LDO's step response between model and schematic.

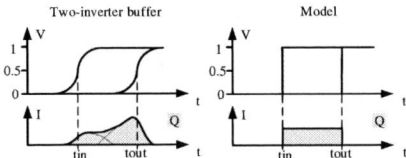

Fig. 11: Buffer current and voltage waveform.

A constant current model is used to estimate the power consumption of the subcircuits. As seen in Fig. 11, it assumes a constant current during the whole time a block is active. The current is chosen such that the total charge is equal to simulations from schematic. Since the circuits are abstracted on a low hierarchical level, even this simple power estimation gives quite good results. As an example, the current waveforms from schematic and model of the output buffer and divider are shown in Fig. 12. The current peaks are captured well by the model, while there are small deviations for low current settings. On average the total current consumption diverges by a mean value of $2\,\text{mA}$, which translates to a static deviation of $0.7\,\text{mV}$ in the supply voltage.

The output phase noise spectrum for an ideal $10\,\text{GHz}$ input source is shown in Fig. 13. As can be seen by comparing the blue line for schematic simulation with ideal supply with the yellow curve from the model, some self-mixing products

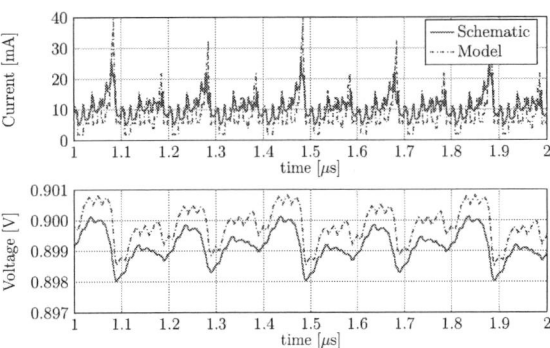

Fig. 12: Transient plot of current and voltage on the output buffer's supply.

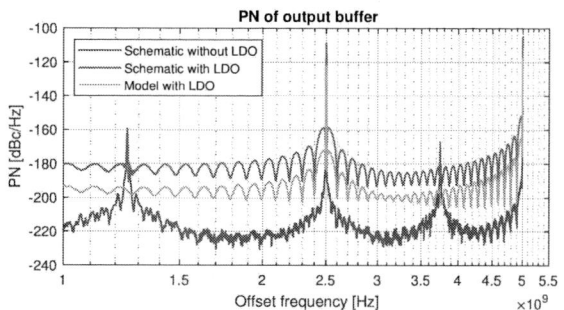

Fig. 13: Phase noise of the output buffer compared.

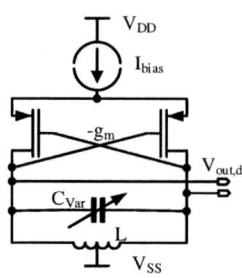

Fig. 14: DCO core.

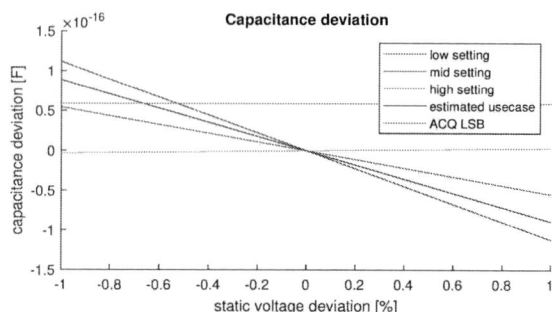

Fig. 15: Capacitor bank deviation for static supply modulation.

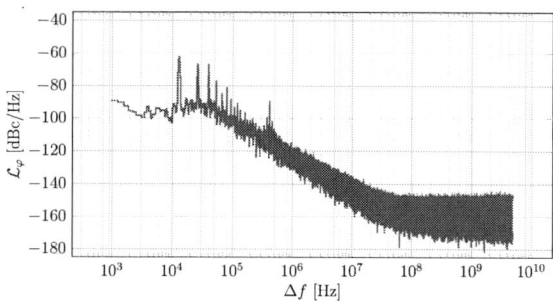

Fig. 16: Output phase noise without considering variation of supply voltage.

at 1.25 GHz and 3.75 GHz are not captured. Nevertheless, the spurs at 2.5 GHz and 5 GHz induced by variation of the supply are well predicted.

C. DCO Model

Figure 14 shows a schematic of the DCO's implementation. The tunable capacitance C_{var} consists of multiple switchable capacitor cells. They are controlled from the 0.9 V domain, e.g. the transistor gates are connected to the analog 0.9 V VDD. Thus, the capacitance value and consequently the DCO's frequency varies with fluctuations on the VDD. Since the effect differs for on and off cells, it is modeled for each cell separately. The capacitance variation for different tuning settings are shown in Fig. 15. The influence of the DCO core from the 1.2 V domain on itself is small. Thus, additionally a static noise model considering the white noise floor and wander noise for the DCO is implemented. The noise levels are extracted from schematic simulations at -155 dBc/Hz for the noise floor and -120 dBc/Hz at 1 MHz offset frequency for the wander noise.

D. Simulation results

The phase noise spectrum of the ADPLL output without considering the influence of supply noise is shown in Fig. 16. The spectrum is simulated for a close-to-integer setting with the frequency control word being one fractional LSB next to

an integer value. The resulting fractional spurs at ~15 kHz offset and its multiples are clearly seen.

Looking at the results in Fig. 17, it can be seen that the divider hardly affects the other circuits. Only the additional spurs at 2.5 GHz and 5 GHz already explained in section V-B are added.

However, considering the current drawn by the TDC in the simulation, changes the output spectrum. Fig. 18 shows the output spectrum of the ADPLL considering currents from the TDC and the divider. Several effects can be identified. First, the fractional spurs are larger, indicating additional non-linearity in the TDC induced by its own current consumption as seen in the green curve. Furthermore, the phase noise

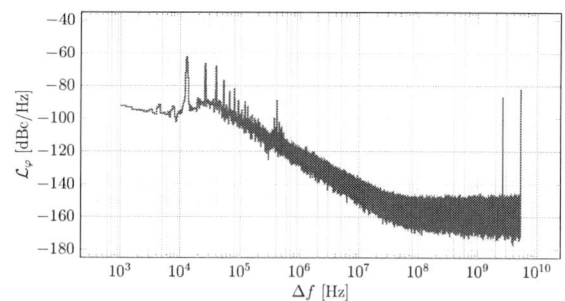

Fig. 17: Output phase noise considering variation of supply voltage caused by the divider.

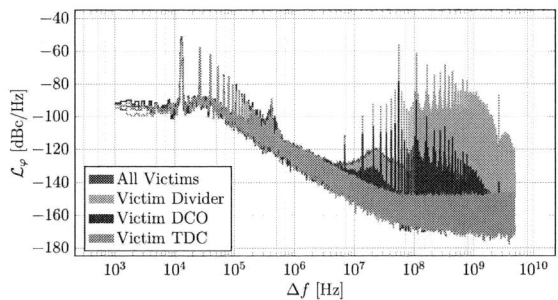

Fig. 18: Output phase noise considering TDC and divider as current loads.

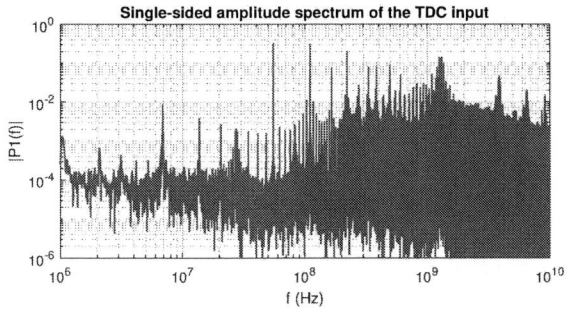

Fig. 19: Spectrum of the TDC input signal.

from the divider (orange) is elevated by the additional current drawn by the TDC. The effect can be clearly assigned to the divider by comparing the orange, green and purple lines. Last, additional spurs at 6.8 MHz and its multiples arise. These can be attributed to the supply modulating the DCO's frequency and the divider, as seen in the purple and orange curves. The TDC's activity is periodic, as once per clock cycle the TDC is calibrated and used to measure the DCO phase. Inspecting the input signal of the TDC, shown in Fig. 19, it can be seen that the Spectrum also contains spurs at 6.8 MHz and its multiples. These translate to the supply and can thus cross-couple to the DCO and the divider.

VI. Conclusion

This paper provides an overview on modeling strategies considering power supply noise in RF SoCs. Recent modeling approaches for block level simulation are reviewed. They can be classified as either providing a variation of internal model parameters or treating the supply as additional input to the blocks transfer function. Considering the application to system-level simulations, the importance of simple, less computational heavy models is highlighted, as a simplification in the net representation can already increase the simulation speed by a factor of 4. Using simple models, only considering power consumption and an α-model for delay variations, a sample SoC consisting of an ADPLL and its LDOs is investigated. These models are already sufficient to

qualitatively highlight major cross-coupling paths in the SoC, giving designers a guideline on which circuits need careful isolation.

References

[1] J. N. Tripathi, V. K. Sharma, and H. Shrimali, "A review on power supply induced jitter," *IEEE Transactions on Components, Packaging and Manufacturing Technology*, vol. 9, no. 3, pp. 511–524, 2019.

[2] I. Galton and C. Weltin-Wu, "Understanding phase error and jitter: Definitions, implications, simulations, and measurement," *IEEE Transactions on Circuits and Systems I: Regular Papers*, vol. 66, no. 1, pp. 1–19, 2019.

[3] X. Wang and A. Martin, "On-die supply-inducecd jitter behavioral modeling," in *IEEE 22nd Conference on Electrical Performance of Electronic Packaging and Systems*, 2013, pp. 147–150.

[4] J. Meier, C. Beyerstedt, F. Speicher, F. Menke, R. Wunderlich, and S. Heinen, "Event - driven modeling of cross-coupling in phase-locked loops through supply paths," in *Kleinheubach Conference*, 2019, pp. 1–4.

[5] M. van Ierssel, H. Yamaguchi, A. Sheikholeslami, H. Tamura, and W. W. Walker, "Event-driven modeling of cdr jitter induced by power-supply noise, finite decision-circuit bandwidth, and channel isi," *IEEE Transactions on Circuits and Systems I: Regular Papers*, vol. 55, no. 5, pp. 1306–1315, 2008.

[6] J. Kim, Y.-C. Lu, and R. Dutton, "Modeling and simulation of jitter in phase-locked loops due to substrate noise," in *Proceedings of the 2005 IEEE International Behavioral Modeling and Simulation Workshop*, 2005, pp. 25–30.

[7] J. Meier, F. Speicher, C. Beyerstedt, T. Saalfeld, G. Boronowsky, R. Wunderlich, and S. Heinen, "Modeling power supply noise effects for system-level simulation of $\Delta\Sigma$ - ADCs," in *16th International Conference on Synthesis, Modeling, Analysis and Simulation Methods and Applications to Circuit Design (SMACD)*, 2019, pp. 265–268.

[8] J. C. Pedro and N. B. Carvalho, *Intermodulation distortion in microwave and wireless circuits*. Artech House, 2003.

[9] G. P. Gibiino, G. Avolio, D. M. M.-P. Schreurs, A. Santarelli, and F. Filicori, "A three-port nonlinear dynamic behavioral model for supply-modulated rf pas," *IEEE Transactions on Microwave Theory and Techniques*, vol. 64, no. 1, pp. 133–147, 2016.

[10] J. Xu, M. Yagoub, R. Ding, and Q.-J. Zhang, "Neural-based dynamic modeling of nonlinear microwave circuits," *IEEE Transactions on Microwave Theory and Techniques*, vol. 50, no. 12, pp. 2769–2780, 2002.

[11] H. Yu, H. Chalamalasetty, and M. Swaminathan, "Modeling of voltage-controlled oscillators including i/o behavior using augmented neural networks," *IEEE Access*, vol. 7, pp. 38 973–38 982, 2019.

[12] H. Yu, J. Shin, T. Michalka, M. Larbi, and M. Swaminathan, "Behavioral modeling of pre-emphasis drivers including power supply noise using neural networks," in *IEEE 10th Latin American Symposium on Circuits Systems (LASCAS)*, 2019, pp. 37–40.

[13] L. Wang, M. Fassbender, M. Scholl, J. Meier, R. Wunderlich, and S. Heinen, "A low-noise low-dropout regulator using a 28-nm technology," in *27th IEEE International Conference on Electronics, Circuits and Systems (ICECS)*, 2020, pp. 1–4.

SMACD / PRIME 2021 | 19 – 22 July 2021, Online Event

High-Performance Flexible and Printed Electronics Based on Inorganic Semiconducting Structures

Abhishek Singh Dahiya, Dhayalan Shakthivel, Yogeenth Kumaresan and Ravinder Dahiya*

Bendable Electronics and Sensing Technologies (BEST) Group, University of Glasgow,
Glasgow G12 8QQ, U.K
*Correspondence to: Ravinder.Dahiya@glasgow.ac.uk

Abstract— The transfer and contact printing methods have recently gained huge attention to integrate nano to chip (macro) scale inorganic structures over soft flexible substrates. This paper gives an insight into the mechanism of printing and development of high-performance flexible electronic devices from assembly of inorganic semiconducting structures of various dimensions using above-mentioned printing techniques, including nano (nanowires (NWs), nanoribbons (NRs), etc.), micro (microwires), and the chip scale (ultra-thin chips (UTCs)). These simple fabrication strategies have opened new avenues for high-performance printed electronic on large area flexible substrates.

Keywords— *Printed flexible electronics, High-performance, Large area electronics, Contact printing, Transfer printing.*

I. INTRODUCTION

The last two decades have witnessed huge progress in the field of flexible and printed electronics. So far, different fabrication technologies have been reported to develop flexible electronics targeting applications across numerous sectors, including wearable systems, robotics, displays, Internet of Things (IoT) and healthcare [2, 4-12]. Fig. 1 qualitatively compares these technologies in terms of fabrication-cost and device-performance along with their advantages and limitations. Currently, organic printed electronics dominate the flexible electronics' market due to advantages such as lower fabrication cost, solution processability and higher flexibility [15]. However, the market for organic based electronics is considered to lie in the low-cost printed circuits aiming at high-volume market segments such as displays, RFIDs etc. where the high performance of conventional electronics is not required [16]. Combining the high-performance of conventional silicon technology with system-in foil applications, ultra-thin chips (UTCs) could provide an alternative means to implement innovative solutions for high-performance flexible electronics with acceptable level of flexibility [18, 19]. However, applications are limited to small and compact areas due to economic reasons and integration related difficulties and is classified as a subtractive method (more material wastage). Carrying the advantages of both above-mentioned technologies i.e., organic printed electronics and UTCs, transfer, and contact printing of high-mobility inorganic nanostructured materials is an additive manufacturing route which can bring innovations in high performance flexible electronics with less material wastage [13].

Employing contact and transfer printing, nano to chip scale inorganic structures of materials such as silicon, carbon, transition-metal dichalcogenides (TMDCs), and metal oxides have been printed over fully flexible substrates. Further, using these printing methods and materials, variety of high performance flexible electronic devices such as transistors, photodetectors, biosensor, radio frequency identification tags (RFIDs), energy harvesters etc. have been developed. This paper provides a thorough understanding of contact and transfer printing technologies for nano to chip scale inorganic semiconducting structures. High-performance flexible electronic devices and circuits from assembly of multitude of printed inorganic structures of various dimensions are presented.

This paper is organized into five sections: In section II, we describe the synthesis methods for inorganic nanostructures. In the section III contact printing of vertically aligned nanoscale structures over large areas are presented. Section IV describes the transfer printing of nano to chip or macro scale inorganic elements. In section V, potential of these printing techniques in the development of electronic skin (e-Skin) is presented.

II. GROWTH/FABRICATION METHODS OF INORGANIC NANOSTRUCTURES

Inorganic nanostructures are promising candidates and holds huge advantage for applications in flexible electronics [20, 21]. The commendable properties of these structures are strongly influenced by the growth and fabrication techniques

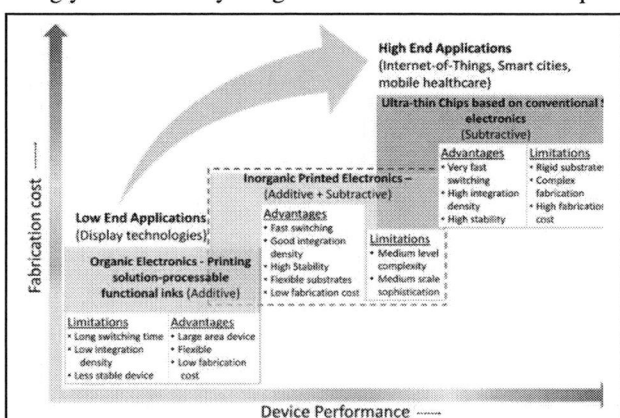

Fig. 1. Different manufacturing routes for the fabrication of large area electronics and the relation between fabrication-cost and device-performance. Inorganic printed electronics could potentially provide the long-term solution for high performance large area electronics. Reproduced with permission from [2].

© VDE VERLAG GMBH · Berlin · Offenbach

employed in the nanomaterials processing. These techniques largely classified into two categories: top-down and bottom-up. The choice of the process and techniques are highly dependent on factors such as material requirement, aspect ratio, doping control, mechanical properties (flexibility, endurance etc.) and targeted applications. Fig. 2 illustrates an overview of the four represented nanostructures preparation techniques which encompasses wide variety of high-quality inorganic materials suitable for high performance flexible electronic applications. These are described briefly in the following section before describing contact printing and transfer printing mechanism and device examples.

A. Top-down synthesis

Top-down fabrication approaches uses bulk single crystalline wafers as starting source material which will be converted into nanostructures using controlled dry plasma or wet chemical etching with aid of lithography patterning [22]. For instance, metal assisted chemical etching (MACE) is a wet etching process (Fig. 2a) for the fabrication of high aspect ratio Si nanowires (NWs) [23]. The process consists of a fabrication of metal nano-mesh (typically Ag or Au) over Si wafers by lithography and selective etching using HF and H_2O_2. The fabrication of metal nano-mesh over the source wafer could be carried out using conventional optical lithography, self-assembled silica nano-spheres (Fig. 1a) and nanoimprint lithography which produces the mesh with the nano-sized holes. MACE holds advantage due to process simplicity, reduced tapering and lateral etching, high aspect ratio NWs, higher etching rate and cost-effectiveness. However, the process is limited to materials which are available in the bulk single crystalline wafer form such as Si, Ge, GaAs etc. Alternative top-down techniques have been developed in the past to fabricate engineered array of nanoribbons (NR) and/or nanomembranes (NMs) of Si and III-V semiconductors. Si NR horizontal arrays are commonly

fabricated using silicon on Insulator (SOI) wafers consists of Si layer (~70 nm) on a sacrificial buried oxide layer (2 µm) over a 600 µm thick bulk wafer (Fig. 2b). The top Si layer is converted into NRs/NMs arrays using lithography followed by dry/wet etching. Further, this technique allows to perform selective doping of the ends of NRs and subsequently, the underneath oxides are etched to create suspended 1D structures anchored at both ends. The suspended NRs are transfer printed (discussed in section IV) over flexible substrates for the device fabrication.

B. Bottom-up synthesis

Bottom-up approaches offers synthesis of wide inorganic NWs and 2D materials with good physical and chemical properties. These methods use solid, liquid, or gaseous sources as initial precursors and build single crystals NWs or 2D monolayers under controlled experimental conditions (temperature, pressure, catalysts etc). Vapour-liquid-solid (VLS) growth process is one of the popular methods to produce semiconducting NWs with wide diameter range of few nms to several microns (Fig. 2c) [24]. The mechanism uses predefined catalyst particles for site specific growth of NWs at temperature in the range of 300-1000°C. For example, Si NWs are commonly grown using Au catalysts by chemical vapour deposition (CVD) with SIH_4 or $SiCl_4$ as precursors. Au nanoparticles in the sub-100 nm diameter range are generally prepared using lithography, self-assembled Au NPs, aerosol and dewetting techniques. The method has been highly successful to produce elemental, binary and ternary NWs such as Si, Ge, CNTs, ZnO, ITO, InGaAs etc in the sub-100 nm diameter range. Additionally, VLS mechanism has been employed to the growth of branched NWs, axial, and core-shell heterostructures, which have potential to realise sensors, optoelectronics, and energy devices. In addition to NWs, bottom-up techniques have shown the advantage of growing high quality 2D materials such as graphene, MoS_2, WS_2, WSe_2, BN etc. Graphene is the first experimentally reported monolayer which subsequently demonstrated to grow over large area using CVD technique (Fig.1d) [25, 26]. It is conveniently grown over Cu foil substrates using CH_4 gaseous source at temperature excess of 1000°C. Large area graphene has been successfully transferred over flexible substrates and has shown interesting applications such as e-Skin [17] (discussed in section V).

III. CONTACT PRINTING

Contact printing has been extensively used to transfer vertically aligned NWs from the donor (growth) substrates over flexible receiver substrate. The printing process mainly carried out in three steps: i) NW bending due to vertical force; ii) alignment of NWs by the applied shear force; and iii) detachment and transfer of NWs to the receiver substrate through surface interactions (Fig. 3a). The in-depth mechanics of the printing system was studied and presented recently [27, 28]. The authors have carried out various printing studies such as influence of applied pressure, sliding velocity etc. were performed to gain a better understanding of the printing dynamics by examining the effects of various

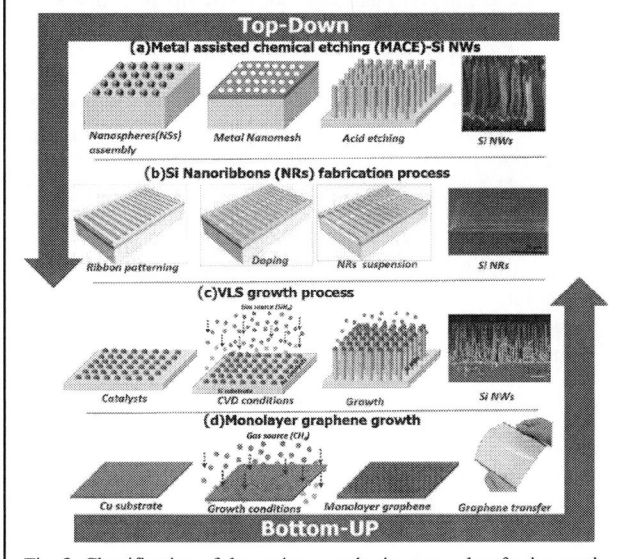

Fig. 2. Classification of the various synthesis approaches for inorganic nanostructures: Top-down and bottom-up methods.

printing parameters [27]. The major advantage of contact printing is that it is a single step dry process with directional alignment of NWs and negligible contamination. The printing process is promising for the fabrication of transistors capable of delivering high ON currents. For instance, ON current of ~6 mA at V_{DS} of 3V was achieved using InAs printed NW based transistor. The device consists of 400 printed NWs in a 3 µm long channel (Fig. 3b-d). Similarly, parallel contact printed ZnO NWs arrays enable fabrication of high performance piezoelectric nanogenerators (PENGs) [29]. The contact printing technique was further employed to form optoelectronic devices such as photodetectors, light-emitting diodes (LEDs) etc. in a mechanically flexible format. For instance, ZnO and Si NWs were contact printed to realized ultraviolet (UV) photodetectors in a Wheatstone bridge (WB) configuration (Fig. 3e-f) [28]. The fabricated devices showed high efficiency, a high photocurrent to dark current ratio (>10^4) and reduced thermal variations because of inherent self-compensation of WB arrangement. The above-mentioned examples are demonstrated in 2D layouts but contact printing could also offer 3D assembly of NWs [30]. Towards this, ten functional device layers of Ge/Si NWs, stacked to form a 3D electronic structure are noteworthy [30]. The fabricated devices showed a high average on-current (4 mA) and minimal variation in the threshold voltage.

IV. TRANFER PRINTING

The transfer printing technique enable the transfer of laterally aligned structures of different length scales (nano to chip scale) from a donor substrate to a receiver substrate generally using soft elastomeric stamp. A generalised mechanism of printing for all three structures in presented in the following section.

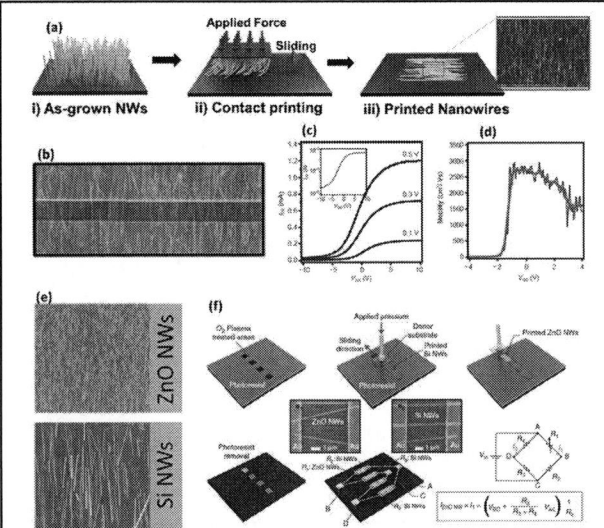

Fig. 3. (a) Schematic illustration for the mechanism of contact printing, (b-d) SEM image and electrical characterisation data of a back-gated FET fabricated on a contact printed, parallel array of InAs nanowires [1], and (e-f) 3D Schema of UV Photodetector fabrication using contact printed ZnO and Si NWs.

A. Tranfer printing mechanism

The printing mechanism can be understood by studying the competing fracture between the stamp/structures interface and the structure/substrate interface. During the first step (retrieval step), the stamp/structure interface must be stronger than the structure/donor interface to retrieve it from the donor substrate. In the second step (transfer step), the stamp/structure interface must be weaker than the structure/target interface to print the structure over donor substrate. These steps are depicted in Fig. 4a. For large area electronics, controllable and reproducible transfer of nano to chip scale structures from the donor to the receiver substrate is required, and hence the precise control of the interface properties of the stamp and substrate is necessary. To this end, control over factors such as surface functionalization, surface morphology modification, temperature and peeling velocity etc. is needed. Accordingly, efforts have been made to address these challenges with modified transfer printing involving the surface morphology [31], interface engineering [32], magnet-controlled [33], and laser driven method [34] etc. These modified transfer printing methods indeed improve the yield and reliability of the process, but also require additional excitation equipment.

The transfer printed structures (nano to chip scale) have shown enormous potential for realising high performance flexible electronic devices and circuits. These are discussed as following.

B. Transfer printing of nanoscale structures

Top-down methods have been used mainly to obtain horizontally aligned nanostructures (see section II). Printing of these nanostructures is performed using different transfer printing methods mentioned in section Iva. The printed nano structures have been employed to develop high-performance flexible electronics. For instance, high performance transistors were successfully developed over fully flexible polyimide (PI) substrates using transfer printed silicon NRs, as shown in Fig. 4b-f [13]. The interesting feature of these devices is that the dielectric (silicon nitride (SiNx)) was deposited at room temperature (RT). The transistors have shown high performance (mobility ≈656 cm²/Vs and on/off ratio >10^6) which is on par with the highest performance of similar devices reported with high-temperature processes, and significantly higher than devices reported with low-temperature processes. The electrical properties of the RT deposited dielectric was also studied. The measured breakdown field strength (>2.2 MVcm^{-1}) confirms its excellent quality. The reported transistor devices are mechanically robust, with the ability to withstand mechanical bending cycles (100 cycles tested) without performance degradation.

C. Transfer printing of microscale structures

Like nanoscale structures, transfer printing has been utilised to print micro scale structures fabricated using top-down approaches. For example, an array of micro-Si plates were transferred on to numerous substrates, including glass, watch, curved surface of earphone etc., by controlling the bending angle of PDMS which enables adhesion less transfer printing process [35]. Likewise, n-type Si based transistors were fabricated using the adhesion less transfer printing process [36]. The devices demonstrated a high mobility of >325 cm^2/Vs. Heterogeneous integration of inorganic light emitting diodes, transistors and stretchable interconnect were performed to fabricate stretchable active matrix displays [37]. The transfer printing was also used to develop flexible micro thermoelectric generators (μ-TEGs) on Poly (ethylene terephthalate) substrate [38]. A TEG module, consisting of an array of 34 alternately doped p-type and n-type Si microwires, is developed using similar top-down technique described in section II. A maximum of 9.3mV open circuit voltage was recorded from the flexible μ-TEG prototype with a temperature difference of 54 °C.

D. Transfer printing of chip or macro scale structures

In case of macro-scale structure (>1mm dimension) transfer process, the entire inorganic or silicon-based devices fabricated using the conventional approach were transferred over flexible substrates [39]. For example, a layer transfer process was explored to physically remove the top thin layer

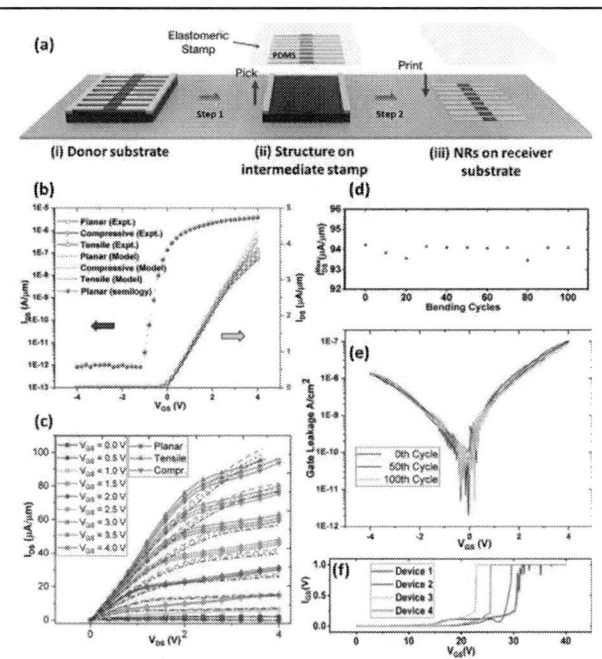

Fig. 4. (a) Schematic illustration of mechanism for transfer printing, (b) Transfer characteristics (experimental (line) versus model (dashed) simulations) and (c) output characteristics of the NR-FET at planar, tensile, and compressive bending conditions (Rc = 40 mm). (d) Variation of the I_{DS} at planar condition after cycles of compressive and tensile bending at $V_{DS} = V_{GS} = 4$ V. (e) Gate dielectric leakage current after subjecting it to cyclic bending. (f) Breakdown voltage characteristics of four randomly chosen devices after subjecting to cyclic bending of 100 cycles. Reproduced from [13].

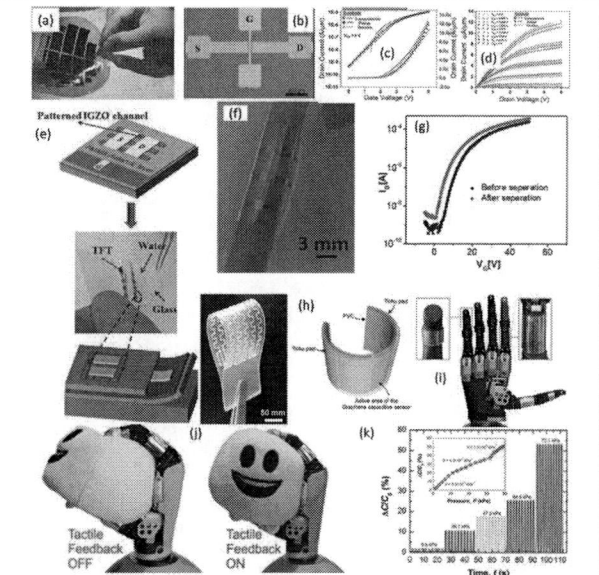

Fig. 5. (a-d) Transfer of UTCs using transfer printing: (a) photomicrograph of flexible UTCs on polyimide, (b) magnified image of n-MOSFET, (c-d) their transfer and output characteristics under various bending condition [3]. (e-g) Transfer printing of TFTs: (e) schematic illustration of separation of electronic device from ridged glass substrate by placing a water at the interface between PMMA and glass slide; (f) optical image of bottom gated IGZO TFT wrapped around cylindrical surface, and (g) their transfer characteristics under different bending radii [14]. (h-k) Large area graphene based capacitive tactile e-Skin: (h) Photograph and 3D schema of a flexible graphene capacitive touch sensor, (i) e-Skin with capacitive sensors integrated onto a robotic hand, (j) Self-powered e-Skin used as tactile feedback for a robotic hand, and (k) touch sensor response vs. applied pressure [17].

of processed silicon wafer along with the active device region through SlimCut process [40]. This SlimCut process is also named as the controlled spalling technique (CST) which uses a thin tensile stressor layer, attached to active device region, to create a stress induced crack under external sheer force. This helps to separate active device as UTCs [40]. Further, the UTCs were easily transferred to desired substrate for potential application. Similarly, PDMS assisted wafer scale transfer process was used to transfer the electronic devices fabricated on 4-inch Si wafer Fig. (5a-d) [3]. Firstly, the high-performance electronic device such as metal oxide semiconductor (MOS) capacitor and MOS field-effect transistor (MOSFET) were fabricated in rigid silicon wafer using conventional process Fig. 5b. Sequentially, the backside of silicon wafer was thinned using a chemical process and transferred to polyimide through PDMS assisted dual transfer process. In Si MOSFETs, the compressive and tensile strain affects the effective mass of the carrier which is reflected in slight variation in the device performance (Fig. 5c-d) with the effective mobility of 384 cm^2/Vs under tensile strain and 333 cm^2/V·s under compressive strain. Similar to Si-based devices, inorganic thin film devices such as thin film transistors (TFTs), gas sensors and photodetectors were transferred to various substrate using water assisted transfer technique [14]. Here, the PMMA was spin coated on the non-treated glass substrate and the electronic devices were fabricated using conventional approach. The weak adhesion between the PMMA to the non-treated glass slide helps to separate the PMMA layer along with the prefabricated device by placing a water-droplet at the interface (Fig. 5e). As shown in Fig. 5f, the fabricated devices were transferred to cylindrical surface and the device demonstrate similar performance after transfer (Fig. 5g).

V. APPLICATION IN ELECTRONIC SKIN DEVELOPMENT

Contact printing has shown potential towards large-scale and heterogeneous integration of printed inorganic elements to realise electronic skin (e-Skin) for application in prosthetics, robotics, etc. Circuit level integration of contact printed NWs that utilizes both the sensory (CdSe NWs) and electronic functionalities (Ge/Si NWs) has been demonstrated [41]. These functionalities are then interfaced to enable an all-NW circuitry with on-chip integration, capable of detecting and amplifying an optical signal with high sensitivity and precision. Large area (49 cm^2) printing of aligned NW arrays were used to develop active-matrix backplane of a flexible pressure-sensor array (18 x 19 pixels) [42]. The integrated sensor array effectively functions as an e-Skin to monitor the applied pressure profiles with high spatial resolution.

Similarly transfer printing of nano to macro scale structures has also led to the e-Skin development. For instance, large area transfer printing of graphene layer on a photovoltaic (PV) cell resulting in energy autonomous tactile sensitive system for soft robotics (Fig. 5h-k) [17]. Hot lamination method was used to transfer 4-inch CVD grown monolayer of graphene from Cu foil to 125-μm-thick poly vinyl chloride (PVC) substrates. The printed graphene sheets were used for the fabrication of large area flexible graphene based capacitive touch sensors. The fabricated sensors showed high sensitivity (4.3 kPa^{-1}) to a wide range of pressures (0.11–80 kPa). One of the key features of the fabricated e-Skin relied on its great transparency, i.e., a sunlight absorption below 5%, which allowed the effective energy harvesting of light energy by a PV cell underneath the e-Skin. This has provided energy autonomy to the fabricated e-Skin. The viability of graphene-based skin sensors is also analysed by means of a dynamic characterization consisting in the grabbing of a soft object. The response obtained from the capacitive sensors was successfully used as tactile feedback in an artificial hand, allowing the manipulation of rigid and soft objects with different shapes.

ACKNOWLEDGMENT

This work was supported by Engineering and Physical Sciences Research Council through Engineering Fellowship for Growth (EP/M002527/1 and EP/R029644/1) and Hetero-print Programme Grant (EP/R03480X/1).

REFERENCES

[1] A. C. Ford, J. C. Ho, Z. Fan, O. Ergen, V. Altoe, S. Aloni, H. Razavi, and A. Javey, "Synthesis, contact printing, and device characterization of Ni-catalyzed, crystalline InAs nanowires," *Nano Research,* vol. 1, no. 1, pp. 32-39, 2008.

[2] A. S. Dahiya, D. Shakthivel, Y. Kumaresan, A. Zumeit, A. Christou, and R. Dahiya, "High-performance printed electronics based on inorganic semiconducting nano to chip scale structures," *Nano Convergence,* vol. 7, no. 1, pp. 33, 2020.

[3] W. T. Navaraj, S. Gupta, L. Lorenzelli, and R. Dahiya, "Wafer Scale Transfer of Ultrathin Silicon Chips on Flexible Substrates for High Performance Bendable Systems," *Adv. Electron. Mater.,* vol. 4, no. 4, pp. 1700277, 2018.

[4] P. Escobedo, M. Ntagios, D. Shakthivel, W. T. Navaraj, and R. Dahiya, "Energy Generating Electronic Skin with Intrinsic Touch Sensing," *IEEE Transactions on Robotics,* vol. 37, no. 2, pp. 683 - 690, 2021.

[5] P. Escobedo, M. Bhattacharjee, F. Nikbakhtnasrabadi, and R. Dahiya, "Smart Bandage with Wireless Strain and Temperature Sensors and Battery-less NFC Tag," *IEEE Internet of Things Journal,* vol. 8, no. 6, pp. 5093 - 5100, 2021.

[6] M. Bhattacharjee, F. Nikbakhtnasrabadi, and R. Dahiya, "Printed Chipless Antenna as Flexible Temperature Sensor," *IEEE Internet of Things Journal,* vol. 8, no. 6, pp. 5101 - 5110, 2021.

[7] M. Soni, M. Bhattacharjee, M. Ntagios, and R. Dahiya, "Printed Temperature Sensor Based on PEDOT: PSS-Graphene Oxide Composite," *IEEE Sensors Journal,* vol. 20, no. 14, pp. 7525-7531, 2020.

[8] Y. Ling, T. An, L. W. Yap, B. Zhu, S. Gong, and W. Cheng, "Disruptive, Soft, Wearable Sensors," *Adv Mater,* vol. 32, no. 18, pp. 1904664, 2020.

[9] A. S. Dahiya, J. Thireau, J. Boudaden, S. Lal, U. Gulzar, Y. Zhang, T. Gil, N. Azemard, P. Ramm, T. Kiessling, C. O'Murchu, F. Sebelius, J. Tilly, C. Glynn, S. Geary, C. O'Dwyer, K. M. Razeeb, A. Lacampagne, B. Charlot, and A. Todri-Sanial, "Review—Energy Autonomous Wearable Sensors for Smart Healthcare: A Review," *J Electrochem Soc,* vol. 167, no. 3, pp. 037516, 2020.

[10] T. Someya, and M. Amagai, "Toward a new generation of smart skins," *Nat Biotechnol,* vol. 37, no. 4, pp. 382-388, 2019.

[11] R. Dahiya, N. Yogeswaran, F. Liu, L. Manjakkal, E. Burdet, V. Hayward, and H. Jörntell, "Large-Area Soft e-Skin: The Challenges Beyond Sensor Designs," *Proceedings of the IEEE,* vol. 107, no. 10, pp. 2016-2033, 2019.

[12] R. Dahiya, D. Akinwande, and J. S. Chang, "Flexible Electronic Skin: From Humanoids to Humans [Scanning the Issue]," *Proceedings of the IEEE,* vol. 107, no. 10, pp. 2011-2015, 2019.

[13] A. Zumeit, W. T. Navaraj, D. Shakthivel, and R. Dahiya, "Nanoribbon-Based Flexible High-Performance Transistors Fabricated at Room Temperature," *Adv Electron Mater,* vol. 6, no. 4, pp. 1901023, 2020.

[14] Y. Kumaresan, R. Lee, N. Lim, Y. Pak, H. Kim, W. Kim, and G.-Y. Jung, "Extremely Flexible Indium-Gallium-Zinc Oxide (IGZO) Based Electronic Devices Placed on an Ultrathin Poly(Methyl Methacrylate) (PMMA) Substrate," *Adv. Electron. Mater.,* vol. 4, no. 7, pp. 1800167, 2018.

[15] L. Manjakkal, A. Pullanchiyodan, N. Yogeswaran, E. S. Hosseini, and R. Dahiya, "A Wearable Supercapacitor Based on Conductive PEDOT:PSS-Coated Cloth and a Sweat Electrolyte," *Adv Mater,* vol. 32, no. 24, pp. 1907254, 2020.

[16] U. Zschieschang, J. W. Borchert, M. Giorgio, M. Caironi, F. Letzkus, J. N. Burghartz, U. Waizmann, J. Weis, S. Ludwigs, and H. Klauk, "Roadmap to Gigahertz Organic Transistors," *Adv Funct Mater,* vol. 30, no. 20, pp. 1903812, 2020.

[17] C. G. Núñez, W. T. Navaraj, E. O. Polat, and R. Dahiya, "Energy-Autonomous, Flexible, and Transparent Tactile Skin," *Adv Funct Mater,* vol. 27, no. 18, pp. 1606287, 2017.

[18] A. Vilouras, A. Christou, L. Manjakkal, and R. Dahiya, "Ultrathin Ion-Sensitive Field-Effect Transistor Chips with Bending-Induced Performance Enhancement," *ACS Appl Electron Mater,* vol. 2, no. 8, pp. 2601-2610, 2020.

[19] S. Gupta, W. T. Navaraj, L. Lorenzelli, and R. Dahiya, "Ultra-thin chips for high-performance flexible electronics," *npj Flex Electron,* vol. 2, no. 1, pp. 8, 2018.

[20] D. Shakthivel, M. Ahmad, M. R. Alenezi, R. Dahiya, and S. R. P. Silva, "1D Semiconducting Nanostructures for Flexible and Large-Area Electronics: Growth Mechanisms and Suitability," Elements in Flexible and Large-Area Electronics, Cambridge University Press, 2019.

[21] C. García Núñez, F. Liu, S. Xu, and R. Dahiya, "Integration Techniques for Micro/Nanostructure-based Large-Area

Electronics," Elements in Flexible and Large-Area Electronics, Cambridge University Press, 2018.

[22] R. S. Dahiya, A. Adami, C. Collini, and L. Lorenzelli, "Fabrication of single crystal silicon micro-/nanostructures and transferring them to flexible substrates," *Microelectronic Engineering*, vol. 98, pp. 502-507, 2012.

[23] Z. Huang, N. Geyer, P. Werner, J. de Boor, and U. Gösele, "Metal-Assisted Chemical Etching of Silicon: A Review," *Adv. Mater.*, vol. 23, no. 2, pp. 285-308, 2011.

[24] D. Shakthivel, W. T. Navaraj, S. Champet, D. H. Gregory, and R. S. Dahiya, "Propagation of amorphous oxide nanowires via the VLS mechanism: growth kinetics," *Nanoscale Adv.*, vol. 1, no. 9, pp. 3568-3578, 2019.

[25] F. Liu, W. T. Navaraj, N. Yogeswaran, D. H. Gregory, and R. Dahiya, "van der Waals Contact Engineering of Graphene Field-Effect Transistors for Large-Area Flexible Electronics," *ACS Nano*, vol. 13, no. 3, pp. 3257-3268, 2019.

[26] E. O. Polat, O. Balci, N. Kakenov, H. B. Uzlu, C. Kocabas, and R. Dahiya, "Synthesis of Large Area Graphene for High Performance in Flexible Optoelectronic Devices," *Sci Rep*, vol. 5, no. 1, pp. 16744, 2015.

[27] A. Christou, F. Liu, and R. Dahiya, "Development of a highly controlled system for large-area, directional printing of quasi-1D nanomaterials," *Microsyst Nanoeng*, 2021.

[28] C. García Núñez, F. Liu, W. T. Navaraj, A. Christou, D. Shakthivel, and R. Dahiya, "Heterogeneous integration of contact-printed semiconductor nanowires for high-performance devices on large areas," *Microsyst Nanoeng*, vol. 4, no. 1, pp. 22, 2018.

[29] G. Zhu, R. Yang, S. Wang, and Z. L. Wang, "Flexible High-Output Nanogenerator Based on Lateral ZnO Nanowire Array," *Nano Lett*, vol. 10, no. 8, pp. 3151-3155, 2010.

[30] A. Javey, Nam, R. S. Friedman, H. Yan, and C. M. Lieber, "Layer-by-Layer Assembly of Nanowires for Three-Dimensional, Multifunctional Electronics," *Nano Lett*, vol. 7, no. 3, pp. 773-777, 2007.

[31] J. W. Jeong, S. R. Yang, Y. H. Hur, S. W. Kim, K. M. Baek, S. Yim, H.-I. Jang, J. H. Park, S. Y. Lee, C.-O. Park, and Y. S. Jung, "High-resolution nanotransfer printing applicable to diverse surfaces via interface-targeted adhesion switching," *Nat Commun*, vol. 5, no. 1, pp. 5387, 2014.

[32] B. Yoo, S. Cho, S. Seo, and J. Lee, "Elastomeric Angled Microflaps with Reversible Adhesion for Transfer-Printing Semiconductor Membranes onto Dry Surfaces," *ACS Appl. Mater. & Interfaces*, vol. 6, no. 21, pp. 19247-19253, 2014.

[33] C. Linghu, C. Wang, N. Cen, J. Wu, Z. Lai, and J. Song, "Rapidly tunable and highly reversible bio-inspired dry adhesion for transfer printing in air and a vacuum," *Soft Matter.*, vol. 15, no. 1, pp. 30-37, 2019.

[34] J. Bian, L. Zhou, X. Wan, C. Zhu, B. Yang, and Y. Huang, "Laser Transfer, Printing, and Assembly Techniques for Flexible Electronics," *Adv. Electron. Mater.*, vol. 5, no. 7, pp. 1800900, 2019.

[35] S. Cho, N. Kim, K. Song, and J. Lee, "Adhesiveless Transfer Printing of Ultrathin Microscale Semiconductor Materials by Controlling the Bending Radius of an Elastomeric Stamp," *Langmuir*, vol. 32, no. 31, pp. 7951-7957, 2016/08/09, 2016.

[36] T.-H. Kim, A. Carlson, J.-H. Ahn, S. M. Won, S. Wang, Y. Huang, and J. A. Rogers, "Kinetically controlled, adhesiveless transfer printing using microstructured stamps," *Applied Physics Letters*, vol. 94, no. 11, pp. 113502, 2009/03/16, 2009.

[37] M. Choi, B. Jang, W. Lee, S. Lee, T. W. Kim, H.-J. Lee, J.-H. Kim, and J.-H. Ahn, "Stretchable Active Matrix Inorganic Light-Emitting Diode Display Enabled by Overlay-Aligned Roll-Transfer Printing," *Advanced Functional Materials*, vol. 27, no. 11, pp. 1606005, 2017/03/01, 2017.

[38] S. Khan, R. S. Dahiya, and L. Lorenzelli, "Flexible thermoelectric generator based on transfer printed Si microwires," *44th European Solid State Device Research Conference (ESSDERC)*, pp. 86-89, 2014.

[39] D. S. Wie, Y. Zhang, M. K. Kim, B. Kim, S. Park, Y.-J. Kim, P. P. Irazoqui, X. Zheng, B. Xu, and C. H. Lee, "Wafer-recyclable, environment-friendly transfer printing for large-scale thin-film nanoelectronics," *Proceedings of the National Academy of Sciences*, vol. 115, no. 31, pp. E7236, 2018.

[40] D. Shahrjerdi, and S. W. Bedell, "Extremely Flexible Nanoscale Ultrathin Body Silicon Integrated Circuits on Plastic," *Nano Lett*, vol. 13, no. 1, pp. 315-320, 2013.

[41] Z. Fan, J. C. Ho, Z. A. Jacobson, H. Razavi, and A. Javey, "Large-scale, heterogeneous integration of nanowire arrays for image sensor circuitry," *Proceedings of the National Academy of Sciences*, vol. 105, no. 32, pp. 11066-11070, 2008.

[42] K. Takei, T. Takahashi, J. C. Ho, H. Ko, A. G. Gillies, P. W. Leu, R. S. Fearing, and A. Javey, "Nanowire active-matrix circuitry for low-voltage macroscale artificial skin," *Nature Mater*, vol. 9, no. 10, pp. 821-826, 2010/10/01, 2010.

© VDE VERLAG GMBH · Berlin · Offenbach

Optimizing Neural Networks for Embedded Hardware

Domenik Helms[†], Karl Amende[‡], Saqib Bukhari[§], Thies de Graaff[†], Alexander Frickenstein[*],
Frank Hafner[§], Tobias Hirscher[‡], Sven Mantowsky[§], Georg Schneider[§], and Manoj-Rohit Vemparala[*]
[*]BMW AG, Munich, Germany, first.last@bmw.de
[†]OFFIS e.V., Oldenburg, Germany, first.last@offis.de
[‡]Valeo, Kronach, Germany, first.last@valeo.com
[§]ZF Friedrichshafen AG, Saarbrücken/Friedrichshafen, Germany, first.last@zf.com

Abstract—**Neural networks are a pervasive technology, which is, however, still held back in the area of embedded systems by the high resource requirements, especially memory size, memory access time and power dissipation. In recent years, several different methods have been proposed to transform given neural networks in such a way that they can get by with much fewer resources while maintaining almost the same accuracy. This work reviews, categorizes and describes the state of the art in adapting and simplifying neural networks to make them better applicable to embedded systems. Even though we developed this study from a purely automotive context, the techniques described are also valid in other areas.**

I. INTRODUCTION

In many embedded application areas, there is a need for high-performance neural network (NN) computation. For highly automated vehicles, for example the perception of a vehicle must be applicable in real time in order to be able to react quickly to changes in the environment. For commercial reasons, however, the recent HW used in the vehicle is subject to stricter limits in terms of computing power and memory than is the case for other application areas of NN technologies. First high performance, low power HW platforms, such as the ZF ProAI are approaching the market. Current NN procedures, however, place quite high demands on the required HW.

To bridge the gap between algorithmic demand and HW capability, methods for performance optimization of NN modules are needed. All recent work in the area of NN simplification can be classified into one of the five categories pruning, quantization, tensor compression, neural architecture search (NAS), and knowledge distillation, which will be presented in the subsequent chapters.

II. PRUNING

Pruning is the process of removing parts of the NN which do not significantly contribute to the final result. A hypothesis was presented in [1] that inside a deep NN a sub-network exists, which when trained in isolation reaches the same performance as the initial NN. This motivates that there are a lot of unnecessary parameters in trained deep NN which are not needed to solve the actual task. The goal of pruning is to identify these parts and remove them to reduce the size of the NN leading to a lower memory consumption and runtime (depending on the technique and HW used).

The first way of categorizing the pruning work in the literature is HW efficiency arising due to regularity. A fast and parallel HW like GPU is inefficient in handling sparse matrices properly, while an embedded CPU may be optimized for sparse data. Even though regular sparsity has some computational advantages, obtaining it might be challenging. When a NN is forced to be pruned in a structured manner, it is highly possible that some important parameters might be deleted from the NN. As a result, a compromise between pruning rate and/or degraded performance must be made. In general, most of the pruning work reported in the literature use GPU for performance evaluation. We obtain sparse matrices as a result of pruning where some of the elements are zero. Based on the composition of zero elements in the pruned matrix, the pruning scheme itself can be classified into four types of pruning methods, namely element weight pruning, channel pruning, filter (or kernel) pruning, and layer pruning.

The second way to categorize pruning methods is based on the algorithmic heuristic to determine the importance of the neuron. These fall into two general categories, handcrafted pruning and automated pruning methods. Handcrafted pruning methods depend on static or pseudo-static rules, which are followed when compressing a given model. Human expertise is needed to formulate these heuristics and to perform layers sensitivity analysis or other fine-tuning of pruning related hyperparameters. Such manual task can be tedious because state-of-the art NN models are typically very deep. Unlike handcrafted methods, automated pruning methods are in principle a NAS problem (rf. Section V), usually formulated as a learning task such that the need for the laborious manual fine-tuning tasks is eliminated. However, there is room for improvement in this direction of research. The work [2], [3] demonstrated that it is difficult to formalize a rule to prune NN and instead leverage a reinforcement learning agent to learn the criteria for pruning, based on a given cost function.

Finally, pruning can also be categorized with respect to when it happens, either at post processing level for a trained model or training and pruning together.

A. Weight Pruning

Handcrafted Pruning: [4] determined the importance of individual elements in the weight matrix based on their magnitude demonstrating the redundancy in deep NN. Pruning individual weights, referred to as irregular pruning, leads to inefficient memory accesses, making it impractical for general purpose computing platforms. Regularity in pruning becomes an important criterion towards accelerator-aware optimization.

[5] use a HW-model to estimate the energy requirements of each layer. The layers with the highest energy contribution present a good starting point for the pruning process, which is based on the L2-norm heuristic. However, the energy estimates do not influence the sparsity ratio directly.

Automated Pruning: [6], [7] proposed a fine-grained pruning, exploiting the flexibility of custom HW.

B. Channel Pruning

Post Processing Pruning: Meta learning [8] allows automatic channel pruning of NN. First, a Meta-Network is trained based on stochastic structure sampling method, which can generate weight parameters for any pruned structure given the target NN. Then, an evolutionary procedure is used to find suitable pruned NN. This method is said to be highly efficient, because weights are generated by the trained Meta-Network, so no fine-tuning is necessary at search time. With a single Meta-Network trained for the target NN, one can search for various pruned NN under different constraints.

Handcrafted Pruning: Achieving better inference times on GPUs requires pruning entire channels instead of individual weights. However, pruning NNs in structured manner might hurt the NN's performance and reduce overall accuracy. Thus, [9] prunes redundant channels by LASSO regression and then minimizes the output error of the remaining feature maps by solving least square minimization. [10] uses geometric median heuristics to assess the filters and those filters that have the smallest values are pruned. Although the filter pruning scheme is useful with respect to HW implementations, it is challenging to remove filters as they directly impact the input channels of the subsequent layer. [11] uses the channel pruning method from [9] a part of an efficient implementation of semantic segmentation. The method was able to reduce the memory consumption by 30% and the number of FLOPs by around 17%. The method used follows a bottom up approach pruning layers one by one using a LASSO regression.

Automated Pruning: [2] used a *reinforce* algorithm to learn pruning filters without using any heuristics. They formulate a non-differentiable policy with an accuracy and efficiency term and train the agent iteratively for each layer to identify redundant filters. The objective for the agent is to maximize the two contrary objectives. [3] demonstrated an auto pruning agent based on an actor-critic deep deterministic policy gradient algorithm. Their agent learns to prune the full model by outputting a sparsity ratio of each layer but without specifying the exact position of filters, for this they fall back to use a heuristic based on L1-norm. The reinforcement learning agent doesn't fine-tune the NN in between the intermediate episodes.

C. Filter (or Kernel) pruning

Post Processing Pruning: Filter Pruning via Geometric Median [10] combines two aspects: First, the minimum of a filter norm should be small and second, the filter norm-deviation should be large. Gate Decorator [12] is a *global* filter pruning method for NN modules, multiplying the output by channel-wise scaling factor sf ($sf \equiv 0$ removes the filter). Taylor expansion is used to estimate the loss when scaling factor is set to zero and use the estimation for a global filter importance ranking. After ranking all filters, the unimportant filters are deleted, and all scaling factors are merged to its original module. Layer-Compensated Pruning [13] combines two common problems: Which and how many filters to prune for each layer, based on a global filter ranking. This improves prior methods by learning to compensate the approximation error incurred in the derivation.

Simultaneous Training and Pruning: [14] simultaneously trains from scratch and prunes the model. After a training epoch, the filters are pruned based on their importance evaluation (lp-norm), where the small filters are pruned. In the next training epoch, the pruned filters can still be updated from zero, so that other filters get pruned, whereas the reconstructed filters remain. This can lead to faster NN with better accuracy.

Handcrafted Pruning: It is demonstrated in [15] that the irregular pruning is HW inefficient despite achieving a high degree of sparsity. Thus, they use evolutionary particle filter to select the fittest combination of connections to prune at three structured level (feature maps, kernels and strided intra-kernel). [16] introduce Dense-Sparse Convolution pruning kernels/filters instead of individual weight elements. They also make use of quantization along with Winograd algorithm to realize an efficient vectorized low latency CPU kernel. They show that the Winograd algorithm when combined with kernel pruning enables acceleration of CNNs with high sparsity ratio.

Automated Pruning: Zhang et al. [17] propose ADAM-ADMM, a unified, systematic framework of structured weight pruning of NNs, that can be employed to induce different types of structured sparsity (filter, channel, kernel-based) on ADMM. They perform pruning during the training with an extra regularization objective which determines the set of redundant weights in the CNN model.

D. Layer Pruning

Automated Pruning: [18] defines a block as a convolution layer together with its normalization and activation. Using a dense NN as linear discriminator, they could identify and remove entire layers from the NN, reducing the execution time on highly parallel devices by $\times 1.4 - 2.2$ for $0.4 - 1.3\%$ accuracy loss.

III. QUANTIZATION

Quantization constraints the numerical precision from a continuous or otherwise large set of values (such as the real numbers) to a discrete or otherwise small set, such as a 16bit float representation, 8bit integers or even binary values. In regards to NN, it is the process of lowering the bitwidth of weights and activations to some defined set with limited range. Quantization can be used to significantly reduce the memory consumption and inference latency of NN to make them fit on embedded HW (if the memory and processor HW yields sufficient support for the reduced bitwidth). In addition, quantization can enable processing of NN on more architectures by e.g. converting floating-point numbers to fixed-point numbers, which can be represented as integers.

The drawback of quantization is the information loss and the subsequent reduction in model performance. The research topic in the field is to reduce this impact on accuracy. Quantization methods can be further subdivided into reduced precision approaches (e.g. float32 to float16 or float to fix-point), clustering (only a few separate constants, each with a high precision), extremely low bitwidth.

A. Reduced precision approaches

[19] presents and evaluates a generic way for quantization to arbitrary bitwidths. A straight-through estimator (STE) was used for training, which is an operation with arbitrary forward and backward function. Any discrete function such as round with a limited range set would introduce zero gradient at most of the solution space. Since this would nullify gradients during training, in the backward path the gradient of the quantization function is replaced by the identity function using STE.

[20] stated that a good parametrisation of the quantizer enables stable training and good final performance. They propose to learn step size and dynamic range for each weight and activation layer independently and infer the bitwidth out of these parameters afterwards, resulting in a mixed precision model. During training they used a penalty function for the model size in order to fulfil specific memory constraints. [21] augmented this idea by using a linear interpolation between two bitwidths, facilitating a differential quantization function. They called their approach FracBits because they use fractional bitwidth in the first of their two training phases. In the second training phase they use final bitwidths derived from the fractional bitwidths and finetune these.

In contrast to the previous publications [22] included the final HW in the design and training process, proposing HW-aware automated quantization which uses RL-based exploration to determine suitable, layer-wise quantization levels for weights and activations resulting in a mixed precision NN model. The reward function, including real HW metrics, is generated by directly executing the inference of a NN model on an FPGA design, which supports on-chip quantization libraries. In contrast, [23] also uses mixed precision approach, wherein only the fully connected layers use binary weights to tackle the limited memory bandwidth problem in mid range FPGA's such as Arria-10. A specialized FPGA based accelerator is developed to accelerate the mentioned mixed precision approach.

B. Clustering

One of the early works was [24] which proposed a unified framework for quantization. The main idea is to divide the input space in sub spaces and later clustering them. A value is then quantized by finding the closed cluster centroid. This also leads to runtime improvements since the dot product between all centroids inside a subspace can be precalculated.

[25] started with pruning a NN, then used weight sharing and then used Huffmann compression to result in an optimal codebook quantization. [26] adds to this by allowing groups of weights to share a common codebook, thus further reducing the memory demand of the overall NN. Sacrificing just 1% of top-1 accuracy, the memory demand can be compressed by almost $\times 20$, outperforming other reference quantization methods such as ABC-Net [27], binary weight network [28] or HW-aware automated quantization [22].

In [29], a hybrid *Mixed Precision* scheme was presented, referred to as Dynamic Fixed Point. The idea is to use fixed point arithmetic and to add a single scaling factor to a group of values. For instance, a (sub-)tensor is represented as INT16 and contains one additional INT8 value to describe it's scaling of all parameters.

C. Vector quantization

In [30], several techniques are presented and compared, reducing the memory footprint of the NN. k-means quantization interprets all (potentially high dimensional) weights of a NN layer as a single linear vector. For these values, a set of s centeroids c_s is found, minimizing the error. [31] adds a permutation step, reordering the sequence of filters in a layer and then applying group-wise vector quantization. Like this, they reduced the accuracy loss by around 2% compared to stand-alone vector quantization. For 1% accuracy loss, they offer a compression ratio of $\times 15 - 20$ and for 4% accuracy loss, they achieve more than an $\times 30$ compression.

D. Extremely low bitwidth

[32] presents methods, improving the accuracy of a post-training low bitwidth (e.g. 8bit) quantization. They propose slightly spreading the representable range from the required minimum to always make 0 an exactly representable number.

In [33], a signal to quantization noise ratio is introduced to represent the quantization losses. They use the ratio to guide an improved parameter rescaling (if an output variable is scaled by factor x, all weights, where this variable is used have to be scaled by $1/x$). Using their approach, they show the top-1 accuracy loss of 8bit quantization can be reduced from several percent to less than one percent.

[19] relaxes the binarization constraint for activations allowing more than one bit but still leverage the binary weights. They explore the quantized weights and activations during the forward pass as well as the backward pass, reducing the complexity of inference as well as training. The authors of [34] propose to replace the usually employed ReLU function by a clipping version of ReLU: $\text{PACT}(x) := max(0, min(a, x))$, where the clipping range is trainable, but bound by an L2 regularization. That way, there is a natural upper bound at the output of each neuron, which helps in avoiding gradient explosions, in extremely low bitwidth quantization.

[35] proposes to learn the step size of a quantizer for weights and activations in conjunction with other NN parameters. They managed to reach the same accuracy as the

baseline model with a 3-bit version of it. [36] improved this approach, also proposing a learnable offset and introducing a new initialization scheme which lowers the variance of performance over several training runs.

E. Binary and ternary

In [37] a binary quantization is used by taking the sign of the value to be quantized. Switching from FP32 to 1bit reduces the memory consumption by $\times 32$. This form of quantization also enables very optimized processing on e.g. CPUs. Instead of calculating the dot product between binary values, the values can be combined into one value e.g. unsigned 32bit integer and then two of these integers are combined using the XNOR operation. After this, the population count operation (popcnt) is calculated by counting the ones in the binary representation of the result. This way the dot product between 32 values can be calculated in a few CPU instructions. In [28] this approach was improved by introducing multiple scaling factors. Estimating the scaling factor for the activations is computationally intense compared to weights, as the distribution of the input features has to be taken into consideration at run time. Yet, the accuracy degradation is still high.

[38] proposes CompactBNN and gets rid of the scale factor for weights. They introduce a higher number of bases to find a better estimate for the activation maps as they observed that the binarization of activations is more challenging compared to the weights. They also replace the standard ReLU function by the parametric ReLU (pReLU). Using binary values in forward propagation, but float32 in backward propagation helps increase training speed, but even then, training rate lacks some $\times 100$ behind.

Ternary quantization [39]–[41] extends the set of possible weights from $\{-1, 1\}$ to $\{-1, 0, 1\}$, thus effectively introducing an option for fine-grained pruning. Ternary quantization greatly reduces the accuracy loss, compared to the binary counterpart. For weights it can be efficiently be realized in FPGAs by fine-grained pruning.

F. Multi-bit networks

Instead of simply reducing the bitwidth to an extreme, always yielding huge accuracy losses, the idea of multi-bit networks (MBN) is to binarize, but to scale up the NN's topology in a schematic way, reducing the accuracy loss, while still saving memory size, memory bandwidth and potentially also instruction counts. MBN starts with the observation, that almost each 32bit machine can do 32 MAC operations in parallel by using bitwise XNOR, popcnt and addition. In memory, 32 weights can be stored in one word and can also be transported simultaneously over the bus. Thus an increase in NN size by 10x (or even higher, if MAC does not compute in a single cycle) will still save computation time and an increase of up to 32x will save memory size and bandwidth.

ABC-Net [27] enables representing weights and activations as multiple single-bit values. Even though the key idea is really powerful, the authors could not bring the base idea to its full potential. Nevertheless, they could show, that when multiplying the NN size by 25 (x5 for the weights and x5 for the activations), the accuracy loss of binarization (effectively assigning 5 single bits to each value) can be kept below 5%, while still enabling effective XNOR multiplications. [42] then significantly reduces both, the final size (i.e. average bitwidth for weights and activations) and the accuracy loss by optimizing the assignment of *bitlayers* to parts of a layer's tensor. This enables binarization and coarse grained pruning at the same time, resulting in below one bit average bitwidth needed. [42] outperforms other pruning proposals in size and accuracy in a mode, where only weights, not activations are binarized. They could show a $\times 13$ weight reduction with an average of 2 bitlayers losing only 3.4% of accuracy.

IV. NUMERICAL ACCELERATION

A. Tensor compression

CNN compression [43] starts with identifying the largest possible linear operation in each CNN layer by shifting the activation function, the pooling, and the repeated application of the operation of the convolution layers into a post processing step, following a huge (usually 4 dimensional) linear tensor operation. For this tensor, an approximate rank can be determined (and minimized, sacrificing some application accuracy) using variational Bayesian matrix factorization [44] or heuristic search [45]. Each layer's kernel tensor can then be reduced in terms of numerical complexity (MAC count) and memory accesses without any further loss of accuracy using Tucker decomposition [46], Candecomp/Parafac decomposition [45] or Tensor Train decomposition [47]. Finally, the parameters of the remaining tensors can be fine-tuned by regular backpropagation, to mitigate some of the accuracy loss from the reduction step. In [43] the accuracy loss was reported $0.2 - 1.7\%$ resulting in $\times 1.4 - 3.7$ speedup and $\times 1.6 - 4.2$ power reduction for mobile devices. For huge, highly parallel accelerators, the speedups are slightly below.

In [48], the authors focus only on a tensor compression of the interface between the last convolutional layer and the first fully connected layer. Here, a low rank filter (either by Tucker or Tensor Train or even by Candecomp/Parafac compression) not only reduces the layer size, but also introduces a powerful regularization technique, avoiding information loss at the vectorization stage.

There are several recent papers on determining the best rank per level to guide the compression step itself. In [49], the optimal rank per level is determined by a reinforcement learning. [50] presents an iterative layer wise rank search. Finally, [45] presents an optimization over all layers at once and [51] presents a much faster version (in terms of host runtime) of this. Final results depend on the compression rate as well as the NN itself. AlexNet, VGG-16 and ResNet-56 all showed between 0.6% better and 0.1% worse accuracy after an $\times 2 - 5$ parameter reduction.

As the resulting tensors are smaller, yet still regular, tensor compression is beneficial for all architectures, even though small embedded architectures tend to profit more than highly parallel accelerators such as GPUs [43].

B. Winograd kernels

A purely numerical acceleration is the Winograd kernel implementation [52]. Here the input and parameter tensors are transformed in a way, that the usual $\mathcal{O}(k^2)$ convolution of one datapoint can instead be done in $\mathcal{O}(d+k-1)$ for d datapoints at once. Though this transformation is very sensible to numerical artefacts, [52] showed that a $\times 4$ speedup is achievable. Due to the numerical sensibility, it is not straightforward to do quantization for the Winograd algorithm. Nevertheless, [53] presents a reduced algorithm in 8bit.

V. Neural Architecture Search

Neural Architecture Search (NAS) is the process of finding an optimal NN topology for a given task and dataset under certain constraints in an automated manner. Hereby, especially the target specific design of NN for edge devices has gained traction over recent years [54]–[57].

As the space of NN topologies is almost infinite, NAS methods commonly define a limited search space to reduce the combinatorial complexity of the NAS. Commonly, the search space contains unary operations such as convolutional layers, activations or pooling and multivariate operations such as addition or concatenation. On top of that, different convolutional layer types, kernel sizes or layer shapes are frequently included in the search space [54], [55], [58]. Concurrently, the search space can also be designed with segments or repeating cells consisting of to a certain extend predefined combinations of unary and multivariate operations [55]. Another approach of designing a search space is to overparametrize a starting NN and reduce it by choice of an ideal subpath [59]. If available, it was found beneficial to incorporate pre-knowledge on a given target HW for the definition of the search space [60]. For instance, [60] focuses on regular convolutions instead of inverted bottleneck layers which are generally predominant in NAS, due to prior knowledge on the HW design of EdgeTPUs.

Next to the search space, an optimization target is needed for NAS. Typical optimization targets are a certain target latency on the target HW or a defined amount of parameters. This is often combined with the goal of a maximum target accuracy [54], [55]. The increased focus on target HW in NAS research is also reflected in a shift of optimization targets from more theoretical metrics like mathematical operations per NN to actually measured latency or memory consumption on the target HW [61].

Given optimization target and search space, NAS is carried out based on a search strategy which is sometimes complemented with a performance estimation strategy [54]. There exists a multitude of search strategies including random search [62], reinforcement learning [58], Bayesian optimization [63] which can be combined with clever performance estimation strategies to avoid retraining every NN topology from scratch. Despite search and performance estimation strategies, the computation cost for NAS remains high and scales with increasing complexity of the design decisions [64].

As a logical consequence, the resulting NN from NAS are tightly coupled with the compute layout of the underlying HW.

For example, [65] found that GPU models tend to be shallower and wider than CPU or mobile models. This reflects the known superiority of GPUs in computations with high operational intensity. Motivated by that, a recent trend in NAS is to investigate the co-design of HW accelerator configurations with NAS for NN topologies [56], [57].

VI. Knowledge Distillation

The idea of knowledge distillation is to use a large, pretrained AI model (the teacher) to better train a much smaller model (the student). Instead of just using the labels, the teacher was trained against, some of the teacher's internal data, i.e. the output of some of the teacher's hidden layers (soft labels), is also used in order to train the student NN. That way, the smaller student model can converge towards a solution, closely to the teacher's solution.

An ensemble of learning algorithms, when trained on the same data, tends to always outperform each of the individual models of the ensemble [66]. The idea of knowledge distillation is based on the observation, that these ensembles can be compressed back into a single, much smaller model [67]. In [68], this idea is extended, introducing soft targets from the teacher NN acting as regularization in the student NN and improving its convergence far beyond the point reachable for the student using only training by avoiding early overtraining.

Based on the result, that the teacher-student efficiency degrades, once the size difference between the teacher and the student NN exceed a certain limit, [69] propose to introduce a third NN, called teacher assistant to step-wise boil down large NN into much smaller, yet accurate student NN.

[70] present a NAS, to search directly for a NN, with flexible channel and layer sizes, in which the knowledge is later transferred by the large NN. The topology is based on minimizing the loss of the NN. The feature map of the pruned NN is an aggregation of K feature map fragments, which are sampled based on the probability. The maximum probability for the size in each distribution serves as the width and depth of the pruned NN, whose parameters are learned by knowledge distillation from original NN.

VII. Conclusion

Some of the techniques presented here have already been worked out very precisely, but some still offer a lot of room for improvement. The combination of different techniques promises a still largely untapped but promising optimisation potential. In particular, NAS-based control of pruning, quantisation and tensor compression seems to be a promising research direction.

Acknowledgment

The research leading to these results is funded by the German Federal Ministry for Economic Affairs and Energy within the project "KI Delta Learning – Development of methods and tools for the efficient expansion and transformation of existing AI modules of autonomous vehicles to new domains." The authors would like to thank the consortium for the successful cooperation.

REFERENCES

[1] J Frankle, M Carbin, *The lottery ticket hypothesis: Finding sparse, trainable neural networks*, arXiv:1803.03635, 2018.

[2] Yihui He et al., *Amc: Automl for model compression and accel-eration on mobile devices*, ECCV, 2018.

[3] Qiangui Huang, Kevin Zhou, Suya You, Ulrich Neumann, *Learning to Prune Filters in Convolutional Neural Networks*, WACV; 2018.

[4] Song Han, Jeff Pool, John Tran, William J. Dally, *Learning both weights and connections for efficient neural networks*, NIPS, 2015.

[5] T Yang et al., *Designing Energy-Efficient Convolutional Neural Networks Using Energy-Aware Pruning*, CVPR, 2017.

[6] Song Han et al., *Eie: Efficient inference engine on compressed deep neural network*, ISCA, 2016.

[7] Angshuman Parashar, et al., *SCNN: An accelerator for compressed-sparse convolutional neural networks*, ISCA, 2017.

[8] Zechun Mu Liu et al. *MetaPruning: Meta Learning for Automatic Neural Network Channel Pruning*, ICCV, 2019.

[9] Yihui He, Xiangyu Zhang, Jian Sun, *Channel pruning for accelerating very deep neural networks*, ICCV, 2017.

[10] Yang Liu He et al., *Filter Pruning via Geometric Median for Deep Convolutional Neural Networks Acceleration*, CVPR, 2019.

[11] Sämann, Timo, et al., *Efficient semantic segmentation for visual bird's-eye view interpretation*, IAS, 2018.

[12] Z You et al. *Gate Decorator: Global Filter Pruning Method for Accelerating Deep Convolutional Neural Networks*, NeurIPS, 2019.

[13] Ting-Wu Chin et al., *Layer-compensated Pruning for Resource-constrained Convolutional Neural Networks*, arXiv:1810.00518, 2018.

[14] Yang He et al. *Soft Filter Pruning for Accelerating Deep Convolutional Neural Networks*, IJCAI, 2018.

[15] Sajid Anwar, Kyuyeon Hwang, Wonyong Sung, *Structured pruning of deep convolutional neural networks*, JETC, 2017.

[16] A Frickenstein et al., *DSC: Dense-sparse convolution for vectorized inference of convolutional neural networks*, CVPR-W 2019.

[17] T Zhang et al., *StructADMM: A Systematic, High-Efficiency Framework of Structured Weight Pruning for DNNs*, arXiv: 1807.11091, 2019.

[18] Wenxiao Wang et al. *DBP: Discrimination Based Block-Level Pruning for Deep Model Acceleration*, arXiv:1912.10178, 2019.

[19] Shuchang Zhou et al., *Dorefa-net: Training low bitwidth convolutional neural networks with low bitwidth gradients*, CoRR; abs/1606.06160.

[20] S Uhlich et al., *Mixed precision DNNs: All you need is a good parametrization*, arXiv:1905.11452, 2019.

[21] L Yang, Q Jin, *FracBits: Mixed Precision Quantization via Fractional Bit-Widths*, arXiv:2007.02017, 2020.

[22] Wang, Kuan Liu, Zhijian Lin, Yujun Lin, Ji Han, Song, *HAQ: Hardware-Aware Automated Quantization With Mixed Precision*, CVPR, 2019.

[23] M R Vemparala, A Frickenstein, W Stechele, *An Efficient FPGA Accelerator Design for Optimized CNNs using OpenCL*, ARCS, 2019.

[24] J Wu et al., *Quantized convolutional neural networks for mobile devices*, CVPR, 2016.

[25] S Han, et al., *Deep compression: Compressing deep neural networks with pruning, trained quantization and Huffman coding*, ICLR, 2016.

[26] Pierre Stock et al., *And the Bit Goes Down: Revisiting the Quantization of Neural Networks*, arXiv:1907.05686, 2019.

[27] Lin, Xiaofan Zhao, Cong Pan, Wei, *Towards Accurate Binary Convolutional Neural Network*, NIPS; 2017.

[28] Mohammad Rastegari et al., *Xnor-net: Imagenet classification using binary convolutional neural networks*, ECCV, 2016.

[29] Dipankar Das et al. , *Mixed Precision Training Of Convolutional Neural Networks Using Integer Operations*, ICLR, 2018.

[30] Yunchao Gong et al., *Compressing deep convolutional networks using vector quantization*, arXiv:1412.6115, 2014.

[31] Julieta Martinez et al., *Permute, Quantize, and Fine-tune: Efficient Compression of Neural Networks*, arXiv:2010.15703, 2020.

[32] B Jacob et al., *Quantization and Training of Neural Networks for Efficient Integer-Arithmetic-Only Inference*, arXiv:1712.05877, 2017.

[33] E Meller et al., *Same, Same But Different - Recovering Neural Network Quantization Error Through Weight Factorization*, arXiv:1902.01917.

[34] Jungwook Choi et al., *PACT: Parameterized Clipping Activation For Quantized Neural Networks*, arXiv:1805.06085. 2018.

[35] Steven K. Esser et al., *Learned step size quantization*, CoRR, arXiv:1902.08153, 2019.

[36] Y Bhalgat et al, *LSQ+: Improving low-bit quantization through learnable offsets and better initialization*, ICCV, 2020.

[37] M Courbariaux, *Binarized neural networks: Training neural networks with weights and activations constrained to +1 or -1.*, 1602.02830.

[38] Wei N Tang Gang Hua Liang Wang, *How to Train a Compact Binary Neural Network with High Accuracy?*, AAAI; 2017.

[39] Z Lin, M Courbariaux, R Memisevic, Y Bengio, *Neural networks with few multiplications*, ICLR, 2015.

[40] Fengfu Li, Bo Zhang, Bin Liu, *Ternary weight networks*, NeurIPS, arXiv:1605.04711, 2016.

[41] Chenzhuo Zhu et al., *Trained ternary quantization*, ICLR, arXiv:1612.01064, 2017.

[42] Zhongnan Qu, Zimu Zhou, Yun Cheng, Lothar Thiele, *Adaptive Loss-aware Quantization for Multi-bit Networks*, arXiv:1912.08883. 2019.

[43] Y D Kim et al., *Compression of Deep Convolutional Neural Networks for Fast and Low Power Mobile Applications*, arxiv:1511.06530, 2016.

[44] Nakajima, Shinichi et al., *Global analytic solution of fully-observed variational bayesian matrix factorization*, JMLR, 14(1):1–37, 2013.

[45] H Kim, C-M Kyung, *Automatic rank selection for high-speed convolutional neural network*, arXiv:1806.10821, 2018.

[46] Y Jieping, *Generalized low rank approximations of matrices*, Machine Learning, 61(1-3):167–191, 2005.

[47] I V Oseledets, *Tensor-train decomposition*, SIAM Journal on Scientific Computing, 33(5):2295–2317, 2011.

[48] X Cao, G Rabusseau, *Tensor Regression Networks with various Low-Rank Tensor Approximations*, arXiv:1712.09520v2, 2018.

[49] Y He, S Han, *Adc: Automated deep compression and acceleration with reinforcement learning*, arXiv:1802.03494, 2018.

[50] X Zhang et al., *Accelerating very deep convolutional networks for classification and detection*, TPAMI, 38(10):1943–1955, 2016.

[51] H Kim, M K Khan, C-M Kyung, *Efficient Neural Network Compression*, arXiv:1811.12781v3, 2020.

[52] Partha Maji et al., *Efficient Winograd or Cook-Toom convolution kernel implementation on widely used mobile CPUs*, CoRR, 2019.

[53] Lingchuan Meng, John Brothers, *Efficient Winograd convolution via integer arithmetic*, CoRR, abs/1901.01965, 2019.

[54] T Elsken et al., *Neural architecture search: A survey*, J. Mach. Learn. Res., 20(55), 1-21; 2019.

[55] M Wistuba et al. *A survey on neural architecture search*, arXiv preprint arXiv:1905.01392; 2019.

[56] Y Zhou et al., *Rethinking Co-design of Neural Architectures and Hardware Accelerators*, arXiv preprint arXiv:2102.08619, 2021

[57] W Jiang et al., *Hardware/software co-exploration of neural architectures*, IEEE TCAD, 39(12), 2020

[58] B Zoph, Q V Le, *Neural architecture search with reinforcement learning*, arXiv preprint arXiv:1611.01578, 2016.

[59] M Tan et al., *MnasNet: Platform-Aware Neural Architecture Search for Mobile*, CVPR; 2019.

[60] Y Xiong et al, *Mobiledets: Searching for object detection architectures for mobile accelerators*, arXiv:2004.14525, 2019.

[61] B Wu et al., *Fbnet: Hardware-aware efficient convnet design via differentiable neural architecture search*, CVPR; 2019.

[62] L Li, A Talwalkar, *Random search and reproducibility for neural architecture search*, PMLR, 2020.

[63] K Kandasamy et al., *Neural architecture search with bayesian optimisation and optimal transport*, arXiv:1802.07191, 2018.

[64] N Wang et al., *NAS-FCOS: Fast neural architecture search for object detection*, CVPR, 2020.

[65] Han Cai Ligeng Zhu Song Han, *ProxylessNAS: Direct Neural Architecture Search on Target Task and Hardware*, ICLR; 2019.

[66] T G Dietterich, *Ensemble methods in machine learning*, Multiple classifier systems, 1–15. Springer, 2000.

[67] C Bucilua, R Caruana, A Niculescu-Mizil, *Model compression*, SIGKDD, 2006.

[68] Geoffrey Hinton, Oriol Vinyals, Jeff Dean, *Distilling the Knowledge in a Neural Network*, arXiv:1503.02531, 2015.

[69] Seyed-Iman Mirzadeh et al., *Improved Knowledge Distillation via Teacher Assistant*, arXiv:1902.03393, 2019.

[70] Xuanyi Dong, Yi Yang, *Network Pruning via Transformable Architecture Search*, NeurIPS, 2019.

The Essential Role of Procedural Approaches in Electronic Design Automation

(Invited Paper)

Daniel Marolt and Jürgen Scheible
Electronics & Drives, Reutlingen University
Reutlingen, Germany
{daniel.marolt, juergen.scheible}@reutlingen-university.de

Göran Jerke and Vinko Marolt
Automotive Electronics, Robert Bosch GmbH
Reutlingen, Germany
{goeran.jerke, vinko.marolt}@de.bosch.com

Abstract—**Electronic design automation approaches can roughly be divided into optimizers and procedures. While the former have enabled highly automated synthesis flows for digital integrated circuits, the latter play a vital (but mostly underestimated role) in the analog domain. This paper describes both automation strategies in comparison, identifying two fundamentally different automation paradigms that reflect the two basic design practices known as "top-down" and "bottom-up". Then, with a focus on the latter, the history of procedural approaches is traced from their early beginnings until today's evolvements and future prospects to underline their practical importance and to accentuate their scientific value, both in itself and in the overall context of EDA.**

Index Terms—**electronic design automation, optimization algorithms, procedures, generators, parameterized cells, analog ICs**

I. Introduction

Electronic design automation (EDA) is a key discipline for the creation of integrated circuits (ICs), having automated the synthesis of digital IC parts via *optimization algorithms*. But apart from a couple of successes (for example in circuit sizing), attempts to apply such *optimizers* to the analog domain have not attained the same industrial acceptance. While verification steps such as simulation, design-rule-check (DRC) and layout-versus-schematic (LVS) are well automated, the creative tasks of analog circuit and layout design still require laborious manual work, supported by rather simple *procedural automatisms*.

Despite their petty abilities, these *procedures* are indispensable in practice today, whereas academia has largely dismissed them for the reason of being too trivial scientifically. However, this misbelief ignores that procedural design approaches have blossomed in a variety of different directions over time. These ongoing developments open up interesting new possibilities, therefore suggesting that procedures might bear a much greater significance for EDA than has been acknowledged so far.

This paper covers the topic of procedural design automation to illuminate its pivotal merits for EDA. First, Section II looks into the origin of the persistent automation gap in the analog domain. Next, Section III compares the traits of the two basic automation strategies (optimizers, procedures) reflective of the two converse design practices called *top-down* and *bottom-up*. Focusing on the latter, Section IV then recounts the historical evolvement of procedural concepts (especially PCells), tools, and use cases. Finally, Section V concludes with a summary and outlines a path towards a *bottom-up meets top-down* flow.

II. The Analog Design Automation Gap

Why are optimizers not as successful in the analog domain as in the digital domain? A closer look at the respective design styles in Section II-A and Section II-B shall give an answer.

A. The Digital Design Style: Standardized

The capabilities of digital IC sections depend on the amount of implemented logic, thus feeding the desire to integrate more and more devices onto an IC. This ongoing trend is well known as *More Moore* [1] and delineates the problem complexity as a matter of *quantity*. Today, digital circuit and layout design has become highly automated via optimizers, but their practical application upon large-scale circuits requires a simplification of the design problem, as achieved by two kinds of abstraction. The first one is to introduce artificial constraints (such as the restriction to signals which are discrete in time and value, as well as the use of fixed-height standard cells that are placed in rows). This enables the second abstraction: handling the design problem as a theoretical model with a focused coverage of the circuit-relevant aspects. The abstraction reduces the degrees of freedom and benefits the optimizer, but entails a loss of design quality, because even an algorithmically optimal solution may be electrically suboptimal in reality. Yet, this *quality gap* can be tolerated due to the discrete nature of digital signals, and so the overall automation strategy sustains itself (Fig. 1, left). In digital design, the design flow determines the design style.

B. The Analog Design Style: Full-Custom

While digital sections represent the "brain" of an IC, analog parts provide it with various voltage supplies and serve as the interface to its continuous-valued environment. The immanent challenges to handle non-linearities, parasitic effects, thermal influences, high voltages, and diverse physical quantities are called *More than Moore* [2] and make the complexity of analog IC design a matter of *quality* rather than quantity. Maintaining signal integrity calls for an optimal (full-custom) design that utilizes the entire spectrum of all available degrees of freedom. This in turn opposes an abstraction of the design problem, since that would cause even the theoretically global optimum to be insufficient for practical application. On the other hand, without abstraction, conventional algorithms can not ensure to optimally solve the NP-hard design problem within practical

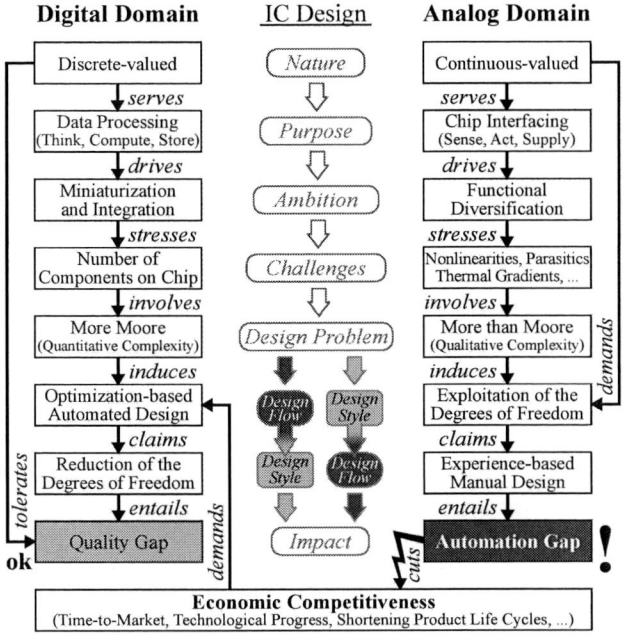

Fig. 1. Implications of IC design in the digital domain and the analog domain.

runtime ranges. This dilemma, first visualized in [3], puts a natural limit on the capabilities of optimizers, in chapter 4 of [4] denoted as *optimization horizon*. Hence, analog design is still done in a manual way, relying on the knowledge, experience, skills, and creativity of human experts. So, economic penalties due to this *automation gap* are tolerated as the lesser of two evils, because design quality does not permit any trade-offs (Fig. 1, right). In analog design, entirely opposite to the digital domain, the design style determines the design flow.

III. TWO DISTINCT AUTOMATION PARADIGMS

While optimizers are rather inadequate for analog IC design, procedures offer at least some relief. This section inspects both strategies in detail (III-A, III-B) and compares them (III-C).

A. Optimizers: Top-Down Automation

In EDA, an optimizer is a relatively universal automatism for a certain task such as topology selection and circuit sizing (in schematic design) or floorplanning, placement, and routing (in layout design). It treats that design task as a mathematical *optimization problem* defined by design restrictions and optimization goals, translating the problem into a theoretical model to find the best *solution* via a repetitive loop of *exploration* (selecting a new solution candidate) and *evaluation* (rating the candidate according to a metric) [5]. Cycling through this loop can require millions of iterations and substantial computation time until a stop criterion is met. Depending on the problem complexity, heuristics and randomization may be employed.

The paradigm behind this strategy fits into top-down design flows (which tackle the design problem from a strategic global perspective) and is called *top-down automation* due to a major strength: the ability to take into account formal expressions of

abstract, high-level design requirements that easily elude the scope of human attention in manual design. On the other hand, top-down automation only optimizes those aspects which are specifically covered by the model used to abstract the problem (hence, "optimized" does not necessarily mean "optimal" [6]).

This abstraction restrains an incorporation of human expertise because that expert knowledge must be *explicitly* expressed in a formal, comprehensive, unambiguous and consistent representation that can be processed by the algorithm [7]. In analog design however, where expert knowledge remains indispensable to achieve proper electrical functioning, it is tremendously intricate to *efficiently* and *sufficiently* describe the full diversity, various impacts and correlated dependencies of all crucial design restrictions and optimization goals in an explicit fashion. This issue is particularly critical when many tightly-linked, contrary, low-level design requirements need to be satisfied concurrently. Therefore, optimizers alone remain far from suitable for handling *More than Moore* complexities.

B. Procedures: Bottom-Up Automation

In contrast to optimizers, a procedure can be regarded as a script-like sequence of successive commands, most of which perform simple design operations (like drawing or moving a shape). A procedure takes a set of user-given input parameters and executes its command sequence (typically in a few seconds or less) to produce a deterministic result (e.g., a schematic or a layout), usually for a very specific design component [8]. That *result* is merely some design output[1] whereas the cognitive *solution* is –contrary to an optimizer– not found at runtime, but was preconceived by the design expert who implemented the procedure, often based on many years of practical experience. The parameters represent degrees of freedom which provide a certain amount of flexibility to customize the result according to the respective requirements and/or the design context.

With the inventive task of devising a solution being left to the human expert's creativity, a procedure merely mimics the expert's laborious needlework. Compared to top-down automation, procedures thus follow a different philosophy, capturing expert knowledge in an informal manner and handling design requirements *implicitly*. This paradigm is denoted as *bottom-up automation* and supports the common practice of bottom-up design, where local details of low-level design entities are addressed before giving rise to more complex components, thereby successively reaching the overarching design intent.

Apart from design automation, one other major approach to increase design productivity is re-use. However, re-using existing designs is impeded by the drawback that a fix design has no degrees of freedom. Procedural approaches resolve this problem with parametrization. They perform re-use not by duplicating singular circuits or layouts but by replicating a human expert's way of engineering such a design. This augments the copy-paste fashion of re-use to a more sophisticated form of

---

[1]The type of output depends on the procedure. For example, if the procedure is a *generator* that creates a layout, its output consists of layout shapes. If it is a schematic generator, its output involves a netlist. If the procedure implements a circuit sizing routine, its output represents a set of (mostly numeric) values.

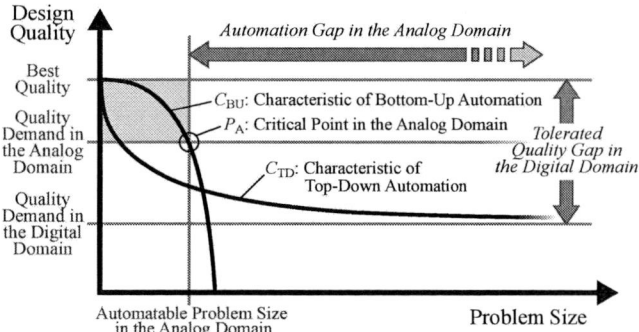

Fig. 2. Quality characteristics of top-down and bottom-up automation [8].

re-using expert knowledge. As a primal note about the basic traits of the underlying strategy, top-down automation requires an *abstraction* of the *problem*, which *lessens* the degrees of freedom, whereas the bottom-up paradigm relies on the *generalization* of a conceived *solution* (by providing parametrical variability), which *increases* the degrees of freedom. In that manner, the bottom-up paradigm naturally combines the two main efficiency-raising measures available in analog IC design, facilitating *automation* through *re-use* (and vice versa).

C. Comparison of Top-Down and Bottom-up Automation

In terms of their respective strengths, top-down automation (explicit consideration of abstract high-level requirements) and bottom-up automation (implicit handling of complex low-level requirements) precisely complement each other. In regard to a design problem's size, the two paradigms exhibit characteristic quality losses compared to the best possible degree of design quality (full-custom) as depicted in Fig. 2 (adapted from [8]).

Top-down automation involves a significant initial loss of design quality due to its abstraction of the problem (shown by curve C_{TD}). Yet, this way of reducing the degrees of freedom allows optimizers to tackle problems of immense quantitative complexity. With increasing problem size, the involved quality decline continues, but becomes more and more marginal. The overall quality gap can be tolerated in the digital domain, but dissatisfies the tight quality requirements of analog IC design.

The bottom-up way of (re-)generating pre-determined design solutions has an inverse characteristic curve (C_{BU}). It is quite feasible for frequently used analog basic circuits (whose design requirements are usually well known from experience). For such small problem sizes, the achievable quality is close to full-custom. However, larger design problems precipitate an ever-increasing quality drop since the design requirements are hard to anticipate in advance. In the analog domain, bottom-up automation can thus ensure the demanded degree of quality up to a certain problem size (P_{A}). The remaining automation gap is continually reduced by improving procedural approaches.

IV. THE EVOLUTION OF PROCEDURAL APPROACHES

Throughout the last 40 years, procedural design automation has evolved from the origin of PCells (IV-A) over their usage in layout design (IV-B) to applications in circuit design (IV-C).

A. Basic Remarks

Parameterized Cells (PCells): The history of procedures dates back to the 1980s, when –originating from simple scripts for layout creation– the groundbreaking concept of *parameterized cells* (PCells) arose. The notion of a *parameterized cell* hints at the fact that a PCell[2] embodies two crucial traits in one single entity: on the one hand it is a parameter-controlled software routine for generating a design, on the other hand it is a library component that can be instantiated like an ordinary (fix, non-parameterized) cell, encapsulating the created design objects. These traits facilitated the so-called *schematic-driven layout* (SDL)[3] flow in the 1990s, which is still fundamental today.

Masters and Instances: When we instantiate a PCell, we get a PCell *instance*. The procedure (an abstract piece of software as opposed to its concrete instantiations in a parent cell) that produces the design objects[4] inside that instance is denoted as the PCell *master*. The term "PCell" can be used to denote both (an instance or its master). Likewise, the term "parameter" is also used ambiguously: it can refer to a parameter *specification* (name, type, etc.) or to a parameter *value*. The former belongs to the PCell master whereas the latter is set for an individual instance of that PCell (typically via a parameter input form). If you *flatten* a PCell instance, its contents are "unpacked" (raised by one level of hierarchy) and the instance vanishes.

Programming Languages: In the early years, implementing a PCell required textual coding. While there have been domain-specific languages such as STICKS [9] (1978), Layla [10] (1985) and MOGLAN [11] (1998), the three IC development frameworks of today's market-leading EDA companies feature their own general-purpose languages for PCell development: Mentor Graphics' *Pyxis Custom IC Design Platform* provides the C-like scripting language AMPLE, the *Galaxy Design Platform* by Synopsys uses Python, and Cadence *Virtuoso* has the proprietary Scheme-based scripting language SKILL.

B. Layout Design

Relative Object Design: Though the very first PCells targeted simple devices, substantial programming effort was required for computing and tracking point coordinates. For that reason, the early years of the 1990s lead to the origin of *relative object design* (ROD). The basic idea of ROD is to provide powerful functions for specifying persistent relationships between PCell shapes, thus making it easier to build and manipulate geometries (e.g., to create a new shape from an existing one).

Stretchline Generators: ROD-driven, the convenient concept of *stretchline generators* arose in the 1990s. It enables PCell developers to draw a *stretchline* (also: *stretch control line*), set a stretch direction (horizontal, vertical, or both), choose PCell shapes that are to be affected by that stretch, and associate the stretch with a PCell parameter (letting PCell users enter their desired stretch amount). This concept, which may also include

---

[2]The remaining text uses *PCell* as a synonym for *procedural generator*, thereby also implicating other vendors' names (PyCells, T-Cells, Magic Cells, user-defined devices, intellectual property, etc.) denoting the same concept.
[3]Another common term for this concept is *connectivity-driven layout*.
[4]A shape inside a PCell instance is subsequently called a *PCell shape*.

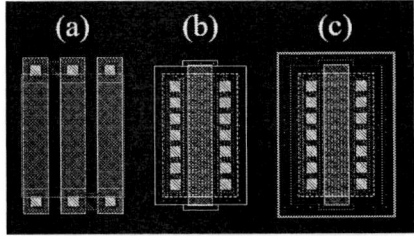 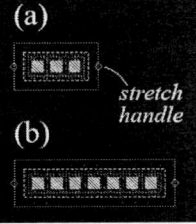

Fig. 3. Left: (a) folded poly resistor, (b) NMOS transistor, (c) PMOS transistor. Right: stretchable substrate contact PCell: (a) default variant, (b) stretched.

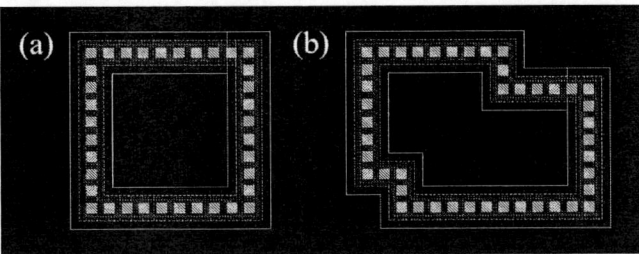

Fig. 4. Fluid guard ring PCell: (a) default variant, (b) stretched and chopped.

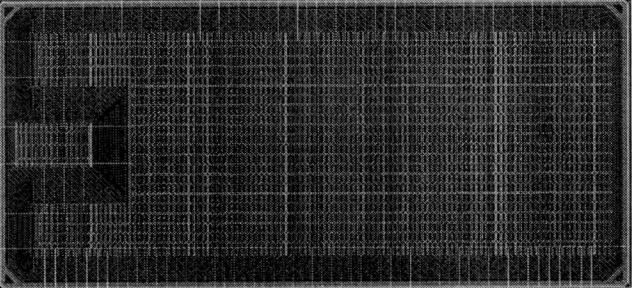

Fig. 5. Hierarchical power MOS PCell using fix cells (corners), PCells (sides), and a ⊐-shaped fluid PCell (inside) around a hierarchical pad PCell (left).

further features (e.g., repetition or conditional inclusion), eased the creation of resistor and transistor PCells (see Fig. 3, left).

Stretchable PCells: The *stretchline* concept above is not to be confused with *stretchable PCells* (also from the 1990s). The former reduces a PCell's implementation effort, whereas the latter enhances its usability. Stretchable PCells are created by specifying *stretch handles* on a PCell shape and assigning them to a numeric PCell parameter. Users can stretch the PCell instance by dragging a stretch handle via mouse (Fig. 3, right). Under the hood, this modifies the respective parameter's value (which has the same effect as editing that value textually).

Fluid PCells: Sooner or later, stretchable PCells have evolved into *fluid PCells*. They can mark internal shapes as being fluid, making them editable from the outside. If such a *fluid shape* (other vendors may use the term *model line*) is modified by the user, its coordinates are turned into a PCell parameter value which is parsed by the PCell to update its contents accordingly. The difference to stretchable PCells (that follow a similar idea) is that the shape-related parameter value holds a point list (not a mere number), which enables fluid PCells based on arbitrary polygons and paths with full editing support (e.g., fluid PCells can not only be stretched but also chopped, as shown in Fig. 4). Main fluid applications are wells, guardrings, and capacitors.

Hierarchical PCells: PCells became infinitely more powerful when they were equipped with the ability to instantiate other PCells. Such *hierarchical PCells* enabled the implementation of compound *modules* (e.g., a current mirror PCell that places many transistors side by side and connects them). Furthermore, hierarchical PCells foster a divide-and-conquer approach of modularity (imagine a power transistor PCell assembling multiple instances of one or more sub-PCells as in Fig. 5). To cope with the intricacies of hierarchy (especially when it involves many levels), auxiliary concepts such as HIPE [12] are helpful.

Wrapper PCells: Hierarchical PCells offered an interesting use case: *wrapper PCells*. In its typical form, a wrapper PCell instantiates a native device PCell from the *process design kit* (PDK) and enhances its layout (e.g., by providing device-level routing). But it may also occur, that the layout of a native PDK PCell is flawed. A *repair PCell* can remedy this situation by first instantiating the PDK PCell, then flattening that internal instance (to eliminate one level of sub-hierarchy), and finally deleting/modifying the faulty PCell shapes that arose from the flattening (note: the repair PCell itself is not flattened hereby).

PCell Development Tools: Beginning in this millennium, the bloom of integrated development environments directed EDA attention to the realization of professional PCell creation tools such as Freescale's *PCell Compiler* [13], IPGEN's *1Stone Developer* [14], SpringSoft's *Laker Custom Layout Automation System* [15], *HiPer DevGen* [16] by Tanner EDA, AnaGlobe's *Geometric Object Layout Formula* [17], Ciranova *PyCell Studio* [18], *Berkeley Analog Generator* (BAG) [19], Fraunhofer's *IIP Framework* [20], and Cadence *PCell Designer* [21].

PCell Development Concepts: Tools like the ones listed above bring along further PCell development concepts. Naming just a few, they include: GUI support (e.g., graphical PCell code editing, layout preview, cross-probing between code and layout), shape import (turning a drawn shape into a shape-creation command call in the PCell code), nestable groups of shapes (for applying a command to a set of shapes or certain subsets), inheritance (deriving a PCell from another PCell, e.g., NMOS from MOS), geometry expressions (to conveniently query and measure geometrical quantities), and technology abstraction (e.g., retrieving PDK-specific values from a lookup table).

Context-Based PCells: The fluid concept (as discussed above) blurs the line between a PCell's internal world and its external context. This idea is driven further by *context-based PCells*, e.g. by *statically context-enhanced PCells* [22] (SCE PCells) where arbitrary context information (not just a single shape) is encoded as a formatted string and passed to the PCell (via a dedicated parameter) which decodes the string again (as [23] does for a routing PCell). But in contrast to fluid PCells, there is no universal method for encoding and decoding arbitrary context data. Furthermore, SCE PCells do not automatically get notified of context changes. Overcoming these limitations would facilitate fully *context-aware PCells*, but no such imple-

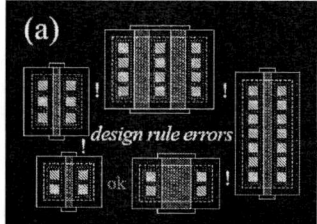 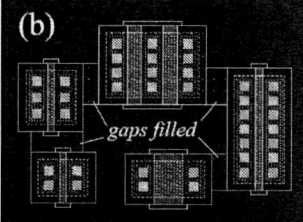

Fig. 6. appCell: five NMOS instances (a) with DRC-violating too-small gaps and (b) with gap-filling shapes directly created in the design by an appCell.

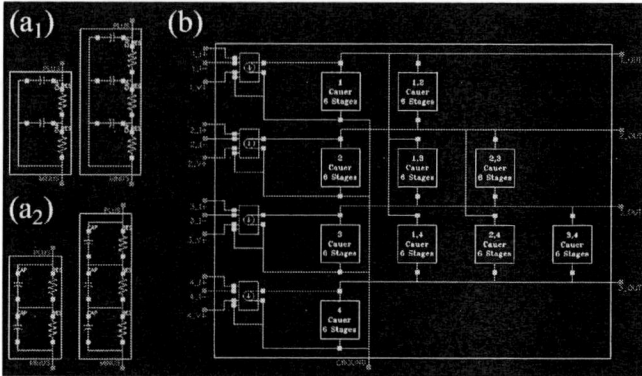

Fig. 7. RC ladder PCell instances with varying number of stages in (a₁) Cauer topology and (a₂) Foster topology, used inside an (b) RC ladder network PCell to evaluate the thermal behavior of IC devices through circuit simulation [35].

mentation is available so far ([31] provides an *interface fabric* as a makeshift solution). Still, PCells with context access are a hot topic now (especially in consideration of increasing neighborhood effects), for example investigated in RF design [24].

appCells: In 2016, the concept of *appCells* was introduced by Cadence PCell Designer [25]. Utilizing the tool's rich PCell development functionalities, appCells can be implemented like conventional PCells. But in contrast to those, appCells are not *instantiated* in a design. Instead, they are *applied* to the design as "apps", driven by parameters and the design content. An appCell can access (read/modify) existing design objects and create new ones directly in the design as in Fig. 6 (thereby not leaving behind any PCell instances). AppCells are deployed as traditional library components and can be easily integrated into the design flow via bindkeys, toolbar icons and menu entries.

Further Works: Further works focused on procedural layout automation are (amongst others): CAPABLE [26] (to ease the utilization of PCells in the design flow), [27][28] (abstract generator descriptions to attain technology-independence), MESH [29] (for generating analog arrays), [30] (facilitating template-driven layout generators), SWARM [31] (which implements an interaction of autonomous PCells to provoke the emergence of overall self-organization) and MEMS+ [32] (a tool able to create a layout PCell for a micro-electro-mechanical design).

C. Circuit Design

Circuit PCells: Although PCells are predominantly utilized in layout design, they can also create schematic diagrams and their symbol views. Such *schematic PCells* and *symbol PCells* are subsumed under *circuit PCells*. Early examples from 2004 and 2005 (done in plain SKILL) are given by [33] and [34], presenting a parameterized transmission line, I/O buffer, and inverter. In analog design, circuit PCells help reduce schematic drawing effort and also maintain SDL functionality above the device level. PCell examples of an OTA, RC ladder, and RC ladder network are demonstrated in [35] (also see Fig. 7).

Circuit PCell Development: In 2007, the procedural circuit description language *Stratus* was presented by [36], targeting VLSI modules. Analog circuits are supported by the already mentioned BAG from 2013, facilitating schematic generators (e.g., [19] shows a variable resistor string). In 2014, the *Parameterized Circuit Description Scheme* (PCDS) [35] introduced several new ideas, for example to create two PCells –schematic and symbol– from one single PCDS description. Driven by

PCDS and gPCDS [37], Cadence PCell Designer also supports the creation of schematic and symbol PCells since 2017 [38].

Procedural Circuit Design: Procedural automatisms are quite adequate for circuit design (topology selection and sizing), and textbooks like [39] readily describe *design procedures* in prose and equations. Early systems with executable design programs (often called *design plans*) are IDAC [40] (1987) and OASYS [41] (1989), followed by many others [42]. The *Procedural Analog Design* (PAD) tool [43] (2008) also treats the design problem with a predefined procedural design sequence (not as a mathematical problem). BAG (2013) supports the creation of sizing scripts via helper classes with template architectures and design routines. In 2017, the *Expert Design Plan* (EDP) approach [44] (which is open to enhancements such as design-space transformation [45] and machine learning [46]) initiated the domain-specific language EDPL [47]. It tries (as potential graphical tools also should) to match the mindset and workflow of human designers as closely as possible, and thus to furnish a convenient bridge allowing those IC experts to create their own custom procedures that fit their respective design environments and circuits while being generic enough to sustain re-usability.

V. SUMMARY AND OUTLOOK

EDA distinguishes optimizers from procedures. The former have highly automated digital design but are not equally suited for analog IC parts since the two domains are entirely different with regard to complexity and design style. Likewise, the two automation strategies follow two distinct paradigms, *top-down* and *bottom-up*, with analog design mostly relying on the latter.

In the last four decades, procedural approaches (especially PCells) have evolved in diverse ways, enriching both layout and circuit design with helpful generators, tools, development methods and usage concepts in different fields (digital, analog, RF, MEMS), stand-alone or combined with other techniques.

In the long run, procedural automatisms have the potential to cover the entire scope of analog basic circuits. Mirroring the essential role of standard cells in digital design, PCells and other procedures could thus provide an abstraction level high enough for a successful utilization of top-down optimizers.

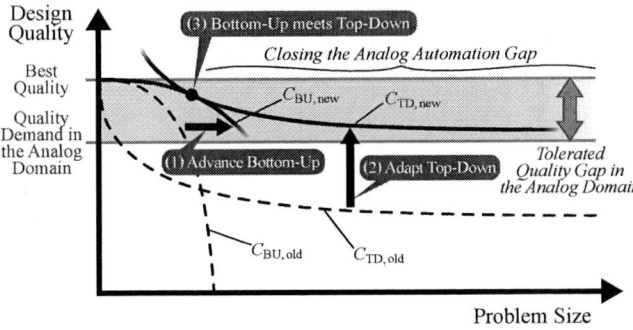

Fig. 8. Milestones leading to a *bottom-up meets top-down* design flow [8].

That idea of teaming bottom-up with top-down automation could ultimately resolve the analog automation gap. Hence, a future mission for EDA is to tap the full potential of both paradigms and drive their convergence toward a novel *bottom-up meets top-down* flow [3]. A roadmap leading to that goal takes five steps [8] visualized in Fig. 8: (1a) provide instruments for capturing expert knowledge implicitly, (1b) turn that implicit expert knowledge into new bottom-up automatisms, (1c) integrate these into present IC design flows in an evolutionary way, then (2) tailor existing top-down algorithms to the new bottom-up automatisms, and finally (3) combine the two paradigms. This accomplishment would be a major breakthrough for EDA.

REFERENCES

[1] Semicond. Industr. Assoc., "Executive Summary", *ITRS*, pp. 1–89, 2005.

[2] W. Arden / M. Brillouët / P. Cogez / M. Graef / B. Huizing / R. Mahnkopf, "'More-than-Moore' White Paper", *ITRS*, pp. 1–31, 2010.

[3] J. Scheible / J. Lienig, "Automation of Analog IC Layout – Challenges and Solutions", *Proc. of ISPD*, pp. 33–40, Mar./Apr. 2015.

[4] J. Lienig / J. Scheible, "Fundamentals of Layout Design for Electronic Circuits", Springer, 2020, DOI: 10.1007/978-3-030-39284-0.

[5] R. A. Rutenbar, "Analog CAD: Not Done Yet", *NSF Workshop*, 2009.

[6] J. Scheible / D. Marolt / M. Schweikardt / H. Habal, "Optimiert ist nicht immer optimal: Automatisierung des Entwurfs analoger ICs – Teil 1", *Elektronik*, no. 6, pp. 46–50, Mar. 2019.

[7] G. Jerke / J. Lienig, "Constraint-Driven Design: The Next Step Towards Analog Design Automation", *Proc. of ISPD*, pp. 75–82, Mar. 2009.

[8] D. Marolt, "Layout Automation in Analog IC Design with Formalized and Nonformalized Expert Knowledge", Ph.D. Thesis, Stuttgart, 2018.

[9] J. D. Williams, "STICKS – A Graphical Compiler for High Level LSI Design", *Proc. of National Computer Conf.*, pp. 289–295, Jun. 1978.

[10] W. E. Cory, "Layla: A VLSI Layout Language", *Proc. of 22nd Design Automation Conference*, pp. 245–251, Jun. 1985.

[11] M. Wolf / U. Kleine / J. Schulze, "New Description Language and Graphical User Interface for Module Generation in Analog Layouts", *Proc. of ISCAS*, vol. 6, pp. 290–293, May/Jun. 1998.

[12] D. Marolt / T. Burdick / G. Jerke / P. Herth / V. Marolt / J. Scheible, "HIPE: Hierarchical Instance Parameter Editing of Parameterized Modules in Analog IC Design", *Proc. edaWorkshop*, pp. 18–23, May 2016.

[13] J. Perez / L. Kasel, "Method and Apparatus for Compiling a Parameterized Cell", *Patent*, Nov. 2007, Patent Number: US 7296248 B2.

[14] D. Friebel, "Automatic Analog IP Generation with 1Stone", *MunEDA User-Group-Meeting Europe*, Nov. 2009.

[15] SpringSoft, Inc., "Laker User Guide and Tutorial", *Laker Custom Layout Automation System*, Jul. 2009, Laker 3.2v4p3.

[16] Tanner, "High Performance Device Generation", *White Paper*, 2010.

[17] AnaGlobe Technology, Inc., "GOLF PCell Designer", *Brochure*, 2010.

[18] Ciranova, Inc., "Ciranova PyCell Studio Tutorial", May 2010.

[19] J. Crossley / A. Puggelli / H.-P. Le / B. Yang / R. Nancollas *et al.*, "BAG: A Designer-Oriented Integrated Framework for the Development of AMS Circuit Generators", *Proc. of ICCAD*, pp. 74–81, Nov. 2013.

[20] B. Prautsch / U. Eichler / S. Rao / B. Zeugmann / A. Puppala / T. Reich / J. Lienig, "IIP Framework: A Tool for Reuse-Centric Analog Circuit Design", *Proc. of SMACD*, pp. 1–4, Jun. 2016.

[21] G. Jerke / V. Marolt / C. Bürzele / P. Herth / T. Burdick, "Visual PCell Programming with Cadence PCell Designer", *CDNLive! EMEA*, 2013.

[22] Cadence Design Systems, https://patents.justia.com/patent/10460069.

[23] K. Sperl, "Ein neuartiger Automatisierungsansatz zur Kombination von Modulgeneratoren mit parametrisierten Verdrahtungs-Zellen im analogen Layoutentwurf", M.Sc. Thesis, Reutlingen University, Mar. 2014.

[24] M. Thoma / D. Marolt / J. Scheible, "Entwurf kontextbasierter PCells für Hochfrequenzanwendungen in modernen CMOS-Technologien", *Proc. of 58th MPC-Workshop*, Reutlingen, Germany, pp. 33–40, Jul. 2017.

[25] G. Jerke / V. Marolt / C. Bürzele / J. Rajpurohit / P. Herth / T. Burdick / G. Cao, "Custom Silicon Design Automation with Cadence PCell Designer", *CDNLive! EMEA*, May 2019, CUS-Techtorial V.

[26] D. Marolt / J. Scheible / G. Jerke / V. Marolt, "CAPABLE: A Layout Automation Framework for Analog IC Design", *Proc. of 54th MPC-Workshop*, Ulm, Germany, vol. 54, pp. 49–59, Jul. 2015.

[27] B. Prautsch / U. Eichler / T. Reich / A. Puppala / J. Lienig, "Abstract Technology Handling for Generator-Based Analog Circuit Design", *Proc. of ZuE, Symposium Reliability by Design*, pp. 56–61, Sep. 2015.

[28] B. Prautsch / U. Eichler / T. Reich / J. Lienig, "Explicit Feature and Edge Insertion for Improved Analog Layout Generators in Advanced Semiconductor Technologies", *Proc. of ANALOG*, pp. 22–27, Sep. 2016.

[29] B. Prautsch / U. Eichler / T. Reich / J. Lienig, "MESH: Explicit and Flexible Generation of Analog Arrays", *Proc. SMACD*, pp. 1–4, 2017.

[30] B. Prautsch / U. Hatnik / U. Eichler / J. Lienig, "Template-Driven Analog Layout Generators for Improved Technology Independence", *Proc. of ANALOG*, pp. 156–161, Sep. 2018.

[31] D. Marolt, "SWARM: A Novel Methodology for Integrated Circuit Layout Automation Based on Principles of Self-organization", *Fortschritt-Berichte VDI*, Reihe 20, Nr. 475, VDI Verlag, Germany, 2020.

[32] Coventor, https://www.coventor.com/products/coventormp/mems-plus/.

[33] P. Bhushan / R. Mitra, "Schematic PCell Implementation in Virtuoso Platform", *Proc. of Internat. Cadence Users Group Conf.*, Sep. 2004.

[34] P. Bhushan / R. Arumugam, "Schematic PCells, Future of Deep Submicron Custom IC Design", *Proc. of CDNLive! Silicon Valley*, Sep. 2005.

[35] D. Marolt / M. Greif / J. Scheible / G. Jerke, "PCDS: A New Approach for the Development of Circuit Generators in Analog IC Design", *Proc. of Austrochip*, pp. 1–6, Oct. 2014.

[36] S. Belloeil / D. Dupuis / C. Masson / J.-P. Chaput / H. Mehrez, "Stratus: A Procedural Circuit Description Language Based Upon Python", *Proc. of ICM*, pp. 1–4, Dec. 2007.

[37] M. Greif / D. Marolt / J. Scheible, "gPCDS: An Interactive Tool for Creating Schematic Module Generators in Analog IC Design", *Proc. of PRIME*, pp. 1–4, Jun. 2016.

[38] G. Jerke / V. Marolt / C. Bürzele / P. Herth / T. Burdick, "Schematic and Symbol PCell Development with Cadence PCell Designer", *CDNLive! EMEA*, May 2017.

[39] P. E. Allen / D. R. Holberg, "CMOS Analog Circuit Design", Third Edition, Oxford University Press, 2012.

[40] M. G. R. Degrauwe / O. Nys / E. Dijkstra / J. Rijmenants / S. Bitz *et al.*, "IDAC: An Interactive Design Tool for Analog CMOS Circuits", *Journal of Solid-State Circuits*, vol. 22, no. 6, pp. 1106–1116, Dec. 1987.

[41] R. Harjani / R. A. Rutenbar / L. R. Carley, "OASYS: A Framework for Analog Circuit Synthesis", *IEEE Transactions on Computer-Aided Design*, vol. 8, no. 12, pp. 1247–1266, Dec. 1989.

[42] G. G. E. Gielen / R. A. Rutenbar, "Computer-Aided Design of Analog and Mixed-Signal Integrated Circuits", *Proceedings of the IEEE*, vol. 88, no. 12, pp. 1825–1852, Dec. 2000.

[43] D. Stefanović / M. Kayal, "Structured Analog CMOS Design", *Analog Circuits and Signal Processing* series, Springer, Netherlands, 2008.

[44] F. Leber / J. Scheible, "A Procedural Approach to Automate the Manual Design Process in Analog Integrated Circuit Design", *Proc. of ANALOG*, pp. 175–180, Sep. 2018.

[45] M. Schweikardt / J. Scheible, "Improvement of Simulation-Based Analog Circuit Sizing using Design-Space Transformation", *Proc. of SMACD*, Jul. 2021, in press.

[46] Y. Uhlmann / M. Essich / M. Schweikardt / J. Scheible / C. Curio, "Machine Learning Based Procedural Circuit Sizing and DC Operating Point Prediction", *Proc. of SMACD*, Jul. 2021, in press.

[47] M. Schweikardt / Y. Uhlmann / F. Leber / J. Scheible / H. Habal, "A Generic Procedural Generator for Sizing of Analog Integrated Circuits", *Proc. of PRIME*, pp. 17–20, Jul. 2019.

© VDE VERLAG GMBH · Berlin · Offenbach

Trash or Treasure? Machine-learning based PCB layout anomaly detection with AnoPCB

Henning Franke[†§], Paul Kucera[†§], Julian Kuners[†§], Tom Reinhold[*],
Martin Grabmann[*], Patrick Mäder[†], Marco Seeland[†], Georg Gläser[*]

[*]IMMS Institut für Mikroelektronik- und Mechatronik-Systeme gemeinnützige GmbH (IMMS GmbH), Ilmenau, Germany
[†]Technische Universität Ilmenau, Germany
[§]*Equal contribution*
Email: {tom.reinhold, martin.grabmann, georg.glaeser}@imms.de,
{henning.franke, paul.kucera, julian.kuners, patrick.maeder, marco.seeland}@tu-ilmenau.de

Abstract—**Designing PCBs requires experience and knowledge about bad designs, e.g. sensitive signals being influenced by aggressively switching ones in close proximity. Those interactions can hardly be detected by formal checks and are usually treated in review processes or by lengthy design iterations. An automated way for identifying potentially harmful regions is still missing. This contribution introduces *AnoPCB*, a tool for automated detection of such potentially harmful regions. Due to the lack of pre-classified data and tremendous potential for bad design choices, AnoPCB is designed as unsupervised method and detects deviations from well-known design practices. In addition, we made AnoPCB freely-available as plugin for the open-source KiCad PCB design environment. Instead of using top-down imagery, AnoPCB processes geometrical relationships and signal properties in terms of layout slices containing category-based signal annotations. After training the anomaly detection on well-functional PCB layouts, our system is able to identify novel and potentially anomalous design patterns in new PCB layouts. We demonstrate our approach using freely available PCB layouts from the HackRF projects and showcase how novel design patterns are detected by AnoPCB.**

I. INTRODUCTION

Transforming design knowledge and experience into usable engineering tools is the major goal of modern design automation. Thus, modern PCB layout design relies on different checking methods to target issues like couplings, signal integrity or electromagnetic interference that are known to cause design malfunction or performance degradation. These checks provided by commercial tools are shown for instance in [1]. Geometry based checks are used to ensure manufacturability or match trace length and impedance. In addition, simulation-based electrical and signal integrity checks determine couplings or current paths. They usually rely on field-solvers or signal-level models [2] that are later included in a circuit simulation. The later hence require simulation models for all components. This result can then be used to optimize a given layout with respect to different properties [3].

Still, these tools are very specifically aiming at known issues. The results are often to be interpreted and only hint whether the layout follows good design practice. In advanced layouts, this leads to the requirement of reviews and consultation of experienced designers. With this contribution, we aim at shortening this design-evaluate-feedback loop as shown in

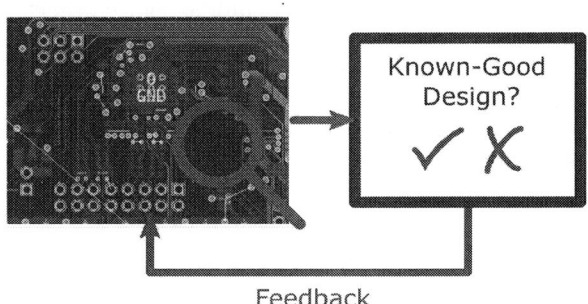

Fig. 1. The design and feedback loop for PCB design.

Fig. 1 by using machine learning to model the *experience* about good design into design tools.

The task of including the *rules* of good design into executable code is therefore the main challenge. The tools and concepts of machine learning target this problem in an approach orthogonal to conventional checks: Instead of programming checks or algorithms, an approximation method is *trained* to realize the wanted check *well enough*. This comes at the cost of a hard-to-explain result and the need for an appropriate preprocessing and ideally labeling [4, p.95f].

In PCB design, we face the situation of existing, unlabeled data. Besides from industrial data of previous products, the open source community forged several designs such as for instance the HackRF [5] board. Even if multiple versions with changes due to electrical effects exist, the areas of good or bad design practice is not marked precisely. Therefore, we assume, that final versions of existing designs are done mostly in a good way. The task of identifying areas following non-optimal design patterns can therefore not be directly done since a data set representing these practices is near impossible to provide. Hence, we map this problem on the challenge of detecting deviations from known-good designs realized by an *anomaly detection* method. For supplying data to the algorithm, we create slices of the PCB layout augmented with category information about the signal properties for marking for instance sensitive signals. This results in slice *pictures* that are *colored* by signal properties. We use this

© VDE VERLAG GMBH · Berlin · Offenbach

data to train a convolutional autoencoder, a type of deep neural network. This network is trained to reconstruct input data after compression to a latent representation of notably reduced size. During training, the network learns to compress and successfully reconstruct layout slices of well-functional layouts. However, the network fails to reconstruct data that differs from training data. Anomaly detection, i.e., detection of novel and potentially harmful layouts, can thus be realized by evaluating the reconstruction error of the network.

This contribution is structured as follows. After introducing the preprocessing of PCB layout data and presenting the autoencoder approach for anomaly detection, we present our KiCAD plugin. This plugin is then trained on open source PCB layouts and used to evaluate modified layouts resulting in unseen data. We conclude with the discussion of the results of this case study.

II. PCB LAYOUT PROCESSING WITH MACHINE LEARNING

To process PCB layouts with machine learning algorithms, an appropriate data format has to be found that allows for the integration of signal properties and geometrical relationships. Usually, layout shapes are given in form of polygons on the different routing layers. Machine learning systems or especially deep neural networks expect the input data to be presented in form of vectors, matrices or, in general, tensors. A first idea is to use rasterized chunks of the layout resulting in 3D tensors with the layers being represented by the third dimension. If these should be annotated with signal information, this would result in 4D tensors exhibiting difficulties in processing as for instance demand for custom network configurations and potentially high memory consumption. Hence, a more lightweight form is to be found that both preserves neighborhood and signal information. For this reason, we use horizontal and vertical layout slices through the signal lines as shown in Fig. 2 that are annotated with signal information. For preserving the geometrical properties, we rasterize with a step width of 4 times the minimum track-width specified in the PCB's design rules. The neighborhood is included by choosing the default width of the slices as 28 times the minimum track-width divided by 4. As in the naive tensor representation, this method also assumes that all layouts use the same number of routing layers. Consider this slice as an 2D image with the colors representing different signal attributes. For encoding these attributes, we categorize the signals by their type as for instance stable analog signals, switching digital signals or power routing. These categories are one-hot encoded and put into the third dimension tensor, i.e. the *color channels* of the slice image.

For detecting anomalous layout configurations, i.e. configurations that are not present in the training set, we use an autoencoder [4, p. 493f] as shown in Fig. 3. This autoencoder is a neural network that learns to create a copy, i.e., reconstruction, of an input image. To make this task non-trivial, autoencoders are implemented with a bottleneck, i.e., an intermediate layer with a significantly reduced dimension. This bottleneck forces the network to first compress the input-data and reconstruct

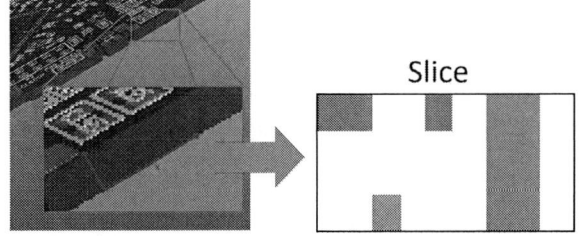

Fig. 2. The slices are taken from the PCB through all layers. Colored areas represent areas with signals. The different colors correspond to different signal-types.

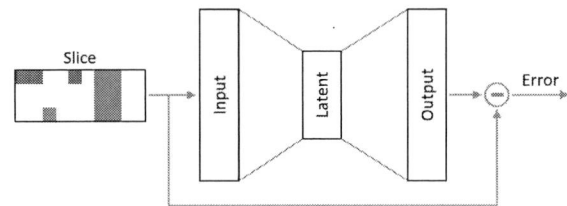

Fig. 3. The slice is fed into the autoencoder. The difference between the autoencoder's output and the original slice is used to compute the error.

afterwards. By training the autoencoder on well designed PCBs we force it to learn good practices typically used by layout designers. In contrast to a basic lookup-table, it also *learns* to reconstruct *similar* configurations to the items present in the training set. When confronted with an input different from training data, the autoencoder will struggle to reconstruct the input. Hence, the reconstruction error representing the difference between the input and the reconstructed version, can be used as a measure for anomaly. In our case, we use the mean-squared error (MSE) for measuring the reconstruction error.

Our autoencoder is composed of six convolution layers and compresses to a latent space of 32 dimensions. This results in approx. 15.000 parameters to be trained which is on the low side but speeds up both the training and the inference process. This topology can be extended in the future to target more complicated configurations or an advanced signal property processing.

III. IMPLEMENTATION AND INTERFACING WITH KICAD

AnoPCB is designed as a plugin for KiCad's pcb editor Pcbnew [6] written in Python. It is implemented in two components: An annotation & evaluation user interface that can be started from a custom toolbar shown in Fig. 4. This toolbar can be used to annotate an open layout with signal properties, create the slices for the machine learning algorithm and evaluate the anomaly detection results. The machine learning algorithm itself is implemented in a server as shown in Fig. 5. This provides two advantages: The computational load can be delegated to a central server with appropriate hardware and the trained model – the *knowledge base* – can be made available to several clients.

After initializing the plugin, the signals on the PCB are to be annotated either by selecting components of the PCB and changing their nets signal type or using a dedicated GUI. To support the user in this process, naming rules can be used for annotation and un-annotated signals can be highlighted. Note that the rules are stored within the project settings and can therefore be used throughout. Instead of differentiating signal types by numerical identifiers, names are used and specified in the preferences, with the standard being the following:

- 1 - digital stable
- 2 - digital switching
- 3 - analog
- 4 - analog sensitive

- 5 - supply
- 6 - reserved
- 7 - reserved
- 8 - unknown

These numbers must not be modified or shuffled between the training and test process since this might result in un-wanted effects. The annotated signals are stored along with configuration options and naming rules in the project folder. For simplicity, this saving process is attached to the normal saving of the actual PCB layout.

After annotation, the training and evaluation processes can be started from within the plugin. The training data is organized in data sets extracted from different projects. In this way, the user can select on which previous PCBs the autoencoder should be trained, i.e. which *experience* should be learned. The training process can be configured by means of a maximum training time or a number of epochs. For stabilizing the training and avoiding over-fitting, dataset augmentation (slice mirroring) and a training/validation split can be enabled. After the training has finished, the loss-trajectory is shown and indicates if the training succeeded. The training result is stored on the server to be available in future evaluations.

For performing PCB analysis, an autoencoder configuration from a previous training has to be chosen. After the plugin transferred the data to the server, the actual anomaly detection method is to be performed. The computation time depends on the PCB size and complexity and the host hardware specification but typically not longer than 10 minutes for designs such as *HackRF one* [5].

After the computation finished, the distribution of the re-construction error is shown to the user. In another dialog the image of the PCB is shown along with slice markers indicating the error, i.e. the *measure of anomaly*. For reducing the shown markers to actual anomalies, they can be limited by setting a threshold for the reconstruction error. The value to be set may depend on the design under review but setting it in a way to see the top five or ten percent of the reconstruction error is reasonable.

For further insight in the actual anomalies, the displayed slice markers can be clustered with K-Means or DBSCAN [7] using their latent representations. This feature allows to group repetitive anomalies together aiming at supporting and speeding-up the actual debugging process.

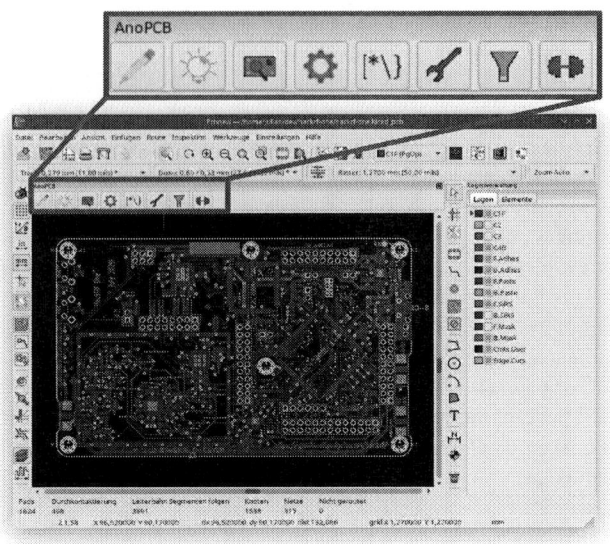

Fig. 4. AnoPCB's toolbar, integrated into Pcbnew's GUI.

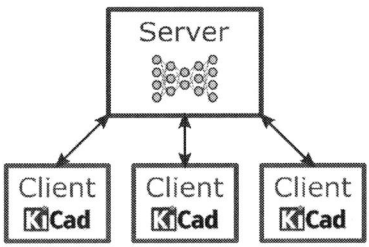

Fig. 5. AnoPCB's client server structure for centralizing the training and evaluation procedure along with the machine learning model management.

IV. CASE STUDY

To evaluate our approach, we train the system by four freely available four-layer PCB designs:

- Ubertooth One [8]
- HackRF One [5]

- Jawbreaker [5]
- Crazyflie 1.0 [9]

For this purpose, we labeled all signals on the layout according to the previously presented categories and used them to initiate the training procedure.

The actual evaluation is done using a modified HackRF One board as shown in Fig. 6. In the shown layout, two types of modifications were made. The modifications marked in red should be detected since they introduce for instance switching signal lines to sensitive analog parts – which is avoided in all layouts in the training set. To the opposite, the greenly marked regions indicate that a variant was introduced that is also present in other regions in the dataset. More specifically, we did the following modifications:

- Create a new trace of the clock signal `clkin`
- Create a trace of `tx_amp` to the digital switching parts
- Change signal category on the lowest layer of

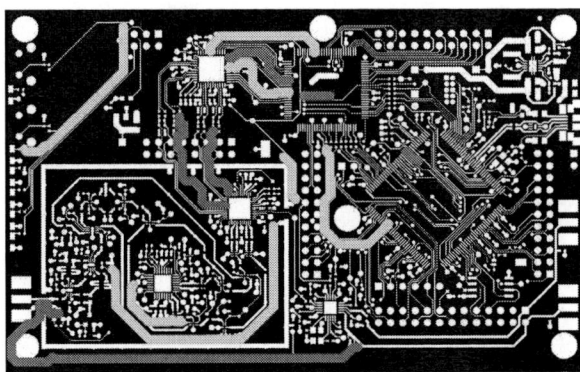
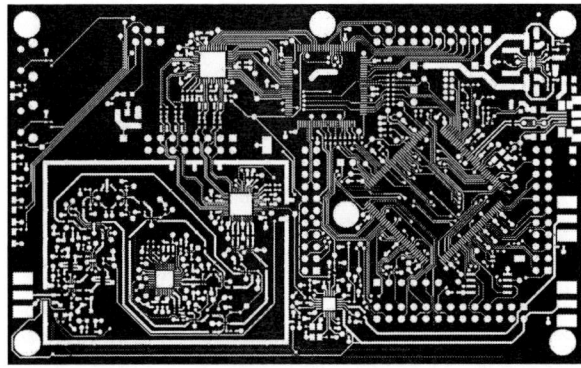

Fig. 6. Upper layer of our modified HackRF One. The red markings denote modifications that have been qualified as critical and have to be found by AnoPCB, the green markings are less critical.

Fig. 7. Upper layer view of the modified HackRF One with highlighted anomalies (marked in red, threshold: 0.009). Along with some false-positives, the critical modifications are mostly highlighted correctly.

`rx_amp_pwr` from *analog sensitive* to *digital switching*

- Change signal category (on the layer above) of `/mcu/usb/power/sd_cmd` and `/mcu/usb/power/sgpio6` from digital switching to *analog sensitive* and *power*
- Change signal category (on the layer above) of `/baseband/exbbq` and `/baseband/rxbbq` from *analog sensitive* to *power* and *digital switching*
- Change signal category (un the upper layer) of `dd0`, `dd8to` and `da4` from *digital stable* to *analog sensitive*, *power* and *analog*
- Change signal category of `/mcu/usb/power/en1v8` from *digital stable* to *digital switching*

As shown in the figure, these changes have been categorized by a PCB designer into critcal (red) and non-critical modifications. On this layout, we performed the anomaly analysis with AnoPCB. The result is shown in Fig. 7. In this result, we discover that the previously modified parts are highlighted correctly along with some false-positives. In some cases, the markings are rather small but can be seen in the dataset. Based on this information, a senior designer can guide his review directly to erroneous parts. In this way, the false-positives can be identified. This paves the way for an incrementally learning system that can guides a review and may learn from its results since the false-positives can be used to extend the dataset to make the overall system more precise.

V. CONCLUSION

We have presented a novel PCB layout analysis method to identify possibly erroneous parts in the design to guide a review process. By extending the training data set with results from the review process, the system's accuracy can be improved. In this way, the review could be also used to label the found anomalies to enable a later identification on previously unseen boards. This might also lead to the fact that the topology of the comprised machine learning system should be extended to increase its capacity. A combination with an automatic routing could provide the potential to create more human-like and optimized layouts. In a more general scope,

an extension of this approach to other layout-driven designs such as for instance, ASICs, MEMS or fluidic devices seems to be promising.

ACKNOWLEDGMENT

This work has been done in the research group IntelligEnt with support of the software project at the Technische Universität Ilmenau. The authors would like to thank Prof. A. Zimmermann and Dr. R. Maschotta for the valuable feedback and support. The initial implementation was also supported by Murad Babayev and Nicholas Heyer.

The IntelligEnt research group is supported by the Free State of Thuringia, Germany, and the European Social Fund under the reference 2018 FGR 0089.

REFERENCES

[1] J. Li and Y. Wu, "Post-layout pcb check and simulations for signal integrity," in *2014 IEEE International Symposium on Electromagnetic Compatibility (EMC)*, 2014, pp. 727–731.

[2] J. Hsu, T. Su, Y. Li, E. Hsiung, K. Xiao, X. Ye, and K. Wu, "Fast signal integrity methodology for pcb pre-layout analysis and layout quality check," in *2013 IEEE 63rd Electronic Components and Technology Conference*, 2013, pp. 2012–2017.

[3] X. Zhao, C. Chen, and J. Lai, "Optimization of pcb layout for 1-mhz high step-up/down llc resonant converters," in *2019 IEEE Applied Power Electronics Conference and Exposition (APEC)*, 2019, pp. 1226–1230.

[4] I. Goodfellow, Y. Bengio, and A. Courville, *Deep Learning*. MIT Press, 2016, http://www.deeplearningbook.org.

[5] *HackRF, open source hardware for software-defined radio*. [Online]. Available: https://github.com/mossmann/hackrf

[6] J.-P. Charras, F. Tappero, and W. Stambaugh, *KiCad Complete Reference Manual*, 2018.

[7] F. Pedregosa, G. Varoquaux, A. Gramfort, V. Michel, B. Thirion, O. Grisel, M. Blondel, P. Prettenhofer, R. Weiss, V. Dubourg, J. Vanderplas, A. Passos, D. Cournapeau, M. Brucher, M. Perrot, and E. Duchesnay, "Scikit-learn: Machine learning in Python," *Journal of Machine Learning Research*, vol. 12, pp. 2825–2830, 2011.

[8] *Project Ubertooth, open source wireless development platform for Bluetooth experimentation*. [Online]. Available: https://github.com/greatscottgadgets/ubertooth

[9] *Crazyflie 1.0 quadcopter*. [Online]. Available: https://github.com/bitcraze/crazyflie-electronics/tree/master/ecad

SMACD / PRIME 2021 | 19 – 22 July 2021, Online Event

A Deep Learning Toolbox for
Analog Integrated Circuit Placement

António Gusmão[1,2], António Canelas[1], Nuno Horta[1,2], Nuno Lourenço[1], Ricardo Martins[1]

[1]Instituto de Telecomunicações, Lisboa, Portugal
[2]Instituto Superior Técnico – Universidade de Lisboa, Lisboa, Portugal
{agusmao; acanelas; nuno.horta; nlourenco; ricmartins}@lx.it.pt

Abstract—This paper presents a deep learning toolbox, DEEPPLACER, to assist designers during the layout design of analog integrated circuits. DEEPPLACER relies on a simple pair-wise device interaction circuit description, i.e., the circuits' topological constraints, to propose valid floorplan solutions for block-level structures, including topologies and deep technology nodes not used for its training, at push-button speed. Despite its automatic functionalities, the toolbox is focused on explainable artificial intelligence, involving the designer in the synthesis flow via filtering and editing options over the candidate floorplan solutions. This constant state of human-machine feedback environment turns the designer aware of the impact of each device's position change and inherent tradeoffs while suggesting subsequent moves, ultimately increasing the designers' productivity in this time-consuming and iterative task. Finally, DEEPPLACER is shown to instantly generate a floorplan with 61% better constraint fulfilment than a human designed solution.

Keywords—Artificial neural networks, Electronic design automation, Explainable artificial intelligence, Physical design.

I. INTRODUCTION

The layout design of analog integrated circuits (ICs) has been defying all automation attempts, and it is still primarily a handcrafting process carried by circuit designers on traditional layout editing frameworks. Notably, in the placement task of the layout design flow, designers must cope with many conflicting topological requirements necessary to produce a robust floorplan against process variations and parasitic layout structures. Simultaneously, a "good" placement solution must also minimize the occupied area, be robust to layout-dependent effects, improve the circuit's routability, and so on. Moreover, changes in the circuit/system topologies, design specifications, or even bias, may annul part of, or all, the previous work.

Different electronic design automation (EDA) tools for automatic analog IC placement were proposed in the literature, which can be distinguished by the number of legacy layouts (LLs) used to generate the new floorplan solution: (1) *generated from topological constraints* [1], where no LLs are used, and every solution is generated by a, usually time-consuming, optimization process; (2) *retargeted from a LL or template* [2], where a previously produced floorplan of the same circuit is migrated using fast compaction techniques to a new technology node or different devices' sizes; and (3) *synthesized by knowledge mining* [3], where a library of LLs is decomposed into sub-blocks, and then, the floorplan of the circuit being generated matches them to its own sub-blocks. Commercial solutions have also evolved in the last decade, providing better user-assisted experiences [4].

However, the advances in machine learning (ML), including deep learning (DL), create new IC design automation opportunities [5]. In the analog spectrum, for automatic circuit sizing, DL models that predict the devices' sizes when requested for specific circuit performances were

This work is funded by FCT/MCTES through national funds and when applicable co-funded EU funds under the project UIDB/50008/2020. Including internal research project LAY(RF)².

recently proposed [6]. Still, DL models that map from devices' sizes to circuit performances, bypassing expensive simulations, are also becoming attractive solutions [7]. Other recent ML applications include hierarchy and building block recognition from circuit netlists [8]; generation of wells during layout design [9]; or even guidance for routing problems [10]. For analog IC placement, fully supervised artificial neural networks (ANNs) that reproduce the layout patterns of a dataset of legacy placement solutions were proposed [11]. But only in [12] it was enhanced with multi-circuit-topology support within the same model, whose distinction between topologies is achieved with a new topological constraint encoding at the ANN's input layer.

In this paper, DEEPPLACER is proposed, an innovative multi-topology technology-independent DL-based toolbox to assist circuit designers in the analog IC placement problem. Its main contributions can be summarized as follows:

- DEEPPLACER's underlying DL model relies on pair-wise device interactions, which can encode any of several topological constraints. This feature results in an enhanced generalization, turning it capable of dealing with **newer circuit topologies** (i.e., not used for its training), unlike approaches of type (2) and (3) [2][3], while simultaneously handling **different technology nodes**;

- Similar to the approaches of type (1) [1], DEEPPLACER produces placement solutions from scratch, fully automatically. However, it increases designers' productivity by proposing tens of valid floorplan solutions for block-level analog structures at **push-button speed** (i.e., bypassing optimization);

- Additionally, DEEPPLACER breaks with previous ML-based analog IC placement approaches [11][12][13] by focusing on explainable artificial intelligence (XAI), where the expert designer is involved in the synthesis flow. This is accomplished through its graphical user interface (GUI), offering filtering and editing options over the several proposed floorplan solutions. As there are several conflicting constraints and objectives in floorplan design, the tool is in a constant state of **human-machine feedback**, presenting the impact of each device's position change and the inherent tradeoffs, while suggesting subsequent moves;

- Experimental results show that DEEPPLACER produces competitive floorplans compared to those human-designed, complementing automation solutions in the industry.

This paper is organized as follows. In Section II, the DEEPPLACER toolbox is fully detailed. Afterward, in Sections III and IV, the experimental results and conclusions are addressed, respectively.

II. DEEPPLACER TOOLBOX

The DEEPPLACER high-level synthesis flow is illustrated in Fig. 1. Its inputs are topological constraint and devices'

© VDE VERLAG GMBH · Berlin · Offenbach

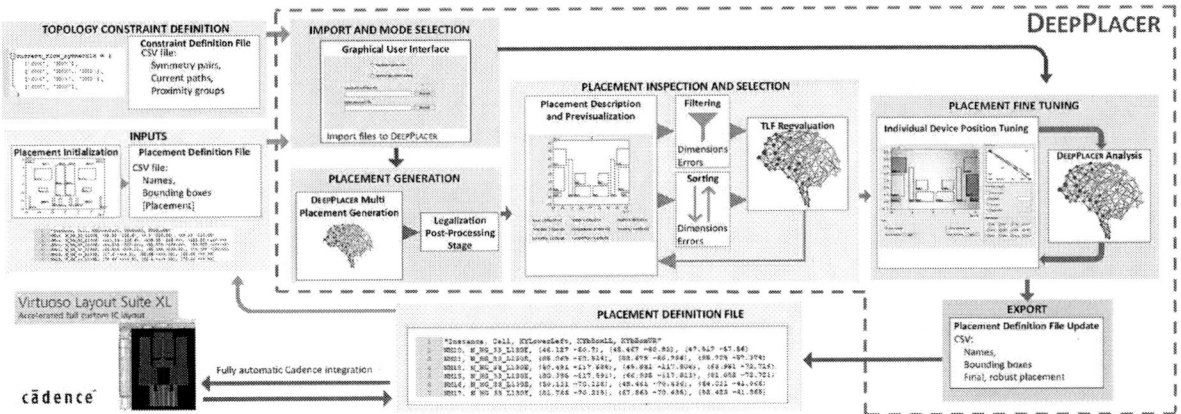

Fig. 1. DEEPPLACER's high-level flow.

dimensions. A DL model, an improved version of the one from [12], with a more abstract constraint codification and graph structure compatible, generates a set of recommended placement solutions that can be inspected and selected. Note that any DL model could be used as long as it supports the input structure, contributing to the tool's modularity. Alternatively, the tool can also input an initial placement. The produced or imported placement can then be analyzed in a user-assisted flow that provides detailed information on constraint satisfaction errors calculated through the developed topological loss function (TLF) [12]: the total error and the discriminated components. In this *placement fine-tuning*, a human-DEEPPLACER feedback environment is established, where the impact of each device's location can be immediately analyzed to understand the tradeoffs of its location in the floorplan, keeping the designers in the loop and aware of the impact of their decisions. Moreover, all constraints are defined as inter-device relations. The error that drives DEEPPLACER can be intuitively visualized in a graph structure that shows the relative position's error contribution between pairs of devices, enabling the designer to identify critical devices and quickly produce an improved solution.

Multi-topology technology-independent placement production is essential to build a competing EDA tool, and thus, the solution adopted was a graph-based description of the constraints that ultimately define the circuit topology in a graph description of the problem. Emphasis is given to interactions between device pairs, resulting in a segmentation of the input into smaller parts. Since graph-encoded features were used, a graph-structured DL model was developed. Additionally, each example's sizing is independently scaled before being solved, keeping only information regarding the devices' relative sizes, turning the tool capable of dealing with multiple technology nodes. Finally, the designed TLF is by nature fully unsupervised, turning the model's training dependent on schematic-level examples only. As sizing solutions are much easier to obtain/produce than floorplans, it grants access to sufficient data to train the model to a competitive level.

A. Complexity of the Problem

Analog IC placement design is heavily reliant on manual labor. Its flow can be described as a feedback loop, where the designer identifies patterns and tests different topological relations between cells, but whose impact on circuit performances is only known after complete routing and extraction. Naturally, experienced designers have developed

intuition in the form of guidelines that should be followed to minimize the layout design risks. These guidelines can be interpreted as placement constraints. In [14], eight different constraints were identified, e.g., symmetry, proximity, alignment, current-flows, etc. However, by considering some of them, a deeply entangled multi-constraint problem whose complexity increases exponentially with the number of devices must be solved.

Consequently, due to the tradeoffs involved, no single *optimal* floorplan solution for a given problem exists. The usage of these guidelines has greatly improved human designer performance, but the development of automatic tools that properly interpret them and simulate human behavior has been a challenging research topic. While they are very intuitive for humans, they are hard to translate to an automatic tool in such a way that they maintain their full abstraction, preserving their high applicability and scalability capabilities. Nonetheless, past approaches have generally considered some of them [1]-[3][12],[13].

B. Level of Automation

DEEPPLACER operates fully automatically, proposing floorplans at push-button speed given the bounding boxes of the circuit's devices and the set of topological constraints that bind them. Due to the constraints' complexity, it is common that no solution meets all constraints, and tradeoffs must be made. DEEPPLACER uses a deep model to produce a set of recommended placements automatically. The criteria used to evaluate the solutions are also shown, explaining the tool's design choices. This feedback mechanism identifies, for each device, its location contribution to the overall cost, providing intuition to the designer on the many constraints that the placement is subject to. The designer can then analyze and select one of the placement solutions out of the many produced by DEEPPLACER. Moreover, the feedback mechanism can also guide manual adjustments to the generated floorplans or imported previous designs, keeping the designer informed of the changes' impact.

C. Designer Interface

The DEEPPLACER toolbox main interface with the designer is done using the GUI shown in Fig. 2. The tool requires a topology-specific file that identifies the devices and their constraints. It also inputs a file with the sizes of the device's bounding boxes (and optionally their location) and outputs the resulting devices' locations. The first needs to be crafted manually, while the others are exported/imported automatically from/to Cadence's Virtuoso layout editor [4].

© VDE VERLAG GMBH · Berlin · Offenbach

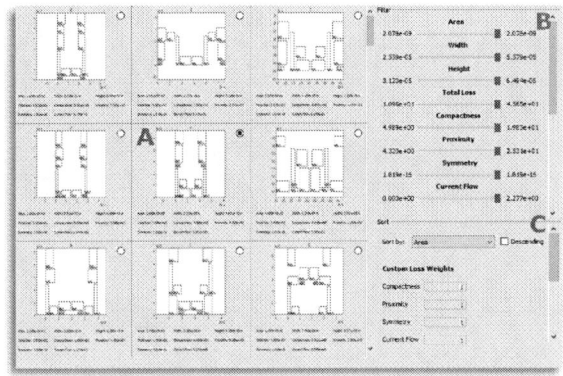 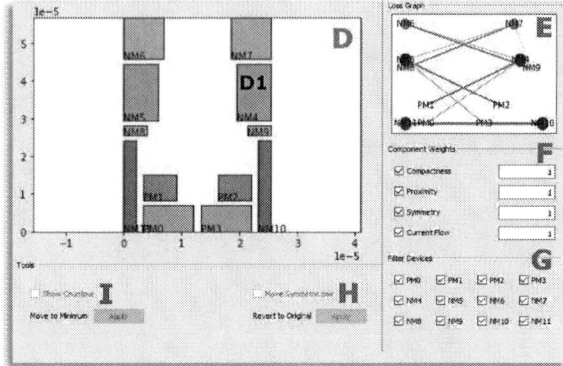

Fig. 2. DEEPPLACER GUI: (a) Placement inspection, selection, interface; (b) Placement analysis and fine tuning interface. Total error shown.

Fig. 2(a) shows the ***placement inspection and selection*** window. Panel **A** shows a card for each floorplan, containing its previsualization and detailed description: floorplan's area, width, and height, as well as the total error and the discriminated topological components. Panel **B** is used for filtering, while panel **C** is used for sorting. The designer can filter for placement solutions that do not meet any wider-scope criteria. The floorplans can be filtered and sorted by any dimension or constraint satisfaction error. Panel **C** allows defining custom TLF weights through which the candidate floorplans will be reevaluated and properly sorted.

Fig. 2(b) shows the ***placement fine-tuning*** window, a human-DEEPPLACER feedback environment. In this environment, the designer can select from the placement representation (panel **D**) one of the devices to fine-tune. To strategize any change, while no device is selected, DEEPPLACER shows the normalized TLF errors of each device as their filling color (**D1**). Since all constraints were defined as inter-device relations, the error can be visualized in a graph. Panel **E** shows the error graph of the circuit. Each node's/edge's size and color codify the device's/interaction's error, enabling the intuitive interpretation of the error. The influence of any device on the error can be toggled in panel **G** to better discern the interaction that is causing the error. Panel **F** enables the visibility of any TLF error components to be toggled, equivalent to setting the component's weight to 0. The option of adjusting each weight is also given. The "move symmetric pair" checkbox (**H**) enables the device's symmetric pair to be moved symmetrically. Once a device has been selected, the errors shown are only those explicitly related to it. For other devices, the error that is shown results from their interaction with the selected device. Meanwhile, panels **F** and **G** still serve their purpose facilitating error interpretation. DEEPPLACER can also superimpose a contour (**I**) on the placement, indicating, for each point in the 2D plane, the expected normalized error if the device is moved to that position.

D. Applicability

DEEPPLACER relies on a carefully designed multi-topology and technology-independent placement model for analog cells. It is unfeasible to train a model with all the existing circuit topologies on all technology nodes, but the graph-based description of the constraints was carefully conceived to reduce inter-topology variability. While the circuit topology ultimately defines the constraints, the model learns to place the boxes subject to the topological constraints, and not to place a transistor or a capacitor of a specific circuit. Analogously, the constraint satisfaction error calculated through the TLF, evaluates the robustness of a placement

through the analysis of these pair-wise device interactions, and moreover, device dimensions are scaled per example (unlike usual ML approaches that scale the entire datasets). These considerations promote generalization, enabling an utterly unknown circuit topology or technology to be successfully decomposed into smaller, known device pair-wise interactions. The tool is highly scalable and versatile, capable of dealing with different topologies composed of a varying number of devices.

E. Integration with Comercial Design Suites

DEEPPLACER's core is fully coded in Python and is fully integrated with mainstream deep learning frameworks (Scikit-learn and PyTorch), allowing state-of-the-art performance in designing, training, and usage of the developed ML models. Additional SKILL scripts bridge the toolbox with Virtuoso, producing DEEPPLACER's inputs directly from the layout editor and using DEEPPLACER's outputs to set the generated placement back into the layout view. As such, the toolbox is fully and transparently integrated into the design flow.

F. Robustness of the Produced Placements

DEEPPLACER relies on an elegant description of intuitive expert-designed guidelines to produce several robust placements at push-button speed. These Placement guidelines, detailed in [14], were generated through years of experience and directly impact the placement quality. DEEPPLACER is capable of reproducing expert designer's patterns that lead to effective placements. This tool's entire design flow is built around the robustness of the solutions and the design tradeoffs' impact on it.

III. RESULTS

Towards demonstrating DEEPPLACER's predictive, interactive, and integration capabilities, predictions were generated for the current starving single-stage amplifier biased by voltage-combiners [15], for a 130-nm technology node, as presented in Fig 3. DEEPPLACER is used to improve the floorplan generated from Virtuoso XL [4], resultant from automatic device instantiation. Results are also compared to the original human-made placement.

Fig. 4 shows the placement solutions and the error graphs associated with the total, proximity, and current-flow errors (constraints considered in the manual design). Table I shows their numeric mean errors. The input placement (generated in Virtuoso XL) shows no current-flow or symmetry errors; however, the placement is far from compact with ~4×

DEEPPLACER's error. Additionally, the pair NM10 and 11 that should be placed in close proximity are placed far away, resulting in the largest proximity error of the floorplans shown, resulting in a proximity error ~6× larger than DEEPPLACER. Note that the graphs of Fig. 4(a) facilitate the identification of the problematic relations. Overall, DEEPPLACER presented the best floorplan: the most compact, with better proximity relations, and the second-best current-flow implementation; with the manual-designed placement presenting 1.6× its error. By analyzing the generated floorplan, it is clear through the devices' colors of Fig. 4(c) that devices NM10 and 11 are the ones with the most error associated. Through the proximity error graph, their relations with devices PM0 and 3 are identified as the most problematic. As such, NM10/11 could be moved downwards on the floorplan. By swapping their places with devices NM21 and 20, the remaining current-flow error would also be solved.

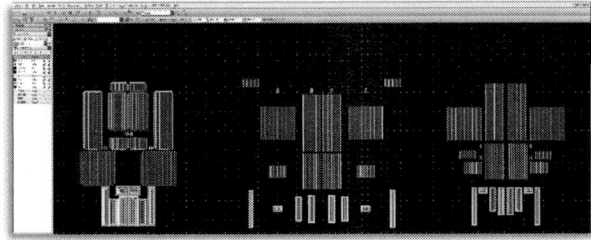

Fig. 3. Generated placements, instantiated in Cadence. From left to right, generated by: DEEPPLACER, Virtuoso XL and expert designer.

TABLE I.   FLOORPLANS' ERROR COMPARISON

Placement	Compactness	Comp.	Proximity	Comp.	Current Flow	Total Error	Comp.
Virtuoso XL	24.77	3.99×	19.42	6.37×	0.00	44.19	4.47×
Human [15]	6.72	1.08×	8.33	2.73×	0.89	15.94	1.61×
DEEPPLACER	6.21	-	3.05	-	0.62	9.89	-

IV. CONCLUSIONS

This paper presented a DL toolbox to assist designers during the layout design of analog ICs, complement automation solutions available in the industry. Its usage flow was thoroughly explained, covering its fully-automatic and user-assisted functionalities, e.g., placement inspection and selection, but also fine-tuning. Interface format from/to Virtuoso layout editor ease its integration into the traditional flow, ultimately increasing designers' productivity.

REFERENCES

[1] A. Patyal, et al., "Analog placement with current flow and symmetry constraints using pcp-sp," in *55th DAC*, pp. 1–6, June 2018.

[2] R. Martins N. Lourenço, N. Horta, "LAYGEN II – Automatic Analog ICs Layout Generator based on a Template Approach" in *Genetic and Evolutionary Computation Conf. (GECCO)*, pp. 1127-1134, Jul. 2012.

[3] P. Wu, M. Lin, T. Chen, C. Yeh, X. Li, and T. Ho, "A novel analog physical synthesis methodology integrating existent design expertise," in *IEEE TCAD*, vol. 34, no. 2, pp. 199–212, Feb. 2015.

[4] "Cadence's Virtuoso Layout L/XL/GXL," http://www.cadence.com.

[5] E. Afacan, N. Lourenço, R. Martins, G. Dündar, "Review: Machine learning techniques in analog/RF integrated circuit design, synthesis, layout, and test," in *Integration, the VLSI Journal*, 77, 113-130, 2021.

[6] N. Lourenço et al., "Using Polynomial Regression and Artificial Neural Networks for Reusable Analog IC Sizing," in *16th Int. Conf. on SMACD*, pp. 13–16, July 2019.

[7] G. Islamoğlu, T. Çakici, E. Afacan, G. Dündar, "Artificial Neural Network Assisted Analog IC Sizing Tool," in *16th Int. Conf. on SMACD*, pp. 9–12, July 2019.

[8] K. Kunal, et al. "GANA: Graph Convolutional Network Based Automated Netlist Annotation for Analog Circuits," in *Proceedings of DATE Conference and Exhibition*, pp. 55–60, Mar. 2020.

[9] B. Xu, et al., "WellGAN: Generative-adversarial-network-guided well generation for analog/mixed-signal circuit layout," in *56th ACM/ESDA/IEEE Design Automation Conference (DAC)*, June 2019.

[10] K. Zhu, et al., "GeniusRoute: A new analog routing paradigm using generative neural network guidance," in *IEEE/ACM International Conference on Computer-Aided Design (ICCAD)*, Nov. 2019.

[11] D. Guerra, A. Canelas, R. Póvoa, N. Horta, N. Lourenço, R. Martins, "Artificial Neural Networks as an Alternative for Automatic Analog IC Placement," in *16th Int. Conf. on SMACD*, pp. 1–4, Jul. 2019.

[12] A. Gusmão, F. Passos, R. Póvoa, N. Horta, N. Lourenço, R. Martins, "Semi-Supervised Artificial Neural Networks towards Analog IC Placement Recommender," in *IEEE ISCAS*, pp. 1–5, Oct. 2020.

[13] R. He and L. Zhang, "Artificial neural network application in analog layout placement design," in *Canadian Conf. ECE*, pp. 954–957, 2009.

[14] M. Lin, Y. Chang, and C. Hung, "Recent research development and new challenges in analog layout synthesis," in *Proceedings of the ASP-DAC*, pp. 617–622, Mar. 2016.

[15] R. Povoa, et al., "Single-Stage OTA Biased by Voltage-Combiners with Enhanced Performance Using Current Starving," in *IEEE Transactions on Circuits and Systems II: Express Briefs*, vol. 65, no. 11, pp. 1599–1603, Nov. 2018

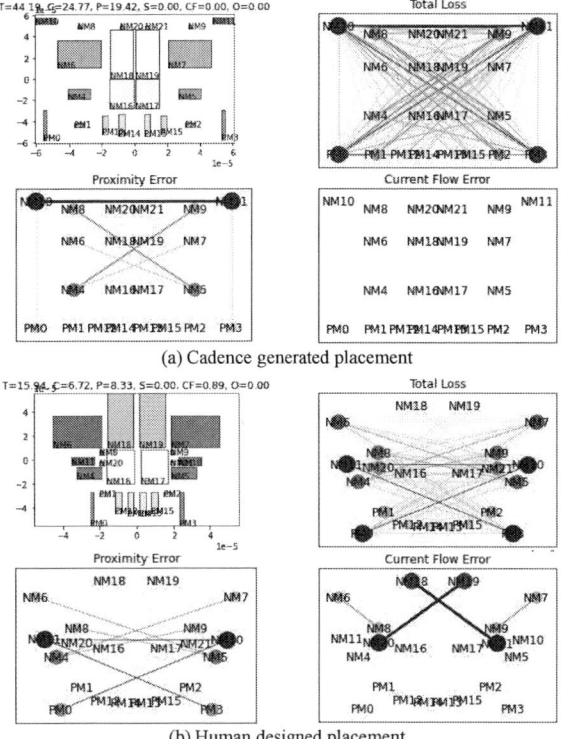

(a) Cadence generated placement

(b) Human designed placement

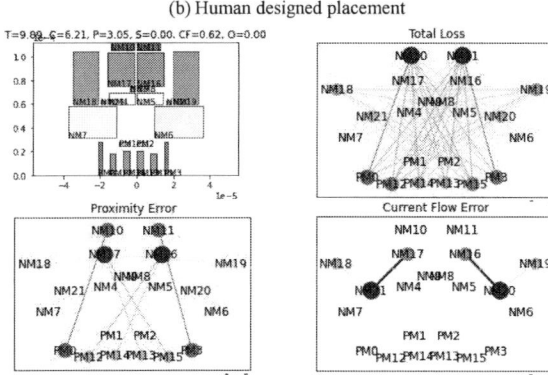

(c) DEEPPLACER generated placement

Fig. 4. Placements generated by Virtuoso XL, a human designer and DEEPPLACER, along with the total, the proximity and the current-flow error graphs.

A Differential Evolution based Methodology for Parameter Extraction of Behavioral Models of Electronic Components

Gazmend Alia
Infineon Technologies AG,
Bundeswehr University Munich
Munich, Germany
alia.external@infineon.com

Andi Buzo
Infineon Technologies AG
Munich, Germany
andi.buzo@infineon.com

Daniel Ludwig
Infineon Technologies AG
Munich, Germany
daniel.ludwig@infineon.com

Linus Maurer
Bundeswehr University Munich
Munich, Germany
linus.maurer@unibw.de

Georg Pelz
Infineon Technologies AG
Munich, Germany
georg.pelz@infineon.com

Abstract—**Behavioral models of electronic components are crucial for system simulation, as they are quick to simulate and yet provide reliable information on the behavior of the original circuit. Parameter extraction of such models, i.e. calibrating the model to match the experimental characteristics of the device, is tedious work, as the number of such parameters can add up to tens of them. There are attempts in the literature to solve this problem with the help of optimization algorithms. However, when put to practice, new challenges arise due to the large number of devices to be calibrated, time restrictions and the wide variety of behavioral models. These challenges require novel techniques that ensure generality, speed and scalability. We address these challenges by proposing a fully automated flow, which includes the following novel features: an evolutionary algorithm, a smart sampling technique for reducing redundancy in the reference data, and a method for making use of the knowledge acquired in previous parameter extraction tasks. We tested the flow with a set of more than 200 Si-diodes and IGBT behavioral models with more than 50 parameters and 30 response curves to be calibrated. The results show that full automation of parameter extraction is possible, i.e. no human intervention is needed. Hundreds of Si-diodes and IGBT behavioral models are calibrated within 48 hours as compared to 1.5 years of manual work.**

Keywords—behavioral models, parameter extraction, calibration, smart sampling, differential evolution, IGBT.

I. Introduction

Behavioral models are very important for simulating and understanding the behavior of electronic systems. They enable fast and accurate simulations that mimic the original devices in their context. The models have usually a large number of parameters and have to be calibrated so that they match the real static and dynamic characteristics of a given device. A human being can usually handle very well a handful of input and output parameters. When it comes to tens of parameters though, it becomes extremely difficult to have an intuitive understanding of what the impact of a certain combination of parameters will be, mainly due to the cross-correlations between different parameters.

The increase in the size and complexity of electronic systems has led to an increase in the complexity and variety of behavioral models. As a result, new techniques are required which enable generic, fast and scalable solutions for parameter extraction. We propose here a fully automatic flow, which fulfills the aforementioned requirements with the following novel features:

o *Evolutionary algorithm.* Differential evolution is used as the calibration algorithm. According to [1], the top 10 optimization algorithms are evolutionary ones. According to our research and experiments, differential evolution proved to be fast, and the calibration converged in all the cases we tested. The great advantage of DE compared to gradient descent approaches is the ability of DE to easily overcome local minima, which is very often the case for models with a large number of parameters.

o *Smart sampling technique for reducing the redundancy of the reference data.* The number of experimental data points are usually in the order of hundreds or thousands. However, data which are close to each other are correlated. We developed a technique that reduces the amount of data to be simulated by a factor of 10 to 20, hence speeding up the calibration routine.

o *Use of previous knowledge.* Differential evolution does not make use of any previous knowledge. We implemented a method, which, based on all previous calibrations, determines a good starting region of model parameters, where it is more probable to find a good solution. This speeds up the calibration by a factor of 5 to 10 times.

The flow is tested with hundreds of Si-diodes and IGBTs. The behavioral models had 51 parameters, 30 response curves and 3 operating conditions, i.e. temperature, gate–emitter voltage and collector-emitter voltage.

The results show that it is possible to implement a fully automatic flow for parameter extraction. As a consequence of our novel techniques, the calibration of more than 200 Si-diodes and IGBTs was done in 48 hours using a parallelization factor of 50, using smart sampling and previous knowledge. The manual work required for such a job is at least 1.5 years for an experienced engineer with good knowledge of the behavioral models.

II. Related Work

There is research in the literature that has addressed aspects of behavioral models for transistors and the parameter extraction procedures using different techniques.

Parameter extraction can be done either analytically or by numerical methods. In the case of analytical methods, for instance [2], the calibration procedure is split into five steps, and in each step a few parameters are calculated that are used in the next step. This mimics more or less the way an engineer does a manual calibration, by determining a few parameters at a time and then using these parameters as the starting point for

© VDE VERLAG GMBH · Berlin · Offenbach

the next iteration. The analytical procedures are very fast and suitable for models with few parameters, and uncorrelated parameters. They do not work when there is moderate correlation between the parameters. In addition, these methods are dedicated to a specific model.

These restrictions have turned people's attention towards numerical methods and the use of machine learning. [3] shows a sequential approach to the calibration of IGBT models by splitting the static characteristics into 3 regions, subthreshold, linear and saturated, and then fitting each part separately. [4] and [5] have used machine learning to calibrate the models. Even though the models used in [4] and [5] have few parameters, respectively 7 and 4, they have shown how one can use swarm-intelligent and genetic algorithms, i.e. Artificial Bee Colony, Queen Bee and Crossing Mates, to extract the model parameters.

To the best of our knowledge, there is no evidence in the literature about applying these techniques to a wide variety of behavioral models, how much time the parameter extraction takes or how the techniques scale up when hundreds of behavioral models are to be calibrated. For this reason it is very hard for us to compare our methodology to other ones given in literature, such as the ones in [3], [4] and [5].

We have implemented a fully automatic routine from the beginning, i.e. reading experimental data, until the end, i.e. determining the values of the model parameters, including verification and documentation of the results. This approach is generic, i.e. it is not limited to a certain type of behavioral model. It is fast, at least 200 times faster than manual work in which hundreds of behavioral models are to be calibrated. Moreover, it scales up very well to hundreds and thousands of behavioral models. To make this possible, human intervention is minimized. A smart sampling technique to reduce data redundancy was developed, and previous calibration knowledge is used when calibrating a new device.

III. PARAMETER EXTRACTION

A. Schematic of the calibration routine

To be efficient, a very general flow for parameter extraction is developed. At each stage of the calibration, the same steps are followed, as depicted in "Fig. 1".

The flow starts with data preparation, where the input data are converted to a standard format. Then the data consistency is checked to make sure that different characteristics of the behavioral model are consistent with each other. After that, the smart sampling technique is applied in order to reduce the amount of data points to be simulated by a factor of 10 to 20, hence reducing the time. At this stage, the reference data and stimuli to the behavioral model are obtained. Now, the calibration loop is entered and in each iteration, the differential evolution algorithm generates the candidate parameter sets. The first parameter set is generated based on the previous calibrations, which sets the calibration flow directly into a good path as compared to a random starting point. This process leads to a speed-up of up to ten fold. The candidates are simulated, and the results are compared to the reference data. The error is calculated and weighted as necessary. This loop continues until a good enough error is obtained. Once we are satisfied with the behavioral model responses, the results are saved in the calibration database, so that the new results are available for the following calibrations.

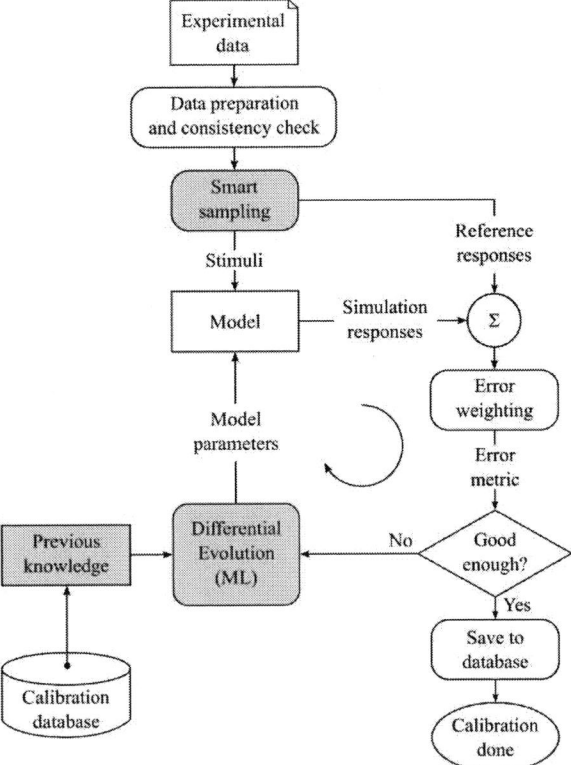

Fig. 1. Flowchart of the calibration process of a single stage.

B. Data preparation and consistency check

This step has the goal of converting the experimental data to a standard format and making sure that the data are consistent, for example, the output and transfer characteristics of a transistor have to match each other. This is a necessary step to ensure successful calibration. In addition, it is used to get important feedback to determine problems in the process of measuring the characteristics of the device. In case an inconsistency is detected, unless it can be corrected, the responses are weighted appropriately to ensure the desired calibration.

C. Smart sampling to reduce data redundancy

One data set usually contains hundreds to thousands of points. In order to gain speed, the least possible number of data points have to be simulated. One has to simulate as few data points as possible to gain speed, but at the same time, has to be careful not to lose accuracy. For this reason, a smart automatic sampling method has to be used in order to sample the data with a variable density depending on the curvature. This means that wherever the data have a high curvature, more samples are needed compared to flat regions. Only 2D data (lines) are considered here, however, the technique can easily be extended to include further dimensions. Here is how the technique works:

First, a smoothening of the data is needed in case the data are noisy, especially when data come from real measurements in the lab. In our case, a "smoothing spline" is used, which for a function $y_i = f(x_i)$ constructs the curve $s(x_i)$ that minimizes the function given in (1). The parameter p determines the tradeoff between the cubic spline and the least square straight line. We have used a p value of 0.95 to 0.99, depending on the noise of the data.

$$p \sum_i |y_i - s(x_i)|^2 + (1-p) \int \left(\frac{d^2 s}{dx^2}\right)^2 dx \qquad (1)$$

The exact formula for the curvature, which is used in the Serret-Frenet formulas [6], is:

$$\kappa = \left| \frac{d^2 y / dx^2}{\left(1 + \left(dy / dx\right)^2\right)^{\frac{3}{2}}} \right| \qquad (2)$$

where κ is the curvature, d^2y/dx^2 and dy/dx are the second and first derivative respectively. The curvature is tightly connected to the derivatives of the data. Fig. 2 shows qualitatively what the intuition behind this relation is.

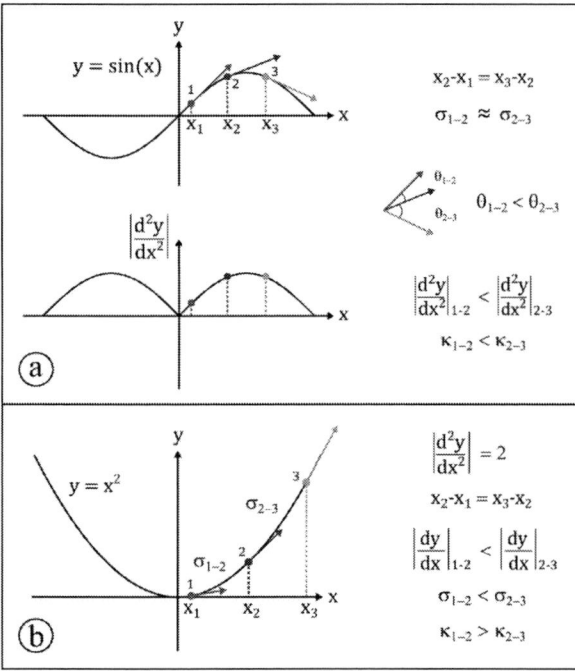

Fig. 2. a) The curvature (κ) of the data is proportional to the second derivative. It shows how much the tangent rotates, and the greater the curvature, the more the tangent rotates. In this case, the curvature in region 1-2 is less than in region 2-3. σ is the arc length. b) The curvature (κ) is inversely proportional to the first derivative. In this case, even though the second derivative is constant (the tangent rotates equally from 1 to 2 to 3), it is obvious that the line is more curved near the origin than further away. This is because the first derivative is smaller closer to the origin. This means that the arc length (σ) 1-2 is smaller than arc length 2-3, which in turn means that the tangent rotates with respect to a smaller distance in 1-2 than in 2-3, hence more curvature in 1-2 than 2-3.

Once the curvature is calculated, the area under the curvature is split into equal areas called split-areas, as illustrated in Fig. 3. The number of split areas is one less than the number of samples. Then, one sample is taken at each point where two adjacent split-areas meet. An example of the application of this smart sampling for the output characteristic of an IGBT is shown in Fig. 3.

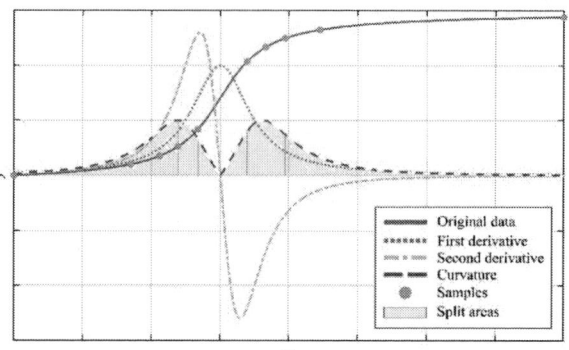

Fig. 3. Example of the smart sampling. The density of sampling is higher in regions of greater curvature.

D. Use of previous knowledge

We have modified the DE by introducing the possibility to use prior calibrations as the starting point for new ones. This modification only, enables a speed-up of 5 to 10 fold compared to random starting points.

Since there is a large number of devices to be calibrated and devices of similar technologies usually have similar characteristics, it is very beneficial to be able to use the previous calibrations as a starting point for the upcoming ones. Fig. 4 shows the difference between a calibration with and without previous knowledge. As shown in the figure, the algorithm reaches small errors much quicker when the previous knowledge is used.

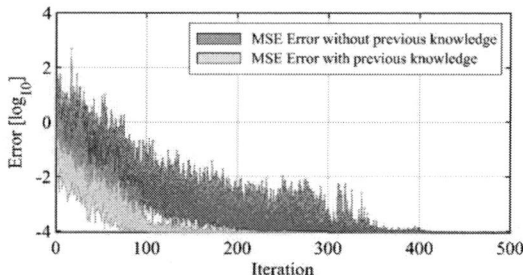

Fig. 4. Comparison of the MSE error evolution between calibration with (green) and without (red) previous knowledge.

There is, however, a risk of directing the algorithm from the beginning towards a very specific region. To avoid this, randomness in the starting point of the algorithm is introduced. Out of 25 starting population members, 15 are random, and only 10 members are selected as the top 10, out of the entire pool of previously calibrated devices. This way, there is enough room for the algorithm to explore regions outside those of the previous devices.

IV. RESULTS

In this section, a few examples of the calibrations are shown that demonstrate how the calibration looks like for some static and dynamic characteristics of the models we tested. Fig. 5 shows the calibration of a single device, i.e. the static characteristic of the Si-diode, the output characteristic of the IGBT, the capacitances of the IGBT and the gate charge of the IGBT. There is an almost perfect match between the experimental data (markers) and the simulated ones (solid lines).

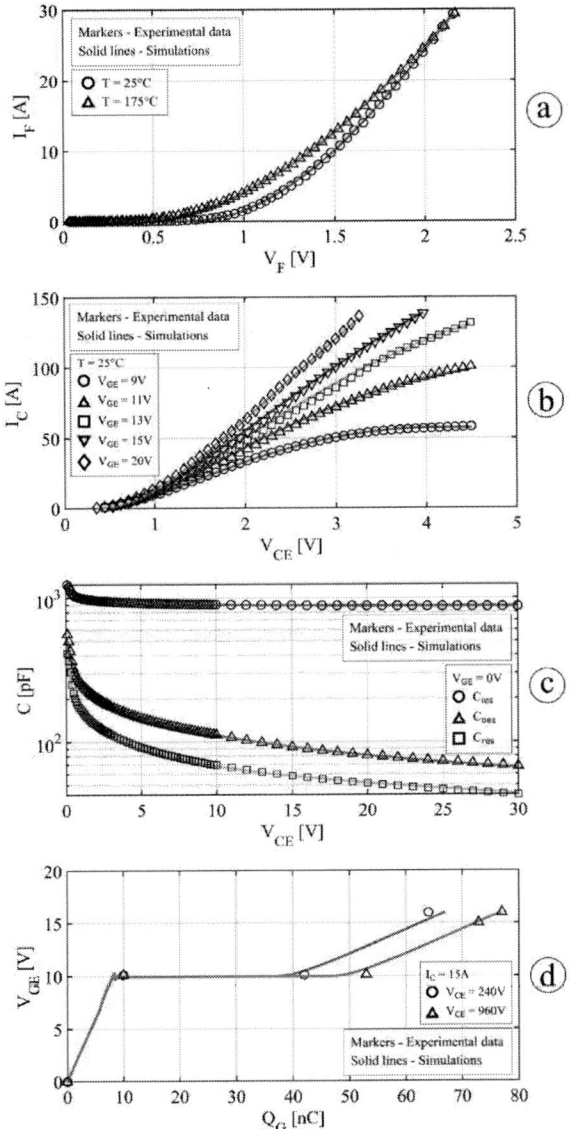

Fig. 5. One example of the quality of the calibration. Experimental data are shown by the markers, whereas the solid lines show the simulation data. a) Diode, forward current vs. forward voltage. b) IGBT, collector current vs. collector-emitter voltage. c) IGBT, capacitances vs. collector-emitter voltage. d) IGBT, gate-emitter voltage vs. gate charge.

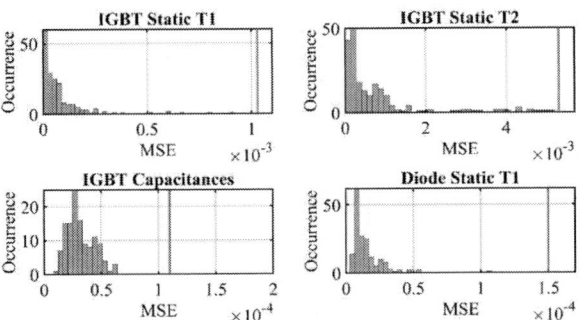

Fig. 6. Histograms of the MSE achieved for more than 200 devices. The red line shows the MSE which is considered by engineers as very good for a manual calibration.

A more comprehensive view is given in Fig. 6, where the histograms of the MSE-s (mean square errors) between the experimental data and the simulations are depicted. In order to give a reference point, hundreds of manual calibrations were analyzed and the red lines in Fig. 6 were extracted by getting the minimum MSE of these manual calibrations. The red lines represent the MSE-s which were considered as very good for the manual calibration. The quality of the calibration is very good for all the characteristics of the devices. The flow is very generic, and once it is in place, it is effortless to apply it to all kinds of behavioral models.

V. CONCLUSION

The paper presents a generic, fast and scalable fully automatic flow for parameter extraction of behavioral models of electronic components using differential evolution as the calibration algorithm. To ensure this, a smart sampling technique based on the curvature of the data is developed. It reduces the data redundancy by a factor of 10 to 20. In addition, the knowledge of previous parameter extraction tasks is used to speed up the parameter extraction of the new models. This leads to a speed-up of 5 to 10 times compared to calibration without the use of previous knowledge. The flow scales up very well. It is tested on more than 200 Si-diode and IGBT behavior models. The entire parameter extraction for all the models took 48 hours using a parallelization factor of 50. This is a gain of 200 in speed compared to the manual work. For the future, we expect to improve three important aspects: automate the measurement of experimental data, modify differential evolution to work with adaptive model parameter ranges and combine differential evolution with gradient descent in order to speed up the calibration process. As a last point, it is important to understand that the entire methodology and flow that we show here is not tightly connected to IGBT and Si-Diode models. It is in principle applicable to other electronic components and even beyond behavioral models. It can be applied to any model that has a large number of input parameters.

REFERENCES

[1] K. Chiba, "Performance comparison of evolutionary algorithms applied to hybrid rocket problem," The 6th International Conference on Soft Computing and Intelligent Systems, and The 13th International Symposium on Advanced Intelligence Systems, Kobe, 2012, pp. 1673-1678.

[2] A. Cerdeira, M. Estrada, R. García, A. Ortiz-Conde and F. García-Sánchez, "New procedure for the extraction of basic aSi:H TFT model parameters in the linear and saturation regions". Solid-state Electronics - SOLID STATE ELECTRON. 45, 2001, pp. 1077-1080.

[3] D. Navarro, T. Sano and Y. Furui, "A Sequential Model Parameter Extraction Technique for Physics-Based IGBT Compact Models," in IEEE Transactions on Electron Devices, vol. 60, no. 2, pp. 580-586, Feb. 2013.

[4] P. Moreno, R. Picos, M. Roca, E. Garcia-Moreno, B. Iniguez and M. Estrada, "Parameter Extraction Method using Genetic Algorithms for an Improved OTFT Compact Model," Spanish Conference on Electron Devices, 2007, pp. 64-67.

[5] N. Akkan, M. Altun and H. Sedef, "Parameter Extraction Method Using Hybrid Artificial Bee Colony Algorithm for an OFET Compact Model," 15th International Conference on Synthesis, Modeling, Analysis and Simulation Methods and Applications to Circuit Design (SMACD), Prague, 2018, pp. 105-108.

[6] F. Frenet "Sur les courbes à double courbure." Thèse. Toulouse, 1847. Abstract in J. de Math. 17, 1852.

[7] R. Storn and K. Price, "Differential Evolution: A Simple and Efficient Adaptive Scheme for Global Optimization Over Continuous Spaces". Journal of Global Optimization. 23, 1997.

Machine Learning in the Analog Circuit Simulation Loop

Petar Tzenov* and Ahmed Sokar[†]

Infineon Technologies AG
Neubiberg, Germany
Email: *petar.tzenov@infineon.com, [†]ahmed.sokar@infineon.com

Abstract—**This paper presents the software coupling of an analog circuit simulator (ACS) to a machine learning (ML) execution engine, in order to enable usage of ML models in circuit simulation context. This is achieved by interfacing Infineon's in-house simulator, TITAN, with the widely accepted machine learning framework TensorFlow (TF), via an easy to use Verilog-A API. Here we introduce the basic characteristics of this interface and present an application example for its usage in analog circuit behavioral modeling.**

Index Terms—**analog circuit simulation, machine learning, tensorflow, verilog-a**

I. INTRODUCTION

Compared to the predominant role which machine learning plays in various other businesses (e.g. robotics, autonomous driving, finance etc.), its productive use for chip design and verification seems to be only in its infancy [1]. This is in part due to strong dependence on EDA-vendor solutions in our industry, whereas the advancement of machine learning has mainly been driven by open source. Furthermore, the introduction of artificial intelligence in the IC design process requires highly skilled experts with knowledge of both chip design and ML methodologies. It is, however, a rarity for a data scientist to understand analog circuit design, or for a circuit designer to understand data science. This suggests that ML-methodologies can only be introduced by a collaborative effort from both camps of experts.

Yet there remains a problem with the tooling, as each industry comes with its own set of software tools and technologies. Whereas a chip designer mainly relies on company specific design flows and develops on scripting and hardware description languages, a data scientist mostly uses open source libraries and develops on high level programming languages.

So the natural question arises, namely how can one seamlessly connect those two worlds, so that neither designers nor data scientists would have to give up on their tools or learn new technologies?

In this work we have endeavored to resolve the above enumerated issues by coupling one of the most widely used machine learning libraries, Google's TensorFlow [2], to our in-house analog circuit simulator TITAN [3]. The interface is realized via an easy to use Verilog-A API, consisting of several basic system tasks for model loading, model evaluation (i.e. performing inference in ML-parlance), model Jacobian computation etc. This coupling is meaningful and presents a

substantial improvement over the state of the art [4]–[6], as it automates the deployment of highly complex mathematical models in the simulation environment. To the best of our knowledge, previous attempts to use machine learning models in circuit simulation have relied either on manual implementation of the model in Verilog-A [4], generation of System Verilog code representing the model [5] or through a co-simulation between e.g. Cadence Virtuoso and Matlab [6]. Our solution hides the complexity of the manual implementation and the generation of HDL code, and constitutes an improvement over the co-simulation approach as it provides internal mechanisms to compute the model's Jacobian. Finally, a direct coupling to TensorFlow has the added benefit that since TF already contains an execution model for heterogeneous compute systems, it can effectively be used for the construction of high performance device models, without having to do any modification of the simulator's internals.

In the following section, Sec. II, we present the interface in question, and in Sec. III, we illustrate an application example of this technology for behavioral modeling via Gaussian process regression.

II. ACS AND TENSORFLOW COUPLING API

In order to ensure fast-prototyping and ease of use, there are several basic requirements that a coupling between a ML-framework and a circuit simulator should satisfy.

- The ML-model should be serializable to a computer readable format. Ideally, this format shall be an industry accepted standard to ensure the portability of the solution across different development environments (e.g. from the model construction stage to the model deployment etc.), as well as different stages of the design process (e.g. system level and transistor level).

- The serialized model should be loadable and executable inside analog circuit simulation context (as opposed to manually implementing the model in a HDL, or generating HDL code implementing the model). This trait is desirable, as it hides the model internals from the engineer, and allows for the model to be tested in and out of simulator context.

- For fast model development, it will be beneficial if the model serialization is also compatible with one or more of the widely accepted machine learning libraries, and can be constructed with a high-level programming language.

© VDE VERLAG GMBH · Berlin · Offenbach

This will enable the model builder to create state-of-the-art models outside of his/hers simulation environment in an "offline" stage, i.e. without having to run an analog simulator to verify the correctness of their ideas.

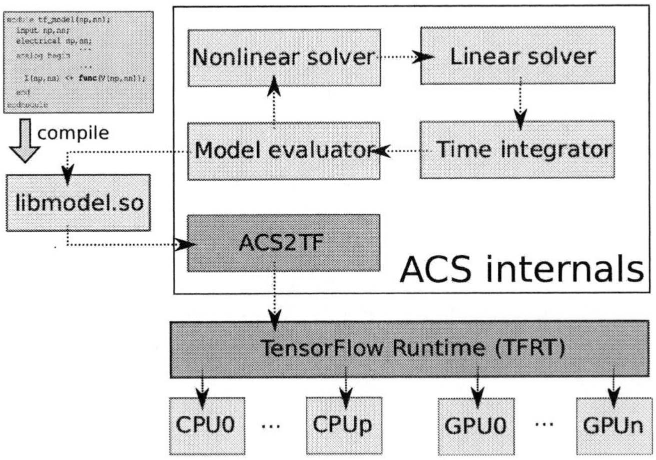

Fig. 1. Schematic representation of the **ACS2TF** coupling architecture.

Guided by the above principles, we chose TensorFlow as it is widely used, actively supported and is completely open source. Models developed with TF can be serialized in various different ways, but we have chosen the Protocol Buffer (PB) format as it is portable across operating systems, has compact binary representation, and there are various software tools that support it.

The coupling with our simulator is realized as depicted in Fig. 1. The internals of a typical analog circuit simulator contain among others, basic modules for device model evaluation, nonlinear and linear solver components and, in the case of time-dependent analyses, a time integration module. The coupling is realized via an additional software module, here denoted as **ACS2TF**, where all the logic of model construction, evaluation, exporting etc. is implemented. Initially, the user provides the simulator with a Verilog-A file, containing the model instantiation and evaluation API-calls, as well as the serialized model or models as a set of Protocol Buffer files. The Verilog-A file is then compiled to a shared library which is dynamically loaded and called during model evaluation stage of the analog simulation. At each evaluation stage, the simulator queries the shared library, which in turn delegates the actual ML-model evaluation to the **ACS2TF** software module. **ACS2TF** then invokes the TensorFlow Runtime, which serves as an execution engine for the model or models in question. As mentioned before, constrained only by their computational resources, the users can distribute the TF-computations over a heterogeneous set of compute nodes.

Broadly speaking, the functions in the API can be divided into two categories. The first one comprises of all API-calls needed for TF-model inference, whereas the second category, encompasses API-calls needed for TF-model construction, training and exportation (to PB file). We briefly present the look and feel of each API, but the reader shall note that we have omitted quite a lot of detail for brevity of presentation.

A. Model inference API

This is a basic API for evaluation of externally constructed TensorFlow models. It includes only three simple Verilog-A system tasks for loading and evaluating a given model.

- `$acs2tf_model(string filename);`
  Load a TensorFlow model from PB file `filename`. If successful, the call returns a non-negative integer uniquely identifying the model, or a negative integer if an error occurred.
- `$acs2tf_execute(integer model_id, [string opname [, real[] opdata]?]*);`
  Execute a TensorFlow model with model id `model_id`, where both input and output operations which need to be executed are specified as `string, real[]` array pairs. The function returns a negative integer if an error occurred or 0 if evaluation succeeded.
- `$acs2tf_evaluate(integer model_id, string in_opname, real[] in_array, string out_opname, real[] out_array, [string opname [, real[] opdata]?]*);`
  Evaluate `out_op` of a TF model with id `model_id`, by feeding its input op with name `in_opname` with the data stored in `in_array` and writing the result in `out_array`. The function returns a negative integer if an error occurred or 0 if evaluation succeeded. This function also automatically performs the computation of the Jacobian of `out_opname` with respect to `in_opname`, using TF's built-in symbolic differentiation engine. Arbitrary number of additional input, output and update operations can (optionally) be provided to the function call, these however are excluded from the Jacobian computation.

B. Model construction and training API

The second set of Verilog-A extensions is targeted to users who wish to construct and train simplistic neural network (NN) models with our interface, directly from within Verilog-A code, i.e. without "leaving" the simulation environment.

The API implementation for construction and training of NN models follows the look and feel of the popular ML-library for python Keras, and as such allows the construction of deep neural networks in a sequential manner, by specifying the characteristics of each "layer" of the model. The so constructed models can be trained, (possibly) during multiple simulator runs, and exported to a Protocol Buffer file for later use in inference. An example of Verilog-A module containing API-calls for model construction, training and exportation is presented in Listing 1.

III. BEHAVRIOAL MODELING OF ANALOG CIRCUITS VIA GAUSSIAN PROCESS REGRESSION

To illustrate the power and ease of use of the **ACS2TF** API, we have used the Gaussian process regression (GPR) method

Listing 1. Example syntax of Verilog-A TensorFlow binding.

```
module tf_model(np,nn);
  input np,nn;
  electrical np,nn;
  integer model, l0, l1, l2;
  real loss; real x[0:1]; real y[0];
  analog initial begin
    // empty model instantiation:
    model = $acs2tf_model();
    // layer structure: 1xinput, 1xhidden and 1xoutput
    l0 = $acs2tf_input_layer(model,2);
    l1 = $acs2tf_dense_layer(model,10,10,"sigmoid");
    l2 = $acs2tf_dense_layer(model,1,11,"identity");
    // 'compile' the model for training. input layer is 'l0'
    // and output layer 'l2'. Loss is 'MeanSquaredError' and
    // training algo is 'Adam'.
    $acs2tf_compile(model,l0,l2, "MSE","ADAM:1e-3");
  end
  analog begin
    x[0] = V(np,nn);
    x[1] = ddt(V(np,nn))
    y[0] = I(np,nn);
    // given input x and target y, fit the model
    loss = $acs2tf_fit(model,x,y);
    @(final_step) begin
      // when done, export the model to file
      $acs2tf_export(model,"my_neural_network.pb");
    end
  end
endmodule
```

to construct a behavioral model of an analog circuit. GPR has recently attracted interest as a methodology to develop surrogate models for circuit optimization [7], as it is known to generate simple and robust regression models, and it has the desirable property that it provides the uncertainty of the prediction, information which can be used to dynamically update the model during simulation. On the downside, GPR has a $O(N^2)$ inference cost, where N is the number of training data points. This issue could in principle be overcome by either developing sparse Gaussian process regression models [8], or by distributing the calculations on a massively parallel many-core compute nodes.

For brevity here we summarize only the most central equations of GP regression methodology, whereas the interested reader is referred to existing literature for more detail [8].

Assume that we have a training dataset $\{X = \{\mathbf{x}_j\}_{j=1}^N, \mathbf{y}\}$, consisting of N input feature vectors $\mathbf{x}_j \in \mathbb{R}^{D_x}$ and N corresponding response values $y_j \in \mathbb{R}$. Given a test datapoint $\mathbf{x}^*$, the conditional posterior distribution of the value $y^* = y(\mathbf{x}^*)$ of the target function at that point is (within the framework of GP regression)

$$y^*|\mathbf{y},X,\mathbf{x}^* \sim \mathcal{N}\Big(K(\mathbf{x}^*,X)K(X,X)^{-1}\mathbf{y}, \quad (1)$$
$$K(\mathbf{x}^*,\mathbf{x}^*) - K(\mathbf{x}^*,X)K(X,X)^{-1}K(X,\mathbf{x}^*)\Big).$$

In Eq. (1) $K(\cdot,\cdot)$ denotes the covariance function kernel, which intuitively is a measure for the relatedness between points in feature space. In this work we have used the *squared exponential kernel*, computed as

$$K(\mathbf{x}_i,\mathbf{x}_j) = \sigma_f^2 \exp\Big(-\frac{d(\mathbf{x}_i,\mathbf{x}_j)^2}{2l^2}\Big), \quad (2)$$

where $d(\cdot,\cdot)$ stands for the Euclidean distance between the points and l and σ_f^2 represent the hyper parameters of the model, encoding a characteristic length scale in feature space

and the maximum covariance between data points, respectively. Finally, we have also abused notation to denote with $K(X,X)$ the matrix with elements $K(X,X)_{i,j} = K(\mathbf{x}_i,\mathbf{x}_j)$ (and similarly for $K(\mathbf{x}^*,X)$).

During inference Eq. (1) can be used to sample from the posterior distribution, or simply to approximate y^* by its mathematical expectation $\mu(\mathbf{x}^*)$. In addition, the same equation also gives us the variance $\sigma(\mathbf{x}^*)^2$ of y^* at the test datapoint, which is a direct measure of how "uncertain" the GP regressor is about the prediction it is about to make.

We have directly implemented this mathematical formalism using TensorFlow's python API, in around 300 lines of code. The model construction is completely decoupled from the analog circuit simulator environment, and it only uses simulation results as its training data. The test circuit we have chosen is a structurally simple four wave bridge rectifier circuit, and was selected due to its analytical tractability and yet highly nonlinear output, which is usually difficult to model accurately.

During the data generation stage, a sinusoidal voltage source, with unit amplitude and 1kHz frequency, is connected to the input of the rectifier. On the other hand, at the output a resistive load with $R = 1\text{k}\Omega$ impedance is connected to the remaining two pins of the circuit. Since we would like to have GPR behavioral model which is a pin-for-pin replacement of the original SPICE level rectifier circuit, this mandates that our model shall have 4 inputs (the node voltages) and 3 outputs (the terminal currents), where the current of one of the terminals is computed as the negative sum of the other three.

The so generated training data is preprocessed with python where, the training set's feature and response values are rescaled to the interval $[0;1]$ and, in order to avoid overfitting, the responses are additionally perturbed with a small additive noise component.

Next, for the computation of the covariance matrix $K(X,X)$, we use an educated guess for σ_f and l and set them both to 0.1. Also, we do not actually invert $K(X,X)$ (since K is usually ill-conditioned), but instead use the Moore-Penrose pseudoinverse K^+ in place of K^{-1} in Eq. (1).

Finally, the computation is also split into two subcomponents and executed in parallel on a single GPU and CPU as depicted in Fig. 3. This way, one can benefit from the hundreds of cores available on modern graphic processors to compute the expensive matrix-vector product in Eq. (1).

After implementing the so described model, it is serialized to a binary Protocol Buffer file and is readily deployed in the circuit simulation environment. As seen from external circuit components, the deployed model is structurally equivalent to the original rectifier circuit. A simple netlist, describing the circuit topology is written and the corresponding Verilog-A module, containing **ACS2TF** API calls for model initialization and evaluation is provided to the simulator. Importantly, the Verilog-A file does not need to include any normalization/denormalization routines, as these are already an internal part of the TensorFlow model (see Fig. 2). Sample HDL code for the deployed model is given in Listing 2.

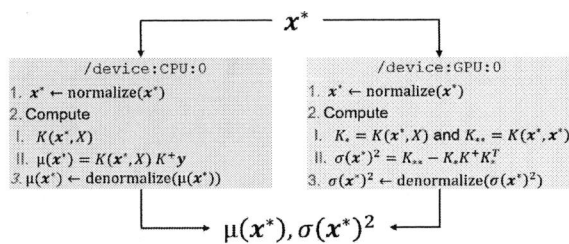

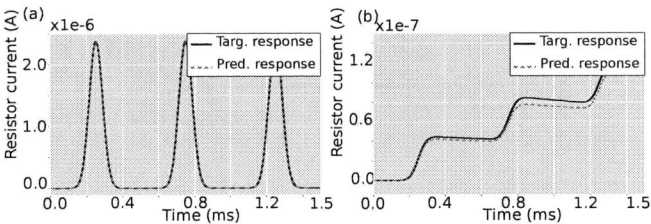

Fig. 2. A schematic representaton of how data and instruction flow are distributed for the computation of Eq. (1). As depicted, the expectation value is calculated on the CPU, whereas the standard deviation on the GPU. All necessary data is replicated to avoid expensive CPU to GPU communication.

Fig. 3. Target and predicted response for the load resistor current where (a) the load is a single $1k\Omega$ resistor and (b) the load comprises of the $1k\Omega$ resistor in parallel with a $C = 5\mu F$ capacitor.

Listing 2. Example model deployment in Verilog-A.

```
module tf_model(p1,p2,p3,p4);
  input p1,p2,p3,p4;
  electrical p1,p2,p3,p4;
  integer model; real x[1:4]; real y[1:3];
  (* desc="Prediction_variance", units="Amps^2" *)
  real var;
  analog initial begin
    model = $acs2tf_model("my_gpr_model.pbtxt");
  end
  analog begin
    x[1] = V(p1); x[2] = V(p2); x[3] = V(p3); x[4] = V(p4);
    $acs2tf_evaluate(model,"x",x,"y",y,"out:var",var);
    I(p1) <+ y[1]; I(p2) <+ y[2]; I(p3) <+ y[3];
    I(p4) <+ -I(p1)-I(p2)-I(p3);
  end
end
endmodule
```

To generate the training data, we performed SPICE-level transient simulations with the rectifier circuit's output connected to a resistive load of $1k\Omega$ impedance. For validation of our model we modified the training cirucit to include also a an additional capacitor with value $C = 5\mu F$, connected in parallel to the resistor. For comparison the GPR model was also executed on CPU only and on the GPU+CPU combination of Fig. 2. The distributed model executes in total of ≈ 2.8 seconds, whereas the CPU-only GPR computation needs around 4 seconds to complete. In both cases the inital TF model and runtime loading takes around 1.5 seconds. As a comparison, the transistor level implementation executes for around a tenth of a second from start to finish, but we expect our approach to scale better with larger circuits. Further investigation on performance are out of scope for this publication.

Comparison of results between the distributed GPR simulations and the transistor level model on from both kinds of simulations are illustrated in Fig. 3. Optically the simulation results between the SPICE representation and the GPR model agree quite well. Root Mean Square (RMS) error for the current entering the resistive load around 1.04% of the maximal supplied current, for the netlist with resistive load only, whereas for the netlist with the $5\mu F$ capacitor added (topolgy unseen during model training), the RMS error is around 3.7%.

Note that these numbers might not be satisfactory for high accuracy and high fidelity applications, but these are deficiencies of the constructed model and *not* of the proposed API. In addition we believe that the errors are mainly due to the fact that the training responses were perturbed from their original simulated values, in order to avoid overfitting. Finally, in the above cited results, the standard deviation of the prediction was consistently under $10nA$.

IV. Conclusion and Final Remarks

We have presented a Verilog-A API, coupling an analog circuit simulator to a machine learning library, in order to enable the easy deployment of state of the art ML models in analog circuit simulation context. As a proof of concept application of the API, we have demonstrated the usage of a Gaussian process regression model in transient simulations. Both SPICE level netlist and GPR model simulation reveal satisfactory agreement. More importantly, the model construction and deployment are completely decoupled, which allows for fast prototyping, easy debugging and creates an avenue for interdisciplinary collaboration. The proposed API can be used productively in several different application areas such as, but not limited to, behavioral and surrogate modeling, IP-obfuscation, signal anomaly detection and safe operating area checks, aging model construction and others. The authors intend to pursue some of the aforementioned areas in future publications.

References

[1] Qi, Weiyi. "IC Design Analysis, Optimization and Reuse via Machine Learning." (2017).

[2] Abadi, Martín, et al. "Tensorflow: A system for large-scale machine learning." 12th USENIX symposium on operating systems design and implementation (OSDI 16). 2016.

[3] Feldmann, Uwe, and Reinhart Schultz. "TITAN: an universal circuit simulator with event control for latency exploitation." ESSCIRC'88: Fourteenth European Solid-State Circuits Conference. IEEE, 1988.

[4] Chen, Zaichen, Maxim Raginsky, and Elyse Rosenbaum. "Verilog-A compatible recurrent neural network model for transient circuit simulation." 2017 IEEE 26th Conference on Electrical Performance of Electronic Packaging and Systems (EPEPS). IEEE, 2017.

[5] Grabmann, Martin, Frank Feldhoff, and Georg Gläser. "Power to the Model: Generating Energy-Aware Mixed-Signal Models using Machine Learning." 2019 16th International Conference on Synthesis, Modeling, Analysis and Simulation Methods and Applications to Circuit Design (SMACD). IEEE, 2019.

[6] Hasani, Ramin M., et al. "Compositional neural-network modeling of complex analog circuits." 2017 International Joint Conference on Neural Networks (IJCNN). IEEE, 2017.

[7] Zhang, Shuhan, et al. "Bayesian optimization approach for analog circuit synthesis using neural network." 2019 Design, Automation & Test in Europe Conference & Exhibition (DATE). IEEE, 2019.

[8] Rasmussen, Carl Edward. "Gaussian processes in machine learning." Summer school on machine learning. Springer, Berlin, Heidelberg, 2003.

© VDE VERLAG GMBH · Berlin · Offenbach

Bringing Structure into Analog IC Placement with Relational Graph Convolutional Networks

António Gusmão[1,2], Nuno Horta[1,2], Nuno Lourenço[1], Ricardo Martins[1]

[1]Instituto de Telecomunicações
[2]Instituto Superior Técnico – Universidade de Lisboa, Lisboa, Portugal
{agusmao; nuno.horta; nlourenco; ricmartins}@lx.it.pt

Abstract—**In this paper, disruptive research using modern embedding techniques and a deep learning (DL) model based on a relational graph convolutional network (R-GCN)** *encoder* **that automates the placement task of analog layout synthesis is conducted. The proposed methodology introduces structure in the input data, drastically reducing the total number of trainable parameters, leading to a smaller and more effective regression model. Moreover, its unsupervised training does not rely on expensive legacy layout data but only on sizing solutions. Experimental results show that the proposed R-GCN deep model generates placement solutions at push-button speed for multiple technology nodes and generalizes to circuit topologies not used in training. Moreover, the model outperforms other dense DL models while being 3000x smaller and producing solutions that compete with highly optimized analog designs.**

Keywords—Analog layout synthesis; Deep learning; Graph convolutional networks; Node Embeddings, Multiplex Graphs.

I. INTRODUCTION

Even though analog integrated circuit (IC) layout synthesis has been intensively studied in the last few decades, there is still no established methodology being used in the industrial environment. Regarding the techniques for automatic device placement, two major lines can be found in the literature: placement optimization with topological constraints [1][2] and placement retargeting from legacy designs [3] or templates [4]. In the first, every solution is generated from scratch. However, even if this time-consuming process is highly constrained, it does not always produce meaningful solutions for the circuit designer. In the second, fast techniques exploit the layout patterns from a single or multiple previously validated legacy layouts or templates. Still, as the placer matches the solution's sub-circuits with those found on a library of legacy data, the acquired knowledge does not generalize beyond training.

Quite recently, machine learning (ML) is starting to show its capabilities in the whole analog IC synthesis, promising to bypass some of the limitations of traditional approaches and revolutionize this field. For example, in the circuit sizing task (prior to layout generation), a two-model chain that maps circuit performances to devices' sizes was recently proposed [5], while in other works, DL models map from device sizes to circuit performances [6]. Other recent alternative applications include the identification of hierarchy levels and building blocks from a circuit netlist [7]; automated generation of wells via a conditional generative adversarial network [8]; or even variational autoencoders that mimic human guidance in routing problems [9]. In the particular case of analog IC placement, a straightforward, fully supervised artificial neural network (ANN) that reproduces the layout patterns of a dataset of legacy placement solutions was proposed [10]. In [11], it was enhanced, providing a multi-circuit-topology model that achieves the distinction between topologies by describing the topological constraints

This work is funded by FCT/MCTES through national funds and when applicable co-funded EU funds under the project UIDB/50008/2020. Including internal research project LAY(RF)².

at the ANN's input layer. Nonetheless, in all the proposed works so far, the description of the circuits' traits on the model's input layer is still a bottleneck, with no apparent solution [12].

Graph representations and graph-based ML techniques that consider the input data's underlying structure have been hot topics in the past few years. Especially as they are, in theory, scalable to graphs of any dimension, and their node-wise (or edge-wise) parameter sharing reduces the model's complexity and increases its generalization capabilities by splitting a problem into the conjunction of smaller problems. This work builds an R-GCN [13] *encoder*, whose architecture is illustrated in Fig. 1, introducing structure in the input data via inter-device relations, drastically reducing the total number of trainable parameters, leading to a smaller and more effective regression model. The novel contributions of this work can be summarized as follows:

- The R-GCN model uses the input graph to exploit the relations between devices. The R-GCN is conceptually more complex than a multilayer perceptron (MLP) or even a convolutional neural network (CNN), but makes an enormous reduction in the number of trainable parameters;

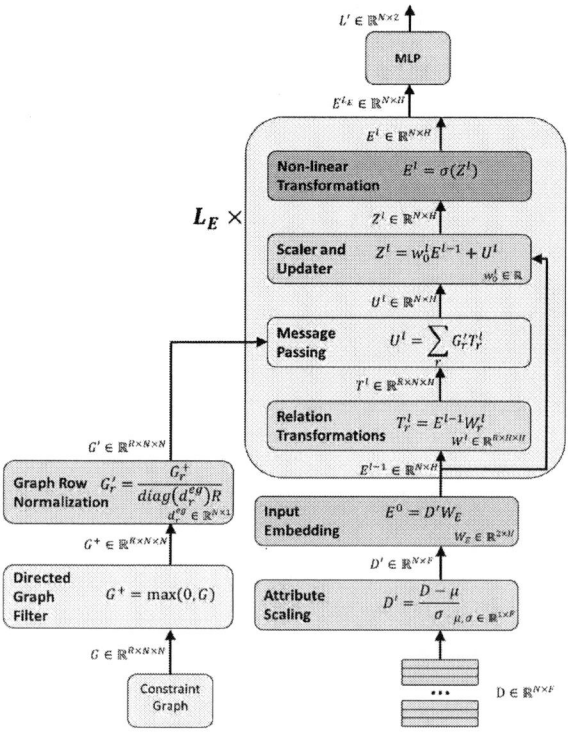

Fig. 1. Proposed relational graph convolutional network *encoder* and embedding generator for analog IC placement synthesis.

- Unlike other graph-based approaches recently applied to analog IC design automation [7][14], this work's input structure is placement specific, i.e., a heterogeneous graph representing the placement constraints to be fulfilled;

- As stated, the topological constraints are described in graphs that encode the relations between pairs of devices. This representation is a significant step forward when compared to the only ML-based placement approach with multi-topology support available in the literature [11], as it removes its dependence from arbitrary parameters, e.g., the length of the current/signal-paths; and enables the use of mature graph-based ML;

- Traditional placement retargeting from legacy designs [3] or ML-based approaches [10] use previously designed placement solutions that contain the knowledge from the expert designers. These layouts are scarce and expensive to obtain, and thus, here, the model's training is focused on complying to the explicitly and efficiently described topological constraints, requiring only sizing data.

This document is organized as follows. In Section 2 the model's input is described. In Sections 3 and 4 the R-GGN *encoder* architecture and the topological loss function, respectively, are detailed. And finally, in Sections 5 and 6 the experimental results and conclusions are addressed.

II. MODEL'S INPUT: HETEROGENEOUS GRAPH DESCRIPTION

The proposed R-GCN [13] *encoder* takes as input a multigraph G defined by the adjacency matrix for each of the types of relations accounted, e.g., this work considers 3 topological constraints: proximity, current-flow, and symmetry. In practice, this is equivalent to defining a constraint graph for each of the constraints, where each node represents one of the circuit's devices, and an edge represents a constraint between them. This feature space results in 3 graphs that share the nodes that constitute them. Purposely defining each constraint at an inter-device level enables the use of efficient, easily parallelized matrixial computations, as well as decreasing its complexity through parameter sharing and enabling the use of established graph-based ML techniques. Other topological constraints can be easily incorporated in the graph description.

A. Proximity Constraints

These describe groups of devices that should be placed closely to account for well sharing or routing. The proximity group description is encoded in the matrix $\mathbf{P}_{N\times N}$, where entries $p_{i,j}$ are set to '1' when device i belongs to the same proximity group as device j, and '0' otherwise. A device can share proximity constraints with any number of devices, but never with itself. As such, for a row of $\mathbf{P}_{N\times N}$ representing a given device, more than one element may be '1' (representing a proximity relation with another device), but not an element in its main diagonal. Fig. 2(a) shows the proximity adjacencies for a generic circuit with 4 devices/nodes.

B. Current-flow Constraints

These constraints are fulfilled with a monotonic routing [15] of current-flows, ultimately reducing the interconnect wirelength and routing-induced parasitics. In Fig. 3, six paths from VDD to VSS on a single-stage amplifier biased with voltage combiners for gain enhancement [15] are highlighted. These correspond to the current-flow constraints encoded by a matrix $\mathbf{C}_{N\times N}$, whose entry $c_{i,j}$ is '1' if device i precedes device j in any of the circuit's current-paths, or '-1' if the device j precedes device i. The result is a directed acyclic

graph where self-connections are also not possible. Note that usually in directed graphs only one direction (and positive elements) is considered, however, by considering the redundant inverse path (represented by the '-1's), if a current-flow constraint fails, both devices produce an amount of error. Hence, $\mathbf{C}_{N\times N}$ is not symmetric, as shown in Fig. 2(b).

C. Symmetry Constraints

Symmetry restricts two devices to a mirrored placement along an axis and reduces the sensitivity to on-die thermal gradients and parasitic mismatches between two identical signal/current-flows. These constraints are encoded in the matrix $\mathbf{S}_{N\times N}$, where the entry $s_{i,j}$ is set to '1' if device i should be placed symmetrically to device j and 0 otherwise. Note that a device can only share symmetry with another single device or itself (self-symmetric device). Finally, devices with no symmetry constraints are also supported, with its row of $\mathbf{S}_{N\times N}$ filled with zeros, as shown in Fig. 2(c). All the previous matrices are unweighted.

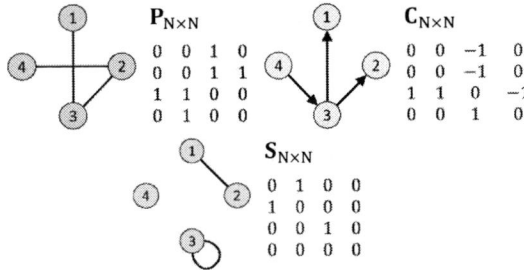

Fig. 2. Generic example of the adjacency matrices that form each constraint of the multigraph.

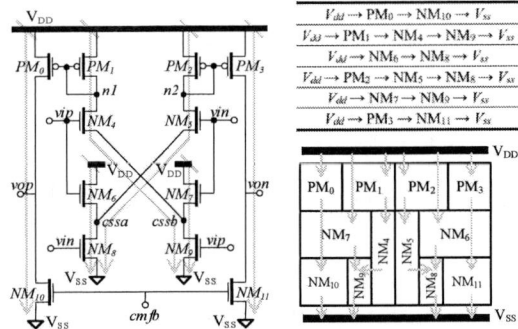

Fig. 3. Schematic of the voltage-combiners biased amplifier (*OpAmp₁*) [15], and, a generic placement with monotonic current-paths.

D. Attribute Vector

Additionally, a node attribute matrix $\mathbf{D}_{N\times F}$, whose row i corresponds to the attribute vector $\mathbf{d}_{i_F}$ for device i, is also considered, where F is the length of the attribute vector. In this work, the *width* and *height* of the devices' bounding boxes, along with a one-hot positional identifier, are the attributes for a total length of $N_{max} + 2$. N_{max} is the maximum number of supported devices. The *widths* and *lengths* of each design are scaled to have zero mean and unitary variance, resulting in the scaled node attribute matrix $\mathbf{D}'_{N\times F}$. Note that this is different from the usual feature standardization, where the features are scaled among all examples. This scaling is suited for placement, which is only sensitive to the devices' relative sizes. Moreover, by scaling each point individually, the model can be generalized for several technology nodes whose dimensions have different magnitude orders.

III. Proposed R-GCN Architecture

Given the adjacency matrix describing the circuit's topological constraints, $\mathbf{G}_{R \times N \times N}$, where R denotes the number of relations considered (3 in this work), and the scaled attribute vectors of each device, $\mathbf{D}'_{N \times F}$, the model must place each device in an $\mathbb{R}^2$ plane such that all the topological constraints are fulfilled, and the placement's area is minimized. As such, the model outputs the devices' positions as an $N \times 2$ matrix $\mathbf{L}'$. The positions of each device are "un"scaled with the σ and μ values used to scale the input attribute vector $\mathbf{D}'$ before being fed to the TLF.

The R-GCN *encoder* of Fig. 1 builds, for each node, a meaningful representation that considers their relationship with the remaining graph's nodes. First, embeddings are generated for each node's scaled attribute vector, using a linear transformation defined by the trainable weight matrix $\mathbf{W}_{E_{F \times H}}$ where H is the hidden size hyperparameter. The embeddings matrix $\mathbf{E}^0_{N \times H}$ is thus generated through the matrixial multiplication $\mathbf{E}^0 = \mathbf{D}' \mathbf{W}_E$. In parallel, the multilayered constraint graph $\mathbf{G}_{R \times N \times N}$ is filtered so that the redundant, negative edges from directed graphs (e.g., the current-flow layer) are eliminated, resulting in the filtered matrix $\mathbf{G}^+_{R \times N \times N}$, which is scaled to be used on message passing. The graph's scaling is done so that the rows of each relation's adjacency matrix sum to R^{-1}, by dividing each row by R times the degree of the node to which it corresponds to, with respect to the relation being considered. The degree of each node with respect to each relation can be determined by the sum of all the elements in the row representing its connections, through $\sum_n g^+_{r,n}$. For a specific relation, the vector $\mathbf{d}^{eg}_{r\ N}$ contains the information regarding each node's degree with respect to relation r. As such, each of the graphs can be scaled by $R^{-1} diag\left(\mathbf{d}^{eg}_r\right)^{-1} \mathbf{G}^+_r$ resulting in the scaled adjacency matrix $\mathbf{G}'_{r_{N \times N}}$. Once all layers are scaled in parallel, the resulting scaled graphs are represented by $\mathbf{G}'_{R \times N \times N}$. The encoding layer is repeated L_E times (hyperparameter) and, composed of four sub-modules:

- *Relation transformation*: the embeddings $\mathbf{E}^0_{N \times H}$ are linearly transformed by R independent, relation and layer-specific, transformations represented by the trainable 3-dimensional weight matrix $\mathbf{W}^l_{R \times H \times H}$, where l denotes the R-GCN layer (out of L_E) to which the transformation matrix corresponds to. Note that each R-GCN layer's parameters are independent from each other. The updated embedding matrix $\mathbf{T}^l_{R \times N \times H}$ is computed through the matrixial multiplication of each of the weight matrix's layers by the embeddings' matrix through: $\mathbf{T}^l_r = \mathbf{E}^{l-1} \mathbf{W}^l_r$, where $\mathbf{E}^{l-1}$ represents the embedding generated by the previous R-GCN layer; for the first layer, it corresponds to $\mathbf{E}^0$;

- *Message passing*: this step computes the aggregation of each node's neighbors transformed embedding resulting in the update matrix $\mathbf{U}_{N \times H}$ which is calculated by summing the $R\ N \times H$ matrices that resulted from the operation $\mathbf{G}'_r \mathbf{T}^l_r$, formally defined as $\mathbf{U}^l = \sum_r \mathbf{G}'_r \mathbf{T}^l_r$. Since the rows of $\mathbf{G}'_r$ sum to R^{-1}, the message passing stage is essentially equivalent to mean pooling R independent message passing procedures;

- *Scaler and updater*: the original embeddings $\mathbf{E}^{l-1}$ (denoted by the skip connection) are added to the updates $\mathbf{U}^l$, that are scaled by a trainable, layer-specific weight, $\mathbf{w}^l_0$;

- *Non-linear transformation*: finally, the embeddings passed to the next R-GCN layer are defined through $\sigma(w^l_0 \mathbf{E}^{l-1} + \mathbf{U})$, where σ represents a non-linearity. After the L_E layers, the final $N \times H$ encoding matrix, $\mathbf{E}^{L_E}_{N \times H}$, is obtained.

IV. Topological Loss Function

A topological loss function (TLF), similar to the one in [11], is used to train the R-GCN *encoder* in a fully unsupervised fashion. It quantifies the constraint fulfillment level in a predicted placement. In addition to the 3 abovementioned topological constraints, the model is also trained to eliminate overlap between devices and minimize the distance between devices (indirectly, minimizing the distance between devices reduces the area; while providing better gradients for the optimization that act on all devices and not only the ones at the edges of the placement). Fully connected graphs describe these relations. Following, the optimization problem was defined through matrix notation, as it reflects the most effective implementation of the computations in state-of-the-art DL platforms. The TLF is defined as the sum of five normalized and weighted cost components: a, the placement's wasted area (or lack of compactness); o, the overlap among devices; c, current-flow constraint violations; s, unsymmetrical devices, and p, distant devices in the same proximity group. The weights associated with each cost component are represented by $\lambda_j, j \in \{a, o, c, s, p\}$ and, for each example in each iteration, are independently sampled from a log-uniform distribution with minimum value 10^{-1} and maximum value of 10^1. Thus, TLF is a function of the matrices $\mathbf{D}$, $\mathbf{C}$, $\mathbf{S}$, and $\mathbf{P}$ as well as the unscaled devices' positions $\mathbf{L}$, is given by:

$$TLF(\mathbf{L}, \mathbf{D}, \mathbf{C}, \mathbf{S}, \mathbf{P}) = \lambda_a a(\mathbf{L}, \mathbf{D}) + \lambda_o o(\mathbf{L}, \mathbf{D}) + \lambda_c c(\mathbf{L}, \mathbf{D}, \mathbf{C}) \\ + \lambda_s s(\mathbf{L}, \mathbf{D}, \mathbf{S}) + \lambda_p p(\mathbf{L}, \mathbf{D}, \mathbf{P}) \quad (1)$$

V. Experimental Results

The methodology was coded in python and demonstrated with a dataset containing sizing solutions (split into training, validation, and test) for different state-of-the-art OpAmps on multiple foundries/nodes: $OpAmp_1$ (umc130) previously shown in Fig. 3, $OpAmp_2$ (umc130 & tsmc65) and $OpAmp_3$ (ams350), as detailed in Table I. Note that $OpAmp_1$ was kept for validation and test only. Additionally, in order to balance the dataset, the sizing solutions of $OpAmp_2$ and $OpAmp_3$ were augmented by shuffling the order of the devices.

TABLE I.   **DATASET DETAILS**

Topology	$OpAmp_1$	$OpAmp_2$		$OpAmp_3$
Technology	umc013	umc013	tsmc65	ams350
#devices	12	15		12
#sizing examples	10422	258	40	220
Train	0%	70%	70%	70%
Validation	50%	15%	15%	15%
#training & validation examples	5211	5160*	5060*	5040*
Test / #test examples	50%/522	15%/13	15%/220	15%/40

*augmented data through shuffling and padding.

The R-GCN test errors on each circuit are compared with a dense DL model, i.e., the MLP [11] of 4 hidden layers (2500/1000/250/100 neurons). The input vector of the MLP is the concatenation of each device's dimension vector along with the flattened adjacency matrices of the constraint graphs to a total vector length of $3N^2 + 2N$. The R-GCN *encoder* hyperparameters are: $L_E = L_D = 4, H = 12$, and has 1502 trainable parameters, about 3000x less than the MLP, that needs 4544380 parameters for comparable performance.

TABLE II. TRAIN, VALIDATION AND TEST ERRORS (TEE) FOR MULTILAYER PERCEPTRON AND PROPOSED R-GCN.

Architecture	Train Error	Validation Error	Test Error (TeE)	$OpAmp_2$ (umc130) TeE	Comp. (%)	$OpAmp_2$ (tsmc65) TeE	Comp. (%)	$OpAmp_3$ TeE	Comp. (%)	$OpAmp_1$ TeE	Comp. (%)	Overall Score
MLP	**0.82**	1.18	1.15	**0.74**	-	0.86	+10.3	0.84	+2.4	1.44	+87.0	75.1
R-GCN	0.92	**0.90**	**0.86**	1.00	+35.1	**0.78**	-	**0.82**	-	**0.77**	-	91.2

Both models were trained for 2000 epochs using Adam optimizer, and its parameters were set to: $\alpha = 0.001$, $\beta_1 = 0.9$ and $\beta_2 = 0.999$. A batch size of 500 samples was adopted, and a dropout with a keep rate of 0.9 was used to prevent overfitting. The used activation function was the exponential linear unit. The models were evaluated using a constant, unitary weighted TLF, i.e., with the weights λ_j , $j \in \{a, o, c, s, p\}$ set to 1. A summary of these results is shown in Table II.

As observable, MLP has a clear bias towards $OpAmp_2$ (0.74 TeE). This is due to the model's invariance to technology changes, as $OpAmp_2$ is twice as common in the training set than $OpAmp_3$. This fact is reinforced with the lowest training error (0.82) between both models, however, its performance is clearly degraded when predicting floorplans for a circuit not included in training ($OpAmp_1$), with 87% higher error than R-GCN. In this dataset, R-GCN presents the best overall score, with the lowest TeE for 3 topologies. Moreover, its generalization potential is observed on $OpAmp_1$ (0.77 TeE), with errors even lower than train and validation circuits.

Additionally, five randomly selected $OpAmp_2$ (tsmc65) solutions were highly optimized using the most-recent multi-objective placement approaches [2], producing a tradeoff between placements' width and height, as illustrated in Fig. 4. R-GCN predictions, after post-processing stage similar to [11], are capable of presenting, at push-button speed, floorplans that compete with the non-dominated fronts produced in several seconds (*camosa*) and up to 10-min (layout-dependent-effects-aware *camosqa*). Post-layout performances of all the R-GCN solutions were automatically validated in Eldo (after Calibre xRC extraction), which are shown in Table III.

TABLE III. POST-LAYOUT PERFORMANCES OF 5 RANDOMLY SELECTED $OPAMP_2$ FLOORPLAN SOLUTIONS PROPOSED R-GCN.

	Pre-layout					Post-layout (R-GCN)				
	Gbw (MHz)	Gdc (dB)	Idd (µA)	PM (°)	FOM	Gbw (MHz)	Gdc (dB)	Idd (µA)	PM (°)	FOM
Sol. 0	56.4	50.9	155	66	2187	56.1	50.9	155	64	2180
Sol. 1	70.9	50.9	140	62	3041	70.2	50.9	140	60	3015
Sol. 2	85.1	50.1	154	62	3324	84.0	50.1	153	59	3287
Sol. 3	66.4	50.8	138	61	2880	65.9	50.8	138	59	2863
Sol. 4	69.0	50.5	201	64	2057	68.7	50.5	201	62	2053

VI. CONCLUSIONS

This work proposed a disruptive approach based on state-of-the-art embedding techniques and a single *multi-topology multi-technology* R-GCN *encoder* for push-button analog IC placement. By introducing structure in the input data, it was possible to develop a model that reduced more than 3000x the number of trainable parameters, when comparing with most-recent DL techniques proposed, while achieving better test sets errors, especially on circuits not included on training. This reduced number of parameters will open the possibility of DL model development capable of dealing with circuits with higher complexity (both in number of devices and interactions among them). Ultimately, the floorplans generated by the R-GCN after post-processing stage, were capable of matching those produced by time-consuming placement optimizers.

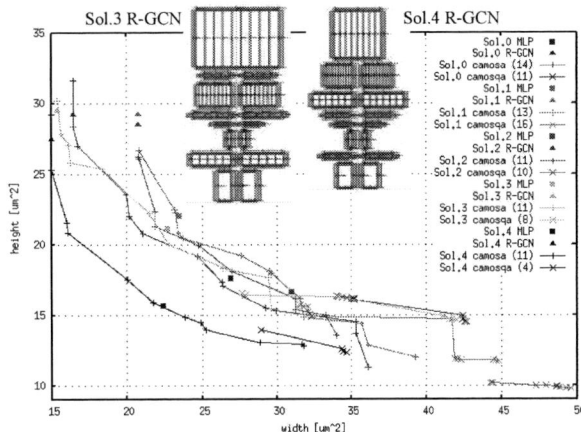

Fig. 4. Tradeoff between placements' height *versus* width (µm²) of the DL models (push-button) and optimization methods (seconds to minutes).

REFERENCES

[1] A. Patyal, et al., "Analog placement with current flow and symmetry constraints using pcp-sp," in *55th DAC*, pp. 1–6, June 2018.

[2] R. Martins, N. Lourenço, R. Póvoa, N. Horta, "Shortening the gap between pre-and post-layout analog IC performance by reducing the LDE-induced variations with multi-objective simulated quantum annealing," *Engineering Applications of Artificial Intelligence*, vol. 98, 104102, Feb. 2021.

[3] P. Pan, et al., "A fast prototyping framework for analog layout migration with planar preservation," in *IEEE TCAD*, vol. 34, no. 9, pp. 1373–1386, Sep. 2015.

[4] R. Martins, N. Lourenço, N. Horta, "LAYGEN II – Automatic Analog ICs Layout Generator based on a Template Approach" in *Genetic and Evolutionary Computation Conf.* (GECCO), pp. 1127-1134, Jul. 2012.

[5] N. Lourenço et al., "Using Polynomial Regression and Artificial Neural Networks for Reusable Analog IC Sizing," in *16th Int. Conf. on SMACD*, pp. 13–16, July 2019.

[6] G. Islamoğlu, T. Çakici, E. Afacan, G. Dündar, "Artificial Neural Network Assisted Analog IC Sizing Tool," in *16th Int. Conf. on SMACD*, pp. 9–12, July 2019.

[7] K. Kunal, et al. "GANA: Graph Convolutional Network Based Automated Netlist Annotation for Analog Circuits," in *Proceedings of DATE Conference and Exhibition*, pp. 55–60, Mar. 2020.

[8] B. Xu, et al., "WellGAN: Generative-adversarial-network-guided well generation for analog/mixed-signal circuit layout," in *56th ACM/ESDA/IEEE Design Automation Conference (DAC)*, June 2019.

[9] K. Zhu, et al., "GeniusRoute: A new analog routing paradigm using generative neural network guidance," in *IEEE/ACM International Conference on Computer-Aided Design (ICCAD)*, Nov. 2019.

[10] D. Guerra, A. Canelas, R. Póvoa, N. Horta, R. Martins, "Artificial Neural Networks as an Alternative for Automatic Analog IC Placement," in *16th Int. Conf. on SMACD*, pp. 1–4, Jul. 2019.

[11] A. Gusmão, F. Passos, R. Póvoa, N. Horta, N. Lourenço, R. Martins, "Semi-Supervised Artificial Neural Networks towards Analog IC Placement Recommender," in *IEEE ISCAS*, pp. 1–5, Oct. 2020.

[12] E. Afacan, N. Lourenço, R. Martins, G. Dündar, "Review: Machine learning techniques in analog/RF integrated circuit design, synthesis, layout, and test," in *Integration, the VLSI Journal*, 77, 113-130, 2021.

[13] M. Schlichtkrull, et al., "Modeling Relational Data with Graph Convolutional Networks," in *Lecture Notes in Computer Science*, 10843, pp. 593–607, 2017.

[14] M. Liu, et al, "S3DET: Detecting System Symmetry Constraints for Analog Circuits with Graph Similarity," in *Proc. of ASP-DAC*, pp. 193–198, Jan. 2020.

[15] R. Martins, R. Póvoa, N. Lourenço, and N. Horta, "Current-flow and current-density-aware multi-objective optimization of analog IC placement," in *Integration, the VLSI*, vol. 55, pp. 295-306, Sept. 2016.

Machine Learning in Charge: Automated Behavioral Modeling of Charge Pump Circuits

Martin Grabmann*, Christian Landrock[†], and Georg Gläser*

*IMMS Institut für Mikroelektronik- und Mechatronik-Systeme gemeinnützige GmbH (IMMS GmbH), Ilmenau, Germany
Email: {martin.grabmann, georg.glaeser}@imms.de
[†]X-FAB Global Services GmbH, Erfurt, Germany Email: christian.landrock@xfab.com

Abstract—Behavior models of Analog/Mixed-Signal (AMS) components are used in today's System-on-Chip (SoC) verification mainly for improving simulation speed. In addition, they can be used to enable verification scenarios including back-box intellectual properties (IP). Writing and maintaining such models is still time consuming and prevents widespread use. One class of essential building blocks are on-chip charge pumps (CP), which enable a variety of different features in SoCs e.g. embedded non-volatile memory solutions. This contribution presents a novel concept for automating the generation of grey-box behavior models of charge pump circuits using a Machine Learning (ML) approach. Compared to classical modeling approaches, it is not necessary to formulate an analytic description specific to the used circuit topology. The applicability of the approach is presented in a case study using an industrial charge pump design.

I. INTRODUCTION

The creation of behavior models of mixed-signal components is a key topic within today's integrated circuit design verification. Models are not only used to improve simulation speed, compared to transistor-level simulation, but also to enable the simulation of scenarios involving black-box IP, where a transistor-level representation is not available. Despite of these advantages, behavior models are not generally available, because engineers fear the amount of time spent in model creation and validation.

Modeling switched-capacitor circuits is of particular interest when dealing with slow system-level simulation performance. Their switching activity limits the achievable step size of the overall transient simulation. If a model is available, it might be created by a design engineer during top-down design. Such models are usually not optimized for the means of system-level verification e.g. inputs are not declared as digital in order to stay in the analog simulation domain. Alternatively, verification engineers might start writing a model if system-level test cases suffer from bad simulation performance. In this bottom-up case, they first need to build up an in-depth understanding of the block's functionality to avoid flawed results. Furthermore, switched-capacitor circuit's behavior is significantly influenced by parasitic effects, which should be considered during modeling.

These points underline the overall demand for more automation in behavior model generation. This contribution presents a novel concept for the generation of grey-box behavior models of charge pump circuits using a Machine Learning approach. They are realized as switched-capacitor circuits

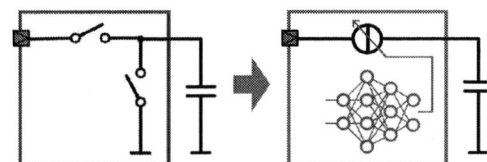

Fig. 1. Modeling switched charge transfers by neural-controlled current sources. The neural net models voltage and current dependencies by observing ports and state quantities.

and are frequently used in on-chip power management units e.g. for wireless sensor solutions or embedded non-volatile memories. The idea is shown in Fig. 1. By introducing neural-controlled current sources, i.e. current sources that are driven by the output of a neural net, we model the charge transfer behavior. In this way, we shift the problem of determining the discrete-time transfer function into the domain of the neural net training. The neural net can be trained from simulation data to reproduce the behavior of the circuit that may include parasitic elements.

The article is organized as follows: After providing a review of related work, the concept of a novel neural-network based charge pump model is introduced. A flow for automating its generation is proposed and evaluated for an application example. Last, the conclusion is drawn and an outlook on future work is provided.

II. RELATED WORK

Several publications, like [1], show how to derive analytic modeling equations for different charge pump topologies, mainly with the focus on calculating and optimizing the efficiency of the design, but not for the purpose of building behavior models. Depending on the used clock frequency, charge pumps can either operate within the slow switching limit (SSL) or the fast switching limit (FSL) [2]. In SSL the RC time constant of the switches and capacitance is much shorter than the clock period as depicted in Fig. 2(a). As long as charge pumps operate in this region, modeling can be done trough RC equivalent circuits [3], [4], [5]. However, state-of-the-art designs are operating within FSL where this assumption does not hold.

In [6] a Verilog-A behavior model is presented for general applicability in both SSL and FSL for different charge pumps with the aim to speed up simulations. It introduces

© VDE VERLAG GMBH · Berlin · Offenbach

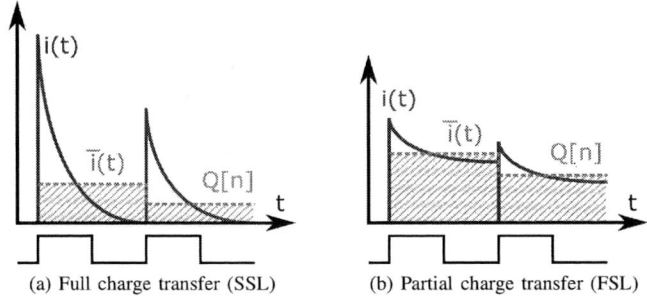

(a) Full charge transfer (SSL)   (b) Partial charge transfer (FSL)

Fig. 2. Charge transfer in different operating regions. $i(t)$ - instantaneous current, $\overline{i}(t)$ - average current per cycle, $Q[n]$ - transfered load per cycle

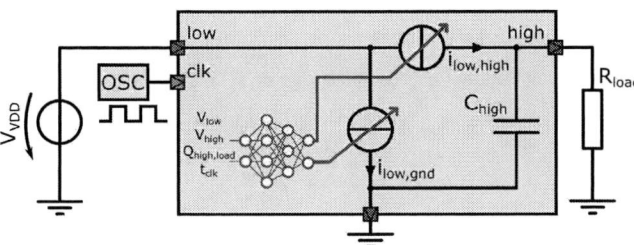

Fig. 3. Example for a implementation of the proposed neural charge transfer model concept.

an equivalent circuit replacing the charge pump and a set of fundamental equations for finding the equivalent circuit's parameters. Technology dependent parameters for the parasitic capacitance as well as voltage drop and resistance of the used switch type have to be defined by the user.

Nowadays, system-level verification is pushed to higher levels of abstraction by modeling analog behavior using Real Number Modeling (RNM) running in the discrete event simulator's domain. In [7] real valued charge pump modeling is discussed and compared to SPICE simulations using the example of a simple voltage doubler. One of the insights is that applying a cycle-averaging approach is very effective in terms of simulation performance. Instead of detailed modeling of the charge transfer mechanism, only the amount of charge that will be transferred from the input to the output per clock cycle is estimated. Models implementing this cycle averaged behavior can be used in digital context and are sufficient for many system-level considerations.

III. NEURAL CHARGE TRANSFER MODEL CONCEPT

We picked up the idea to model charge transfers between nodes. By using a machine-learning regressor to estimate the amount of charge, the implementation of specific charge transfer equations is bypassed. Since time discretization is inherent to switched cap circuits, we assume that it is sufficient to only update the system on each start of a new transfer cycle, denoted by an edge of the switching clock. In an analog context the charge transfer can be represented by current sources between each pair of nodes with a current value hold constant for the duration of one clock cycle.

A. Problem statement

The current state of the system is defined by the charges that are stored on internal capacitors, represented by the node voltages. Furthermore, the charge transfer depends on the applied voltage at the input ports and the current that is flowing to an attached load. To fit into the concept, the output current is also discretized by integration over the last clock cycle. These relations lead to the following equations for the structure defined in Fig. 3.

$$Q_{\text{low,high}}[n+1] =$$
$$f_{\text{low,high}}(Q_{\text{high,load}}[n], V_{\text{low}}[n], V_{\text{high}}[n], t_{\text{clk}}[n]) \quad (1)$$

$$Q_{\text{low,gnd}}[n+1] =$$
$$f_{\text{low,gnd}}(Q_{\text{high,load}}[n], V_{\text{low}}[n], V_{\text{high}}[n], t_{\text{clk}}[n]) \quad (2)$$

The generalisation of these expressions to switched cap converters with multiple-inputs, multiple-outputs and additional internal nodes is shown in (3).

$$\mathbf{Q}_{\text{SW}}[n+1] = \mathbf{f}_{\text{QTNN}}(\mathbf{Q}_{\text{OUT}}[n], \mathbf{V}[n], t_{clk}[n]) \quad (3)$$

with

- $\mathbf{Q}_{\text{SW}}$ representing all charges transferred between switched caps by the switching process
- $\mathbf{Q}_{\text{OUT}}$ representing charges that are passed to an attached load and flow in a non-switched manner
- $\mathbf{V}$ denoting the operating point
- t_{clk} operating frequency

B. Neural network approach

The non-linear function $\mathbf{f}_{\text{QTNN}}$, is unknown and needs to be approximated in order to predict charge transfers for the next switching cycle. Since we are aiming for a universal approach valid for a wide range of possible charge pump implementations, it is unfavorable to design a specific function and fit free parameters to it. Therefore, we use a machine learning algorithm to train a function approximator without explicitly defining a template for $\mathbf{f}_{\text{QTNN}}$. Among different machine-learning algorithms, we selected feed forward neural networks, also called multi-layer perceptrons (MLP) for this task, because of their good scalability and tool support.

IV. GENERATION FLOW

This section describes our proposed workflow for the implementation of models following the concept presented in the last section.

A. Generating training data

First, training data has to be generated by performing transistor-level simulations. Even though neural network regressors have a good potential for interpolation between training data, extrapolation to completely unknown scenarios is not reliable. Therefore, the relevant behavior of the DUT, including the application scenario of the model, has to be captured. With (3) in mind, sweeps on the clock period, input voltage and attached load must be performed. Transient simulations should show the charge-up until reaching the steady state for each sweep point. To generate a balanced data set, including a similar amount of samples for each sweep point, the stop time should be adjusted to the settling time.

B. Data pre-processing

The simulation data is imported to Python and several pre-processing steps are performed in order to extract relevant features for the training. Therefore, the simulation traces are merged into a data frame on which operations from a self-written Python toolbox are performed. These operations can be chained in order to generate a pre-processing pipeline. First, the clock period t_{clk} is extracted from consecutive edges t_{n} and $t_{\text{n}+1}$ of the analog CLK input. Next, the flows between nodes and through ports of the DUT are integrated in the interval $[t_{\text{n}}, t_{\text{n}+1}]$ resulting in a charge value $Q[n]$ per cycle for each pair of nodes. Since the input data originates from the analog solver, the stepwidth of the raw data is not fixed and integration is done using the trapezoidal rule. Last, a value table is assembled containing all inputs of the desired function f_{QTNN} as features and the corresponding outputs as targets.

C. ML training

The readily available modules from scikit-learn [8] can be used to construct the MLP regressor and perform it's training. In the present case of more than one target variables, the library splits the task into a seperate MLP network for each target output. The number and size of the layers need to be well-chosen to prevent the regressor from overfitting. Therefore, it is best-practice to perform a hyperparameter search to find the minimum possible layer topology with optimal score on training and validation data.

D. Code generation and validation

Last, the model algorithm must be translated to behavioral Verilog code that can be simulated in a mixed-signal simulation. Therefore, we use the a templating approach based on [9] to generate Verilog building blocks for the pre-processing steps and the MLP network. These building blocks are automatically instantiated and connected within a target VerilogAMS view, which can contain additional user-defined functionality. Finally, the new cell view should be validated in block-level simulations, in order to rate the resulting model error. This is not possible during training, because a prediction error in the charge transfers propagates to an inaccurate output voltage, that is used as input in subsequent time steps. Afterwards, it can be rolled out to system-level simulations.

V. CASE STUDY

We demonstrate the ability of our proposed flow using the example of a three-stage cross-coupled charge pump. The topology, which was implemented in a standard 180-nm CMOS technology, is shown in Fig. 4. Following the procedure described in Section IV, the model shown in Fig. 5 was implemented.

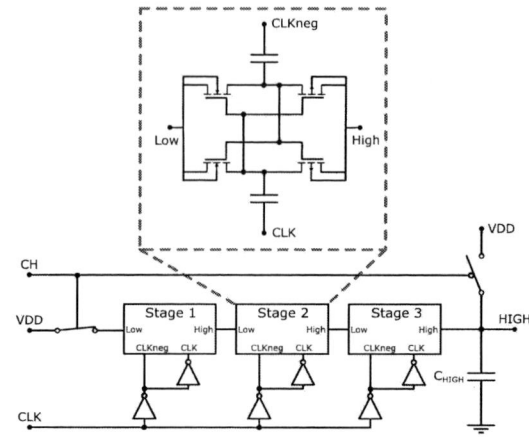

Fig. 4. Simplified scheme of the three-stage cross-coupled charge pump used in the case study.

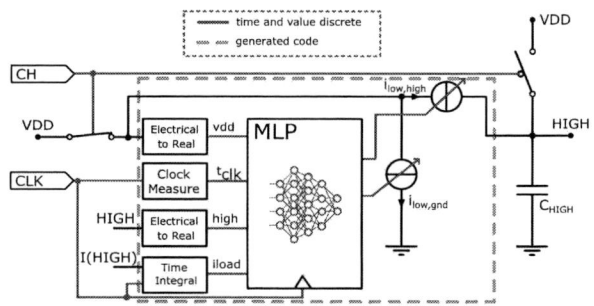

Fig. 5. Structure of the generated model from the case study.

A. Model generation

First, training simulations for the parameter set from Table I have been generated for the schematic view, whereat an analog extracted view could be used in the same way. To generate a balanced data set in which the charge-up process is equally well represented as the steady state, the simulations are stopped when the output voltage changes not more than $10\,\text{mV}$ but not earlier than after 15 clock cycles. After pre-processing, the training data set contained 6870 samples in total. The neural network topology was selected as the result of a coarse hyper-parameter search to be build up onto 3 hidden layers, with 50 neurons in the first, 20 in the second and 10 in the last layer. The neural activation function has been selected to *tanh* and the optimizer *adam* [8] has been used. The time for pre-processing as well as training of the network took $9\,\text{min}$ by making use of multi-threading.

The generated model has been evaluated in validation simulations with the second parameter set from Table I including additional parameter combinations. Fig. 6 shows the output voltage generated by the model simulation compared to a schematic simulation. The simulation time was significantly reduced as shown in Table II while a good accuracy was maintained. Simulations results for all validation parameter sets are summarized in Table III. The mean error over the

TABLE I
SIMULATED PARAMETER SETS FOR TRAINING AND VALIDATION DATA.

Parameter	Training	Validation
V_{dd}	lin [1.2 : 0.6 : 3.6] V	lin [1.2 : 0.3 : 3.6] V
R_{load}	log [100k : 10 : 1M] Ω	log [100k : 20 : 1M] Ω
f_{clk}	lin [10M : 10M : 50M] Hz	lin [10M : 5M : 50M] Hz
N	250	1620

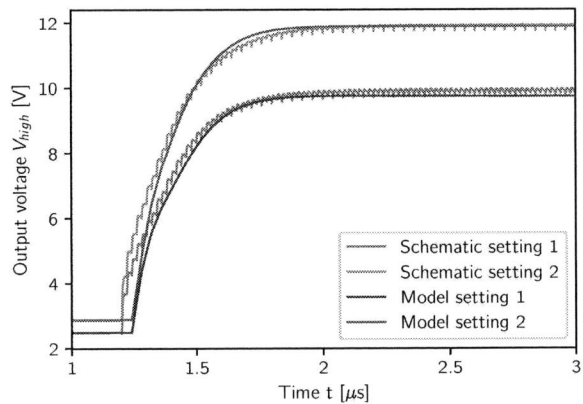

Fig. 6. Comparison of simulation results between schematic reference and implemented model for two exemplary parameter settings.
Setting 1: R_{load}=200 kΩ V_{low}=3 V f_{clk}=20 MHz
Setting 2: R_{load}=1 MΩ V_{low}=3.3 V f_{clk}=20 MHz
Simulation metrics are explained in Table II.

TABLE II
COMPARISON OF SIMULATION METRICS FOR TWO PARAMETER SETS.

	Setting 1		Setting 2	
	Schematic	Model	Schematic	Model
CPU time	7.77 s	0.25 s	8.3 s	0.19 s
CPU time gain	31.1		43.7	
Error $V_{high}(t_{final})$	0.4%		0.95%	

TABLE III
ERROR STATISTIC OVER ALL VALIDATION SIMULATIONS.

	Min	Max	Mean
Error $V_{high}(t)$	65.6 mV	1.23 V	0.44 V
Error $V_{high}(t_{final})$	4.1 µV	279 mV	57.1 mV
Error $V_{high}(t_{final})$ in %	0.1	9.6	0.8

be further investigated to target the open challenge of a formal stability proof and corresponding regularization schemes for the generated non-linear difference equations.

ACKNOWLEDGMENT

The presented results are based on the work of the IntelligEnt research group which is now carried on by the project VE-VIDES. The IntelligEnt research group was supported by the Free State of Thuringia, Germany, and the European Social Fund under the reference 2018 FGR 0089. The project VE-VIDES is funded by the German Federal Ministry of Education and Research (BMBF) under grant no. 16ME0246.

whole trajectory and the steady state is small enough for the requirements of system-level simulations.

B. Discussion

The presented case study shows a first implementation of the proposed modeling concept. As shown, the results seem to be of acceptable accuracy. Still, we experienced issues during different phases of the process that have to be addressed in future work. First, the training set has to be balanced in the way that it exhibits the full behavior that should be modelled. For instance, when the switch-on behavior is only represented in a small subset of the training set, the final result may show artifacts in this phase. Furthermore, we find that scenarios represented at the border of the training set may demonstrate errors or oscillations in the resulting trained model. This leads to the fact, that the training set should cover the application scenario with a safety margin. The training process itself still requires user-defined neural network layer sizes which should be targeted by automated hyperparameter search in the next step.

VI. CONCLUSION

We have shown the automated model generation for a charge pump circuit based on the neural charge transfer model concept. The short computation times for training and execution promise the potential for more advanced models. Hence, extensions towards regulated systems with more complex charge transfers controlled by state machines are to be taken as a next step. Also, the relations to state-space averaging methods is to

REFERENCES

[1] M. H. Eid and E. Rodriguez-Villegas, "Analysis and design of cross-coupled charge pump for low power on chip applications," *Microelectronics Journal*, vol. 66, pp. 9–17, Aug. 2017.

[2] T. Tanzawa, "On the output impedance and an output current–power efficiency relationship of dickson charge pump circuits," *IEEE Transactions on Circuits and Systems II: Express Briefs*, vol. 65, no. 11, pp. 1664–1667, Nov. 2018.

[3] D. El-Ebiary, M. Fikry, M. Dessouky, and H. Ghitani, "Average behavioral modeling technique for switched-capacitor voltage converters," in *2006 IEEE International Behavioral Modeling and Simulation Workshop*. IEEE, Sep. 2006.

[4] I. Boccuni, R. Gulino, and G. Palumbo, "Behavioral model of analog circuits for nonvolatile memories with vhdl-ams," *Analog Integr. Circuits Signal Process.*, vol. 33, no. 1, pp. 19–28, Oct. 2002. [Online]. Available: https://doi.org/10.1023/A:1020328928180

[5] R. Mita, G. Palumbo, and M. Pennisi, "Behavioral model of charge pumps with VHDL," in *2005 12th IEEE International Conference on Electronics, Circuits and Systems*. IEEE, Dec. 2005.

[6] A. Ballo, M. Bottaro, A. D. Grasso, and G. Palumbo, "A general behavioral model of charge pump DC-DC converters," in *2020 International Conference on Electrical, Communication, and Computer Engineering (ICECCE)*. IEEE, Jun. 2020.

[7] "Using EEnet to perform Electrical Equivalent Modeling in SystemVerilog," Cadence Design Systems, Inc., Tech. Rep.

[8] F. Pedregosa, G. Varoquaux, A. Gramfort, V. Michel, B. Thirion, O. Grisel, M. Blondel, P. Prettenhofer, R. Weiss, V. Dubourg, J. Vanderplas, A. Passos, D. Cournapeau, M. Brucher, M. Perrot, and E. Duchesnay, "Scikit-learn: Machine learning in Python," *Journal of Machine Learning Research*, vol. 12, pp. 2825–2830, 2011.

[9] *Mako Templates for Python*. [Online]. Available: http://www.makotemplates.org/

© VDE VERLAG GMBH · Berlin · Offenbach

Frequency-Limited Reduction of RLCK Circuits via Second-Order Balanced Truncation

Olympia Axelou, Dimitrios Garyfallou, George Floros

Department of Electrical and Computer Engineering, University of Thessaly, Volos, Greece
{oaxelou, digaryfa, gefloros}@e-ce.uth.gr

Abstract—Second-order formulation using susceptance elements has become very effective in modeling on-chip inductive couplings. Several prior works have proposed model order reduction techniques for RLCK circuits, mostly based on balanced truncation (BT) and moment matching, providing reduced-order models (ROMs) that can be simulated over the whole frequency range. However, in most applications, the ROMs are simulated only at specific frequency windows, which means that established methods usually provide models that may become unnecessarily large to achieve approximation over all frequencies. In this paper, we present a second-order frequency-limited approach for RLCK circuits, which may be combined with efficient low-rank Lyapunov solvers, leading to ROMs which are either smaller or exhibit better accuracy compared to an established second-order BT method. Experimental results on interconnect bus structures demonstrate the advantages of the proposed method.

Index Terms—model order reduction (MOR), balanced truncation (BT), circuit simulation, second-order systems.

I. INTRODUCTION

As integrated circuits move to advanced technology nodes, 2.5D and 3D chiplet-based architectures have become key enablers for overcoming the related manufacturing challenges [1]. In this type of integration, the challenge of efficient parasitic modeling is completely different compared to monolithic 2D design, since inductive effects have to be taken into account as well. These effects are more pronounced in power delivery and clock networks, as well as at long and wide bus structures, and can affect the power and signal integrity of the chiplet. The electrical models of the above are usually formulated as RLCK circuits in second-order form, since many desired properties of the susceptance matrix can be preserved in this form [2]. However, the large and dense inductance matrix, mostly due to mutual inductance, hinders the application of numerical simulation methods. Model order reduction (MOR) provides efficient techniques to reduce the model complexity by replacing the original model with a much smaller one, while achieving accurate approximation of the input–output port behavior.

MOR methods that have been applied in second-order systems are divided into two main categories. Moment matching (MM) techniques [3], [4] are well established due to their computational efficiency in producing reduced-order models (ROMs). Their drawback is that the ROM depends only on the number of matching moments and the quality of the produced Krylov subspace. On the other hand, system theoretic techniques, such as balanced truncation (BT) [5], [6],

provide very satisfactory and reliable bounds for the approximation error. However, BT techniques require the solution of Lyapunov matrix equations which are very computationally expensive, and also involve storage of dense matrices, even if the system matrices are sparse. In order to make such techniques amenable to large circuit models, low-rank solution methods, such as the extended Krylov subspace (EKS), have been developed [7].

The majority of the aforementioned methods focus on approximating the original model over the whole frequency range (from DC to infinity). In most practical applications, however, we are only interested in a specific finite frequency range. Frequency-limited BT methods have been proposed in the past [8], [9], where a user-specified frequency range is given in order to obtain solutions of Lyapunov matrix equations that improve the accuracy of the ROM in this particular range. The problem is that the existing frequency-limited BT techniques are only applied in first-order systems and cannot handle RLCK circuits, where the susceptance matrix is utilized.

In this paper, we introduce a frequency-limited second-order BT method for RLCK circuits, which extends [9] in order to handle second-order Gramians, as defined in [10]. In contrast to the first-order frequency-limited BT, the proposed methodology produces ROMs which preserve the structure information that is inherent to the RLCK circuits. Finally, we evaluate our methodology on actual multi-line bus examples, and we demonstrate that frequency-limited second-order BT may produce ROMs with either smaller size or superior accuracy compared to standard BT in a specific frequency range.

The rest of the paper is organized as follows. Section II presents the theoretical background of second-order BT methods for the reduction of RLCK circuit models. Section III presents our main contributions in the application of the frequency-limited framework to second-order BT methods. Section IV presents our experimental results, while conclusions are drawn in Section V.

II. SECOND-ORDER BT FOR RLCK CIRCUITS

Consider the second-order LTI system:

$$\mathbf{M}\ddot{\mathbf{q}}(t) + \mathbf{D}\dot{\mathbf{q}}(t) + \mathbf{K}\mathbf{q}(t) = \mathbf{B}_1\mathbf{u}(t)$$
$$\mathbf{y}(t) = \mathbf{L}_1\mathbf{q}(t) + \mathbf{L}_2\dot{\mathbf{q}}(t) \tag{1}$$

where $\mathbf{M}, \mathbf{D}, \mathbf{K} \in \mathbb{R}^{n \times n}$, $\mathbf{B}_1 \in \mathbb{R}^{n \times p}$, $\mathbf{L}_1, \mathbf{L}_2 \in \mathbb{R}^{q \times n}$, $\mathbf{q} \in \mathbb{R}^n$, $\mathbf{u} \in \mathbb{R}^p$, $\mathbf{y} \in \mathbb{R}^q$ and in which $\mathbf{M}$ consider to be nonsingular. The objective of MOR is to produce a ROM:

$$\tilde{\mathbf{M}}\ddot{\mathbf{q}}_r(t) + \tilde{\mathbf{D}}\dot{\mathbf{q}}_r(t) + \tilde{\mathbf{K}}\mathbf{q}_r(t) = \tilde{\mathbf{B}}_1\mathbf{u}(t)$$
$$\mathbf{y}(t) = \tilde{\mathbf{L}}_1\mathbf{q}_r(t) + \tilde{\mathbf{L}}_2\dot{\mathbf{q}}_r(t) \tag{2}$$

where $\tilde{\mathbf{M}}, \tilde{\mathbf{D}}, \tilde{\mathbf{K}} \in \mathbb{R}^{r \times r}$, $\tilde{\mathbf{B}}_1 \in \mathbb{R}^{r \times p}$, $\tilde{\mathbf{L}}_1, \tilde{\mathbf{L}}_2 \in \mathbb{R}^{q \times r}$, $\mathbf{q}_r \in \mathbb{R}^r$, and in which the order $r << N$ and the output error is bounded as $||\tilde{\mathbf{y}}(t) - \mathbf{y}(t)||_2 < \varepsilon ||\mathbf{u}(t)||_2$ for given input $\mathbf{u}(t)$ and given small ε. The bound in the output error can be equivalently written in the frequency domain as $||\tilde{\mathbf{y}}(s) - \mathbf{y}(s)||_2 < \varepsilon ||\mathbf{u}(s)||_2$ via Plancherel's theorem [11]. If

$$\mathbf{H}(s) = (\mathbf{L}_1 + s\mathbf{L}_2)(s^2\mathbf{M} + s\mathbf{D} + \mathbf{K})^{-1}\mathbf{B}_1$$
$$\tilde{\mathbf{H}}(s) = (\tilde{\mathbf{L}}_1 + s\tilde{\mathbf{L}}_2)(s^2\tilde{\mathbf{M}} + s\tilde{\mathbf{D}} + \tilde{\mathbf{K}})^{-1}\tilde{\mathbf{B}}_1$$

are the transfer functions of the original model and the ROM, then the output error in the frequency domain is:

$$||\tilde{\mathbf{y}}(s) - \mathbf{y}(s)||_2 = ||\tilde{\mathbf{H}}(s)\mathbf{u}(s) - \mathbf{H}(s)\mathbf{u}(s)||_2$$
$$\leq ||\tilde{\mathbf{H}}(s) - \mathbf{H}(s)||_\infty ||\mathbf{u}(s)||_2 \tag{3}$$

where $||.||_\infty$ is the induced $\mathcal{L}_2$ matrix norm, or $\mathcal{H}_\infty$ norm of a rational transfer function. Therefore, in order to bound the output error, we need to bound the distance between the transfer functions as $||\tilde{\mathbf{H}}(s) - \mathbf{H}(s)||_\infty < \varepsilon$.

In order to directly apply BT and reduce the second-order system of (1), the basic idea is to first transform the second-order system into an equivalent first-order form as

$$\mathbf{E}\frac{d\mathbf{x}(t)}{dt} = \mathbf{A}\mathbf{x}(t) + \mathbf{B}\mathbf{u}(t)$$
$$\mathbf{y}(t) = \mathbf{L}\mathbf{x}(t) \tag{4}$$

and then obtain the balancing matrices by a standard BT procedure. To this end, the second-order Gramians, which are introduced in [10], are formed based on the first-order realization of the state-space formulation of (1) with $2n$ dimensions, $\mathbf{x}^T = [\mathbf{q}, \dot{\mathbf{q}}]$, and

$$\mathbf{A} \equiv -\begin{pmatrix} \mathbf{0} & -\mathbf{K} \\ -\mathbf{K} & \mathbf{D} \end{pmatrix}, \quad \mathbf{E} \equiv \begin{pmatrix} -\mathbf{K} & \mathbf{0} \\ \mathbf{0} & \mathbf{M} \end{pmatrix},$$
$$\mathbf{B} \equiv \begin{pmatrix} \mathbf{0} \\ \mathbf{B}_1 \end{pmatrix}, \quad \mathbf{L} \equiv (\mathbf{L}_1 \quad \mathbf{L}_2) \tag{5}$$

The first-order realization of (4) has the same input–output behavior as the second-order system of (1). In order to reduce the second-order system, first we need to compute the Gramians for the first-order realization, which are derived by the solution of the Lyapunov matrix equations [12]:

$$\mathbf{E}\mathbf{P}\mathbf{A}^T + \mathbf{A}\mathbf{P}\mathbf{E}^T = -\mathbf{B}\mathbf{B}^T$$
$$\mathbf{E}^T\mathbf{Q}\mathbf{A} + \mathbf{A}^T\mathbf{Q}\mathbf{E} = -\mathbf{L}^T\mathbf{L} \tag{6}$$

If we conformally partition the computed first-order Gramians of (6) as defined in [10], we obtain

$$\mathbf{P} = -\begin{pmatrix} \mathbf{R} & \mathbf{S} \\ \mathbf{S}^T & \mathbf{O} \end{pmatrix}, \quad \mathbf{Q} = \begin{pmatrix} \mathbf{U} & \mathbf{X} \\ \mathbf{X}^T & \mathbf{H} \end{pmatrix} \tag{7}$$

In the above, $\mathbf{R}, \mathbf{U} \in \mathbb{R}^{n \times n}$ submatrices are the second-order controllability and observability matrices, respectively. Finally,

there exists a similarity transformation matrix $\mathbf{T}$, which makes the eigenvalues of the Gramians product invariant such as:

$$\mathbf{T}^{-1}\mathbf{R}\mathbf{U}\mathbf{T} = diag(\sigma_1^2, \sigma_2^2, ..., \sigma_n^2) \tag{8}$$

where $\sigma_1, \sigma_2 ..., \sigma_n$ are the Hankel singular values (HSVs) of the model. The most controllable and observable states correspond to the largest HSVs, and r of them are kept by truncating the $n-r$ states corresponding to the smallest HSVs. The BT method provides an error bound with respect to the $\mathcal{H}_\infty$ norm such as:

$$||\tilde{\mathbf{H}}(s) - \mathbf{H}(s)||_\infty \leq 2(\sigma_{r+1} + \sigma_{r+2} + ... + \sigma_n) \tag{9}$$

Finally, in order to compute the projection matrix, the singular value decomposition (SVD) of $\mathbf{U}\mathbf{\Sigma}\mathbf{V} = \mathbf{Z}_R\mathbf{K}\mathbf{Z}_U$ has to be computed, where $\mathbf{Z}_R$ and $\mathbf{Z}_U$ are the Cholesky factors of $\mathbf{R}$ and $\mathbf{U}$, respectively, i.e., $\mathbf{R} = \mathbf{Z}_R\mathbf{Z}_R$, $\mathbf{U} = \mathbf{Z}_U\mathbf{Z}_U$. The projection matrices onto a lower dimensional subspace are defined as $\mathbf{W}_l = \mathbf{W}_r = \mathbf{Z}_R\mathbf{U}_{n \times r}\mathbf{\Sigma}_{r \times r}^{-1/2}$. Then the reduced-order matrices are:

$$\tilde{\mathbf{M}} = \mathbf{W}_l^T\mathbf{M}\mathbf{W}_r, \quad \tilde{\mathbf{D}} = \mathbf{W}_l^T\mathbf{D}\mathbf{W}_r, \quad \tilde{\mathbf{K}} = \mathbf{W}_l^T\mathbf{K}\mathbf{W}_r$$
$$\tilde{\mathbf{B}}_1 = \mathbf{W}_l^T\mathbf{B}_1, \quad \tilde{\mathbf{L}}_1 = \mathbf{L}_1\mathbf{W}_r, \quad \tilde{\mathbf{L}}_2 = \mathbf{L}_2\mathbf{W}_r \tag{10}$$

Regarding the RLCK circuits, they can be formulated as a second-order system of (1) using nodal analysis (NA). By setting $\mathbf{M} \equiv \mathbf{C}$, $\mathbf{D} \equiv \mathbf{G}$, $\mathbf{K} \equiv \mathbf{\Gamma}$, $\mathbf{L}_1 \equiv \mathbf{0}$, and $\mathbf{L}_2 \equiv \mathbf{B}_1^T$, the second-order formulation of an RLCK circuit is the following:

$$\mathbf{C}\ddot{\mathbf{q}}(t) + \mathbf{G}\dot{\mathbf{q}}(t) + \mathbf{\Gamma}\mathbf{q}(t) = \mathbf{B}_1\mathbf{u}(t)$$
$$\mathbf{y}(t) = \mathbf{B}_1^T\dot{\mathbf{q}}(t) \tag{11}$$

where $\mathbf{u}(t), \mathbf{y}(t) \in \mathbb{R}^p$ are the input currents and output voltages, $\dot{\mathbf{q}}(t) \in \mathbb{R}^n$ are the nodal voltages, and $\mathbf{C}, \mathbf{G}, \mathbf{\Gamma} \in \mathbb{R}^{n \times n}$ are the capacitance, conductance, and susceptance matrices, respectively. The capacitance matrix is considered nonsingular.

III. Second-Order BT for RLCK circuits in Limited Frequency Intervals

The solution of the Gramians that appear in (6) consider the entire frequency interval, i.e., $(-\infty, \infty)$. If we restrict the interval to a certain range, i.e., $[-\omega_2, -\omega_1] \cup [\omega_1, \omega_2]$, we can obtain the frequency-limited Gramians, $\mathbf{P}_\omega$ and $\mathbf{Q}_\omega$, which may be derived by the solution of the following modified Lyapunov equations [13]:

$$\mathbf{A}\mathbf{P}_\omega\mathbf{E}^T + \mathbf{E}\mathbf{P}_\omega\mathbf{A}^T = -(\mathbf{E}\mathbf{F}\mathbf{B}\mathbf{B}^T + (\mathbf{E}\mathbf{F}\mathbf{B}\mathbf{B}^T)^T)$$
$$\mathbf{A}^T\mathbf{Q}'_\omega\mathbf{E} + \mathbf{E}^T\mathbf{Q}'_\omega\mathbf{A} = -((\mathbf{L}^T\mathbf{L}\mathbf{F}\mathbf{E})^T + \mathbf{L}^T\mathbf{L}\mathbf{F}\mathbf{E}) \tag{12}$$

where

$$\mathbf{F} = \frac{1}{2\pi}\left(\int_{-\omega_2}^{-\omega_1}(i\omega\mathbf{E} - \mathbf{A})^{-1}d\omega + \int_{\omega_1}^{\omega_2}(i\omega\mathbf{E} - \mathbf{A})^{-1}d\omega\right) \tag{13}$$

The above matrix integral can be evaluated as:

$$\mathbf{F} = \frac{1}{\pi}Re\left(\int_{\omega_1}^{\omega_2}(i\omega\mathbf{E} - \mathbf{A})^{-1}d\omega\right)$$
$$= Re(\frac{i}{\pi}ln((\mathbf{A} + i\omega_1\mathbf{E})^{-1}(\mathbf{A} + i\omega_2\mathbf{E})))\mathbf{E}^{-1} \tag{14}$$
$$= \mathbf{E}^{-1}Re(\frac{i}{\pi}ln((\mathbf{A} + i\omega_2\mathbf{E})(\mathbf{A} + i\omega_1\mathbf{E})^{-1}))$$

where $ln()$ is the function of the matrix logarithm, which represents the inverse of the matrix exponential, i.e., a matrix $\mathbf{X}$ such that $exp(\mathbf{X}) = \mathbf{Y}$.

The frequency-limited Gramians characterize the controllability and observability of the model in the selected frequency range, and the process of balancing and truncation eliminates states that are difficult to control and observe inside this frequency range. This means that more states can be eliminated for a given tolerance in (9), which would not have been eliminated otherwise (e.g., states which are easily controllable and observable in other frequencies), leading to lower ROM order r (or alternatively, lower error in the frequency range for a given order r). In order to compute $\mathbf{P}_\omega$, $\mathbf{Q}_\omega$ by solving (12), we have to deal with the two different right-hand sides (RHS) of frequency-limited Lyapunov equations, which are in the forms of $-(\mathbf{B}_\omega \mathbf{B}^T + \mathbf{B} \mathbf{B}_\omega^T)$ and $-(\mathbf{L}_\omega^T \mathbf{L} + \mathbf{L}^T \mathbf{L}_\omega)$ (where $\mathbf{B}_\omega \equiv \mathbf{EFB}$ and $\mathbf{L}_\omega \equiv \mathbf{LFE}$), instead of the standard forms $-\mathbf{B}\mathbf{B}^T$ and $-\mathbf{L}^T\mathbf{L}$ of (6). However, efficient Lyapunov solvers, which can efficiently deal with the matrix logarithm and the modified RHS, have been already presented in the literature [9]. The main steps of the proposed frequency-limited BT for second-order RLCK models are summarized in Algorithm 1.

Algorithm 1: Second-order frequency-limited reduction of RLCK circuits by BT

Input: System matrices $\mathbf{C}$, $\mathbf{G}$, $\boldsymbol{\Gamma}$, $\mathbf{B}_1$, and frequency range $[\omega_1, \omega_2]$
Output: ROM matrices $\tilde{\mathbf{C}}, \tilde{\mathbf{G}}, \tilde{\boldsymbol{\Gamma}}, \tilde{\mathbf{B}}_1$
Function so_freq_lim_BT $(\mathbf{G}, \mathbf{C}, \boldsymbol{\Gamma}, \mathbf{B}_1, [\omega_1, \omega_2])$:

  1) Solve the frequency-limited Lyapunov equations of (12) to obtain the Gramian matrices $\mathbf{P}_\omega$, $\mathbf{Q}_\omega$ with respect to the first-order realization matrices of (5).
  2) Partition the Cholesky factors, $\mathbf{R} = \mathbf{Z}_R \mathbf{Z}_R$, $\mathbf{U} = \mathbf{Z}_U \mathbf{Z}_U$, according to (7).
  3) Compute the singular value decomposition (SVD) of the product of the Cholesky factors, i.e., $\mathbf{U}\boldsymbol{\Sigma}\mathbf{V} = \mathbf{Z}_R \boldsymbol{\Gamma} \mathbf{Z}_U$.
  4) Compute the truncated part of the balancing transformations using $\mathbf{W}_l = \mathbf{W}_r = \mathbf{Z}_R \mathbf{U}_{n \times r} \boldsymbol{\Sigma}_{r \times r}^{-1/2}$.
  5) Compute the corresponding ROM matrices as:
$$\tilde{\mathbf{C}} = \mathbf{W}_l^T \mathbf{C} \mathbf{W}_r, \qquad \tilde{\mathbf{G}} = \mathbf{W}_l^T \mathbf{G} \mathbf{W}_r,$$
$$\tilde{\boldsymbol{\Gamma}} = \mathbf{W}_l^T \boldsymbol{\Gamma} \mathbf{W}_r, \qquad \tilde{\mathbf{B}}_1 = \mathbf{W}_l^T \mathbf{B}_1.$$
  return $\tilde{\mathbf{C}}, \tilde{\mathbf{G}}, \tilde{\boldsymbol{\Gamma}}, \tilde{\mathbf{B}}_1$
End Function

Finally, it should be noted that BT-type MOR methods do not generally preserve the passivity of the original model (due to the similarity transformation involved in the balancing before truncation). However, instead of provably passive models, MOR techniques have been focused on passivity enforcement after efficient reduction. A wealth of passivity enforcement techniques, such as [14], have been developed to assure passivity of the ROMs obtained by frequency-limited second-order BT.

IV. EXPERIMENTAL RESULTS

For the experimental evaluation of the proposed methodology, we created 5 benchmarks of 3D geometry bus structures using FastHenry [15], which consist of layers of parallel wires. More specifically, different layers are planes across the z-axis, different wires are line segments across the y-axis, and each wire extends on the x-axis. The characteristics of these benchmarks are shown in Table I, where *#layers* is the number of layers of each benchmark, *#wires per layer* represents the number of wires that each layer has, and *#filaments per wire* denotes the number of filaments that each wire is broken into. The wires composed of multiple filaments are split along their width. In all benchmarks, the wires have length $1mm$ and cross section $1\mu m^2$. The distance between successive layers is $2\mu m$ and the distance between wires of the same layer is $1\mu m$.

TABLE I
CHARACTERISTICS OF THE EVALUATION BENCHMARKS

Benchmark	#layers	#wires per layer	#filaments per wire
interc1	9	91	1
interc2	3	273	2
interc3	3	410	1
interc4	3	410	2
interc5	3	546	1

The second-order frequency-limited BT was implemented with the procedure of Algorithm 1 for the frequency range of $[\omega_1, \omega_2] = [10^5, 10^9]$, and was compared against SBPOR [5]. In both methods, the Lyapunov equations were solved with the method presented in [9]. The ROMs of SBPOR and second-order frequency-limited BT were compared with respect to both their order for given tolerance ε and their accuracy for given ROM order. In the first case, the error tolerance was chosen as $\varepsilon = 10^{-2}$ for both methods, while in the second case, the order that resulted from the execution of SBPOR was reused for the truncation of the HSVs of the frequency-limited Gramians. All experiments were executed with MATLAB R2021a on a Linux workstation, having a 3.6GHz Intel Core i7 processor with 32GB memory.

Our experimental results are reported in Table II, where *Max error* refers to the maximum error between the infinity norms of the transfer functions of the original model and the ROM in the selected frequency range, i.e., $||\mathbf{H}(s) - \tilde{\mathbf{H}}(s)||_\infty$, *Time* refers to the computational time (in seconds) needed to generate the ROMs, while *Reduction percentage* refers to the percentage of ROM reduction for the same error bound between SBPOR and second-order frequency-limited BT. As can be seen, the proposed second-order frequency-limited BT can produce ROMs which, in the selected frequency range, exhibit either smaller size for given error, or smaller error for given order in comparison to SBPOR. The proposed method provides ROMs which have up to 50% less states with respect to SBPOR in interc1 benchmark, while achieving up to $2615\times$ less error in interc5 benchmark. Note that the execution time of second-order frequency-limited BT was slightly larger than SBPOR due to the matrix logarithm calculation.

TABLE II
REDUCTION RESULTS OF FREQUENCY-LIMITED BT VERSUS SBPOR IN THE $[10^5, 10^9]$ FREQUENCY INTERVAL

Benchmark	#nodes	#ports	Second-order BT (SBPOR)			Second-order frequency-limited BT (Proposed)			
			ROM order	Max error	Time (s)	ROM order for same error	Reduction percentage	Max error for same order	Time (s)
interc1	4095	32	64	1.9e-03	35.48	32	50%	6.9e-05	40.74
interc2	4095	64	98	4.6e-02	41.23	70	28.57%	7.0e-05	47.85
interc3	6150	96	197	3.3e-02	144.56	152	22.84%	2.2e-04	186.21
interc4	6150	192	480	3.4e-02	312.09	312	35%	1.3e-05	354.52
interc5	8190	256	704	6.4e-02	653.90	430	38.92%	9.2e-05	733.88

To demonstrate the accuracy of our method, we compare the transfer functions of the original model and the ROMs generated by the second-order frequency-limited BT and SBPOR. Fig. 1 presents the transfer functions of ROMs produced by second-order frequency-limited BT and SBPOR for the interc2 benchmark, in the frequency interval $[10^5, 10^9]$, along with the absolute errors induced over the original model for the selected benchmark in the same frequency range. As can be seen, the response of second-order frequency-limited BT ROM is performing very close to the original model, while the response of SBPOR exhibits a clear deviation.

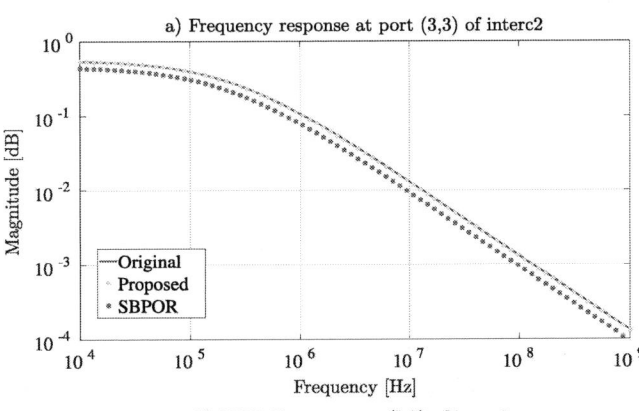

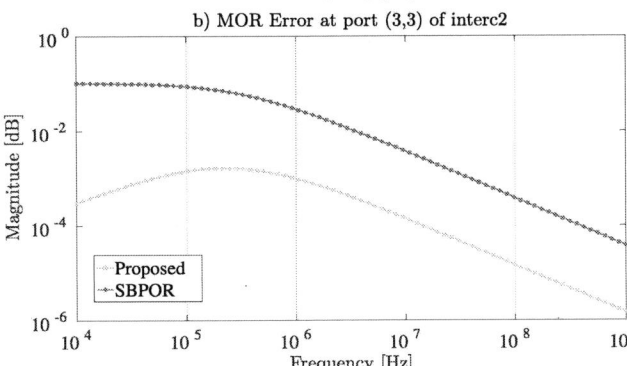

Fig. 1. Comparison of ROM transfer functions and absolute error magnitudes obtained by the proposed second-order frequency-limited BT and SBPOR in the range $[10^5, 10^9]$ at port (3,3) of interc2 benchmark.

V. CONCLUSION

In this paper, we presented a frequency-limited methodology for reducing second-order systems arising in the modelling of RLCK circuits, which requires only the specification of the end frequencies. Experimental results indicate that our approach provides clear improvements in model accuracy or size with respect to SBPOR, while retaining its benefits of specified error bounds, introducing only a small overhead in the reduction process.

REFERENCES

[1] M. Liu, "1.1 Unleashing the Future of Innovation," in *Proc. of the IEEE International Solid-State Circuits Conference*, pp. 9–16, 2021.

[2] Hui Zheng and Lawrence T. Pileggi, "Robust and passive model order reduction for circuits containing susceptance elements," in *Proc. of the IEEE/ACM International Conference on Computer Aided Design*, pp. 761–766, 2002.

[3] B. N. Sheehan, "ENOR: model order reduction of RLC circuits using nodal equations for efficient factorization," in *Proc. of the Design Automation Conference*, pp. 17–21, 1999.

[4] Yangfeng Su, Jian Wang, Xuan Zeng, Zhaojun Bai, C. Chiang and D. Zhou, "SAPOR: second-order Arnoldi method for passive order reduction of RCS circuits," in *Proc. of the International Conference on Computer Aided Design*, pp. 74–79, 2004.

[5] B. Yan, S. X. -. Tan, P. Liu and B. McGaughy, "SBPOR: Second-Order Balanced Truncation for Passive Order Reduction of RLC Circuits," in *Proc. of the Design Automation Conference*, pp. 158–161, 2007.

[6] B. Yan, S. X. -. Tan and B. McGaughy, "Second-Order Balanced Truncation for Passive-Order Reduction of RLCK Circuits," *IEEE Trans. on Circuits and Systems II: Express Briefs*, vol. 55, no. 9, pp. 942–946, 2008.

[7] V. Simoncini, "A New Iterative Method for Solving Large-Scale Lyapunov Matrix Equations," *SIAM Journal on Scientific Computing*, vol. 29, no. 3, pp. 1268–1288, 2007.

[8] G. Floros, N. Evmorfonoulos and G. Stamoulis, "Efficient Circuit Reduction in Limited Frequency Windows," in *Proc. of the International Conference on Synthesis, Modeling, Analysis and Simulation Methods and Applications to Circuit Design*, pp. 129–132, 2019.

[9] G. Floros, N. Evmorfopoulos and G. Stamoulis, "Frequency-Limited Reduction of Regular and Singular Circuit Models Via Extended Krylov Subspace Method," *IEEE Trans. on Very Large Scale Integration (VLSI) Systems*, vol. 28, no. 7, pp. 1610–1620, 2020.

[10] D. G. Meyer and S. Srinivasan, "Balancing and model reduction for second-order form linear systems," *IEEE Trans. on Automatic Control*, vol. 41, no. 11, pp. 1632–1644, 1996.

[11] K. Gröchenig, *Foundations of time-frequency analysis*. Springer Science & Business Media, 2001.

[12] S. Gugercin and A. C. Antoulas, "A survey of model reduction by balanced truncation and some new results," *International Journal of Control*, vol. 77, no. 8, pp. 748–766, 2004.

[13] P. Benner, P. Kurschner and J. Saak, "Frequency-limited balanced truncation with low-rank approximations," *SIAM Journal on Scientific Computing*, vol. 38, no. 1, pp. 471–499, 2016.

[14] S. G. Talocia and A. Ubolli, "A comparative study of passivity enforcement schemes for linear lumped macromodels," *IEEE Trans. on Advanced Packaging*, vol. 31, no. 4, pp. 673–683, 2008.

[15] M. Kamon, M. J. Tsuk and J. K. White, "FASTHENRY: a multipole-accelerated 3-D inductance extraction program," *IEEE Trans. on Microwave Theory and Techniques*, vol. 42, no. 9, pp. 1750–1758, 1994.

A Mixed Time-Frequency RF Simulation Technique Based on Numerical Time-Slot Partitioning

Jorge F. Oliveira [1,2]

[1] *School of Technology and Management*
Polytechnic of Leiria
Leiria, Portugal
jorge.oliveira@ipleiria

[2] *Instituto de Telecomunicações*
University of Aveiro
Aveiro, Portugal
ORCID: 0000-0002-8525-3888

Abstract—The increasing complexity of radio frequency (RF) architectures and the continuous push to profit from digital signal-processing techniques, have been addressing new challenges to circuit-level simulation. This paper describes the most relevant details of a time-frequency simulation technique specially conceived for the efficient numerical simulation of circuits whose stimuli are turned on and off for unknown periods of time, as is the case of RF circuits managing signals coded in some on-off digital scheme. The proposed technique is based on a time-slot partition stratagem with automatic switching between different numerical schemes (multitime envelope transient harmonic balance and time-step integration) along the simulation process, according to the on-off state of the circuits' stimuli. Simulation tests performed in an illustrative application example, an on-off amplitude shift keying (ASK/OOK) transmitter used in low-power applications, as radio frequency identification (RFID) or biomedical imaging, revealed significant gains in computational speed over commercial computer-aided design tools.

Keywords—*circuit simulation, numerical simulation, partial differential equations, radio frequency.*

I. INTRODUCTION

Noteworthy improvements have been made in RF and microwave numerical circuit-level simulation in the last two decades. Clear examples of that are the efficiency gains brought by innovative algorithms based on multivariate formulations [1]-[2], artificial frequency mapping schemes [3], subset circuit division (circuit block partitioning) techniques [4]-[7], enhanced adaptations of traditional time-domain [8] and frequency domain [9] techniques, etc. Regardless of the significant accomplishments that have resulted from such innovative techniques, some unresolved challenges still remain in the RF computer-aided design research field. For instance, evaluating the numerical solution of RF circuits driven by stimuli randomly switched on and off for undetermined time slots is an example of practical interest where an efficient simulation tool is still missing. Amplitude-burst wireless transmitters [10], or on-off amplitude shift keying transceivers [11]-[14], are illustrative examples of practical application in which this on-off situation takes place. When the stimuli are turned off the circuits operate in their natural response regimes, i.e., in their autonomous behaviour. In such situations only conventional initial value solvers (SPICE-like simulation schemes) are appropriated to compute the numerical solution of the circuits. However, SPICE numerical schemes are generally very ineffective if utilized to simulate RF circuits. In most cases they incur high computational costs in terms of time and memory consumption.

In order to attempt to address the above described scenario, two innovative simulation techniques based on a time-slot partitioning approach have already been proposed in [15] and [16]. The one described in [15] is a technique that operates in a pure time-domain framework. It combines several time-step integration (TSI) schemes with an envelope transient over shooting engine. In the same way as in the approach now discussed in this paper, the decision of which method to use depends on the state of the excitation. Since it operates strictly in the time domain, the technique described in [15] is not able to compute smooth narrowband waveforms in the frequency domain. The method now discussed in this research work is based on the ideas previously advanced in [16], by proposing a mixed time-frequency method (operating in both time and frequency domains) combining an envelope transient harmonic balance (ETHB [7]) engine (used for the time slots in which the circuits' stimuli are turned on) with a time-step integration engine (used for the time slots in which there are no stimuli). A special focus will now be given to the BDF2 and Runge-Kutta Bogacki-Shampine methods, considered here as the preferred time-step integration engines. In what concerns to the switching between numerical schemes, a slightly different approach will also be taken.

II. MATHEMATICAL MODEL OF AN ELECTRONIC CIRCUIT

A. Univariate Formulation

Under the quasi-static assumption the behaviour of an electronic circuit can in general be described with a system of *differential algebraic equations* (DAEs) of the form

$$\mathbf{p}\left(\mathbf{y}\left(t\right)\right) + \frac{d\mathbf{q}\left(\mathbf{y}\left(t\right)\right)}{dt} = \mathbf{x}\left(t\right), \qquad (1)$$

where $\mathbf{x}\left(t\right) \in \mathbb{R}^{N}$ represents the set of excitations of the circuit (independent voltage or current sources), whereas $\mathbf{y}\left(t\right) \in \mathbb{R}^{N}$ stands for the state-variable vector (node voltages and brunch currents). $\mathbf{p}\left(\mathbf{y}\left(t\right)\right)$ represents the memoryless linear or nonlinear elements (resistors, nonlinear voltage controlled current sources, etc.), while $\mathbf{q}\left(\mathbf{y}\left(t\right)\right)$ models all dynamic linear or nonlinear elements (capacitors or inductors, represented as voltage-dependent electric charges, or current-dependent magnetic fluxes, respectively). N is the number of unknowns (state variables) of the circuit.

B. Multitime Formulation

Electronic circuits containing voltages and currents that evolve at two, or more, widely separated time scales (which is typically the case of RF systems) can be represented much more efficiently if their behaviour is defined as a function of multiple time variables, i.e., as a function of multiple artificial time scales. With this subterfuge all the signals (forcing functions and state variables) will be characterized as multitime entities, implying that the circuits' dynamic behaviour will be described by systems of partial differential algebraic equations and not by systems of ordinary differential algebraic equations.

Since in this case we are interested in circuits running on two different time scales (the periodic carrier fast time, t_2, and the aperiodic slow baseband time scale, t_1), the following description is assumed for $\mathbf{x}(t)$ and $\mathbf{y}(t)$ in the DAE system of (1). t is substituted by t_1 in the slowly varying entities and t is substituted by t_2 in the fast-varying ones. The system of (1) is then converted into the following *multirate partial differential algebraic equations'* (MPDAE) system [1]-[2]

$$\mathbf{p}\left(\hat{\mathbf{y}}\left(t_1, t_2\right)\right) + \frac{\partial \mathbf{q}\left(\hat{\mathbf{y}}\left(t_1, t_2\right)\right)}{\partial t_1} + \frac{\partial \mathbf{q}\left(\hat{\mathbf{y}}\left(t_1, t_2\right)\right)}{\partial t_2} \\ = \hat{\mathbf{x}}\left(t_1, t_2\right). \quad (2)$$

The mathematical relation between (1) and (2) establishes that, if $\hat{\mathbf{x}}\left(t_1, t_2\right)$ and $\hat{\mathbf{y}}\left(t_1, t_2\right)$ satisfy (2), then the univariate forms $\mathbf{x}(t) = \hat{\mathbf{x}}(t, t)$ and $\mathbf{y}(t) = \hat{\mathbf{y}}(t, t)$ satisfy (1) [1]. Consequently, if we want to obtain the univariate solution in the generic $\left[0, t_{Final}\right]$ interval, due to the periodicity of the problem in the t_2 fast carrier dimension we will have

$$\mathbf{y}(t) = \hat{\mathbf{y}}\left(t, t \bmod T_2\right) \quad (3)$$

on the rectangular domain $\left[0, t_{Final}\right] \times \left[0, T_2\right]$, where $t \bmod T_2$ represents the remainder of division of t by T_2.

III. INNOVATIVE SIMULATION METHOD

A. Basic Slot Partitioning Strategy

For achieving an intuitive explanation of the simulation method proposed in this paper, let us consider an example of a circuit driven by a signal like the one depicted in Fig. 1. The method discussed in this paper will compute the solution of the circuit using a purely time-domain TSI scheme and a mixed time-frequency ETHB engine. The TSI engine will be mostly employed for the time intervals where the stimulus is off, i.e., the intervals where the circuit presents its transient response due to the energy stored in the reactive elements. The ETHB engine will be employed for

certain segments of the time intervals in which the stimulus is switched on, i.e., where the circuit exhibits a forced response.

B. Numerical Schemes

For the time slots where the solution of the solution has to be evaluated with a time-step integration engine (typically when the circuit is not excited by any independent stimulus), initial-value solvers have to be employed. To do that one can make use of Runge-Kutta (RK) schemes [17],

$$\mathbf{k}_i = \mathbf{f}\left(t_0 + c_i h, \mathbf{y}_n + h\sum_{j=1}^{s} a_{ij}\mathbf{k}_j\right), \quad i = 1, 2, \ldots, s \\ \mathbf{y}_{n+1} = \mathbf{y}_n + h\sum_{i=1}^{s} b_i \mathbf{k}_i, \quad (4)$$

or linear multistep methods [17], as is the case of the popular backward differentiation formulas (BDF)

$$\sum_{j=0}^{s} \alpha_j \mathbf{y}_{n+j} = h\beta\mathbf{f}\left(t_{n+s}, \mathbf{y}_{n+s}\right), \quad (5)$$

where $\mathbf{f}$ is defined according to (1) as

$$\mathbf{f}\left(t, \mathbf{y}(t)\right) = \left[\frac{d\mathbf{q}}{d\mathbf{y}}\right]^{-1}\left[\mathbf{x}(t) - \mathbf{p}\left(\mathbf{y}(t)\right)\right]. \quad (6)$$

$\mathbf{y}_n$, $\mathbf{y}_{n+1}$, $\mathbf{y}_{n+2}$, ..., denote the numerical solution at the time instants t_n, t_{n+1}, t_{n+2}, ..., respectively, and the parameter $h = t_{n+1} - t_n = t_{n+2} - t_{n+1} = \cdots$ is the time-step integration size. In the experiments reported in Section IV it was considered the second-order backward differentiation formula (BDF2)

$$\mathbf{y}_{n+2} - \frac{4}{3}\mathbf{y}_{n+1} + \frac{1}{3}\mathbf{y}_n = \frac{2}{3}h\mathbf{f}\left(t_{n+2}, \mathbf{y}_{n+2}\right) \quad (7)$$

and also the Bogacki-Shampine RK method of order 2.

For the time slots where the stimulus is switched on it is possible to make use of ETHB to evaluate the solution of the circuit. Considering the condition $\hat{\mathbf{y}}\left(t_1, 0\right) = \hat{\mathbf{y}}\left(t_1, T_2\right)$, expressing the periodic regime in t_2, where T_2 is the carrier period, and using a backward differentiation formula (e.g., the BDF2) to approximate the derivatives of (2) in t_1, the following boundary value problem with periodic boundary conditions is obtained for each slow time instant:

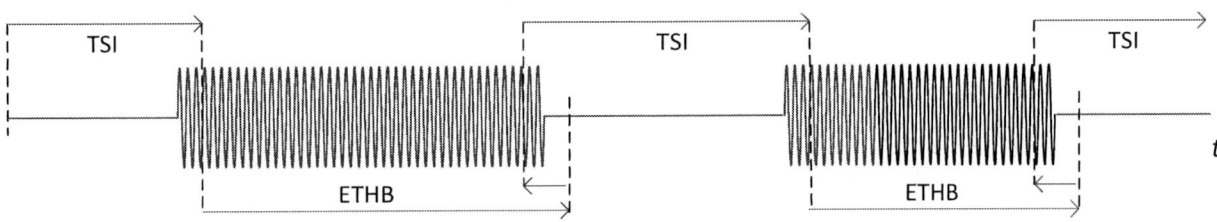

Fig. 1. Time-slot partitioning technique for switching between the transient response computed with TSI and the forced response calculated with ETHB.

$$\mathbf{q}\left(\hat{\mathbf{y}}_{n+2}\left(t_2\right)\right)-\frac{4}{3}\mathbf{q}\left(\hat{\mathbf{y}}_{n+1}\left(t_2\right)\right)+\frac{1}{3}\mathbf{q}\left(\hat{\mathbf{y}}_n\left(t_2\right)\right)$$
$$+\frac{2}{3}h_1\frac{d\mathbf{q}\left(\hat{\mathbf{y}}_{n+2}\left(t_2\right)\right)}{dt_2}=\frac{2}{3}h_1\left[\hat{\mathbf{x}}\left(t_{1,n+2},t_2\right)-\mathbf{p}\left(\hat{\mathbf{y}}_{n+2}\left(t_2\right)\right)\right],\quad(8)$$
$$\hat{\mathbf{y}}_{n+2}\left(0\right)=\hat{\mathbf{y}}_{n+2}\left(T_2\right).$$

With ETHB, the solution of (8) is evaluated by harmonic balance (HB). The corresponding HB system for each slow time instant $t_{1,i}$ is the $N\left(2K+1\right)$ set of algebraic equations given by

$$\mathbf{Q}\left(\hat{\mathbf{Y}}\left(t_{1,n+2}\right)\right)-\frac{4}{3}\mathbf{Q}\left(\hat{\mathbf{Y}}\left(t_{1,n+1}\right)\right)+\frac{1}{3}\mathbf{Q}\left(\hat{\mathbf{Y}}\left(t_{1,n}\right)\right)$$
$$+\frac{2h_1}{3}j\mathbf{\Omega}\,\mathbf{Q}\left(\hat{\mathbf{Y}}\left(t_{1,n+2}\right)\right)+\frac{2h_1}{3}\left[\mathbf{P}\left(\hat{\mathbf{Y}}\left(t_{1,n+2}\right)\right)-\hat{\mathbf{X}}\left(t_{1,n+2}\right)\right]\quad(9)$$
$$=0$$

where $\hat{\mathbf{X}}\left(t_{1,i}\right)$ and $\hat{\mathbf{Y}}\left(t_{1,i}\right)$ are the vectors containing the Fourier coefficients of the excitation and the solution (the state variables), respectively, and $j\mathbf{\Omega}$ is a diagonal matrix whose entries are $[-jK\omega_0,\ldots,0,\ldots,jK\omega_0]$. K is the order of the adopted harmonic truncation, and $\omega_0=2\pi f_C$ is the carrier frequency.

C. Transition Between Numerical Schemes

Fig. 1 shows that the stimulus off-on transition does not coincide with the TSI stop instant. This is because the simulation algorithm automatically obliges a postponement in the TSI to ETHB switching process. Indeed, after the excitation is turned on the simulator continues to use the TSI engine for a certain number of carrier cycles. Then, it computes both the TSI and ETHB solutions for a small period of time and compares the solutions with one another. If $\left\|\mathbf{y}_{TSI}-\mathbf{y}_{ETHB}\right\|<\delta$, where δ is a pre-defined maximum allowed quantity, then the simulator switches to ETHB. Otherwise it will carry on computing both the TSI and ETHB solutions until such condition is satisfied.

Fig. 1 also illustrates that the stimulus on-off transition does not match the ETHB stop instant. Indeed, when the stimulus is turned off at some time instant between $t_{1,i-1}$ and $t_{1,i}$, the simulator has to go back to the previous slow time instant $t_{1,i-1}$ and stops using ETHB. The reason for that is because ETHB is only applicable when the circuit is driven by a periodic excitation. It must be noted that the h_1 time-step length is considerably longer than the carrier period. Therefore, the solution obtained with ETHB for the $[t_{1,i-1},t_{1,i}]$ interval will be not considered and the TSI engine will be used to perform the numerical computations from $t_{1,i-1}$.

It must be noted that with this proposed method the solution of the circuit will be computed in the 1-D time t for some time slots (by using a TSI engine), and in the 2-D $\left(t_1,t_2\right)$ framework for the other time slots (by using the multitime ETHB engine). Indeed, when performing the inverse Fourier transform of the mixed time-frequency solution computed with ETHB, we obtain the corresponding bivariate solution $\hat{\mathbf{y}}\left(t_1,t_2\right)$ in the $\left(t_1,t_2\right)$ framework, which have to be converted to the 1-D time t according to (3).

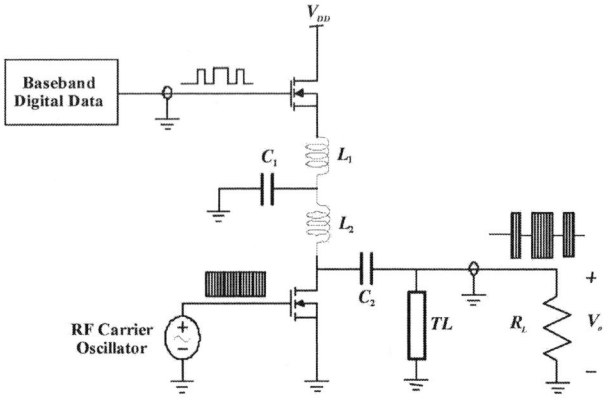

Fig. 2. ASK/OOK (on-off amplitude shift keying) wireless transmitter [15]

IV. ILLUSTRATIVE APPLICATION EXAMPLE

With the intention of demonstrating the capabilities of the method described in this paper the ASK/OOK transmitter depicted in Fig. 2 was considered as the sample application example. This circuit can be used in low-power applications, as RFID (RF identification), or biomedical imaging, and is basically composed of a MOS switch followed by a low-pass filter (needed to control the drain bias of the output drain modulator at the digital data rate, but providing the necessary high impedance for the RF carrier) and an output modulator constituted by an RF MOSFET and a high-Q $\lambda/4$ resonator at the carrier frequency of 2 GHz (TL) represented by its parallel L-C lumped equivalent circuit.

The circuit was simulated in MATLAB running on an Intel Core i7-2640M CPU @ 2.80 GHz with 16 GB RAM computer, with the proposed method versus conventional SPICE. A dynamic step size h starting from 5 ps was assigned to the t univariate time scale. A step length $h_1=50$ ns was used for the t_1 time scale. $K=15$ was chosen to be the highest harmonic order for the harmonic balance calculations. The numerical computation times required to carry out the simulations of the circuit in the [0, 2 μs] time interval are presented in Tables I and II. With the purpose of assessing the impact of the "on" and "off" periods in the efficacy of the method, the baseband digital data input has been considered to be a PWM signal whose duty cycle was manually adjustable. Additionally, in order to attest the accuracy of the numerical solution achieved with the proposed method, two normalized mean square errors (NMSE) are given. NMSE 1 stands for the largest of the NMSE values of all variables in the circuit. NMSE 2 denotes the average of these NMSE values. The NMSE assesses the global discrepancies between the numerical solution computed with the proposed method, when compared to the solution evaluated with conventional SPICE. NMSE values summarized in Tables I and II seem to indicate that the accuracy of the proposed method is very acceptable.

Tables I and II also show that there is an increase in the speedup gain as the input duty cycle is raised, i.e., as the number of '1's tends to be greater than the number of '0's. It is worth to point out that even in the case of equal probability between '1's and '0's the method discussed in this paper is more effective than conventional SPICE (a speedup gain of approx. 2 times was achieved).

TABLE I

NUMERICAL RESULTS FOR SIMULATIONS OF THE CIRCUIT SHOWN IN FIG. 2 USING SPICE IMPLEMENTATIONS BASED ON THE BDF2 SCHEME

CPU time (h:min:sec) and accuracy	Input duty cycle				
	10%	30%	50%	70%	90%
SPICE (BDF2)	01:27:22	01:31:17	01:37:16	01:39:51	01:46:11
Proposed method	01:20:14	01:07:19	00:45:53	00:32:57	00:23:22
Speedup	1.09	1.36	2.12	3.03	4.54
NMSE 1	3.81×10^{-6}	3.70×10^{-6}	3.66×10^{-6}	3.61×10^{-6}	3.42×10^{-6}
NMSE 2	2.55×10^{-6}	2.43×10^{-6}	2.39×10^{-6}	2.29×10^{-6}	2.03×10^{-6}

TABLE II

NUMERICAL RESULTS FOR SIMULATIONS OF THE CIRCUIT SHOWN IN FIG. 2 USING SPICE IMPLEMENTATIONS BASED ON THE RK SCHEME

CPU time (h:min:sec) and accuracy	Input duty cycle				
	10%	30%	50%	70%	90%
SPICE (RK)	01:29:52	01:33:09	01:40:41	01:43:19	01:48:55
Proposed method	01:24:16	01:09:34	00:49:45	00:34:18	00:25:09
Speedup	1.07	1.34	2.05	3.01	4.33
NMSE 1	4.05×10^{-6}	3.92×10^{-6}	3.81×10^{-6}	3.78×10^{-6}	3.56×10^{-6}
NMSE 2	2.69×10^{-6}	2.57×10^{-6}	2.48×10^{-6}	2.40×10^{-6}	2.29×10^{-6}

V. CONCLUSIONS

An efficient mixed time-frequency simulation method has been discussed and tested in this study. The method operates in univariate and bivariate frameworks and automatically switches between two very distinct numerical schemes, the time-step integration engine and the multitime envelope transient harmonic balance engine, along the simulation process.

The method is particularly suitable for computing the solution of RF circuits containing periodic forcing functions being successively switched on and off for unspecified time slots. With this method considerable savings on the computational costs can be achieved, without noticeable loss in the accuracy of the results.

It has also been demonstrated that the speedup advantage of the the method, when compared to conventional SPICE, depends on the discrepancy between the duration of the on and off periods. Thus, even though the effectiveness of the method has been proved to be quite considerable for the tested application example, it is expected that this efficacy will become progressively more noticeable for circuits in which the discrepancy between the duration of the on and off time slots is increased.

ACKNOWLEDGMENT

This work is funded by National Funds through FCT – Fundação para a Ciência e Tecnologia, under the project UIDB/50008/2020-UIDP/50008/2020.

REFERENCES

[1] T. Mei, J. Roychowdhury, T. Coffey, S. Hutchinson, and D. Day, "Robust stable time-domain methods for solving MPDEs of fast/slow systems," *IEEE Trans. on Computer-Aided Design of Integrated Circuits and Systems*, vol. 24, no. 2, pp. 226-239, Feb. 2005.

[2] J. F. Oliveira, "Radio frequency and microwave numerical simulation techniques based on multivariate formulations", *Applied Mathematics and Computation*, vol. 294, pp. 238-252, Feb. 2017.

[3] N. B. Carvalho, J. C. Pedro, W. Jang, and M. B. Steer, "Nonlinear RF circuits and systems simulation when driven by several modulated signals," *IEEE Trans. on Microwave Theory and Tech.*, , vol.54, no.2, pp.572,579, Feb. 2006.

[4] V. Rizzoli et al., "Domain-decomposition harmonic balance with block-wise constant spectrum,", in Proc. *IEEE MTT-S International Microwave Symp. Digest*, San Francisco, CA, Jun. 2006, pp. 860-863.

[5] J. F. Oliveira and J. C. Pedro, "A multiple-line double multirate shooting technique for the simulation of heterogeneous RF circuits," *IEEE Trans. on Microwave Theory and Tech.*, vol. 57, no. 2, pp. 421-429, Feb. 2009.

[6] J. F. Oliveira and J. C. Pedro, "A new mixed time-frequency simulation method for nonlinear heterogeneous multirate RF circuits," in *Proc. IEEE MTT-S International Microwave Symp. Digest*, Anaheim, CA, May 2010, pp. 548-551.

[7] J. Oliveira and J. C. Pedro, "Efficient RF circuit simulation using an innovative mixed time-frequency method", *IEEE Trans. on Microwave Theory and Tech.*, vol. 59, no. 4, pp. 827-836, Apr. 2011.

[8] M.A Farhan, E. Gad, M.S. Nakhla, and R. Achar, "Fast Simulation of Microwave Circuits With Nonlinear Terminations Using High-Order Stable Methods," *IEEE Trans. on Microwave Theory and Tech.*, vol. 61, no. 1, pp. 360-371, Jan. 2013.

[9] F. Bizzarri, A. Brambilla and L. Codecasa, Harmonic balance based on two-step Galerkin method, in IEEE Transactions on Circuits and Systems I: Regular Papers, 63 (9) (2016) 1476-1486.

[10] S. C. Pires, P. M. Cabral and J. C. Pedro, "Radio frequency carrier amplitude-burst transmitters - from architecture to circuit," in *IET Microwaves, Antennas & Propagation*, vol. 9, no. 3, pp. 271-280, Feb. 2015.

[11] J. Hsieh, Y. Huang, P. Kuo, T. Wang and S. Lu, "A 0.45-V Low-Power OOK/FSK RF receiver in 0.18μm CMOS technology for implantable medical applications," *IEEE Transactions on Circuits and Systems I: Regular Papers*, vol. 63, no. 8, pp. 1123-1130, Aug. 2016.

[12] S. J. Kim, C. S. Park and S. Lee, "A 2.4-GHz ternary sequence spread spectrum OOK transceiver for reliable and ultra-low power sensor network applications," *IEEE Transactions on Circuits and Systems I: Regular Papers*, vol. 64, no. 11, pp. 2976-2987, Nov. 2017.

[13] Hien-Hua Jung and Kea-Tiong Tang, "A 0.9-V 2.36-GHz MedRadio-band 10-Mbps low-power OOK modulator for neural implants," in *Proc. International Symposium on VLSI Design, Automation and Test (VLSI-DAT)*, Hsinchu, 2017, pp. 1-4.

[14] H. Zhu, K. Goi and K. Ogawa, "All-Silicon waveguide photodetection for low-bias power monitoring and 20-km 28-Gb/s NRZ-OOK signal transmission," *IEEE Journal of Selected Topics in Quantum Electronics*, vol. 24, no. 2, pp. 1-7, March-April 2018.

[15] D. Ferreira, J. F. Oliveira and J. C. Pedro, "A novel time-domain CAD technique based on automatic time-slot division for the numerical simulation of highly nonlinear RF circuits", *IEEE Trans. on Microwave Theory and Tech.*, vol. 62, no. 1, pp. 18-27, Jan. 2014.

[16] J. F. Oliveira, "On the usage of a time-frequency switch mode algorithm to efficiently simulate RF circuits", *AEU - International Journal of Electronics and Communications*, vol. 127, Dec. 2020,

[17] E. Hairer and G. Wanner, *Solving Ordinary Differential Equations II: Stiff and Differential Algebraic Problems*. Springer Series in Computational Mathematics, 1996.

© VDE VERLAG GMBH · Berlin · Offenbach

Application of Asymmetric Crosstalk Harnessed Signaling on 3D Hexagonal Interconnect Arrays

Daniel Iparraguirre
Intel Corporation
Hillsboro, OR, USA
daniel.iparraguirre@intel.com

José Delgado-Frías
Washington State University
Pullman, WA, USA
delgado-frias@wsu.edu

Abstract—This paper describes a novel application of the Asymmetric Crosstalk Harnessed Signaling (ACHS) on 3D routing environments, where interconnect wires are distributed across the two dimensions perpendicular to the routing direction. The ACHS scheme completely eliminates the common encoding mode that is part of the original CHS scheme. A fully 3D multi-stripline stackup containing a wiring bundle in a hexagonal wire arrangement is applied where channels are placed together. Simulation results, that include crosstalk and jitter effects, have shown a significant performance improvement of ACHS over the original CHS scheme.

Keywords—CHS, signal integrity, 3D routing, multi-stripline stackup, cross-section, jitter.

I. Introduction

Next-generation computer systems are expected to have increased demands for highly parallel data access with low latencies, thereby requiring very high-speed interfaces with high interconnect density. Current technologies allow for high-density 2D single-ended wiring environments with top/bottom general ground or power reference layers, which allow for high-speed communication targets when coupled with transmitter/receiver equalization and crosstalk cancellation circuitries [5 – 7]. Higher performance targets, however, are reaching the limit of what technology can offer in terms of wiring density with the traditional 2D routing scheme, therefore requiring engineering teams to explore basic 3D routing solutions like dual stripline, where two stacked signaling layers share a general top and bottom reference layer set. A further step includes an intensive 3D routing scheme where multiple signal layers are stacked together, allowing interface wiring to be distributed across two dimensions perpendicular to the routing direction. This scheme brings a significant challenge regarding the higher crosstalk levels to be handled. Crosstalk Harnessed Signaling (CHS) has been proposed as a very promising alternative for very high-speed parallel interfaces [1]. CHS takes advantage of the wire coupling relationships inside a bus and sending the information encoded by them, instead of dealing with the wire coupling as an undesirable phenomenon. A further study has also shown the effectiveness of CHS in full 3D routing stackups [2], as also made evident its weakness related to the common-mode noise sensitivity, which ultimately hurts the overall bus performance and requires the application of additional inter-channel spacing or ground shielding, or particular encoding solutions [4].

Asymmetric Crosstalk Harnessed Signaling (ACHS) is being introduced as a very promising modification over the originally defined CHS [3]. In ACHS, the Hadamard encoding matrix is transformed so that the common-mode vector is eliminated or differentially complemented by either removing the common-mode row altogether, or a combination of adding columns and diagonally concatenating the elementary encoding matrices. The conversion is aimed towards eliminating the common encoding mode in order to provide a comparable crosstalk rejection and performance across all the encoded bits in a channel with a small penalty in terms of a larger wiring for the encoded bus, making it therefore highly tolerant to crosstalk coming from neighboring channels in the interface.

This study shows the application of ACHS on 3D routing bundles with cross-sections organized in hexagonal patterns, so that every wire is potentially dealing with 6 immediate crosstalk aggressors, hence crosstalk effects have a higher impact on the unprotected common encoding mode inside the original CHS, than in traditional 2D routing wire sets with 2 immediate aggressors per wire. ACHS offers a comparable performance across the encoded data bits with a small routing overhead, therefore provided a much better bus performance than the original CHS scheme in terms of worst eye height and width across the encoded bus. The study also shows how a particular arrangement of encoded wire bundles affects the internal coupling relationships essential to the success of CHS, resulting in different eye openings per wire bundle.

This paper is organized as follows. The Asymmetric CHS scheme and its application to a generic high-speed 16-bit parallel source-synchronous interface are introduced in Section II. Section III describes the hexagonal 3D routing and wire arrangement for CHS and ACHS encoded channels (including how bytes, nibbles and strobes are encoded spatially across the bus wiring cross section). Also, this section includes how the impedance, insertion loss and crosstalk effects are modeled in a 3D multi-stripline stackup for a full-bus dataflow simulation. Section IV describes the simulation results and performance comparisons between CHS and ACHS for all the data bits in the 16-bit interface. Finally, Section V presents some concluding remarks from this study.

II. Asymmetric Crosstalk-Harnessed Signaling (ACHS)

Crosstalk-Harnessed Signaling (CHS) is defined as the signaling that makes use of the existing wire coupling as a basic building block for the information transmission, instead of a noise component to be removed or minimized. This signaling scheme is accomplished by the application of a square eigenvector matrix that encodes the original data into orthogonal vector components to be carried by the wire coupling elements in the bus in multiple voltage levels. The Hadamard matrix,

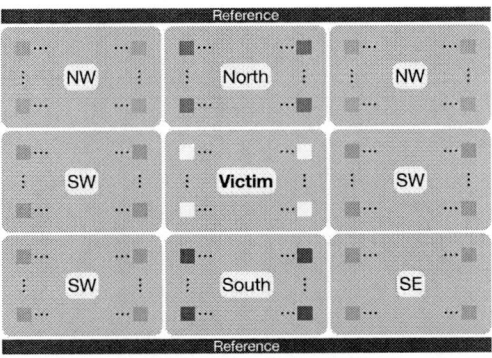

Fig. 1. Cross-section of a 3D-routed parallel bus with general top and bottom ground reference layers, with a central Victim channel surrounded by 8 Aggressor channels. * *Implementation for the purposes of this research, with no relationship to any actual product.*

eigenvector matrix, has been successfully used for this purpose; for a 4-bit data encoding it becomes:

$$W_{CHS} = \frac{1}{4}\begin{bmatrix} 1 & -1 & 1 & -1 \\ 1 & -1 & -1 & 1 \\ 1 & 1 & -1 & -1 \\ 1 & 1 & 1 & 1 \end{bmatrix} \quad \dots\dots\dots\dots (1)$$

This basic encoding scheme offers 4 orthogonal encoding modes for the 4 data bits and distributed across 4 interconnect wires in the bus through 5 encoded voltage levels. Three of these modes are differential and therefore take great advantage of the internal wire coupling for the safe transmission of data with high crosstalk rejection (this crosstalk coming from outside the 4-wire bundle); the fourth mode instead is common (non-differential) and thus highly sensitive to any sort of crosstalk or noise in general.

The novel ACHS scheme is presented as a matrix modification aimed to either eliminate the common encoding mode or turn it differential, by means of either: i) removing the common encoding mode row, ii) adding columns to the matrix, or iii) diagonally concatenating Hadamard matrices [3]. For a 16-bit data bus with a differential strobe pair, it is possible to build a non-square 20x17 encoding matrix, containing a set of concatenated 4x4 and 2x2 matrices in order to encode together data and strobe bits:

$$W_{ACHS} = \frac{1}{4}\begin{bmatrix} 1 & -1 & 1 & -1 & 0 & 0 & 0 & 0 & 0 & 0 & 0 & 0 & 0 & 0 & 0 & 0 & 0 & 0 & 0 & 0 \\ 1 & -1 & -1 & 1 & 0 & 0 & 0 & 0 & 0 & 0 & 0 & 0 & 0 & 0 & 0 & 0 & 0 & 0 & 0 & 0 \\ 1 & 1 & -1 & -1 & 0 & 0 & 0 & 0 & 0 & 0 & 0 & 0 & 0 & 0 & 0 & 0 & 0 & 0 & 0 & 0 \\ 1 & 1 & 1 & 1 & -1 & -1 & -1 & -1 & 0 & 0 & 0 & 0 & 0 & 0 & 0 & 0 & 0 & 0 & 0 & 0 \\ 0 & 0 & 0 & 0 & -1 & -1 & 1 & 1 & 0 & 0 & 0 & 0 & 0 & 0 & 0 & 0 & 0 & 0 & 0 & 0 \\ 0 & 0 & 0 & 0 & 1 & -1 & -1 & 1 & 0 & 0 & 0 & 0 & 0 & 0 & 0 & 0 & 0 & 0 & 0 & 0 \\ 0 & 0 & 0 & 0 & -1 & 1 & -1 & 1 & 0 & 0 & 0 & 0 & 0 & 0 & 0 & 0 & 0 & 0 & 0 & 0 \\ 0 & 0 & 0 & 0 & 0 & 0 & 0 & 2 & -2 & 0 & 0 & 0 & 0 & 0 & 0 & 0 & 0 & 0 & 0 & 0 \\ 0 & 0 & 0 & 0 & 0 & 0 & 0 & 2 & 2 & -2 & -2 & 0 & 0 & 0 & 0 & 0 & 0 & 0 & 0 & 0 \\ 0 & 0 & 0 & 0 & 0 & 0 & 0 & 0 & 0 & -2 & 2 & 0 & 0 & 0 & 0 & 0 & 0 & 0 & 0 & 0 \\ 0 & 0 & 0 & 0 & 0 & 0 & 0 & 0 & 0 & 0 & 1 & -1 & 1 & -1 & 0 & 0 & 0 & 0 & 0 & 0 \\ 0 & 0 & 0 & 0 & 0 & 0 & 0 & 0 & 0 & 0 & 1 & -1 & -1 & 1 & 0 & 0 & 0 & 0 & 0 & 0 \\ 0 & 0 & 0 & 0 & 0 & 0 & 0 & 0 & 0 & 0 & 1 & 1 & -1 & -1 & 0 & 0 & 0 & 0 & 0 & 0 \\ 0 & 0 & 0 & 0 & 0 & 0 & 0 & 0 & 0 & 0 & 1 & 1 & 1 & 1 & -1 & -1 & -1 & -1 \\ 0 & 0 & 0 & 0 & 0 & 0 & 0 & 0 & 0 & 0 & 0 & 0 & 0 & 0 & -1 & -1 & 1 & 1 \\ 0 & 0 & 0 & 0 & 0 & 0 & 0 & 0 & 0 & 0 & 0 & 0 & 0 & 0 & 1 & -1 & -1 & 1 \\ 0 & 0 & 0 & 0 & 0 & 0 & 0 & 0 & 0 & 0 & 0 & 0 & 0 & 0 & -1 & 1 & -1 & 1 \end{bmatrix} \dots(2)$$

In this scheme, two pairs of 4x4 Hadamard matrices are concatenated along the big matrix diagonal, so that the respective common-mode rows get aligned and complemented to each other, turning two common modes into one differential mode; the result is 2 groups of 7 bits encoded into 16 ACHS encoded signals to be transmitted. Similarly, two 2x2 Hadamard matrices get concatenated in a same fashion in order to encode 3 bits (2 data bits and a strobe) into 4 ACHS signals. Therefore,

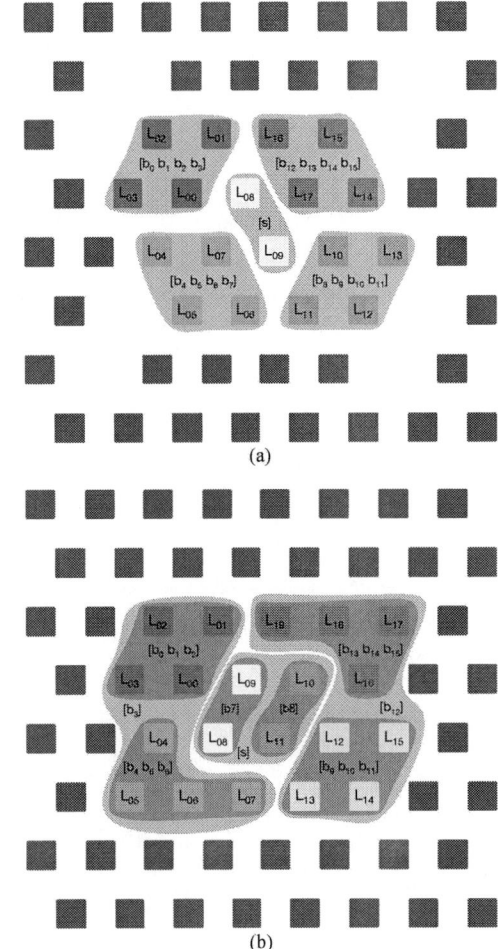

(a)

(b)

Fig. 2. Cross-sections of: (a) A CHS-encoded 16-bit data channel with differential strobe pair into an 18-wire bundle, where all data nibbles are independently encoded by 4x4 CHS matrices. (b) An ACHS-encoded 16-bit data channel plus strobe, into a 20-wire set, through the encoding matrix described in (2); data bits are encoded differently, some of them into 4 wires, some others into 8 wires and others encoded together with the strobe pair. * *Implementation for the purposes of this research, with no relationship to any actual product.*

an original 17-bit bus (16 data bits plus 1 strobe) is encoded into a 20-wire ACHS set. Given the fact that the strobe is generally differential in a source-synchronous bus design practice, the routing overhead (relative additional wiring related to the original CHS encoding) becomes (20 - 18)/18 = 11.11%.

III. 3D HEXAGONAL INTERCONNECT ARRAYS

A. CHS and ACHS Interconnect Arrangements

In 3D routing, interconnects are distributed across the plane perpendicular to the routing direction. In a parallel single-ended channel design, wires can be allocated in 2 dimensions forming different patterns and sharing a common general ground or power reference somewhere on top or bottom of the routing stackup. Fig. 1 shows the cross-section of a general multi-channel bus 3D routing representation, in which the central "victim" channel is surrounded by 8 channels on left/right and up/down directions. In addition, every channel has a number of

SMACD / PRIME 2021 | 19 – 22 July 2021, Online Event

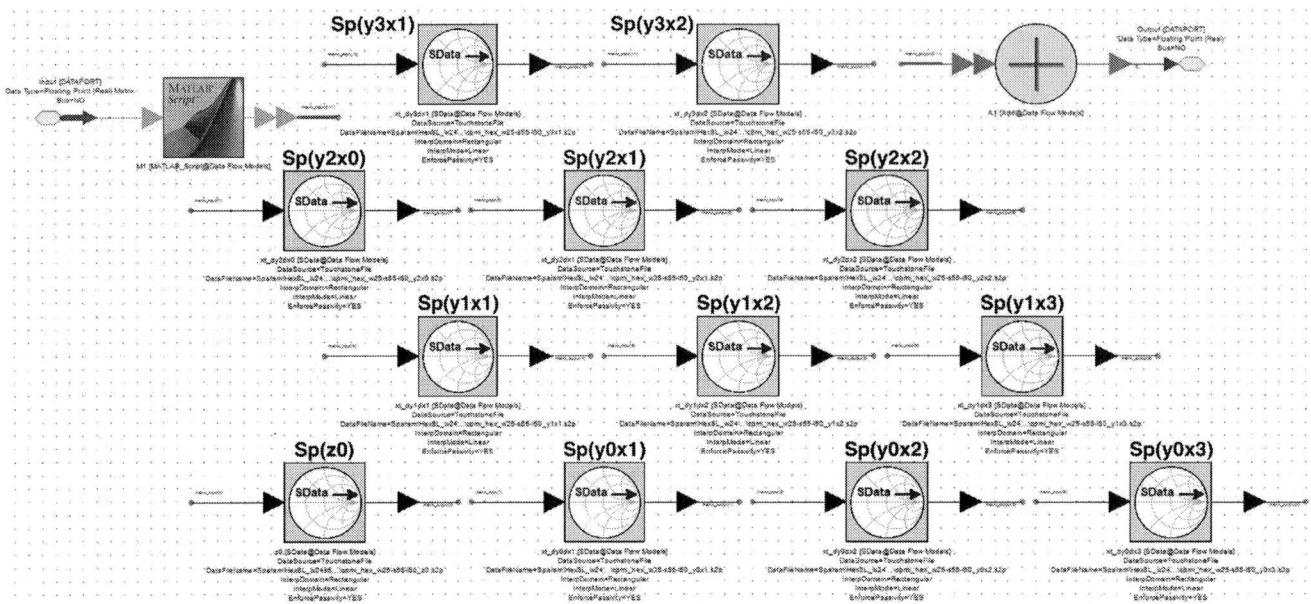

Fig. 3. SystemVue® dataflow implementation of the 3D interconnect array response, represented by the S-parameter convolutions for the victim's insertion loss and FEXT interactions from neighboring aggressors. The involved S-parameter responses are extracted from the implemented multi-stripline stackup shown in Fig. 4. *Implementation for the purposes of this research, with no relationship to any actual product.*

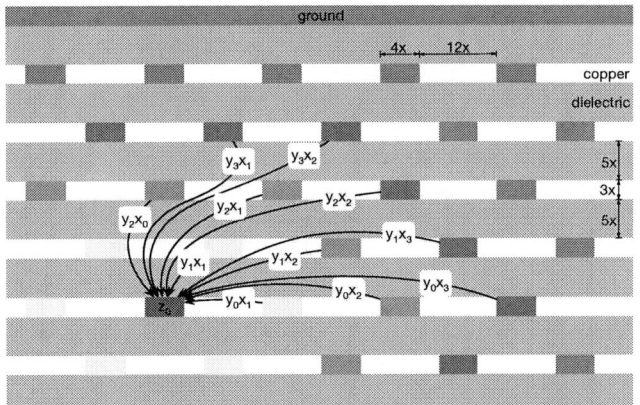

Fig. 4. Partial cross-section view of the basic multi-stripline stackup employed for the S-parameter extractions, pertaining to the victim's insertion loss (z_0) as well as the FEXT components from every neighboring wire in the 3-hexagon region (36 aggressors in total). A quadrilateral approximation is assumed by extracting the x_iy_j components from the shown wire subset. General bottom ground layer is not displayed. *Implementation for the purposes of this research, with no relationship to any actual product.*

wires distributed in a variety of arrangements. In this work, a hexagonal wire arrangement is studied so a 20-wire ACHS channel is organized in a 5x4 array; the equivalent CHS counterpart employs only 18 wires per channel, therefore a collection of CHS channels can benefit itself from additional wire spacing that provides some relief in terms of crosstalk reduction. Fig. 2(a) shows a CHS-encoded 16-bit channel in an 18-wire bundle, in which every nibble is encoded independently in a 4-wire set, while the differential strobe pair in the middle of the array still employs 2 wires; every nibble has one bit encoded through common mode (bits b_3, b_7, b_8 and b_{12}). Fig. 2(b) shows the same 16-bit bus with strobe encoded through ACHS with the encoding matrix from (2). Following the scheme described in

Section II, bit groups [b_0, b_1, b_2], [b_4, b_5, b_6], [b_7], [b_8], [b_9, b_{10}, b_{11}], [b_{13}, b_{14}, b_{15}] are encoded into independent wire sets, while each of bits b_3, b_{12} and the strobe get encoded across 2 of the aforementioned sets. Unlike CHS, encoding groups take clearly different shapes across the channel cross-section, therefore all participant bits get encoded differently and are exposed to different crosstalk sources; this gets reflected in the simulation results, which ultimately reveal which bits get the worst crosstalk conditions are therefore are the bus performance limiters.

B. Hexagonal Interconnect Modeling

In order to properly compare the performances of CHS and ACHS routing implementation, a full-link dataflow-based simulation environment with generic transmitter and receiver electrical characteristics has been implemented in Keysight® SystemVue®. The environment performs a full-transient simulation of a random-generated data sequence on the 9 channels belonging to the bus structure from Fig. 1, working at 20 gigatransfers per second. Fig. 3 shows the dataflow module implementing the 3D interconnect response processing, in which the crosstalk effects are added individually by modeling the S-parameters from a general multi-stripline stackup shown in Fig 4. A hexagonal wire pattern is implemented in Ansys® HFSS® Layout tool using generic conductor and dielectric materials, so the victim's insertion loss as well as the individual FEXT effects from every surrounding aggressor are extracted. Since some data bits from the wire distribution in Fig. 2(b) are encoded across 4 signal layers, the FEXT modeling has to include aggressors separated up to 3 layers from the victim. Fig. 4 shows a subset of the included aggressors across 3 concentric hexagons around the victim, making up for a total of 36 aggressors. A 11-aggressor subset is then selected for the individual FEXT S-parameter extraction, then assuming a quadrilateral symmetry approximation in order to cover for the

© VDE VERLAG GMBH · Berlin · Offenbach

TABLE I
CHS AND ACHS PERFORMANCE RESULTS
TRANSFER RATE = 20GTPS, R_{JITTER} = 2PS

Bit	Eye Width (ps)			Eye Height (mV)		
	CHS	ACHS	Δ	CHS	ACHS	Δ
0	32.35	33.40	1.05	342.6	350.8	8.2
1	31.26	32.10	0.84	315.2	319.8	4.6
2	32.29	32.58	0.29	326.2	342.0	15.8
3	12.02	32.18	20.16	99.8	307.0	207.2
4	33.00	29.77	-3.23	356.2	309.4	-46.8
5	32.02	31.70	-0.32	307.2	327.8	20.6
6	32.36	33.03	0.67	338.0	345.4	7.4
7	1.94	34.22	32.28	62.2	357.2	295.0
8	15.12	33.19	18.07	134.0	330.4	196.4
9	32.77	33.63	0.86	341.6	353.0	11.4
10	32.73	31.71	-1.02	320.0	307.0	-13.0
11	33.66	33.12	-0.54	356.2	348.4	-7.8
12	19.62	29.51	9.89	124.0	290.6	166.6
13	32.58	31.17	-1.41	335.6	298.2	-37.4
14	32.04	31.26	-0.78	303.0	330.4	27.4
15	32.18	31.71	-0.47	352.8	345.8	-7.0
Worst	1.94	29.51	27.57	62.2	290.6	228.4
μ	27.37	32.14	4.77	275.9	329.0	53.0
σ	9.68	1.32	-8.36	104.1	21.3	-82.8

* No relationship to any actual product.

entire vicinity. Fig. 3 shows the S-parameter convolution components (SData) laid out in SystemVue®, for the victim's insertion loss as well as for each one of the 11 designated FEXT aggressors. Previous Matlab® blocks in the dataflow simulator create the corresponding wire neighborhood matrices for every bit inside the victim channel.

IV. SIMULATION RESULTS AND COMPARISON

Both CHS and ACHS performances were assessed by using the previously described full-link dataflow environment, with 9 channels arranged according to Fig. 1; this means, a Victim channel being exposed to crosstalk coming from 8 surrounding channels. Generic silicon buffers on the transmitter side were modeled with a slew-rate of 30V/ns and a random jitter of 2ps, sending random 1024 random bit patterns at 20 gigatransfers per second. The performance results are presented in Table I, in terms of eye width and eye height. CHS presents an eye collapse in almost all the common-mode encoding instances (i.e. bits b_3, b_7, b_8 and b_{12} from Fig. 2(a) cross-section). Thus, the common mode encoding brings down the entire bus performance, rendering it inoperable with the employed design features. On the contrary, ACHS offers very comparable eye opening across the 16 data bits. On the other hand, for ACHS bits b_3 and b_{12} which pertain to the compensated modes from (1) and which are encoded across 8 wires according to Fig. 2(b) happen not to be

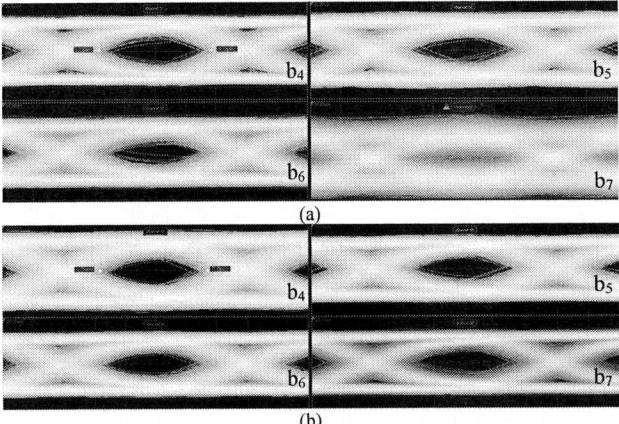

(a)

(b)

Fig. 5. Simulation eye diagrams for simulated bits b_4 (yellow), b_5 (blue), b_6 (green) and b_7 (pink) for the 16-bit hexagonal 3D routing arrangement with (a) CHS, with total eye collapse on bit b_7; and (b) ACHS, offering similar performance across all bits. ** No relationship to any actual product.*

the performance limiters in terms of eye height. This in turn is a good indicator of the degree of success of ACHS in order to overcome the common-mode encoding issues inherent to the original CHS scheme. Finally, Fig. 5(a) and 5(b) show the comparison between CHS and ACHS eye openings respectively.

V. CONCLUDING REMARKS

An application of ACHS on 3D hexagonal interconnect arrays has been presented, showing the significant advantages in terms of performance with respect to the original CHS encoding scheme for very high wiring integration levels, due to the elimination of the common mode which is highly sensitive to crosstalk noise, with a small routing overhead penalty. These results show that ACHS provides an excellent performance for all the encoded bits in a channel, without the need for additional inter-channel clearance, which is usual practice in high-speed parallel bus implementations. This in turn helps design teams to envision novel computer systems with higher wiring integration levels, taking full advantage of 3D routing using ACHS in order to achieve both high transfer speed and access parallelism.

REFERENCES

[1] C. Sreerama, S. H. Hall, P. G. Huray, A. Zenteno, M. W. Leddige, O.B. Oluwafemi, A. A. Aziz and J. Wight, "A Crosstalk-Friendly Signaling Method," *IEEE T. Comp., Pack. Manuf. Techn.*, vol. 8, no. 9, pp. 1621–1632, Sep. 2018.

[2] A. A. Aziz, P. G. Huray and C. Sreerama, "CHS Application on 3D Novel High-Bandwidth Signaling," in *Proc. CSPA 2018*, Penang, Malaysia, Mar. 2018, pp. 82–87.

[3] D. Iparraguirre, J. Delgado-Frías, H. Heck, "A Crosstalk-Harnessed Signaling Enhancement that Eliminates Common-Mode Encoding," *Proc. ISCAS'2021*, Daegu, South Korea, May 22-28, 2021. *To appear.*

[4] M. W. Leddige, S. H. Hall, C. Sreerama, O. B. Oluwafemi, A. Zenteno and M. C. Falconer, "Mode Selective Balanced Encoded Interconnect," U.S. Patent Appl. 20160026597, Jan. 28, 2016.

[5] S. H. Hall and H. L. Heck, *Advanced Signal Integrity for High-Speed Digital Designs*, 1st ed., Hoboken, NJ, USA: Wiley, 2009.

[6] S. H. Hall, G. W. Hall and J. A. McCall, *High-Speed Digital System Design: A Handbook of Interconnect Theory and Design Practices*, 1st ed., Hoboken, NJ, Wiley-IEEE Press, 2000.

[7] C. Sreerama, "Novel Crosstalk Mitigation Solutions for High-Speed Interconnects to Maximize Bus Band-Width and Density," in *8th Annual Signal Integrity Symp.*, Penn State Harrisburg, PA, Apr. 4, 2014, pp. 1.

© VDE VERLAG GMBH · Berlin · Offenbach

Adaptive Test Bench Generation, Simulation and Parameter Extraction for AMS Circuitry

Alexander Meyer, Léon Weihs, Ralf Wunderlich, Stefan Heinen

Integrated Analog Circuits and RF Systems Laboratory RWTH Aachen University
Kopernikusstr. 16, D-52074 Aachen, Germany
E-Mail: mailbox@ias.rwth-aachen.de

Abstract—This paper presents a novel semi-automatic test bench generation and parameter extraction workflow for analog-mixed signal circuitry. Specific test benches for a given circuit are automatically generated, simulated, and model parameters are extracted. By relying on a modular test bench structure, changes in the circuit's design or port list can be automatically considered, facilitating the overall verification process and avoiding errors during the parameter extraction phase. Hence, precise and pin-accurate models can be generated faster and more reliably. A first implementation of this workflow using Cadence's SKILL language is shown.

Index Terms—Test Bench Generation, Parameter Extraction, Model Generation, SKILL, SKILL++, OCEAN, VERILOGA

I. INTRODUCTION

With the ever-increasing complexity of analog-mixed signal circuitry, even more sophisticated models for verification are required to keep simulation time in check while also correctly depicting the underlying circuitry's behavior [1], [2]. Fulfilling these requirements not only depends on a model's mathematical foundation but also relies on the correct and standardized extraction of its parameters. This relation, which is well understood for digital systems [3], is however rarely addressed in analog circuits.

Extracting the required parameters for a model can be tedious, often resulting in a convoluted cluster of different slightly varying test benches, and the device under test (DUT) might change throughout the design and verification process. Also, dozens of parameters might be needed for the model depending on the desired level of complexity, requiring various types of analyses and post-processing capabilities during the extraction process [4]. If any parameters are wrongly determined or simply outdated, the model will not depict the correct behavior of the DUT [5]. Additionally, the verification must guarantee pin-accuracy to ensure the reliability and re-usability of the model [6].

Using a structured test bench generation and parameter extraction approach, misconceptions about parameters and their extraction can be avoided, and verification models can be generated with the same principle every time. Instead of referring to a test bench as a single, rigid entity, this work views a test bench as a composition of different, standardized test modules. Depending on the DUT, the required test modules are chosen while maintaining the same and easily recognizable test bench structure.

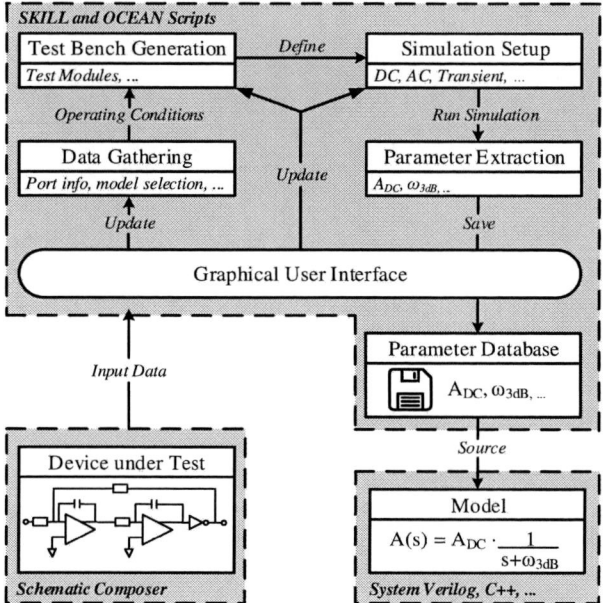

Fig. 1. Flow for test bench generation, simulation and parameter extraction.

Utilizing Cadence's native programming language SKILL, the test benches and the corresponding simulation setup can be generated automatically [7]. Furthermore, simulation control and parameter extraction can also be performed with Cadence's OCEAN and MATLAB's post-processing capabilities. Consequently, the otherwise error-prone and time-consuming parameter extraction process can be significantly facilitated, paving the way for a more robust verification flow.

The paper is structured as follows. Section II introduces the proposed workflow, highlighting necessary steps to extract the model parameters for a given DUT automatically. Section III focuses on the implementation of a software tool capable of providing the necessary functionality. Here, particular focus is laid upon a systematic object-oriented approach to generate and evaluate test benches. Finally, the results are discussed, and a conclusion is drawn in Section IV.

II. CONCEPT

Fig. 1 visualizes the proposed workflow and consists of four steps. These include the gathering of DUT data, test bench

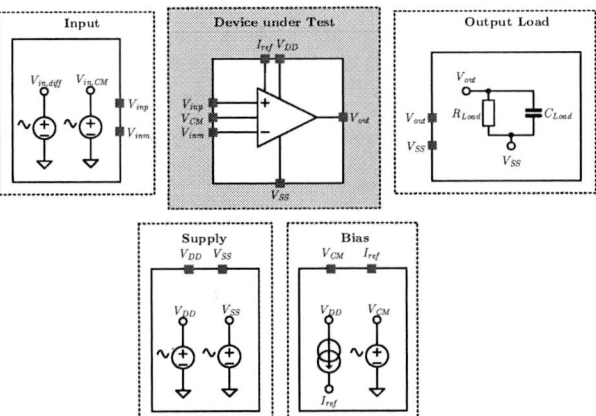

Fig. 2. Exemplary modular test bench setup for an opamp.

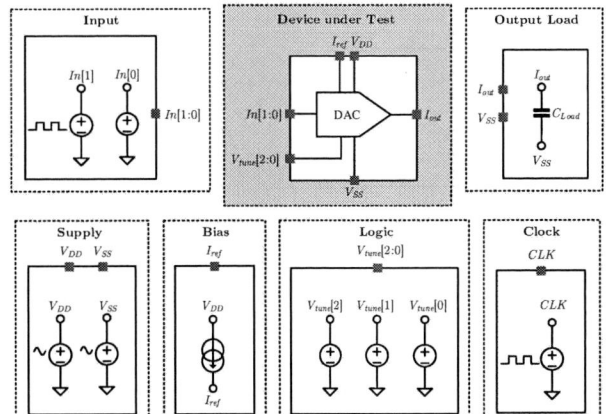

Fig. 3. Exemplary modular test bench setup for a DAC.

generation, simulation run control and parameter extraction. A Graphical User Interface (GUI) links the different steps and guides the process.

A. Data Gathering

In the first step, specific DUT data is collected. This includes all port information regarding the DUT as well as the selection of a corresponding verification model. Only then can pin-accurate models be generated, as well as the correct test benches and simulations be set up.

B. Test Bench Generation

Afterward, the test bench setup process is triggered. Depending on the DUT, several different test benches may be needed, such as an open and closed-loop test bench in the case of an operational amplifier (opamp). Multiple test benches are generated iteratively, utilizing a novel approach for the generation of each test bench itself. Instead of creating one large stimulus block for the DUT, seven different test modules are available and are selected depending on the test bench/DUT. Table I lists these different test modules with corresponding exemplary use-cases. To increase the level of automation, each test module is implemented using VerilogA.

To motivate the idea of a modular test bench structure, the conceptual test benches of an opamp and a Digital-to-Analog Converter (DAC) are shown in Fig. 2 and Fig. 3. Both test benches require input sources, loads, bias voltages, and currents. Additionally, the DAC requires a sampling clock and digital tuning words, whereas the opamp does not. Such differences in the test benches can be easily considered following a modular test bench approach by either adding or omitting a clock and logic test module. Test modules can be replaced by transistor-level circuitry, such as a Bandgap reference source as a bias generator, if a more accurate simulation with the actual bias voltages is needed. Such adaptations are achievable due to the flexibility of the modular implementation and would not be possible using one large stimulus block. Furthermore, keeping the individual test bench module's complexity relatively low

TABLE I
OVERVIEW OF THE DIFFERENT TYPES OF TEST MODULES

Test Module:	Examples:
INPUT	Sinusoidal, dc sweep or rectangular waveform
LOAD (in/out)	Resistive, capacitive or inductive load
SUPPLY	Analog V_{DD} or V_{SS}
BIAS	Reference current, Bandgap voltage
LOGIC	Digital tuning words for capacitors, resistors
CLOCK	Reference clock, local oscillator

allows for much better code readability during manual inspection of the generated test benches.

C. Simulation Setup and Execution

As the standard analog design environments provide only basic automation functionality [4], a custom-made solution is needed. Utilizing Cadence's native OCEAN language, this problem can be overcome. Like the previously discussed modular test bench generation, all the different simulations needed e.g., dc, ac or transient, are set up individually. A master simulation file then coordinates the execution of all simulations as well as providing shared simulation settings such as model libraries and standard simulator settings.

D. Parameter Extraction

The final step before the model generation is the extraction of suitable parameters. For this purpose, SKILL's extensive logging capabilities and specialized OCEAN functions for simulation data processing are utilized [4], [7]. Also, MATLAB's vast data processing capabilities can be exploited due to native integration with Cadence. The obtained parameter extraction results are then written to a dedicated database which the verification engineer can access. Hence, the verification engineer only has to source the model parameters from the database. The model generation itself is then entirely independent of the parameter extraction process and can utilize the verification

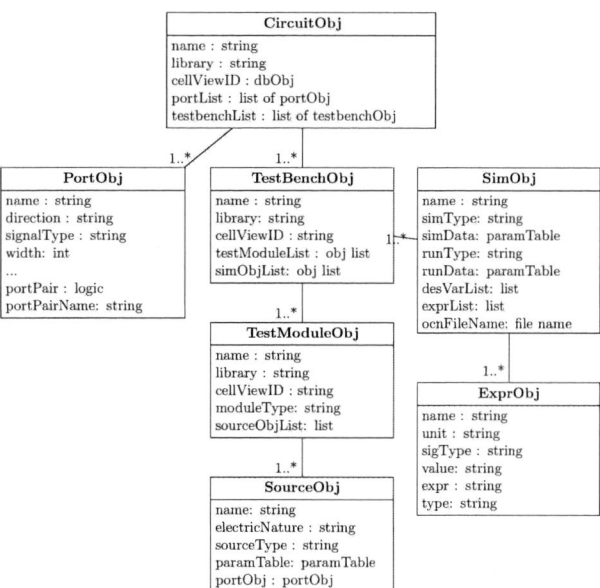

Fig. 4. Simplified UML diagram of the software prototype.

engineer's preferred verification method whether it be System Verilog, C++, etc. [1], [4], [5], [6].

III. IMPLEMENTATION

A. Code Structure

Fig. 4 presents a simplified unified modeling language (UML) class diagram of the implementation, omitting details that do not aid in understanding its structure. Here, an object-oriented programming paradigm was chosen for the implementation. Its features, namely easy manipulation of an object's attributes, the generation of multiple objects with the same properties, and the definition of unique methods for different classes, are decisive in regard to the creation of numerous test benches, simulation setups, and parameter extraction.

Each DUT is abstracted as a circuit object (circuitObj) which comprises all required test bench, simulation, and parameter extraction settings. Due to the object-oriented nature, all attributes of the circuitObj are individually and easily addressable.

As mentioned in Section II-A, initially, all port data of the DUT must be gathered to correctly assign all ports in the test bench to the corresponding test modules. This data includes the port name, its direction, signal type, and width. It also comprises any affiliation with other ports, such as the differential input pair of an operational amplifier visualized in Fig. 2. Each port is mapped to a portObj, where all its related properties are stored. All portObjs of the DUT are then stored in a list which is accessible to the circuitObj.

Accurate modeling of the DUT may require many different parameters to be extracted, each in turn requiring different test benches and simulations. Therefore, each test bench is implemented as a test bench object (testbenchObj), forming a

list of testbenchObjs, which, similarly to the list of portObjs, is accessible to the circuitObj.

A testbenchObj contains all test modules that are needed for proper simulation of the DUT. The test modules are entities themselves (testModuleObj) and hold all ports assigned to them, database handles (cellViewID) and the parameters for the different types of signal sources needed.

More precisely, each signal source is treated as an individual signal source object (sourceObj). Due to the vast differences between the different types of signal sources several sub-classes of sourceObj exist. Table II depicts these sub-classes and their properties in more detail. Common properties such as electric nature (current or voltage signal) or the corresponding port are inherited from the sourceObj class.

Besides a list of testModuleObjs, each testbenchObj also incorporates all data required for simulation control. Each simulation type such as dc, ac, or transient is regarded as an individual simulation object (simObj). A simObj stores the value of sweep variables in case a parametric sweep is required as well as all expressions that are needed for proper parameter extraction using that specific simulation.

The required expressions are abstracted and initialized as expression objects (exprObj) that hold the name, unit, signal type, and value of the dedicated expression, along with its respective evaluation result. An exprObj also references the underlying OCEAN or MATLAB expression for the post-processing.

B. Showcase

The first version of a software tool capable of offering all previously discussed functionalities has been implemented and tested. The tool itself is implemented using SKILL and can be launched directly from Cadence Virtuoso's Command Interpreter Window. Hence, no programming interface or temporary storage of data is required to bridge the gap between Virtuoso and a proprietary third-party tool or programming language. However, due to the implementation's flexibility, third-party software can be used for post-processing tasks if desired.

Extracting the port information of a given DUT can be done in a semi-automatic fashion as shown in Fig. 5. If the circuit designer sticks to certain naming conventions, this process can be significantly facilitated as the first-time-right guessing of port properties becomes possible and no user input is required.

TABLE II
THE DIFFERENT SOURCEOBJ CLASSES AND THEIR ATTRIBUTES.

Class:	Class Attributes:
DC	amplitude
AC	amplitude, phase
SINE	period, dc-level, amplitude, phase
PULSE	period, amplitude, delay, rise/fall time, waveform shape
XF	amplitude, phase
LOGIC	high level, low level, value, encoding (hex, bin, decimal)

Port Specification

Specify the Port Information for cellView: `MEKOWA_LF_OP1`

Pin Name	Direction	Signal Type	Width	Port Pair	Port Pair Name	Test Module Affiliation
inp	input	supply	1	diff_p	inm	INPUT
pon	input	supply	1	se	nil	BIAS
AVDO	inputOutput	supply	1	se	nil	SUPPLY
outp	output	signal	1	diff_p	outm	OUTPUT
AVSS	inputOutput	ground	1	se	nil	SUPPLY
outm	output	signal	1	diff_n	outp	OUTPUT
iref_5u	input	supply	1	se	nil	BIAS
vcm	input	supply	1	se	nil	BIAS
inm	input	supply	1	diff_n	inp	INPUT

Fig. 5. Port Specification with automatic recognition of some port parameters.

```
'include "constants.vams"
'include "discilpines.vams"
module OP_OL_TB_Bias( vcm, I_BIAS_5u, ...)
// Port and Bus Declaration
output vcm;
output I_BIAS_5u;
...
electrical vcm;
electrical I_BIAS_5u;
// Parameter Definition
real parameter dc_value_vcm = 600m;
real parameter dc_value_I_BIAS_5u = 5u;
...
// Analog Behavior:
analog begin
V(vcm) <+ dc_value_vcm;
I(I_BIAS_5u) <+ dc_vlaue_I_BIAS_5u;
...
end
endmodule
```

Fig. 6. Automatically generated VerilogA Test Module Code Sample.

Although automation is the tool's primary target, flexibility for the user is maintained by allowing various settings to be set manually. While all source, tuning words and load settings are set automatically, user input determines the operating conditions of the test benches. An automatically generated bias test module's VerilogA code is shown as an exemplary snippet in Listing 6. With respect to the VerilogA conventions, the module's ports are declared first, all parameters are initialized, and the analog behavior is specified for each port.

Focusing on the simulation setup, manual input by the user is inevitable as can be seen in Fig. 7. Although all simulations, parametric sweeps, and expressions can be automatically initialized, specific parameters, such as ac frequency range or parametric sweep values, have to be manually specified. Extraction of known parameters can be bypassed by manual entry if known beforehand, potentially cutting down on simulation time. Upon confirmation, OCEAN files are generated, executed, and the model parameters are extracted.

IV. CONCLUSION

An innovative approach to automatic test bench generation, simulation run, and parameter extraction has been presented, offering a faster and more reliable generation of verification models. Depending on the DUT, suitable test benches are automatically generated using a modular approach, corresponding

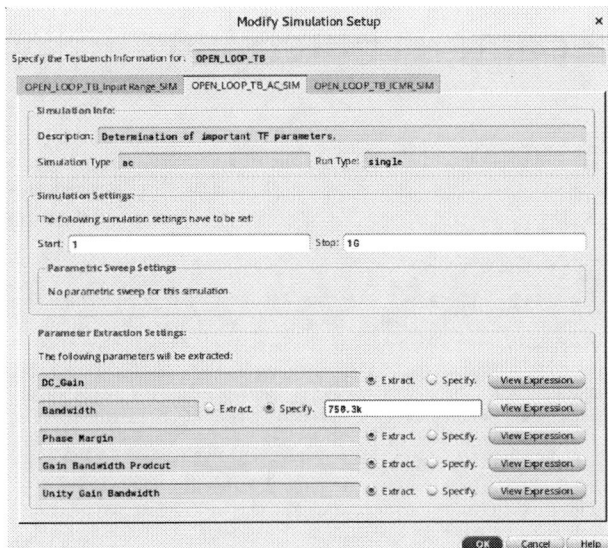

Fig. 7. Simulation Setup and Parameter Extraction Overview.

simulations are set up, and required model parameters are extracted.

First results of the implemented software tool, including a simplified UML class diagram, have been shown. Port specification, test bench, and simulation set-up were implemented and tested for non-hierarchical circuits. Further development will focus on improving automatic test bench generation, simulation run, and parameter extraction for hierarchical circuitry, where multiple sub-blocks have to be modeled individually.

ACKNOWLEDGMENT

The authors acknowledge the financial support of the German Federal Ministry of Education and Research (FKZ 16ES08353).

REFERENCES

[1] F. Speicher et al., "Advanced Modeling Methodology for expedient RF SoC Verification and Performance Estimation", 2018 15th International Conference on Synthesis, Modeling, Analysis and Simulation Methods and Applications to Circuit Design (SMACD), Prague, Czech Republic, 2018

[2] J. E. Chen, "A Modeling Methodology for Verifying Functionality of a Wireless Chip", 2009 IEEE Behavioral Modeling and Simulation Workshop, San Jose, CA, USA, 2009

[3] Y.-N. Yun et al., "Beyond UVM for practical SoC verificiation", 2011 International SoC Design Conference, Jeju, Korea (South), 2011

[4] M. Schleyer et al. "Automated Parameter Extraction Framework Functional Verification of Mixed-Signal SoCs", 8th Conference on Ph.D. Research in Microelectronics & Electronics (PRIME), Aachen, Germany, 2012

[5] Y. Wang et al., "Hierarchical generation of pin accurate SystemC models based on RF circuit schematics", 2010 IEEE International Behavioral Modeling and Simulation Workshop, San Jose, CA, USA, 2010

[6] C. Beyerstedt et al., "A Fast and Accurate True Event-driven Phase Locked Loop Model," 2020 27th IEEE International Conference on Electronics, Circuits and Systems (ICECS), Glasgow, UK, 2020

[7] T. J. Barnes, "SKILL: A CAD System Extension Language," 27th ACM/IEEE Design Automation Conference, Orlando, FL, USA, 1990

Monitoring Analog Circuit Performance using Adaptive Filters and RSM-based Behavioral Models

Maike Taddiken, Steffen Paul, Dagmar Peters-Drolshagen
Institute of Electrodynamics and Microelectronics (ITEM.me)
University of Bremen, Germany, +49(0)421/218-62551
{taddiken, steffen.paul, peters}@me.uni-bremen.de

Abstract—This paper presents a concept for monitoring the performance of an analog circuit by using adaptive filters in combination with behavioral models. Digital calibration techniques based on adaptive filters are applied to monitor the circuit performance. The reference signal representing the ideal error-free circuit performance required by the filter is provided by a behavioral model based on Response Surface Modeling (RSM). Comparing the output signal of the circuit to the ideal reference signal results in a number of filter coefficients that are used to estimate the circuit performance. The shift of the filter coefficients over time is used to monitor the degradation of performance caused by aging. The monitoring approach is analyzed on a simulative basis with an amplifier in a 28nm process.

Index Terms—monitoring, degradation, adaptive filter, behavioral model

I. INTRODUCTION

Modern analog integrated circuits are used in many safety-critical applications which require the circuits to operate within given specifications throughout their lifetime. However, the behavior and functionality of these circuits is influenced by various effects such as temperature and voltage fluctuations as well as aging which become more pronounced in small technology nodes. In oder to guarantee the correct functionality of a circuit or system, a method for determining the current state of performance and analyzing whether the circuit is degrading is needed.

Some work has already been published concerning aging or health monitoring of circuits. The focus is mainly on the degradation of digital circuits, where the comparison between stressed and unstressed ring oscillator behavior [1] or monitoring of critical paths is used [2]. However, these approaches allow only an indirect conclusion about the state of the circuit and do not allow to monitor the actual performance. Additionally, the test structures may receive different workloads than the actual circuit, thus the degradation also differs [3]. For analog circuits, sensors observing the threshold voltage shift caused by aging [4] or dedicated monitoring of the NBTI effect [5] have been proposed. These monitors however are very specialized and do not always allow a direct conclusion about the shift of the circuits performance values.

In [6], [7], digital calibration techniques based on the application of digital adaptive filters are used for on-line monitoring and correction of the error caused in the analog part in a sensor interface circuit. This enables the correction of any occurring linear errors independent of their origin in the digital domain, which reduces the need for compensation and high design requirements in the analog domain.

Based on these techniques, a first approach of a monitoring concept is developed in this work that uses the filter coefficients of the adaptive filters to determine the current performances of a dedicated analog circuit. The adaptive filter requires a reference signal that represents the error-free behavior of the circuit under test.

Response Surface Modeling (RSM) is widely used to model the behavior of both analog and digital integrated circuits, especially to enable the analysis of process variation and/or aging [8], [9], [10]. In many cases these models are used to speed up simulation and enable the analysis of certain physical effects on higher levels of abstraction which then enables system level design and verification.

In this approach, the reference signal required by the filter is provided by a behavioral model which uses RSMs to describe the ideal circuit behavior. The deviation from this ideal performance behavior is represented by the resulting filter coefficients and enables a backward-estimation of the actual performance values that are monitored by the filter. Observing the shift of the filter coefficients over time also enables the detection of degradation.

In section II, reliability aspects and the application of adaptive filters for error detection in circuits is discussed. The monitoring approach is presented in section III. In section IV, results from the simulative analysis of an example amplifier circuit are discussed and a conclusion is given in section V.

II. BACKGROUND

A. Reliability

Influences like process variation, which occurs during the manufacturing of integrated circuits as well as changing operating conditions like temperature and voltage cause shifts in the behavior of transistors and thus alter the performance of a circuit in comparison to its nominal specifications. In addition, aging effects like Bias Temperature Instability (BTI) and Hot Carrier Injection (HCI) lead to a gradual change of transistor parameters and, subsequently, circuit performances over time. The degree of degradation heavily depends on the stress profile consisting of temperature, supply voltage and input signals [11]. Thus, the performances of a circuit fulfilling the specifications at time zero can shift so much that the functionality of the system is in jeopardy.

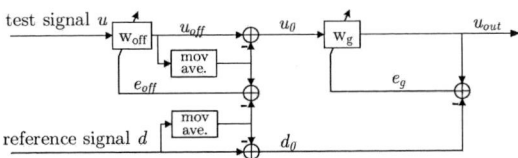

Fig. 1: Adaptive filter structure as proposed in [7].

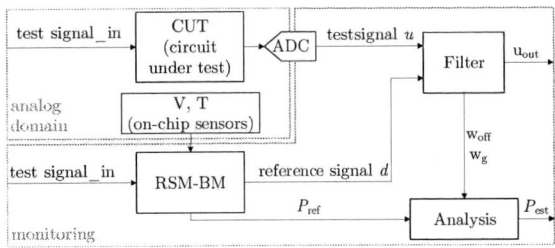

Fig. 2: Overview of the monitoring concept.

B. On-line error detection with adaptive filters

An on-line error detection and correction method based on digital calibration using adaptive filters with the least-mean-squares (LMS) algorithm has been presented in [6], [7] for the correction of a sensor interface signal. The adaptive filter requires a reference signal which represents the error-free ideal output signal of the circuit under test (CUT), whose origin is not stated in [7].

The structure of the adaptive filter used for the error detection is shown in fig. 1 and taken from [7]. In contrast to the work in [6], the error detection does not have to be done continuously because degradation due to aging is only recognizable on a larger timescale. It is assumed that a test cycle is carried out for the circuit at predefined times where a known test signal is used as an input signal, resulting in the output test signal u.

Two types of errors can be directly determined with the proposed filter: offset and gain errors. The error determination is divided into two steps: first the offset error of the DC output level is determined and corrected with an adaptive filter W_{off}. Afterwards, the gain error is analyzed with a second filter W_g. In this work, the order of the filters is set to $N = 0$, resulting in one filter coefficient for each filter.

As the offset error is additive, the corrected signal results in

$$u_{off}[n] = u[n] + \sum_{k=0}^{N} w_{off,k}[n] \stackrel{N=0}{=} u[n] + w_{off,0}. \quad (1)$$

LMS algorithm is used to determine the filter coefficients:

$$\mathbf{w}_{off}[n+1] = \mathbf{w}_{off}[n] + sign(\overline{u})\mu_{off}\mathbf{u}[n]e_{off}[n] \quad (2)$$

where μ_{off} is the step size, $u[n]$ is the current unfiltered input test signal and the error is given by $e_{off}[n] = \overline{u}[n] - \overline{d}[n]$, with $\overline{\square}$ being the moving averages of the test u and reference signal d.

In order to get an average-free signal for the determination of the gain error, the moving averages are subtracted from both the test and reference signal. The average-free test signal u_0 and reference signal d_0 are subsequently used to determine the filter coefficient using the LMS algorithm

$$\mathbf{w}_g[n+1] = \mathbf{w}_g[n] + \mu_g\mathbf{u}_0[n]e_g[n] \quad (3)$$

with $e_g[n] = u_{out}[n] - d_0[n]$

$$u_{out}[n] = \mathbf{w}_g[n]\mathbf{u}_0 = \sum_{k=0}^{N} w_{g,k}[n]u_0[n] \stackrel{N=0}{=} w_{g,0}[n]u_0[n] \quad (4)$$

III. MONITORING

In this work, the filter coefficients resulting from a comparable digital on-line correction are analyzed to monitor the performance and ultimately the degradation state of an analog circuit. An overview of the monitoring concept is shown in fig. 2. The error-free reference signal d is provided by a behavioral model of the CUT, which is based on the use of RSMs as performance models (RSM-BM). The model includes the key performances of the circuit that are monitored by the filter and their dependence on other operational parameters such as temperature and voltage. The behavioral model does not include the deviation of the performances due to aging. The values of the operational parameters are provided by additional on-chip sensors. Thus, the reference signal d represents the ideal response of the circuit for a given set of operating conditions at time zero.

Analyzing the resulting filter coefficients and the performance values P_{ref} calculated with the RSMs enables the estimation of the current performance values of the CUT at the time of the monitoring. Due to the chosen filter order of $N = 0$, there is only one filter coefficient for each filter. Therefore, the filter coefficients are directly related to the offset and gain error and can be used to estimate the circuit performance values P_{est} using

$$DC_{offset,est} = DC_{offset,ref} - w_{off} \quad (5)$$

$$gain_{est} = \frac{gain_{ref}}{w_g} \quad (6)$$

where $\square_{ref}$ are the performances calculated by the RSMs of the behavioral model.

Taking into account the shift over time of the filter coefficients, they can be used to determine the performance degradation of the circuit by comparing to the initial coefficients at time zero. Ideally, the reference and test signal should be equal at time zero. However, an error is to be expected caused by the modeling error and inaccuracies of the RSMs and the ideal behavioral model in comparison to the real CUT. Additionally, process variation is not captured by the model. This means that even at time zero, an initial model error will occur and result in initial filter coefficients that can be used to estimate the 'real' performance of the CUT at tim zero. Any deviation additional to this initial model error measured at a point later in time can then be attributed to aging.

© VDE VERLAG GMBH · Berlin · Offenbach

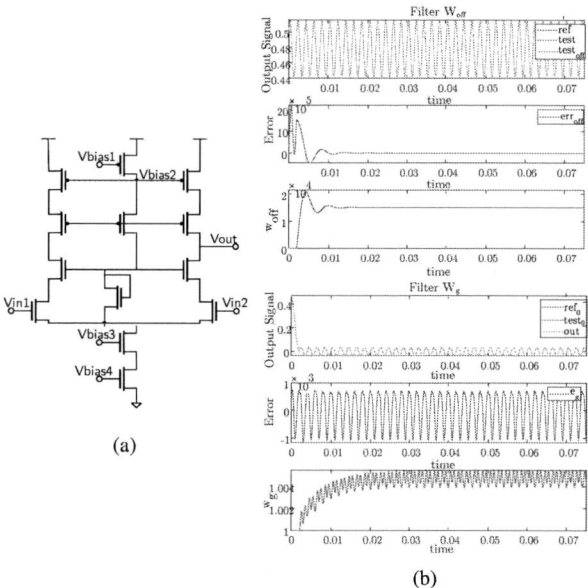

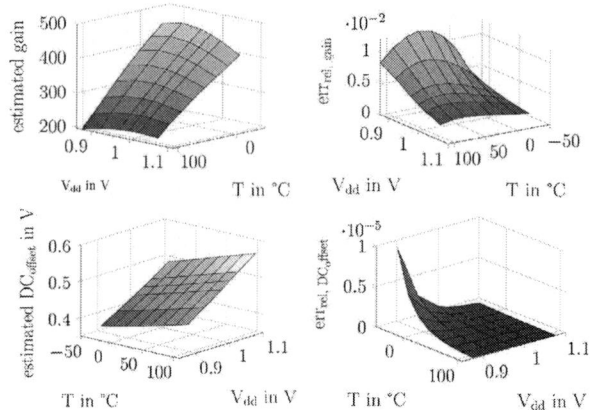

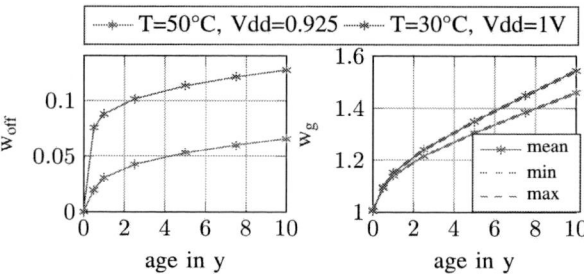

Fig. 4: Estimated gain and DC offset and relative error compared to spectre simulations at $age = 0$.

Fig. 3: (a) schematic of the cascode differential amplifier and (b) transient filter output for $T = 30°C$ and $V_{dd} = 1V$ at $age = 0$yrs.

Additionally to the performance monitoring, the filter coefficients can be used to correct the error in the output signal of the amplifier, as shown in [6].

IV. RESULTS

A cascode differential amplifier in open-loop configuration as shown in fig. 3(a) is used as a test circuit. The 28nm FDSOI process includes HCI and NBTI aging models provided by the manufacturer for Cadence® Spectre reliability simulator.

The behavior of the amplifier is analyzed for temperatures between -40°C and 120°C and the supply voltage V_{dd} ranging between 0.85V and 1.1V with a nominal voltage of $V_{dd} = 1V$ over a range of 10 years (315360000 sec).

Since the filters are able to correct offset and gain errors, these two performance values are included in the behavioral model. The RSMs for the linear gain and DC offset are constructed in dependence on the temperature T and supply voltage V_{dd} based on simulations. The behavioral model describes an ideal amplifier with linear gain and DC offset modeled by the RSMs. Frequency response and slew rates or delays are not modeled.

A sinusoidal signal at 500Hz is used as test signal u. The signals in the filter structure as well as the error signals and resulting filter coefficients are shown in fig. 3(b) at time zero $age = 0$yrs. The initial model error causes a small deviation between the test and reference signal which is corrected by the adaptive filter. It can be seen that in contrast to the offset error, the gain error is not static and a periodic ringing can be observed in the resulting filter coefficient w_g. This is due to a small time difference between the test and reference

Fig. 5: Shift of filter coefficients w_{off} and w_g over age.

signal because the ideal behavioral model does not include any delay or slew rate. To deal with this ringing, the mean of the coefficient w_g, $\overline{w_g}$, is calculated once the filter has ramped up to a steady state of periodical ringing and used for estimating the gain.

$$gain_{est} = \frac{gain_{RSM}}{\overline{w_g}} \qquad (7)$$

The estimation is analyzed at time zero $age = 0$ yrs at different conditions for T and V_{dd} that differ from the sample points that were used to create the RSMs for the behavioral model. The estimation error is calculated by comparing P_{est} to the performance values from spectre circuit simulations. The results are shown in fig. 4. It can be seen that the initial model error causes only a little error in the performance estimation at $age = 0$ yrs for the gain caused by the periodical ringing and subsequent usage of the mean for the estimation. The estimation error for the DC offset is negligible which is explained by the linearity of the error in the first place.

To monitor degradation, the shift of the filter coefficients over time is observed as shown in fig. 5. For w_g, the mean as well as the minimum and maximum is shown. It can be seen that the variation of the filter coefficient gets larger over age, which is due to the fact that aging causes an additional

SMACD / PRIME 2021 | 19 – 22 July 2021, Online Event

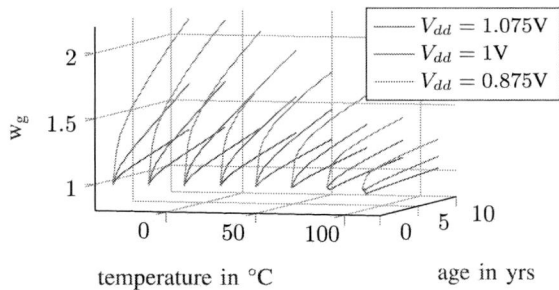

Fig. 6: Shift of filter coefficient $\overline{w_g}$ over age at different temperatures and V_{dd}.

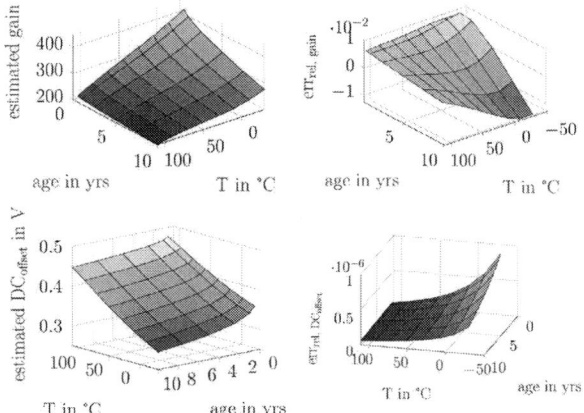

Fig. 7: Estimation af aged performances over temperature at $V_{dd} = 1$V and relative error.

non-linearity in the output signal due to the different effects on PMOS and NMOS transistors.

The shift of filter coefficient $\overline{w_g}$ over age at different temperatures and V_{dd} is shown in fig. 6. It can be seen that the shift due to degradation is usually a lot greater than the initial model error. This enables the analysis of the degradation, provided that the sensor measurements for temperature and voltage are accurate. It is also apparent that the influence of aging is greater for smaller temperatures where the initial gain was higher compared to high temperatures where the gain was initially lower.

This also highlights why the dependence on the operating conditions has to be included in the behavioral model: to differentiate between different causes for the shift in filter coefficients a point of reference has to be known. If only the nominal performances were included, it would not be possible to conclude whether a shift in performance and thus in the filter coefficients is caused by aging or a change in operating conditions.

The resulting estimation of the aged performances for the gain and DC offset is shown in fig. 7. Again, the relative error compared to actually simulated performances is small, showing that a good estimation of the circuit performances is possible.

V. Conclusion

In this paper an approach for monitoring the performance of analog circuits has been presented. It exploits the capabilities of digital calibration and on-line error detection using adaptive digital filters and combines it with the usage of RSMs for behavioral modeling to estimate the performance values of an analog circuit at the time of monitoring. The estimation results in a very good accuracy in a simulative validation of the approach. By observing the shift of the filter coefficients over time, the degradation of the circuit and its performances can be monitored. Further work is needed to evaluate how process variation can be considered and to determine the impact of inaccuracies of the sensors providing the input data for the behavioral model. Also, the application of more sophisticated behavioral models and test signals should be analyzed to include additional performance values such as frequency behavior as well as the possibility to estimate the remaining lifetime based on the shift of the filter coefficients and the complexity of the digital implementation.

Acknowledgment

This work is supported by the German research Foundation, DFG, in the research project LEMON (project number 354944200).

References

[1] K. K. Kim, W. Wang, and K. Choi, "On-chip aging sensor circuits for reliable nanometer MOSFET digital circuits," *IEEE Transactions on Circuits and Systems II: Express Briefs*, vol. 57, no. 10, pp. 798–802, 2010.

[2] A. F. Gomez and V. Champac, "Early Selection of Critical Paths for Reliable NBTI Aging-Delay Monitoring," *IEEE Transactions on Very Large Scale Integration (VLSI) Systems*, vol. 24, no. 7, pp. 2438–2448, 2016.

[3] M. Agarwal, B. C. Paul, M. Zhang, and S. Mitra, "Circuit Failure Prediction and Its Application to Transistor Aging," in *25th IEEE VLSI Test Symposium (VTS'07)*, 2007, pp. 277–286.

[4] B. L. Ji, D. J. Pearson, I. Lauer, F. Stellari, D. J. Frank, L. Chang, and M. B. Ketchen, "Operational Amplifier Based Test Structure for Quantifying Transistor Threshold Voltage Variation," *IEEE Transactions on Semiconductor Manufacturing*, vol. 22, no. 1, pp. 51–58, 2009.

[5] M. B. Yelten, P. D. Franzon, and M. B. Steer, "Analog Negative-Bias-Temperature-Instability Monitoring Circuit," *IEEE Transactions on Device and Materials Reliability*, vol. 12, no. 1, pp. 177–179, 2012.

[6] S. Heinssen, T. Hillebrand, M. Taddiken, S. Paul, and D. Peters-Drolshagen, "On-line monitoring and error correction in sensor interface circuits using digital calibration techniques," in *2018 IEEE 36th VLSI Test Symposium (VTS)*. IEEE, 2018, pp. 1–6.

[7] ——, "On-Line Error Correction in Sensor Interface Circuits by Using Adaptive Filtering and Digital Calibration," in *Multidisciplinary Digital Publishing Institute Proceedings*, vol. 2, no. 13, 2018, p. 963.

[8] W. Daems, G. Gielen, and W. Sansen, "Simulation-based generation of posynomial performance models for the sizing of analog integrated circuits," *IEEE Transactions on Computer-Aided Design of Integrated Circuits and Systems*, vol. 22, no. 5, pp. 517–534, 2003.

[9] E. Maricau and G. Gielen, "Efficient variability-aware NBTI and hot carrier circuit reliability analysis," *IEEE Transactions on Computer-Aided Design of Integrated Circuits and Systems*, vol. 29, no. 12, pp. 1884–1893, 2010.

[10] M. Taddiken, S. Paul, and D. Peters-Drolshagen, "ReSeMBleD-Methods for Response Surface Model Behavioral Description," in *2018 15th International Conference on Synthesis, Modeling, Analysis and Simulation Methods and Applications to Circuit Design (SMACD)*. IEEE, 2018, pp. 157–160.

[11] G. G. Elie Mariceau, *Analog IC Reliability in Nanometer CMOS*. Springer, 2013.

© VDE VERLAG GMBH · Berlin · Offenbach

A Compact Model for Scalable MTJ Simulation

Fernando García-Redondo
Arm Ltd, Cambridge, UK
fernando.garciaredondo@arm.com

Pranay Prabhat
Arm Ltd, Cambridge, UK
pranay.prabhat@arm.com

Mudit Bhargava
Arm Inc, Austin, USA
mudit.bhargava@arm.com

Cyrille Dray
Arm Ltd, La Paros, France
cyrille.dray@arm.com

Abstract—This paper presents a physics-based modeling framework for the analysis and transient simulation of circuits containing Spin-Transfer Torque (STT) Magnetic Tunnel Junction (MTJ) devices. The framework provides the tools to analyze the stochastic behavior of MTJs and to generate `Verilog-A` compact models for their simulation in large VLSI designs, addressing the need for an industry-ready model accounting for real-world reliability and scalability requirements. Device dynamics are described by the Landau-Lifshitz-Gilbert-Slonczewsky (s-LLGS) stochastic magnetization considering Voltage-Controlled Magnetic Anisotropy (VCMA) and the non-negligible statistical effects caused by thermal noise. Model behavior is validated against the `OOMMF` magnetic simulator and its performance is characterized on a 1-Mb 28 nm Magnetoresistive-RAM (MRAM) memory product.

Index Terms—STT-MRAM, MTJ, Compact modeling, s-LLGS

I. INTRODUCTION

Recent advances in MTJ devices [1]–[4] open the path to the next generation of system architectures, from embedded battery-less edge devices where off-chip Flash will be replaced by embedded MRAM, to future power-efficient caches in HPC systems. STT-MRAM is being actively developed by foundries and integrated into 28 nm generation CMOS Process Design Kits (PDKs) [4].

The complex multi-layered MTJ structures have heavily non-linear magnetic and electric behaviors that depend on the device structure, the thermal and external magnetic environment, and the applied electrical stress (Figure 1). On top of this complex system of equations, the evolution of MTJ magnetization m shows a stochastic behavior caused by the random magnetic field induced by thermal fluctuations.

Many MTJ models have been presented, from micro-magnetic [5], [6] approaches to macro-magnetic `SPICE` and `Verilog-A` compact models for circuit simulations [1]–[3], [7]. Compact models present in the literature account for different behavioral aspects: a better temperature dependence [1], a more accurate anisotropy formulation for particular MTJ structures [1]–[3], and the integration of Spin-Orbit Torque (SOT) for three-terminal devices [2], [3]. These prior works focus on the ability to model a specific new behavior for the simulation of a single or a few MTJ devices in a small circuit. In these approaches, the simulation of device stochasticity imposes a high computational load on the simulation, complicating the simulation of larger VLSI circuits. To address the need to simulate thousands of MTJ devices in a single memory macro, circuit designers may have to oversimplify

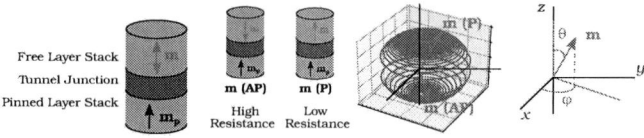

Fig. 1. Basic MTJ structure, magnetization vector switching trajectory [10] from Parallel (P) to Anti-Parallel (AP) states, and reference coordinate system.

MTJ switching to a basic interpolated behaviour, and design for worst-case fixed voltage and current targets. This hinders the design of optimized circuits.

Therefore, there is a clear need for an accurate and modular compact model accounting for the magnetization dynamics of MTJ devices yet scalable enough for the simulation of large MRAM memory macros. In this paper we present three key contributions: Section II presents a study of reliability issues in MTJ model simulation. The proposed modeling framework is presented in Section III, including a modular multi-threaded s-LLGS solver validated against `OOMMF` [5], an efficient `Verilog-A` compact model for MRAM macro simulation and a solution to model thermal noise effects. Finally, in Section IV we present a case study with a 1-Mb 28 *nm* MRAM macro.

II. MTJ COMPUTATIONAL AND CONVERGENCE ISSUES

From early behavioral prototypes to mature, product-ready PDKs, the simulation of emerging devices is a process that requires not only research contributions capturing novel device behavior [1]–[3], [7], but also optimization stages enabling circuit design [8]. For STT-MRAM, circuit design needs the complex device dynamics to be incorporated into the standard `SPICE` and `Cadence® Spectre` [9] solvers. In this section we address the challenges of reliable and efficient simulation of MTJ devices. All results are generated using `Spectre` 17.1.0.583 with default settings unless noted otherwise.

The s-LLGS system resolution can easily lead to artificial oscillations if solved in a Cartesian reference system [1] when using the Euler, Gear and Trapezoidal methods provided by circuit solvers. In [10] it is studied how MTJ devices benefit from resolution in spherical coordinates, specially when using implicit midpoint and explicit RK45 methods. Additionally, Cartesian methods require [1] a non-physical normalization stage for the magnetization vector [10]. Some further issues arising from solving in a spherical reference system are detailed below.

© VDE VERLAG GMBH · Berlin · Offenbach

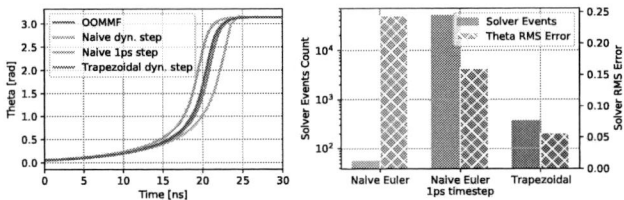

Fig. 2. MTJ simulation comparing a naive Euler integration against `Spectre` trapezoidal solver. The naive method leads to larger RMSE (vs. the `OOMMF` reference) even under $1ps$ timestep with $100\times$ computational load.

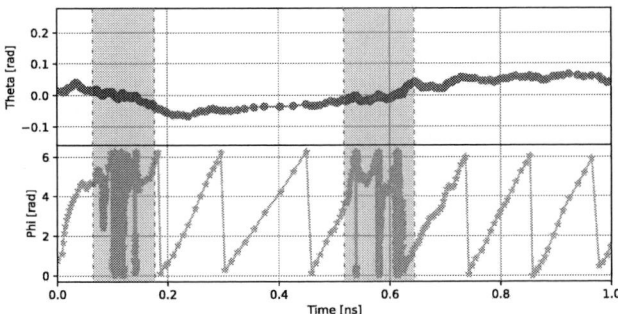

Fig. 3. MTJ s-LLGS simulation under no external current/field, considering thermal $\boldsymbol{H_{th}}$ effects. The evolution of $\boldsymbol{m_\phi}$ and $\boldsymbol{m_\theta}$ over time highlights the asymptote on $\frac{d}{dt}\boldsymbol{m_\phi}$ at $\theta = 0$. The trapezoidal solver under relaxed tolerance successfully provides the desired solution while using a dynamic time step.

A. $\boldsymbol{m_\phi}$ Wrapping

Prior work on MTJ `Verilog-A` models uses a naive Euler method to integrate the magnetization vector [1], [3]

$$\boldsymbol{m}(t) = \boldsymbol{m}(t - dt) + \boldsymbol{dm}dt. \qquad (1)$$

However, Equation 1 causes $\boldsymbol{m_\phi}$ to exceed the $[0, 2\pi)$ physically allowed range, and grow indefinitely over time, leading to unnatural voltages that eventually prevent simulator convergence. Our solution is to use the circular integration `idtmod` function provided by `Spectre` [8], [9] which wraps $\boldsymbol{m_\phi}$ at 2π and prevents it from increasing indefinitely.

B. $\boldsymbol{m_\theta}$ Accuracy and Minimum Timestep

Vector $\boldsymbol{m_\theta}$ represents the binary state of the MTJ data bit and is the most critical component to accurately model MTJ switching. The naive integration of $\boldsymbol{m_\theta}$ directly encoding Equation 1 in the `Verilog-A` description incurs a substantial error during a switching event, as shown in Figure 2. Reducing the timestep improves the RMS error but at the cost of a large increase in computational load. Our solution is to use the `idt` function for the circuit solvers provided by `Spectre` [9], which adapt the integration based on multiple evaluations of the differential terms at different timesteps, leading to better accuracy. Figure 2 describes the accuracy RMSE when using a naive Euler method and the `idt` trapezoidal solver, when compared against `OOMMF` reference. The graph highlights how even with a smaller timestep involving $100\times$ more computed events, the naive integration leads to larger errors.

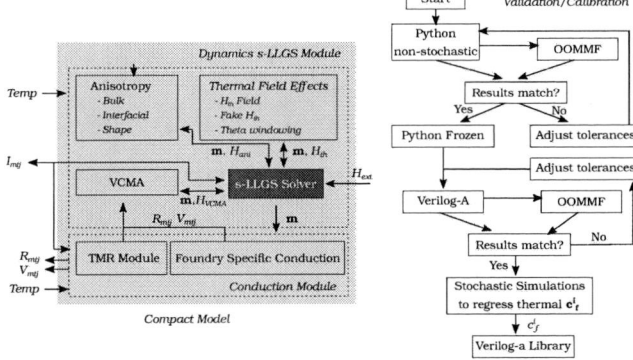

Fig. 4. Proposed MTJ modeling framework. The *Conduction* and *Dynamics* modules determine the device conduction and the magnetization based on the external, device anisotropy, STT and thermal induced fields. The `Python` and `Verilog-A` models validate the MTJ behavior against `OOMMFC`.

C. m_ϕ acceleration

If a larger circuit is to be simulated and fixed timesteps banned, to avoid instabilities it is essential to manage the tolerance/timestep scheme [10], especially during the computation of the $\boldsymbol{m_\phi}$ component. The asymptote $\frac{d}{dt}\boldsymbol{m_\phi} \to \infty$ at $\boldsymbol{m_\theta} = 0$ accelerates the precession mechanism [10]. As described by Figure 3, this requires the solver to accordingly increase its resolution. A bounded or fixed time step would fail either to provide sufficient resolution or sufficient performance.

III. PROPOSED FRAMEWORK

In this section we describe the implementation and validation of the proposed MTJ compact model addressing the challenges described in Section II, emphasizing its *dynamics module* computing the MTJ magnetization and proposing solutions to incorporate thermal noise effects.

A. Compact Model Structure

Figure 4 describes the implemented compact model, composed of two modules: the *Conduction* and *Dynamics* modules. The conduction scheme describing the instantaneous MTJ resistance is dependent on the foundry engineered stack. Our modular approach allows foundry-specific conduction mechanisms to complement the basic Tunnel-Magneto-Resistance (TMR) scheme [1]–[3], [11]. The *Dynamics* module describes the temporal evolution of the MTJ magnetization m as a monodomain nanomagnet influenced by external and anisotropy fields, thermal noise and STT [5], [10], described by the s-LLGS equations [5] in I.S.U.

$$\frac{d\boldsymbol{m}}{dt} = -\gamma'\boldsymbol{m} \times \boldsymbol{H_{eff}} + \alpha\gamma'\boldsymbol{m} \times \frac{d\boldsymbol{m}}{dt}$$
$$+\gamma'\beta\epsilon(\boldsymbol{m} \times \boldsymbol{m_p} \times \boldsymbol{m}) - \gamma'\beta\epsilon'(\boldsymbol{m} \times \boldsymbol{m_p})$$

$$\beta = |\frac{\hbar}{\mu_0 e}|\frac{I}{VM_s}, \epsilon = \frac{P\Lambda^2}{(\Lambda^2 + 1) + (\Lambda^2 - 1)(\boldsymbol{m} \cdot \boldsymbol{m_p})} \qquad (2)$$

where $alpha$ and γ are the Gilbert damping factor and gyromagnetic ratio respectively, related by $\gamma' = \frac{\gamma\mu_0}{1+\alpha^2}$, P is

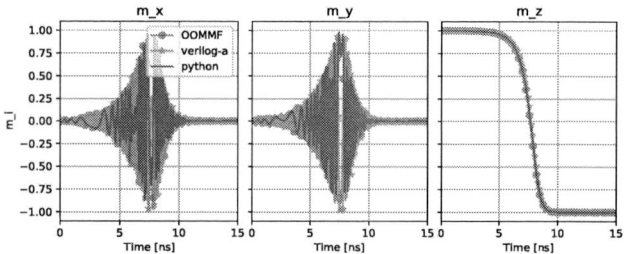

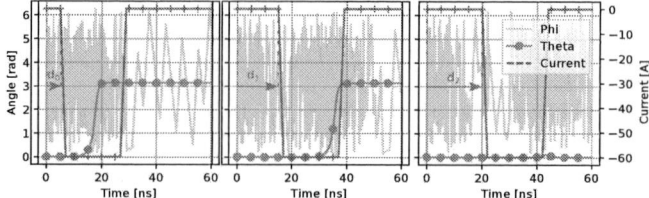

Fig. 5. Validation against OOMMF Framework. A cylindrical MTJ with 50 nm diameter, 1 nm thickness, $K_i = 1e-3\ Jm^{-2}$, $P = 0.75$, $\alpha = 0.01$, $\gamma = 1.76e11\ rad/s/T$, $M_s = 1.2\ Am^{-1}$ switches by a current of 35uA.

Fig. 6. Damping effect on m_θ under H_{th} field absence. The absence of STT-current (during the d_2 delay) collapses m_θ preventing MTJ switching.

the polarization factor, M_s is the magnetization saturation, I is the current flowing through the MTJ volume V, m_p is the pinned-layer unitary polarization direction and Λ and ϵ' set the primary and secondary spin transfer terms respectively. The effective magnetic field for a Perpendicular Magnetic Anisotropy (PMA) is defined by the anisotropy field, the external field and the thermal induced field $H_{eff} = H_{ani} + H_{ext} + H_{th}$. The anisotropy field is composed of the PMA uniaxial term, the shape anisotropy demagnetization field, and the VCMA field [1], [3], becoming the H_{eff} vector

$$
\begin{aligned}
H_{eff} =\ & H_{ext} + H_{uni} + H_{demag} - H_{VCMA} + H_{th} \\
=\ & H_{ext} + \frac{2K_i}{t_{fl}\mu_0 M_s} m_z - M_s N \cdot m \\
& - \frac{2\xi I R_{mtj}}{t_{fl} t_{ox} \mu_0 M_s} m_z z + \mathcal{N}(0,1)\sqrt{\frac{2K_B T\alpha}{\gamma' M_s V \Delta_t}}, \quad (3)
\end{aligned}
$$

where K_i is the interfacial energy constant, t_{fl} and t_{ox} are the free layer and oxide thicknesses, N is the shape anisotropy demagnetization factor, ξ the VCMA coefficient, K_B the Boltzmann constant and $\mathcal{N}(0,1)$ a Gaussian random vector with components in x, y, z meeting [10], [12] conditions.

The compact model has been implemented in `Python` by adapting the `Scipy solve_ivp` [13] solvers to support the H_{th} simulation as a pure Wiener process. The parallel `Python` engine enables MC and statistical studies. The `Verilog-A` implementation uses `idt /idtmod` integration schemes with parameterizable integration tolerances.

B. Validation Against OOMMF

The compact model is validated against the NIST OOMMF MicroMagnetic Framework [5], testing our results against its OOMMFC interface [6]. Following the method from [10], we use `SpinXferEvolve` to simulate a single magnet under an induced spin current. Figure 5 shows that `Python` and `Verilog-A` implementations compare well with OOMMF with adaptive tolerance driven computations.

C. Thermal noise and MTJ stochasticity

The random magnetic field H_{th} caused by thermal fluctuations induces stochastic MTJ behaviour resulting in a non-zero Write Error Rate (WER) upon switching events. H_{th} follows a Wiener process [10], [12], in which each x, y, z random component is independent of each other and previous

states. This implies that the computation of H_{th} requires large *independent* variations between steps, hindering the solver's attempts to guarantee signal continuity under small tolerances. The scenario shown in Figure 3 leads to computational errors under default solver tolerances, and excessive computational load under 1 ps bounded time steps.

Three solutions have been proposed in the literature. First, to emulate the random field by using an external current or resistor-like noise source [7]. However, SPICE simulators impose a minimum capacitance on these nodes filtering the randomness response, therefore preventing a true Wiener process from being simulated. Second, to bound a fixed small timestep to the solver [1], [2], but as described before this is not feasible for large circuits. Third, to only consider scenarios where the field generated by the writing current is much larger than H_{th}, forcing $H_{th} = 0$ [1], [3]. This has strong implications when moving from single to multiple successive switching event simulations. Under no H_{th} thermal field, the magnetization damping collapses m_θ to either 0 or π. s-LLGS dynamics imply that the smaller the m_θ the harder it is for the cell to switch, and if completely collapsed, it is impossible. This artificial effect, depicted in Figure 6, does not have an equivalent in reality, as $H_{th} \neq 0$ imposes a random angle. Design, validation and signoff for large memory blocks with integrated periphery and control circuits requires the simulation of sequences of read and write operations, with each operation showing identical predictable and switching characteristics. However, the damping artifact discussed above prevents or slows down subsequent switches after the first event, since the subsequent events see an initial θ value vanishingly close to zero. Two solutions are proposed below.

1) Windowing Function Approach: Our first objective is to provide a mechanism for m_θ to saturate naturally to the equilibrium value given by H_{th} during switching events. By redefining the evolution of m on its θ component we are able to saturate its angle at θ_0, the second moment of the Maxwell-Boltzmann distribution of m_θ under no external field [1]–[3], [7]. The new derivative function $\frac{d}{dt}m'_\theta$ uses a Tukey window with the form $\frac{d}{dt}m'_\theta = w_{tukey}(m_\theta, \theta_0)\frac{d}{dt}m_\theta$ where

$$
w(m_\theta) = \begin{cases} 0 & \text{if} \quad m_\theta < \theta'_0 \\ 0.5 - \frac{cos}{2}\left(\frac{4\pi(m_\theta - \theta'_0)}{\theta'_0}\right) & \text{if} \quad m_\theta - \theta'_0 < \frac{\theta'_0}{4} \\ 1 & \text{otherwise} \end{cases}
$$

(4)

SMACD / PRIME 2021 | 19 – 22 July 2021, Online Event

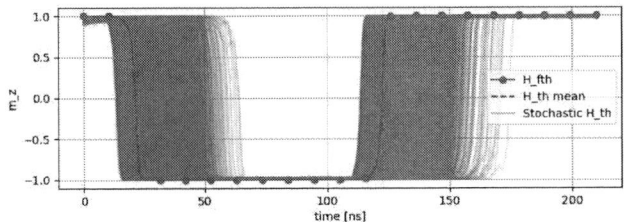

Fig. 7. 10^5 stochastic simulations for the calibration of H_{fth}.

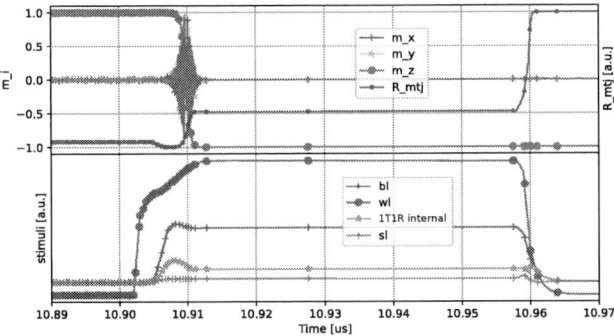

Fig. 8. Memory macro simulation showing MTJ switching.

defined for $m_\theta \in [0, \frac{\pi}{2}]$ and defined symmetrically ($w'(m_\theta) = w(\pi - m_\theta)$) for $m_\theta \in [\frac{\pi}{2}, \pi]$. This function slows down $\frac{d}{dt} m'_\theta$ when reaching the angle $\theta'_0 = c_w \theta_0$, therefore saturating m_θ and avoiding m from collapsing over z. Moreover, by using c_w^{mean} we are able to define the angle that statistically follows the mean stochastic MTJ behavior. Similarly, by simply using the set of parameters $[c_w^{worst}, c_w^{best}, c_w^{WER_0}, ..., c_w^{WER_i}]$ we can simulate the worst, best, WER_i behaviors, analyzing how a given circuit instantiating that MTJ device would behave statistically with negligible simulation performance degradation.

2) Emulated Magnetic Term: The windowing function approach prevents artificial m_θ saturation at the end of switching events, but still does not capture the mean effect of H_{th} during switching or under low-current excitation, such as during read events. Our next objective is to address this without performing a large ensemble of random transient simulations. The expansion of Equation 3 after its expression in spherical coordinates describes m_θ evolution as proportional to $H_{eff\phi} + \alpha H_{eff\theta}$, leaving $\frac{d}{dt} m_\theta \simeq \frac{\gamma'}{1+\alpha^2} H_{eff\phi}$. We propose to add a fictitious $H_{fth\phi}$ term H_{eff} with the purpose of emulating the mean/best/worst statistical H_{th} contribution that generates θ_0 [1]–[3], [7]. By defining

$$H_{fth} = c_f \sqrt{\frac{2K_B T \alpha}{\gamma' M_s V \Delta_t}} \phi \qquad (5)$$

we are able to efficiently model – by simply using $[c_f^{worst}, c_f^{best}, c_f^{WER_0}, ..., c_f^{WER_i}]$ set of parameters – the statistical behavior caused by the thermal effects while avoiding their inherent simulation disadvantages.

The process of calibrating a particular compact model instance is described in Figure 4, and ensures the validation of `Python` and `Verilog-A` instances against OOMMF, before the regression of c_f^i or c_w^i coefficients takes place with a one-off stochastic simulation step as shown in Figure 7.

IV. 1-Mb MRAM Macro Benchmark

To validate scalability on a commercial product, the model is instantiated into the the 64×4 memory top block of the extracted netlist from a 1-Mb 28 nm MRAM macro [4], and simulated with macro-specific tolerance settings. The emulated magnetic term from Section III-C2 enables the previously impossible capability of simulating successive writes with identical transition times due to non-desired over-damping. Figure 8 –resistance/voltage units omitted for confidentiality– describes a writing operation 10 μs after power-on sequence. We combine the s-LLGS OOMMF validated dynamics with foundry-given thermal/voltage conductance dependence, providing the accurate resistance response over time. Compared to using fixed resistors, there is an overhead of $3.1\times$ CPU time and $1.5\times$ RAM usage. In return, circuit designers can observe accurate transient switching behaviour and read disturbs.

V. Conclusions

This work presents an MTJ modeling approach for large VLSI circuits. We analyze MTJ modeling challenges and propose solutions to accurately capture stochastic thermal noise effects. Accuracy is validated against OOMMF , and scalability shown by incorporation into a 1-Mb 28 nm MRAM macro. The framework code is available upon request.

References

[1] H. Lee *et al.*, "Analysis and Compact Modeling of Magnetic Tunnel Junctions Utilizing Voltage-Controlled Magnetic Anisotropy," *IEEE Trans. Magn.*, vol. 54, no. 4, 2018.
[2] M. M. Torunbalci *et al.*, "Modular Compact Modeling of MTJ Devices," *IEEE Trans. Electron Devices*, vol. 65, no. 10, 2018.
[3] K. Zhang *et al.*, "Compact Modeling and Analysis of Voltage-Gated Spin-Orbit Torque MTJ," *IEEE Access*, vol. 8, 2020.
[4] E. M. Boujamaa *et al.*, "A 14.7Mb/mm2 28nm FDSOI STT-MRAM with Current Starved Read Path, 52Ω/Sigma Offset Voltage Sense Amplifier and Fully Trimmable CTAT Reference," *IEEE Symp. VLSI Circuits, Dig. Tech. Pap.*, vol. 2020-June, 2020.
[5] M. J. Donahue *et al.*, "OOMMF user's guide, v1.0," National Institute of Standards and Technology, Gaithersburg, MD, Tech. Rep., 1999.
[6] M. Beg *et al.*, "User interfaces for computational science: A domain specific language for OOMMF embedded in Python," *AIP Adv.*, vol. 7, no. 5, may 2017.
[7] G. D. Panagopoulos *et al.*, "Physics-based SPICE-compatible compact model for simulating hybrid MTJ/CMOS circuits," *IEEE Trans. Electron Devices*, vol. 60, no. 9, 2013.
[8] F. Garcia-Redondo *et al.*, "Building Memristor Applications: From Device Model to Circuit Design," *IEEE Trans. Nanotechnol.*, vol. 13, no. 6, nov 2014.
[9] Cadence, "Spectre Accelerated Parallel Simulator."
[10] S. Ament *et al.*, "Solving the stochastic Landau-Lifshitz-Gilbert-Slonczewski equation for monodomain nanomagnets : A survey and analysis of numerical techniques," 2016.
[11] Y.-J. Tsou *et al.*, "Write Margin Analysis of Spin–Orbit Torque Switching Using Field-Assisted Method," *IEEE J. Explor. Solid-State Comput. Devices Circuits*, vol. 5, no. 2, dec 2019.
[12] W. F. Brown, "Thermal Fluctuations of a Single-Domain Particle," *Phys. Rev.*, vol. 130, no. 5, jun 1963.
[13] P. Virtanen *et al.*, "SciPy 1.0: fundamental algorithms for scientific computing in Python," *Nat. Methods*, vol. 17, no. 3, mar 2020.

© VDE VERLAG GMBH · Berlin · Offenbach

A Quantitive Analysis of the Recovery Effect in Batteries from Datasheets

Alberto Bocca

Yukai Chen

Wenlong Wang

Alberto Macii

Enrico Macii

Massimo Poncino

Department of Control and Computer Engineering
Politecnico di Torino
Torino, Italy
{name.surname}@polito.it

Abstract—Over the past decade, battery modeling using datasheets has been intensively researched due to the growing number of battery-powered devices. One of the typical non-ideal discharge behaviors of certain batteries is the partial recovery of their energy after a current pulse; this is known as *recovery effect*. Consequently, the battery runtime is generally longer in the case of a battery operating at pulsed current than at constant current of the same magnitude.

This work demonstrates how to analyze the recovery effect simply by considering the manufacturer's data. The proposed method was applied to the Energizer E91 alkaline cell. The results show that, as additional available capacity, the absolute recovered energy increases as the battery current magnitude becomes greater. In this case, the additional battery capacity can reach up to about 30% for medium currents. Direct experimental measurements validate the proposed methodology.

Index Terms—Battery modeling, datasheet, energy, recovery effect

I. INTRODUCTION

In a modern world based on information and communication technology (ICT), the ever greater increase of mobile devices and Wireless Sensors Networks (WSN) has attracted much research on data communication and batteries on optimizing the use of energy to maximize the lifetime of such systems [1], [2]. Both the task scheduling and frequency of data transmission of battery-powered devices generally affect the runtime and longevity of batteries, which are divided into two main categories: primary (i.e., non-rechargeable) and secondary (i.e., rechargeable) batteries. For instance, nowadays, consumer electronics (e.g., sensor-based monitoring devices, portable radio/CD players) often use primary batteries, especially alkaline type. In fact, they currently dominate more than 70% of the battery market (more than 80% of the primary battery market). They are also the preferred battery for wireless and sensory applications due to low cost/maintenance, low self-discharge, and high capacity [3]. On the other hand, most mobile applications (e.g., smartphones, electric vehicles) generally use secondary batteries based on lithium-ion technology [4]. In this context, one of the most important non-ideal characteristics of certain electrochemical energy storage systems is known as *recovery effect*. It refers to the capability of a battery to recover part of its energy in the rest time

following each current pulse. In this way, the total available capacity (i.e., energy) of a battery is more significant in the case of a pulsed rather than a continuous discharge current at the same rate [1], [5].

The main purpose of this article is to present a method for analyzing this feature simply by considering the product data of a battery. Although a primary alkaline cell is considered here, the method can also be used for secondary batteries. This is, to the best of our knowledge, the first work on modeling the recovery effect in batteries from product datasheet.

II. BACKGROUND

Battery modeling from datasheet has had a notable development in recent years [6], [7]. Most of the recently proposed methodologies concern lithium-based batteries, and especially rechargeable cells, due to the rapid development of electric vehicles [8] and also applications for storing energy in the systems related to renewable sources [9]. These models focus primarily on battery runtime. However, nowadays they are also developed for the analysis of life degradation in battery cells [10]. The accuracy of these models depends on the quality and quantity of manufacturer's data [11].

The recovery effect in batteries mainly depends on load current magnitude, the length of recovery period and depth of discharge (DOD) [4], [12]. For instance, the measurements reported in [4] show that the recovery effect is positively correlated with these three parameters in lithium sulfur (Li_2S) batteries. In general, this capacity recovery property is critical in many applications such as wireless sensor nodes for intermittent data sampling and transmission [13], and automotive systems [14]. However, only a few battery models include the recovery effect, which is not linear with respect to the discharge current magnitude and state-of-charge (SOC) of a battery [15]. For example, an equivalent circuit model including recovery effect was proposed in [16], whereas a stochastic model was proposed in [17] where the battery behavior is described through a Markov process. The models proposed so far in the literature are generally populated by direct experimental results.

The *rated capacity effect* is another non-linear characteristic of electrochemical batteries, whose total capacity generally

© VDE VERLAG GMBH · Berlin · Offenbach

depends on the continuous discharging current rates. Peukert's equation describes it in the following way [18]:

$$C_P = I^k \cdot t \qquad (1)$$

In (1), C_P is nominal capacity, I is continuous discharge current, t is battery runtime, and k is the Peukert number, a coefficient which is generally near to 1. Although this relationship was initially being proposed for lead-acid batteries, it is also useful for other chemistries, sometimes with some precaution in the correct use of it [19].

III. METHODOLOGY

To illustrate the proposed methodology more simply, this section reports directly an example for the performance analysis of Energizer E91 alkaline cells (zinc-manganese dioxide, Zn/MnO_2) from their datasheet [20].

Firstly, it was necessary to analyze the rated capacity effect of this cell for different continuous discharge currents. The datasheet reports in a bar graph the battery's total capacity when reaching the cut-off voltage of 0.8 V at four different currents: 25, 100, 250, and 500 mA. From these published data, we extracted the Peukert number for this cell, that is 1.1152 for small currents, in order to analyze the total capacity also for a continuous current of 50 mA. In this way, a comparison was made possible with the total capacity of a pulse current of 50 mA (radio/clock), as reported in the datasheet. After applying the Peukert coefficient to (1), the rated capacity at a continuous discharge of 50 mA was found to be approximately 2,816 mAh for 56.32 service hours. Table I reports the extracted rated capacity for each of the considered continuous currents.

TABLE I
THE RATED CAPACITY FOR VARIOUS CONTINUOUS CURRENTS FOR THE ENERGIZER E91 CELL.

Current [mA]	Capacity [mAh]
25	3050
50	2816
100	2600
250	1900
500	1550

On the other hand, Table II reports the total capacity for various pulse currents after considering the total service hours as extracted from the manufacturer's data, which are included in the datasheet in *voltage vs. service time* plot characteristics. In this case, three concern applications at small and medium pulse currents (i.e., 50, 100, and 250 mA), whereas the others mainly concern applications of constant loads.

Figure 1 shows a comparison between the cell's total capacity under test when considering continuous and pulse currents extracted from the datasheet. The difference in total capacity increases as the discharge current also increases. In other terms, the absolute recovered energy increases as the battery current magnitude becomes greater. Furthermore, in these application cases, the rest times of the battery are always

TABLE II
THE TOTAL CAPACITY OF THE ENERGIZER E91 CELL FOR VARIOUS PULSE CURRENTS.

Current [mA]	Capacity [mAh]	Note
50	2856	1 hour ON / 7 hours OFF
100	2712	1 hour ON / 23 hours OFF
250	2450	1 hour ON / 23 hours OFF

several hours for pulsed currents (see column three in Table II). Consequently, the recovery effect is mostly completed at each rest time, before the next current pulse. For this reason, we can directly compare the discharge currents in continuous and pulsed modes. Conversely, we should consider the average current, which depends on the duty cycle of pulse time with respect to the total period including rest time.

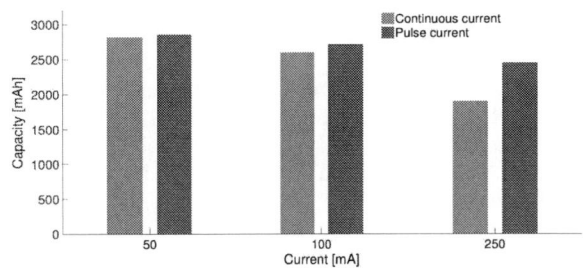

Fig. 1. A comparison of the total capacity for different currents.

By considering the differences of the extracted capacity values, for continuous and pulsed currents at 50, 100, and 250 mA, we found the following fitting function of the additional capacity C_{add} of the battery in the case of a pulsed current I_p instead of a continuous one:

$$C_{add} = 4 \cdot 10^{-06} \cdot I_p^3 + 0.0058 \cdot I_p^2 + 0.5 \cdot I_p \qquad (2)$$

This function refers to the additional capacity for very long rest times only (see Table II). In fact, it is assumed that the shorter is the rest time, the less the additional capacity will generally be. Figure 2 shows this characteristic, which is however expected to be different for high current rates, so that the gain margin in the use of pulsed currents instead of continuous currents will no longer exactly follow this function.

In addition, we also searched for the effect of energy recovery from a temporal point of view, so that *time* could be included in (2). Unfortunately, this information is generally not available in datasheets and, therefore, it cannot be extracted from these documents alone. For this reason, some direct experimental data could help to accomplish the goal of a more comprehensive analysis of this phenomenon and, especially, to validate the proposed methodology.

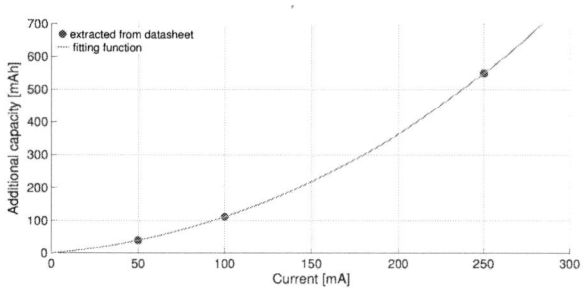

Fig. 2. The additional capacity by pulse currents to the rated capacity by continuous currents.

IV. EXPERIMENTAL RESULTS

A. Experiment design

Three different groups of experiments were planned to discharge fresh Energizer E91 batteries at various currents: 50 mA, 100 mA, and 250 mA, respectively, with a cycle of 1-hour pulse of discharging current and 1-hour pause until the cell was exhausted. In this way we explored with a first timing analysis how the recovery effect acts on battery capacity during the pulse-off period. As the pause time reported in the datasheet is instead very large (7 or 23 hours), the comparison of the results obtained by us and those published is mostly of a qualitative nature. However, this analysis is sufficient to validate the method, as demonstrated below.

B. Experiment setup

The laboratory equipment includes, as reported in Table III, a DL3021 electronic load produced by RIGOL for discharging the batteries under test and an HP 34401A digital multimeter to measure the battery voltage during discharge. The measured data were transmitted to a PC via a serial connection.

TABLE III
DEVICES USED FOR THE TESTS

Device	Type	Role in the experiment
Energizer E91	Primary battery	The deployed battery in the experiments
HP 34401A	Digital multimeter	Monitoring and recording battery voltage
RIGOL DL3021	Electronic load	Discharging battery cells

C. Experimental results

Since the battery cut-off voltage is not clearly mentioned in the product data [20], we considered the discharge termination point indicated in the datasheet (i.e., 0.8 V). The experimental results are summarized in Table IV, which reports the mean values of the total test time and total capacity of the cells for various pulse currents.

Fig. 3 shows the recorded plot for one of the experiments at 100 mA. It is worth noting that the cut-off voltage is confirmed to be 800 mV, although the cell was eventually fully depleted.

TABLE IV
SUMMARY OF THE MEAN TIME OF ALL EXPERIMENTS TO REACH CUT-OFF
VOLTAGE (1 HOUR ON / 1 HOUR OFF)

Pulse-on current [mA]	50	100	250
Time to reach 800 mV [h]	101.40	46.63	14.74
Total capacity [mAh]	2535	2331	1842

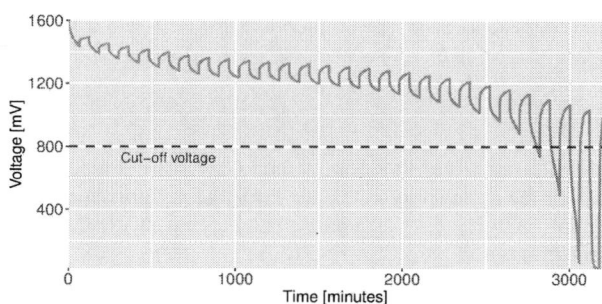

Fig. 3. Battery discharge voltage profile at pulse-on current $I_p = 100$ mA.

Figure 4 shows both the total capacity extracted from the datasheet and that obtained from our experiments. In addition, it includes the respective fitting functions for a comparison purpose.

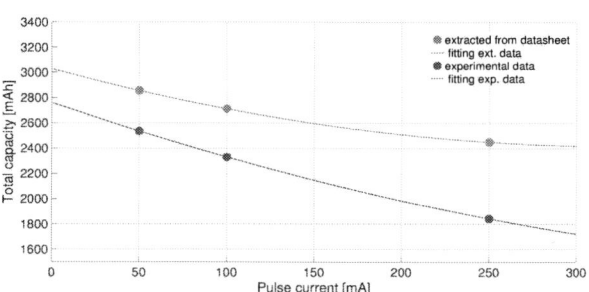

Fig. 4. The total capacity of the E91 cell for various pulse currents but at different rest time.

Firstly, the experimental results confirm the general energy recovery trend for the battery considered here, as the two fitting functions plotted in Fig. 4 have a very similar trend. However, they do not overlap, even partially; this is mainly due to the different rest time between product data and our experiments. In fact, in this case the total time for the recovery transient to reach the asymptotic expected value is certainly greater than one hour. In a preliminary analysis, this is depicted in Fig. 5, which includes a zoom view of the pulses in Fig. 3. In this graph, Δt and Δv are the missing time and voltage, respectively, to accomplish a full energy recovery.

Therefore, the idle period length in pulsed currents plays a critical role for recovery effect as well as discharge current magnitude. Adding this time information in a comprehensive battery model accounting for recovery effect will be addressed in future.

© VDE VERLAG GMBH · Berlin · Offenbach

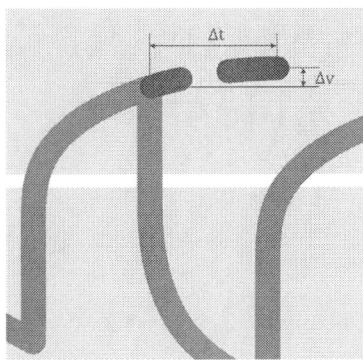

Fig. 5. A detail of the plot in Fig. 3 with a possible transient that would be expected for a full recovery.

Secondly, we can draw a few other conclusions from all these results:

- Our experimental data could possibly be obtained from cells with some degree of aging due to their being in stock for a time before being sold. In this case, an accurate analysis would not be so easy unless battery degradation was known at the start of testing.
- Although the values reported in Table IV are mean values, part of the expected error is due to process variation, so that results, for different cells of the same brand and type, usually differ from each other. For this reason, an extensive analysis of more experimental data should be preferred, in order to minimize this error.

Finally, the difference of the total capacity between the values extracted from the datasheet and those obtained from the experiments carried out at 50, 100 and 250 mA, is 11%, 14% and 25%, respectively. Nevertheless, this difference is expected to decrease considerably with a full energy recovery. In the overall analysis, the total capacity of the battery operating at pulsed current is generally greater than that at continuous current, up to approximately 30% for medium currents.

V. CONCLUSION

This work presents a simple method for analyzing the *recovery effect* in batteries, using the product data published by manufacturers and applied to the Energizer E91 battery. The total capacity of the battery operating at pulsed current, which involves a recovery of energy during rest times, is in fact greater than at continuous current. The proposed method allows a preliminary analysis of the limits of battery use through the extraction of a function that characterizes this effect. The results indicate that as the pulse current magnitude increases, the absolute amount of energy that can be recovered becomes greater. A more detailed analysis of energy recovery time will be addressed in a future work.

REFERENCES

[1] S. Narayanaswamy, S. Schlueter, S. Steinhorst, M. Lukasiewycz, S. Chakraborty, and H. E. Hoster, "On battery recovery effect in wireless sensor nodes," *ACM Transactions on Design Automation of Electronic Systems (TODAES)*, vol. 21, no. 4, pp. 1–28, 2016.

[2] Y. Chen, D. Jahier Pagliari, E. Macii, and M. Poncino, "Battery-aware design exploration of scheduling policies for multi-sensor devices," in *Proceedings of the 2018 on Great Lakes Symposium on VLSI*, 2018, pp. 201–206.

[3] M. Radfar, A. Nakhlestani, H. Le Viet, and A. Desai, "Battery management technique to reduce standby energy consumption in ultra-low power iot and sensory applications," *IEEE Transactions on Circuits and Systems I: Regular Papers*, vol. 67, no. 1, pp. 336–345, 2020.

[4] C. Maurer, W. Commerell, A. Hintennach, and A. Jossen, "Capacity recovery effect in lithium sulfur batteries for electric vehicles," *World Electric Vehicle Journal*, vol. 9, no. 2, p. 34, 2018.

[5] L. Benini, D. Bruni, A. Macii, E. Macii, and M. Poncino, "Discharge current steering for battery lifetime optimization," *IEEE Transactions on Computers*, vol. 52, no. 8, pp. 985–995, August 2003.

[6] M. Petricca, D. Shin, A. Bocca, A. Macii, E. Macii, and M. Poncino, "An automated framework for generating variable-accuracy battery models from datasheet information," in *Proc. International Symposium on Low Power Design (ISLPED)*, 2013, pp. 365–370.

[7] J. V. Barreras *et al.*, "Datasheet-based modeling of Li-Ion batteries," in *Proc. IEEE Vehicle Power and Propulsion Conference (VPPC)*, 2012, pp. 830–835.

[8] X. Zeng, M. Li, D. Abd El-Hady, W. Alshitari, A. S. Al-Bogami, J. Lu, and K. Amine, "Commercialization of lithium battery technologies for electric vehicles," *Advanced Energy Materials*, vol. 9, no. 27, p. 1900161, 2019.

[9] J. Hoppmann, J. Volland, T. S. Schmidt, and V. H. Hoffmann, "The economic viability of battery storage for residential solar photovoltaic systems–a review and a simulation model," *Renewable and Sustainable Energy Reviews*, vol. 39, pp. 1101–1118, 2014.

[10] M. Petricca, D. Shin, A. Bocca, A. Macii, E. Macii, and M. Poncino, "Automated generation of battery aging models from datasheets," in *Proc. 32nd IEEE International Conference on Computer Design (ICCD)*, 2014, pp. 483–488.

[11] A. Bocca, A. Macii, E. Macii, and M. Poncino, "Composable battery model templates based on manufacturers' data," *IEEE Design & Test*, vol. 35, no. 3, pp. 66–72, 2018.

[12] D. Rakhmatov, S. Vrudhula, and D. A. Wallach, "A model for battery lifetime analysis for organizing applications on a pocket computer," *IEEE transactions on very large scale integration (VLSI) systems*, vol. 11, no. 6, pp. 1019–1030, 2003.

[13] C.-K. Chau, F. Qin, S. Sayed, M. H. Wahab, and Y. Yang, "Harnessing battery recovery effect in wireless sensor networks: Experiments and analysis," *IEEE Journal on Selected Areas in Communications*, vol. 28, no. 7, pp. 1222–1232, 2010.

[14] A. Baumgardt, F. Bachheibl, and D. Gerling, "Utilization of the battery recovery effect in hybrid and electric vehicle applications," in *2014 17th International Conference on Electrical Machines and Systems (ICEMS)*. IEEE, 2014, pp. 254–260.

[15] M. R. Jongerden and B. R. Haverkort, "Which battery model to use?" *IET software*, vol. 3, no. 6, pp. 445–457, 2009.

[16] J. Zhang, S. Ci, H. Sharif, and M. Alahmad, "An enhanced circuit-based model for single-cell battery," in *2010 Twenty-Fifth Annual IEEE Applied Power Electronics Conference and Exposition (APEC)*. IEEE, 2010, pp. 672–675.

[17] C.-F. Chiasserini and R. R. Rao, "A model for battery pulsed discharge with recovery effect," in *WCNC. 1999 IEEE Wireless Communications and Networking Conference (Cat. No. 99TH8466)*, vol. 2. IEEE, 1999, pp. 636–639.

[18] W. Peukert, "Über die Abhängigkeit der Kapazität von der Entladestromstärke bei Bleiakkumulatoren," in *Elektrotechnische Zeitschrift*, 1897, p. 20.

[19] N. Omar, P. V. d. Bossche, T. Coosemans, and J. V. Mierlo, "Peukert revisited–Critical appraisal and need for modification for lithium-ion batteries," *Energies*, vol. 6, no. 11, pp. 5625–5641, 2013.

[20] Energizer, "Energizer E91 - Product Datasheet," [Online]. Available: https://data.energizer.com/pdfs/e91.pdf [Accessed March 3, 2021].

© VDE VERLAG GMBH · Berlin · Offenbach

SMACD / PRIME 2021 | 19 – 22 July 2021, Online Event

A Phase Error Correction Algorithm for RF Energy Harvesters Using Two Antennas

Ali Doğuş Güngördü, Didem Erol, Alican Çağlar, Mustafa Berke Yelten

Istanbul Technical University

Electronics and Communications Engineering

Istanbul, Turkey

gungordua@itu.edu.tr, erold@itu.edu.tr, caglara@itu.edu.tr, yeltenm@itu.edu.tr

Abstract—In this study, a phase error correction algorithm is proposed for radio frequency (RF) energy harvesting through a fully differential RF-to-DC converter with two antennas. The system keeps one of the inputs as the reference and adjusts the phase of the other input. A 2-bit resolution is used to demonstrate how the concept works. The system aims to operate without an external battery. System-level simulation results prove that through the proposed concept, the efficiency of the harvester can be considerably boosted for an RF harvester with two antennas.

Index Terms—RF energy harvester, fully differential RF-to-DC converter, phase error correction, finite state machines

I. INTRODUCTION

In recent years, there is a growing interest in remote-controlled systems known as passive-powered devices [1]. Consequently, harvesting the ambient energy to partially or fully deliver the required power for the operation of portable electronic devices has become increasingly more popular [2]. An internal power supply is not required for passive-powered devices to extract energy from the sunlight, electromagnetic waves, or mechanical vibration [3]. Elimination of batteries benefits the applications where battery replacement is impractical or expensive, including biomedical implants, radio-frequency identification (RFID), environmental monitoring in inaccessible locations, or disaster recovery [4]. Energy extraction from radio-frequency (RF) electromagnetic waves, referred to as radio-frequency energy harvesting is preferred over other sources of energy [5]. RF energy harvesting enables powering low-power wireless sensors and it has a potentially widespread accessibility in urban areas [6]. An RF energy harvester converts the AC power into DC power [7].

Various applications, such as biomedical and healthcare devices, RFIDs, and remote sensors have made the wireless power transfer (WPT) important. The frequency range of transfer extends from low Megahertz to ultra-high frequency (UHF) bands depending on the application [8]. The industrial, scientific, and medical (ISM) band at 902–928 MHz is suitable for RF energy harvesting [9]. Moreover, RF sources provide strong and reliable power at 868 MHz European ISM, which allows maximum of 3.28 W effective isotropic radiated power (EIRP) [10]. The sensitivity of an RF energy harvester defined

This work was supported by the Department of Scientific Research Projects, Istanbul Technical University, under the Project 41030.

as the minimum required input power to generate a useful output DC voltage within the network coverage area for a given source EIRP [8] - [9]. The sensitivity of an RF harvester depends on the threshold voltage of rectifying devices (diodes or transistors) in a multi-stage rectifier [9]. Dickson and full-bridge rectifiers are not suitable for the ambient WPT, where the RF power is low since a high dropout voltage results in poor sensitivity and reduced overall performance [8]. There are many low threshold schemes to overcome such limitations, which depend on either technology solutions or suitable compensation circuits [9]. The fully cross-coupled rectifier is a power-efficient technique to solve the sensitivity issue by utilizing the RF power deferentially to four rectifying transistors [8].

The main problem of the topologies consisting of two antennas is that the possible phase difference (other than 180°) between the antennas, causes the received signals to fade. In the fully differential architecture, energy can be harvested with maximum efficiency, if there is a 180° phase difference between the input signals. Thus, the purpose of this paper is to harvest RF energy by keeping the phase difference between input signals at 180°. To that end, a phase detection circuit that can find the optimum phase shift to be applied on two input signals to achieve maximum energy harvesting has been proposed.

The organization of this paper is given as follows: Section II briefly describes the system and the algorithm of phase correction, thereby discussing the energy consumption of active blocks, and the external battery issue. Section III demonstrates a scenario, in which the initial phase difference between the inputs are 120°, in conjunction with the system-level simulation results for the overall system. Finally, conclusions are drawn in Section IV.

II. SYSTEM DESCRIPTION

A. Overview

The proposed system is given in Fig. 1. To provide the maximum efficiency for a fully differential RF-DC converter, there must be a 180° phase difference between the input signals. However, the phase difference between the inputs may vary due to factors such as antenna directivity and system nonidealities. The purpose of the system is to arrange a 180° phase difference by adjusting the phase of Input-2. Input-1 can

© VDE VERLAG GMBH · Berlin · Offenbach

be seen as a reference signal whereas Input-2 is the signal that must be adjusted. To show that the concept operates correctly, the resolution of the system is chosen as two bits between 0-180°, and the corresponding phase shift amounts are 0°, 60°, 120°, and 180°. The system has two programmable phase shifters to modify the phase of Input-2. An appropriate phase shifter design can be found in [11]. SW1 and SW2 are the switch signals to program the phase shifters. At the beginning, when the RESET signal is on, the Finite State Machine (FSM) sets the initial phase shift amount to 0° and 60° at phase shifter 1 and phase shifter 2, respectively.

When the clock is activated, the comparison stage starts. Outputs of the phase shifters arrive to the passive mixer. The function of the passive mixer can be represented by $f(x, y, \alpha)$ where x and y are the amplitudes of Input 1 and Input 2, respectively. Moreover, α is the phase difference between the inputs. The mixer produces $f_1 + f_2$ and $f_1 - f_2$ at the output where f_1 and f_2 are the frequencies of the input signals. If f_1 and f_2 are equal, the mixer will produce a DC signal (a function of α) and it can be easily translated by an ultra-low power analog comparator to a logic signal. The comparator produces a logic signal and accordingly, the FSM produces the new phase shift values. After the comparison stage, the FSM lets the controller to activate the phase adjuster according to the final state of FSM, and the RF-DC converter starts to operate with the final two inputs.

For a more precise description of the system operation, the algorithmic flow of the system can be reviewed in Fig. 2. When the comparison starts with the clock signal, a two-bit counter starts to count. The comparison stage finishes after three cycles. When the counter is equal to three, the phase adjusting function is activated and the adjusted Input-2 signal goes to the RF-DC conversion block.

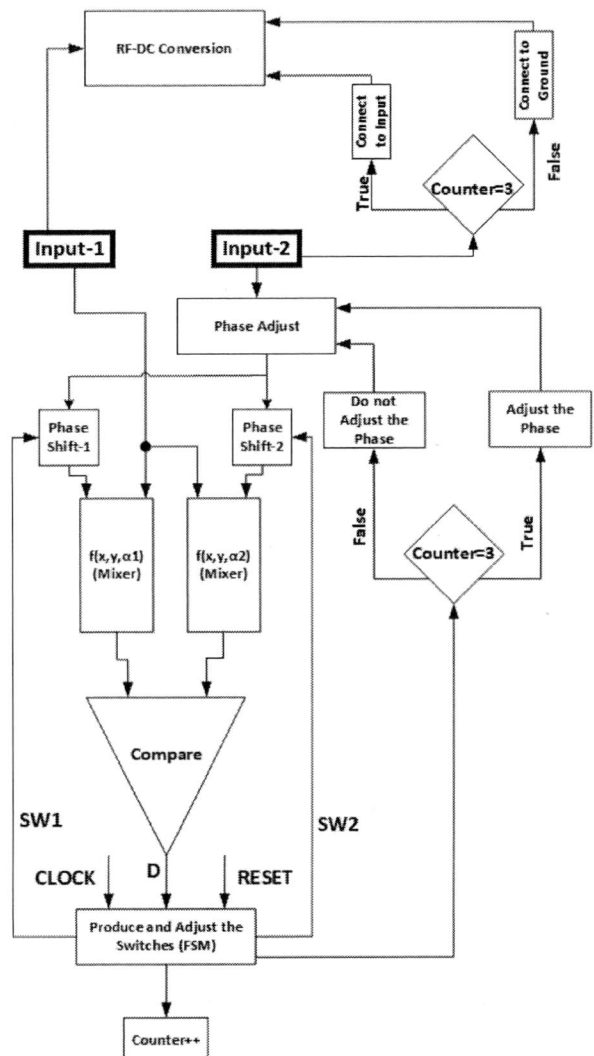

Fig. 2. The algorithmic flow of phase correction.

B. Finite State Machine

The details of phase shifting can be understood with the FSM state diagram, shown in Fig. 3. It is important to note that the state diagram should be eventually implemented in hardware; hence, it is designed and synthesized accordingly. Here, D represents the logic of the output of the comparator, whereas $P1$ and $P2$ stand for the phase shift amounts of the two phase shifters in Fig. 1. Finally, there are seven different states, $S0$-$S6$. In $S0$, 0° and 60° are compared. If $D = 0$, it means that 60° wins, whereas if $D = 1$, 0° overcomes. The logic continues that way for the next three cycles. The final decision on the phase shift amount is provided with one of the states, $S3$ through $S6$, each indicating to a different phase shift value. $P1$ stores the final result. For example, after three cycles, if the state is $S3$, then the FSM sends $P1$ data to the phase adjusting block to provide a 60° of phase shift. Another example would be, when the final state is $S6$, then the

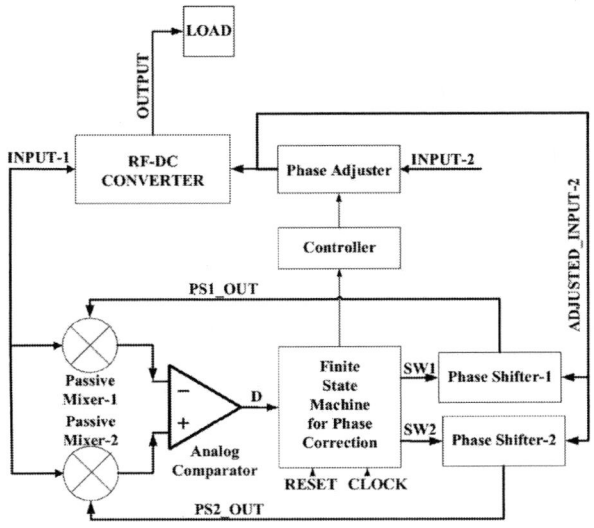

Fig. 1. The structure of the designed system.

adjusting block obtains a 120° phase shift amount since $P1$ is guided by $S6$. Moreover, final phase shift amounts of 180° and 0° can be acquired through the states $S4$ and $S5$, respectively. In summary, the FSM is at the heart of the whole system, processes the comparison data, arranges the switches of the phase shifters, and makes the final phase shift adjustment for Input-2.

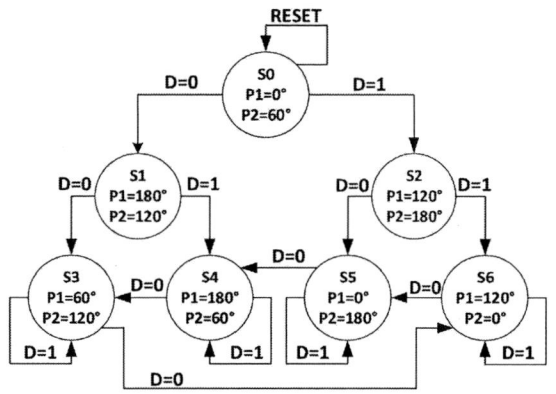

Fig. 3. The FSM State Diagram.

C. Controller

The controller can be seen in Fig. 1, which is fed by the FSM. Since the FSM completes the comparison stage in three cycles, a two-bit counter and a D flip-flop, which keeps the final phase shift amount for phase adjustment, constitute the controller block. Initially, the memory is set to 0° phase shift. After the comparison stage ends, the memory writes the $P1$ data to the phase adjusting block.

D. Power Consumption

It is beneficial to give an insight about the power consumption of the components. The static power of the digital blocks (FSM and controller) only includes the power due leakage currents in the standard CMOS technology. The phase shifter and the mixer do not consume power. However, the analog comparator consumes some power as it is an active circuit. Assuming that the frequencies of the input signals are equal, the mixer produces a DC signal at the output. Thus, an ultra-low-power analog comparator (close to 1 μW) should be employed in the system. The dynamic power of the digital blocks heavily depends on the frequency of the clock signal. The phase correction side of the system only works for three cycles. Thus, the frequency can be chosen arbitrarily low to minimize the dynamic power consumption.

E. External Battery

An ideal RF energy harvester should not employ an external battery. The system is designed based on this constraint. When the reset is on, the RF-to-DC converter harvests only the signal from Input-1 as the phase of Input-2 is not ready. After this initial energy harvesting, the accumulated energy is sent to the digital blocks and the analog comparator. The harvester

should provide the necessary power for the phase correction. After the comparison stage finishes, the system is ready to deliver power to the load. Now, the RF harvester can utilize its two inputs simultaneously to achieve the maximum power efficiency.

III. SIMULATION RESULTS

To demonstrate the proposed concept, system-level simulation results are provided in this section. All of the blocks, including the digital part in the system, are designed at schematic level. For the phase shifter, ideal capacitors, inductors and switches are used. The simulations have been performed in Cadence Design Environment. A 50 Ω antenna along with a -13 dBm electromagnetic power is assumed since it is suggested in the literature that it is a common power level for a realistic environment [12]. The passive voltage boosting is assumed to be $\times 4.3$, thus it corresponds to a 300 mV voltage amplitude for both input signals. The design in [10] has been utilized as the basis of the RF harvester architecture. At the output, a 400 $k\Omega$ resistor is used to emulate the load.

As an example, consider the scenario where the phase difference between inputs is 120°. The operation of the circuit can be demonstrated as follows: Initially, when the reset is logic 0 (it is active-low zero), the phase adjuster (see Fig. 1) is set for 0°. The phase adjuster remains at 0° phase shift until the counter is equal to three. The FSM is at the $S0$ state where $P1$ is set to 0° and $P2$ is set to 60° (see the state diagram in Fig. 3). After the clock signal is activated, and the reset signal is deactivated, the counter starts to count, the FSM starts to compare, and accordingly change the state. In Fig. 4, the clock and reset signals, output data from the comparator (represented by D), switches for the phase shifter-1 and the phase shifter-2, the switch of the phase adjusting block, and the output data of the counter are shown for three cycles of comparison. 8-bit switches are employed in all phase shifters; thus, the decimal equivalents of 8-bit data are provided in Fig. 4. Here, 3 (00000011) corresponds to a 0° phase shift, 6 (00000110) corresponds to a 60° phase shift, 52 (00011100) corresponds to a 120° phase shift, and 248 (11111000) corresponds to a 180° phase shift. As observed in Fig. 4, when the reset is logic 0 (this is the reset state as the system uses an active-low reset), phase shifter-1 is equal to 3, phase shifter-2 is equal to 6, and the phase adjusting block output is equal to 3. To adjust the 120° phase shift amount, the comparator produces "001". However, until the counter is equal to 3, the phase adjusting block stays at 3, and after the comparison is done and the counter being equal to 3, the phase adjusting block changes its state to 6, which corresponds to a 60° phase shift.

To demonstrate the improvement on the harvester, the outputs with and without phase adjustment are given in Fig. 5. Using the phase adjustment algorithm, the output voltage is around 2.9 V, whereas without using the algorithm, it is equal to 2.3 V. Since the power conversion efficiency (PCE) is proportional to the square of the output voltage, it can be inferred that the RF-to-DC power conversion efficiency factor drops by 1.6×. Moreover, this is not even the worst scenario.

© VDE VERLAG GMBH · Berlin · Offenbach

For example, if the inputs would have the same phases (0° phase difference), no reasonable output voltage could have been obtained as the contribution from both inputs would cancel each other.

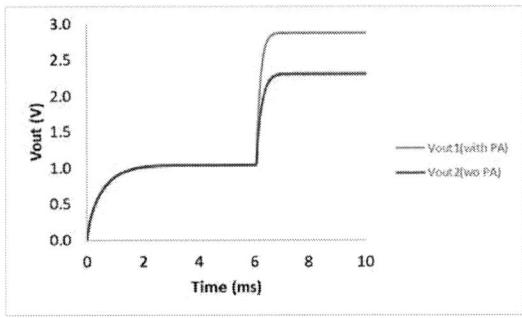

Fig. 5. Comparison of the output voltages with and without the phase adjustment.

feeding antennas. The system has a 2-bit phase resolution. The resolution can be increased by updating the phase shifters and the digital parts without changing the other blocks. Simulation results reveal that the system corrects the phase difference and boosts the efficiency of the harvester automatically. The next step would be to take this concept and implement its corresponding hardware via a chip prototype.

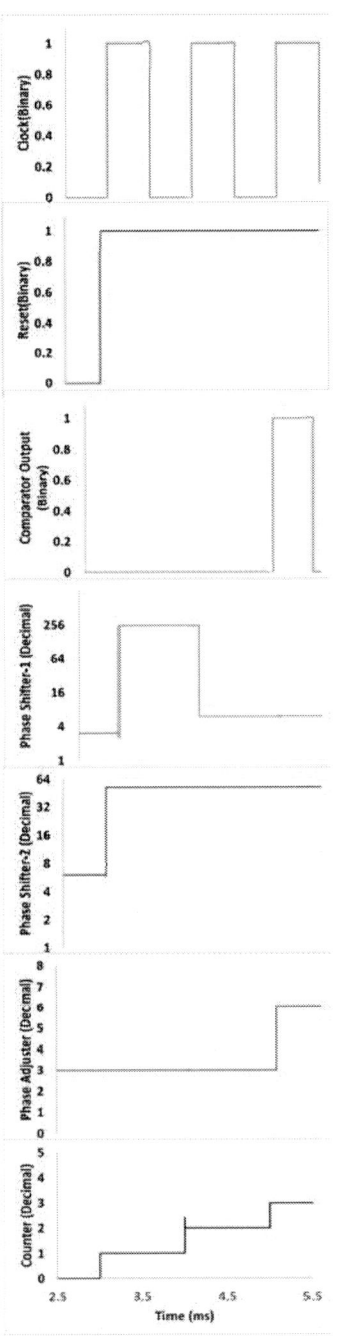

Fig. 4. Digital outputs obtained in the system.

IV. CONCLUSION

In this study, a phase correction algorithm is demonstrated at the system-level for fully differential RF harvesters with two

REFERENCES

[1] J. A. Paradiso and T. Starner, "Energy scavenging for mobile and wireless electronics," *IEEE Pervasive computing*, vol. 4, no. 1, pp. 18–27, 2005.

[2] V. Bhatnagar and P. Owende, "Energy harvesting for assistive and mobile applications," *Energy Science & Engineering*, vol. 3, no. 3, pp. 153–173, 2015.

[3] P.-H. Hsieh, C.-H. Chou, and T. Chiang, "An rf energy harvester with 44.1% pce at input available power of-12 dbm," *IEEE Transactions on Circuits and Systems I: Regular Papers*, vol. 62, no. 6, pp. 1528–1537, 2015.

[4] J. A. Hagerty, F. B. Helmbrecht, W. H. McCalpin, R. Zane, and Z. B. Popovic, "Recycling ambient microwave energy with broadband rectenna arrays," *IEEE Transactions on Microwave Theory and Techniques*, vol. 52, no. 3, pp. 1014–1024, 2004.

[5] S. O'Driscoll, A. S. Poon, and T. H. Meng, "A mm-sized implantable power receiver with adaptive link compensation," in *2009 IEEE International Solid-State Circuits Conference-Digest of Technical Papers*, pp. 294–295, IEEE, 2009.

[6] E. Davut, O. Kazanci, A. Caglar, D. Altinel, M. B. Yelten, and G. K. Kurt, "A test-bed based guideline for multi-source energy harvesting," in *2017 10th International Conference on Electrical and Electronics Engineering (ELECO)*, pp. 1267–1271, 2017.

[7] B. Li, X. Shao, N. Shahshahan, N. Goldsman, T. Salter, and G. M. Metze, "An antenna co-design dual band rf energy harvester," *IEEE Transactions on Circuits and Systems I: Regular Papers*, vol. 60, no. 12, pp. 3256–3266, 2013.

[8] A. S. Almansouri, M. H. Ouda, and K. N. Salama, "A cmos rf-to-dc power converter with 86% efficiency and- 19.2-dbm sensitivity," *IEEE Transactions on Microwave Theory and Techniques*, vol. 66, no. 5, pp. 2409–2415, 2018.

[9] G. Papotto, F. Carrara, and G. Palmisano, "A 90-nm cmos threshold-compensated rf energy harvester," *IEEE Journal of Solid-State Circuits*, vol. 46, no. 9, pp. 1985–1997, 2011.

[10] M. Stoopman, S. Keyrouz, H. J. Visser, K. Philips, and W. A. Serdijn, "Co-design of a cmos rectifier and small loop antenna for highly sensitive rf energy harvesters," *IEEE Journal of Solid-State Circuits*, vol. 49, no. 3, pp. 622–634, 2014.

[11] M. Cook and J. W. Rogers, "A highly compact 2.4-ghz passive 6-bit phase shifter with ambidextrous quadrant selector," *IEEE Transactions on Circuits and Systems II: Express Briefs*, vol. 64, no. 2, pp. 131–135, 2016.

[12] W. Serdijn, A. Mansano, and M. Stoopman, "Introduction to rf energy harvesting," in *Wearable Sensors*, pp. 299–322, Elsevier, 2014.

Robust Design Methodology for RF LNA including Corner Analysis

Antonio D. Martinez-Perez
Group of Electronic Design (GDE)
Universidad de Zaragoza
Zaragoza, Spain
adimar@unizar.es

Francisco Aznar
Group of Electronic Design (GDE)
Centro Universitario de la Defensa
Zaragoza, Spain
faznar@unizar.es

Guillermo Royo
Group of Electronic Design (GDE)
Universidad de Zaragoza
Zaragoza, Spain
royo@unizar.es

Pedro A. Martinez-Martinez
Group of Electronic Design (GDE)
Universidad de Zaragoza
Zaragoza, Spain
pemar2@unizar.es

Santiago Celma
Group of Electronic Design (GDE)
Universidad de Zaragoza
Zaragoza, Spain
scelma@unizar.es

Abstract—This work presents a design methodology of a competitive inductorless single-ended LNA in 65-nm standard CMOS technology. Instead of relying only on typical conditions, the method uses corners to find the optimal device sizing, anticipating variations on the implemented circuit, and hence, the design significantly improves its reliability, being able to fulfil specifications in a wider range of process deviations. Also, in addition to its robustness, the designed circuit is very effective, achieving a Noise Figure of 2.9 GHz at 5 GHz with a simple g_m-enhanced common-gate amplifier thanks to a careful design window selection. The paper also describes the tradeoff-oriented design-window methodology that accomplishes the demanding specifications. Moreover, the authors provide a biasing strategy to offset the high process variability of the technology and statistical simulations show a 50 % reduction of failed samples due to process variations. The paper includes figures and statistical results from Montecarlo analysis for a more detailed description of the effect of the method and strategy employed on the complete design. Finally, a table compares the results with other similar circuits in the state-of-art.

Index Terms—CMOS analog design, design strategy, inductorless, low-noise amplifier, process variations

I. INTRODUCTION

Typically, the first stage in an RF receiver is a Low-Noise Amplifier (LNA). Thus, its operation and restrictions have a notable impact on the global system performance. In other words, the LNA is a critical part of an electronic communication system. Moreover, desired specifications in RF amplifiers are opposite: improving one of them normally will cause one or several of the others to degrade. Hence, achieving an optimum trade-off, instead of independent-specification optimization, is essential to obtain competitive results [1].

Due to the importance of this element for the system performance, errors and deviations in its operation could propagate the effect along all the receiver chain. For example, a lower gain (G) will imply that the next stage must work with a lower

This paper has been supported by MINECO-FEDER (TEC2017-85867-R) and DGA fellowship to Antonio D. Martinez-Perez

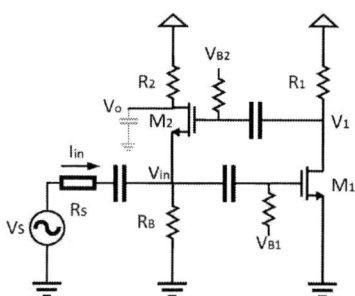

Fig. 1: Topology of the proposed LNA.

signal amplitude; or a worse noise figure (NF) will mean a decreased signal-to-noise ratio for all the following blocks.

In order to reduce to the minimum the impact of all these inevitable variances, the LNA presented in this paper has been designed from optimizing the trade-off between specifications for competitive results, while evaluating worst cases to guarantee the achieved trade-off is valid when deviations take effect.

This paper is organized as follows. Section II presents the selected LNA topology and the design specifications. Section III describes the employed design methodology, the optimization process and their limitations. Section IV provides a biasing strategy to mitigate the severe effects of process variations over the circuit. Finally, conclusions are drawn in Section V.

II. PROPOSED LNA

The selected topology is a g_m-enhanced common-gate amplifier [2] (see Fig. 1). This topology presents two relevant advantages. First, it does not require inductors, which suppose a considerable area consumption and has severe limitations in their quality factor and parasitic effects [3]. And second, it is a very effective topology despite being simple. The reduced number of elements favours low noise and applying more-precise optimization methodologies [4].

The necessity of adapting input impedance to source impedance defines the LNA topologies. The common-gate (CG) stage presents a finite input impedance in a single-stage amplifier without the gain constraint of resistor feedback. This impedance depends on transistor transconductance (g_{m-CG}); however, noise analysis reveals that also noise is dependent on g_{m-CG}. Thus, a CG stage with a g_{m-CG} large enough to present adequate input impedance will also imply too high noise.

The g_m-enhanced technique improve the trade-off between input adaptation and noise [5]. If an auxiliary amplifier increases the global transconductance, the transistor from the CG stage can reduce its g_m2 to obtain the desired performance. Thus, the circuit achieves the desired result, but it does with a more beneficial noise level. Following the strategy of simplicity mentioned in the first section, a common-source (CS) stage works as the auxiliary amplifier.

In Fig. 1, the M_2 transistor and the R_2 resistor compound the CG stage, while the M_1 transistor and the R_1 resistor do the CS amplifier. The R_B resistor is mandatory for an adequate polarization of the CG stage, as it works as a current source if R_s is negligible by comparison. Besides, as usual in RF circuits, there are pairs of biasing resistors and coupling capacitors to introduce the appropriate bias voltages (V_{b1} and V_{b2}).

Although higher-order effects have notable importance to achieve competitive results, the first-order analytical expressions provide a useful approach to the values of the design variables and the complex relationships among these and the desired specifications. These expressions can be found in [4].

In a first-order approach, R_2/R_S defines G. Thus, R_2 should be increased; however, a high value of R_2 will cause a pole with the parasitic load capacitances and will severely limit cut-off frequency (f_c). R_B must be at least an order of magnitude greater than R_S to be neglected but for the DC domain. On the part of R_1, its reduction implies an improvement of NF, but only if the transconductance of M_1 can be increased to compensate it without decreasing the gain of the CS stage. The bias voltages (V_{B1} and V_{B2}) must place the transistors on the region of strong or moderate inversion. Finally, bias resistors and coupling capacitors have high enough values to not interfere in the signal domain, but they must be implementable in the technology.

Due to transistors being implemented with minimum length to maximize f_c [6], the only free variables for the design window are the transistor width.

III. Design Methodology

The different design variables cannot be considered independent, and a change of any of them can alter the result of most, if not all, specifications. This implies a multidimensional optimization problem. Thus, each design variable requires multiplying the resources to find the solution, and having

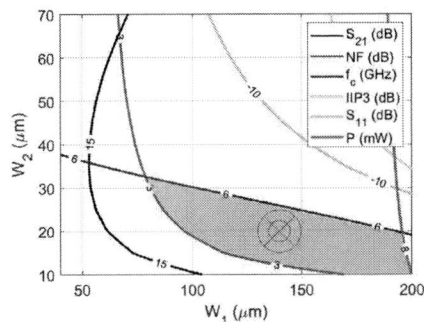

Fig. 2: Isolines of the desired specifications, each one indicates its value. G and NF restrict lower-left corner of the figure, while P, $IIP3$, and S_{11} do the same on the opposite side. f_c limits the upper part. The shaded area is the resulting design window. The design point is 140 μm and 20 μm for W_1 and W_2, respectively.

TABLE I: Montecarlo results (512 samples) without bias compensation

	Mean	Std Dev	Spec	Yield
$NF @ 5\ GHz$	2.7 dB	0.16 dB	3 dB	94.9 %
$S_{11} @ 5\ GHz$	−13.47 dB	0.95 dB	−10 dB	100 %
$G @ 5\ GHz$	16.23 dB	0.49 dB	15 dB	97.46 %
f_c	6.4 GHz	550 MHz	6 GHz	77 %
$IIP3$	−6.3 dBm	0.99 dBm	−7.5 dBm	99.02 %
P	6.82 mW	1.01 mW	8 mW	87 %

All specification yield: 60.74 %

more than two dimensions makes hardly representable the optimization in a graph.

However, the problem can become manageable if just two main variables are selected. Each of these variables must have a strong influence over each stage of the system. The rest of the variables will be the operation point of the system. Hence, an isograph with the main variables as axes can show the results of a specification. Furthermore, this allows determinate the area composed of the W1-W2 values that fulfil all desired specification. In other words, the design window is graphically visualized.

The design is implemented in 65-nm standard CMOS technology, and all presented figures and results emerge from simulations employing complete 4.6 BSIM models provided by the manufacturer. Also, the design considers relevant external parasitics effects, including in simulation the 50-fF load capacitance from the following stage and represented in Fig. 1 in grey.

Thanks to the mentioned methodology, specifications are accomplished on a large design window. The size of the design window is enough to guarantee performance against eventual mismatch issues. Fig. 2 shows the design window.

Nevertheless, the unavoidable simplification for an understandable evaluation of the problem implies the loss of some information about the variables of the operation point. Fig. 2 clearly shows that the circuit can accomplish the desired specification even though both W_1 and W_2 could change in ± 5 μm. However, the allowed range of variation in resistors

SMACD / PRIME 2021 | 19 – 22 July 2021, Online Event

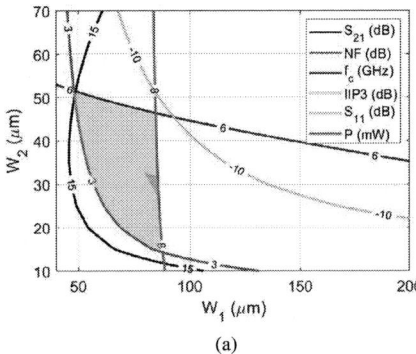

(a)

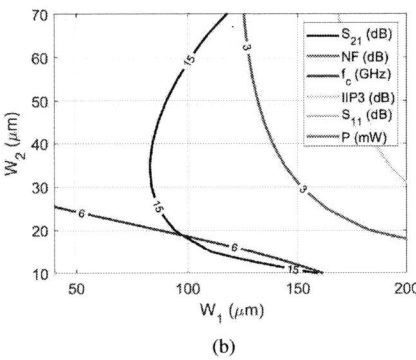

(b)

Fig. 3: Design window for different conditions: (a) ff corner and (b) ss corner. Design window (red shaded area) moves to the left side of the figure for ff corner due to the P constraint. As a result, there is almost no overlap with nominal-case design window (green shaded area). In ss corner, there is no valid design window as NF and f_c restriction never meet.

or bias voltage is unknown due to the assumptions done for solving the optimization problem.

That missing information might still have a notable impact. When the simulation applies process variations, and the design faces extreme conditions while evaluating corners, the design window will suffer significant changes. Nevertheless, transistor width must remain the same along with all corners and typical case, so the real design window is the intersection among the design window for each corner.

This issue has its impact on the results of a Montecarlo analysis with random process variation of the circuit devices (using technology models for statistical simulations). Table I shows the results. The yield, i.e., the percentage of valid samples among all studied ones on the analysis, has an important role to arrive at conclusions. Despite the wide design window, almost one sample in four will not achieve the desired 6-GHz f_c and other samples will demand higher consumption power than 8 mW. Even some samples do not fulfil gain or linearity requirements although the limits seem to be significantly far from the design point in Fig. 2. Besides, the yield of samples fulfilling all specifications falls to 60.74 %. The reason for this low number is that deviations rarely cause failing in more than one specification; the variance displaces the design out of the design window falling out of one of its borders.

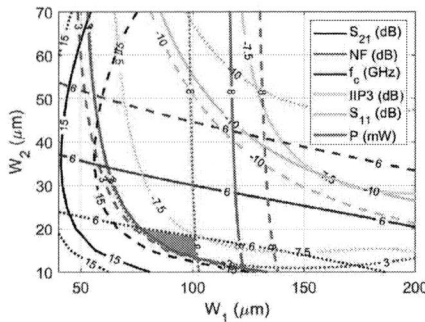

Fig. 4: Design window compensating process variations by bias voltage. Results from nominal case (solid line), ss corner (dot-dash line) and ff corner (dashed line) are shown. The point marks the final values for transistor widths.

To summarize, the previous methodology provides a solid design in the typical conditions, but the assumption for the optimization ignores important process variations. Thus, a method for evaluating the impact of process variability on the operation point variables is essential for robust optimization.

IV. CORNER ANALYSIS

An ss and ff corners are defined to evaluate the circuit against process variations. As the three-sigma extreme values are too restrictive and not a close representation of expectable variations, $\pm$ 10 % deviations from typical value are applied to resistors (+10 % on ss and -10 % on ff). In the case of transistors, the ff and ss transistor models are directly used on respective corners, as process variations modify internal device variables. The initial design is evaluated in these corners as shown in Fig. 3. The graphs clearly show that corners cause an enormous change.

Besides, the design variables cannot easily change in an integrated circuit to compensate for the variations, as they correspond to the physical sizing of the devices. Bias voltages are the exception. These voltage values can be externally controlled for a manual fix or by means of an adaptive bias control.

This method can deeply relax restrictions imposed by the corners to reach a common design window between corners. In other words, tuning the bias voltage provides a freedom degree to align ff and ss design windows.

Nevertheless, the method still requires evaluating simultaneously all specifications in both corners. As the bias voltage can radically change the shape of the window, the most advantageous biasing cannot be obtained from studying a single specification, neither observing the normally most critical for the corner; the designer must consider all the parameters. On the other hand,if a pair of transistor widths are valid in both corners with adequate bias voltages, the continuity of the effect of variations will guarantee that the pair is also inside of a typical design window. Besides, the biasing voltage in typical conditions is between ss and ff voltage bias values.

Fig. 4 shows the design window in the three corners: ss (dot-dashed), ff (dashed) and typical (solid). The common design

© VDE VERLAG GMBH · Berlin · Offenbach

TABLE II: Montecarlo results (512 samples) with bias compensation

	Mean	Std Dev	Spec	Yield
NF @ 5 GHz	2.88 dB	0.13 dB	3 dB	83.8 %
S_{11} @ 5 GHz	−15.94 dB	0.30 dB	−10 dB	100 %
G @ 5 GHz	15.91 dB	0.53 dB	15 dB	93.4 %
f_c	7.08 GHz	650 MHz	6 GHz	97.8 %
$IIP3$	−6.03 dBm	0.67 dBm	−7.5 dBm	100 %
P	7.77 mW	0.13 mW	8 mW	100 %

All specification yield: 81.64 %

TABLE III: Comparison of LNA performance

Spec.	This work	[2]	[5]	[6]	[3]
Tech. (nm)	65	180 (SiGe)	130	65	65
V_{DD} (V)	1.2	1.8	1.2	1.2	1.2
NF (dB)	2.88	4	4.0	3.5	5
G (dB)	15.9*	18**	20**	15.6**	20**
f_c (GHz)	7	9.6	2.7	5.2	7
S_{11} (dB)	-15.9	-10	-10	-10	-6
$IIP3$ (dBm)	-6	-11.3	-12	0	2
P (mW)	7.77	10	1.32	14	3.84

* Single-ended output
** Differential output

window is remarked in blue. In all corners V_{B2} is 1.1 V to keep M$_2$ in strong inversion, while V_{B1} varies to offset the process variation impact: 550 mV for the ff corner; 650 mV for the typical case; and 750 mV for ss corner. The ss f_c restriction noticeable constraints the design window because of the increment on the already high R_2 value. Note that it cannot decrease or the gain in ff corner will be drastically reduced and hence the NF. According to this figure, the optimal values for W_1 and W_2 are 100 μm and 15 μm, respectively. Paradoxically, despite having a significantly smaller area (and much closer borders), the design window from Fig. 4 is more robust against variations than Fig. 2.

For a fair comparison with that initial sizing and evaluating the advantage of the proposed method, Table II shows the results from a 512-samples Montecarlo analysis under the same conditions. Note that although NF yield slightly decreases, that sacrifice allows a noticeable improvement of f_c yield, and hence, there is an important increment in the number of samples that fulfil all specifications. The design of Table I has a yield of 60.74 % while the proposed optimization achieves an 81.64 % yield, i.e., the number of out-of-specification samples is halved thanks to the method.

Moreover, the yield estimation shown in Table II is pessimistic. The simulation process requires defining the concrete values for bias voltage. Thus, in the simulation V_{B1} cannot be tuned as an analog variable but in fixed steps (0.12 V due to computational constraints). Automated selection is done according to P as the closest to 8 mW that does not exceed the limit. The existence of this discretization of V_{B1} range can be observed in P results, as the mean is not 8 mW and it presents some deviation. Without that subtle loss of available P, some failed samples due to G or NF can reach respective limits and the change will improve the yield.

Table III compares the final results with similar designs in the literature. Although there are not experimental results from the proposed circuit, the extreme simulations with the complete manufacturer models provide a detailed insight into the system performance and capabilities. Note that a significant difference exists in gain comparison because of the different nature of outputs (differential/single-ended). If the application could use a differential LNA, transforming the proposed circuit to differential could duplicate G (+6 dB).

V. CONCLUSIONS

A robust, simple and competitive LNA in 65-nm standard CMOS technology is achieved. Besides, the design also accomplishes specifications in the defined corners. This is possible thanks to mitigating the highly adverse effects of process variations by a biasing strategy.

Due to the demanding nature of LNA, obtaining a design window or optimizing the circuit raises a series of issues. The methodology described in [1] allows for graphical representation of the design window at the cost of not considering some effects that can be evaluated by corners.

The effects of these extreme conditions cannot be neglected in design. They have a huge impact on LNA performance, dramatically altering the design window to the point it can disappear. Thus, a fixed set of design variables that accomplishes target specifications along all the corners is extremely impractical.

However, a variable biasing voltage can offset the process variations of the corners. Tuning V_{B1} in a 200 mV range significantly decreases the effect of process variability to the point that failed samples on Montecarlo analyses are halved. Thanks to the proposed methodology, the yield is increased from 60 % to beyond the 80 % mark in highly demanding specification circumstances.

The circuit has been recently sent to manufacture. Thus, in the conference, the authors will provide experimental results.

REFERENCES

[1] A. D. Martinez-Perez, C. Gimeno, D. Flandre, F. Aznar, G. Royo and C. Sanchez-Azqueta, *Methodology for Performance Optimization in Noise- and Distortion-Canceling LNA*, 16th International Conference on Synthesis, Modeling, Analysis and Simulation Methods and Applications to Circuit Design (SMACD), July 2019.

[2] I. R. Chamas and S. Raman, *Analysis, Design, and X-Band Implementation of a Self-Biased Active Feedback Gm-Boosted Common Gate CMOS LNA*, IEEE Transactions on Microwave Theory and Technique, Vol. 57, No. 3, pp. 542-551, March 2009.

[3] T. Chen, S. Rodriguez, J. Akerman and A. Rusu, *An Inductorless Wideband Balun-LNA For Spin Torque Oscillator-Based Field Sensing* IEEE International Conference on Electronics Circuits and Systens (ICECS), pp. 36-39, December 2014.

[4] A. D. Martinez-Perez, P. A. Martinez-Martinez, G. Royo, F. Aznar, G. Royo and S. Celma, *A New Approach to the Design of CMOS Inductorless Common-gate Low-noise Amplifiers*, 24th European Conference on Circuit Theory and Design (ECCTD), September 2020.

[5] F. Belmas, F. Hameau and J. M. Fournier, *A Low Power Inductorless LNA With Double Gm Enhancement in 130 nm CMOS*, IEEE Journal of Solid-State Circuits, Vol. 47, No. 5, pp. 1094-1103, May 2012.

[6] S. C. Blaakmeer, E. A. M. Klumperink, D. M. W. Leenaerts, and B. Nauta, *Wideband Balun-LNA With Simultaneous Output Balancing, Noise-Canceling and Distortion-Canceling* IEEE Journal of Solid-State Circuits, Vol. 43, No. 6, pp. 1341-1350, June 2008.

SMACD / PRIME 2021 | 19 – 22 July 2021, Online Event

Event-Driven Modeling and Simulation of 5G NR-Band RF Transceiver in SystemVerilog

Chan Young Park and Jaeha Kim

Elelctrical and Computer Engineering Department, Seoul National University, Seoul, Korea

chanyoung_park@mics.snu.ac.kr, jaeha@snu.ac.kr

Abstract—While baseband-equivalent real-number models (RNMs) are the current state-of-the-art for modeling RF transceivers in SystemVerilog, but their simulation speeds and accuracies are not adequate for predicting performance degradation due to DC offsets or high-order harmonic effects. This paper presents the models for a multi-standard, direct-conversion RF transceiver using XMODEL, for evaluating its system-level performance as well as verifying its digital controllers. The simulation results indicate that the presented models, including the digital configuration/calibration logic for the 5G sub-6GHz-band and mmWave-band transceiver, can deliver 30–1800× higher speeds than the baseband-equivalent RNMs while estimating the quadrature amplitude modulation signal constellation and error vector magnitude in the presence of non-idealities such as non-linearities, DC offsets, and I/Q imbalances.

Keywords—RF circuits, Modeling, SystemVerilog, XMODEL.

I. INTRODUCTION

With the growing interaction between the RF analog front-end and the digital calibration/selection logic in 5G multi-standard RF transceivers (TRX), an efficient simulation solution that is entirely based on SystemVerilog is required to verify these transceivers' functionality and to evaluate their performance. In RF transceivers that support various legacy bands and carrier aggregation, over 5,000 configuration bus bits are controlled by large complex digital logic to select a certain frequency band of operation, and various digital calibration and signal processing techniques are employed to improve their communication performance in the presence of non-idealities [1]. To verify the functionality and evaluate performance metrics, such as error vector magnitude (EVM), time-domain simulations including both analog/RF and digital sub-systems are necessary. In this study, to address the slow simulation speed of SPICE or SPICE-HDL co-simulation, multi-standard RF transceiver models are proposed that can run entirely within SystemVerilog and deliver 30–1800× faster speeds than those of the baseband-equivalent real-number models (RNMs) [2,3], leveraging the event-driven simulation of XMODEL [4].

Currently, there are mainly two approaches for modeling RF systems using RNMs: fast-but-inaccurate baseband-equivalent (BBEQ) modeling and accurate-but-slow passband modeling [2,3]. To express high-frequency RF signals that are modulated by low-frequency data, the baseband-equivalent models assume that the RF signals have a fixed-frequency carrier and only express its magnitude and phase information or, equivalently, the in-phase (I) and quadrature-phase (Q) information with a small number of events [3]. However, when these signals have to include the passband information, for example, to model frequency tones far from the carrier frequency such as the DC or high-order harmonic components, signals must be sampled with a sufficiently fine time-step to avoid aliasing, and the key benefits of the BBEQ modeling are lost. In these cases, the passband models that

```
module sin_gen #(parameter freq = 3.5e9,parameter tsp  = 10e-9)(
    input reg clk,                    // sampling clock
    input reg [11:0] ctrl_amp,        // control amplitude
    output real sin_out
);
real amp, t;
assign amp = 2 * (real'(ctrl_amp)) / 4096;
always @(posedge clk) begin
    t = $realtime;
    sin_out = amp * $sin(2*`M_PI*freq*t*tsp);
end
endmodule
```
(a) Example of the conventional RNM model.

```
module sin_gen #(parameter freq = 3.5e9)(
    input reg [11:0] ctrl_amp,        // control amplitude
    output real sin_out
);
xreal amp, sin_unit;
dac      #(.num_bit(12), .min(0.0), .max(2.0))
         XP0 (.in(ctrl_amp), .out(amp));
sin_gen #(.freq(freq)) XP1 (sin_unit);
multiply XP2 (.in(amp, sin_unit), .out(sin_out));
endmodule
```
(b) Example of the proposed XMODEL model.

Fig. 1. Signal representation.

express the RF signals may as well be used for their direct computations. However, the required large number of events typically slows down the simulations and limit the duration of the simulation to a few symbols, which is not sufficient to evaluate EVM and collect signal constellation, requiring at least 1,000 symbols [2].

In comparison, the event-driven signal representation and simulation algorithm used by XMODEL [4] can be an effective solution to address the aforementioned challenges. XMODEL expresses the continuous-time waveform of an analog signal $x(t)$ using the following functional expression, which also has its counterpart $X(s)$ in the Laplace domain [4]:

$$x(t) = \sum_i c_i t^{m_i-1} e^{-a_i t} u(t) \rightarrow X(s) = \sum_i \frac{c_i}{(s+a_i)^{m_i}} \quad (1)$$

In other words, each event during the simulation updates the values of the coefficients c_i's, m_i's, and a_i's, which collectively describe how the signal varies with time according to (1). The key difference between this approach and the RNM-based approaches is that the former does not rely on a large number of events to express a time-varying RF/analog signal. As the expression in (1) can contain an arbitrary number of terms, it can include additional frequency components without triggering additional events. Furthermore, the XMODEL primitives, such as *sin_gen* and *multiply*, each of which performs the operation suggested by its name, makes it easy to compose models simply by connecting them together [5].

Fig. 1 compares the RNM and XMODEL models for generating a sinusoidal signal of which amplitude is controlled by a digital code (*ctrl_amp*). The RNM model in Fig. 1(a) needs to evaluate a $sin()$ function at a constant time-step interval, which is set by the period of the clock (*clk*). This time-step interval will greatly determine the speed and accuracy of the simulation. On the other hand, the

© VDE VERLAG GMBH · Berlin · Offenbach

108

XMODEL model in Fig. 1(b) is described using a set of primitives and its simulation triggers events only when there is a change in the digital code (*ctrl_amp*). It is because (1) can express a sinusoidal function directly, without compromising accuracy.

This paper showcases a multi-standard 5G RF transceiver model using XMODEL. First, Section 2 addresses the key challenges in modeling the direct-conversion RF transceiver. Section 3 then describes the presented RF transceiver models in detail. Section 4 discusses the verification and simulation results for the 5G sub-6 GHz and mmWave-band operations. And finally, Section 5 concludes the paper.

II. DIRECT-CONVERSION RF TRANSCEIVER

While the direct-conversion architecture has the lower implementation costs and simpler frequency plan that can cater to multiple standards when compared to its heterodyne counterparts [6], it may be susceptible to the degradations in the communication performance due to the DC offsets and I/Q imbalance issues . For instance, the LO signal that leaks into the mixer input can cause self-mixing and create a DC offset in the mixer output. On the other hand, the gain/phase mismatch can cause an imbalance between the I/Q phases, which can destroy the orthogonality of the quadrature amplitude modulation (QAM) symbols. Consequently, most modern transceivers employ digital calibration loops to compensate for such imbalances.

III. PROPOSED TRANSCEIVER MODELS

This section describes the proposed SystemVerilog models for the direct-conversion RF transceiver. Fig. 2 shows the overall testbench configuration for the system. The system comprises mainly an analog/digital transceiver model described in SystemVerilog and Python scripts that can generate control codes and compute system performance from the simulated results. The analog parts of the transceiver are modeled using XMODEL primitives, whereas the digital parts are described in pure Verilog, which can be synthesized into gate-level descriptions after their functionalities are verified. The RF TRX model is designed based on a direct-conversion structure. The performance of the system is predicted through the chain simulation of sending modulated symbols, such as the QAM of OFDM on the transmitter side and restoring the symbols to data on the receiver side.

A. Mixer Model

The mixer block performs frequency conversion by multiplying the RF and carrier signals in time domain, and the modeling aims to reflect the nonlinear factors of the mixer that degrade the performance of the system. Any mismatch and leakage can cause a DC offset or gain/phase mismatch between the in-phase and quadrature-phase paths. The proposed method can simulate the passband signal in a fully event-driven method; thus, the output signal can be calculated by multiplying the actual RF/carrier signals reflecting the amplitude/phase error and leakage in the time domain. Fig. 3 shows the pseudo code of the proposed mixer model and illustrations of the calibration methods, respectively. First, for the RF signals ($IN_{I/Q}$) and carrier signals ($LO_{I/Q}$) applied to the mixer model, the gain and phase values are distorted by means of externally assigned control values, and any difference between these values may reflect the gain/phase

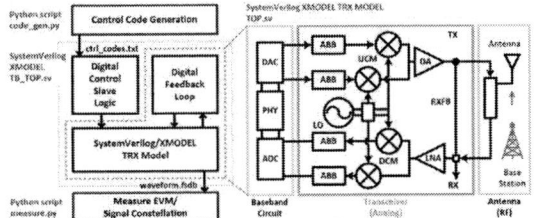

Fig. 2. Overall test bench organization.

```
module mixer (
    input reg [11:0] ctrl_XXXs, …      // control signals
    input xreal IN_I, IN_Q,            // input RF signals
    input xreal LO_I, LO_Q,            // input LO signals
    output xreal OUT_I, OUT_Q          // output RF signals
);
xreal g_IN_I, g_IN_Q, g_LO_I, g_LO_Q, …;

dac      #(.num_bit(12), .min(0.0), .max(2.0))
         XD0 (.in(ctrl_IN_AM_I), .out(g_IN_I));
multiply XM0 (.in(g_IN_I,IN_I), .out(s_IN_I));
dac      #(.num_bit(12), .min(0.0), .max(2.0))
         XD1 (.in(ctrl_IN_PM_I), .out(p_IN_I));
delay    XE1 (.delay(p_IN_I), .in(s_IN_I), out(d_IN_I));
…        // same for d_IN_Q, d_LO_I, d_LO_Q

// LO leakage signals added to input RF signals
dac      #(.num_bit(12), .min(0.0), .max(2.0))
         XD8 (.in(ctrl_LOL_AM_I), .out(g_LOL_I));
multiply XM4 (.in(g_LOL_I,LOL_I), .out(s_LOL_I));
dac      #(.num_bit(12), .min(0.0), .max(2.0))
         XD9 (.in(ctrl_LOL_PM_I), .out(p_LOL_I));
delay    XE3 (.delay(p_LOL_I), .in(s_LOL_I), out(d_LOL_I));
add      XA0 (.in(d_IN_I,d_LOL_I), .out(MIX_IN_I));
…        // same for MIX_IN_Q

// Mixing operation
multiply XM6 (.in(MIX_IN_I,d_LO_I), .out(MIX_OUT_I));
multiply XM7 (.in(MIX_IN_Q,d_LO_Q), .out(MIX_OUT_Q));

// DC offset calibration
dac      #(.num_bit(12), .min(0.0), .max(2.0))
         XD12 (.in(DCOC_I), .out(DC_offset_I));
add      XA3 (.in(MIX_OUT_I, DC_offset_I), .out(OUT_I));
…        // same for OUT_Q
endmodule
```

(a) Proposed pseudocode for the mixer model.

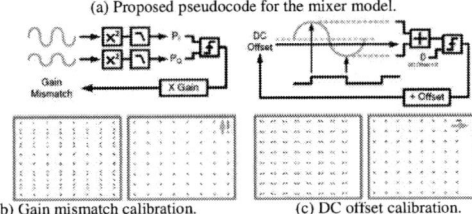

(b) Gain mismatch calibration.   (c) DC offset calibration.

Fig. 3. Mixer model and constellation diagrams.

mismatch between paths I and Q. The 12-bit gain/phase control codes are converted into an analog gain/phase value using the *dac* primitive, and the input signals are amplified by the gain value using the *multiply* primitive and delayed by the phase value using the *delay_var* primitive thereafter. Furthermore, some carrier signals (LO leakages) are leaked in the same manner and are added to the input signal using the *add* primitive. The input signals of the combined mixer are multiplied by the carrier signal, thus multiplying the actual mixing behavior as well.

Figs. 3 (b) and (c) show the gain mismatch calibration (GMC) method, DC offset calibration (DCOC) method, and signal constellation diagrams before and after the calibration. When the GMC calibration is turned on, the RX GMC block measures the power of each down-converted signal in paths I and Q, and then compares the magnitudes of the two signals. If the Q-path gain is greater than the I-path gain, it is lowered through the feedback loop, and vice versa. The DC offset value is calculated from the RX ADC's DCOC block and then

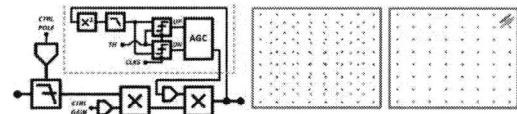

Fig. 4. Block diagram of the ABB model and constellation diagram.

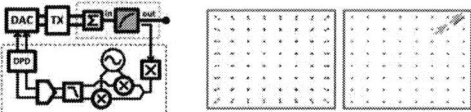

Fig. 5. Block diagram of the PA model and constellation diagram.

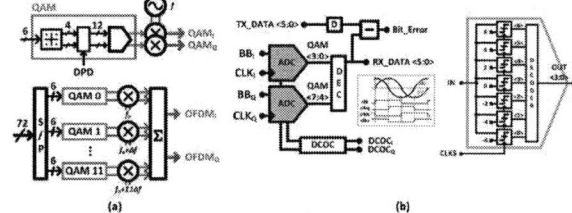

Fig. 6. Block diagrams of (a) the QAM/OFDM symbol generation model and (b) the ADC and ADC unit model.

assigned to the mixer through a feedback loop. These 12-bit offset values are converted to analog offset values via the *adc* primitive and then subtracted from each path using the *add* primitive. The DCOC block uses a set of four clock phases to sample the signals of the paths I and Q twice each, and searches for the offset values that makes the absolute difference between the two sample values equal to zero.

B. Analog Baseband Model

The analog baseband model (ABB) filters the high-frequency components that are generated during the modulation process on the TX and RX sides and automatically controls the signal power to fit within a defined range. As shown in Fig. 4, the ABB model consists of a filter block and a variable-gain amplifier block. First, the filter model is designed with a Chebyshev type-2 structure using the channel bandwidth value as the cut-off frequency. As the characteristics of the filter are determined by the parameters (gain, poles, and zeros) of the *filter* primitives, the values of these parameters must be changed according to the digital control code to design a variable-bandwidth filter. Furthermore, the variable-gain amplifier of the ABB maintains the amplitude of the output signal within a certain range through a feedback loop that automatically controls the amplifier gain. The input signals are amplified by the gain that is determined by the 12-bit external bus (*gain_dac*) and the gain generated through the AGC feedback loop (*gain_agc*). In the AGC loop, first, the power is calculated from the output signal (*out*) as in GMC, and this value is compared with a reference value (*th*) to ensure that the level of the output signal is within the reference level.

C. Power Amplifier Model

The power amplifier (PA) block amplifies large power input signals outside of the linear region owing to the transconductance of the transistor. Consequently, the non-linearities should be compensated to avoid signal distortion. The gain-compression relationship between the input and output signal can be defined by the polynomial function ($y = gain \cdot x - gain \cdot k \cdot x^3$) and it can be modeled by assigning

the coefficient of each term to the parameter of the *poly_func* primitive, as shown in Fig. 5.

One method used to compensate the nonlinearity of the PA is the digital predistortion (DPD) [7], which measures the distortion of the output signal and pre-distorts the input signal from the DAC model through a feedback loop so that the output signal achieves the desired linearity. In the DPD model (dotted box), the output signal is first scaled by the reciprocal of the gain value and then multiplied by the LO signal to separate the paths I and Q. The harmonic components of each signal are filtered, and the amplitude level of the remaining baseband signal is converted into a 12-bit bus signal ($DPD_{I/Q}$) using the *adc* primitive. In digital DPD logic, this 12-bit measured amplitude is compared to the reference level, and the offset predistortion value is calculated and assigned to the DAC model. The signal constellation diagrams show the symbol locations without and with the DPD enabled.

D. Baseband Circuit Model

The baseband circuit (BB) block models the function used in the transceiver modem to process signals on both the TX and RX sides. The TX side performs a DAC function that generates pseudo-random binary sequence data and converts it to a QAM/OFDM symbol, and the RX side performs the reverse function of the transmitter to restore the data, that is, the ADC function.

Fig. 6 (a) shows the block diagrams of the QAM and OFDM symbol generation models. When the "*qam_gen*" block converts the 6-bit data into I/Q QAM level which reflects the DPD information, the "cosine" and "sine" signals, each consisting of a sub-carrier frequency, are multiplied and sent to the TX side. In addition, the OFDM generation block models the behavior of the inverse fast Fourier transform, which then multiplies the signals generated from 12 QAM generation blocks by each sub-carrier signal and combines all these signals.

Fig. 6 (b) describes the ADC model, which samples the output signal of the RX ABB and restores this information to the original data sequences. The QAM and DC offset information can be obtained by sampling these two signals with a 4-phase clock. First, the "*clk*" clock signal with period "*time_sym*" is delayed by T/4, T/2, and 3T/4 to generate *clkq*, *clkb*, and *clks*, respectively. When the clock (*clkq* or *clkb*) is triggered, each input signal ($BB_{I/Q}$) is sampled by the *sample* primitive and converted into 4-bit QAM level information in the "*adc_unit*" model. Finally, an 8-bit QAM signal (4 bits in each I/Q path) is demapped to a 6-bit original data sequence.

IV. SIMULATION RESULTS

The EVM metric measures the distance of the QAM constellation points from the reference locations, and is computed as a root-mean-squared magnitude of these error vectors normalized to the ideal signal level. Fig. 7 shows the signal constellation diagrams and EVM values, obtained by post-processing the simulated results in Python. For each constellation diagram, the blue and red dots represent the I/Q reference and I/Q received symbols, respectively. Figs. 7(a) and (b) show the signal constellations when the transceiver model is operating at 5G n78 (sub-6GHz) and n257

© VDE VERLAG GMBH · Berlin · Offenbach

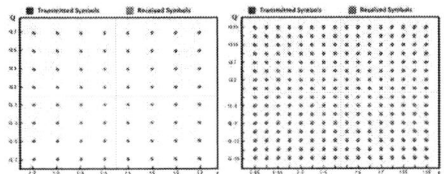

Fig. 7. Signal constellation diagrams : (a) 64-QAM and (b) 256-QAM.

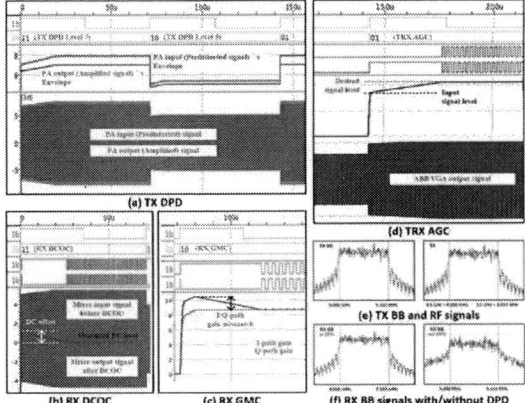

Fig. 8. (a-d) Simulated waveforms and (e-f) FFT analysis results.

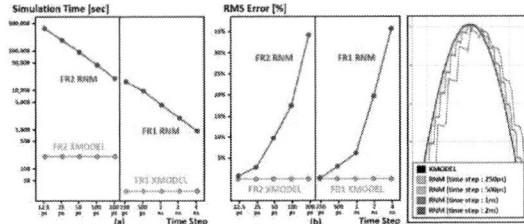

Fig. 9. Simulation performances : (a) run time and (b) RMS error.

(mmWave) band, and the simulated EVMs are 0.78% and 1.23%, respectively.

Figs. 8(a) to (d) show the simulated waveforms assuming a non-ideal case, where the DC offset of the I-path is -0.5, and the gain of the Q-path is 20% greater than that of the I-path. Figs. 8(e) and (f) show the FFT analysis results of OFDM symbols with 15 kHz spacing from 9 to 9.165 MHz and 3.5 GHz up/down-link: the OFDM symbol, the up-converted signal, and RX signals with and without DPD calibration in the PA. Without DPD, a spectral regrowth due to the intermodulation components deteriorates the RX sensitivity.

To evaluate the speed of simulation of the proposed model, we created a baseband-equivalent RNM that could perform the same operation. In RNM, only the frequency components of the baseband-equivalent signal near the carrier frequency are modeled, and the high-frequency harmonics and DC components are ignored. The non-idealities and non-linearities, caused due to the mismatch between the gain and phase, are equally reflected in the models. In addition, the filters are modeled in the z-domain through bilinear transformation. Both the RNM and XMODEL models share the same pure Verilog digital control/calibration loop, thus creating a model that performs exactly the equivalent operations.

The RNM witnesses a trade-off between the simulation performance and accuracy over time-step. The simulation

speed and accuracy of the RNM models are dependent on the simulation time-step of SystemVerilog, which determines the time spacing between the adjacent points on the RF signal waveforms. Figs. 9(a) and (b) show the speed and accuracy of the simulation results when the simulation is run for 1 s under various time-step conditions. For the FR1 band (3.5 GHz carrier and 10 MHz bandwidth), the runtime of the RNM decreases from 16,367 to 910 s when the simulation is performed while sweeping time-steps from 250 ps to 4 ns, whereas the RMS percentage error increases from 0.2 to 35.69%. Moreover, for the FR2 band (28 GHz carrier and 200 MHz bandwidth), the runtime decreases from 366,660 to 19,587 s, whereas the RMS error increases from 0.7 to 34.25 with sweeping time-steps from 12.5 ps to 200 ps. However, the proposed XMODEL-based model delivers a constant runtime and accuracy regardless of the time-steps (27 s in the FR1 band and 203 s in the FR2 band). For both FR1 and FR2 bands, the XMODEL-based model runs 30–1,800 times faster while transmitting high-frequency passband signals without sacrificing the accuracy, as shown in Fig. 9(b).

V. CONCLUSION

In this study, the SystemVerilog models for a multi-standard, direct-conversion RF transceiver that enables efficient event-driven simulation were presented. The models can estimate the performance metrics of the systems in the presence of various non-ideality conditions, such as DC offsets and I/Q imbalances, and verify the operation of the digital configuration/calibration controllers. The proposed models can serve the roles of the high-level MATLAB models for performance evaluation and digital RTL models for digital verification, and SPICE netlists for analog simulation while delivering fast speed entirely within SystemVerilog. The proposed models can be used as a simulation platform for exploring various RF transceiver architectures before IC design and as a verification testbed for checking the digital configuration/ calibration controllers before sign-off.

ACKNOWLEDGMENT

This work was supported by Samsung Electronics Co., Ltd, through the project SLSI-202005GE002S. The EDA tools were supported by the IC Design Education Center and Scientific Analog, Inc.

REFERENCES

[1] J. Lee, et al., "21.6 A Sub-6GHz 5G New Radio RF Transceiver Supporting EN-DC with 3.15Gb/s DL and 1.27Gb/s UL in 14nm FinFET CMOS," in IEEE ISSCC Dig. Tech. Papers, Feb. 2019.

[2] J. E. Chen, "A Modeling Methodology for Verifying Functionality of a Wireless Chip," in IEEE Behavioral Modeling and Simulation Workshop (BMAS), Sep. 2009.

[3] J. He, et al., "System-Level Time-Domain Behavioral Modeling for a Mobile WiMax Transceiver," in IEEE Behavioral Modeling and Simulation Workshop (BMAS), Sep. 2006.

[4] J. Jang, et al., "True Event-Driven Simulation of Analog/Mixed-Signal Behaviors in SystemVerilog: A Decision-Feedback Equalizing (DFE) Receiver Example," in IEEE Custom Integr. Circuits Conf., Sep. 2012.

[5] XMODEL Reference Manual, Release 2020.05, Scientific Analog, Inc., 2020.

[6] Mak, et al., "Transceiver Architecture Selection: Review, State-of-the-art Survey and Case Study," in IEEE Circuits Syst. Mag., Sep. 2007.

[7] Y. Liu, et al., "A General Digital Predistortion Architecture Using Constrained Feedback Bandwidth for Wideband Power Amplifiers," IEEE Trans. Microw. Theory Techn., May 2015.

Sensitivity analysis in dynamic WPT systems based on non-intrusive stochastic methods

P. Lagouanelle[*†], G. Di Capua[‡], N. Femia[§], F. Freschi[†], A. Maffucci[‡], L. Pichon[*], S. Ventre[‡]

[*]Group of Electrical Eng., Paris, CNRS, CentraleSupélec, Université Paris-Saclay, Gif-sur-Yvette, France
Group of Electrical Engineering-Paris, CNRS, Sorbonne Université, Paris, France
[†]Department of Energy, Politecnico of Torino, Turin, TO, Italy
[‡]Department of Electrical and Information Eng., University of Cassino and Southern Lazio, Cassino, FR, Italy
[§]Department of Information and Electrical Eng. and Applied Math., University of Salerno, Fisciano, SA, Italy

Abstract—The analysis of coil pairs mutual inductance is of great interest in the characterization, design and optimization of dynamic Wireless Power Transfer (WPT) systems for automotive applications. The objective of this paper is to show the use of non-intrusive stochastic methods to build accurate predictors of the mutual inductance. These methods are based on a polynomial-Chaos-Kriging metamodeling approach, which enables an accurate sensitivity analysis at system level. This approach is here applied to study the most influential spatial parameters in a dynamic WPT system, given different trajectories of the vehicle during its motion.

Index Terms—Inductive power transfer, mutual inductance, non-intrusive stochastic methods, sensitivity analysis.

I. INTRODUCTION

Dynamic battery-charging applications for Wireless Power Transfer (WPT) systems are crucial for the development of the overall mobility of electric vehicles [1]. The knowledge of the coupling between the receiving coil (RX) at the vehicle side and the transmitting coil (TX) at the ground side, is the key for maximizing the efficiency of the WPT system. In particular, the mutual inductance M is the key parameter influencing the overall performance of WPT systems, in terms of efficiency, power transfer, harmonic distortion, etc. In fact, the optimal design of such systems would require the knowledge of the mutual inductance in the variability range of the geometrical and physical parameters.

In real-world WPT systems, the coil pairs consist of complicated 3D geometries, including metallic and ferrite parts for the magnetic flux lines confinement. Analytical formulas of the mutual inductance are not easily computable, and its evaluation requires experimental measurements and/or numerical solutions of a magneto-quasi-static problem. Finite Element Method (FEM)-based analysis are in fact commonly adopted [2]. Recently, an analytical model of the mutual inductance of coil pairs in static WPT systems has been proposed in [3]. Similarly, in [4] an analytical model of the mutual inductance between coupled coils in dynamic WPT system has been presented. These models are obtained by using evolutionary algorithms. In both cases, several FEM simulations with different reciprocal positions between

the coils are performed by means of a commercial 3D FEM solver (Ansys Maxwell) and an in-house solver (CARIDDI code, [5]).

Different geometrical, physical and/or spatial parameters can be taken into account for the identification of analytical behavioral models of the mutual inductance in WPT applications. In this paper, we aim to perform the sensitivity analysis of spatial parameters of interest for dynamic WPT systems. In particular, the goal is to show the usefulness of non-intrusive methods by combining polynomial chaos expansions with Kriging metamodels in assessing the sensitivity of the electromagnetic problem under discussion to the lateral and the longitudinal displacements of the RX coil with respect to the ground TX coils. Using the behavioral model derived in [6], it was possible to map the mutual inductance in a large set of coils reciprocal positions, so providing the results used in this work as the input for the proposed methodology. Specifically, they are here used to perform the Sobol index sensitivity analysis [7] at a low computation cost. Such tools, implemented by using the UQLab framework [8], have been successfully used in the past for the determination of specific absorption rate in biological tissues due to mobile phones at microwaves frequencies [9], [10]. The same goes for an automotive WPT system with a simplified 3D model, where Polynomial chaos and Kriging methods have been really efficient [11].

The case study presented in this paper refers to a real WPT system, realized by the Politecnico di Torino, Italy, and analyzed in the frame of the project "Metrology for Inductive Charging of Electric Vehicles" (MICEV) [12].

II. THE MUTUAL INDUCTANCE MODEL

A. The input model

Fig. 1 shows the reference geometry in the plane (y, z), where the vehicle trajectory lies. The nominal trajectory is given by the red arrow in Fig. 1, representing the RX coil moving along the y-axis, with no lateral displacement ($\Delta z = 0$). Any other trajectory can be represented by coordinates ($\Delta y, \Delta z$). In this paper, we consider the behavioral model given in [6], which

© VDE VERLAG GMBH · Berlin · Offenbach

describes the mutual inductance as a function of Δy and Δz, as given in (1):

$$M_{tot} = M_{RX-TX1} + M_{RX-TX2} =$$
$$= M_{tx1,bhv}(\Delta y, \Delta z) + M_{tx1,bhv}(\Delta y - 2\Delta y_{mid}, \Delta z) \tag{1}$$

where $M_{tx1,bhv}$ is the analytical model for the $RX - TX1$ coil pair, and $2\Delta y_{mid} = 2.104$ m is the longitudinal displacement between the center of the two TX coils. The inductance $M_{tx1,bhv}$, expressed in µH, is given by (2):

$$M_{tx1,bhv} = p_0 tanh[p_1(\Delta y^2 + p_2)] + p_3 atan(|p_4\Delta y|^{p_5}) + p_6 \tag{2}$$

where the coefficients p_i (for $i = 0, .., 6$) are:

$$p_i = a_{i0} atan[a_{i1}(|\Delta z| - a_{i2})] + a_{i3} \tag{3}$$

with values of the fitting coefficients listed in Table I.

B. Polynomial-Chaos-Kriging metamodelling

Kriging is a stochastic interpolation algorithm that interpolates the local variations of the output $\mathbf{M}$ as a function of the neighboring experimental design points, whereas Polynomial-Chaos expansion approximates well the global behavior of $\mathbf{M}$. By combining the global and local approximation, a more accurate stochastic process can be achieved. Polynomial-Chaos-Kriging (PCK) is defined as a universal Kriging model the trend of which consists of a set of orthonormal polynomials. Given an input X of the parameters, the output $M(X)$ can be estimated by (4):

$$\widehat{M}(X) = \sum_{\alpha \in \mathcal{A}} y_\alpha \psi_\alpha(X) + \sigma^2 Z(X, \omega) \tag{4}$$

where $\sum_{\alpha \in \mathcal{A}} y_\alpha \psi_\alpha(X)$ is a weighted sum of orthonormal polynomials describing the trend of the PCK model, σ^2 and $Z(X, \omega)$ denote the variance and the zero mean, unit variance, stationary Gaussian process, respectively. Hence, PCK can be interpreted as a universal Kriging model with a specific trend.

Consistency of the metamodel: let's consider a set $\{(X_1, M_1), \ldots, (X_n, M_n)\}$ of n datapoints, where X_i and M_i for $i = 1, .., n$ are the input and their corresponding outputs. Using this set, one can build a metamodel $\widehat{M}(X)$ with PCK. The accuracy of the metamodel is calculated using the mean Leave-One-Out error (LOO), given in (5):

$$LOO = \frac{1}{n} \sum_{i=1}^{n} \left(\frac{\widehat{M}_{/i}(X_i) - M_i}{M_i} \right)^2 \tag{5}$$

where $\widehat{M}_{/i}$ is the mean predictor trained using all (X, Y) but (X_i, M_i). The LOO allows to evaluate the consistency of the metamodel. If the LOO is close to 1, the metamodel is highly modified if one datapoint is missing.

Accuracy of the metamodel: if one aims to build a metamodel $\widehat{M}_k(X)$ using a subset of k datapoints out of the aforementioned n datapoints, the accuracy of the predictor on the $(n - k)$ remaining points $\{(X_1, M_1), \ldots, (X_{n-k}, M_{n-k})\}$ can be calculated using the Out-of-Sample-Error (OSE), given in (6):

$$OSE = \frac{1}{n - k} \sum_{i=1}^{n-k} \left(\frac{\widehat{M}_k(X_i) - M_i}{M_i} \right)^2 \tag{6}$$

If the OSE for k datapoints is extremely small, it means that, at the non-sampled points, there is almost no difference between the predictor and the real value.

III. CASE STUDIES AND NUMERICAL RESULTS

A. Training datasets

To train our metamodels, three different datasets have been considered, referred to different trajectories of the vehicle during its motion (see Fig. 1):

- dataset 1, given by trajectories parallel to the nominal one, with only lateral displacement (blue arrow);
- dataset 2, given by a trajectory that moves diagonally with respect to the nominal one (green arrow);
- dataset 3, given by various trajectories for any possible misalignment of the coils: we sampled different points in the plane (y, z) to build our metamodel.

B. Dynamic charging applications with one TX coil

The aforementioned metamodel has been first run on a coil pair system made by only one TX coil. The mutual inductance values for the pair RX-TX1 against the longitudinal displacement Δy and the lateral misalignment Δz for any trajectory is shown in Fig. 2. It can be observed that the mutual inductance is also reaching its maximum in this area. The points sampled by our algorithm are shown in Fig. 3: a higher number of samples fall in the

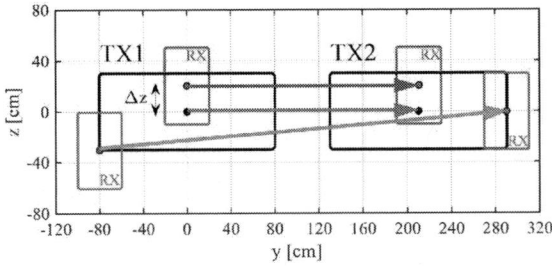

Fig. 1: Different trajectories of the RX coil moving along two TX coils. The nominal trajectory is represented by a red arrow.

TABLE I: Coefficient values for the model given in Eqs. (2)(3)

coefficient	a_{i0}	a_{i1}	a_{i2}	a_{i3}
p_0	$1.33 \cdot 10^1$	7.35	$1.90 \cdot 10^{-1}$	$-1.77 \cdot 10^1$
p_1	$1.36 \cdot 10^1$	$2.02 \cdot 10^1$	$2.57 \cdot 10^{-1}$	2.93
p_2	$-5.01 \cdot 10^{-2}$	8.40	$2.34 \cdot 10^{-1}$	$-4.84 \cdot 10^{-1}$
p_3	9.92	7.32	$1.87 \cdot 10^{-1}$	$-1.40 \cdot 10^1$
p_4	$1.20 \cdot 10^{-1}$	8.46	$2.63 \cdot 10^{-1}$	-1.50
p_5	1.08	7.28	$3.23 \cdot 10^{-1}$	-2.73
p_6	$-1.32 \cdot 10^1$	7.40	$1.89 \cdot 10^{-1}$	$1.79 \cdot 10^1$

range $\Delta z \in [40, 40]$cm and $\Delta y \in [80, 80]$cm, whereas less samples can be found outside these ranges. The three computed metamodels (one for each dataset) have been used to perform the sensitivity analysis, whose results are given in Table II. First, the three metamodels are extremely consistent with themselves ($LOO < 10^{-5}$), which ensures three independent but accurate sensitivity analysis indices. Then, the three indices are almost giving the same result for the trajectories: the longitudinal displacement Δy is the most relevant parameter. Indeed, the lateral displacement Δz realizes only a minor shift when the vehicle is moving along the charging lane.

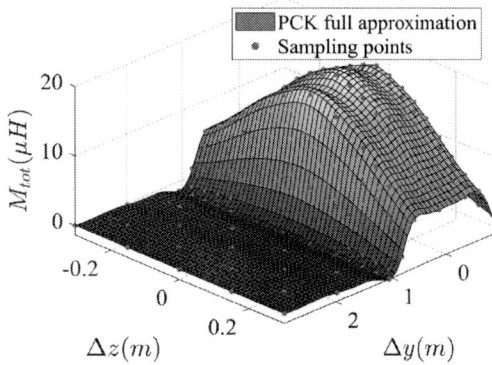

Fig. 2: Mutual inductance values for a single pair RX-TX against the longitudinal displacement Δy and the lateral misalignment Δz for any trajectory.

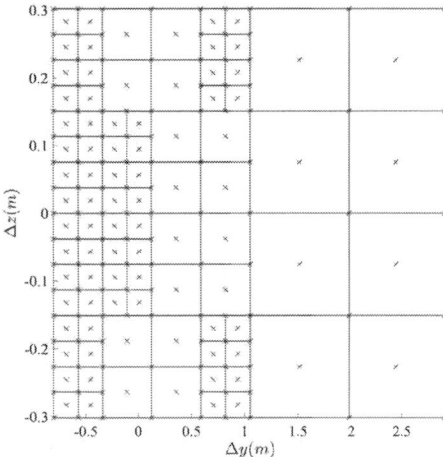

Fig. 3: Parameter domains and relevant samples used to build the metamodel for the mutual inductance for a single pair RX-TX against the longitudinal displacement Δy and the lateral misalignment Δz for any trajectory.

C. Dynamic charging applications with two TX coils

For two consecutive TX coils, as shown in Fig. 1, we consider the total mutual inductance M_{tot}. The area of

TABLE II: Sobol index analysis of the mutual inductance for a single pair RX-TX, against Δy and Δz.

Trajectory	Total $S_{\Delta y}$	Total $S_{\Delta z}$	LOO
dataset 1 (parallel trajectory)	0.943	0.132	$7.69 \cdot 10^{-6}$
dataset 2 (diagonal trajectory)	0.935	0.133	$1.26 \cdot 10^{-6}$
dataset 3 (any trajectory)	0.935	0.135	$1.56 \cdot 10^{-5}$

interest ranges from the position where the RX coil is on the top of the TX1 coil to that where it is on the top of the TX2 coil, which corresponds to $\Delta z \in [40, 40]$cm and $\Delta y \in [80, 80] \cup [130, 290]$cm. From Figs. 4 and 5 it can be seen that our algorithm sampled more points in these domains for various trajectories, and that the mutual inductance is also reaching its maximum in this area. As the spacing between the two TX coils is big enough, their areas of effect are not overlapped. As a consequence, the maximum value of the mutual inductance is the same one achieved for the model with only one TX coil. The three computed metamodels (one for each dataset) have been used to perform the sensitivity analysis, whose results are given in Table III. The three metamodels are still consistent with themselves ($LOO < 10^{-5}$), but the three analysis indices are not giving the same results. For a parallel trajectory, the RX coil is only seeing the effect of the second coil at the end of the trajectory which is not affecting the mutual inductance compared to the previous analysis. Conversely, for a diagonal trajectory and for any other trajectory, the RX coil cannot avoid the effect of the second TX coil. Therefore, the longitudinal displacement is not the most important parameter anymore. This means that a car, moving forward over a series of TX coils but not along a trajectory parallel to the nominal one, realizes a mutual inductance that is now much more dependent on its lateral misalignment. However, given the TX coil dimensions and the motion of the car along the y-axis, the longitudinal displacement remains the most influential parameter ($S_{\Delta y} > S_{\Delta z}$), as suggested by the results listed in Table III.

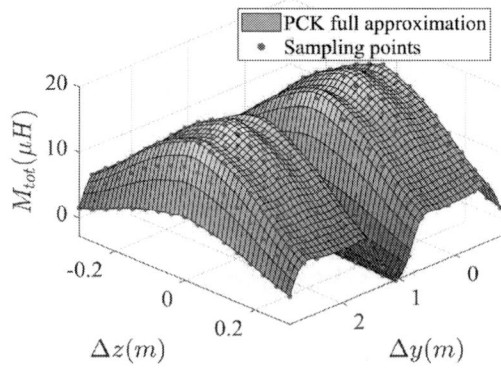

Fig. 4: Mutual inductance values for a single RX coil and two TX coils against the longitudinal displacement Δy and the lateral misalignment Δz for any trajectory.

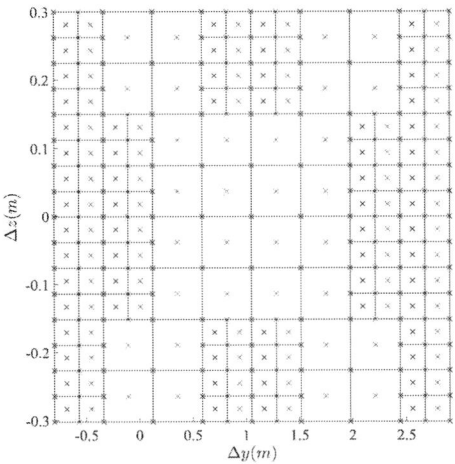

Fig. 5: Parameter domains and relevant samples used to build the metamodel for the mutual inductance for a single RX coil and two TX coils against the longitudinal displacement Δy and the lateral misalignment Δz for any trajectory.

TABLE III: Sobol index analysis of total mutual inductance over two TX coils, against Δy and Δz for various trajectories.

Trajectory	Total $S_{\Delta y}$	Total $S_{\Delta z}$	LOO
dataset 1 (parallel trajectory)	0.908	0.166	$1.60 \cdot 10^{-6}$
dataset 2 (diagonal trajectory)	0.656	0.375	$1.27 \cdot 10^{-6}$
dataset 3 (any trajectory)	0.684	0.349	$5.34 \cdot 10^{-6}$

D. Best sensitivity analysis

Three different metamodels have been computed so far, giving us three different sensitivity analysis. To find which metamodel is the best at predicting the behavior of the total mutual inductance, we tried to predict the values from the dataset 1 using the values of the dataset 2 and the other way around, while predicting both datasets with the metamodel build with any trajectory. By evaluating the OSE, we understand that the first two datasets are only able to predict the total mutual inductance behavior within their own validity range for the parameters values, as suggested by the results listed in Table IV. On the contrary, the third metamodel ensures quite low OSE values for both datasets, and can predict values of the mutual inductance in a wider range for the input parameters. As expected the metamodel for any trajectory is providing the best metamodel to work with, and its sensitivity analysis is the one to be taken into account.

TABLE IV: OSE on the values from datasets 1 and 2, against the metamodel predictors built with various trajectories.

Metamodel	dataset 1	dataset 2
dataset 1 (parallel trajectory)	$1.77 \cdot 10^{-2}$	2.28
dataset 2 (diagonal trajectory)	1.09	$5.29 \cdot 10^{-4}$
dataset 3 (any trajectory)	$3.04 \cdot 10^{-4}$	$4.20 \cdot 10^{-5}$

CONCLUSIONS

By using a Polynomial-Chaos-Kriging algorithm, we computed an accurate and consistent estimator for the mutual inductance in Wireless Power Transfer (WPT) system for dynamic charging applications. Using this predictor, an accurate sensitivity analysis has been performed to examine the effects of different spatial parameters at a low computation cost. Even if the effect of lateral misalignment between the coil pair on the resulting mutual inductance cannot be totally neglected for the design of WPT systems, the longitudinal displacement remains the most influential parameter in dynamic charging applications, which has to be taken into account in WPT systems analysis and design.

ACKNOWLEDGMENT

The results here presented have been developed in the framework of the EMPIR 16ENG08 MICEV Project. The EMPIR initiative is co-funded by the European Union's Horizon 2020 research and innovation program and the EMPIR participating States.

REFERENCES

[1] V. Cirimele, M. Diana, F. Freschi, and M. Mitolo, "Inductive power transfer for automotive applications: State-of-the-art and future trends," *IEEE Transactions on Industry Applications*, vol. 54, no. 5, pp. 4069–4079, 2018.

[2] B. Olukotun, J. Partridge, and R. W. Bucknall, "Optimal finite element modelling and 3d parametric analysis of strong coupled resonant coils for bidirectional wireless power transfer," in *2018 53rd Int. Universities Power Engineering Conf. (UPEC)*, 2018.

[3] G. Di Capua *et al.*, "Mutual inductance behavioral modeling for wireless power transfer system coils," *IEEE Transactions on Industrial Electronics*, vol. 68, no. 3, pp. 2196–2206, 2020.

[4] ——, "Analysis of dynamic wireless power transfer systems based on behavioral modeling of mutual inductance," *Sustainability*, vol. 13, no. 5, March 2021.

[5] R. Albanese and G. Rubinacci, "Finite element methods for the solution of 3d eddy current problems," *Advances in Imaging and Electron Physics*, vol. 102, pp. 1–86, 1997.

[6] K. Stoyka *et al.*, "Behavioral models for the analysis of dynamic wireless charging systems for electrical vehicles," in *2020 IEEE Int. Symp. on Circuits and Systems (ISCAS)*, Oct. 2020.

[7] I. M. Sobol, "Global sensitivity indices for nonlinear mathematical models and their monte carlo estimates," *Mathematics and computers in simulation*, vol. 55, no. 1-3, pp. 271–280, 2001.

[8] S. Marelli and B. Sudret, "Uqlab: A framework for uncertainty quantification in matlab," in *Vulnerability, uncertainty, and risk: quantification, mitigation, and management*, 2014, pp. 2554–2563.

[9] D. Voyer, F. Musy, L. Nicolas, and R. Perrussel, "Probabilistic methods applied to 2d electromagnetic numerical dosimetry," *COMPEL-The International Journal for Computation and Mathematics in Electrical and Electronic Engineering*, 2008.

[10] Silly-Carette *et al.*, "Variability on the propagation of a plane wave using stochastic collocation methods in a bio electromagnetic application," *IEEE microwave and wireless components letters*, vol. 19, no. 4, pp. 185–187, 2009.

[11] P. Lagouanelle, V.-L. Krauth, and L. Pichon, "Uncertainty quantification in the assessment of human exposure near wireless power transfer systems in automotive applications," in *2019 AEIT Int. Conf. of Electrical and Electronic Technologies for Automotive (AEIT AUTOMOTIVE)*, 2019.

[12] M. Zucca *et al.*, "Metrology for inductive charging of electric vehicles (micev)," in *2019 AEIT Int. Conf. of Electrical and Electronic Technologies for Automotive (AEIT AUTOMOTIVE)*, 2019.

© VDE VERLAG GMBH · Berlin · Offenbach

Performance Analysis of IPT Systems for Electric Vehicles Dynamic Battery Charging

Giulia Di Capua[1], Luca De Guglielmo[2] and Nicola Femia[2]

[1] Department of Electrical and Information Eng., University of Cassino and Southern Lazio, Cassino, FR, Italy
[2] Department of Information and Electrical Eng. and Applied Math., University of Salerno, Fisciano, SA, Italy
E-mails: {giulia.dicapua@unicas.it; ldeguglielmo@unisa.it; femia@unisa.it}

Abstract - **This paper investigates the joined influence of the vehicle trajectory, the coils mutual inductance, and the control strategy of transmitter and receiver power converter stages on the energy and efficiency performances of Inductive Power Transfer (IPT) systems for electric vehicles dynamic battery charging. A model based on normalized electric and magnetic variables is adopted to verify the maximum charge and energy efficiency of the IPT system, during the vehicle motion. The performances with respect to the control action implemented at the inverter and post-regulator stages and the vehicle nominal trajectory are also analyzed.**

Keywords – Control, Inductive Power Transfer Systems, Power and Efficiency Performance.

I. INTRODUCTION

Wireless Power Transfer Systems (WPTSs) allow to overcome the limitations of electric vehicle battery charging based on wired systems, mainly connected to the diversity of plugs and to charging points distribution [1]. Static WPTS have landed to a good technological maturity level in terms of components and architectures, and their installation can retrofit and/or integrate plug-in charging stations based on existing grid infrastructure and parking areas [2]. The development of dynamic WPTSs is still in progress, and several commercial and laboratory prototypes implementing different coupling coils, geometries, power ratings, and control strategies are available [3]. The H2020 *"Metrology for Inductive Charging of Electric Vehicles"* (MICEV) project [4] has been funded in a recent European Metrology Programme for Innovation and Research (EMPIR) call, to specifically investigate in a unitary frame the manifold WPTS issues about metrological techniques, models, methods and tools for human exposure assessment, and models and methods for power/efficiency performance analysis and design optimization. All these topics play an important role, especially for the possibility of performing reliable system-level investigations by means of analytical models and studying the combined effect of inherent coils characteristics and power electronics control features [5].

A key parameter affecting WPTS performances is the mutual inductance (M) between transmitting (TX) coil and receiving (RX) coil, which varies during the vehicle motion according to the relative instantaneous positions of the coil pair. The trajectory followed by a human driver is non-deterministic and likely deviates from the nominal one along TX coil longitudinal symmetry axis, which provides maximum coupling [3]. If the mutual inductance M is known as analytical function of the relative spatial position of the coils, an adaptive and predictive control of the TX and RX power converter stages, matching the trajectory of the motion, can be developed to achieve optimal WPTS performances.

Recently, a behavioral modeling approach has been profitably applied to obtain an analytical model of the mutual inductance between couples of TX-RX coil pairs comprising ferrite cores and metallic shields [6][7]. In particular, it has been proved that these behavioral models provide very simple analytical formulas that enable quite reliable WPTS performance predictions, both in static and in dynamic charging, starting from a reduced set of coils experimental measurements and/or numerical simulation results as input data to suitable evolutionary algorithms. In general, the behavioral modeling is a reliable and effective approach to solve performance analysis problems where the inherent complexity of the component or system under investigation does not allow for the formulation of a mathematical model based on physical laws and equations, mainly because of its intrinsic non-linearity.

This paper integrates the mutual inductance behavioral model of a coil pair and the main control system equations of a WPTS to predict the joint effect of coils characteristics and power electronics control constraints on the overall performances of a dynamic battery-charging system.

II. WPTS NORMALIZED MODELING

Fig. 1 shows a series-series resonant WPTS battery charger, with a Pulse-Width Modulated (PWM) inverter feeding the TX coil, a diode bridge rectifying the RX coil current, and a PWM DC-DC boost converter managing the power delivered to the battery under charge. The TX and RX total resistances R_1 and R_2 are determined by the coils and the resonant capacitors. The first harmonics model of the WPTS at resonance is given by (1):

$$\overline{V_1} = R_1 \overline{I_1} - j\omega_o M \overline{I_2} \qquad (1.a)$$

$$\overline{V_2} = R_{ac} \overline{I_2} = j\omega_o M \overline{I_1} - R_2 \overline{I_2} \qquad (1.b)$$

where ω_o is the resonance frequency, $R_{ac} = 8\,R_L/\pi^2$ is the equivalent first harmonic resistance at rectifier input, and R_L is the equivalent resistance at the boost converter input [8]. The main quantities of interest for a WPTS energetic performance analysis are the battery voltage and current V_{dc} and I_{dc}, and the DC source voltage and current V_{in} and I_{in}, which determine the input power, output power and power efficiency ratings, respectively given by $P_{in} = V_{in} I_{in}$, $P_{dc} = V_{dc} I_{dc}$, and $\eta_P = P_{dc} / P_{in}$.

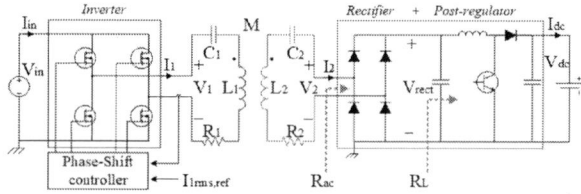

Fig. 1. Series-series resonant WPTS battery charger.

© VDE VERLAG GMBH · Berlin · Offenbach

The power efficiency η_P varies with the instantaneous position of the vehicle over each TX coil along the charging lane. Beyond the power efficiency, the energy efficiency η_U and the final energy U_{dc} transferred to the battery are also of interest, both varying during the vehicle motion. Finally, the *rms* value of TX coil current I_1 must be considered too, because it is subjected to the inverter current limit setting and determines the current I_{in} injected in the input DC bus. In fact, the inverter current limit and the post-regulator control setup heavily influence the overall WPTS performance. In [8] it was shown that some improvement in terms of power and energy transferred to the battery can be achieved by means of the post-regulator, if an optimal value of the equivalent load resistance R_L (or equivalently, R_{ac}) is adopted. In this paper, an enhancement of that control optimization is investigated, by exploiting the mutual inductance behavioral model discussed in [9] that provides M as function of the reciprocal TX-RX coils position.

The starting point of the study is the DC battery voltage. The value of V_{dc} can range from 100 V to 200 V for hybrid/plug-in hybrid vehicles and from 400 V to 800 V (and higher) for electric-only vehicles. The battery operation is managed by the boost post-regulator of Fig. 1, ensuring the limits for fast and safe battery charging while preventing a fast decay of the battery health. In static charging, the post-regulator control drives the duty-cycle D_p to feed the battery current I_{dc} until $V_{dc} = V_{max}$. Then, the duty-cycle D_p is driven so that I_{dc} decays over the time, until the full battery charge. In dynamic charging, instead, the post-regulator control can be designed to pursue Maximum Power (MP), Maximum Charge (MC) or Maximum Efficiency (ME) goal during the transit of the vehicle over the charging lane, while guaranteeing that $V_{dc} \leq V_{dc,max}$ and $I_{dc} \leq I_{dc,max}$. Meanwhile, the inverter control ensures that $I_1 \leq I_{1,max}$. Solving (1) yields:

$$I_1 = n_i \frac{4V_{in}}{\pi R_1} \frac{(1+r_{ac})}{(1+m+r_{ac})}; \quad I_{in} = n_i \frac{2}{\pi} I_1 \qquad (2)$$

$$I_2 = n_i \frac{4V_{in}}{\pi \sqrt{R_1 R_2}} \frac{\sqrt{m}}{(1+m+r_{ac})}; \quad I_{dc} = n_p \frac{2}{\pi} I_2 \qquad (3)$$

$$n_i = \sin\left(\frac{\pi}{2} D_i\right) = \frac{\pi}{4} \frac{V_1}{V_{in}}; \quad n_p = \frac{V_{rect}}{V_{dc}} = 1 - D_p \qquad (4)$$

$$m = \frac{\omega_o^2 M^2}{R_1 R_2}; \quad r_{ac} = \frac{R_{ac}}{R_2} = \frac{8V_{dc}}{\pi^2 R_2 I_{dc}} n_p^2 \qquad (5)$$

where D_i is the phase-shifted H-bridge inverter duty-cycle and D_p is the post-regulator boost converter duty-cycle. It can be demonstrated that:

$$\eta_P = \frac{P_{dc}}{P_{in}} = \eta_{inv}\eta_{rec} \frac{r_{ac}\, m}{(1+r_{ac})(1+m+r_{ac})} \qquad (6)$$

η_{inv} and η_{rec} being the inverter and rectifier efficiencies, respectively. For sake of simplicity, we consider ideal power efficiency conditions for the inverter and rectifier stages ($\eta_{inv} = \eta_{rec} = 1$). Based on (2)-(5), for any vehicle trajectory, the WPTS performances are then determined by D_i and D_p controls. Accordingly, an MP, MC or ME control can be alternatively implemented.

The vehicle trajectory and speed determine the mutual inductance time evolution $M(x_t, y_t)$, where (x_t, y_t) are the coordinates in the plane (x, y) of the trajectory followed by the RX coil with respect to the TX coil during the time t. The non-linearity resulting from ferrites and shields surrounding the TX-RX coils does not allow for an

analytical calculation of the value of M by means of the Maxwell formula and of the coils geometry and position. Experimental measurements or electromagnetic numerical calculations are then needed to obtain the value of M during the vehicle motion. A behavioral model of the coil pair mutual inductance has been recently discussed in [9], enabling easier performance analysis of dynamic WPTSs. Given any TR-RX coils setup, such a model allows to obtain a simple yet reliable analytical model of $M(x_t, y_t)$, including all non-linearity effects. This paper shows how the $M(x_t, y_t)$ model can be used to optimize the WPTS control in order to achieve MC and ME goals.

Given $m(x_t, y_t)$, let us suppose that the WPTS inverter control imposes a fixed nominal current operation of TX coil, with $I_1 = I_{1,ref}$. The simplest post-regulator control setting can be based on a target load resistance $R_{L,n}$, corresponding to a nominal value of $R_{ac,n}$, ensuring the best achievement of the MC-ME control goal. Accordingly, from (2)(3)(5) we get:

$$n_i = \frac{\pi R_1 I_{1,ref}}{4V_{in}}\left(\frac{1+r_{ac,n}+m}{1+r_{ac,n}}\right) \qquad (7)$$

$$n_p = \frac{\pi \sqrt{R_1 R_2} I_{1,ref}}{4V_{dc}} \frac{r_{ac,n}\sqrt{m}}{1+r_{ac,n}} \qquad (8)$$

Based on (7)(8), the input and output currents are:

$$I_{in} = \frac{R_1 I_{1,ref}^2}{2V_{in}}\left(\frac{1+r_{ac,n}+m}{1+r_{ac,n}}\right) \qquad (9)$$

$$I_{dc} = \frac{R_1 I_{1,ref}^2}{2V_{dc}} \frac{r_{ac,n} m}{(1+r_{ac,n})^2} \qquad (10)$$

and the inverter and post-regulator duty cycles are:

$$D_i = \frac{2}{\pi}\arcsin\left[\frac{\pi R_1 I_{1,ref}}{4V_{in}}\left(\frac{1+r_{ac,n}+m}{1+r_{ac,n}}\right)\right] \qquad (11)$$

$$D_p = 1 - \frac{\pi \sqrt{R_1 R_2} I_{1,ref}}{4V_{dc}} \frac{r_{ac,n}\sqrt{m}}{1+r_{ac,n}} \qquad (12)$$

subjected to the constraints $0 \leq D_i \leq 1$, $0 \leq D_p \leq 1$. Given $I_{1,ref}$, V_{in}, and V_{dc}, the inverter duty-cycle D_i saturates to 1 if:

$$m > \left(\frac{4V_{in}}{\pi R_1 I_{1,ref}} - 1\right)(1+r_{ac,n}) = m_i \qquad (13)$$

whereas the post-regulator duty-cycle D_p saturates to 0 if:

$$m > \left[\frac{4V_{dc}}{\pi \sqrt{R_1 R_2} I_{1,ref}}\left(\frac{1+r_{ac,n}+m}{1+r_{ac,n}}\right)\right]^2 = m_p \qquad (14)$$

Eqs. (13) and (14) allow to investigate the impact of the nominal value of $r_{ac,n}$ on the WPTS operation and performances, based on the shape of m determined by the vehicle trajectory. In particular, a desirable goal is to set $r_{ac,n}$ so that the inverter and post-regulator duty-cycles do not cross their own saturation thresholds during the transit of the vehicle over the TX coils, so that the system performances can be predicted analytically. Accordingly, we consider the operation of the WPTS such that neither the inverter nor the post-regulator duty cycles saturate at 1 and 0. This condition is satisfied if the peak value of normalized mutual inductance (m_{pk}) along the trajectory is always lower than the values of m_i and m_p. From (13) and (14), we get:

$$r_{ac,n} > \frac{m_{pk}}{4V_{in}/(\pi R_1 I_{1,ref})-1} - 1 = r_{ac,\min i} \quad (15)$$

$$r_{ac,n} > \frac{1}{\pi I_{1,ref}\sqrt{R_1 R_2}\sqrt{m_{pk}}/(4V_{dc})-1} = r_{ac,\min p} \quad (16)$$

Given the nominal RX current $I_{2,nom}$, from (3) the maximum DC battery current can be evaluated as:

$$I_{dc,\max} = I_{2,nom}\frac{\sqrt{R_1 R_2}\, I_{1,ref}}{2V_{dc}}\frac{r_{ac,n}\sqrt{m}}{1+r_{ac,n}} \quad (17)$$

From (10) and (17) we get a further minimum for $r_{ac,n}$:

$$I_{dc} < I_{dc,\max} \Rightarrow r_{ac,n} > \frac{I_{1,ref}}{2I_{2,nom}r_2}\sqrt{m_{pk}}-1 = r_{ac,\min dc} \quad (18)$$

The maximum value for $r_{ac,n}$ can be obtained by calculating the maximum power efficiency from (6):

$$\frac{\partial \eta_p}{\partial r_{ac,n}} = 0 \Rightarrow r_{ac,n\max} = \sqrt{1+m_{pk}} \quad (19)$$

Consequently, the $r_{ac,n}$ value must fulfill the condition (20):

$$\max\left\{r_{ac,n\min i}, r_{ac,n\min p}, r_{ac,n\min dc}\right\} < r_{ac,n} < r_{ac,n\max} \quad (20)$$

Finally, from Eqs. (9) and (10) and given $r_{ac,n}$, it can be demonstrated that the charge transferred to the battery and the resulting energy efficiency while the vehicle transits over each TX coil are given by:

$$Q = \int_{t_1}^{t_2} I_{dc}(t)dt = \frac{R_1 I_{1,ref}^2 \Delta t}{2V_{dc}}\frac{r_{ac,n}\, m_{av}}{(1+r_{ac,n})^2} \quad (21)$$

$$\eta_U = \frac{\int_{t_1}^{t_2} V_{dc}I_{dc}(t)dt}{\int_{t_1}^{t_2} V_{in}I_{in}(t)dt} = \frac{r_{ac,n}\, m_{av}}{(1+r_{ac,n})(1+r_{ac,n}+m_{av})} \quad (22)$$

where $\Delta t = (t_2 - t_1)$ is the time interval during which the RX coil moves crossing the TX coil, entering in $t = t_1$ and exiting in $t = t_2$, and m_{av} is the average value of the normalized squared mutual inductance m over the vehicle trajectory.

III. CASE STUDY AND DISCUSSION

As reference case study, we consider the WPTS dynamic battery charger discussed in [8], adopted as test bench in the frame of the European H2020-EMPIR MICEV project [4]. Nonetheless, the approach proposed in this paper is of general validity, independently of the specific TX-RX coils. In [8], the following nominal values of the parameters have been considered: $V_{in,n} = 500$ V, $I_{1,ref} = 40$ A, $I_{2,n}=50$ A $V_{dc,n} = 370$ V, $R_1 = 0.78\,\Omega$, $R_2 = 0.53\,\Omega$, $L_1 = 281.4\,\mu H$, $L_2 = 119.8\,\mu H$, $\omega_o = 534$ krad/s.

The coil pair of reference is shown in Fig. 2, with a list of the main geometrical parameters. Under low-speed conditions, we can consider the behavioral model of the mutual inductance valid for this coil pair as suggested in [9], which describes the value of M as a function of the longitudinal and lateral displacements between the center of the RX and TX coil, expressed as point coordinates (x_t, y_t), as given in (26)(27):

$$M(x_t, y_t) =$$
$$= p_0 \tanh\left[p_1(x_t^2 + p_2)\right] + p_3 \text{atan}(|p_4 x_t|^{p_5}) + p_6 \quad (26)$$

$$p_i = a_{i,0}\,\text{atan}(a_{i,1}(|y_t|-a_{i,2})) + a_{i,3}, \quad i = 0,...,6 \quad (27)$$

where x_t and y_t are in meters, M is in microhenries, and the coefficients $\{a_{i,0}, a_{i,1}, a_{i,2}, a_{i,3}\}$, for $i = 0,...,6$, are listed in Table II for a given vertical distance of 20 cm between the RX and TX coils. Based on Eqs. (26)(27), Fig. 3 shows the resulting plots of m while the vehicle is moving along the nominal trajectory the RX coil (black arrow in Fig. 3(a)). In particular, we can observe that the RX coil enters the TX coil with coordinates ($x_t = -100$ cm, $y_t = 0$ cm) and exits the TX coil with coordinates ($x_t = 100$ cm, $y_t = 0$ cm). From (26)(27), we have:

$$m_{pk} = \frac{\omega_o^2 M(0,0)^2}{R_1 R_2} = \frac{\omega_o^2}{R_1 R_2}\left[p_0 \tanh(p_1 p_2) + p_3\frac{\pi}{2} + p_6\right]^2 \quad (28)$$

$$p_i = a_{i,0}\,\text{atan}(-a_{i,1}a_{i,2}) + a_{i,3}, \quad i = 0,1,2,6 \quad (29)$$

where $M(0,0)$ is the value of the mutual inductance when the two coils are perfectly aligned in the center of their symmetry axis, thus ensuring the highest value of M.

TABLE II. COEFFICIENT VALUES FOR THE BEHAVIORAL MODEL IN (26)(27)

Coefficients	$a_{i,0}$	$a_{i,1}$	$a_{i,2}$	$a_{i,3}$
p_0	1.33E+01	7.35E+00	1.90E-01	-1.77E+01
p_1	1.36E-01	2.02E+01	2.57E-01	2.93E+00
p_2	-5.01E-02	8.40E+00	2.34E-01	-4.84E-01
p_3	9.92E+00	7.32E+00	1.87E-01	-1.40E+01
p_4	1.20E-01	8.46E+00	2.63E-01	-1.50E+00
p_5	1.08E+00	7.28E+00	3.23E-01	-2.73E+00
p_6	-1.32E+01	7.40E+00	1.89E-01	1.79E+01

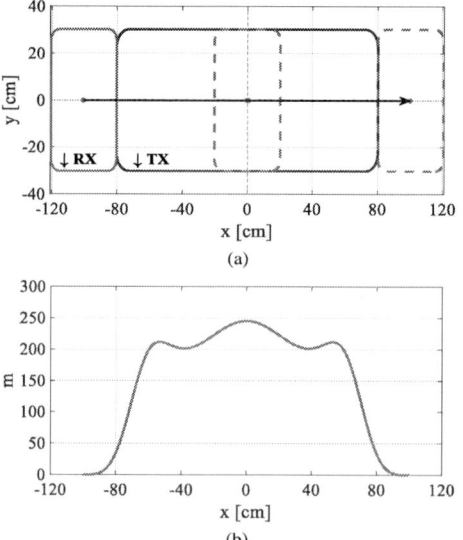

TX wire cross-section = 28 mm²
RX wire cross-section = 28 mm²
TX inner dimensions = 1.5m x 0.5m
RX inner dimensions = 0.3m x 0.5m
TX number of turns = 10
RX number of turns = 10

Fig. 2. Analyzed coil pair, with aluminum shield (grey) and ferrites (blue).

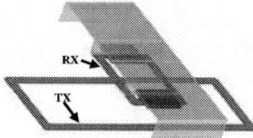

Fig. 3. (a) Nominal trajectory of the RX coil (red rectangle) over one TX coil (blue rectangle). (b) Normalized mutual inductance for the nominal trajectory of the RX coil moving along the TX coil longitudinal axis.

For the nominal trajectory considered in this paper, at the constant speed of 4 km/h, we have $m_{pk} = 245$. From the analysis of the minimum and maximum value of $r_{ac,n}$ discussed in Section II, we get: $r_{ac,n\,min} = r_{ac,n\,mini} = 12.3$ and $r_{ac,n\,max} = 15.7$. Fig. 4 shows the plot of the duty-cycle values of the inverter stage (blue lines) and of the post-regulator stage (red lines), for $r_{ac,n} = 1.05\,r_{ac,n\,min}$ (solid lines) and for $r_{ac,n} = r_{ac,n\,max}$ (dashed lines), proving that neither the inverter nor the post-regulator control saturates. Fig. 5 shows the system input current at the inverter stage (blue lines) and the output current of the post-regulator stage (red lines), for $r_{ac,n} = 1.05\,r_{ac,n\,min}$ (solid line) and for $r_{ac,n} = r_{ac,n\,max}$ (dashed line).

Fig. 6 shows the charge transferred to the battery, which is 34.5 C for $r_{ac,n} = 1.05\,r_{ac,n\,min}$ (solid lines) and 27.0 C for $r_{ac,n} = r_{ac,n\,max}$ (dashed lines), while Fig. 7 shows the energy delivered by the source (blue lines) and to the battery (red lines) for $r_{ac,n} = 1.05\,r_{ac,n\,min}$ (solid lines) and for $r_{ac,n} = r_{ac,n\,max}$ (dashed lines). At $t = t_2$ (when $x_t = 100$ cm and $y_t = 0$ cm), the RX coil is exiting the TX coil and the following values of energy are achieved: $U_{in} = 15.0$ kJ and $U_{dc} = 12.8$ kJ (when $r_{ac,n} = r_{ac,n\,min}$), $U_{in} = 11.8$ kJ and $U_{dc} = 10.0$ kJ (when $r_{ac,n} = r_{ac,n\,max}$). It follows that the energy efficiency are $\eta_U = 85.3\%$ and $\eta_U = 84.8\%$, respectively.

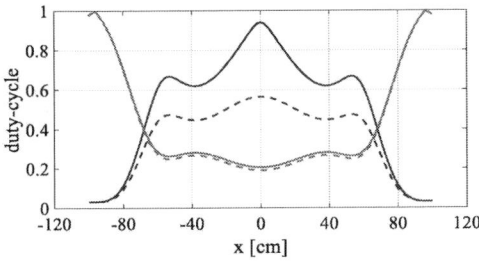

Fig. 4. Duty-cycle values of the inverter stage (blue lines) and of the post-regulator stage (red lines) for $r_{ac,n} = 1.05\,r_{ac,n\,min}$ (continuous lines) and $r_{ac,n} = r_{ac,n\,max}$ (dashed lines).

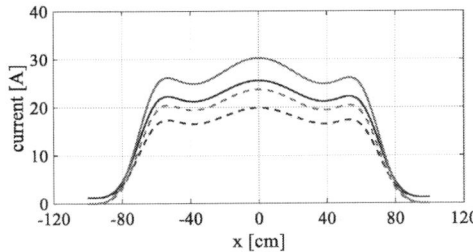

Fig. 5. Input current at the inverter stage (blue line) and output current at the post-regulator stage (red lines) for $r_{ac,n} = 1.05\,r_{ac,n\,min}$ (continuous lines) and $r_{ac,n} = r_{ac,n\,max}$ (dashed lines).

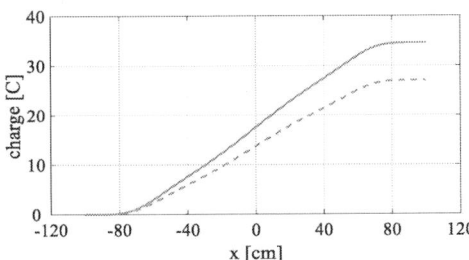

Fig. 6. Charge transferred to the battery for $r_{ac,n} = 1.05\,r_{ac,n\,min}$ (continuous line) and $r_{ac,n} = r_{ac,n\,max}$ (dashed line).

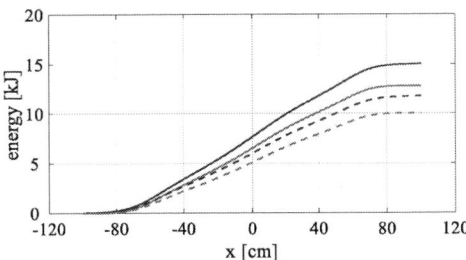

Fig. 7. Input energy (blue lines) and battery delivered energy (red lines) for $r_{ac,n} = 1.05\,r_{ac,n\,min}$ (continuous lines) and $r_{ac,n} = r_{ac,n\,max}$ (dashed lines).

CONCLUSIONS

This paper provides a novel analytical modeling approach that allows the setup of inverter and post-regulator controls in dynamic inductive power transfer systems for electric vehicle battery charging, fulfilling the requirements of maximum charge transferred to the battery and maximum energy efficiency. A behavioral model of the mutual inductance of coils has been adopted to predict the optimal value of the equivalent ac resistance to be set at receiver coil in order to achieve the given charge and efficiency goals, based on a reference trajectory of the vehicle transiting over a transmitter coil. The proposed model shows that the value of the ac resistance providing optimal performance depends on the vehicle trajectory, and allows to achieve a global performance optimization considering the trajectory uncertainty and tolerance, the reference values of the transmitter and receiver currents, and the dc input and battery voltages.

ACKNOWLEDGMENTS

The work was partially supported by the University of Salerno, through the project "Sistemi di Carica Induttiva di Veicoli Elettrici" (300638FRB18DICAPUA).

REFERENCES

[1] A. Ahmad, M. S. Alam, R. Chabaan, "A comprehensive review of wireless charging technologies for electric vehicles", *IEEE Trans. Transp. Electrif.*, vol. 4, pp. 38-63, 2018.

[2] D. Patil, M. K. McDonough, J. M. Miller, B. Fahimi, P. T. Balsara, "Wireless power transfer for vehicular applications: Overview and challenges", *IEEE Trans. Transp. Electrif.*, vol. 4, pp. 3–37. 2018.

[3] S. Laporte *et al.*, "Dynamic wireless power transfer charging infrastructure for future EVs: from experimental track to real circulated roads demonstrations", *World Electric Vehicle Journal*, vol. 10, n. 4, Nov. 2019.

[4] Project EU-H2020, MICEV—Metrology for Inductive Charging of Electric Vehicles, Available online: www.micev.eu (accessed on 15 March 2021).

[5] M. Zucca *et al.*, "Metrology for Inductive Charging of Electric Vehicles (MICEV)", *2019 AEIT Intern. Conf. of Electrical and Electronic Technologies for Automotive* (AEIT AUTOMOTIVE), Torino, Italy, July 2019.

[6] G. Di Capua *et al.*, "Mutual Inductance Behavioral Modeling for Wireless Power Transfer System Coils", *IEEE Trans. Ind. Electron.*, vol. 68, n. 3, pp. 2196–2206, March 2021.

[7] G. Di Capua *et al.*, "Analysis of Dynamic Wireless Power Transfer Systems Based on Behavioral Modeling of Mutual Inductance", *Sustainability*, vol. 13, n. 5, Feb. 2021.

[8] R. Ruffo et al., "Sensorless Control of the Charging Process of a Dynamic Inductive Power Transfer System with an Interleaved Nine-Phase Boost Converter", *IEEE Trans. Ind. Electron.*, vol. 65, n. 10, pp. 7630–7639, Oct. 2018.

[9] K. Stoyka *et al.*, "Behavioral Models for the Analysis of Dynamic Wireless Charging Systems for Electrical Vehicles", *2020 IEEE Int. Symp. on Circuits and Systems* (ISCAS), Seville, Spain, Oct. 2020.

© VDE VERLAG GMBH · Berlin · Offenbach

Coil Geometry Modeling and Optimization for a Bidirectional Wireless Power Transfer System

Simon Nigsch
Institute for Energy Systems
Eastern Switzerland University OST
Buchs SG, Switzerland
simon.nigsch@ost.ch

Falk Kyburz
Institute for Energy Systems
Eastern Switzerland University OST
Buchs SG, Switzerland
falk.kyburz@ost.ch

Kurt Schenk
Institute for Energy Systems
Eastern Switzerland University OST
Buchs SG, Switzerland
kurt.schenk@ost.ch

Abstract—The coil geometry and improvement of misalignment tolerance is a key challenge for the optimization of wireless power transfer (WPT) systems. In this paper a double-D shaped coil geometry is modeled and optimized for a bidirectional wireless power transfer system for electrical vehicles (EVs) with vehicle to grid (V2G) capabilities. With the new proposed winding configuration, the magnetic coupling is less sensitive to parking misalignments, which is useful for a highly efficient battery charging system. Further efforts to reduce the amount of ferrite material helps to lower the weight and cost of the system without affecting the performance. In order to verify the effect of the winding geometry experimentally, a 3.6 kW hardware prototype was built.

Keywords—*Wireless power transfer system, coil geometry optimization, bidirectional wireless charging*

I. INTRODUCTION

Contactless charging of electrical vehicle batteries with wireless power transfer (WPT) offers unique advantages compared with conventional conductive charging. Due to the absence of a galvanic connection, the charging progress requires no user interaction and no moving of mechanical components. This provides a more convenient and safer way to charge a car. It also enables the charging process to be integrated into regular vehicle operations by charging at bus stops, taxi stands or traffic lights along the route. Due to the shorter dwell times when charging in the depot, operators can reduce the number of fleet vehicles and the associated operating costs. In addition, the more frequent recharging reduces the depth of discharge of the battery. This extends the battery life and allows the electric vehicle to be dimensioned with a smaller storage capacity on board and consequently with less investment. Furthermore, wireless charging enables the possibility to use the EV as a local storage to provide grid stabilization and backup power since an average car is parked most of the time during the day. By extending a wireless charger with V2G capabilities, the availability of charging stations and different services is improved. However, in every wireless power transfer system, the power transfer efficiency is related to the coupling between the coils [1-3]. In the case of an EV charger this is variable as the different EVs consist of different charging pad size and ground clearance. Furthermore, the magnetic coupling is dependent on the alignment of the ground pad module (GPM) and car pad module (CPM). Therefore, a different type of double-D coil geometry has been developed to optimize the magnetic coupling under consideration of the given lateral and longitudinal misalignments. In this paper a magnetic coupler for a bidirectional wireless charging system has been simulated and optimized to achieve highest efficiencies while limiting cost and weight of the system. Starting from the conventional double-D winding design, this paper shows how the magnetic coupling variation can be reduced for different ground clearances and parking misalignments. With an adapted winding structure, the magnetic coupling is noticeably less sensitive to lateral parking misalignment, which helps to improve the magnetic coupling and the system efficiency. Additionally, the goal was to optimized the ferrite core geometry to reduce the weight and cost of the system while keeping or even increasing the magnetic coupling. Different types of core geometries have been proposed in the literature to maximize the coupling coefficient. The focus in those designs was to maximize the magnetic coupling without limiting the amount of core material. However, setting up a limit on the core loss with a systematic minimization of the core weight and loss were mostly ignored in those designs [5, 6]. Hence, the core design principles were realized only to enhance the mutual coupling. In this work the coupler and core geometry were optimized with the help of a 3D-FEM tool to reduce the size, weight and cost of the magnetic coupler without impacting the efficiency of the system.

II. SYSTEM DESIGN SPECIFICATIONS

The design of the magnetic coupler is based for a bidirectional WPT system for EV charging [9]. The nominal power for charging and discharging the car battery is 3.6kW. The system is meant to charge 400V batteries with a battery voltage in the range of 330 V – 440 V. This allows to use a 230 V / 16 A outlet to charge an EV within a few hours. The specifications of the coupler geometry and ground clearance meet the IEC 61980 and SAE J2954 standards. The designed system fulfills the power class WPT1 with the maximum power requirements specified in SAE J1772 for AC Level 1 charging. The vertical distance over which the power must be transferred is an important parameter for WPT system specifications. This distance is limited to 150 mm as given in TABLE I. The system must operate over a range in longitudinal and lateral direction to account for expected misalignments from the optimal parking position. A parking accuracy of 75 mm in longitudinal and 100 mm in lateral direction is supported by the designed system.

The specification of the primary and secondary power inverter for a bidirectional power flow as well as the compensation circuit is not described in this paper. This work focuses only on the design and optimization of the magnetic coupler.

TABLE I. GEOMETRY SPECIFICATIONS

GPM dimensions	520 mm x 650 mm
CPM dimensions	400 mm x 450 mm
Ground clearance distance	100 – 150 mm
Misalignment x / y	±75 mm / ±100 mm

© VDE VERLAG GMBH · Berlin · Offenbach

III. Coil Geometry Optimization

The coil-to-coil power transfer efficiency of an inductively coupled system depends on the quality factor Q of the WPT coils and the magnetic coupling k. The quality factor describes the power losses in the winding, the ferrite cores and aluminum shielding plate. The magnetic coupling is highly dependent on the coil geometry, size and misalignment of the coils. The maximum efficiency of an inductive power transfer system with series resonant compensation is given by the magnetic coupling and quality factor of the GPM and CPM coils. This relationship is stated in [4] and is given by

$$\eta_{max} = \frac{(kQ)^2}{\left(1 + \sqrt{1 + (kQ)^2}\right)^2} \approx 1 - \frac{2}{kQ} \tag{1}$$

From (1) it can be seen that the maximum transfer efficiency can be optimized by increasing the magnetic coupling. To reach the maximum efficiency for different misalignments of the system, the magnetic coupling among all parking positions should be as high as possible and the variation low. To optimize this, a 3-D FEM tool was used in the frequency domain. Starting from a standard double-D coil geometry shown in Fig. 1, the magnetic coupling for the given ground pad module (GPM) and car pad module (CPM) dimensions were simulated. As can be seen in Fig. 2, the magnetic coupling of this geometry is strongly dependent on the ground clearance and the lateral misalignment. The difference between the best and worst coupling within the same ground clearance is in the range of 200 %. This variation aggravates a highly efficient design as the compensation of the coil stray-inductances with the resonant capacitors cannot be designed efficiently. It can be shown that less variation for the coupling increases the overall performance of the system.

To improve this behavior, several geometry optimizations and core structure variations have been published. Various papers state an improvement of the coupling behavior by adjusting its design, but most of them focus on a single parking position or add more copper and core material to enhance the magnetic coupling. Additionally, none of them optimized the coil geometry to counteract for a high variation of magnetic coupling for different parking positions [7, 8].

In this work, an adapted double-D coil geometry as shown in Fig. 3 is presented, which shows an almost constant magnetic coupling over the entire lateral and longitudinal misalignment range. Compared to the coil geometry shown in Fig. 1, the double-D windings in the center of the transmitter coil are separated. To compare the performance and magnetic coupling of the GPM, the outer dimensions are kept constant. This adaption of the coil geometry results in a more homogenous magnetic field distribution but reduces the magnetic coupling to about 25 % for ideal positioned coils. However, it increases also the magnetic coupling for maximum misalignment to achieve a near constant magnetic coupling with a variation of less than 5 % at a defined ground clearance. The simulation results of this coil arrangement are shown in Fig. 4. At the same time, to optimize the compensation network, an almost constant self and leakage inductance helps to reduce the reactive power on the inverter stage. This lowers the losses and increase the overall efficiency of the WPT system for the given specifications.

Fig. 1 Simulation model of a standartd double-D coil for the given assembly dimensions and ferrite bars to limit the core weight.

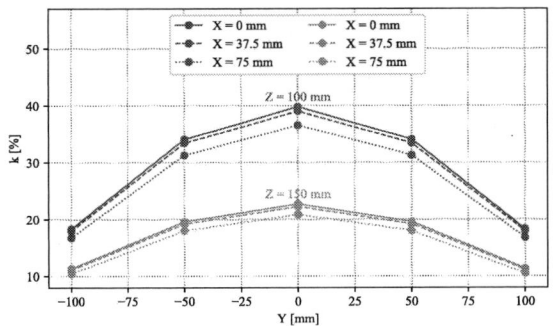

Fig. 2 Simulation of the magnetic coupling with a standard double-D coil for lateral and longitudinal misalignment with two different airgaps according to SAE J2954.

Fig. 3 Simulation model of the adapted double-D coil for the given assembly dimensions and ferrite bars to limit the core weight.

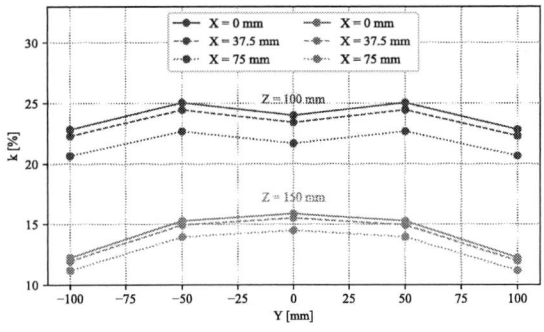

Fig. 4 Simulation of the magnetic coupling with the optimized double-D coil for lateral and longitudinal misalignment with different airgaps according to SAE J2954.

IV. Core Structure Optimization

For a WPT system, the ferrite core of a GPM and CPM partially provides a low reluctance path for the magnetic flux. The overall effect of this core in the flux path can be represented as a reduction of the reluctance. Therefore, it increases both the self and mutual inductance of the GPM and CPM. From Fig. 1 and Fig. 3, it can be seen that the size of the magnetic ferrite plates covers the whole double-D windings. By reducing the length of those ferrite bars, as shown in Fig. 6, the magnetic coupling is increased. By reducing the length of the ferrite bars, the magnetic flux is guided more through the coil, which increases the mutual flux. At the same time, the leakage flux is reduced since less flux is guided to the outside of the double-D coil caused by the lack of ferrite plates. This effect is visible when comparing the simulation results of the magnetic flux for both couplers shown in Fig. 6. A further reduction of the ferrite bar length and width will decrease the coupling. The simulation showed that the optimum ferrite size covers the area showed in Fig. 5 and applies to both couplers, the GPM and CPM.

This optimization of the ferrite bars results in a higher magnetic coupling compared to the initial design of Fig. 3. The improvement of the magnetic coupling is in the range of 4 % and does not depend on the misalignment or ground clearance. Additionally, the 3D-FEM simulation shows slightly lower core losses as the flux density in the bars stays the same. The saving of ferrite weight and cost for this design is in the range of 20 %.

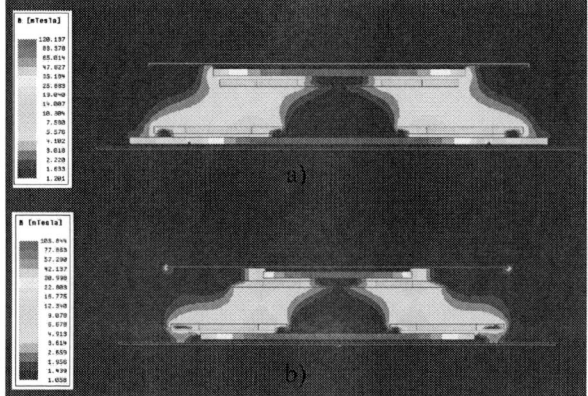

Fig. 6 3D-FEM simulation of the coupler, comparing the magnetic flux density of GPM and CPM. (a) with ferrite bars covering the whole winding and (b) optimized ferrite size.

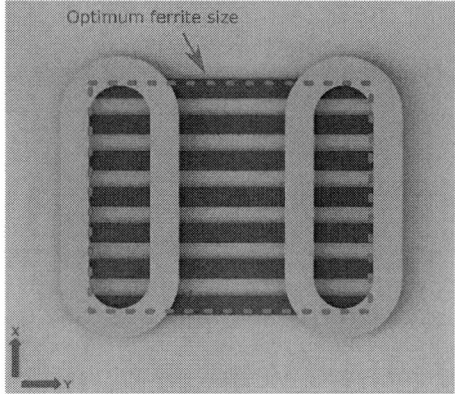

Fig. 7 Simulation model of the adapted double-D coil for the given assembly dimensions and ferrite bars to limit the core weight.

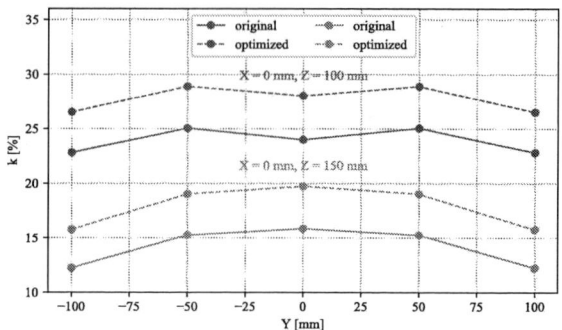

Fig. 8 Simulation of the magnetic coupling with the original and optimized double-D coil for lateral misalignment with different airgaps according to SAE J2954.

V. EXPERIMENTAL VERIFICATION

For experimental verification, a physical prototype was built according to the simulation model and characterized using measurements. Fig. 8 shows a photo of the lab prototype GPM coil. The dimensions are kept true to the model as close as possible to reduce the introduction of errors caused by geometry differences. Materials used in the prototype are 3 mm and 5 mm transparent PMMA, 1575 x 0.071 mm copper litz wire, DMR44 MnZn 30.5 mm x 60.5 mm ferrite plates, aluminum sheet metal and PA glass filled bolts. Using non-conducting materials for the construction of the prototype is required to prevent losses caused by the ac magnetic fields.

Fig. 9 Photograph of the investigated magnetic coupler with the new double-D coil winding for an optimized coupling behaviour.

A difference between simulation model and the prototype is the construction of the winding, which is modelled as a solid one turn winding in the simulation but built using several turns of litz wire with the same exterior dimensions as the rectangular single winding. The wires are kept in the rectangular shape with plastic coil formers, where the litz wire is snapped into slots at a regular distances. A comparison of the solid and litz wire version in a simulation without external magnetic fields resulted in negligible difference. Besides the geometric requirements, the coil former also ensures a minimum distance between the wires, which is required for the high voltage applied during operation. Unfortunately, the start and the end of the coil windings are not at the same location, which results in partial turns which violate the rectangular coil geometry slightly and can be a source of error.

To compare the simulation to the prototype, the characteristic inductances are used, which describe the wireless charging coupler as a regular transformer using the equivalent T model shown in Fig. 9. The FEM simulation outputs an inductance matrix of the form

$$\begin{bmatrix} u_1 \\ u_2 \end{bmatrix} = \begin{bmatrix} L_1 & M_{12} \\ M_{12} & L_2 \end{bmatrix} \begin{bmatrix} i'_1 \\ i'_2 \end{bmatrix} \tag{2}$$

which can easily be mapped to the T model. L_1 and L_2 are the inductances measured on the primary and secondary where the opposite winding has an open circuit. M is the mutual inductance.

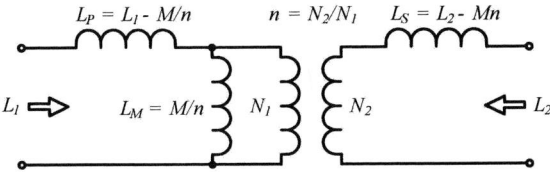

Fig. 10 Equivalent T model for the transformer with conversion formula from the inductance matrix.

Since the simulation model uses a single turn winding, the inductance values are scaled to the actual number of turns used in the prototype.

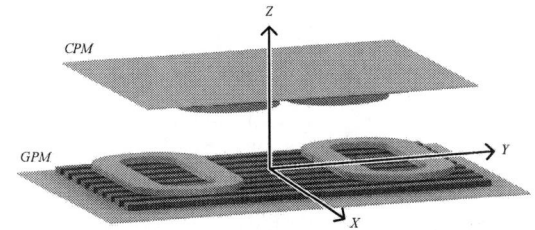

Fig. 11 Axis orientation and misalignment in X, Y and Z direction.

The coil arrangement definition an axis orientation of the longitudinal (x) and lateral (y) misalignment is shown in Fig. 10. The magnetic coupling was determined by measuring the open-circuit and short-circuit inductance with an impedance analyzer. The comparison of prototype measurements and simulation results for four different parking positions are summarized in TABLE II. As expected in a wireless power transfer system, the leakage inductances of the primary and secondary coils are much larger than the magnetizing inductance due to the low flux coupling. The four positions shown in the table are each two ground clearance variants at the center and at maximum misalignment. Comparing the minimum and maximum coupling their ratio is 2.14. On further inspection of the comparison, one can see that the errors are acceptable and the FEM model can be considered valid. Regarding the source of the errors, multiple issues are brought up:

- Parasitic capacitance in the primary and secondary windings of the prototype, which does not affect the transformer model, but is significant during operation.

- Inaccuracies in the way the windings are built in the prototype.

- Partial turns are necessary to due to start and end of the winding, which can lead to asymmetry.

- Measurable asymmetry in the prototype between misalignment quadrants.

For the purpose of the current work, the relative error of the self inductance is in the range of 4 % and therefore sufficient.

TABLE II.   COMPARISON OF FEM AND MEASURED DATA

Position [mm]				Inductance [µH]			Coupling [%]
X	Y	Z		L_P	L_S	L_M	k
0	0	100	FEM	317.2	213.8	74.0	24.0
			Meas.	304.7	215.5	62.8	21.5
0	0	150	FEM	311.6	227.9	45.0	15.9
			Meas.	305.4	227.0	40.7	14.8
75	100	100	FEM	293.8	230.6	60.6	20.7
			Meas.	289.8	228.1	53.2	18.8
75	100	150	FEM	313.2	245.1	31.1	11.2
			Meas.	308.7	243.2	28.2	10.4

VI. CONCLUSIONS

In this paper it is shown that the variation in the coupling coefficient in the standard double-D coil geometry can be improved by spacing the two inner parts of the winding apart. Reducing the variation in the characteristic inductances of the coupler simplifies the over-all system design and the average conversion efficiency is increased. The modification also improves the problem of high flux density in the ferrite at the center of the double-D coil. Another objective is the reduction of the ferrite volume required under the coils. The improved positioning of the ferrite bars reduces the volume of ferrite required by approximately 50% and increases the average coupling coefficient at the same time. In the new scheme the ferrite is only placed at areas where the coil requires a low reluctance flux path. The simulations are verified on a prototype built to the dimensions of the simulation model and the measurements are compared to the simulation output, which results in a good accuracy for practical applications.

REFERENCES

[1] Bi, Zicheng & Kan, Tianze & Mi, Chris & Zhang, Yiming & Zhao, Zhengming & Keoleian, Gregory. (2016). A review of wireless power transfer for electric vehicles: Prospects to enhance sustainable mobility. Applied Energy. 179. 413-425. 10.1016.

[2] J. Marquart, F. Kyburz, C. Mathis and K. Schenk, "FEA assisted design and optimization for a highly efficient 22 kW inductive charging system for electric vehicles with large air gap and output voltage variation," 2017 IEEE Applied Power Electronics Conference and Exposition (APEC), Tampa, FL, 2017, pp. 3640-3647.

[3] R. Haldi and K. Schenk, "A 3.5 kW wireless charger for electric vehicles with ultra high efficiency," 2014 IEEE Energy Conversion Congress and Exposition (ECCE), Pittsburgh, PA, 2014, pp. 668-674.

[4] R. Bosshard, J. W. Kolar, J. Mühlethaler, I. Stevanović, B. Wunsch and F. Canales, "Modeling and η - α -Pareto Optimization of Inductive Power Transfer Coils for Electric Vehicles," in IEEE Journal of Emerging and Selected Topics in Power Electronics, vol. 3, no. 1, pp. 50-64, March 2015.

[5] M. Budhia, J. T. Boys, G. A. Covic, and C.-Y. Huang, "Development of a single-sided flux magnetic coupler for electric vehicle IPT charging systems," IEEE Trans. Ind. Electron., vol. 60, no. 1, pp. 318–328, Jan. 2013.

[6] K. Aditya and S. S. Williamson, "Design considerations for loosely coupled inductive power transfer (IPT) system for electric vehicle battery charging," 2014 IEEE Transportation Electrification Conference and Expo (ITEC), Dearborn, MI, 2014, pp. 1-6.

[7] M. Mohammad and S. Choi, "Optimization of ferrite core to reduce the core loss in double-D pad of wireless charging system for electric vehicles," 2018 IEEE Applied Power Electronics Conference and Exposition (APEC), San Antonio, TX, 2018, pp. 1350-1356.

[8] M. Mohammad, S. Choi, Z. Islam, S. Kwak and J. Baek, "Core Design and Optimization for Better Misalignment Tolerance and Higher Range of Wireless Charging of PHEV," in IEEE Transactions on Transportation Electrification, vol. 3, no. 2, pp. 445-453, June 2017

[9] S. Nigsch, F. Kyburz and K. Schenk, "Bidirectional Wireless Power Transfer System with Optimized Coil Geometry," 2020 IEEE PELS Workshop on Emerging Technologies: Wireless Power Transfer (WoW), 2020, pp. 135-140.

© VDE VERLAG GMBH · Berlin · Offenbach

SMACD / PRIME 2021 | 19 – 22 July 2021, Online Event

Impact of the Pad Geometry on System-Level Performance Indicators for IPT Systems in Electrical Vehicles

Antonio Maffucci[1], Salvatore Ventre[1]

[1] Dep. of Electrical and Information Engineering
Univ. of Cassino and Southern Lazio
Cassino, Italy
maffucci@unicas.it, ventre@unicas.it

Alberto Delgado Exposito[2]

[2] Centre of Industrial Electronics (CEI)
Universidad Politécnica de Madrid
Madrid, Spain
a.delgado@upm.es

Abstract— **This paper deals with the choice of the shapes of the coils of a resonant Inductive Power Transfer (IPT) system for dynamic recharging of electrical vehicles. It investigates the impact of the coil shapes on some relevant system-level performance indicators, such as the voltage gain and the resonant frequency, associated to the variation of the coil coupling factor along the vehicle's trajectory. A trade-off is found between maximizing the coupling factor and minimizing its variations along the trajectory. The case study here is a dynamic real 85-kHz resonant IPT system, whose reference coil geometry is rectangular. Alternative coil shapes (circular, rounded edges) are investigated, highlighting what are the best options for the specific design optimization targets.**

Keywords— *electrical vehicle, inductive coupling, mutual inductance, wireless power transfer.*

I. INTRODUCTION

Developing easy and efficient battery recharging systems is one of the main challenges of the Electrical Vehicles (EV) mobility. One of the most promising recharging technologies is based on the Inductive Power Transfer (IPT), belonging to the more general class of the Wireless Power Transfer (WPT) technologies [1]-[2]. In such systems, the energy is transmitted via inductive coupling between transmitting (TX) coils embedded in the road and receiving coils (RX) located on the EV. The overall system, made by the coil pair and all the electronic subsystems at TX and RX sides (e.g., converters, compensation networks,..) is working in resonance condition. Thus, to guarantee the optimal performance of the IPT system, a parametric optimization is required during the design, and a proper control strategy must be implemented during the operation.

The geometry of the so-called "pad" (composed by coil pair, magnetic core and shields) is known to have a major impact on the system-level performance, and therefore its optimal configuration has been intensively investigated over the years [3]-[6]. However, so far a standard for the geometry has not been yet set, since the possible choices can have different impact depending on the performance indicators that the design aim to optimize. Indeed, a system-level optimization can address targets associated to different performance indicators (e.g., maximum transmitted power, power efficiency, total harmonic distortion, stability of gain and resonance frequency, human exposure safety, etc..), and so usually a trade-off compromise must be chosen, coming from a multi-objective optimization.

In this paper, we investigate two geometrical kinds of shapes for designing the coils. Specifically, the impact of these geometrical shapes is referred to the coupling factor k of the coils, that in turns affects two system-level performance indicators of resonant IPT systems, such as the voltage gain and the resonant frequency.

In Section II, these performance indicators are introduced with reference to different possible topologies of resonant IPT systems. In addition, their relation to the coupling factor k is also discussed. In Section III, starting from a real IPT system, with rectangular coils, a FEM analysis is carried out to investigate the impact of the shape of coils on the variation of coupling factor along the vehicle trajectory. Conclusions are drawn in Section IV.

II. SYSTEM-LEVEL PERFORMANCE INDICATORS

Figure 1 shows the general schematic of the series-series topologies for IPT applications where C_i ($i = TX, RX$) represents the primary and secondary capacitors places to resonate with the inductances L_i from transmitter or receiver side. Although this topology overview does not illustrate any dependence on the coupling factor k, it has large effects on different aspects such as gain or resonant frequency, requiring the implementation of a sophisticated regulator.

For instance, the series-series resonant topology operating as a current source, explained in [7]-[8], necessitates primary and secondary capacitors C_i tuned to each self-inductance L_i at the resonant frequency ω_{sf}:

$$C_i = \frac{1}{\omega_{sf}^2 L_i} \quad i = TX, RX \tag{1}$$

Then, the converter presents a particular current gain (input voltage V_{in} to output current I_{out}) that depends on the coupling factor k as follows:

$$\frac{I_{out}}{V_{in}} = \frac{1j}{\omega_{sf} k n_T L_{RX}} \tag{2}$$

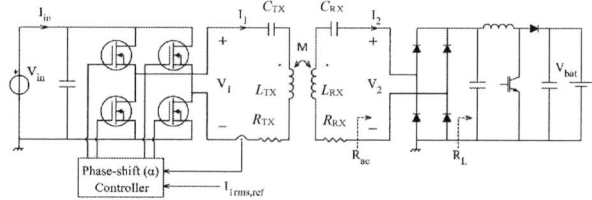

Fig. 1 Schematic of the series-series resonant IPT system studied here.

© VDE VERLAG GMBH · Berlin · Offenbach

being n_T defined as follows:

$$n_T = \sqrt{L_{TX}/L_{RX}}.\tag{3}$$

If TX and RX coils have the same geometry, with a number of turns equal to N_{TX} and N_{RX}, it is $n_T = N_{TX}/N_{RX}$.

Assuming that the air-gap variances has almost no effects on the self-inductance, we have that for the whole range of vertical distances the resonance is maintained. However, even if the air gap is fixed, in the case of dynamic charging the coupling factor can have large changes associated to the different reciprocal positions assumed by TX and RX coils along the vehicle trajectory. As a consequence, the gain described by equation (2) will have a large variation, inversely proportional to the coupling factor. It means that we must implement a very reliable regulation for the converter.

On the other hand, for the specific case of series-series resonant topology analyzed here, it is possible to obtain voltage source behavior [7] by changing the capacitor value (or the switching frequency):

$$C_i = \frac{1}{\omega_{sf}^2(1-k)L_i}\tag{4}$$

By doing this, as long as the resonance is kept, the converter shows a constant voltage gain that only depends on the turn ratio selected between transmitter and receiver:

$$\frac{V_{out}}{V_{in}} = \frac{1}{n_T}.\tag{5}$$

Note that (5) is valid since the series-series resonant compensation adopted here is using a leakage cancelation.

In this particular case, from eq. (4) one can notice that if the coupling factor varies the resonance will be compromised, so the constant output voltage behavior will be lost. Once again, this leads to the request of implementing a regulator that must be able to regulate the dynamic charging.

Therefore, it is very crucial to understand the importance of designing a link that is the least sensitive to coupling fluctuations, and for this reason, in this work we will study which geometrical shape is the best from the point of view of k variations.

III. GEOMETRY IMPACT ON INDUCTANCES

The reference geometry analyzed in this paper is given by a real IPT system for the dynamic EV charging, realized by the Politecnico di Torino, Italy, and analyzed in the project MICEV [9]. The system is composed of 50 TX coils embedded in the road, each of them able to deliver a power up to 20 kW, with a maximum vehicle speed of 100 km/h. The details of the systems are given in [10]. The geometry of the pad is shown in Fig. 2, where the TX and RX coils are depicted in red, the aluminum shielding structure in grey and the ferrite magnetic core in blue. The geometrical and physical parameters are summarized in Table I.

The analysis of this system has been performed in [11] for a static recharging condition by using an in-house full 3D

magneto-quasi-static (MQS) solver, Cariddi, based on integral formulation. This approach is well-tailored for such problems, since it only requires meshing the conducting and magnetic regions, avoiding to mesh air and impose fictitious boundaries to close the problem. The analysis has been extended to the dynamic case in [12], where the coil parameters are again obtained by using the MQS model. In both static and dynamic cases, the numerical results obtained with the MQS solver Cariddi have been successfully validated against the experimental data [11]-[12].

In Fig. 1 the circuit representation of overall system is provided, including the power electronics related to the TX and RX sides. The reference geometry for the dynamic charging systems is given in Fig. 2: we are here considering the effect of the motion of the vehicle over two adjacent transmitting coils, put at a distance of 50 cm. The other dimensions and parameters are given in Table 1. The RX coil is moving along the longitudinal axis following the nominal trajectory plotted in red in Fig. 3. Starting from this reference problem, that has been studied in [11], in this paper two different shapes of the TX coils are considered: rectangular (reference case) and rounded corner shapes. As for the RX coils, the considered shapes are rectangular or full circular.

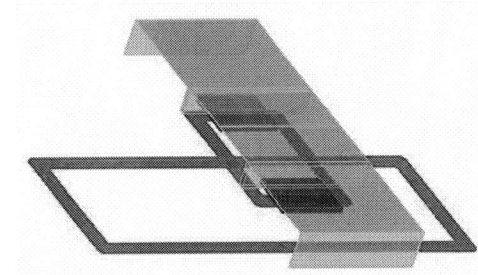

Fig. 2 Geometry of the pad of the reference IPT system [9]-[10]: coils are in red, shield in grey and ferrites in blue.

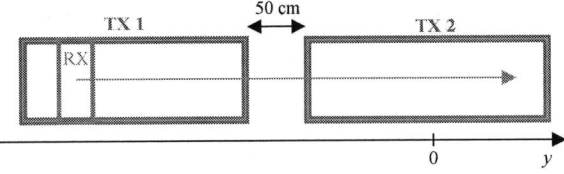

Fig. 3 Reference geometry for the dynamic charging system Polito IPT: nominal trajectory of a rectangular RX coil over two adjacent TX rectangular coils. The distance between the two TX coils is 50 cm. All the other dimensions are given in Table 1.

TABLE I - PARAMETERS OF THE REFERENCE IPT SYSTEM COIL PAIR [10]

Parameter	Value
Frequency	85 kHz
TX and RX wire cross-section	28 mm²
TX number of turns, and inner dimensions	10, 1.5 m x 0.5 m
RX number of turns, and inner dimensions	10, 0.3 m x 0.5 m
Air gap (vertical distance between coils), Δx	20 cm
Chassis conductivity (Aluminum)	33.4 MS/m
Ferrite relative permeability	2000

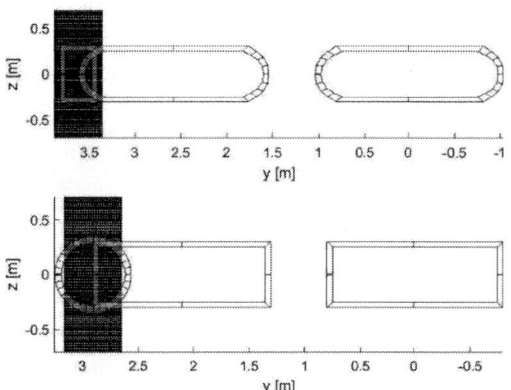

Fig. 4 Two shape combinations analyzed here: (a) C-R, circular edge TX, rectangular RX; (b) R-C, rectangular TX, circular RX.

TABLE II – COMBINATIONS OF THE COIL SHAPES ANALYZED IN THIS PAPER

Combination	Description
R-R	rectangular TX, rectangular RX (reference)
C-R	circular edge TX, rectangular RX
R-C	rectangular TX, circular RX
C-C	circular edge TX, circular RX

TABLE III – RESISTANCES AND INDUCTANCE VALUES OF THE COIL PAIRS IN THE REFERENCE IPT SYSTEM [10]-[11]

MQS values	R_{TX} (mΩ)	R_{RX} (mΩ)	L_{TX} (μH)	L_{RX} (μH)	M (μH)
R-R	359	128	281.4	119.8	18.3

All the four combinations of coil shapes in Table II have been analyzed: the combinations C-R and R-C are shown in Fig. 4. The choice adopted here is to maintain the same maximum size for the TX coil and the same maximum size in the z-direction for the RX coil.

In order to study the dynamic system, we have preliminarily checked the conditions under which the MQS model may be used to derive the inductances associated to each position along the trajectory. As set in [12], this is possible if the following condition holds:

$$\frac{4\,\tau_M}{t_f} \ll 1 \qquad (6)$$

where t_f is the flight time, that is the time needed by the vehicle to cover the distance of a TX coil at the given velociti, amd τ_M is the magnetic time constant of the system, defined as:

$$\tau_M = max\,[eig\,(\mathbf{R^{-1}L})] \qquad (7)$$

being **R** and **L** the resistance and inductance matrices describing the coils in the MQS limit and in the nominal position (RX coil placed at $y = 0$, see Fig. 3). For the reference problem analyzed here, these parameters are listed in Table III. Given these parameters, the time constant is estimated to be $\tau_M \approx 0.99$ ms, and the flight time at a speed of 100 km/h is $t_f \approx 54$ ms, hence condition (7) is satisfied and the MQS model can be used to analyse our case.

As expected, the main change occuring along the trajectory is related to the mutual inductance, whereas the self inductances are only slightly influenced by the position: for instance, Fig. 5 reports the spatial distribution of the mutual and self inductances along the trajectory for the case C-R. As clearly shown, the relative variation of the mutual inductance is high, with the maxima located in the positions where the RX coils stay in the center of the TX ones, and a minimum corresponding to the positions between the two coils. The relative variations on the self inductances, related to the eddy currents excited in the coils and in the shield, are below 0.5% for the TX coils and of 3% for the RX coil. Similar behavior is found for the other cases in Table II.

Starting from the knowledge of the inductance values, we have computed the coupling factor, defined as:

$$k = M/\sqrt{L_{TX}L_{RX}}. \qquad (8)$$

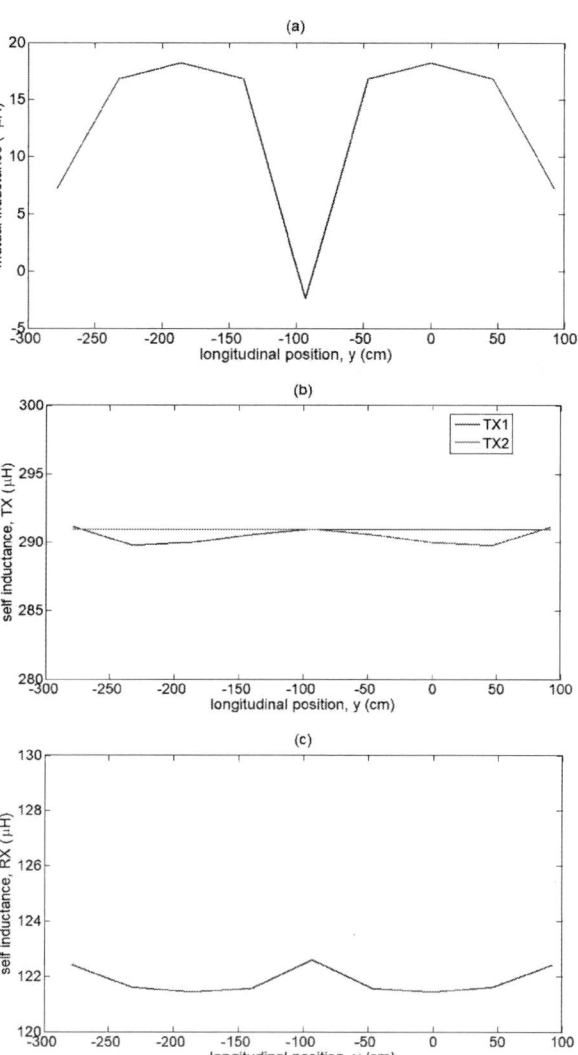

Fig. 5 Spatial distribution of the inductances along the trajectory for the case C-R: (a) mutual inductance; (b) self-inductances, TX; (c) self-inductance, RX.

In Fig. 6 the spatial variation of k along the vehicle trajectory is plotted, referred to the four combination of coil shapes. This variation impacts on the actual values of the resonant frequency along the trajectory. Indeed, the values of the capacitances imposing the resonant behavior are chosen from (4) by using the value of k in the nominal position of the coils. Thus, changing k would provide a shift in the actual value of the resonant frequency, as shown in Fig. 7. This, in turns, would lead to a shift of the gain from its ideal value (5).

By comparing the shapes, it is evident that the two combinations with rectangular RX coils (R-R and C-R) provide smaller relative variations of k along the trajectory, hence a more stable resonant frequency (whose relative variations are below 5%, see Fig. 7). In addition, by comparing these two choices, it results that R-R profile is flatter in the trajectory tracts where the RX coil is above one of the TX coils, and changes with slower transitions in the tract where the RX coil is outside the TX coils. This behavior is desirable to easier the action of the controller.

From another point of view, if the system designed is aimed at maximizing the value of k, then the best choices are given by configurations associated to circular TX shapes (C-C and R-C), ase clearly show in Fig. 6. Among these two choices, C-C provides a flatter profile of k when the RX coils are inside the TX ones, and a smoother spatial transition outside these coils.

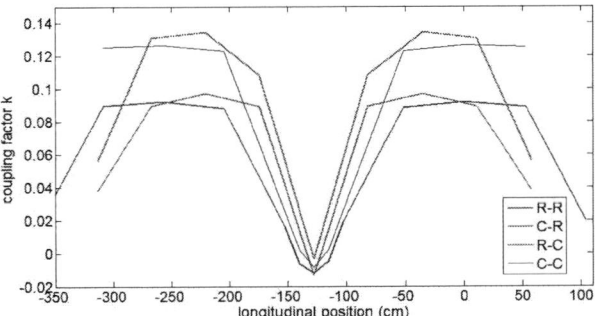

Fig. 6 Spatial distribution of the coupling factor k, for the four coil shapes (as in Table II).

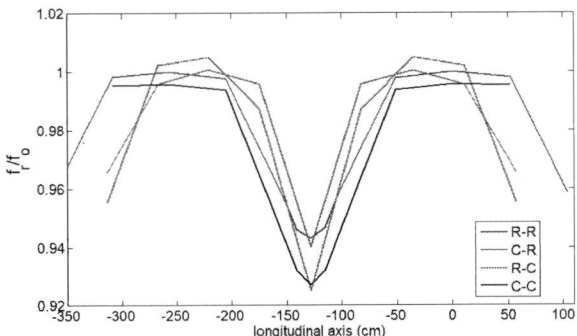

Fig. 7 Spatial distribution of the actual resonant frequency normalized to the nominal value of 85 kHz, for the four coil shapes (Table II).

IV. CONCLUSIONS

In this paper, we have shown how the shapes of the coils influence the resonance frequency and the voltage gain of the considered class of IPT systems, through the variation of the coupling factor k along the vehicle trajectory. Specifically, a trade-off is found between the requests of maximizing k and to minimize its variations along the trajectory. In the first case, the best option is that of choosing circular shapes of TX and RX coils. Instead, choosing the rectangular shape for the TX coil minimizes the variation k. With these TX coils, a rectangular RX coil would also help the action of the controller, by providing smoother spatial variations.

ACKNOWLEDGMENTS

The results here presented have been developed in the framework of the EMPIR 16ENG08 MICEV Project. The EMPIR initiative is co-funded by the European Union's Horizon 2020 research and innovation program and the EMPIR participating States.

REFERENCES

[1] Z. Bi, T. Kan, C. Mi, Y. Zhang, Z. Zhao, G. A. Keoleian, "A review of wireless power transfer for electrical vehicles: prospect to enhance sustainable mobility", Applied Energy, vol. 179, pp. 413-425, 2016.

[2] V. Cirimele, M. Diana, F. Freschi, M. Mitolo, "Inductive Power Transfer for Automotive Applications: State-of-the-Art and Future Trends", IEEE Trans. on Industry Appl., vol. 54, pp. 4069-4079, 2018.

[3] Z. Luo and X. Wei, "Analysis of Square and Circular Planar Spiral Coils in Wireless Power Transfer System for Electric Vehicles," in IEEE Trans. on Industrial Electronics, vol. 65, pp. 331-341, 2018.

[4] U. Castillo-Zamora, P. S. Huynh, D. Vincent, F. J. Perez-Pinal, M. A. Rodriguez-Licea and S. S. Williamson, "Hexagonal Geometry Coil for a WPT High-Power Fast Charging Application," in IEEE Trans. on Transportation Electrification, vol. 5, no. 4, pp. 946-956, Dec. 2019.

[5] A. Delgado, D. Schoenberger, J. Á. Oliver, P. Alou and J. A. Cobos, "Design Guidelines of Inductive Coils Using a Polymer Bonded Magnetic Composite for Inductive Power Transfer Systems in Electric Vehicles," in IEEE Trans. on Power Electronics, vol. 35, no. 8, pp. 7884-7893, Aug. 2020.

[6] M. Kim, D. Joo and B. K. Lee, "Design and Control of Inductive Power Transfer System for Electric Vehicles Considering Wide Variation of Output Voltage and Coupling Coefficient," in IEEE Trans. on Power Electronics, vol. 34, no. 2, pp. 1197-1208, Feb. 2019.

[7] A. Delgado, N. A. Requena, R. Ramos, J. A. Oliver, P. Alou and J. A. Cobos, "Design of Inductive Power Transfer System With a Behavior of Voltage Source in Open-Loop Considering Wide Mutual Inductance Variation," in IEEE Trans. on Power Electronics, vol. 35, no. 11, pp. 11453-11462, Nov. 2020.

[8] Y. H. Sohn, B. H. Choi, G. Cho, C. T. Rim, "Gyrator-Based Analysis of Resonant Circuits in Inductive Power Transfer Systems," in IEEE Trans. on Power Electronics, vol. 31, no. 10, pp. 6824-6843, Oct. 2016.

[9] M. Zucca, et al., "Metrology for Inductive Charging of Electric Vehicles (MICEV)," Proc. of 2019 AEIT International Conference of Electrical and Electronic Technologies for Automotive (AEIT AUTOMOTIVE), Torino, Italy, 2-4- July 2019.

[10] R. Ruffo, V. Cirimele, M. Diana, M. Khalilian, A. L. Ganga, P. Guglielmi, "Sensorless Control of the Charging Process of a Dynamic Inductive Power Transfer System with an Interleaved Nine-Phase Boost Converter", IEEE Trans. on Industrial Electronics, vol. 65, no. 10, pp. 7630-7639, Oct. 2018.

[11] Di Capua, G.; Femia, N.; Stoyka, K.; Di Mambro, G.; Maffucci, A.; Ventre, S.; Villone, F. Mutual Inductance Behavioral Modeling for Wireless Power Transfer System Coils. IEEE Trans. on Industrial Electronics, Vol. 68, pp. 2196 – 2206, 2021.

[12] G. Di Capua, A. Maffucci, K. Stoyka, G. Di Mambro, S. Ventre, V. Cirimele, F. Freschi, F. Villone, and N. Femia, "Analysis of Dynamic Wireless Power Transfer Systems Based on Behavioral Modeling of Mutual Inductance," Sustainability, vol. 13, no. 5, p. 2556, Feb. 2021.

A Probe Placement Method for Efficient Electromagnetic Attacks

Minmin Jiang* and Vasilis F. Pavlidis[†]

Advanced Processor Technologies group, Department of Computer Science, University of Manchester
Email: *minmin.jiang@manchester.ac.uk, [†]vasileios.pavlidis@manchester.ac.uk

Abstract—**Electromagnetic (EM) emissions have been explored as an effective means for non-invasive side-channel attacks. The leaked EM field from the memory bus when the data is loaded from the on-chip memory has received considerable attention in literature. Meanwhile, off-chip memory buses gradually become the new attack target due to the relative ease of access in the modern system in package technologies, such as 2.5-D integration where processing and memory chips are integrated, for example, on a silicon interposer. This paper, therefore, investigates EM snooping attacks on interposer-based off-chip memory buses. A gradient-search algorithm is proposed to locate fast (i.e. $O(N)$) the most efficient attack point. The effectiveness of the search algorithm and attack efficiency is evaluated on a 64-bit bus. It is demonstrated that at the optimal attack point, EM attacks can succeed with more than $10\times$ fewer traces, compared to placing the probe to sub-optimal locations.**

Index Terms—**interposer, electromagnetic emission, side-channel attack, near-field probing, pearson correlation**

I. INTRODUCTION

A side channel is a physical communication channel, which allows attackers to infer secret information from a system through, for instance, power supply lines, EM leaks, and timing or acoustic noise [1]. Among these side channels, the power side-channel attacks (SCAs) have greatly been explored, and are therefore regarded as the major threat to the security of integrated circuits (ICs) [2]. Compared to the semi-invasive characteristic of the power SCA, such as modifying or intervening with the circuits to measure power, the EM SCA is regarded as non-invasive and, hence, easier to launch. EM SCAs utilize a probe to couple the EM field and can attack the system faster than power SCAs [3].

Noise generated by the switching activity of the logic cells during the encryption process can also be leaked from an EM side channel. The fluctuations of the EM field can be received by near-field probing. By analyzing the spatial localized emanations, the EM emissions originating from the encryption modules rather than the non-encryption modules are mainly captured. Furthermore, there is another EM SCA called far-field EM attack. In this type of attack, the switching noise can also couple into the analog part of the chip through the substrate [4]. This noise can be amplified and transmitted by the analog wireless communication circuits along with the useful radio frequency. By intercepting it with a radio receiver, the attackers can recover the sensitive messages.

EM attacks on memory blocks on-chip have also been performed. In this case, the adversary can not physically access

This work is funded by European Commission under the Horizon 2020 Framework Programme for Research and Innovation through the EuroExa project under Grant 754337.

the internals of the memory circuit, and, thus, has to rely on EM emissions for extracting the desired information (e.g., secret key). However, EM emissions in this scenario are also produced from adjacent circuit blocks, hindering significantly the efficiency of an on-chip EM attack. On the other hand, the large physical size of the off-chip memory buses facilitates EM attacks as EM emissions from neighbouring components can be weaker due to the larger physical distance. Dayeol *et al.* [5] perform an off-chip side channel attack, named *MEMBUSTER*, on the memory bus between the CPU and the off-chip DRAM. A Dual In-line Memory Module (DIMM) interposer, as stated in [5], inserted between the processor and the DRAM, captures the memory bus signals and finally sends the signals to an analyzer for the attack. However, with the development of new packaging solutions, the processor and memory chips can be integrated on the same substrate, which enables improved interconnections among these components [6]. To the best of our knowledge, no exploration has been performed to show how best to attack the interconnections implemented in such an interposer. A gradient-search algorithm [7], adapted in this paper, helps to determine the optimal probe position and retrieve the sensitive message on the memory bus with the minimal measurements to disclosure (MTD).

This paper is structured as follows. Preliminaries about the AES algorithm and EM correlation attack are presented in Section II and the investigated system is discussed in Section III. The algorithm used for the probe position search is described in Section IV. The simulation results are analyzed in Section V. Finally, conclusions are drawn in Section VI.

II. PRELIMINARIES

In this section, a description of the targeted encryption algorithm, the correlation based EM attack, and SNR, commonly used to evaluate information leakage, are introduced.

A. Advanced encryption standard (AES)

AES is the most commonly targeted algorithm in side-channel attacks due to its widespread application. It is a symmetric block cipher, which encrypts messages segmented into blocks [8]. The encryption is symmetric because the key used for decryption is the same as that used for the encryption.

When implemented for blocks of 128 bits and a 128-bit (or 256-bit) key, the algorithm normally has ten rounds of processing. Within each round processing module except for the last one, there is a combination of 4 processes: substitution, transposition, substitution, and XORing with the sub-key.

© VDE VERLAG GMBH · Berlin · Offenbach

B. Correlation based EM attack

Correlation based EM attacks use Pearson's correlation coefficient to recover the most probable key by collecting a sufficient number of EM traces [3]. By linking the measured EM traces with a leakage model, the correlation coefficients of the traces are calculated to extract the key. The Hamming Distance (HD) model is employed as the leakage model, which assumes that the number of transitions ($0 \rightarrow 1$ or $1 \rightarrow 0$) predicts the magnitude of EM field.

If AES encryption is repeated N times with known plaintexts and a fixed unknown key, the collected magnetic traces are denoted as $\mathcal{L} = \{l_1, ..., l_N\}$, where l_m ($m = 1, ..., N$) is a magnetic trace generated by a certain plaintext. Each trace is a time series with s sampling points, $l_m = \{l_{m,1}, ..., l_{m,s}\}$. $\mathcal{F}$ is a function of an intermediate value χ, which depends on the plaintext p and the key k, denoted as $\mathcal{F} = \psi(\chi) = \psi(p, k)$. $\mathcal{F}$ is a subset of m-bit χ, whose value is the Hamming Weight of χ. The correlation between the traces $\mathcal{L}$ and the function $\mathcal{F}$ for each guessed key is described by

$$\rho = \frac{E\left[(\mathcal{F} - E(\mathcal{F}))(\mathcal{L} - E(\mathcal{L}))\right]}{\sqrt{D(\mathcal{F}) D(\mathcal{L})}}, \qquad (1)$$

where $E\left[(\mathcal{F} - E(\mathcal{F}))(\mathcal{L} - E(\mathcal{L}))\right]$ is the co-variance between them, and $D(\mathcal{F})$, $D(\mathcal{L})$ are, respectively, their individual variances. The position where the highest ρ appears is the best sub-key candidate.

C. Signal-to-noise ratio (SNR)

In side-channel attacks, SNR is characterized as an essential parameter to describe the leakage in side channels, which affects the attack capability. The greater the SNR, the fewer traces are needed to disclose the secret key [9]. In side-channel measurements, the SNR of the gathered traces is [9]

$$SNR = \frac{D(V_{data})}{D(V_{noise})}, \qquad (2)$$

where $D(V_{data})$, $D(V_{noise})$ is the variance of coupled voltages originating from the useful data and noise, respectively.

III. SYSTEM UNDER EM ATTACK

S-box is a critical non-linear operation of substitution of AES. It is typically implemented on hardware by complex logic gates or a look-up table (LUT). In 2.5-D ICs, to reduce the dynamic power of the encryption chip, a LUT based S-box can be implemented on a custom-designed off-chip read-only ROM [10], as shown in Fig. 1. The ROM receives the address and sends the substitution data through the memory bus routed with the redistribution layers (RDL) of the interposer.

In this paper, the register value is selected as χ, and the first round of AES is preferred to be attacked. This choice is because the intermediate result of AES is stored in χ at the end of each round. Thus, the sub-key used in the first round can be revealed by the EM leakage generated at the edge of the clock when χ updates its value. Meanwhile, χ is the memory

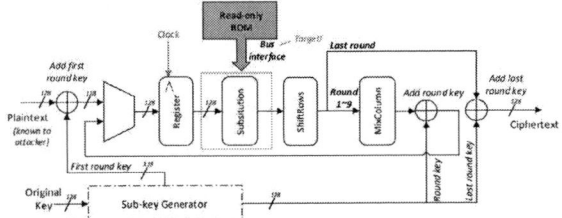

Fig. 1. Block diagram of AES encryption with ROM LUT based S-box.

address, being sent to ROM to find the correct swapped data. The leakage generated by the memory address (L_a) is

$$L_a = \psi(\chi) = \psi(PT, k_1) = HW(PT \oplus k_1), \qquad (3)$$

where PT is the plaintext, k_1 is the sub-key byte used in the first round, and HW represents the Hamming Weight. Based on (1) and (3), the off-chip memory address can be snooped to recover the sensitive key.

IV. EFFICIENT PROBE PLACEMENT

High spatial resolution can help maximize the coupling for near-field probing [11]. However, this high resolution leads to an extremely large space that must be searched to determine the best probe location for the attack. When the most effective probe location is determined, the target system is attacked fast. If the probe is positioned far from optimal places, MTD can require 4.3 times more traces or fail to attack [12]. Consequently, a gradient-search approach is introduced to facilitate the attack over brute-force search across the overall system area. SNR is used to evaluate the search efficiency of the algorithm in this paper.

A. Probe placement algorithm

When the signal current flows along the bus lines, the amplitude of the magnetic emissions in different y positions has negligible differences due to the voltage drop. However, the EM field varies significantly along x and z directions. Consequently, the search space reduces from three dimensions (x, y, z) to two dimensions (x-z plane). In the search across the x-z plane, the normalized standard deviation (NSD) of the emissions is selected to evaluate information leakage.

The evaluation function f of the gradient-search algorithm is denoted as $f(x, y, z) = h_{NSD}$, where h_{NSD} is the value of normalized standard deviation at each point. The search starts from a random position (x_1, y_1, z_1), where NSD is measured at the closest grid cell corresponding to this position. Next, NSD is measured, separately, in four adjacent grid cells to the location (x_1, y_1, z_1). The difference of measured NSD between the tested cell and the current cell is regarded as the magnitude of the vector, and the direction of search is pointing from the tested cell to the current cell. The addition of the four vectors is the gradient. The next location to be measured is ($x_1 - \Delta \cdot \nabla f$, y_1, $z_1 - \Delta \cdot \nabla f$), where Δ is the step size and ∇f is the calculated gradient. NSD can be measured in the new grid cell which is mapped at the new position. In the next iteration, the monitored grid cell is shifted between adjacent grid cells until the edge of *x-z* plane or NSD reaches the maximum.

B. Search area reduction

In this section, the 8-bit bus described in Fig. 2(a), is taken as an example to verify the quality of gradient-search algorithm. This bus width is chosen as, typically, EM attacks are launched on parts of wide buses to mitigate the exponential increase in the number of vectors that need to be analyzed. Rather a byte by byte attack is preferred without considerably degrading the success rate of attacks [12]. As shown in Fig. 2(a), the sweeping range for *x-z* plane is 40 μm × 40 μm and the probe is assumed to be placed vertically over the *y-z* plane of the bus, depicted in Fig. 2(b). When the *x-z* plane is divided into a 8 × 8 grid, the NSD map is shown in Fig. 3(a). MTD is normally inversely proportional to SNR^2, denoted as $MTD = \frac{k1}{SNR^2} = \frac{k2}{NSD^4}$ [13], where k_1 and k_2 are empirical parameters chosen to match the measurements. In this 8-bit bus model, when NSD equals 1.112, MTD is 30, in that case, k_2 is taken as 45.87. As depicted in Fig. 3(b), as the NSD increases, the number of traces needed for the EM attack decreases.

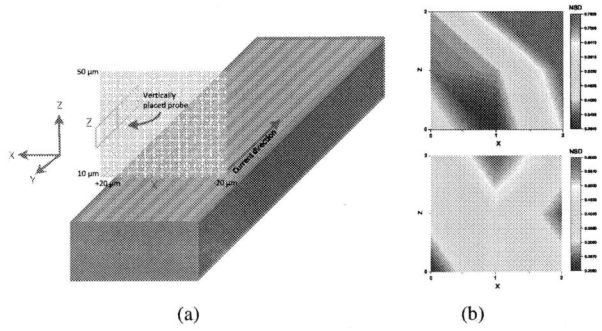

Fig. 2. (a) Discretized *x-z* plane where leakage measurements are performed, and (b) a rough scan (3 × 3) of NSD distribution in *x-z* plane when probe is placed vertically (above) and horizontally (below). In the maps, the vertical orientation exhibits a more significant NSD than the other.

Different grid scales are used to verify the algorithm. As shown in Fig. 3(c), for a grid of $N \times N$, the gradient-search algorithm can reduce the NSD measurements needed to reach minimal MTD from N^2 (brute-force measurements) to approximate *N*. Furthermore, for the 16 × 16 grid, the effect of step size Δ on the number of iterations is demonstrated in Fig. 3(d). If the step size is too small, for example one grid cell (2.5 μm), the search can be trapped at local minimum. If the step size is too large, for example 4 grid cells (10 μm), the search might miss the optimal location. Thus, given a reasonable step size, the algorithm reduces the search space over the exhaustive brute-force search.

V. EFFECTIVENESS OF THE ALGORITHM

The cross-section of the investigated stacked structure of an 8-bit bus on an interposer is shown in Fig. 4(a) along with related geometry parameters. The data rate of the bus is 1 *Gbps* and the sampling rate of the trace is set to 10 *Gbps*. In EM simulations, the probe is modeled as a single

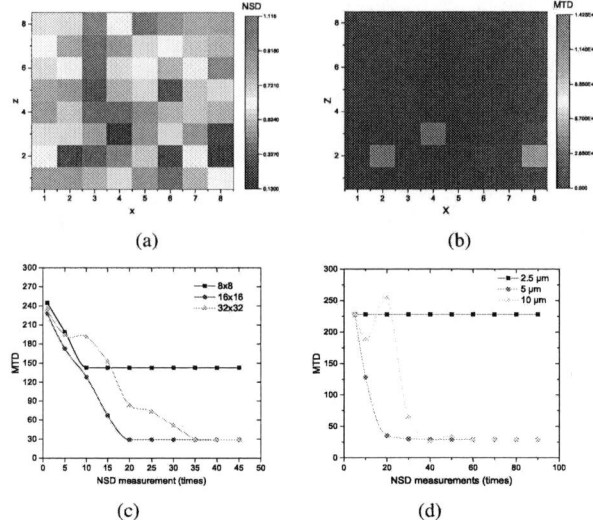

Fig. 3. (a) NSD map for an 8 × 8 scan, (b) Heatmap of MTD. Higher NSD means fewer traces are needed for a successful attack, (c) The search time complexity is reduced from $O(N^2)$ to $O(N)$ for a $N \times N$ grid, and (d) Effect of step size on the convergence of the search.

turn rectangle coil with 200 μm length and width equal to a quarter of the length, which exhibits the maximum NSD, and is placed vertically over the bus. The step size is set to 5 μm.

Based on the gradient-search algorithm presented in Section IV-A, the optimal EM attack location for this 8-bit bus is found at $(-2.5, 700, 45)$, where the value for y-axis can be randomly chosen as close as possible to the near-end of the bus. When the probe is placed at this point, frequency sweeping is performed with *ANSYS HFSS* [14]. The S-parameters generated at the frequency domain are exported from *HFSS* and imported into *Spectre* for transient analysis in the time domain. When the plaintext is swept from 0 to 255, 256 HD values are recorded in addition to the coupled voltage on the probe. When plotting all 256 peak voltage values for the 256 different plaintexts according to their HD values (0 to 8), as in Fig. 4(b), the captured EM trace demonstrates a very good linear correlation with the Hamming Distance that depends on the plaintext and the key.

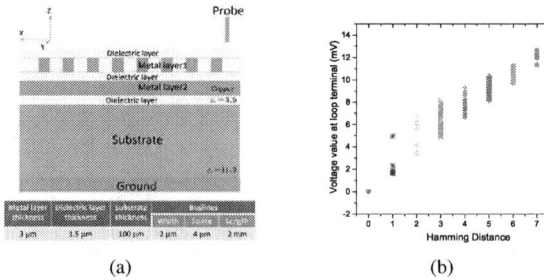

Fig. 4. (a) Stacking structure of the interposer-based bus, and (b) Correlation between the coupled voltage and HD at the optimal position.

In order to demonstrate that the key can be extracted with this method, the AES encryption process is repeated 256 times for each 8-bit plaintext and an 8-bit fixed key. As shown in Fig. 5(a), among the correlation coefficients of all the guessed keys (256 keys), the position with the highest correlation corresponds, indeed, to the right key (165). There is another similar peak obtained at the position where the symmetric key (90), called the "shadow key", appears. This is because the XOR operation is symmetric. Moreover, the number of traces needed for this successful attack is illustrated in Fig. 5(b), where each of the 256 lines corresponds to the probability of the corresponding 8-bit value to be the correct key. The line with the highest correlation corresponds to the correct key.

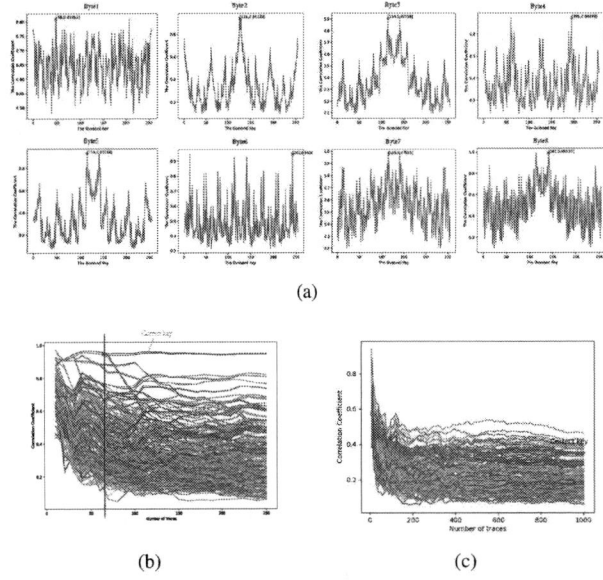

(a)

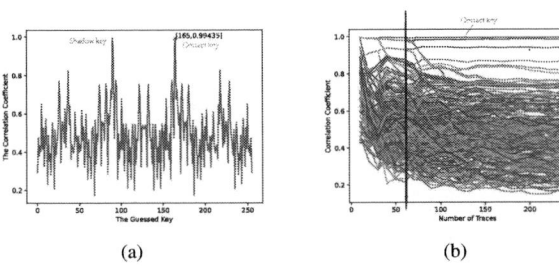

(a)   (b)

Fig. 5. Attack results of the 8-bit bus. (a) Correlation based EM attack uses a HD leakage model to attack the sub-key used in the first round of AES, and (b) the correct key is distinguished in less than 80 traces.

After the successful attack on 8-bit bus, the bus width is extended to 64 bits to verify the effectiveness of the algorithm on wider buses. If each byte of the 64-bit key can be attacked individually, the measurements for attacking the whole key can be highly reduced from $2^{64} = 1.84467 \times 10^{19}$ to $8 \times 2^8 = 2048$. This number can significantly increase (decrease) if the EM is sub-optimally (optimally) placed. Far from optimal probe locations completely fail to retrieve even a single sub-key. However, using the proposed algorithm, the correct 64-bit key is generated byte by byte with fewer than 256 traces per byte by placing the EM probe at the optimal position as determined for attacking each byte. The results of the correlation attack for all the sub-keys are shown in Fig. 6(a).

Furthermore, the EM attack results on the subkey Byte4 for two different measurement configurations are shown in Fig. 6(b) and Fig. 6(c). When the probe is placed at the optimum location, the sub-key can be efficiently recovered with fewer than 100 traces (MTD=70), while at the non-optimal position, more than $10\times$ traces are recorded and yet the correct key (that corresponds to the thick line in Fig. 6(c)) exhibits a low correlation coefficient and is not detected.

VI. CONCLUSION

In this paper, a fast and efficient SCA on an interposer-based off-chip memory bus is introduced. The core idea is to exploit high spatial EM field resolution to determine the the best attack location quickly. An algorithm based on gradient-search is developed to provide a scanning strategy, which reduces the brute-force search of $N \times N$ space to N. NSD is preferred as the metric for leakage, such that the attack hotspot can be precisely determined for EM attacks with considerably low

(b)   (c)

Fig. 6. (a) Byte-by-byte attack results of the 64-bit bus. EM attack results for Byte4 where the probe is (b) optimally placed as determined by the gradient-search algorithm, and (c) placed 10 μm away from the optimal position.

MTD. For the off-chip 64-bit bus scenario, $10\times$ fewer traces are needed for sub-key attacks where the probe is placed at the optimum location. The proposed search algorithm is scalable as it is shown to apply to different bus widths.

REFERENCES

[1] S. François-Xavier, "Introduction to side-channel attacks," *Proc. of Secure integrated circuits and systems conference.* pp. 27–42, Dec. 2010.
[2] K. Paul, J. Jaffe, and B. Jun, "Differential power analysis," *Proceedings of Advances in Cryptology conference*, pp. 388–397, Dec. 1999.
[3] A. Dakshi *et al.*, "The EM side channel (s)," *International workshop on cryptographic hardware and embedded systems*, pp. 29–45, Aug. 2002.
[4] C. Giovanni *et al.*, "Screaming channels: When electromagnetic side channels meet radio transceivers," *Proceedings of the ACM Conference on Computer and Communications Security*, pp. 163–177, Oct. 2018.
[5] L. Dayeol *et al.*. "An off-chip attack on hardware enclaves via the memory bus," *Proceedings of USENIX Security Symposium*, 2020.
[6] V. F. Pavlidis, I. Savidis, and E. G. Friedman, *Three-Dimensional integrated circuit design*, 2nd Ed. Morgan Kaufmann Publishers, 2017.
[7] R. Sebastian, "An overview of gradient descent optimization algorithms," arXiv, preprint, Sep. 2016.
[8] D. Joan and V. Rijmen, "AES proposal: Rijndael," 1999.
[9] O. Changhai *et al.*, "SNR-Centric Power Trace Extractors for Side-Channel Attacks," *IEEE Transactions on Computer-Aided Design of Integrated Circuits and Systems*, Jun. 2020.
[10] T. Craig, M. Bhargava, and K. Mai, "Side-channel attack resistant ROM-based AES S-Box," *Proceedings of IEEE International Symposium on Hardware-Oriented Security and Trust*, pp. 124–129, Jun. 2010.
[11] S. Laurent, S. Guilley, and Y. Mathieu, "Electromagnetic radiations of fpgas: High spatial resolution cartography and attack on a cryptographic module," *ACM Transactions on Reconfigurable Technology and Systems*, Vol. 2, No. 1, pp. 1–24, Mar. 2009.
[12] I. Vishnuvardhan and A. E. Yilmaz, "An adaptive acquisition approach to localize electromagnetic information leakage from cryptographic modules," *IEEE Texas Symposium on Wireless and Microwave Circuits and Systems*, pp. 11-6, Mar. 2019.
[13] M. Stefan, "Hardware countermeasures against DPA–a statistical analysis of their effectiveness," *Proceedings of In Cryptographers' Track at the RSA Conference*, pp. 222–235, Feb. 2004.
[14] https://www.ansys.com/products/electronics/ansys-hfss

© VDE VERLAG GMBH · Berlin · Offenbach

SMACD / PRIME 2021 | 19 – 22 July 2021, Online Event

Dealing with hierarchical partitioning in bottom-up design methodologies

F. Passos[1,2], P. Saraza-Canflanca[1], R. Castro-Lopez[1], E. Roca[1] and F.V. Fernandez[1]

[1]Instituto de Microelectrónica de Sevilla, IMSE-CNM (CSIC/Universidad de Sevilla), Sevilla, Spain
francisco.fernandez@imse-cnm.csic.es

[2]Instituto de Telecomunicações, Lisboa, Portugal

Abstract—**This paper deals with the expertise blend of circuit design and design methodology development required to successfully address hierarchical partitioning of analog, radio-frequency and mm-Wave circuits in bottom-up design methodologies. A set of guidelines for the optimal configuration of the bottom-up process is discussed. Two case studies are used to demonstrate that these guidelines yield sound design results.**

I. INTRODUCTION

Analog, mixed-signal, radiofrequency (RF) and millimeter-wave (mm-Wave) design constitute a major bottleneck in modern electronic circuits and systems. Traditionally, the design of complex electronic systems has been addressed by performing a hierarchical decomposition followed by a top-down design optimization flow, mapping top-level requirements into lower-level blocks specifications [1]. However, this procedure has some limitations: (a) some specifications of some blocks may be impossible to achieve; (b) when the whole system is verified, it may end up not fulfilling the specifications; or (c) the design may be globally suboptimal due to inappropriate specification transmission. These problems may incur unpredictable and costly iterations of the whole design flow.

A promising alternative follows just the opposite direction. After properly decomposing the system into two or more hierarchical levels, the design starts at the lowest level by finding the optimal trade-offs between circuit performances, i.e., their Pareto front, and composing the results up the hierarchy until reaching the highest level, as illustrated in Fig. 1 [2]-[4]. Whereas in top-down design methodologies the design of each block at each hierarchical level is posed as a constrained single-objective optimization problem, in bottom-up design methodologies, block design at each level is posed as a multi-objective optimization problem whose results will exhibit the best trade-offs among the circuit performances of that particular block. Major efforts to date have concentrated in efficient algorithms for multi-objective optimization, appropriate performance evaluation strategies at each level and bottom-up composition using indexation techniques.

Usually, reported approaches have been demonstrated in experiments in which the system was already partitioned, the hierarchical levels and the blocks at each level were well-defined, and so did the topologies of each block. It has been assumed that this has been previously provided by the user. But

what has not been extensively discussed in the literature is the required expertise for this task. Design expertise is obviously needed but the expertise required is usually well beyond that of a conventional designer. This paper addresses some issues related to the hierarchical partitioning by expanding on the basic considerations that must be taken into account, with emphasis on those aspects beyond a conventional design expertise. In particular, the topology selection will not be addressed for two reasons: (a) it is an issue that designers decide on comfortably to the extent that they even do not wish someone/something else selecting the topology; and (b) alternative topologies for the same functionality can be easily handled by bottom-up approaches.

Section II in this paper details the basic guidelines that hierarchical partitioning in a bottom-up design methodology must consider and Section III shows two case studies that illustrate them. Section IV provides some concluding remarks.

II. DESIGN AND DESIGN METHODOLOGY GUIDELINES

An essential aspect in bottom-up design methodologies is how should the system be hierarchically partitioned, i.e., how many hierarchical levels, which blocks at each level and which optimization objectives and constraints should be considered. To a large extent, the partitioning strategy imitates the designer, because the divide-and-conquer strategies applied by designers are usually the smartest way to proceed. However, this design expertise must be matched to some other considerations linked to the bottom-up design strategy:

1) Properly identify the performances. Each block at each hierarchical level should be characterized by a set of performances for which it should be clear if they can be handled as objectives and/or constraints.

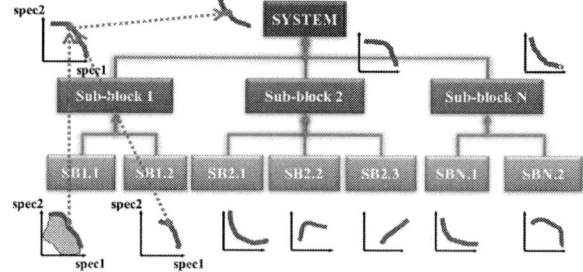

Fig. 1. Illustrating bottom-up design methodologies.

This work was supported by the VIGILANT Project (PID2019-103869RB / AEI / 10.13039/501100011033). Pablo Sarazá Canflanca acknowledges MICINN for supporting his research activity through the predoctoral grant BES- 2017-080160.

© VDE VERLAG GMBH · Berlin · Offenbach

2) All relevant performances must be included in the optimization process. If any performance of a block at a given level has a relevant influence in any performance characteristic at higher hierarchical levels it must be considered appropriately, either as an objective or constraint. A typical example can be the area occupation. Assume that area minimization is an objective of a block at a given level and that the performances of this block are obtained by composition of Pareto fronts of lower-level blocks. Then, the area must also be an objective of the lower-level block. Otherwise, the design of the higher-level block would be using design points of the lower-level block that have not been optimized in area, and, therefore, area minimization at the higher level is not guaranteed.

3) The number of minimization/maximization objectives should be reasonably limited. A first reason is that Pareto fronts are generated using multi-objective optimization algorithms. Although many-objective optimization algorithms constitute a hot area of research [5], most popular algorithms find difficulties beyond four objectives. These objectives should also be conflicting among them; otherwise, we would be wasting resources exploring inexistent trade-offs between objectives. On the other hand, consider that the Pareto front used in the bottom-up composition is just a set of samples, hopefully closed to, the true Pareto front. The quality of the design space exploration at a higher level is good if the Pareto front sampling is sufficiently dense. Here, 'sufficiently dense' means achieving the right balance between the number of objectives and the computational effort to generate the Pareto front since, when the number of objectives increases either the sampling density decreases or the computational effort to keep the density increases exponentially.

4) Impose convenient constraints. Multi-objective optimization algorithms try to find the optimal trade-offs of all conflicting objectives, e.g., giving two minimization objectives it will allow arbitrarily large values of one of them as long as the other can be further minimized. Many of those solutions may lack interest from the design point of view, e.g., improving the processing speed of a circuit may lead to uninteresting high values of power consumption. Preventing the exploration of such solutions by imposing appropriate constraints reduces the computational burden and promotes a larger density of samples of the Pareto fronts in the regions of interest.

5) Select appropriate objectives for composition of Pareto fronts. When composing Pareto fronts up the hierarchy, optimization at higher levels is performed by using samples of the Pareto fronts at lower levels as search space. Correct composition implies that the samples of the Pareto front at the higher level can only be obtained by using the samples of the lower-level Pareto fronts, i.e., that the samples of the Pareto front at the higher level cannot be obtained from dominated solutions at the lower level.

6) Check the dependence of surrounding circuitry. Bottom-up approaches are based on the generation of optimal trade-offs between circuit performances. Some of these performances may change with surrounding circuitry, a circuitry that can be unknown a priori. This problem and its solution may take different forms. For instance, a mutual common-mode voltage of different sub-blocks of a $\Delta\Sigma$ converter is used in [6]. Another example is the dependence of some performances of operational amplifiers with their resistive and capacitive load, which can be accounted for by introducing a Pareto front transformation technique in [7].

7) Check the reduction of complexity. For every block considered at any intermediate level, it is possible to decide to keep it at that level or promote it to the immediately upper hierarchical level. The bottom-up composition of an intermediate block only makes sense if the complexity is considerably reduced.

In some cases, these guidelines may be conflicting among them. Hence, they must be appropriately balanced on a case-by-case basis.

III. CASE STUDIES

Two practical case studies in this section highlight some of the considerations that must be accounted for in the application of bottom-up design methodologies.

A. RF frontend

The RF frontend in Fig. 2 will be considered for the Bluetooth standard (2.4-2.5 GHz ISM band), resulting in the receiver frontend specifications in Table I. As it can be observed, it is desired to obtain the trade-off between area and power consumption. An upper constraint on power consumption is imposed as solutions with larger power lack interest for our application. The topologies in Fig. 3 are used for the sub-blocks.

If the design problem is solved in a bottom-up way, objectives and constraints for the LNA, VCO and mixer must be formulated, as shown in Table II. Several of the guidelines above can be observed in this table. Power and area are objectives at the receiver level and, hence, they are also objectives for the LNA, VCO and mixer. There are not more

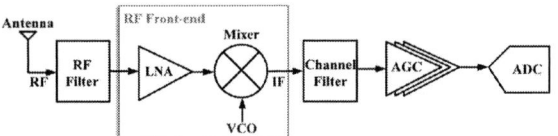

Fig. 2. Complete RF receiver signal chain with focus on the RF front-end (LNA, VCO and Mixer).

TABLE I
RECEIVER FRONT-END SPECIFICATIONS.

Front-end Performance	Bluetooth standard
CG @ down-frequency	> 12 dB
CG @ up-frequency	> 12 dB
P_{DC}	**Minimize** < 40 mW
NF @ down-frequency	< 8.79 dB
NF @ up-frequency	< 8.79 dB
IIP_3	> -10.35 dBm
Area (μm^2)	**Minimize**

TABLE II
OBJECTIVES AND CONSTRAINTS FOR THE LNA, VCO AND MIXER

LNA Performance	LNA Specifications	VCO Performance	VCO Specifications	Mixer Performance	Mixer Specifications
S_{11} @ 2.45; 2.5; 2.55 GHz	< -12 dB	f_{osc}	> 2.45 GHz	CG @ 10 MHz	> 5 dB
S_{22} @ 2.45; 2.5; 2.55 GHz	< -12 dB	f_{osc}	< 2.55 GHz	CG @ 40 MHz	> 5 dB
S_{21} @ 2.45; 2.5; 2.55 GHz	**Maximize** (>7 dB)	PN @ 1MHz offset	**Minimize** (< -110 dBc/Hz)	P_{DC}	**Minimize** (< 20 mW)
k	> 1	P_{DC}	**Minimize** (< 20 mW)	NF @ 10 MHz	< 20 dB
NF @ 2.45; 2.5; 2.55 GHz	**Minimize**	P_{OUT}	**Maximize** (> -2 dBm)	NF @ 40 MHz	**Minimize** (< 20 dB)
P_{DC}	**Minimize** (< 20 mW)			IIP_3	**Maximize** (> -15 dBm)
IIP_3	> -15 dBm			Port-to-Port Isolation	>30 dB
Area (μm^2)	**Minimize**	Area (μm^2)	**Minimize**	Area (μm^2)	**Minimize**

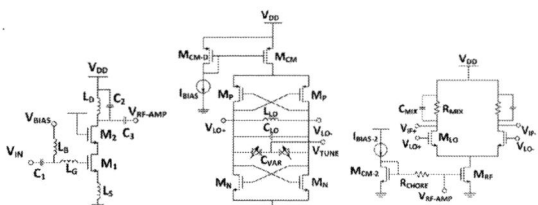

Fig. 3. Schematics of LNA, VCO and Mixer.

than four objectives for each block and, in some of them, a reasonable constraint has been imposed to improve the quality of the fronts, e.g., if the received has a power constraint of 40 mW it is reasonable that the lower-level blocks also have a power constraint. In other cases, reasonable constraints are given, e.g., for LNA matching it is enough to impose constraints to S_{11} and S_{22}; considering them as objectives would bring little benefit and would lead to a very complex 6-objective optimization problem.

It is also interesting to highlight the advantages of using a bottom-up approach. Fig. 4 shows the comparison of the optimization results of the receiver using the bottom-up approach described above vs. the design of the receiver front-end as a single, flat optimization process. The figure shows three fronts for a different number of generations of the flat approach. It can be concluded that the bottom-up approach contributes to a much better exploration of the design space.

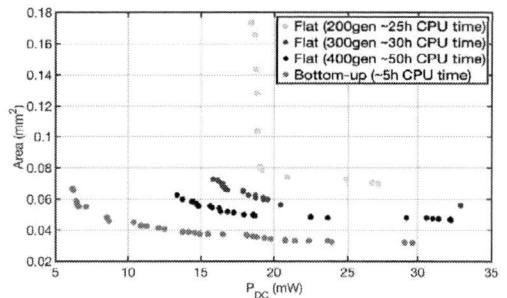

Fig. 4. Comparison of Pareto fronts with the bottom-up and the flat approaches for the Bluetooth standard.

B. mm-Wave receiver

The second case study is the mm-Wave receiver (RX) for radar applications (76-81 GHz band) in Fig. 5, composed of an LNA, a mixer (MIX) and a mixer driver (MD). Its integration into monolithic microwave integrated circuits (MMIC) have paved the way to automotive radars on a chip [8].

A priori it could be considered that two hierarchical levels should be defined: at the lower level we have the three blocks: MD, MIX and LNA, and at the upper level we have the receiver. Although this strategy is not impossible to implement, there are some associated difficulties that make other alternatives addressing the aspects in Section II more advisable, in particular, considering the MD and LNA at the lower level and promoting the MIX to the upper level.

In this case, we are considering minimization of power consumption, area and noise figure, as shown in Table III. Notice that area has been assimilated to length since the floorplan for this kind of circuit has been previously established

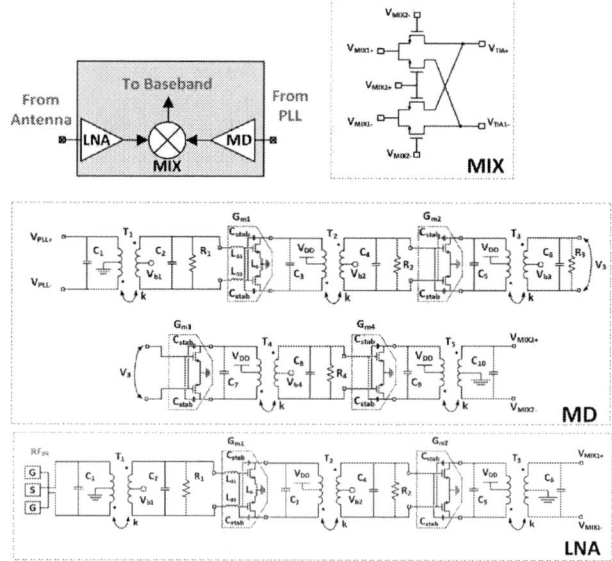

Fig. 5. Schematics of a mm-Wave Rx MMIC.

TABLE III
DESIRED SPECIFICATIONS FOR THE Rx MMIC OPTIMIZATION.

Rx MMIC Performance	Rx MMIC Specifications
CG @ 65-85 GHz	$25 < CG < 35$ dB
P_{DC}	Minimize (< 60 mW)
NF	Minimize (<10 dB)
Area (length)	Minimize
$IIP3$	> 0 dBm
V_{out} (V_{MIX2+}-V_{MIX2-})	$0.9 < V_{out} < 1.1$ V

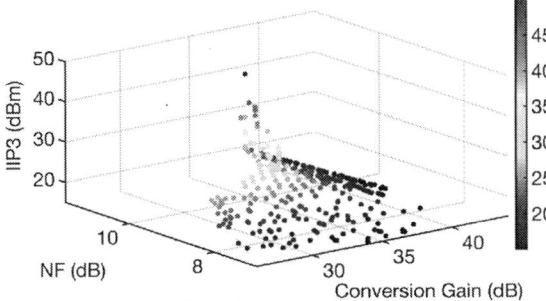

Fig. 6.   Pareto front of the mixer.

TABLE IV
OBJECTIVES AND CONSTRAINTS FOR THE LNA AND MD.

LNA performance	LNA specs	MD Performance	MD specs
S_{11} @ 65-85 GHz	< -12 dB	S_{11} @ 65-85 GHz	< -12 dB
S_{21} @ 65-85 GHz	Maximize (>7 dB)	S_{21} @ 65-85 GHz	Maximize (>15 dB)
k	>1	k	>1
P_{DC}	Minimize (< 40 mW)	P_{DC}	Minimize (< 70 mW)
Area (length)	Minimize	PAE	$>15\%$
$S_{21@65GHz}$ - $S_{21@75GHz}$	<1.5dB	Area (length)	Minimize
$S_{21@85GHz}$ - $S_{21@75GHz}$	<1.5dB	$S_{21@65GHz}$ - $S_{21@75GHz}$	<1.5dB
$S_{21@70GHz}$ - $S_{21@75GHz}$	<0.75dB	$S_{21@85GHz}$ - $S_{21@75GHz}$	<1.5dB
$S_{21@80GHz}$ - $S_{21@75GHz}$	<0.75dB	$S_{21@70GHz}$ - $S_{21@75GHz}$	<0.75dB
NF	Minimize (<7dB)	$S_{21@80GHz}$ - $S_{21@75GHz}$	<0.75dB
IIP_3	>-5 dBm	V_{out}	$0.9 < V_{out} < 1.1$

kind of methodologies. Its successful implementation requires a combination of design expertise and experience in the development of design methodologies that hardly can be automated and that hardly can be found in a single person. This paper elaborates on this combination and reports two design examples where it has been put to practice.

and the only area-related relevant parameter is length since height is dominated by the passives. Since the Rx MMIC is part of a larger system, it must be designed with a set of specifications such that it complies with the full system needs (e.g., the gain of the Rx cannot be just maximized due to linearity issues; therefore, an upper constraint is established, as shown for the conversion gain, CG, row in Table III). If the MIX were to be optimized at the circuit level, a mandatory objective would be minimization of the noise figure (as it is also an objective at the higher level). However, NF minimization in the MIX also maximizes the gain. The consequence is that the Pareto front of the MIX would contain solutions with too high of a gain so as to meet the Rx MMIC specification in Table III, as shown in the mixer front in Fig. 6. Although this could be handled, at least partially, by introducing constraints, it is not trivial, and might require redesign iterations. If, on the other hand, we consider that due to the reduced number of independent design variables of the MIX, there is not a significant advantage in generating a MIX Pareto front, the decision to handle the MIX directly at the upper level arises naturally.

By following this approach, only Pareto fronts of the LNA and MD are generated according to the specifications in Table IV and following similar criteria to those in Section III.A. Notice that in addition to gain maximization, several constraints are imposed so that the gain is sufficiently flat in the band of interest. Then, the Pareto front of the Rx MMIC in Fig. 7 is obtained.

IV. CONCLUSIONS

Hierarchical partitioning in bottom-up design methodologies has been usually skipped in papers reporting this

REFERENCES

[1] H. Chang et al., A top-down constraint-driven design methodology for analog integrated circuits. Boston: Kluwer Academic Publishers, 1997.

[2] T. Eeckelaert, T. McConaghy, and G. Gielen, "Efficient multiobjective synthesis of analog circuits using hierarchical Pareto-optimal performance hypersurfaces," in Proc. Design, Automation and Test in Europe Conference, 2005, pp. 1070-1075 Vol. 2.

[3] R. Gonzalez-Echevarria, et al., "An automated design methodology of RF circuits by using Pareto-optimal fronts of EM-simulated inductors," in IEEE Trans. Comp.-Aided Design of Integr. Cir. Sys., vol. 36, no. 1, pp. 15-26, Jan. 2017.

[4] F. Passos, et al. "A multilevel bottom-up optimization methodology for the automated synthesis of RF systems." IEEE Trans. Comp.-Aided Design of Integr. Cir. Sys, vol. 39, no.3, pp. 560-571, 2020.

[5] C. Coello, S. Gonzalez, J. Figueroa, M. Castillo and R. Hernandez, "Evolutionary multiobjective optimization: open research areas and some challenges lying ahead," in Complex & Intelligent Systems, 2020.

[6] T. Eeckelaert et al., "Hierarchical gottom-up analog optimization methodology by a delta-sigma A/D converter design for the 802.11a/b/g standard," in Proc. IEEE Design Automation Conf., 2006, pp. 25-30.

[7] E. Roca, M. Velasco, R. Castro-Lopez and F.V. Fernandez, "Context-dependent transformation of Pareto-optimal performance fronts of operational amplifiers," Analog Integrated Circuits and Signal Proc., vol. 73, pp 65-76, 2012.

[8] C. Cui et al., "A 77-GHz FMCW radar system using on-chip waveguide feeders in 65-nm CMOS," in IEEE Trans. Micro. The. and Techn., vol. 63, no. 11, pp. 3736-3746, 2015.

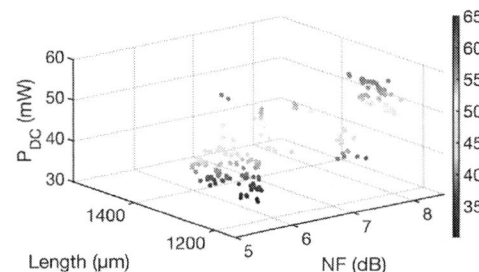

Fig. 7.   Pareto front of the Rx MMIC.

SMACD / PRIME 2021 | 19 – 22 July 2021, Online Event

Modeling and Optimization of Supply Sensitivity for a Time-Domain Temperature Sensor

Jun Tan, *Member, IEEE*

IMMS Institut für Mikroelektronik- und Mechatronik-Systeme gemeinnützige GmbH (IMMS GmbH), Ilmenau, Germany
jun.tan@imms.de

Abstract—**This paper presents a methodology for modeling and optimizing the supply sensitivity for a time-domain temperature sensor. Many modern System-on-Chip (SoC) designs integrate multiple functional blocks, such as sensors, power management units (PMU), and wired/wireless communication interfaces. Interference on the supply line between such blocks becomes critical, since it has a significant impact on sensor performance. Many state-of-the-art designs utilize supply sensitivity as a sensor parameter to describe the immunity to supply interference. Therefore, modeling and analysis of this parameter to optimize sensor performance is needed. This paper begins with the modeling of the common time-domain temperature sensor, while an instance of the model is precisely created with extracted parameters. Additionally, verification is performed by comparing the model with a transistor-level design. Finally, the optimization methodology for supply sensitivity improvement is discussed. The results show that the DC supply sensitivity is significantly reduced by a factor of 41. With the proposed methodologies, a further AC supply sensitivity can be analyzed and optimized, so that sensor performance in SoC designs can be improved.**

I. INTRODUCTION

Nowadays, system-on-chip (SoC) becomes more important and practical, since devices need to integrate more functions and be miniaturized at the same time. Low-power SoC solutions, powered by batteries or passively harvested from environment [1], have attracted more attentions for applications, such as Wireless Sensor Networks (WSN) and the Internet of Things (IoT).

Temperature sensing, which is required for many applications, is an attractive sensor choice in many low-power SoCs. However, such sensors suffers from fluctuating supply voltages, since Power Management Unit (PMU) of such devices is often limited due to lack of good passive components or low power budgets. This results in a significant degradation of sensor performance, which limits the performance of the overall system. The quantitative characterization of the sensor error caused by supply noise is defined as supply sensitivity, which is usually written by the ratio of the error variation and the variation of the input supply voltage (°C/V). In recent years, much research has focused on optimizing supply sensitivity. In [2], the supply sensitivity of ADC-based temperature sensors achieves 0.28 °C/V and 0.45 °C/V, respectively. In [3] a VCO-based temperature sensor reaches 34 °C/V, while another VCO-based sensor [4] achieves only 0.36 °C/V.

So far, these results of supply sensitivity are obtained either with no optimization or with simple optimization. However, systematic analysis is still lacking, making it difficult to push

the immunity of supply noise to the limit, which is still not revealed. This paper presents a common approach to model the supply sensitivity of common low-power time-domain temperature sensors, while it optimizes the supply sensitivity utilizing a typical sensor design. However, the optimization technique can also be applied to other time-domain temperature sensors. Section II introduces the modeling of the time-domain temperature sensor. Section III verifies the modeling, while section IV presents the improvement method. Finally, conclusions are drawn in Section V.

II. MODELING OF A TIME-DOMAIN TEMPERATURE SENSOR

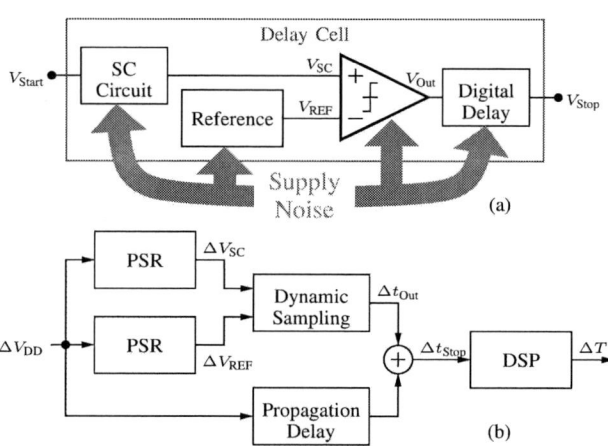

Fig. 1. (a) The block diagram summarizes the delay cell of time-domain temperature sensors and (b) the signal flow to be analyzed in this paper.

A. Block diagram of a common delay cell topology

Most state-of-the-art time-domain temperature sensors convert the temperature into a timing signal using a delay cell, which is shown in Fig. 1 (a). The delay cell consists of four basic components. A switched-capacitor (SC) circuit and a reference, which are normally analog circuits, output two voltages V_{SC} and V_{REF}, respectively. A comparator flips the output, when $V_{SC} \geq V_{REF}$, while the output also shifts a small optional digital delay. The entire delay cell is triggered by a start signal V_{Start} and generates the stop signal V_{Stop} for the next stage. All components can be used for temperature sensing and are sensitive to supply noise. A VCO-based

© VDE VERLAG GMBH · Berlin · Offenbach

topology [3], [4] can be seen as a temperature dependent digital delay, since the inverter is a comparator that compares the input voltage with its own threshold voltage. For an SC-circuit-based topology [5], [6], the analog parts are used to produce temperature dependent voltages, which are converted into temperature dependent delay by the comparator. The final measured delay is evaluated in the DSP, so that the measured temperature can be obtained (Fig. 1 (b)).

B. Analog Frontend

The simplified analog circuit in Fig. 2 (a) is analyzed for power supply rejection (PSR), which is used to describe the behaviour from supply voltage to output. On the left side, a typical current reference is built via $Q_{1,2}$, R_1, $M_{1,2}$, and A_1. The current reference generates a proportional-to-absolute-temperature (PTAT) current for the SC circuit, which is shown on the right side. The base-emitter voltage (V_{BE}) of Q_1 is utilized for the reference voltage, so this block is reused. In Fig. 2 (b), the signal flow graph [7] is drawn to analyze the transfer function from ΔV_{DD} to the three output voltages: ΔV_{REF}, ΔV_{SC1}, and ΔV_{SC2}. Compared to [7], the current I_{SC} is added and is not onlz regulated by V_X (g_m) but also by the supply voltage (g_{ds}). Finally, it flows through two different SC configurations to generate V_{SC1} and V_{SC2}.

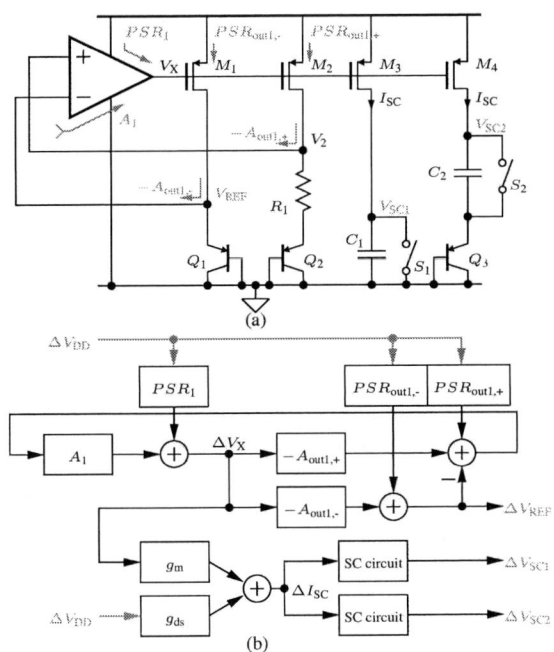

Fig. 2. (a) The block diagram of the simplified analog circuit can be modeled in a (b) signal flow graph.

C. Timing Generation Scheme

The periodic sampling (Fig. 3 (a)) utilizes a simple comparator topology (Fig. 3 (b)), whose output is buffered by an inverter. The comparator samples the crossing of the two input voltages and outputs a pulse as an event. If V_{REF} and V_{SC} are

stable, the comparator outputs a fixed periodic sampling time. However, with supply noise the sampling becomes dynamic so that it generates a dynamic timing. Besides that, the supply voltage also affects the timing of the comparator output by shifting the operational point of V_O and the threshold voltage of the inverter stage. This delay is combined with the delay from digital logic to form to the digital propagation delay. The modeling of this part is achieved by a simple Verilog-A code that detects the cross between the two input voltage and adds a propagation delay, which represents the original comparator delay.

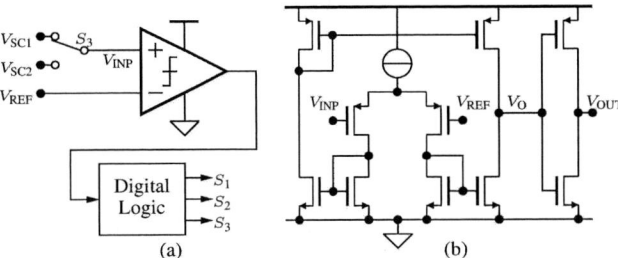

Fig. 3. (a) The circuit for sampling the voltage and generating the continuous control signal and (b) the schematic of the comparator

D. Digital Signal Processing (DSP)

In Fig. 4 (a), the positive input voltage V_{INP} is switched between V_{SC1} and V_{SC2} (Fig. 4 (b)). The control signals S_1 and S_2 are generated by the digital logic, so that the timing of S_1 is digitized by the time-to-digital converter (TDC). In [5], [1], the duty cycle of S_1 (Fig. 4 (c)) is linear to the absolute temperature, if the ratio of C_2/C_1 and Q_2/Q_1 are designed delicately. Therefore, the temperature can be simply calculated by a linear function of the duty cycle (Fig. 4 (d)).

In this modeling, the temperature output can be expressed by

$$T(D) = 554 \cdot D - 280.4, \qquad (1)$$

where D is the measured duty cycle and the resulted temperature is in degree Celsius.

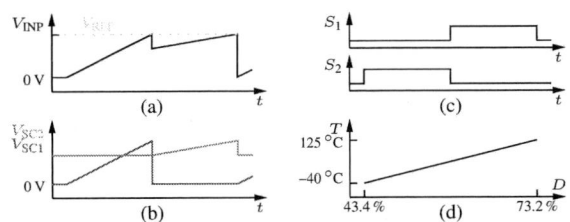

Fig. 4. The input signals (a) V_{INP}, (b) V_{SC1}, and V_{SC2} generate the timing signals (c) S_1 and S_2. (d) The DSP evaluates the duty cycle and generates the final measured value.

III. MODELING VERIFICATION

To verify the modeling, the simulation results of the transistor-level circuit and the modeling are compared. The

TABLE I
EXTRACTED PARAMETERS OF A TYPICAL LOW-POWER TEMPERATURE SENSOR DESIGN

Parameter	Extracted value	Parameter	Extracted value
$A_{\text{out1+,DC}}$	1.9340	$A_{1,\text{DC}}$	1446.5
$PSR_{\text{out1+,DC}}$	1.9352	$PSR_{1,\text{DC}}$	0.971
$f_{\text{out1+}}$ ($\omega_{\text{out1}}/2\pi$)	995 kHz	f_1 ($\omega_1/2\pi$)	1.4 Hz
$A_{\text{out1-,DC}}$	0.6295	$g_m^{(1)}$	500 nΩ^{-1}
$PSR_{\text{out1-,DC}}$	0.6299	$g_{ds}^{(1)}$	70 pΩ^{-1}
$f_{\text{out1-}}$ ($\omega_{\text{out1}}/2\pi$)	24 kHz	C_1	40 pF
R_{off} of $S_{1,2,3}$	>1 GΩ	C_2	496 pF

(1) g_m and g_{ds} are obtained by a cascode current mirror, which is simplified to $M_1..M_4$ in Fig. 2 (a).

parameter in Fig. 2 (b) is listed in Table I. Please notice that the parameters A_1, $A_{\text{out1,+}}$, $A_{\text{out1,-}}$, PSR_1, and PSR_{out1} are modeled using first-order system [8] to extract a simple but sufficiently accurate frequency response.

The results of the PSR simulation in the frequency domain are shown in Fig. 5. The PSR of V_{REF} has the typical outlook with one zero and two poles as regulated voltage references in [8]. However, the PSR curves of V_{SC1} and V_{SC2} are totally different, since they have multiple zeros and poles. The both PSR curves start at approximately the same DC PSR of 1.5 dB, while they reach different PSR from 1 Hz to 100 Hz. From 1 kHz the PSR of V_{SC1} donates a pole so that it reduces with 20 dB/Decade. The PSR of V_{SC2} increases to another flat region at 10 kHz, and decreases with 20 dB/Decade beyond 100 kHz. For the three PSR curves, the results of transistor circuit and model overlap, so that the models correctly describe the system behaviour.

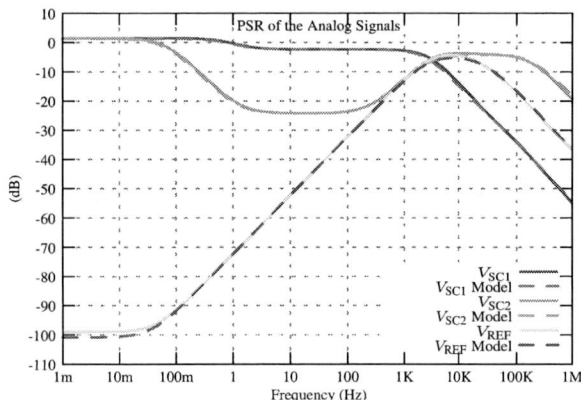

Fig. 5. PSR simulation results of the three output voltages between the transistor circuit and the model

Finally, the system modeling is compared with the transistor design in transient simulation in Fig. 6. In Fig. 6 (a) the comparator input signals V_{INP} and V_{REF} overlap, so that their timing outputs are approximately the same as well. Since the modeling can represent the analog circuit at the nominal operational point (V_{DD}=1.2 V and T=25 °C), the impact of DC supply shift is simulated and shown in Fig. 6 (b). The temperature errors of transistor circuit and modeling are starting

from −1.21 °C and −1.29 °C at zero DC shift, respectively. With increasing DC supply shift from 0 to 200 mV, the error increases linearly to 0.16 °C and 0.08 °C with the error shift of 1365.6 mK and 1362.5 mK (Table II), respectively. The DC supply sensitivity is calculated by the error shift over the supply shift, which in this case are 6.83 °C/V and 6.81 °C/V of transistor and model, respectively. Please notice that the nominal error at 25 °C is not zero, because the second-order error of the temperature sensing still exists and is not compensated from the equation 1.

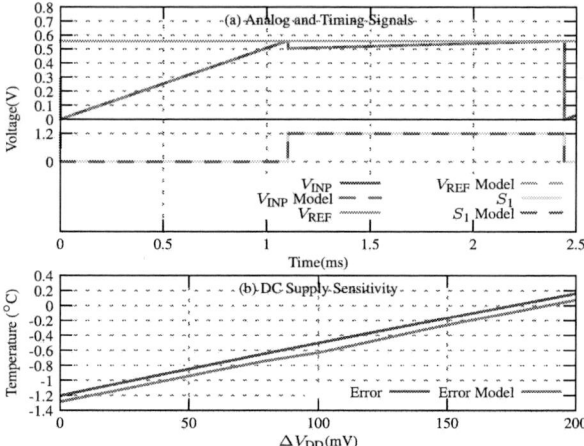

Fig. 6. (a) The comparison of the analog signals and timing signals, (b) the DC supply sensitivity can be calculated by the error shift over the supply shift.

IV. IMPROVEMENT OF THE DC SUPPLY SENSITIVITY

To analyze the strategy for efficient improvement, each error is isolated from other sources, so that it can be tested alone for its total error contribution. Since $V_{\text{SC1,2}}$ in Fig. 2 (a) are generated by a current I_{SC}, this current is analyzed instead of $V_{\text{SC1,2}}$. The results in Table II show that the dominant source comes from the comparator, while V_{REF} is totally unimportant. The reason is that V_{REF} already achieves approximately −100 dB DC PSR, while PSR of $V_{\text{SC1,2}}$ is above 0 dB. The reason of the dominance of the comparator is that the comparator is sensitive to the slope of the input voltage. Since the slopes of C_1 and C_2 are different in Fig. 6 and Fig. 4, the comparator adds different delay to the pulse and gap of S_1. Please notice that, since both timing signals are added with additional delays, the resulting overall error is smoothed by the ratio-metric measurement, while a single pulse or a gap of S_1 brings more error [6].

To reduce the error of the comparator, different optimization strategies can be implemented. Firstly, the bias current of the comparator can be increased to reduce the propagation delay, but the current consumption is increased. Furthermore, several native NMOS transistors can be inserted into the supply wire [4] to reduce the propagation delay, but it will increase the technology dependence. In this paper, a simple switch is added to the reference path of the comparator, so that the propagation delay can be completely removed.

TABLE II
EXTRACTED CONTRIBUTION OF EACH ERROR SOURCE FROM THE
MODELING FOR ΔV_{DD}=0 V TO 0.2 V

Source	Contributed Error	Percentage	DC Supply Sensitivity
Total	1362.5 mK	100 %	6.81 °C/V
V_{REF}	4.0 mK	0.3 %	
I_{SC}	150.2 mK	11.0 %	
Delay	1209.9 mK	88.8 %	

TABLE III
DC SUPPLY SENSITIVITY OF OPTIMIZED MODEL AND TRANSISTOR
CIRCUIT FOR ΔV_{DD}=0 V TO 0.2 V

	Source	Contributed Error	Percentage	DC Supply Sensitivity
Model	Total	31.6 mK	100 %	0.158 °C/V
	V_{REF}	4.6 mK	14.6 %	
	I_{SC}	27.0 mK	85.4 %	
	Delay	<0.1 mK	<1 %	
Transistor Circuit	Total	33.1 mK	100 %	0.165 °C/V

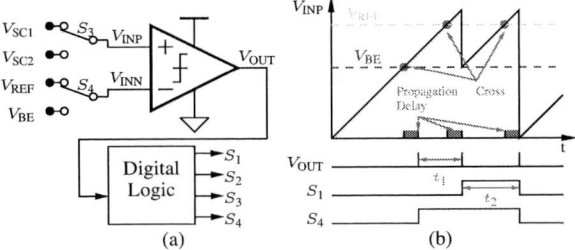

Fig. 7. (a) The improved timing generation scheme utilizes an additional switch (S_4) and makes $C_1 = C_2$ (Table I) to generate (b) there phases to calculate the propagation delay.

In Fig. 7 (a), the modified block generates two crosses on a rising edge to cancel the propagation delay (Fig. 7 (b)). The timing t_1 is without propagation delay, while the next phase t_2 contains a propagation delay. By calculating $t_2 - t_1$, the propagation delay can be obtained, so that the DSP can systematically remove the delay and achieve better supply sensitivity.

ΔI_{SC} can be reduced by manipulating the transconductance g_m and g_{ds} in Fig. 2 (b), since V_{REF} maintains the lowest supply dependency. To achieve this, the source-degenerated current mirror of $M_1 .. M_4$ in Fig. 2 (a) is utilized, so that the g_m and g_{ds} of these PMOS transistors are significantly reduced by the added serial resistors. While the optimization is applied to both the model and the transistor circuit, the DC supply sensitivity is analyzed and shown in Table III. The errors of the model and the transistor circuit are significantly reduced to 31.6 mK and 33.1 mK, respectively. The new timing generation scheme eliminates the error from the propagation delay, while the source-degenerated current mirror reduces the error from I_{SC} to 29 mK. The final optimized DC supply sensitivity achieves 0.158 °C/V and 0.165 °C/V from the model and the transistor circuit, respectively. Thus, the optimization improves the DC supply sensitivity by a factor of approximately 41.

V. CONCLUSION

The modeling and optimization of a time-domain temperature sensor are presented to achieve a better DC supply sensitivity. The topology of the temperature sensor is modeled for an analog frontend, a timing generation block, and a signal processing algorithm. The modeling of the supply path reveals the system behavior without going into details of transistor circuits. The modeling is verified by comparing the DC supply sensitivity to the transistor-level circuit, while the results match between them. The modeling is optimized for two major error source, namely the propagation delay of the comparator and the charging current of the SC load. For the first source, the timing generation scheme is modified, so that the propagation delay can be completely removed during data processing. For the charging current, the source-degenerated current mirror is utilized to reduce g_m and g_{ds}, so that less current error is generated. The optimization significantly improves the DC supply sensitivity from 6.83 °C/V to 0.165 °C/V by a factor of approximately 41. With the achieved modeling and optimization methodology, the AC supply sensitivity can be further analyzed and optimized in the future work.

ACKNOWLEDGMENT

The RoMulus project is supported within the Research Program ICT 2020 by the German Federal Ministry of Education and Research (BMBF) under the reference 16ES0362. The BICCell project is funded by DECHEMA (Gesellschaft für Chemische Technik und Biotechnologie e.V.) via the AiF (Arbeitsgemeinschaft industrieller Forschungsvereinigungen) as a joint industrial research project (IGF) by the Federal Ministry of Economics and Energy (BMWi) by resolution of the German Parliament under the reference 21174 BR/2.

REFERENCES

[1] J. Tan and et al., "A Fully Passive RFID Temperature Sensor SoC with an Accuracy of ±0.4 °C (3σ) from 0 °C to 125 °C," *IEEE Journal of Radio Frequency Identification*, vol. 3, no. 1, pp. 35–45, March 2019.

[2] K. Souri, Y. Chae, F. Thus, and K. Makinwa, " A 0.85 V 600 nW All-CMOS Temperature Sensor with an Inaccuracy of ±0.4 °C (3σ) from −40 to 125 °C," in *2014 ISSCC*, 2014, pp. 222–223.

[3] T. Anand, K. A. A. Makinwa, and P. K. Hanumolu, "A VCO Based Highly Digital Temperature Sensor With 0.034 °C/mV Supply Sensitivity," *IEEE Journal of Solid-State Circuits*, vol. 51, no. 11, pp. 2651–2663, Nov 2016.

[4] K. Yang and et al., "A 0.6 nJ −0.22/0.19 °C Inaccuracy Temperature Sensor Using Exponential Subthreshold Oscillation Dependence," in *2017 ISSCC*, Feb 2017, pp. 160–161.

[5] J. Tan, A. Rolapp, and E. Hennig, "A Low-Voltage Low-Power CMOS Time-Domain Temperature Sensor Accurate To Within [-0.1,+0.5] °C From −40 °C To 125 °C," in *2014 APCCAS*. IEEE, 2014, pp. 463–466.

[6] M. K. Law, A. Bermak, and H. C. Luong, "A Sub-μW Embedded CMOS Temperature Sensor for RFID Food Monitoring Application," *IEEE Journal of Solid-State Circuits*, vol. 45, pp. 1246–1255, 2010.

[7] J. Tan and R. Sommer, "Analysis and Optimization of Power Supply Rejection for Power Management Unit Design in RFID Sensor Applications," in *2019 SMACD*, July 2019, pp. 181–184.

[8] ——, "Modeling of Low-dropout Regulator to Optimize Power Supply Rejection in System-on-Chip Applications," in *2019 SMACD*, July 2019, pp. 113–116.

Noise Performance in Current Mirror Circuit based on CNTFET and MOSFET

Roberto Marani
National Research Council of Italy (CNR), Institute of Intelligent Industrial Technologies and Systems for Advanced Manufacturing (STIIMA)
Bari, Italy
https://orcid.org/0000-0002-5599-903X

Anna Gina Perri
Department of Electrical and Information Engineering, Electronic Devices Laboratory, Polytechnic University of Bari
Bari, Italy
https://orcid.org/0000-0003-4949-987X

Abstract—**In this paper we present a comparative analysis of noise performance of Carbon Nanotube Field Effect Transistors (CNTFETs) and MOSFET, through the design of a basic current mirror. For reference current of 1 μA and 10 μA the output static and dynamic characteristics are better in the case of CNTFET, but for all cases the output noise current is always higher for the CNTFET than for the MOS. The software used is Advanced Design System (ADS) which is compatible with the Verilog A programming language.**

Keywords—**CNTFET, MOSFET, Modelling, Current Mirror Design, Advanced Design System (ADS)**

I. INTRODUCTION

A promising candidate which allows further scaling down is the Carbon Nanotube Field Effect Transistor (CNTFET). CNTFET is a new kind of molecular device, using a carbon nanotube as channel [1-4]. Among carbon nanotube FETs, conventional CNTFET (also denoted as C-CNTFET), with heavily doped source and drain contacts, is utilized for high-performance and low-power memory designs, also because this device has a significantly smaller off current which greatly reduces the power consumed at off state of CNTFET [5-6].

For this device we have already proposed a compact, semi-empirical model [1-7], in which we introduced some improvements to allow an easy implementation both in SPICE and in Verilog-A. Then our model has been implemented to carry out static and dynamic analysis of A/D circuits [8-13].

In this paper we present a comparative analysis of noise behavior of CNTFETs and MOSFET devices through the design of a basic current mirror. To have comparable results, we refer to a C-CNTFET and a MOSFET in 32 nm technology.

The software used is Advanced Design System (ADS) which is compatible with the Verilog A programming language. The simulation results allow to show the differences between CNTFET and MOS technology and the advantages of the first for analogue VLSI circuits.

II. A BRIEF REVIEW OF CNTFET AND MOSFET MODEL

An exhaustive description of our I-V CNTFET model is in [11-15]. Therefore we suggest the reader to consult these References.

It is a compact, semi-empirical model directly and easily implementable in simulation software to design analog and digital circuits: in fact the most complex part of the model is contained in Verilog A.

With the hypothesis that each sub-band decreases by the same quantity along the whole channel length, the total drain current can be expressed as [2]:

$$I_{DS} = \frac{4qkT}{h} \sum_{p} \left[\ln\left(1 + \exp\xi_{Sp}\right) - \ln\left(1 + \exp\xi_{Dp}\right) \right] \quad (1)$$

where k is the Boltzmann constant, T is the absolute temperature, h is the Planck constant, p is the number of sub-bands, while ξ_{Sp} and ξ_{Dp}, depending on temperature through the sub-bands energy gap, have the expressions reported in [4-6].

Regarding the C-V model, an exhaustive description of our C-V model is widely described in [6] and therefore the reader is requested to consult it, in which the following expressions of quantum capacitances C_{GD} and C_{GS} are explained:

$$\begin{cases} C_{GD} = q \sum_{p} \frac{\partial n_{Dp}}{\partial V_{GS}} = q \sum_{p} \frac{\partial n_{Dp}}{\partial \xi_{Dp}} \frac{\partial \xi_{Dp}}{\partial V_{CNT}} \frac{\partial V_{CNT}}{\partial V_{GS}} \\ C_{GS} = q \sum_{p} \frac{\partial n_{Sp}}{\partial V_{GS}} = q \sum_{p} \frac{\partial n_{Sp}}{\partial \xi_{Sp}} \frac{\partial \xi_{Sp}}{\partial V_{CNT}} \frac{\partial V_{CNT}}{\partial V_{GS}} \end{cases} \quad (2)$$

being V_{CNT} the surface potential.

In order to simulate correctly the CNTFET behavior, we needed to estimate parasitic capacitances and inductances as well as the drain and source contact resistances.

In this paper we have achieved this goal using an empirical method [1-2] more suitable for simulations in CAD environment. This method requires the extraction of the previous parasitic elements comparing the device characteristics with the measured ones. In this way all elements of the equivalent circuit can be determined [1-2].

Fig. 1 shows our model, in which we have reported the values of circuital elements, while Fig. 2 shows its symbol, that we will use in the following simulations.

© VDE VERLAG GMBH · Berlin · Offenbach

SMACD / PRIME 2021 | 19 – 22 July 2021, Online Event

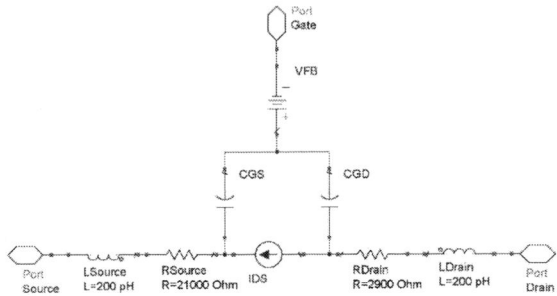

Fig. 1. Equivalent circuit of n-type CNTFET.

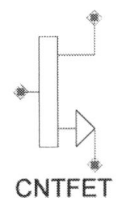

Fig. 2. CNTFET symbol.

For the MOSFET model we use the BSIM4 model of ADS library.

BSIM (Berkeley Short-channel IGFET Model) [16] refers to a family of MOSFETs for integrated circuit design. It also refers to the BSIM group located in the Department of Electrical Engineering and Computer Sciences (EECS) at the University of California, Berkeley, that develops these models. In this work BSIM4 has been used for the 32 nm technology nodes.

Moreover the MOSFET parameters, obtained using an evolution of previous Berkeley Predictive Technology Model (BPTM), are improved by us through parametric simulations to obtain performance of the MOSFET model comparable to the CNTFET one.

Fig. 3 shows the MOSFET symbol, which refers to BSIM4 model.

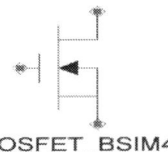

Fig. 3. MOSFET symbol.

Regarding noise model, in [17] we have proposed a compact noise model of CNTFET, and therefore we suggest the reader to consult this paper.

In particular we have considered the main noise sources, which are:

Thermal noise of R_G
Thermal noise of R_S and R_D
Channel thermal noise and shot noise
Flicker noise
Channel-induced gate noise.

For an exhaustive analytical description of our noise model we suggest the reader to consult our Reference [17].

III. BASIC CURRENT MIRROR CIRCUIT

The basic current mirror is the keystone of the current mirror circuits and it consists in just two active component as shown in Fig. 4.

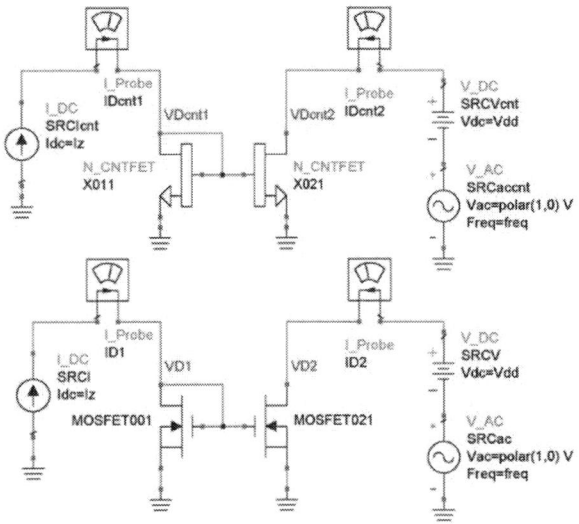

Fig. 4. Basic current mirror: on top the CNTFET version, on bottom the MOS version.

In this circuit the reference current is injected by the current generator on the far left side, the two active devices mirror this current on the output, on the right side. The constant voltage generator is plugged in series to the output to force the output voltage. The AC voltage generator is used to obtain the output resistance at various frequencies.

We ran first a static simulation to measure the output behavior of the circuit when the output is forced at various voltages. We first plot ratio of the output current to the input current, since in a current mirror it should be 1.

In Fig. 5 we present the obtained results for CNTFET and MOS circuit having considered three input currents, 1 uA (in red), 10 uA (in blue) and 100 uA (in violet). This scheme will be repeated in all following analysis and therefore we will no more stress this.

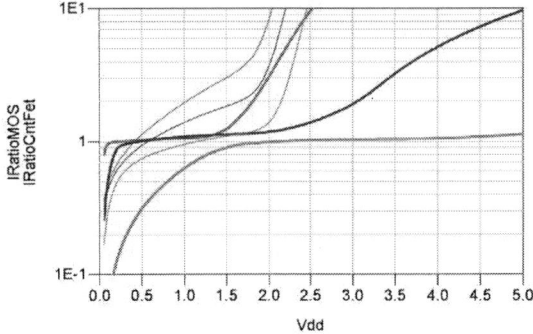

Fig. 5. Ratio of the output current divided to the input current. For the CNTFET circuit the curves are bold lines, while for the MOS circuit the curves are thin lines.

© VDE VERLAG GMBH · Berlin · Offenbach

We note that for the CNTFET the ratio is near to 1 for a larger interval of output voltage values, indeed for the MOS circuit the ratio is not stable near the value of 1.

In the CNTFET case the 1 uA current the ratio is stable from 0.1 V to 1.5 V, the 10 uA has a good output voltage interval between 0.3 V and 2.5 V, while the 100 uA current has a ratio near 1 at voltages greater than 1.5 V.

Fig. 6 shows the differential output resistance of the circuit, defined as the derivative of the I-V output static curves.

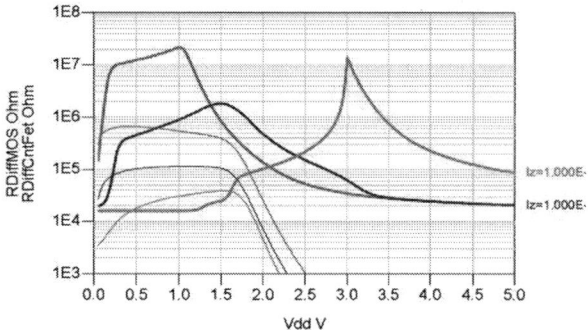

Fig. 6. Differential output resistance of the current mirror circuits. Lines as in Fig. 5.

Here we observe that the higher resistance values are in the same intervals where the current ratio is near 1, as expected. Moreover we observe that the CNTFET circuit has almost always an higher output resistance than the MOS circuit.

In the second simulation we studied the frequency behaviour of the output admittance for small signal in linearized approximation and the obtained results are shown in Figs 7 and 8. The output voltage is held constant at 1 V for all currents.

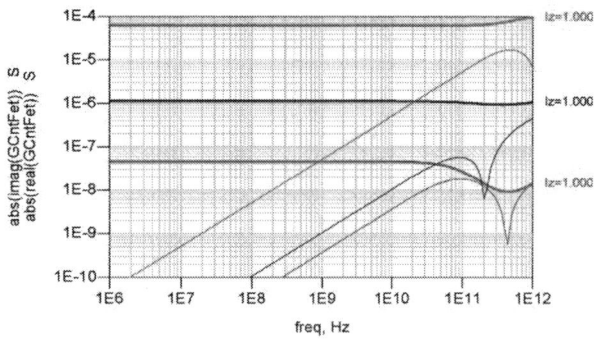

Fig. 7. Differential output admittance for the CNTFET circuit, values in Siemens. The real part is in bold lines, the imaginary part in thin lines.

We can observe that while the real part is almost independent from the frequency, the imaginary part (at frequency lower than 100 GHz) is just proportional to the frequency, so that, at lower frequency, the differential output circuit could be represented as a parallel of a resistor, a capacitor and the current generator.

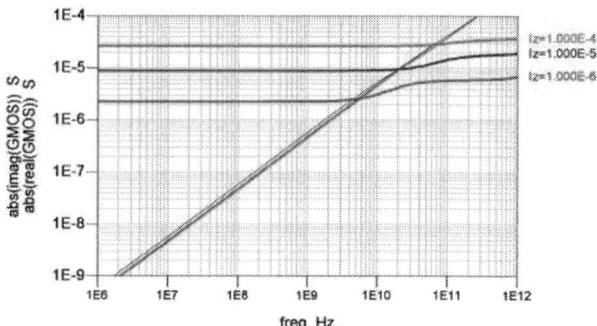

Fig. 8. Differential output admittance for the MOS circuit, values in Siemens. The real part is in bold lines, the imaginary part in thin lines.

Fig. 9 shows the spectral density of the output noise current for both CNTFET and MOS at different currents at output voltage 1 V.

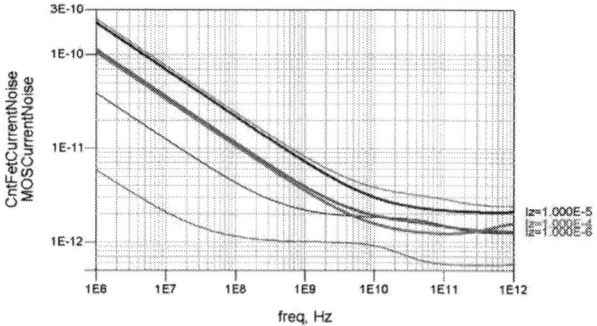

Fig. 9. Spectral density of the output noise current for the CNTFET and for the MOS circuit, values in A Hz$^{-1/2}$. Lines as in Fig. 5.

We can see higher noise in the CNTFET circuit in almost all cases, but over 1 GHz it is no more than three times higher, as shown in Fig. 10.

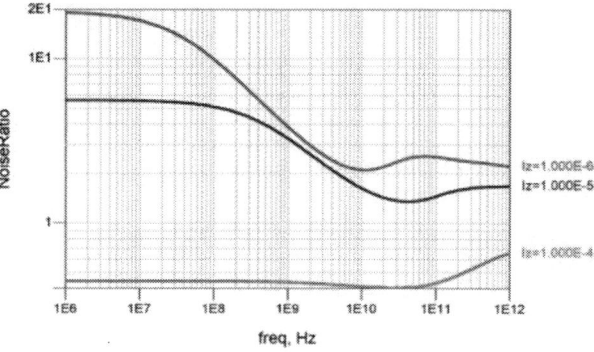

Fig. 10. Ratio of the noise current spectral density of the CNTFET circuit divided by the noise current spectral density of the MOS circuit.

In the case of 100 uA current, the noise is lower for the CNTFET circuit.

We also calculate the spectral density of noise current for the component coming from the flicker, the shot and the thermal noise coming from input and output devices for the CNTFET circuit. Except the case of the 100 μA current, devices contribute evenly to the output noise in the lower frequency range. Since the main idea of the current mirror is just mirroring current, it was easy to understand how the noise current on the input branch is mirrored in the output branch when the circuit is properly working.

In the CNTFET, compared to BSIM4 MOS model, for frequencies below 10 GHz the main contribution is the flicker noise which largely dominates for its 1/f dependence. For frequency over 10 GHz the main contributor are the shot noise and the thermal noise. Since the thermal noise is proportional to the resistances, its reduction requests a better control of the various parasitic resistances, always considering that the limit for channel resistance is the quantum limit.

IV. CONCLUSION

We have presented a simulation study of a basic current mirror based on CNTFET, comparing it with the same using MOS device. For reference current of 1 µA and 10 µA the output static and dynamic characteristics are better in the case of CNTFET, but for all cases the output noise current is always higher for the CNTFET than for the MOS. The output noise for CNTFET is no more than three times higher (10 dB) than for the MOS, but at some frequency and current we foresee no more than two (6 dB) times higher.

Currently we are studying the effect of temperature [18-19] in CNTFET based circuits and analyzing more thoroughly the effects of parasitic elements of interconnection lines in CNT embedded integrated circuits [20] and the impact of technology on CNTFET-based circuits performance [21].

REFERENCES

[1] A.G. Perri, R. Marani, *CNTFET Electronics: Design Principles*, Progedit Editor, Bari, Italy, ISBN: 978-88-6194-307-0, (2017).

[2] G. Gelao, R. Marani, R. Diana, A.G. Perri: A Semi-Empirical SPICE Model for n-type Conventional CNTFETs, *IEEE Transactions on Nanotechnology*, **10**(3), 506-512, (2011).

[3] R. Marani, A.G. Perri: A Compact, Semi-empirical Model of Carbon Nanotube Field Effect Transistors oriented to Simulation Software, *Current Nanoscience*, 7(2), 245-253, (2011).

[4] R. Marani, A.G. Perri: A DC Model of Carbon Nanotube Field Effect Transistor for CAD Applications, *International Journal of Electronics*,. **99**(3), 427-444, (2012).

[5] R. Marani, G. Gelao, A.G. Perri: Comparison of ABM SPICE library with Verilog-A for Compact CNTFET model implementation, *Current Nanoscience*, **8**(4), 556-565, (2012).

[6] R. Marani, G. Gelao, A.G. Perri: Modelling of Carbon Nanotube Field Effect Transistors oriented to SPICE software for A/D circuit design. *Microelectronics Journal*, 44(1), 33-39, (2013).

[7] R. Marani, A.G. Perri: Analysis of CNTFETs Operating in SubThreshold Region for Low Power Digital Applications. *ECS Journal of Solid State Science and Technology*, 5(2), M1-M4, (2016).

[8] R. Marani, A.G. Perri: A De-Embedding Procedure to Determine the Equivalent Circuit Parameters of RF CNTFETs, *ECS Journal of Solid State Science and Technology*, 5(5), M31-M34, (2016).

[9] G. Gelao, R. Marani, L. Pizzulli, A.G. Perri: A Model to Improve Analysis of CNTFET Logic Gates in Verilog-A-Part I: Static Analysis, *Current Nanoscience*, **11**(4), 515-526, (2015).

[10] G. Gelao, R. Marani, L. Pizzulli, A.G. Perri: A Model to Improve Analysis of CNTFET Logic Gates in Verilog-A-Part I: Dynamic Analysis, *Current Nanoscience*, **11**(6), p. 770-783, (2015).

[11] R. Marani, A.G. Perri: A Simulation Study of Analogue and Logic Circuits with CNTFETs, *ECS Journal of Solid State Science and Technology*, 5(6), M38-M43, (2016).

[12] R. Marani, A.G. Perri: A Comparison of CNTFET Models through the Design of a SRAM Cell, *ECS Journal of Solid State Science and Technology*, 5(10), M118-M1, (2016).

[13] R. Marani, A.G. Perri: CNTFET-based Design of Current Mirror in Comparison with MOS Technology, *ECS Journal of Solid State Science and Technology*, 6(5), M60-M68, (2017).

[14] R. Marani, A.G. Perri: Design and Simulation Study of Full Adder Circuit based on CNTFET and CMOS technology by ADS, *ECS Journal of Solid State Science and Technology*, 7(6), M108-M122, (2018).

[15] R. Marani, A.G. Perri: A Review on the Study of Temperature Effects in the Design of A/D Circuits based on CNTFET, *Current Nanoscience*, 15(5), p. 471-480, (2019).

[16] http://bsim.berkeley.edu/models/bsim4/

[17] R. Marani, G. Gelao, A.G. Perri: A Compact Noise Model for C-CNTFETs, *ECS Journal of Solid State Science and Technology*, 6(4), pp. M118–M126, (2017).

[18] R. Marani, A.G. Perri: A Review on the study of Temperature Effects in the Design of A/D Circuits based on CNTFET. *Current Nanoscience*, 15, pp. 471-480, (2019),.

[19] R. Marani, A.G. Perri: Temperature Dependence of I-V Characteristics in CNTFET Models: A Comparison. *International Journal of Nanoscience and Nanotechnology*. 17(1), pp. 33-39, (2021)..

[20] R. Marani, A.G. Perri: Effects of Parasitic Elements of Interconnection Lines in CNT Embedded Integrated Circuits. *ECS Journal of Solid State Science and Technology*, 9, (2020).

[21] R. Marani, A.G. Perri: Impact of Technology on CNTFET-based Circuits Performance. *ECS Journal of Solid State Science and Technology*, 9, (2020).

SMACD / PRIME 2021 | 19 – 22 July 2021, Online Event

A g_m/I_D Sizing Method for High-speed Multi-stage Operational Amplifiers with Feedforward-only Compensation[1]

Qixu Xie and Guoyong Shi and Yaoyao Ye

Dept of Micro/Nano Electronics, Shanghai Jiao Tong University, Shanghai, China

{smallxie,shiguoyong,yeyaoyao}@sjtu.edu.cn

Abstract—We present a g_m/I_D-based sizing method for multi-stage operational amplifiers (Op Amps) with feedforward-only compensation. For this class of Op Amps, we have to take into account of the dominant parasitic capacitances in pole-zero analysis. The presented design method combines analytical design equations with the g_m/I_D tables to facilitate the calculation of the sizing parameters. To enhance the sizing accuracy, we propose to use a three-dimensional lookup table of g_{ds}/I_D versus both g_m/I_D and V_{DS} and the current-normalized parasitic capacitance lookup tables, i.e., $C_{p,q}/I_D$ ($p,q = D, G, S$) versus g_m/I_D in sizing calculations. We verify by a case study that a high-speed three-stage Op Amp (attaining a GBW beyond 3 GHz) can be quickly sized using a TSMC 65nm CMOS technology.

Index Terms—feedforward compensation, high-speed operational amplifier (Op Amp), g_m/I_D method, transistor sizing

I. Introduction

In recent years research on the high-speed data converter design is receiving rising attentions [1]–[3]. Implementation of this class of data converters often requires the design of high-speed operational amplifiers (Op Amps) with the gain-bandwidth product (GBW) beyond gigahertz (GHz) while maintaining a sufficient gain at a high operating frequency. Designers would tend to adopt a multi-stage design to cope with such requirements. However, the requirement of gigahertz GBW prevents the designers from adopting any Miller compensation method in a multi-stage design as the existence of Miller capacitors would result in low frequency dominant poles, against the goal of achieving gigahertz GBW. One feasible design option is to apply feedforward-only compensations [4] to place the dominant poles at higher frequencies. However, since the location of the dominant poles and zeros are mainly influenced by the capacitive parasitics, it necessitates a detailed analysis on the poles and zeros with explicit dependence on the parasitics. Recently, Gebreyohannes et al. [3] proposed a g_m/I_D-based heuristic searching method to synthesize the device sizing of this type of Op Amps, but the computational cost seems high due to lack of properly using design equations.

The design flow adopted in the paper [3] is summarized in Fig. 1(a). Due to lack of using design equations, the heuristic search conducted in that work seemed blind, hence costs more runtime. With the help of analytical design equations, as we follow in this paper, more deterministic sizing steps can be deduced. Further with the help of customized device-level gm/ID lookup tables, quick sizing (even without laborious heuristic searches) can be realized.

The proposed design flow is depicted in Fig. 1(b), where one of the most important aspects is on the use of analytical design equations derivable from the transfer function analysis.

[1]This research was supported in part by the National Key R&D Program of China No. 2019YFB2205002 and the National Natural Science Foundation of China (NSFC) grant No. 61974087.

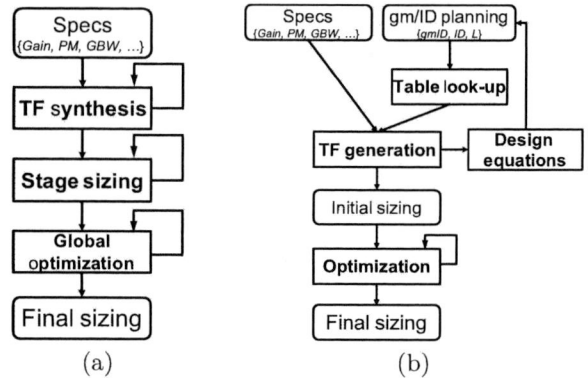

Fig. 1: Comparison of the design flows: (a) The design flow proposed in the paper [3]. (b) The proposed design flow.

In addition to that, properly prepared gm/ID lookup tables also play an important role, as we shall explain in the sequel.

The effectiveness of the g_m/I_D design method has been extensively validated in many publications recently [5]–[8]. In particular, it is confirmed by Shi [8] that combining g_m/I_D table lookup with design equations is a more powerful sizing method that can save a great deal of searching work. We note that the sizing method based on the combination of lookup tables and design equations is only targeted at an initial sizing. It does not rule out further sizing optimization procedures.

In this paper we shall emphasize the characterization of the parasitic capacitances $C_{p,q}$, ($p,q = D, G, S$) (D, G, S refer to the drain, gate, and source terminals of a MOSFET) normalized to the drain current I_D ([7]). It turns out to be helpful for planning the poles and zeros because the MOS device sizes affect the parasitic capacitances.

The rest of this paper is organized as follows. In section II we present the necessary gm/ID lookup tables to be used in the proposed sizing flow. In section III we present the pole-zero analysis on a three-stage feedforward-compensated Op Amp targeted at gigahertz operation. We customize the design equations in the forms of explicit gm/ID design variables for sizing calculations. Then a simulation-based verification on the achieved performance by sizing is reported in section IV. Section V concludes this paper.

II. gm/ID Lookup Tables

We introduce the gm/ID lookup tables to be used in sizing. The traditionally used gm/ID table construction methods have been described in the references such as [5], [6]. Due to the specific needs in our sizing tasks, customized use of the gm/ID tables is quite necessary. In this work we are interested in using the lookup tables involving $I_D/(W/L)$, g_{ds}/I_D and C_{gg}/I_D and

© VDE VERLAG GMBH · Berlin · Offenbach

their relations to the sweeping variables V_{GS}, V_{DS}, and g_m/I_D, where $C_{gg} = C_{gd} + C_{gs} + C_{gb}$ (called the gate parasitics).

A. g_{ds}/I_D Versus g_m/I_D and V_{DS}

In the traditional g_m/I_D method, V_{DS} of a MOSFET is assumed fixed when measuring the g_m/I_D curve sets [5], [6], [8]. However, due to the uncertainty of V_{DS} in sizing, g_{ds} (the MOSFET channel conductance) is undetermined during sizing, causing uncertainty in determining the dc gain of a gain stage.

In this work, we recommend to use a three-dimensional (3D) curve set capturing the dependence of g_{ds}/I_D on the variations of both g_m/I_D and V_{DS} as illustrated in Fig. 2, where a fixed channel length $L = 1$ µm was chosen (in a TSMC 65 nm technology). In practice, it is recommended to collect a set of such surfaces by sweeping over a grid of the transistor lengths, which can be used to look up a channel length of a MOS device in sizing.

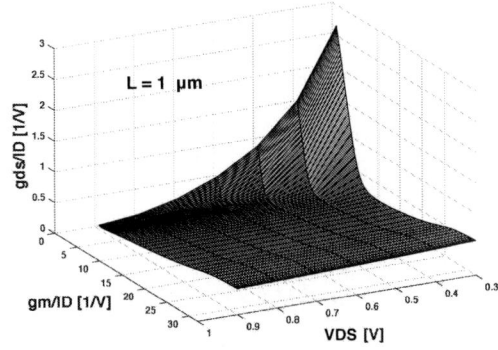

Fig. 2: Dependence of g_{ds}/I_D on g_m/I_D and V_{DS} for an N-type MOS with the TSMC 65 nm technology. L is fixed at 1 µm.

B. C_{gg}/I_D Versus g_m/I_D and L

As discussed earlier, parasitic capacitances of MOSFETs must be taken into account in the pole-zero analysis of a feedforward-compensated Op Amp. Although in general the capacitive parasitics of a MOSFET are dependent on the device dimension, the normalized quantity C_{gg}/I_D is virtually independent of the transistor channel width W when the channel length L and the biasing voltages are fixed. This property is particularly useful in sizing due to the fact that the normalization weakens the dependence on the device size. This property was also effectively made use of in Liao and Zhang [7] for sizing while considering the layout effect.

An example of the C_{gg}/I_D-versus-g_m/I_d curves is shown in Fig. 3 for an NMOS device in the TSMC 65 nm technology. With the selection of a g_m/I_D value, an I_D, and an L, we can determine the value of C_{gg} approximately.

Other gm/ID curve sets used in sizing are: V_{GS}-versus-g_m/I_D and V_{DSAT}-versus-g_m/I_D. We fixed V_{DS} at the half of $V_{DD} = 1.2$ V when sweeping these curves.

III. Sizing with Design Equations and gm/ID Tables

We focus on sizing a three-stage Op Amp with two feedforward compensation paths as shown in Fig. 4. The sizing method is general in that it can be applied to a class of analogous multistage Op Amps.

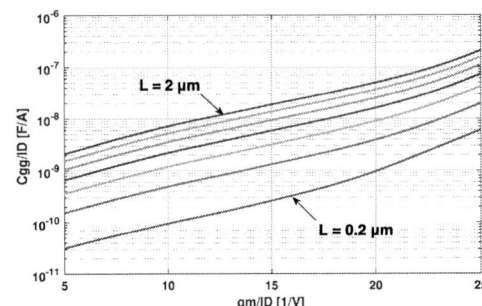

Fig. 3: C_{gg}/I_D versus g_m/I_D for an N-type MOS in the TSMC 65nm technology.

The Op Amp consists of three identical differential source-coupled stages. The input differential pairs of the succeeding stages are used as the feedforward g_{mf} elements while the loading transistors play the role stage transconductances.

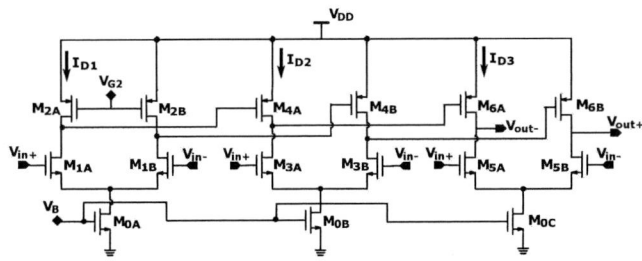

Fig. 4: Fully differential three-stage Op Amp with feedforward compensations.

A. Small-signal Analysis

Frequency domain analysis is important for designing high-speed Op Amps. To proceed, we perform the small-signal analysis on the macromodel drawn in Fig. 5. The symbols G_{mi}, R_{oi}, and C_{oi} ($i = 1, 2, 3$) represent the equivalent transconductance, output resistance, and parasitic capacitance of the ith stage, respectively.

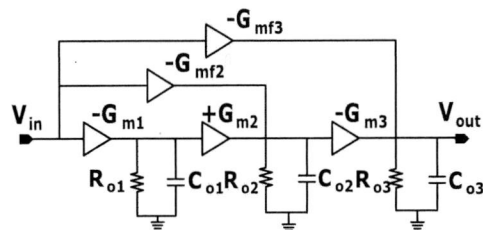

Fig. 5: Macromodel of the three-stage Op Amp with feedforward compensation.

The transfer function (TF) of this circuit is derived as (after ignoring the minor terms)

$$TF(s) \approx \frac{A_v \left(1 + \frac{g_{mf2}C_{o1}}{g_{m1}g_{m2}}s + \frac{g_{mf3}C_{o1}C_{o2}}{g_{m1}g_{m2}g_{m3}}s^2\right)}{\prod_{i=1}^{3}(1 + R_{oi}C_{oi}s)}. \quad (1)$$

The key design equations derivable from this TF are discussed below.

1) DC Gain: The dc gain of the three-stage Op Amp is given by

$$A_v = \prod_{i=1}^{3} A_{oi} = \prod_{i=1}^{3}(g_{mi}R_{oi}). \quad (2)$$

By our proposed method, we pre-choose the g_m/I_D and L values for all transistors and determine the transistor sizes stage by stage.

Using the pre-selected g_m/I_D and L values of of all transistors, the V_{GS} of transistors can be found according to the precomputed V_{GS}-to-g_m/I_D curves. Due to the interconnection of the successive stages, the V_{DS} of a preceding stage PMOS is equal to V_{GS} of a succeeding stage PMOS. In addition, noting that the drain voltages of all tail transistors, M_{0A}, M_{0B}, M_{0C}, are quite free, we can choose the V_{DS} values of the source-coupled differential pairs of all stages to be anywhere $V_{DSAT,M_n} \leq V_{DS} \leq V_{DD} - V_{DS,M_p} - V_{DSAT,M0}$, where M_n stands for an NMOS input transistor and M_p a loading PMOS in each stage. Then the g_{ds}/I_D of the M_n and M_p pair in each stage can be determined from the 3D surface in Fig. 2 according to the prescribed values of g_m/I_D, V_{DS}, and L.

Then the dc gain of each stage can be calculated by

$$A_{oi} = \frac{(g_m/I_D)_{M_{n,i}}}{(g_{ds}/I_D)_{M_{n,i}} + (g_{ds}/I_D)_{M_{p,i}}}. \quad (3)$$

2) Poles and Zeros: To ensure that the two zeroes are real and located in the left-half plane (LHP) of the complex plane, we require that (referring to the numerator of (1))

$$\left(\frac{g_{mf2}C_{o1}}{g_{m1}g_{m2}}\right)^2 - \frac{4g_{mf3}C_{o1}C_{o2}}{g_{m1}g_{m2}g_{m3}} \geq 0. \quad (4)$$

Equivalently, it can be written in the following gm/ID variant

$$\frac{\left(\frac{C_{o1}}{I_{D1}}\right)\left(\frac{g_{mf2}}{I_{D2}}\right)^2\left(\frac{g_{m3}}{I_{D3}}\right)}{4\left(\frac{g_{m1}}{I_{D1}}\right)\left(\frac{g_{m2}}{I_{D2}}\right)\left(\frac{C_{o2}}{I_{D2}}\right)\left(\frac{g_{mf3}}{I_{D3}}\right)} \geq 1, \quad (5)$$

where I_{Di} are the branch currents pre-allocated, as marked in Fig. 4.

Note that the capacitances C_{o1} and C_{o2} are due to the parasitics. They are approximately given by

$$C_{o1} \approx C_{gs4} + C_{gd4} \times A_{o2}, \quad C_{o2} \approx C_{gs6} + C_{gd6} \times A_{o3} \quad (6)$$

after considering the Miller effect. Furthermore, we may write the terms C_{oi}/I_{Di} in (5) as

$$\frac{C_{o1}}{I_{D1}} = \frac{I_{D2}}{I_{D1}} \times \left(\frac{C_{gs4}}{I_{D2}} + \frac{C_{gd4}}{I_{D2}} \times A_{o2}\right) \quad (7a)$$

$$\frac{C_{o2}}{I_{D2}} = \frac{I_{D3}}{I_{D2}} \times \left(\frac{C_{gs6}}{I_{D3}} + \frac{C_{gd6}}{I_{D3}} \times A_{o3}\right) \quad (7b)$$

and find $C_{gs,gd}/I_D$ from the $C_{gs,gd}/I_D$-versus-g_m/I_D curves given the values of g_m/I_D and L.

Assuming that the two zeros are widely separated, we write them approximately as

$$z_1 = \frac{g_{m1}g_{m2}}{g_{mf2}C_{o1}}, \quad z_2 = \frac{g_{m3}g_{mf2}}{g_{mf3}C_{o2}}. \quad (8)$$

Then written in the g_m/I_D variants, they are

$$z_1 = \left(\frac{g_{m1}/I_{D1}}{C_{o1}/I_{D1}}\right) \times \left(\frac{g_{m2}/I_{D2}}{g_{mf2}/I_{D2}}\right), \quad (9a)$$

$$z_2 = \left(\frac{g_{m2}/I_{D2}}{C_{o2}/I_{D2}}\right) \times \left(\frac{g_{m3}/I_{D3}}{g_{mf3}/I_{D3}}\right). \quad (9b)$$

The poles are easily derived from (1) as

$$p_i = \frac{1}{R_{oi}C_{oi}} \quad (i = 1, 2, 3), \quad (10)$$

which can be written equivalently as

$$p_i = \frac{(g_{ds,p} + g_{ds,n})_i/I_{Di}}{C_{oi}/I_{Di}} \quad (i = 1, 2, 3), \quad (11)$$

where $g_{ds,p}$ and $g_{ds,n}$ are the respective channel conductances of the P- and N-type transistors of each stage.

It follows that the pole-zero (PZ) ratios of z_1-to-p_2 and z_2-to-p_3 are

$$z_1/p_2 = \frac{A_{o2}\left(\frac{g_{m1}}{I_{D1}}\right)\left(\frac{C_{o2}}{I_{D2}}\right)}{\left(\frac{g_{mf2}}{I_{D2}}\right)\left(\frac{C_{o1}}{I_{D1}}\right)}, \quad (12a)$$

$$z_2/p_3 = \frac{A_{o3}\left(\frac{g_{mf2}}{I_{D2}}\right)\left(\frac{C_{o3}}{I_{D3}}\right)}{\left(\frac{g_{mf3}}{I_{D3}}\right)\left(\frac{C_{o2}}{I_{D2}}\right)}. \quad (12b)$$

These expressions bring us insight on the pole-zero placement. In order to attain gigahertz GBW, we shall restrict these PZ ratios in the interval of [0.2, 5] so that nearly cancelling zero-pole pairs are formed. Under these constraints, the frequency attenuation before reaching the GBW frequency can be slowed down. This is a critical design tactic for high-speed operation.

IV. Verification by Simulation

The proposed sizing procedure was applied to the three-stage Op Amp shown in Fig. 4. We used a TSMC 65 nm technology in simulation. The final sizes of the transistors are listed in Table I. Other numerical settings are $V_{in,dc} = 651.9$ mV, $V_{G2} = 628.6$ mV, and $V_B = 432.6$ mV.

TABLE I: Transistor sizes.

Device	W/L (µm/µm)	Device	W/L (µm/µm)
M_{0A}	3.8/0.3	$M_{3A,B}$	20.0/0.3
M_{0B}	190/0.3	$M_{4A,B}$	140.8/0.3
M_{0C}	266/0.3	$M_{5A,B}$	133/0.3
$M_{1A,B}$	0.4/0.3	$M_{6A,B}$	127/0.3
$M_{2A,B}$	0.75/0.3		

After sizing, we ran the operating point (OP) simulation to verify the sizing accuracy by the gm/ID lookup table and design equation based calculations. Table II shows that the simulated parameter values of the sized circuit are in good agreement to the pre-chosen parameter values during our sizing planning phase.

The simulated open-loop ac response is plotted in Fig. 6. This circuit achieved a dc gain of 75.3 dB and a high-frequency gain of 37.3 dB at 100 MHz and a phase margin of 78 degrees (when driving a 100 fF load, refer to [3]). Other simulated performance numbers are listed in Table III with a comparison to the target numbers. It achieved a figure-of-merit (FoM)

$$\left(\frac{f_u C_L}{Power}\right) = 272 \frac{\text{MHz} \cdot \text{pF}}{\text{mW}}.$$

In Fig. 6 we also plotted the planned transfer function (dotted lines). It shows that the gm/ID-based pole-zero planning predicts reliably the final ac response result.

TABLE II: Cross check of the design parameters between their preset and simulated values.

Name listing	Parameter	Preset value	Simulation
Inversion Coefficient	$(g_m/I_D)_{M1}$	8 V^{-1}	9 V^{-1}
	$(g_m/I_D)_{M2}$	8 V^{-1}	8.2 V^{-1}
	$(g_m/I_D)_{M3}$	8 V^{-1}	9.5 V^{-1}
	$(g_m/I_D)_{M4}$	14 V^{-1}	14.8 V^{-1}
	$(g_m/I_D)_{M5}$	15 V^{-1}	16.1 V^{-1}
	$(g_m/I_D)_{M6}$	12 V^{-1}	12.1 V^{-1}
Poles and zeroes	p_1	1.05 MHz	0.72 MHz
	p_2	93.8 MHz	99.8 MHz
	p_3	838.1 MHz	800.5 MHz
	z_1	29.6 MHz	42.3 MHz
	z_2	0.98 GHz	1.26 GHz

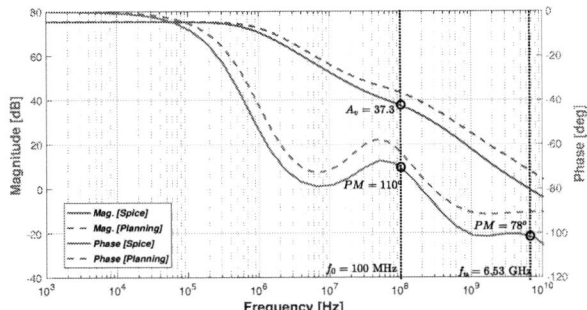

Fig. 6: Open-loop ac response of the three-stage Op Amp in the TSMC 65 nm technology.

TABLE III: Simulated circuit performance.

Spec name	Spec value	Simulation
Power Supply	1.2 V	\
Load Capacitance	100 fF	\
Bias Current (I_{D1})	10 µA	8.38 µA
Bias Current (I_{D2})	500 µA	392.1 µA
Bias Current (I_{D3})	700 µA	614.6 µA
Total Current (I_{DD})	2.42 mA	2.0 mA
Power	2.88 mW	2.4 mW
Gain	> 60 dB	75.32 dB
GBW (f_u)	> 1 GHz	6.53 GHz
Phase Margin	> 45 °	78 °

V. Conclusion

We have presented a gm/ID-based transistor sizing method for application to multiple-stage Op Amps with feedforward-only compensation. We have customized the gm/ID sizing method by incorporating design equations and extending the gm/ID lookup tables to accommodate the influence of the parasitic capacitances. A notable benefit of the proposed sizing method is a substantial reduction of simulation-based searching efforts to arrive at an initial sizing result. The proposed method can be integrated in a sizing automation tool.

References

[1] H. Shibata, R. Schreier, et al., "A DC-to-1 GHz tunable RF $\Delta\Sigma$ ADC achieving DR= 74 dB and BW= 150 MHz at $f_0 = 450$ MHz using 550 mW," IEEE Journal of Solid-State Circuits, vol. 47, no. 12, pp. 2888–2897, 2012.

[2] X. Yang and H. Lee, "Design of a 4th-order multi-stage feedforward operational amplifier for continuous-time bandpass Delta Sigma modulators," Proc. IEEE International Symposium on Circuits and Systems (ISCAS), pp. 1058–1061, 2016.

[3] F. T. Gebreyohannes, J. Porte, M. M. Louërat, and H. Aboushady, "A g_m/I_D methodology based data-driven search algorithm for the design of multistage multipath feed-forward-compensated amplifiers targeting high speed continuous-time $\Sigma\Delta$-modulators," IEEE Transactions on Computer-Aided Design of Integrated Circuits and Systems, vol. 39, no. 12, pp. 4311–4324, 2020.

[4] B. K. Thandri and J. Silva-Martinez, "A robust feedforward compensation scheme for multistage operational transconductance amplifiers with no Miller capacitors," IEEE Journal of Solid-State Circuits, vol. 38, no. 2, pp. 237–243, 2003.

[5] P. G. A. Jespers and B. Murmann, Systematic Design of Analog CMOS Circuits Using Pre-computed Lookup Tables. Cambridge, UK: Cambridge University Press, 2017.

[6] M. N. Sabry, H. Omran, and M. Dessouky, "Systematic design and optimization of operational transconductance amplifier using g_m/I_D design methodology," Microelectronics Journal, vol. 75, pp. 87–96, May 2018.

[7] T. Liao and L. Zhang, "Efficient parasitic-aware gm/ID-based hybrid sizing methodology for analog and RF integrated circuits," ACM Trans. Des. Autom. Electron. Syst., vol. 26, no. 2, p. 31, 2020.

[8] G. Shi, "Sizing of multi-stage Op Amps by combining design equations with the gm/ID method," Integration, the VLSI Journal, vol. 79, pp. 48–60, 2021.

Hybrid Capacitor-less LDO with Switched-Mode Dead-Zone Control

Nellie Laleni, Andreas Tsiougkos, Vasilis Pavlidis

Department of Electrical and Computer Engineering, Aristotle University of Thessaloniki, Greece

Abstract—**A hybrid capacitor-less low-dropout regulator that includes switched-mode dead-zone control is proposed. Differently from the prior art, the combination of one digital and two analog low-dropout regulators effectively mitigates the oscillations of the output voltage within the dead-zone and reduces the overall power consumption. The digital part consists of an 1-bit comparator, shift-registers and the PMOS array, and the analog part is constructed by an error amplifier and a pass transistor. In addition, a single-stage high gain amplifier without extra circuitry or an output capacitor is included ensuring that the hybrid LDO operates for load currents between 10 μA - 10 mA, offering significant savings in area. The digital part is modeled using Verilog-A/AMS and the analog part and the output PMOS array of the digital part are designed using 180 nm X-fab CMOS technology.**

Index Terms—**low-dropout regulator, capacitor-less, analog-assistance, switched-mode control, dead-zone control**

I. Introduction

The ongoing demand for increased functionality and the related computational power on battery-supplied devices, make power management a critical issue. A power management unit typically includes DC-DC converters that generate different DC voltages for diverse loads according to the application requirements. For a cleaner supply and resilience from the noise on the supply lines, circuits called low-dropout (LDO) regulators are utilised.

Analog low-dropout (ALDO) regulators consist of an analog error amplifier and a pass element (mostly a power PMOS). ALDO regulators are widely used due to their effective line and load regulation. However, in advanced technologies due to the lower supply voltages, digital low-dropout (DLDO) regulators are preferred [1], [2]. In the DLDO, the analog amplifier is replaced by a digital controller with a quantization error and, instead of the power pass element, a power switch array is used (Fig. 1). A digital controller drives the output array according to the quantized error in order to stabilize the output at the desired voltage.

Steady-state oscillations can also be caused due to the quantized nature of the DLDO. There are two main approaches to suppress these oscillations. Firstly, a separate analog loop can be constructed [3]. However, the performance of the analog loop is restricted by the technology since ensuring high gain and high power supply rejection ratio (PSRR) in low supply voltages is challenging. As a result, the steady-state oscillations are poorly suppressed, especially for high load conditions. Another approach is to use a dead-zone control, *i.e.* disabling the regulation within a voltage range [4], [5]. The

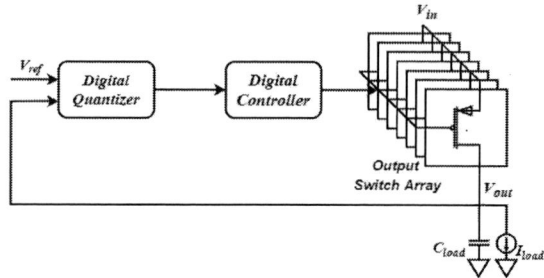

Fig. 1: A high level concept of a typical DLDO.

challenge of using a dead-zone type control is to determine the dead-zone range and settle the output voltage inside the dead-zone with respect to the reference voltage [4].

In this paper, a hybrid capacitor-less LDO with a dead-zone switched-mode control is introduced to overcome all aforementioned limitations of the recent DLDO circuits (Fig 2a). In this topology, the digital part is responsible for the high load currents, the analog part drives the smaller loads and helps the transient response across the whole operation. No output capacitance is required, thereby reducing both area and power. One additional analog LDO, is used in the dead zone to avoid the steady-state ripples [6].

The paper is organised as follows. In Section II, the working principle of the topology is described. The implementation details of the proposed digital LDO are provided in Section III. The simulation results and the conclusion of this paper are presented in Section IV and Section V, respectively.

II. Working Principle of the Proposed LDO

The proposed topology consists of a shift register DLDO and two identical ALDOs with a folded cascode error amplifier and a PMOS pass transistor. Analog assistance provides faster response and overcomes the need for a faster responding DLDO, which subsequently requires a higher sampling frequency or a complex design of the DLDO. In addition, having an analog loop provides high PSRR and increases the output accuracy in the voltage regulation. These high performance characteristics of the LDO allow the removal of the large output capacitance, since the PSRR is high due to the analog gain and no filter output capacitence is needed, thereby considerably decreasing the area. Moreover, the ALDOs with Miller capacitors operate within a small load range providing support to this hybrid LDO (HLDO) and keeping it stable. ALDOs eliminate the steady-state ripples with the help of the dead-zone control. One of the ALDOs

© VDE VERLAG GMBH · Berlin · Offenbach

SMACD / PRIME 2021 | 19 – 22 July 2021, Online Event

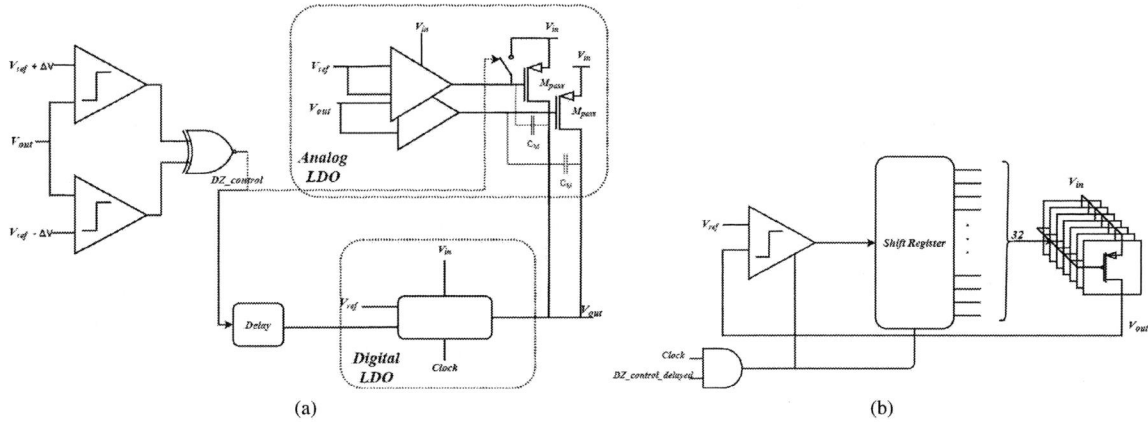

Fig. 2: Block diagram of the proposed LDO, where (a) is the complete system, and (b) only the shift-register DLDO

is active in the entire range of operation, whereas the other ALDO is active only in the dead-zone of the DLDO, such that the power consumed by the analog part of the HLDO is decreased.

Therefore, the proposed hybrid LDO has two modes of operation: analog-digital (AD) and analog-analog (AA). AD stands for the operation where one of the ALDO and the DLDO simultaneously supply current. AD occurs outside of the dead-zone. Within the dead-zone, the DLDO is paused and the second ALDO switches on for the regulation, leading to the AA mode. The two ALDOs operate in parallel to produce a precise output voltage during this mode.

In the AD mode, the main current supplier is the DLDO. The output voltage and the reference voltage are compared using an 1-bit comparator and the output switch array is driven using a 5-bit thermometer code bi-directional shift register triggered at the rising clock edges, in order to regulate the output voltage. The ALDO during the AD mode, operates synergistically with the DLDO improving the response time. As shown in Fig. 2a, one of the comparators outputs '1' whenever the output voltage higher (lower) than $V_{ref} - \Delta V$ ($V_{ref} + \Delta V$). The XNOR drives the $DZ_{control}$ signal to high, enabling the operation of the DLDO after a certain delay. This delay ensures the stability of the operation during the transition between modes. Within the dead-zone, the two ALDOs drive the output voltage to the required level. With a change in the load current, if the output voltage exceeds the dead-zone, the DLDO starts supplying current again.

The dead-zone is implemented to eliminate unwanted steady-state ripples [1]. However, in the absence of the ALDOs, the output can settle to different voltages within the range of the dead-zone. The always-on ALDO, provides the necessary current to settle the desired output voltage as in [3]. However, in the proposed LDO, the ALDO does not provide high current. Therefore, in the dead-zone, the second ALDO assists the operation by providing additional current and leverages the regulation collectively with the other ALDO. Two small ALDOs are more power efficient than one large ripple-cancellation amplifier [4]. The proposed LDO usefully

combines the ripple cancellation technique of [3] using the benefits of two analog amplifiers as in [4], offering a superior topology.

III. IMPLEMENTATION

The digital LDO consists of an 1-bit comparator, a 5-bit thermometer code bi-directional shift register, and 32 PMOS switches driving the output (Fig. 2b). The clock frequency is selected as 100 MHz. A clock gating circuit for the dead-zone is also included. All the digital part is modeled in Verilog-A/Verilog-AMS. The output switch is designed in 180 nm X-fab CMOS technology with maximum deliverable current 10 mA.

The dead-zone occurs when the output voltage lies within a region, ΔV, around the reference voltage level *i.e.*, $V_{out} \pm \Delta V$. Determining ΔV is critical and includes a trade-off among the level of steady-state ripples, the response time, and the value of V_{out} (with respect to V_{ref}) [4]. In this implementation, ΔV has been chosen such that the dead-zone corresponds to 10% of the reference voltage symmetrically around the V_{ref} ($V_{out} = V_{ref} + / - 10\% * V_{ref}$).

Two identical ALDOs are constructed to provide correct operation within the dead-zone as shown in Fig. 2a. The folded-cascode amplifier of the ALDO is designed to have high gain and maximum output swing. High gain provides better PSRR and also better transient performance. Maximum output swing improves the operating range of the amplifier. A Miller capacitor of 2 pF shown in Fig. 2a is added to the output of the error amplifier to improve stability. This capacitor also provides smoother switching for the second amplifier. The magnitude and phase responses of the ALDO are illustrated in Fig. 3a with different load currents. The ALDO yields 40 dB DC gain with 64.7° phase margin for the worst condition (minimum load current). In Fig. 3b, it can be seen that if the Miller capacitor is not used, the ALDO is not stable even for the maximum load current.

In many analog or hybrid LDO designs, the analog part is designed using two-stage OTA to obtain reasonable DC gain [4]. This approach, however, makes the stability

© VDE VERLAG GMBH · Berlin · Offenbach

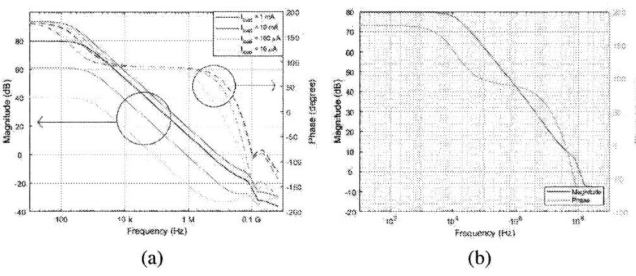

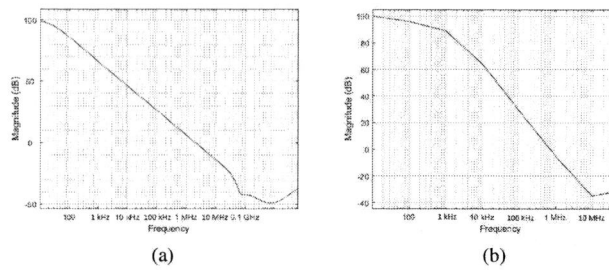

| (a) | (b) | (a) | (b) |

Fig. 3: Frequency response of the ALDO, where (a) for different load currents, and (b) for the maximum load without a Miller capacitor.

Fig. 4: (a) PSRR of the ALDO, and (b) PSRR of the hybrid LDO for minimum load current.

challenging especially for high load conditions. The identical folded-cascode amplifiers are one-stage amplifiers and provide easier control of the stability. Moreover, there are designs with one stage amplifier and additional circuitry to provide higher gain [3] or stability by splitting the second dominant pole into two smaller poles [6]. The proposed ALDO offers simplicity as an one-stage folded cascode amplifier, stability, and high gain during operation.

For the load currents lower than the LSB of the DLDO, ALDO provides all the current for the regulation. The second ALDO improves the steady-state performance of the LDO and provides high precision for the output voltage. The total bias current of the two ALDOs is 30 μA and the total quiescent current (I_Q) is estimated as 50 μA. Also, for advanced technologies, where V_{supply} is low, the specifications of high-performance analog error amplifiers are strict and the design becomes challenging. As a result, extra gain-boosting techniques and/or stages are needed to provide high gain and high PSRR for the whole circuit. Using two one stage-amplifiers instead of one with more stages, reduces the design complexity and doubles the provided current, without degrading the bandwidth [4].

IV. SIMULATION RESULTS

In this section, the steady-state and the transient performance are analyzed. The steady-state performance of the LDO is dominated by the ALDOs because of the dead-zone. Therefore, the simulations for the PSRR are investigated for the ALDO and the overall HLDO. The simulation results are shown in Fig. 4. The ALDO's PSRR is simulated via AC analysis using spectre, whereas the PSRR of the HLDO is extracted from the system's transient response for input disturbance of 100 mV for different frequencies. Around DC, the output PMOS transistor improves the gain and the PSRR. The maximum demonstrated PSRR is 98 dB and remains above 70 dB around 1 kHz. The HLDO follows the PSRR of the ALDO as mentioned previously. In higher frequencies, where the ALDO has poor PSRR, the hybrid LDO exhibits better performance as the DLDO supports the operation of the system (whenever the output voltage lies outside the dead-zone).

In the transient response simulations, the current step rise and fall times are equal to 300 ns and the C_{load} is 20 pF.

The LDO transient response is supported by the ALDOs for load currents smaller than the LSB of the DLDO. The output voltage for the step between 10 μA and 0.4 mA, where only the ALDO operates, is plotted in Fig. 5. As the step becomes larger, the undershoot deepens. However, the settling time of the LDO decreases when the DLDO switches on and quickly recovers the V_{out} even if the undershoot is greater.

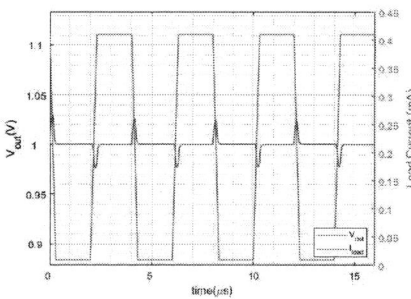

Fig. 5: Output voltage for the current step 10 μA to 0.4 mA.

The circuit performance for the maximum load step is critical for the stable operation of the LDO. The change in the output voltage for the maximum load step, is drawn in Fig. 6. For the load increasing from 10 μA to 10 mA, the output drops by 80 mV. The maximum overshoot level is 120 mV. The output voltage settles back after the maximum load step in less than 0.5 μs, which is comparable to state of the art LDOs [4], [7]. Moreover, in conventional LDOs, an output capacitor is added to filter out the V_{droop} and to enhance the PSRR [1]. However, this output capacitance adds considerable area to the LDO. For example, in [4], the switched-mode hybrid LDO includes 500 pF capacitance which constitutes 30% of the total area of 0.081 mm^2 and in [5], the DLDO has an integrated capacitance of 1 nF which consumes 40% of the total area (0.1 mm^2). However, topologies as in [7] (0.034 mm^2) designed without a capacitance, are demonstrated with around half of the total area of the LDO in [4]. This approach means that LDOs without capacitance as the proposed LDO, can save area up to 35-40% compared to prior art.

The transient response of the overall system and DLDO only, are plotted, respectively, in Fig. 7a. The current step in this operation is 1 mA to 10 mA such that the ALDO and the DLDO operate in tandem. Without the ALDOs, the steady-state ripples are visible (100 mV peak to peak) and the settling time is longer. With the assistance of the ALDOs, the ripples are eliminated and V_{out} settles faster.

Environmental variations, such as temperature can also affect the response of the LDO. As the temperature increases, the levels of the overshoot and undershoot decrease. The simulation results for the maximum load step across a wide range of temperatures is depicted in Fig. 7b. As can be seen in this figure, the maximum change at the output deviation is around 5% for the temperatures between $-20°C$ and $90°C$ and the settling time is not affected, which shows insensitivity to changes in temperature.

Other state-of-the-art designs are compared with this work in Table I including both simulation and measurement results from other works. The supply voltage of the proposed LDO is comparably low with respect to the supply voltages of the state-of-the-art. This means that the analog part of the proposed LDO works properly under low supply conditions, which are similar to DLDO supplies. Moreover, stable low load current drive with very small headroom voltage is achieved, which is superior to the designs listed in Table I. Finally, the low V_{droop} is produced without any off-chip output capacitance, which further reinforces the usefulness of this new LDO. To compare the performance of the LDOs two figures of merit (FOMs) are calculated from [8]. FOM_1 is used for the dynamic performance and FOM_2 is used for static performance (smaller values are better) where the proposed LDO is competitive with state-of-the-art circuits.

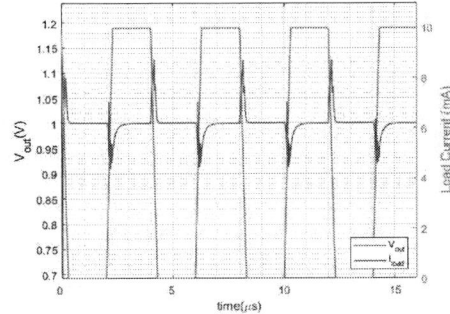

Fig. 6: Output voltage for the current step 10 μA to 10 mA.

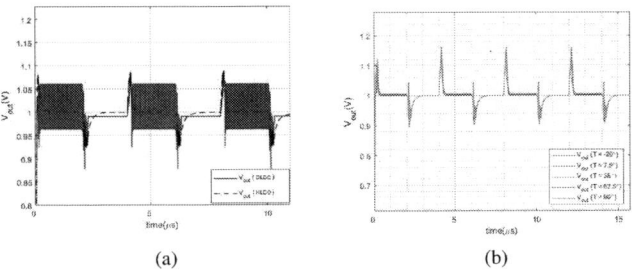

(a)   (b)

Fig. 7: Transient performance of the LDO, (a) Comparison between all-digital and hybrid for the current step 1 mA to 10 mA, and (b) maximum load step for different temperatures.

TABLE I: Comparison between the proposed LDO and prior art

	This Work	[9]	[10]	[8]	[11]
LDO Type	Hybrid	Analog	Analog	Hybrid	Analog
Process	180 nm	28 nm	65 nm	65 nm	65 nm
Input Voltage	1.2 V	1.2-2.5 V	1.2 V	1.8 V	1.2 V
Output Voltage	1 V	0.7-0.9 V	0.7-0.8 V	0.75-0.95 V	0.8-1.1 V
Minimum Load Current	10 μA	0 A	1 μA	10 μA	100 μA
Maximum Load Current	10 mA	50 mA	10 mA	40 mA	5 mA
Load Capacitance	20 pF	80 pF	70 pF	Cap-less	1 pF
V_{droop} (mV) @ I_{max} (mA)	80 @ 9.99	190 @ 50	96 @ 10	10 @ 98	62 @ 4.9
I_Q (μA)	50*	308	60	120	11.5
FOM_1 (μA.ps)	7.99	N.A.	8.232	15	2.91
FOM_2 (ps)	0.8	1.87	8.232	0.0188	0.02852

$FOM_1 = \frac{V_{droop}}{CurrentStep} \times C_{load} \times \frac{I_{MIN}}{I_{MAX}} \times I_Q$,

$FOM_2 = V_{droop} \times C_{load} \times \frac{I_Q}{I_{MAX}^2}$,

* estimated value.

V. CONCLUSION

The hybrid capacitor-less LDO presented in this work provides stable operation for the load range between 10 μA to 10 mA. The fast response, the compatibility for low supply voltages, the design simplicity of the digital LDOs, and the good steady-state performance of the analog LDOs are combined in this topology. Thanks to the analog-assistance, the steady-state ripples are eliminated and the PSRR of the overall LDO is improved. The proposed hybrid LDO is a strong candidate for cutting-edge digitally dominated, low-power systems, demonstrating appropriate regulation at low power supplies.

REFERENCES

[1] Y. Okuma, K. Ishida, Y. Ryu, X. Zhang, P.-H. Chen, K. Watanabe, M. Takamiya, and T. Sakurai, "0.5-V input digital LDO with 98.7% current efficiency and 2.7-μA quiescent current in 65 nm CMOS," in *IEEE Custom Integrated Circuits Conference*, 2010, pp. 1–4.

[2] S. B. Nasir, S. Gangopadhyay, and A. Raychowdhury, "A 0.13 μm fully digital low-dropout regulator with adaptive control and reduced dynamic stability for ultra-wide dynamic range," in *IEEE International Solid-State Circuits Conference*, 2015, pp. 1–3.

[3] M. Cheah, D. Mandal, B. Bakkaloglu, and S. Kiaei, "A 100-mA, 99.11% current efficiency, 2-mVpp ripple digitally controlled LDO with active ripple suppression," *IEEE Transactions on Very Large Scale Integration (VLSI) Systems*, vol. 25, no. 2, pp. 696–704, 2017.

[4] S. B. Nasir, S. Sen, and A. Raychowdhury, "Switched-mode-control based hybrid LDO for fine-grain power management of digital load circuits," *IEEE Journal of Solid-State Circuits*, vol. 53, no. 2, pp. 569–581, 2018.

[5] M. Huang, Y. Lu, S.-W. Sin, U. Seng-Pan, and R. P. Martins, "A fully integrated digital LDO with coarse–fine-tuning and burst-mode operation," *IEEE Transactions on Circuits and Systems II: Express Briefs*, vol. 63, no. 7, pp. 683–687, 2016.

[6] S. B. Nasir, S. Sen, and A. Raychowdhury, "A 130nm hybrid low dropout regulator based on switched mode control for digital load circuits," *IEEE International Solid-State Circuits Conference*, no. 2, pp. 1–3, 2015.

[7] M. Huang, Y. Lu, S.-P. U, and R. P. Martins, "An analog-assisted tri-loop digital low-dropout regulator," *IEEE Journal of Solid-State Circuits*, vol. 53, no. 1, pp. 20–34, 2018.

[8] Y. Zhang, H. Song, R. Zhou, W. Rhee, I. Shim, and Z. Wang, "A capacitor-less ripple-less hybrid LDO with exponential ratio array and 4000x load current range," *IEEE Transactions on Circuits and Systems II: Express Briefs*, vol. 66, no. 1, pp. 36–40, 2019.

[9] W. Wang and B. Chi, "A wideband high PSRR capacitor-less LDO with adaptive DC level shift and bulk-driven feed-forward techniques in 28nm CMOS," in *IEEE International Symposium on Circuits and Systems*, 2019, pp. 1–5.

[10] S. Gweon, J. Lee, K. Kim, and H. Yoo, "93.8% current efficiency and 0.672 ns transient response reconfigurable LDO for wireless sensor network systems," in *IEEE International Symposium on Circuits and Systems*, 2019, pp. 1–5.

[11] A. Santra, A. De Carmine, G. V. S. Rao, and Q. A. Khan, "A highly scalable, time-based capless low-dropout regulator using master-slave domino control," *IEEE International Symposium on Circuits and Systems*, pp. 1–4, 2019.

© VDE VERLAG GMBH · Berlin · Offenbach

Verilog-A model development of a DC–DC boost controller with autonomous optimization

1st Davide Severin
Infineon Technologies - Italy
Davide.Severin-EE@infineon.com

2nd Capodivacca Giovanni
Infineon Technologies - Italy
Giovanni.Capodivacca@infineon.com

3rd Tchodjie Tchamabe Bernard Blaise
Infineon Technologies - Italy
BernardBlaise.TchodjieTchamabe@infineon.com

4th Buzo Andi
Infineon Technologies - Germany
Andi.Buzo@infineon.com

5th Diaconu Cristian-Vasile
Infineon Technologies - Romania
Cristian-Vasile.Diaconu@infineon.com

Abstract—**Simulation time of wide schematic is often the root-cause of time-to-market worsening, especially during the design verification of large switchable circuit. To speed CPU-time up equivalent models can be used in place of the original modules, but they are often created for a specific project and requires a manual tuning of the specification. This contribution provides a methodology to develop Verilog-A equivalent models whose implementation is general, maintaining the physical coherence and parameterizing the internal properties. This large set of parameters is tuned with an autonomous Artificial Intelligence algorithm (GDE3) so that to achieve the required Mean Percentage Error (MPE) but reducing the human effort in optimization from some weeks down to a couple of days. The entire methodology has been applied to a DC–DC boost converter, decreasing the simulation time of an order of magnitude.**

I. INTRODUCTION

Over the years, transistor-level complexity in analog and digital circuits has increased, led by the Moore's law. Although the computational power of computers has grown as well, the simulation of very wide electrical designs is still burdensome and time-consuming. The computation time is one of the major factor of time-to-market worsening. A DC–DC switching converter is a typical example of a complex circuit, which consists of of a large number of functional sub-blocks.

To reduce simulation time during the early stages of the top schematic verification and to aid designers to investigate the most suitable specification for each sub-blocks, a Verilog-A behavioural model of each basic building block is developed. The model implementation is physically coherent, so that the resulting simulation is consistent with the transistor-level behaviour [1] and the debug is facilitated. Moreover, the characteristics are parametric and the same model is a valid representation of different version of the same basic block. The manual tuning of a large set of parameters is resource-intensive, especially if a high accuracy is required. Therefore, an Artificial Intelligence technique, based on the Generalized Differential Evolution algorithm, is employed to tune autonomously the parameters, given the reference circuit. Differently from the IGBT standalone optimization [2], in this case the algorithm is applied on a more complex circuit, i.e. a controller of a DC–DC converter. Moreover, the optimization is performed along a continuous waveform, i.e. the transient responses of each sub-modules, considering different stimuli. The application of the GDE brings many advantages in terms of time: the amount of working hours is lowered from some weeks down to a couple of day and the final response is the most accurate possible.

In next sections the methodology used is presented: in Section II the boost controller is introduced and the use-case Current Sense model is explained in details. In Section III the validation and optimization process is presented. In Section IV the results are listed, in particular an overview of the working hours saved using the AI algorithm and the overall Mean Percentage Error (MPE).

II. DEVELOPMENT OF THE MODELS

A. Peak Current-Mode in DC–DC Boost

The use-case project in which the models were developed is a DC–DC switching boost with a Peak Current-Mode (PCM) control, but the modeling methodology must be as general as possible so that it can be re-used for every future project. The DC–DC switching boost is an high-efficiency step-up voltage regulator, which is typically stabilized with a negative voltage feedback to prevent the output voltage from over-shooting or dropping. The Current-Mode Control (CMC) is a multi-loop control strategy in which both the output voltage and the inductor's current are feed-backed, to determine the duty-cycle of the Pulse-Width-Modulated ignition signal of the main switch. Since the inductor's current value is meaningful for the control loop only within the on-time, it is sensed in series with the power switch. A current compensation slope is added to prevent the system from the sub-harmonic oscillations while the duty-cycle is greater than 50% [3]. The DC–DC Boost controller developed in Infineon and given as reference circuit for the Verilog-A modeling, implements a PCM control. An overview of the system designed is shown in Fig. 1.

B. Modeling description

Verilog–A is a very flexible modeling language for analog circuits and it is largely employed nowadays to realize behavioural models of circuits and devices, starting from basic

© VDE VERLAG GMBH · Berlin · Offenbach

SMACD / PRIME 2021 | 19 – 22 July 2021, Online Event

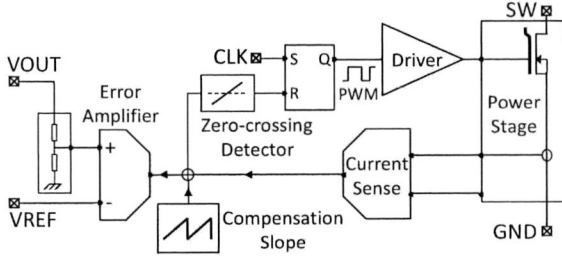

Fig. 1. Overview of the Infineon controller for a DC–DC Boost

components up to high-level description of fully-functioning chips. The controller was therefore modeled in Verilog-A: each basic building block was developed, tested and optimized independently and then combined to evaluate the performance. For the sake of simplicity, only the current sense model is presented. In Fig. 2 it is shown an outline of the transistor-level implementation of the power stage and the current sense [4]. The power stage implements six nMOS transistors, forced to work in ohmic region during the on-time:

- M1 is the main switch, which the inductor current flows through during the on-time.
- M2 and M3 stand in series and the same gate voltage make them operate in the ohmic region. The current flowing through M2 is proportional to I_1 by a factor set to N, obtained scaling the $\frac{W}{L}$ ratio of M2. Similarly, M3 is scaled down by a factor M.
- M4 and M5 match M1, M2 and M3 in at least one electrical characteristics. Their $\frac{W}{L}$ is scaled by a factor K with respect to M1. M3 and M4 are driven by the same gate voltage of M3.

In the ohmic region, transistors may be approximated with their on-resistance R_{on}: M2 and M3's are respectively N and M times bigger with respect to M1's. On-resistances of M4 and M5 are K times bigger. The current sense is a transconductive stage with a current feedback. The input stage is an amplifier that enhances the V_d by a factor of A, and turns it into a driving voltage for the gate of a pMOS M6, whose current I_6 is feed-backed toward M5. An output current mirror constituted by M6 and M7 turns the I_6 into the desired I_{sense} which flows toward the output terminal: the mirror ratio is user-defined. A

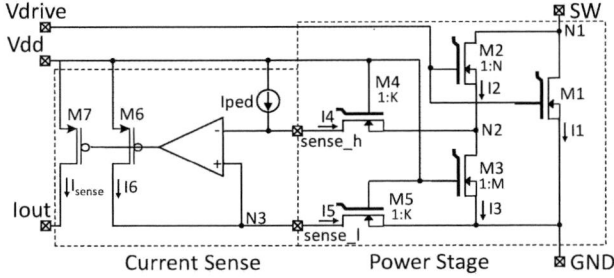

Fig. 2. Schematic of the power stage and the current sense of the Infineon controller for the DC–DC Boost

current I_{ped} is added to bias the input amplifier: I_{ped} flows through M4 and M3 toward ground. It is designed so that its value is significantly smaller than I_2. In steady state, the system provides an I_{sense} current equal to:

$$I_{sense} = I_{ped} + \frac{M}{K\,(M+N+1)} \cdot I_L \qquad (1)$$

The relationship between I_L and I_{sense} is proportional if $I_{ped} = 0$, otherwise it is still linear: the circuit is effectively used within a PCM control to sense the inductor's current.

The Verilog–A model of the Power Stage is physically coherent with the schematic: an nMOS is modeled using the theoretical relationships that link the voltages across the terminals and the drain-source current. Capacitances are added across the terminals. This nMOS is instantiated five times with the same topology: N, M and K factors are parameters of the system.

The model of the Current Sense is instead shown in Fig. 3. The basic idea is to keep the same pin-out and topology of the original circuit. Thus it is perfectly plug-&-play and the parameters reflect the physical quantities that are found in the schematic. The Operational Transconductive Amplifier (OTA)

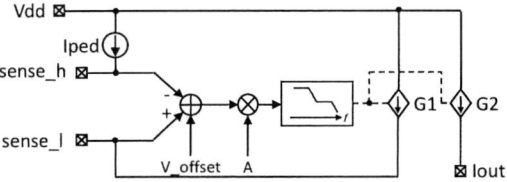

Fig. 3. Overview of the proposed model for the Current Sense

is modeled as a differential amplifier, indeed it can be broken down into three stages:

1) A differential stage, wherein the two sensing inputs are subtracted. An additional voltage offset v_offset parameter is inserted into the operation.
2) An amplification stage, with the gain parameter A.
3) A frequency response adapter, which introduces poles and zeros in the open-loop transfer function, as the reference OTA does. This module is short-circuited in DC-simulations.

The OTA output stage and the pMOS M6 are modeled by means of a Voltage Controlled Current Source (VCCS) G1 which withdraws from the power supply a current proportional to the control node toward the $sense\_l$ node, similarly to the original circuit. The transconductance of the controlled source is a parameter of the system. The current mirror is easily replicated with a parallel VCCS G2 connected from the power supply to the I_{out} pin. An additional independent current source is connected from the Vdd to $sense\_h$, which pushes the pedestal current toward the power stage. Although the nominal value of some parameters (e.g. the I_{ped}) is already fixed by design specifications, all the parameters are meant to be optimized by the algorithm mainly for two reasons:

© VDE VERLAG GMBH · Berlin · Offenbach

1) Albeit the fixed nominal value, the actual magnitude may be slightly different.
2) The model is meant to be re-used in other projects, wherein the same parameter may assume different values.

III. OPTIMIZATION PROCEDURE

Once a Verilog-A module is developed, the optimization of the parameters is performed in order not only to fit the output response to the reference circuit, but primarily to validate the thoroughness of the model. A scheme of the validation and optimization procedure is presented in Fig. 4. The optimizer is the Generalized Differential Evolution (GDE) algorithm and it is a global maximum or minimum finder, with a reduced number of control settings. The usage of an autonomous routine is the key-point because the manual tuning of complex models with tens of parameters, is extremely complicated and time-consuming and the optimum is rarely found. With the AI only the setup is up to the user, with an enhanced saving of working hours. Since the GDE evolves heuristically within a pre-defined number of iterations, if the optimum is not reached, a larger number of cycles is required. If the algorithm still returns a non-satisfying optimization, a re-modeling is suggested to introduce new elements. Conversely, if the module has been over-modeled, the redundancies may be removed in favor of a lighter and time-saving model. Once the single model is optimized, it is connected with the other Verilog-A modules to evaluate the overall performances.

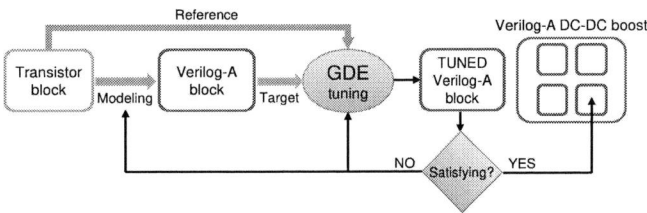

Fig. 4. Overview of the modeling flow

A. Generalized Differential Evolution

The *Evolutionary Computation* is a family of algorithms for global optimization. One of the most famous subset are the *Evolutionary Algorithms*. This class of algorithms is inspired by the biological evolution of the species, that naturally tends to preserve the strongest individual to the detriment of the most unfit to live. Among these, the GDE is a an heuristic population-based algorithm for global optimization. Unlike its predecessor, the Differential Evolution [5], the GDE is capable of optimizing more than one Objective Function (OF), i.e. target cost function, in parallel [6], without the need to establish *a priori* the mutual priorities among them. This leads to a wider solution space and, hopefully, a better result.

B. Setup and Optimization

The procedure to optimize a model is hereinafter described. The setup requires a test-bench in which both the reference circuit and the Verilog-A model are instantiated and tested with

the same set of stimuli. The analysis and the configuration options are set on a simulation state, in which are also loaded all the Verilog-A parameters so that they can be directly set within the simulator environment. Typically the analysis is a transient simulation and the response is an output continuous waveform, both for the transistor-level circuit and the Verilog-A model. The algorithm is a global minimum finder and cannot manage a continuous waveform: therefore some key-measurements are performed on both the responses. From the two values the percentage error is evaluated, which is a perfect Objective Function for the GDE algorithm:

$$Objective\ Function = \left| \frac{x^{ref} - x^{mod}}{x^{ref}} \right| \qquad (2)$$

In Fig. 5, $x \in \{t_{rise}, V_{over1}, \dots\}$, which are the rise and fall-time, the overshoot peaks and the halfway step measured on the output current of the modeled current sense. In general, these key-measurements are specific for each block according to its application. Measurements are often differential since, for this application, a little time delay is acceptable: measuring the difference between the two waveforms at every time point is not significant. The GDE algorithm will minimize all the errors in parallel. Eventually, the choice of the best Pareto-solution is up to the user according to the designer's need and it may not necessarily be the one with minimum MPE. The more measurements are performed all over the waveforms, the more similar the two will be, but the computation time and the human effort will increase.

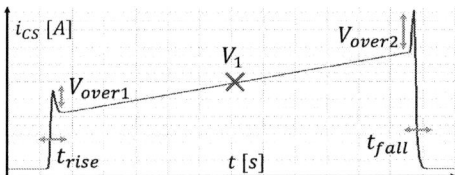

Fig. 5. Example of key-measurements performed on transient response of the Current Sense output current during the on-phase of the main switch

IV. RESULTS

A. Top simulation results

The Current Sense model along with the other blocks that compose the boost controller are inserted in a top schematic in place of the transistor-level ones. A transient simulation is performed with a load jump and a battery step to evaluate not only the steady-state but also the model's dynamic behaviour. In addition, the system is tested at different corners, for the battery voltage, the Bill Of Material and the load current, to validate the model as an eligible substitute of the original circuit for a rapid verification plan. At the end of the simulations, the output voltage obtained from the model is compared with the one in the same corner but with the transistor-level schematic. Some interesting waveforms are shown in Fig. 6. The comparison is very promising: the responses are consistent in the majority of the situations and sometimes almost

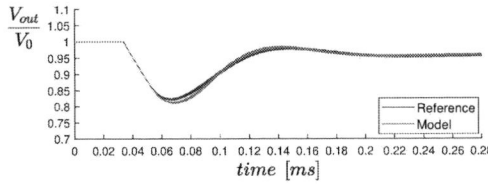

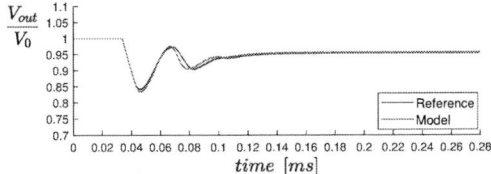

Fig. 6. Output voltage response of the DC–DC switching boost to a positive load jump for two different corners at 27°C - Comparison between the original response and the Verilog-A model one with normalized voltage axis

superimposable. The computation time required to run the transistor-level plan is compared to the Verilog-A one. Results are shown in Table I, along with the improvement calculated as the difference $t_{ref} - t_{model}$ over the transistor-level time t_{ref}. Results are collected from simulations run at different corners: the worst, the best and the mean improvements are listed. As expected, they show an order of magnitude of improvement in using the models instead of the original blocks.

TABLE I
TIME IMPROVEMENT TABLE FOR THE BATTERY STEP AND LOAD JUMP TRANSIENT SIMULATION IN THE TOP SCHEMATIC OF THE DC–DC BOOST

Time Improvement	t_{ref} [s]	t_{model} [s]	CPU–time Improvement
Best value	53.2k	6.41k	88.0%
Worst value	25.7k	5.12k	80.0%
Mean value	36.0k	5.97k	83.4%

B. Autonomous Optimization Advantages

The application of the GDE autonomous algorithm is meant to reduce the human working time while optimizing the parameter set of each sub-module and to achieve the best MPE possible, according to the model limitations.
In the DC–DC controller, 11 blocks were developed. Each was manually optimized on a dedicated testbench and then with the GDE routine, to evaluate the improvement. The terms of the comparison, whose results are listed in Table II, are:

- Human working time: the number of working hours required to achieve an *as fit as possible* model. The autonomous machine-time is not counted.
- MPE: Mean Percentage Error of the fit output response.

All in all, the usage of the GDE algorithm leads to a working hours reduction by a factor of $1/6$ ($\sim$ 65 hours saved) and an MPE improvement by a factor of 2 on average. In Table (II) a complete summary is shown.

V. CONCLUSIONS

A new Verilog-A model of a DC–DC boost controller has been developed and its response has been fit using an AI

TABLE II
COMPARISON BETWEEN THE MANUAL AND THE GDE OPTIMIZATION FOR THE VERILOG-A PARAMETERS

	Manual Opt.	GDE Opt.
Working Hours per block	7h[a]	1h 15m[b]
Working Hours in total	$\sim$ 80h	$\sim$ 15h
MPE	< 20%	< 10%

[a] 1h for the test-bench, 6h for the optimization
[b] 1h for the test-bench, 15m for the algorithm setup

routine. The model has been tested and compared with the reference circuit to prove the validity of the code and the optimization: the output response of the regulator is always consistent and reveals the same stability issues.
The methodology developed allows the user to fit an entire continuous waveform, performing distributed measurements along it, employing a Multi-Objective Algorithm to handle all of them without introducing *a priori* weighting factors. Working hours are spared thanks to the AI routine, reducing the human intervention at least by a factor of 6, achieving the best MPE possible. Indeed, the deviation is always below 10%. The top simulations comparison is very promising: the responses are consistent in the majority of the situations and sometimes almost perfectly superimposable. Finally, the overall saving in terms of CPU–computation time is almost an order of magnitude. This new model will allow the preliminary verification phase to be reduced from many weeks down to a couple of days. It is moreover flexible and can be re-used for many other projects which implement the same topology structure: engineers will be aided starting from the concept phase up to the pre-tapeout stages.

ACKNOWLEDGMENT

This work was partially supported by the projects iDev40. The project iDev40 has received funding from the ECSEL Joint Undertaking under grant agreement No. 783163. The JU receives support from the European Union's Horizon 2020 research and innovation program. It is co-funded by the consortium members, grants from Austria, Germany, Belgium, Italy, Spain and Romania. The information and results set out in this publication are those of the authors and do not necessarily reflect the opinion of the ECSEL Joint Undertaking.

REFERENCES

[1] P. Wilson and R. Wilcock, "Behavioural modeling and simulation of a switched-current phase locked loop," 2005 IEEE International Symposium on Circuits and Systems, Kobe, 2005, pp. 5174-5177 Vol. 5, doi: 10.1109/ISCAS.2005.1465800.
[2] D. Ludwig, G. Alia, A. Biswas and M. Cotorogea, "Behavioral compact models of IGBTs and Si-diodes for data sheet simulations using a machine learning based calibration strategy," PCIM Europe digital days 2020; International Exhibition and Conference for Power Electronics, Intelligent Motion, Renewable Energy and Energy Management, 2020, pp. 1-8.
[3] R. D. Middlebrook, "Topics in Multiple-Loop Regulators and Current-Mode Programming," in IEEE Transactions on Power Electronics, vol. PE-2, no. 2, pp. 109-124, April 1987, doi: 10.1109/TPEL.1987.4766345.
[4] Infineon Technologies AG, D. Vacca Cavalotto, E. Orietti, "Circuit Arrangement," U.S. Patent US 13 899 626, Nov.27, 2014.
[5] R. Storn and K. Price, "Differential Evolution - A Simple and Efficient Heuristic for Global Optimization over Continuous Spaces," Journal of Global Optimization, 11.341-359.10.1023/A:1008202821328, 1997.
[6] S. Kukkonen and J. Lampinen, "Generalized Differential Evolution for General Non-Linear Optimization," 10.1007/978-3-7908-2084-3_38, 2008.

© VDE VERLAG GMBH · Berlin · Offenbach

SMACD / PRIME 2021 | 19 – 22 July 2021, Online Event

Analog Circuit Abstraction to SystemC-AMS Secured by Affine Forms

Ahmad Tarraf
Institute for Computer Science
Goethe University Frankfurt, Germany
tarraf@em.cs.uni-frankfurt.de

Lars Hedrich
Institute for Computer Science
Goethe University Frankfurt, Germany
hedrich@em.cs.uni-frankfurt.de

Abstract—Formal verification of analog circuits still suffers from a lot of challenges. The continuous nature of the variables along with the well-known state space explosion problem obstruct most verification approaches. Using behavioral abstraction, a circuit can be verified to some extent, however, the quality of the results is limited to the accuracy of the abstract model. To solve these problems, we propose an automated abstraction methodology that generates from transistor-level Spice netlists accurate models at the system level in SystemC-AMS. The models are deployed as hybrid automatons (HAs) with reduced complexities and linear system equations. Further, to compensate for modeling errors resulting from the abstraction technique, an extended version of the approach utilizes affine forms to generate models that produce bounded results during symbolic simulations.

Index Terms—abstraction, hybrid automaton, behavioral modeling, SystemC-AMS

I. INTRODUCTION

During the design process of analog and mixed-signal systems, behavioral modeling is often a mandatory step. SystemC-AMS [1] offers a fast, universal, and flexible description for mixed-signal systems at high-level. However, a challenge is to incorporate transistor-level circuits at the system-level in SystemC-AMS. Previous automated modeling approaches result in little speedups and ignore the modeling error in the high-level simulations. Additionally, manual written models suffer from being too abstract, often neglecting non-idealities like limiting behavior. Moreover, they lack update capabilities if the underlying transistor level circuit changes.

With the extension of affine arithmetic decision diagrams (AADDs) to SystemC-AMS [2], a system can be described using affine forms to enclose analog behaviors. This allows for the generation of models that include all deviations from the modeling process as well as opens the doors towards modeling with parameter deviations. Considering the first aspect, a transient analysis of a model with affine forms results in safe upper and lower bounds during a SystemC-AMS simulation.

In this paper, we propose the following contributions:

- Automatically generate accurate SystemC-AMS models from Spice netlists with BSIM accuracy.
- Propose four methods for the dynamic behavioral description of the models. All models obtained exhibit large speedup factors compared to their Spice netlists.
- Extended the basic approach utilizing AADDs [2], to enclose errors and deviations from the modeling approach.

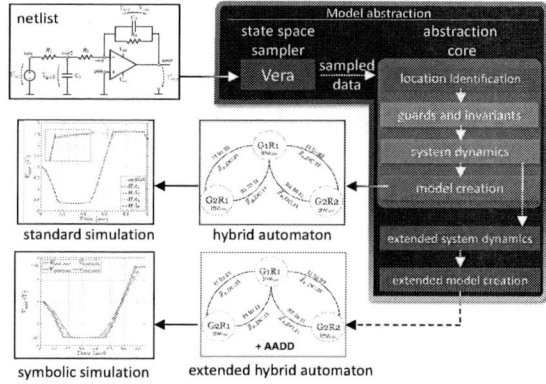

Fig. 1: Overview of the presented methodology.

Fig. 1 shows an overview of the approach. After sampling the circuit with *Vera* (Sec. III-A), a HA is modeled for standard simulations (Sec. III-B-III-D). Alternatively, a HA can be constructed to enclose the whole sampled behavior (Sec. IV).

II. PREVIOUS WORK

Analog behavioral abstraction has a long research history [3]. An automated abstraction process is presented in [4], which mainly improves simulation speed but is not targeting HAs. A symbolic simulation-based method for nonlinear behavioral models is presented in [5], while [6] presents an approach for piecewise linear models. The method in [7] models on the fly the underlying differential-algebraic equations (DAEs) of electrical networks using piecewise linear regions for each nonlinear element. It suffers from using an abstract transistor-model and is limited to the number of transistors to be verified. Other approaches use numeric approximation such as radial basis functions with neural networks techniques [8] or general kernels models in support vector machines [9]. Unfortunately, these data-driven methods suffer from lower accuracy and difficulties in obtaining significant speedups. In [10], an automated abstraction approach is described that abstracts a circuit to a HA in Verilog-A. Unlike [10], the approach here focuses on the generation of SystemC-AMS models with extension capabilities for transient enclosures considering the whole sampled system behavior.

© VDE VERLAG GMBH · Berlin · Offenbach

III. Automatic Behavioral Abstraction

The HA used in this context is as defined in [11] with some restrictions on the jumps and guards. The invariants are only used to find the guards as shown later.

A. Sampling the State Space

The approach starts by numerically sampling the transistor level netlist using *Vera* [12]. A circuit can be described by a DAE system, for simplicity here as a SISO system:

$$f(x(t), \dot{x}(t), u(t)) \qquad (1)$$

With $x \in \mathbb{R}^n$ denoting the vector of unknown voltages and currents in the original state space $\mathcal{S}_o$ of the system. *Vera* steps through the state space, linearizing the system locally and calculating the reachability of the sampled points of the overall nonlinear system. The result is a data set containing the DC points $x_{DC} \in \mathbb{R}^n$, the conduction and capacitance matrices G and C, and the input vector b. The system is transformed to a Kronecker form using the transformation matrices F and E. By performing a dominant pole reduction as in [9] on the obtained matrices, the large dynamic order resulting from parasitic poles is reduced to m i.e. the functional needed. In particular, using $\Delta x = x - x_{DC}$ and:

$$\Delta x = F \Delta x_s, \qquad (2)$$

this results in:

$$s \underbrace{\begin{bmatrix} I_\lambda & 0 \\ 0 & 0 \end{bmatrix}}_{ECF} \begin{bmatrix} \Delta x_{s,\lambda} \\ \Delta x_{s,\infty} \end{bmatrix} + \underbrace{\begin{bmatrix} -\Lambda & 0 \\ 0 & -I_\infty \end{bmatrix}}_{EGF} \begin{bmatrix} \Delta x_{s,\lambda} \\ \Delta x_{s,\infty} \end{bmatrix} = \underbrace{\begin{bmatrix} \tilde{b}_\lambda \\ \tilde{b}_\infty \end{bmatrix}}_{Eb} \Delta u, \qquad (3)$$

with the identity matrix I and the diagonal (or band-diagonal) matrix $\Lambda \in \mathbb{R}^{m \times m}$ filled with the m finite eigenvalues from the underlying generalized eigenproblem. Note that transformed vectors are marked with a tilde (̃). Eq. (3) can be split into a dynamic part with the state vector $x_{s,\lambda} \in \mathbb{R}^m$ in the reduced state space $\mathcal{S}_\lambda$, and an algebraic part with the state vector $x_{\infty,\lambda} \in \mathbb{R}^{(n-m)}$ in the $\mathcal{S}_\infty$ space. For simplicity, $x_{s,\lambda}$ will be denoted as x_λ. Eq. (3) shows the strength of our approach: To calculate x using Eq. (2), only x_λ needs to be determined, as $x_{s,\infty}$ is directly given from the second row of Eq. (3). However, Eq. (3) is pointwise valid. Hence, our objective is to create a HA with a set of locations in which a generalized form of Eq. (3) is valid for a set of sample points.

B. Locations, Invariants, and Guards of the Hybrid Automaton

The next step of the abstraction process is to find the finite set of locations $loc_i \in Loc$. Using the finite m eigenvalues of the linearized system, groups (g) are found by clustering the eigenvalues using an extended version of k-means. Next, groups are split into regions (r). This is necessary in case several disjoint portions in the $\mathcal{S}_\lambda$ space belong to the same group. Regions are found by examining the neighborhood information of sample points from the same group (see [10]). Finally, a location is specified as $loc_i = g_j r_k \in Loc$ where j is the group and k the region index. Considering for example a neuron with a ReLU activation function, the found locations are illustrated in Fig. 2a. Two locations were found: a linear location ($g2r1$) and a limiting location ($g1r1$).

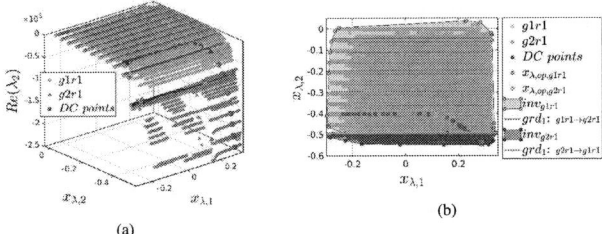

(a)

(b)

Fig. 2: $\mathcal{S}_\lambda$ space illustrated (a) against the real part of the second eigenvalue of the reduced system and (b) of the generated HA with two locations. The large red points represent the DC points.

For each identified location, the invariant ($inv_{g_j r_k}$) is found by hulling the underlying sample points in the $\mathcal{S}_\lambda$ space using convex shapes. In Fig. 2b, polytopes are used.

Between the invariants, guards are identified by examining the edges (facet in general) of the current invariant for the presence of points from the neighbor locations. In case the number of points found is above a specified tolerance value, the edge is taken as a guard. Although different representation forms exist, guards are modeled here as halfspaces as shown in Fig. 2b. A guard from location loc_i to loc_t is denoted as:

$$grd_k : loc_i \rightarrow loc_t \qquad | \ loc_i, loc_t \in Loc \qquad (4)$$

A guard allows for a specific location switch. Once a guard is taken, its corresponding jump condition is applied:

$$x_{\lambda, new} = I x_{\lambda, old} - x_{\lambda, op, loc_t} \qquad (5)$$

Where $x_{\lambda, op, loc_t} \in \mathbb{R}^m$ is the operating point of the target location (Sec. III-C), and $x_{\lambda, new}$ and $x_{\lambda, old}$ are the state vectors after and before the transition, respectively.

C. Dynamics of the Hybrid Automaton

Using the linear state space representation, the system behavior in each location of the HA is described. For that, the operating point in each location (x_{λ, op, loc_i}) is chosen from the DC points (see Fig. 2) calculated by *Vera*. In particular, the operating points are chosen as the DC points closest to the guards as shown in Fig. 2b. The system behavior in a location $loc_i \in Loc$ of the HA is then described as:

$$\Delta \dot{x}_\lambda = A_{loc_i} \Delta x_\lambda + B_{loc_i} \Delta u, \qquad (6)$$

such that:

$$\Delta x_\lambda = x_\lambda - x_{\lambda, op, loc_i} \qquad \Delta u = u - u_{op, loc_i} \qquad (7a)$$

$$A_{loc_i} = \bar{\Lambda}_{loc_i} \qquad B_{loc_i} = \bar{E}_{\lambda, loc_i} \bar{b}_{\lambda, loc_i} \qquad (7b)$$

Matrices with subscript λ correspond to sub-matrices belonging to x_λ (see Eq. (3)). A and B are computed from the mean values of the matrices of the sampled points belonging to a location. Using Eq. (6), x_λ is calculated in the $\mathcal{S}_\lambda$ space. To obtain the values of the nodal voltages and currents in the original state space $\mathcal{S}_o$, x must be calculated. This is done by a back-transformation obtained by using Eq. (2) with Eq. (3):

$$x = x_{op, loc_i} + \bar{F}_{\lambda, loc_i} \Delta x_\lambda - \bar{F}_{\infty, loc_i} \bar{E}_{\infty, loc_i} \bar{b}_{\infty, loc_i} \Delta u \quad (8)$$

D. Hybrid Automaton in SystemC-AMS

The last step of the abstraction process is to deploy the HA in SystemC-AMS using the Time-Data-Flow model of computation (TDF-MoC). For a standard simulation, the invariants of the HA are neglected and the guards are expressed using if conditions. Hence, in case a guard becomes active, a location switch occurs immediately. Thus, the current location loc_i has a unique value at each time step. As the further aim is to generate models that use AADDs to perform range computations, the integration method used for solving Eq. (6) is of special importance. Hence, four methods we deployed: state space function provided by System-AMS (*ss*), backward Euler (*eu*), Runge-Kutta (*rk*), and discretization of the linear state space (*di*). The *ss* method uses the SystemC-AMS TDF solver for a state space description (sca_tdf::sca_ss). The remaining methods use a constant predefined sample time (T). The *eu* method finds the current value $\boldsymbol{x}_\lambda(iT)$ by:

$$\Delta \boldsymbol{x}_\lambda(iT) = (\boldsymbol{I} - T\boldsymbol{A}_{loc_i})^{-1}[\boldsymbol{x}_\lambda((i-1)T) - T\boldsymbol{A}_{loc_i}\boldsymbol{x}_{\lambda,op,loc_i} \\ + T\boldsymbol{B}_{loc_i}(u((i-1)T) - u_{op,loc_i})] \tag{9}$$

As the *rk* (Runge-Kutta) method is similarly described, the explanation is skipped. The *di* method discretizes the continuous solution of the linear state space in the $\mathcal{S}_\lambda$ space to yield:

$$\Delta \boldsymbol{x}_\lambda(iT) = e^{\boldsymbol{A}_{loc_i}T}(\boldsymbol{x}_\lambda((i-1)T) - \boldsymbol{x}_{\lambda,op,loc_i}) \\ + \boldsymbol{A}_{loc_i}^{-1}(e^{\boldsymbol{A}_{loc_i}T} - \boldsymbol{I})\boldsymbol{B}_{loc_i}(u(i-1)T) - u_{op,loc_i}) \tag{10}$$

The matrix operations in Eqs. (9, 10) are precomputed in Matlab and the resulting matrices are written to the SystemC-AMS model. Hence, T must be set before the simulation. At each time step of the simulation, the obtained values of $\boldsymbol{x}_\lambda(iT)$ are transformed back into the $\mathcal{S}_o$ space using Eq. (8).

IV. MODELING THE SYSTEM USING AFFINE FORMS

The models can be extended for symbolic simulations using the AADD library [2]. The library combines ordered binary decision diagrams with affine arithmetic forms which implement a range arithmetic similar to interval arithmetic or zonotopes to model uncertainties in an efficient (symbolic) way. The library extends the SystemC-AMS data structure with data types that allow symbolic executions: doubleS, floatS, etc. Moreover, the control statements can be adjusted with ifS, elseS, and endS. Thus, during simulation both condition cases are added to the AADD as terminal nodes. For more details see [2]. The AADD library represents continuous propagated uncertainties by affine arithmetic forms [2], [13]:

$$\{x = x_c + \sum_{i=1}^{n} x_i\beta_i \mid \beta_i \in [-1,1]\}, \qquad x, x_i, x_c \in \mathbb{R} \tag{11}$$

Each variable x_i models the sensitivity of x to the basic uncertainty $\beta_i \in [-1, 1]$, which is a symbolic variable whose exact value is unknown. Each β_i is scaled by the corresponding partial deviation x_i. The variables are then traced with the opt_sol class to capture its minimum and maximum values. Currently, the library does not support the TDF state space (sca_ss). Hence, the methods *di*, *eu*, and *rk* were developed, however, we will focus only on the *di* method. The elements of the matrices in Eqs. (8, 10) can be modeled as affine forms (*Afm*). For simplicity, affine description are limited to single symbolic variables allowing for the calculation of

elements in Eqs. (7b, 8, 10) using the Taylor series and interval arithmetic. If several uncertainties were considered per element, the particular solution in Eq. (10) is replaced by discretizing the integral in linear state space solution using the Taylor series along with Simpson's numerical integration.

Moreover, by modeling loc_i as an affine form and adjusting the guard conditions to affine conditions (*Afc*), several locations can be in a reachable set at each time point.

V. RESULTS

For demonstration, we present three examples. The first example is a 2nd order low-pass filter implemented using a CMOS 350nm technology with 17 transistors. Besides the wanted linear low-pass behavior, the circuit has a limiting behavior at the output $V_{out} \in [-1.65, +1.65]$V. In Fig. 3, four HAs of the circuit that only vary in their integration method were compared to the netlist based on a simulation for 2s with a time step of $0.1\mu s$ and an input $V_{in} = 4sin(2\pi t)$. The figure holds the output error δ_y (first row) and output voltage V_{out} (second row) plotted against the time. As shown in Fig. 3 and presented in Table I, all models have significant speedups while maintaining acceptable output deviations. Note that the *di* method exhibits the largest speedup with the smallest deviations and is thus used as the standard method. For all models, the Spice netlist was sampled in 130.1s by *Vera* and imported into Matlab's memory in 2.55s. The modeling time in Table I measures the time consumed by the abstraction process only, while $\delta_{y,r}$ stands for the relative output error which divides δ_y over the output range of both systems.

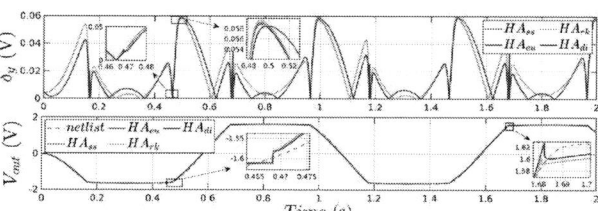

Fig. 3: Simulation results of the 4 SystemC-AMS models and the Spice netlist.

The second example shows the scalability of the approach. Consider the GmC filter from Fig. 4. The circuit consists of

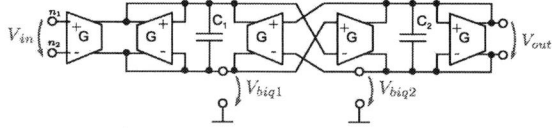

Fig. 4: Schematic of a Spice netlist describing a GmC filter.

46 transistors and has $n = 45$ nodal variables. The circuit dynamic order is 26 which is reduced by *Vera* to $m = 2$. In 243.41s, *Vera* sampled 27191 points and in 0.57s the data was imported to Matlab. The abstraction processes consumed 5.48s to build a HA with 5 locations using the *di* method. In Fig. 6a, the simulation results of the Spice netlist and the HA for a sine wave of amplitude 0.5V and frequency of 1KHz at the input V_{in} is shown. The output voltages V_{out}, internal voltages V_{biq1} and V_{biq2}, and the current location of the HA (loc_i) are shown

in the first, second, and third row of this figure, respectively. The vertical purple lines indicate a location switch. As shown, the output of the HA exhibits slight deviations from the output of the netlist. This can be traced back mainly to the matrices in Eqs. (7b, 8) which were calculated by mean values. For the internal voltages, V_{biq1} and V_{biq2}, strong deviations to the netlist exist. However, the difference between these signals is nearly the same, preserving the differential behavior of the architecture of the filter.

To model the whole sampled behavior, affine forms can be used. For simplicity, interval hulls ($\mathcal{I}$) are used, which correspond to an affine description with a single symbolic variable. In Fig. 5, the hulled eigenvalues are shown, illustrating the elements of the extended system matrix A_{loc_i}. Similarly, this is done to the remaining matrices resulting in

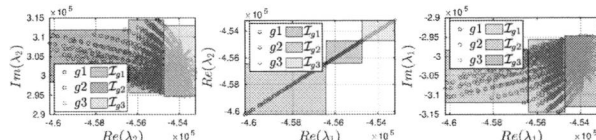

Fig. 5: Hulled eigenvalues of the GmC filter.

range computations in Eqs. (7b, 8, 10). This encloses the whole sampled behavior, however, presents undesired fluctuations in the results, especially in the location variable loc_i which affects the runtime. To eliminate these fluctuations, loc_i is modeled as an affine form and the guard conditions are ajdusted (*Afc*). As shown in Fig. 6b, the location variable loc_i can attain various values in a single time step. This encloses the behavior of the Spice netlist with little over approximations (see Table I), however, worsens the runtime.

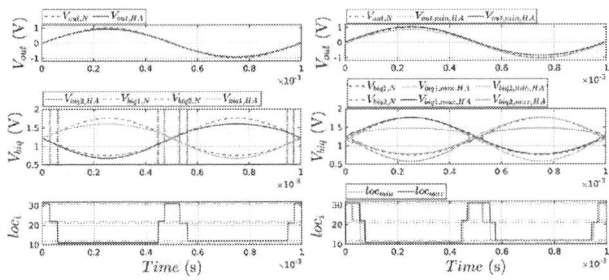

(a) Standard simulation.   (b) Symbolic simulation.

Fig. 6: Results of (a) a standard simulation of a HA (*di*) and (b) a symbolic simulation of a HA with affine forms (*di+Afm+Afc*) with 5 location compared to the Spice netlist of the GmC filter.

The third example in Table I is an analog neuron with a nonlinear ReLU activation function. Fig. 2b shows the HA.

VI. CONCLUSION

This paper presents a fully automated abstraction approach that results in a behavioral model in SystemC-AMS syntax at the system level. The obtained models exhibit large speedups and have reduced complexities compared to the original Spice netlists. An extension to the approach allows for symbolic simulations using AADDs to enclose the whole sampled behavior

TABLE I: Comparison of transistor level netlists with the generated behavioral models

	Model	Model. time (s)	Compile time (s)	Runtime (s)	Speed-up	$\hat{\delta}_{\dot{y}}$ (V)	$\hat{\delta}_{\dot{y},r}$ (%)
Low-pass	Netlist (17 tr)	-	-	15.28	-	-	-
	HA *ss*	8.33	1.56	0.15	101.8	0.058	3.58
	eu	8.24	1.58	0.119	128.4	0.059	3.60
	rk	8.27	1.56	0.125	122.2	0.059	3.63
	di	8.32	1.57	0.109	140.1	0.058	3.56
	di+Afm	8.40	1.66	0.176	86.81	-	-
	di+Afm+Afc	8.40	1.67	262.9	0.058	encl.	encl.
GmC	Netlist (46 tr)	-	-	1.884	-	-	-
	HA *di*	5.48	1.61	0.012	157	0.082	8.2
	di+Afm	5.58	2.04	2.4	0.785	-	-
	di+Afm+Afc	5.58	2.05	5.295	0.355	encl.	encl.
Neuron	Netlist (18 tr)	-	-	4.49	-	-	-
	HA *ss*	3.01	1.68	0.101	44.45	0.12	6
	di	3.07	1.67	0.08	56.12	0.11	5.5

$\hat{\delta}$ stands for the maximum value of δ, *Afm* for affine forms in the matrices, *Afc* for AADD extended if conditions, and tr for transistors

of the transistor level circuit. In particular, this generates over-approximative SystemC-AMS models that incorporate abstraction errors. Future work will focus on decreasing the runtime of the affine models and incorporating technology-dependent process variations. This can replace extensive Monte Carlo simulations by a single simulation with affine forms.

REFERENCES

[1] M. Barnasconi, G. Christoph, K. Einwich, M. Damm, M.-M. Louëra, T. Maehne, F. Pecheux, and A. Vachoux, "Standard SystemC AMS Language Reference Manual," 2010.

[2] C. Radojicic, C. Grimm, A. Jantsch, and M. Rathmair, "Towards verification of uncertain cyber-physical systems," in *Proceedings 3rd International Workshop on Symbolic and Numerical Methods for Reachability Analysis (SNR@ETAPS 2017), Uppsala, Sweden*, pp. 1–17, 2017.

[3] G. G. Gielen and J. R. Phillips, "Simulation and modeling for analog and mixed-signal integrated circuits," *Electronic Design Automation for IC Implementation, Circuit Design, and Process Technology*, 2016.

[4] W. Zheng, Y. Feng, X. Huang, and H. Mantooth, "Ascend: Automatic bottom-up behavioral modeling tool for analog circuits," in *Circuits and Systems (ISCAS). IEEE International Symposium On*, May 2005.

[5] C. Borchers, "Symbolic Behavioral Model Generation of Nonlinear Analog Circuits," *IEEE Transactions on Circuits and Systems II: Analog & Digital Signal Processing*, vol. 45, no. 10, pp. 1362–1371, 1998.

[6] F. Fernandez, B. Perez-Verdu, and A. Rodriguez- Vazquez, "Behavioral Modeling of PWL Analog Circuits Using Symbolic Analysis," *IEEE International Symposium on Circuits and Systems (ISCAS)*, 1998.

[7] H. S. L. Lee, M. Althoff, S. Hoelldampf, M. Olbrich, and E. Barke, "Automated generation of hybrid system models for reachability analysis of nonlinear analog circuits," in *Design Automation Conference (ASP-DAC), 2015 20th Asia and South Pacific*, pp. 725–730, Jan. 2015.

[8] M. Isaksson, D. Wisell, and D. Ronnow, "Wide-band dynamic modeling of power amplifiers using radial-basis function neural networks," *IEEE Transactions on Microwave Theory and Techniques*, vol. 53, 2005.

[9] J. Phillips, J. Afonso, A. Oliveira, and L. M. Silveira, "Analog macromodeling using kernel methods," in *Proceedings of the 2003 IEEE/ACM International Conference on Computer-Aided Design*, p. 446, IEEE Computer Society, 2003.

[10] A. Tarraf and L. Hedrich, "Behavioral Modeling of Transistor-Level Circuits using Automatic Abstractionto Hybrid Automata," in *Design Automation and Testin Europe (DATE)*, (Florence), 2019.

[11] O. Stursberg and B. H. Krogh, "Efficient representation and computation of reachable sets for hybrid systems," in *Hybrid Systems: Computation and Control*, Springer Berlin Heidelberg, 2003.

[12] S. Steinhorst and L. Hedrich, "Advanced methods for equivalence checking of analog circuits with strong nonlinearities," *Formal Methods in System Design*, vol. 36, no. 2, pp. 131–147, 2010.

[13] L. H. de Figueiredo and J. Stolfi, "Affine Arithmetic: Concepts and Applications," *Numerical Algorithms*, vol. 37, pp. 147–158, Dec. 2004.

Simulating the impact of Random Telegraph Noise on integrated circuits

P. Saraza-Canflanca[1], E. Camacho-Ruiz[1], R. Castro-Lopez[1], E. Roca[1] J. Martin-Martínez[2], R. Rodriguez[2], M. Nafria[2] and F.V. Fernandez[1]

[1]Instituto de Microelectrónica de Sevilla, IMSE-CNM (CSIC/Universidad de Sevilla), Sevilla, Spain
[2]Electronic Engineering Department (REDEC) group, Universitat Autònoma de Barcelona (UAB) Barcelona, Spain

Abstract— **This paper addresses the statistical simulation of integrated circuits affected by Random Telegraph Noise (RTN). For that, the statistical distributions of the parameters of a defect-centric model for RTN are experimentally determined from a purposely designed integrated circuit with CMOS transistor arrays. Then, these distribution functions are used in a statistical simulation methodology that, taking into account transistor sizes, biasing conditions and time, can assess the impact of RTN in the performance of an integrated circuit. Simulation results of a simple circuit are shown together with experimental measurements of a circuit with the same characteristics implemented in the same CMOS technology.**

Keywords—RTN, CMOS, Simulation, Transistor, Characterization

I. Introduction

Random Telegraph Noise (RTN) has become a subject of increasing concern in deeply-scaled CMOS technologies [1], due to its role as a source of device and circuit performance variability [2], [3]. At device level, RTN is observed as random and sudden discrete jumps of the drain current, which are caused by threshold voltage shifts associated to stochastic charge trapping/de-trapping events in/from device defects [4]. The impact of RTN has been reported for a wide variety of circuits, such as SRAMs or Ring Oscillators [3].

Fig. 1 shows an example of a current trace measured on a PMOS transistor with $|V_{gs}| = 1.0$V and $|V_{ds}| = 0.1$V clearly showing the RTN effect. The parameters that characterize the RTN phenomenon are the number of defects in the transistor, the amplitude of the current shifts (or, analogously, the amplitude of the threshold voltage shifts) associated to each of these defects, and their time constants, which may depend on the bias and temperature conditions. These time constants are the capture time (τ_c), i.e., the average time that an empty defect takes to capture a charge carrier, and the emission time (τ_e), i.e.,

the average time that an occupied defect takes to emit the charge carrier. All of these are stochastic parameters that can be modeled by distribution functions; hence, the impact of RTN on circuit performances must be studied statistically. To obtain the characteristic parameters of such distribution functions, an integrated circuit containing thousands of devices was designed on a 65-nm CMOS technology [5] and characterized with the experimental setup [6] shown in Fig. 2.

This paper presents the complete methodology for statistical simulation of RTN effects. To that end, Section II presents the distribution functions used. Section III presents the simulation methodology and Section IV shows the experimental results on a practical circuit.

II. RTN Characterization

The experimental setup allows to obtain thousands of current traces under different biasing and temperature conditions. From the current traces, the TiDeVa tool [7], [8], identifies the current levels by applying a maximum likelihood estimation (MLE)-based method. From the current levels, the number of defects, their associated current shifts, and the times at which captures and emissions take place can be identified [7].

These data must be modelled with a mathematical formulation that can be exploited for circuit simulation. The characteristic parameters of RTN are random variables, and, therefore, they are modeled with probability distribution functions.

A. Distribution of the number of defects

Frequently, the number of defects in a transistor has been modeled as a Poisson distribution [9], [10]. A discrete random

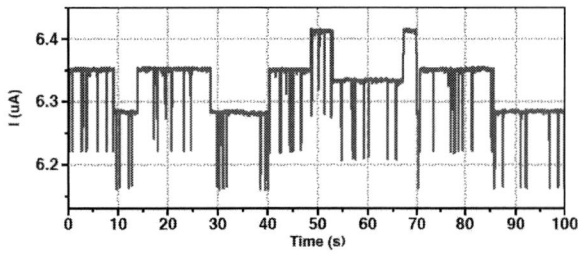

Fig. 1. Example of a current trace measured for a PMOS device displaying RTN-induced current shifts.

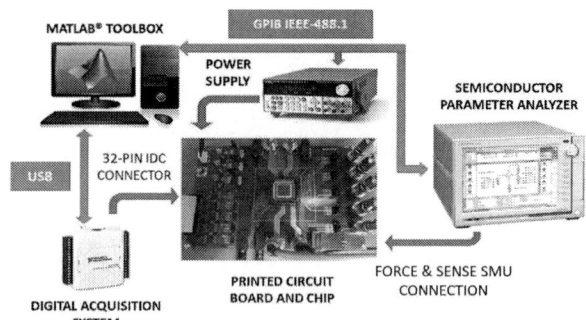

Fig. 2. Schematic representation of the experimental setup used in this work.

This work was supported by the VIGILANT Project (PID2019-103869RB / AEI / 10.13039/501100011033). Pablo Sarazá Canflanca also acknowledges MICINN for supporting his research activity through the predoctoral grant BES-2017-080160.

© VDE VERLAG GMBH · Berlin · Offenbach

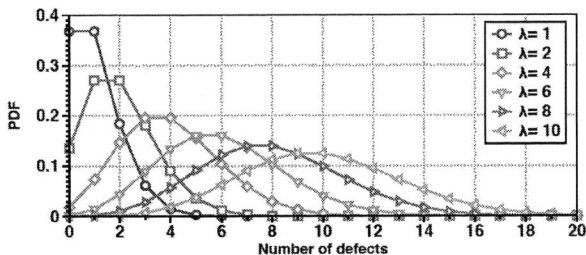

Fig. 3. Examples of Poisson distribution for several values of the mean number of defects λ.

variable X is said to have a Poisson distribution with parameter λ ($\lambda \in \mathbb{R}$, $\lambda \geq 0$) if the probability mass function of X follows:

$$f(k; \lambda) = \Pr(X = k) = \frac{\lambda^k e^{-\lambda}}{k!} \tag{1}$$

For the RTN problem, this probability mass function represents the probability that, for transistors having an average number of defects λ, a given transistor contains exactly k defects. The equation can be easily adapted if instead of the average number of defects in the device, the average density of defects in the technology is considered:

$$\Pr(k \text{ defects in device}) = \frac{(\lambda A)^k e^{-\lambda A}}{k!} \tag{2}$$

where A is the device area and λ is now the average number of defects per unit area (defect density). Fig. 3 shows some examples of a Poisson distribution for several values of λ.

B. Distribution of the capture and emission times

The joint probability density function of the emission and capture times is formulated as a bivariate log-normal function [4],[11]:

$$P_{def}(\tau_e, \tau_c) = \frac{1}{\tau_e \tau_c 2\pi \sigma_{\tau_e} \sigma_{\tau_c} \sqrt{1 - \rho^2}} \cdot$$
$$\cdot e^{-\frac{1}{2\sqrt{1-\rho^2}} \left[\frac{(l\tau_e - \mu_{\tau_e})^2}{\sigma_{\tau_e}^2} + \frac{(l\tau_c - \mu_{\tau_c})^2}{\sigma_{\tau_c}^2} + \frac{2\rho(l\tau_e - \mu_{\tau_e})(l\tau_c - \mu_{\tau_c})}{\sigma_{\tau_e} \sigma_{\tau_c}} \right]} \tag{3}$$

where $l\tau_e = \log(\tau_e)$ and $l\tau_c = \log(\tau_c)$, μ_{τ_e} and μ_{τ_c} are the mean value of the emission and capture times, σ_{τ_e} and σ_{τ_c} are their standard deviations and ρ is the correlation coefficient. Distribution parameters are fitted to match the capture and emission events observed experimentally, resulting in the plot of Fig. 4.

It must be remarked that emission and capture times in (3) are in fact mean values of random variables corresponding to the real time at which particular emission or capture events occur. The emission and capture events are commonly modeled by a Markov process, yielding the probability that a defect with emission/capture time τ is emitted/captured at time instant t as:

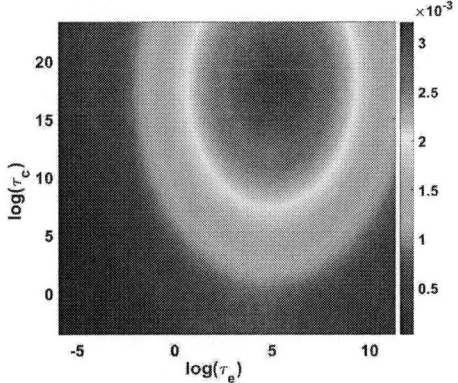

Fig. 4. Bivariate log-normal distribution of capture and emission times.

$$P(t) = \frac{1}{\tau} e^{-t/\tau} \tag{4}$$

C. Distribution of the threshold voltage and current shifts

The current shifts are extracted from the RTN current traces. Fig. 5 shows the histogram of the current shifts for 400 PMOS devices with $W/L = 80\text{nm}/60\text{nm}$ for $|V_{gs}| = 1.0\text{V}$ and $|V_{ds}| = 0.1\text{V}$. It can be concluded that a two-lognormal seems a good approximation for the current shifts δI:

$$f(\delta I) = \frac{R}{\delta I \sqrt{2\pi} \sigma_l} e^{-\frac{(\log(\delta I) - \mu_l)^2}{\sigma_l^2}} + \frac{(1 - R)}{\delta I \sqrt{2\pi} \sigma_u} e^{-\frac{(\log(\delta I) - \mu_u)^2}{\sigma_u^2}} \tag{5}$$

where μ_l, μ_u, σ_l and σ_u, represent the mean and standard deviation of the lower and upper lognormal distribution and R represents the relative amplitude of both distributions.

Fitting the two-lognormal distribution in (5) to the experimental results in Fig. 5 yields the associated distribution function, displayed in red. Equivalently, a similar fitting process can be performed for the threshold voltage shift distribution.

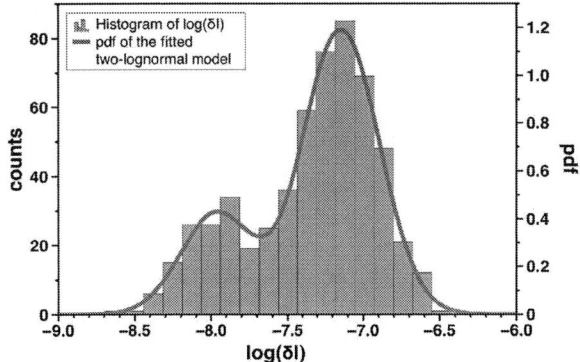

Fig. 5. Histogram of the experimentally extracted amplitudes for the RTN-induced current shifts (in blue), together with their associated pdf (in red).

III. SIMULATION METHODOLOGY

Once the probability density functions that govern RTN defects are established, a methodology to simulate circuits can be developed. Charge capture and emission in/from defects induce changes in the threshold voltages of the transistors and, hence, shifts of their drain current along time. As circuit simulators can obtain the drain current from a given threshold voltage (and operating conditions) modeling the shifts of the latter ones will enable obtaining the changes of the former ones.

Basically there are two possible approaches to incorporate this threshold voltage variation into circuit simulation [12]. The first one is to incorporate such variations into the transistor model, and the second one is to modify the circuit netlist to add a controlled voltage source to each transistor gate to model the threshold voltage shift, as illustrated in Fig. 6. Both approaches have benefits and downsides but the RTN mathematical framework shown before can be used with either approach. In the examples shown below, however, the circuit netlist modification approach has been used.

Being RTN a stochastic effect, circuit simulations should be also statistical (e.g., Monte-Carlo simulations). For that, hundreds or thousands of instances like that in Fig. 6(b) must be generated. The generation of instances in the form of a transient voltage along a certain simulation time T_{sim} proceeds through the following steps for each device:

Step 1: Generate the number of defects N_{def} for this device according to Poisson distribution. A simple algorithm to generate samples of this distribution can be found in [13]. If $N_{def} = 0$, the process is completed for this device.

Step 2: Generate emission τ_e and capture time τ_c of each defect according to the bivariate lognormal distribution in (3).

Step 3: Generate a sample of the associated threshold voltage shift ΔV_{th} for that defect. The governing probability density function, analogous to that for current shifts in eq. (5), can be efficiently sampled by: (1) deciding if the first or the second term in (5) becomes active by comparing the relative amplitude R with a sample of the commonly available uniform distribution; and (2) generating a sample of the selected log-normal distribution by applying the Box-Muller method using also two samples of uniform distributions [14].

Step 4: Generate the initial state $(t = 0)$ of each defect

according to the occupation probability: $P_{occ} = \frac{\tau_e}{\tau_e + \tau_c}$

Step 5: If the defect is occupied (empty), generate a sample of the time instant t_{ec} in which the charge will be emitted/captured. The probability, as given in (4), uses the time in the decimal scale. As the defects have been sampled from a bivariate log-normal distribution, it is necessary to transform (4) to logarithmic scale:

$$P(t) = \frac{t}{\tau} e^{-t/\tau} \tag{6}$$

The probability density function in (6) can be efficiently sampled by generating its cumulative density function and selecting a sample from a uniform distribution.

Step 6: Increase t by t_{ec}. Increase (decrease) threshold voltage shift at time t by ΔV_{th} if the defect has been captured (emitted).

Step 7: If $t < T_{sim}$ go to Step 5.

Step 8: Increase defect counter n. If $n \leq N_{def}$ go to Step 2.

Step 9: Sort the times t_{ec} corresponding to emission and capture events of all defects, and calculate the time evolution of their combined threshold voltage shift.

IV. SIMULATION AND EXPERIMENTAL RESULTS

The methodology in Section III was applied to the generation of hundreds of traces of threshold voltage shifts vs. time. For illustration's sake, four of these traces are shown in Fig. 7. By comparing these traces with the current trace in Fig. 1, it can be noticed that they show the same kind of random behavior, but Fig. 1 is a noisy one. This is because: (a) the voltage source in Fig. 6(b) must not contain noise as the noise is already included in the transistor model, and (b) the experimental trace in Fig. 1 contains noise intrinsic to the measurement setup that must not be included in a circuit simulation. The generated traces were used to simulate the impact of RTN on a basic integrated circuit: the simple current mirror in the inset of Fig. 8. An example of such a simulated output current for a constant input current is shown in Fig. 8(a), together with the associated copy factor in (b).

To experimentally validate the simulation methodology, the current mirror was integrated in the same 65-nm CMOS technology [15]. When biased in the same conditions, a typical behavior of the output current for a constant input current and the corresponding copy factor of the current mirror from one of the samples is shown in Fig. 9. It can be observed that the RTN directly impacts the copy factor of the current mirror, as expected from the simulation results displayed in Fig. 8.

V. CONCLUSIONS

In this paper, a complete methodology to simulate the impact of RTN on circuits has been presented. Since the parameters that characterize RTN are stochastic, a statistical

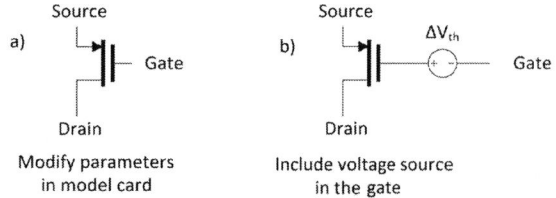

Fig. 6. Modeling RTN for circuit simulation: (a) modifying the parameters of the model card; (b) including a voltage source at the transistor gate.

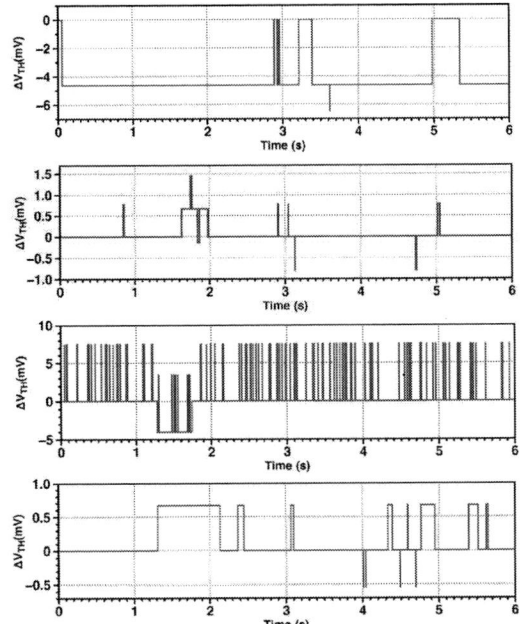

Fig. 7. Examples of ΔV_{th} traces generated for four different devices by using the simulation methodology presented in this work.

characterization of the phenomenon has been performed to extract the corresponding parameter distributions. Then, by sampling these distributions, the statistical simulation of integrated circuits affected by RTN can be performed. To do this, the effect of RTN on the devices V_{th} is included as a variation in the gate voltage of the device, which is incorporated through a voltage source on the transistor gate. Finally, this simulation methodology has been experimentally validated.

VI. References

[1] N. Tega et al., "Increasing threshold voltage variation due to random telegraph noise in FETs as gate lengths scale to 20 nm", in Symp. on VLSI Technology, IEEE, pp. 50, 2009.

[2] N. Tega et al., "Impact of threshold voltage fluctuation due to random telegraph noise on scaled-down SRAM", in Int. Rel. Phys. Symp. (IRPS), IEEE, pp. 541-546, 2008.

[3] M. Luo, R. Wang, S. Guo, J. Wang, J. Zou, R. Huang, "Impacts of random telegraph noise (RTN) on digital circuits", IEEE Trans. Electron Devices, vol. 62, pp. 1725-1732, 2015.

[4] T. Grasser et al., "Switching oxide traps as the missing link between negative bias temperature instability and random telegraph noise," in Int. Electron Devices Meet. (IEDM), IEEE, pp. 1-4, 2009.

[5] J. Diaz-Fortuny et al., "A versatile CMOS transistor array IC for the statistical characterization of time-zero variability, RTN, BTI and HCI", IEEE J. Solid-State Circuits, vol. 54, no 2, pp. 476-488, 2018.

[6] J. Diaz-Fortuny et al., "Flexible setup for the measurement of CMOS time-dependent variability with array-based integrated circuits", IEEE Trans. Instrum. Meas., vol. 69, no. 3, pp. 853-864, 2019.

[7] P. Saraza-Canflanca et al., "TiDeVa: a toolbox for the automated and robust analysis of Time-Dependent Variability at transistor level", Proc. of Synthesis, Modeling, Analysis and Simulation Methods and Applications to Circuit Design (SMACD), pp. 1-4, 2018.

[8] P. Saraza-Canflanca et al., "A robust and automated methodology for the analisis of time-dependent variability at transistor level," in Integration, the VLSI Journal, Vol. 72, pp. 13-20, 2020.

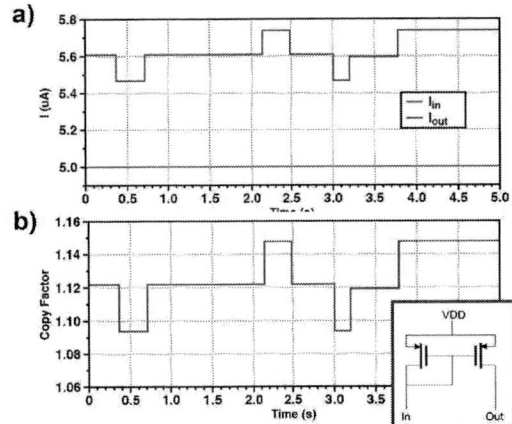

Fig. 8. a) Simulated I_{out} for a constant I_{in} displaying the impact of RTN, and b) the corresponding copy factor, for the current mirror in the inset.

[9] K. Takeuchi et al., "Direct observation of RTN-induced SRAM failure by accelerated testing and its application to product reliability assessment," Proc. Symp. VLSI Technology, pp. 189-190, 2010.

[10] T. Nagumo et al., "New analysis method for comprehensive understanding of random telegraph noise," IEEE Int. Electron Devices Meet., (IEDM) 2009.

[11] T. Grasser et al., "Analytical modeling of the bias temperature instability using capture/emission time maps," in Int. Electron Devices Meet. (IEDM), IEEE, pp. 27.4.1-27.4.4, 2011.

[12] A. Lange, F. A. V. Gonzalez, I. Lahbib and S. Crocoll, "Comparison of modeling approaches for transistor degradation: model card adaptations vs subcircuits," 49th European Solid-State Device Research Conference (ESSDERC), Cracow, Poland, pp. 186-189, 2019.

[13] D. Knuth, The Art of Computer Programming, Vol. 2: Seminumerical Algorithms, 3rd edition. Addison Wesley, 1997.

[14] G. Box and M.E. Muller, "A Note on the generation of random normal deviates," in the Annals of Mathematical Statistics, Vol. 29, No. 2, pp. 610-611, 1958.

[15] P. Martin-Lloret et al., "An IC array for the statistical characterization of time-dependent variability of basic circuit blocks", Proc. of Synthesis, Modeling, Analysis and Simulation Methods and Applications to Circuit Design (SMACD), pp. 241-244, 2019.

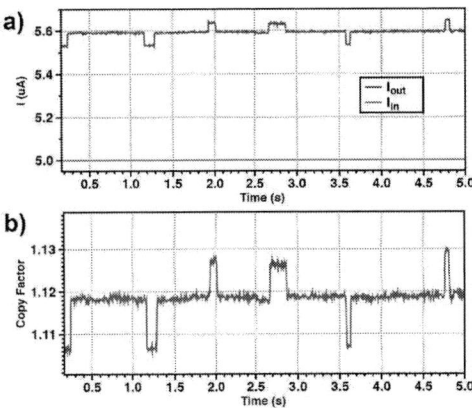

Fig. 9. a) Measured I_{out} for a constant I_{in} displaying the impact of RTN on the current mirror in the inset of Fig. 8 fabricated in the same 65nm CMOS technology [15], and b) the corresponding copy factor.

Connecting Energy Storages from Tool Independent, Signal-flow Oriented FMUs

Meik Ehlert
Scientific Automation
Fraunhofer Institute for
Mechatronic Systems
Design IEM
Paderborn, Germany
meik.ehlert@iem.fraunhof
er.de

Jan Michael
Scientific Automation
Fraunhofer Institute for
Mechatronic Systems
Design IEM
Paderborn, Germany
jan.michael@iem.fraunhof
er.de

Christian Henke
Scientific Automation
Fraunhofer Institute for
Mechatronic Systems
Design IEM
Paderborn, Germany
christian.henke@iem.fraun
hofer.de

Ansgar Trächtler
Heinz-Nixdorf-Institute
University of Paderborn
Paderborn, Germany
ansgar.traechtler@hni.uni-
paderborn.de

Matthias Kalla
Institute for Drive Systems
and Power Electronics
Leibniz Universität
Hannover
Hannover, Germany
matthias.kalla@ial.uni-
hannover.de

Bakr Bagaber
Institute for Drive Systems
and Power Electronics
Leibniz Universität
Hannover
Hannover, Germany
bakr.bagaber@ial.uni-
hannover.de

Bernd Ponick
Institute for Drive Systems
and Power Electronics
Leibniz Universität
Hannover
Hannover, Germany
ponick@ial.uni-
hannover.de

Axel Mertens
Institute for Drive Systems
and Power Electronics
Leibniz Universität
Hannover
Hannover, Germany
mertens@ial.uni-
hannover.de

Abstract— In cross-domain system simulation, models from a wide variety of tools are used. The FMI standard provides the possibility to connect these models as Functional Mockup Units. These so called FMUs have signal-flow oriented interfaces to each other. The interfaces can often be inconsistent due to missing knowledge of what kinds of energy storage are present in each model. The inconsistency leads to false simulation results. This article therefore presents design guidelines according to how energy storages can be distributed to different models of the electric drive technology. The goal here is to ensure correct numerical simulations. To validate these guidelines, a drive system is built from several FMUs and is compared against a topology-oriented reference.

Keywords—energy stores, signal flow, interfaces, FMI, FMU

I. INTRODUCTION

Complex mechatronic systems, such as electric drive systems, require holistic system simulation throughout the entire development process [1][2]. Thus, errors can be detected at an early stage, which reduces cost. Using the simulation results, better prototypes can be built. This also leads to cost and time savings, which results in overall higher quality in development.

The challenge conducting simulations of complex mechatronic systems is the diversity of different engineering disciplines. Different tools are used for modelling different subcomponents. Application engineers are facing the major challenge of coupling the models from these tools with each other. The options for this are often proprietary and vary from manufacturer to manufacturer. The Functional Mockup Interface is a widely used standard that makes it possible to couple cross-tool models for an overall system simulation [3]. However, this standard does not provide any guidelines for the design of models to ensure a physically correct and consistent simulation.

Particularly when creating separate models for individual subcomponents of a drive system, the system borders for model coupling can be very unfavourable when cutting the connection between two energy storages. Energy storage denotes physical elements like capacitors or masses, whose state variable is directly related to the stored energy. When two masses are connected, they represent the same physical state. However, the states of the two masses become independent when they are located in different models. In course of the same torque, two different speeds would occur for the masses. To avoid this, algebraic loops have to be involved in the system [4]. However, the algebraic loops result in numerical difficulties as the FMUs don't provide the full equations of their systems.

This article therefore aims to present design guidelines according to how tool independent models for electric drive systems are to be created in order to ensure the correct physical simulation. The aim is to avoid connections of similar energy storages in different models. For this purpose, approaches for the coupling of energy storages are shown. Subsequently, a reference architecture for electric drive systems is given. For this, a procedure is presented to treat energy storages in a generally valid way. Finally, a summary and an outlook are given.

II. STATE OF THE ART

A. Functional Mockup Interface

The Functional Mock-Up Interface (FMI) is a standardized interface agreed upon by various tool vendors to ensure interoperability of models in simulation environments [3]. For this purpose, code is generated from models which implements the standardized interface. These so called FMUs can be connected via signal-flows. In [4] the FMI function *getDirectionalDerivatives* from 2.0 version is presented, which addresses the problem of connecting similar energy storages. However, this function is only applicable if the interfaces of the FMUs respect algebraic constraints. Furthermore, the function is optional and is not supported by all tools. The specification of the pre-released FMI 3.0 standard deals with terminals, which enable topology-oriented modelling [5]. However, this feature does not solve the described problem, as terminals just describe a collection of signal-flows. The equations of the FMUs are still compiled and cannot be changed.

B. Modeling Guidelines

In signal-flow oriented modelling, the MAB guideline from the MathWorks Advisory Board is of particular importance. This guideline gives recommendations for modelling in Matlab, Simulink and StateFlow. It intends to increase the maintainability of models and thus their reusability [6]. However, these refer to general modelling methods within a model and do not make any specific statements about energy storage coupling.

In [7], a catalogue of measures was presented to enable efficient model coupling for distributed simulations. Here, a higher-level coupling model is proposed to ensure consistent interfaces. However, no concrete interface definition is given, which is based on an energy storage distribution.

C. Energy storages in signal-flow oriented circuits

In this article, the electrical and mechanical domains are considered for electrical drive systems. For these domains, the energy storages inductor, capacitor, mass and spring exist. Table 1 summarizes their differential equations, where L denotes the constant for inductivity, C for capacity, m for mass and c for spring stiffness [8]. The variables u and i describe the electrical system quantities voltage and current. v and F represent the mechanical quantities velocity and force. In order to avoid numerical problems to solve these equations, only an integration of the potential and flux quantities is considered. This results in the input and output quantities given in table 1.

TABLE 1   EQUATIONS OF ENERGY STORAGES

Energy Storage	Equation	Input	Output
Inductor	$\dot{\imath} = \frac{1}{L} \cdot u$	u	i
Capacitor	$\dot{u} = \frac{1}{C} \cdot i$	i	u
Mass	$\dot{v} = \frac{1}{m} \cdot F$	F	v
Spring	$\dot{F} = c \cdot \dot{x} = c \cdot v$	v	F

Based on the input and the output variables, it can be seen that the connection of similar energy storages leads to inconsistent interfaces. That is shown on the example of connecting two masses with following equations

$$\dot{v}_1 = \frac{1}{m_1}F_1 \qquad \dot{v}_2 = \frac{1}{m_2}F_2 \qquad (1)$$

To realize a connection, the potential quantities must be the same and the sum of the flux quantities must result in zero [9].

$$0 = F_1 - F_2 \qquad (2)$$
$$v_1 = v_2 \qquad (3)$$

Applying equation (2) in equation (1) results in

$$\dot{v}_1 = \frac{1}{m_1}F_1 \qquad \dot{v}_2 = \frac{1}{m_2}F_1 \qquad (4)$$

For $m_1 \neq m_2$ the potential quantity constraints from equation (3) are broken. Thus, a signal-flow oriented connection between two similar energy storages cannot be performed.

D. Coupling methods for energy storages

In the following, three possible solutions are shown that can be used to solve the interface problem.

1) Forming of substitute variables: This method is often the standard procedure in modelling when several identical energy storages occur. The quantities are summerized to a substitute variable. Subsequently, the numerical integration takes place with only one energy storage to be considered. The example of the two masses would result in the following equations:

$$m_{total} = m_1 + m_2 \qquad (5)$$
$$\rightarrow \dot{v} = \frac{1}{m_{total}}F \qquad (6)$$

The problem with cross-tool simulation, however, is that the system equations of the FMUs cannot be changed after the model has been compiled.

2) Dynamic coupling with fictitious coupling elements: Another solution to solve the problem of connecting two identical energy storages is to use a fictious energy storage as coupling element. In the example of the two masses, a fictious spring can be used, to convert the inconsistent interfaces of the masses in consistent interfaces. The first mass would tension the spring, which drives the second mass. This results in the following equations, where F_c denotes the force of the spring:

$$\dot{v}_1 = \frac{1}{m_1} \cdot (F_1 - F_c) \qquad (7)$$
$$\dot{F}_c = c \cdot (v_1 - v_2) \qquad (8)$$
$$\dot{v}_2 = \frac{1}{m_2} \cdot (F_2 - F_c) \qquad (9)$$

The stiffer the spring is, the smaller is the error resulting from the fictitious spring. However, stiff systems result in small time constants, which significantly increases the simulation time.

3) Definition of algebraic constraints: Instead of a signal-flow oriented interconnection, the model coupling can also be carried out via the definition of algebraic constraints. The subsystems would first be arranged independently of each other in a state space model. After that, the coupling constraints from equation (2) and (3) are inserted. These conditions result in a DAE system as follows

$$\begin{bmatrix} \dot{v}_1 \\ \dot{v}_2 \\ 0 \\ 0 \end{bmatrix} = \begin{bmatrix} 0 & 0 & 0 & 0 \\ 0 & 0 & 0 & 0 \\ 1 & -1 & 0 & 0 \\ 0 & 0 & 0 & 0 \end{bmatrix} \cdot \begin{bmatrix} v_1 \\ v_2 \\ 0 \\ 0 \end{bmatrix} + \begin{bmatrix} \frac{1}{m_1} & 0 \\ 0 & \frac{1}{m_2} \\ 0 & 0 \\ 1 & -1 \end{bmatrix} \cdot \begin{bmatrix} F_1 \\ F_2 \end{bmatrix} \qquad (10)$$

This can be solved for example by index reduction. With this approach, however, it must be noted that the system equations of the models must be known in order to solve the algebraic constraints. This is not the case with the FMI standard. Thus, this method is not applicable for a design guideline.

For the style guide presented here, the coupling method *1) Forming substitute variables* is used. This enables a simulation with accurate results that does not involve fictitious dynamics. Likewise, this method is applicable with common cross-tool coupling options, such as the FMI standard, since the system equations do not need to be known at the time of simulation. Nevertheless, the locations of the energy storages have to be declared at the modelling stage.

For this purpose, the guidelines are presented in the following.

III. DESIGN OF THE REFERENCE ARCHITECTURE

By reviewing simulation models from various research projects, typical components of an electric drive system were identified. These are compiled in the reference architecture in Figure 1. The interfaces of the signal-flows are derived from the energy storage distribution in table 2.

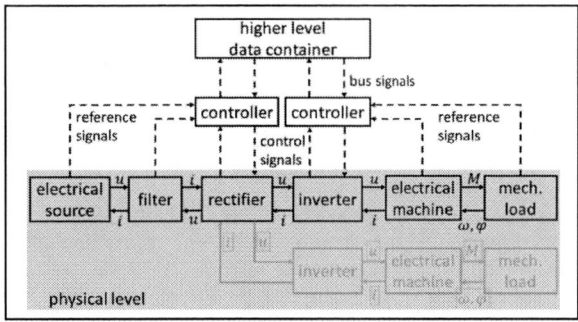

Figure 1: Reference architecture of an electrical drive system

In addition to the physical level, the architecture includes two information signal levels for controllers and higher level data containers. The actual connection of the energy storages takes place on the physical level. Here, it should be possible to model a drive system from the electrical supply to the mechanical load. The signal flows refer to the potential and the flux variables. The quantities required by the numerical integration are forwarded. The resulting quantities are fed back. Here, the interface definition depends on which energy storages are considered in which component. Table 2 shows how the substitute quantities of the energy storages are distributed in the architecture. It should be noted that a substitute variable not only refers to the component in which it is modelled, but also depends on the connected components.

TABLE 2   DISTRIBUTION OF ENERGY STORAGES

Components	Energy Storages
Electrical source	
Filter	Filter inductance Filter capacity Source inductance
Rectifier	DC link capacitor
Inverter	
Electrical machine	Machine inductance Inverter inductance
Mechanical load	Motor inertia Load inertia

The electrical energy supply is assumed to be an ideal voltage source. No energy storage is provided here. Possible inductances are considered in the filter together with the filter inductance. As a result, several components must not be branched off from the supply in order to ensure energy conservation. If several motors are used in the drive system, the branching must be provided in the DC link. The filter can also include a filter capacity.

The DC link is spanned by the rectifier and the inverter components. Since several capacitances can be directly connected here, the total capacitance of the DC link must be taken into account in the rectifier. This also includes capacitances from different inverters. Each motor is fed from one inverter. A possible inverter inductance is considered in the motor model with the motor inductance. The motor can be rigidly coupled to a mechanical load. The total mass inertia of motor and load is thereby provided in the load model.

IV. VALIDATION OF THE REFERENCE ARCHITECTURE

For the validation, the created reference architecture is characterized with a simple drive system, shown in Figure 2. This is modelled purely topology oriented in Modelica to form a reference result. The system consists of a single-phase line voltage source with an LR filter. This voltage is rectified and feeds two DC motors. To keep the model simple, no inverter or other voltage regulators are used. Each motor drives a mechanical load. The motors and their loads have different parameterizations. It should be noted that the mechanical connection deals with rotational components. Thus, the energy storages consider an inertia θ. To validate the design guidelines for the connection of energy storages, only the physical level of the reference architecture is of interest. Controllers are not considered in this example.

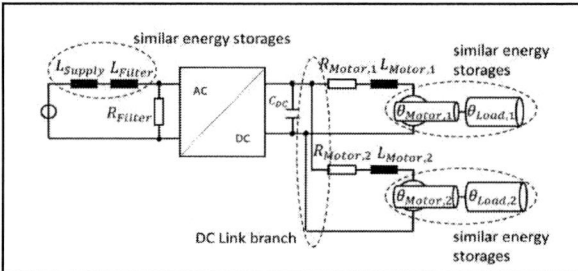

Figure 2: Example of a drive system

The places where the guidelines are taken into account are marked with dotted circles. The design guidelines have to be considered in the connection of the supply and the LR filter as well as in the connection of the motors and their loads. The supply includes a small inductance. This inductance has to be considered in the filter component together with the filter inductance. In the connection of motor and load, there are two masses directly coupled. Here, the overall inertia has to be considered in the load model. Furthermore, the reference architecture of the design guidelines specifies that the two motors are connected to the same rectifier and not to the same supply with an own rectifier. This is necessary since the supply model doesn't consider an energy storage and the energy conservation would be violated.

The system is subsequently partitioned to the components from the reference architecture. To generate FMUs from these components, the input and the output variables have to be declared. This process is called casualization [10]. Sensors and sources are used to generate signal-flows that form the interfaces between the components. Figure 3 shows the causalized system. After that, the generation process of the FMUs can be started. The drive system is then simulated by connecting the FMUs. Figure 4 shows the comparison of the simulation results between the reference model and the FMU

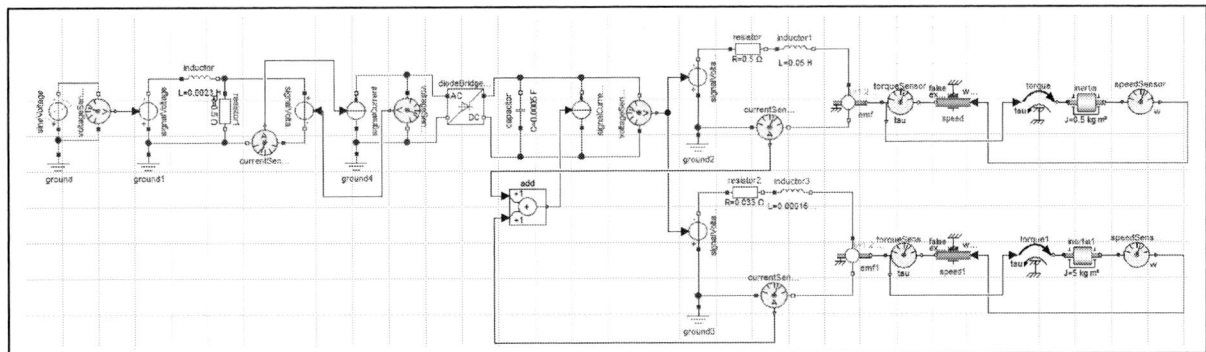

Figure 3: casualized drive system

connection. The speeds of the two motors are considered. From these quantities, it can be seen that the results of the signal-flow-oriented FMU connection match those of the topology-oriented reference model.

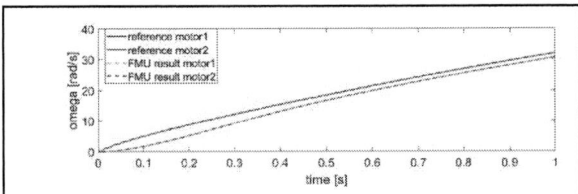

Figure 4: Simulation results of the drive system

V. SUMMARY

In this article, a reference architecture with design guidelines was presented to connect energy storages from different models. First, the FMI standard was introduced as a possibility for cross-tool model coupling. Since such couplings typically have signal-flow oriented interfaces, special attention must be paid to the energy storage distribution in the overall system. It was shown that the connection of similar energy storages results in inconsistent interfaces. To avoid such connections, a reference architecture was given, which defines how energy storage substitute variables have to be deployed in an overall drive system. This reference architecture was validated on an example system. This system was first modelled topology oriented in Modelica. The system was split into individual components, which were exported as FMUs and connected via signal-flows. A simulation provided the same results as the reference model.

VI. FUTURE WORK

In the next steps, the reference architecture should be validated with other examples. Use cases such as the DC link control of a drive system or the start-up of an induction machine are conceivable for this purpose. The creation of a topology-oriented reference model can be a particular challenge if the system to be considered was created with different modelling tools.

Currently, the parameterization of the energy storage substitute variables must be performed at the time of modelling. In the future, this parameterization could be done automatically at runtime. Here, it is conceivable that the

FMUs get a special tag, which specifies which energy storage values are to be considered for the model. An environment which simulates these FMUs can read these values and parameterize the corresponding substitute variables in the intended components. Additional information, such as the tag for energy storage values, could be held in a higher level coupling model or in a digital twin.

Another possibility for resolving inconsistent interfaces lies in the system identification of the overall system. Here, the FMUs of the individual components are not used to be connected to the overall system. Rather, they are used to identify the mathematical system equations of the components. These are then used to form an overall system of equations. The connections between the individual components are implemented using constraints that relate to the potential and the flux quantities. This approach has the advantage that models can be used which have not considered the design guidelines.

VII. REFERENCES

[1] Michael, J.; Holtkötter, J.; Henke, C.; Trächtler, A.: Modellbildung und Simulation im Kontext des Systems Engineering. In: ASIM-Treffen STS/GMMS 2016, S. 174-179, March. 2016 ASIM

[2] Popp, M..; Mathis, W.; John, M.; Korolova, O.; Mertens, A.; Ponick, B.: A Modified CCM Approach for Simulating Hierarchical Interconnected Dynamical Systems, 2017

[3] Blochwitz, T., Otter, M.: The Functional Mockup Interface for Tool independent Exchange of Simulation Models, 2011

[4] Blochwitz, T.; Otter, M.; Akesson, J.; Arnold, M.; Clauß, C.; Elmquist, H.; Friedrich, M.; Junghans, A.; Mauss, J.; Neumerkel, D.; Olsson, H.; Viel, A.: Functional Mockup Interface 2.0: The Standard for Tool independent Exchange of Simulation Models, 9th International Modelica Conference, 2012

[5] Modelica Association: FMI-Standard 3.0 beta, https://fmi-standard.org/docs/3.0-dev/, 05.06.2021

[6] MathWorks Advisory Board: Cobtrol Algorithm Modeling Guidelines Using MATLAB, Simulink and Stateflow, Version 5.0, 2020

[7] BERNHARD, J.; WENZEL, S.: Verteilte Simulationsmodelle für produktionslogistische Anwendungen - Anleitung zur effizienten Umsetzung, Proceedings Simulation und Visualisierung, 2006

[8] Isermann, R.: Mechatronische Systeme, Springer Berlin, 2007, ISBN 978-3-540-32336-5

[9] Elmqvist, H.; Mattson, S.; Otter, M.: Modelica – The new object-oriented Modeling Language, 12th European Simulation Multiconference, 98

[10] Mehlhase, A.: Konzepte für die Modellierung und Simulation strukturvariabler Modelle, Technische Universität Berlin, Fakultät für Elektrotechnik und Informatik, Berlin, 2015

Adaptive Simulation with HDL Control Module for Frequency Converting Circuits

Zoltán Tibenszky, Martin Kreißig, Corrado Carta, Frank Ellinger
Chair of Circuit Design and Network Theory
Technische Universität Dresden, 01069 Dresden, Germany
Email: zoltan.tibenszky@tu-dresden.de

Abstract—An adaptive simulator-independent method is presented, which reduces the simulation and post-processing time of transient simulations through the use of an intelligent sweep control implemented as a behavioural model. It is capable to effectively reduce the points to be simulated for determining the operation regions, or to find the local extrema of a continuous value performance parameter. It controls the variable sweeps based on real time simulation results, and delivers the circuit's performance parameters to be determined in a text file. This eliminates the often time consuming additional post-processing steps. The effectiveness of the proposed method is demonstrated for the case of a frequency divider verification, where it has led to a 10 times reduction in the simulation time, and eliminated any additional post-processing steps.

I. Introduction

As the manufacturing technologies overcome ever newer challenges in order to control the fabrication processes of the most advanced technologies, the number of factors affecting the circuit performance increase. Consequently, device models have become computationally more complicated by incorporating these effects. This increases both the required design and verification times. The problem is even more pronounced in analog circuit designs, where time domain verification is often needed. Circuits with frequency conversion functions – mixers, frequency dividers and multipliers – must be simulated with a time step small enough for the high frequency components of the signals involved, but the simulation time has to be long enough to cover the initial settling time and at least a few periods of the lowest frequency component.

Harmonic balance methods provide a computationally efficient way to simulate mildly non-linear circuits in the frequency domain, but their effectiveness is greatly reduced for rapidly changing signals, such as digital edges. In these cases time based solvers are preferred for accurate timing values [1]. Time domain simulators could adaptively adjust their time step according the signal behaviour and are less susceptible to convergence problems for strongly non-linear circuits [1].

The aim of the presented method is to increase the verification efficiency with means readily available to the circuit designers. It solely relies on common tools, such as a behavioural analog module, without sophisticated

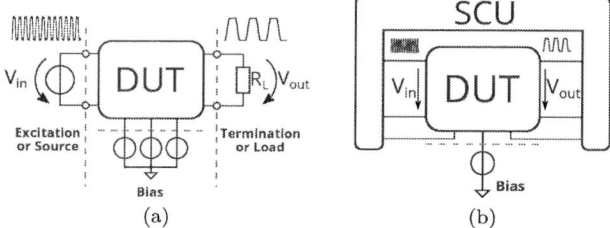

Fig. 1. Simulation methods: (a) Conventional analog simulation method: one excitation per run, and (b) Using a module to intelligently control the simulation sweeps and obtain the results.

or dedicated software packages. Verilog-A behavioural description is supported in most of the simulators. Some device models are also implemented with a it [2]. Behavioral models are a simple way to implement experimental models [3], [4] or to verify models and the contribution of their elements used in analytical calculations. Open-source analog and mixed signal simulators are also available with support of behavioural models [5]. In this paper we will use frequency dividers as an example to illustrate the method and its advantages, but the approach is sufficiently universal to be applied to other circuit types. The presented simulation method was used to simulate the performance of the frequency dividers reported in [6], [7].

II. Problem Description

Table I lists the specifications of a true single-phase clock (TSPC) frequency divider. This list of specifications is generally applicable for a variety of circuits, and it would be similar for mixers, current mode logic (CML) or Miller dividers. The backgate often serves as a tuning terminal to compensate for process, voltage and temperature (PVT) variations or mismatches, and therefore it was added to the table as well [8]. Its importance has increased with the proliferation of silicon on insulator (SOI) processes allowing forward body bias (FBB), which increased its effectiveness. The meaning of the majority of the listed specification is well known. Self-resonance frequency of a frequency divider is defined as the input frequency where it requires the smallest input power. Such circuits could oscillate at this frequency under given conditions even if no input signal is present. The input sensitivity curve defines the operating region in terms of required input amplitude

TABLE I
LIST OF SPECIFICATIONS OF A FREQUENCY DIVIDER

specification		symbol	unit
frequency	self resonance	f_{SR}	GHz
	highest operating	$f_{in,high}$	GHz
	lowest operating	$f_{in,low}$	GHz
power	required input	A_{in}	V or dBm
	consumption	P_{DC}	mW or dBm
	output	P_{out}	Hz
sensitivity	supply Voltage	V_{DD}	V
	backgate voltage	V_{BG}	V
	DC input voltage	$V_{in,DC}$	V
division ratio		N	—
area		—	μm^2
temperature		T	°C
input sensitivity curve		—	—

TABLE II
VERIFICATION PARAMETER SPACE

	Unit	from	to	case 1		case 2	
				step	n	step	n
V_{DD}	V	0.5	0.9	0.05	9	0.1	5
V_{BG}	V	0	3	0.5	7	1	4
$V_{in,DC}$	V	$\frac{3V_{DD}}{10}$	$\frac{7V_{DD}}{10}$	$\frac{V_{DD}}{40}$	17	$\frac{V_{DD}}{10}$	5
f_{in}	GHz	1	70	0.5	141	2	35
P_{in}	dBm	-40	4	1.5	31	3	16
Number of points:				$4.7 \cdot 10^6$		56000	

at a given frequency, and is considered the most important performance parameter of a frequency divider.

Table II shows an example parameter space for two different accuracy levels, without temperature or corner variations. The last two rows contain the parameters related to the input signal. It is clear that they dominate the total number of the simulations to be run. The presented example reduces the amplitudes values, but it can be extended to reduce the frequency points as well.

The large number of runs will increase the time required to design, debug and run the post-processing scripts. It is especially the case for transient simulations, where timing related information is to be obtained. The storage of long transient waveforms from multiple simulation runs requires non-negligible storage space, and access time. Their post-processing could be computationally classified as a *big data* problem, which a design or verification engineer might not be an expert in. Despite the huge volume of obtained data, very often only a few scalar values or functions with significantly reduced number of points are enough to characterize and verify a design. The conventional approach of determining one scalar performance parameter involves the loading and the mathematical processing of one or even multiple a signal waveforms. Significant time saving is possible if the parameters and functions to be obtained are generated in simulation time, and are stored into a file at the end of the simulation.

Another consideration is the ease of comparison between simulation and measurement results. Typically automated measurement results are stored in a coma-separated values (CSV) file, which can be generated from a manually filled spreadsheet as well. Storing all simulation results in the same or similar format would ease their comparison against measurements.

In cases where no performance parameter matrix, but an operation range is to be determined, a reduction of the number of points to be simulated is possible through the assumptions of continuity and smooth boundaries. The former assumption allows a reduction of the parameter space only to the operating range boundary, while latter assumption allows to efficiently reuse the boundary of the current simulation point, as an initial guess for the next one.

For adaptive simulation control an assessment of the circuit performance under the current conditions is carried out, and the next sweep value is determined. At the same time the obtained performance parameters can be written into a CSV file, without the computational overhead of conventional waveform post-processing.

III. PROPOSED METHOD

All the desired functionality listed in the previous section could be accomplished with circuit blocks written in a hardware description language (HDL), such as Verilog-A, placed into the simulation testbench. Many IC designers routinely use behavioral HDL models for circuit blocks interfacing their designs to increase simulation speed of higher level system simulations, and thus they usually posses the knowledge required to develop, or modify such a simulation controller. We will refer to this control circuit as simulation control unit (SCU) in this paper.

The proposed testbench architecture is depicted in Fig. 1 (b). In contrast to the conventional approach of Fig. 1 (a), the SCU provides all input signals and some bias inputs to the device under test (DUT). Another advantage of this method is the concentration of most simulation parameters in just one instance. It provides a better overview than many sources, which control only one or two parameters each.

The SCU used in our work provided the input signal defined by its frequency, amplitude and DC voltage, as well as the supply voltages of the circuit core and its buffer. It also assesses the average power of the provided bias signals, as well as the output power of the DUT through a programmable internal load resistance, and saves them into a CSV file.

It lays within the freedom of the designer to determine which input parameters to be set and monitored by the SCU circuit, and which is swept by the simulator. Even though the supply voltages and the input DC voltage was provided by the SCU, it was not swept by it. This provided the same data structure, which was used by our automatic measurement setup. Due to the adaptive nature of the simulation, the sweep limits can be set wide enough for

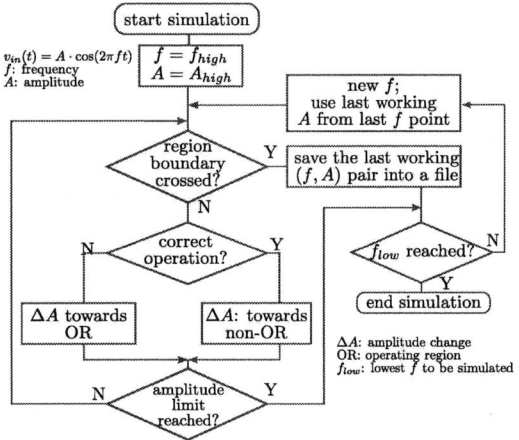

Fig. 2. High level operation of the proposed algorithm for the case of a frequency divider.

all simulation conditions without significant penalty in the simulation time.

IV. Control Algorithm

The operation of the algorithm used to simulate the frequency dividers is summarized in Fig. 2. The calculation and the saving of the performance parameters of interest were omitted for clarity. If circuit parameters, such as power consumption, or output power, are to be acquired at a given amplitude level for better comparison, then this additional amplitude level can be added to every simulated frequency point in a straightforward way.

Circuits based on digital logic cells form the core of many frequency dividers, such as CML and TSPC dividers. These circuits act on one of the transitions of their input signal. Even in injection locked frequency dividers, the injection signal affects the circuit at its peak value the most, while being less effective at other time points in the period. Thus the salient input property of these circuits is time, and not power or voltage. They need only an input signal strong enough to make the detection of the signal's threshold crossing possible. If these circuits function correctly at a given input amplitude, then their functionality are kept at higher input amplitudes as well in the majority of cases. The algorithm depicted in Fig. 2 uses this assumption by looking for only the lower amplitude boundary of the operation range. The extension of the algorithm for the more general case of both lower and upper amplitude boundary searches is conceptually straightforward.

Utilizing the assumptions of continuous operating range with a smooth boundary and only lower amplitude boundary the algorithm will start the search for reliable operation with the maximal amplitude (A_{high}) at the highest simulation frequency (f_{high}). If no operation is detected the frequency will be reduced. Once reliable operation was detected at A_{high}, the amplitude is reduced towards the boundary of the operational and non-operational regions until the circuit does not operate reliable any more. Then the frequency is changed to the next value, and the amplitude of the previous reliable operation is applied. If the circuit is in the operational region, then the amplitude is reduced to reach the operation region boundary. Otherwise the amplitude is increased for the same reason. The lower end of the operating range is reached when the circuit will not operate any more after operation at previous frequencies. At this point the whole operating region has been determined, and the simulation will be stopped by the SCU to save simulation time. Therefore stop time of the transient simulation does not influence the actual simulation time.

In order to detect the boundary of the operational region, the boundary has to be crossed, which sets the minimum number of simulations points to two for every frequency. The error of the results are limited to the amplitude and frequency steps.

A high operating frequency is usually less straightforward to achieve than a low operation frequency. Starting the sweep at f_{high} allows the designer to spot early both inaccurate f_{high} setting and unsatisfactory circuit performance.

The SCU relies on time measurement between output transitions of the DUT. Reliable operation is defined as a programmable number of consecutive output periods, which deviate from the ideal division ratio (N) less than the specified tolerance ($\Delta N/N$). Non-operation at frequencies where the DUT does not generate any signal transitions are detected with a watchdog, whose timer is set at every output transition of the DUT.

In our case we have used an AC coupling capacitor between the DUT output and the monitoring port of the SCU. This allowed to keep the edge detection threshold at 0 V. The SCU was written to be reusable for many frequency dividers to provide a unified test setup and to reduce the testbench development time. The following options can be set through netlist parameters of the SCU: the DC voltage of three bias outputs, the range and the step size of the frequency and amplitude sweeps, the name of the CSV output file, the desired division ratio, length of the initial startup transient, the load resistance connected to the DUT's output terminal, and the criteria of reliable operations, such as the relative tolerance of the output period error, and the number of consecutive correct output periods. The Verilog-A code of the used module is available at [9] together with a more detailed flowchart using the same variable names as the Verilog-A code.

It is possible to reach additional improvement at the cost of increased code complexity and thus development time. An adaptive frequency sweep could be used instead of fix frequency steps. Nested parameter sweeps where a small change in the parameter value is expected to result in a similar sensitivity curve could be another possibility.

A similar adaptive simulation method could be useful

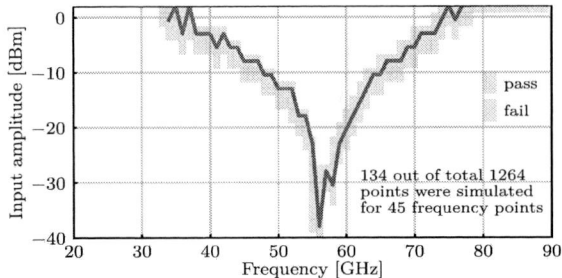

Fig. 3. The simulated sensitivity curve of a frequency divider. Every simulated point is highlighted according to the operational status of the DUT in the given point. The width and height of the highlight are set to the step size of the frequency and amplitude sweep.

to find the highest and lowest performance values in an operating condition range to identify the strengths and weaknesses of the circuit under extreme conditions. This problem definition is very similar to many optimization problems and assuming continuous functions, algorithms such as gradient descent could be used [10].

The simulated amplitude-frequency pairs and pertinent operational status are shown with colored highlights in Fig. 3. In the plotted sensitivity curve example the amplitude was swept between 4 and −40 dBm with 2.5 dB steps, which corresponds to −25 % degrease in the voltage amplitude. The frequency was swept from 90 GHz down to 20 GHz with 1 GHz steps. In a conventional sweep approach it would be equivalent to 18×71 = 1278 simulation points. The proposed method required only 134 point to determine the sensitivity curve, which is an order of magnitude improvement. The benefits of the method become even stronger for a higher amplitude resolution.

Operation were detected at 45 frequency points in our example, which corresponds to less than 3 amplitudes per frequency point. The simulation time improvement is visualized as the ratio of the highlighted and white graph area. Additional benefits are expected through the elimination of the post-processing steps.

V. CONCLUSION

A simulation methodology reducing the number of the required number of parametric transient simulation runs through intelligent sweep control has been presented for binary pass/fail outcomes, such as operational range. Beyond the simulation time improvement, it does not require the saving of any signal waveforms to the disk, or any post-processing steps, which leads to additional verification time improvements. In average less than three amplitude values were simulated for every frequency point in the presented frequency divider example. This is equivalent to one order of magnitude improvement of simulation time compared to the conventional nested amplitude and frequency sweeps.

ACKNOWLEDGMENT

This work was funded by SMWK and SAB through project PROSECCO.

REFERENCES

[1] K. S. Kundert, "Introduction to RF Simulation and Its Application," *IEEE J. Solid-State Circuits*, vol. 34, no. 9, pp. 1298–1319, sep 1999. doi: 10.1109/4.782091

[2] W. Grabinski *et al.*, "FOSS EKV2.6 Verilog-A Compact MOSFET Model," in *ESSDERC 2019 - 49th European Solid-State Device Research Conference (ESSDERC)*, Sep. 2019. doi: 10.1109/ESSDERC.2019.8901822. ISSN 2378-6558 pp. 190–193, iSSN: 2378-6558.

[3] H. E. Dawale, L. Sibeud, S. Regord, G. Jourdan, S. Hentz, and F. Badets, "Compact Modeling and Behavioral Simulation of an Optomechanical Sensor in Verilog A," *IEEE Trans. Electron Devices*, vol. 67, no. 11, pp. 4677–4681, Nov. 2020. doi: 10.1109/TED.2020.3024477

[4] N. Sharma, A. Marshall, and J. Bird, "VerilogA based compact model of a three-terminal ME-MTJ device," in *2016 IEEE 16th International Conference on Nanotechnology (IEEE-NANO)*, Aug. 2016. doi: 10.1109/NANO.2016.7751550 pp. 145–148.

[5] M. Brinson, "The Qucs/QucsStudio and Qucs-S Graphical User Interface: An Evolving "White-Board" for Compact Device Modeling and Circuit Simulation in the Current Era: Invited Paper," in *2020 27th International Conference on Mixed Design of Integrated Circuits and System (MIXDES)*, Jun. 2020, pp. 23–32.

[6] Z. Tibenszky, C. Carta, and F. Ellinger, "A 0.35 mW 70 GHz divide-by-4 TSPC frequency divider on 22 nm FD-SOI CMOS technology," in *2020 IEEE Radio Frequency Integrated Circuits Symposium (RFIC)*, Aug. 2020. doi: 10.1109/RFIC49505.2020.9218362. ISSN 2375-0995 pp. 243–246, iSSN: 2375-0995.

[7] M. Kreißig, S. Buhr, M. El-Shennawy, and F. Ellinger, "A 0.1–13 GHz wide range multi-modulus divider with adaptive sensitivity for broad band operation," in *2018 IEEE 61st International Midwest Symposium on Circuits and Systems (MWSCAS)*, Aug. 2018. doi: 10.1109/MWSCAS.2018.8624050. ISSN 1558-3899 pp. 901–904, iSSN: 1558-3899.

[8] E. Wittenhagen, M. Runge, N. Lotfi, H. Ghafarian, Y. Tian, and F. Gerfers, "Advanced Mixed Signal Concepts Exploiting the Strong Body-Bias Effect in CMOS 22FDX®," *IEEE Trans. Circuits Syst. I*, vol. 68, no. 1, pp. 57–66, Jan. 2021. doi: 10.1109/TCSI.2020.3023077

[9] Z. Tibenszky. (2021, Feb.) horror-vacui/freq_div_meas. Zenodo. Doi:10.5281/zenodo.4589072. [Online]. Available: https://doi.org/10.5281/zenodo.4589072

[10] S. Ruder, "An overview of gradient descent optimization algorithms," *arXiv preprint arXiv:1609.04747*, 2016.

Step Size Determination Approach for Aging Simulations in Analog ICs

Engin Afacan

Department of Electronics Engineering, Gebze Technical University

Abstract—**Simulation of time-dependent variations is quite complicated since the degradation is a function of time, where the time step directly affects the accuracy and the efficiency of the analysis. Commercial tools use a constant step count during simulations, in which choosing a large step count may degrade the efficiency whereas keeping it small may result in accuracy problems. To overcome this problem, a couple of different adaptive time-step approaches have been proposed in the literature. Nevertheless, they suffer from the initial workload during step count determination or some other accuracy problems. In this study, we present a two-level step count determination approach. At the first level, the step count induced estimation error can be promptly determined via an effective simulation strategy. At the second level, the error is fitted into a saturated power law model; thus, the efficient step count can be determined without any simulation effort. The proposed approach provides a remarkable save in computation time and can be used for all analog circuits without loss of generality.**

I. INTRODUCTION

CMOS analog circuits suffer from several time-independent and -dependent variability problems, where the adverse effect of such problems on circuit performance has been risen due to aggressive and continuous down-scaling of feature size [1], [2]. Process, temperature, and voltage (PVT) variations are the well-known time-independent variability problems, which have been studied for many years and numerous accomplished models, efficient simulation techniques, and circuit design approaches robust to those variations have been developed. On the other hand, modeling of aging mechanisms is quite complicated due to the their time-dependency. Since it would be highly inefficient to measure the aging effects under nominal conditions, semi-empirical aging models are constructed via the acceleration aging tests (AAT) [3]–[5]. Simulation of time-dependent degradation mechanisms is another challenging problem. Namely, aging mechanisms such as Hot Carrier Injection (HCI) and Bias Temperature Instability (BTI) cause an increase in the threshold voltage over time, where the change in V_{th} is a function of time, channel length, temperature, and the drain-source and source-gate voltages [6], [7]. Among all these variables, transistor terminal voltages change over time (age), which means that the stress conditions should be updated within certain periods during aging simulation. Conventionally, the total simulation time is divided into sub-periods by a certain step count in commercial tools such as MOSRA (Synopsys) [8], UDRM (Mentor) [9], and RelXpert (Cadence) [10]. However, using a constant step count may lead to a demanding

trade-off between the accuracy and the efficiency of the analysis. Choosing a small step count results in inaccurate estimations whereas keeping the step count larger causes a dramatic increase in the computation time. This problem becomes very critical especially for variability-aware design automation tools [11], [12], in which excessive number of simulations are performed through optimization. This study proposes a novel approach to determine the step count of aging simulations, which consists of two levels. At the first level, a projection of estimation error is obtained with an efficient strategy by running just a few simulations. At the second level, the obtained error data is fitted into the power-law model and the efficient step count is determined without running any simulations. Remainder of the paper is as follows. In Section II, the related work in the literature is summarized. In Section III, the proposed step count determination approach is introduced and explained in detail. The case study circuit and simulation results are provided and discussed in Section VI. Finally, the paper is concluded in Section V.

II. RELATED WORK

Commercial aging simulators use a constant step count, where the user determines the step count at the beginning and the total simulation is divided into sub-periods. However, this is an impractical way to simulate aging effects, which may result in either waste of computation time or inaccurate estimations. A couple of different step count determination approaches have been proposed in the literature. In [10], the step count is dynamically determined by considering the variation on the circuit performance at each time step. The next step count is determined by using a formula when any considerable change occurs in the performance; otherwise, the simulation proceeds with the former step count. In this approach, unlike the discussion in [13], the accuracy has the priority over the efficiency since the change in the performance is considered rather than the change in the transistor parameters. Even though the claim in [13] was missing of some important stress condition updates, that is not the fact. Actually, considering V_{th} change may result in mis-evaluation since even small changes in transistors' parameters may degrade the circuit performance due to non-linear nature of analog circuits. Furthermore, being out of transistors' nominal operation regime will directly affect the circuit performance, so such change can also be captured via performance based evaluation. Two adaptive step-count approaches are presented in [13]. In the first approach, the step count is dynamically determined considering the time to update stress conditions, where the critical value for acceptable variation is determined by the user. In the second approach, a fixed number of step count is allowed and distributed over time. One step calculation is performed for the whole circuit and the transistor exposed to worst case degradation is only considered. The allowable variation is determined with dividing this worst case degradation by the fixed step count. In both approaches, the decision to update time is made by considering the variation in the V_{th}, which may result in accuracy problems as aforementioned. Furthermore, no simulation is performed at each intermediate step to avoid computation workload; however, this may lead to accuracy problems in some applications. Another step count determination approach is proposed in [14]. Unlike the approaches proposed in [13], the efficient step count is determined via SPICE simulations, at which an initial phase of the lifetime is used for the step count determination. Here, this initial phase is evaluated as the whole lifetime duration, where a reliable simulation is performed by using a relatively large step count. Then, the same simulation is repeated for different step counts starting from the minimum number. Finally, the efficient step count is determined considering the estimation error between the reliable and the instant simulations. This loop ends when the error decreases to a predetermined level and the last step count value is used for the remaining lifetime simulation. However, the number simulations performed at the initial phase may degrade the time efficiency due to the repeated considerations of the same time duration.

III. THE PROPOSED APPROACH

A general flow of the proposed reliability simulation approach is provided in Fig. 1. The approach is based on extracting the error trajectory by performing simulations with different step counts; thus, an error model is obtained for each design under test (DUT) circuit. A simulation based aging simulator is used in this study, in which both BTI and HCI are considered. The aging simulation starts with a fresh simulation to obtain the prior stress conditions. By using this data, degradation amount of each device is calculated via the semi-empirical model developed in [5]. Then, the transistor model file is updated and aging simulation is performed for $T_{age} = T_{final}/N_{step}$. This loop is repeated until $T_{final} = T_{age}$, where the stress conditions are updated at each intermediate step.

The proposed approach involves two different phases. The first phase starts with initial description of simulation duration (T_{final}), the initial step count ($N_{initial} = 2$), and the allowable degradation error in the worst case circuit performance (ε_{allow}). $N_{initial}$ is chosen as 2 to provide common intercept points between the iterations. The first aging simulation is performed with these initial parameters to obtain the first reference point. As the next step, the simulation duration is halved by keeping $N_{initial} = 2$ and the aging simulation is called again. Here, it should be noted that dividing the time duration by two means increasing of the step count in a logarithmic manner. The relative error is calculated for circuit performances of the final point of the last and the intermediate point of the former analysis. Then, the calculated error (ε_i) is compared with the maximum allowable error, where if (ε_i) is greater than ε_{allow}, one

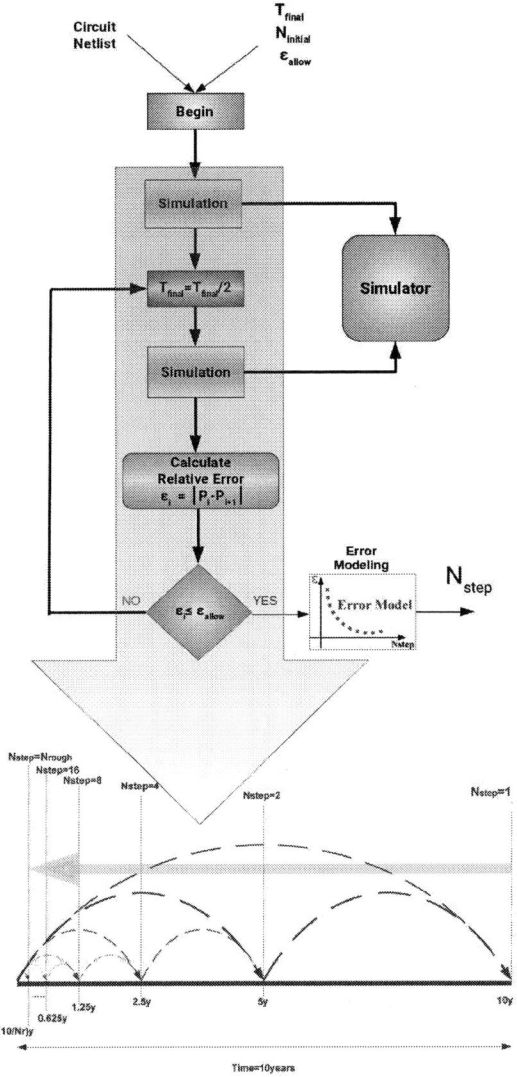

Figure 1. A general flow of the propoped approach.

more simulation is performed by dividing the time duration into half (increasing N_{step} by 2). This procedure is visually illustrated in Fig. 1. Let us consider a 10 years simulation. At first, the total time is divided by $N_{initial} = 2$, so there will be data for 5 and 10 years for this initial simulation. Then, the total time is divided by 2 again; hence, we will have 2.5 and 5 years aging data. In the next step, the error is calculated according to the difference between the first 5 years (one-step) and the second 5 years (two-step) data. This iteration is repeated until the error is less than the targeted error tolerance. Otherwise, the calculated error data is transmitted to the modeling part, in which the error points are fitted into a model. The saturated power-law model given in Equation 1, where a and b are model coefficients, is used during the modeling phase since the aging exhibits a power-law behavior in its nature.

$$N_{step} = a(1 + \epsilon_{allow})^b \qquad (1)$$

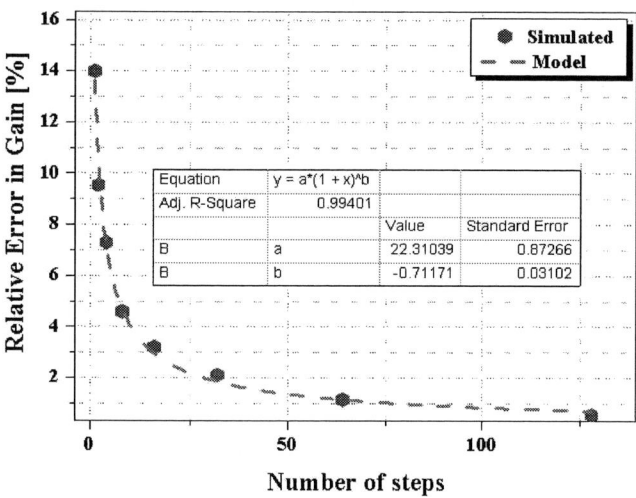

Figure 2. Estimation Error vs number of steps.

Here, one should consider that the last step count value is not the optimal one since the step count is increased in a logarithmic manner. Therefore, the actual optimal step count is a value between the last and the one former points. To demonstrate this situation, the obtained error data for a single stage amplifier is provided in Fig. 2. The ϵ_{allow} was selected as 1%, which means the simulation phase is stopped when the estimation error in the worst case performance metric is less than 1%. As seen from the figure, the relative error in the gain decreased down to 1% somewhere between 64 and 128. However, the simulation data just have two points there and it is impossible to determine that point. Here, the model takes the role and accurately localizes the optimal point ($N_{step} = 82$) without any simulation effort. The initial workload for this example is only 12 simulations. Furthermore, the designer have the flexibility to determine the error level by only alternation of the ϵ_{allow} without performing any additional simulation. A list of error levels for different step counts is provided in Table I.

Table I. ERROR LEVEL VS STEP COUNT.

Error Level (E_{allow})	Step Count (N_{step})
0.1%	198
1%	82
3%	21
5%	8

IV. CASE STUDY: FOLDED CASCODE AMPLIFIER

To demonstrate the proposed approach, a folded cascode amplifier, whose schematic is given in Fig. 3, was designed using 130nm CMOS technology parameters. The open-loop gain, the 3dB bandwidth, and the phase margin were considered as the major design performances. According to the fresh simulation results, design performances were measured as $62dB$, $65kHz$, and 55°, respectively. User defined variables, t_{final}, $N_{initial}$, and ϵ_{allow} were determined as 10 years, 2, and 1%, respectively. Here ϵ_{allow} is the maximum allowable estimation error in the worst degraded specification.

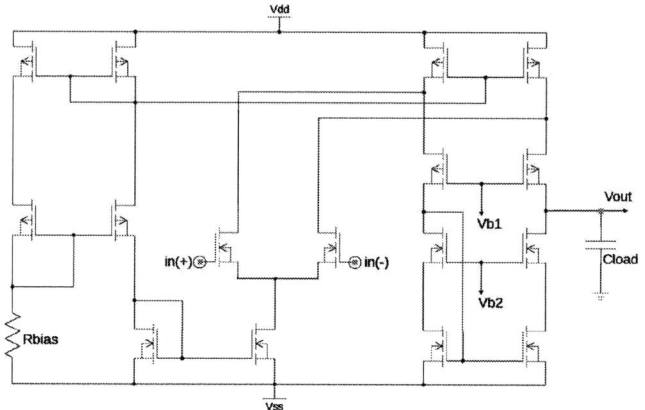

Figure 3. Schematic of the folded cascode amplifier.

The circuit was simulated by using both the fixed step count approach ($N_{step} = 10$, 20, 50, and 1000) and the proposed approach. The efficient step count was estimated as 162 by the proposed approach, where 14 initial simulations were performed. Threshold voltage degradation results

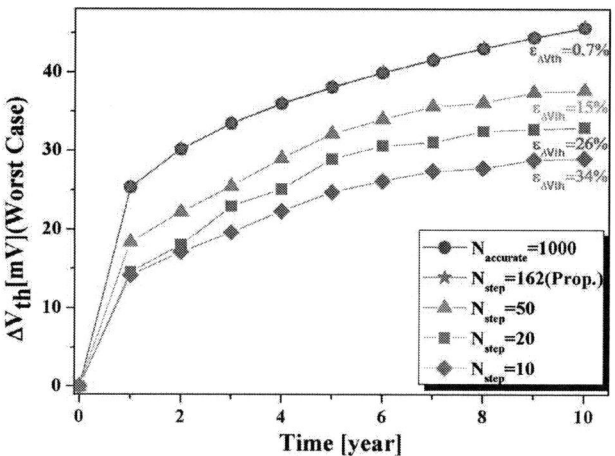

Figure 4. V_{th} degradation for different step counts.

for the worst case device are given in Fig. 4. According to the results, as expected, the estimation error considerably decreases through increasing step counts. The estimation error of the determined step count is 0.7% at the end of the lifetime simulation (10 years). The response of the circuit to this error is provided in Fig. 5, which is also used in the step count decision mechanism since the 3dB bandwidth

© VDE VERLAG GMBH · Berlin · Offenbach

is the most degraded specification. The estimation error was found 0.95%. To demonstrate the efficiency of the tool, three different solutions for the folded cascode amplifier circuit were simulated by the proposed tool and the tool presented in [15] and the results are listed in Table II. As can be seen from the table, the proposed work provides a remarkable decrease in the initial simulation work, which ultimately results in a save in the simulation time up to % 67 without any loss in accuracy. This save shoots up when the proposed aging simulator is used in automatic synthesis of robust circuits, where reliability simulations are performed for several hundreds, even thousands, of candidate circuits.

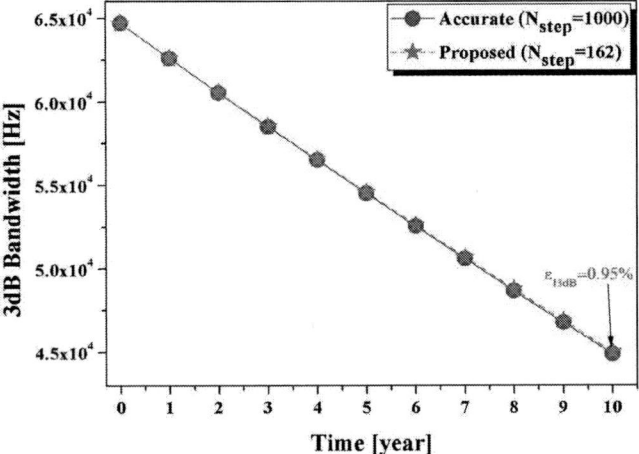

Figure 5. 3dB bandwidth simulation results.

Table II.   COMPARISON RESULTS.

		N_{step}	$N_{accurate}$	Initial Work [# of sim.]	$\epsilon_{\Delta V_{th}}$ [%]	$\epsilon_{f_{3dB}}$ [%]	Time [s]
This Work	1	89	1000	12	1.02	0.9	86
	2	123	1000	12	1.2	0.84	115
	3	135	1000	14	0.95	0.45	130
Tool in [15]	1	100	1000	50	0.85	0.95	128
	2	140	1000	78	0.96	1.4	194
	3	160	1000	84	0.78	1.2	208

V.   CONCLUSION

This study proposes an efficient and reliable way of determination of the efficient step count for reliability simulations. To increase the reliability, the efficient step count is determined via SPICE simulations, in which degradation in performance specifications are considered rather than the V_{th} degradation. Furthermore, to increase the efficiency, a two-phase step count determination approach was constructed. At the first phase, estimation errors are obtained via SPICE simulations by increasing the step count in a logarithmic manner. At the second phase, these errors are fitted into a saturated power-law model; thus, the efficient step count is determined according to the accuracy constraint without performing any simulations. The proposed approach can be used for all analog circuits without loss of generality. Simulation results indicated that the proposed approach saves the computation time up to 67%.

REFERENCES

[1] W. Weikang *et al.*, "Line-edge roughness induced single event transient variation in SOI FinFETs," *Journal of Semiconductors*, vol. 36, no. 11, p. 114001, 2015.

[2] S. Kiamehr *et al.*, "The impact of process variation and stochastic aging in nanoscale VLSI," in *2016 IEEE International Reliability Physics Symposium (IRPS)*. IEEE, 2016, pp. CR–1.

[3] C.-N. Shen *et al.*, "The study of activation energy (ea) by aging and high temperature storage for quartz resonator's life evaluation," in *Proceedings of the 2010 Symposium on Piezoelectricity, Acoustic Waves and Device Applications*. IEEE, 2010, pp. 118–122.

[4] A. Bravaix *et al.*, "Hot-carrier acceleration factors for low power management in DC-AC stressed 40nm NMOS node at high temperature," in *2009 IEEE International Reliability Physics Symposium*. IEEE, 2009, pp. 531–548.

[5] E. Afacan *et al.*, "Semi-empirical aging model development via accelerated aging test," in *2016 13th International Conference on Synthesis, Modeling, Analysis and Simulation Methods and Applications to Circuit Design (SMACD)*. IEEE, 2016, pp. 1–4.

[6] N. G. Junior and S. Barraud, "Experimental analysis of negative temperature bias instabilities degradation in junctionless nanowire transistors," in *2018 33rd Symposium on Microelectronics Technology and Devices (SBMicro)*. IEEE, 2018, pp. 1–4.

[7] G. Thareja *et al.*, "NBTI reliability of strained SOI MOSFETs," in *International Symposium for Testing and Failure Analysis*, vol. 32. ASM International; 1998, 2006, p. 423.

[8] B. Tudor *et al.*, "MOSRA: An efficient and versatile MOS aging modeling and reliability analysis solution for 45nm and below," in *2010 10th IEEE International Conference on Solid-State and Integrated Circuit Technology*. IEEE, 2010, pp. 1645–1647.

[9] S. M. Graphics, "Reliability simulation in CMOS 90nm design using Eldo."

[10] G. Keith, F. Mu, G. Kapila, and V. Reddy, "Simulation of circuit reliability with relxpert," *Texas Instrum. Cadence*, 2005.

[11] E. Afacan *et al.*, "A lifetime-aware analog circuit sizing tool," *Integration*, vol. 55, pp. 349–356, 2016.

[12] P. M. Ferreira, H. Cai, and L. Naviner, "Reliability aware AMS/RF performance optimization," in *Performance Optimization Techniques in Analog, Mixed-Signal, and Radio-Frequency Circuit Design*. IGI Global, 2015, pp. 28–54.

[13] P. Martín-Lloret, A. Toro-Frías, J. Martín-Martínez, R. Castro-López, E. Roca, R. Rodríguez, M. Nafría, and F. V. Fernández, "A size-adaptive time-step algorithm for accurate simulation of aging in analog ics," in *2017 IEEE International Symposium on Circuits and Systems (ISCAS)*. IEEE, 2017, pp. 1–4.

[14] E. Afacan, G. Berkol, G. Dündar, A. E. Pusane, and F. Başkaya, "A deterministic aging simulator and an analog circuit sizing tool robust to aging phenomena," in *2015 International Conference on Synthesis, Modeling, Analysis and Simulation Methods and Applications to Circuit Design (SMACD)*. IEEE, 2015, pp. 1–4.

[15] E. Afacan *et al.*, "A deterministic aging simulator and an analog circuit sizing tool robust to aging phenomena," in *Synthesis, Modeling, Analysis and Simulation Methods and Applications to Circuit Design (SMACD), 2015 International Conference on*. IEEE, 2015, pp. 1–4.

Schematic Generation of Programmable Analog Neural Networks for Signal Proccessing

Florian Aul*, Nikoletta Katsaouni[†], Lukas Krischker*, Sascha Schmalhofer*, Marcel H. Schulz[†], Lars Hedrich*

*Institute for Computer Science, Goethe University Frankfurt, Germany
aul/schmalhofer/hedrich@em.cs.uni-frankfurt.de
[†]Institute for Cardiovascular Regeneration, Goethe University Frankfurt, Germany
katsaouni/marcel.schulz@em.uni-frankfurt.de

Abstract—This paper presents a methodology for generating analog neural networks (ANNs) from trained TensorFlow models using an intermediate description of the network. All needed circuitry and helper circuits are available in a block library provided for the operation in an energy efficient low power region. The weights of the ANN are programmable by the corresponding bitstream, which is also generated by the proposed approach. The feasibility of the approach is demonstrated with two examples, the larger being an ANN for arrhythmia detection of electrocardiograms (ECGs) with 2,570 neurons and 10,042 weights.

Index Terms—analog, neural networks, generation, power-efficiency

I. INTRODUCTION

Many neuronal networks (NNs) suffer from large power consumption being executed on standard CPUs or GPUs. An interesting alternative is the use of analog circuits for the basic operations (multiplication, sum), termed analog neural network (ANN). The multiplication with a weight can be realized with several techniques like a dedicated multiplier/mixer or by having programmable gain stages. In this paper we want to automatize the generation of ANNs.

We focus on an implementation of pretrained nets. The training is done before the realization/programming. The generated ANNs are not capable of training themselves, which definitely makes things easier – and more power efficient.

In Fig. 1 an overview of the proposed workflow to generate an ANN is shown. We start with the training of an arbitrarily selected architecture in TensorFlow. This model is able to use quantized weights. The number of weights and neurons are held small to reduce the energy and the chip area. From the trained model a model file is generated. The ANN generator itself finally generates a schematic, a symbol and a bitstream of the programmable ANN.

II. PREVIOUS WORK

The use of analog circuits for the implementation of neural networks has been proposed in several different works [1, 2]. Some approaches use mixed signal blocks for energy reduction [3]. These methods often have a kernel block for the analog operation but do not implement a full analog circuit for the neural network. A synthesis of analog neurons is proposed in [4], while the number of neurons and weights is limited. [5] implements a full ANN to detect numbers (MNIST) with 1.5k neurons and 17k synapses. However, this approach is not programmable and operates on a fixed task. In [6] a wafer

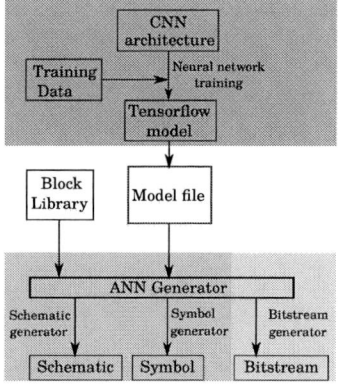

Fig. 1. Workflow of generation of analog neural net (ANN).

scaled implementation of ANNs was proposed. This approach is trainable but is expensive in terms of chip area.

Over the last years, Convolutional Neural Networks (CNNs) have attracted considerable interest from the scientific community with numerous and diverse applications in various fields such as speech recognition [7, 8], biosignals [9, 10, 11] and medical imaging [12].

The architecture of a CNN is inspired by the structure and functionality of the human brain and the information flow between the artificial neurons is resembling the neurons of the visual cortex. A CNN consists mainly of convolutional layers followed by a non-linear activation function and a pooling operation. Each convolutional layer extracts important features from the input signal and generates feature maps by applying filters whose parameters are learned during the training. Contrary to the conventional fully connected neural networks, the parameter sharing, as a result of the convolution operations, allows the network to be scalable to long signals while maintaining a low number of parameters. Furthermore, the number and size of filters and layers can be adjusted to increase the network energy efficiency.

In this contribution we train and design the CNNs in TensorFlow as sketched in Fig. 1. The trained model is written into a model file (see Sec. V-A).

III. ARCHITECTURE OVERVIEW

The proposed methodology implements a pure ANN with programmable weights and biases. Mainly ReLU activation

© VDE VERLAG GMBH · Berlin · Offenbach

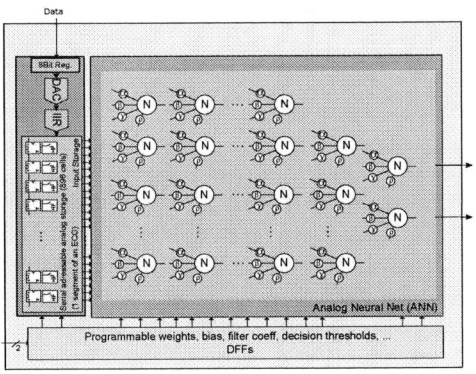

Fig. 2. Blockdiagram of ANN with supporting input and signal processing units. All blocks are programmable.

functions are used, but tanh-like activation functions are also available. In the current implementation one-dimensional input layers of arbitrary size are supported. In our example case, we use electrocardiogram (ECG) signals in the time domain. The signal can either come from directly connected sensor data or be provided by a streaming interface. Additionally, the sequential streaming data enables the use of signal preprocessing like filtering with an infinite impulse response filter (IIR).

The architectural overview in Fig. 2 shows this structure. The ANN itself is programmable and can consist of several different layers as explained in the following sections. We use 6-bit programming cells for the weights and 8-bit programming cells for the bias generators. Accordingly, the whole network is programmed by a bitstream.

IV. BLOCK LIBRARY

For the wanted implementation of a CNN we developed analog basic blocks in a low power region to be implemented in a 130nm BiCMOS process. We decided to take a current signal in the range of -200nA – +200nA in the subthreshold region of the MOS transistors as the main signal representation. The ReLU implementation is based on current-mirrors (see Fig. 3). The frontend of the current mirror implements the sum. The backend of the current mirror implements a 6-bit programmable weight. The real transistor implementation has additional cascode transistors for improving the linearity. In Fig. 4 all the necessary blocks, with the above mentioned signal range, for the generation of the ANN are listed.

The analog cells are biased with a balanced tree of current mirrors and bias voltage generators. The number of cells connected to one bias cell and to one ReLU frontend and Tanh frontend has to be restricted due to leakage currents into the gates. If the leakage currents load the frontend cells too much, the accuracy of the backend cells is decreased. This biasing tree is also used for the power down of the ANN. However, a better power-down strategy could further reduce the power-up time.

The signal streamed in is converted to the analog domain with a 8-bit current steering digital-to-analog converter (DAC, see Fig. 2). Subsequently, the analog signal is bandpass filtered by 7 programmable second order IIR filter stages to remove the DC offset and the high frequency noise and stored in capacitive analog storage cells.

V. SCHEMATIC GENERATION

The principle workflow of the schematic generation is described in Fig. 1. The schematic generator uses a layer based approach to create the neural network. This was necessary to keep the complexity manageable without introducing restrictions for the created network.

The programming cells for the bias and weights are implemented as serially connected D-Flip-flops (DFFs). This connection allows to (re-)program the network with a bitstream. If a programming cell is connected to more than ten weight cells, the schematic generator inserts buffer cells.

The neuron (ReLU-frontend) and weight cells (ReLU-backend) need bias voltages to work properly. These biases will be called *circuit bias*. The circuit bias is layer dependant and each bias cell gets a 5nA biasing current. The circuit which provides the bias current is implemented as a balanced tree to keep the number of mirror stages low and equal, and thus the mismatch error low.

A. Model file

The model file is used as a human-readable intermediate file format to transfer the neural network model to the schematic generator. It contains the structural and numeric information

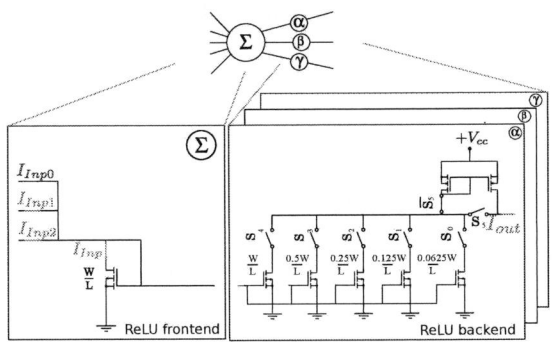

Fig. 3. Principle of ReLU neuron with 6-bit programmable weights (S_5=sign).

Schematics & VerilogA models		Recipes	
Basic blocks	**Programming**	**Layer generation**	**Programming**
ReLU frontend	6-bit programming	Input layer	6-bit programming
ReLU backend	8-bit programming	Dense layer	8-bit programming
Tanh frontend	6-bit buffer	Convolutuional layer	6-bit buffer
Tanh backend	8-bit buffer	Pooling layer	8-bit buffer
Inputcells frontend	**Signal Processing**	Output layer	**Circuit bias**
Inputcells backend	IIR cells		Circuit bias cells
Poooling cells (3 - 6)	DAC		

Fig. 4. Block library: Blocks used for the ANN schematic generation (left) and recipes for generating the layers with these blocks (right).

of the network and the used layers needed by the generator. The extraction of the model file from TensorFlow is done automatically.

Accordingly to the layer based approach, the model file holds layer dependant information, e.g. filter size or pooling function.

```
# HIDDENLAYER #
# LAYERTYPE CONVOLUTION
# NODECOUNT 289
# ACTIVATIONFUNCTION relu
# FILTERCOUNT 2
# KERNELSIZE 8 1
# STRIDES 1
# BIAS TRUE
0.0625   0.
# WEIGHTS
-1.0625  -1.0625  -1.0625  -0.1875  -0.1875   0.      0.      0.0625
-0.1875   0.125   0.3125   0.375   -0.0625   1.375   0.     -1.25
# HIDDENLAYEREND #

# HIDDENLAYER #
# LAYERTYPE POOLING
# NODECOUNT 96
# POOLINGFUNCTION avg
# ACTIVATIONFUNCTION relu
# POOLINGCOUNT 3
# STRIDES 3
# CHANNELS 2
# HIDDENLAYEREND #
```

Fig. 5. Excerpt of the *model file* showing the description for two layers. The first layer is a convolutional layer with two filters, the weights and a layer bias. The second layer is the following pooling layer with the average pooling function of input size three.

The model file is also used to create the neural network part of the programming bitstream. Due to the well defined order, in which the programming cells are connected, the corresponding bitstream is created directly from the model file.

Algorithm 1: Create Schematic

input : model file
output: schematic
createInputConnections(schematic, neuralNet);
foreach *layer in layers* **do**
    updatePrefixes();
    createBiasInstances(schematic, layer);
    createWeights(schematic, layer);
    if *hasNetworkBias(layer)* **then**
        | buildNetworkBias(schematic, layer);
    buildNeuronFrontends(schematic, layer);
createOutputConnections(schematic, neuralNet);
createBiasNetwork(schematic);

B. Algorithms

The model file is parsed once and used for both the symbol and the schematic creation. The symbol is connecting the schematic to the environment through predefined ports which only differ in the bus width of the selected ports. For both representations the portnames are hardcoded and the bus width is taken from the parsed model file in order to avoid connection problems. The layer-based approach is taken from the object oriented programming paradigm and abstracts the generation of a layer from the implementation details of the basic cells (see Alg. 1). The arrangement of the cells in each layer is

handled by the given recipe for the layer type. With small gaps in between each layer, they are visually separated. The connections between the cells of different network layers are not fully drawn to keep the schematic clear. Instead, we use labeled wire snippets to connect the instances. The naming of the instances and connections is based on the actual layer and the position inside the layer. This makes the relation between the implementation and the model visible.

VI. RESULTS

We used the approach on two examples. All results are simulation results on transistor level with a 130nm design kit. The first example is a fully connected neural net that classifies an arbitrary signal, based on whether it contains a 10Hz sine. The network has 11 input neurons, one hidden layer with 18 neurons and one output neuron. Overall it has 216 weights. The schematic generation process starting from the *model file* for this example needs 0.47s runtime. In Table I the accuracy of the classification process is reported. Due to the inaccuracy of the transistor level implementation the accuracy is going down. For one evaluation the ANN needs 1.13ms system time and consumes 1.03nJ. The simulation time for one evaluation is 81s.

TABLE I
ACCURACY MEASUREMENT OF 10Hz DETECTION NET

Model	Accuracy	Power	Energy
TensorFlow Model	81.5%	-	-
Schematic with VerilogA	79.25%	-	-
Schematic with transistors	77.0%	$0.862\mu W$	$1.03nJ$

The second example is a more complex, larger CNN with 2,570 neurons and 10,042 weights for arrhythmia detection in ECGs. The architecture and structure of this example are shown in Fig. 6. For this application, we read segments of 7s original time, sampled at 128 Hz, filtered and stored in the input cells (see Fig. 2).

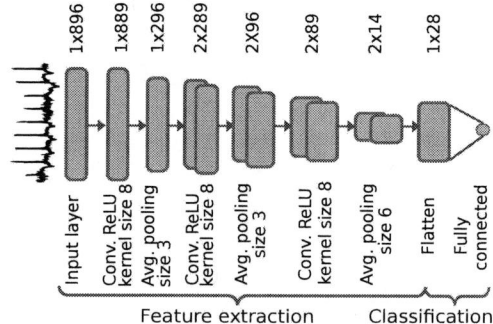

Fig. 6. Architecture of CNN for arrythmia detection.

The schematic generation process has a runtime of 54.5s. The full resulting schematic is shown in Fig. 7. Besides the above mentioned number of neurons and weights the schematic contains 1,645 bias cells for the neurons, 516 pooling cells and 2,326 cells for duplicating the bias quiescent currents and generating the biasing voltages for the analog circuits.

© VDE VERLAG GMBH · Berlin · Offenbach

The power consumption of the different blocks in the overall design is shown in Table II. We have two different phases to evaluate the streamed signals:

- READ_SEGMENT: Read the signal, filter it with the IIR Filter and fill the input cells for the ANN.
- EVALUATE_SEGMENT: Evaluate the actual input with the ANN.

Due to the streaming nature of the approach, most of the time the READ_SEGMENT phase is used. It reads and filters the input signal at 200kHz clock frequency. The reading of one segment needs 19.2ms system time and a current of 63 μA.

TABLE II

SIMULATED CURRENT CONSUMPTION FOR THE MAIN BLOCKS IN THE PHASES OF THE WHOLE DESIGN ON TRANSISTOR LEVEL

Cell / Phase	READ_SEGMENT	EVALUATE_SEGM
CNN	27.1 μA	395.5 μA
IIR	29.8 μA	29.4 μA
Input storage	5.5 μA	51.0 μA
DAC	19.8 nA	20.6 nA
Programming cells	479.3 nA	479.3 nA
Other blocks	69.5 nA	64.5 nA
Sum	63.0 μA	476.5 μA

The decision computation using the ANN in the EVALUATE_SEGM phase from power up to evaluation and power down needs 2ms but has a much larger current consumption of overall 476.5 μA. Hence, the reading filtering and evaluating of one segment takes 2.595 μJ at 1.2 V supply voltage. The main contributors to the energy consumption according to Table II is the CNN with about 64% followed by the IIR filter with about 25 % and the input storage cells with about 10%. The other cells e.g. the DAC and controller logic are in sum below 2% and hence negligible. A variant without the IIR filter can be implemented with digital storage cells and simple DACs at each input reducing the power of the overall chip by 25% - 30%.

The simulation for one segment (21ms system time, 896 inputs) needs about 2 days to run as the overall circuit has 1.5 million transistors. Hence, an accuracy measurement on a big test set is not feasible. We measured the accuracy of the implementation of the ReLU neurons, exhibiting an absolut deviation of 0.5 nA max. near 0 input and a relative deviation of 2.5% in the positive range of the function.

As all weights and signal bias cells as well as the IRR-filter coefficients are programmable, different tasks can be performed with a fabricated chip.

The estimated active area needed under the gates sums up to 7.4mm^2. Hence it can be implemented in reasonable chip size. However, as we do not optimize for area, a lot of improvement could be expected.

VII. CONCLUSION

In this paper we presented an automatic generation approach for implementing programmable analog neural networks on chip. The method uses a model file which can automatically be derived from a TensorFlow implementation of the desired NN. All needed neural blocks and the supporting blocks like

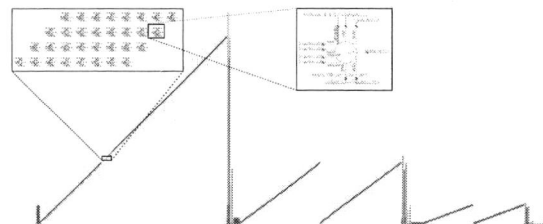

Fig. 7. Generated schematic for the arrhythmia detection NN. The diagonal segments are the weight blocks for the convolutional kernels. A weight block is magnified.

biasing and programming cells are generated automatically. The presented example has about 10,000 weights and needs 1.5 million transistors.

Future work concerns two main parts for a fully automatic solution: an automated debugging framework to come up with the very long simulation times of the transistor netlist and an automatic layout generation of these regular topologies.

VIII. ACKNOWLEDGMENT

The authors acknowledge the financial support by the Federal Ministry of Education and Research of Germany in the ENERGICS project (project number 16ES1141).

REFERENCES

[1] H. P. Graf and L. D. Jackel, "Analog electronic neural network circuits," *IEEE Circuits and Devices Magazine*, vol. 5, no. 4, 1989.
[2] J. Misra and I. Saha, "Artificial neural networks in hardware: A survey of two decades of progress," *Neurocomputing*, vol. 74, no. 1-3, 2010.
[3] D. Bankman, L. Yang, B. Moons, M. Verhelst, and B. Murmann, "An always-on 3.8 μj 86% CIFAR-10 mixed-signal binary CNN processor with all memory on chip in 28-nm CMOS," *IEEE Journal of Solid-State Circuits*, vol. 54, no. 1, 2018.
[4] F. Aul, N. Katsaouni, M. H. Schulz, and L. Hedrich, "Synthesis of Power-Efficient Analog Neural Networks for Signal Processing," in *Analog 2020: 17. ITG/GMM-Fachtagung Analog*, 2020.
[5] K. Jia, Z. Liu, F. Qiao, X. Liu, Q. Wei, and H. Yang, "AICNN: Implementing Typical CNN Algorithms with Analog-to-Information Conversion Architecture," in *2017 IEEE Computer Society Annual Symposium on VLSI (ISVLSI)*, 2017.
[6] J. Schemmel, J. Fieres, and K. Meier, "Wafer-scale integration of analog neural networks," in *2008 IEEE International Joint Conference on Neural Networks (IEEE World Congress on Computational Intelligence)*. IEEE, 2008.
[7] S. Kwon *et al.*, "A CNN-assisted enhanced audio signal processing for speech emotion recognition," *Sensors*, vol. 20, no. 1, 2020.
[8] Y. Zhao, X. Jin, and X. Hu, "Recurrent convolutional neural network for speech processing," in *2017 IEEE International Conference on Acoustics, Speech and Signal Processing (ICASSP)*. IEEE, 2017.
[9] U. R. Acharya, S. L. Oh, Y. Hagiwara, J. H. Tan, and H. Adeli, "Deep convolutional neural network for the automated detection and diagnosis of seizure using EEG signals," *Computers in biology and medicine*, vol. 100, 2018.
[10] Ö. Yıldırım, U. B. Baloglu, and U. R. Acharya, "A deep convolutional neural network model for automated identification of abnormal EEG signals," *Neural Computing and Applications*, 2018.
[11] S. L. Oh, Y. Hagiwara, U. Raghavendra, R. Yuvaraj, N. Arunkumar, M. Murugappan, and U. R. Acharya, "A deep learning approach for Parkinson's disease diagnosis from EEG signals," *Neural Computing and Applications*, 2018.
[12] R. Zeleznik, B. Foldyna, P. Eslami, J. Weiss, I. Alexander, J. Taron, C. Parmar, R. M. Alvi, D. Banerji, M. Uno *et al.*, "Deep convolutional neural networks to predict cardiovascular risk from computed tomography," *Nature communications*, vol. 12, no. 1, 2021.

SMACD / PRIME 2021 | 19 – 22 July 2021, Online Event

Generators, Templates, and Code Generation for Flexible Automation of Array-Style Layouts

Benjamin Prautsch[*], Reimund Wittmann[†], Uwe Eichler[*], Uwe Hatnik[*], Jens Lienig[§]

[*]Fraunhofer IIS/EAS, Institute for Integrated Circuits, Division Engineering of Adaptive Systems, Dresden, Germany
{Benjamin.Prautsch, Uwe.Eichler, Uwe.Hatnik}@eas.iis.fraunhofer.de
[†]IMST GmbH, Kamp-Lintfort, Germany, reimund.wittmann@imst.de
[§]Dresden University of Technology, Dresden, Germany; jens@ieee.org

Abstract—**The design of integrated circuits from the specification onward aims at the successful validation by silicon measurements. One key milestone in this process is the completion of the layout. This, however, is very challenging as many iterations are usually necessary due to parasitic effects. In order to address this challenge in analog layout design, our work extends procedural generator-based automation. A declarative array template is embedded into the common generator structure. Following this structure, generator code is automatically generated with a schematic as the input. Using this approach, a flexible generator is created immediately that allows automatic design of array-style layouts with template-based flexibility and at generator-based execution speed. In addition, the template enables early and fast parasitic estimates. Our combined approach contributes to analog layout automation by bridging the gap between generators and templates.**

Keywords—*Analog automation, generator, template, array layout, reuse, design migration, analog layout*

I. INTRODUCTION

Analog integrated circuit design is a very sophisticated task that requires much expertise and consumes considerable efforts. Contrary to digital design, analog suffers from a lack of automation which results in long design time and limited reuse. As technologies evolve and become more advanced while cycle times are getting ever shorter, analog design engineers must meet increasingly challenging schedules. Thus, several approaches are followed to accelerate design.

Most pragmatically, design engineers concentrate on few types of designs so that they become very familiar with them. This increases productivity as they can then take known best-practice decisions (e.g., hierarchical organization or additional unconnected layout devices). However, such steps are all manually done. Thus, in the last decades several additional features were included into common design environments in order to improve productivity. Well-known examples are schematic-driven layout (SDL), placement tools for matching structures, visually-assisted automation, or instant design rule check (DRC). These environments are of great help, however, still most parts of analog are implemented manually.

Lately, tools that support more automation through reuse and generation of cells begin to find their ways into design environments. For layout reuse, flexible templates are used [1, 2]. They usually represent the layout arrangement in an abstract way and guide layout creation through optimization approaches [3, 4, 5]. Subsequently, the actual layout generation step follows which, depending on the capability of the generation engine, might support multiple process technologies. At this level, procedural layout *generators* come into picture [6, 7, 8, 9, 10] that automate generation of devices

and building blocks. The advantage of generators is that they provide fast and parameterizable solutions to known problems employing expert knowledge [11]. However, their drawback is that they often lack flexibility and hardly allow structural changes as adapting their source code often requires significant development time. *Templates*, on the other hand, allow much flexibility in layout definition; they also help considering (geometric) constraints that are essential for analog layout automation [12, 13, 14].

We believe that the combination of templates and generators is very promising to tackle the aforementioned shortcomings. In [15], generators are combined with optimization based on swarm intelligence showcasing the advantage of combining flexible techniques with generator-based approaches. More generally, [11] proposes a *bottom-up meets top-down design flow* approach besides a continuous layout design flow in order to tackle analog layout automation.

In this work, we combine the bottom-up generator approach with the top-down template approach and add automatic schematic-to-code creation, which, to the best of our knowledge, has not been done before. This way, generators are automatically derived from an input schematic while incorporating the known flexibility of a template for user-defined placement. As the result, generator-based automation with more flexibility and faster generator development becomes possible. Our approach combines the following contributions into a single flow (Fig. 1):

- Generator framework that is technology-agnostic [16],
- Template approach for flexible placement and routing,
- Automated source code creation for fast generator availability (that can also be extended manually).

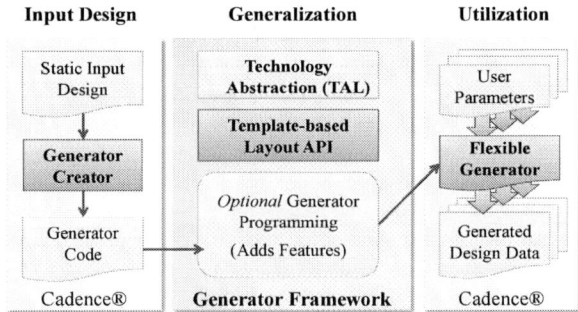

Figure 1 Illustration of the flow for automated generator code creation and utilization (with our contributions highlighted in blue). First, a static schematic is analyzed by the Generator Creator. All PDK devices are then mapped through technology abstraction (TAL [16]) and assigned to initial positions of the array template in the generator code. The resulting generator can be executed by the user. It allows to (re)arrange and adapt the layout generated for appropriate results across sizing and PDKs. Optionally, the generator code can be extended in order to implement additional functionality.

© VDE VERLAG GMBH · Berlin · Offenbach

180

II. OUR DESIGN FLOW APPROACH

When utilizing generators, the analog design flow is often extended by an additional hierarchy of (generated) building blocks. Instead of designing static blocks, procedural generators describe and create the required views flexibly. As designers might be required to programing code, it has even been called a "paradigm shift in AMS circuit design" [17]. So far, however, generator programming has been the bottleneck.

A. Automated Generator Creation

As designers are usually not into programming and as every design differs, automated generator code creation is desirable. Therefore, we follow an automated approach that creates flexible generators for schematic, symbol, and layout automatically based on a static input design. Our approach standardizes the code structure, thus, simplifies generator development and code creation. At the same time, valuable expert knowledge (being a key aspect of generators [11]) is stored in an executable way in order to accelerate both design and reuse.

In Fig. 1, the automatic process of generator creation is shown. Using our *Generator Creator*, an existing schematic (and symbol) is converted into parameterizable generator code. It can exactly replicate the input schematic through a default parameter set, creates a similar symbol (or a standard symbol), and assigns the instances found in the schematic to the abstract array-style template (see Section II.B). Moreover, a technology abstraction layer (TAL) [16] is used in order to map particular technology-dependent parameters to a generic representation (e.g. adaptation of the name of the width parameter of a transistor or the type of a transistor). The generator code created is the basis to include additional functionality into the generator. Programming the initial code manually would consume significant development time.

Once the generator is created, it can be parameterized and executed in order to create variants of building blocks of a design including the related array-style layouts. As the result, the created generator can be used for designs with different specifications or process technologies. This, in turn, enables fast layout prototyping and design.

B. Template-based Layout

Templates are widely used in optimization-based approaches in order to define floorplanning and partly placement and/or routing. They are abstract definitions of the target layout. In contrast to generators, they are not executable. Examples are LDS [1], dynamically generated templates in AIDA [18], as well as methods that use templates for porting like IPRAIL [4] or the fast prototyping approach in [5]. Templates have different appearances. They can be described in a flexible way [1, 19] or represent regular structures, such as arrays [20] or "streets" [21]. Some EDA approaches also use templates for both placement and routing [21, 22].

Inspired by the template approach, this work follows the idea of structuring procedural generator code using flexible layout templates for the layout description (currently limited to array-styles). This is beneficial as programming structural/topological variants, flexible sizing, placement options, and routing at the same time is hard to maintain in procedural code. Instead, our approach first describes the abstract layout arrangement declaratively using a template (that has to be implemented before). In the next step, this description is analyzed and translated into positions for each element of the template.

Based on a given placement, the following routing strategy is applied to the array. The regions between instance columns form the routing channels. Their size is not static but an adaptive result from the routing contained (implemented as composite pattern). The routing approach assumes unit devices but is not limited to it and connects two nets each. The method first prioritizes cohesive unit devices available on either side of a routing channel. The algorithm prevents occupying neighboring routing channels (which means that only two in three routing channels will be used to connect similar devices). Then, the remaining devices are collected and assigned to the routing channels in the same way while starting with the least filled ones. Subsequently, vertical wires are drawn to each unit device (from bottom to the most distant logical device) and a horizontal trunks plus vias follow. Finally, a procedural command sequence translates all these abstract elements defined in the template to "real" instances, wires, and vias in the actual layout.

As a result, layout description and program flow are structured separately which improves flexibility and eases generator development. Utilization of the *Generator Creator* also reduces faults significantly as the code is generated automatically and as checks for logical errors are included using callbacks (see III.B). This means that it is possible to promptly detect fatal LVS errors such as missing instances in case of wrong user inputs or faulty (manual) code updates. A valid LVS finalizes the verification of a generated building block.

As the template controls the first part of the layout sequence (see III.B), it can afterwards be combined with procedural code. This way, much effort for placement can be automated in order to reduce overall code development time of custom generators. Additional generator code can, among others, include details of the layout generation such as additional shapes, vias, well contacts, or routing details.

III. APPLICATION IN OUR GENERATOR FRAMEWORK

This work extends the functionality of our generator framework in [9]. The *Layout API* (application programming interface) is included into the generator structure adding array-style template flexibility. Our *Generator Creator* has been adapted in order to automatically create generator code that controls this template via the generator GUI.

A. Layout API

The *Layout API* is an extension to the generator API that adds template capability. It allows to use declarative and procedural code within the same generator. In this work, it uses (among others) a grid-based approach comparable to the approach in [20], meaning the layout area is segmented into rows and columns. Each row's height can vary as well as each column's width. Fields of the grid can be defined to represent a device or a hierarchical template recursively. The generator that utilizes this API requires only few code lines to define and interpret a template in order to create the actual layout at the level of the design environment.

B. Programming Interface and Generator Structure

Generators implemented with our method apply a similar code structure using class inheritance. Particular tasks are separated in specialized methods which are then executed by the base class for each generator (Fig. 2). First, the code defines user parameters which are shown in a parameter mask (for interaction with the designer) or hierarchically passed through the generator hierarchy. Second, callbacks are defined that automatically consider cross-dependencies between parameters in order to validate user inputs. Third, common properties of all views of the generated cell(s) are defined in the *prepare()* method in order to ensure consistency. Finally, respective methods are implemented to describe the circuit representations (views) for schematic, layout, and symbol.

The template has been included into this structure in order to enable user configuration (*param_spec()*), run checks (*param_check()*), and control the layout process *layout()*.

```
class Generator(gen.HierBlock):
    def param_spec(self): # parameters and constraints
        self.template = Template(…)
    def param_check(self):# parameter updates & callbacks
        self.template.update(…)
    def prepare(self):      # common data for all views
        self.template.evaluate(…)
    def schematic(self, cv):
        # procedural schematic description
    def layout(self, cv):
        self.template.draw(cv) # template execution
        myShape = cv.create_shape(…) # optional code
```

Figure 2 Basic structure of (hierarchical) generator code. The names of the class methods are predefined in the abstract base class that implements the general execution order of all methods when run. In addition to schematic and layout, also methods for symbol and testbench creation can be defined in the generator code.

C. Design Reuse with the Generator Creator

Our Generator Framework [9] interfaces with the Cadence® Virtuoso® design environment where the *Generator Creator* can be run in order to translate existing schematics into generator code. The *Generator Creator* executes the following steps: (1) the schematic is read and all elements of it are stored, (2) all data fetched is converted into a PDK-independent description, (3) the generic data is translated into generator code fragments that represent the generator structure given in Section III.B., and (4) the fragments are merged to a complete generator file. This method tremendously accelerates generator development as thousands of lines of code are created within seconds.

Each created generator can be run immediately or whenever required. As the generator code links parameter mask and *Layout API*, entry fields allow to modify the template. As default, an (almost) quadratic array is defined with the devices assigned. Based on user input, the arrangement can be adapted (e.g. number of matrix rows or the placement algorithm). All parameters entered into the generator GUI can be saved, too, in order to reuse them across projects and even PDKs along with the generator. For example, the placement definition can be reused this way.

IV. Design Experiment

We validated our presented approach on a capacitor arrangement that has a regular structure based on unit devices. It is part of a SAR ADC architecture [23, 24] that we intend to build for very complex matching constraints. Such structures cannot efficiently be designed manually by following best practices (see Section I). This prevents *close to optimal* unit capacitor placement with minimal influence of the interconnects. The main unit capacitor array of the popular differential split-cap architecture consists of 2 sub arrays, each again split into two arrays for MSB and LSB (most and least significant bit) values. So, in total, 4 sub DACs have to be combined into a single layout for high sub DAC linearity, high MSB accuracy, N channel and P channel symmetry, and matching requirements. The capacitor ratios have to match in presence of inherent layout parasitics in order to meet the SAR target resolution requirements. It is not easy to find a well-working layout arrangement, as the unit capacitor size and total capacitor array configuration strongly influence both random and systematic errors of the circuit.

Using our combined template and generator approach, we are able to fast try placement options and estimate the related parasitic effects. While manual placement would be very time-consuming, the *Generator Creator* combined with the template approach results in a flexible generator to allow fast early analyses. An example schematic with 1024 capacitors (twice 256 for LSB and 252 for MSB plus eight coupling capacitors) was defined. They represent the capacitances of the block diagram shown in Fig. 3. In order to achieve a well-matched placement, the logical devices must be split into unit devices which are arranged close and with sufficient size [25]. At the same time, a common-centroid pattern compensates process gradients [26]. In order to achieve a good arrangement, we adapted the common-centroid algorithm from [27] which creates a highly dispersed placement.

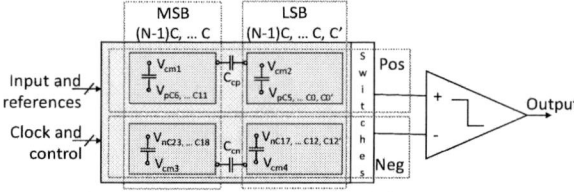

Figure 3 Coarse block diagram of the SAR ADC core. The capacitor array instantiates the capacitors C0 to C23 (and dummies C0' and C12') for MSB and LSB on both positive (Pos) and negative (Neg) channel. The split cap coupling capacitors are indicated, too.

Additionally, systematic errors from the routing are to be considered, too, as even a perfectly matched placement will be degraded significantly by parasitic routing capacitances [28], especially with very small capacitances [29].

The overall time from schematic import to initial layout generation is only a few minutes, with an overall layout generation run time of about 70 seconds. A reusable generator is created, too. This way, large parts of the layout can be generated much faster than manually possible, which enables optimization over parameter variants and arrangements. A generated capacitor arrangement is shown in Fig. 4.

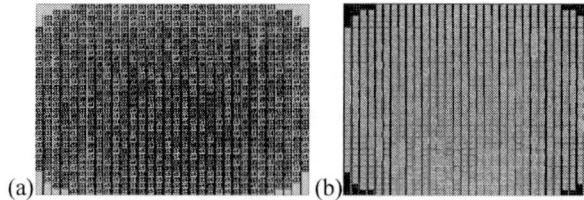

Figure 4 Generated capacitor arrangement. (a) shows the abstract layout template with each logical device colored identically among unit devices and (b) depicts the generated layout with the capacitor instances of logical device C4, C10, C16, and C22 being marked.

In order to support design decisions, we included a method into the template that quickly estimates the effective capacitance ratio of the logical capacitors with routing. Based on the areas of both unit devices and routing, estimates of the effective capacitance ratio are calculated (Fig. 5). The influence of routing is estimated by the template using the cumulative routing area per logical device over substrate. In the PDK used, both capacitor area and routing area contribute by about the same amount. Thus, the curves of both device area and routing area over logical capacitor (that should be congruent) can be added. The overall ratio is already acceptable, especially when considering the fast generation speed. Some extensions of the routing options might still be included in order to further improve the result.

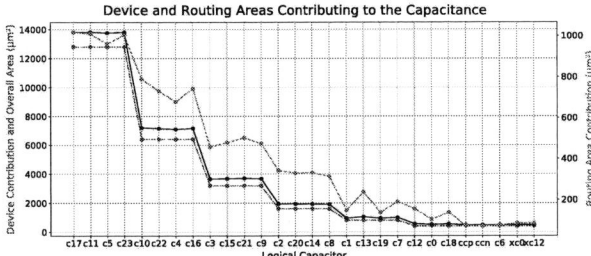

Figure 5 Areas of both capacitors (dash-dotted) and routing (dashed) that contribute to the capacitance ratio (solid). While the capacitors realize the exact target ratio, the routing causes deviations. Still, a good overall ratio is achieved quickly.

V. Conclusion and Outlook

Our presented approach enables accelerated and more flexible generator-based layout creation. It combines a template-based *Layout API* with generators which, in addition, are created automatically through a *Generator Creator*. This novel combination allows flexible array-style layout automation that is user-driven as designers have full control over the placement. In addition, our approach eases generator-based reuse across design projects and process technologies. We validated our methodology using a capacitor arrangement for later use in a SAR ADC. With this, a flexible, template-driven generator is automatically created which both helps to accelerate layout design and eases parasitic estimation (that would not be possible this fast manually). We believe that the combination of templates and generators is a valuable automation approach to meet both flexibility and fast generation speed in a user-driven way.

As next step, we will extend the *Layout API* to support more template types. In addition, we will include more routing options in order to provide higher flexibility and to further extend parasitic trade-off estimations.

Acknowledgment

This work was enabled by the project AnastASICA (grant no. 16ES0990 and 16ES0989) which is funded by BMBF. Many thanks to Christian Albrecht who significantly improved and extended the Generator Creator and many thanks to the reviewers for providing detailed feedback.

References

[1] A. Unutulmaz, G. Dündar and F. V. Fernández, "LDS - A Description Script for Layout Templates," *2011 20th European Conf. on Circuit Theory and Design (ECCTD)*, pp. 857-860, 2011.

[2] R. Castro-López, O. Guerra, E. Roca and F. V. Fernández, "An Integrated Layout-Synthesis Approach for Analog ICs," *IEEE Trans. on Computer-Aided Design of Integrated Circuits and Systems*, vol. 27, no. 7, pp. 1179-1189, 2008.

[3] R. Martins et al., "AIDA: Robust Layout-Aware Synthesis of Analog ICs Including Sizing and Layout Generation," *2015 Int. Conf. on Synthesis, Modeling, Analysis and Simulation Methods and Applications to Circuit Design (SMACD)*, pp. 1-4, 2015.

[4] N. Jangkrajarng et al., "IPRAIL—Intellectual Property Reuse-based Analog IC Layout Automation," *Integration, the VLSI Journal*, vol. 36, pp. 237-262, 2003.

[5] P. Pan et al., "A Fast Prototyping Framework for Analog Layout Migration With Planar Preservation," *IEEE Trans. on Computer-Aided Design of Integrated Circuits and Systems*, vol. 34, no. 9, pp. 1373-1386, 2015.

[6] A. Graupner, R. Jancke and R. Wittmann, "Generator Based Approach for Analog Circuit and Layout Design and Optimization," *2011 Design, Automation & Test in Europe (DATE)*, pp. 1-6, 2011.

[7] D. Payne, "A Review of an Analog Layout Tool Called HiPer DevGen," Nov. 28 2011. [Online]. Available: https://semiwiki.com/x-subscriber/tanner-eda/885-a-review-of-an-analog-layout-tool-called-hiper-devgen/. [Accessed 14 05 2021].

[8] E. Chang et al., "BAG2: A Process-Portable Framework for Generator-Based AMS Circuit Design," *2018 IEEE Custom Integrated Circuits Conf. (CICC)*, pp. 1-8, 2018.

[9] B. Prautsch, U. Eichler, S. Rao, B. Zeugmann, A. Puppala, T. Reich and J. Lienig, "IIP Framework: A Tool for Reuse-Centric Analog Circuit Design," *13th Int. Conf. on Synthesis, Modeling, Analysis and Simulation Methods and Applications to Circuit Design (SMACD 2016)*, pp. 1-4, 2016.

[10] T. Reich, U. Eichler, K.-H. Rooch and R. Buhl, "Design of a 12-bit Cyclic RSD ADC Sensor Interface IC Using the Intelligent Analog IP Library," *ANALOG 2013 - Entwicklung von Analogschaltungen mit CAE-Methoden*, 2013.

[11] J. Scheible und J. Lienig, „Automation of Analog IC Layout – Challenges and Solutions," *Proc. of Int. Symp. on Physical Design (ISPD'15)*, pp. 33-40, 2015.

[12] A. Krinke, M. Mittag, G. Jerke and J. Lienig, "Extended Constraint Management for Analog and Mixed-Signal IC Design," *2013 European Conf. on Circuit Theory and Design (ECCTD)*, pp. 1-4, 2013.

[13] A. Krinke, "Constraint Propagation for Analog and Mixed-Signal Integrated Circuit Design," *Fortschritt-Berichte VDI*, vol. 20, no. 474, Dissertation. 2020.

[14] A. Nassaj, J. Lienig and G. Jerke, "A New Methodology for Constraint-Driven Layout Design of Analog Circuits," *Proc. of the 16th IEEE Int. Conf. on Electronics, Circuits and Systems (ICECS 2009)*, pp. 996-999, 2009.

[15] D. Marolt, Layout Automation in Analog IC Design with Formalized and Nonformalized Expert Knowledge, Dissertation. Stuttgart, 2018.

[16] B. Prautsch, U. Eichler, T. Reich, A. Puppala and J. Lienig, "Abstract Technology Handling for Generator-Based Analog Circuit Design," *GMM-Fachbericht 83, Reliability by Design (ZuE 2015)*, VDE Verlag, pp. 56-61, 2015.

[17] J. Crossley et al., "BAG: A Designer-Oriented Integrated Framework for the Development of AMS Circuit Generators," *Computer-Aided Design (ICCAD), 2013 IEEE/ACM Int. Conf.*, pp. 74-81, 2013.

[18] R. Martins, A. Canelas, N. Lourenço and N. Horta, "On-the-fly Exploration of Placement Templates for Analog IC Layout-Aware Sizing Methodologies," *2016 13th Int. Conf. on Synthesis, Modeling, Analysis and Simulation Methods and Applications to Circuit Design (SMACD)*, pp. 1-4, 2016.

[19] B. Prautsch, U. Hatnik, U. Eichler and J. Lienig, "Template-Driven Analog Layout Generators for Improved Technology Independence," *Proc. of ANALOG 2018*, pp. 156-161, 2018.

[20] B. Prautsch, U. Eichler, T. Reich and J. Lienig, "MESH: Explicit and Flexible Generation of Analog Arrays," *2017 14th Int. Conf. on Synthesis, Modeling, Analysis and Simulation Methods and Applications to Circuit Design (SMACD)*, pp. 1-4, 2017.

[21] A. C. Kammara and A. König, "Absynth: A Comprehensive Approach for Full Front to Back Analog Design Automation," *2018 15th Int. Conf. on Synthesis, Modeling, Analysis and Simulation Methods and Applications to Circuit Design (SMACD)*, pp. 165-168, 2018.

[22] A. Unutulmaz, G. Dündar and F. V. Fernández, "A Template Router," *2011 20th European Conf. on Circuit Theory and Design (ECCTD)*, pp. 334-337, 2011.

[23] L. Sun, Q. Dai, C. Lee and G. Qiao, "The Analysis on the Parasitic Capacitors Effect of the Fully Differential Architecture of SAR ADC," *Applied Mechanics and Materials*, vol. 20–23, pp. 342-345, 2010.

[24] Y. Zhu, U.-F. Chio, H.-G. Wei, S.-W. Sin, S.-P. U and R. P. Martins, "Linearity Analysis on a Series-Split Capacitor Array for High-Speed SAR ADCs," *VLSI Design*, 2010.

[25] M. J. M. Pelgrom, A. C. J. Duinmaijer and A. P. G. Welbers, "Matching Properties of MOS Transistors," *IEEE Journal of Solid-State Circuits*, vol. 24, no. 5, pp. 1433-1439, 1989.

[26] A. Hastings, The Art of Analog Layout, 2. ed., Pearson Prentice Hall, 2006.

[27] J. Chen, P. Luo and C. Wey, "Placement Optimization for Yield Improvement of Switched-Capacitor Analog Integrated Circuits," *IEEE Trans. on Computer-Aided Design of Integrated Circuits and Systems*, vol. 29, no. 2, pp. 313-318, 2010.

[28] M. P. Lin, Y. He, V. W. Hsiao, R. Chang and S. Lee, "Common-Centroid Capacitor Layout Generation Considering Device Matching and Parasitic Minimization," *IEEE Trans. on Computer-Aided Design of Integrated Circuits and Systems*, pp. 991-1002, 2013.

[29] H. Omran, H. Alahmadi and K. N. Salama, "Matching Properties of Femtofarad and Sub-Femtofarad MOM Capacitors," *IEEE Trans. on Circuits and Systems I: Regular Papers*, vol. 63, no. 6, pp. 763-772, 2016.

© VDE VERLAG GMBH · Berlin · Offenbach

Improvement of Simulation-Based Analog Circuit Sizing using Design-Space Transformation

Matthias Schweikardt and Jürgen Scheible
Electronics & Drives, Reutlingen University
Reutlingen, Germany
{matthias.schweikardt, juergen.scheible}@reutlingen-university.de

Abstract—**This paper presents an improvement in usability and integrity of simulation-based analog circuit sizing. Instead of using geometrical sizing parameters (width, length), a transformed design-space, consisting exclusively of electrical parameters (branch currents, efficiencies and speed) is utilized. This design-space is explored more efficiently by optimizers. Moreover, this design-space can be reduced without affecting the quality of the result. The method is illustrated on two application examples, a symmetrical and a miller operational amplifier. Sizing the circuits using the transformed design-space showed significant reduction in required circuit simulations (up to 11x faster), better convergence, without loss in quality.**

Index Terms—**analog circuit sizing, gm over id, look-up table, optimization, downhill-simplex, simulation**

I. INTRODUCTION

Since the complexity of analog integrated circuit design increases much faster than the design productivity, there is urgent need for automation [1]. One major task in the analog development cycle is sizing, i.e. provide valid geometrical parameters (e.g. widths, lengths) to the devices in the circuit. The two main paradigms for automated circuit sizing are to use either knowledge-based (procedural) or optimization-based strategies [2], both are covered extensively in research. In procedural automation, the manual, experience-based sizing strategy of circuit designers is captured in a procedure [3], which promises same quality as manual design and short execution time. Its major disadvantage is, that initial effort must be spent to capture the knowledge, s.t. it can be reused.

The main principle of optimization-based sizing is illustrated in Fig. 1. It consists of an *optimization engine* and an *evaluation engine* (terminology from [4]). The optimization engine provides a candidate $\mathbf{y}$ to the evaluation engine, which will return the performance in $\mathbf{z}_1$ and operating point information in $\mathbf{z}_2$. The candidate vector $\mathbf{y}$ contains the sizing parameters, i.e. geometrical parameters. In contrast to procedural sizing, there is less initial effort needed, but with the disadvantage that optimization can diverge or return physically infeasible results. As a consequence, e.g. constraint observers must be introduced, which will observe the operating points (e.g. $V_{GS} - V_{TH}$) of the devices in the circuit (conf. Fig. 1).

There are two different implementations for the evaluation engine, using equations or a simulator (Fig. 1) [2]. Evaluating

This work was sponsored by the German Federal Ministry of Education and Research (BMBF) within the project PLASMA under Ref. No. 13FH197PX8.

equations is fast, but they cannot model all performances accurately, and one must spend an initial effort in deriving the equations for every new topology. Circuit simulations take time, but the result will be always accurate. Anyhow, every equation-based sized circuit is verified at the end using simulation. Various research of optimization-based sizing was carried out in the past, e.g. on the implications of different algorithms (overview presented in [2]) on the runtime.

At this point it becomes clear, that one challenge in (optimization-based) circuit sizing is to reduce the number of required simulations, without affecting quality or integrity. We refer to the optimization principle in Fig. 1 in the following as *Variant 1*.

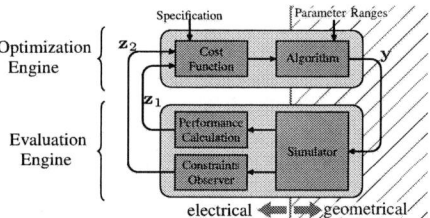

Fig. 1: Classical Simulation-Based Optimization Principle

II. OUR CONTRIBUTION

We propose, that both usability and performance of optimization-based sizing using simulation can be improved using a transformed design-space (concept is shown in Fig. 2).

Instead of using geometrical parameters of transistors in the candidate vector, electrical parameters, in fact *branch currents*, *efficiencies* and *speed* (vector $\mathbf{x}$ in Fig. 2), are used here as parameters. They must be transformed back in geometrical parameters, as input for circuit simulation. As a consequence, the optimization engine operates in an environment, spanned exclusively by electrical parameters, since both inputs $\mathbf{z}_1$, $\mathbf{z}_2$ and output $\mathbf{x}$ comprise electrical characteristics. We refer to this alternative in the following as *Variant 2*. The transformation is described in the following subsections II-A and II-B in detail. We do not focus on passive devices, because this can be covered easily using formulas.

A. Branch Currents

Most analog circuits consist of building blocks, with a dedicated functionality each. Two of them, a differential pair and a

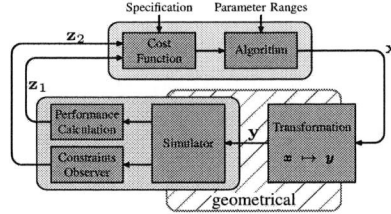

Fig. 2: Optimization Principle using Design-Space Transformation

current mirror, are shown in Fig. 3, more are presented in [5]. In equilibrium, only a dedicated number of distinct terminal

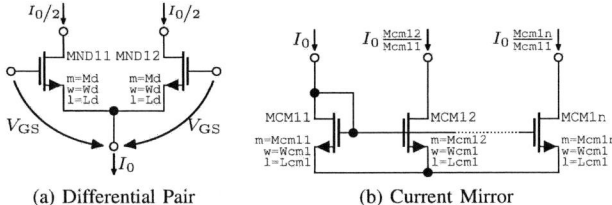

(a) Differential Pair   (b) Current Mirror

Fig. 3: Examples of typical functional blocks

currents can be in contact with the block. Only one distinct current can be specified for the differential pair, the other two terminal currents are directly related (conf. Fig. 3a). For a current mirror, the number of terminal currents corresponds directly to the number of terminals (conf. Fig. 3b). A more complex circuit is build up from these blocks by connecting them. Since the sum of currents at a node is zero, the terminal currents of different blocks depend on each other in a circuit.

As a consequence, a given analog circuit can comprise only a dedicated number of independent choosable currents. We refer them as *branch currents* $I_{B,i}$. These branch currents are one subset of the new candidate vector $\mathbf{x}$. The multipliers in current mirrors can be directly calculated from the connected branch currents (conf. Fig. 3b). When a biasing current mirror is used, typically one branch current (in this paper $I_0 = 3\,\mu\text{A}$) is fixed.

B. Efficiency and Speed

In the last decades, an enhancement to equation-based sizing was introduced, the g_m/I_D-method [1] [7]. The difference to classical equation-based design is, that the characteristics of a single transistor are initially evaluated using circuit simulation and stored in a Look-Up Table (LUT). The contents of the LUT is then inserted in the equations and used for sizing, which provides more accurate results than text-book transistor equations (e.g. square-law).

Here, these LUTs are used for transformation. Two characteristics, representing the *efficiency* (g_m/I_D) of the operating/biasing point and *speed* (unity-gain frequency, f_{UG}) are utilized from the LUT. We chose these parameters, because there is little impact of body-bias on these characteristics. The contents of a LUT for a NMOS transistor is shown in

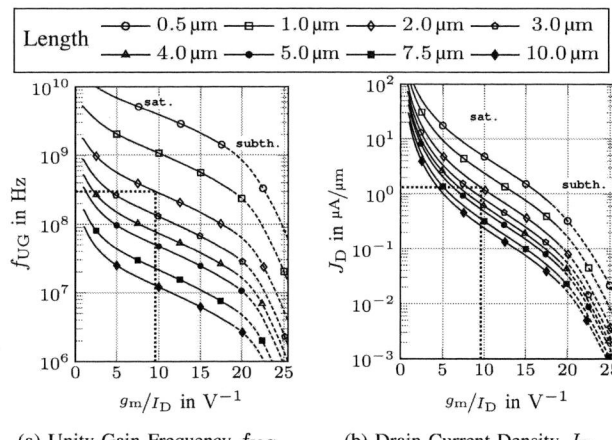

(a) Unity Gain Frequency f_{UG} (b) Drain Current Density J_D

Fig. 4: Extracting Length l and Drain Current Density J_D for a given Unity Gain Frequency $f_{UG} = 300\,\text{MHz}$ and efficiency $g_m/I_D = 9.6\,\text{V}^{-1}$ of a NMOS-Transistor

Fig. 4. The sizing parameters width and length can be directly calculated from a given g_m/I_D and f_{UG}. In most cases, all transistors in a building block share the same efficiency and speed, therefore width and length of every transistor in the block can be determined simultaneously. It should be noted here, that the multipliers M are derived from the branch currents (conf. II-A).

First, the LUT from Fig. 4a is used to extract the length of the transistor (here 2 μm). When a desired point is not available in the LUT, the two neighbouring lengths are interpolated. Secondly, the LUT from Fig. 4b is used to extract the drain current density J_D (here 1.3 μA/μm). From the branch current and drain current density, the width can be calculated directly. The time, needed for evaluating the transformation is negligible ($\ll 0.1\,\%$ of a single circuit simulation).

Improving the sizing using a LUT was carried out in other research [6], but the here presented method has several benefits. In [6], the voltages and (branch) currents in the circuit are used as candidate vector, which has the disadvantage, that the LUT must be regenerated every iteration. Moreover, the here chosen characteristics allow to reduce the design-space (see following section II-C).

C. Design-Space Reduction using Expert Knowledge

In the previous subsections, the utilized transformation was described in detail. One difficulty of every optimization method is, that beforehand suitable parameter ranges must be provided. This can be difficult, because too tight ranges lead to invalid results, while too broad ranges can provoke non-convergent behaviour. Circuit designers tend to have their own strategy when manually designing a circuit, based on empirical knowledge. This expert knowledge typically relates to electrical and not geometrical parameters. Hence, when electrical parameters are used as design-space, this setup can be carried out more easily.

One strategy of experienced designers is to utilize a certain g_m/I_D for sizing of their devices. When electrical parameters are used as candidate vector, this parameter can be specified directly, resulting in a reduced design-space.

Another strategy to reduce the design space is to utilize equations (when available), e.g. from literature, which map electrical parameters in the circuit to performances. Since there is still a simulator in the loop, there is no harm when these equations are inaccurate. In this case, too, the design-space can be reduced (*Variant 3*).

III. APPLICATION EXAMPLES

To prove the validity of the presented method, we consider the sizing of two common operational amplifier topologies. The goal of the optimization is to size Unity Gain Bandwidth (UGBW), Phase-Margin (PM), rising and falling Slew-Rate (SR_r, SR_f), DC-Loopgain (A_0), mismatch-induced offset (V_{off}), Common-Mode Rejection-Ratio (CMRR), Power-Supply Rejection-Ratio (PSRR), the input ($V_{IL} - V_{IH}$) and output voltage range ($V_{OL} - V_{OH}$) for a given load capacitance C_L while minimizing the (estimated) layout area A_e. After optimization, the Figure of Merit FOM = UGBW $\cdot$ C_L/I_{DD} is calculated for comparison, using the supply current I_{DD}.

For optimization the well know Downhill-Simplex Algorithm [8] is used. This simple algorithm was chosen to illustrate that the here presented method is capable to improve optimization inherently, without the need to utilize newer, more advanced algorithms. The cost function is an unweighted sum of the relative deviations from specification. When $V_{DS} - V_{DSAT}$ and/or $V_{GS} - V_{TH}$ of any transistor in the circuit is less than $50\,mV$, the relative deviation is added to the cost function as well (Constraints Observer in Fig. 1).

For classical geometrical parameter-based sizing (*Variant 1*), a range from $1\,\mu m$ to $6\,\mu m$ for the length and a range from $2\,\mu m$ to $20\,\mu m$ for the width is applied. For *Variant 2*, a f_{UG}-range from $20\,MHz$ to $2\,GHz$ and from $5\,V^{-1}$ to $15\,V^{-1}$ in $|g_m/I_D|$ (center of the saturation region) is used individually for each building block.

The initial simplex is generated by randomly sampling the design-space (all Variants). The algorithm is run for 10 different initial simplexes. A maximum of 500 simulations is permitted for each run, terminating the algorithm in case of non-convergence. The best result wrt. number of simulations is shown in the tables I and II.

A. Symmetrical Operational Amplifier

First, the Symmetrical Operational Amplifier (Fig. 5) is considered. The circuit contains 5 functional blocks, 2 of them are identical (MPCM2*). A total of 10 parameters must be found for a valid sizing (Wcm1, Lcm1, Wd, Ld, Wcm2, Lcm2, Mcm11, Mcm12, Mcm21, Mcm22). The parameters Md, Mcm31 and Mcm32 are set to 2, due to symmetry. When geometrical parameters are used for sizing (*Variant 1*) it takes 168 simulations, when electrical parameters are used it takes only 77 simulations (see Tab. I). Here, the design space

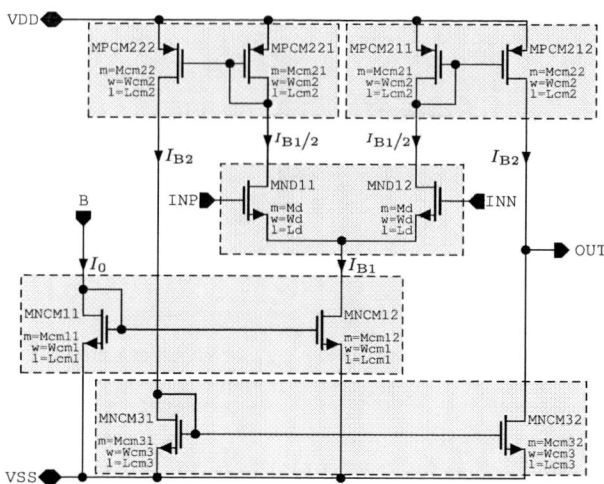

Fig. 5: Symmetrical Operational Amplifier (Body of NMOS is connected to VSS, Body of PMOS is connected to VDD)

consists of g_m/I_D and f_{UG} of the four blocks ($2 \cdot 4$) and the two branch currents I_{B1}, I_{B2}.

From the design, the correlation

$$|SR_i| \approx 2 \cdot \frac{I_{B2}}{C_L} \qquad (1)$$

can be extracted easily. By considering internal capacitances, the branch current $I_{B2} = 12\,\mu A$ can be approximated directly from the specification. Furthermore, the $|g_m/I_D|$ of every transistor in the circuit is set to $10\,V^{-1}$. Using this expert knowledge (conf. II-C), the design space is reduced to 5 electrical parameters (*Variant 3*).

When optimization with this reduced design-space is utilized, the number of needed simulations reduces to 31. All three variants provide a result within specification, but *Variant 2* is best, both with respect to FOM and A_e. The result provided by *Variant 3* is only slightly worse than *Variant 2*.

Another interesting characteristic is, that optimization converged always using *Variant 3*, in seven out of ten cases when using *Variant 2* and only in three out of ten cases when using *Variant 1*.

B. Miller Operational Amplifier

Secondly, the Miller Operational Amplifier (Fig. 6) is considered. The circuit contains 4 functional blocks, but with an additional resistor and capacitor for compensation. A total of 12 parameters must be determined for a proper sizing (Md = Mcm21 = Mcm22 = 2). Here also, *Variant 2* takes much less simulations than *Variant 1* (75 compared to 147), with much better FOM for the same area. For *Variant 2*, the design space consists of the four g_m/I_D and f_{UG} plus I_{B1}, I_{B2}, C_c and R_c.

From [9] the correspondences

$$C_c \approx \frac{I_{B1}}{SR_r} \qquad (2)$$

SMACD / PRIME 2021 | 19 – 22 July 2021, Online Event

TABLE I: Results of the Symmetrical Operational Amplifier

Objective	Unit	Spec.	Results		
			Variant 1	*Variant 2*	*Variant 3*
No. of Parameters (Design-Space)			10	10	5
UGBW	MHz	3.5 - 4	3.92	3.93	3.67
PM	°	> 65	70.5	71.14	71.1
SR_r	V/µs	3.5 - 4	3.53	3.56	3.59
$|SR_f|$	V/µs	3.5 - 4	3.54	3.53	3.59
A_0	dB	≥ 55	57.5	57.35	55.3
$V_{off}(1\sigma)$	mV	≤ 3	3.0	2.9	2.9
CMRR	dB	≥ 80	95.0	95.0	92.8
PSRR	dB	≥ 80	124.3	106.1	102.3
V_{IL}	V	≤ 0.9	0.58	0.56	0.8
V_{IH}	V	≥ 3.2	3.26	3.26	3.27
V_{OL}	V	≤ 0.1	0.08	0.09	0.07
V_{OH}	V	≥ 3.2	3.21	3.22	3.22
A_e	µm^2	≤ 550	533	472	497
C_L	pF		5		
FOM	MHz·pF/mA	-	609	651	607
No. of Simulations			**168**	**77**	**31**
Improvement regarding Variant 1			-	**x 2.18**	**x 5.42**

TABLE II: Results of the Miller Operational Amplifier

Objective	Unit	Spec.	Results		
			Variant 1	*Variant 2*	*Variant 3*
No. of Parameters (Design-Space)			12	12	6
UGBW	MHz	2.5 - 3	2.9	2.88	2.61
PM	°	> 80	81.8	87	81.4
SR_r	V/µs	2.5 - 3	2.74	2.57	2.51
$|SR_f|$	V/µs	2.5 - 3	2.75	2.8	2.57
A_0	dB	≥ 100	107.8	102.4	105.6
$V_{off}(1\sigma)$	mV	≤ 3	1.4	2.9	2.8
CMRR	dB	≥ 80	105.3	122.7	109.9
PSRR	dB	≥ 80	115.1	105.9	105.6
V_{IL}	V	≤ 0.9	0.76	0.85	0.79
V_{IH}	V	≥ 2.7	2.8	2.77	2.72
V_{OL}	V	≤ 0.1	0.04	0.05	0.05
V_{OH}	V	≥ 2.7	2.74	2.73	2.73
A_e	µm^2	≤ 5500	5442	5434	4813
C_L	pF		15		
FOM	MHz·pF/mA	-	449	576	522
No. of Simulations			**147**	**75**	**13**
Improvement regarding Variant 1			-	**x 1.96**	**x 11.3**

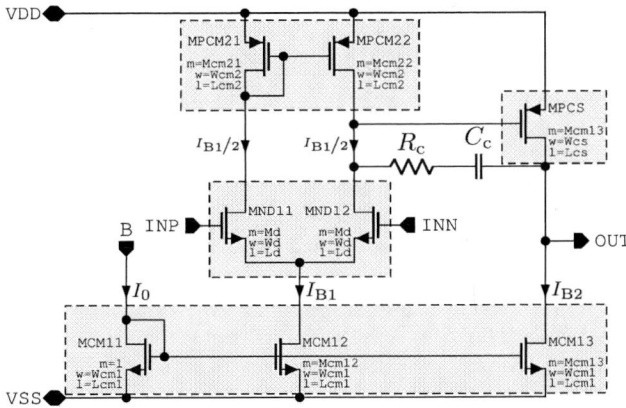

Fig. 6: Miller Operational Amplifier (Body of NMOS is connected to VSS, Body of PMOS is connected to VDD)

and

$$R_c \approx \frac{1}{g_{m,CS}} \cdot \frac{C_c + C_L}{C_c} = \frac{1}{|g_m/I_D|_{CS} \cdot I_{B2}} \cdot \frac{C_c + C_L}{C_c} \quad (3)$$

are extracted for the circuit. When these equations are utilized, the design-space can be reduced by two parameters, because R_c and C_c are calculated directly from other electrical parameters. Again, the $|g_m/I_D|$ of every transistor is set to $10\,\text{V}^{-1}$. Summarized, the design-space reduces to 6 parameters (*Variant 3*). This variant requires only 13 simulations, with slightly worse FOM, but significantly less area A_e. Similar than for the previous example, *Variant 3* converged always, *Variant 2* in 9 out of 10 cases and *Variant 1* only in 7 out of 10 cases.

IV. SUMMARY AND OUTLOOK

The here presented results show, that optimization using a transformed design-space improves convergence, runtime and quality of the resulting circuits. Moreover, the setup of an optimization is simplified, since electrical parameter ranges can specified more easily. Additionally, the design-space can be reduced, with little effect on the quality of the result. This transformation also can improve procedural design [10], since the transformation matches circuit designers way of thinking.

However, there is still room for improvement. First, more advanced optimization algorithms than Downhill-Simplex can be utilized. Moreover, other LUT values (e.g. self-gain) could be used for transformation, which could result in even better results. Since *Variant 3* turned out to converge in every case, it seems reasonable to first fix the $|g_m/I_D|$ of every transistor in optimization and later increase the design-space to fine-tune the circuit.

REFERENCES

[1] Y. Abdelrahman, B. Murmann an H. Omran. (2020). Analog IC Design Using Precomputed Lookup Tables: Challenges and Solutions. IEEE Access. PP. 1-1. 10.1109/ACCESS.2020.3010875.

[2] Nuno Lourenço, Ricardo Martins and Nuno Horta, "Automatic Analog IC Sizing and Optimization Constrained with PVT Corners and Layout Effects",Springer Int. Publishing Switzerland 2017.

[3] D. Stefanovic, M. Kayal, M. Pastre and V. B. Litovski, "Procedural analog design (PAD) tool," 4th Int. Symposium on Quality Electronic Design, 2003. Proceedings., San Jose, CA, USA, 2003, pp. 313-318.

[4] J. Scheible and J. Lienig, "Automation of Analog IC Layout – Challenges and Solutions"; ISPD 2015, Monterey, CA, USA, pp. 33-40.

[5] T. Massier, H. Graeb and U. Schlichtmann, "The Sizing Rules Method for CMOS and Bipolar Analog Integrated Circuit Synthesis," in IEEE Transactions on CAD of Integrated Circuits and Systems, vol. 27, no. 12, pp. 2209-2222, Dec. 2008.

[6] Cheng-Wu Lin, Pin-Dai Sue, Ya-Ting Shyu and Soon-Jyh Chang, "A bias-driven approach for automated design of operational amplifiers," 2009 Int. Symposium on VLSI Design, Automation and Test, Hsinchu, Taiwan, 2009, pp. 118-121.

[7] P. Jespers and B. Murmann, Systematic design of analog CMOS circuits, Cambridge University Press, 2017.

[8] J. A. Nelder and R. Mead, "A Simplex Method for Function Minimization," The Comp. Journ., Volume 7, Issue 4, Jan. 1965, pp. 308–313

[9] P. E. Allen and D. Holberg, "CMOS Analog Circuit Design," Third Edition, Oxford University Press, 2012

[10] Y. Uhlmann et al., "Machine Learning Based Procedural Circuit Sizing and DC Operating Point Prediction," Int. Conference on Synthesis, Modeling, Analysis and Simulation Methods and Applications to Circuit Design (SMACD), 2021

© VDE VERLAG GMBH · Berlin · Offenbach

Machine Learning Based Procedural Circuit Sizing and DC Operating Point Prediction

Yannick Uhlmann[*] , Michael Essich[†] , Matthias Schweikardt[*] , Jürgen Scheible[*] and Cristóbal Curio[†]

[*] Electronics & Drives, Reutlingen University

Email: {yannick.uhlmann, matthias.schweikardt, juergen.scheible}@reutlingen-university.de

[†] Cognitive Systems, Reutlingen University

Email: {michael.essich, cristobal.curio}@reutlingen-university.de

Abstract—This paper presents a machine learning powered, procedural sizing methodology based on pre-computed look-up tables containing operating point characteristics of primitive devices. Several Neural Networks are trained for 90 nm and 45 nm technologies, mapping different electrical parameters to the corresponding dimensions of a primitive device. This transforms the geometric sizing problem into the domain of circuit design experts, where the desired electrical characteristics are now inputs to the model. Analog building blocks or entire circuits are expressed as a sequence of model evaluations, capturing the sizing strategy and intention of the designer in a procedure, which is reusable across different technology nodes. The methodology is employed for the sizing of two operational amplifiers, and evaluated for two technology nodes, showing the versatility and efficiency of this approach.

Index Terms—machine learning, neural networks, gm over id, analog ic sizing

I. INTRODUCTION

Despite previous research regarding the design automation of analog Integrated Circuits (ICs), in terms of topology synthesis [1], design- [2] or yield-optimization [3] this domain has still not caught up with its digital counterpart [4].

Recently, the $g_\mathrm{m}/i_\mathrm{d}$-method [5], based on pre-computed Look-Up Tables (LUTs), has been gaining popularity among analog circuit designers. This approach combines the accuracy of simulation based methods with the execution time of equation-based methods [6]. Additionally, sizing strategies are made reusable by capturing them in executable form.

Formalizing expert knowledge lays the foundation for *procedural* automation approaches, which have been successfully employed in the physical design domain before [4], [7]. These procedures are primarily used to generate layouts for smaller, frequently reoccurring circuit structures. Recently, similar approaches have made advances in the circuit design domain as well [8]. The work presented here, aims to extend this approach with Machine Learning (ML) models, based on a remapping of the LUTs, for the purpose of procedurally describing a sizing strategy in terms of electrical characteristics.

Related work shows that Artificial Neural Networks (ANNs) can model the behaviour of active devices around the operating point [9]. Furthermore, Multilayer Perceptrons (MLPs) have

This work was sponsored by the German Federal Ministry of Education and Research (BMBF) within the project PLASMA under Ref. No. 13FH197PX8.

been employed to map building block specific performances to device dimensions [10], [11].

Similar to equation based design approaches, this work focuses on the transistor level, where sizing parameters *propagate* from a *reference device* to functionally related devices [12]. For smaller technology nodes however, where simplified equations break down [5], ML models trained on simulation data remain accurate. Therefore, previously conceived sizing procedures can be seamlessly reused across technology nodes by exchanging the correspondingly trained device models.

The basis for this approach are LUTs, characterising primitive devices around the operating point. Section II describes how data is sampled and how it is used to train Deep Neural Networks (DNNs) for different mappings and technologies. These mappings and the propagation of sizing parameters to functionally related devices are detailed in Section III. Section IV shows the process of composing a sizing procedure and the work flow for reuse with different technology models. Finally, Section V concludes this contribution.

II. DATA SAMPLING AND TRAINING

As mentioned in the previous Section, the foundation for this approach are LUTs containing the operating point parameters of a transistor [5]. ANNs can be trained to learn a mapping of this data, approximating the simulation model. However, previous work has shown, that it can be difficult for ANN predictions to converge over the entire input space [10]. Instead of binning the input space and training a correction model, in this approach, the dataset is sampled [13] and transformed [14], changing the distribution of the outputs, as shown in Fig. 1, such that training a single model is sufficient.

All of the mappings presented in this work are trained using the same network architecture and training algorithm, which was found by iteratively extending the architectures of previous works [10], [11]. The result is a DNN with 5 inputs, 7 fully connected hidden layers and 7 outputs, where rectified linear unit (ReLU) [15] is used as activation for all hidden layers. These hidden layers consist of 128, 256, 512, 1024, 512, 256 and 128 neurons respectively. The Adam optimizer [16] is used to minimize the Mean Squared Error (MSE) between the LUT and predictions. Training on a dataset of 4 million samples for 100 epochs with a batch size of 2000,

© VDE VERLAG GMBH · Berlin · Offenbach

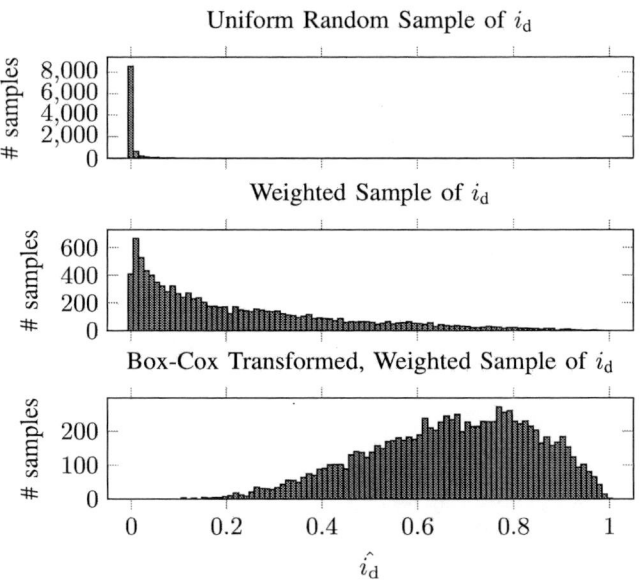

Fig. 1: Histograms of scaled drain current $\hat{i}_d \in [0, 1]$, showing how the distribution changes for different sampling techniques.

takes ca. 15 min on an NVIDIA® TITAN RTX™. Fig. 2 shows how well the predictions agree with the LUT afterwards.

The training data is obtained by characterizing 45 nm and 90 nm bsim4v5 [17] Predictive Technology Model (PTM) [18] devices over a range of *terminal voltages* (V_{ds}, V_{ds}, V_{ds}) and *geometries* (W, L).

III. FUNCTION MAPPINGS FOR CIRCUIT SIZING

The equations in the underlying simulation model map the terminal voltages and device geometry to the operating point parameters. While there are additional, model internal parameters, such as charge carrier mobility, the analysis of process variations is beyond the scope of this work.

When sizing a circuit however, most of the input parameters of this mapping are not known, while some of the outputs are either known a priori or desired for certain design objectives. This would require function optimization to find *geometric* parameters, which result in desired *electrical* behaviour. The search space, spanned by possible combinations of widths and lengths for multiple devices in a circuit is considerable and not easily constrained. It would be much more intuitive for a circuit designer however, to constrain the *speed* (f_{ug}) of a device, or outright fix the *efficiency* (g_m/i_d) to a specific value, greatly reducing the search space. Rearranging these equations would therefore transform the problem into the domain of analog circuit design experts [19]. However, instead of analytically inverting the equations of the transistor model, in this approach, non-linear regression models are trained to *learn* and approximate the mappings shown in Tab. I.

Related work [20] shows that breaking a circuit into *building blocks*, such as the *current mirror* or the *differential pair* which are highlighted in Fig. 3 and Fig. 4, reduces the design space

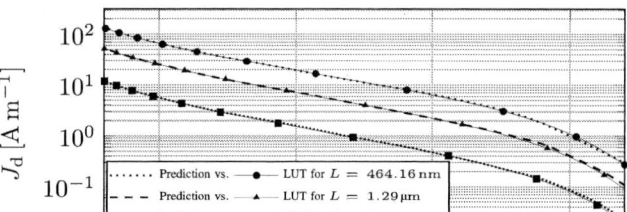

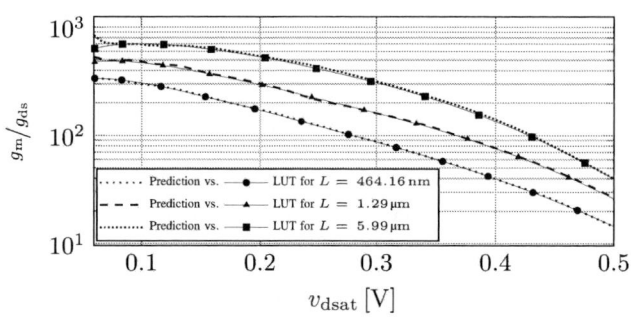

Fig. 2: The 90 nm NMOS models compared to the LUT for length $L \in \{464.16\,\text{nm}, 1.29\,\mu\text{m}, 5.99\,\mu\text{m}\}$.

TABLE I: Function Mappings represented by Trained DNNs.

DNN	Mapping
ν_L :	$[L, v_{dsat}, i_d, V_{ds}, V_{bs}]^\top \mapsto [J_d, g_m, g_{ds}, V_{gs}, v_{th}, (g_m/i_d), f_{ug}]^\top$
γ_L :	$[L, (g_m/i_d), i_d, V_{ds}, V_{bs}]^\top \mapsto [J_d, g_m, g_{ds}, V_{gs}, v_{th}, v_{dsat}, f_{ug}]^\top$
ν_f :	$[f_{ug}, v_{dsat}, i_d, V_{ds}, V_{bs}]^\top \mapsto [J_d, g_m, g_{ds}, V_{gs}, v_{th}, (g_m/i_d), L]^\top$
γ_f :	$[f_{ug}, (g_m/i_d), i_d, V_{ds}, V_{bs}]^\top \mapsto [J_d, g_m, g_{ds}, V_{gs}, v_{th}, v_{dsat}, L]^\top$

even further. While DNNs can be trained to approximate a mapping from building block specific performances to individual sizing parameters for each device therein [11], this work focuses on an alternative approach. It is shown, that sizing a single *reference* device and simply *propagating* the sizing parameters to related devices within a building block is sufficient [12]. This only requires training data for a single device and implicitly ensures functionally related devices are matched. Hence, the sizing of a building block is reduced to a single DNN evaluation, which will be elaborated on during the example in the following Section.

IV. PROCEDURAL SIZING EXAMPLE

For demonstrating the viability of the methodology presented in this work, the symmetrical operational amplifier shown in Fig. 3, as well as the miller operational amplifier seen in Fig. 4, are sized to meet a specification. The strategy for both is initially formulated using the same models trained for the 90 nm technology, and captured in an executable procedure. Subsequently, this is *reused* with the models for the

TABLE II: Specification for both Sizing Examples.

Parameter	V_{DD}	$V_{\mathrm{in,cm}}$	$V_{\mathrm{out,cm}}$	I_{B0}	C_{L}
Specification	1.20 V	600 mV	600 mV	10 µA	10 pF

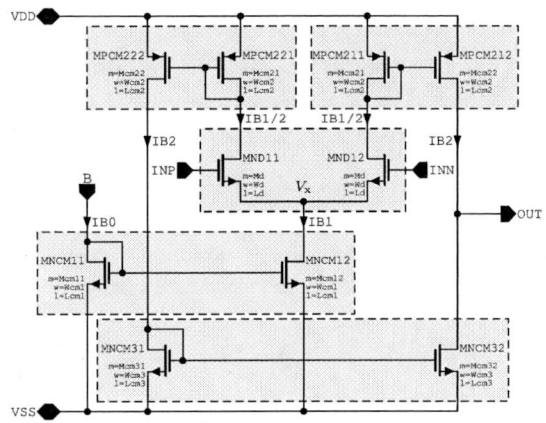

Fig. 3: Symmetrical operational amplifier, where bulk potentials of NMOS and PMOS are VSS and VDD respectively.

45 nm technology, sizing the circuit for the same specification and achieving similar performance.

A. Sizing Procedure for a Symmetrical Operational Amplifier

Analytically, the *gain* (A_0) and *cutoff frequency* (f_0) are approximated with (1) and (2) commonly found in literature [21], where the load (C_{L}) is part of the testbench, connected to OUT.

$$A_{0,\mathrm{sym}} \approx M \cdot \frac{g_{\mathrm{m,d}}}{g_{\mathrm{ds,cm22}} + g_{\mathrm{ds,cm32}}} \qquad (1)$$

$$f_{0,\mathrm{sym}} \approx \frac{g_{\mathrm{ds,cm22}} + g_{\mathrm{ds,cm32}}}{2 \cdot \pi \cdot C_{\mathrm{L}}} \qquad (2)$$

Beforehand, prior knowledge is considered by observing the specification, given in Tab. II, and deciding on a biasing current $I_{\mathrm{B1}} = I_{\mathrm{B0}}/2$. Additionally, the ratio $M = 1:4$ of the PMOS current mirrors MPCM2 is defined. Usually, this is chosen to balance power consumption and phase margin. Since this has to be analyzed separately by simulation, starting values $M_{\mathrm{cm21}} = 1$ and $M_{\mathrm{cm22}} = 4$ are selected.

Thus, a sizing strategy is expressed by sequentially selecting devices from the circuit and choosing a corresponding model from Tab. I, depending on prior knowledge and previously obtained parameters. This procedure, which is reminiscent of the manual work flow with analytical equations, is illustrated in detail hereafter and captured in executable form.

First, since the common mode output voltage $V_{\mathrm{out,cm}}$ is known, current mirror MNCM3 is considered with (3).

$$\nu_{f,\mathrm{cm32}}(f_{\mathrm{ug,cm3}}, v_{\mathrm{dsat,cm3}}, I_{\mathrm{B2}}V_{\mathrm{out,cm}}, 0.0)$$
$$\Rightarrow [J_{\mathrm{d,cm32}}, V_{\mathrm{gs,cm3}}, g_{\mathrm{ds,cm32}}, \dots]^{\mathsf{T}} \quad (3)$$

This defines the width $W_{\mathrm{cm3}} = I_{\mathrm{B2}}/J_{\mathrm{d,cm32}}$ in terms of $f_{\mathrm{ug,cm3}}$, and constitutes $M_{\mathrm{cm31}} = M_{\mathrm{cm32}} = 2$ for a $1:1$ ratio.

Next, the current mirrors MPCM21 and MPCM22 are considered with the single model evaluation given in (4), thereby sizing them identically in terms of the same $f_{\mathrm{ug,cm2}}$.

$$\nu_{f,\mathrm{cm2}}(f_{\mathrm{ug,cm2}}, v_{\mathrm{dsat,cm2}}, I_{\mathrm{B2}}, (V_{\mathrm{DD}} - V_{\mathrm{out,cm}}), 0.0)$$
$$\Rightarrow [J_{\mathrm{d,cm22}}, V_{\mathrm{gs,cm2}}, g_{\mathrm{ds,cm22}}, \dots]^{\mathsf{T}}, \quad (4)$$

where $I_{\mathrm{B2}} = M_{\mathrm{cm22}} \cdot I_{\mathrm{B1}}/2$. Let $V_{\mathrm{y}} = V_{\mathrm{DD}} - V_{\mathrm{gs,cm2}}$, then the differential pair MND1 is defined in terms of $f_{\mathrm{ug,d}}$ with (5).

$$\gamma_{f,\mathrm{d}}(f_{\mathrm{ug,d}}, (g_{\mathrm{m}}/i_{\mathrm{d}})_{\mathrm{d}}, (I_{\mathrm{B1}}/2), (V_{\mathrm{y}} - V_{\mathrm{x}}), -V_{\mathrm{x}})$$
$$\Rightarrow [J_{\mathrm{d,d}}, V_{\mathrm{gs,d}}, g_{\mathrm{m,d}}, \dots]^{\mathsf{T}}, \quad (5)$$

where $V_{\mathrm{x}} = \arg\min_{V_{\mathrm{x}}} |V_{\mathrm{gs,d}} - |V_{\mathrm{in,cm}} - V_{\mathrm{x}}||$. Finally, (6) defines the sizing for MNCM1 in terms of the length L_{cm1}.

$$\nu_{L,\mathrm{cm1}}(L_{\mathrm{cm1}}, v_{\mathrm{dsat,cm1}}, I_{\mathrm{B1}}, V_{\mathrm{x}}, 0.0)$$
$$\Rightarrow [J_{\mathrm{d,cm12}}, V_{\mathrm{gs,cm1}}, \dots]^{\mathsf{T}} \quad (6)$$

Special care is taken with this biasing current mirror, since it might not be in saturation. Therefore, W_{cm1} has to be increased, accounting for lower V_{x}.

Consequently, the sizing procedure p_{sym} for this entire circuit, consisting of 10 devices, is wholly expressed by a sequence of 4 DNN evaluations, and defines the strategy in terms of electrical characteristics and L_{cm1}, according to (7).

$$p_{\mathrm{sym}} : \begin{bmatrix} f_{\mathrm{ug,d}} \\ L_{\mathrm{cm1}} \\ f_{\mathrm{ug,cm2}} \\ f_{\mathrm{ug,cm3}} \\ (g_{\mathrm{m}}/i_{\mathrm{d}})_{\mathrm{d}} \\ v_{\mathrm{dsat,cm1}} \\ v_{\mathrm{dsat,cm2}} \\ v_{\mathrm{dsat,cm3}} \end{bmatrix} \mapsto \begin{bmatrix} A_{0,\mathrm{sym}} \\ f_{0,\mathrm{sym}} \end{bmatrix} \quad (7)$$

For this example, a target gain of $A_{0,\mathrm{t}} = 48\,\mathrm{dB}$ is chosen. Then, $(g_{\mathrm{m}}/i_{\mathrm{d}})_{\mathrm{d}} = 15\,\mathrm{V}^{-1}$, $v_{\mathrm{dsat,cm1}} = 0.2\,\mathrm{V}$, $v_{\mathrm{dsat,cm2}} = 0.1\,\mathrm{V}$, $v_{\mathrm{dsat,cm3}} = 0.2\,\mathrm{V}$ and $L_{\mathrm{cm1}} = 800\,\mathrm{nm}$ are fixed, further reducing the problem to (8).

$$\arg\min_{f_{\mathrm{sym}}} |A_{0,\mathrm{t}} - A_{0,\mathrm{sym}}| \quad (8)$$

Wherein $f_{\mathrm{sym}} = [f_{\mathrm{ug,d}}, f_{\mathrm{ug,cm2}}, f_{\mathrm{ug,cm3}}]^{\mathsf{T}}$, for which an optimizer quickly finds a solution. The obtained sizing is simulated and the results are compared to the prediction in Tab. III.

Sizing the circuit with the same procedure in another technology, only requires the correspondingly trained ML models. For the 45 nm technology, the only *geometric* parameter $L_{\mathrm{cm1}} = 500\,\mathrm{nm}$ is adjusted, while *all* other parameters, including the found f_{sym}, remain identical. Simulating the results yields comparable performances, as shown in Tab. III.

B. Sizing Procedure for a Miller Operational Amplifier

The second example is sized similarly, starting from the output with MPCS. Then, let $V_{\mathrm{ds,cm2}} = V_{\mathrm{gs,cs}}$, for sizing MPCM2. Systematic matching is considered, by ensuring

TABLE III: Prediction vs. Simulation Results for the Symmetrical Operational Amplifier in 90 nm and 45 nm Technology.

Technology	90 nm		45 nm	
Parameter	Prediction	Simulation	Prediction	Simulation
A_0	48.00 dB	48.65 dB	48.01 dB	48.26 dB
f_0	9.35 kHz	8.90 kHz	9.39 kHz	8.08 kHz
$(g_m/i_d)_d$	15.00 V^{-1}	15.15 V^{-1}	15.00 V^{-1}	15.38 V^{-1}
$v_{\mathrm{dsat,cm1}}$	100 mV	96.22 mV	100 mV	88.92 mV
$v_{\mathrm{dsat,cm2}}$	200 mV	195.62 mV	200 mV	188.10 mV
$v_{\mathrm{dsat,cm3}}$	200 mV	204.75 mV	200 mV	188.14 mV

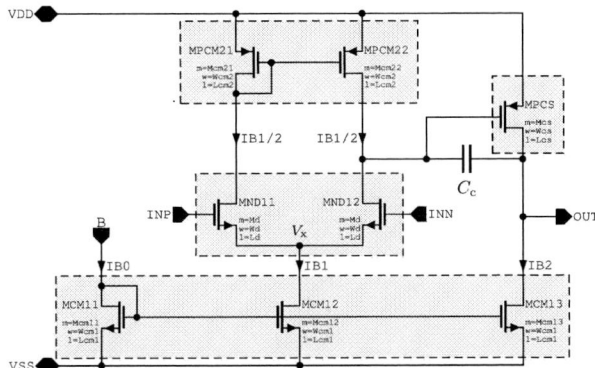

Fig. 4: Miller operational amplifier, with bulk potentials of NMOS and PMOS being `VSS` and `VDD` respectively.

$V_{\mathrm{ds,cm21}} = V_{\mathrm{ds,cm22}}$ [21]. V_{x} is obtained as in the previous example, for sizing `MND1` and subsequently, `MCM1`.

For this example a target gain $A_{0,\mathrm{t}} = 84$ dB is chosen. Initially, the procedure is composed with the 90 nm models and then executed again with the same inputs, but the 45 nm models. The results and comparisons are shown in Tab. IV.

V. CONCLUSION

A ML based, procedural approach for circuit sizing and DC operating point approximation is presented. The formulated procedures express sizing strategies in terms of electrical parameters, transforming this geometric problem into the domain of circuit design experts [19]. With an execution time and work flow comparable to equation based sizing methods, the results reasonably approximate the simulation, even for smaller technology nodes. While the initial setup involving the generation of training data, training the models and composing the procedure demands considerable effort, the reuse value is significant, especially for technology migrations.

TABLE IV: Prediction vs. Simulation Results for the Miller Operational Amplifier in 90 nm and 45 nm Technology.

Technology	90 nm		45 nm	
Parameter	Prediction	Simulation	Prediction	Simulation
A_0	84.00 dB	84.62 dB	84.19 dB	84.39 dB
f_0	233.87 Hz	225.09 Hz	249.66 Hz	202.10 Hz
$(g_m/i_d)_d$	15.00 V^{-1}	15.26 V^{-1}	15.00 V^{-1}	15.47 V^{-1}
$v_{\mathrm{dsat,cm2,cs}}$	200 mV	195.14 mV	200 mV	185.16 mV
$v_{\mathrm{dsat,cm1}}$	100 mV	96.42 mV	100 mV	89.87 mV

REFERENCES

[1] M. G. R. Degrauwe, O. Nys, E. Dijkstra, J. Rijmenants, S. Bitz, B. L. A. G. Goffart, E. A. Vittoz, S. Cserveny, C. Meixenberger, G. van der Stappen, and H. J. Oguey, "Idac: an interactive design tool for analog cmos circuits," *IEEE Journal of Solid-State Circuits*, vol. 22, no. 6, pp. 1106–1116, 1987.

[2] W. Nye, D. C. Riley, A. Sangiovanni-Vincentelli, and A. L. Tits, "Delight.spice: an optimization-based system for the design of integrated circuits," *IEEE Transactions on Computer-Aided Design of Integrated Circuits and Systems*, vol. 7, no. 4, pp. 501–519, 1988.

[3] K. Antreich and R. Koblitz, "Design centering by yield prediction," *IEEE Transactions on Circuits and Systems*, vol. 29, no. 2, pp. 88–96, 1982.

[4] J. Scheible and J. Lienig, "Automation of analog ic layout: Challenges and solutions," in *Proceedings of the 2015 Symposium on International Symposium on Physical Design*, ser. ISPD '15. New York, NY, USA: Association for Computing Machinery, 2015, p. 33–40. [Online]. Available: https://doi.org/10.1145/2717764.2717781

[5] P. G. A. Jespers and B. Murmann, *Systematic Design of Analog CMOS Circuits: Using Pre-Computed Lookup Tables*. Cambridge University Press, 2017.

[6] F. Silveira, D. Flandre, and P. G. A. Jespers, "A gm/id based methodology for the design of cmos analog circuits and its application to the synthesis of a silicon-on-insulator micropower ota," *IEEE Journal of Solid-State Circuits*, vol. 31, no. 9, pp. 1314–1319, 1996.

[7] D. Marolt, J. Scheible, and G. Jerke, "A practical layout module pcell concept for analog ic design," *Proceedings of CDNLive EMEA*, 2013.

[8] M. Schweikardt, Y. Uhlmann, F. Leber, J. Scheible, and H. Habal, "A generic procedural generator for sizing of analog integrated circuits," in *2019 15th Conference on Ph.D Research in Microelectronics and Electronics (PRIME)*, 2019, pp. 17–20.

[9] J. Xu and D. E. Root, "Artificial neural networks for compound semiconductor device modeling and characterization," in *IEEE Compound Semiconductor Integrated Circuit Symposium (CSICS)*, 2017, pp. 1–4.

[10] K. Mendhurwar, H. Sundani, P. Aggarwal, R. Raut, and V. Devabhaktuni, "A new approach to sizing analog cmos building blocks using precompiled neural network models," *Analog Integrated Circuits and Signal Processing - ANALOG INTEGR CIRCUIT SIGNAL*, vol. 70, 03 2012.

[11] N. Kahraman and T. Yildirim, "Technology independent circuit sizing for fundamental analog circuits using artificial neural networks," in *PRIME - 2008 PhD Research in Microelectronics and Electronics, Proceedings*, 06 2008, pp. 1 – 4.

[12] R. Iskander, M.-M. Louërat, and A. Kaiser, "Automatic dc operating point computation and design plan generation for analog ips," in *Analog Integrated Circuits and Signal Processing*, vol. 56, 2008, pp. 93–105. [Online]. Available: https://doi.org/10.1007/s10470-007-9075-3

[13] P. Efraimidis and P. Spirakis, *Weighted Random Sampling*. Boston, MA: Springer US, 2008, pp. 1024–1027.

[14] G. Box and D. Cox, "An analysis of transformations," in *Journal of the Royal Statistical Society. Series B (Methodological)*, vol. 26(2), 1964, pp. 211–252. [Online]. Available: http://www.jstor.org/stable/2984418

[15] P. Ramachandran, B. Zoph, and Q. V. Le, "Searching for activation functions," 2017. [Online]. Available: https://arxiv.org/abs/1710.05941

[16] D. P. Kingma and J. Ba, "Adam: A method for stochastic optimization," in *3rd International Conference on Learning Representations, ICLR 2015, San Diego, CA, USA, May 7-9, 2015, Conference Track Proceedings*, 2015. [Online]. Available: http://arxiv.org/abs/1412.6980

[17] W. Liu, K. Cao, X. Jin, and C. Hu, "Bsim 4.0.0 technical notes," EECS Department, University of California, Berkeley, Tech. Rep. UCB/ERL M00/39, Aug 2000. [Online]. Available: http://www2.eecs.berkeley.edu/Pubs/TechRpts/2000/3863.html

[18] Y. K. Cao, "What is predictive technology model (ptm)?" *SIGDA Newsl.*, vol. 39, no. 3, p. 1, Mar. 2009. [Online]. Available: https://doi.org/10.1145/1862891.1862892

[19] M. Schweikardt and J. Scheible, "Improvement of simulation-based analog circuit sizing using design-space transformation," in *2021 17th International Conference on Synthesis, Modeling, Analysis and Simulation Methods and Applications to Circuit Design (SMACD)*, 2021.

[20] H. Graeb, S. Zizala, J. Eckmueller, and K. Antreich, "The sizing rules method for analog integrated circuit design," in *IEEE/ACM International Conference on Computer Aided Design. ICCAD 2001. IEEE/ACM Digest of Technical Papers (Cat. No.01CH37281)*, 2001, pp. 343–349.

[21] P. E. Allen and D. R. Holberg, *CMOS Analog Circuit Design*. Oxford University Press, USA, 1995-06.

SMACD / PRIME 2021 | 19 – 22 July 2021, Online Event

Surrogate-Assisted Multi-objective Differential Evolution based on Gaussian Process for Analog Circuit Synthesis

Sen Yin, Wenfei Hu, Ruitao Wang, Zhikai Wang, Jian Zhang, Yan Wang
Institute of Microelectronics, Tsinghua University, China.
yins18@mails.tsinghua.edu.cn, wangy46@tsinghua.edu.cn

Abstract—In this paper, a surrogate-assisted multi-objective differential evolution based on Gaussian process is proposed for analog circuit synthesis. NSGA-II-DE is used as the multi-objective optimizer and online Gaussian process surrogate model is constructed to prescreen the best two trial vectors according to non-dominated sorting and modified crowding distance. Only two instead of multiple designs are simulated by HSPICE in one generation. The efficiency of proposed approach is verified on two real-world circuits. Compared with two state-of-the-art multi-objective evolutionary algorithms, our method can achieve better Pareto front (lowest I_H^-) with much less number of simulations.

Keywords—**Multi-objective evolutionary algorithm, Gaussian process, differential evolution, analog circuit synthesis**

I. INTRODUCTION

The automatic synthesis of analog circuits remains in the immature stage compared with digital circuits. Traditional methodology focuses on high-level system specification tools, where the performance are distributed among the circuits [1] by behavior models constituting the system. The inappropriate distribution of performances for circuits challenges the designers or leads to over-design. The bottom-up design methodology [2] has received much attention in academic community in recent years, in which the complete system is decomposed into several hierarchical levels. Each low-level circuit design space is reduced to Pareto front (PF) by multi-objective optimization algorithms and the obtained PFs are directly passed to the upper level via a set of coordinates. During the optimization of system level, all low-level modules are available and there is no need to redesign.

Multi-objective optimization is of vital importance in bottom-up design methodology. Most existing multi-objective sizing methods can be classified into two categories: model-based and simulation-based approaches. Model-based methods [3] firstly sample points from the simulation of circuits, then construct models for each circuit performance. Finally, multi-objective evolutionary algorithms are followed to solve the problem via computationally cheap models. Circuit simulation is directly driven by multi-objective evolutionary algorithms such as NSGA-II [4] and MOEA/D [5] in simulation-based methods. NSGA-II redefines the ranking rules by non-dominated sorting and crowding distance to handle multi-objectives. MOEA/D decomposes the problem into several single-objective problems. A large number of simulations is needed in simulation-based methods and the accuracy of model determines the optimization results in model-based methods. To enhance the efficiency of simulation-based methods while maintaining accuracy, surrogate-assisted evolutionary algorithms (SAEA) are adopted. Gaussian process (GP) is used to substitute the simulation in part in [6]. Bayesian optimization is extended to multi-objective optimization in [7].

In this paper, a surrogate-assisted multi-objective differential evolution (SAMODE) is proposed for analog circuit synthesis. SAMODE is a novel SAEA where GP is combined with the framework of NSGA-II-DE [8], which aims to prescreen the best two trial vectors according to non-dominated sorting and modified crowding distance. Only two instead of multiple designs are simulated in each generation. The modification of crowding distance concerns more about boundary solutions, preventing the stagnation of the population. The main contribution of the work are: (1) the development of the new framework in which any accurate model can be introduced into the multi-objective optimization, (2) the modification of crowding distance, guaranteeing the reliability of optimization results, (3) much less prediction times of GP compared with Bayesian optimization in [7], and (4) the development of SAMODE, showing the combined advantages of high efficiency.

The rest of paper is organized as follows. In Section II, the basic technique used in this paper is presented. In Section III, the proposed SAMODE is elaborated. In Section IV, the efficiency of proposed algorithm is verified by two real-world circuits. We conclude the paper in Section V.

II. BACKGROUNDS

A. Gaussian Process Regression

Gaussian process regression [9] is a non-parametric and interpolate model. The kernel function is of vital importance in GP, which is defined as (1) in this paper.

$$k(\boldsymbol{x_i}, \boldsymbol{x_j}) = \sigma_f^2 exp(-\frac{1}{2}(\boldsymbol{x_i} - \boldsymbol{x_j})^T \Lambda^{-1}(\boldsymbol{x_i} - \boldsymbol{x_j})) \quad (1)$$

where Λ and σ_f are hyper-parameters of the kernel function.

GP is a collection of random variables, any finite number of which have a joint Gaussian distribution. Given training set $\{\boldsymbol{X}, \boldsymbol{y}\}$, where $\boldsymbol{X} = \{\boldsymbol{x_1}, \boldsymbol{x_2}, ..., \boldsymbol{x_N}\}$ and $\boldsymbol{y} = \{y_1, ..., y_N\}$, The log likelihood of training set can be expressed as

$$\log p(\boldsymbol{y}|\boldsymbol{X}, \theta) \propto -\frac{1}{2}(\boldsymbol{y} - \mu_0)^T K_\theta^{-1}(\boldsymbol{y} - \mu_0) - \frac{1}{2}\log |K_\theta| \quad (2)$$

© VDE VERLAG GMBH · Berlin · Offenbach

where μ_0 is the prior mean function, $K_\theta(i,j) = k(\boldsymbol{x_i}, \boldsymbol{x_j})$ and k is the kernel function. θ is denoted to the hyper-parameters of kernel function, which is determined by maximize the log likelihood estimation (MLE).

Given new data points $\boldsymbol{x^*}$, the posterior distribution of $f(\boldsymbol{x^*})$ can be calculated as follows

$$f(\boldsymbol{x^*}) \sim N(\mu(\boldsymbol{x^*}), \sigma^2(\boldsymbol{x^*})) \qquad (3)$$

$$\begin{cases} \mu(\boldsymbol{x^*}) = \mu_0 + k(\boldsymbol{x^*}, \boldsymbol{X})K_\theta^{-1}(\boldsymbol{y} - \mu_0) \\ \sigma^2(\boldsymbol{x^*}) = k(\boldsymbol{x^*}, \boldsymbol{x^*}) - k(\boldsymbol{x^*}, \boldsymbol{X})K_\theta^{-1}k(\boldsymbol{X}, \boldsymbol{x^*}) \end{cases} \qquad (4)$$

The constant mean function and squared exponential covariance function are used in this paper. The mean function values $\mu(\boldsymbol{x^*})$ are the predictive values of GP.

B. NSGA-II-DE

NSGA-II [4] is one of the most popular multi-objective evolutionary algorithms, in which offspring are generated following three operators, tournament selection, simulated binary crossover (SBX) and polynomial mutation. Differential evolution (DE) operator often outperforms other operators in single objective optimization. In light of this, NSGA-II-DE [8] was proposed by Hui Li et al, which is almost the same as NSGA-II except that the SBX operator is replaced by DE operator. Several modifications are made on NSGA-II-DE in this work, focusing on ranking rules. The DE operator used in this work is defined as listed in (5).

$$v_i = x_i + F \cdot (x_{r1} - x_{r2}) \qquad (5)$$

where x_i is the i-th individual of the population, $r1$, $r2$ are distinct integers uniformly chosen from the set $\{1, 2, ..., N\}\backslash\{i\}$, $F \in (0,2]$ is a control parameter called scaling factor. F is set to a rand number uniformly chosen from $(0.5, 1)$ in this work.

III. THE PROPOSED SAMODE APPROACH

The details of proposed SAMODE method are demonstrated in Algorithm 1. λ stands for the size of evolving population. $nfev$ and $maxnfev$ are denoted to the number and the maximum number of circuit simulations. At first, Latin hypercube sampling (LHS) [10] is used to sample λ designs between the lower bound $\boldsymbol{lb}$ and upper bound $\boldsymbol{ub}$ of design space and perform circuit simulations on initial designs. All designs and corresponding fitness values are added to the database and $nfev$ is set to λ. Secondly, it jumps out a loop until $nfev$ is bigger than $maxnfev$. Then the population is formed by selecting the λ best designs according to the ranking rules in Section III-A. Thirdly, unlike DE where only DE operator is used to generate next generation, DE operator is followed by polynomial mutation to generate λ trial vectors, which guarantees the diversity of the population. In the next stage, GP is constructed for each objective by all samples in the database. The λ trial vectors are combined with all designs in the database to form Q. We sort Q according to the ranking rules in Section III-A, the best 2 trial vectors x_{best1} and x_{best2} are chosen in the light of the ranking number. It is noticed that the fitness values of trial vectors are predicted by GP in the

---

Algorithm 1 Pseudo code of SAMODE

1: Use LHS to sample λ designs from $[\boldsymbol{lb}, \boldsymbol{ub}]$. Perform circuit simulations by HSPICE to all of these designs to form initial population. Add them to the database;
2: $nfev = \lambda$;
3: **while** $nfev < maxnfev$ **do**
4:     Select the λ best candidate designs from the database to form population P in terms of simulation results according to the ranking rules in Section III-A;
5:     Apply DE operator and polynomial mutation on P to generate λ trial vectors;
6:     Build global surrogates $\hat{f}_g$ for each objective by Gaussian process using all samples in the database;
7:     Combine λ trial vectors with all samples in the database to form Q;
8:     Sort Q according to the ranking rules in Section III-A and select the best two trial vectors, x_{best1} and x_{best2} in the light of ranking number;
9:     Perform circuit simulations by HSPICE to x_{best1} and x_{best2};
10:     Add x_{best1}, x_{best2} and corresponding objective values to the database;
11:     $nfev = nfev + 2$;
12: **end while**

---

process of sorting. Finally, x_{best1} and x_{best2} are performed circuit simulation by HSPICE to obtain the specification and added to the database. $nfev$ increases by 2. The reason why two designs instead of one are selected is to prevent premature.

A. Ranking Rules

To extend genetic algorithm to handle with multi-objective problems, NSGA-II redefines the ranking rules based on non-dominated sorting and crowding distance, which shows high efficiency in practice. If the ranking rules of NSGA-II are directly adopted in SAMODE, the population is easy to fall into stagnation. As a result, we make several modifications on crowding distance.

Firstly, the crowding distance is defined as the summation of distances of two adjacent individuals of all objectives in NSGA-II, while in SAMODE, the crowding distance is calculated by the multiplication of distances, which is scale-dependent. Secondly, there are two steps using the ranking rules, selecting designs to form P and sorting Q. The crowding distance of boundary solutions (solutions with smallest and largest fitness values) defines differently in these two steps. In the first step, the boundary solutions are assigned an infinite distance value like in NSGA-II. In the second step, the crowding distances of boundary solutions are calculated by the multiplication of the nearest distance in Q of all objectives and then multiplied by 2^m, where m is the number of objectives.

If an infinite distance value is assigned on boundary solutions in sorting Q, the best two trial vectors tend to choose the boundary designs instead of other promising designs in the PF, which degenerates the diversity of the population.

IV. EXPERIMENTAL RESULTS

In this section, the proposed SAMODE is compared with two state-of-the-art multi-objective evolutionary algorithms, NSGA-II and MOEA/D. The efficiency of proposed method is verified by two real-world circuits. All experiments were conducted on a Linux operating system with dual Intel 1.90GHz Xeon CPUs and 64G memory. All experiments are carried out 10 times to average out the random fluctuations.

To compare the performance of different multi-objective algorithms, the hypervolume difference metric I_H^- [11] is used in this work. The I_H^- is defined as follows:

$$I_H^-(P) = HV(P_*) - HV(P) \qquad (6)$$

where HV(P) is the hypervolume of the obtained PF, all simulated data of different algorithms are combined to extract a true PF P_*. Besides, λ is set to 100 in all examples. We draw a reference point using the worst values of all sample points in ten runs and three algorithms.

A. X-band Low Noise Amplifier

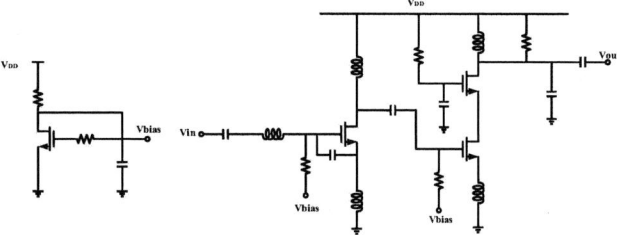

Fig. 1. schematic of the X-band low noise amplifier

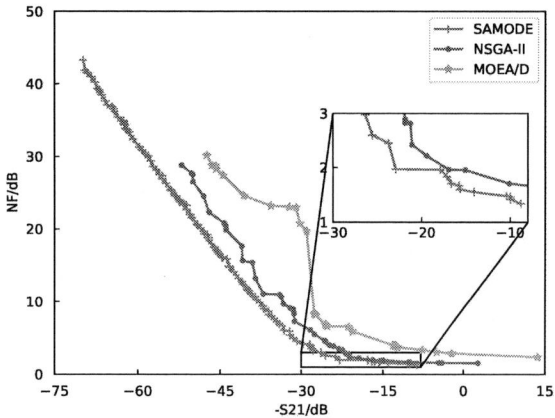

Fig. 2. PFs generated by three algorithms for the X-band low noise amplifier

The first example is an X-band low noise amplifier (LNA) as shown in Figure 1. The circuit is designed based on 130nm process with 14 design variables including the lengths and widths of transistors and the values of capacitance, inductance

and resistance. the pass band of LNA is from 8.5GHz to 10.5GHz. The goal is to maximize the minimum S21 and minimize the maximum noise factor (NF) in the pass band, both are as listed in (7).

$$\text{minimize } \{-S21, NF\} \qquad (7)$$

Designers of analog circuits can not optimize two conflicting circuit performances at the same time. The trade-off between NF and S21 offered by multi-objective optimization facilitates the design of circuits. The $maxnfev$ of proposed SAMODE is set to 400, including 100 initial sample points, while the maximum number of simulations is 800 both for NSGA-II and MOEA/D.

TABLE I
STATISTICS OF X-BAND LNA RESULTS

Algorithm	SAMODE	NSGA-II	MOEA/D
Maximum S21(dB)	**72.58**	52.07	47.45
Minimum NF(dB)	**1.12**	1.30	1.65
mean I_H^-	**108.18**	640.69	1122
median I_H^-	**112.22**	661.69	1107
max I_H^-	**168.30**	763.3	1417
min I_H^-	**39.46**	466.2	879

The PFs with minimum I_H^- values are illustrated in Figure 2 and the statistic results of I_H^- are listed in Table I. The points in PF of SAMODE are more widely distributed than that of NSGA-II and MOEA/D and get closer to the ideal PF. As we can see from Table I, NSGA-II performs better than MOEA/D in terms of I_H^- and the average I_H^- of SAMODE is 17% that of NSGA-II with half the number of simulations, which shows the efficiency of SAMODE.

B. Fully Differential Operational Amplifier

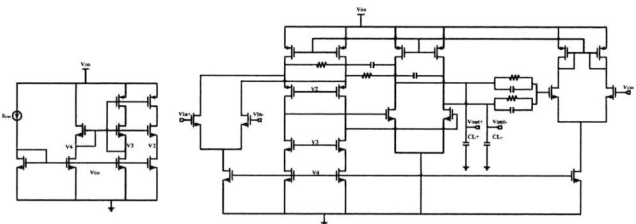

Fig. 3. schematic of the fully differential operational amplifier

The second example is a fully differential operational amplifier with 21 design variables implemented in 65nm process as shown in Figure 3. In order to test the ability of dealing with three-objective optimization of SAMODE, two problems are considered in this example. One is to maximize gain and maximize the gain-bandwidth product (GBW). The other one is to maximize gain, maximize the gain-bandwidth product and minimize the total current (Itot) in the circuit. Two questions are listed as follows:

$$\text{minimize } \{-Gain, -GBW\} \qquad (8)$$

© VDE VERLAG GMBH · Berlin · Offenbach

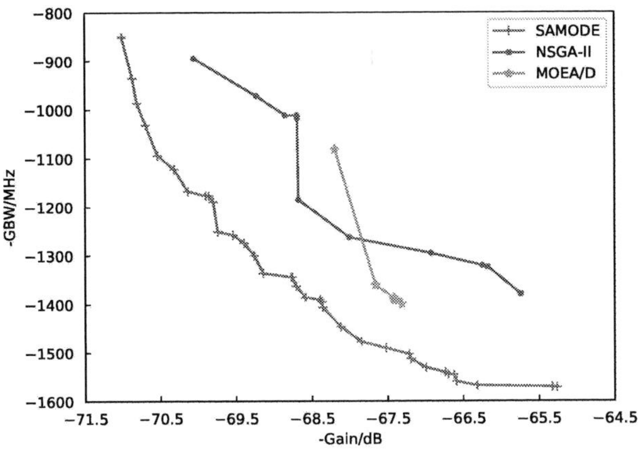

Fig. 4. PFs generated by three algorithms for the fully differential operational amplifier with two objectives

TABLE II
STATISTICS OF FULLY DIFFERENTIAL OPERATIONAL AMPLIFIER RESULTS
WITH TWO OBJECTIVES

Algorithm	SAMODE	NSGA-II	MOEA/D
Maximum Gain(dB)	**71.0**	70.1	68.2
Maximum GBW(MHz)	**1607**	1464	1546
mean I_H^-	**987.9**	3689.0	6034.8
median I_H^-	**1002.0**	3665.2	6090.0
max I_H^-	**2315.5**	4177.9	7633.8
min I_H^-	**269.2**	3327.4	4569.8

$$\text{minimize } \{-\text{Gain}, -\text{GBW}, \text{Itot}\} \qquad (9)$$

We limit $maxnfev$ to 400 for SAMODE with 100 initial samples. 800 simulations are performed for NSGA-II and MOEA/D. Both problems share the same parameter setting.

The optimization results for two-objective problem are detailed in Table II. According to Table II, we can conclude that SAMODE has achieved much lower average I_H^- than NSGA-II and MOEA/D. The PFs with minimum I_H^- values for fully differential operational amplifier are pictured in Figure 4. we can see that all points in the PF of NSGA-II and MOEA/D are dominated by the PF of SAMODE. The proposed method searches for a better PF with less number of simulations.

TABLE III
STATISTICS OF FULLY DIFFERENTIAL OPERATIONAL AMPLIFIER RESULTS
WITH THREE OBJECTIVES

Algorithm	SAMODE	NSGA-II	MOEA/D
Maximum Gain(dB)	**70.40**	70.05	68.21
Maximum GBW(MHz)	**1603**	1398	1545
Minimum Itot(A)	**0.0013**	0.0015	0.0022
mean I_H^-	**13.87**	22.19	45.52
median I_H^-	**14.67**	21.66	45.19
max I_H^-	**20.38**	26.68	55.21
min I_H^-	**4.02**	18.59	37.25

The statistic results of fully differential operational amplifier results with three objectives are presented in Table III. As can be seen, SAMODE achieves better Gain, GBW and Itot than other two algorithms. The lowest I_H^- is also achieved by SAMODE, which verifies the ability of proposed approach to handle with three-objective optimization problems.

On the basis of results of fully differential operational amplifier with two and threes objectives, SAMODE is capable to achieve better PF with much lower number of simulations compared with NSGA-II and MOEA/D.

V. CONCLUSION

In this paper, we propose a surrogate-assisted multi-objective differential evolution, SAMODE for analog circuit synthesis based on Gaussian process. Offspring are generated following the flowchart of NSGA-II-DE and GP is applied to prescreen the best two individuals in offspring. We make several modifications on the ranking rules, mainly about boundary solutions in PF to guarantee the diversity of the population. The efficiency of proposed SAMODE is verified by two real-world circuits and three multi-objective problems. Compared with NSGA-II, we reduce up to 83% of I_H^- with half the number of simulations. Significantly better results can be achieved with less number of simulations compared with two state-of-the-art multi-objective evolutionary algorithms.

REFERENCES

[1] Gang Zhang, Aykut Dengi, and L Richard Carley. Automatic synthesis of a 2.1 GHz SiGe low noise amplifier. In *2002 IEEE Radio Frequency Integrated Circuits (RFIC) Symposium. Digest of Papers (Cat. No. 02CH37280)*, pages 125–128. IEEE, 2002.

[2] Fábio Passos, Elisenda Roca, Javier Sieiro, Rafaella Fiorelli, Rafael Castro-López, José María López-Villegas, and Francisco V Fernández. A multilevel bottom-up optimization methodology for the automated synthesis of RF systems. *IEEE Transactions on Computer-Aided Design of Integrated Circuits and Systems*, 39(3):560–571, 2019.

[3] Prakash Kumar Rout and Debiprasad Priyabrata Acharya. Design of CMOS ring oscillator using CMODE. In *2011 international conference on energy, automation and signal*, pages 1–6. IEEE, 2011.

[4] Kalyanmoy Deb, Amrit Pratap, Sameer Agarwal, and TAMT Meyarivan. A fast and elitist multiobjective genetic algorithm: NSGA-II. *IEEE transactions on evolutionary computation*, 6(2):182–197, 2002.

[5] Qingfu Zhang and Hui Li. MOEA/D: A multiobjective evolutionary algorithm based on decomposition. *IEEE Transactions on evolutionary computation*, 11(6):712–731, 2007.

[6] Bo Liu, Hadi Aliakbarian, Soheil Radiom, Guy AE Vandenbosch, and Georges Gielen. Efficient multi-objective synthesis for microwave components based on computational intelligence techniques. In *DAC Design Automation Conference 2012*, pages 542–548. IEEE, 2012.

[7] Wenlong Lyu, Fan Yang, Changhao Yan, Dian Zhou, and Xuan Zeng. Multi-objective Bayesian optimization for analog/RF circuit synthesis. In *Proceedings of the 55th Annual Design Automation Conference*, pages 1–6, 2018.

[8] Hui Li and Qingfu Zhang. Multiobjective optimization problems with complicated Pareto sets, MOEA/D and NSGA-II. *IEEE transactions on evolutionary computation*, 13(2):284–302, 2008.

[9] Carl Edward Rasmussen. Gaussian processes in machine learning. In *Summer School on Machine Learning*, pages 63–71. Springer, 2003.

[10] Michael Stein. Large sample properties of simulations using latin hypercube sampling. *Technometrics*, 29(2):143–151, 1987.

[11] Shi Cheng, Yuhui Shi, and Quande Qin. On the performance metrics of multiobjective optimization. In *International Conference in Swarm Intelligence*, pages 504–512. Springer, 2012.

SMACD / PRIME 2021 | 19 – 22 July 2021, Online Event

A fast Structural Synthesis Algorithm for Op-Amps based on Multi-Threading Strategies

Inga Abel, Clara Kowalsky, and Helmut Graeb,

Technical University of Munich, Chair of Electronic Design Automation, {first name}.{last name}@tum.de

Abstract—**This paper presents a method to speed up the structural synthesis of op-amps presented in [1]. Op-Amp topologies are created, sized and evaluated according to a given set of specifications. Three different types of op-amps are supported: single-output, fully-differential and complementary op-amps. The sizing and evaluation process is parallelized using multi-threading techniques. On a common computer the new algorithm needs less than eight hours to find all suitable topologies for a given set of specifications out of thousands of topologies compared to 22 h in [1]. The experimental results present the speed-up of the algorithm on seven different specification sets.**

Index Terms—**analog circuit design, synthesis, CMOS, operational amplifiers**

I. INTRODUCTION

Operational amplifiers are the fundamental building block of analog/mixed-signal circuits. While for digital circuits, fundamental building blocks are designed almost fully automatically, op-amps are still designed mostly manual till now.

In this paper, we propose a parallelization of the structural synthesis method in [1]. Compared to the state of the art, which creates a huge set of variants and evaluates them by numerical simulation [2]–[6], in [1] a creation of only design-relevant topology candidates and an equation-based sizing and evaluation process was presented. While [1] already provided a significant reduction in CPU time, this paper introduces a multi-threading-based parallel evaluation process, which achieves a speed-up up to a factor of five compared to [1] on a machine with 16 threads.

Table I gives an overview of state-of-the-art structural synthesis algorithms. It shows the number of supported topologies, the runtime of the algorithms and the used type of sizing. State-of-the-art synthesis approaches which use equation-based methods to evaluate circuit topologies like [7], [8] support a small number of topologies, as the set up of the equation-based description is not automated. By using a hierarchical functional block composition method based on [9] to create op-amp topologies and formalizing the knowledge of these functional blocks to describe the circuit behavior [10], the set-up of the equation-based circuit model is fully automated in [1] such that several thousand topologies are supported. The hierarchical functional block composition graph omits many redundant and design-irrelevant topologies, compared to the state of the art. Please note that the search duration for [2], [3], [5] was obtained on advanced hardware, while the maximum search duration for this algorithm is obtained on a common desktop computer.

Work	# supported topologies	search duration	Type of Sizing
[7]	64	< 1 h	equation-based
[8]	~70	< 0.2 h	equation-based
[11]	24	< 0.5 h	numerical
[2]	> 10 000	2 d	numerical
[3]	~3500	15 h	numerical
[4]	> 10 000	n.a.	numerical
[5]	> 10 000	8 h	numerical
[6]	~4000	secs. (w/o sizing)	numerical
[1]	~2000	0.5 - 22 h	equation-based
this work	~2000	0.2 - 7 h	equation-based

TABLE I: Comparison of state-of-the-art structural synthesis algorithms

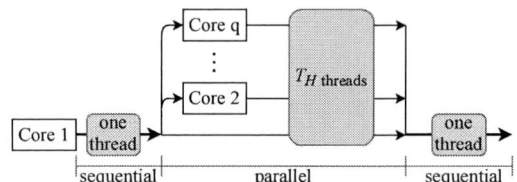

Fig. 1: Sequential and parallel tasks

In the following, an overview of multi-threading is given (Sec. II). Sec. III presents an outline of the structural synthesis algorithm and shows how we use multi-threading to parallelize the sizing and evaluation of multiple op-amp topologies. The experimental results (Sec. IV) show the major speed-up obtained by using multi-threading in the synthesis algorithm.

II. MULTI-THREADING

Many programming languages as C++ or Python allow to distribute tasks to multiple threads which run in parallel during program execution (Fig. 1). This is called multi-threading. Depending on the hardware, T_H threads can be executed in parallel. T_H does not have to be identical to the number of cores q as there exist cores which can run multiple threads in parallel. Running tasks in parallel can result in a significant speed-up. Not all tasks in a program are suited to run in parallel. The tasks running in parallel should be as independent as possible, needing only read access to common variables. Write access to common variables might lead to race conditions, a variable might exceed its limit, as two tasks write on it in the same time. Thus, a program can be divided into sequential and parallelized parts. In structural synthesis, the sizing of different circuits is a task suitable for parallel programming, as the sizing of one circuit is independent of the sizing of another circuit.

© VDE VERLAG GMBH · Berlin · Offenbach

HL 5	op-amp (single-output, complementary, fully-differential)
HL 4	first stage, feedback stage, second stage, bias
HL 3	simple transconductance, feedback transconductance, complementary transconductance, stage bias, load
HL 2	voltage bias, current bias, differential pair
HL 1	normal transistor, diode transistor, capacitor

Fig. 2: Functional blocks in op-amps

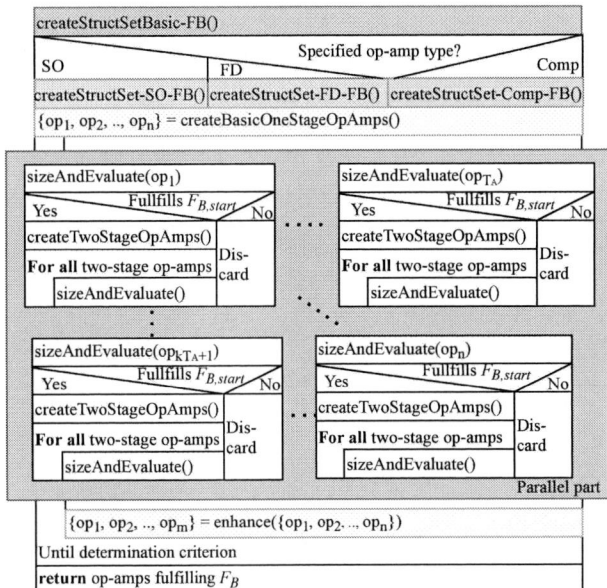

Fig. 3: Structural synthesis algorithm

III. STRUCTURAL OP-AMP SYNTHESIS BASED ON MULTI-THREADING

The structural synthesis algorithm is based on the algorithm in [1]. The algorithm has as input the circuit specifications and the op-amp type to be created. Three different types are supported: single-output (SO, e.g. Fig. 4a), fully-differential (FD, e.g. Fig. 4b) and complementary (Comp, Fig. 4c). Complementary op-amps are defined as op-amps having a pmos and nmos differential pair forming the input stage. In the following, we will give a brief overview how topologies are created based on functional blocks and describe the general algorithm. We explain how we use multi-threading to parallelize the sizing and evaluation of different op-amp topologies.

A. Creation of Op-Amp Topologies

The structural synthesis algorithm creates topologies based on functional blocks (FBs). Fig. 2 gives an overview of functional blocks in op-amps. They are hierarchically structured. On the lowest level are devices like normal and diode transistor, e.g., Fig 4a, N_3; P_3. On the second level are basic analog structures like voltage and current bias, e.g., Fig. 4b, N_8; N_9. The third level contains amplification stage subblocks like transconductance, load, and stage bias, e.g., Fig. 4a, P_1 - P_2; N_1 - N_2; P_3. The fourth level consists of amplification stages like first and second stage, e.g., Fig. 4c, P_1 - P_7, N_1 - N_7, and the op-amp bias. The top level contains complete op-amp topologies. Structural implementations of a functional block on level i are created based on structural implementations of functional blocks on level i - 1 having characteristic pin connections. Thus, complete op-amp topologies are hierarchically constructed.

To reduce the memory usage, it is differentiated between three groups of functional blocks: basic functional blocks (violet), functional blocks corresponding to a specific op-amp type (orange: single-output, blue: fully-differential, green: complementary) and functional blocks only generated if they are part of the created op-amp topology (yellow). The basic functional blocks and the op-amp type-specific functional block are created at the beginning of the structural synthesis algorithm (Fig. 3). They are used in many other functional blocks and do not have many structural representatives. The remaining functional blocks have many structural implementations and are created on demand when the op-amp topology is created (Fig. 3, yellow). This is a good compromise in terms of runtime and memory usage [1].

Two-stage op-amps are defined as one-stage op-amps to which a second stage is added. They are only created if its one-stage version fulfills certain requirements during the sizing and evaluation process discussed in Sec. III-B. This reduces the search space inherently.

B. General Algorithm

The synthesis algorithm (Fig. 3) starts by creating one-stage op-amps with a small number of transistors (basic op-amps). The set is sized and evaluated using the equation-based sizing method in [10], [12]. It uses analytical equations instead of SPICE-like simulations. As in [1], the performance specifications F_B are split into two groups: $F_{B,start}$ denotes all performance specifications which degrade or are not influenced if another stage is added to the op-amp, e.g., phase margin. $F_B \setminus F_{B,start}$ denotes all performance specifications which are or might be positively influenced if another stage is added to the op-amp, as open-loop gain. If a one-stage op-amp fulfills $F_{B,start}$, its corresponding two-stage topologies are created, sized and evaluated. For the sizing and evaluation, approximately one minute is needed per op-amp. The deviations between the performance values calculated using analytical equation and simulation results are within designers expectations and less than 30 %.

After evaluating all one-stage op-amps in the set and evaluating the corresponding two-stage op-amps, the one-stage op-amps are enhanced e.g., by changing the stage bias or load. The new set of one-stage op-amps is then again sized, evaluated, and enhanced. The algorithm terminates if the most simple op-amps fulfilling the specification are found or if all op-amps fulfilling the specifications are found. That depends on the user preferences.

Specs #	1	2	3	4	5	6	7
Op-amp type	Single-output			Fully-differential		Complementary	
Bias current (μA)	10	100	10	100	100	100	100
Load Capacity (pF)	20	20	20	20	20	20	20
Supply voltage (V)	5	5	5	5	5	5	5
Gate-area ($10^3 \mu m^2$)	≤ 15	≤ 5	≤ 5	≤ 50	≤ 20	≤ 15	≤ 5
Quiescent power (mW)	≤ 15	≤ 8	≤ 5	≤ 25	≤ 15	≤ 10	≤ 5
Phase Margin (°)	≥ 60	≥ 60	≥ 80	≥ 60	≥ 60	≥ 60	≥ 60
CMRR (dB)	≥ 70	≥ 70	≥ 70	≥ 80	≥ 80	≥ 80	≥ 80
Max. common-mode input volt. (V)	≥ 3	≥ 3.5	≥ 3.5	≥ 3	≥ 3	-	-
Min. common-mode input volt. (V)	≤ 2	≤ 1.5	≤ 1.5	≤ 2	≤ 2	-	-
Open-loop gain (dB)	≥ 80	≥ 70	≥ 45	≥ 70	≥ 60	≥ 70	≥ 70
Unity-gain bandwidth (MHz)	≥ 2.5	≥ 10	≥ 10	≥ 2.5	≥ 10	≥ 2.5	≥ 10
Slew rate ($\frac{V}{\mu s}$)	≥ 3.5	≥ 20	≥ 20	≥ 3.5	≥ 15	≥ 3.5	≥ 15
Max. output volt. (V)	≥ 3.5	≥ 3.5	≥ 4	≥ 3	≥ 3.5	≥ 3.5	≥ 4
Min. output volt. (V)	≤ 1.5	≤ 1.5	≤ 1	≤ 2	≤ 1.5	≤ 1.5	≤ 1

TABLE II: Specifications sets 1, 4, 5: general; Specification sets 2, 3, 5, 7: challenging

C. Parallelizing the Sizing and Evaluation Process

Different to [1], the sizing and evaluation of a set of topologies now runs in parallel (Sec. II). For this task, T_A - 1 threads are started in addition to the main thread in which the program runs (Fig. 1). T_A depends on the number of threads the used hardware supports, T_H, and the number of created op-amps n. It is calculated by:

$$T_A = \begin{cases} T_H, & if \ n \geq T_H \\ n, & if \ n < T_H \end{cases} \quad (1)$$

Each thread t_i, $i \in \{1, .., T_A\}$, is assigned a set of k_i op-amps to be sized and evaluated. The number of topologies k_i depends on the number of created threads T_A and the number of created op-amp topologies n:

$$k_i = \begin{cases} \lfloor n/T_A \rfloor, & \text{for } t_i, i \in \{1, .., T_A - (n \mod T_A)\} \\ \lceil n/T_A \rceil, & \text{for } t_i, i \in \{T_A - (n \mod T_A) + 1, .., T_A\} \end{cases} \quad (2)$$

The main thread t_1 also runs the sequential parts of the program. It has a smaller number of assigned circuits if the circuits cannot be evenly distributed, as it also controls the starting and closing of the threads $t_2, ..., t_{T_A}$ created for the parallel execution of the program.

The sizing and evaluation of two-stage op-amps still runs sequentially, which lessens the effect of the inherent reduction of the search space when two-stage op-amps are not created if the one-stage op-amp did not fulfill $F_{B,start}$. Some threads running in parallel may have more one-stage op-amps for which two-stage op-amps are created than others. Future work remains in parallelizing the sizing and evaluation of two-stage op-amps to distribute tasks more equally.

IV. EXPERIMENTAL RESULTS

The synthesis algorithm was tested using seven different specification sets (Table II). *Specs 1* to *Specs 3* specify single-output as op-amp type. *Specs 1* are general specifications, many different topologies fulfill them. *Specs 2* and *Specs 3* are

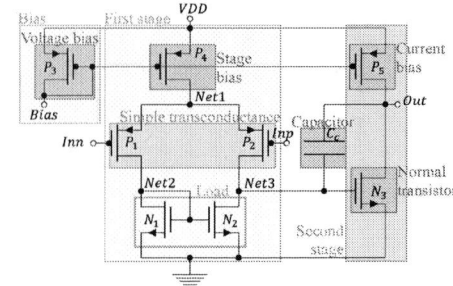

(a) Single-output op-amp, *Specs 1* and *Specs 2*

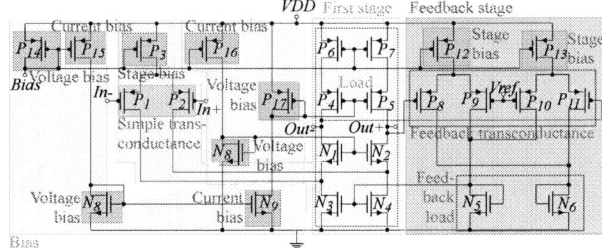

(b) Fully-differential op-amp, *Specs 4* and *Specs 5*

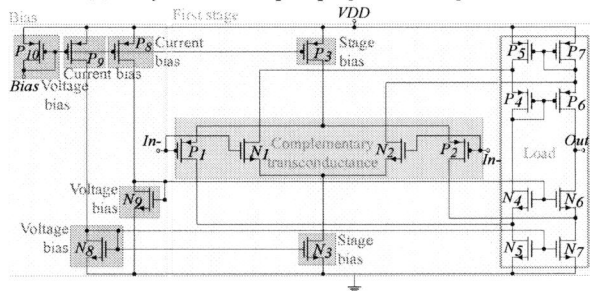

(c) Complementary op-amp, *Specs 6* and *Specs 7*

Fig. 4: Example topologies for each op-amp type

Specs #	1	2	3	4	5	6	7
# one-stage op-amps fulfilling $F_{B,start}$	144 (210)	96 (210)	81 (210)	39 (72)	37 (72)	36 (36)	36 (36)
# created op-amp topologies	1728 (2940)	1152 (2940)	972 (2940)	468 (936)	444 (936)	36 (36)	36 (36)
# op-amps fulfilling F_B	228 (2940)	71 (2940)	54 (2940)	34 (936)	2 (936)	10 (36)	6 (36)

TABLE III: Overview of the created topologies and their performance; brackets: maximum # supported topologies

more challenging specifications, varying in the gain and phase margin requirement. *Specs 4* and *Specs 5* comprises a fully-differential op-amp as type. *Specs 4* are general specifications and *Specs 5* are more demanding. *Specs 6* and *Specs 7* includes a complementary op-amp as type. *Specs 5* are general and *Specs 6* are more strict. For all specification sets, all topologies fulfilling the specifications were created.

Table III gives an overview of the created topologies. Not all one-stage op-amps fulfilled the specifications in $F_{B,start}$ such that the corresponding two-stage op-amps were not created. The number of op-amps fulfilling the more challenging specification sets (*Specs* 2, 3, 5, 7) is less than the number of op-amps fulfilling the more general specifications (*Specs* 1,

Specs #	1	2	3	4	5	6	7
Sequential algorithm [1] (1)	21 h	16 h	14.5 h	8 h	8 h	35 min	30 min
Intel® Core™ i5-7500 CPU@3.4GHz (4)	7 h [33%]	5.5 h [34%]	5.5 h [38%]	3 h [38%]	3.5 h [44%]	15 min [43%]	14 min [47%]
2 x Intel® Xeon® CPU E5-2695 @2.4GHz (2 x 8)	4 h [19%]	3.2 h [20%]	3.2 h [22%]	1.6 h [20%]	1.6 h [20%]	11 min [31%]	9 min [30%]
2 x Intel® Xeon® Gold 6126 CPU @2.6GHz (2 x 24)	3.5 h [17%]	3 h [19%]	3 h [21%]	1.2 h [15%]	1.2 h [15%]	7.5 min [21%]	6 min [20%]

TABLE IV: Runtime comparison on different CPUs; (m): max. # of threads; [n] runtime in percent compared to sequential algorithm

4, 6), which corresponds to the expectations. Fig. 4 shows example topologies for the different specification sets. The two-stage op-amp with single output in Fig. 4a fulfills *Specs* 1 and *Specs* 2 with different dimension sets. Because of the high demands on the phase margin, it does not fulfill *Specs* 3. Differently sized versions of the op-amp in Fig. 4b are part of the solution set of *Specs* 4 and *Specs* 5. The op-amp in Fig. 4c fulfills *Specs* 6 and *Specs* 7. Please note that for complementary op-amps only one-stage op-amps are currently supported.

Table IV shows a runtime comparison of the proposed parallel synthesis algorithm and its sequential counterpart in [1]. The synthesis algorithm was tested on three different CPUs. The Intel®Core™ i5-7500 CPU@3.4GHz is a common processor for desktop computers able to run 4 threads in parallel. The Intel® Xeon® CPU E5-2695@2.4GHz and Intel® Xeon® Gold 6126 CPU@2.6GHz are two server CPUs supporting 8, respectively 24, threads in parallel. Even on a small desktop CPU, the runtime of the algorithm is reduced to 40% of the sequential runtime, needing less than 8 h to find all possible topologies for general specifications. On more advanced hardware only 20% of the sequential runtime is needed, reducing the runtime for general specifications to half a work day.

Regarding *Specs* 2, 3, the runtime of the sequential algorithm differs by 1.5 h as fewer two-stage op-amps are created for *Specs* 3. This runtime reduction cannot be seen for the parallelized algorithms. Both specification sets have identical runtimes. The reason for this lies in the missing parallelization of the sizing and evaluation of two-stage op-amps. For *Specs* 3, the additional one-stage op-amps not fulfilling $F_{B,start}$ are unevenly distributed among threads, several of the parallel running threads have more one-stage op-amps for which two-stage op-amps are created.

The synthesis algorithm (Fig. 3) creates a fixed number of op-amp topologies before the sizing and evaluation step. For CPUs supporting a high number of threads T_H, this can lead to unused threads, as the number of created op-amp topologies n is smaller than T_H. In many runs on the machine with the two Intel® Xeon® Gold CPUs with a total of 48 threads, not all threads are used. This leads to similar runtimes as for the Intel® Xeon® E5 processor. However, creating as many op-

amp topologies as number of threads T_H might exceed the memory. It remains future work to enhance the algorithm such that the number of op-amp topologies created at the same time corresponds to the amount of threads being able to run in parallel on the CPU and the assigned RAM storage to the program.

V. CONCLUSION

This paper presented structural synthesis algorithm which is able to size and evaluate 3000 topologies in less than a work day on a common desktop computer. The short runtime is obtained by using multi-threading to enable a parallel sizing and evaluation of multiple op-amp topologies. The experimental results show that using a more advanced hardware further reduces the runtime to 20% of the runtime of the sequential version of the algorithm.

ACKNOWLEDGMENT

The authors would like to thank the Cusanuswerk for partly funding this work.

REFERENCES

[1] I. Abel and H. Graeb, "Structure Synthesis of Op-Amps by Functional Block Composition," 2021, submitted to IEEE Transactions on Computer-Aided Design of Integrated Circuits and Systems. [Online]. Available: http://arxiv.org/abs/2101.07517

[2] J. R. Koza, F. H. Bennett, D. Andre, M. A. Keane, and F. Dunlap, "Automated synthesis of analog electrical circuits by means of genetic programming," *IEEE Transactions on Evolutionary Computation*, vol. 1, no. 2, pp. 109–128, 1997.

[3] T. McConaghy, P. Palmers, M. Steyaert, and G. G. E. Gielen, "Variation-Aware Structural Synthesis of Analog Circuits via Hierarchical Building Blocks and Structural Homotopy," *IEEE Transactions on Computer-Aided Design of Integrated Circuits and Systems*, vol. 28, no. 9, pp. 1281–1294, 2009.

[4] A. Das and R. Vemuri, "A graph grammar based approach to automated multi-objective analog circuit design," in *2009 Design, Automation Test in Europe Conference Exhibition*, 2009, pp. 700–705.

[5] M. Meissner and L. Hedrich, "FEATS: Framework for Explorative Analog Topology Synthesis," *IEEE Transactions on Computer-Aided Design of Integrated Circuits and Systems*, 2015.

[6] Z. Zhao and L. Zhang, "An Automated Topology Synthesis Framework for Analog Integrated Circuits," *IEEE Transactions on Computer-Aided Design of Integrated Circuits and Systems*, vol. 39, no. 12, pp. 4325–4337, 2020.

[7] P. C. Maulik, L. R. Carley, and R. A. Rutenbar, "Integer programming based topology selection of cell-level analog circuits," *IEEE Transactions on Computer-Aided Design of Integrated Circuits and Systems*, vol. 14, no. 4, pp. 401–412, April 1995.

[8] R. Harjani, R. A. Rutenbar, and L. Carley, "OASYS: A Framework for Analog Circuit Synthesis," *IEEE Transactions on Computer-Aided Design of Integrated Circuits and Systems*, 1989.

[9] I. Abel, M. Neuner, and H. Graeb, "A Functional Block Decomposition Method for Automatic Op-Amp Design," 2020, . [Online]. Available: http://arxiv.org/abs/2012.09051

[10] ——, "A Hierarchical Performance Equation Library for Op-Amp Design," 2020, submitted to IEEE Transactions on Computer-Aided Design of Integrated Circuits and Systems. [Online]. Available: http://arxiv.org/abs/2012.09088

[11] A. Gerlach, J. Scheible, T. Rosahl, and F. Eitrich, "A generic topology selection method for analog circuits with embedded circuit sizing demonstrated on the OTA example," in *Design, Automation Test in Europe Conference Exhibition (DATE), 2017*, 2017, pp. 898–901.

[12] I. Abel, M. Neuner, and H. Graeb, "COPRICSI: COnstraint-PRogrammed Initial Circuit SIzing," *Integration*, vol. 76, pp. 148 – 158, 2021.

An Essay on the Next Generation of Performance-driven Analog/RF IC EDA Tools: The Role of Simulation-based Layout Optimization

Ricardo Martins[1], António Gusmão[1,2], António Canelas[1], Fábio Passos[1,3], Nuno Lourenço[1], Nuno Horta[1,2]

[1]Instituto de Telecomunicações, Lisboa, Portugal
[2]Instituto Superior Técnico, Universidade de Lisboa, Portugal
[3]Dialog Semiconductors, Lisboa, Portugal
{ricmartins}@lx.it.pt

Abstract—Despite the fact that analog and radio-frequency (A/RF) integrated circuit (IC) design automation has been intensively studied in the last few decades, only automatic circuit-level sizing methodologies have achieved a satisfactory level of maturity. Layout and its countless issues have challenged all automation attempts, and two limiting factors must be addressed to force their way into the industrial environment: plug-and-play capabilities and accurate assessment of post-layout performance degradation. This paper brainstorms around the idea of developing the ultimate fully automatic "performance-driven" A/RF IC synthesis by incorporating simulation-based layout optimization concepts in the flow. The essay is carried the PONDEROUS tool, a novel and highly integrated, but exceptionally computationally intensive, placement A/RF IC optimizer.

I. INTRODUCTION

Block-level A/RF IC design automation has been an intensive topic in the past few decades. However, while circuit sizing has achieved a reasonably well-established concept for its automation, i.e., simulation-based optimization, layout has challenged all automation attempts and there is still no consensual methodology applied in the industrial environment. Mainly, all automation attempts failed at two points: plug-and-play capabilities and accurately assessing the performance degradation post-layout. Additionally, as A/RF IC design dives into smaller technology nodes, the increasing topological requirements, additional process design kit (PDK)'s design rules, and parasitic structures or layout-dependent effects arising from the physical layout description are continuously defying previous approaches. This paper digs into whether it could be possible to apply the successful concept of circuit sizing automation directly to the layout, i.e., *Are modern workstations (WS) ready for the computational requirements of a simulation-based layout optimization?* The answer is pursued with the PONDEROUS tool, a placement optimizer that imports PDK's PCells, and simulates every candidate floorplan towards optimal post-layout solutions.

This paper is organized as follows: in Section II, the literature review is conducted; in Section III, the PONDEROUS tool is described; and finally, in Sections IV and V, the preliminary results and conclusions, respectively, are drawn.

II. LITERATURE REVIEW AND DISCUSSION

This Section starts by overviewing the existing sizing tools, their "performance-drive" nature, afterwards the existing layout tools, and finally, the layout-aware sizing optimization concept.

This work is funded by FCT/MCTES through national funds and when applicable co-funded EU funds under the project UIDB/EEA/50008/2020, including internal research project LAY(RF)^2 (X-0002-LX-20).

A. "Performance-driven" Sizing Automation: from Knowledge-based to Simulation-based

A/RF IC sizing task, and its automation, have been ever since performance-driven. Whether using old-fashioned knowledge-based design plans that capture designers' expertise and are followed to produce the component sizes that meet the performance requirements [1]. Alternatively, by using optimization techniques embedded with different evaluation methods: (1) model-based, where simple equations, posynomial models [2] or more sophisticated models, e.g., artificial neural networks [3], are used to estimate the circuit performance; or (2) simulation-based [4], where an off-the-shelf simulator is used to accurately assess the circuit performance, as illustrated in Fig. 1. However, since the circuit simulator is the ultimate tool to validate an A/RF IC's performance, simulation-based optimization has been cementing over the years as the most widely accepted approach, even in commercial EDA solutions. Nonetheless, model-based evaluation is still relevant in situations where a single solution's simulation time is prohibitive, e.g., components that require computationally intensive electromagnetic evaluations [5] or at system-level [6].

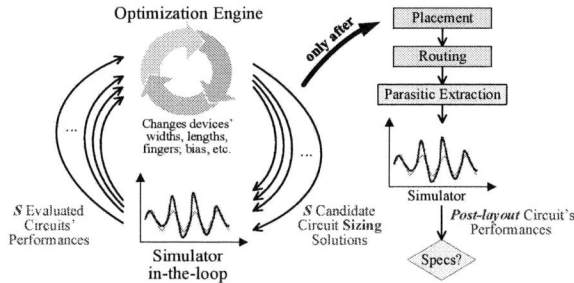

Fig. 1. Simulation-based sizing optimization concept, followed by placement, routing and parasitic extraction of one sizing solution. S is the size of the population of the evolutionary algorithm, or 1 in a single-element metaheuristic.

In retrospect, knowledge-based sizing, despite inaccurate, was proposed at a time where common WS could simply not handle computationally intensive optimizations. With the advance of computing power, sizing optimizations progressively moved from simpler models to simulation-based, although devices' models used for circuit simulation, e.g., BSIM [7], also followed a tremendous complexity increase as the nanometer integration technologies were scaled down. Due to the enormous computational capabilities available nowadays, simulation-based sizing optimization of A/RF ICs is no longer used only to fine-tune manual designs or center a solution within a set of PVT corners or yield, but being used, instead, to solve from scratch sizing problems beyond human capabilities. Nevertheless, these sizing optimizations are still challenging modern WS in complex circuits, taking from days to months [4].

B. "Aesthetics-driven" Placement Automation

The automatic A/RF IC layout generation tools usually follow the traditional design flow, i.e., first placement and, only after, routing, as the complexity of simultaneous place & route grows exponentially. The initial optimization-based approaches were based on the macro-cell placer in absolute coordinates [8], where each cell is represented by its coordinates on a grid-less plane. However, as a huge solution space of both feasible and unfeasible solutions is explored by the optimization kernel, and due to the limitations of old WS, relative representations [9], where the optimizer only perturbs an encoding structure, and it is packed into an admissible placement solution, became a strong research topic during the last three decades [10]. Nonetheless, absolute representations were recently revived [11], as they are the most natural approach to satisfy all topological constraints and can describe any possible floorplan.

In both cases, to reduce the optimization processes' unpredictability and produce a solution meaningful for the designers, a high amount of topological constraints is set. These constraints span from symmetry, common centroid, proximity, regularity, etc., with the ultimate objective of reducing the parasitic mismatch between groups of devices or sensitivity to process variations [12]. Nonetheless, most of these constraints are implemented as insurance and, for "aesthetics" purposes, as their impact on the post-layout circuit performance cannot always be easily demonstrated. However, a perfectly packed and constraint-correct floorplan can still present: (1) a deficient routability, and thus, placement tools have tried to include routing-related data during this process, e.g., with current-flow/density considerations [13] or by minimizing estimated wiring topologies [14]. Still, while there are signs of correlation between routing quality/interconnect length and circuit's performances, as shown in [15], placements with longer wiring topologies may even lead to better circuit performances; or, (2) a high-impact of layout dependent-effects [16].

Alternatively, placement can be automatically generated by retargeting [17] or templates [18], which may incorporate pseudo-post-simulation as criteria in the selection of the sub-partitions [19]. Still, they require a library of specific legacy data to be available, and thus, far from plug-and-play. Ultimately, all of the overviewed approaches failed at accurately assessing the precise performance degradation due to the layout parasitic or layout-dependent effects, and only a final verification step may prove it, as illustrated in Fig. 1. A survey of the routing mechanisms applied to A/RF is remitted to the "Routing Analog Circuits" chapter of [20].

C. Bringing "Performance-Driven" Criteria to Layout Automation: the Layout-aware Sizing Loop

In order to bypass the lack of "performance-driven" criteria in A/RF IC layout synthesis, layout-aware sizing approaches were proposed. These approaches start with the simulation-based sizing concept and perform explicit layout generation in-the-loop for each candidate sizing solution, as illustrated in Fig. 2. The layout generators used can be based on procedures [21], templates [22], mixed-optimization/template [23], or even enumeration of the hierarchical floorplan [24]. These layouts'

parasitics are then estimated or extracted in off-the-shelf tools, providing the simulation-based loop precise post-layout performances to guide the optimization process. While this approach is a major step towards closing the gap between electric and physical design, its convergence is strongly dependent on the quality of the automatic layout generator selected. Clarifying, none of the layouts generated in-the-loop are actually produced towards *optimal* performances but, instead, used as a means to adapt the sizing according to the layout generator's peculiarities or deficiencies. In summary, while Fig. 1 moves towards *optimal* pre-layout performances, Fig. 2 moves towards *sub-optimal* post-layout performances. The lack of accurate and thorough performance-aware layout generators is currently compromising the optimality of a fully automatic A/RF IC flow.

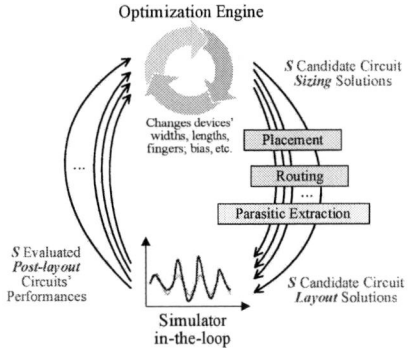

Fig. 2. Layout-aware sizing optimization concept. Automatic placement, routing, and extraction are carried in-the-loop for each candidate sizing.

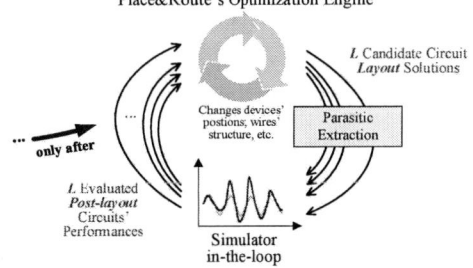

Fig. 3. Simulation-based layout optimization concept. L is the size of the population of the evolutionary algorithm, or 1 in a single-element metaheuristic.

III. Simulation-Based A/RF IC Placement Optimization

The proposed simulation-based A/RF IC layout optimization is a direct analogy from the simulation-based sizing. Instead of changing the dimensions of the devices, bias voltage, etc., the sizing is fixed, and the optimization engine changes the place & route (P&R) layout details, e.g., locations or rotations of the devices for placement, and locations or layers of the interconnects for routing, as shown in Fig. 3. Ultimately, by using a multi-objective metaheuristic, each fixed sizing solution will generate multiple layout alternatives, each with its own tradeoffs. As a side note, the authors are aware that routing interconnects are major culprits for post-layout performance degradation. Still, for the sake of simplicity and demonstrability, the essay conducted on this paper, without loss of generalization, is carried on the placement task only using the Ponderous tool.

A. Problem Formulation

The simulation-based placement optimization of A/RF ICs addressed in this paper is described by a set of MOSFET devices

$M = \{m_1(w_1, h_1), ..., m_{|M|}(w_{|M|}, h_{|M|})\}$, each one defined by an effective layout implementation with a width w and height h; a set of symmetry pairs $SP = \{sp_1(m_a, m_b), ..., sp_{|SP|}(m_a, m_b)\}$ that share a common symmetry axis in the floorplan; a set of axis-aligned self symmetric cells $SS = \{ss_1(m_a), ..., ss_{|SS|}(m_a)\}$; and, a set of monotonic current-flows $MCF = \{mcf_1, ..., mcf_{|CF|}\}$, linking any number of devices M. The many-objective optimization problem is set to find x, the bottom left coordinates of each device's location in the layout, such that the placements' *area* are minimized; but also, the optimization objectives previously set for the sizing optimization. For example, suppose the sizing of the MOSFET-only operational transconductance amplifier for biomedical and healthcare (OTABH) applications of Fig. 4, optimized pre-layout towards maximum gain and minimum *Idd*. In that case, the simulation-based placement optimization objectives are set to maximize post-layout *gain* and minimize post-layout *Idd*, placements' *area*. The same applies to the specifications of Fig. 4.

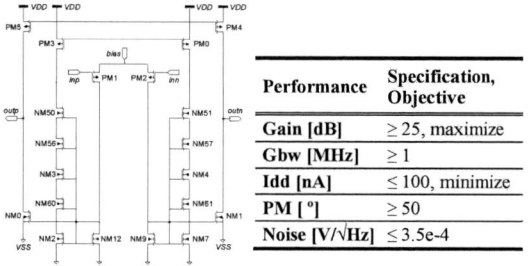

Fig. 4. Schematic of the OTABH with current-paths highlighted [25], and, its specifications and objectives.

Performance	Specification, Objective
Gain [dB]	≥ 25, maximize
Gbw [MHz]	≥ 1
Idd [nA]	≤ 100, minimize
PM [º]	≥ 50
Noise [V/√Hz]	≤ 3.5e-4

B. Placement Representation

For a complete exploration of the placement design space, the optimizer is allowed to move the cells explicitly, i.e., each cell is represented through its absolute coordinates on a plane whose grid is equivalent to the manufacturing grid, e.g., 5-nm. Therefore, cells are allowed to overlap each other, and thus, the following constraint must be considered:

$$g_0(x) \geq 0 \quad g_0(x) = -overlap(x) \quad (1)$$

The overlap is calculated with the interception of the bounding box of each device. Its value must be driven to zero in order to obtain a feasible solution, i.e., $g_0(x) \geq 0$. Nonetheless, in order to decrease the number of design variables and force the desired symmetry pairs SP, in a symmetry-pair, only the bottom-left coordinates of one element are moved, and the other is found deterministically from the first.

C. Ponderous Architecture

To test the simulation-based layout, a novel PONDEROUS tool, developed in Java™, as illustrated in Fig. 4, is composed of the following modules:

1) Optimization Engine: The engine adopted is based on the constrained multi-objective simulated annealing (MOSA) algorithm proposed in [25]. The adapted MOSA is based on the concepts of the amount of objective and constraint dominance and uses an archive to store the set of non-dominated solutions.

2) Instantiator and Compiler: While the previous sizing step fixes the sizing, the optimization engine is allowed to change the number of fingers of the MOSFET devices, and thus, different devices are needed as each new floorplan candidate

arrives. PONDEROUS' *instantiator* automatically generates SKILL scripts that are used to open a Cadence session, instantiate the desired PDK's PCell, and then save it as a GDSII file. These files are imported to the tool and the *compiler* places each cell according to the values of x and computes the area and $g_0(x)$.

3) Port Mapping and GDSII Exporter: For placement solutions with $g_0(x) \geq 0$, and since no routing is available at this stage, the circuit's netlist is analyzed, and each cells' terminals are labeled with the correct ports of each net. Afterward, the GSDII file containing the placement solution is created. Floorplan solutions with overlaps are fed directly to the optimizer without post-layout performances.

4) Off-the-shelf Extraction and Simulation: Automatic calls are performed, first, to the Mentor Graphics' nmLVS to extract the complete geometries required for layout-dependent effects calculations and produce the LVS report file; and after, to the Mentor Graphics' Calibre to extract resistive and ground, self and mutual capacitance parasitics. This extracted netlist is then simulated in the same testbench setup used for pre-layout optimization, such as the methodology of [4], and thus, no additional effort is required.

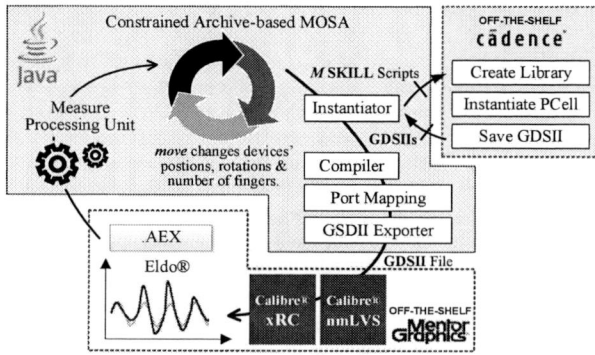

Fig. 5. PONDEROUS architecture: performance-driven A/RF IC placement optimizer with off-the-shelf tools' integration.

IV. EXPERIMENTAL RESULTS

The experiments are conducted using the OTABH of Fig. 4 for a 65-nm technology node. A sizing optimization was carried with the objectives and constraints of Fig. 5 using the methodology from [4]. As shown in [11], this circuit is particularly susceptible to the impact of layout-dependent effects (LDEs), specially to the length of oxide diffusion (LOD). The pre-layout performances of a sizing solution are presented in the *Pre-Layout* line of Table I. The layout was produced using a template-based approach, reproducing the designer's guidelines, and, its post-LDE performances are detailed in the *Template 1^lde* line, where some performance figures not converged to feasible values due to the LDE-impact. A template correction was made by increasing the length of the oxide diffusions of the MOSFET devices, where the noise constraint is still marginally violated post-LDE (*Template 2^lde* line). The optimization-based methodology proposed in [11] was then used to minimize the LDEs-impact, resulting in a competitive solution in terms of performance figures and area (*LDE-aware^lde* line). Still, when computing the complete self and mutual layout parasitics (*LDE-aware^rcc* line), its post-layout performances diverge significantly from the post-LDE only, with *Idd* increasing ~17% and failing constraint. This reveals that by even considering accurate in-the-loop LDE variations, the layout generator is still inaccurately assessing the performance degradation post-layout.

© VDE VERLAG GMBH · Berlin · Offenbach

TABLE I.   POST-LDE AND POST-LAYOUT PERFORMANCES

	Gain [dB]	Gbw [MHz]	Idd [nA]	PM [°]	Noise [V/√Hz]	Area [μm²]	Time
Specification	≥25	≥8	≤100	≥50	≤3.5e-4	n/d	n/d
Target	Max.		Min.			Min.	
Pre-layout	25.14	8.97	99.1	51.67	3.4e-4	n/d	n/d
Template 1^{lde} [18]	25.14	n/c	91.8	n/c	3.4e-4	193.7	Push-button
Template 2^{lde} [18]	25.20	8.80	88.9	56.17	3.7e-4[fail]	212.6	
LDE-aware^{lde} [11]	25.00	8.68	91.4	55.65	3.5e-4	166.5	~10min
LDE-aware^{rce} [11]	25.16	10.57	106.6[fail]	51.32	3.3e-4		
PONDEROUS a^{rce} (comp. w/[11]^{rce})	25.12 / −0.04	8.51	98.3 / −8.3	56.43	3.4e-4	**227.7** / +61.2	
PONDEROUS b^{rce} (comp. w/[11]^{rce})	25.08 / −0.08	8.17	**90.4** / −16.2	57.04	3.4e-4	479.5 / +313	~5hrs
PONDEROUS c^{rce} (comp. w/[11]^{rce})	**25.67** / +0.51	8.93	98.4 / −8.2	52.26	3.5e-4	392.2 / +226	

^{lde}The complete LDE parameters were computed using Mentor Graphics' Calibre PEX with "Extraction Type" set to "No R/C";

^{rce}The complete LDE parameters and self/mutual layout parasitics were computed using Mentor Graphics' Calibre PEX with "Extraction Type" set to "R,C,CC";

PONDEROUS was then used to optimize the floorplan, where a time budget of 5 hours was allowed in a CentOS machine with an 8-core CPU and 16 GB of RAM. The solutions with best *area*, *Idd* and *gain* are detailed in Table I. To meet all the performance constraints after R, C, CC extraction there is an increase in the placement area when comparing with [11]. The obtained post-layout performance tradeoff for the *gain* and *idd* is shown in Fig. 6, showing that by varying the placement and number of fingers of its devices, about 8-nA of *idd* and 0.55-dB of *gain* are traded post-layout, insights which are not commonly available to the circuit designer.

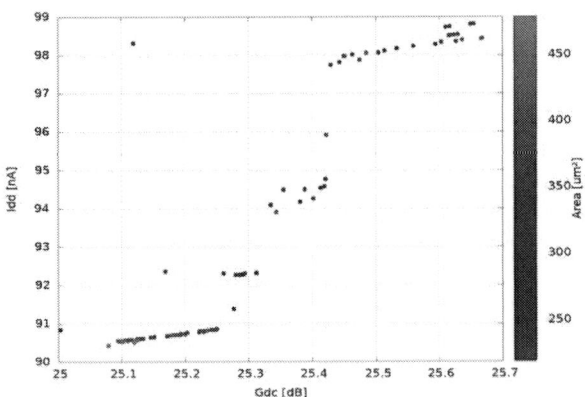

Fig. 6. PONDEROUS solutions ilustrating the performance variation due to placement.

V. CONCLUSIONS & FUTURE RESEARCH DIRECTIONS

Layout design of A/RF ICs has been heavily constrained ever since as an insurance for correctness, however, its real impact is only known after complete post-layout extraction and simulation. This paper explored the possibility of applying the successful concept of circuit sizing automation directly to the layout, i.e., developing a simulation-based layout generator. The essay was carried on placement task with the PONDEROUS tool, set towards maximum post-layout accuracy by importing PDK's PCells, optimizing the cells' coordinates in an absolute representation and number of fingers, and, simulating every candidate floorplan. While it addressed the two major limiting

factors of existent layout generators, i.e., plug-and-play capabilities and accurate assessment of post-layout performance degradation, it is exceptionally computationally intensive. Ultimately, this is the first tool in the literature that provides the designer with the complete post-layout performance and area tradeoffs. As future research directions, the authors plan to extend this essay to simulation-based routing, as well as the possibility of incorporating these concepts into the layout-aware sizing optimization of Fig. 2.

REFERENCES

[1] M. Degrauwe, et. al., "IDAC: an interactive design tool for analog CMOS circuits," in *IEEE JSSC*, vol. 22, no. 6, pp. 1106-1116, Dec. 1987.

[2] M. Hershenson, S. Boyd, and T. Lee, "GPCAD: a tool for CMOS op-amp synthesis," in *Proc. of the IEEE/ACM ICCAD*, pp. 296-303, Nov. 1998.

[3] H. Liu, et al., "Remembrance of circuits past: macromodeling by data mining in large analog design spaces," in *DAC*, pp. 437-442, June 2002.

[4] R. Martins, et al., "Many-Objective Sizing Optimization of a Class-C/D VCO for Ultralow-Power IoT and Ultralow-Phase-Noise Cellular Applications," in *IEEE TVLSI*, vol. 27, no. 1, pp. 69 – 82, Jan. 2019.

[5] B. Liu, et al., "An Efficient High-Frequency Linear RF Amplifier Synthesis Method Based on Evolutionary Computation and ML Techniques," in *IEEE TCAD*, vol. 31, no. 7, pp. 981-993, July 2012.

[6] M. Velasco-Jiménez, et al., "Design space exploration using hierarchical composition of performance models," in *Proc. of IEEE ISCAS*, May 2015

[7] BSIM Group. BSIM4 Model. (2017). http://bsim.berkeley.edu/models/.

[8] D. Jepsen, and C. Gelatt, "Macro placement by Monte Carlo annealing," in *IEEE International Conference on Computer Design*, 495–498, 1983.

[9] D. Wong, and L. Liu, "A new algorithm for floorplan design," in *ACM/IEEE Design Automation Conference (DAC)*, 101–107, 1986.

[10] R. Martins, N. Lourenço, and N. Horta, "Analog Integrated Circuit Design Automation – Placement, Routing and Parasitic Extraction Techniques," Springer, 2017. ISBN 978-3-319-34060-9

[11] R. Martins, et al., "Shortening the gap between pre-and post-layout analog IC performance by reducing the LDE-induced variations with multi-objective simulated quantum annealing," in *Engineering Applications of Artificial Intelligence*, vol. 98, 104102, Feb. 2021.

[12] M. Lin, Y. Chang, C. Hung, "Recent Research Development and New Challenges in Analog Layout Synthesis," *ASP-DAC*, pp. 617–622, 2016.

[13] P.-H. Wu, et al., "Exploring Feasibilities of Symmetry Islands and Monotonic Current Paths in Slicing Trees for Analog Placement," in *IEEE TCAD*, vol. 33, no.6, pp. 879 - 892, June 2014.

[14] L. Zhang and Y. Jiang, "Global-Routing Driven Placement Strategy in Analog VLSI Physical Designs," in *MWCAS*, pp. 1239-1242, Aug. 2005.

[15] P.-H. Wu, et al., "Performance-driven Analog Placement Considering Monotonic Current Paths," *IEEE/ACM ICCAD*, pp. 613–619, Nov. 2012.

[16] H.-C. Ou, et. al, "Layout-dependent-effects-aware analytical analog placement," in *IEEE TCAD*, no. 35, vol. 8, pp. 1243−1254, Aug. 2016.

[17] P. H. Wu, et. al. "A Novel Analog Physical Synthesis Methodology Integrating Existent Design Expertise," in *IEEE Trans. Comput.-Aided Des. Integr. Circuits Syst.*, vol. 34, no. 2, pp. 199–212, Feb. 2015.

[18] R. Martins, N. Lourenço, and N. Horta, "LAYGEN II – Automatic Analog ICs Layout Generator based on a Template Approach" in *Genetic and Evolutionary Computation Conf.* (GECCO), pp. 1127-1134, Jul. 2012.

[19] P.-C. Pan, et al., "On Closing the Gap Between Pre-simulation and Post-simulation Results in Nanometer Analog Layouts," in *International Conference on SMACD*, pp. 181-185, Jul. 2018.

[20] H. E. Graeb (Editor), "Analog Layout Synthesis: A Survey of Topological Approaches," Springer, 2010.

[21] C. De Ranter, et al., "CYCLONE: Automated Design and Layout of RF LC-Oscillators," *IEEE TCAD*, vol. 21, no. 10, pp. 1161 – 1170, Oct. 2002.

[22] F. Passos, et al., "Enhanced systematic design of a voltage controlled oscillator using a two-step optimization methodology," in *Integration, the VLSI*, vol. 63, pp. 351 – 361, Sep. 2018.

[23] R. Martins, A. Canelas, N. Lourenço, and N. Horta, "Stochastic-based Placement Template Generator for Analog IC Layout-aware Synthesis", in *Integration, the VLSI Journal*, vol. 58, pp. 485–495, June 2017.

[24] H. Habal, H. Graeb, "Constraint-based Layout-driven Sizing of Analog Circuits," in *IEEE TCAD*, vol. 30, no. 8, pp. 1089-1102, Aug. 2011.

[25] R. Martins, R. Póvoa, and N. Horta, "Current-flow and current-density-aware multi-objective optimization of analog IC placement," in *Integration, the VLSI*, vol. 55, pp. 295-306, Sept. 2016.

[26] A. Gusmão, et al., "Semi-Supervised Artificial Neural Networks towards Analog IC Placement Recommender", in *IEEE ISCAS*, pp. 1-5, Oct. 2020.

An Efficient Modeling Approach for Large Ring Oscillator Based Ising Machines

Markus Graber, Nico Angeli, Klaus Hofmann

Integrated Electronic Systems Lab, Technical University of Darmstadt

Darmstadt, Germany

{Markus.Graber, Nico.Angeli, Klaus.Hofmann}@ies.tu-darmstadt.de

Abstract—Using the Ising model for computation of optimization problems is getting more and more popular. A very promising approach is the use of electrical oscillators, often called Oscillator-based Ising machine (OIM), integrated on a single silicon die. A new modeling approach to handle the complexity of large networks during the design phase is proposed. The simulation runtime is significantly reduced while valuable design insight is provided. It helps to understand the relation between circuit level properties and Ising machine behaviour. It is easy to use and has minimal requirements on the circuit topology. All relevant properties used for the proposed modeling approach are extracted from a transistor-based implementation by a circuit simulator. Systems with large number of nodes can be fast and accurately analyzed and optimized for their computation ability.

Index Terms—Ring Oscillator, Ising computing, Impulse Sensitivity Function (ISF), Oscillator based Ising machine (OIM)

I. Introduction

Today's demand for high computing power is steadily increasing. The solution of optimization problems is of key interest in multiple disciplines. Traditional CMOS based computing struggles solving NP-hard problems efficiently. Approaching optimization problems using the Ising model rises more interests. Implementations based on various physical oscillator principles like lasers or nano-oscillators have been proposed [1], [2]. Recently, the use of electrical oscillator has been demonstrated. Based on its theoretical analysis, it shows a high potential and working hardware prototypes were developed [3]. An implementation on a single silicon die with thousands of nodes is a next step towards competing with today's digital processors. Since mass production of modern CMOS technology is well established, an integrated Ising machine can be easily manufactured in high volume and simply interfaced with existing computers. For the design of such systems, we identify two goals: Firstly, a large number of nodes with versatile connectivity to flexible represent optimization problems shall be achieved. Secondly, a high quality solutions, close to the ground state, with fast execution time shall be obtained. Challenges for the development of such systems are discussed in II, a simulation model is presented in III and its potential shown in IV.

II. Simulation Challenges

The previously demonstrated hardware prototype is based on coupled LC-oscillators [3]. The high area consumption of a single inductor makes this unfavorable for an integrated implementation. By just using transistors, avoiding any passive elements, ring oscillators offer a significant area advantage and are well suited for this application. A first integrated ring oscillator Ising machine prototype has already been presented, where an unit cell occupies roughly $950\,\mu m^2$ [4].

Unfortunately, the oscillators do not always settle to their ground state, the optimal solution. Experimental results show, that the optimal solution can be reached, but the outcome of a single run is statistically distributed [4], [5]. A similar behaviour is observed in simulations using the here proposed model. While small problems are usually solved with the optimal solution in a single run, this is not the case for difficult problems. Also the impact of unavoidable physical effects like device mismatch and parasitic elements need to be considered. Thus, the analysis of circuit properties to increase the solution quality is of key interest and arises a need for a fast and accurate simulation model.

Small quantities of oscillators and couplers can be simulated on transistor level. Large scale simulation of oscillator networks with hundreds to thousands of nodes get unfeasible due to long run-time. Locking of LC-based oscillators can be analytically described using Adler's equation [6]. Even enhanced versions are available, which are applicable for ring oscillators [7]. These are based on simplifications using ideal elements, introducing an error varying with circuit topology and technology. Hence, we propose an easy to use simulation & modeling approach, which is based on numerical extraction by a circuit simulator. With exception of a load replica, the oscillator is treated as a black-box.

III. Simulation Model

For the mathematical description, we define a network for an Ising machine as shown in Fig. 1: It consists of n oscillators with phase ϕ_i, each representing a discrete variable of the Ising model. The oscillators are connected with each other through m couplers with adjustable strength $K_{ij} = K_{ji}$, which is determined by the optimization problem. While phase differences between oscillators are in general continuous, the Ising model only uses two discrete states. Thus, before evaluation an additional sync clock is applied to force alignment of all oscillators in either of two possible phases.

An oscillator is observed at three ports. An *out* port drives its couplers. An *in* port is manipulated by other nodes through

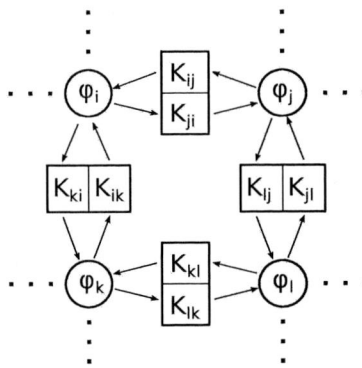

Fig. 1: General structure of the Ising machine network. An oscillator is indicated by its phase ϕ_i and the coupling strength between two oscillators is defined by K_{ij}.

couplers. Additionally, a *sync* port for discretization is included. Fig. 2 shows an example of an oscillator, a four-stage differential architecture with an additional output buffer.

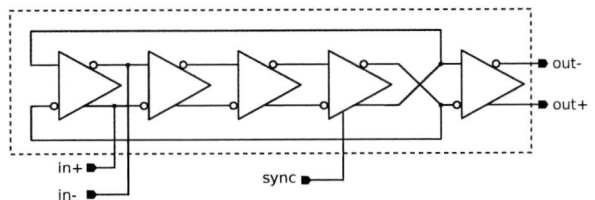

Fig. 2: Example oscillator with an *in*, *out* and *sync* port.

A. Oscillator behaviour

The Ising model is represented by the phase difference between oscillators. Therefore, amplitude perturbations as well as the spectrum of the oscillator are not considered. Thus, we approximate the phase ϕ_i of the i-th oscillator as:

$$\frac{\mathrm{d}\phi_i}{\mathrm{d}t} = \omega_i^* + \sum_{j=1, j \neq i}^{n} f(\phi_i, \phi_j, K_{ij}, t) + f(\phi_i, \phi_{sync}, t) \quad (1)$$

Where ω_i^* is the natural frequency, an initial phase ϕ_i^* can be added as boundary condition. The influence on the phase due to coupling is a function $f(\phi_i, \phi_j, K_{ij}, t)$ of oscillator phases, coupling strength K_{ij} between the i-th and j-th node as well as time. Similar, the phase manipulation from the sync clock is expressed as $f(\phi_i, \phi_{sync}, t)$ with respect to the sync clock phase ϕ_{sync}. These two functions need to be calculated based on circuit properties of the oscillators and couplers. We break these functions down into the contribution of the oscillator and coupler: The phase response of an oscillator on an injected current is described as ISF_{cpl} and ISF_{sync} by the concept of an Impulse Sensitivity Function (ISF), which is handled in section III-B. A coupler forces a current I_{ij} into the oscillator based on the phases, analyzed in section III-C. The oscillator time behaviour is characterized by a delay t_{del} as discussed in section III-D. This allows to replace the general functions of

equation 1 based on circuit extraction, where we now separate between the phase observed at the input $\phi_{in,i}$ and the delayed output $\phi_{out,i} = \phi_{in,i}(t - t_{del}) + \phi_{offset}$:

$$\frac{\mathrm{d}\phi_{in,i}}{\mathrm{d}t} = \omega_i^* + \sum_{j=1, i \neq j}^{n} ISF_{cpl}(\phi_{in,i}) * K_{ij} * I(\phi_{in,i}, \phi_{out,j})$$

$$+ ISF_{sync}(\phi_{in,i}) * i_{sync}(t) \quad (2)$$

ϕ_{offset} preserves the oscillator phase difference $\phi_{in} - \phi_{out}$ in the unperturbed state, enabling any t_{del}. i_{sync} is the current injected by the global sync clock. Jitter can be considered by adding a random, zero-mean variable.

B. Impulse Sensitivity Function

Predicting the excessive oscillator phase, which is the additional phase change on top of the natural oscillation, as a response to a coupling current input is essential. We use the concept of an ISF, which was initially introduced for calculating noise conversion in oscillators [8]. Motivated by the current modelling of the coupler, we use the modified ISF definition $ISF(\phi) = \dfrac{\mathrm{d}\delta\phi}{\mathrm{d}q_{inj}}$ presented in [9]. This 2π-periodic function describes the excessive phase $\delta\phi$ for an injected charge q_{inj}.

The ISF has a linear relationship between excessive phase and injected charge. In general, coupling currents are several orders of magnitude larger than noise current, requiring the analysis of non-linearity effects. We use a Periodic Steady State (PSS) analysis based extraction as linear reference [9]. This is compared with ISF calculation based on transient simulation with varying intensities to evaluate non-linear effects [8]. For our circuit, the observed error from non-linearities is small. If non-linearities get more relevant for other designs, the ISF could be extended with the non-linearity data obtained from transient simulation. ISF functions for the example circuit are shown in Fig. 3, where it is separated between *in* port for coupling and the *sync* port.

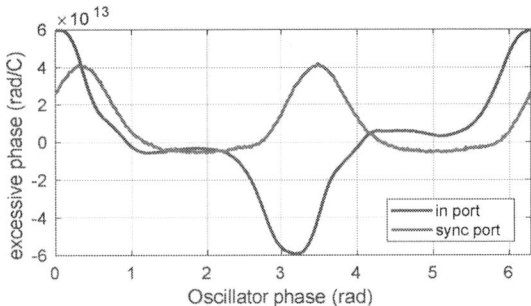

Fig. 3: Example of an oscillator ISF function. The *in* and *sync* port have different characteristics.

C. Coupler Extraction

Forced by coupling, the network of oscillators will naturally align towards one average frequency (assuming all oscillators are locking). Small temporary frequency variations will occur as response to coupling between nodes. Thus, the coupler will operate at a fixed frequency with small perturbations. Therefore, the current I_{ij} generated by the coupler is approximated by a function of the oscillator phases ϕ_i and ϕ_j, allowing interpretation as ϕ_i and the phase difference $\Delta\phi_{ij} = \phi_i - \phi_j$.

$$I_{ij} = f(\phi_i, \phi_j) = \widetilde{f}(\phi_i, \Delta\phi_{ij}) \tag{3}$$

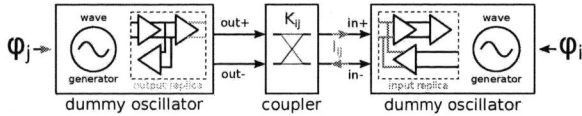

Fig. 4: Testbench for simulating the current I_{ij} injected by the coupler into the oscillator. Replica shown in Fig. 5.

Fig. 4 shows the principle of extraction. The coupler is driven by the *out* port with phase ϕ_j and injects the differential current I_{ij} into the *in* port of the oscillator with phase ϕ_i. Based on the periodic nature, I_{ij} is recorded over a full oscillation period for various phase differences $\Delta\phi_{ij} \in [0, 2\pi[$. Transistor based oscillators can not be used, because they settle towards their ground state of $\Delta\phi_{ij} = 0$. Thus, we use a novel dummy oscillator scheme shown in Fig. 5 allowing precise control of the phase difference while maintaining the output impedance. A recorded waveform of an oscillator with an identical load is replayed with variable phase shift by the wave generator. To replicate the *in/out* port impedance, an "unrolled" oscillator is added. For this "unrolling", the oscillator feedback loop is cut and a copy of the circuit attached. The exact recorded waveform travels through the circuit elements, giving it an accurate output impedance. The loading of the "unrolled" oscillator and the conditions of the recorded waveform must be precisely matched using load dummys. Any differences will cause unwanted phase shift and deviations of the replicated waveform.

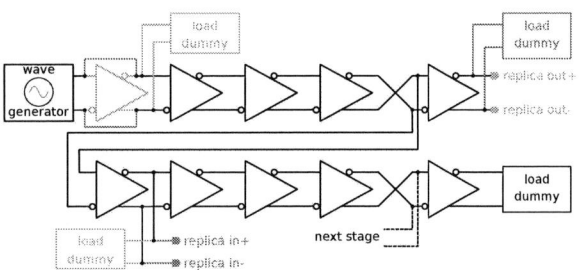

Fig. 5: Dummy oscillator principle with "unrolling". The phase is precisely controllable by the wave generator, while an accurate output impedance is maintained at the replica ports. **red:** for *in* port only, **blue:** for *out* port only, **black:** common

D. Transient behaviour

While the ISF function predicts the (asymptotic) changes of phase of the oscillator it is commonly used with a unit-step response, leading to an instantaneous jump in phase [8],

[9]. The coupling nature with multiple feedback paths from coupling between oscillators make the system highly sensitive to time delays. Due to different *in* and *out* ports, any input propagates through multiple stages until it reaches the output, which is approximated by a delay. This is obtained by injecting a current pulse into the *in* port for different oscillator phases. The delay is measured by the time difference between the rising edge of the current pulse and a noticeable deviation of the oscillator *out* port compared to an identical unperturbed oscillator. The delay appeared to be almost constant across different oscillator phases, although deviations at phases around the low intensity sections of the ISF are hardly identifiable. The coupler delay is small compared to the oscillator delay and hence neglected.

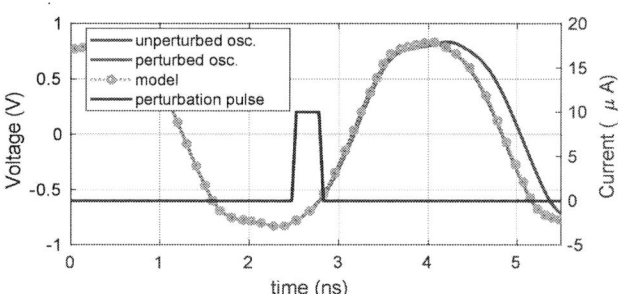

Fig. 6: Transient response of the oscillator model. Comparison of the model with the transistor based circuit. An unperturbed oscillator is shown for reference.

IV. Model application & verification

For verification, equation 2 is implemented in VerilogA enabling a direct comparison to the transistor based circuit in a 65 nm node. Fig. 6 compares the response to a current pulse input. On close inspection, small differences directly after the impulse can be seen and amplitude perturbations are present around 3.7 ns. These settle and a close fit of the model is achieved. Fig. 7 compares the coupling of two oscillators with an initial phase difference of $\frac{\pi}{2}$. Coupling is activated at $t = 0$, forcing both oscillators to the same phase. A qualitative similar settling behaviour is achieved by the model. Small differences are present due to circuit specific effects, which are only relevant for the off-state and turn-on of the coupler.

A. Simulation example

An example for a large system simulation based on a Matlab implementation is shown in Fig. 8. A random generated 10x10 node max-cut problem is simulated over 250 nominal oscillation cycles, equivalent to 1 µs. The phase difference between the individual oscillators is shown in Fig. 8 (a). The discretization during sync cycles around 125 and 225 nominal oscillation cycles is clearly visible by the alignment of the nodes to integer multiple of π differences. Based on the phase representation, the solution for a graph in Fig. 8 (b) is obtained.

Instead of using circuit extraction to simulate the Ising machine, the design methodology can be reversed. Arbitrary

© VDE VERLAG GMBH · Berlin · Offenbach

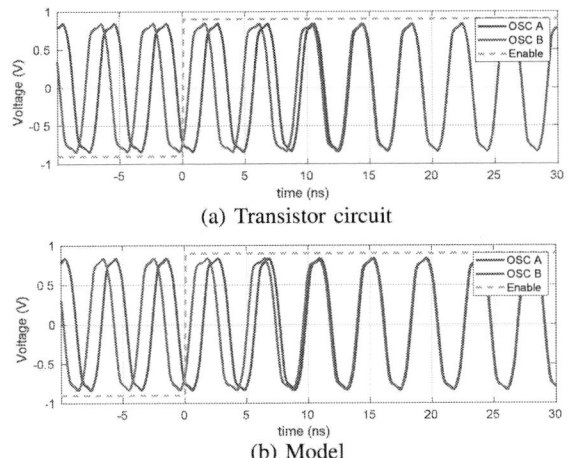

(a) Transistor circuit

(b) Model

Fig. 7: Comparison of the settling behaviour for two coupled oscillators. After a short free-running period, the coupler is enabled at $t = 0$.

ISF and coupler functions can be quickly evaluated to analyse the impact on the optimization problem solution. Based on these high level simulations, specifications can be derived to guide the circuit design process.

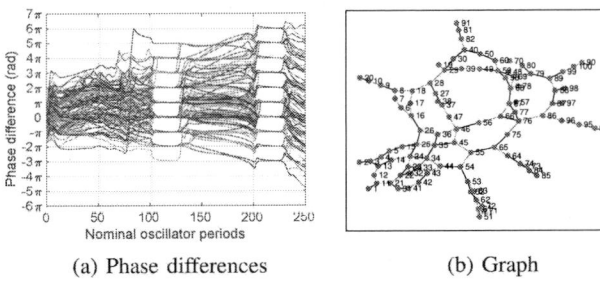

(a) Phase differences   (b) Graph

Fig. 8: Example of a 10x10 node max-cut problem with 120 random positive and negative edges. (a) shows the phase difference between the oscillator nodes, which translates to the graph solution in (b). Blue: in phase, Red: opposite phase

B. Simulation Performance

Using the proposed model, simulation speed is significant improved. Already a transistor level simulation of a 20x20 node problem takes multiple hours, while the model takes only minutes. TABLE I compares the simulation runtime of the model and the transistor level circuit for random generated system of various sizes.

Practical implementation of equation 2 is based on interpolation of the ISF and the coupler current. Assuming a constant number of couplers per node, the runtime for a fixed number of cycles increases linear $\mathcal{O}(n)$ with the number of nodes, while a fully connected network achieves quadratic scaling $\mathcal{O}(n^2)$. The scaling for solving optimization problems is worse, since it might require more cycles with increasing number of nodes.

TABLE I: Simulation runtime for random generated problems with Matlab on a i7-8086K with 64 GB RAM over 1000 oscillation cycles. Setup and post-processing excluded.

Problem	No. nodes	No. couplers	Model Sim. runtime	Transistor level Sim. runtime
5x5	25	35	26.3 s	6 min 9 s
10x10	100	120	43.8 s	36 min 2 s
20x20	400	500	1 min 38 s	5 h 18 min
25x25	625	900	2 min 56 s	8 h 24 min[1]
30x30	900	1400	5 min 50 s	12 h 17 min[1]
40x40	1600	2400	17 min 46 s	-
50x50	2500	3800	37 min 38 s	-
75x75	5625	8400	3 h 52 min	[1]approximated

Due to inefficiencies in our Matlab-based implementation, a worse scaling is observed.

V. CONCLUSION

An intuitive, easy-to-use modeling and simulation concept is proposed. It is based on the oscillator's ISF and uses a replica oscillator scheme to accurately estimate coupling current in a periodic steady state. All relevant functions can be extracted using automated circuit simulation without the need for analytic calculations. Compared to transistor level circuit simulations the runtime is reduced from days and hours to minutes. Based on the individual phases of all oscillator nodes, it provides design insight and allows system and circuit optimization for high quality solutions with fast settling. Using the presented mode, a relation between circuit properties and overall Ising machine can be established.

REFERENCES

[1] P. L. McMahon *et al.*, "A fully programmable 100-spin coherent ising machine with all-to-all connections," *Science*, vol. 354, no. 6312, pp. 614 617, 2016.

[2] S. Dutta, A. Khanna, J. Gomez, K. Ni, Z. Toroczkai, and S. Datta, "Experimental demonstration of phase transition nano-oscillator based ising machine," in *2019 IEEE International Electron Devices Meeting (IEDM)*, 2019, pp. 37.8.1–37.8.4.

[3] T. Wang, L. Wu, and J. Roychowdhury, "Late breaking results: New computational results and hardware prototypes for oscillator-based ising machines," in *2019 56th ACM/IEEE Design Automation Conference (DAC)*, 2019, pp. 1–2.

[4] I. Ahmed, P. W. Chiu, and C. H. Kim, "A probabilistic self-annealing compute fabric based on 560 hexagonally coupled ring oscillators for solving combinatorial optimization problems," in *2020 IEEE Symposium on VLSI Circuits*, 2020, pp. 1–2.

[5] M. K. Bashar, A. Mallick, D. S. Truesdell, B. H. Calhoun, S. Joshi, and N. Shukla, "Experimental demonstration of a reconfigurable coupled oscillator platform to solve the max-cut problem," *IEEE Journal on Exploratory Solid-State Computational Devices and Circuits*, vol. 6, no. 2, pp. 116–121, 2020.

[6] R. Adler, "A study of locking phenomena in oscillators," *Proceedings of the IRE*, vol. 34, no. 6, pp. 351–357, 1946.

[7] P. Bhansali and J. Roychowdhury, "Gen-adler: The generalized adler's equation for injection locking analysis in oscillators," in *2009 Asia and South Pacific Design Automation Conference*, 2009, pp. 522–527.

[8] A. Hajimiri and T. H. Lee, "A general theory of phase noise in electrical oscillators," *IEEE Journal of Solid-State Circuits*, vol. 33, no. 2, pp. 179–194, 1998.

[9] F. Pepe *et al.*, "An efficient linear-time variant simulation technique of oscillator phase sensitivity function," in *2012 International Conference on Synthesis, Modeling, Analysis and Simulation Methods and Applications to Circuit Design (SMACD)*, 2012, pp. 17–20.

The Merging Technique to Simulate Synchronization Mode of Coupled Oscillators

Mark M. Gourary,
IPPM RAS, Russian Academy of Sciences
Moscow, Russia
gourary@yandex.ru

Sergey G. Rusakov, *Senior Member IEEE*
IPPM RAS, Russian Academy of Sciences
Moscow, Russia
rusakov@ippm.ru

Abstract—New approach is applied to the analysis of synchronized Kuramoto oscillators. The approach is based on the technique of merging the oscillator models. This new technique provides computational efficiency due to the developed transformation of the pair of synchronized oscillators into single Kuramoto oscillator. Analytical expressions are derived that define the characteristics of merged oscillators, the transform of coupling functions and the sequence of merging operations. The results of numerical experiments are presented.

Keywords—: coupled oscillators, Kuramoto model, phase macromodels, synchronization, coupling functions.

I. INTRODUCTION

The application of synchronized oscillatory ensembles to the development of oscillatory neural networks [1-3] and oscillatory computing systems [4-7] is a promising direction in design of information processing systems.

The state variables in models of the oscillatory ensembles are represented by the phases of oscillators with respect to the phase of some reference oscillation. The effective description of the oscillatory ensembles is provided by widely used Kuramoto model (KM) [8]. KM is given by the system of ordinary differential equations (ODE). The transient curves of ensemble phases can be obtained by solving this ODE system. But the behavior of the synchronized ensemble is determined by the steady-state phases of the KM and the transient analysis is not necessary in this case.

The steady-state phases can be determined by solving KM algebraic system. However, the features of KM lead to numerical difficulties that arise due to nonmonotonic dependencies of right hand sides of algebraic equations on oscillator phases. This leads to multiple solutions of KM algebraic systems that are usually obtained by the homotopy methods [9]. The special purpose homotopy based approaches to solve steady-state KM have been developed [10-13].

The goal of this paper is to avoid the complicated homotopy based approaches while steady-state analysis with KM and to construct a more economical algorithm to solve this problem. We propose an approach based on sequential replacing the model of pair of synchronized oscillators in an ensemble by the single oscillator model. We called this operation as the merging of oscillators. The paper presents the algorithms for estimating parameters of the merged oscillator and for determining the merging sequence, as well as the algorithms to obtain the phases of the original oscillators.

The rest of the paper is organized as follows. Section II presents basic model expressions for description of coupled Kuramoto oscillators. The proposed approach based on considering a pair of synchronized oscillators in an ensemble as a separate single oscillator is given in Section III. The

algorithm of merging oscillator models is discussed in a short form in Section IV. Some results of applying the considered merging procedure are given in Section V. The conclusions are drawn in section VI.

II. BASIC EXPRESSIONS OF THE KURAMOTO MODEL

A. General Form of KM

Kuramoto model (KM) describes a system of N coupled oscillators defined by their natural frequencies (fundamentals) ω_m and the time-varying instantaneous phases $\theta_m(t)$. KM is represented in the most general form as ODE system [8]:

$$\frac{d\theta_m}{dt} = \omega_m + \sum_{n=1}^{N} u_{mn}(\theta_n - \theta_m), m = 1\ldots N. \quad (1)$$

Here $u_{mn}(\theta_{mn})$ are 2π-periodic coupling functions, $\theta_{mn} = \theta_n - \theta_m$. The most common form of KM uses sinusoidal coupling functions with coupling factors A_{mn} and phase shifts α_{mn}.

$$u_{mn}(\theta_{mn}) = A_{mn}\sin(\theta_{mn} + \alpha_{mn}). \quad (2)$$

The coupling functions of the KM general form (1) can be often approximated with desired accuracy by truncated Fourier series with given number of terms:

$$u_{mn}(\theta_{mn}) = \sum_{k=1}^{K_m} A_{mn}^{(k)} sin(k\theta_{mn} + \alpha_{mn}^{(k)}). \quad (3)$$

Constant term in (3) is omitted because its nonzero value is equivalent to the change of the fundamental ω_m in (1). Similarly, the diagonal couplings $u_{mm}(0)$ are also omitted.

The oscillators are synchronized [8] at sufficiently strong couplings. All oscillators in the fully synchronized ensemble share common frequency $\omega = d\theta_m/dt$ (for all m) and instantaneous phases $\theta_m(t) = \omega t + \varphi_m$ with relative phases φ_m. After the substitution of $\theta_m(t)$ into (1) we have

$$\omega = \omega_m + \sum_{n=1}^{N} u_{mn}(\varphi_n - \varphi_m), \qquad m = 1\ldots N. \quad (4)$$

The set of synchronized oscillators produces free-running oscillations with an arbitrary phase shift, so the phase of one oscillator can be arbitrary set, e.g. $\varphi_N = 0$. Then (4) represents N algebraic equations with N variables: $\omega, \varphi_1, \ldots, \varphi_{N-1}$.

B. The Solution of KM Equations for Pair of Oscillators

In the simplest case of two oscillators (p, q) with unidirectional coupling the synchronization frequency

© VDE VERLAG GMBH · Berlin · Offenbach

coincides with the fundamental of exciting (q-th) oscillator ($\omega = \omega_q$). The relative phase ($\varphi_{pq} = \varphi_q - \varphi_p$) of the excited ($q$-th) oscillator satisfies the equation obtained from (4)

$$\omega_{pq} - u_{pq}(\varphi_{pq}) = 0, \qquad (5)$$

Here $\omega_{pq} = \omega_q - \omega_p$. The solution of (5) exists if the frequency deviation is located within the range of values of the function u_{pq} (locking range)

$$\min_{0 \le \varphi < 2\pi} u_{pq}(\varphi) \le \omega_{pq} \le \max_{0 \le \varphi < 2\pi} u_{pq}(\varphi). \qquad (6)$$

Inequalities (6) define the synchronization condition for the given coupling function $u_{pq}(\varphi)$. The minimal deviation of the frequencies mismatch ω_{pq} from the nearest bound of the locking range define the synchronization depth [14]:

$$d_{pq} = \min\left(\omega_{pq} - \min_{\varphi} u_{pq}(\varphi), \max_{\varphi} u_{pq}(\varphi) - \omega_{pq}\right). (7)$$

The condition (6) can be presented as $d_{pq} \ge 0$.

The equations (5), (6), (7) are illustrated in Fig. 1.

It can be seen that there are at least two solutions (5) for the frequency deviation within the locking range. The complete ODE system (1) applied to a pair of unidirectionally coupled oscillators allows to obtain the stability condition for solution φ_{pq} of (6) in the following form [15, 16]:

$$u'_{pq}(\varphi_{pq}) < 0, \text{where} \quad u'_{pq}(\varphi_{pq}) = \frac{\partial u_{pq}(\varphi)}{\partial \varphi}\bigg|_{\varphi = \varphi_{pq}} \quad (8)$$

Equation (5) can be numerically solved by sweeping φ_{pq} in sufficiently dense set of points in $[0, 2\pi]$ and finding approximate solution that satisfies the stability condition (8).

For mutually (bidirectionally) coupled two oscillators (p, q) system (4) is presented in the form:

$$\begin{aligned}\omega &= \omega_p + u_{pq}(\varphi_{pq}), \\ \omega &= \omega_q + u_{qp}(\varphi_{qp}).\end{aligned} \qquad (9)$$

Denote phase difference of oscillators in the isolated pair as $\overline{\varphi} = \varphi_{pq} = -\varphi_{qp}$. After subtracting the second equation in (8) from the first one we can obtain

$$\omega_{pq} - v_{pq}(\overline{\varphi}) = 0, \qquad (10)$$

where $v_{pq}(\overline{\varphi}) = u_{pq}(\overline{\varphi}) - u_{qp}(-\overline{\varphi})$.

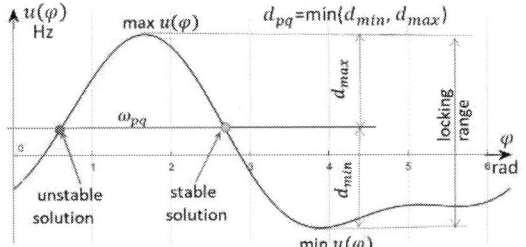

Fig. 1. Solutions of the phase equation for the externally excited oscillator.

Equation (10) has the form (5). Therefore, the synchronization conditions (6) and the stability condition (8) are also valid for (9) after the replacement of u_{pq} by v_{pq}. The solution of (9) can also be determined similarly to Fig. 1, and after finding the phase φ_{pq} the synchronization frequency ω can be obtained through any of the equations (8).

III. MERGING OF SYNCHRONIZED OSCILLATORS WITH EXTERNAL COUPLINGS

Here we consider the problem of replacing a pair of synchronized Kuramoto oscillators with external couplings by a single oscillator with equivalent couplings. The process of forming the oscillator parameters can be called pair merging.

We assume that the pair can be approximately presented as an oscillator with the fundamental equal to the frequency of the synchronized pair (8) after solving (1)

$$\omega_M = \omega_p + v_{pq}(\overline{\varphi}) \qquad (11)$$

Here M is the index of the merged oscillator. We define its phase as the phase of p-th oscillator ($\varphi_M = \varphi_p$). The phase of q-th oscillator is shifted by $\overline{\varphi}$:

$$\varphi_p = \varphi_M, \qquad \varphi_q = \varphi_M - \overline{\varphi}. \qquad (12)$$

Let's now consider the case when each oscillator of the pair is additionally connected to any r-th oscillator:

$$\begin{aligned}\omega &= \omega_p + u_{pq}(\varphi_{pq}) + u_{pr}(\varphi_{pr}), \\ \omega &= \omega_q + u_{qp}(-\varphi_{pq}) + u_{qr}(\varphi_{qr}).\end{aligned} \qquad (13)$$

After subtracting the second equation in (13) from the first one and considering $\varphi_{qr} = \varphi_{pr} - \varphi_{pq}$ the following expressions are obtained:

$$\omega_{pq} + v_{pq}(\varphi_{pq}) = u_{qr}(\varphi_{pr} - \varphi_{pq}) - u_{pr}(\varphi_{pr}), \qquad (14)$$

$$\omega = \omega_p + u_{pq}(\varphi_{pq}) + u_{pr}(\varphi_{pr}) = 0, \qquad (15)$$

Supposing couplings u_{pr}, u_{qr} are small we present the solution φ_{pq} of (14), (15) as the small deviation $\delta\varphi$ from the solution $\overline{\varphi}$ of the uncoupled pair (9) $\varphi_{pq} = \overline{\varphi} + \delta\varphi$. Thus we can linearize functions $v_{pq}(\varphi_{pq})$, $u_{pq}(\varphi_{pq})$, $u_{qr}(\varphi_{pr} - \varphi_{pq})$ using the first order Taylor expansion:

$$v_{pq}(\varphi_{pq}) = v_{pq}(\overline{\varphi} + \delta\varphi) = v_{pq}(\overline{\varphi}) + v'_{pq}(\overline{\varphi})\delta\varphi. \qquad (16)$$

$$\begin{aligned}u_{qr}(\varphi_{pr} - \varphi_{pq}) &= u_{qr}(\varphi_{pr} - \overline{\varphi}) - u'_{qr}(\varphi_{pr} - \overline{\varphi})\delta\varphi \approx \\ &\approx u_{qr}(\varphi_{pr} - \overline{\varphi}).\end{aligned} \qquad (17)$$

$$u_{pq}(\varphi_{pq}) = u_{pq}(\overline{\varphi} + \delta\varphi) = u_{pq}(\overline{\varphi}) + u'_{pq}(\overline{\varphi})\delta\varphi \qquad (18)$$

Here the derivatives are defined similarly to (8). We neglected the term $u'_{qr}(\varphi_{pr} - \overline{\varphi})\delta\varphi$ in (17) because it has the second order of smallness.

After substituting (16), (17) into (14), and considering (10) the equation $v'_{pq}(\overline{\varphi})\delta\varphi = u_{qr}(\varphi_{pr} - \overline{\varphi}) - u_{pr}(\varphi_{pr})$ can be obtained which can be solved with respect to $\delta\varphi$ as:

$$\delta\varphi = \frac{u_{qr}(\varphi_{pr} - \overline{\varphi}) - u_{pr}(\varphi_{pr})}{v'_{pq}(\overline{\varphi})}. \tag{19}$$

The substitution of (18) into (15) considering (11) leads to:

$$\omega = \omega_M + u'_{pq}(\overline{\varphi})\delta\varphi + u_{pr}(\varphi_{pr}). \tag{20}$$

We substitute then $\delta\varphi$ (19) into (20) and replace φ_{pr} by φ_{Mr} from (12). This results in the following:

$$\omega = \omega_M + a \cdot u_{qr}(\varphi_{Mr} - \overline{\varphi}) + b \cdot u_{pr}(\varphi_{Mr}), \tag{21}$$

where $a = u'_{pq}(\overline{\varphi})/v'_{pq}(\overline{\varphi}), b = 1 - a$. $\tag{22}$

Equation (21) represents in KM the merged pair of oscillators p, q. The standard form of the M-th equation of (4) for the oscillator coupled with the only r-th one is following:

$$\omega = \omega_M + u_{Mr}(\varphi_{Mr}), \tag{23}$$

The comparison (21) with (23) shows that the coupling function from r-th oscillator to the merged pair is defined through initial coupling functions as from the r-th generator to the combined pair is defined through the initial coupling functions as follows

$$u_{Mr}(\varphi_{Mr}) = \alpha \cdot u_{qr}(\varphi_{Mr} - \overline{\varphi}) + \beta \cdot u_{pr}(\varphi_{Mr}) \tag{24}$$

The coupling function from the merged pair to r-th oscillator (u_{rM}) is defined by couplings from p-th and q-th couplings (u_{rp}, u_{rq}). Considering (12) we obtain:

$$u_{rM}(\varphi_{rM}) = u_{rq}(\varphi_{rM} + \overline{\varphi}) + u_{rp}(\varphi_{rM}) \tag{25}$$

Expressions (24), (25) fully define couplings after merging. If initial coupling functions are defined by Fourier series (4) then resulting functions also can be presented in the similar form. It is required to perform (24), (25) for each oscillator outside the pair to evaluate all the couplings.

IV. The Sequence of Merge Operations

The algorithm to solve KM equations for synchronized oscillators is based on the sequential merging of oscillator models. The sequence of merge operations and selection of oscillators for merging at each step are the main problems in the development of the considered algorithm.

The merging process is not accurate due to only first order Taylor expansion in (17), (18). We propose to select merging oscillators by the minimal value of the estimated linearization error that increases with the growth of $|\delta\varphi|$ (19). The estimate of $|\delta\varphi|$ can be found by estimating the magnitudes of the numerator and the denominator in (19).

The numerator in (19) can be measured by the power of couplings. We define the power of one mn coupling (P_{mn}) and the input power of m-th oscillator (P_m) by

$$P_{mn} = \int_0^{2\pi} u_{mn}^2(\varphi)\, d\varphi = \frac{1}{2}\sum_{k=1}^{K_m}\left(A_{mn}^{(k)}\right)^2, P_m = \sum_{k=1}^{N} P_{mn}$$

The input power of mn pair ($P^{(mn)}$) and its rms of pair power ($R^{(mn)}$) have the following view:

$$P^{(mn)} = P_m + P_n - P_{mn} - P_{nm}, R^{(mn)} = \sqrt{P^{(mn)}}.$$

Thus, the numerator in (19) is estimated by rms $R^{(mn)}$.

To estimate the denominator (19) we use the synchronization depth (7) with the replacement u_{pq} by v_{mn} for the mutually coupled oscillators (10)

$$d_{mn} = min\left(\omega_{mn} - \min_{\varphi} v_{mn}(\varphi), \max_{\varphi} v_{mn}(\varphi) - \omega_{mn}\right).$$

The value of d_{mn} characterizes the derivative $v'_{mn}(\overline{\varphi})$ due to the approximately direct dependence of the derivative v'_{pq} on the synchronization depth. Therefore, the pair providing the minimum error can be found by minimizing the criterion $S^{(mn)} = d_{mn}/R^{(mn)}$ among all synchronized pairs (m, n):

$$(p, q) = \operatorname*{argmin}_{m,n}\left(\frac{d_{mn}}{R^{(mn)}}\middle| d_{mn} > 0\right). \tag{26}$$

If for $d_{mn} > 0$ the pair (p, q) does not exist then the algorithm is completed. Otherwise, the merge of p-th and q-th oscillators is performed by applying (10), (11), (22), (24), (25). If the result of the algorithm operation is a single oscillator with index L, then the frequency ω_L is equal to the frequency of a fully synchronized ensemble of oscillators.

To determine the phase states of the original oscillators the backward process is performed after the merge of oscillators.

V. Numerical Examples

The proposed method is illustrated below by its application to the simulation of coupled nonlinear oscillators for computing [4–6]. We consider encoding information bits (0/1) using phases of oscillator signals (0/π).

We applied the developed approach to two oscillator based Boolean logic configurations presented in Fig. 2 a,b.

The ideal chain of inverters (Fig. 2a) is excited by the reference oscillator with $f_1 = 100$Hz. Its phase φ_1 is considered as $\varphi_1 = 0$. If natural frequencies of all oscillators are equal to f_1 ($f_1 = f_2 = f_3 = f_4 = 100$Hz) then oscillators' phases are represented by alternating values 0, π corresponding with Boolean '0', '1'. Coupling factors in Fig 1 are the same: $A_{21} = A_{32} = A_{43} = 10$Hz.

Series of numerical experiments was performed to analyze impact of imperfect characteristics of oscillators on deviations of phases. Each experiment was defined by the deviation of oscillators' natural frequencies from the natural frequency of the reference oscillator. For a given frequency deviation Δf of the second oscillator ($f_2 = f_1 + \Delta f$) the deviations for other oscillators are defined by $f_3 = f_1 - \Delta f$, $f_4 = f_1 + 2\Delta f$.

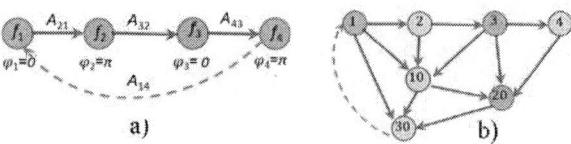

Fig. 2. Oscillator-based Boolean logic networks a) chain of inverters, b) three majority gates with the output inverting.

The first row of the Table 1 ($\Delta f = 0$) contains results for the ideal chain. The next three rows present results for $\Delta f = 1, 2, 4$ Hz correspondingly. Note that at $\Delta f > 5$ the synchronization is not achieved. Values of oscillators' phases evaluated by merging approach are presented in columns 3, 5, 7 and are marked by M. The errors are estimated as the difference from the numerical steady-state solution of full Kuramoto ODE system (1). They are given in columns 4, 6. 8 and are marked with the letter K. One can see that deviations from ideal phases increase with the growth of frequency deviations. Errors in the merging method demonstrate similar dependence. The last two rows of Table 1 contain the results obtained under the presence of the parasitic feedback coupling of the 4-th oscillator with the reference oscillator (dashed line in Fig. 2a). It is seen that even under small coupling factors ($A_{14} = 10^{-3}$, 10^{-2}Hz) errors increase but save relatively small values.

Parasitic couplings are analyzed in the example of three majority gates 10, 20, 30 excited by the chain of inverters 1,2,3,4 (Fig. 2b). Natural frequencies of oscillators 1- 4 are equal to the reference frequency $f_4 = f_3 = f_2 = f_1 = 100$Hz. The natural frequencies of majority gates are defined by the parameter Δf: $f_{10} = f_1 + \Delta f$, $f_{20} = f_1 - \Delta f$, $f_{30} = f_1 - 0.5 \cdot \Delta f$.

Results of numerical experiments performed for feedback coupling factor $A_{1,30} = 0.01$ are presented in Table 2. Table 2 contains phases of majority gates 10, 20, 30 evaluated by merging method (columns 2, 4, 6) and their errors with respect to full Kuramoto ODE system (columns 3, 5, 7).

Table 2 also includes the comparison of the deviation of the synchronization frequency from the reference frequency ($\Delta f_s = f_{syn} - f_1$, mHz) evaluated by merging method and by the conventional Kuramoto ODE solution. (columns 8, 9).

The results in Table 2 show the increase of the deviations from ideal phases with the growth of the frequencies discrepancies and similar increase in the merging errors.

VI. CONCLUSIONS

The method of sequential merging of oscillators models was developed to find steady state solution of Kuramoto equations for coupled oscillators. The proposed technique is an effective alternative to homotopy-based methods for analyzing ensembles of synchronized oscillators.

TABLE I. THE ANALYSIS OF A CHAIN OF INVERTERS WITH DIFFERENT PARAMETERS

Δf Hz	A_{14} Hz	$\varphi_2/2\pi$ M	$E_2/2\pi$ K	$\varphi_3/2\pi$ M	$E_3/2\pi$ K	$\varphi_4/2\pi$ M	$E_4/2\pi$ K
1	2	3	4	5	6	7	8
0	-	0.5	0	0	0	0.5	0
1	-	0.484	$7 \cdot 10^{-6}$	$1 \cdot 10^{-6}$	$1 \cdot 10^{-6}$	0.468	$3 \cdot 10^{-5}$
2	-	0.468	$1 \cdot 10^{-5}$	$1 \cdot 10^{-5}$	$1 \cdot 10^{-5}$	0.434	$5 \cdot 10^{-5}$
4	-	0.434	$33 \cdot 10^{-6}$	$34 \cdot 10^{-6}$	$34 \cdot 10^{-6}$	0.352	$15 \cdot 10^{-5}$
4	10^{-3}	0.434	$14 \cdot 10^{-4}$	$34 \cdot 10^{-6}$	$27 \cdot 10^{-4}$	0.354	$28 \cdot 10^{-4}$
4	10^{-2}	0.434	$11 \cdot 10^{-3}$	$34 \cdot 10^{-6}$	0.022	0.370	$20 \cdot 10^{-3}$

TABLE II. ANALYSIS OF MAJORITY GATES WITH DIFFERENT PARAMETERS

Δf	φ_{10} M	$E_{10}/2\pi$ K	φ_{20} M	$E_{20}/2\pi$ K	φ_{30} M	$E_{30}/2\pi$ K	Δf_s M	Δf_s K
1	2	3	4	5	6	7	8	9
1		$73 \cdot 10^{-7}$		$24 \cdot 10^{-4}$		$11 \cdot 10^{-4}$	45	38
2		$11 \cdot 10^{-4}$		$40 \cdot 10^{-4}$		$17 \cdot 10^{-4}$	83	62
4	0.484	$34 \cdot 10^{-4}$	$73 \cdot 10^{-7}$	$58 \cdot 10^{-4}$	0.493	$21 \cdot 10^{-4}$	142	90
6		$29 \cdot 10^{-4}$		$68 \cdot 10^{-4}$		$21 \cdot 10^{-4}$	187	106
8		$37 \cdot 10^{-4}$		$75 \cdot 10^{-4}$		$21 \cdot 10^{-4}$	221	116

ACKNOWLEDGMENT

The reported study was funded by RFBR, project number 19-29-03012.

REFERENCES

[1] C. Bick, M. Goodfellow, C.R. Laing, et al. "Understanding the dynamics of biological and neural oscillator networks through exact mean-field reductions: a review," J. Math. Neurosc. **10**, 9, 2020. https://doi.org/10.1186/s13408-020-00086-9

[2] P. Ashwin, S. Coombes, R.J. Nicks, "Mathematical Frameworks for Oscillatory Network Dynamics in Neuroscience," Journal of Mathematical Neuroscience 6(2), 2016, pp.1-92.

[3] M. Bonnin, F. Corinto and M. Gilli, "Periodic Oscillations in Weakly Connected Cellular Nonlinear Networks," IEEE Transactions on Circuits and Systems I: Regular Papers, vol. 55, no. 6, 2008, pp. 1671-1684. doi: 10.1109/TCSI.2008.916460.

[4] Y. Fang, V. V. Yashin, D. M. Chiarulli and S. P. Levitan, "A Simplified Phase Model for Oscillator Based Computing," 2015 IEEE Computer Society Annual Symposium on VLSI, Montpellier, pp. 231-236, 2015, doi: 10.1109/ISVLSI.2015.44.

[5] M. Bonnin, F. Bonani and F. L. Traversa, "Logic Gates Implementation with Coupled Oscillators," 2018 IEEE Workshop on Complexity in Engineering (COMPENG), Florence, 2018, pp. 1-4, doi: 10.1109/CompEng.2018.8536222.

[6] T. Wang and J. Roychowdhury, "Design tools for oscillator-based computing systems," 2015 52nd ACM/EDAC/IEEE Design Automation Conference (DAC), San Francisco, CA, 2015, pp. 1-6, doi: 10.1145/2744769.2744818.

[7] G. Csaba, A. Raychowdhury, S. Datta and W. Porod, "Computing with Coupled Oscillators: Theory, Devices, and Applications," 2018 IEEE International Symposium on Circuits and Systems (ISCAS), Florence, 2018, pp. 1-5, doi: 10.1109/ISCAS.2018.8351664.

[8] J. A. Acebrón, et al. "The Kuramoto model: A simple paradigm for synchronization phenomena," Reviews of Modern Physics 77(1), 2005, pp.137 – 185 doi: 77. 10.1103/RevModPhys.77.137.

[9] J.M. Ortega; W.C. Pheinboldt. Iterative solution of equations in several variables. Academic Press, New York, 1970

[10] E. Y. Huang, S. Jafarpour and F. Bullo, "Synchronization of Coupled Oscillators: The Taylor Expansion of the Inverse Kuramoto Map," 2018 IEEE Conference on Decision and Control (CDC), Miami Beach, FL, 2018, pp. 5340-5345, doi: 10.1109/CDC.2018.8619559.

[11] S. Jafarpour, E. Y. Huang, and F. Bullo "Synchronization of Kuramoto Oscillators: Inverse Taylor Expansions," SIAM Journal on Control and Optimization Volume 57, Issue 5 pp. 3101-3602, 2019, doi: 10.1137/18M1216262

[12] S. Jafarpour and F. Bullo, "Synchronization of Kuramoto Oscillators via Cutset Projections," in IEEE Transactions on Automatic Control, vol. 64, no. 7, pp. 2830-2844, July 2019, doi: 10.1109/TAC.2018.2876786.

[13] A. Savostyanov, A. Shapoval, M. Shnirman, "The inverse problem for the Kuramoto model of two nonlinear coupled oscillators driven by applications to solar activity," Physica D: Nonlinear Phenomena, Vol. 401, 132160, 2020, https://doi.org/10.1016/j.physd.2019.132160.

[14] X. M. Zhang, J. B. Zhang, A. Q. Liu, F. Chollet and J. Z. Hao, "Study of injection-locking phenomenon using MEMS tunable laser," 18th IEEE International Conference on Micro Electro Mechanical Systems, 2005. MEMS 2005. Miami Beach, FL, USA, 2005, pp. 80-83, doi: 10.1109/MEMSYS.2005.1453872.

[15] M.M. Gourary, S.G. Rusakov, et al. "Smoothed Form of Nonlinear Phase Macromodel for Oscillators," In: IEEE/ACM Int. Conf. on Comp.-Aided Design, 2008, pp. 807 - 814. doi: 10.1109/ICCAD.2008.4681669.

[16] M. M. Gourary, S. G. Rusakov, S. L. Ulyanov, "Nonlinear Phase Macromodel for the Analysis of Oscillator circuits," In Problems of Perspective Micro- and Nanoelectronic Systems Development - 2014. Proceedings / edited by A. Stempkovsky, Moscow, IPPM RAS, 2014. Part 1. pp. 77-82.

RTL Implementation of MCMC-based Constraints Solver

Moemen Ahmed[1], Youssef Ahmed[1], Younan Nagy[1], Manar Adbel-Rahman[1],
Khaled Salah[2], M. Watheq El-Kharashi[1], Ayub Khan[2]

[1]Department of computer and systems, faculty of engineering, Ain Shams University, Cairo, Egypt.
[2]Siemens Digital Industries Software, Fremont, USA.
Email: [2]Khaled_mohamed@mentor.com, [2]Ayub_khan@mentor.com

Abstract – **Functional hardware verification is one of the most challenging areas in the hardware design cycle. With the increase in the complexity and size of the design, the time needed for verification becomes the largest part of the total design time. The most recent technique in practical verification is constrained random simulation. This method needs a solver to produce random input stimuli that satisfies a pre-defined set of input constraints. The efficiency of the overall verification process depends critically on the speed of the constraints solver and the distribution of the generated solutions. In this paper, we propose hardware acceleration for RTL constraints solver integrated with SystemVerilog. The solver is based on Markov Chains Monte Carlo methods.**

Index Terms - Hardware acceleration, hardware constraint solver, constrained random verification, SystemVerilog.

I. INTRODUCTION

Functional hardware verification is commonly recognized as the hardware design cycle's bottleneck. Due to the big number of stimuli needed to cover the scenarios indicated in the verification plan, constrained random verification is now the most common used technique. In constrained random verification, the testbench for the design under test (DUT) involves a generator of random stimuli, but to avoid generating invalid stimuli, the generator must generate stimuli that satisfy some constraints [1], [8]. Therefore, there is a need for effective constraints solver to produce stimuli that meet input constraints. There are two classifications for constraints solvers: searching algorithms and probability distribution sampling algorithms. Searching algorithms explore exhaustively the whole search space until finding a valid solution. One of the most important searching algorithms is **tree search backtracking** [2].

Probability distribution sampling algorithms such as **Binary Decision Diagrams (BDD)**, **Davis-Putnam-Logemann Loveland (DPLL)** and **Markov chain Monte Carlo (MCMC)** use the set of constraints to construct a probability distribution space which is suitable for sampling solutions or samples, then check the satisfiability of these samples to the pre-defined set of constraints [3]-[7].

Markov chain Monte Carlo (MCMC), According to the empirical results in [3], this algorithm is better than the previous mentioned ones. MCMC is a method that is based on "metropolis-hasting" algorithm. In this algorithm a sequence of Markov-chain transitions is taken to approximate a target distribution using a proposal distribution and a cost function to guide these transitions. "Gibbs" sampling is used to get that proposal distribution. And the cost function is the number of violated constraints. There are many software-based constraints solvers. For hardware-based constraints solvers, only few works exist. Authors in [1] focus on creating a big number of

solutions per run, not the same as our need to produce a single solution as rapidly as possible.

In this paper, we propose a hardware constraints solver based on MCMC algorithm that is integrated with SystemVerilog by a GUI-based parser. The proposed solver allows the user to write and edit System Verilog constraints [9] and then converts these constraints to a synthesizable format and generate text files to be downloaded on the FPGA memory to be solved by Hardware. The architecture of the proposed hardware constraints solver is based on MCMC sampling that is discussed in [1]. The rest of the paper is organized as follows. We describe the preliminaries in section II, then we describe the overall proposed methodology in section III. After that, we describe the proposed GUI-based parser in section IV, then the proposed hardware architecture in section V. We test our software and hardware solvers and provide the experimental results in section VI. Finally, the conclusion is given in section VII.

II. PRELIMINARIES

A. Constraints

Let $x = (x_1, \ldots, x_m)$ be the boolean input variables and $y = (y_1, \ldots, y_n)$ be the integer input variables. The goal is to find a valid solution (x, y) that satisfies the given constraints set. We use the mixed boolean/integer constraints normal form (MBINF) to describe the set of constraints as a conjunction (Λ) of clauses as shown in Equations (1)-(5).

$$MBINF \coloneqq \Lambda\, C_i \tag{1}$$
$$C_i = V\, l_{ij} \tag{2}$$
$$l_{ij} \in \{x_k, \neg x_k, R\} \tag{3}$$
$$R = [\textstyle\sum a_v y_v + b \le 0] \tag{4}$$
$$a_v, b \in \mathbb{Z}. \tag{5}$$

Each clause c_i is a disjunction (V) of literals. A literal l_{ij} can be a boolean literal or an integer literal R, where i is the index of the clause and j is the index of the literal. A boolean literal can be either a boolean variable x_k or its negation $\neg x_k$. An integer literal R is a linear inequality over the integer variables y, where a_v is the coefficient of the integer variable y with index v, b is the bias of the equation and $\mathbb{Z}$ is the integer domain.

B. MCMC Sampler

Algorithm 1 provides the pseudocode for the MCMC sampler, which is the main algorithm of the sampler that performs two types of moves, probabilistic search move (which is shown in the upcoming algorithm 2) and stochastic search move (which is shown in the upcoming algorithm 3). The algorithm starts with assigning random initial values to the variables, then performs a Probabilistic-Search-Move to get a new assignment (x, y), if this assignment does not satisfy the given set of constraints, it performs a sequence of recovery moves until it finds a valid one. These recovery moves can be either a Probabilistic-Search-Move or a Stochastic-Search-Move. The decision of which move to be performed is made randomly and controlled by parameter p_{ss} (probability of stochastic search).

Algorithm 2 provides the pseudocode for the Probabilistic-Search-Move. The algorithm starts with choosing a variable at random. If this variable is boolean, it just flips the assignment of that variable. If it is integer, it proposes a new value from the probability distribution space of the variable that is constructed from the given set of constraints [3]. After that, the new assignment is compared to the current assignment by making some calculations to know if the new one is closer to the valid solution space or not. If it is closer, the move is accepted.

Algorithm 3 provides the pseudocode for the Stochastic-Search-Move (greedy algorithm). This algorithm starts with choosing an unsatisfied clause at random, then it loops on all active variables in that clause. At each cycle, it proposes a new assignment by changing the value of only one variable, then it computes the number of satisfied clauses of the new assignment. At the end of the algorithm, only the assignment that have the largest number of satisfied clauses is accepted by the move and all the others are discarded. Note that in this algorithm changing the value of variables is performed with the same way described in **Algorithm 2**.

Algorithm 1 MCMC Sampler

1: {Given: formula $f(x, y)$; parameter p_{ss}}
2: (x, y): = random assignment
3: (x, y): = PROBABILISTIC_SEARCH_MOVE (x, y)
4: **while** $\neg f(x, y)$ do
5:    **with** probability p_{ss} **do**
6:       (x, y): = STOCHASTIC_SEARCH_MOVE (x, y)
7:    **else**
8:       (x, y): = PROBABILISTIC_SEARCH_MOVE (x, y)
9: **end while**
10: output (x, y)

Algorithm 2 Probabilistic-Search-Move

1: {Given: assignment$(x, y) = (x_1, ..., x_m; y_1, ..., y_n)$}
2: select variable v from $(x_1, ..., x_m; y_1, ..., y_n)$ uniformly
   at random
3: **if** v is Boolean x_k **then**
4:    (x', y'): = $(x_1, ..., x_{k-1}, \neg x_k, x_{k+1}, ..., x_m; y)$
5: **else** v is integer y_k
6:    (x', y') :=PROPOSE (y_k, x, y)
7:    U: = # clauses unsatisfied under (x, y)
8:    U':= # clauses unsatisfied under (x, y)
9:    p: = min $\{1, e^{-(U'-U)}\}$
10: **with** probability p **do**
11:    **return** (x', y')
12: **else**
13:    **return** (x, y)

Algorithm 3 Stochastic-Search-Move

1: {Given: formula f; assignment $(x, y) = (x_1, ..., x_m; y_1, ..., y_n)$}
2: select unsatisfied clause $c_i \in$ f uniformly at random
3: **for each** variable v $\in c_i$ **do**

4:    **if** v is Boolean x_k **then**
5:       (x^j, y^j): = $(x_1, ..., x_{k-1}, \neg x_k, x_{k+1}, ..., x_m; y)$
6:    **else** v is integer y_k
7:       (x^j, y^j) :=PROPOSE (y_k, x, y)
8:    **end if**
9:    U^j := # clauses unsatisfied under (x^j, y^j)
10: **end for**
11: $J^* := \arg(min_j U^j)$
12: **return** (x^{J^*}, y^{J^*})

III. THE PROPOSED METHODOLOGY

The proposed methodology is shown in Fig.1. We provide a graphical user interface to allow the user to write SystemVerilog constraints or upload a SystemVerilog constraints file, also the user can choose how to solve the constraints: using software solver or hardware solver.

IV. THE PROPOSED GUI-BASED PARSER

We provide a GUI-based parser and solver for our methodology in Python and Tkinter. In this section, we illustrate the implementation of the GUI-based parser. The parsing process starts from a modified subset of regular expressions and grammar rules that describe SystemVerilog syntax, then we use PyPEG(a parse tree generator) to generate the parse tree of the input SystemVerilog syntax. We use a modified subset of SystemVerilog grammar rules as we are interested only in the data declaration and constraints parts in SystemVerilog.

The next step is to understand the generated parse tree and convert it to the required format. The GUI-Based parser allows the user to write real SystemVerilog code even with hexadecimal numbers, as we support all types of numbers as in real SystemVerilog codes i.e. binary, octal, decimal, and hexadecimal. In section IV-A, we talk about our supported SystemVerilog syntax and examples about it. In section IV-B, we talk about how to convert these SystemVerilog constraints into a synthesizable format. And finally, in section IV-C, we show some implementation techniques in our proposed system.

A. Supported SystemVerilog Syntax

We support only the constraint part of SystemVerilog. An example of supported SystemVerilog syntax is shown in Listing 1. We support the most important constraints types in the constraint part of SystemVerilog such as **linear inequalities, implication, if statements, inside expressions, and unpacked arrays.**

B. The GUI-Based Parser: Synthesizable Constraints

We need to convert each type of these constraints into a format that can be solved either by the software solver or the hardware solver. First, the linear inequalities: we convert any inequality to the standard format shown in Equation (6):

$$a_1 y_1 + a_2 y_2 + a_3 y_3 + ... + a_n y_n \leq \text{bias} \qquad (6)$$

Where y is integer variable, and a is coefficient. So, any inequality is mapped only to an array of coefficients $[a_1, a_2, ..., a_n]$. Second, the implication constraints as shown in Listing 1 line 14. An implication constraint consists of two parts: condition part and body part. The condition part is mapped to boolean variables which are ORed with integer constraints in the body part as shown in (7). We support only equality constraint for one variable in the condition part.

$$(z \text{ OR } x+1>=0) \text{ AND } (z \text{ OR } x+2>=0) \tag{7}$$

Where z is a boolean variable and x is an integer variable. And then each clause is mapped as its boolean and integer coefficients.

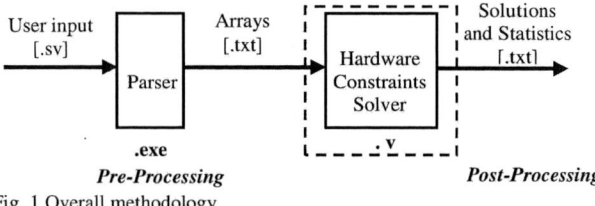

Fig. 1 Overall methodology.

Listing 1 SystemVerilog code

```
1: class c;
2:     rand bit [15 :0] x;
3:     rand integer y;
4:     rand bit [4 :0] arr [12];
5:     rand bit [1:0] z;
6:     constraint legal {
7:         foreach (arr[i]) {arr[i]+10 <=0;}
8:         y inside {-20, [-30: -10]};
9:         x+8>=y;
10:        -5*y>0;
11:        x>0;
12:        z==3 -> x+1<=0;
13:        if(z==2) {x+3<=0;}
14:        z==0 -> {x+1>=0; x+2>=0;}
15:        }
16: endclass
```

And for the unpacked arrays: the array is converted by the parser GUI tool into multiple variables, the number of the variables equals the array size. At the end of parsing the tool encode all of coefficients into binary numbers and write them into files as a memory source for our Verilog codes.

V. THE PROPOSED HARDWARE ARCHITECTURE

In this section we talk about our proposed hardware implementation and architecture. In section V-A, we describe the most hardware implementation features of our proposed hardware methodology. In section V-B, we describe the state machines and the hardware architectures.

A. Implementation Features

Our methodology of solving constraints and generating valid stimuli has many important and useful features and implementation techniques for constrained random verification and simulation field. The most important features and implementation techniques are:

- **The random number generator**: in our algorithm, we need randomization in a specific range, however the normal Linear Feedback Shift Register (LFSR) generates random numbers within the bit-width range not user-specified range as from 4 to 20. So, we map the output of the LFSR to a value in that range using Equation (8). In case the output of LFSR is negative, the mapper uses Equation (9), but if the result of applying modulo is zero, the mapper uses Equation (10).

$$Random\ out = ((LFSR\ out)\ \%\ (max+1- min) + min \tag{8}$$
$$Random\ out = ((LFSR\ out)\ \%\ (max+1- min) + max+1 \tag{9}$$
$$Random\ out = ((LFSR\ out)\ \%\ (max+1- min) \tag{10}$$

- **Output consistency**: which is controlled by a seed value which is determined by the user. So, whenever the user enters the same constraints set and the same seed, he gets the same solution.

- **Constraints conflict detection**: our program counts the number of moves taken from the initial assignment to reach a solution, so when this counter exceeds a specific limit, a conflict detection is announced.

B. Hardware Architecture

The high-level architecture of our proposed RTL constraints solver, shown in Fig.2, has four primary components, solver control unit, probabilistic search move, stochastic search move, and checker.

Solver control unit manages the high-level architecture. it decides between enabling two other blocks, probabilistic search block and stochastic search block, with input probability P_{ss} i.e. probability of stochastic search.

Probabilistic search move architecture, as shown in fig.3, is the hardware implementation of Algorithm 2. The algorithm decides to accept or reject the new value according to probability calculated with Equation (11), where U and V is the number of violated constraints before and after.

$$Probabiltity = \begin{cases} 1, & U-V < 0 \\ 2^{U-V}, & U-V \geq 0 \end{cases} \tag{11}$$

Stochastic search move architecture, as shown in fig.4, is the hardware implementation of algorithm 3. We use a parallel implementation of the stochastic search move that proposes a new assignment for each variable and evaluates it in parallel, which guarantees that the stochastic search move takes only one clock cycle.

The used sampling technique in both moves is based on "Gibbs-Sampling". The detailed block diagram for this technique is shown in Fig.5. This part gets new values for input variables by changing the value of only one variable selected uniformly at random depending on the type of that variable. If it is Boolean, its value is inverted, if it is integer, we pick a new value from a probability distribution which is constructed from the given constraint set. Some integer variables are constrained with discrete values, so in this case we get one of these stored values uniformly at random.

© VDE VERLAG GMBH · Berlin · Offenbach

VI. EXPERIMENTAL RESULTS

We show experimental results for our proposed hardware solver compared with cloud-based tool [7], and the hardware solver proposed in [1]. We compare in terms of time and number of clock cycles. Our proposed hardware solver shows significant improvement. The complete comparison is shown in table I.

Table I Comparison with cloud-based tool EDA.

No. Variables	No. Constraints	Cloud-based Tool [7]	MCMC hardware solver [1]		The Proposed Hardware Solver	
		Time (s)	Time (us)	No. Cycles	Time (us)	No. Cycles
10	20	0.28	0.44	44	0.25	25
12	12	0.28	0.44	44	0.34	34
32	32	0.3	1.12	112	0.92	92
64	64	0.3	2.4	240	1.42	142

VII. CONCLUSIONS

In this paper, we propose an enhanced hardware architecture for RTL constraints solver. Any overhead for the constraints solver has a significant cost on the overall verification efficiency. The proposed accelerated solver speeds up the runtime and it is also integrated with SystemVerilog by a GUI-based parser. In the future work, we will generalize the parser to other HDLs as well and not restricted to SystemVerilog alone.

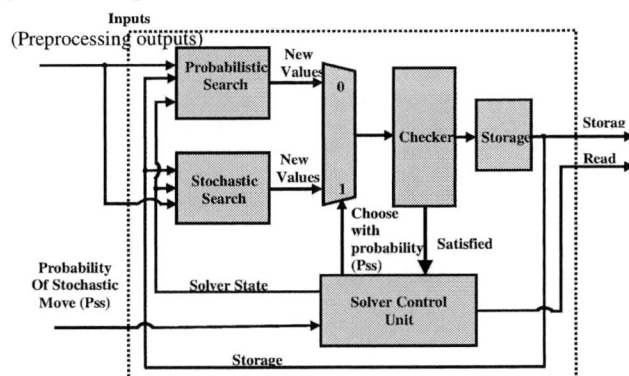

Fig. 2 The constraints solver hardware architecture.

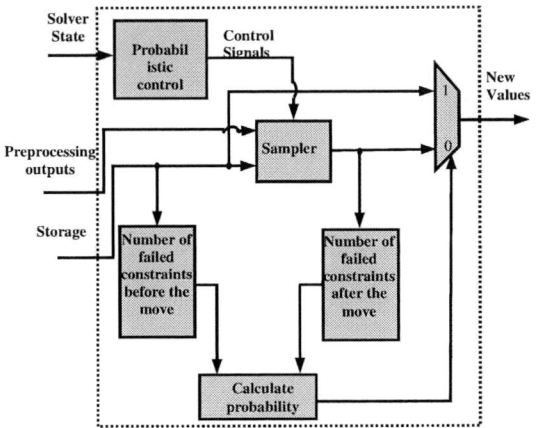

Fig. 3 Probabilistic search hardware architecture.

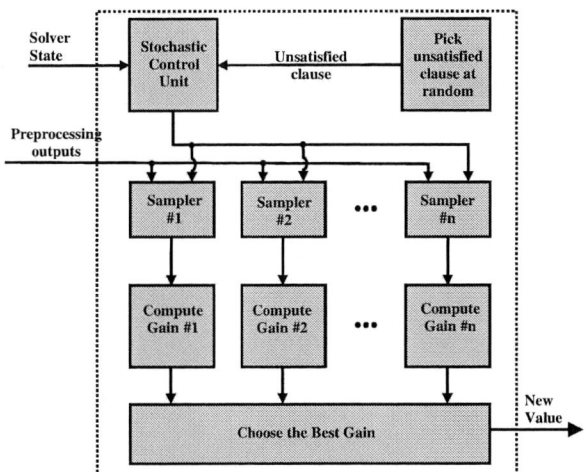

Fig. 4 Stochastic search architecture.

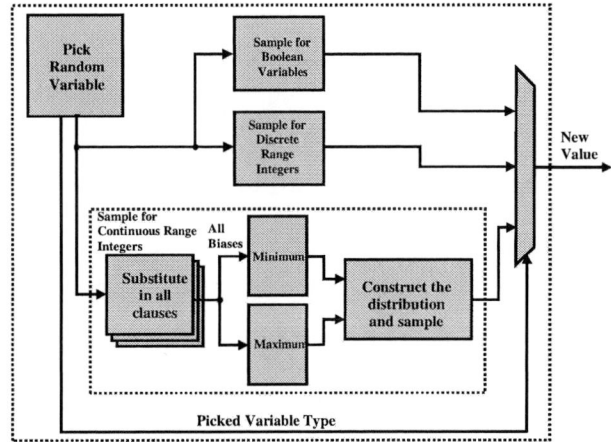

Fig. 5 The sampler architecture.

REFERENCES

[1] T. Welp, N. Kitchen and A. Kuehlmann, "Hardware Acceleration for Constraint Solving for Random Simulation," in IEEE Transactions on Computer-Aided Design of Integrated Circuits and Systems, Vol. 31, no. 5, pp. 779-789, May 2012.

[2] N. Kitchen and A. Kuehlmann, "Stimulus generation for constrained random simulation," 2007 IEEE/ACM International Conference on Computer-Aided Design, San Jose, CA, 2007

[3] M. Hosseini, R. Islam, A. Kulkarni and T. Mohsenin, "A Scalable FPGA-Based Accelerator for High-Throughput MCMC Algorithms," IEEE 25th Annual International Symposium on Field Programmable Custom Computing Machines (FCCM), Napa, CA, 2017.

[4] J. Yuan, A. Aziz, C. Pixley, and K. Albin, "Simplifying Boolean constraint solving for random simulation-vector generation," IEEE Trans. Comput.-Aided Des. Integr. Circuits Syst., vol. 23, no. 3, Mar. 2004.

[5] I. Skliarova and A. de Brito, "Reconfigurable hardware SAT solvers: A survey of systems," IEEE Trans. Comput., vol. 53, no. 11, Nov. 2004.

[6] H. Kim, H. Jin, K. Ravi, P. Spacek, J. Pierce, B. Kurshan, and F. Somenzi, "Application of formal word-level analysis to constrained random simulation," in Proc. Comput.-Aided Verif., Jul. 2008.

[7] https://www.edaplayground.com/

[8] Fathy, Khaled, and Khaled Salah. "An efficient scenario based testing methodology using uvm." 2016 17th International Workshop on Microprocessor and SOC Test and Verification (MTV), IEEE, 2016.

[9] Salah, Khaled. "A Unified UVM Architecture for Flash-Based Memory." 2017 18th International Workshop on Microprocessor and SOC Test and Verification (MTV). IEEE, 2017.

SMACD / PRIME 2021 | 19 – 22 July 2021, Online Event

A study of SRAM PUFs reliability using the Static Noise Margin

E. Camacho-Ruiz*, P. Saraza-Canflanca, R. Castro-Lopez, E. Roca, P. Brox and F. V. Fernandez
Instituto de Microelectrónica de Sevilla, IMSE-CNM (CSIC and Universidad de Sevilla), Sevilla, Spain
*Corresponding author: camacho@imse-cnm.csic.es

Abstract— The use of SRAM cells as key elements in a Physical Unclonable Function (PUF) has been widely reported. An essential characteristic the SRAM cell must feature for a reliable PUF is stability, i.e., it must power up consistently to the same value. Different techniques to measure this stability (and thus improve the PUF reliability) have been reported, such as the Multiple Evaluation method and, more recently, the Maximum Trip Supply Voltage method, the latter using the Data Retention Voltage (DRV) concept. While experimental results have been reported, this paper sheds some light from a different perspective: simulation. In this sense, and using well-known concepts like butterfly curves, static noise margin and voltage-transfer curves, an analysis is provided on why and how stability originates in the cell. Moreover, by simulating the butterfly curve behavior when the supply voltage scales down, it is possible to correlate DRV with stability, thereby confirming the correct theoretical foundation of the MTSV method.

Keywords— SRAM, PUF, reliability, SNM, DRV

I. INTRODUCTION

In recent years, there has been a trend to escalate security in electronic devices. In this sense, Physical Unclonable Functions or PUFs are increasingly being deployed as an alternative to traditional non-volatile memories for their superiority in metrics like attack resiliency, flexibility, and cost [1]. PUFs compute a function which is unique for each device: a challenge is received by the PUF and it generates a response that will be used to form a particular key. A secure PUF should feature these two characteristics: uniqueness (different PUF instances should return different responses for the same challenge) and unpredictability (not even the designer can anticipate the response) [2]. That is why the variability of the manufacturing process (unique and unpredictable) of integrated circuits is often used as leverage to develop these traits. The PUF should be also reliable (the same response should be obtained for the same challenge applied to the same instance); this depends on the PUF's implementation, its behavior when impeding factors like noise are present and the response post-processing [3].

Different PUF implementations have been developed [2] and this work is focused on the use of Static Random Access Memory (SRAM) cells as building components of the PUF [4]. This type of memories is very popular in electronic devices as fingerprint identifier because they can take advantage of general-purpose SRAM arrays, reducing the implementation area dedicated to the PUF itself. One of the most used topologies is the 6T cell, the schematic of which is shown in Fig. 1. Two data-holding cross-coupled inverters (transistors M_1-M_4) form the SRAM core (with Q and $\overline{Q}$ as input/output nodes). Another two transistors, M_5 and M_6, allow access (to write and read information) through the bit-lines (BL and $\overline{BL}$) when the word-line (WL) is active.

The response of an SRAM PUF, which is built from a collection of SRAM cells, is given by the power-up voltage

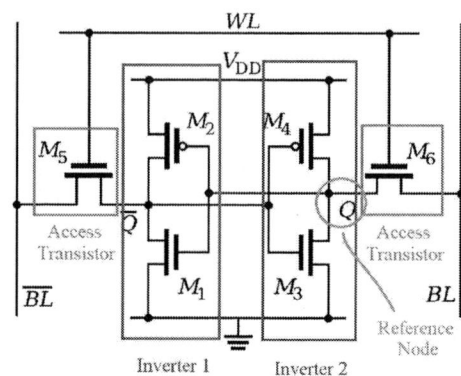

Fig. 1. Schematic of a 6T SRAM cell.

values of these cells. Due to process variations during fabrication, an unbalance in the threshold voltages of the core transistors usually occur. When this happens, node Q sets at either a logic value "1" or "0" when the cell is powered-up. The challenge then is the powering up of the cells and the response is the collective power-up values. Ideally, the cells within an SRAM PUF would be stable. In this context, a stable SRAM cell means that it always powers up to the same state. However, although factors like electrical noise can change the power-up value from one iteration to another, cells are not equally affected by these factors. There are different techniques to improve the PUF reliability that consist in characterizing the stability of each cell so that only the most stable cells are used. The Multiple Evaluation method, for instance, uses multiple power-ups [5][6] to discriminate between stable and unstable cells. Other methods [7][8] rely on the concept of Data Retention Voltage (DRV) to identify stable cells. That value is the minimum supply voltage in a SRAM cell for which the logic values can be stored safely and reliably. More specifically, the Maximum Trip Supply Voltage (MTSV) method [7] correlates the value of DRV to the degree of stability of each SRAM cell.

The goal of this paper is to corroborate the experimental findings in [7] but using simulation this time around: i.e., to demonstrate that the MTSV method can indeed identify stability and thus improve the PUF reliability. This will be done using the concept of Static Noise Margin (SNM) [9]-[11] of the two data-holding inverters. The SNM can be interpreted as the maximum amount of noise that the SRAM can tolerate without changing its state (from "1" to "0", or vice versa). The SNMs are calculated from the butterfly curve generated from the voltage transfer characteristic (VTC) of each inverter. As V_{DD} is raised from 0V to its nominal value during the power-up, the butterfly curve evolves (as it will be shown later) and so does the value of the SNMs. This can be used to relate the DRV used by the MTSV metric with the SNM.

To achieve that goal, in Section II the basic concepts to be used in the simulations (butterfly curve and SNM) are first introduced. Then, in Section III, some representative scenarios

This work has been supported by the PID2019-103869RB-C31/AEI/10.13039/501100011033 project. Pablo Sarazá Canflanca acknowledges MICINN for supporting his research activity through the predoctoral grant BES-2017-080160.

© VDE VERLAG GMBH · Berlin · Offenbach

that will help in examining a variety of SRAM cells (in terms of their core transistors' threshold voltages) are proposed. These scenarios, defined as combinations of the VTCs of the SRAM core inverters, will be simulated at nominal V_{DD} values and the correlation between SNMs and stability described. These results will be used in Section IV where the cell power-up will be carefully simulated and then, the correlation between stability and DRV will be established. Finally, Section V presents the concluding remarks.

II. THE BUTTERFLY CURVE AND THE SNMs

As depicted in Fig. 2, a SRAM butterfly curve is formed by the VTCs of Inverter 1 (input Q vs output $\overline{Q}$) and Inverter 2 (input $\overline{Q}$, in the vertical axis, vs output Q). There are two lobes in the curve, each one with its own associated SNM value, defined as the length of the side of the largest square that can be embedded inside that lobe [11]. The SNM of the left lobe, SNM_0, corresponds to the logic "0" while the SNM of the right lobe, SNM_1, corresponds to the logic "1".

The logic value at node Q after a power-up ("1" or "0") will directly depend on the specific differences in the threshold voltages of the inverters' transistors. For the sake of illustration, consider that the NMOS M_1 and M_3 have the same threshold voltage; then, the power-up state of the cell would only depend on the mismatch between the PMOS M_2 and M_4. With $\Delta Vth = |Vth_2| - |Vth_4|$, the cell should set at "1" in node Q for $\Delta Vth > 0$ and to "0" for $\Delta Vth < 0$. As a consequence, when the actual threshold voltages change from the nominal value due to manufacturing variations, the VTC of the inverter shifts to the left or to the right with respect to the nominal VTC. As illustration, Fig. 2 shows a leftwards shift of the VTC of inverter 1 (Vth_2 has increased due to variability) while the VTC of inverter 2 remains the same (Vth_4 has still a nominal value). Accordingly, SNM_1 increases (to SNM_1*) while SNM_0 decreases (to SNM_0*). In general, the stability of the cell will depend on the mismatches and unbalances in the threshold voltages of the four core transistors. If these unbalances are not significantly large, the cell may frequently change its power-up values at different moments, due to, for instance, the presence of noise [5]. As illustrated before, the unbalances may also shift the magnitudes of the SNMs; so, a correlation arises between the power-up trend to a certain state and the relative magnitudes of SNM_1 and SNM_0. This fact is key in the simulations that will be presented here.

Thanks to the variety of scenarios (i.e., specific unbalances in the threshold voltages) that can be simulated, the relationship between SNM and stability can be analyzed. Furthermore, simulation can be used to inspect the detailed evolution of the butterfly curve and the SNM along the supply voltages experienced during a power-up and, in this way, extract the DRV and relate it with the cell stability. The technology used for the simulations has been a commercial 65-nm CMOS process, with a nominal supply voltage of 1.2V. The size of the NMOS access transistors and the PMOS core transistors is W=80nm and L=60nm, while the size of the NMOS core transistors is W=160nm and L=60nm. The electrical simulation tool used was Spectre®.

III. PREDICTION OF CELL STABILITY

As pointed out before, it is possible to predict the stability of the SRAM cell with the butterfly curve. The SNM of each lobe will show the tendency in stability: the lobe with the higher SNM will be more likely to determine the cell power-

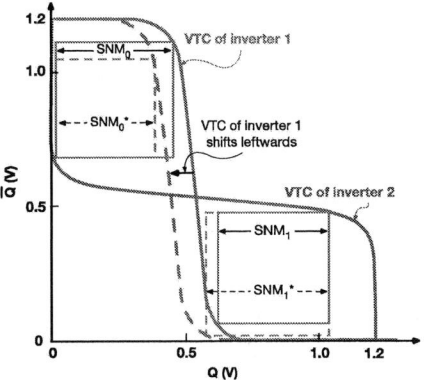

Fig. 2. Butterfly curve representation of the SRAM cell.

up value. While, in the absence of noise, the SRAM will always power up to the value determined by the lobe with the largest SNM, this may not be the case when noise is present. In this case, and during consecutive power-ups, the cell may set at a different value. Therefore, the stability of those cells with very similar SNM_1 and SNM_0 values will dramatically depend on the noise present. However, for those cells with one of their two SNM values sufficiently larger than the other, the cell will power up to the state corresponding to the largest lobe (i.e., larger SNM) even in the presence of noise.

Instead of simulating the infinite combinations of threshold voltages when mismatch is present (each one yielding a specific difference between SNM_1 and SNM_0 and, therefore, a specific power-up trend), this work considers nine representative combinations that accounts for any possible difference between SNM_1 and SNM_0 (and thus any power-up trend). These combinations or scenarios yield SNM differences ranging from zero to maximum values. Therefore, any other combination of threshold voltages outside these nine scenarios will yield a SNM difference equal or smaller than the ones already attained. The scenarios stem from considering, in addition to the nominal situation, two extremal shifts for each inverter's VTC (i.e., leftwards or rightwards), with each shift resulting from the transistors having one of three possible values for its threshold voltage, the nominal threshold voltage value μ_{Vth}, and $\mu_{Vth} \pm 3\sigma_{Vth}$:

- *NOMINAL*: This would be the situation in which both transistors have nominal threshold voltages (e.g., the solid-line VTC of inverter 1 in Fig. 2).

- *RIGHT*: the VTC is shifted right with respect to the NOMINAL one. This is because the variations of the threshold voltages in both transistors shift the VTC to the right (e.g., $Vth_{NMOS} = \mu_{Vth_{NMOS}} + 3\sigma_{Vth_{NMOS}}$, $|Vth_{PMOS}| = |\mu_{Vth_{PMOS}}| - 3\sigma_{Vth_{PMOS}}$).

- *LEFT*: The VTC is shifted to the left with respect to the NOMINAL one (e.g., $Vth_{NMOS} = \mu_{Vt\,NMOS} - 3\sigma_{Vth_{NMOS}}$, $|Vth_{PMOS}| = |\mu_{Vt\,PMOS}| + 3\sigma_{Vt\,PMOS}$, as with the dashed-line VTC of inverter 1 in Fig. 2).

Each inverter will then have one of these three representative VTCs (i.e., NOMINAL, RIGHT or LEFT). This yields nine combinations or scenarios, each one resulting in a specific butterfly curve with a particular lobe asymmetry. The scenarios range from the three ones with equal lobes (and thus equal SNMs, e.g., RIGHT-RIGHT), to the two extreme ones with the largest asymmetry (and, thus, largest difference in SNMs, e.g., RIGHT-LEFT). The other four scenarios (e.g., RIGHT-NOMINAL) present SNM differences in between the two previous groups. Noise may cause the cell to power-up to a different state in scenarios with similar SNMs. Otherwise, the stability of the cell will very much depend on how different SNMs are in relation to the amount of noise present.

The nine scenarios are detailed in Table I, showing the specific threshold voltages values that were used to simulate each scenario. The values of SNM_1 and SNM_0 that result from the simulations (obtained by alternatively sweeping Q and $\overline{Q}$ voltages from 0V to 1.2V with DC analyses at V_{DD} = 1.2V) are shown in columns 3 and 4 of Table II. In the "Stability" column, and after analyzing the SNM values, it is indicated whereas the SRAM cell is deemed unstable or stable, and, in the latest case, to which logic state will most likely power up, which corresponds to the one with the largest SNM value. Those scenarios with two similar VTCs (e.g., Scenario 1) are expected to be less stable than the other ones. On the other hand, those scenarios with two completely different VTCs will correspond to the most stable cells (e.g., Scenario 5). The rest of scenarios (e.g., Scenario 7) are in between those two stability extremes. The butterfly curves of Scenarios 1, 5, 7 and 9 are shown in Fig. 3.

The butterfly curves of Scenarios 5 and 9 have one lobe bigger than the other. This difference in the lobe size, or, equivalently, in the values of SNM, is expected to make the cell stable at a certain value ("1" for Scenario 5 and "0" for Scenario 9) even when noise appears in the cell core. However, Scenario 7, although it is more stable at "0" (i.e., $SNM_0 > SNM_1$), it is expected to be less stable than Scenario 9. This is because the amount of noise needed to change its state is smaller in Scenario 7 than in Scenario 9 and because Scenario 7 has more similar lobes (and, therefore, more equal SNMs). In Scenario 1, since the threshold voltages of the SRAM core transistors are balanced, the lobe sizes are the same and, therefore, they have the same SNM. Any noise present in the cell will produce different power-up values on consecutive power-ups, thus resulting in an unstable cell.

IV. POWER-UP AND DRV SIMULATION

Once the tendency of stability has been predicted with the use of SNM data and DC simulations at V_{DD}=1.2V, this section will detail power-up simulations for each scenario with a two-fold goal: quantify the stability of each scenario and inspect how the butterfly curve evolves along the supply voltages experienced during a power-up, to measure the DRV. Power-up simulations will require a transient analysis, while DRV measurement involves a parametric DC analysis.

A. Power-up simulations

These transient simulations (with noise included) are aimed at evaluating the relationship between the SNMs and the power-up stability. In the process of a power-up, the supply voltage follows a ramp evolution as shown in Fig. 4b.

One hundred power-ups were simulated for each scenario. The results are reported in the fifth and sixth columns of Table II. For the sake of illustration, Fig. 4a shows the power-up of Scenario 1 in the first three seconds, with 15 power-ups in total. As it can be seen, there are 7 power-ups at "1" and 8 at "0". This scenario was one of the most unstable ones: this is because the SNM of each lobe was very close to each other and, consequently, any noise could make the cell evolve into one state or another during a power-up. This fact was

TABLE I. SCENARIOS AND CORRESPONDING THRESHOLD VOLTAGES

SCEN.	Inv. 1	Inv. 2	Vth_{M1} (mV)	Vth_{M2} (mV)	Vth_{M3} (mV)	Vth_{M4} (mV)
1	NOM.	NOM.	374.3	-311.1	374.3	-311.1
2	NOM.	RIGHT	374.3	-311.1	461.3	-223.9
3	NOM.	LEFT	374.3	-311.1	287.3	-398.3
4	LEFT	NOM.	287.3	-398.3	374.3	-311.1
5	LEFT	RIGHT	287.3	-398.3	461.3	-223.9
6	LEFT	LEFT	287.3	-398.3	287.3	-398.3
7	RIGHT	NOM.	461.3	-223.9	374.3	-311.1
8	RIGHT	RIGHT	461.3	-223.9	461.3	-223.9
9	RIGHT	LEFT	461.3	-223.9	287.3	-398.3

TABLE II. SNM VALUES, NUMBER OF POWER-UPS AND DRV SIMULATION RESULTS

SCEN.	Stability	SNM_0	SNM_1	No. of 0s	No. of 1s	DRV
1	Unstable	0.4355	0.4355	51	49	-
2	1	0.4006	0.4963	0	100	0.19
3	0	0.4601	0.3644	100	0	0.22
4	1	0.3644	0.4601	0	100	0.22
5	1	0.3399	0.5315	0	100	0.27
6	Unstable	0.3808	0.3808	56	44	-
7	0	0.4963	0.4006	100	0	0.19
8	Unstable	0.4485	0.4485	48	52	-
9	0	0.5315	0.3399	100	0	0.27

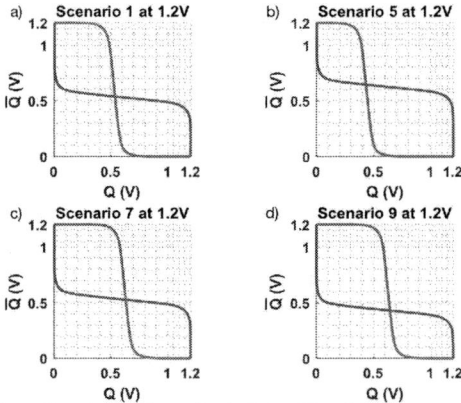

Fig. 3. Butterfly curves of Scenarios 1, 5, 7 and 9 at 1.2V.

predicted by the SNM value of each lobe in Table I and these power-up simulations confirm it.

As expected, unstable cells obtain around half of the power-ups to "0" (and thus half to "1"), as listed in columns five and six of Table II. For those cells labelled as stable by the SNM-based prediction, the percentage of power-ups to the same logic value out of the two possible ones is 100%.

B. DRV simulations

The MTSV method predicts stability according to the experimentally-measured value of a cell's DRV. With DRV being the maximum supply voltage at which the bit flip occurs after writing the non-preferred value, those cells with a lower DRV will be less stable than those with a higher

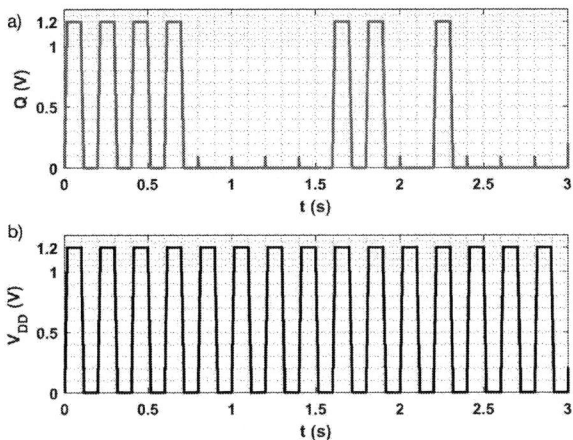

Fig. 4. Power-up of Scenario 1: (a) Evolution of the voltage at node Q on consecutive power-up; (b) Evolution of the supply voltage V_{DD}.

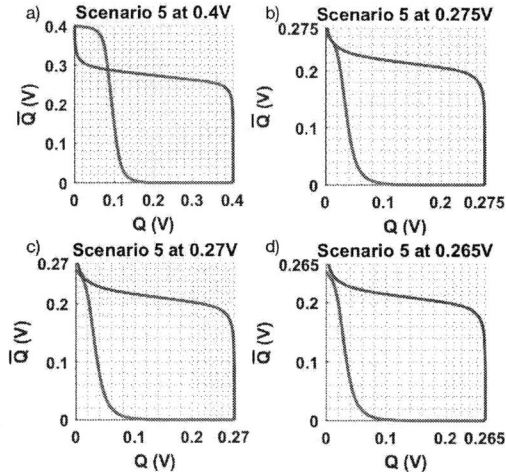

Fig. 5. Butterfly curves of Scenario 5 at 0.4V, 0.275V, 0.27V and 0.265V.

DRV. Since the noise in the cell core is a factor directly responsible for the bit-flipping and the SNM can be interpreted as the maximum amount of noise that the SRAM can tolerate without changing its state, it follows that there is a correlation between DRV and SNM. However, this correlation should not be examined when the cell is fully powered up (i.e., V_{DD}=1.2V), but during the power-up itself since (1) DRV is measured lowering V_{DD} and, (2), the final cell state is decided early during the power-up [12]. The DRV manifests in the butterfly curves when V_{DD} is scaled down from 1.2V to 0V. In doing so, the VTCs degrade to such level that one of the SNMs reduces to zero, i.e., the butterfly curve loses one of its lobes or, in other words, the midpoint at which the two inverters' VTC intersect disappear [13]. The maximum V_{DD} voltage at which this happens is the DRV.

The simulations consisted in generating the butterfly curves of each scenario at different values of V_{DD}, from 1.2V to 0V, in 0.005V increments. For each value of V_{DD}, a butterfly curve is obtained with a DC analysis. As an example, Fig. 5 shows the evolution of the butterfly curve of Scenario 5 around the found DRV value (0.27V). At 0.4V, the curve still retains its original shape (shown in Fig. 3b) and two lobes can still be clearly observed. As V_{DD} is scaled down, one of the lobes decreases until it disappears. The voltage at which that happens is the cell's DRV. Below this voltage, the butterfly curve no longer presents two clearly differentiated lobes. The last column in Table II shows the DRV value of each scenario. As it was predicted by the SNM value at 1.2V, cells with larger DRV, which the MTSV method associates better stability, are cells with much more different SNM values. For unstable scenarios, no value of V_{DD} was found for which a lobe disappears.

V. Conclusions

In this paper, a set of simulations has been presented to analyze the stability of SRAM cells in PUFs and to confirm the validity of the MTSV method presented in [7]. Emulating the process variability, a set of characteristic scenarios was defined and, with the use of DC simulations to obtain butterfly curves, it was possible to set a relation between SNM and stability. Then, consecutive power-ups and detailed V_{DD} scaling down simulations for DRV measurement were performed. With transient analyses it was possible to assess the stability of each scenario on multiple power-ups. For DRV measurement, the evolution and degradation of butterfly curves were simulated when V_{DD} is lowered. Those scenarios where SRAM cells had larger differences between SNM_0 and SNM_1, predicted as the most stable cells, attained higher values of DRV. This confirms, via simulation, that the MTSV method, experimentally demonstrated elsewhere, has a solid foundation for improving the reliability of SRAM PUFs.

References

[1] M. Alioto, "Trends in hardware-security: from basics to ASICs," IEEE Solid-State Circuits Magazine, vol. 11, no 3, pp. 56–74, 2019.

[2] G. E. Suh, S. Devadas, "Physical unclonable functions for device authentication and secret key generation," *in Proc. of Design Automation Conference,* pp. 9-14, 2007.

[3] Intrinsic ID, "The reliability of SRAM PUF," 2017. [Online]. Available: http://www.intrinsic-id.com/. [Accessed: Mar. 17, 2021].

[4] D. E. Holcomb, W. P. Burleson, K. Fu, "Power-up SRAM state as an identifying fingerprint and source of true random numbers," IEEE Transactions on Computers, vol. 58, no 9, pp. 1198-1210, 2008.

[5] I. Baturone, M. A. Prada-Delgado, S. Eiroa, "Improved generation of identifiers, secret keys, and random numbers from SRAMs," IEEE Transactions on Information Forensics and Security, vol. 10, no 12, pp. 2653-2668, 2015.

[6] J. Lee, D. Jee, D. Jeon, "Power-up control techniques for reliable SRAM PUF," IEICE Electronics Express, vol. 16, no. 13, pp. 1-6, 2016.

[7] P. Saraza-Canflanca *et al.,* "Improving the reliability of SRAM-based PUFs under varying operation conditions and aging degradation," Microelectronics Reliability, vol. 118, 2021.

[8] X. Xu *et al.,* "Reliable physical unclonable functions using data retention voltage of SRAM cells," IEEE Transactions on Computer-Aided Design of Integrated Circuits and Systems, vol. 34, no. 6, pp. 903–914, 2015.

[9] E. Seevinck, F. J. List, J. Lohstroh, "Static-noise margin analysis of MOS SRAM cells," IEEE Journal of Solid-State Circuits, vol. 22, no. 5, pp. 748-754, 1987.

[10] B. H. Calhoun, A. P. Chandrakasan, "Static noise margin variation for sub-threshold SRAM in 65-nm CMOS," IEEE Journal of Solid-State Circuits, vol. 41, no. 7, pp. 1673-1679, 2006.

[11] G. Arora, Poonam, A. Singh. "SNM Analysis of SRAM Cells at 45nm, 32nm and 22nm Technology," International Journal of Engineering Research and General Science, Vol. 2, Issue 4, 2014.

[12] A. Alheyasat *et al.,* "Selection of SRAM cells to improve reliable PUF implementation using cell mismatch metric," *in Proc. of Conference on Design of Circuits and Integrated Systems,* pp. 1-6, 2020.

[13] Hulfang Qin *et al.,* "SRAM leakage suppression by minimizing standby supply voltage," *in Proc. of International Symposium on Signals, Circuits and Systems,* pp. 55-60, 2004.

© VDE VERLAG GMBH · Berlin · Offenbach

SMACD / PRIME 2021 | 19 – 22 July 2021, Online Event

Design and Optimization of a Control Algorithm for a Digital Low-Dropout Regulator in System-on-Chip Applications

Benedikt Ohse*, Jun Tan[†], *Member, IEEE*
*Ernst-Abbe-Hochschule Jena, Jena, Germany
benedikt.ohse@eah-jena.de
[†]IMMS Institut für Mikroelektronik- und Mechatronik-Systeme gemeinnützige GmbH (IMMS GmbH), Ilmenau, Germany
jun.tan@imms.de

Abstract—The *System-on-Chip* (SoC) solutions have gained more importance in the field of microelectronics in the past years. The aim of this work is to develop an algorithm for a digital low-dropout regulator (DLDO) for a near-threshold / low supply voltage application. The developed LDO is fully synthesisable and has a high robustness.

In the work, simulation models were developed for a combination of a 9-bit-Successive-Approximation-Register (SAR) and a proportional–integral–derivative (PID) controller. In addition, research and optimizations were carried out regarding the phase shift for better control of the PID controller, the switching between the two algorithms and the stability of the digital LDO.

The design is implemented in a commercial CMOS technology, with the input supply voltage from 0.85 V to 1.8 V. The input voltage range of the developed design ranges from 0.9 V to 1.8 V and the power consumption is smaller as 10 μW at an operating voltage of 0.9 V. In order to achieve a trade-off between a fast settling for large transient and a low ripple at DC, a combined algorithm of a SAR and a PID controller is implemented. We have also optimised the performance for a wide input voltage range (0.9 V to 1.8 V), fast settling time (50 μs at 4 MHz, 1.2 V) and stable setting (0.79 V to 0.84 V at 4 MHz, 1.2 V). The clock and the power consumption are also reduced (average <10 μW with 0.9 V VDC and 4 MHz, while the output power is between 10 μA and 320 μA using constant load.)

I. INTRODUCTION

Modern System-on-Chips (SoC) implement multiple voltage domains in power management unit (PMU) to achieve high performance and ultra-low-power [1]. The digital circuit, which typically consumes the highest power in SoC [2], is normally powered by near/subthreshold supply voltage. Analog low-dropout regulator (ALDO) [3] can achieve low-quiescent current, good power supply rejection (PSR) and fast transient, making it suitable for noise sensitive analog/RF blocks. On the other hand, the ALDO error amplifier is difficult to operate in the near/subthreshold range due to the lack of voltage headroom.

Recently, it has been found that digital LDO (DLDO) fits into this scenario due to its low-voltage capability, high robustness and process scalability. The most common DLDO architecture utilizes a voltage comparator followed by shift registers to enable the control of the pass transistor array [1].

However, the simple shift registers are not suitable for fast transient, since they generate +1/-1 control at every clock. Many state-of-the-art papers optimize the control algorithm for fast transient and better stability [1], [4]. It is noticeable that in other LDOs, the operating voltage and the output voltage are much closer together. This was also noticed during development,as the high dynamic range proved to be particularly demanding. For example, the voltage drop in the publications of [1] and [5] is significantly lower (200 mV and 50 mV). Also, in some cases, as in the publication by [6], the current load (max. 100 mA) is significantly higher for the same load capacitance, which means that the efficiency is higher and the control simpler than with lower currents.

In this paper, we present a combined adaptive control algorithm to enhance the transient response with a better settling stability. The paper is organized as follows. Section II introduces the idea and the implementation of this combined PID and SAR control algorithm. Section III verifies the algorithm by mixed-signal verification. Finally, conclusions are drawn in Section IV.

II. IMPLEMENTATION OF THE COMBINED SAR AND PID ALGORITHM

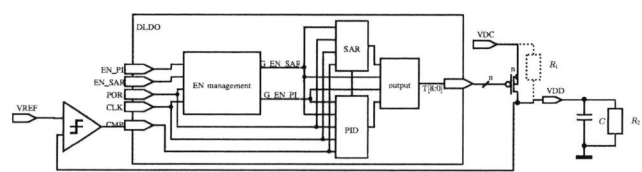

Fig. 1. **Block diagram of the combined design.** The DLDO consists of a logic (EN management) that decides which algorithm is best suited for the current situation. The SAR and PID algorithms are implemented in the modules of the same name. Which code is given to the PMOS array (T) is decided in the output module. R_1 is the spare resistor of the PMOS array. C and R_2 are the load capacitance and the load resistance.

Fig. 1 shows the combined design that consists of a PID and a SAR controller. These can either be selected explicitly by

© VDE VERLAG GMBH · Berlin · Offenbach

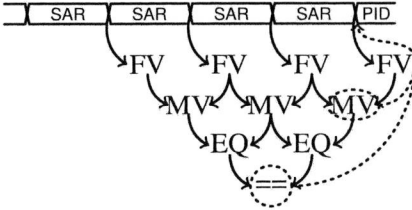

Fig. 2. Controller sequence when changing from SAR controller to PID controller The first line shows which control algorithm is used. At the end of a SAR cycle, a mean value (MV) is calculated from two final values (FV). If two of these consecutive mean values are equal (EQ) or similar, a *flag* is set. If two consecutive *flags* are equal, a transition to the PID controller takes place and the last mean value is passed.

the enable PINs or, if both are selected, they can be operated in an automatic mode. For this switchability, the module **EN management** is required and so that the output is occupied accordingly, the module **output**. The individual controllers, the switching of these and the optimisation of the controllers are described in more detail below

A. PID

The controlled system contains both the transistor array and the load. The step response of the transistors is so fast that its time influence can be neglected, which means they are considered adjustable resistors. These are grouped together as a single resistor (R_1) that can be adjusted by the controller. The load can be assumed to be a parallel connection of a capacitance C and a resistor R_2 connected in series to the transistor array simplified as a resistor. This leads to a PT_1 element as a controlled system with the transfer function:

$$G_S(s) = \frac{\frac{R_2}{R_1+R_2}}{1 + \frac{R_1 R_2 C}{R_1+R_2} \cdot s} = \frac{K_P}{1 + T_1 \cdot s}$$

,where K_P is the amplification and T_1 the time constant. The disturbances due to the variable operating voltage and variable load are not considered as individual disturbance variables, but as a change in the load resistance.

Due to the fact that a comparator is used in the LDO (like in Fig. 1), the P component can only set two output values. This problem also affects the D component, which can only set three values. Only the I component can produce more values. Therefore, the PID controller is reduced to an I controller. In summary, the controller parameters K_{PC} and K_{DC} are 0 and K_{IC} is initially set to 2. The T_C is set to 0.5, which leads to the conclusion that $K_{IC} \cdot T_C$ results is 1.

B. SAR

The advantage of the standard SAR algorithm is that the output voltage is set within approximately as many clock cycles as there are transistors in the transistor array. The disadvantage is that the output voltage is not approximated with small changes in value, but large step changes are made.

C. Switch condition

1) SAR → PID: The Fig. 2 illustrates the approach from SAR to PID graphically. At this point the term "final value" is introduced. This is the value that the SAR controller has set when it has completed a full run. If a final value is stored, then it can be offset against the new one to form an average value before this is overwritten. After this mean value has been saved, it can be compared with the next mean value and if these two do not differ too much, a *flag* (EQ in Fig. 2) is set. The transition to the PID controller occurs when this *flag* has been set twice in succession without interruption. For the transition, the last mean value is then used as "transfer value".

2) PID → SAR: The output of the comparator (Fig. 1) is monitored. When it flips regularly, no switching from PID to SAR is triggered. On the other hand, if it remains the same in a predefined time, the switching to SAR is generated.

D. Optimizations

1) PID: The only parameter that can be changed in the I controller for an improvement is the K_I. This is increased or decreased by means of a counter. A higher K_I can reduce the time until the target voltage is reached, but it increases the overshoot. This can be counteracted by reducing the K_I so that the oscillation in the trapped area is also reduced.

2) SAR: One way to improve the SAR algorithm is to attenuate the transient amplitude of the output voltage during large jumps. In the design, this can be achieved by increasing the load capacitance. Thus, a relative small output voltage swing is achieved, instead of a large dynamic voltage swing, which could potentially ruin the performance of the load.

3) Phase shift: The output capacitance leads to a phase shift of up to 90° between code and output voltage, as can be seen in the Fig. 3 (a). In the publication of [4] it was explained that once the output voltage is at its maximum or minimum, it is approximately the correct output value of the controller. For the detection of the maximum or minimum, an analogue module was developed in the mentioned publication. The assumption of this work is that the output voltage oscillates with a constant frequency. Because then, theoretically, the correct output value of the controller lies in the middle between the two extremes. The Fig. 3 (b) serves as a principle sketch for a better understanding. In this figure, the stored extrema are shown in addition to the code and the output voltage. Various conditions have been defined for saving the extrema, which are not described in more detail here.

III. VERIFICATION

For all simulations a transistor level synthesis was used.

A. Operation point simulation

First, it is checked whether the design sets the output voltage correctly for different reference voltages, operating voltages and current loads. For the Fig. 4, the load was set to $226\,\mu A$ and a point was plotted in the diagram for each simulated reference voltage. To make it easier to see deviations, the points were connected horizontally. It can be seen that the voltages are not only set correctly, but are also very close to the desired point.

© VDE VERLAG GMBH · Berlin · Offenbach

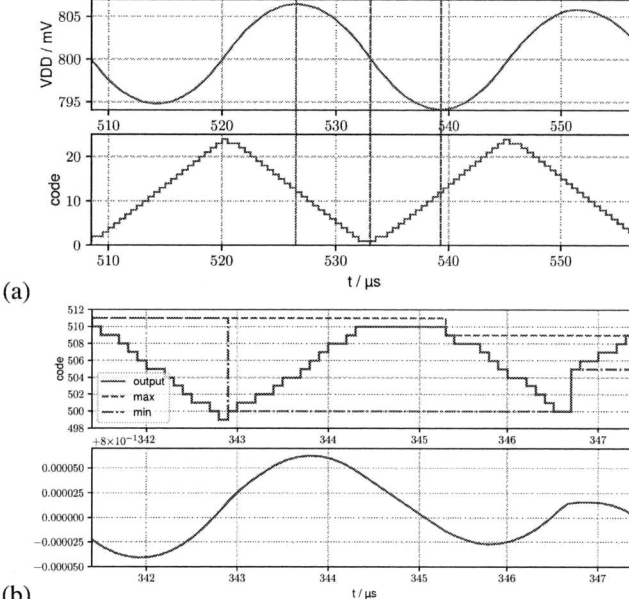

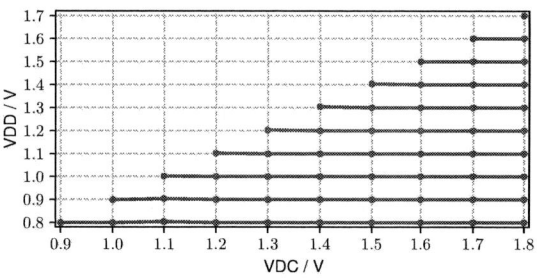

(a)

(b)

Fig. 3. **(a) Phase shift between output voltage and output value** A phase shift of up to 90° occurs between the output value of the controller (code) and the output voltage VDD of the controller. **(b) Reduction of the influence of the phase shift** The partial image shows the output voltage VDD, the code and the stored minimum and maximum values.

Fig. 4. **Operation point simulation of the combined controller** Shown is the output voltage VDD over the operating voltage VDC at different reference voltages VREF. Each point corresponds to a reference voltage. These points were connected in order to better detect rises and falls. 100 nF (discrete) was used as load capacitance.

B. Dynamic simulation

In addition to checking the regulation of load changes, a simulation was also carried out in which a current source was used as the load. This simulates a real circuit. The data for this came from a simulation of a circuit from a previous application. Likewise, areas were marked in the first row and enlarged in the second row. With the lower operating voltage, significantly more time is needed until the voltage is set. At the higher voltage, the amplitude of the oscillation is greater. In general, the voltage is well regulated to the 0.8 V and the voltage is maintained as desired.

C. Power simulation

At the same voltage, using the higher frequency is more energy-consuming because most of the energy is consumed at

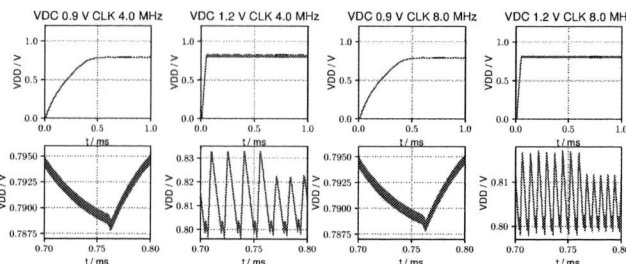

Fig. 5. **Simulation course of the combined controller with realistic current load** The first row shows four simulation curves at different operating voltages and frequencies. The marked sections are shown in more detail in the second row. 100 nF (discrete) was used as the load capacitance.

each edge change. And with a higher frequency, more edge changes occur within a period of time. At the same frequency, the power consumption increases with the voltage.

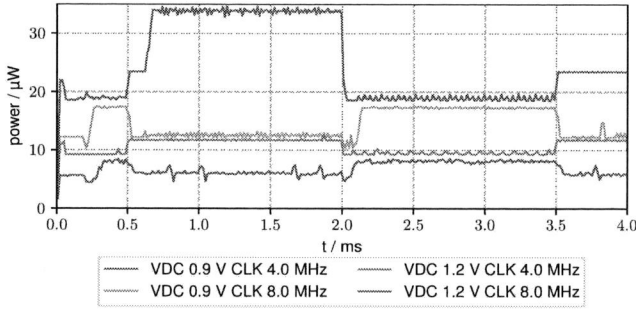

Fig. 6. **Power consumption of the combined controller** The curves show the power consumption of the controller in four different situations. For the calculation, the consumed current was divided into 300 blocks and the integral was calculated in each case, each block was divided by its duration and multiplied by the operating voltage. The current load is changed between 10 μA and 320 μA at 0.5 ms, 2 ms, and 3.5 ms with 100 nF (discrete) as load capacitance

D. Stability

For the determination of the pole and zero points, an AC simulation is usually carried out in which the frequency is swept. Since this simulation uses a functional description for the comparator, this simulation cannot be used. Therefore, 3000 transient simulations were carried out with a clock of 8 MHz, automatic mode, 100 nF (discrete), a current load of 200 μA and a voltage of 1.2 V. The frequency of the input signal was distributed logarithmically up to 4 MHz. The reference voltage was used as a sine wave with the function $VREF = 0.1\,V \cdot sin(2\pi ft) + 0.8\,V$ as the input variable for the system. A DFT (Digital Fourier Transformation) was performed, this signal at the input frequency divided by the input signal and split into real and imaginary parts. This was then used for the Nyquist plot and for the amplitude and phase response in the bottom diagram in the Fig. 7. The Nyquist plot suggests that the system behaves like a PT_2. It follows that the system must have no zero but two pole points. This statement about the stability due to the position of the poles only applies to this one operating voltage and exactly this load - A general statement is not possible. For the system to be unstable, the amplitude of the output signal would have to increase. This is only achieved when the damping of the system is < 0. From the amplitude response it can be read whether the damping $d = 0, > 0$ or $0 < d < 1$. If the amplitude response has no elevation, as is the case in the Fig. 7, the damping of the system is greater than zero and thus the system is stable. [7]

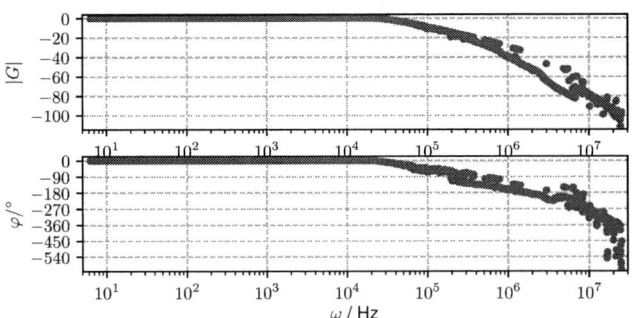

Fig. 7. **Bode plot of the overall system** The Bode plot of the entire system shown consists of the upper graph, which shows the amplitude response of the transfer function G, and the phase response of the system in the lower graph.

In Table I, the comparison with state-of-the-art DLDO paper is shown. This proposed DLDO provides largest supply voltage range and requires relative low clock frequency.

TABLE I
COMPARISON OF VARIOUS PUBLICATIONS

	voltage drop	use ADC	clk / MHz	I Load (max)	algorithm
[8]	50 mV	no	1 - 10	200 µA	shift register
[9]	0.05 - 0.75 V	no	400	4.6 mA	PID
[10]	0.05 - 0.95 V	no	10	1.5 mA	SAR
[6]	0.1 - 0.7 V	no	500	100 mA	dual-loop shift register
[5]	50 mV	no	-	33.2 mA	dual-loop shift register
[11]	200 mV	no	500	100 mA	dual-loop shift register
[4]	0.05 - 0.7 V	no	100	2 mA	SAR + PD + PWM
this work	50 mV - 1 V	no	8	200 µA	PID + SAR

IV. CONCLUSION

In this paper we present a combination of a SAR and a PID controller, the associated switching between the two controllers, a research of the phase shift for a better control of the PID controller and a stability consideration of the digital LDO. The switching from SAR to PID is triggered, when the averaged middle value in three detection cycle outside a defined range. The switching from PID to SAR is triggered, when the output voltage remains unstable for longer than necessary clock cycles. To reduce output swing amplitude, the output code is directly assigned with the calculated optimized

value in order to minimize the phase shift. The bode diagram of the proposed digital LDO is plotted as well, so that the stability of the digital LDO can be revealed and compared to the traditional analog LDO. With the proposed digital LDO, the input voltage range is achieved from 0.9 V to 1.8 V, while the settling time is achieved for 50 µs at 4 MHz clock and 1.2 V VDC. The ripple is limited only in the range of 0.79 V to 0.84 V. When the typical clock is 4 MHz, the power consumption is as low as $<10\,\mu W$ with 0.9 V VDC, while the output current can be supported between 10 µA and 320 µA.

ACKNOWLEDGEMENT

The BICCell project is funded by DECHEMA (Gesellschaft für Chemische Technik und Biotechnologie e.V.) via the AiF (Arbeitsgemeinschaft industrieller Forschungsvereinigungen) as a joint industrial research project (IGF) by the Federal Ministry of Economics and Energy (BMWi) by resolution of the German Parliament under the reference 21174 BR/2.

REFERENCES

[1] Y.-J. Lee, W. Qu, S. Singh, D.-Y. Kim, K.-H. Kim, S.-H. Kim, J.-J. Park, and G.-H. Cho, "A 200-mA Digital Low Drop-Out Regulator With Coarse-Fine Dual Loop in Mobile Application Processor," *IEEE Journal of Solid-State Circuits*, vol. 52, Jan. 2017.

[2] X. Liu, J. Zhou, Y. Yang, B. Wang, J. Lan, C. Wang, J. Luo, W. L. Goh, T. T. Kim, and M. Je, "A 457 nw near-threshold cognitive multi-functional ecg processor for long-term cardiac monitoring," *IEEE Journal of Solid-State Circuits*, vol. 49, no. 11, pp. 2422–2434, Nov 2014.

[3] Y. Lu, Y. Wang, Q. Pan, W. Ki, and C. P. Yue, "A Fully-Integrated Low-Dropout Regulator With Full-Spectrum Power Supply Rejection," *IEEE Transactions on Circuits and Systems I: Regular Papers*, vol. 62, no. 3, pp. 707–716, March 2015.

[4] L. G. Salem, J. Warchall, and P. P. Mercier, "A Successive Approximation Recursive Digital Low-Dropout Voltage Regulator With PD Compensation and Sub-LSB Duty Control," *IEEE Journal of Solid-State Circuits*, vol. 53, Jan. 2018.

[5] Y. Huang, Y. Lu, F. Maloberti, and R. P. Martins, "A Dual-Loop Digital LDO Regulator with Asynchronous-Flash Binary Coarse Tuning," in *2018 IEEE International Symposium on Circuits and Systems (ISCAS)*. Florence: IEEE, 2018.

[6] M. Huang, Y. Lu, S.-W. Sin, U. Seng-Pan, and R. P. Martins, "A Fully Integrated Digital LDO With Coarse–Fine-Tuning and Burst-Mode Operation," *IEEE Transactions on Circuits and Systems II: Express Briefs*, vol. 63, no. 7, Jul. 2016.

[7] S. Zacher and M. Reuter, *Regelungstechnik für Ingenieure*. Wiesbaden: Springer Fachmedien Wiesbaden, 2017.

[8] Yasuyuki Okuma, Koichi Ishida, Yoshikatsu Ryu, Xin Zhang, Po-Hung Chen, Kazunori Watanabe, Makoto Takamiya, and Takayasu Sakurai, "0.5-V input digital LDO with 98.7% current efficiency and 2.7-µa quiescent current in 65nm CMOS," in *IEEE Custom Integrated Circuits Conference 2010*. San Jose, CA, USA: IEEE, Sep. 2010.

[9] S. B. Nasir, S. Gangopadhyay, and A. Raychowdhury, "5.6 A 0.13µm fully digital low-dropout regulator with adaptive control and reduced dynamic stability for ultra-wide dynamic range," in *2015 IEEE International Solid-State Circuits Conference - (ISSCC) Digest of Technical Papers*. San Francisco, CA, USA: IEEE, Feb. 2015.

[10] Y. Li, X. Zhang, Z. Zhang, and Y. Lian, "A 0.45-to-1.2-V Fully Digital Low-Dropout Voltage Regulator With Fast-Transient Controller for Near/Subthreshold Circuits," *IEEE Transactions on Power Electronics*, vol. 31, no. 9, Sep. 2016.

[11] K.-C. Woo, T.-W. Kim, S.-K. Hwang, M.-J. Kim, and B.-D. Yang, "A fast-transient digital LDO using a double edge-triggered comparator with a completion signal," in *2018 International Conference on Electronics, Information, and Communication (ICEIC)*. Honolulu, HI: IEEE, Jan. 2018.

© VDE VERLAG GMBH · Berlin · Offenbach

A Differential Public PUF Design for Lightweight Authentication

Shengyu Duan[1,2,*], Gaole Sai[3,4]

[1]School of Computer Engineering and Science, Shanghai University, Shanghai 200444, China
[2]State Key Laboratory of Computer Architecture, Institute of Computing Technology, Chinese Academy of Sciences, China
[3]Guangdong Provincial Key Lab of Robotics and Intelligent System, Shenzhen Institutes of Advanced Technology, Chinese Academy of Sciences, China
[4]CAS Key Laboratory of Human-Machine Intelligence-Synergy Systems, Shenzhen Institutes of Advanced Technology, China

Abstract—**Physical Unclonable Functions (PUFs) have emerged as a promising primitive to provide a hardware keyless security mechanism for integrated circuit applications. Public PUFs (PPUFs) address the crucial PUF vulnerability of requiring databases to store the secrete reference information. This paper investigates one of the PPUF families, differential PPUF (dPPUF), which exploits complexly interacted logic to create time gap between actual execution and simulation, and to prevent simulation attacks. Conventional dPPUFs have high costs. We observe the repressers have large impact on simulation time, but take a great proportion in current dPPUF structure. We thereby present a represser network with specially designed interconnection and shared logic, to maintain a high circuit complexity and to reduce the area cost. We show the proposed dPPUF has similar resilience over simulation attacks as the conventional design. Area cost can be reduced by up to 44%, and thus the proposed design can be used for lightweight authentication.**

Index Terms—**Physical Unclonable Function (PUF); Public PUF (PPUF); Lightweight authentication; Hardware security**

I. INTRODUCTION

The exponential increase of Internet of Things devices leads to major security concerns. It is crucial to provide fundamental security services like authentication. Physical Unclonable Functions (PUFs) emerged as a promising candidate for keyless security mechanisms and resource-constrained applications. PUFs map a set of challenges to a set of responses, known as challenge-response pairs (CRPs), by leveraging inherent and unclonable process variations and uniquely identifying silicon chips in integrated circuits. There is currently a variety of PUF implementations, such as delay-based [1], [2], memory-based [3], [4], mixed-signal PUFs [5].

For the most PUF-based authentication protocols, a verifier has prior access to a database storing a subset of CRPs or a model of the PUF, but this poses vulnerabilities as the integrity of some verifier nodes may get compromised [6], [7]. One approach to counter this issue is public PUF (PPUF), where the mathematical model of a PPUF is publicly accessible and emulatable, but the emulation takes significantly longer than initiating the actual PPUF, and can be noticed in the protocol [8], [9]. A few PPUF architectures have been proposed such as differential PPUF (dPPUF) [10], matched PPUF [11] and digital PPUF [12].

In this work, we investigate dPPUF for lightweight authentication. We evaluate the security and cost for different structure of dPPUF, and present a design for dPPUF repressers to reduce the overall area, while maintaining the security and quality.

II. BACKGROUND

A. Differential PPUF Architecture

Figure 1 shows the overall architecture of a typical dPPUF [10], which has two structurally identical blocks, each composed by boosters (*i.e.*, XOR gates) and repressers. The number of cells at each booster/represser stage is determined by the width of the challenge, and each stage provides highly shared mixing signals. Each two stages are interacted by a partly maximally connected and partly random interconnection.

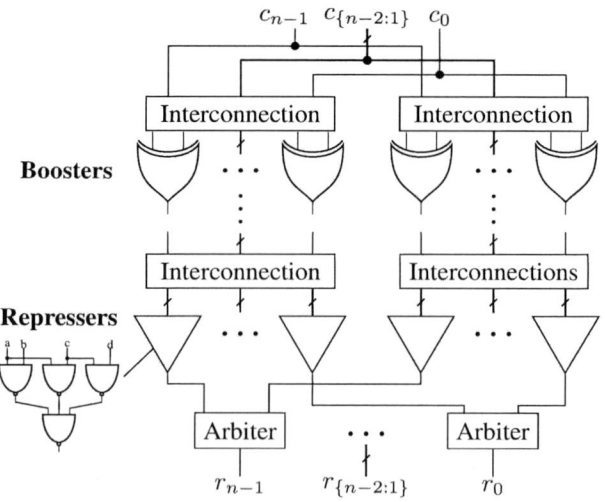

Fig. 1. Circuit architecture of conventional differential PPUF [10]

---

*Corresponding author (email: sduan@shu.edu.cn).
The work was supported by State Key Laboratory of Computer Architecture (ICT, CAS) under Grant No. CARCH201909.

For an authentication scheme, a verifier challenges a dPPUF by triggering rise transitions on some of the inputs, which propagate to the boosters. In a booster stage, each transition at the input causes an output switching, and some transitions fan out due to the interconnection. Thus, the amount of transitions exponentially increases through multiple interconnection and boosters, until the propagated signals arrive at the repressers.

A represser cell eliminates some arrived transitions, because a transition is propagated only when multiple and specific inputs are switched, as can be seen from the represser circuit in Fig. 1. Response of a dPPUF is determined by the frontier signals, which are the ones that firstly cause transitions on the outputs. For booster cells, the frontier signals can be easily simulated, as the gate characteristics of a PPUF are public. For represser cells, simulations of the frontier signals are highly costly, because any transitions might be repressed, depending on the arrival time. So an attacker has to precisely examine a large number of transitions at the represser inputs, which are exponentially increased through the previous booster stages.

Signals from the two nominally identical but physically unique (*i.e.*, due to process variations) dPPUF blocks race to the arbiters. An arbiter produces a response bit based on the first arrived transition from the two blocks, and locks the response bit. In this work, we design an arbiter circuit by using a SR latch and an edge detector.

B. PPUF Security Attacks and Susceptibility

There are several potential security attacks for dPPUFs and other PPUF architectures. Firstly, the primitive of PPUF assumes the simulation to reveal the relationship of a CRP cannot be completed in a reasonable amount of time, compared with the execution time on the actual hardware. A large enough execution-simulation gap (ESG) is therefore expected to reduce the susceptibility to simulation attacks [9]. For dPPUFs, ESG can be widened by increasing the height and thereby increasing the circuit complexity [10].

Guessing attacks are another type of attack, which exploit the statistics of CRPs [7], [10]. For instance, the probability of each response bit is desired to be 0.5, avoiding a consistent output. The correlation between any challenge and response bits, or any two response bits, or any intermediate and response bits, is expected to be 0.5, preventing a simple guessing for the response from the relevant information. The probability of a response bit switching for the challenges with hamming distances of 1, also known as avalanche effect property, is expected to be 0.5 as well, so that an attack cannot predict a CRP from a previous verified and similar input challenge. One can choose to only verify those CRPs having desired statistical metrics, or the ones not causing avalanche effect, to improve the resilience over guessing attacks.

Few works have investigated the susceptibility of PPUF to machine learning-based attacks. To the best of the authors' understanding, PPUFs generally have high resilience against machine learning-based attacks. Firstly, it is exhaustive to collect enough CRP samples for training due to the extremely complex structure of PPUFs. Secondly, any model that cannot

infer the response in a comparable time as the actual chip would be considered as an unsuccessful attempt.

III. DIFFERENTIAL PPUF FOR LIGHTWEIGHT AUTHENTICATION

A. ESG and Cost Evaluation

The security primitive of dPPUFs relies on a substantial ESG to prevent a straightforward simulation attack. It has been reported a greater simulation effort will be taken for a dPPUF with a greater height [10], which increases the complexity in both circuit structure and circuit timing. We further investigate the relationship between ESG and the heights of booster and represser stages. We denote the heights of booster and represser stages as h_b and h_r, respectively, and simulate the dPPUFs with different h_b and h_r.

For a preliminary experiment, we simulate a 8-bit dPPUF with h_r equaling 2, and increase h_b from 3 to 7. Fig. 2 (a) demonstrates the simulation time for different h_b. The simulation time is averaged from 10 CRP simulations and normalized. As can be seen, a linear increase on simulation time is caused when a greater h_b is applied. According to Fig. 1, each signal delivered into a booster layer would have a fan-out of 2 on average, which ideally doubles the number of signal transitions through each booster stage. However, for a transient simulation, an attacker only needs to simulate until the first transitions arrive at all arbiters, which lock the arbiters. On the other words, although the total number of candidate transitions might be doubled through a booster layer, it is not necessary to simulate all transitions propagated to the outputs, especially for those late arrived signals. As a result shown in Figure 2 (a), the observation limits the increase of ESG when implement more booster layers into a dPPUF.

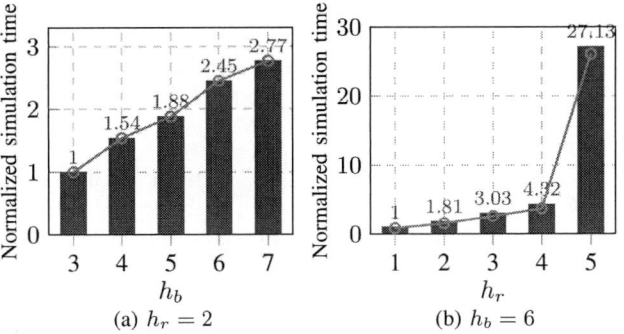

(a) $h_r = 2$ (b) $h_b = 6$

Fig. 2. Simulation efforts for 8-bit dPPUFs

The relationship between simulation effort and the height of represser stages is shown in Fig. 2 (b), where h_b is set to be 6 and h_r is increased from 1 to 5. Compared with the results of Fig. 2 (a), a more significant increase of simulation effort can be observed in Fig. 2 (b). Some early arrived transitions might be eliminated by the represser, unless multiple and the specific inputs are switched, so more candidate transitions caused by the previous stages need be simulated, increasing the simulation effort. The exact simulation time may certainly be varied due to different simulation methods, so we hereby only consider the comparison between the increases of simulation time caused by different h_b and h_r. According to Fig. 2, the simulation time can be more significantly increased by increasing the height of represser stages. This fact is independent of the simulation methods, because it is determined by overall structure of a dPPUF. Implementing more repressers not only enlarges the circuit scale, but increases the complexity of circuit timing as well.

Note that there are other considerations when adjust the height of booster and represser layers. Specifically, it should be avoided to implement too many represser stages into a dPPUF, as this may repress most propagated transitions, and the responses would be thereby largely determined by the process variations of the arbiters, independently of the boosters and repressers. This work will only focus on the resilience of dPPUFs over simulation attacks.

Generally, a greater ESG can be realized by increasing the height of represser stages. However, repressers induce greater costs, compared with the other components of a dPPUF. In Fig. 3, we show the transistor counts and the proportions of different components for the dPPUFs of Fig. 2 (b). As can be seen, nearly 20% area overhead would be induced by increasing one stage of repressers in this case. Repressers make a significantly large proportion in a dPPUF with high security.

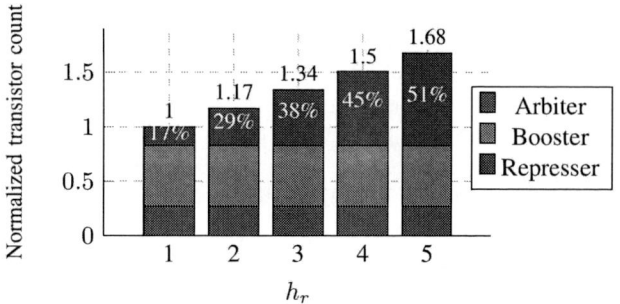

Fig. 3. Transistor counts of different components in 8-bit dPPUFs ($h_b = 6$)

B. Proposed Represser Network

In order to provide a simulation attack-resilient and lightweight authentication, the cost of repressers of dPPUFs has to be reduced. We observe some represser cells located at the same layer may partly share some inputs, and thus some logic cells can also be shared to reduce the area. Based on the above observation, we propose the following technique to design an interconnection interacting two represser stages or booster and represser stages:

Assume O_i^j is an output of a represser located at jth layer in an n-bit dPPUF (i.e., $i \in \{0, 1, ..., n-1\}$). We give Equation (1) to indicate the relationship between O_i^j and the output of the previous stage, O_i^{j-1}.

$$O_i^j = \mathcal{R}(O_i^{j-1}) \qquad (1)$$

where

$$O_i^{j-1} = \{O_{i0}^{j-1}, O_{i1}^{j-1}, O_{i2}^{j-1}, O_{i3}^{j-1}\},$$
$$i0, i1, i2, i3 \in \{0, 1, ..., n-1\}$$

We then give Equation (2) to assign different sets of O_i^{j-1}, to maximally share the logic cells, where the cardinality of the intersection of each two adjacent O_i^{j-1} is set to be 3, so that each represser would have three inputs shared by the one in neighbor. This allows the repressers receiving different signal combinations from the outputs of the previous stage, while having the biggest intersection to reduce the cost.

$$\begin{cases} |O_i^{j-1} \cap O_{n-1}^{j-1}| = 3, & i = 0 \\ |O_i^{j-1} \cap O_{i-1}^{j-1}| = 3, & i \neq 0 \end{cases} \qquad (2)$$

Following the above presented interconnection method, we demonstrate an n-bit represser stage with the interconnection interacted to the previous stage, In Fig. 4. As can be seen, the proposed represser cell has the same circuit structure as the conventional design (Fig. 1), to maintain the function, but each two adjacent repressers have two shared NAND gates. Compared with the conventional design, area of the repressers, measured by transistor count, can be reduced by 44.4%. It should be noted the outputs from the previous stage, O_i^{j-1}, do not have to be interacted to the repressers in the exact connections as shown in Fig. 4, as long as Equation (2) is satisfied. One can shuffle O_i^{j-1} for each represser stage to create more randomly interacted connections.

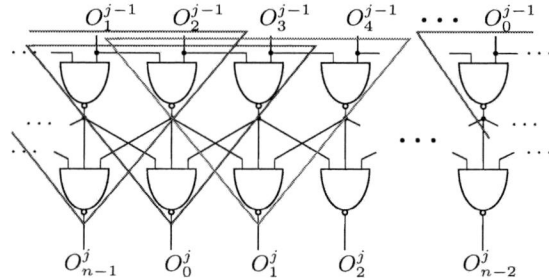

Fig. 4. The proposed represser stage with shared logic

IV. SIMULATION RESULTS

We use the technique described in Section III-B to design dPPUFs for lightweight authentication, and compare them with the conventional design. A 65-nm CMOS technology is applied to implement the circuits, and a SPICE model is used for the simulations and evaluations.

We firstly implement 8-bit dPPUFs with the conventional and proposed repressers, where h_b is set to be 6 and h_r is increased from 1 to 6, to capture the trends of simulation time, in Fig. 5. As can be seen, for all the values of h_r, the simulation time of the proposed dPPUF is very close to that of the conventional design. In both cases, the trends of the simulation time when increase h_r are similar, indicating the proposed dPPUF maintain the similar structural complexity as the conventional design, to provide a high resilience over simulation attacks. This can be expected, because the proposed repressers have exactly the same structure as the one of the previous work in Fig. 1, but only share some logic to reduce the overall cost, so the circuit complexity is not affected.

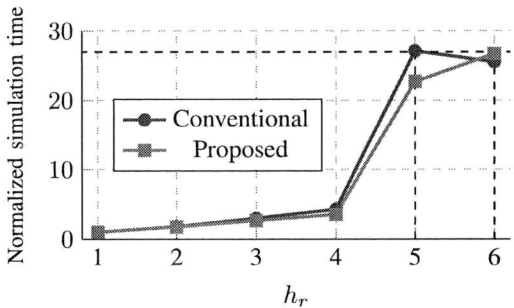

Fig. 5. Simulation time comparisons of conventional and proposed dPPUFs

The cost reduction of the proposed dPPUF is related to the proportion of repressers in the entire chip, and is generally independent of the width of the dPPUF. We evaluate the costs considering different ratio of h_b to h_r (h_b/h_r), without considering the area of arbiters, which are generally small compared with boosters and repressers in a dPPUF, according to Fig. 3. Fig. 6 demonstrates the area reduction of our proposed method, based on transistor count. As our method specifically reduces the cost of repressers, the maximal area reduction can be realized when h_b/h_r is close to 0, which indicates repressers take the most proportion in the entire dPPUF. In such a case, our method leads to a 44% around area reduction. The area reduction becomes smaller when fewer represser stages are used, compared with booster stages.

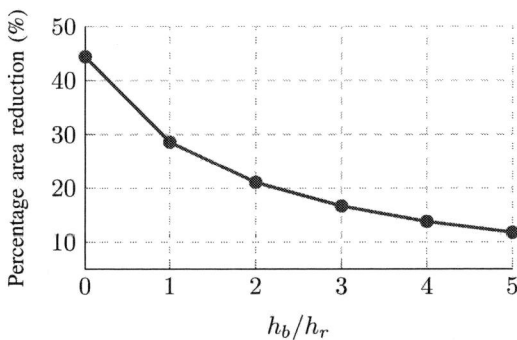

Fig. 6. Area reductions for the proposed dPPUFs

Finally, we quantify the quality of our presented dPPUF. A practical security protocol may require a key of at least 64-bit, but the simulation for the dPPUFs with such a width is costly. We design a 16-bit dPPUF with 7 stages of boosters and 2 stages of repressers, to complete the simulations in a reasonable time. Some quality metrics of the proposed dPPUF, including uniformity, uniqueness and reliability, are compared with those of the traditional design, shown in Table I. For uniformity and uniqueness evaluations, we examine 5 dPPUF chips, each producing 1,000 CRPs. For reliability evaluation, the temperature is varied within the range of -40°C to 85°C and the supply voltage of 1.2V varied by ±10%. As can be seen from Table I, the proposed design method leads to negligible change for uniformity, compared with the conventional design. Uniqueness and reliability is decreased by 12.12% and increased by 6.27%, respectively. This can be explained as follows. Due to the shard logic in the proposed

repressers, fewer unique logic cells are included in the paths from the primary inputs to the primary outputs, and fewer process variations are accumulated in each path. This results in more consistent responses for different dPPUFs or for a dPPUF running at different temperature and voltage corners, and thus leads to reduced uniqueness and increased reliability.

TABLE I
QUALITY METRICS OF CONVENTIONAL AND PROPOSED DPPUFS

	Uniformity	Uniqueness	Reliability
Conventional	0.614	0.297	0.750
Proposed	0.624	0.261	0.797
Percentage change	+1.63%	-12.12%	+6.27%

V. CONCLUSION

PPUF is a recently emerging hardware security primitive. This paper investigates one of the PPUF families, dPPUF, to realize a lightweight authentication. By evaluating the ESGs and the costs of conventional dPPUFs, we find the ESG can be largely increased by implementing more represser stages, but this significantly increases the cost. We thereby propose a new represser network, including a specially designed interconnection and the represser stage with shared logic. We show the proposed design exhibits a similar structural complexity, to maintain a great ESG, while the area can be reduced by up to 44%, compared with the conventional design, so that it is more applicable for lightweight authentication. Besides, the proposed dPPUF design causes acceptable changes for uniformity, uniqueness and reliability.

REFERENCES

[1] G. E. Suh and S. Devadas, "Physical unclonable functions for device authentication and secret key generation," in *44th ACM/IEEE Design Automat. Conf.*, 2007, pp. 9–14.

[2] J. W. Lee *et al.*, "A technique to build a secret key in integrated circuits for identification and authentication applications," in *Symp. VLSI Circuits. Dig. of Tech. Papers*, 2004, pp. 176–179.

[3] M. S. Mispan, S. Duan, B. Halak, and M. Zwolinski, "A reliable PUF in a dual function SRAM," *Integration*, vol. 68, pp. 12–21, 2019.

[4] J. Miskelly and M. O'Neill, "Fast DRAM PUFs on commodity devices," *IEEE Trans. Comput.-Aided Design Integr. Circuits Syst.*, vol. 39, no. 11, pp. 3566–3576, 2020.

[5] H. Su, B. Halak, and M. Zwolinski, "Two-stage architectures for resilient lightweight PUFs," in *IEEE 4th Int. Verification Secur. Workshop (IVSW)*, 2019, pp. 19–24.

[6] U. Chatterjee *et al.*, "Building PUF based authentication and key exchange protocol for IoT without explicit CRPs in verifier database," *IEEE Trans. Dependable Secure Comput.*, vol. 16, no. 3, pp. 424–437, 2019.

[7] C. Wachsmann and A. R. Sadeghi, *Physically Unclonable Functions (PUFs): Applications, Models, and Future Directions*, Morgan and Claypool, 2014.

[8] M. Potkonjak and V. Goudar, "Public physical unclonable functions," *Proceedings of the IEEE*, vol. 102, no. 8, pp. 1142–1156, 2014.

[9] U. Rührmair *et al.*, "Towards electrical, integrated implementations of SIMPL systems," in *IFIP Int. Workshop Inf. Secur. Theory Practices*. Springer, 2010, pp. 277–292.

[10] M. Potkonjak, S. Meguerdichian, A. Nahapetian, and S. Wei, "Differential public physically unclonable functions: Architecture and applications," in *48th ACM/EDAC/IEEE Design Autom. Conf. (DAC)*, 2011, pp. 242–247.

[11] S. Meguerdichian and M. Potkonjak, "Matched public PUF: Ultra low energy security platform," in *IEEE/ACM Int. Symp. Low Power Electronics Design*, 2011, pp. 45–50.

[12] T. Xu and M. Potkonjak, "Lightweight digital hardware random number generators," in *IEEE SENSORS*, 2013, pp. 1–4.

© VDE VERLAG GMBH · Berlin · Offenbach

Organic Transistor Parameter Estimation and Accurate Modeling for Process Optimization

Rosalba Liguori, Gian Domenico Licciardo, Luigi Di Benedetto
Department of Industrial Engineering
University of Salerno
Fisciano (SA), Italy
email: rliguori@unisa.it

Abstract—**The development of accurate tools and models able to analyze and predict the electronic device properties is compulsory to promote the technological advancement and particularly to improve fabrication process in the case of unconventional materials. In this work the model proposed for the interpretation of the data obtained with the admittance spectroscopy technique is useful for investigating the causes of the property differences in organic thin film transistors fabricated with various gate dielectrics. The extracted parameters show that the use of the UV-cured copolymer PVP-co-PMMA as dielectric layer in a pentacene-based OTFT, instead of the as-deposited copolymer or the standard polymer PVP, reduces the ion diffusion through gate insulator and the density of trap states at the insulator-semiconductor interface, while improving the semiconductor structural order, therefore producing devices with lower and stable threshold voltage and lessened hysteresis.**

Keywords—*admittance, interfaces, modeling, OTFT, performance, processing*

I. INTRODUCTION

The rapid evolution in consumer electronics market, in the last years increasingly interested in wearable, flexible and lightweight devices, is favoring the development of technologies based on unconventional materials such as organic electronics [1]. Great research efforts have been devoted on the synthesis of novel materials with desirable electrical and optical characteristics [2],[3]. Although organic materials offer the possibility to adjust their properties by changing their microstructure [4],[5], the realization of stable and reliable devices to be employed in real applications is only allowed by the proper arrangement of different layers and the right manufacturing processes. In this context, the use of numerical simulations and the development of accurate analytical model are fundamental to understand the physical mechanisms underlying the device operation and predict the behavior of novel devices [6]-[9]. In this work, the objective is to investigate the properties of organic thin film transistors (OTFTs) fabricated with different polymer gate dielectrics and was achieved through the development of an accurate model of the equivalent metal-insulator-semiconductor (MIS) capacitors, with its bases in physical theory and usable in circuit simulators.

II. EXPERIMENTALS

A. Device fabrication

OTFT devices (Fig. 1) were fabricated in the bottom-gate top-contact geometry on three glass substrates. The gate electrodes consist of 50 nm thick aluminum strips deposited by thermal evaporation. The insulator layers, all having a thickness of about 400 nm, were obtained with three different approaches: one by spin-coating a solution of Poly(4-vinylphenol) (PVP) and two by a solution of Poly(4-vinylphenol-co-methyl methacrylate) (PVP-co-PMMA), one of which was then irradiated with UV light at a wavelength of 250 nm for 15 minutes. In the order, they were indicated as device A, B and C. A pentacene active layer and gold source and drain electrodes, with a thickness of 50 nm, were evaporated on the top through shadow masks. For each sample, the equivalent MIS device with a single gold contact (referred to as source) and a MIM (metal-insulator-metal) junction were fabricated for the admittance measurements.

B. Measurement results

OTFT transfer characteristics (Fig. 2) were collected at room temperature in ambient air and parameters such as onset voltage and channel mobility were estimated. Results show that the use of the copolymer as dielectric layer increases the mobility and reduces hysteresis loop width, compared to the OTFT based on the standard polymer, whereas adding the UV-curing treatment favors the reduction of threshold voltage and further improves hysteresis. The admittance of MIS devices was measured using an Agilent E4980A LCR meter, by superimposing a small signal of 10 mV and frequency between 200 Hz and 2 MHz to the dc bias (V_{GS}), between the gate and source terminals, which was swept from -40 to 40 V. The complex admittance $Y(\omega)$ was represented according to the parallel model through the capacitance C_P and the loss L_P, that is, the conductance G_P divided by the angular frequency ω taking account of the energy dissipated by the device:

$$Y(\omega) = j\omega C_P(\omega) + G_P(\omega) = j\omega[C_P(\omega) - jL_P(\omega)]. \quad (1)$$

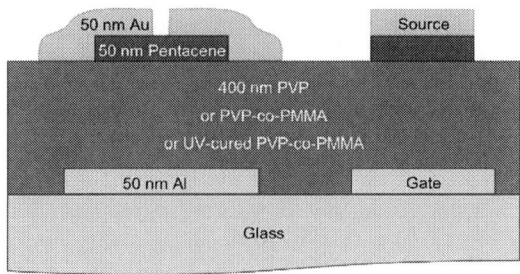

Fig. 1. Cross section of the fabricated OTFTs and MIS capacitors.

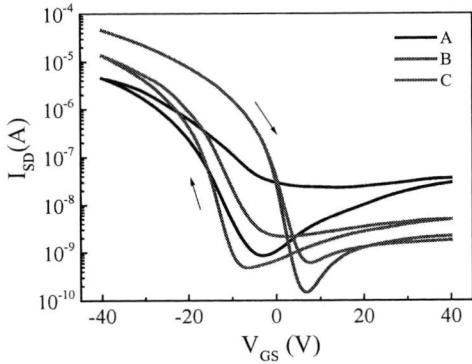

Fig. 2. Typical transfer characteristics of the fabricated pentacene-based OTFTs with three different dielectrics at $V_{DS}=-20$ V.

Experimental results in all cases (square symbols in Fig. 3) show the p-type behavior of pentacene, since at low frequency C_P goes from the maximum value measured in accumulation regime, matching the insulator layer capacitance C_i, to lower values in depletion regime, corresponding to the series sum of C_i and the depletion layer capacitance C_D, whose width rises with positive biases. As frequency increases, two main dispersion phenomena are distinguishable from the simultaneous capacitance reduction and loss peak appearance. The first peak shifts with bias proving the presence of trapping states at the interface between semiconductor and insulator. The second one is nearly constant in frequency but decreases in amplitude with bias, showing a correlation with the dispersive transport in the organic semiconductor bulk.

III. THEORETICAL MODEL

Admittance spectroscopy turns out to be a powerful characterization technique, which needs simple measurements but which, in order to provide a wide range of information, requires a suitable equivalent model to correctly interpret the experimental data. Several models based on an equivalent circuit with different numbers of elements have been used to describe organic devices for various applications [10]-[15]. The model proposed in this work (Fig. 4) accurately describes the experimental data, by associating an element of the equivalent circuit to each of the underlying physical processes occurring in organic MIS devices. It allows to extract the fundamental OTFT parameters, distinguishing bulk and interface properties, and thus to correlate the electrical performance to the processing.

In parallel to the capacitor $C_i=\varepsilon_i/t_i$, describing the gate insulator layer, where ε_i and t_i are its permittivity and thickness, the constant phase element (CPE) Y_δ was placed for the presence of a weak dispersion visible at the lowest frequencies, which was attributed to the anomalous diffusion of slow ions through the insulator [16], presumably water molecules:

$$Y_\delta = A_{diff}(j\omega)^{1-\delta}. \qquad (2)$$

Here, A_{diff} depends on the diffusing particle density and $0<\delta<0.5$ defines the continuous time random walk ($\delta=0.5$ corresponds to normal diffusion).

The semiconductor depletion region is represented by the capacitor $C_D=\varepsilon_{osc}/w_D$, where ε_{osc} and w_D are the semiconductor permittivity and depletion width, whereas the interface trap states, which cannot be described by the single time constant model, are defined with an equivalent admittance given by C_{it} and G_{it} as in [17], according to the model for a distribution of single trap levels with density D_{it}:

$$Y_D = j\omega\left[C_D + \frac{qD_{it}}{\omega\tau_{it}}tan^{-1}(\omega\tau_{it})\right] + \frac{qD_{it}}{2\tau_{it}}ln[1 + (\omega\tau_{it})^2], \quad (3)$$

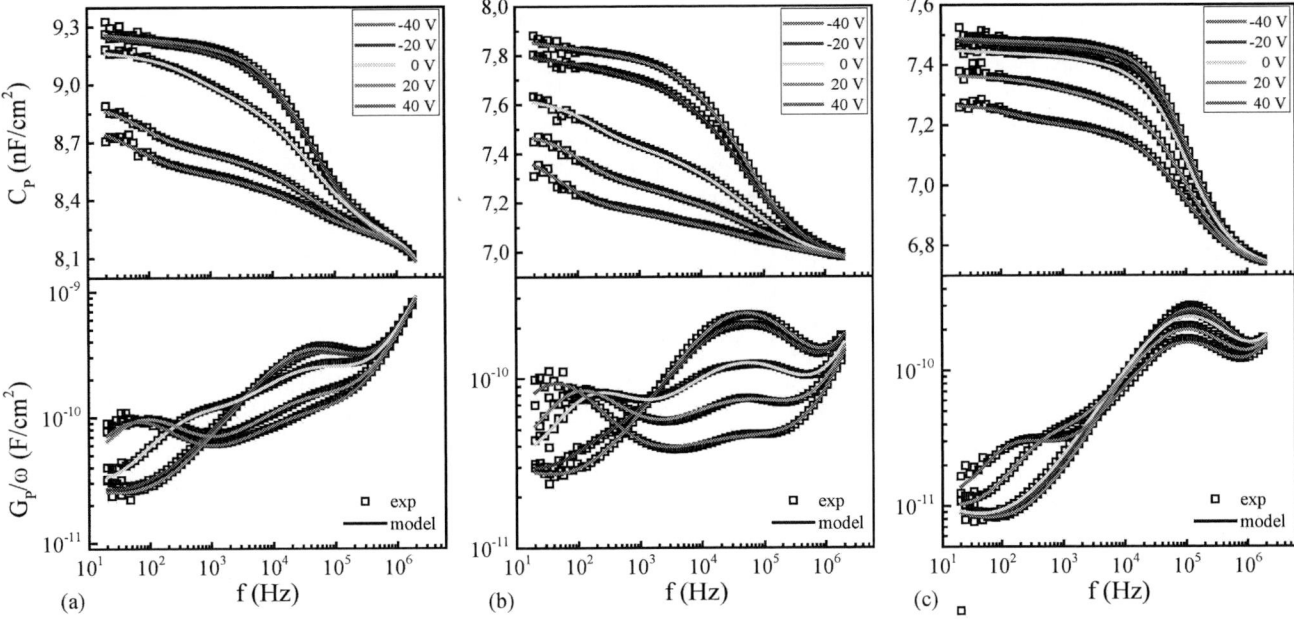

Fig. 3. Frequency dependent capacitance and loss of MIS devices based on PVP (a), as-deposited PVP-co-PMMA (b) and UV-treated PVP-co-PMMA (c).

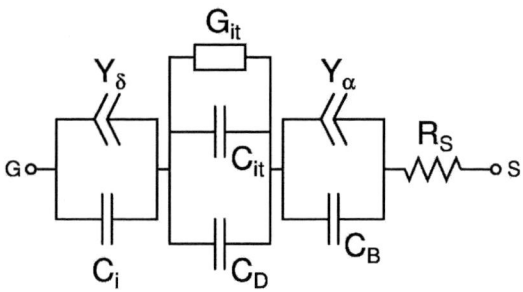

Fig. 4. Organic MIS capacitor equivalent circuit.

where τ_{it} is the relaxation time constant. The effect of depletion and interface trap states is visible at low frequency.

The non-depleted semiconductor region is characterized by a dispersive transport due to bulk structural disorder, which causes an upper limit to the frequency at which the injected holes are able to follow the ac signal. Following the Cole-Cole theory, the bulk admittance was expressed as follows:

$$Y_B = j\omega C_B[1 + (j\omega\tau_B)^{\alpha-1}], \qquad (4)$$

and thus it was represented through the capacitance C_B and a CPE in the form $Y_\alpha = G_B(j\omega\tau_B)^\alpha$, where τ_B is the pentacene relaxation time and $0 < \alpha < 1$ is the dispersion parameter [18].

Finally, a back contact resistance R_S, responsible for the high frequency dispersion, describes the quality of the interface between the semiconductor and the metal contact.

IV. DISCUSSION

The model was simulated for the three MIS capacitors, starting from the relevant known physical and geometrical constants of the devices, whereas all the other parameters were extracted by fitting the simulated curves with experimental data. The model curves are plotted in Fig. 3, superimposed on the measures. They were obtained keeping constant C_i and the semiconductor film capacitance, given by the series of C_D and C_B, reaching a good fit with an average R^2 value of 0.9998.

As first step, the width of depletion layer was extracted from C_D, obtaining the values reported in Fig. 5, and then the onset

voltage V_{ON} and the semiconductor doping density N_A were deduced using the following formula:

$$w_D = \frac{\varepsilon_{osc}}{c_i}\left[\sqrt{1 + \frac{2c_i^2}{qN_A\varepsilon_{osc}}(V_{GS} - V_{ON})} - 1\right]. \qquad (5)$$

For devices A, B and C, V_{ON} is equal to −13 V, −15 V and 5 V, respectively, in accordance with the OTFT characteristics in Fig. 2, whereas N_A is about $8.5 \cdot 10^{17}$ cm⁻³, $5 \cdot 10^{17}$ cm⁻³ and $7 \cdot 10^{17}$ cm⁻³.

Interface trap densities were calculated, along with the corresponding time constants, from C_{it} and G_{it} and reported in Fig. 6. Here, the surface potential φ_S, which is the band bending needed to access interface trap at energy E_T, was estimated as:

$$\varphi_S = E_T - E_F = \frac{qN_Aw_D^2}{2\varepsilon_{osc}}, \qquad (6)$$

where E_F is the bulk Fermi level. The obtained values of D_{it}, showing an approximately exponential distribution, are higher in device A and lower in device C, in accordance with the OTFT charge carrier mobility, demonstrating that interface traps are the main element affecting performance [19].

As regards the semiconductor bulk, the average resistivity ρ_{osc} of the non-depleted region at different distance from the channel was evaluated from the relaxation time constant ($\tau_B = \varepsilon_{osc}\rho_{osc}$) as bias varies. The values reported in Fig. 7 indicate an improvement of conductivity from sample A to sample C, as expected, whereas the dependence on the depth proves that the semiconductor morphology is affected not only by the insulator surface quality, but also by the back contact deposition causing structural disorder. Sample C also shows a reduced transport dispersion, with an average α of 0.24, versus 0.38 of device A and 0.46 of device B. Furthermore, back contact resistance is about 0.2 Ω·cm² in devices based on copolymer and about 1 Ω·cm² for the PVP-based capacitor.

Finally, all dielectrics are characterized by a dispersion parameter δ equal to about 0.2, proving the presence of anomalous diffusion, while showing different A_{diff}: it was estimated to be about $1.8 \cdot 10^{-10}$, $1.2 \cdot 10^{-10}$ and 10^{-10} F(rad/s)^δ from sample A to sample C. The main model parameters extracted for the three devices are listed in Table I.

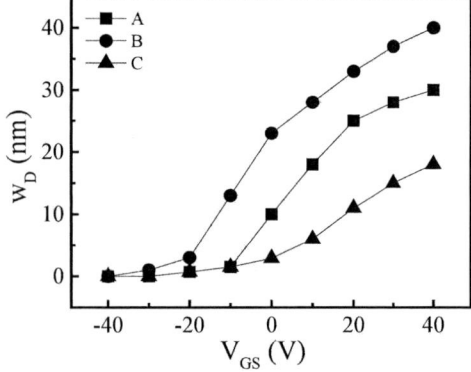

Fig. 5. Depletion width extracted from experimental admittance data.

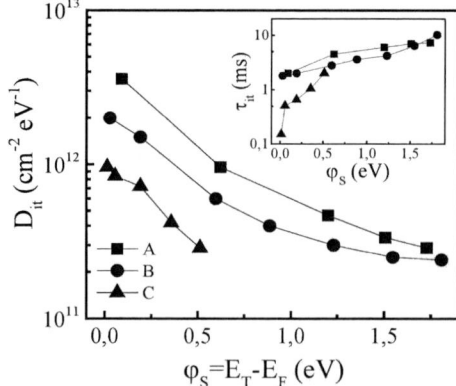

Fig. 6. Interface density of states D_{it} and corresponding time constants τ_{it}.

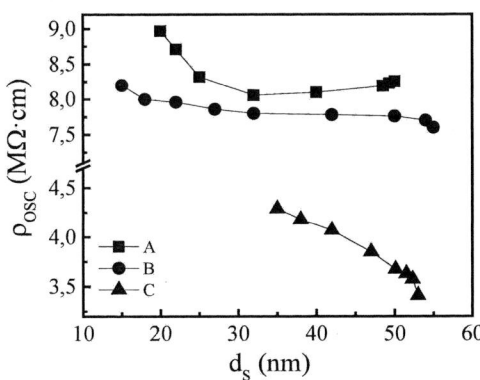

Fig. 7. Average resistivity of the bulk pentacene layer of thickness d_s next to the gold back contact.

TABLE I.   MODEL PARAMETERS

Dev.	V_{ON} (V)	N_A (cm^{-3})	A_{diff} (F(rad/s)$^\delta$)	D_{it}^{max} (cm^{-2}eV^{-1})	ρ_{osc}^{min} (MΩ·cm)	α	R_s (Ω·cm^2)
A	−13	$8.5\cdot10^{17}$	$1.8\cdot10^{-10}$	$3.6\cdot10^{12}$	8	0.38	1
B	−15	$5\cdot10^{17}$	$1.3\cdot10^{-10}$	$2\cdot10^{12}$	7.6	0.46	0.2
C	5	$7\cdot10^{17}$	10^{-10}	$9.6\cdot10^{11}$	3.4	0.24	0.2

The obtained parameters relative to diffusion through the dielectric layer demonstrate the efficacy of methyl methacryl groups contained in PVP-co-PMMA in reducing water adsorption at polymer surface and diffusion into the gate dielectric [20], thus obtaining more stable devices with reduced hysteresis. On the contrary, the hydroxyl groups in PVP favor slow polarization, justifying the reduced performance of the sample A, which also shows the highest gate leakage current. The difference between N_A values previously reported proves the influence of polymer surface energy on the semiconductor microstructure: indeed, a higher doping density can be associated to the presence of wider grain boundaries that arise when semiconductor layer grows forming large grains. This occurs on hydrophilic surfaces, as in the cases of PVP or a UV-treated surface. However, the reduced semiconductor resistivity (Fig. 7) exhibited by the UV-treated sample demonstrates an enhanced semiconductor structural order induced by an improved insulator surface, which, along with a lower density of interface states (Fig. 6), contributes to a higher OTFT charge carrier mobility and a lower threshold voltage.

V. CONCLUSION

Since the study of device physical phenomena is necessary to improve performance and stability, an electrical model is proposed to extract the significant parameters of organic MIS devices characterized through admittance spectroscopy and to identify the correlations between process and electrical characteristics. The dynamics of OTFTs based on three different gate dielectric polymers have been fully described, showing the critical role played by the interface between insulator and semiconductor layers.

REFERENCES

[1] E. Ragonese, M. Fattori, and E. Cantatore, "Printed organic electronics on flexible foil: Circuit design and emerging applications," IEEE T. Circuits-II, vol. 68, pp. 42-48, Jan 2021.

[2] S. Fusco et al., "Novel DPP derivates functionalized with auxiliary electron-acceptor groups and characterized by narrow bandgap and ambipolar charge transport properties," Dyes Pigments, vol. 186, 109026, Feb 2021.

[3] S. Fusco et al., "N-rich fused heterocycluc systems: Synthesis, structure, optical and electrochemical characterization," Eur. J. Org. Chem., vol. 2016, pp. 1772-1780, Mar 2016.

[4] R. Liguori, A. Botta, A. Rubino, S. Pragliola, and V. Venditto, "Stereoregular polymers with pendant carbazolyl groups: Synthesis, properties and optoelectronic applications," Synthetic Met., vol. 246, pp. 185-194, Dec 2018.

[5] A. Botta et al., "Optoelectronic properties of poly(N-alkenyl-carbazole)s driven by polymer stereoregularity," J. Polym. Sci. Pol. Chem., vol. 56, pp. 242-251, Jan 2018.

[6] A. Falco et al., "Simulation and fabrication of polarized organic photodiodes," Proc. IEEE Sensors, 7808585, Jan 2017.

[7] R. Liguori, W.C. Sheet, A. Facchetti, and A. Rubino, "Light- and bias induced effects in pentacene-based thin film phototransistors with a photocurable polymer dielectric," Org. Electron., vol. 28, pp. 147-154, Jan 2016.

[8] L. Di Benedetto et al., "A model of electric field distribution in gate oxide and JFET-region of 4H-SiC DMOSFETs," IEEE T. Electron Dev., vol. 63, pp. 3795-3799, Sep 2016.

[9] R. Liguori, and A. Rubino, "Metastable light induced effects in pentacene," Org. Electron., vol. 15, pp. 1928-1935, Sep 2014.

[10] A. Arnal et al., "DC characterization and fast small-signal parameter extraction of organic thin film transistors with different geometries," IEEE Electron Dev. Lett., vol. 41, pp. 1512-1515, Oct 2020

[11] P. K. Manda, L. Karunakaran, and S. Dutta, "Modeling the frequency response of organic metal-insulator-semiconductor capacitors," Mater. Today Proc., volume 19, pp. 53-57, 2019

[12] E. Bezzeccheri, A. Femia, R. Liguori, and A. Rubino, "Interface trap state characterization of metal-insulator-semiconductor structures based on photosensitive organic materials," Mater. Today Proc., vol. 4, pp. 5045-5052, 2017.

[13] M. R. Fiorillo et al., "Analysis of HMDS self-assembled monolayer effect on trap density in PC70BM n-type thin film transistors through admittance studies," Mater. Today Proc., vol. 4, pp. 5053-5059, 2017

[14] H. Hirwa, S. Pittner, and V. Wagner, "Interface analysis by impedence spectroscopy and transient current spectroscopy on semiconducting polymers based metal-insulator-semiconductor capacitors," Org. Electron., vol. 24, pp. 303-314, May 2015.

[15] A. T. Marin, K. P. Musselman, and J. L. MacManus-Driscoll, "Accurate determination of interface trap state parameters by admittance spectroscopy in the presence of a Schottky barrier contact: Application to ZnO-based solar cells," J. Appl. Phys., vol. 113, pp. 144502-1-144502-10, Apr 2013.

[16] R. Ledru et al., "Low frequency dielectric loss of metal/insulator/organic semiconductor junctions in ambient conditions," Org. Electron., vol. 13, pp. 1916-1924, Oct 2012.

[17] J. Li, Z. Chen, Y. Qu, and R. Zhang, "Traps around Ge Schottky junction interface: Quantitative characterization and impact on the electrical properties of Ge MOS Devices," IEEE J. Electron Devi., vol. 8, pp. 350-357, Mar 2020.

[18] L. Drewniak, and S. Kochowski, "The origin of constant phase element in equivalent circuit of MIS (n) GaAs structures," J. Mater. Sci. Mater. Electron., vol. 31, pp. 19106-19118, Sep 2020.

[19] R. Liguori et al., "Insights into interface treatments in p-channel organic thin-film transisotrs based on a novel molecular semiconductor," IEEE T. Electron Dev., vol. 64, pp. 2338-2344, May 2017.

[20] S. Baang et al., "Influence of the surface properties of polymeric insulators on the electrical stability of 6,13-bis(triisopropylsilylethynyl)-pentacene thin-film transistors," J. Korean Phys. Soc., vol. 67, pp. 2124-2130, Dec 2015.

© VDE VERLAG GMBH · Berlin · Offenbach

Bias Temperature Instability Characterization and Modeling for 0.18um CMOS Under Extreme Thermal Stress Conditions

Yen Tran
Etudes et Production Schlumberger
Clamart, France
Univ. Bordeaux, Bordeaux INP, UMR CNRS 5218, IMS Laboratory,
Talence, France
Ttran50@slb.com

Toshihiro Nomura
Etudes et Production Schlumberger
Clamart, France
TNomura@slb.com

Mohamed Salim Cherchali
Etudes et Production Schlumberger
Clamart, France
MCherchali@slb.com

Claire Tassin
Etudes et Production Schlumberger
Clamart, France
CTassin@slb.com

Yann Deval
Univ. Bordeaux, Bordeaux INP, UMR CNRS 5218, IMS Laboratory,
Talence, France
yann.deval@ims-bordeaux.fr

Cristell Maneux
Univ. Bordeaux, Bordeaux INP, UMR CNRS 5218, IMS Laboratory,
Talence, France
cristell.maneux@u-bordeaux.fr

Abstract—**We investigated the significance of 0.18um CMOS (Complementary Metal-Oxide-Semiconductor) degradation due to bias temperature instability (BTI) under extreme temperature operations (150°C and 210°C). The transistors have been applied dedicated DC bias and temperature conditions to investigate the wear-out mechanism in specific severe environment for oilfield applications. The aging tests have been monitored up to 1,000 hours. These results are preliminarily used to develop equations reflecting aging laws to be included in commercial software tool for further investigation at logic circuit level.**

Keywords—Reliability, oilfield, 0.18μm CMOS, extreme temperature, DC characterization, BTI

I. INTRODUCTION

CMOS transistor reliability is becoming more and more critical in integrated circuit (IC) technology domain due to high-density and mixed-signal integration in the context of extreme temperature application. Although the device failure mechanisms have been intensively investigated for many years, the range of investigations in term of operating conditions is usually limited to conventional application domains involving mass markets. Therefore, the application domain involving specific extreme temperatures are seldom studied even though the reliability of the associate electronic circuits turned out to be crucial for the application. For oilfield applications, downhole tools require reliable and stable performance of electronic devices during several years at high temperature. Therefore, the focus of this work is to develop a compact model including ageing laws for 0.18um CMOS technology to optimize circuit design for a targeted lifetime under extreme temperatures. For that purpose, we have conducted an intensive aging test campaign for both nMOS and pMOS featuring two different gate lengths. Two most important intrinsic wear-out mechanisms namely BTI and HCI (Hot Carrier Injection) were characterized and modeled under operating voltage biases and elevated stress temperatures. However, for the sake of paper length, we only report here the BTI degradation and modeling for 0.18um CMOS.

According to the literature, correct modelling of BTI is quite difficult due to the fact that the nature of BTI damage as well as its experimental results are not consensus in scientific community [1-3]. Furthermore, a distinctive feature of BTI is the recovery effect when the stress conditions are reduced or removed making the BTI modelization become more complicated.

It is well accepted that BTI induce threshold voltage shift and this shift is usually considered as the best monitor of these degradation [2-5]. Therefore, in this study, the threshold voltage shift was investigated through its extraction and fitted to a power-law model based on experimental results. After that, this model was implemented into the simulator using tool-specific application programming interfaces (APIs) provided by electronic design automation (EDA) commercial software. Finally, simulation results were compared to measurements to validate the accuracy of this BTI and HCI models for extreme temperatures for the specific technology under investigation.

BTI characterization and modelling has been extensively studied so far; however, most of these research works were conducted at elevated voltages beyond the safe operating conditions to activate the degradation within a short time. However, this approach presents the risk of over-accelerating other different wear-out mechanisms that are not supposed to be experienced in normal operating voltages. Therefore, our experiment is a long-term test, in which we characterized BTI degradation at the operating voltages with the stress time up to 1,000 hours. These experiment results are expected to reflect the real wear-out in practical application.

The second highlight of this study is that BTI tests were performed under severe temperatures (150°C and 210°C) to simulate the oilfield application in extreme thermal conditions. Note that the stress temperatures of already published works on BTI degradation are commonly achieved in the range between 125°C to 175°C [1-7].

© VDE VERLAG GMBH · Berlin · Offenbach

II. AGING TEST DESCRIPTION

A. Device Under Test

The technology under test is a CMOS 0.18 um qualified for high temperatures. The devices under test are 1.8 V isolated n-type and p-type MOSFETs (Metal-Oxide-Semiconductor Field-Effect Transistor) with W/L= 20um/0.18um. The gate oxide thickness (typical value) is t_{ox} = 4.1nm for nMOS and t_{ox} = 3.9nm for pMOS. At least 10 devices were characterized at wafer level to evaluate the technology dispersion. The average dispersion of 5% indicates the maturity of the technology. All dies were packaged into ceramic DIL-40 allowing performing the aging tests at elevated temperature.

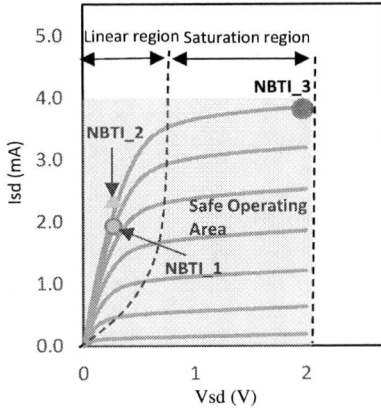

Figure 1: *Output characteristics of a 0.18um pMOS with the location of different bias stress points. The values of V_{sg} are 0.8V, 1.0V, 1.2V, 1.4V, 1.6V, 1.8V and 2V, respectively.*

B. Stress Bias and Temperature Conditions

The characterization includes periodical measurements of the DC characteristics. A high-level programing language (Python) was used to control the whole system providing advanced capabilities, such as controlling time delay between stress and measurement that allows a limited amount of relaxation; performing automatic measurement of multiple devices at the given stress time. In this aging test campaign, 2 devices were used for each stress bias conditions. The threshold voltage shift ΔV_{th} has been chosen to capture the BTI degradation and the shifts of the threshold voltage were extracted from I_d-V_g curves of MOSFETs biased in saturation region based on procedure described in [9]. According to aging test results, the average dispersion of V_{th} degradation under each stress bias was observed below 4%. BTI recovery modeling is not considered in this paper and will be considered in future work. The aging test conditions are listed in Table 1. The body is connected to the source and the substrate is grounded in all cases to reflect the real practical applications. The aging tests were performed at two different temperatures: 150°C and 210°C.

Table 1: *Bias and temperature stress conditions.*

Type	$|V_{ds}|$(V)	$|V_{gs}|$(V)	Bias conditions	T°C
nMOS	0.05	2.0	PBTI	
pMOS	0.05	1.8	NBTI_1	150°C
	0.05	2.0	NBTI_2	210°C
	2.0	2.0	NBTI_3	

For BTI characterization, NBTI_1 and NBTI_2 correspond to two different $|V_{gs}|$ values (1.8V and 2.0V, respectively) and have been considered to activate uniform BTI degradation in which $|V_{ds}|$ is kept quite small (0.05V). More aggressive $|V_{ds}|$ would have triggered unwanted HCI causing the overestimation of the BTI degradation. NBTI_3 bias condition with high $|V_{ds}|$ (2V) has been applied to the aging test in order to investigate the contribution of high $|V_{ds}|$ to NBTI degradation. The output characteristics of a 0.18μm pMOS and corresponding bias stress points are presented in Figure 1.

III. EXPERIMENTAL RESULTS AND DISCUSSION

A. Negative Bias Temperature Instability (NBTI) in pMOS

NBTI occurs in pMOS at high temperature in inverted channel with or without carrier conduction. Three configurations NBTI_1, NBTI_2, and NBTI_3 bias conditions were applied at 150°C and 210°C to characterize NBTI degradation. According to our results, NBTI in pMOS is considered as the most serious reliability issue of 0.18um CMOS technology cause the threshold voltage shifts due to NBTI at 210°C can achieve up to 9% after 1,000 hours applying stress.

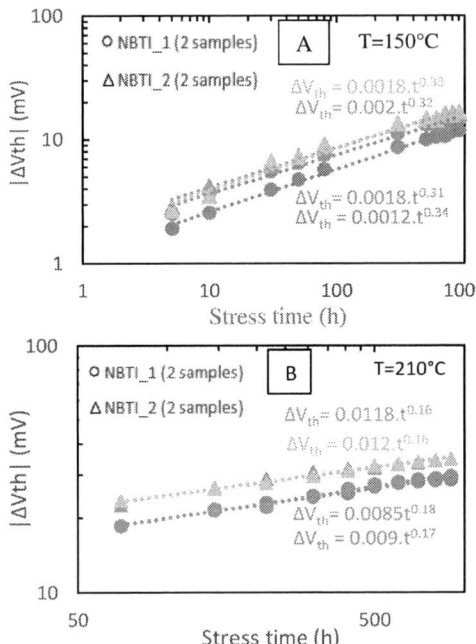

Figure 2: *Threshold voltage shift ΔVth associated with NBTI degradation under NBTI_1 and NBTI_2 bias conditions at 150°C (A) and 210°C (B) of a 0.18μm pMOS. The dotted lines are fitted to the results.*

NBTI degradation is clearly observed as the shift of I_{sd}-V_{sg} curve over stress time. The threshold voltage shifts due to NBTI_1 and NBTI_2 bias voltages have been observed following a power-law as function of stress time (Figure 2):

$$\Delta Vth = A \times t^n \tag{1}$$

where A is the prefactor and n is the time exponent parameter. At 150°C, the value of n is in the range of 0.31-0.34 regardless of different V_{sg} bias conditions. However, this value decreases significantly down to 0.16-0.17 under 210°C with the similar stress voltages applied. Our values of n are presented in the Table 2.

According to Table 2, the values of n do not depend on bias stress conditions. However, it seems that n has close relation to stress temperature, i.e, n~ 0.3 at 150°C and n~0.18 at 210°C.

Table 2: *Value of n extracted from experiment.*

T°C	Bias type	n value
150	NBTI_1	0.31 – 0.34
	NBTI_2	0.32 – 0.33
	NBTI_3	0.30
	PBTI	Not clear
210	NBTI_1	0.17 – 0.18
	NBTI_2	0.16
	NBTI_3	0.18
	PBTI	0.18 – 0.23

Referring to published works [4,10,11] on NBTI degradation at T=100°C - 200°C, a various values of $n \approx 0.16$-0.5 are reported depending on the diffusion species inside the oxide layer, i.e, H_2 or/and H^0 or/and H^+ using reaction-diffusion (R-D) model. In these works, the values of n have been considered independent with temperature, but n varies with stress time due to the creation and/or recombination of hydrogen ions during the test. Table 3 presents the possible value of n depending on the type of mobile species inside dielectric layer.

Table 3: *Effect of different species on time exponent n [11]*

Model	Reaction	Mobile Species	n value
I	$SiH + H^+ \Leftrightarrow Si^+ + H$	H	0.25
II	$SiH + H^+ \Leftrightarrow Si^+ + 0.5\,H_2$	H_2	0.165
III	Model I and II	H, H_2	0.165-0.25
IV	$SiH + 2H^+ \Leftrightarrow Si^+ + H^+$	H^+	0.25-0.5
V	Models I, II and III	H^0, H^+, H_2	0.165-0.5
VI	Model I and trapping of H	H	> 0.25
VII	Model I and release of H	H	< 0.25
VIII	Model II and trapping of H_2	H_2	>0.165
IX	Model II and release of H_2	H_2	< 0.165

From the experimental results, our proposal hypothesis here is that different temperatures can trigger different types of mobile species resulting in the large discrepancy of n.

The magnitude of NBTI degradations were observed to depend on the gate voltage, more precisely, on the oxide electric field (E_{ox}). Therefore, the prefactor A typically follows a power law of the applied oxide electric field [4,7]:

$$A \propto E_{ox}^{\gamma} \approx \left(\frac{|V_{gs} - V_{th0}|}{t_{ox}} \right)^{\gamma} \qquad (2)$$

Furthermore, the temperature dependence of the threshold voltage can be expressed as an Arrhenius law [7]:

$$\Delta V_{th} \propto exp \left(\frac{-E_a}{k_B T} \right) \qquad (3)$$

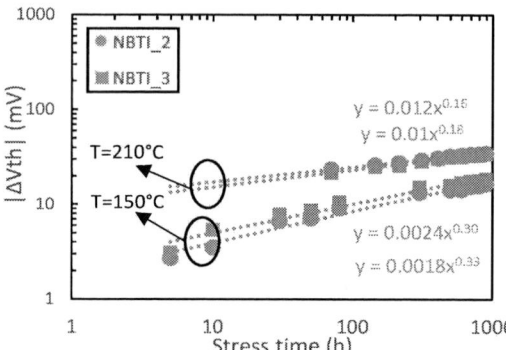

Figure 3: *Threshold voltage shift ΔVth associated with NBTI degradation under NBTI_2 and NBTI_3 bias conditions at 150°C and 210°C of a 0.18µm pMOS.*

Where k_B is the Boltzmann constant (k_B= 8,617.10^{-5} eV/K), T is the stress temperature in Kelvin, and E_a is the apparent activation energy which is typically reported in the range of 60-80 meV [4].

To investigate the role of $|V_{ds}|$ in NBTI degradation, we performed the aging test with NBTI_3 bias condition and compared the ΔV_{th} with one suffering NBTI_2 bias condition under 150°C and 210°C. The high $|V_{ds}|$ in NBTI_3 bias condition can activate both NBTI and HCI. In our case, NBTI degradation is not dependent on $|V_{ds}|$ where $|V_{ds}| \leq 2V$. The ΔV_{th} under NBTI_2 and NBTI_3 are quite similar implying that HCI under NBTI_3 is negligible and $|V_{ds}|$ does not have a significant impact on NBTI degradation (Figure 3). With $|V_{ds}| \leq 2V$, it is expected that holes have less chance to gain the energy up to 4.5eV to become channel hot carriers [1,8].

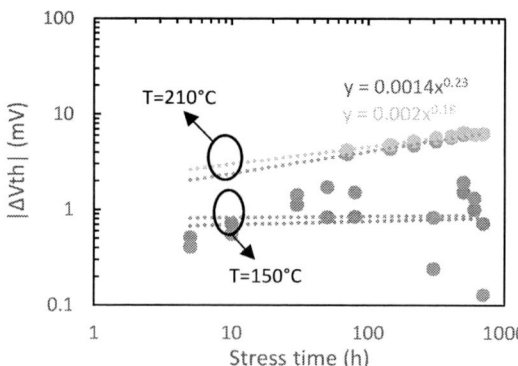

Figure 4: *Threshold voltage shift ΔVth associated with PBTI degradation under PBTI bias conditions at 150°C and 210°C of a 0.18µm nMOS.*

Following these standpoints, the dependence of threshold voltage shifts due to BTI degradation on temperature, gate bias voltage and stress time can be summarized as follows:

$$\Delta V_{th}(t) = C \times \left(\frac{|V_{gs} - V_{th0}|}{t_{ox}} \right)^{\gamma} \times exp \left(\frac{-E_a}{k_B T} \right) \times t^{n(T)} \qquad (4)$$

Where C, γ, E_a, and n are fitted parameters and they are related to the specific CMOS technology; $n(T)$ implies the dependence of n on temperature.

B. Positive Bias Temperature Instability (PBTI) in nMOS

PBTI degradation in nMOS is usually neglected due to the insignificant degradation. Our results after 1,000h stress show that PBTI degradation does not represent an important impact at 150°C since ΔV_{th} is below 0.5%. However, PBTI becomes more pronounced at 210°C in which ΔV_{th} shift is around 2.5% after the same stress period and bias conditions as the PBTI test at 150°C. PBTI degradation at 210°C also follows power-law with stress time and the prefactor $A \sim$ 0.01-0.02 and $n \sim$ 0.18-0.23 as shown in Figure 4.

Although the parameter degradation values of nMOS are different from those of pMOS, the PBTI model for nMOS can be written:

$$\Delta V_{th} = C' \times \left(\frac{|V_{gs} - V_{th0}|}{t_{ox(Nmos)}}\right)^{\omega} \times exp\left(\frac{-E_{a(Nmos)}}{k_B T}\right) \times t^{m(T)}$$

(5)

Where C', ω, $E_{a(Nmos)}$, and m are fitted parameters related to technology under investigation.

IV. INTEGRATION OF AGING MODELS INTO CIRCUIT SIMULATOR

The aging models have been proposed in previous section to simulate the degradation behavior observed from experiments. Equation (4) and (5) was adapted to the underlying BSIM3 parameter, V_{th0}, with the fitted parameters C, γ, E_a and n extracted from experimental results. The degraded V_{th0} in BSIM3 is defined as:

$$V_{th0}(aged) = V_{th0} - \Delta V_{th0}$$

(5)

Where V_{th0} is the threshold voltage at zero-volt substrate bias and $V_{th0} < 0$ for pMOS.

In this paper, Cadence unified reliability interface (URI) was used to implement BTI model for the reliability circuit simulation.

The compact model provided by the foundry just ensures DC characteristics accuracy lower than T=175°C. Beyond this critical temperature, there is a large difference between simulation and measurement (~14%). Therefore, before performing aging simulation, some BSIM3 parameters (W, L, V_{th0}) have been slightly modified to obtain the good agreement between measurement and simulation of fresh device.

Reliability simulations of a single pMOS at NBTI_2 were performed to validate the proposed NBTI model. We obtained very good accuracy compared to experimental results (Figure 5) even at extreme temperature close to 210°C.

V. CONCLUSION

In this study, we have demonstrated the aging models for BTI degradation and implemented it into a commercial software tool. The simulated DC characteristics were compared with measurements at different bias and temperature conditions. A good agreement between simulation and experimental results implies that the extracted parameter value combined with BSIM3 model ones can

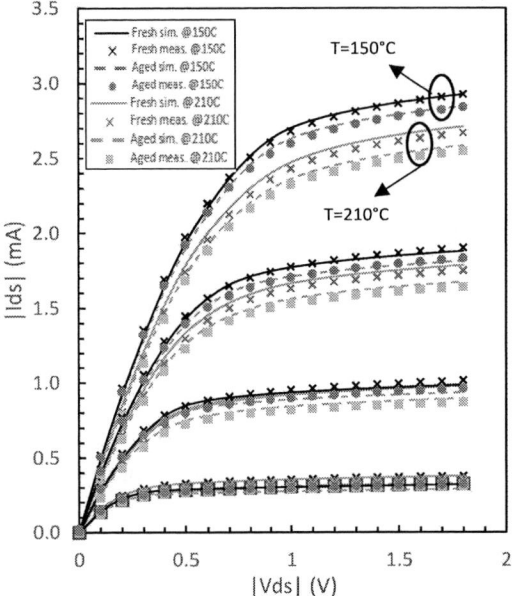

Figure 5: *Comparison of aged simulations and aged measurements of a 0.18um pMOS under NBTI_2 bias condition at 150°C and 210°C after 900h stress. The values of |Vgs| are 0.9V, 1.2V, 1.5V and 1.8V, respectively.*

provide a good description of BTI degradation. The future research will further focus on circuit-level aging simulation [12] to verify the compact model including aging laws.

REFERENCES

[1] Alvin W. Strong et al., ed. (2009), "Reliability Wearout Mechanisms in Advanced CMOS Technologies". Wiley-IEEE Press.

[2] D. Schroder et al., "NBTI: Road to cross in deep submicron silicon semiconductor manufacturing". J. Appl. Phys., 94(1)1, 1-18, 2003.

[3] X. Li et al., "Compact modeling of MOSFET Wearout Mechanisms for Circuit-Reliability Simulation". IEEE Transactions on device and materials reliability, Vol. 8, No.1, March 2008

[4] V. Huard, M. Denais, C. R. Parthasarathy. "NBTI degradation from physical mechanisms to modeling". Microelectronics Reliability, 46: 2006, pp 1–23

[5] G. Chen et al. "Dynamic NBTI of PMOS transistors". Proc. IEEE Reliab. Phys. Symp 2003, pp 196–202.

[6] Schroder et al., "NBTI: What do we understand". Microelectron. Reliab. 2007,47,841-852.

[7] J. Franco, B. Kaczer, G. Groeseneken, "Reliability of High Mobility SiGe Channel MOSFETs for Future CMOS Applications". Springer Series in Advanced Microelectronics.

[8] S. Rangan et al, " Universal recovery behavior of negative bias temperature instibility". IEDM Tech, Dig., 2003, pp. 341-344

[9] Ortiz-Conde et al., "A review of recent MOSFET threshold voltage extraction methods". Microelectronics and Reliability, 42(4), 583-596, 2002.

[10] Ogawa et al., " Generalized diffusion-reaction model for the low-field charge-buildup instibility at the Si-SiO₂ interface". Physical Teview B, 51(7), 4218-423

[11] S. Chakravarthi et al.,"A comprehensive framework for predictive modeling of NBTI". 2004 IEEE International Reliability Physics Symposium, 2004, pp. 273-282.

[12] M. B. Yelten et al "Surrogate-Model-Based Analysis of Analog Circuits—Part II: Reliability Analysis," in IEEE Transactions on Device and Materials Reliability, vol. 11- 3, pp. 466-473, Sept. 2011.

Run-Time Adaptive Hardware Accelerator for Convolutional Neural Networks

Cristian Sestito, Fanny Spagnolo, Pasquale Corsonello
Department of Informatics, Modeling, Electronics and System Engineering
University of Calabria
Arcavacata di Rende, Italy
cristian.sestito@unical.it, f.spagnolo@dimes.unical.it,
p.corsonello@unical.it

Stefania Perri
Department of Mechanical, Energy and Management Engineering
University of Calabria
Arcavacata di Rende, Italy
s.perri@unical.it

Abstract—State-of-the-art Convolutional Neural Networks are characterized by heterogeneous convolutional layers to proper balance accuracy and computational complexity. Run-time adaptive convolution architectures able to process feature maps with kernels of various sizes and strides are highly desirable to achieve a favorable speed/power dissipation balance. This paper presents the design of an adaptive architecture able to manage efficiently convolutional layers with different running parameters. In order to guarantee high resources utilization for all the supported kernel sizes and strides, in contrast with existing competitors, the proposed design combines non-uniform basic blocks differently customized from each other. As a further nice characteristic, the hardware architecture here presented efficiently manages both odd and even kernel sizes, useful in models also requiring transposed convolutional layers. When accommodated within a Xilinx XC7Z045 FPGA SoC device, the proposed engine reaches a peak throughput of 217.2 GOPS and dissipates about 2.75 W at the 150 MHz clock frequency.

Keywords—*Convolutional neural networks (CNN), reconfigurable hardware architecture, field programmable gate array (FPGA), heterogeneous embedded systems.*

I. INTRODUCTION

Convolutional Neural Networks (CNNs) are nowadays widely used to perform computer vision tasks, including image classification, object detection and semantic segmentation. Typically, CNNs consist of a feature extractor, made of Convolutional Layers (CONVs) followed by non-linear activation functions and pooling layers, and a task-oriented section, which depends on the specific application. For instance, Fully-Connected Layers (FCs) comply with the classification, whereas up-sampling decoders are adopted for the segmentation. Anyway, the generic CONV performs 3D convolutions between M channels of $H_I \times W_I$ input feature maps (*ifmaps*) and N filter kernels of $M \times K \times K$ coefficients. Such a layer provides N channels of $H_O \times W_O$ output feature maps (*ofmaps*) containing different features extracted from the inputs.

Unfortunately, in practical cases, this process requires a huge number of Multiply-Accumulations (MACs): as an example, the well-known VGG-16 model [1] performs about 15.5 GMACs to classify objects within 224×224 images. For this reason, proper acceleration techniques and architectures are necessary.

While Graphics Processing Units (GPUs) are the most suitable solution to face up the computational complexity within performance-constrained environments, dedicated hardware accelerators are usually considered to cope also with low-power applications, such as Internet-of-Things and edge computing. In these contexts, recent FPGA-based Systems-on-Chips (SoCs) are widely adopted since they offer a reasonable speed-power trade-off, as well as the possibility to implement flexible parallel architectures. As discussed in the comprehensive survey of FPGA-based accelerator architectures for CNNs presented in [2], many attempts have been done in the last few years to efficiently manage variable kernel sizes (K) at different strides (S).

This paper presents an innovative efficient adaptive convolution architecture that complies with the above-mentioned behavior. In contrast with its direct competitors [3-8], the novel Reconfigurable Convolution Processing Element (RCPE) supports both odd and even kernel sizes by combining non-uniform computational blocks. Thanks to this approach, for a given CNN model, the proposed hardware accelerator guarantees either a higher resources utilization to be achieved or a cheaper implementation platform to be used. Furthermore, the possibility of also managing even-sized kernels makes the approach proposed here very suitable for CNN models that exploit transposed convolutional layers [9]. The RCPE can be integrated within modern FPGA-based SoCs that merge the powerfulness of a dedicated Processing System (PS) to the customizability of the programmable fabric. When implemented within the Xilinx XC7Z020 chip, the proposed design, made able to operate on K=1, 3, 4, 5, 7, 9 with S=1, 2, exhibits a peak throughput of 42.7 GOPS at the 118 MHz running frequency, which grows up to 217.2 GOPS when the wider XC7Z045 device is used, thus becoming a redoubtable competitor for recent state-of-the-art counterparts.

II. RELATED WORKS

In last years, several efficient FPGA-based reconfigurable hardware architectures have been proposed for deep CNNs [3-8, 10]. Taking into account that state-of-the-art models frequently use 3×3 kernels, the convolution engine proposed in [3] combines multiple 3×3 basic blocks to manage different

---

This work was supported by "POR Calabria FSE/FESR 2014-2020 International Mobility of PhD students and research grants/type A Researchers". Actions 10.5.6 and 10.5.12, actuated by Regione Calabria, Italy.

© VDE VERLAG GMBH · Berlin · Offenbach

kernel sizes. Unfortunately, this approach causes a non-negligible resource underutilization. For example, to support 5×5 CONVs, four 3×3 blocks are employed, with ~30.5% of the computing elements unused.

The uniform 4×256 MAC array proposed in [4] alleviates this issue and supports up to 16×16 CONVs and 256×256 *fmaps* with a remarkable scalability. However, the management of both *ifmaps* and coefficients buffering requires a quite complex auxiliary control logic, which significantly affects the resources requirements.

The reconfigurable architecture presented in [5] accelerates 3×3, 5×5 and 7×7 convolutions with a very high resources utilization. However, this design was mainly tailored for high-end powerful platforms, thus it reaches a quite high throughput at expenses of the power consumption.

As demonstrated in [6], reconfigurable structures based on systolic arrays reach good computational times, but they require extra buffering resources to accommodate data batches and show a relatively high energy dissipation.

The approach adopted in [7] to accelerate the YOLO model supports only 1×1 and 3×3 kernels at the stride 1. Moreover, even though the parallelism adopted makes this strategy devoted to high-end devices, it achieves limited peak throughput.

In [8], Multiple Convolutional Layer Processors (M-CLP) are used to perform different CONVs on several *fmaps* in parallel. Given that faster CLPs stay in idle until slower ones end their computations, this approach cannot utilize the resources as efficiently as possible.

The Adaptive and Hierarchical CNN proposed in [10] uses resources much more efficiently by exploiting the partial reconfiguration of FPGAs, but the reconfiguration time (tens of milliseconds) significantly reduces the actual throughput.

III. THE PROPOSED DESIGN

Unlike most of the competitors [3-7], the novel design uses non-uniform computing units that allow increasing the resources utilization. The top-level architecture of the novel RCPE is depicted in Fig. 1. The accelerator complies with the Advanced eXtensible Interface (AXI4) protocol to be easily integrated within modern heterogeneous embedded systems. The *Configuration & Control Unit* (CCU) receives the protocol signals and proper control signals that allow self-adapting the computation to different *fmaps*, kernel sizes and strides. The CCU dispatches the configuration signals to the reconfigurable modules. On the basis of this information, the *ifmap Buffer* (IFB) reads multiple *ifmaps* and arranges proper convolution windows. Meanwhile, the *Weights Buffer* (WB) stores the needed filters. Then, the *Compute Unit* (CU) executes several convolutions in parallel by using multiple *Convolution Engines* (CEs). The *ofmap Buffers* (OFBs) store these intermediate results and perform accumulations to compose the final *ofmaps*.

As illustrated in Fig. 2, the IFB consists of three parts: the $K_M{\times}K_M$ *Register Matrix*, the *Line Shift Buffer* and the *Padding Unit*. It must be noted that K_M depends on the maximum supported kernel size. As an example, Fig. 2 shows the buffer architecture tailored for $K_M{=}10$. In this case, due to the multiplexing logic interleaving the $K_M{\times}K_M$ registers of the *Register Matrix*, the IFB can arrange input pixels within different types of convolution windows: (a) one 9×9 window, (b) one 7×7 window, (c) three 5×5 windows, and (d) eight 3×3

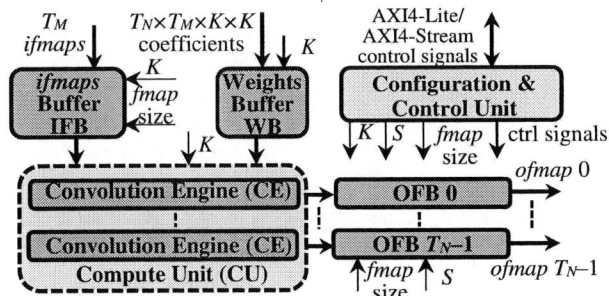

Fig. 1. Top-level architecture of the proposed Reconfigurable Convolution Processing Element.

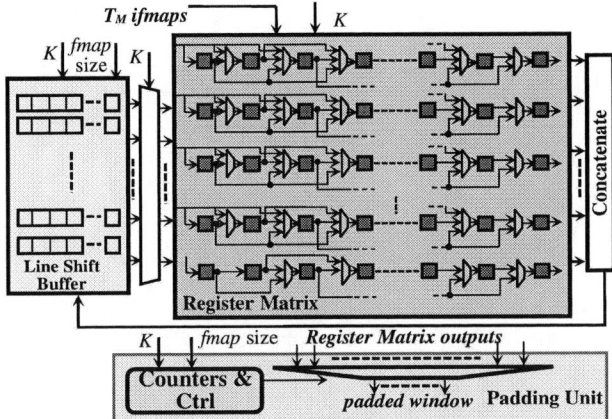

Fig. 2. Example schematic of the *ifmap* Buffer when $K_M{=}10$.

windows. The top leftmost registers are responsible for storing T_M input pixels, each of them belonging to a different channel, whereas the others shift the incoming data and make them available for the next processing. The rightmost registers are also responsible for transmitting pixels to the *Line Shift Buffer*. The latter consists of embedded Block RAMs, running as $K_M{-}1$ First-In-First-Out memories with depth *fmap_size–K*, and temporarily stores pixels, thus providing expected input data to the leftmost registers of the *Register Matrix*. Finally, the *Padding Unit* performs proper zero-padding over the composed windows. Such a module consists of $K_M{\times}K_M$ banks of multiplexers, which output either the current pixels or zero in the case of padding. Two counters, that indicate the position of the current anchor point within the *fmap*, manage the selectors of the multiplexers.

As depicted in Fig. 1, the CU includes T_N CEs, with T_N being the number of *ofmaps* provided in parallel. According to the windows provided by the IFB, each CE performs up to T_M convolutions and accumulates their results to furnish a provisional *ofmap* value. Among the possible non-uniformly sized tiles solutions, we chose the CE architecture illustrated in Fig. 3, because it well fits the Zynq target platforms and exploits almost completely their available DSPs. Of course, following the same approach, several other configurations could be conceived on the basis of the target device. The CE consists of two types of *Processing Elements* (PEs): the MAC units, hereby indicated as type-A PEs (PE-As), and the simple adders, named type-B PEs (PE-Bs). Both PE types are carefully implemented by means of embedded Digital Signal Processing (DSP) slices, assuring that critical signals are routed through fast dedicated

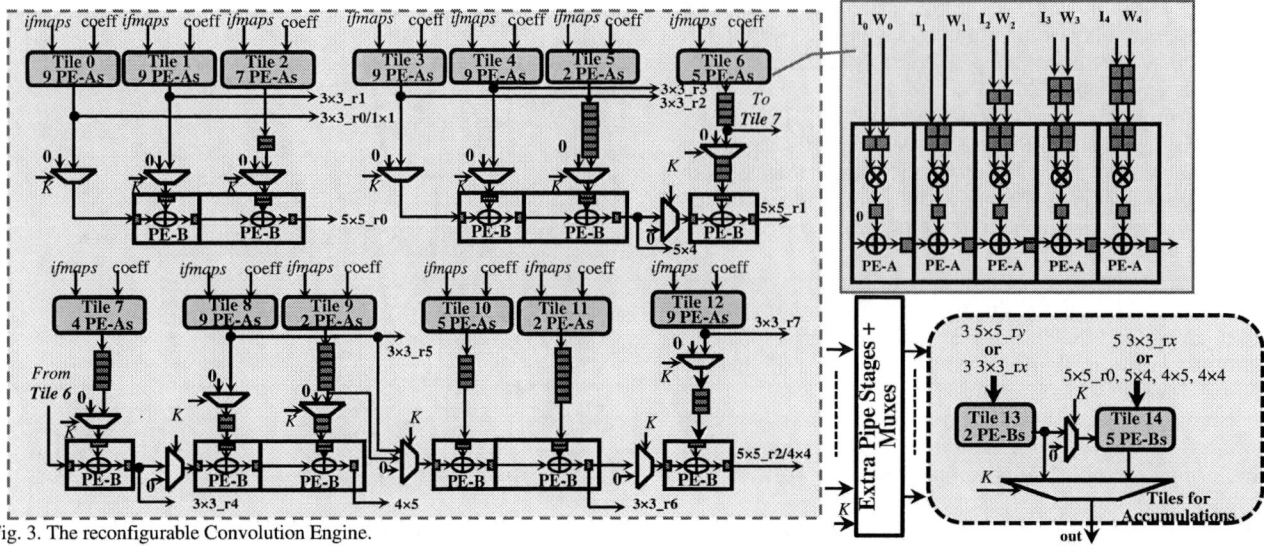

Fig. 3. The reconfigurable Convolution Engine.

interconnections and adaptive paths between PEs are properly activated by the current K. As an example, the inset of Fig. 3 depicts the arrangement of 5 PE-As within the *Tile 6*. The CE consists of 81 PE-As and 18 PE-Bs that allow performing different types of convolutions. As shown in Fig. 3, all the PE-As and the respective PE-Bs can perform (a) eight 3×3 convolutions, (b) three 5×5 convolutions, (c) 5×4, 4×5, 4×4 convolutions, (d) eight 1×1 convolutions in parallel. Pipeline stages, depicted as gray rectangles, and the *Extra Pipe Stages+Muxes* unit ensure the parallel results provided in the various configurations by the PE-As/PE-Bs tiles to be correctly time aligned. The *Extra Pipe Stages+Muxes* unit also selects data for the subsequent accumulations. Indeed, the last 7 PE-Bs employed in the *Tiles for Accumulations* module sum either the eight 3×3 results, indicated as *3×3_rx* (with $x=0,...,7$), or the three 5×5 results, indicated as *5×5_ry* (with $y=0,...,2$). In addition, the latter unit can be exploited to provide other types of convolution results. For instance, both the *5×5_r0* and the *5×5_r1* can be accumulated by the *Tile 13* to comply with a 7×7 convolution. Alternatively, the *5×5_r0*, the 5×4 and 4×5 outputs, and the 4×4 result can be supplied to the *Tile 14* to compose a 9×9 convolution output.

The T_N OFBs visible in Fig. 1 receive the results from the T_N CEs and accumulate them to the T_N results produced at the previous steps. These provisional results are stored within simple dual-port BRAMs adapted to the current *ofmaps* size. The OFBs also take care of managing the stride S. Indeed, when $S=1$, all the incoming results are stored within the BRAMs. Conversely, when $S=2$, BRAMs buffer ¼ of the outputs.

IV. IMPLEMENTATION AND RESULTS

The proposed RCPE was designed by using the Very High-Speed Integrated Circuits Hardware Description Language (VHDL) at the Register-Transfer-Level abstraction and, for purposes of validation, it was integrated within the heterogeneous Embedded System (ES) depicted in Fig. 4. The Processing System (PS) configures and controls the custom circuits implemented within the Programmable Logic (PL) that hosts the RCPE and the Direct Memory Access (DMA) units.

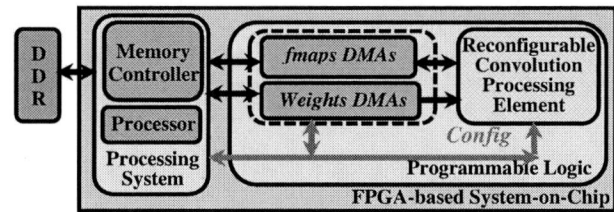

Fig. 4. The referred embedded system architecture.

The latter manage the data transfers from/to an external Double Data Rate memory (DDR). The *fmaps DMAs* provide the RCPE with up to T_M input pixels at a time. Conversely, at the completion of the entire processing, the RCPE drives the *fmaps DMAs* with up to T_N output pixels. Meanwhile, the *Weights DMAs* input the filters for each convolutional step. The DMA units and the RCPE receive configuration signals from the PS by means of AXI4-Lite transactions.

Two versions of the proposed ES have been implemented within the Xilinx Zynq 7020 and 7045 SoCs using the 2019.2 Vivado Design Suite. Table I summarizes the results obtained by the comparison with several state-of-the-art counterparts. Apart the resource utilization, the clock frequency and the power consumption, Table I shows the supported kernel sizes and strides, the maximum parallelism achievable ($T_{Mmax}×T_{Nmax}$) and the data precision. Performance metrics are also considered, including the throughput, expressed in terms of both the peak and the effective Giga Operations per Second (GOPS), and the CONVs execution time related to different CNN models. In this regard, it must be noted that, while the effective GOPS depend on the specific CNN model, the peak GOPS rely on the computational capability of the implemented architecture.

The cheapest version of the proposed design, implemented within the Zynq-7020 device, supports 8-bit fixed-point representation and performs up to 8×2 parallel convolutions at the 118 MHz running frequency. At a parity of the supported data precision, our design exhibits appreciable advantages in terms of resource utilization and power consumption over the competitors [3, 5, 6]. Indeed, the new reconfigurable architecture, purposely designed to maximize the portion of the global computational load sustained by DSPs slices, saves up to ~87%, ~84.1% and ~91.3% of Look-Up Tables (LUTs), Flip-Flops (FFs) and Block RAMs (BRAMs), respectively. Among the compared 8-bit designs, at the parity of CNN model, both those described in [3] and [5] exhibit better effective-throughput/parallelism ratio. This is because, while the design in [3] significantly relies on a clock frequency ~1.8× higher, that demonstrated in [5] is based on a more powerful platform that, in turn, offers more DSPs and consumes more energy. However, both [3] and [5] support much less kernel sizes and strides configuration, thus limiting their flexibility.

When the Zynq-7045 device is used, the novel accelerator performs up to 8×8 parallel convolutions and, compared to the 16- and 18-bit counterparts [3, 4, 7, 8], it exhibits the most favorable effective-throughput/parallelism ratio. This behavior is further confirmed if the above ratio is correlated to the dissipated power. Referring to the VGG-16 and the YOLO

models, the proposed architecture achieves an effective throughput only ~1.2× lower than that reported by [3] and [7], albeit exhibiting halved parallelism and clock frequency, respectively.

The architectures introduced in [4] and [8] reach, in turn, the highest flexibility and the most relevant parallelism. However, such results require a noticeable amount of DSPs, actually among the highest as reported in Table I.

Finally, it is worth mentioning the behavior of the proposed 16-bit architecture when performing the VGG-S model. The latter consists of CONVs of different K, ranging from 3 to 7, and S, ranging from 1 to 2. In such a case, the effective throughput reaches 140.7 GOPS, thus confirming the performance efficiency of the proposed adaptive scheme.

Results summarized in Table I allow us to conclude that the proposed approach overcomes alternative implementations especially in power constrained design environments and in presence of higher flexibility requirements.

TABLE I.   COMPARISON RESULTS

	[3]	[5]	[6]	New				[3]	[4]	[7]	[8]	New			
Device	7Z020	ZU9EG	7Z035	7Z020				7Z045	7VX485T	ZU9EG	7VX690T	7Z045			
LUTs	29867	49022	75241	9772				182616	78318	95136	133854	28743			
FFs	35489	85340	91444	14544				127653	96929	90589	161411	57775			
BRAMs [Mb]	3	13	10.97	1.13				17.09	12.5	8.63	19.48	7.49			
DSPs	190	1048	758	200				780	1034	609	3494	800			
Freq. [MHz]	214	195	100	118				150	150	300	170	150			
Peak GOPS	-	350.4		42.7				-	300.0	289	-	217.2			
Precision	8-bit	8-bit	8-bit	8-bit				16-bit	18-bit	16-bit	16-bit	16-bit			
K (S)	1,3,5(1)	3,5,7(1)	1,3,5,7 (1,2)	1,3,4,5,7,9 (1,2)				1,3,5(1)	≤16(1)	1,3 (1)	1,3,7 (1,2)	1,3,4,5,7,9 (1,2)			
$T_{Mmax} \times T_{Nmax}$	16×2	8×8	-	8×2				64×2	-	8×64	5×256	8×8			
Power [W]	3.50	4.80	3.44	1.98				9.63	18	11.80	7.20	2.75			
CNN	VGG16	VGG16	TinyYOLO[1]	VGG16	YOLO[2]	TinyYOLO[2]	VGGS	VGG16	Custom	YOLO[1]	SqueezeNet	VGG16	YOLO[2]	TinyYOLO[2]	VGGS
Eff. GOPS	84.3	214	125	32	17.1	19.7	28.6	187.8	129.7	102	909.7	153.6	82.3	91.9	140.7
GOPS/DSP	0.444	0.204	0.165	0.160	0.086	0.099	0.143	0.241	0.125	0.167	0.260	0.192	0.103	0.115	0.176
Time [ms]	364	140	53	958.3	325.5	79.5	241.9	163	0.3	288	0.8	199.8	67.8	17	49.2

[1] 416×416 input images   [2] 224×224 input images

V.   CONCLUSION

This paper presented a novel reconfigurable architecture able to run-time adapt itself to different kernels, strides and *fmap* sizes. Even-sized kernels, required by CNN models that exploit both convolutional and transposed convolutional layers, are also supported. When implemented within the Xilinx XC7Z045 device, the novel convolution engine dissipates only ~2.75 W at the 150 MHz running frequency and occupies ~13.1%, ~13.2%, ~39% and ~88.9% of the available LUTs, FFs, BRAMs and DSPs, respectively. When compared to several state-of-the-art counterparts, the proposed architecture exhibits the lowest resource requirement and power consumption with the most favorable effective-throughput/parallelism ratio.

REFERENCES

[1] K. Symonyan and A. Zisserman, "Very deep convolutional networks for large-scale image recognition", Proc. of the 3rd International Conference on Learning Representations, ICLR 2015, San Diego, California, USA, May 2015.

[2] S. Mittal, "A survey of FPGA-based accelerators for convolutional neural networks", *Neural Comput. Appl.*, vol. 32, no. 4, pp. 1–31, 2018.

[3] K. Guo et al., "Angel-Eye: A Complete Design Flow for Mapping CNN Onto Embedded FPGA", *IEEE Transactions on Computer-Aided Design of Integrated Circuits and Systems*, vol. 37, no. 1, pp. 35-47, Jan. 2018.

[4] N. Shah, P. Chaudhari and K. Varghese, "Runtime Programmable and Memory Bandwidth Optimized FPGA-Based Coprocessor for Deep Convolutional Neural Network", *IEEE Transactions on Neural Networks and Learning Systems*, vol. 29, no. 12, pp. 5922-5934, Dec. 2018.

[5] F. Spagnolo, S. Perri, F. Frustaci and P. Corsonello, "Reconfigurable Convolution Architecture for Heterogeneous Systems-on-Chip", 2020 9th Mediterranean Conference on Embedded Computing (MECO), Budva, Montenegro, 2020, pp. 1-5.

[6] Y. Li, S. Lu, J. Luo, W. Pang and H. Liu, "High-performance Convolutional Neural Network Accelerator Based on Systolic Arrays and Quantization", 2019 IEEE 4th International Conference on Signal and Image Processing (ICSIP), Wuxi, China, 2019, pp. 335-339.

[7] S. Zhang, J. Cao, Q. Zhang, Q. Zhang, Y. Zhang and Y. Wang, "An FPGA-Based Reconfigurable CNN Accelerator for YOLO", 2020 IEEE 3rd International Conference on Electronics Technology (ICET), Chengdu, China, 2020, pp. 74-78.

[8] Y. Shen, M. Ferdman and P. Milder, "Maximizing CNN Accelerator Efficiency through Resource Partitioning", Proc. of ACM/IEEE 44th Annual International Symposium on Computer Architecture (ISCA), Toronto, Canada, June 2017.

[9] S. Perri, C. Sestito, F. Spagnolo and P. Corsonello, "Efficient Deconvolution Architecture for Heterogeneous Systems-on-Chip", *J. Imaging*, vol. 6, no. 85, pp. 1-17, Sept. 2020.

[10] M. Farhadi, M. Ghasemi and Y. Yang, "A Novel Design of Adaptive and Hierarchical Convolutional Neural Networks using Partial Reconfiguration on FPGA", 2019 IEEE High Performance Extreme Computing Conference (HPEC), Waltham, MA, USA, 2019, pp.1-7.

SMACD / PRIME 2021 | 19 – 22 July 2021, Online Event

Design and Analysis of a Leading One Detector-based Approximate Multiplier on FPGA

Salvatore Scarfone
Department of Informatics, Modeling,
Electronics and System Engineering
University of Calabria
Arcavacata di Rende, Italy
scrsvt95b07c352b@studenti.unical.it

Fabio Frustaci
Department of Informatics, Modeling,
Electronics and System Engineering
University of Calabria
Arcavacata di Rende, Italy
f.frustaci@dimes.unical.it

Stefania Perri
Department of Mechanical, Energy and
Management Engineering
University of Calabria
Arcavacata di Rende, Italy
s.perri@unical.it

Abstract— **In the context of error-tolerant applications, several approximate multipliers have been proposed to trade the energy consumption with the result accuracy. Unlikely, most of them are conceived for Application Specific Integrated Circuits and they can not be implemented on Field Programmable Gate Arrays due to their unique hardware structure. Among the others, the Leading One Detector-based approximate multipliers have attracted a lot of interest due to their efficiency. Nevertheless, a complete characterization of this kind of multipliers on Field Programmable Gate Arrays is still missing. This paper presents a thorough analysis of the approximate multiplier known as Dynamic Range Unbiased Multiplier when implemented on Field Programmable Gate Arrays, and it provides useful design guidelines to get the optimum energy-quality trade-off. Moreover, a simple approximation strategy is proposed to further increase the multiplier efficiency, leading to an energy reduction of up to 34% and a quality increase of up to 43% with respect to the conventional structure. Finally, the possibility of enhancing the referred approximate multiplier with a dynamically tuning of the energy-quality trade-off is analysed.**

Keywords— Approximate Computing, FPGA, multipliers.

I. INTRODUCTION

In the last few years, approximate computing has emerged as one of the most effective techniques consisting in relaxing the constraint of an exact computation of error-tolerant applications in order to trade the quality of the result with speed, area and power consumption [1]-[2]. The research interest has been devoted to the design of approximate fundamentals computational blocks, such as multipliers. Many approximate multipliers have been proposed in the last few years [3]-[5]. However, the vast majority of these designs is specifically conceived to be used in an Application Specific Integrated Circuit (ASIC) implementation exploiting low-level techniques such as power gating, standard-cell based technology mapping and transistor-level circuital modification. Unfortunately, such techniques cannot be directly applied also to design approximate multipliers on Field Programmable Gate Arrays (FPGAs). Indeed, FPGA chips have a unique structure made of a reduced set of hardware primitives, such as Look-Up Tables (LUTs), dedicated interconnections, Block Rams and Digital Signal Processors (DSPs). Hence, due to the underlying architectural differences between ASICs and FPGAs, trying to replicate an ASIC-based approximate design on an FPGA can be either impossible or inefficient [6]. Among the different proposed approximate multipliers, a special attention has been paid to the one based on the concept of Leading One Detector (LOD). This approach pre-processes the operands to detect the position of the first most significant bit (MSB) equal to one. It was firstly proposed in [7] and named DRUM (Dynamic

Range Unbiased Multiplier). Although such an approach has been targeted for ASIC, its principle can be applied also on FPGA platforms. However, the few proposed FPGA implementations of LOD-based approximate multipliers do not furnish a complete analysis of the quality-energy and quality-resource trade-off that such a kind of multiplier can achieve on the modern FPGA devices [8].

In this paper, we present a thorough analysis of the basic DRUM approximate multiplier when implemented on FPGAs and we provide some design guidelines to get the optimum trade-off between energy, quality and used hardware resources. Moreover, we propose a simple yet effective approximation strategy that is able to reduce (increase) the DRUM multiplier energy (quality) at the parity of output quality (energy) by up to 34% (43%). Finally, we investigate the possibility to enhance the DRUM multiplier with the capability of tuning its energy-quality trade-off at run time.

The remainder of the paper is organized as follows. Section II provides a brief overview of the basic DRUM multiplier. Section III deals with a thorough analysis of the quality-energy-resource trade-off achievable by the selected multiplier when implemented on the Xilinx Artix-7 xc7a100tcsg324-1. Moreover, the proposed approximation strategy applied to the DRUM multiplier is described and analysed. In Section IV, results about enhancing the DRUM multiplier with a dynamic energy-quality trade-off are presented. Finally, some conclusions are drawn in Section V.

II. THE DRUM LOD-BASED APPROXIMATE MULTIPLIER

The Dynamic Range Unbiased Multiplier (DRUM) has been firstly introduced in [7] and it can be considered as the main reference of the LOD-based approximate multipliers. Its scheme is depicted in Fig. 1a.

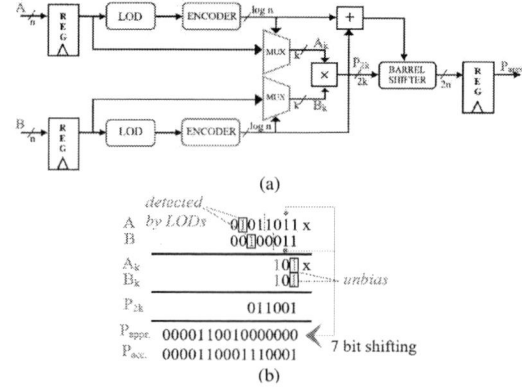

Fig. 1 The DRUM multiplier: (a) the scheme of [7]; (b) an example of operation

© VDE VERLAG GMBH · Berlin · Offenbach

The working principle is based on the idea of extracting two k-bit sub-words, A_k and B_k, from the two n-bit inputs, A and B, respectively, with $k<n$. Each sub-word starts from the position detected by the corresponding LOD, in which the first non-zero MSB is positioned. Then, A_k and B_k are multiplied by an exact $k{\times}k$ multiplier and eventually a barrel shifter left-shifts the result P_{2k} to obtain the final $2n$-bit product $P_{appr.}$. Fig. 1b shows an example of unsigned multiplication with $n = 8$ and $k = 3$. The bits detected by the LODs on inputs A and B are highlighted in red. The last bit of the extracted sub-words A_k and B_k (highlighted in green) is always set to 1 in order to unbias the approximation. The product P_{2k} is finally left-shifted by 7 bit positions to obtain the 16-bit approximate result P_{appr}. In the described example, P_{appr} differs from the accurate product P_{corr} by less than 1%. The value of k is the main parameter and it is set at the design time determining the achievable energy-quality trade-off. In terms of energy dissipation, a low value of k is preferable to reduce the size of the exact used multiplier. However, the lower the value of k, the higher the inaccuracy of the result since a lower number of bits of the inputs A and B are considered in the multiplication. As it is visible in Fig. 1a, the DRUM multiplier employs several auxiliary components in addition to the $k{\times}k$ exact multiplier: two LODs, two Encoders, two Muxes, one Adder and one Barrel Shifter. It follows that the energy-quality trade-off is greatly influenced by the energy dissipation of such auxiliary circuits, whose design complexity is also dependent on k. The purpose of the following Section is to analyse the impact of the auxiliary circuits on the global performance of the DRUM multiplier, when implemented on a FPGA device.

III. ENERGY-QUALITY-RESOURCE TRADE-OFF OF THE FPGA-BASED IMPLEMENTATION OF THE DRUM MULTIPLIER

Taking into account the peculiar hardware structure of an FPGA device with respect to ASIC designs, we implemented the DRUM multiplier on a FPGA for different value of k and we analysed its performances in terms of power, delay, energy, resources occupancy and quality of the result. We focus our analysis on the unsigned multiplier with $n = 8$ since such an input bit width is typical for many error-tolerant applications, such as multimedia processing [5]. All the designs have been implemented on the Xilinx Artix-7 xc7a100tcsg324-1 device using Xilinx Vivado 18.3. The LUTs of the used FPGA chip can be configured either as one Boolean function of up to 6 inputs or two Boolean functions of up to 5 inputs if these two functions share common inputs.

A. Power analysis

Preliminarily, the DRUM multiplier of Fig. 1a has been described in VHDL for different values of k and implemented with a loose clock timing constraint of 10ns. In all the implementations, the Xilinx Vivado's multiplier IP has been used for the exact $k{\times}k$ multiplier. The same IP has been used to implement the reference $8{\times}8$ exact multiplier.

```
// LODin is the input to the LOD, index is the LOD's output
process (LODin)
 begin
  if (LODin (7) = '1') then index <= 7;
   elsif (LODin (6) = '1') then index <= 6;
   elsif (LODin (5) = '1') then index <= 5;
   elsif (LODin (4) = '1') then index <= 4;
   elsif (LODin (3) = '1') then index <= 3;
   elsif (LODin (2) = '1') then index <= 2;
   elsif (LODin (1) = '1') then index <= 1;
   else index <= 0;
  end if;
end process;
```

Fig. 2 The VHDL description of the LOD module

	LUT	FF	Power (mW)	Slack (ns)
Accurate	52	32	3.08	2.65
DRUM_k5	74	32	8.50	1.06
DRUM_k4	59	32	5.80	1.56
DRUM_k3	55	32	3.87	3.32
DRUM_k2	35	32	2.21	3.78

TABLE I.   IMPLEMENTATION RESULTS FOR $T_{CLK}=10NS$

Fig. 2 depicts an excerpt of the VHDL code describing the LOD module. Table I summarizes the obtained implementation results. The power dissipation has been extracted by the Xilinx Vivado Report Power tool on the basis of the actual switching activity of the internal nodes. The latter has been found by annotated post-implementation simulations in which the multiplier has been fed by a sequence of 10,000 random input operands. The power consumptions reported in Table I include the power dissipated by the signals, the logics and the clock network. As expected, the design complexity of the DRUM multiplier increases for higher values of k. Indeed, the utilized LUTs in the case $k = 5$ are more than doubled with respect to the case $k = 2$. More interestingly, Table I reveals that the power dissipation of the accurate multiplier is lower than the one showed by almost all the DRUM configurations. The only exception is represented by the DRUM configuration with $k = 2$, whose power dissipation is 28% lower compared to the one of the accurate multiplier. Therefore, the presented analysis clearly shows that implementing the DRUM multiplier as shown in the scheme of Fig. 1a on a FPGA device leads to inefficient results.

B. Power-quality analysis of the pipelined DRUM multiplier

An accurate analysis of the results of the annotated post-implementation simulations have revealed that the power inefficiency of the DRUM multiplier, shown in the previous sub-section, is due to the high switching activity of the internal nodes. This is a consequence of the high number of glitches generated by the auxiliary circuitry when the multiplier is implemented according to the scheme of Fig. 1a. Therefore, in order to reduce the switching activity of the internal nodes, we investigated the possibility to insert a pipeline stage, as depicted in Fig. 3. The latter shows that the exact $k{\times}k$ multiplier is in a pipeline stage. Therefore, the glitches coming out from the two MUXs cannot propagate towards the multiplier. Similarly, the pipeline registers after the adder prevent the glitches to propagate towards the barrel shifter. This scheme has been implemented and characterized and the implementation results obtained at the clock timing constraint of 10ns are collected in Table II. Moreover, the accuracy achieved by each DRUM configuration has been analysed considering 100,000 couples of uniformly distributed random operands. The multiplier accuracy has been measured in terms of the well-known Mean Absolute Error (MAE) metric [3], whose expression is given in Equation (1).

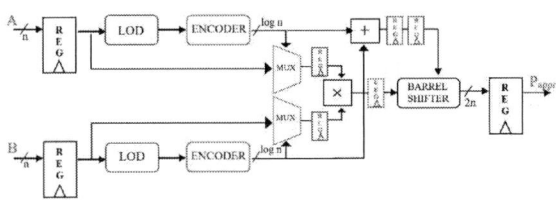

Fig. 3 Pipelined version of the DRUM multiplier

TABLE II. RESULTS FOR THE PIPELINED DRUM MULTIPLIER (T_{CLK}=10NS)

	LUT	FF	Power (mW)	Slack (ns)	MAE
Accurate	32	52	3.08	2.65	0
DRUM_k5_pipe	66	60	3.27	5.34	529
DRUM_k4_pipe	52	56	2.68	5.66	948
DRUM_k3_pipe	46	52	2.42	6.07	1836
DRUM_k2_pipe	31	48	1.64	6.54	3649

$$MAE = \frac{\sum_{i=1}^{N}|P_{appr,i} - P_{corr,i}|}{N} \qquad (1)$$

where N is the number of input operands (A_i, B_i) fed to the multiplier, and $P_{approx,i}$ and $P_{corr,i}$ are the i-th approximate and correct result, respectively.

It is worth noting that the total amount of used resources (i.e. Flip-Flops and LUTs) obviously increases with respect to the multiplier implementation without pipeline. In particular, the number of FFs increases due to the additional pipeline stages. Interestingly, the pipeline allows the synthesizer to perform a more efficient synthesis, therefore the number of LUT decreases with respect to the multiplier implementation without pipeline. Nevertheless, the higher amount of total used resources is a reasonable price to pay for reducing the power dissipation by up to 53%. It is interesting to note that, with $k \leq 4$, the proposed circuit dissipates less power than the accurate 8×8 multiplier. As expected, the lower the value of k the higher the value of the *MAE*, that means a lower accuracy. As a result of the presented analysis, we can conclude that the insertion of a pipeline stage is an effective strategy to make the LOD-based DRUM multiplier suitable for a FPGA-based implementation. In other words, pipelining allows the DRUM multiplier to achieve an effective power-quality trade-off.

C. Proposed approximation strategy to improve the performance of the DRUM multiplier on FPGA

In order to further improve the performance of the DRUM multiplier, we propose a simple approximation strategy that can be easily integrated in the design depicted in Fig. 3. The idea is to enhance the DRUM multiplier with another knob that can be used to expand the energy-quality trade-off. It consists in simplifying the k-bit sub-word A_k (B_k) when the LOD detects the first non-zero MSB within one of the first k LSB positions of the operand A (B). For each configuration of the DRUM multiplier (i.e. for each value of k), let us introduce the parameter $h \leq k$. It is worth noting that h and k allow a configuration of the DRUM multiplier to be defined, in which h LSBs of the sub-words A_k (B_k) are set to 0 when the first non-zero MSB of the operand A (B) lies within its first k LSB positions. As an example, with the configuration (2_5) the DRUM multiplier uses h=2 and k=5 and, when the leading one bit is one of the first five LSBs of the operand A (B), two LSBs of the 5-bit sub-word A_5 (B_5) extracted from A (B) are set to 0.

As well as k, the value of the new parameter h is chosen at the design time. With such a further approximation, the design of the auxiliary circuits of the DRUM multiplier can be simplified. As an example, Fig. 4 depicts the VHDL code of the MUX used to extract the k-bit sub-word from each of the two input operands. Moreover, by varying the value of h, it is possible also to tune the energy dissipation of the exact $k \times k$ multiplier. The design space have been thoroughly analysed by implementing all the possible pipelined multiplier configurations assuming $k \in [2, 5]$ and $h \in [0, 5]$, with the constraint $h \leq k$. The configurations with $h = 0$ are equivalent to the original DRUM multiplier. All the designs have been

synthesized and implemented at their maximum clock frequency. Fig. 5 depicts the obtained trade-off between the achieved quality and the average energy per operation of the multipliers.

Some interesting considerations can be drawn from the analysis of Fig. 5. It is evident that only the configurations with $k \leq 4$ show an energy dissipation significantly lower than the energy dissipated by the exact multiplier (some configurations with k=5 have even a higher energy consumption). Moreover, the proposed approximation strategy is able to make the energy-quality trade-off more graceful. Indeed, we can observe that, for a fixed value of k, the result quality does not significantly decrease as the value of h changes. Conversely, the energy consumption of the multiplier considerably decreases. As an example, the configuration (2_3) achieves a MAE only 1% higher than the standard DRUM multiplier with k=3, but its energy consumption is 34% lower. Similarly, the proposed design strategy can improve the result quality at the parity of the energy consumption. As an example, the configuration (4_4) dissipates almost the same energy of the standard DRUM multiplier with $k = 3$, but its MAE is 43% lower.

Table III reports the implementation results obtained for the multiplier depicted in Fig. 3 with the design configurations leading to the optimum energy-quality trade-off. It is worth noting that, at the parity of the parameter k, the proposed design strategy can reduce the resources utilization with a negligible decrease of the maximum operating frequency. As an example, when compared to the standard DRUM configuration with $k = 4$, the proposed DRUM configuration (4_4) saves 18% of LUTs and 11% of FFs and it is only ~4% slower.

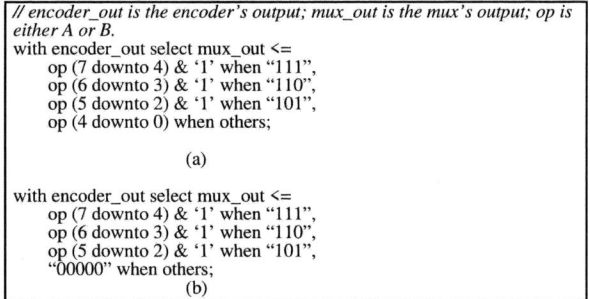

Fig. 4 The VHDL description of the MUX module for k=5: a) conventional DRUM; b) proposed DRUM with h=5

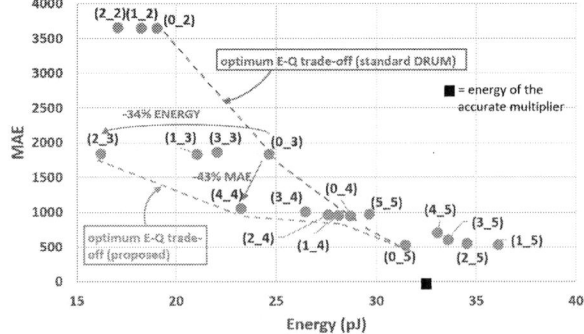

Fig. 5 Energy-quality trade-off of the conventional and proposed DRUM multiplier

TABLE III. IMPLEMENTATION RESULTS OF THE CONVENTIONAL AND PROPOSED DRUM MULTIPLIER

Conventional DRUM				Proposed			
(h_k)	LUT	FF	Freq. (MHz)	(h_k)	LUT	FF	Freq. (MHz)
(0_2)	40	48	401	(2_3)	56	50	380
(0_3)	58	52	394	(2_4)	53	54	274
(0_4)	55	56	278	(4_4)	45	50	267
(0_5)	72	60	250	(0_5)	72	60	250

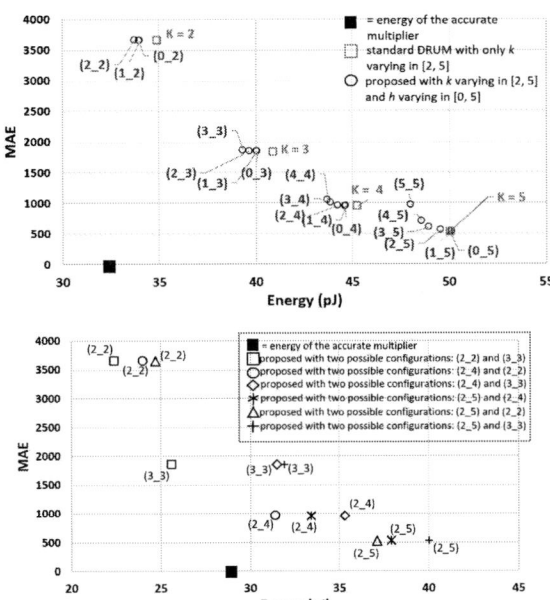

Fig. 7 Energy-quality trade-off obtained with just two tuning levels.

IV. ANALYSIS OF THE DYNAMIC DRUM MULTIPLIER

As a further novelty, we analyzed the performance of the DRUM multiplier when enhanced with the capability of tuning k and h at run time. This is strongly needed when the required energy-quality trade-off is not fixed at the design time but it may change during the application running [5]. Obviously, the dynamic configurability entails a more complex design since the multiplier should be able to adapt itself to different configurations. Fig. 6 shows the energy-quality trade-off achieved with the dynamic tuning. In all the designs, the values of k and h are supposed to be stored into registers that can be set by external control signals. All the designs have been synthetized and implemented at their maximum clock frequency. We firstly analyzed the performances of the standard DRUM multiplier and of the proposed design implemented with the capability to vary k and h in the whole range [2, 5] and [0, 5], respectively. It is worth noting that the proposed approximation strategy leads to better energy results compared to the standard DRUM multiplier, where only k can be dynamically changed. Nevertheless, due to the increased design complexity, all the designs dissipate more energy than the accurate multiplier. Therefore, we concluded that the DRUM multiplier, with a full dynamic configuration of its parameters, is not suitable for an FPGA-based implementation.

Since the increased design complexity is a consequence of the configurability range, we analysed several versions of the proposed DRUM multiplier with only two possible quality configurations, i.e. k can vary between just two different values. The parameter h has been set to the value assuring the best energy-quality trade-off. Fig. 7 depicts the performance of just the designs with the most favourable combination of h and k. It is worth noting that only the case with $k \in [2, 3]$ is able to tune the energy-quality trade-off at run time and to reduce the energy dissipation with respect to the accurate multiplier.

V. CONCLUSIONS

This paper has presented a thorough analysis of the approximate LOD-based DRUM multiplier and useful design guidelines have been drawn to make such a kind of multiplier suitable for an FPGA implementation. We demonstrated that the energy consumption due to the required auxiliary circuits invalidates the traditional DRUM design performances. We showed also that pipelining is an effective strategy to enable a significant energy-quality trade-off. Moreover, we proposed a simple yet efficient approximation technique that is able to make the energy-quality trade-off more graceful. Compared to the traditional DRUM multiplier, the proposed design reduces (increases) the energy consumption (output quality) at the parity of output quality (energy) by up to 34% (43%). Finally, we investigated the performance of the DRUM multiplier when enhanced with the capability to dynamically tune the energy-quality trade-off at run time. We showed that, when implemented on FPGAs, such a kind of multiplier exhibits a favourable behaviour only when the energy-quality trade-off can be tuned between just two different levels.

REFERENCES

[1] J. Han and M. Orshansky, "Approximate computing: An emerging paradigm for energy-efficient design," 2013 18th IEEE European Test Symposium (ETS), Avignon, France, 2013, pp. 1-6.

[2] M. Alioto, V. De and A. Marongiu, "Guest Editorial Energy-Quality Scalable Circuits and Systems for Sensing and Computing: From Approximate to Communication-Inspired and Learning-Based," in IEEE Journal on Emerging and Selected Topics in Circuits and Systems, vol. 8, no. 3, pp. 361-368.

[3] F. Frustaci, S. Perri, P. Corsonello and M. Alioto, "Approximate Multipliers With Dynamic Truncation for Energy Reduction via Graceful Quality Degradation," in IEEE Transactions on Circuits and Systems II: Express Briefs, vol. 67, no. 12, pp. 3427-3431, Dec. 2020.

[4] M. H. S. Javadi, M. H. Yalame and H. R. Mahdiani, "Small Constant Mean-Error Imprecise Adder/Multiplier for Efficient VLSI Implementation of MAC-Based Applications," in IEEE Transactions on Computers, vol. 69, no. 9, pp. 1376-1387, 1 Sept. 2020.

[5] H. Pei, X. Yi, H. Zhou and Y. He, "Design of Ultra-Low Power Consumption Approximate 4–2 Compressors Based on the Compensation Characteristic," in IEEE Transactions on Circuits and Systems II: Express Briefs, vol. 68, no. 1, pp. 461-465, Jan. 2021.

[6] B. S. Prabakaran et al., "DeMAS: An efficient design methodology for building approximate adders for FPGA-based systems," 2018 Design, Automation & Test in Europe Conference & Exhibition (DATE), Dresden, Germany, 2018, pp. 917-920.

[7] S. Hashemi, R. I. Bahar and S. Reda, "DRUM: A Dynamic Range Unbiased Multiplier for approximate applications," 2015 IEEE/ACM International Conference on Computer-Aided Design (ICCAD), Austin, TX, USA, 2015, pp. 418-425.

[8] S. Gandhi, M. S. Ansari, B. F. Cockburn and J. Han, "Approximate Leading One Detector Design for a Hardware-Efficient Mitchell Multiplier," 2019 IEEE Canadian Conference of Electrical and Computer Engineering (CCECE), Edmonton, AB, Canada, 2019, pp. 1-4.

Extending a RISC-V core with an AES hardware accelerator to meet IOT constraints

Anthony ZGHEIB, Olivier POTIN, Jean-Baptiste RIGAUD, Jean-Max DUTERTRE

Mines Saint-Etienne, CEA-Tech Centre CMP, F - 13541 Gardanne, France
zgheib@emse.fr, olivier.potin@emse.fr, rigaud@emse.fr, dutertre@emse.fr

Abstract—Internet of Things devices and applications are subject to strong constraints in terms of cost, code size and power consumption. This leads to difficulties in using resource-hungry encryption algorithms to ensure the confidentiality of the exchanged data. In this paper, we extend with a custom instruction the RISC-V open source Instruction Set Architecture (ISA) and integrate an Advanced Encryption Standard (AES) hardware accelerator to an IBEX RISC-V core. This is achieved for the sake of reducing its energy consumption, encryption time and code size with respect to purely AES software solutions. We consider a Field Programmable Gate Array implementation and ascertain its relevance for an Electrocardiography use case.

Index Terms—IOT, AES, RISC-V, ISA, ECG, power consumption.

I. INTRODUCTION

The Internet of Things (IOT) consists in embedded systems driving physical objects (e.g. sensors, actuators) connected to the internet. Through the internet, these physical objects communicate and exchange information between each others or with servers. Designers aim at improving the performance of the IOT applications in terms of code size, speed, power consumption and security. For a secure data exchange in IOT applications, and from privacy concerns, the exchanged data need to be encrypted. The Advanced Encryption Standard (AES) is a well known algorithm used for data blocks' encryption and decryption [1]. Its implementation can be either hardware or software. However, the use of AES software encryption algorithms requires high energy consumption and encryption time. These characteristics do not meet IOT constraints. In this work, we describe a hardware-based 128-bit AES implementation to an IBEX [2] RISC-V processor [3] to solve this problem. We extend the RISC-V open source Instruction Set Architecture (ISA) with a custom instruction to encrypt the data to be exchanged. Furthermore, we compare our approach to two software-based implementations: the TinyAES [4] and the OpenSSL AES [5] to reflect the performance of our solution. We also consider an Electrocardiography (ECG) application use case. Our paper is organized as follows: Section II details the design of the hardware AES Intellectual Property (IP) and its connection with the RISC-V processor. Section III describes the simulations and the circuit characteristics. Section IV leads to the comparison between the software and hardware AES models. Furthermore, Section V mentions a ciphering example of ECG signals. Finally, Section VI presents our conclusion.

II. HARDWARE AES RISC-V CORE EXTENSION

A. Hardware AES IP design

Our 128-bit AES IP is based on the Federal Information Processing Standard (FIPS) 197 standards [1]. It computes the transformations of one AES round, from the initial round 0 through rounds 1 to 9 up to the final round 10 of the complete AES Rijndael design. The round ciphering depends on the round and the block cipher mode. We integrate four block cipher modes of operation: Electronic codebook (ECB), Cipher Block Chaining (CBC), Cipher Feedback (CFB) and Output Feedback (OFB) [6]. Fig. 1 shows the architecture of the AES IP. It has four 128-bit registers (drawn in green) for storing the data to be encrypted (plaintext), the encryption key, the initial vector (IV) or the encrypted result and the last register for storing the round key. The AES IP receives from the circuit: the clock and reset signals. Similarly, it receives from the RISC-V core the following input signals: the operating cipher mode (mode_i), the calculation round number (round_i), the enable round computing signal (en_rnd_cpt_i), the plaintext, key or IV (data_i), the writing address (wr_addr_i) and its enable signal (wr_en_i). When the encryption is done, the plaintext, key or encrypted result could be read with the read address (rd_addr_i) and read enable (rd_en_i) signals. The input and output signals (data_i and data_o) are 32-bit buses for easier compatibility while loading or storing values from or to the 32-bit RISC-V core registers. Hence, four store operations are needed to store a 128-bit plaintext, key or IV.

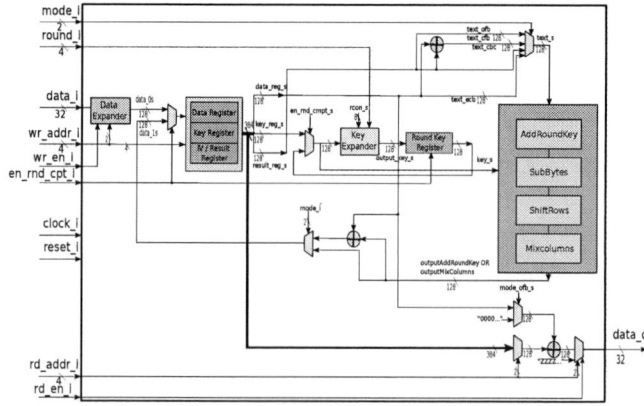

Fig. 1. Architecture and interface signals of the AES IP.

© VDE VERLAG GMBH · Berlin · Offenbach

B. Extending the RISC-V ISA

RISC-V cores have a free open source ISA designed for extensibility in comparison to other cores like x86 and ARM [7]. In our work, we exploit the RISC-V ISA extension's capability to integrate the AES hardware accelerator efficiently. We extended the RISC-V ISA with a new custom instruction in order to communicate with and to control the operations of the AES IP through the IBEX core. The use of a custom dedicated instruction reduces the number of assembly instructions used to perform an encryption. In addition, it brings an improvement in regard of the AES encryption speed. This is also demonstrated by [8] with their AES implementation experiments based on a SPARC V8-compatible LEON-2 processor for the S-BOX calculation. Our AES custom instruction has to be able to perform the three following operations:

- A store operation (moving data from the IBEX registers to the AES registers).
- A round operation (executing the AES round calculations).
- A load operation (moving data from the AES registers to the IBEX registers).

A single RISC-V 32-bit R-type instruction (as a register to register operation is called) permits us to operate our IP in all its functionalities. Its format is given in Fig. 2. Funct7 is used to code the round number, the specified chain mode and the required operation. Table I enumerates the values of funct7 we used for this purpose. R_3, R_2, R_1 and R_0 represent the 4 bits of the round number. For our AES implementation, the funct3 and opcode labels are fixed to "111" and "0001011" having a compatibility with the RISC-V funct3 and opcodes required format [9]. Rs1 and rd represent respectively the addresses of the source and destination registers (the second source register rs2 is not used). Their values depend on the instruction operation, they could point to IBEX or AES registers.

C. AES IP - IBEX connection

The IBEX core is a 32-bit open source RISC-V central processing unit (CPU) core. It is a low power and small processor suitable for IOT applications. It has a 2-stage pipeline: Instruction Fetch (IF) and Intruction Decode and Execute (ID/EX) [2]. IBEX is used in the opentitan project provided by lowrisc [10]. This project also implements a hardware AES IP, for which the communication between the processor and the IP is made using a set of control and status registers (CSRs). This allows the ciphering/deciphering processes to be achieved in parallel to the processor activity. In our case, we communicate with the AES IP and execute its calculation using the new custom instruction which prevent a parallel execution.

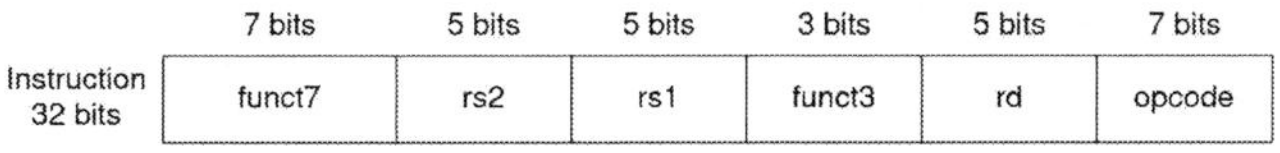

Instruction 32 bits	7 bits	5 bits	5 bits	3 bits	5 bits	7 bits
	funct7	rs2	rs1	funct3	rd	opcode

Fig. 2. The RISC-V 32-bit R-type instruction format.

TABLE I
FUNCT7 VALUES AND DESCRIPTIONS

Funct7 value	Description
XXXX001	Store instruction
$R_3 R_2 R_1 R_0 100$	Round Instruction - Mode ECB
$R_3 R_2 R_1 R_0 101$	Round Instruction - Mode CBC
$R_3 R_2 R_1 R_0 110$	Round Instruction - Mode CFB
$R_3 R_2 R_1 R_0 111$	Round Instruction - Mode OFB
XXXX000	Load instruction-Modes ECB,CBC and CFB
XXXX011	Load instruction-Mode OFB

We made modifications to the IBEX core, especially to its Decode (ID) stage to adapt it to the new custom AES instruction. A simple version of the connection between the IBEX's ID stage and the AES IP is illustrated in Fig. 3. As an instruction arrives from the Fetch Stage to the Decode Stage, it is decoded to various signals. In case of an AES instruction, dedicated signals will be used to activate and control the AES IP. Referring to Fig. 3, the bus data_i is used to transfer either the plaintext to be ciphered, the key, or the IV used in AES chain modes. These values are sent from the IBEX registers. Similarly, data_o can be used to transfer the plaintext, the key, or the ciphertext to be stored in the IBEX registers.

D. Application code modification and compilation

In order to avoid any modification of the compiler toolchain, we used the new AES instruction as a macro instruction to control the AES IP. An AES encryption process involves the use of the AES macro instruction several times, using different funct7 values as mentioned in Table I, in order to sequentially store the plaintext and key in the AES registers, then to launch the 11 AES round calculations and, at the end to store the ciphertext in the IBEX registers. The macro instruction is defined using the inline assembly (ASM) code. An instance of its define format is represented as follows:
#define AES_macro_insn(_funct7, _rs1, _rd) /
r_type_insn(_funct7, 0b00000, _rs1, 0b111, _rd, 0b0001011)
The AES macro instruction is a R-type instruction with fixed values for the rs2, funct3 and opcode fields corresponding to the AES R-type instruction fields. Consequently, we can replace the funct7, rs1 and rd fields with the required values in the application code. By compiling the application code, we obtain the virtual memory (VMEM) file as one of the output files. This file contains the binary instructions interfaced with the RISC-V core to be executed. The VMEM code size of our AES instructions dedicated for an ECB encryption with our AES IP is equal to 0.6 KBytes.

III. SIMULATION RESULTS AND CIRCUIT CHARACTERISTICS

A. Simulation results

Using Vivado 2018.1 tools, we have successfully simulated the place & root design of the IBEX core with the AES IP for all the ciphering modes and their chain processes. As an illustration, we report the execution time of a test code that executes a simple AES encryption in the ECB mode. It takes 140 clock cycles for the whole test code to be executed, from

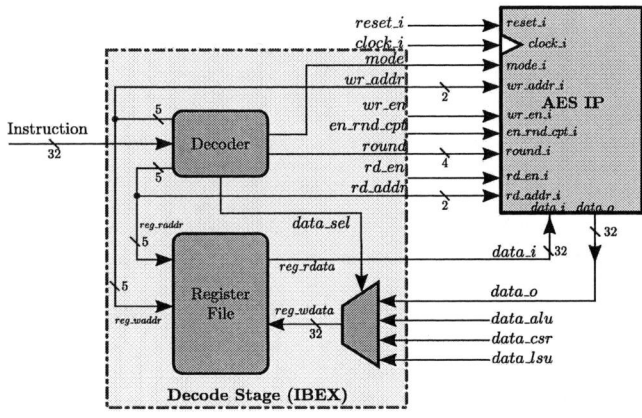

Fig. 3. The connection between the IBEX's decode stage and the AES IP.

the initialization of the plaintext and key values in the memory to the end of the program. From the 140 cycles, the AES IP consumes 60 from the first store instruction to its registers till the last load to the IBEX registers (ciphertext). And from these 60 cycles, 23 correspond to AES instructions execution: there are 4 for the storage of the plaintext and the same for the key, 11 for the AES rounds and 4 to store the ciphertext into the IBEX registers. The instructions that consume the most in terms of clock cycles are the Load Word (LW) operations from the memory to the IBEX registers. These operations are needed before transferring data from the core to the AES registers.

B. Circuit characteristics

1) Area: All our implementations targeted a Nexys Artix-7 Field Programmable Gate Array (FPGA) board, a 18nm CMOS technology. This board contains 33,650 logic slices. Each slice is composed of four 6-input LUTs (LookUp Tables), 8 flip-flops, multiplexers and carry units. As a first approach, we placed and routed the AES IP. Its design requires 547 slices referring to the Vivado utilization report. Also, we did the same for the IBEX core which needed 765 slices. The top level design containing the IBEX core with the clock generator and the random-access memory (RAM) component needs 886 slices. However, when implementing the AES IP to the IBEX core, their utilization requirements in terms of slices were adjusted to adapt the connection between their input and output signals. The AES IP and IBEX core require respectively 609 and 757 slices when placed in a top level design with a clock generator and RAM. In total, the design needs 1,479 slices. For the current AES model and comparing the two top levels, adding the AES IP makes the circuit bigger by 40% in terms of slices. As mentioned before, the IP contains 4 block cipher modes. By modifying it and just conserving the ECB mode, we can gain approximately 5% in the AES area. Depending on the application, a compromise could be made between the area gain or the block cipher modes requirements.

2) Power: We estimated the power consumption of our place & root design using the Vivado report power utility. For

an accurate measure, we took into consideration the Switching Activity Interchange Format (SAIF) file which plots all the signals activity information for one ECB ciphering process simulation. The total power consumed for one ECB ciphering by the IBEX core and AES IP is equal to 397 mW.

IV. COMPARISON WITH 2 SOFTWARE AES ALGORITHMS

A. Description of the software algorithms

The 8-bit TinyAES is a small and portable implementation of the AES. It has the ECB, CounTeR (CTR) and CBC encryption chain modes [4]. The 32-bit OpenSSL AES uses LUTs containing intermediate results of an AES ciphering process and could operate under ECB, CBC, CFB and OFB chain modes [5]. Both codes contain functions for encryption and decryption. The unused functions like the decryption ones have been removed from our test code to have a fair comparison between the two algorithms and our hardware implementation. Also, only the ECB mode have been activated for the simulation. We give in table II, the test code size (VMEM size) dedicated to perform a single ECB encryption either by using the software algorithms (TinyAES or OpenSSL AES), or by using our hardware AES IP with their necessary time, power and encryption energy. The total power consumption for one encryption is equal to 380 mW for the TinyAES and 378 mW for the OpenSSL AES.

B. Encryption energy interpretation

Referring to Table II, the AES IP implementation consumes less energy in µJ than the TinyAES and OpenSSL AES. The largest encryption time for having one ECB encryption corresponds to the TinyAES (70,036*20ns) where 20 ns is the system's clock period. By taking the AES chain modes into consideration, the encryption time for the following encryption processes requires less clock cycles since the key has already been defined for the TinyAES and OpenSSL AES or has been loaded at the first encryption process in the hardware AES key register. Table III shows the chain encryption time with the required energy for the three AES solutions. We can check that, for a single ECB encryption with the TinyAES, 17 data encryption with the OpenSSL AES and 847 data encryption with our AES IP could be made with the same energy budget.

V. ECG USE CASE SCENARIO

Several IOT applications require the confidentiality to be considered as a major aspect while exchanging data. For instance, in the medical field, patients' private medical data need to be protected. In this section, we refer to an AES ciphering process to cipher ECG signals. The ECG is known as a method to record the electric heart muscles' activity.

TABLE II
COMPARISON BETWEEN AES SOLUTIONS FOR A SINGLE ECB CIPHERING

AES Type	Vmem(kB)	Clock cycles	Power(mW)	Energy(µJ)
AES IP	0.6	140	397	1.11
TinyAES	5.7	70,036	380	532.27
OpenSSL	31.6	6,445	378	48.59

TABLE III
CHAIN ENCRYPTION DATA WITH THE SAME KEY

AES Type	Clock cycles	Energy(μJ)
AES IP	76	0.627
TinyAES	57,856	437.39
OpenSSL	3,874	29.13

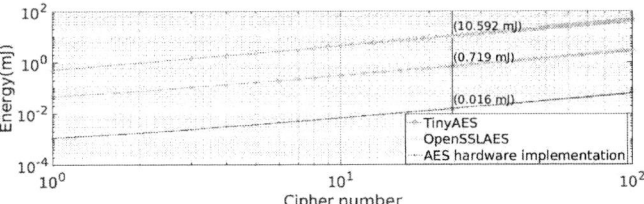

Fig. 4. The energy consumption for an ECB ciphering with the AES hardware implementation, TinyAES and OpenSSL AES via an ECG use case.

We consider a Holter device which is a continuous ECG monitoring medical device. It monitors ECG signals from 24 to 48 hours as a short term or from 1 to 2 weeks as a long term [11]. Referring to [12] and [13], a database named "SHAREE" contains ECG Holter signals of 139 hypertensive patients recorded for 24h in order to be exploited in the context of medical research. The recordings were based on 3 ECG signals that were sampled at 128 samples per second with an 8-bit precision. This leads to 3,072 bits/second which is equivalent to 384 Bytes/second. We consider the case of sending the ECG signals gradually as the Holter device receives the patient heart's signals to the medical unit via an IOT platform. Using the AES encryption model, a message of 128 bits (16 Bytes) could be encrypted. Hence, on the base of this ECG use case, 24 encryption processes are required. Fig. 4, represented on a log-log scale, shows the encryption energy required for the three AES considered solutions. For the ECG application, the ciphering process using the AES hardware implementation requires 0.016 mJ compared to 10.592 mJ and 0.719 mJ for the TinyAES and OpenSSL AES. Our AES IP respectively consumes 662 and 44.9 times less energy than the TinyAES and OpenSSL AES. This reflects the low energy aspect of our IP in favor to be used in IOT applications. If the Holter system's total energy consumption also depends on the energy used for the communication process, the ECG signals filtering, recording and more, this energy will be approximately the same for the three encryption solutions.

VI. CONCLUSION

In this work, we report the implementation of an AES hardware accelerator for the purpose of reducing the energy consumption, encryption time and code size of a circuit designed for IOT applications. We also illustrate the potential of RISC-V cores through the use of their open-source ISA to improve their performances in IOT applications. This is achieved by extending the RISC-V ISA to implement our AES accelerator to an IBEX RISC-V core. We show the benefit of adding a new instruction to the RISC-V ISA by comparing our AES hardware implementation to two software algorithms: the TinyAES and OpenSSL AES. As a result, the application code size (VMEM size) dedicated to encrypt a message with the AES IP is small and have a size gain of 89.47% and 98.1% with respect to the TinyAES and OpenSSL AES. Furthermore, the AES IP requires 140 clock cycles to obtain an ECB ciphertext compared to 70,036 and 6,445 for the TinyAES and OpenSSL AES codes respectively. Alternatively, the AES IP adds 40% in terms of slices to the final circuit. However, more optimizations could be made to our IP to gain in area. Finally, we demonstrate in an ECG use case that our AES accelerator

respectively consumes 662 and 44.9 times less energy than the TinyAES and OpenSSL AES. This shows the relevance of our solution and promotes its use in IOT applications. Another perspective is to investigate the LightWeight Cryptography (LWC) candidates of the National Institute of Standards and Technology (NIST) contest like ASCON [14] or already existing LWC algorithms (e.g. PRESENT [15]) to reduce further the size and energy overheads of the IBEX core. A similar work to our AES implementation could be found in [16], where they implemented ASCON and ISAP accelerators to a RI5CY core while extending the RISC-V ISA. Comparing to ASCON, ISAP offers in addition protection against various types of fault attacks.

REFERENCES

[1] NIST-FIPS Standard, "Announcing the advanced encryption standard (aes)," FIPS Publication, 197, pp.1-51, Nov 2001.
[2] Lowrisc, IBEX documentation, Oct 2020, https://ibex-core.readthedocs.io/en/latest.
[3] D. Patterson and A. Waterman, "The RISC-V Reader: An Open Architecture Atlas," Strawberry Canyon, 1st ed., 2017.
[4] kokke, "Small portable AES128/192/256 in C," https://github.com/kokke/tiny-AES-c, 2016.
[5] The OpenSSL Project, "OpenSSL: The open source toolkit for SSL/TLS," http://www.openssl.org, 2019.
[6] M. Dworkin, "Recommendation for block cipher modes of operation: Methods and techniques," Technical report, USA, 2001.
[7] K. Asanović and D. Patterson, "Instruction sets should be free: The case for risc-v," Technical report, UCB/EECS-2014-146, EECS Department, University of California, Berkeley, Aug 2014.
[8] S. Tillich, J. Großschädl, and A. Szekely, "An instruction set extension for fast and memory-efficient aes implementation," in Communications and Multimedia Security, vol. 3677, J. Dittmann, S. Katzenbeisser, and A. Uhl, Eds. Berlin, Springer Berlin Heidelberg, pp. 11–21, 2005.
[9] The RISC-V International com., RISC-V custom opcodes, Nov 2013.
[10] Opentitan, AES HWIP Technical Specification, https://docs.opentitan.org/hw/ip/aes/doc, 2021
[11] P. Zimetbaum and A. Goldman, "Ambulatory arrhythmia monitoring," Circulation, vol. 122, no. 16, pp. 1629-1636, Oct 2010.
[12] P. Melillo et al., "Automatic prediction of cardiovascular and cerebrovascular events using heart rate variability analysis," PLOS ONE, vol. 10, no. 3, pp.1-14, Mar 2015.
[13] A. Goldberger et al., "Physiobank, physiotoolkit, and physionet: components of a new research resource for complex physiologic signals," Circulation, vol.101, no.23, pp. e215–e220, 2000.
[14] C. Dobraunig, M. Eichlseder, F. Mendel, and M. Schläffer, "Ascon v1.2", NIST Round 2 Candidate, https://csrc.nist.gov/Projects/Lightweight-Cryptography/Round-2-Candidates, 2019.
[15] A. Bogdanov et al., "PRESENT: An ultra-lightweight block cipher," International workshop on cryptographic hardware and embedded systems, Springer, pp. 450–466, 2007.
[16] S. Steinegger and R. Primas, "A Fast and Compact RISC-V Accelerator for Ascon and Friends," in CARDIS 2020: 19th Smart Card Research and Advanced Application Conference, 2020.

Memristive Logic-in-Memory Implementations: A Comparison

Pietro Inglese, Elena Ioana Vatajelu, Giorgio Di Natale

Univ. Grenoble Alpes, CNRS, Grenoble INP*, TIMA, Grenoble, France
{pietro.inglese,ioana.vatajelu,giorgio.di-natale}@univ-grenoble-alpes.fr

Abstract — The technology evolution addresses the demand for faster computers. Despite the achieved speed-up in terms of memory and computation performances, the communication between the memories and the processor remains a bottleneck of today's computers. The Computation in Memory (CiM) paradigm aims at solving this problem by moving the computation directly inside the memory, eliminating thus the need for data transfer between memory and processor. Among the available CiM implementations, this study focuses on the Logic-in-Memory (LiM) solutions, i.e., digital operations to accelerate Boolean Logic. This work provides a comparison among the most prominent LiM solutions in terms of required memory resources (i.e., number of memristors) and number of operations.

Keywords — Emerging technologies, Logic-in-Memory, Memristors, Non-Volatile memories

I. INTRODUCTION

Among the most important challenges faced by today's computing systems are the memory wall (caused by the uneven evolution of processing speed and memory access times and bus data transfer) and the energy efficiency. Emerging non-volatile memories are widely studied today as means to maximize energy efficiency mainly due their ability to reduce the static power consumption. These memories include Resistive Random-Access Memories (RRAMs), Phase Change Memories (PCMs), and Spin-Transfer Torque Magnetic RAMs (STT-MRAMs). Another perceived advantage of emerging non-volatile memories resides in their physical capabilities which can be exploited to perform logic or arithmetic operations directly inside the memory array, therefore providing a solution to bypass the memory wall issue.

Several solutions to mitigate the memory wall by reducing the data movement between memory and CPU have been proposed. They can be classified in 3 main categories: (i) Computation Near Memory [1], where the memory core is placed as close as possible to the CPU, permitting to have a shorter bus and therefore decrease the latency, (ii) Computation via LUT, that exploits a Look-Up Table storing pre-calculated operations, and the (iii) Computation in Memory (CiM) which exploits memory technologies (both classical and emerging) to perform calculations directly within the memory array. The latter is considered the most efficient, since it is more flexible than the Computation via LUT and completely eliminates the need for data transfer via buses.
Depending on the exploited physical memory device characteristic, CiM can allow to perform analog or digital computations. Analog computations are mainly used to perform additions and matrix multiplications mainly to design accelerators for machine learning. Digital computations are used to accelerate Boolean logic. In this scenario, a part or the whole classical Arithmetic Logic Unit (ALU) embedded in the processor is actually implemented directly within memory. This is referred to as Logic-in-Memory (LiM) and it is the main focus of this paper.

In the last five years, the number of research papers dealing with this topic on different levels of abstraction has increased exponentially. In this context, research is focused on the design of LiM architectures, the development of LiM-compatible instructions set, the methods for system integration and development of the programming model for LiM integration in computing systems. Nevertheless, the actual status of the research is fragmented and the reproduction of the reported results, along with the choice of an implementation to be adopted, are not trivial. For instance, [2] presents a review on in-Memory Computing, focusing on the memories enabling it and on its applications, [3] offers a classification of the in-Memory Computing solutions, while [4] describes several adder implementations. Besides the fragmentation, the use of existing electrical models and their parameters is not always supported by physical measurements on real devices.

While these comparative surveys show the characteristics of existing solutions, a thorough analysis of the implementations of Boolean functions is still missing. Therefore, in this work we present a study of the existing LiM implementations in order to perform a fair comparison in terms of resources and efficiency. More in particular, we present a thorough study of simple Boolean functions implemented in-memory resulting in a comprehensive comparison of LiM solutions in terms of required number of memristors and number of operations.

The remainder of this paper is organized as follows. Section II presents the Logic-in-Memory solutions and describes the basic functioning and the primitive logics enabled. Section III presents an analysis of the described solutions through the basic Boolean logic blocks and a full adder implementation. Finally, Section IV concludes the paper.

II. BACKGROUND

This section describes basic memory array modifications to enable LiM and describes some selected LiM solutions.

In the traditional use of a memory array, a memory cell is selected by means of address decoders, it is written in by the write driver and read from with the help of the sense amplifiers. The voltage levels required to enable the operations on the memory cell are set by the voltage regulators. One memory array communicates with the processor or other memory arrays by means of bus connections. In order to enable the LiM operations, several changes need to be implemented to the memory array or/and to its peripheral circuitry. In this context, the peripheral circuitry consists of standard memory periphery (sense amplifier, write driver, address decoders). Moreover, in some instances, additional logic is added to enable computation.

Several LiM proposals exist in literature, some are general, and can be used with any memory technology, others take advantage of the device physics and are only suitable to be implemented in a specific technology. In addition, some of the

---

* Institute of Engineering Univ. Grenoble Alpes

© VDE VERLAG GMBH · Berlin · Offenbach

existing LiM solutions are designed to implement specific logic functions [5]–[11] henceforth called "primitive operations", while others propose solutions for the implementation of any Boolean function [12].

The existing LiM solutions can be classified depending on the way the inputs are stored (the memory content, i.e., stateful logic, or an electrical signal, i.e., non-stateful logic) and depending on where the operations are performed (in the memory array, or in the periphery). In this context, three main LiM classes can be distinguished. They are described in the following and their characteristics are summarized in Figure 1.

A. *Stateful Logic in LiM Array*

The operations are performed within the memory array and the data are stored as memory content. This solution is proposed only for memristive crossbar (1R-RRAM) arrays. The input data are stored within the memory array, and the output (computation result) is obtained as memory content within the memory array. Inputs and outputs are coded as the resistive states of the storing memristors.

In order to enable LiM operations within the memory array, several conditions need to be respected: (1) the memory cells containing the input data and the memory cells to store the result of the computation must share the same row (column); (2) access to multiple memory cells should be enabled; (3) specific control voltages (different than the memory read/write voltages) to be applied for the completion of logic operations. As a consequence, the write driver, the voltage regulator and the address decoders of standard memory array have to be modified to enable LiM.

LiM solutions pertaining to this class include: *Memristor-Aided Logic - MAGIC* [5], with NOT and NOR as primitive operations, *FELIX* [6], [7] with OR, NAND and MIN as primitive operations, *IMPLY* [8], [9] with Boolean implication as primitive operation, and Stateful Three-Input Logic [10] with ORNOR3 (i.e., input1 OR (input2 NOR input3)) as primitive operation.

B. *Stateful Logic in LiM Array and its Periphery*

The operations are performed within the memory array periphery or by means of additional logic and the input data are stored as memory content. This solution can be used with any memory technology. The input data are stored within the memory array, while the output (computation result) is obtained as a voltage (or current) outside of the memory array.

In order to enable LiM operations within the periphery of the memory array, several conditions need to be respected: (1) the memory cells containing the input data must share the same column; (2) access to multiple memory cells should be enabled by modifying the address decoders; (3) the sense amplifiers should be modified such that different references are allowed. We refer to this class of solutions as Logic in Periphery (LIP).

The *MRIMA* architecture [12] is based on the Logic In Periphery: it exploits re-configurable Sense Amplifiers (SAs) to perform arithmetic and logic operations on STT-MRAM. All Boolean functions can be implemented with this solution by resorting to additional combinational gates.

C. *Non-Stateful Logic in LiM Array and its Periphery*

The operations are performed within the memory array and by using additional logic, and the data are coded partially as memory content and partially as voltage levels. This solution can be used with resistive technology only. It uses two types of input data: (1) memory content, (2) voltage level, while the output (computation result) is obtained as memory content within the memory array.

In order to enable this type of LiM operation several conditions need to be respected: (1) specific control voltages (different from the memory read/write voltages) to be applied for the completion of logic operations, (2) specific registers to store the inputs to be given as voltage levels. As a consequence, the write driver, the voltage regulator and the address decoders of standard memory array have to be modified to enable LiM.

An implementation of this solution is *PLiM* [11], which implements, as primitive operation, a special case of majority voter, where one of the inputs is negated (a.k.a., *Resistive majority*).

	CiM Solution	#mem pts	Gate	Remarks	Technology	Operations
LIM Array Stateful Logic	MAGIC (NOR2)	3	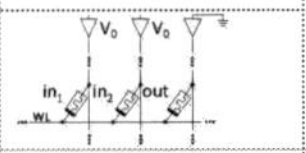	input non-destructive, parallelizable	Memristive crossbar (1R-RRAM)	1. Initialize R_{out} 2. Apply proper voltages to R_{in}, R_{out} 3. Obtain output in R_{in_out}
	IMPLY A→B	2	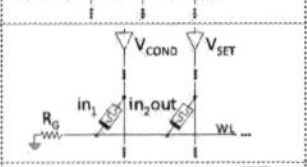	input destructive, parallelizable		
	Stateful Three-Input Logic (ORNOR3)	3	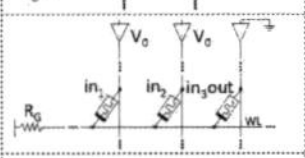	input destructive		
LIM Array + Periphery – Non Stateful Logic (Hybrid inputs)	PLiM (RMAJ3)	3 (1 in the array, 2 external)	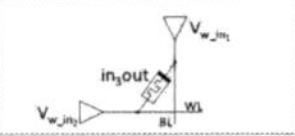	input destructive, requires input pre-processing	Memristive crossbar (1R-RRAM)	1. Read inputs R_{in} and convert to voltages 2. Apply the voltages to R_{in_out} 3. Obtain output in R_{in_out}
LIM Array + Periphery Stateful Logic	Logic in Periphery	2	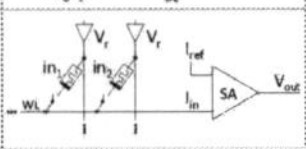	input non-destructive, additional step to store output in memory	SRAM 1T1R-RRAM STT-MRAM	1. Apply voltage on multiple rows 2. Obtain output (via SA) 3. [Store output in memory]

Figure 1 – Primitive logic gates: Column 2 (CiM Solution) lists the considered LiM solutions and the corresponding primitive operations. The number of memory cells needed to implement a 2-input (1-bit) primitive operation is summarized in Column 3 (#mem pts), while the schematic of the "primitive operation" gate for each solution is illustrated in Column 4 (Gate). Column 7 (Operations) lists the algorithm executed to obtain the result of the primitive operation. The executed operation can be input-destructive or not (see column 5, Remarks). An input-destructive operation is an operation that changes the value of the inputs after it is executed.

III. PROPOSED METHOD AND RESULTS

The goal of this work is to provide a comprehensive comparison of existing LiM solutions and understand their implementation complexity. The analysis has been performed on basic Boolean functions, in order to be as generic as possible and to provide the designer an indication of implementation complexity and cost of each LiM solution. In addition, this study can give an indication of which LiM solution is more suitable for a target application, depending on the most used Boolean functions.

In order to achieve a fair comparison among all solutions, we mapped all the 0-input logic functions (TRUE and FALSE), 1-input logic functions (COPY, NOT), 2-input logic functions (NOR, OR, NAND, AND, XNOR, XOR, NIMPLY, IMPLY) and the Full Adder as 3-input logic function, by using the primitive operations of *MAGIC* (and its extensions), *IMPLY* (and *ORNOR3*) and *PLiM* solutions. The full adder has as inputs A, B and C_{IN} while the outputs are S and C_{OUT}.

$$C_{OUT} = ((A + B)' + (B + C)' + (C + A)')'$$
$$S = (((A' + B' + C')' + ((A + B + C)' + C_{OUT})')')'$$

Tables I and II summarize the results of this study. Table I shows the mapping of all considered Boolean functions on LiM primitive operations. For each LiM implementation, the primitive operations are written in blue. Each row contains the mapping of the Boolean function defined in the first column.

Primitives	MAGIC-based		Stateful Logic		IMPLY-based	LIM Array + Periphery / Non Stateful Logic (Hybrid inputs)
	MAGIC	FELIX	IMPLY	IMPLY (non-destructive)	ORNOR3	RMAJ
Primitives	MAGIC_NOT(in1,out) MAGIC_NOR(in1,in2,out)	FELIX_OR(in1,in2,out) FELIX_NAND(in1,in2,out) FELIX_MIN(in1,in2,in3out)	IMPLY_IMPLY(in1,in2out)		ORNOR3(in1,in2,in3)	RMAJ(in1,in2,in3out)
COPY	MAGIC_COPY(in1,f1,out): 1-2) MAGIC_NOT(in1,f1) 3-4) MAGIC_NOT(f1,out)	FELIX_COPY(in1,f1,out): 1) Write0(f1) 2-3) FELIX_OR(in1,f1,out)	IMPLY_COPY(in1,f1,out): 1-2) IMPLY_NOT(in1,f1) 3-4) IMPLY_NOT(f1,out)			RMAJ_COPY(in1,out): 1) Write0(out) r1) Read(in1) 2) RMAJ(in1,0,out)
NOT	MAGIC_NOT(in1,out): 1) Write0(out) 2) MAGIC_NOT(in1,out)		IMPLY_NOT(in,out): 1) Write0(out) 2) IMPLY(in,out)			RMAJ_NOT(in1,out): 1) Write0(out) r1) Read(in1) 2) RMAJ(1,in1,out)
NOR	MAGIC_NOR(in1,in2,out): 1) Write0(out) 2) MAGIC_NOR(in1,in2,out)		IMPLY_NOR(in1,f1,out): 1-3) IMPLY_OR(in1,out,in2) 8-9) IMPLY_NOT(f1,out)	IMPLY_NOR_ND(in1,in2,f1,out): 1-7) IMPLY_OR_ND(in1,in2,out,f1) 8-9) IMPLY_NOT(f1,out)	ORNOR3_NOR(in1,in2,out): 1) Write0(out) 2) ORNOR3(out,in1,in2)	RMAJ_NOR(in1,in2,f1,out): 1-4,r3) RMAJ_AND(in1,in2,out,f1) 5-6,r4) RMAJ_NOT(f1,out)
OR	MAGIC_OR(in1,in2,f1,out): 1-2) MAGIC_NOR(in1,in2,f1) 3-4) MAGIC_NOT(f1,out)	FELIX_OR(in1,in2,out): 1) Write0(out) 2) FELIX_OR(in1,in2,out)	IMPLY_OR(in1,f1,in2out): 1-4) IMPLY_NOT(in1,f1) 5-6) IMPLY_NOT(in2,f1) 3) IMPLY_IMP(f1,in2out)	IMPLY_OR_ND(in1,in2,f1,out): 1-4) IMPLY_COPY(in2,f1,out) 5-6) IMPLY_NOT(in1,f1) 7) IMPLY(f1,out)	ORNOR3_OR_ND(in1,in2,out): 1-2) ORNOR3_NOR(in1,in2,f1) 3-4) IMPLY_NOT(f1,out)	RMAJ_OR(in1,in2,f1,out): 1) Write0(out) 2-3,r1) RMAJ_NOT(in2,f1) r2-r3) Read(in1,f1) 4) RMAJ(in1,f1,out)
NAND	MAGIC_NAND(in1,in2,f1,f2,out): 1-8) MAGIC_AND(in1,in2,out,f1,f2) 9-10) MAGIC_NOT(f2,out)	FELIX_NAND(in1,in2,out): 1) Write1(out) 2) FELIX(in1,in2,out)	IMPLY_NAND(in1,in2,out): 1-2) IMPLY_NOT(in1,f1) 3) IMPLY(in1,in2_out)			RMAJ_NAND(in1,in2,f1,out): 1-4,r4) RMAJ_AND(in1,in2,out,f1) 5-6,r5) RMAJ_NOT(f1,out)
AND	MAGIC_AND(in1,in2,f1,f2,out): 1-2) MAGIC_NOR(in1,f1) 3-4) MAGIC_NOR(in2,f2) 5-6) MAGIC_NOR(f1,f2,out)	FELIX_AND(in1,in2,f1,out): 1-2) FELIX_NAND(in1,in2,f1) 3-4) MAGIC_NOT(f1,out)	IMPLY_AND(in1,in2,f1,out): 1-3) IMPLY_NAND(in1,in2,f1) 4-5) IMPLY_NOT(f1,out)			RMAJ_AND(in1,in2,f1,out): 1-2,r1) RMAJ_NOT(in1,f1) 2-3,r1) RMAJ_NOT(in2,f1) r2-r3) Read(in1,f1) 4) Read(in1,f1,out)
IMP	MAGIC_IMPLY(in1,in2,f1,out): 1-2) MAGIC_NOT(in1,f1) 3-6) MAGIC_OR(out,in2,f1,out)	FELIX_IMPLY(in1,in2,f1,out): 1-2) MAGIC_NOT(in1,f1) 3-4) FELIX_OR(f1,in2,out)	IMPLY_IMPLY(in1,in2out): 1) IMPLY(in1,in2out)	IMPLY_IMPLY_ND(in1,in2,f1,out): 1-4) IMPLY_COPY(in2,f1,out) 5) IMPLY(in1,out)		RMAJ_IMPLY(in1,in2,out): r1-r2) Read(in1,in2) 2) RMAJ(1,in1,in2)
NIMP	MAGIC_NIMPLY(in1,in2,f1,f2,out): 1-2) MAGIC_NOT(in1,f1) 3-4) MAGIC_NOR(in1,in2,out)		IMPLY_NIMPLY(in1,in2,out): 1) Write0(out) 2-3) IMPLY_NOT(in2,out)	IMPLY_NIMPLY_ND(in1,in2,f1,out): 1-5) IMPLY_IMPLY(in1,in2,out,f1) 6-7) IMPLY_NOT(f1,out)		RMAJ_NIMPLY(in1,in2,f1,out): 1-2,r1) RMAJ_COPY(in1,f1) r2) RMAJ(in2) 3) RMAJ(0,in2,f1) 4-5,r2) RMAJ_COPY(in2,f1) r3) Read(in1) r4) Read(f1)
XOR	MAGIC_XOR(in1,in2,f2,out): 1-2) MAGIC_NOR(in1,f1) 3-4) MAGIC_NOR(in2,out) 5-6) MAGIC_NOR(f1,out,f2) 7-8) MAGIC_NOR(f1,out,f2) 9-10) MAGIC_NOR(f2,out)	FELIX_XOR(in1,in2,out): 1-2) FELIX_OR(in1,in2,f1) 3) FELIX_COPY(cin,cout,f1) 4-5) MAGIC_NOT(f1,out)	IMPLY_XOR(in1,in2,f1,f2,out): 1-4) IMPLY_OR(in1,in2,f1) 5-8) IMPLY_COPY(in1,f2) 9) IMPLY(in1,f2) 10) IMPLY(in2,f2) 11-12) IMPLY_NOT(f1,out) 13) IMPLY(f2,out)	IMPLY_XOR_ND(in1,in2,f1,f2,out): 1-2) IMPLY_NOT(in1,f1) 3-4) IMPLY_COPY(in1,f2) 5-6) ORNOR3_NOR(f1,out,f2) 7-8) ORNOR3_NOR(in2,out) 9-10) ORNOR3_NOR(f1,f2,out)	ORNOR3_XOR(in1,in2,f1,f2,out): 1-2) IMPLY_NOT(in1,f1) 3-4) IMPLY_COPY(in1,f2) 5-6) ORNOR3_NOR(f1,out,f2) 7-8) ORNOR3_NOR(in2,out) 9-10) ORNOR3_NOR(f1,f2,out)	RMAJ_XOR(in1,in2,f1,out): 1-7,r4) RMAJ_XOR(in1,in2,out,f1) 8-9,r5) RMAJ_NOT(f1,out)
XNOR	MAGIC_XNOR(in1,in2,f1,f2,out): 1-10) MAGIC_XOR(in1,in2,out,f1,f2) 11-12) MAGIC_NOT(f2,out)	FELIX_XNOR(in1,in2,out): 1-3) FELIX_XOR(in1,in2,f1) 4-5) MAGIC_NOT(f1,out)	IMPLY_XNOR(in1,in2,out): 1-13) IMPLY_XOR(in1,in2,out,f1,f2) 14-15) IMPLY_NOT(f1,out)		ORNOR3_XNOR(in1,in2,f1,f2,out): 1-10) ORNOR3_XOR(in1,in2,out,f1,f2) 11-12) IMPLY_NOT(f2,out)	RMAJ_XNOR(in1,in2,f1,out): 1-7,r4) RMAJ_XOR(in1,in2,out,f1) 8-9,r5) RMAJ_NOT(f1,out)
Full Adder	MAGIC_FA(in1,in2,cin,f1,f2,f3,f4,s,cout): 1-2) MAGIC_NOR(in1,in2,f1) 3-4) MAGIC_XOR(in1,in2,cin,s) 5-6) MAGIC_NOR(in2,cin,s) 7-9) MAGIC_NOR(f1,s,cout) 10) MAGIC_NOR(cin,ini,s) 11-12) MAGIC_NOR(cin,in1,s) 13-14) MAGIC_NOR(in1,in2,f1) 15-16) MAGIC_NOR(f1,s) 17-18) MAGIC_NOR(s,cin,f1) 19-20) MAGIC_NOR(in2,cin,f1) 21-22) MAGIC_NOR(f2,cout) 23-24) MAGIC_NOR(s,f3,s) 25-26) MAGIC_NOR(f3,f2) 27-28) MAGIC_NOR(s,f3) 29-30) MAGIC_NOR(f2,f3) 31-32) MAGIC_NOR(f3,f4,f1) 33-34) MAGIC_NOR(f2,f4,f1) 35-36) MAGIC_NOT(f1,s)		IMPLY_FA_SUM(in1,in2,cin,f1,f2,out): 1-2) Write0(f2), Write1(out) 3-11) IMPLY_OR(in1,f2), IMPLY_OR(in2,f1,f2), OR(cin,f1,f2) 12-16) IMPLY_AND(f2,out,f1,out) 17) Write0(f2) 18-22) IMPLY_OR(in1,f1,f2), IMPLY_OR(in2,f1,f2), IMPLY(cin,f2) 23-27) IMPLY_AND(f2,out,f1,out) 28) Write0(f2) 29-33) IMPLY(in1,f2), IMPLY(in2,f2), IMPLY_OR(cin,f1,f2) 34-38) IMPLY_AND(f2,out,f1,out) 39) Write0(f2) 40-44) IMPLY(in1,f2), IMPLY(cin,f2), IMPLY_OR(cin,f1,f2) 45-49) IMPLY_AND(f2,cout,f1,cout) IMPLY_FA_COUT(in1,in2,cin,f1,f2,cout): 1-2) Write0(f2), Write1(cout) 3-8) IMPLY_OR(in1,f2), IMPLY_OR(in2,f1,f2), IMPLY_OR(cin,f1,f2) 9-13) IMPLY_AND(f2,cout,f1,cout) 14) Write0(f2) 15-20) IMPLY_OR(in1,f1,f2), IMPLY_OR(in2,f1,f2), IMPLY_OR(cin,f1,f2) 21-25) IMPLY_AND(f2,cout,f1,cout) 26) Write0(f2) 27-32) IMPLY(in1,f2), IMPLY(in2,f2), IMPLY_OR(cin,f1,f2) 33-37) IMPLY_AND(f2,cout,f1,cout)		ORNOR3_FA(in1,in2,f1,f2,f3,f4,f5,sum,cin,cout): 1-7) Write0(sum,cout,f1,f2,f3,f4,f5) 8) IMPLY(in2,f1) 9) IMPLY(in2,f2) 10) ORNOR3(cout,f1,f2) 11-12) Write0(f1,f2) 13) ORNOR3(sum,in1,in2) 14) ORNOR3(f1,sum,cout) 15) IMPLY(cin,f2) 16) ORNOR3(cout,f1,sum) 17) Write0(sum) 18) IMPLY(f2,f3) 19) IMPLY(f2,f4) 20) ORNOR3(f5,f3,f2) 21-22) Write0(f1,f3) 23) ORNOR3(f1,f2,f4) 24) ORNOR3(sum,f5,f1)	RMAJ_FA(in1,in2,cin,f1,sum,cout): 1-2,r1) RMAJ_COPY(cin,cout) 3-4,r2) NOT(cin,f1) r3-r4) Read(in1,f1) 5)- RMAJ(in1,in3out) 6-7,r5) RMAJ_COPY(cin,sum) r6,r7) Read(in1) 8) RMAJ(in1,in2,sum) r8) Read(cin) 9) RMAJ(in2,cin,sum)

Table I – Mapping of all considered Boolean functions on LIM primitive operations

	Input			#memristors						#steps					
2 bit	0 0 1 1			MAGIC	FELIX	IMPLY	IMPLY destr	ORNOR3	RMAJ3	MAGIC	FELIX	IMPLY	IMPLY destr	ORNOR3	RMAJ3
	0 1 0 1														
Gate	Output	#in	#out												
TRUE (write 1)	1 1 1 1	0	1	1 + 0	1 + 0	1 + 0	1 + 0	1 + 0	1 + 0	1	1	1	1	1	1
FALSE (write 0)	0 0 0 0	0	1	1 + 0	1 + 0	1 + 0	1 + 0	1 + 0	1 + 0	1	1	1	1	1	1
$in1$ (COPY)	0 0 1 1	1	1	2 + 1	2 + 1	2 + 1	2 + 1	2 + 1	2 + 0	4	3	4	4	4	2 + 1
NOT $in1$	1 1 0 0	1	1	2 + 0	2 + 0	2 + 0	2 + 0	2 + 0	2 + 0	2	2	2	2	2	2 + 1
$in1$ NOR $in2$	1 0 0 0	2	1	3 + 0	3 + 0	3 + 1	3 + 0	3 + 0	3 + 1	2	2	9	5	2	6 + 4
$in1$ OR $in2$	0 1 1 1	2	1	3 + 1	3 + 0	3 + 1	2 + 1	3 + 1	3 + 1	4	2	7	3	4	4 + 3
$in1$ NAND $in2$	1 1 1 0	2	1	3 + 2	3 + 0	3 + 0	3 + 0	3 + 0	3 + 1	10	2	3	3	3	6 + 5
$in1$ AND $in2$	0 0 0 1	2	1	3 + 2	3 + 1	3 + 1	3 + 1	3 + 1	3 + 1	6	4	5	5	5	4 + 3
$in2$ IMP $in1$	1 0 1 1	2	1	3 + 1	3 + 1	3 + 1	2 + 0	3 + 1	3 + 0	6	4	5	1	1	2 + 2
$in2$ NIMP $in1$	0 1 0 0	2	1	3 + 1	3 + 1	3 + 1	3 + 0	3 + 1	3 + 1	4	4	7	3	3	4 + 3
$in1$ XOR $in2$	0 1 1 0	2	1	3 + 2	3 + 0	3 + 2	3 + 2	3 + 2	3 + 1	10	3	13	13	10	7 + 4
$in1$ EQUAL $in2$ (XNOR)	1 0 0 1	2	1	3 + 2	3 + 1	3 + 2	3 + 2	3 + 2	3 + 1	12	5	15	15	12	9 + 5
3 bit	**Input**														
	0 0 0 0 1 1 1 1														
	0 0 1 1 0 0 1 1														
	0 1 0 1 0 1 0 1														
	Output														
FA (sum)	0 1 1 0 1 0 0 1	3	2	5 + 4	5 + 1	5 + 2	5 + 2	5+5	5 + 1	36	12	49+37	49+37	24	9 + 10
FA (c_out)	0 0 0 1 0 1 1 1														

Table II – number of memristors and number of operations per Boolean function

For clarity, a unique syntax is used for all cells of Table I:

```
LiM_Boolean (used memory cells)
i-j)  LiM_Boolean(used memory cells)
```

where the first line defines the name of the Boolean function implemented in a specific LiM, together with the used memory cells for inputs and outputs; the following lines describe the algorithm used to map that function on primitive operations, underlying the number of steps required for its execution (*i-j*). In the case of majority voter requiring additional registers, the algorithm contains extra read operations marked as r_i.

For each function, a number of memristors are used to store the inputs (in_i), the output (*out*), and intermediate results (f_i). The intermediate results are used to solve complex mapping algorithm where several operations are executed in sequence, and they are stored in so-called functional memristors. In case of destructive operations (i.e., *IMPLY*), the content of one of the input memristors is overwritten by the output (noted as in_iout in the table).

To validate the solutions, we have developed a script which checks for the correctness of each Boolean function mapped on LiM primitive operations.

Table II summarizes, for each Boolean function:

its truth table (column 2);

number of inputs and outputs of the function (columns 3 and 4);

for each LiM solution: number of used memristors, in the form #(input and output) + #functional (columns from 5 to 10);

for each LiM solution: number of operations needed to perform the computation (columns from 11 to 16). For the *PLiM* implementation, the steps are indicated as the memory cycles + the reading operations.

IV. CONCLUSIONS

In this paper we have presented an extensive study of the most prominent LiM solutions and provided a comparison in terms of required memory resources (i.e., number of memristors) and number of operations to implement basic Boolean functions.

The obtained results show big discrepancies among LiM solutions in the number of steps to execute the operations. For instance, the XOR requires many more steps if implemented with *IMPLY* logic compared to *FELIX*. These results reflect the complexity of each operation but do not directly translate into an estimation of the actual execution time. This is due to the fact that, due to physical and electrical characteristics of the memristive devices, the timing of each operation can vary significantly.

REFERENCES

[1] M. Gokhale *et al.*, "Processing in memory: the Terasys massively parallel PIM array," *Computer*, vol. 28, no. 4, pp. 23–31, Apr. 1995.

[2] A. Sebastian *et al.*, "Memory devices and applications for in-memory computing," *Nature Nanotechnology*, pp. 1–16, Mar. 2020.

[3] H. A. D. Nguyen *et al.*, "A Classification of Memory-Centric Computing," *J. Emerg. Technol. Comput. Syst.*, vol. 16, no. 2, p. 13:1-13:26, Jan. 2020.

[4] P. L. Thangkhiew *et al.*, "Efficient implementation of adder circuits in memristive crossbar array," in *TENCON 2017 - 2017 IEEE Region 10 Conference*, Nov. 2017, pp. 207–212.

[5] S. Kvatinsky *et al.*, "MAGIC—Memristor-Aided Logic," *IEEE Trans. Circuits Syst. II*, vol. 61, no. 11, pp. 895–899, Nov. 2014.

[6] N. Peled *et al.*, "X-MAGIC: Enhancing PIM Using Input Overwriting Capabilities," in *2020 IFIP/IEEE 28th International Conference on Very Large Scale Integration (VLSI-SOC)*, Oct. 2020, pp. 64–69.

[7] S. Gupta *et al.*, "FELIX: fast and energy-efficient logic in memory," in *Proceedings of the International Conference on Computer-Aided Design*, San Diego California, Nov. 2018, pp. 1–7.

[8] E. Lehtonen and M. Laiho, "Stateful implication logic with memristors," in *2009 IEEE/ACM International Symposium on Nanoscale Architectures*, San Francisco, CA, USA, Jul. 2009, pp. 33–36.

[9] S. Kvatinsky *et al.*, "Memristor-Based Material Implication (IMPLY) Logic: Design Principles and Methodologies," *IEEE Transactions on Very Large Scale Integration (VLSI) Systems*, vol. 22, no. 10, pp. 2054–2066, Oct. 2014.

[10] A. Siemon *et al.*, "Stateful Three-Input Logic with Memristive Switches," *Sci Rep*, vol. 9, no. 1, pp. 1–13, Oct. 2019.

[11] P.-E. Gaillardon *et al.*, "The Programmable Logic-in-Memory (PLiM) Computer," in *Proceedings of the 2016 Design, Automation & Test in Europe Conference & Exhibition (DATE)*, 2016, pp. 427–432.

[12] S. Angizi *et al.*, "MRIMA: An MRAM-based In-Memory Accelerator," *IEEE Trans. Comput.-Aided Des. Integr. Circuits Syst.*, pp. 1–1, 2019.

© VDE VERLAG GMBH · Berlin · Offenbach

A 12-bit 100 MHz SAR ADC in 110-nm CMOS for MAPSs

Silvia Tedesco

Department of Electrical, Electronics and Communications Engineering
Politecnico di Torino
Italian Institute of Nuclear Physics - Sezione di Torino
Turin, Italy
Email: silvia.tedesco@to.infn.it

Abstract—This paper presents a fully differential 12-bit SAR ADC developed for high-voltage CMOS sensors. The converter has been designed in compliance with low power consumption, high resolution and low material budget requirements.

A merged capacitor switching method is employed to decrease power consumption and the capacitor array has been split up into two sub-DACs in order to reduce the area. The prototype has been implemented in a 110-nm CMOS technology. With a power supply of 1.2 V and a 100 MHz clock, simulations show an ENOB 9.87 of and a SFDR of 73.42 dB. The power consumption of the ADC is 513 μW while the Figure of Merit (FOM) results 54.8 fJ/conv-step. The final chip includes also a calibration engine to minimize the capacitor mismatch effect thus further improving the resolution.

Index Terms—SAR ADC, split capacitor, low power

I. INTRODUCTION

Monolithic active pixel sensors (MAPSs) are becoming more and more popular in High Energy Physics (HEP) experiments due to their higher performance compared to hybrid pixels. This technology provides reduced pixel pitch, low power consumption and smaller area. Therefore, MAPSs can be of interest also in other fields such as X-ray imaging, medical physics and space instrumentation. The use of high voltage bias allows to deplete the substrate and collect the charge by drift, improving device speed and radiation tolerance. In the technology adopted for this work, a back-bias voltage of about 200 V allows to obtain an active substrate up to 300 μm [1].

In some applications, it can be of interest to measure the collected charge with high resolution. In this case, the analog information is stored in a capacitor in the pixel and then transferred to the periphery, where it is digitized by column-parallel ADCs. A SAR ADC can provide in this case an optimal compromise between conversion speed, power consumption and area.

In this paper, a power efficient, high radiation tolerant and high resolution converter with a target resolution of 12 bit is presented. This makes the ADC also suitable as a monitoring ADC to measure the on-chip parameters that need to be tracked, such as bias voltages or temperature.

One of the most important source of power consumption in a SAR ADC is the switching operation in the capacitor array. Considering that the single-ended architecture based on the charge redistribution technique has not a power efficient

switching algorithm, a fully differential topology with merged capacitor switching method has been selected.

A first prototype has been developed in a 110 nm CMOS technology. The final layout of the ADC has a size of $(371.7 \times 229.3) \mu m^2$. In simulations, the converter shows a 8.3 MS/s operation speed with 12-bit resolution. However, it can achieve up to 40 MS/s with a resolution of 10 bits, which is an adequate value for most applications. Simulations with 1.2 V supply and 100 MHz clock frequency show an Effective Number Of Bits (ENOB) of 9.87 and a Spurious Free Dynamic Range (SFDR) of 73.42 dB before calibration. Moreover, the power consumption of the ADC results equal to 513 μW which corresponds to a 54.8 fJ/conv-step FoM. A calibration engine has been embedded on chip. Thanks to its features, this device can be suitable to be embedded in readout circuits. Furthermore, if a higher sampling rate is needed, these converters can be organized into a time interleaved configuration thus increasing the operational speed.

The paper is organized as follows. Section II describes the architecture of the ADC. Section III shows the guidelines for the physical implementation. The simulation results are shown in Section IV while first experimental results are depicted in Section V. Finally, Section VI draws the conclusions.

II. ADC ARCHITECTURE

The classical topology adopted for SAR ADC is the charge redistribution technique [2]- [3]. However, the performance of the converter can be affected by external noise due to the single ended architecture. Moreover, the switching algorithm is not power efficient. Therefore, a fully differential architecture has been chosen. In this case, the difference between the two inputs is digitized making the circuit robust to noise. As a matter of fact, common disturbances on both lines are rejected.

The merged capacitor switching method has been selected [4]. During the sampling phase, the input signal is sent to the top plates of the DAC while the bottom plates are connected to the common mode voltage, V_{cm}. For the evaluation of the Most Significant Bit (MSB), the difference between the two inputs is checked without switching any capacitor. If the result is positive, the bit is set to 1 by switching the bottom plate of the largest capacitor in the top array from V_{cm} to the reference voltage, V_{ref}. At the same time, the largest capacitor in the

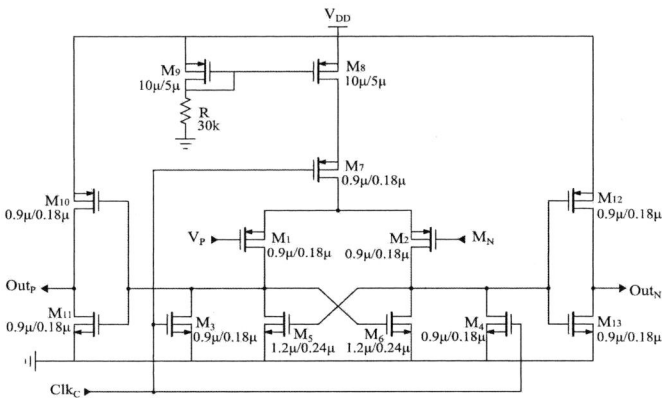

Fig. 1. Dynamic latched comparator schematic.

bottom array is switched to GND. In this way, a voltage equal to $V_{ref}/4$ is subtracted to V_+ and added to V_-. On the other hand, if the difference between the two inputs is negative, the MSB is set to 0 and the opposite operation is performed. The same procedure is carried out to determine the other bits.

A further advantage of this architecture is that the common mode voltage at the input of the comparator is constant in contrast to the step-down switching method presented in [5]. This allows to not affect the behaviour of the comparator.

The main building blocks of the converter are a dynamic latched comparator, a binary weighted Digital-to-Analog Converter (DAC) and a control logic. Moreover, an offset injection circuit, which is useful for the calibration engine, has been embedded. These blocks will be described in the next sections.

A. Comparator architecture

Fig. 1 shows the schematic of the dynamic latched comparator which employs a positive feedback to generate two digital levels [5]. The load consists into a latched stage composed by transistors M5 and M6 in the figure. Moreover, two inverters are embedded in the output stage in order to further speed up the regeneration of the amplifier output into a full swing digital CMOS signal.

Two NMOS switches (M3 and M4 in the figure) are connected in parallel with the load to disable the positive feedback. Furthermore, a PMOS switch (called M7) is connected in series to the output branch of the current mirror in order to avoid static power consumption when the comparator is in the reset phase. For this reason, power is consumed only in the transition between the two logic levels thus making the design suitable for low-power applications.

During the reset phase, the path between the two power rails is disconnected and both the inputs are held to V_{DD}. On the other hand, by enabling the differential pair, the input transistors start working. If V_p is greater than V_n, V_{outp} is driven to 0 faster than V_{outn}, so the positive feedback pulls V_{outp} up and V_{outn} down.

B. DAC architecture

The schematic of the proposed ADC is shown in Fig. 2.

In the SAR ADC, a binary weighted DAC is employed both as a sample and hold block and to provide fraction of the reference voltage. In the layout, most of the area is taken by the DAC. As a matter of fact, every bit of resolution added doubles the DAC occupied area. Moreover, this causes the increase of the load at the input of the comparator. For these reasons, the capacitor array has been split up into two sub-DACs connected by using a coupling capacitor [6]- [7].

The value of the coupling capacitor is such that the capacitance which is seen by the main-DAC is equal to the unit capacitance C_{min}.

In the converter, the bits have been split up into a main-DAC of 9 bits and a sub-DAC of 3 bits. Since $C_{min} = 2fF$, the total capacitance C_{TOT} for the segmented DAC is equal to 1040 fF. On the other hand, for a classical DAC having the same minimum capacitance, C_{TOT} results equal to 8192 fF.

In order to reduce the mismatch, in the sub-DAC the value of the minimum capacitance has been doubled (4 fF). Therefore, the total capacitance has a slight increase going from 1040 fF to 1056 fF. Nevertheless, the segmentation offers the best choice.

C. SAR control logic

The SAR ADC needs a control logic for storing the value of the bits after the conversion result. This circuit is composed of a shift register and D-type flip-flops in which the bits are memorized [5].

Two control signals called *sample* and *convert* are used to manage the SAR logic operation. When both of them are high, the sampling phase is enabled, so the shift register and the flip-flops are in the reset state. When the conversion phase starts (*sample* = 0 and *convert* = 1), a logic "1" is shifted through the register at every clock cycle. The outputs of the shift register are used to sequentially pick out the flip-flops in which the comparison results would be stored. Hence, the value of the bit is stored in the selected flip-flop. The last output of the shift register is the *End of Conversion* (EoC) signal which is used to points out the end of the conversion.

D. Offset injection circuit

Due to the high resolution required, a perturbation-based digital calibration algorithm is embedded to correct non-linearities of the converter. The Offset Double Conversion (ODC) technique has been chosen [8].

In the ODC technique, each analog sample which has been digitized has to be converted also injecting two analog offsets ($+\Delta_a$ and $-\Delta_a$). These offsets have the same absolute value but opposite sign. Hence, two raw codes, D_+ and D_-, are provided to the digital calibration. The offsets can be removed from the output codes and, in case of an ideal converter, the results should be equal to the digitization of the analog sample to which no offset has been added. In case of mismatches, the calibration engine evaluates the error and corrects the bit weights.

Therefore, an offset injection circuit has been embedded. It is composed of the capacitance C_{cal} which can be connected

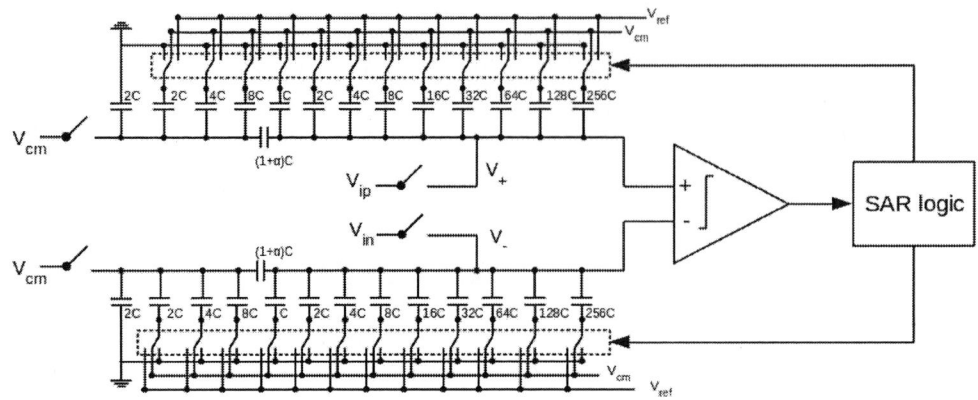

Fig. 2. Proposed SAR ADC with segmented capacitor array.

to the top plate of the main-DAC through a switch . The other plate of the capacitor can be connected to V_{cm}, V_{ref} or ground. The same circuit is embedded both in the top and in the bottom array.

During the sampling phase, no offset is injected, so the calibration capacitance is connected between V_{cm} and the input voltage. Hence, the input signal is sampled and converted without any added offset.

Then, without carrying out a new sampling, a positive offset is injected by connecting the calibration capacitance of the top array to V_{ref} while the one in the bottom array to GND. Thus, the new value equal to $V_{in} + \Delta_a$ is digitized.

Eventually, the opposite operation is performed for injecting a negative offset. As a result, $V_{in} - \Delta_a$ is converted.

The value of the calibration capacitance has been chosen equal to 10 fF in order to inject a voltage corresponding to 40 LSB. Since $V_{ref} = 1\,V$ and $V_{cm} = 0.5\,mV$, the injection is $\sim 9.6\,mV$ on each array that corresponds to $\sim 19.2\,mV$ peak-to-peak.

III. PHYSICAL IMPLEMENTATION

Since crossing particles can degrade the circuit operations, the layout design had to be carried out considering the radiation tolerance of the device. As a matter of fact, transients in the substrate can lead to latch-up. In order to prevent latch-up, a high number of substrate contacts has been added thus reducing its resistance. Moreover, guardrings have been placed around the blocks.

The supplies for the analog blocks and for the digital ones are kept separated in order to protect the analog circuits from digital noise. Furthermore, a dedicated power supply bus has been embedded for the comparator which is both sensitive and noisy. Hence, a total of three power supply busses are present.

In the DAC, a proper number of unit capacitors connected in parallel are used instead of employing a single capacitor for every bit in order to obtain different capacitance values.

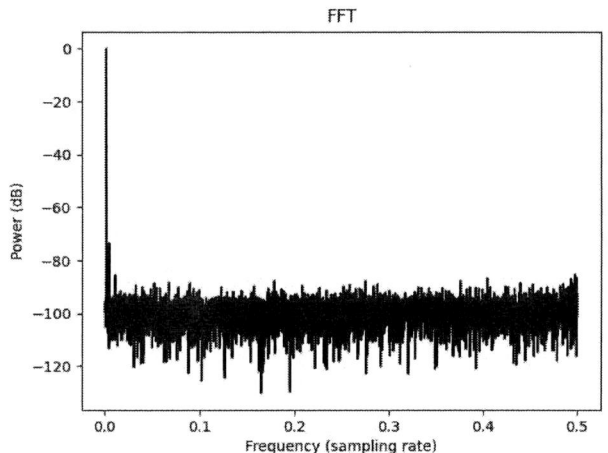

Fig. 3. FFT at 100 MHz clock with input sinusoid of 15.6 kHz.

This allows to improve the capacitor matching and reduce the non-linearities.

The total area occupied by the ADC is $(371.7 \times 229.3)\mu m^2$. The final chip has a size of $(3 \times 1.5)mm^2$ with a core of $(2.4 \times 0.9)mm^2$ and includes both the converter and the calibration engine.

IV. SIMULATION RESULTS

For the simulations, a 1.2V power supply has been used. The converter has been simulated without including the digital calibration circuit. The input sinusoid had a frequency of 15.9 kHz and an amplitude equal to the reference voltage (1V). The clock frequency was equal to 100 MHz and the sampling was 8.3 MS/s. A number of 8192 samples has been taken and the Fast Fourier Transform (FFT) is reported in Fig. 3.

© VDE VERLAG GMBH · Berlin · Offenbach

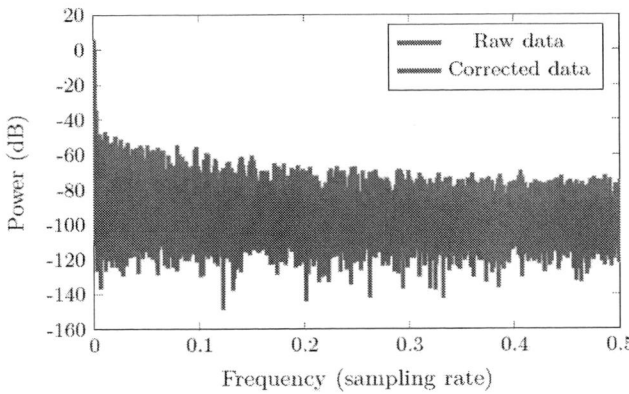

Fig. 4. FFT of a 3.015 kHz sinusoid before (red) and after the calibration (green).

From the FFT, the main parameters to evaluate the dynamic performance of the ADC can be calculated. The Signal to Noise and Distortion (SINAD) is evaluated as follows:

$$SINAD = 20\log\sqrt{\frac{A_f^2}{\sum_{k=0}^{f-1} A_k^2 + \sum_{k=f+1}^{N/2} A_k^2}} \qquad (1)$$

From the SINAD, the ENOB can be calculated:

$$ENOB = \frac{SINAD - 1.76}{6.02} \qquad (2)$$

For the presented ADC, the SINAD results equal to 61.16 dB and the ENOB is 9.87. The SFDR is calculated as:

$$SFDR = 20\log\sqrt{\frac{A_f^2}{A_{spurious}^2}} \qquad (3)$$

From the simulation results that SFDR = 73.42 dB. The power consumption of the converter is 513 μW. Hence, the FOM of the ADC can be evaluate as:

$$FOM = \frac{P}{f_s \times 2^{ENOB}} \qquad (4)$$

where P is the power consumption and f_s the sampling frequency. The FOM of this converter corresponds to 54.8 fJ/conv-step.

V. First experimental results

The ADC has been tested by using a 3.015 kHz sinusoidal waveform as input signal and a 100 MHz clock frequency corresponding to a 8.3 MS/s. The FFT has been evaluated by collecting 2^{15} samples. The plot is shown in Fig. 4.

The raw data coming from the ADC shows a high amount of missing codes. These errors can be well reproduced by assuming a significant mismatch among the unit capacitors. For the chosen capacitors, Monte-Carlo models were not available at the time of the design and the statistical mismatch could not be properly accounted for. However, a mean of 3

bits can be retrieved thanks to the calibration engine. Hence, a ENOB of 8.5 bits, which is equivalent to a 9 bit ADC, is achieved.

VI. Conclusion

In this paper, a 12-bit fully differential SAR ADC was presented. The prototype has been implemented by using 110 nm CMOS technology. The design has been optimized by employing the merged capacitor switching method, which is a power efficient switching procedure in order to reduce power consumption. Moreover, the total area has been decreased by using a segmented capacitor DAC. Another key feature of this converter is the high radiation tolerance.

The simulations show a ENOB of 9.87 and a SFDR of 73.42 dB. This resolution is suitable for most applications. However, it can be improved by using a digital calibration circuit that has been integrated in the final chip. The power consumption is equal to 513 μW which is appropriate for low power MAPS sensors.

First experimental results shows that some codes are missing due to the mismatch between the capacitors of the DAC. An ENOB of 8.5 bits has been achieved after digital correction.

References

[1] Pancheri, Lucio, et al. "Fully Depleted MAPS in 110-nm CMOS Process With 100–300-um Active Substrate." IEEE Transactions on Electron Devices 67.6 (2020): 2393-2399.

[2] James L. McCreary and Paul R. Gray, "All-MOS charge redistribution analog-to-digital conversion techniques - Part I", IEEE J. Solid-State Circuits, vol. SC-10, pp. 371-379, Dec. 1975.

[3] Angelo Rivetti, Giovanni Anelli, Francis Aghinolfi, Giovanni Mazza, Francesco Rotondo, "A low-power 10 bit ADC in a 0.25-μ m CMOS: design considerations and test results". In: *2000 IEEE Nuclear Science Symposium. Conference Record (Cat. No. 00CH37149)*. IEEE 2000, pp 9-15.

[4] V. Hariprasath, J. Guerber, S.-H. Lee, and U.-K. Moon. "Merged capacitor switching based SAR ADC with highest switching energy-efficiency". Electronics Letters, 46, 2010.

[5] Chun-Cheng Liu et al., "A 10-bit 50-MS/s SAR ADC with a monotonic capacitor switching procedure". In: *IEEE Journal of Solid-State Circuits* 45.4 (2010), pp. 731–740.

[6] Y. S. Yee, L. M. Terman, and L. G. Heller, "Two-Stage Weighted Capacitor Network for D/A-A/D Conversio", IEEE J. Solid-State Circuits, vol. SC-14, no.4, pp. 778-781, Aug. 1979.

[7] Wei Tung and Shu-Chuan Huang, *An Energy-Efficient 11-bit 10-MS/s SAR ADC with Monotonic Switching Split Capacitor Array*. In: *2018 IEEE International Symposium on Circuits and Systems (ISCAS)*. IEEE. 2018, pp 1-5.

[8] Wenbo Liu, Pingli Huang, and Yun Chiu. "A 12-bit, 45-MS/s, 3-mW redundant successive-approximation-register analog-to- digital converter with digital calibration". In: IEEE Journal of Solid-State Circuits 46.11 (2011), pp. 2661–2672.

SMACD / PRIME 2021 | 19 – 22 July 2021, Online Event

A Timing Skew Correction Technique in Time-Interleaved ADCs Based on a $\Delta\Sigma$ Digital-to-Time Converter

Gabriele Bè, Mario Mercandelli, Luca Bertulessi

Dipartimento di Elettronica, Informazione e Bioingegneria
Politecnico di Milano
P.zza Leonardo da Vinci 32, 20133 Milano, Italy
gabriele.be@polimi.it

Abstract—This paper reviews state-of-the-art skew correction methods in time-interleaved (TI) analog-to-digital converters (ADCs) and introduces a novel mixed-signal skew correction technique based on dithering the control word of a digital-to-time converter (DTC), which significantly relaxes the stringent DTC resolution requirements. Simulation results of a 10 GS/s 8–bit 16-core TI-ADC reveal that the proposed technique reduces the required DTC resolution by a factor of 4x and improves the spurious-free dynamic range (SFDR) by up to 10 dB with a negligible noise penalty.

Index Terms—Analog-to-digital converters (ADCs), calibrations, time-interleaving, timing skew, mixed-signal, digital-to-time converters (DTCs), dithering

I. INTRODUCTION

Analog-to-digital converters (ADCs) need to satisfy the stringent requirements of modern applications while keeping reasonable power consumption. In recent years, successive-approximation register (SAR) ADCs gained increased popularity thanks to their scaling-friendly implementation, which benefits from CMOS technology scaling, and their reduced power consumption [1]. However, their intrinsically serial operation leads to lower conversion rates than other ADC architectures.

To overcome the speed limitation, and at the same time retain all the benefits of SAR ADCs, *time-interleaving* (TI) has been successfully employed [2]–[4]. Introduced by Black and Hodges in 1980 [5], time-interleaved converters enabled sampling rates beyond the values achievable at that time by a single ADC. As shown in Fig. 1, the underlying idea is to parallelize more cores that cyclically process the input signal. By employing M channels, the overall sampling rate increases to Mf_s, where f_s is the sampling rate of each of the sub-ADC and M is the interleaving factor.

Mismatches in dc offset, static gain, and timing skew among cores produce artifacts in the output spectrum that limit the achievable signal-to-noise plus distortion ratio (SNDR) and spurious-free dynamic range (SFDR) [5]–[7]. Offset mismatch leads to fixed tones at multiples of the core sampling rate, while gain mismatch and timing skew give amplitude and phase modulation images, respectively [8]. In particular, timing skew is caused by the static difference between the ideal

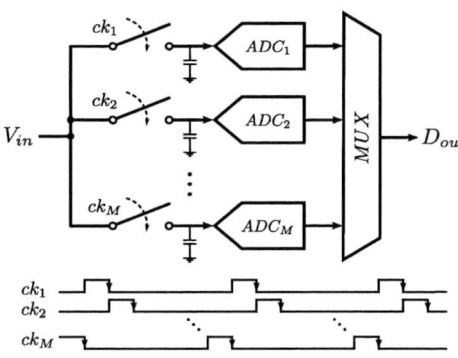

Figure 1. Block diagram of an M-channels TI-ADC.

sampling instants of each core and the actual ones, and its impact becomes more severe as the frequency of the input signal increases. While offset and gain mismatches can be corrected in the digital domain without much effort [9], how to efficiently cope with timing skew is still an unresolved challenge.

This paper reviews existing skew correction methods and introduces a novel mixed-signal technique based on a digital-to-time converter (DTC) with a $\Delta\Sigma$ modulator dithering its control word. The proposed technique relaxes the DTC required resolution, leading to a design with reduced area occupation and power consumption, achieving the same SFDR. A theoretical analysis of the impact of DTC dithering on the ADC total noise that fits with simulation results is also presented. The paper is organized as follows. Section II compares fully-digital and mixed-signal timing-skew correction methods, highlighting the pros and cons of the two approaches. Section III introduces the proposed mixed-signal technique, while section IV reports simulation results. Finally, section V draws the conclusion.

II. SKEW CORRECTION METHODS

Timing skew can be mitigated by introducing a front-end track-and-hold (T&H) at the TI-ADC input [10]. The additional T&Hs inside each channel do not suffer from

© VDE VERLAG GMBH · Berlin · Offenbach

skew mismatches since the sampling instants are set by the first T&H. However, this hierarchical sampling scheme poses significant system-level challenges since the front-end T&H must run at the full interleaving speed [9].

Alternatively, skew calibration algorithms have been proposed in the literature. Skew calibrations can be classified as analog, digital, or mixed-signal methods. In the analog approach, identification and error correction are carried out in the analog domain. However, the resulting performance is intrinsically bounded to the accuracy of analog components [9].

Digital corrections aim at recovering the skew-free samples from the core outputs by knowing the timing skew. Digital interpolators [11], adaptive FIR filters [12], and derivative-based linear reconstructions [3] have been proposed. All of them require digital filters with many taps, which significantly increase latency, area occupation, and power consumption of an on-chip implementation. Due to these limitations, only the derivative-based method has been successfully integrated, so far in a 1.62 GS/s TI-ADC [3]. Due to the periodic nature of digital filters, the correction accuracy is limited near the Nyquist frequency, and extending the input frequency towards Nyquist requires more filter taps [9], [12]. Moreover, it only takes care of linear error terms, and the derivative is computed based on the skewed samples instead than on the input signal itself, not to mention that the digital correction accounts for 21% (19.7 mW) of the overall power consumption of the design reported in [3], [13].

To overcome these issues, mixed-signal approaches have been considered, thus combining the power efficiency of analog corrections with the accuracy of the digital error identification. This approach relies on DTCs, which tune the edges of the sampling clock. They achieve lower power consumption and area occupation than digital methods [14], [15], are not limited in terms of input frequency, remove the error before quantization, and retain the portability advantages of digital algorithms to advanced technology nodes, requiring only a mild porting of the DTC circuit. Consequently, mixed-signal methods are the natural choice to pursue an area- and energy-efficient implementation. However, the design of the analog delay line should be carefully addressed to achieve the required resolution and to limit the introduction of extra jitter.

III. PROPOSED DTC DITHERING TECHNIQUE

Figure 2 shows a popular DTC topology implemented in CMOS logic. The output of the first inverter is loaded by a bank of digitally-switched capacitors that tune the crossing-point of the threshold of the second inverter, providing the required delay. The second inverter is necessary to restore sharp transitions at the output [16].

TI-ADCs in the GS/s range require from a few fs to some tens of fs of DTC resolution to satisfy the matching specifications set by medium-to-high resolution converters. Moreover, they need a tuning range of some ps to cover all possible process, voltage, and temperature (PVT) variations [14]. The required resolutions, along with the wide tuning range, lead to a high number of control bits, which, in turn, makes more

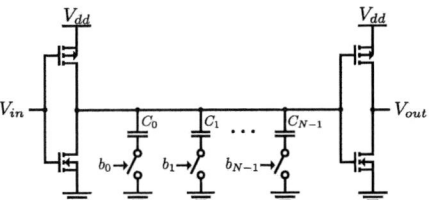

Figure 2. Circuit implementation of a CMOS-logic DTC.

challenging the design of the DTC itself and increases its area occupation and power consumption.

The idea behind the proposed correction method can be understood by looking at the nature of interleaving errors. In the presence of mismatches, the overall TI-ADC with a stationary input becomes a *cyclostationary* system since the same error source appears periodically to the signal path whenever the relative core processes the input signal, and spurs appear in the output spectrum, degrading the SFDR. A coarse DTC resolution would result in only a partial attenuation of these spurious tones, leading to a sub-optimal SFDR.

Figure 3 shows the proposed architecture, which aims at breaking the periodicity coming from the residual skew errors after a coarse correction. A DTC per channel is used to tune the edges of the sampling clock, and its control word is estimated by employing a background algorithm like the one proposed in [9]. In nominal conditions, the correlation between adjacent channels should be equal since their timing distance is constant. Consequently, the difference in the correlations is caused by timing skew, and by minimizing them, the skew is compensated.

The skew correction coefficient is divided into an integer and a fractional word. The former is directly applied to the DTC and compensates for the skew mismatch at a resolution set by the least significant bit (LSB) of the DTC itself. The latter goes through a MASH-1 $\Delta\Sigma$ modulator, which is easily implemented by a modulo-N accumulator. The modulator's single-bit output is given by the carry bit of the accumulator, and it represents a modulated version of its input. The control word is finally obtained by the sum between the integer word and the modulation bit. The same scheme is then employed in all the channels but the first one, which serves as a global reference for the skew identification algorithm [9]. Anyway, a DTC is also employed in the first channel, with its control word fixed at half full-scale range, to nominally compensate the DTC fixed delay and allow both positive and negative skew corrections on the other channels.

The modulo, N, of the accumulator controls the equivalent correction resolution. On average, the resolution is determined by the fractional word instead of the DTC's LSB, while the maximum instantaneous error is still set by the DTC. The effect of the $\Delta\Sigma$ modulator is to break the periodicity of the residual skew—present because of the coarse DTC resolution—and convert it into broadband noise. This will improve the SFDR but increase the noise level, slightly reducing the SNDR.

© VDE VERLAG GMBH · Berlin · Offenbach

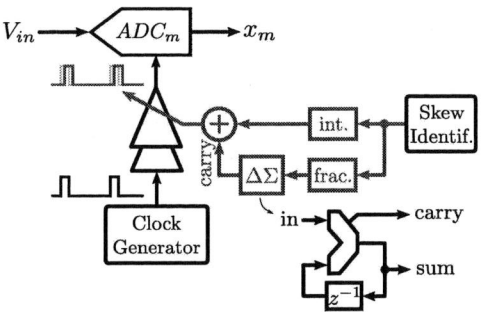

Figure 3. Channel architecture of the proposed mixed-signal skew correction technique. The DTC control word is obtained by the sum between the integer word and the single-bit modulated version of the fractional word.

The noise contribution due to the finite DTC resolution can be estimated by applying the well-known quantization theory of data converters to DTCs. The noise on the sampling time due to the finite DTC resolution is given by:

$$N_\tau = \frac{\Delta_\tau^2}{12} \qquad (1)$$

where N_τ is the quantization noise power, and Δ_τ is the LSB of the DTC. A conversion factor is needed to link the time domain to the voltage domain to quantify the reduction of SNDR due to the finite DTC resolution.

To this aim, a generic timing error, Δ_t, results in a voltage error obtained by approximating the input signal with its first-order Taylor expansion around the ideal sampling instant. By considering a sinusoidal signal, the instantaneous voltage error is given by:

$$\epsilon_\tau = \omega_0 \cdot A \cos\left(\omega_0 t + \phi_0\right) \cdot \Delta_t \qquad (2)$$

where A is the amplitude, ω_0 the frequency, and ϕ_0 the initial phase of the sinusoidal tone. The conversion factor between noise on the sampling time and voltage noise is obtained by the mean squared value of the above equation:

$$N_V = \frac{1}{2}\omega_0^2 A^2 N_\tau \qquad (3)$$

where N_V and N_τ are the voltage and sampling time noise, respectively. By substituting N_τ in the previous equation with (1), the voltage noise contribution due to the finite DTC resolution is thus given by:

$$N_V = \frac{\Delta_\tau^2}{12} \cdot \frac{1}{2}\omega_0^2 A^2 \qquad (4)$$

By taking into account also ADC quantization noise and clock jitter, the overall SNDR results in:

$$\text{SNDR} = \frac{\frac{1}{2}A^2}{\frac{\text{FSR}^2}{2^{2B}\cdot 12} + \frac{\Delta_\tau^2}{12} \cdot \frac{1}{2}\omega_0^2 A^2 + \frac{1}{2}\omega_0^2 A^2 \sigma_j^2} \qquad (5)$$

where B is the number of bits, FSR the full-scale range, and σ_j^2 is the jitter variance. The jitter contribution is obtained by a linear model similar to the skew one. In the case of $\Delta\Sigma$ modulation, the DTC noise power integrated throughout the whole bandwidth doubles due to the noise shaping.

The increasing of the total noise due to the $\Delta\Sigma$ modulation depends on the relative weight of the three contributions. Based on that, the proposed method gives a negligible increase of noise if timing skew is not the dominant noise contributor.

The proposed modulation technique is fully digital and only requires an accumulator and an adder per channel. The modulator, shown in Fig. 3, consists of an accumulator of the fractional bits, where the output is the carry of the accumulator itself. It has reduced area occupation and power consumption since the required number of bits is limited. Compared to previous art, where high-resolution DTCs are used, the proposed method is virtually costless, it greatly benefits from technology scaling, and adds a negligible implementation overhead.

IV. SIMULATION RESULTS

The proposed skew correction method has been applied to a 10 GS/s 8–bit 16-core TI-ADC. The ADC architecture has been simulated in the Matlab environment. The behavioral model includes jitter and timing skew with a standard deviation of 50 fs rms and 3 ps rms, respectively. Figure 4(a) shows the results of a parametric simulation on the DTC resolution. The resolution is swept from 10 fs to 500 fs, the input frequency is set to 4.948 GHz, the input amplitude is equal to −1 dBFS, and the modulo of the $\Delta\Sigma$ modulator is 10 bits. Each point is the average of 100 Monte Carlo runs, and all metrics are computed from 2^{16} samples.

The dashed lines in Fig. 4(a) refer to the case where the DTC dithering is disabled. It can be seen that by increasing the LSB of the DTC, both SNDR and SFDR decrease since the skew compensation is only partial. By decreasing the precision of the skew correction, residual spurs start to rise from the noise floor. For instance, when the SNDR is reduced by 3 dB compared to the 10 fs resolution case, the SFDR is already decreased by more than 15 dB. This outcome is highly undesirable since spurious tones severely degrade system-level performance.

The solid lines in Fig. 4(a) show that when the DTC dithering is enabled, the SFDR stays constant up to 200 fs of DTC resolution. The possibility to increase the DTC resolution up to 200 fs—or even more, based on the system requirements—significantly simplifies the DTC design and reduces its implementation overhead. As a downside, the degradation of SNDR due to the increased noise floor must be carefully considered. The difference between the two lines in Fig. 4(a) is at most 2.2 dB at 500 fs resolution. In the 200 fs case, where the SFDR is still at its maximum value, the reduction is 0.78 dB. Fig. 4(a) also reports the theoretical prediction of the SNDR obtained by applying (5). The model fits with simulation results both with and without considering the $\Delta\Sigma$ modulator. These results demonstrate the effectiveness of the proposed technique in keeping the maximum SFDR while relaxing the DTC design specification and, at the same time, limiting the introduction of extra noise in applications where the SFDR is a critical parameter.

Figure 4(b) shows the output spectrum of a realization of the previous TI-ADC after timing skew correction with a

SMACD / PRIME 2021 | 19 – 22 July 2021, Online Event

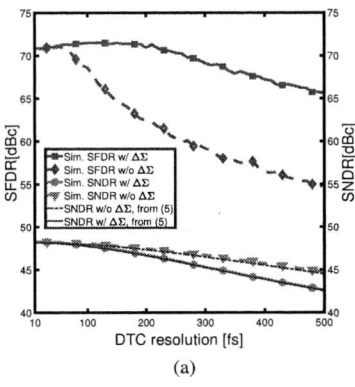 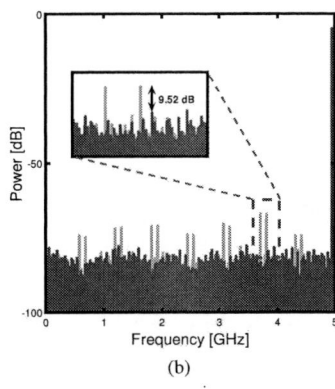 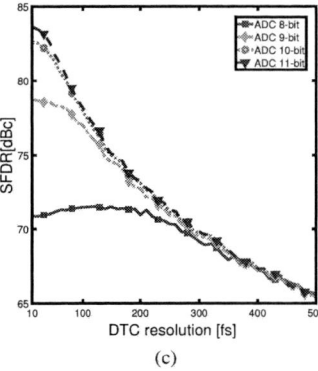

(a) (b) (c)

Figure 4. (a) SNDR and SFDR vs. skew correction resolution. The dashed lines refer to the unmodulated control word, while the solid ones are obtained when the proposed DTC dithering technique is enabled. The black lines refer to the theoretical prediction of the SNDR given by (5), which is consistent with the simulation results. (b) Output spectrum obtained when the DTC modulation is disabled (grey) vs. when it is enabled (blue). (c) SFDR vs. DTC resolution parametrized to the ADC resolution when DTC dithering is enabled.

DTC resolution of 200 fs. The figure superimposes the outputs obtained by disabling (grey) and enabling (blue) the $\Delta\Sigma$ modulator. The SFDR improves by slightly less than 10 dB, from 61.73 dB to 71.25 dB, while the SNDR decreases from 47.34 dB to 46.69 dB.

Since DTC dithering spreads the spur power over the spectrum, slightly increasing the noise level, it is best suited for medium-resolution ADCs. Figure 4(c) shows the results of a Monte Carlo parametric simulation where the proposed technique is applied to different ADC resolutions. It is evident how in the 10– and 11–bit cases, the SFDR falls rapidly with the DTC resolutions, indicating that the interval where the $\Delta\Sigma$ modulation is most effective, i.e., the range in which the SFDR stays almost constant, is for much lower resolutions than 50 fs. This behavior suggests that an accurate DTC design is still needed, and the proposed method gives less benefit compared to the 8-bit case. Consequently, DTC dithering can be effectively applied to medium-resolution SAR-based TI-ADCs, improving their already good energy efficiency.

V. CONCLUSION

This paper presents an improved correction of timing skew in TI-ADCs, relaxing the DTC requirements on medium-resolution ADCs without compromising the spectral purity. Simulation results of a 10 GS/s 8–bit 16-core TI-ADC show an SFDR improvement of 10 dB when the LSB of the correction DTCs has been increased by a factor of $4\times$ compared to the unmodulated case. Since, typically, the resolution required to reduce the interleaving spurs below the noise level is more stringent than the SNDR specification, the proposed technique allows to achieve optimal system-level performances in applications where the SFDR is of utmost importance, such as in communication systems.

REFERENCES

[1] B. Murmann, "The successive approximation register ADC: a versatile building block for ultra-low- power to ultra-high-speed applications," *IEEE Commun. Mag.*, vol. 54, no. 4, pp. 78–83, 2016.

[2] L. Kull, T. Toifl, M. Schmatz, P. A. Francese, C. Menolfi, M. Braendli *et al.*, "22.1 A 90GS/s 8b 667mw 64x interleaved SAR ADC in 32nm digital SOI CMOS," in *IEEE ISSCC Dig. Tech. Papers*, 2014, pp. 378–379.

[3] N. Le Dortz, J. Blanc, T. Simon, S. Verhaeren, E. Rouat, P. Urard *et al.*, "22.5 A 1.62GS/s time-interleaved SAR ADC with digital background mismatch calibration achieving interleaving spurs below 70dBFS," in *IEEE ISSCC Dig. Tech. Papers*, 2014, pp. 386–388.

[4] Y.-Z. Lin, C.-H. Tsai, S.-C. Tsou, and C.-H. Lu, "A 8.2-mW 10-b 1.6-GS/s 4x TI SAR ADC with fast reference charge neutralization and background timing-skew calibration in 16-nm CMOS," in *Proc. IEEE Symp. VLSI Circuits (VLSI)*, 2016, pp. 1–2.

[5] W. C. Black and D. A. Hodges, "Time interleaved converter arrays," *IEEE J. Solid-State Circuits*, vol. 15, no. 6, pp. 1022–1029, 1980.

[6] Y.-C. Jenq, "Digital spectra of nonuniformly sampled signals: fundamentals and high-speed waveform digitizers," *IEEE Trans. Instrum. Meas.*, vol. 37, no. 2, pp. 245–251, 1988.

[7] A. Petraglia and S. K. Mitra, "Analysis of mismatch effects among A/D converters in a time-interleaved waveform digitizer," *IEEE Trans. Instrum. Meas.*, vol. 40, no. 5, pp. 831–835, 1991.

[8] N. Kurosawa, H. Kobayashi, K. Maruyama, H. Sugawara, and K. Kobayashi, "Explicit analysis of channel mismatch effects in time-interleaved ADC systems," *IEEE Trans. Circuits Syst. I*, vol. 48, no. 3, pp. 261–271, 2001.

[9] B. Razavi, "Design considerations for Interleaved ADCs," *IEEE J. Solid-State Circuits*, vol. 48, no. 8, pp. 1806–1817, 2013.

[10] K. Poulton, J. J. Corcoran, and T. Hornak, "A 1-GHz 6-bit ADC system," *IEEE J. Solid-State Circuits*, vol. 22, no. 6, pp. 962–970, 1987.

[11] H. Jin and E. K. F. Lee, "A digital-background calibration technique for minimizing timing-error effects in time-interleaved ADCs," *IEEE Trans. Circuits Syst. II*, vol. 47, no. 7, pp. 603–613, 2000.

[12] S. M. Jamal, D. Fu, N.-J. Chang, P. J. Hurst, and S. H. Lewis, "A 10-b 120-Msample/s time-interleaved analog-to-digital converter with digital background calibration," *IEEE J. Solid-State Circuits*, vol. 37, no. 12, pp. 1618–1627, 2002.

[13] N. Le Dortz, "Digital mismatch calibration of Time-Interleaved Analog-to-Digital Converters," PhD theses, CentraleSupélec, Jan. 2015. [Online]. Available: https://tel.archives-ouvertes.fr/tel-01331558

[14] M. Straayer, J. Bales, D. Birdsall, D. Daly, P. Elliott, B. Foley *et al.*, "27.5 A 4GS/s time-interleaved RF ADC in 65nm CMOS with 4GHz input bandwidth," in *IEEE ISSCC Dig. Tech. Papers*, 2016, pp. 464–465.

[15] S. Devarajan, L. Singer, D. Kelly, T. Pan, J. Silva, J. Brunsilius *et al.*, "A 12-b 10-GS/s Interleaved Pipeline ADC in 28-nm CMOS Technology," *IEEE J. Solid-State Circuits*, vol. 52, no. 12, pp. 3204–3218, 2017.

[16] A. Santiccioli, C. Samori, A. L. Lacaita, and S. Levantino, "Power-jitter trade-off analysis in digital-to-time converters," *IEEE Electron. Lett.*, vol. 53, no. 5, pp. 306–308, 2017.

© VDE VERLAG GMBH · Berlin · Offenbach

A low-noise high-speed comparator for a 12-bit 200-MSps SAR ADC in a 28-nm CMOS process

Luca Ricci, Luca Bertulessi, Andrea Bonfanti
Dipartimento di Elettronica, Informazione e Bioingegneria (DEIB)
Politecnico di Milano
Milan, Italy
luca.ricci@polimi.it, luca.bertulessi@polimi.it, andrea.bonfanti@polimi.it

Abstract—This paper presents a high-speed and low-noise comparator implemented in a 28-nm bulk CMOS technology with a 0.9-V supply voltage. The comparator is designed for a 12-bit 200-MSps successive-approximation-register (SAR) analog-to-digital converter (ADC). Simulations show an input-referred noise of 163 μV and a reset-out delay of 110-ps for an input differential voltage of 100 μV. The energy per conversion is 595 fJ/conv and the Figure-of-Merit is 15.8 nJμV^2, better than the state of the art.

Index Terms—comparator, low-noise, analog-to-digital converter(ADC), successive-approximation-register (SAR), dynamic

I. INTRODUCTION

Comparators are ubiquitous in mixed-signal systems. In particular, they are present in analog-to-digital converters (ADCs), such as flash, interpolating, $\Sigma-\Delta$ or successive-approximation-register (SAR) ADCs. The latter is a popular solution for high-speed medium-resolution converters due to its power efficiency. For SAR ADCs featuring a number of bits larger than 10, the thermal noise due to sampler and comparator can limit the resolution [1,2]. The former noise contribution can be reduced increasing the sampling capacitance or with noise cancellation techniques [1], whereas the latter needs to be addressed by designing a low-noise comparator. However, power consumption and noise trade off and approximately half of the power consumption of SAR ADCs comes from the comparator [1,3]. Therefore, the challenge for high-speed medium-resolution SAR ADCs is to design a fast and low-noise comparator that is relatively energy efficient.

This paper deals with the design of a high-speed comparator for a 12-bit 200-MSps SAR ADC in a 28-nm bulk CMOS technology with a 0.9-V power supply. The structure of this paper is the following. Section II discusses state-of-the-art comparators to justify the chosen architecture. Section III describes the design of the comparator, whereas the simulation results are shown in Section IV. Section V summarizes the work and draws the conclusions.

II. STATE-OF-THE-ART

The strongARM comparator [4], shown in Fig. 1, is a popular choice in ADCs for its good performance in terms of speed and power consumption. In this topology, pre-amplification and regeneration of the input is carried out by M_2-M_3 and M_4-M_7, respectively, embedded in one stage. It works in two

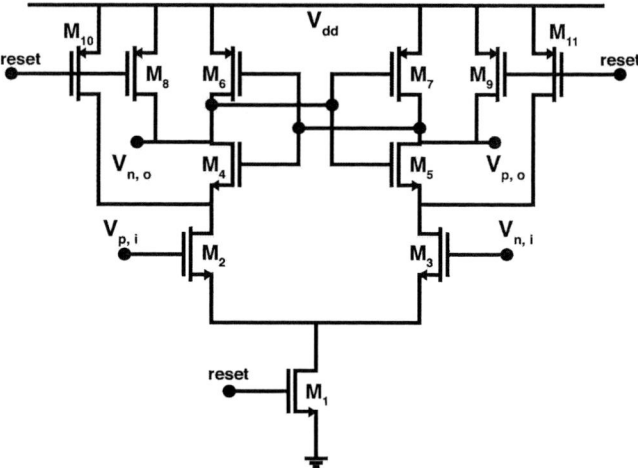

Fig. 1: Schematic of the strongARM comparator.

phases, one for reset and one for comparison. When *reset* is low, the tail transistor is off and the switches M_8-M_{11} are turned on to reset the internal nodes of the comparator to V_{dd}. When *reset* is pulled high, the reset switches are turned off and the tail transistor is turned on starting the comparison. It is a fast and efficient comparator featuring a nil static current. However, it suffers from kickback noise because its inputs are poorly isolated from the voltage transients at the drain and source of the input pair transistors during the reset and comparison phases. Moreover, it is not a topology suitable for advanced technology nodes because four stacked transistors between the power rails require a large voltage headroom.

A dynamic comparator better suited for low supply voltages is the double-tail latch-type sense amplifier [5], shown in Fig. 2a. Here, the input differential signal is amplified by a dynamic amplifier and than latched. The input nodes are shielded from the outputs, thus kickback noise is greatly reduced. Moreover, it requires less voltage headroom, because there are only three stacked transistors. Finally, the addition of the second tail adds a degree of freedom in the design.

The double-tail sense amplifier is designed for high speeds, rather than energy efficiency [3]. Therefore, efforts to improve the energy efficiency of this topology has been carried out in the last decade. In [3], the input transistors of the sec-

© VDE VERLAG GMBH · Berlin · Offenbach

SMACD / PRIME 2021 | 19 – 22 July 2021, Online Event

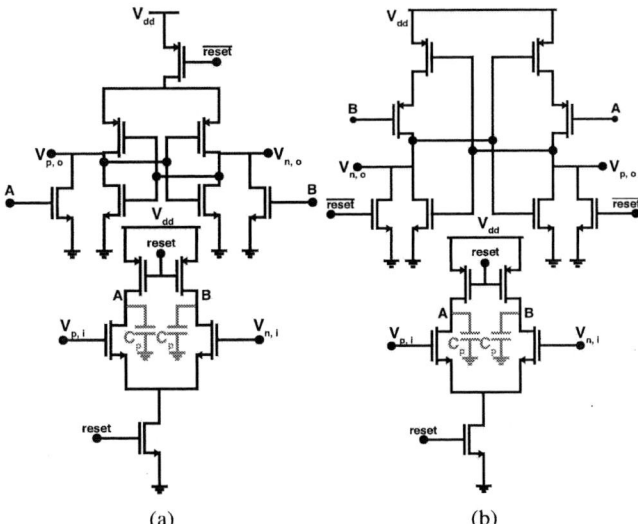

Fig. 2: (a) Double-tail latch-type sense amplifier [5] and (b) the more energy-efficient version in [3].

ond stage are replaced by PMOS transistors and moved in-between the inverter transistors (see Fig. 2b). This decreases the cross-conduction current in the second stage because the PMOS transistors turn on only when the input voltage goes approximately below $V_{dd} - V_{th}$. In [6], the introduction of dynamic biasing reduces the amount of charge taken from the capacitors at the pre-amplifier outputs during the comparison phase. Therefore, less charge is drawn from the supply to reset C_p to V_{dd}. Finally, an energy-efficient comparator has been proposed in [7] featuring a dynamic floating inverter preamplifier followed by a strongARM latch. These topologies [6,7] therefore are a good choice for a low-energy comparator. The energy reduction comes, however, at the expense of comparator speed. In fact, the pre-amplifier transistors are pushed towards the weak inversion region by means of a dynamic bias to reduce the current and increase the g_m/I_D ratio. Consequently, this implies a larger integration time slowing down the comparator. Other topologies have been proposed to enhance the comparison speed worsening the noise performance, as in [8], where high speed is achieved cascading three half-latches.

III. DESIGN OF THE COMPARATOR

From the qualitative description of the state-of-the-art, it is possible to conclude that:

- The strongARM compatarator is fast and features low power consumption but kickback noise reduces the con-verter resolution;
- The double-tail comparator [5] solves the drawbacks of the strongARM topology, even though it is less energy efficient;
- Efforts [6,7] to reduce the comparator energy consump-tion traded off energy with comparison speed.

Based on these reasons the proposed comparator (see Fig. 3) features a dynamic pre-amplifier followed by a strongARM latch. With respect to the solution in [3], the PMOS input pair in the second stage is not embedded in the latch, but directly tied to the supply rail through a switch, M_6 in Fig. 3. This allows to increase the amplification of the pre-amplifier output signal. The comparator speed benefits from the larger gain for small input signals. Moreover, the input-referred noise due to the second stage is further reduced.

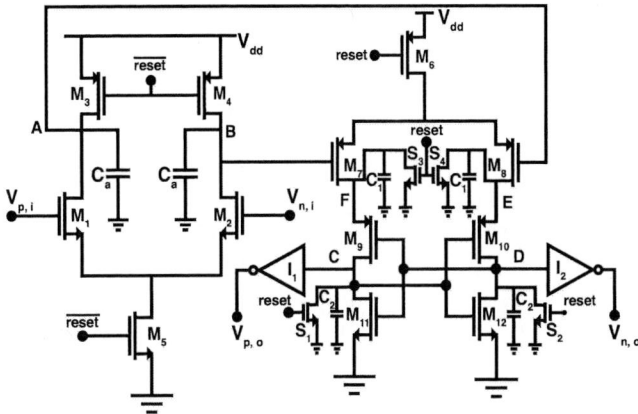

Fig. 3: Schematic of the proposed comparator.

In the reset phase ($reset = 1$), the nodes A and B are charged to V_{dd} by M_3-M_4, whereas the internal nodes (C, D, E and F) of the second stage are discharged to ground by S_1-S_4. The tail transistors M_5 and M_6 are off in this phase. In the comparison phase, the reset transistors M_3-M_4 and the switches S_1-S_4 are turned off, whereas the tail transistors are turned on. They work in linear region during comparison tying the sources of the input pair transistors of the first and second stage to ground and V_{dd}, respectively. During the comparison, M_1 and M_2 discharge the nodes A and B at a different rate based on the input signal. Thus, a differential signal builds up at the input of the following stage. When $reset$ goes low, the voltage at the nodes A and B, V_A and V_B respectively, is V_{dd}. Thus, the second stage does not immediately start to work, because transistors M_7-M_8 are off. Once either V_A or V_B goes below $V_{dd} - V_{th}$, a current starts to flow in the second stage. At this point the nodes C, D, E and F begin to be charged. As the voltage at the nodes C and D increases, M_{11}-M_{12} turn on slowing this process. Based on the input differential voltage, one of the two is able to bring C or D to ground, whereas the other node is driven high. In other words, the latch implemented by the cross-coupled inverters regenerates the amplified signal.

The capacitors C_a at the output of the pre-amplifier reduce the input-referred noise of the comparator. Other capacitors, C_1 and C_2, are implemented in the second stage to reduce its noise contribution. Switches S_1-S_4 are needed in the latch to avoid memory effects during the comparisons. Finally, inverters I_1 and I_2 increase the driving strength at the output of the comparator.

© VDE VERLAG GMBH · Berlin · Offenbach

It is worth pointing out that the comparator requirements in terms of noise and speed are set by the application. In this work, the comparator is designed for a 12-bit 200-MSps SAR ADC with 0.9-V supply voltage. Therefore, the least-significant bit (LSB) is approximately 440 μV. The input-referred noise is targeted to be LSB/3, which guarantees a signal-to-noise ratio degradation due to the comparator noise lower than 3.6 dB, if the sampler noise (i.e., kT/C) is negligible. Assuming a 50% duty cycle for the sampling clock, 12 conversions must be carried out in 2.5 ns, thus the comparator must compare and be reset in approximately 200 ps. This is not necessarily true for asynchronous SAR ADCs, but the SAR logic is assumed to be synchronous in this work for the sake of simplicity. Thus, the same time slot is allocated for each bit [9].

The noise requirement sets the value of the capacitance C_a at the output of the pre-amplifier and consequently the power consumption. Indeed, the input-referred noise of the comparator can be approximated to [10]:

$$\sigma_n^2 = \frac{2kT\gamma}{C_a}\frac{V_{ov}}{V_{th,p}} + \frac{kT}{2C_a}\left(\frac{V_{ov}}{V_{th,p}}\right)^2, \quad (1)$$

where k is the Boltzmann constant, T is the absolute temperature, V_{ov} is the overdrive voltage of M_1-M_2 and $V_{th,p}$ is the threshold voltage of the PMOS transistor. Considering a temperature of $100°$ C, an overdrive voltage of 150mV and a threshold voltage of 350mV, the minimum capacitance for an input-referred noise lower than LSB/3 is 227 fF. In this design, the capacitance is 328 fF to account for the second-stage noise contributions neglected in (1). Therefore, the energy consumption per conversion of the first stage is 531 fJ/conv. Finally, C_1 and C_2 have been set to 22 fF to limit the noise contribution of the second stage to less than LSB/6.

The proposed comparator has been compared to the topology in Fig. 2b sized to have equal input-referred noise and energy per comparison. The pre-amplifier is the same, therefore only the second stage was changed to have the same input-referred noise. The circuits have been simulated computing the delay between the reset and output signals. The result is shown in Fig. 4 (black lines). For large signals, the delay is dominated

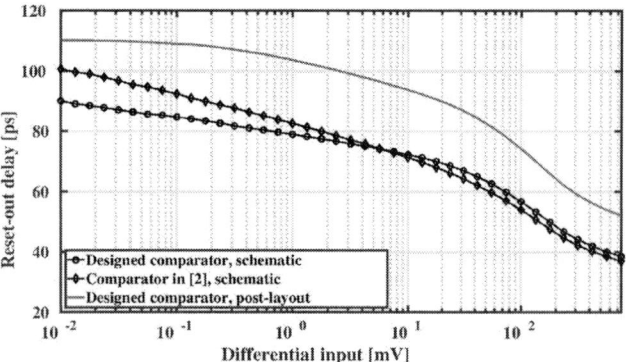

Fig. 4: Simulated delay as a function of the input voltage for the proposed comparator and the topology in [3].

by the first stage, because the latch is fast to regenerate the signal coming from the dynamic pre-amplifier. Therefore, the delay is similar in the two topologies. However, the delay is approximately 10% better in the proposed comparator than in the topology in [3] for small input signals. This can be explained by the fact that the gain in the designed strongARM stage is larger than in the latch with the embedded input pair. The overdrive voltage of the input transistors remains constant in the strongARM, whereas it slowly decreases in the comparator in [3] as the comparison is being carried out. Therefore, a lower current flows into the latch increasing the regeneration time.

IV. RESULTS

The proposed comparator has been implemented in a 28-nm bulk CMOS technology with a 0.9-V voltage supply. Its layout is shown in Fig. 5. All the simulations have been carried out at a temperature of 100 $°C$.

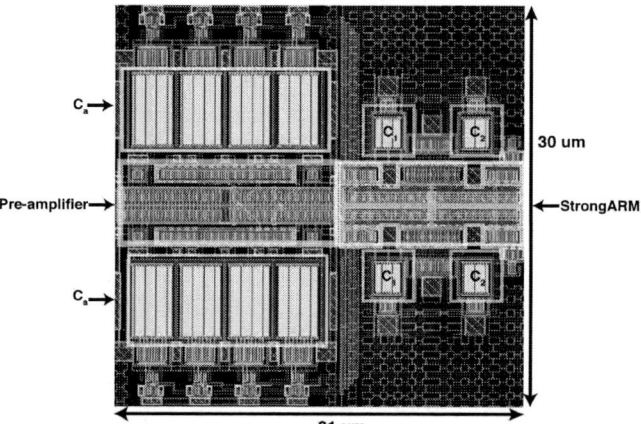

Fig. 5: Layout of the proposed comparator in 28-nm CMOS.

The result of the post-layout simulation of the reset-out delay as a function of the input differential voltage is also shown in Fig. 4 (red line). The delay slope is approximately 6 ps/decade when the latch regeneration time dominates, i.e., for input voltages smaller than 1 mV.

The comparator noise has been computed with the cumulative distribution function (CDF) [14]. A transient noise simulation is run with a constant input signal observing 5000 comparator decisions. This simulation is repeated for different values of the input voltage. The CDF is then computed as the percentage of the '1' decisions with respect to the total number of comparisons. The result is shown in Fig. 6. Fitting the CDF with a Gaussian cumulative distribution, the standard deviation of the input-referred noise is 163 μV, close to the expected value of 140 μV considering (1) and the second-stage contribution. In typical conditions, the standard deviation of the input-referred offset is 2 mV, from 1000 Monte-Carlo simulations. The comparator can be clocked with a 5-GHz reset signal resulting in a power consumption of 2.9 mW. This corresponds to an energy per conversion equal to 595 fJ/conv,

© VDE VERLAG GMBH · Berlin · Offenbach

SMACD / PRIME 2021 | 19 – 22 July 2021, Online Event

	This work	[8]	[5]	[11]	[12]	[13]
Technology [nm]	**28**	28	90	90	90	65
V_{dd} [V]	**0.9**	1	1.2	1.2	1	1.2
Area [μm^2]	**930**	78	82.5	120	543*	1394
Noise [mV_{rms}]	**0.16**	$\approx$ 1	1.5	0.39	N.A.	N.A.
Delay slope [ps/dec]	**6**	6.4	44	16	24.3	N.A.
f_{clk} [GHz]	**5**	13.5	1	1.5	1	5
Power [mW]@f_{clk}	**2.9**	2.2	0.113	0.0915	0.04	2.88
E/conv [fJ/conv]@f_{clk}	**595**	163	113	N.A.	20	N.A.
FoM [nJ μV^2]	**15.8**	163	254	N.A.	N.A.	N.A.

* Estimated from data   FoM = Energy/conv * Noise power

TABLE I: Summary of the performance and comparison with state-of-the-art

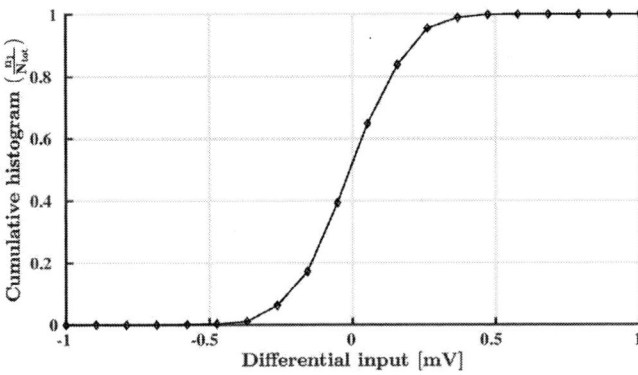

Fig. 6: Simulated CDF.

mostly due to the reset of the pre-amplifier capacitors C_a, being $2 \cdot C_a V_{dd}^2 = 531$ fJ/conv.

In order to compare the performance of the comparator with other works, the Figure-of-Merit (FoM) in [7] has been chosen to account for both energy per conversion and noise. It is defined as $E\sigma_n^2$, where E is the energy per conversion and the second term is the noise power. A summary of the performance of the proposed comparator and a comparison with other relevant works at high clock frequency ($\geq$ 1 GHz) is shown in Table I. The proposed comparator has the lowest input-referred noise among the reported works. The low noise performance results in the highest reported power consumption and energy per conversion. However, the FoM is one order of magnitude lower with respect to [8] and [5]. The area is 30 μm x 31 μm, mostly due the 328-fF capacitances at the output of the pre-amplifier. In terms of maximum operating frequency, it favourably compares to the other works in Table I except for [8], which, however, has an input-referred noise 6 times larger.

V. CONCLUSIONS

This paper presents a comparator designed for a 12-bit 200-MSps SAR ADC in 28-nm technology with a 0.9-V supply. There are two stages, a dynamic pre-amplifier and a strongARM latch with a PMOS input pair. Its power consumption is 2.9 mW at 5 GHz and the energy per conversion is 595 fJ/conv. At the cost of area and power consumption, the input-referred noise is 163 μV that makes the comparator suitable for a

medium resolution SAR ADCs. This comparator shows good efficiency in terms of noise and energy consumption compared to other works operating in the GHz range.

REFERENCES

[1] J. Liu, X. Tang, W. Zhao, L. Shen, and N. Sun, "A 13-bit 0.005-mm2 40-ms/s sar adc with kt/c noise cancellation," *IEEE Journal of Solid-State Circuits*, vol. 55, no. 12, pp. 3260–3270, 2020.

[2] V. Giannini, P. Nuzzo, V. Chironi, A. Baschirotto, G. Van der Plas, and J. Craninckx, "An 820μw 9b 40ms/s noise-tolerant dynamic-sar adc in 90nm digital cmos," in *2008 IEEE International Solid-State Circuits Conference - Digest of Technical Papers*, 2008, pp. 238–610.

[3] M. van Elzakker, E. van Tuijl, P. Geraedts, D. Schinkel, E. A. M. Klumperink, and B. Nauta, "A 10-bit charge-redistribution adc consuming 1.9 μw at 1 ms/s," *IEEE Journal of Solid-State Circuits*, vol. 45, no. 5, pp. 1007–1015, 2010.

[4] B. Razavi, "The strongarm latch [a circuit for all seasons]," *IEEE Solid-State Circuits Magazine*, vol. 7, no. 2, pp. 12–17, 2015.

[5] D. Schinkel, E. Mensink, E. Klumperink, E. van Tuijl, and B. Nauta, "A double-tail latch-type voltage sense amplifier with 18ps setup+hold time," in *2007 IEEE International Solid-State Circuits Conference. Digest of Technical Papers*, 2007, pp. 314–605.

[6] H. S. Bindra, C. E. Lokin, D. Schinkel, A. Annema, and B. Nauta, "A 1.2-v dynamic bias latch-type comparator in 65-nm cmos with 0.4-mv input noise," *IEEE Journal of Solid-State Circuits*, vol. 53, no. 7, pp. 1902–1912, 2018.

[7] X. Tang, L. Shen, B. Kasap, X. Yang, W. Shi, A. Mukherjee, D. Z. Pan, and N. Sun, "An energy-efficient comparator with dynamic floating inverter amplifier," *IEEE Journal of Solid-State Circuits*, vol. 55, no. 4, pp. 1011–1022, 2020.

[8] A. T. Ramkaj, M. S. J. Steyaert, and F. Tavernier, "A 13.5-gb/s 5-mv-sensitivity 26.8-ps-clk–out delay triple-latch feedforward dynamic comparator in 28-nm cmos," in *ESSCIRC 2019 - IEEE 45th European Solid State Circuits Conference (ESSCIRC)*, 2019, pp. 167–170.

[9] A. Yu, D. Bankman, K. Zheng, and B. Murmann, "Understanding metastability in sar adcs: Part ii: Asynchronous," *IEEE Solid-State Circuits Magazine*, vol. 11, no. 3, pp. 16–32, 2019.

[10] P. Nuzzo, F. De Bernardinis, P. Terreni, and G. Van der Plas, "Noise analysis of regenerative comparators for reconfigurable adc architectures," *IEEE Transactions on Circuits and Systems I: Regular Papers*, vol. 55, no. 6, pp. 1441–1454, 2008.

[11] C. Chan, Y. Zhu, U. Chio, S. Sin, U. Seng-Pan, and R. P. Martins, "A reconfigurable low-noise dynamic comparator with offset calibration in 90nm cmos," in *IEEE Asian Solid-State Circuits Conference 2011*, 2011, pp. 233–236.

[12] Masaya Miyahara, Yusuke Asada, Daehwa Paik, and Akira Matsuzawa, "A low-noise self-calibrating dynamic comparator for high-speed adcs," in *2008 IEEE Asian Solid-State Circuits Conference*, 2008, pp. 269–272.

[13] B. Goll and H. Zimmermann, "A comparator with reduced delay time in 65-nm cmos for supply voltages down to 0.65 v," *IEEE Transactions on Circuits and Systems II: Express Briefs*, vol. 56, no. 11, pp. 810–814, 2009.

[14] I. Opris, "Noise estimation in strobed comparators," *Electronics Letters*, vol. 33, no. 15, pp. 1273–1274, 1997.

© VDE VERLAG GMBH · Berlin · Offenbach

SMACD / PRIME 2021 | 19 – 22 July 2021, Online Event

A 2GS/s 10-bit Time-Interleaved Capacitive DAC for Self-Interference-Cancellation Application

Mazyar Abedinkhan Eslami, Danilo Manstretta, Rinaldo Castello

Microelectronics Laboratory, University of Pavia, Pavia, Italy

mazyar.abedinkhan01@universitadipavia.it

Abstract—This article presents a 2-GS/s time-interleaved (TI) 10-bit capacitive digital-to-analog converter (CDAC) for self-interference-cancellation (SIC) application. It is also capable of working as a non-TI & stand-alone CDAC with 1-GS/S clock frequency. By taking advantage low parasitic capacitance and equivalent parasitic capacitance at bottom and top plate of MIM capacitor, the split-capacitor technique is used without significant degradation in the linearity. The special architecture of the designed layout also relieves the local and radial oxide gradient error. The CDAC is designed in 28nm CMOS technology. If the CDAC works in stand-alone mode with 1-GS/s clock frequency, followed by an additional anti-aliasing filter and the baseband input frequency equals 10.74 MHz, the ENOB, SFDR and THD at the output of the filter is equal to11.3-bit,76 dB and 76dB, respectively.

Keywords— digital-to-analog converter (DAC), capacitive DAC (CDAC), split capacitor, self-interference-cancellation (SIC)

I. INTRODUCTION

Communication systems play a key role, prompting universities and companies to try to increase the speed and quality of the transmitted information. Hence, hardware specifications need to be boosted to preserve performance. In the circuits, linearity (e.g. SFDR and IIP3), noise (e.g. SNR and BER), power consumption and area are critical. Since most computing happens in digital, it is necessary to go from analog to digital and vice versa. To this end, high performance analog-to-digital (ADC) and digital-to-analog converter (DAC) are required. In this article, we focus on the latter. DAC has vast area of application. For example, they are used within ADCs like SAR and ΣΔ, or part of an RF transceiver.

The DAC presented here is meant for self-interference-cancellation (SIC) and particularly for the case of a full-duplex (FD) communication system. FD improves spectral efficiency by a factor of two, since both transmitter (TX) and receiver (RX) operate concurrently in the same band. The major drawback of FD is the TX signal leaking on top of the RX one. Since the TX is a much stronger signal compared to RX, the receiver chain will either clip or need an extremely high dynamic range (DR). This is especially challenging to meet for the ADC which becomes very power-hungry or even unfeasible. Since the TX signal is available in digital it can potentially be cancelled in RF and/or in base band (BB) (either in the analog or in the digital domain). In reference [1], the cancellation is done in RF with a current DAC which introduces noise penalty being located before signal amplification. If the DAC performs SIC in BB, proposed here, it can benefit from the front-end amplification, minimizing noise degradation. Moreover, connecting the DAC to the virtual ground of the trans-impedance amplifier (TIA) immediately after the down-conversion mixer, non-linear effects are minimized, and minimal additional power is required.

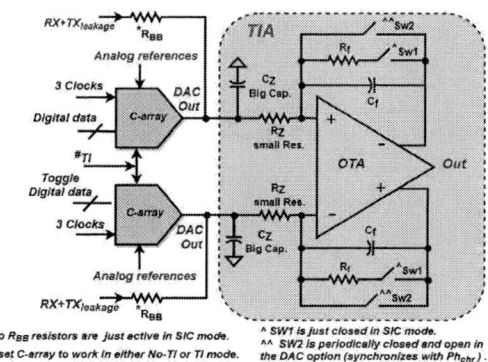

Fig.1 Block diagram of the design.

This article is structured as follows. Section II describes the DAC architecture and its four main sub-blocks. Section III elaborates on the key points of each of these four block. Noise analysis is performed in section IV, Section V shows simulation results both in DAC and SIC mode, finally some conclusions are drowned in section V.

II. BLOCK ARCHITECTURE

The block was designed to operate in two working modes: 1) Stand-alone DAC as shown in Fig. 1. In this case the charge stored in the capacitive array is transferred during the first half of the clock cycle to a capacitor connected in feedback around an OTA. During the second half of the clock a reset switch discharges the feedback capacitor while the capacitor array is pre-charged by the reference voltages according to the digital code. The maximum DAC refresh rate coincide with the clock frequency which is *1 GHz*. 2) The DAC in SIC mode as shown in Fig. 2. In this case two 2KΩ resistors are connected to the OTA virtual ground to model the down converted SI signal. A resistor R_f and a capacitor C_f is connected in feedback to the OTA and no rest switch is used. In this mode, the DAC is operated in time- interleaved mode giving an equivalent refresh rate of *2 GHz*. Having the goal to demonstrate how to cancel the TX interference, the DAC specifications are mainly determined by SIC parameters. First, we find the required signal-to-noise ratio (SNR) according to the typical data of a FD system given in Table I below. Target TX and RX parameters are similar to a SAW-less frequency-division duplex receiver as in [2]. The RX noise floor ($RX_{noise\ floor}$) is equal to -174 dBm/Hz x 10 log (BW_{Cancl}) + NF = -87 dBm. The TX leakage (L_{TXtoRX}) is given by the TX power (20 dBm) minus the TX-RX attenuation (assumed to be 40 dB). Hence, L_{TXtoRX} is -20 dBm and the necessary SNR of the DAC (SNR_{DAC}) equals L_{TXtoRX} - ($RX_{noise\ floor}$) = 67 dB. To minimize reciprocal interference, the operation of the DAC should be synchronous with the down-conversion mixers. Hence, the working frequency of the DAC (F_S) is chosen to be twice the 1 GHz mixer clock i.e. 2 GHz. Thanks to over-sampling, if we

© VDE VERLAG GMBH · Berlin · Offenbach

SMACD / PRIME 2021 | 19 – 22 July 2021, Online Event

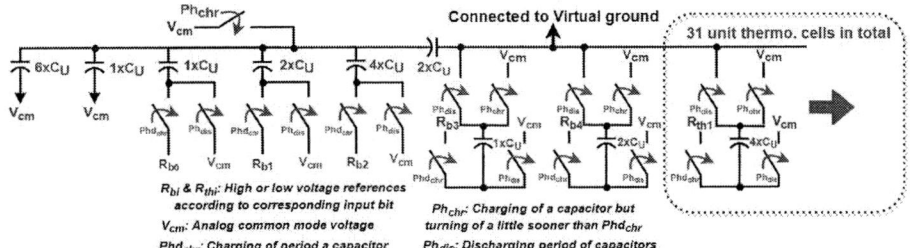

Fig.2 The CA made of the binary and thermometric coded cells along with a split-capacitor.

consider only quantization noise, the required SNR_{DAC} can be achieved with less bits, as expressed in (1), where N is the number of bits and OSR is the oversampling ratio [$F_S/(2×BW_{signal})$].

$$SNR_{DAC}= 6.02× N+ 1.76+ 10×log10 \ (OSR) \quad (1)$$

From the data, in Table I, the minimum N that can satisfy the spec. is 9. Considering DAC non-idealities like static and dynamic non-linearity we have added an extra bit making N=10. In principle we can use noise shaping and mismatch to reduce N. However, given the very high clock the power consumption of the digital pre-processing can be very significant. To minimize power consumption, we avoided any noise shaping.

The next step is to choose the architecture of the DAC. A capacitive DAC (CDAC) could be a good candidate for less than 12-bit resolution because, compared to a current DAC, it does not consume DC power, it can operate with a lower voltage supply and is much less sensitive to jitter [3].

The last point in the definition of the CDAC parameters is the value of the total capacitance of the CDAC array (C_T). In mainly determined form KT/C noise and matching criteria. In this case the seize of C_T is dictated by the fact that the CDAC should provide the cancellation current for the maximum expected SIC, which sets its full-scale current. The current generated by the CDAC is shown below.

$$I_{Cancel}= I_{DAC} \ (Average) = Q_{DAC}/T_S= C_{Total} ×V_{Ref} × F_S \quad (2)$$

Where Q_{DAC}, V_{Ref} and T_S are the maximum charge to be stored in the CDAC, the array reference voltage and the charging time C_T, respectively. It is reasonable to assume for the maximum value of I_{Cacel} between 0.5 to 1 mA [2].

TABLE I.   CANCELLATION PARAMETERS [2]

P_{TX} dBm	TX to RX Attenuation (L_{TXtoRX}) dB	Cancellation BW (BW_{cancl}) MHz	Noise Figure dB	Margin dB
20	40	80 (in RF)	8	6

For V_{Ref} = 0.35 V and a 2 GHz clock, the corresponding C_T is between 0.7 to 1.4 pF. A value of 1 pF was chosen here. The DAC uses a *fully differential topology* and its block diagram is shown in Fig. 1. Its four different sub-blocks are explained in the following sections.

III.   DAC BUILDING BLOCKS

A. CapacitiveArray Architectore

The binary-weighted approach is the simplest method to implement the capacitive array (CA) but it suffers from accuracy limitations when targeting resolution above 8-12 bits [3]. This is because when the ratio between the most-significant-bit (MSB) and the least-significant-bit (LSB) becomes high, many critical limitations arise like the impossibility of scaling the switch size down to the LSB and the increasing role of the parasitic for the smaller capacitors. To mitigate these issues, the capacitor split-array (SCA) configuration is chosen. SCA has two critical issues i.e., fractional value and non-equal bottom and top parasitic capacitance of the split-capacitor. We have addressed these issues in the following way. First, by adding a dummy capacitor in the LSB-side of SCA according to the value given in [4], the size of the split-capacitor was made equal to an integer multiple of the unit capacitor (C_U). The new configuration of CA is shown in Fig. 2. Although this method causes a small gain error, it improves linearity by removing the mismatch associated with the use of a fractionally valued attenuator capacitance. This is favorable trade-off for the target application. In fact, the gain error is automatically compensated for at the system level by the (adaptive loop) that scales amplitude and phase of the DAC input in the digital domain. On the other hand, compensating the distortion terms would require an additional non-linear adaptation algorithm both much more complex and more expensive in terms of power and area. The values of the dummy capacitance in the LSB-side and of the split-capacitor are $6×C_U$ and $2×C_U$, respectively. In this way the equivalent capacitance seen from the MSB side towards the LSB side is C_U. This method changes a 10-bit SCA structure with unit capacitance C_U to an equivalent binary-weighted array with the same number of bits but a unit capacitance of $C_U/8$. Regarding the second issue, we notice that the metal-insulator-metal (MIM) capacitor available in the design kit has identical parasitic capacitance at its bottom and top plates (both ends are in the same layer). Furthermore, the ratio between parasitic and main capacitance is less than 0.6%. Due to both these features, the non-linearity caused by parasitic capacitors is drastically reduced. The segmentation technique is also utilized in CA to further enhance its linearity and high-frequency performance. The SCA is divided into three parts. A three-bit binary-weighted array for the LSB part on the left-side of the split capacitor ($1×C_U+ 2×C_U+ 4×C_U$), two binary-weighted middle bits at the right side of split-capacitor ($1×C_U+ 2×C_U$) and five thermometric-coded MSB ($31×4C_U$) [5]. The five-lower binary-weighted and five upper unary arrays give a lower chip area and lower dynamic error respectively Because the CA just consist of multiple (1, 2 and 4) of C_U, matching is optimized [5]. Choosing C_U equal to 8fF a very low thermal noise and excellent matching is achieved and the effect of the parasitic capacitance of routing is negligible. Using an array reference voltage of 350mV the maximum I_{DAC} for a 1 GHz clock is 350uA, according to (1). Using a time-interleaved (TI) architecture with two identical CA, the equivalent clock

© VDE VERLAG GMBH · Berlin · Offenbach

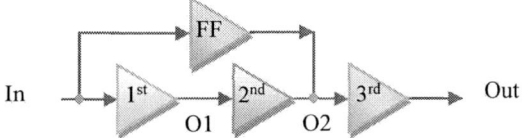

Fig. 3   3-stages Amplifier with FF stage.

becomes 2GHz and $I_{DAC-TI} = 2 \times 350uA = 700uA$. Based on the above observations no calibration was implemented.

The switches are another important part of the SCA, that could contribute distortion. In our design they are built with low voltage (LV) transistors. Thanks to the use of an SCA C_U is equal to 8 fF allowing to use a minimum size switch with minimum effect from its parasitic capacitances. All other switches are scaled up proportionally to the size of the capacitance to which they connect to. This is possible without requiring an excessive area thanks the SCA. Furthermore, both the middle bit and the MSB sub-array use parasitic insensitive switching to mitigate the effects of parasitic components. The switching sequence is shown in Fig. 2. To limit the voltage spike when connecting the array to the TIA an 9 pF capacitance C_Z, is connected in parallel to the virtual ground, it also filter the out-of-band frequency in the RX path. In addition, a resistor proportional to the switch on-conductance is placed in series to each switch. The RC time constant of such a connection is about 40ps which is just a bit smaller than the close loop time constant of the TIA. In this way the TIA settling time is affected in a negligible way, while the distortion due to the loading of the resistors connected to the virtual ground in cancellation mode is strongly reduced.

B. Transimpedance Amplifier Design

When operating in SIC mode, the transimpedance amplifier (TIA) implements a first-order filter which suppresses out-of-band components and allows gain control. It includes three different parts.

1) Operational transconductance amplifier (OTA): The OTA defines the settling time of the DAC and the performance of the RX path. A three stages OTA is a good candidate considering gain, power consumption and complexity [6]. Three stages amplifers have three poles (one at the output of each stage) plus the high-frequency poles of the feedback network. Pole-zero compensation is chosen for stability instead of Miller compensation. If these pole-zero doublets are located at high-frequency, their negative effect of the associated doublet on the settling time of the DAC are negligible. One high-frequency zero is created by a no-capacitor feed-forward (NCFF) method [6] while another one is created in the input network (to be described later). The block diagram of the OTA is shown in Fig. 3. The class-AB feed-forward (FF), shown in Fig. 4, is a modified version of the one developed in [7] whose effective Gm is multiplied by 1.5 in this design. The OTA DC gain, unity-gain-loop-bandwidth, and phase margin are 83dB, 3.8 GHz, and 44°, respectively. The above specifications allow to settle within 1LSB at 10-bit accuracy in 450ps, i.e. in about 10% less than the on-time of the clock. This corresponds to $T_S/2$ for a 1 GHz clock and to T_S for the equivalent 2 GHz clock when operating in TI mode (10% margin is taken to compensate for process and temperature variations).

2) Feedback network: The feedback network is shown in Fig. 1. Rf and Cf have a value of 2 KΩ and 2 pF respectively to give a bandwidth equal to 40MHz. The value of Rf is chosen such that it allows to significantly amplify the RX signal in SIC mode without any clipping at the TIA output assuming at least 20 dB of SI cancellation [2].

3) Input network: This part includs an 8pF capacitor (C_Z) and a 10 Ω resistor (R_Z) as shown in Fig. 1. C_Z is used to absorb the current spikes that occur when discharging the CA capaciatance toward virtual grownd. This drastically reduces the voltage spikes at the virtual grownd node preventing current saturation in the input stage of the OTA. In this way no large signal slewing occurs in the step response even for a full scale input improving linearity. Furthermore C_Z woud be required when connectig the mixer to the TIA to filter the harmonic of the clock. The combination of C_Z and R_Z also creates a high-frequency zero in the loop response of the TIA improving the phase margin.

C. Digital part

This block is made of two parts, i.e. the timing generator and a structure for testing it. Parasitic insensitive switching needs three different clocks to work properly. They are produced starting from a 2 GHz reference clock by a timing generator. Two sets of clocks are generated for No-TI and TI mode. To test the CDAC, it is necessary to internally generate 10-bit digital data synchronized with the CDAC clock.

IV.   NOISE ANALYSIS

TIA's switches and feedback resistor are the three main noise contributors and being uncorrelated, they can be added up as shown in (4).

$$\overline{V^2}_{Out} = \overline{V^2}_{SWchr} + \overline{V^2}_{SWdis} + \overline{V^2}_{Rf} + \overline{V^2}_{OTA} \quad (4)$$

Where $\overline{V^2}_{Out}$ is the total noise at the output of the TIA, $\overline{V^2}_{SWchr}$, is the switch noise in the charging phase $\overline{V^2}_{SWdis}$, is the switch noise in the discharging phase $\overline{V^2}_{Rf}$ is the feedback resistor thermal noise and $\overline{V^2}_{OTA}$ is the noise of the TIA. A single ended topology is first considered (as it shown in Fig. 4) then a 3 dB correction factor is applied to take into consideration the fully differential nature of the circuit [8].

Consider first the switch noise. When the CA is charged, as shown in Fig. 4 (a), the mean square value of the thermal noise sampled charge is equal to $KT \times C_T$, where C_T is the capacitance of the full array. In the discharge phase, as shown in Fig. 4 (b), the sampled noise charge is transferred to the output of the TIA together with another $KT \times C_T$ noise charge packet [8]. Assuming a white noise spectrum for the sampled noise, the noise power at the TIA output integrated up to the band edge. The third term in (4) is the output referred in band

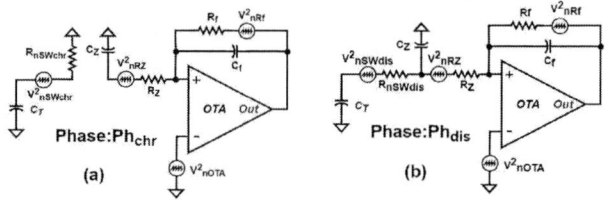

Fig.4 The single ended noise model of the design in (a) charging phase (b) delivering the charge to the circuit.

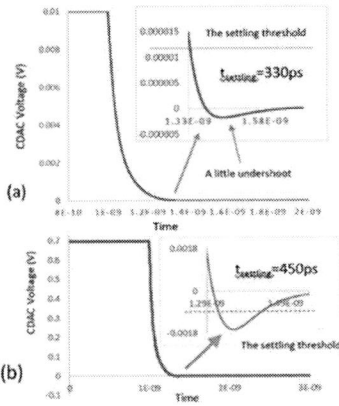

(a)

(b)

Fig. 5 The settling time (a) small signal (b) large signal.

noise power due to R_f. Finally, the last term gives the output referred in band noise power due to the OTA which has an equivalent input noise resistance of 150. Combining the 4 terms we get 40μ total output noise. For a clock of 2 GHz, C_T and R_f of 2KΩ of 2 pF respectively and V_{Ref} of 375 mV the RMS output signal is a bit more than 1 V RMS. This corresponds to a SNR of 88 dB. Taking into account the differential nature of the circuit the SNR becomes 91 dB. Since the SNR due to the thermal noise is at least 20 dB higher than the SNDR of the DAC in both stand alone and SIC mode of operation, we have decided not to include any thermal nose contribution in the simulations reported below. This speeds-up simulations without significantly affecting the final results.

V. SIMULATION RESULTS

The design is in 28nm CMOS technology. A dual supply voltage, 1V and 1.3 V for the digital part and the TIA, respectively, was used. To evaluate the setting time of the CDAC, ach input of the TIA is connected to a capacitor in series with a resistor equal to total CDAC capacitance and total switches resistance, respectively. The two capacitors are charged to V_{CM}+5mV and VCM-5m in one case and to V_{CM}+350mV and VCM-350mV to evaluate the small-signal and large-signal settling-time, respectively. Two capacitors are connected to the TIA at t=1ns. The differential voltage across the two capacitors is shown in Fig. 5. The settling time to 0.1% (equals 10-bit) of the final value is around 330ps and 450ps in the two cases and the shape of the response is very similar indicating a very linear behavior. The input signal of the CDAC is a digitized sinusoidal waveform, whose frequency, chosen according to Coherent sampling, equals 10.74MHz. Two sets of simulations were launched for stand-alone DAC and SIC mode. After the TIA an ideal second order filter with a 40 MHz bandwidth has been included in the simulation. This is intended to represent the anti-alias filter that will be placed between the TIA and the ADC that will follow it. *All reported results are taken after such a filter*. In this way the effect of the oversampling in the DAC SNDR is considered.

A. DAC simulation

For this simulation the DAC is operated in No-TIA mode and the input code range is ±460, 1 dB bellow full scale, so that rail-to-rail swing is reached at the TIA output. The resulting spectrum is shown in Fig. 6. An ENOB of 11.3 bits and SNR of 70.6 dB are demonstrated, i.e. below the

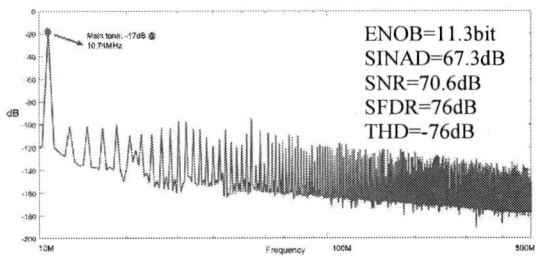

ENOB=11.3bit
SINAD=67.3dB
SNR=70.6dB
SFDR=76dB
THD=-76dB

Fig. 6 Output spectrum of the CDAC in No-TI mode.

maximum theoretical value considering the 12X oversampling ratio and the -1 dB full scale signal.

B. Simulation result of SIC mode

In Fig. 1, the baseband's TX leakage is modeled injecting a voltage to two R_{BB} resistor with the DAC operates in TI mode. Cancellation is defined as the difference between the output before and after cancellation in dB. The 2.6Vpeak SI signal, would give a 5.3dBV signal at the output without cancellation (TIA gain =1). The achieved cancellation is 41.1+5.3=46.4dB. SNDR$_{equiv}$ is 46.4+14.5=60.9dB and gives the SNDR after cancellation referred to the signal before cancellation. The simulation results are shown in Table II.

TABLE II. SIC SIMULATION RESULTS

Input	Output	Cancel.	SNR	SNDR	SFDR	THD	SNDR$_{equiv}$
V	dBV	dB	dB	dB	dBc	dB	dB
2.6	-41.1	46.4	15.6	14.5	21.3	-20.1	60.9

VI. CONCLUSION

The DAC could reproduce the leakage signal of the TX to the RX path and cancel the self-interfere in the wireless communication systems. For performing the cancellation, a 10-bit capacitive DAC is presented in this article which works with frequency of 2GS/s in time-interleaved. For a Sinusoidal input 1 dB below full scale and a baseband frequency equals 10.74 MHz, the ENOB and SFDR of the CDAC followed by an anti-aliasing filter equal 11.3 bit and 76 dB, respectively.

REFERENCES

[1] L. Calderin, et al., "Analysis and design of integrated active cancellation transceiver for frequency division duplex systems," IEEE J. Solid-State Circuits, vol. 52, no. 8, pp. 2038–2054, Aug. 2017.

[2] D. Montanari et al., "An FDD Wireless Diversity Receiver With Transmitter Leakage Cancellation in Transmit and Receive Bands," IEEE JSSC, vol. 53, no. 7, pp. 1945-1959, July 2018.

[3] M. Rajabzadeh et al., "Comparison Study of DAC Realizations in Current Input CT Modulators" IEEE Transactions on Circuits and Systems II: Express Briefs, vol, 68, no. 1,pp. 111, Jan. 2021.

[4] A. H. Chang, H. S. Lee, and D. Boning, "A 12b 50MS/s 2.1mW SAR ADC with redundancy and digital background calibration," IEEE ESSCIRC, pp. 109–112, Sept. 2013.

[5] C. Lin and K. Bult, "A 10-b, 500-Msample/s CMOS DAC in 0.6 mm ," IEEE J. Solid-State Circuits, vol. 33, pp. 1948–1958, Dec. 1998.

[6] B. K. Thandri and J. Silva-Martinez, "A robust feedforward compensation scheme for multistage operational transconductance amplifiers with no Miller capacitors," IEEE Journal Solid-State Circuits, vol. 38, no. 2, pp. 237–243, Feb. 2003.

[7] V. Peluso et. , "A 900-mV low-power ΔΣ A/D converter with 77-dB dynamic range," IEEE Journal of Solid-State Circuits, vol. 33, pp. 1887–1897, December 1998.

[8] J. Xu, et al., "Noise Optimization Techniques for Switched-Capacitor Based Neural Interfaces" IEEE Transactions on Biomedical Circuits and Systems, vol. 14, no. 5, pp. 1024-1035, Oct. 2020.

© VDE VERLAG GMBH · Berlin · Offenbach

SMACD / PRIME 2021 | 19 – 22 July 2021, Online Event

Implementation of a Low Power Decimation Filter in a 180 nm HV-CMOS Technology for a Neural Recording Front-End

Markus Sporer[*], Nicolas Graber[*], Steffen Moll[*], Stefan Reich[*] and Maurits Ortmanns[*]
[*]Institute of Microelectronics, University of Ulm,
Albert-Einstein-Allee 43, Ulm, Germany, markus.sporer@uni-ulm.de

Abstract—The design of decimation filters for Delta Sigma Converters is a rarely discussed topic though these filters are necessary for $\Delta\Sigma$ modulator ADCs. In this work, we present a low-power decimation filter for a neural recording front-end. We show the applied design criteria and compare different filter structures to find the most efficient implementation. The decimation filter has been implemented and simulated in a 180 nm HV-CMOS process. It achieves a low consumption of 6.4 µW at a supply voltage of 1.2 V. The filter has a 13 bit output running at a sampling rate of 22 kHz and requires an area of 460 µm x 210 µm. The input signal's SNDR degradation due to the filter is less than 2 dB.

Index Terms—Decimation, Neural Recording, Low Power, Biomedical Implant, CIC Filter

I. INTRODUCTION

Decimation filters are used in integrated circuits that require alias-free down-sampling, such as $\Delta\Sigma$ ADCs and DSPs. In order to minimize the digital filter's power consumption, a low supply voltage and an advanced process node are preferable. However, there are applications such as neural recording with additional high-voltage stimulation [1] where it is required to use a less scaled, high-voltage process node. Although $\Delta\Sigma$ modulator based neural recording front-ends are thoroughly researched [2]–[4], significantly less attention was given on the corresponding decimation filter. Even though the neural recording frequency range is close to audio applications, decimation filters in the audio frequency range typically occupy a large area and can consume tens of mW [5]. Since biomedical implants such as neural recorders are strictly power and area limited, it is not feasible to simply use these filters.

In this paper, we present a low-power decimation filter especially designed for use within a neural recording front-end. Moreover, we discuss the applied design criteria to achieve a reasonable power and area consumption compared to the modulator. The designed filter decimates a 2.816 MHz single bit output of a second order $\Delta\Sigma$ modulator by a factor of 128 with a low power consumption of 6.4 µW at a supply voltage of 1.2 V and a size of 460 µm x 210 µm. The design has been implemented in a 180nm HV-CMOS technology. In Section II the filter's design criteria are described. Section III shows the comparison of different filter structures and their implementation. Subsequently post-layout simulation results

and the layout of the complete decimation filter are presented in Section IV. Finally the paper is concluded in Section V.

II. DECIMATION FILTER ARCHITECTURE

The neural recorder front-end consists of a single-bit second order continuous-time $\Delta\Sigma$ modulator with an OSR of 128. The modulator achieves an SNDR of 65.6 dB for a 10 kHz inband. Since the modulator contains a chopped input stage, special care must be taken to avoid aliasing of the large out-of-band 25 kHz chopper signal.

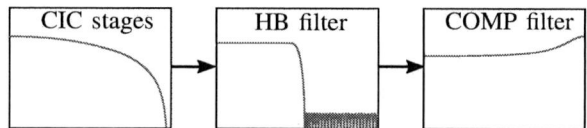

Fig. 1: Block diagram of a typical decimation filter.

Fig. 1 shows a block diagram of a typical decimation filter. Due to their highly efficient design cascaded-integrator-comb (CIC) filters are usually used for the first decimation stages [6]. This type of filter though has a rather poor cut-off and is therefore not able to sufficiently suppress aliasing if used as last filter stage. Hence, a more complex halfband filter is required to ensure a proper cut-off and prevent artifacts due to aliasing. However it is not feasible to compensate the CIC filter's droop by using the halfband filter due to its flat passband [7]. Often, an extra compensation filter is used to compensate the passband droop.

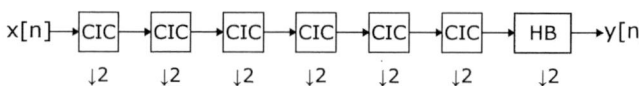

Fig. 2: Structure of the decimation filter with six cascaded CIC filters for high rate- and a halfband filter for final rate reduction.

To minimize the resource demand, no specialized compensation filter was included in this design. Instead, emphasis was put on a minimized passband droop within the CIC stages. The reason is that neural signals are usually post-processed

© VDE VERLAG GMBH · Berlin · Offenbach

for spike sorting and filtering [8] where undesired passband droop can be compensated much more efficiently than on-chip. Moreover instead of a single CIC filter with a high decimation factor, six cascaded CIC filters with a fixed rate decimation of two as shown in Fig. 2 were used. In this design, the decision allows a large part of the circuit to run at lower clock rates and also keeps the filter's internal bitwidths small which otherwise can grow significantly in a single stage design. An increased CIC filter order leads to a higher stopband attenuation but also increases the filter's passband droop and the computational effort. Therefore we used an iterative approach to keep the orders of the single CIC stages as low as possible. As a starting point, the orders of all stages were set to two. To determine a power efficient configuration, the order of the last filter was increased and the raised order was moved towards the first CIC stage until the maximum filter performance in terms of inband noise was achieved. This procedure was repeated until the increase of inband noise due to aliasing between the $\Delta\Sigma$ modulator and the CIC filter chain's output was less than about 0.5 dB.

This threshold was selected since less inband noise due to aliasing would have been possible but also inefficient due to the sharply increasing hardware demand. Table I shows the order of each CIC stage and the resulting bitwidth using this approach.

TABLE I: Determined CIC stage orders and corresponding bitwidths.

Stage	1	2	3	4	5	6
Order N	2	3	3	3	2	2
Rate R	2	2	2	2	2	2
Bitwidth b	3	6	9	12	14	16

Due to their poor passband cut-off, CIC filters are not suitable as last decimation stage. Instead, an elliptic IIR halfband filter combining low passband ripple and a precise cut-off was chosen. Approximately half of the halfband filter's coefficients are zero which reduces the hardware effort greatly. Since a non-linear phase is of no interest for neural recording, an IIR filter with much less computational effort compared to an FIR filter was used. The halfband filter was implemented with a minimum stopband attenuation of 50 dB and a narrow transition width of 1.2 kHz.

Fig. 3 shows the decimation filter's transfer function. No noticeable ripple is occuring due to the use of an elliptic IIR halfband filter architecture. However the cascaded CIC filters cause a maximum passband droop of 1.5 dB at 10 kHz.

III. FILTER IMPLEMENTATION

To find the most economic filter implementation, the second and third order CIC filters' area and power consumption were evaluated for a recursive and a non-recursive structure. Since the non-recursive structure reduces the number of additions by half, it requires less standard cells and has a lower power

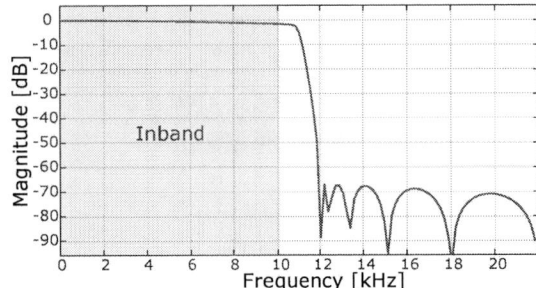

Fig. 3: Magnitude response of the decimation filter.

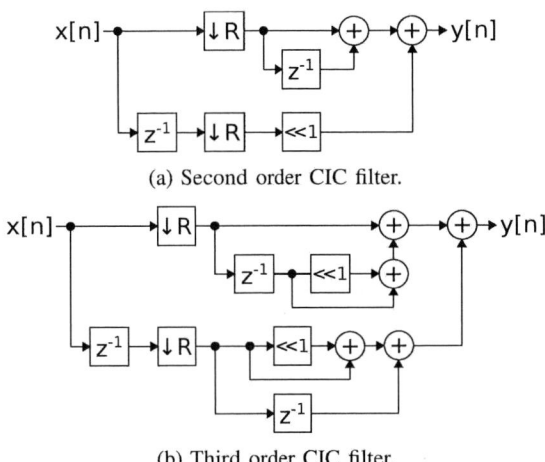

(a) Second order CIC filter.

(b) Third order CIC filter.

Fig. 4: Structure of the implemented non-recursive polyphase CIC filters.

consumption than the recursive implementation. Additionally polyphase decomposition was applied on the non-recursive structure as depicted in Fig. 4. The different input delay of the two branches was realized by triggering on both rising and falling edge of the clock signal, which enables to run all standard cells at the lower output clock rate. This resulted in a significantly reduced power consumption, making the non-recursive structure with polyphase decomposition the most power and area efficient implementation of a second and third order CIC filter with a decimation rate of two in the chosen 180 nm technology. Different implementations of the six-stage CIC filter are compared in terms of power consumption and standard cell usage in Fig. 5.

The halfband filter was implemented in parallel form to benefit from the advantage that there is no possibility of internal filter overflow. Moreover the benefits of a polyphase implementation can be applied enabling the complete halfband filter to run at the lower output clock rate. The internal bitwidth was set to 15 bit with 12 bit coefficients to reduce the filter's complexity. It has a 15 bit input and 13 bit output, consumes 1.1 µW of power and requires 706 standard cells. Fig. 6 shows the structure of the 11th-order elliptic IIR halfband filter.

SMACD / PRIME 2021 | 19 – 22 July 2021, Online Event

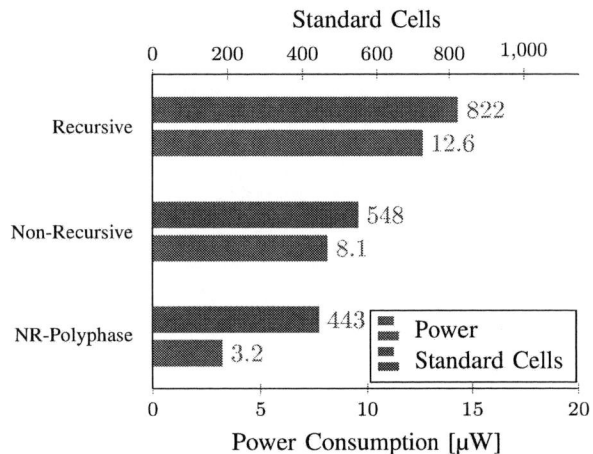

Fig. 5: Comparison of different CIC filter implementations.

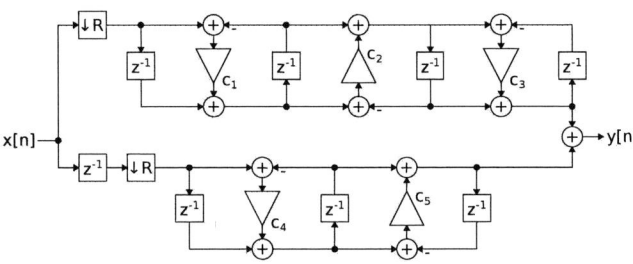

Fig. 6: Parallel structure implementation of the 11th-order IIR halfband filter.

Since this design is a multirate filter, different clock rates are needed. An asynchronous counter was implemented to derive all required clock rates from a single clock input. The asynchronous implementation is more power efficient than a synchronous counter and consumes only 0.33 µW.

IV. POST-LAYOUT SIMULATION

The decimation filter was placed and routed to perform post-layout simulation and to compare the area and power consumption with the preceding $\Delta\Sigma$ modulator. The filter consumes 6.4 µW of power which is about half of the $\Delta\Sigma$ modulator's power consumption of 12 µW. Note, that a neural recorder ADC has to fulfill about 40-60 dB signal amplification, accommodate thereby few µV to a few mV AC input signals, while being able to reject up to 50 mV electrode DC offset [9], which makes it hardly comparable to a standard ADC and consumes more power given ENOB and Nyquist rate f_{NYQ}. Fig. 7a and Fig. 7b show power spectral densities of the modulator's output before and after decimation, respectively. The SNDR degradation is smaller than 0.5 dB at 1 kHz and therefore within the specified range of 2 dB or less. Moreover, no aliasing of the strong 25 kHz chopper signal can be noticed.

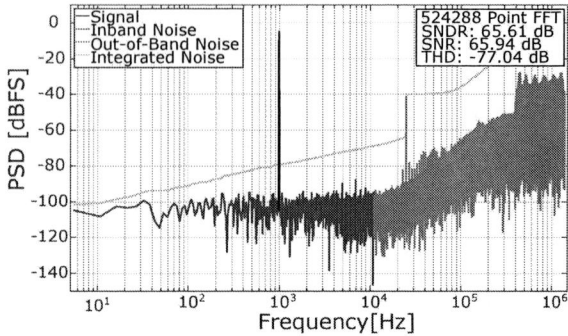

(a) PSD of the raw 1 bit $\Delta\Sigma$ modulator output.

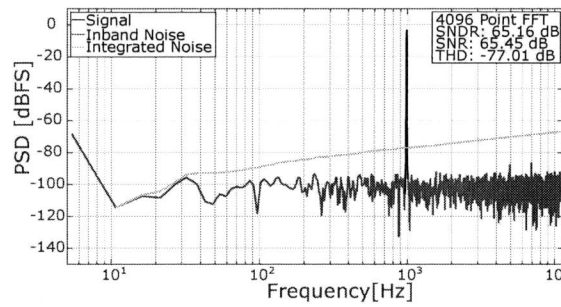

(b) PSD of the decimated $\Delta\Sigma$ modulator output.

Fig. 7: Comparison of the raw and decimated $\Delta\Sigma$ modulator output signal.

Fig. 8 shows the neural recorder's layout consisting of the decimation filter and the GmC based continuous-time $\Delta\Sigma$ modulator which is not further discussed in this paper. The modulator and the filter require an area of 20,250 µm² and 96,600 µm², respectively. Although the decimation filter is tremendously smaller than an audio range decimator, it still requires almost five times more area than the modulator.

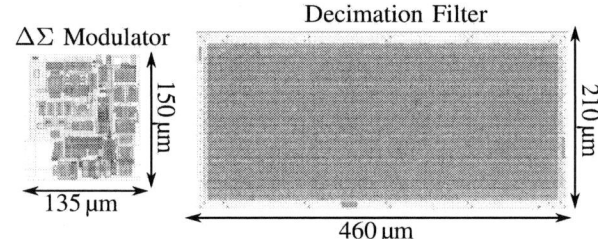

Fig. 8: Neural recorder's layout and dimensions.

V. CONCLUSION

The design challenges of decimation filters for neural implants are rarely addressed. An inadequate designed filter can consume a significant amount of power and could shatter all power-saving efforts which were made in the preceding stages. Moreover, the use of advanced process nodes is often not possible due to the requirement of a HV-CMOS technology

© VDE VERLAG GMBH · Berlin · Offenbach

making an efficient filter design even more important. Our proposed decimation filter is designed for the use within a neural recorder and achieves a very low power consumption of only 6.4 µW. The comparison of different CIC filter architectures enabled us to determine the non-recursive implementation with polyphase decomposition as the most efficient structure in terms of power and area. Allowing a certain amount of non-critical passband droop allowed us to forgo the compensation filter and keep the standard cell usage low. It has been shown over simulations, that the decimator's performance is suitable for the use within a $\Delta\Sigma$ modulator based neural recorder.

ACKNOWLEDGMENT

This work was funded by the German National Science Foundation DFG under grant number EI 1078/1-1.

REFERENCES

[1] Michael Haas, Ulrich Bihr, Jens Anders, and Maurits Ortmanns. A bidirectional neural interface ic with high voltage compliance and spectral separation. In *2016 IEEE International Symposium on Circuits and Systems (ISCAS)*, page 2743–2746, 2016.

[2] Antonios Nikas, Sreenivas Jambunathan, Leonhard Klein, Matthias Voelker, and Maurits Ortmanns. A continuous-time delta-sigma modulator using a modified instrumentation amplifier and current reuse DAC for neural recording. *IEEE Journal of Solid-State Circuits*, 54(10):2879–2891, oct 2019.

[3] Hariprasad Chandrakumar and Dejan Markovic. A 15.2-ENOB 5-kHz BW 4.5-μW chopped CT $\Delta\Sigma$-ADC for artifact-tolerant neural recording front ends. *IEEE Journal of Solid-State Circuits*, 53(12):3470–3483, dec 2018.

[4] Changuk Lee, Taejune Jeon, Moonhyung Jang, Sanggeon Park, Jejung Kim, Jeongsik Lim, Jong-Hyun Ahn, Yeowool Huh, and Youngcheol Chae. A 6.5-μW 10kHz BW 80.4-dB SNDR Gm-C-Based CT $\Delta\Sigma$ modulator with a feedback-assisted gm linearization for artifact-tolerant neural recording. *IEEE Journal of Solid-State Circuits*, 55(11):2889–2901, nov 2020.

[5] YuQing Yang, Terry Sculley, and Jacob Abraham. A single-die 124 dB stereo audio delta-sigma ADC with 111 dB THD. *IEEE Journal of Solid-State Circuits*, 43(7):1657–1665, jul 2008.

[6] E. Hogenauer. An economical class of digital filters for decimation and interpolation. *IEEE Transactions on Acoustics, Speech, and Signal Processing*, 29(2):155–162, apr 1981.

[7] B.P. Brandt and B.A. Wooley. A low-power, area-efficient digital filter for decimation and interpolation. *IEEE Journal of Solid-State Circuits*, 29(6):679–687, jun 1994.

[8] Hernan Gonzalo Rey, Carlos Pedreira, and Rodrigo Quian Quiroga. Past, present and future of spike sorting techniques. *Brain Research Bulletin*, 119:106–117, oct 2015.

[9] Rikky Muller, Simone Gambini, and Jan M. Rabaey. A 0.013 μm^2, 5μW, DC-coupled neural signal acquisition IC with 0.5 v supply. *IEEE Journal of Solid-State Circuits*, 47(1):232–243, jan 2012.

Analog Baseband Filter and Variable-gain Amplifier for Automotive Radars in 22 nm FD-SOI CMOS

Andres Seidel[#1], Songhui Li[#], Laszlo Szilagyi[#], Corrado Carta[#], Jens Wagner[#,$] and Frank Ellinger[#,$]

#Chair for Circuit Design and Network Theory, TU Dresden, Dresden, Germany
$ CeTi, Center for Tactile Internet, TU Dresden, Dresden, Germany
[1]andres.seidel@tu-dresden.de

Abstract—This paper presents the design of an analog baseband for a 77 GHz automotive radar in 22 nm fully-depleted silicon on insulator (FD-SOI) CMOS. The baseband ranges from 0.13 MHz to 14 MHz and requires a steep low-pass filter to avoid aliasing. This is realized with an 8th order Butterworth low-pass filter. In order to achieve the specified gain requirements, a variable gain amplifier (VGA) with DC-offset cancellation is applied before the filter. Its dynamic range was maximized, the optimization strategy is presented. The performance of the baseband components was verified with laboratory measurements. The filter achieves a suppression of ≥ 35 dB at 28 MHz. A dynamic range of 45 dB and a maximum gain of 41 dB is measured for the VGA. In addition, a good agreement between simulation and measurement is obtained. With an area-optimized design of 0.032 mm², the baseband components are a fraction of the size compared to chains of similar radar systems.

Index Terms—Analog baseband, VGA, Butterworth filter, automotive radar, compact layout, Silicon-on-insulator

I. INTRODUCTION

It is an indispensable endeavor of automotive development to ensure greater safety and comfort. Today sophisticated driver assistance systems are becoming more and more standard and experience an even greater precision. For radar applications in the mm-wave range, III/V technologies offer advantages in the design of the analog high frequency front end but fail to integrate high-performance digital signal processing within one chip. For complex applications with extensive process levels, strongly-scaled CMOS processes are still preferred, which benefit from low costs especially at high volumes.

As an essential part of a 77 GHz frequency-modulated continuous-wave (FMCW) radar [1], this paper presents the design of the baseband components in a receiver front end. Fig. 1 illustrates a simplified overview of the baseband chain. In [2] and [3] the analog basebands of similar radar systems are implemented with three programmable gain amplifier (PGA) stages, where each stage has its own DC-offset cancellation. The combined PGAs achieve a high gain performance, but the design is area intense. In addition, such an approach needs an integrated serial peripheral interface (SPI) to adjust the gain.

In order to realize a more compact design, this work focuses on a variable gain amplifier (VGA), which can be adjusted by a single control voltage. It must have a dynamic range of 35 dB, a maximum gain of 40 dB and high linearity to ensure radar performance. In addition, possible interfering signals and

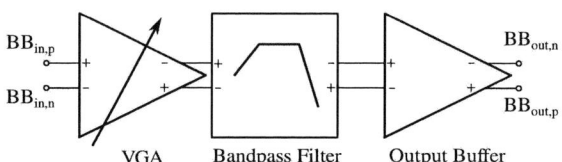

Fig. 1. Block diagram of the baseband chain.

unwanted modulation products have to be filtered in the analog baseband to avoid aliasing. The specifications according to the radar scenario require a steep aliasing low-pass filter with a slope of about 100 dB per decade. An output buffer stage provides a fast recharging of the ADC's input capacitor during the sampling sequence.

The following section focuses on the design of the implemented baseband bandpass filter as well as VGA. The output buffer will be addressed in another publication, since no test chip has been manufactured yet. The results of the laboratory measurements are presented in section 3 and finally compared with the state of the art in section 4.

II. CIRCUIT DESIGN

A. Baseband Filter

A low- and a high-pass filter were cascaded in order to form the baseband filter with bandpass characteristic. To design the low-pass filter, the MATLAB *Analog Filter Design Toolbox* provided a convenient initial point with basic filter classes implemented [4]. The Butterworth filter is selected due to its flat passband and low group delay characteristic. To meet specifications, the 3-dB cutoff frequency is set at 14 MHz, followed by a steep-slope stopband to guarantee the required minimum suppression of 30 dB at 28 MHz. The desired frequency response could be achieved with a Butterworth filter of order 6, but some margin should be considered due to process variations. Therefore an 8th order filter is utilized. Four stages in multiple feedback topology are used for the circuit synthesis as shown in Fig. 2. Since the baseband signals are fully differential, this topology offers the advantage that only one fully differential operational amplifier is required per stage. Furthermore, C_1 is half the size compared to a single ended topology. The op-amp is designed as a two-stage topology with Miller compensation and common-mode control, achieving an

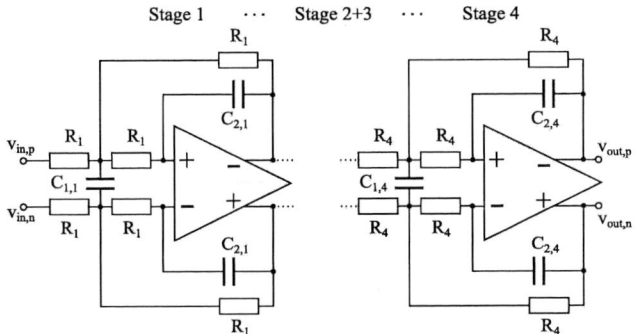

Fig. 2. Schematic of the 8th order Butterworth low-pass filter.

TABLE I

RESISTANCE AND CAPACITANCE VALUES OF THE 8TH ORDER
BUTTERWORTH FILTER ACCORDING TO FIG. 2.

Stage N	1	2	3	4
R_N	13.3 kΩ	11.1 kΩ	15.2 kΩ	12.6 kΩ
$C_{1,N}$	555 fF	815 fF	900 fF	2.85 pF
$C_{2,N}$	500 fF	500 fF	250 fF	100 fF

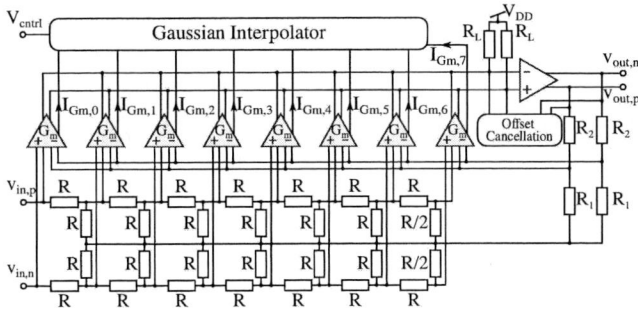

Fig. 3. Schematic of the variable gain amplifier.

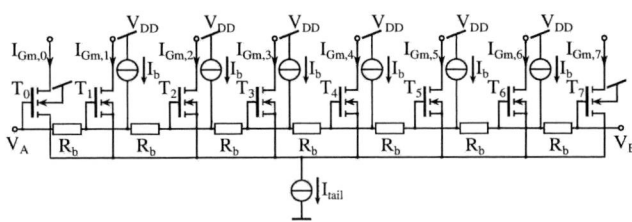

Fig. 4. Schematic of the Gaussian Interpolator for the proposed VGA.

open-loop DC-gain of 64 dB and a gain–bandwidth product of 1 GHz. It draws 360 µA from a 1.2 V supply source. The values of RC components are set trading off the size of the capacitors against the noise of the resistors. This is realized by means of sizing the feedback capacitor C_2. The resistance and capacitance values of the Butterworth low-pass filter are shown in Tab. I.

The high-pass filter is required to suppress DC-signals and has a 3 dB-cutoff frequency of 130 kHz. The implementation is done with a simple RC network ($R_{\text{HP}} = 200$ kΩ and $C_{\text{HP}} = 21$ pF) with a buffer stage output, which has a fixed gain of 6 dB. The complete bandpass filter draws 1.7 mA from the 1.2 V DC-supply.

B. Variable Gain Amplifier

The VGA design depicted in Fig. 3 is based on the topology of [5], which offers high linearity and low noise performance. Following [6], a 7-stepped resistor-resistor (R-R) ladder with 8.4 dB attenuation per step is used to achieve a theoretical dynamic range of 58.5 dB. Before and after each R-R ladder step a transconductance-cell (G$_{\text{m}}$-cell) is connected. A Gaussian interpolator is used to smoothly control the bias current $I_{\text{Gm,x}}$ for each G$_{\text{m}}$-cell in form of a Gaussian distribution curve. Thus, only specific G$_{\text{m}}$-cells are active for the signal to pass.

The principle and the design of the used Gaussian interpolator, shown in Fig. 4, is detailed in [7]. The internal control voltages V_{A} and V_{B} are in differential mode with an offset of 500 mV. They depend on a single-ended control voltage V_{cntrl}, which ranges from 0 V to 1 V.

$$V_{\text{A}} = V_{\text{cntrl}}, \qquad V_{\text{B}} = 1 \text{ V} - V_{\text{cntrl}} \qquad (1)$$

This allows the gain to be varied with only one tuning voltage. Thus, only one pad is needed.

To adjust the gate voltages of the transistors T$_0$ – T$_7$ in the range of V_{cntrl}, I_{b} is set to 5 µA and R_{b} to 12 kΩ. A tail current I_{tail} of 80 µA defines the overall provided bias current for all G$_{\text{m}}$-cells.

Since the circuit is implemented in a CMOS process, n-MOS transistors are used as current sources instead of bipolar transistors. However, the low threshold voltage V_{th} of the available devices and the limited headroom of the control voltages in such a low-voltage technology leads to an overlapping of the Gaussian-shaped bias currents. In order to guarantee a high dynamic range, it would be necessary to ensure that the tail current I_{tail} only flows through T$_0$ (for maximum gain) and T$_7$ (for minimum gain), respectively. If T$_0$ – T$_7$ have the same V_{th}, this is not the case, because the neighboring transistor is still active and draws current. To compensate this behavior, low V_{th} devices are used for T$_0$ and T$_7$ while T$_1$ – T$_6$ are implemented with a high V_{th} transistors. V_{th} of T$_0$ and T$_7$ is further lowered by connecting their back-gates to V_{DD}. Fig. 5 shows the trajectories of the resulting bias currents. In particular, $I_{\text{Gm,0}}$ and $I_{\text{Gm,7}}$ of the external G$_{\text{m}}$-cells are much steeper and reach higher maximum of 53 µA. Thus, a sufficient dynamic range of the VGA of 44 dB is achieved, 15 dB higher as if identical transistors would have been used for the Gaussian interpolator.

The load resistors $R_{\text{L}} = 2$ kΩ are essential for the input biasing of the op amp. R_1 and R_2, which provide a higher linearity via feedback, are 200 Ω and 20 kΩ, respectively. For offset cancellation caused by input referred offset voltages, the differential integrator technique of [7] is used. It needs to be implemented fully integrated without any external capacitor. For the offset cancellation sub-circuit the integrator components $C_{\text{int}} = 45$ pF and $R_{\text{int}} = 280$ kΩ are used. This

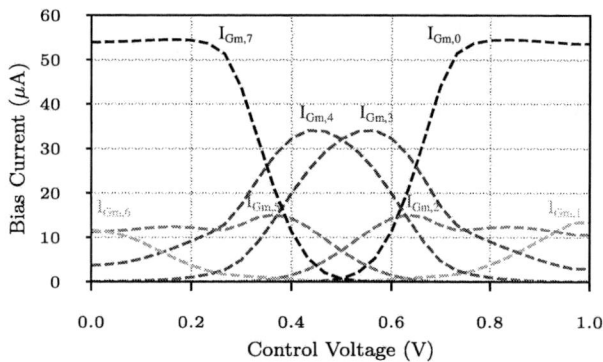

Fig. 5. Bias currents generated within the Gaussion interpolator with enhanced dynamic range versus control voltage V_{cntrl}.

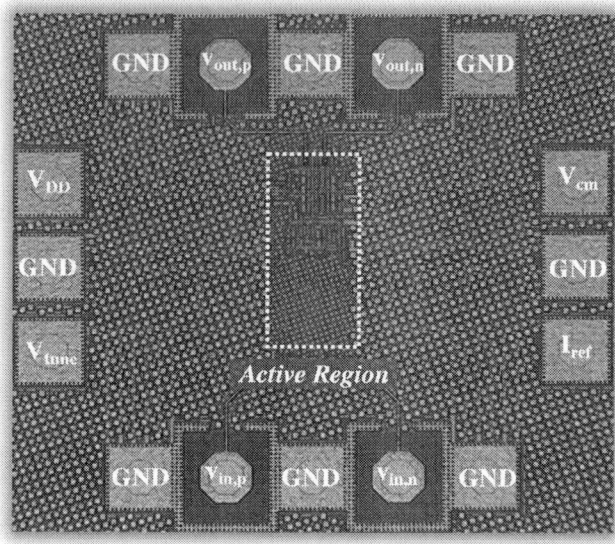

Fig. 6. Chip photograph of the manufactured baseband bandpass filter with an active region area of $0.017\,mm^2$ ($80\,\mu m \times 205\,\mu m$).

causes a high-pass characteristic with a cutoff frequency at 50 kHz, irrelevant for the baseband chain, since the high-pass characteristic of the bandpass filter is higher (130 kHz). The VGA draws 2.65 mA from $V_{DD} = 1.2\,V$.

III. MEASUREMENT RESULTS

The baseband bandpass filter and VGA were manufactured on standalone ASICs for individual characterization before being integrated in a radar system in GLOBALFOUNDRIES 22FDX®, a 22 nm FD-SOI technology. A photograph of the bandpass filter is displayed in Fig. 6, which has an active region area of $0.017\,mm^2$ ($80\,\mu m \times 205\,\mu m$). The active area of the VGA has a similar area of $0.015\,mm^2$ ($85\,\mu m \times 175\,\mu m$). The chips were characterized with a R&S® ZVA67 network analyzer under laboratory conditions on a wafer prober. S-parameter measurements were performed in a $50\,\Omega$ environment and transformed afterwards into a high impedance domain.

A. Bandpass Filter

The filter was measured in two different modes. The first two samples were measured in a two-port setup using BALUNs for differential to single-ended conversion at the in- and output. As shown in Fig. 7, the measured voltage gain in the the passband is about 2.5 dB lower than simulated, but overall the measured characteristics are in good agreement with the simulation. The second measurement mode in the four-port setup of sample number 03 demonstrates very good match between simulation and measurement. A suppression of 35 dB at sample frequency (28 MHz) was achieved. The measured high-pass also corresponds to the simulated curve, which indicates a high reliability of the technology models.

B. Variable Gain Amplifier

The dynamic range of the VGA was measured with the two-port setup. In the $50\,\Omega$ S-parameter measurement the dynamic range of 45 dB for the forward transmission gain is already demonstrated, as shown in Fig. 8. By converting those measurements into a high impedance environment, the

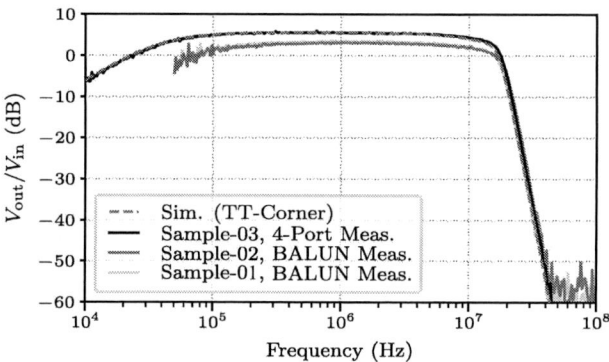

Fig. 7. Simulated and measured frequency response for different samples of the baseband bandpass filter using an 8_{th}-order Butterworth low-pass.

voltages gain ranges from -4 dB to 41 dB. An exemplary frequency characteristic for the maximum gain within the high impedance environment is displayed in Fig. 9. Due to the transformation and the de-embedding of the bandwidth-limited BALUNs some ripples occur but the consistency between simulation and measurement is clearly observable.

IV. CONCLUSION

Due to the different approaches for designing a mm-wave automotive radar, the specifications for the baseband chain differ from system to system. To classify the components presented in this work, Tab. II compares them with baseband chains of similar radar systems. The required bandwidth of this baseband chain is 2.8-times than in [2]. In addition, the slope of the 8th order low-pass filter is 1.6-times steeper than in [2], [3]. Nevertheless, the power consumption of 5.22 mW

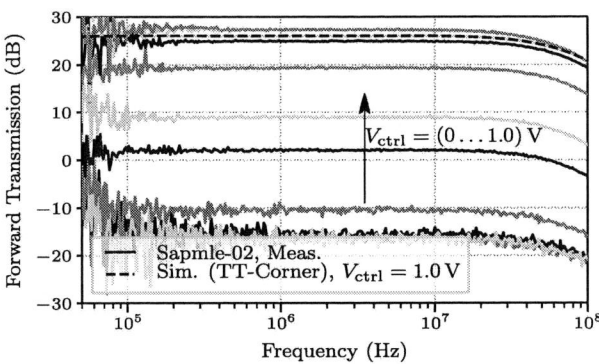

Fig. 8. Measured forward transmission of the VGA within a 50 Ω environment while sweeping the control voltage V_{cntrl} and simulated performance at $V_{\mathrm{cntrl}} = 1$ V.

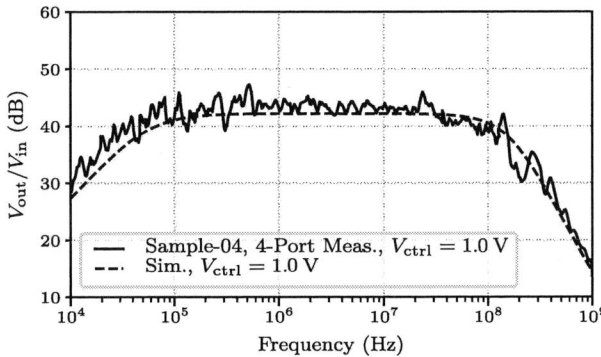

Fig. 9. Simulated and measured frequency response in a high impedance environment at maximum control voltage $V_{\mathrm{cntrl}} = 1$ V.

is similarly low. By means of the proposed dynamic range enhancement technique, the analog controllable VGA achieves a very high dynamic range of 45 dB. All other references have three PGAs stages implemented, which achieve slightly higher gain ranges of 51.4 dB to 54 dB. Although the requirements for the baseband components are similar or in some cases higher, they need only a fraction of the chip area. Since only $0.1\,\mathrm{mm}^2$ of area is available in the final radar chip, a compact design is essential and was a priority for this work.

The presented baseband components comply with the specifications of the FMCW mm-wave radar system. The circuit topologies used prove to be suitable and correspond to the behavior simulated in advance. Based on 22 nm FD-SOI technology, it was also possible to realize an energy efficient and an extremely compact design.

ACKNOWLEDGMENT

This work was funded via subcontract from GLOBAL-FOUNDRIES Dresden Module One within the framework Important Project of Common European Interest (IPCEI), by the Federal Ministry for Economics and Energy (BMWi) and by the State of Saxony. We would like to thank the colleagues

TABLE II
STATE OF THE ART COMPARISON OF ANALOG BASEBAND CHAINS FOR AUTOMOTIVE RADAR SYSTEMS.

Ref.	This	[2]	[3]	[8]
Process	**22 nm**	65 nm	65 nm	65 nm
Power (mW)	**5.22**	5.6	17.1	4.3
Stages	**1 VGA + 8th filter**	3 PGAs + 5th filter	3 PGAs + 5th filter	3 PGAs + 4th filter
Gain range (dB)	**2 – 47**	18 – 72	16.8 – 68.2	-3 – 51
Bandwidth (MHz)	**0.13 – 14 (3 dB)**	0.1 – 5 (1 dB)	0.5 – 5 (3 dB)	0.07 – 0.77 (3 dB)
Area (mm²)	**0.032**	≈0.28*	≈0.68**	0.52

*estimated from chip photograph in [9]; **estimated from chip photograph

from the Center for Tactile Internet with Human-in the-Loop (CeTI) for exchange of experiences in circuit design.

REFERENCES

[1] S. Kolodinski, C. Mart, W. Weinreich, V. Sessi, J. Trommer, T. Chohan, H. Mulaosmanovic, W. M. Weber, S. Slesazeck, B. Peng, C. Esposito, Y. Zimmermann, M. Schröter, X. Xu, P. V. Testa, C. Carta, F. Ellinger, S. Lehmann, M. Drescher, and M. Wiatr, "IPCEI subcontracts contributing to 22-FDX Add-On Functionalities at GF," in *ESSDERC 2019 - 49th European Solid-State Device Research Conference (ESSDERC)*, Sep. 2019, pp. 74–77.

[2] W. Wu, L. Zhang, L. Zhang, and Y. Wang, "A temperature-robust analog baseband chain for automotive radar," in *2017 International Conference on Electron Devices and Solid-State Circuits (EDSSC)*, Oct. 2017, pp. 1–2.

[3] H. Jia, L. Kuang, W. Zhu, Z. Wang, F. Ma, Z. Wang, and B. Chi, "A 77 GHz Frequency Doubling Two-Path Phased-Array FMCW Transceiver for Automotive Radar," *IEEE J. Solid-State Circuits*, vol. 51, no. 10, pp. 2299–2311, Oct. 2016.

[4] J. Squire, "Analog Filter Design Toolbox," 2021. [Online]. Available: https://www.mathworks.com/matlabcentral/fileexchange/9458-analog-filter-design-toolbox

[5] B. Gilbert, "A Low-noise Wideband Variable-gain Amplifier Using An Interpolated Ladder Attenuator," in *1991 IEEE International Solid-State Circuits Conference. Digest of Technical Papers*, Feb. 1991, pp. 280–281.

[6] M. El-Shennawy, N. Joram, and F. Ellinger, "A 55 dB range gain interpolating variable gain amplifier with improved offset cancellation," in *2016 12th Conference on Ph.D. Research in Microelectronics and Electronics (PRIME)*, Jun. 2016, pp. 1–4.

[7] ——, "Techniques for maximizing input handling and improving linearity of gain interpolating VGAs," in *2015 11th Conference on Ph.D. Research in Microelectronics and Electronics (PRIME)*, Jun. 2015, pp. 1–4.

[8] Y. Kim, S. Lee, and Y. Eo, "A 65 nm CMOS base band filter for 77 GHz automotive radar compensating path loss difference," in *2012 Asia Pacific Microwave Conference Proceedings*, Dec. 2012, pp. 688–690.

[9] L. Chen, L. Zhang, W. Wu, L. Zhang, and Y. Wang, "A Compact 76-81 GHz 3TX/4RX Transceiver for FMCW Radar Applications in 65-nm CMOS Technology," in *2019 IEEE Radio Frequency Integrated Circuits Symposium (RFIC)*, Jun. 2019, pp. 311–314.

SMACD / PRIME 2021 | 19 – 22 July 2021, Online Event

A Highly Linear High-Voltage Compliant Current Output Stage for Arbitrary Waveform Generation

Felix Schwarze, Florian Protze, Christian Matthus and Frank Ellinger
Chair for Circuit Design and Network Theory
Technische Universität Dresden, Germany
felix.schwarze@tu-dresden.de

Abstract—**In this paper, a highly linear high-voltage compliant current output stage is presented. The circuit enables the utilization of a conventional 8 bit low-voltage current-steering digital-to-analog converter in a high-voltage environment. Based on the improved active-feedback cascode current mirror topology, several adaptions were implemented to optimize the circuit towards high linearity and high bandwidth. The proposed circuit provides 167 MHz bandwidth, which to the authors' knowledge is the highest reported bandwidth for high-voltage compliant current mirrors. Moreover, with 0.49 LSB at 8 bit resolution it has, to the authors' knowledge, the highest reported linearity to date. Additionally, it is high-voltage compatible for output voltages of up to 60 V and provides the widest output current range of 10 mA maximum output current. At the same time, the power consumption of the utilized cascode control loop was reduced by 92 % compared to the original topology.**

Index Terms—**CMOS analog design, current mirror, high-voltage, Automotive Ethernet, PoDL, DAC**

I. INTRODUCTION

The current mirror is one of the most frequently used basic building blocks in analog and mixed-signal systems. In the vast majority of cases, current mirrors are utilized in biasing circuits. Here, their DC characteristics such as high accuracy and large output impedance are mandatory. Current mirrors can, however, also be utilized as output stages in various signal processing and conditioning units that require an accurate current output. If the required output current is non-static, the design has to fulfill both bandwidth and accuracy requirements that are usually contradictory. Advances in power electronics and high-voltage (HV) CMOS technologies during the last decade enabled the utilization of mixed-signal circuits in high-voltage environments. Corresponding applications are manifold and range from HV compliant implantable neural stimulators [1]–[4] to ultrasonic transducers for noninvasive material evaluation [5] and reconfigurable antenna arrays that are controlled by high-voltage digital-to-analog converters (HVDAC) [6]. For such applications, highly accurate current output stages are required. The deployment in high-voltage environments does however introduce HV compliance as additional design task that complicates the implementation of suitable current mirrors.

Parts of this research were supported by the BMBF (Bundesministerium für Bildung und Forschung, Germany), in the project fast power which is part of the FAST research cluster and the Zwanzig20 funding program.

A novel application is the utilization of a HV compliant current mirror as compensation current source in a mixed-signal noise-cancelling system for Power over Data Lines (PoDL) in Automotive Ethernet. A PoDL system enables the simultaneous transmission of electrical energy via differential DC voltages of typically 48 V and differential data signals over only one unshielded twisted pair of copper cables. Major benefits of PoDL systems are their extremely low cost and low weight as well as high data rates and network reconfigurability. To combine the high DC transmission voltage and the data signal, an inductive coupling network can be used. In the proposed use case, the output current stage is placed within the coupling network and digitally controlled by an adaptive finite impulse response (FIR) filter. Tracking the power supply noise, the FIR filter computes the current that is required to cancel the noise over the coupling network and controls the current source via a current-steering digital-to-analog converter (DAC) accordingly. While being connected to the differential DC transmission voltage, the current mirror output node has to withstand voltages of up to 60 V. The overall system concept is thoroughly described in [7].

In this paper, a HV compliant current output stage manufactured in an 0.18 μm HV CMOS technology is presented. The circuit is described in Section II along with a discussion of the corresponding system requirements. In Section III, measurement results including both DC and AC characteristics as well as a comparison with other related works are presented. The paper is summarized with a conclusion in Section IV.

II. CIRCUIT DESCRIPTION

The presented current mirror was designed to serve as a current output stage of a current-steering DAC that was implemented in the same 0.18 μm HV CMOS technology utilizing only low-voltage CMOS devices at a supply voltage of 1.8 V. The DAC is supposed to have a resolution of 8 bit and a maximum output current of 10 mA. The current output stage thus has to transfer this current into the high-voltage domain at a ratio of 1:1 without degrading the linearity of the DAC output. As the DAC is clocked at a maximum rate of 100 MHz, the bandwidth of the output stage should be in a similar frequency range and was set to exceed 150 MHz.

While HV compliance is a key design parameter, the circuit should still be reasonable small which implies the utilization of as few HV devices as possible. A well-suited topology is the

© VDE VERLAG GMBH · Berlin · Offenbach

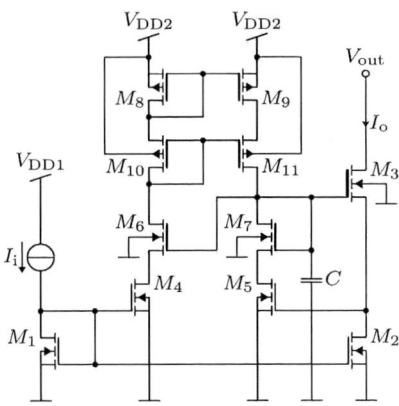

Fig. 1. Transistor-level schematic of proposed high-voltage compliant implementation of improved active-feedback cascode current mirror with $V_{DD1} = 1.8\,V$, $V_{DD2} = 5\,V$ and $V_{out} \leq 60\,V$.

improved active-feedback cascode current mirror proposed in [8]. In the presented work, the original topology was modified as described in detail below to suit the given requirements and is depicted in Fig. 1. The core of the current mirror comprises the low-voltage transistors M_1 and M_2 that are scaled equally and laid out in a two-column common centroid pattern to achieve maximum linearity at minimum sensitivity against manufacturing variances. To enhance the linearity and output resistance, the HV transistor M_3 is placed in a cascode configuration. As was proposed in [9], a HV device is used in place of a standard low-voltage device to enable HV compliance at the output node. The utilization of only a single HV device reduces additional layout efforts and area demands to a minimum. As can be derived from the chip photograph in Fig. 2, the HV transistor takes up nearly half of the overall circuit area, which emphasizes the need to keep the number of HV devices at a minimum.

The remaining transistors M_4-M_{11} form a control loop that significantly enhances the circuit's linearity. The control loop regulates the gate voltage of transistor M_3 taking the input and output current into account to keep the drain-source voltages of the transistors M_1 and M_2 equal. This prevents output current deviations due to channel length variations. As the HV transistor requires gate voltages that exceed $V_{DD1} = 1.8\,V$, the control loop is supplied with $V_{DD2} = 5\,V$. The transistors M_6-M_{11} therefore are low-voltage 5 V-devices while transistors

M_1, M_2, M_4 and M_5 are regular low-voltage 1.8 V-devices.

In the previous works regarding this circuit topology, the transistors M_4 and M_5 were scaled equal to transistor M_1, implying a control loop current that is twice the input current. However, for the proposed application in this work, the bandwidth requirements allow for a different approach. It was observed that a reduction of the control loop current mostly affects the circuit's bandwidth. Hence, the control loop transistor dimensions were optimized to enhance the circuit's efficiency while still achieving the required bandwidth of 150 MHz. By optimum scaling of the associated transistors, the control loop current was reduced by 92 %.

Additionally, a cascode pMOS current mirror (M_8 - M_{11}) was utilized in the control loop instead of a regular pMOS current mirror to increase the circuit's accuracy. This was possible due to the supply voltage headroom at $V_{DD2} = 5\,V$. Further, a capacitor C was placed at the gate of M_3 to prevent large overshoots at steep input current transients.

III. Measurement Results

The measurements presented below were carried out with two different circuit samples, referred to as A and B. Both chips were applied to a measurement printed circuit board (PCB) as shown in Fig. 3. Besides the chip and several connectors, a number of capacitors that were utilized to ensure low noise of the DC-supply voltages and a high precision high frequency resistor at the output node of the circuit were placed on the PCB. Unless stated otherwise, all of the following results were obtained at an output voltage of $V_{out} = 48\,V$. The presented measurements were also performed at different output voltage levels but are not shown due to insignificant deviations in the results achieved.

The DC transfer characteristics of the circuit were obtained from 0.1 mA to 10 mA with a step size of 100 µA and are shown in Fig. 4. As can be seen, the output current I_{out} strictly follows the input current I_{in} at a ratio of 1:1. A slight deviation can be obtained for circuit sample B towards higher currents. Here, the mirror ratio seems to be reduced due to manufacturing variations. These deviations are however not critical for the given application because both circuit samples provide a very high linearity. As can be seen in Fig. 5, the integral non-linearity (INL) of circuit sample A is less than or equal to 0.49 LSB. The INL of circuit sample B is less than or

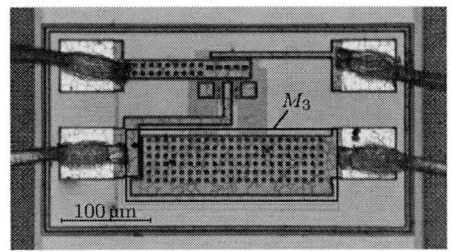

Fig. 2. Chip photograph of assembled circuit die on PCB. Transistor M_3 is the only HV device used in the circuit.

Fig. 3. Measurement PCB including test chip (pink frame).

© VDE VERLAG GMBH · Berlin · Offenbach

SMACD / PRIME 2021 | 19 – 22 July 2021, Online Event

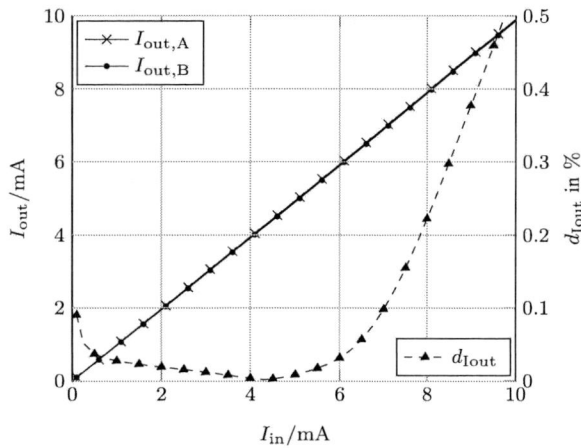

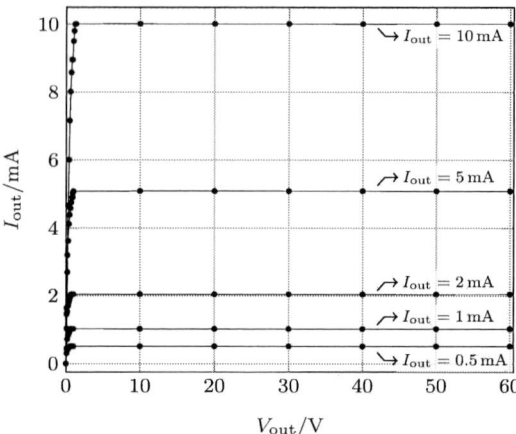

Fig. 6. Measured output characteristics $I_{out}(V_{out})$ of circuit sample A.

Fig. 4. Measured DC transfer function $I_{out}(I_{in})$ of circuit samples A and B acquired at $V_{out} = 48\,\mathrm{V}$ and simulated relative temperature-induced drift of output current d_{Iout} at constant input current levels I_{in} within a temperature range of $-40\,°\mathrm{C}$ to $150\,°\mathrm{C}$ (AEC-Q100 Automotive Grade 0) as defined in Eq. 1.

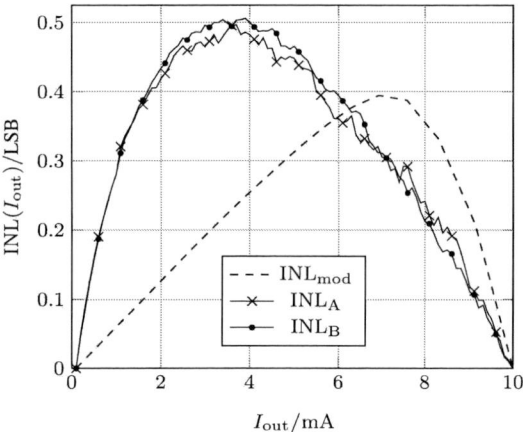

Fig. 5. Modelled and measured integral non-linearity $INL(I_{out})$ of circuit samples A and B acquired at $V_{out} = 48\,\mathrm{V}$ for an output current range of $10\,\mathrm{mA}$ and an assumed resolution of 8 bit, i.e a LSB step size of $I_{LSB} = 39\,\mu\mathrm{A}$.

equal to $0.51\,\mathrm{LSB}$. Given a resolution of 8 bit, this translates to a current deviation of less than $20\,\mu\mathrm{A}$ over the entire output current range. The achieved results indicate that the presented circuit is very well suited for the given application in terms of linearity. Fig. 5 does also show the expected INL_{mod} that was simulated with a circuit model including layout parasitics. Even though the courses of the graphs differ, the measured INL meet the expectations.

A reason for the varying courses of the INL graphs can be derived from the temperature dependence of the circuit that is illustrated in Fig. 4. Here, the simulated relative temperature-induced drift of the circuit's output current d_{Iout} defined by

$$d_{Iout} = \frac{I_{out}(I_{in}, T = -40\,°\mathrm{C}) - I_{out}(I_{in}, T = 150\,°\mathrm{C})}{I_{in}} \quad (1)$$

is shown. The circuit is very insensitive towards temperature

changes and thus well suited for Automotive applications. The maximum relative drift of the circuit's output current at constant input current levels is only $0.5\,\%$ over the whole AEC-Q100 Automotive Grade 0 temperature range of $-40\,°\mathrm{C}$ to $150\,°\mathrm{C}$. The figure also shows an absolute minimum around an input current of $4\,\mathrm{mA}$ and an increase of the relative drift for higher current values. This temperature variation due to self-heating is not considered within the INL simulation.

The aforementioned insensitivity towards changes in the output voltage can be derived from the measured output characteristics of circuit sample A shown in Fig. 6. The current stage turns on fully at an output voltage of $1\,\mathrm{V}$ and does not show any significant output current variations over the entire output voltage range of $1\,\mathrm{V}$ to $60\,\mathrm{V}$.

For the bandwidth measurements of the current output stage, a bias tee was utilized at the input node to simultaneously set the DC operating point and apply a sinusoidal signal at various frequencies. For the generation of this signal, a ROHDE&SCHWARZ SMA100-B signal generator was used. The current variation at the output node was measured as differential voltage over a sense resistor of $R_{sense} = 50\,\Omega$ with a ROHDE&SCHWARZ RTO-1044 digital oscilloscope. A sinusoidal input voltage was applied to produce a sinusoidal input current with an amplitude of $\hat{I}_{in} = 0.5\,\mathrm{mA}$ at any given frequency point above $50\,\mathrm{kHz}$. This lower frequency boundary results from the utilized bias tee and its application range between $50\,\mathrm{kHz}$ and $16\,\mathrm{GHz}$. In order to compensate any frequency dependent deviations of the sense resistor, an initial calibration measurement was performed without the current stage. The AC transfer characteristic $H_i(f)$ of the circuit was derived following the equation

$$H_i(f) = \frac{\hat{I}_{out}(f)}{\hat{I}_{in}(f)}, \quad (2)$$

with the measured output current amplitude $\hat{I}_{out}$. The bandwidth measurement was carried out at different DC input currents and output voltage levels yielding no significant variations. The presented measurement results were obtained

© VDE VERLAG GMBH · Berlin · Offenbach

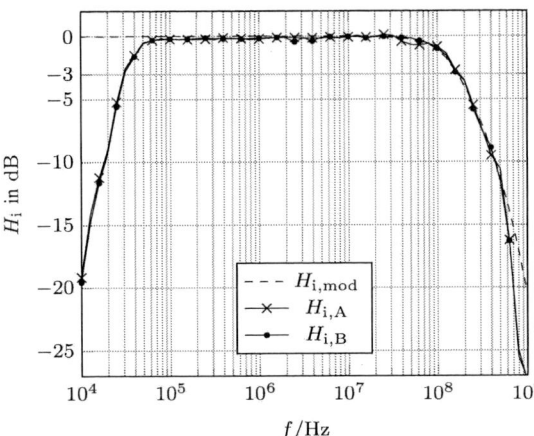

Fig. 7. Simulated transfer function $H_{i,mod}(f)$ and measured transfer functions $H_{i,A}(f)$ and $H_{i,B}(f)$ of proposed circuit acquired at an input current amplitude of $\hat{I}_{in} = 0.5\,\mathrm{mA}$ with DC operating conditions of $I_{out} = 5\,\mathrm{mA}$ and $V_{out} = 48\,\mathrm{V}$. Low-frequency deviations result from high-pass behavior of utilized bias tee.

at a DC input current of $I_{out} = 5\,\mathrm{mA}$ and an output voltage of $V_{out} = 48\,\mathrm{V}$ and are shown in Fig. 7. While the high-pass effect due to the utilized bias tee is visible for lower frequencies, both circuit samples exceed the set bandwidth requirements with a 3 dB-bandwidth of 167 MHz. It is worth emphasizing that both circuit samples not only behave almost identically, but do also agree very well with the simulation results $H_{i,mod}$ shown in the figure that predicted a 3 dB-bandwidth of 170 MHz.

From the simulation model, the output resistance could also be calculated over the entire frequency range. At a frequency of 100 kHz, the output stage delivers an output resistance of more than 1.7 MΩ, which is why an output resistance measurement could not be performed due to accuracy limitations in the measurement setup. The simulated output resistance is however very high over the entire frequency range, emphasizing the suitability of the circuit for the aforementioned noise-cancellation system.

A comparison of the demonstrated current output stage with related HV compliant current mirrors is shown in Tab. I. The presented circuit achieves the highest linearity while at the same time providing the widest output current range. The measured bandwidth considerably exceeds the previously reported works.

IV. CONCLUSION

The design of a highly linear HV compliant current output stage manufactured in an 0.18 μm HV CMOS technology was presented. It is utilized as compensation current source in a novel mixed-signal noise-cancelling system for Power over Data Lines (PoDL) in Automotive Ethernet. The design enables the utilization of a conventional low-voltage current-steering DAC in a HV environment without degrading the linearity of the 8 bit DAC output. Based on the improved active-feedback cascode current mirror topology, several adaptions were implemented to optimize the circuit for the given

application. Considering the system requirements imposed by the noise-cancellation system, the proposed circuit was optimized towards high linearity and bandwidth, while at the same time the power consumption of the cascode control loop could be reduced by 92 %. Compared to related works, the proposed circuit exceeds previously published HV compliant current mirrors in terms of output current range, linearity and bandwidth. In a future work, the presented current output stage will be combined with an 8 bit DAC on one chip to be utilized as digitally controlled current source in the aforementioned noise-cancelling system for PoDL systems.

TABLE I
COMPARISON OF PROPOSED CIRCUIT WITH RELATED WORKS

	[8]	[9]	[10]	This work
Voltage compliance	5 V	300 V	11.5 V	60 V
Maximum output current	500 μA	2 mA	1 mA	10 mA
INL (@8 bit)	0.77 LSB*	0.72 LSB*	0.8 LSB*	0.49 LSB
Bandwidth	35 MHz	-	-	167 MHz
Area	-	-	0.3 mm²	0.127 mm²
Technology	-	0.8 μm HV CMOS	180 nm CMOS	180 nm HV CMOS

*) Parameter calculated from data given in corresponding paper.

REFERENCES

[1] D. Osipov, S. Paul, S. Strokov, A. K. Kreiter, A. Schander, T. Tessmann, and W. Lang, "Current driver with read-out HV protection for neural stimulation," in *2016 IEEE Nordic Circuits and Syst. Conf. (NORCAS)*, 2016, pp. 1–4.

[2] E. Pepin, J. Uehlin, D. Micheletti, S. I. Perlmutter, and J. C. Rudell, "A high-voltage compliant, electrode-invariant neural stimulator front-end in 65nm bulk-CMOS," in *42nd IEEE European Solid-State Circ. Conf. (ESSCIRC)*, 2016, pp. 229–232.

[3] V. N. Tuan and H. Cha, "A standard CMOS neural stimulator IC with high voltage compliant output current driver," in *2017 Int. SoC Design Conf. (ISOCC)*, 2017, pp. 316–317.

[4] J. Uehlin, W. A. Smith, V. Rajesh Pamula, S. Perlmutter, V. Sathe, and J. C. Rudell, "A bidirectional brain computer interface with 64-channel recording, resonant stimulation and artifact suppression in standard 65nm CMOS," in *45th IEEE European Solid State Circ. Conf. (ESSCIRC)*, 2019, pp. 77–80.

[5] J. Borg and J. Johansson, "An ultrasonic transducer interface IC with integrated push-pull 40 Vpp, 400 mA current output, 8-bit DAC and integrated HV multiplexer," *IEEE Journal of Solid-State Circuits*, vol. 46, no. 2, pp. 475–484, 2011.

[6] J. Ning and K. Hofmann, "A 120 V high voltage DAC array for a tunable antenna in communication system," in *17th Int. Symp. on Design and Diagnostics of Elect. Circ. Syst. (DDECS)*, 2014, pp. 65–70.

[7] F. Schwarze, F. Protze, M. Kreissig, and F. Ellinger, "Design and latency analyses of an enhanced mixed-signal coupling network with adaptive noise cancellation for power over data lines," *IEEE Transactions on Vehicular Technology*, 2021, doi = 10.1109/TVT.2021.3060913.

[8] A. Zeki and H. Kuntman, "Accurate and high output impedance current mirror suitable for CMOS current output stages," *Electronics letters*, vol. 33, no. 12, pp. 1042–1043, 1997.

[9] E. Shoukry, M. Mony, and D. V. Plant, "Design of a fully integrated array of high-voltage digital-to-analog converters," in *2005 IEEE Int. Symp. on Circ. and Syst. (ISCAS)*, 2005, pp. 372–375.

[10] V. N. Tuan and H. Cha, "A standard CMOS neural stimulator IC with high voltage compliant output current driver," in *2017 Int. SoC Design Conf. (ISOCC)*, 2017, pp. 316–317.

A RISC-V-based System on Chip for High-Speed Control in Safety-Critical 650 V GaN-Applications

[1]Mike Richter, [1]André Lüdecke, [1]Yoon-Cue Lee, [1]Alexander Stanitzki, [1]Alexander Utz,
[2]Günter Grau, [1]Holger Kappert, [1,3]Rainer Kokozinski
[1]Fraunhofer Institute for Microelectronic Circuits and Systems (IMS), Finkenstr. 61, 47057 Duisburg, Germany
[2]advICo microelectronics GmbH, Münsterstr. 13-15, 45657 Recklinghausen, Germany
[3]Department of Electronic Components and Circuits, University of Duisburg-Essen, Bismarckstr. 81, 47057 Duisburg, Germany
E-Mail: Mike.Richter@ims.fraunhofer.de

Abstract— **In power electronics applications, Gallium Nitride (GaN)-based transistors generally offer benefits in efficiency and switching speed, but come at the cost of harder controllability and thus more complex control circuits. As a contribution towards easier and safer control for GaN-devices, we present a new system-on-chip (SoC) specifically dedicated to the control of GaN-based power modules. The main processing unit is compliant with the RISC-V specification and implements the RV32IMC instruction set. Major parts of a control algorithm may be performed by peripheral modules that reduce workload on the processor and enable low-latency reactions towards safety-critical events in the GaN-circuit. The output is generated by a PWM-unit that supports sub-nanosecond resolution. The processor achieves a benchmark score of 2.03 CoreMark per MHz. The SoC reacts to threshold violations on measurement inputs within 25 ns. Future steps will involve expanding the SoC with more safety features and the coupling of several SoCs for multiphase synchronization.**

Keywords— *RISC-V, Gallium Nitride (GaN), Power Module, System on Chip, Functional Safety*

I. INTRODUCTION

Due to the ever-rising requirements for energy efficiency, state of the art power electronics systems rely more and more on Gallium Nitride (GaN)-based high electron mobility transistors (HEMT). These components offer a high electric breakdown field [1, 2] along with an electron mobility significantly higher than their silicon-based relatives [1–3], making them preferable in high-frequency as well as high-efficiency contexts. However, technical limitations such as parasitic inductance [1] or the tendency towards unintentional turn-on when exposed to cross-talk phenomena [4, 5] complicate the electronics design, especially in applications where functional safety is a critical constraint. At the same time, the development of application-oriented processing units is seeing a tremendous updraft ever since the RISC-V movement started to provide open-source access to sophisticated instruction set specifications and thereby enabled the formation of an entirely new processor ecosystem [6]. As specifications, compilers and debugging tools are already available for free, system designers face way less hurdles when developing systems with their own integrated processors. Conclusively, state of the art GaN power devices have potential for improvement with respect to ease of application and risk limitation, which the conjunction with dedicated processing hardware may provide. Hence, we contribute a novel system on chip (SoC) architecture that brings open RISC-V hardware standards into the domain of 650 V GaN-based power devices. The SoC is outfitted with extensive analog and digital peripherals dedicated to the demands of GaN-based control circuits, allowing modular integration into a plethora of application circuits. In addition

to the regular control circuit, we implement a fast comparator-based decision logic that provides further options to implement circuit surveillance and ensure functional safety of the application system. The decision logic can be configured individually for each measurement signal, with actions ranging from raising a processor interrupt to full shutdown of the output signals towards the driver circuit. With this, we demonstrate the benefits of dedicated processors in GaN-applications. This paper is structured as follows: in chapter II, we summarize related work of the recent years. In chapter III, we give a broad overview of the system. Chapter IV describes the AIRI5C, our RISC-V microprocessor implementation. The digital control circuitry is explained in chapter V. In chapter VI, we present elements for an application control circuit. namely analog frontend, compensator and PWM generator. In chapter VII, we present information on synthesis and layout as well as performance characteristics. Chapter VIII concludes this contribution and gives an outlook to future plans with respect to the system.

II. RELATED WORK

In the research domain of efficient DC-DC converters, several publications propose ways to efficiently handle voltage regulation by a digital PWM controller [7–9]. The publications propose different architectures of digital control circuits to regulate small supply voltages below 10 V. Differences to our approach are, aside from different target operation voltages, the specialization on a certain conversion topology as well as the lack of an on-chip processor. A more similar approach to ours has been presented by Yu et al. [10]. The proposed smart gate driver implements a digital PWM controller as well as the gate driver to actuate a 600 V GaN HEMT co-packaged with the controller. A small, stack-based processing unit is used for setting the internal configuration bits on-the-fly. Major additions that we present with respect to this approach are (1) the integration of a RISC-V based microprocessor for higher computational versatility, (2) an extended analog frontend for measurements, (3) a trip control to react on threshold violations that may be critical to the system's safety and (4) a broadcast mechanism to switch all configuration bits synchronously.

III. SYSTEM OVERVIEW

A detailed block diagram of the SoC is depicted in Fig. 1. As hinted by the background color, the functional elements are assigned to different system domains on the chip. The Analog Frontend contains input buffers as well as fast comparators to indicate threshold violations on critical parameters. Values can be measured by three different ADCs. The Feedback ADC allows direct forwarding to the Compensator, which may be used to generate a manipulation value in a control circuit. The Compensator is a third-order Infinite Impulse Response (IIR)

This work received funding by the German Bundesministerium für Bildung und Forschung under grant no. 16ME0078K and is part of the international "19009 GaNext" project of the PENTA EUREKA cluster.

© VDE VERLAG GMBH · Berlin · Offenbach

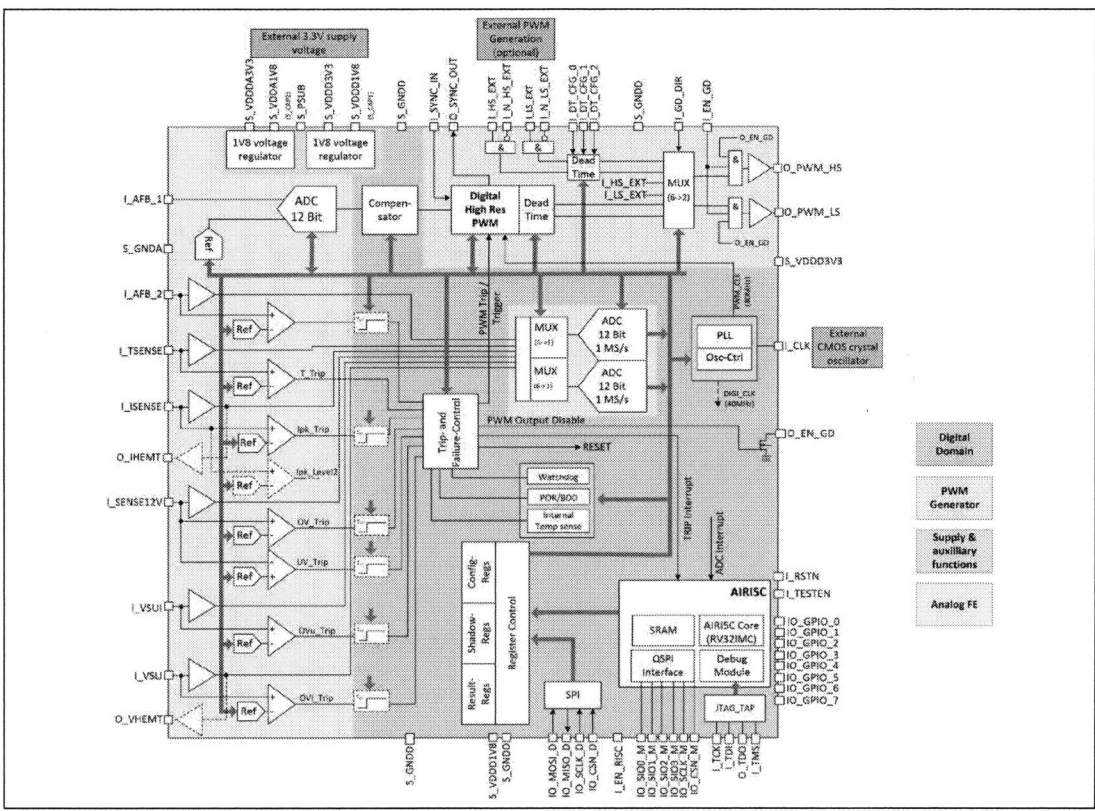

Fig. 1.   System Diagram of the GaN-Control SoC.

filter that provides the PWM Generator with the next duty cycle. The Digital Control Unit involves all entirely digital elements not belonging to the Processor or the PWM-Block. It facilitates the configuration registers of the system as well as the trip control, which takes safety measures in case of alarming signals from internal and external sources. Both the internal AIRI5C Processor and external microprocessors connected via SPI may access the digital control. The on-chip bus knows its master implicitly by the state of the EN_RISC input pin. The output generation is performed in the PWM-Block in the upper right corner. Auxiliary functions are found in light red boxes across Fig. 1 and feature the clock generation, internal voltage regulation, reset functionality, internal temperature surveillance and surveillance for clock failures (clock watchdog).

IV. AIRI5C MICROPROCESSOR

The AIRI5C is a single-core in-order microprocessor that implements the RV32IMC instruction set of the RISC-V Instruction Set Architecture (ISA). For debugging purposes, the processor features a debug module that can be accessed from the chip's JTAG port. Using a commercial debugger such as the *Olimex ARM-USB-Tiny-h*, the processor will be accessible by *OpenOCD*. The main part of the processor consists of four separate pipeline stages: (1) Decompression, (2) Instruction Fetch, (3) Decode/Execute and (4) Write Back. Fig. 2 illustrates the overall structure of the processor pipeline. For better clarity, the control and status registers (CSR), the control logic and the debug module have been omitted. In the Decompression stage, data from the memory interface is evaluated to detect compressed instructions in accordance to the RISC-V C-Extension. The Instruction Fetch stage is where

the actual decompression is performed, i. e. a compressed instruction is replaced by its 32-Bit equivalent and enters the Decode/Execute stage. RV32I-instructions can be executed by the processor's arithmetic-logical unit (ALU) in a single cycle. For instructions of the M-extension subset, an additional hardware multiplier as well as a divider may be accessed using a Pico Co-Processor Interface as seen in the open-source processor PicoRV32 [11]. A multiplication may stall the pipeline for five clock cycles while a division or remainder operation takes 32 clock cycles to complete. In the Write Back stage, data writes are performed into the register file or the data memory. As the AHB-Lite based on-chip bus to memory and peripherals needs two clock cycles to perform a write, a store instruction imposes a stall of one clock cycle onto the pipeline. Memory accesses can generally lead to three different storage units: QSPI-Flash memory (QSPI), SRAM and peripherals. The QSPI is meant to serve as a large non-volatile memory which holds e.g. a bootloader, the application program code and initialization data. The expectable drawback of the QSPI is the high latency (especially for write accesses). Hence, the actual program execution is supposed to happen in the Dual Port SRAM. Here, one instruction can be accessed per clock cycle while data accesses happen on the second port. The peripheral module registers as the third memory unit are a designated part of the memory map. The corresponding hardware is located in the digital control block outside the processor core. The registers are used to configure and read data from the analog frontend, PWM, safety features and others. To fully make use of the SoC architecture, the processor supports external interrupts from ADCs as well as the trip control logic described in chapter V.

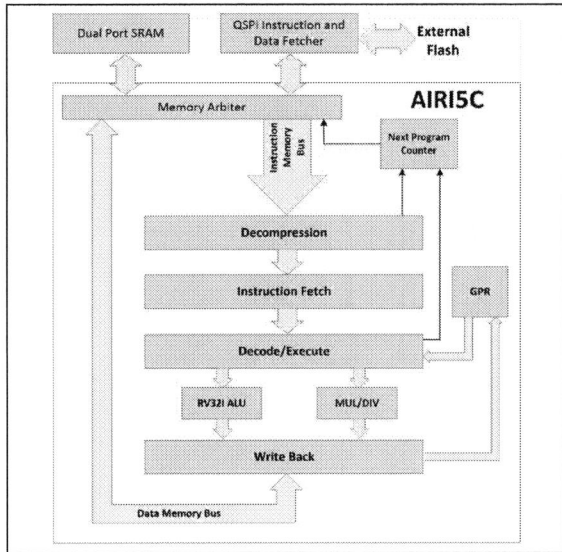

Fig. 2.   Pipeline structure of the internal AIRI5C processor.

V. DIGITAL CONTROL

The Digital Control Block comprises all registers for the configuration of the PWM Generator and analog devices and the implemented functions for trip control, GPIO input/output selection and ADC control and readout. The registers can be accessed through the on-chip bus using either the AIRI5C Processor or an external microprocessor connected to the SPI interface. To prevent concurrency issues when accessing registers, the peripheral on-chip bus is switched between "only SPI interface" and "only RISC-V" depending on the EN_RISC input. As an alternative to direct write accesses, all registers can be set to a shadow mode in which changes to their value do not directly take effect. New configuration values are instead saved in shadow registers. When writing to a certain broadcast address within the peripheral memory map, all registers are synchronously updated with their respective shadow register value. This feature is meant to improve initialization and may also be used to seamlessly change between different modes of operation with minimal risk of failure inducing temporary system states. Another crucial safety element is the trip control which evaluates the signal comparators of the analog frontend. Based on the configuration the trip control will react to a trip signal with six possibilities:

1. Reset of the SoC
2. PWM Output Disable
3. Assertion of a PWM trip signal (critical)
4. Assertion of a PWM trigger signal (non-critical)
5. Assertion of an interrupt flag for the AIRI5C
6. Start of an ADC conversion

There is a total of eleven possible trip signals that can be configured to assert any of the trip options.

VI. CONTROL CIRCUIT ELEMENTS

A. Analog Frontend

The analog frontend has two main tasks. The first task is to prepare the analog input signals for the optionally subsequent digital processing. Secondly, it must detect external error conditions in the shortest possible time and signal them to the trip control unit. If necessary, a predefined reaction can then be carried out for the respective error case. The analog frontend has a total of seven analog input pads. For six of these, the signal processing consists of one or two fast signal comparators and the option to multiplex the input signal to one of two 12-bit ADCs. The signal comparators have a maximum response time of 10 ns. The compare voltage can be configured individually for each comparator using an associated DAC. Two of the inputs have two independently configurable comparators providing the option to monitor two different thresholds. A comparator's output can be blanked for a certain period of time after the occurrence of a PWM edge via a programmable timer. This way, the trip control can ignore a comparator when an uncritical temporary threshold violation due to e.g. commutation noise occurs. Further processing of the resulting comparator signals is user-configurable as described in chapter V. For further digital signal processing the input signals can be converted by two 12-bit ADCs. Both the number of signals to be converted and the trigger of a conversion (successive, specific event) can be freely configured for each ADC. For example, it is possible to use one of the two ADCs to convert only one input signal at a time for a specific event in order to realize fast closed loop control. The other ADC may convert several other input signals or the same input signal at a different event. For the seventh analog input, the difference of the input voltage to an adjustable reference voltage level is converted by a third 12-bit ADC. The ADC in this path is reserved exclusively for this input to implement fast closed loop control. The execution of an ADC conversion is again configurable by the user in the same way as for the other two ADCs. The resulting output signal of the ADC can then be processed either by the integrated RISC-V or the following internal Compensator.

B. Compensator

Optionally, an integrated digital Compensator can be used as part of the closed-loop control for setting the PWM on-time to compensate the error voltage using a digital filter. For this, the output signal of the upstream ADC is used as the input value of the compensator, which in this case provides the error voltage between the configured reference voltage value and the resulting voltage value of the actual PWM cycle. The digital Compensator consists of an input multiplier, with which the 12-bit digital input value can be multiplied by a 16-bit scaling factor, followed by the 3rd order IIR filter, and an output multiplier, which in turn allows scaling of the 18-bit output value. The required number of poles and zeros in the digital filter depends on the application. The 3rd order IIR filter is based on second order sections (SOS structure), a second order IIR filter section is cascaded with a first order IIR filter section. In addition to the 16-bit scaling coefficients for the input and output values, the total of eight 16-bit filter coefficients is also user programmable, with the most significant bit being the sign bit.

C. PWM Generator

A digital PWM Generator has been developed by *advICo microelectronics GmbH*. The module is designed for a PWM frequency of up to 8 MHz and can be configured to drive either double-N switches or inverter style switches. It uses differential buffers to achieve duty cycles with resolutions below 200 ps and current-starved inverters for a sub-ns programmable interlock time to allow optimum efficiency without shoot-through in the power devices. For general use cases, the PWM allows the configuration of independent dead times between high-side and low-side outputs for each transition to support soft-switching and resonant switching applications. The PWM Generator can be configured by setting registers via the on-chip bus. The duty cycle to be used can be provided by the internal Compensator.

VII. RESULTS

The ASIC was designed and layouted in *X-FAB*'s 180 nm technology *XH018*. Digital parts were synthesized and routed using the *Cadence* Toolchain utilizing *Genus* for logic synthesis and *Innovus* for physical implementation (place & route). The digital part comprises 9,000 Flip-Flops and 21,000 logic gates. The overall chip-area is 18 mm^2 with a dimension of 4.5 mm x 4 mm. Fig. 3 shows the layout of the final ASIC. At the time of submission of this paper the chip is under fabrication at *X-FAB*. To assess the performance of the processor, the benchmarking software *CoreMark* has been performed on the simulation model as well as FPGA implementations of the core. Due to the limited size of the on-chip SRAM, the simulated benchmark has been performed using an ideal memory model of greater size. The access latency of both the ideal memory and the actual SRAM was one clock cycle, thus the score is not distorted by varying memory access latencies. Using the -o3 compiler flag, the benchmark score was found to be 2.03 CoreMark/MHz. As functional safety is an important concern for the SoC, we have analyzed the required time to react on threshold violations. Therefor we estimated the time until a violating internal or external signal has caused the desired trip signal to be risen. The six different trip signals have been listed in chapter V. We based our estimations on simulated delays of the analog design as well as the latency of the digital circuit. As illustrated in Table I, the delay of a trip signal within the digital domain is expected to be around 15 ns for internally measured signals and 25 ns for external signals. The additional 10 ns arise from the propagation delay we expect of the analog comparators. For both internal and external signals, we expect 15 ns of worst-case digital latency to correctly synchronize the signals into the digital clock domain. With respect to the processor's branch delay we estimate the worst-case latency between the execution of the first instruction within an Interrupt Service Routine (ISR) and internal and external signal violations to be 115 ns and 125 ns respectively.

VIII. SUMMARY AND OUTLOOK

A system architecture for a PWM-Control in safety-critical GaN applications has been proposed. The system extends comparable approaches by implementing a versatile RISC-V microprocessor as well as trip signals to quickly react on threshold violations within the GaN-circuit. Using shadow registers, the control system can be reconfigured synchronously. Key data of the synthesis results, processor benchmark and estimated delay of trip signals have been provided. Future work on the SoC architecture will investigate the power and area overhead for extending the on-chip processor into a dual core lockstep design. Based on a safe-island approach, the most critical features of the SoC will be modified to increase their reliability and error resilience. With respect to the system's versatility, our plans for the next iteration of the SoC also involve coupling and synchronizing PWM outputs of multiple SoCs to improve support of multiphase applications.

Fig. 3. Layout of the System-on-Chip.

TABLE I.   ESTIMATED WORST-CASE DELAY OF TRIP SIGNALS.

Source	Worst-case delay Time [ns]	
	Trip Signal	*ISR*
Internal (e.g. On-chip Temperature)	15	115
External analog input	25	125

REFERENCES

[1] K. J. Chen *et al.*, "GaN-on-Si Power Technology: Devices and Applications," *IEEE Trans. Electron Devices*, vol. 64, no. 3, pp. 779–795, 2017, doi: 10.1109/TED.2017.2657579.

[2] T. Oka, "Recent development of vertical GaN power devices," *Jpn. J. Appl. Phys.*, vol. 58, SB, SB0805, 2019, doi: 10.7567/1347-4065/ab02e7.

[3] M. A. Khan, J. M. van Hove, J. N. Kuznia, and D. T. Olson, "High electron mobility GaN/Al x Ga 1– x N heterostructures grown by low-pressure metalorganic chemical vapor deposition," *Appl. Phys. Lett.*, vol. 58, no. 21, pp. 2408–2410, 1991, doi: 10.1063/1.104886.

[4] E. A. Jones, F. F. Wang, and D. Costinett, "Review of Commercial GaN Power Devices and GaN-Based Converter Design Challenges," *IEEE J. Emerg. Sel. Topics Power Electron.*, vol. 4, no. 3, pp. 707–719, 2016, doi: 10.1109/JESTPE.2016.2582685.

[5] J. Wang, D. Liu, H. C. P. Dymond, J. J. O. Dalton, and B. H. Stark, "Crosstalk suppression in a 650-V GaN FET bridgeleg converter using 6.7-GHz active gate driver," in *ECCE 2017: IEEE Energy Conversion Congress & Expo : Cincinnati, Ohio, October 1-5*, Cincinnati, OH, 2017, pp. 1955–1960.

[6] RISC-V International, *RISC-V*. [Online]. Available: https://riscv.org/ (accessed: Mar. 5 2021).

[7] E. Abramov, T. Vekslender, O. Kirshenboim, and M. M. Peretz, "Fully Integrated Digital Average Current-Mode Control Voltage Regulator Module IC," *IEEE J. Emerg. Sel. Topics Power Electron.*, vol. 6, no. 2, pp. 485–499, 2018, doi: 10.1109/JESTPE.2017.2771949.

[8] S.-Y. Kim *et al.*, "Design of a High Efficiency DC–DC Buck Converter With Two-Step Digital PWM and Low Power Self-Tracking Zero Current Detector for IoT Applications," *IEEE Trans. Power Electron.*, vol. 33, no. 2, pp. 1428–1439, 2018, doi: 10.1109/TPEL.2017.2688387.

[9] F. G. R. Ramos, T. C. Pimenta, and L. H. C. Ferreira, "Design of a low-cost and high-performance digital PWM controller for DC-DC converters," in *LASCAS 2018: 9th IEEE Latin American Symposium on Circuits and Systems : Puerto Vallarta, Mexico : February 25-28 : proceedings*, Puerto Vallarta, 2018, pp. 1–4.

[10] J. Yu, W. J. Zhang, A. Shorten, R. Li, and W. T. Ng, "A smart gate driver IC for GaN power transistors," in *Proceedings of the 30th International Symposium on Power Semiconductor Devices and ICs: May 13-17, 2018, Chicago, USA*, Chicago, IL, 2018, pp. 84–87.

[11] GitHub, *cliffordwolf/picorv32*. [Online]. Available: https://github.com /cliffordwolf/picorv32#pico-co-processor-interface-pcpi (accessed: Mar. 2 2021).

© VDE VERLAG GMBH · Berlin · Offenbach

An Approach to Online Wear Out Monitoring of PCB Interconnects in Safety-Critical Systems

Saeid Yazdani
dept. of ETIT
TU Chemnitz
Chemnitz, Germany
saeid.yazdani@etit.tu-chemnitz.de

Werner Wolz
dept. of ETIT
TU Chemnitz
Chemnitz, Germany
werner.wolz@etit.tu-chemnitz.de

Rainer Engelhardt
Steinbeis Center of Design and Test
Steinbeis GmbH
Chemnitz, Germany
rainer.engelhardt@stw.de

Christian Schott
dept. of ETIT
TU Chemnitz
Chemnitz, Germany
christian.schott@etit.tu-chemnitz.de

Ulrich Heinkel
dept. of ETIT
TU Chemnitz
Chemnitz, Germany
ulrich.heinkel@etit.tu-chemnitz.de

Daniel Kriesten
dept. of Engineering
HS Mittweida
Mittweida, Germany
kriesten@hs-mittweida.de

Abstract —**Online monitoring inside electronic control units (ECU) is mandatory for all safety-critical systems. Dedicated *quality and safety monitoring structures* will be described that support the early detection of health problems. Interconnect wear-out and delamination are major failure causes in printed circuit boards, and sensitive detector arrangements allow for the estimation of near-future fail situations. Countermeasures like pre-emptive service actions can keep the system safe without running into emergency situations (i.e. activating redundancy), and the number of system emergency states can be reduced. This paper suggests methods for online monitoring of wear-out detectors to increase reliability and safety.**

Keywords— *Safety, PCB wear-out, FR4, wire, state-of-health, delamination, interconnect failure, via fatigue, defect prediction.*

I. INTRODUCTION

Safety and reliability are key requirements for all automotive or airborne electronics. The printed circuit board PCB is by far the component with the largest impact on quality, reliability, and safety: Their interconnect quality should offer long-term stability, and wiring must not show any fatigue or wear-out under extreme environmental conditions (mission profile). Varying air pressure in airplanes as well as shocks from road potholes must not lead to a degradation. The growing desire for increased power density leads to smaller dimensions, and wires or via contacts must stay reliable even at miniature size. High-speed designs for radar and lidar applications have introduced high-precision waveguides, and geometrical variations of board parameter's lower reliability and safety. Therefore, board parameters should be monitored to detect wear-out of via contacts or delamination of sheets that could introduce signal distortion until functionality is no longer available. We suggest the introduction of PCB state-of-health monitors that enable wear-out detection far from the end of life of all operational parts.

II. PCB FAILURE CAUSES AND MODES

A. Failure Causes

The ever-increasing complexity of printed circuit boards due to availability of high pin count integrated circuit in smaller packages forces the layout designers to use thinner traces and small diameter through hole vias as well as blind and buried vias extensively.

Present-day designs have the tendency to get smaller in footprint as a requirement usually passed from industrial engineers to layout engineers. The trend in mobile phone industry clearly indicates the move to high numbers of layers

in PCBs and the use of blind and buried vias result in HDI (High Density Interconnect) printed boards and emerge of related standards such as IPC-2226.

These features are not trivial to manufacture and once produced successfully, are susceptible to degradation induced by stress. Therefore, each added layer in a PCB can increase the chances of delamination and blistering.

The main causes of interconnect failure can be expressed in the following categories:

- Environmental stress such as extended durations or multiple repetitions of temperature and humidity cycles.

- Mechanically induced stresses like vibration and shock.

- Electrical stress such as voltage spikes and current surges.

The above-mentioned failure causes manifest as cracks in vias and traces and copper separation from substrate as shown in Fig. 1 and Fig .2. In extreme cases, these failures can lead to a total break-down of vias and traces, resulting in total breakdown of the electronic system. Fig. 1 and Fig.2 shows examples of such failures manifesting as cracks in vias and delamination on surface of a PCB.

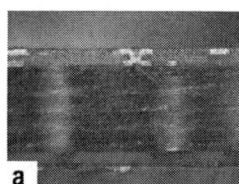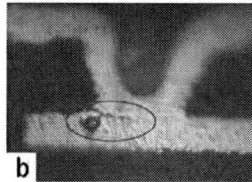

Fig. 1. Via cracking due to stress as illustrated in [1, Fig. 4]. (a) appearance of the via and (b) 40 times maginification.

Fig. 2. Copper separation as illustrated in [2, Fig. 7].

© VDE VERLAG GMBH · Berlin · Offenbach

In case of environmental stress, the most dominant factor is the difference in CTE (Coefficient of Thermal Expansion) between insulator and conductor materials of a printed board. Copper alloy, as the most prominent conductor, has a CTE of 18 ppm/K whereas for typical FR4 laminates this number is between 14 to 17 ppm/K in X and Y axis and up to 60 ppm/K in Z axis.

Conversely, epoxy resin used as binding material could have a CTE of 30 ppm/K on the lower end and 40 ppm/K on the upper end. While the manufacturing techniques are seemed to be mature enough to account for these mismatching CTE numbers (e.g. usage of reinforcing woven glass on the resin) risks of uneven expansion of materials exist and to make it worse, the amount of expansion in X, Y and Z axis are not usually uniform.

In case of PTH (Plated Through Holes) vias and other variants of vias (e.g. blind and buried vias), the main cause of failures are significant differences in the magnitude of thermal expansions between copper and surrounding glass epoxy [3]. CTE of a few selected common base materials used in PCB production are shown in Tab. 1.

TABLE I.   CTE of common PCB base materials

Material	Tg ª (°C)	CTE (ppm/K)		
		X	Y	Z
FR4 135	135	17	17	55
FR4 150	150	17	17	52
FR4 170	170	15	15	45
ISOLA FR406	170	13	13	60
ISOLA FR408	180	13	13	60
Arlon AD250C	250	16	16	50
Taconic RF-35	259	9	9	120
Rogers 4003	280	11	14	46

ª· Glass Transition Temperature.

Fig. 3 Depicts cracks in the barrel of a PTH via as a direct result of mismatching CTE between copper, base material and epoxy resin after prolonged exposure to thermal cycling.

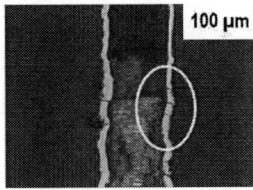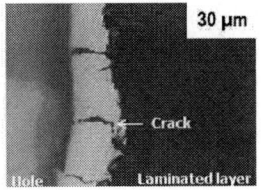

Fig. 3. Copper crack in a via barrel [3, Fig. 4].

B. Failure Modes

In printed circuit boards, the interconnect failures initially propagate as a resistive path between two traces or planes [4] and if left unattended, the resistance will increase over time by the continuation or repetition of the root cause, resulting in abnormally high resistance values and finally can resolve in a totally open circuit of the conductor under stress.

As it can be seen in Eq. 1, resistance (R) of a conductor in a simple form is directly proportional to its resistivity (ρ), length (L) and cross section area (A).

$$R = \frac{\rho L}{A} \text{ [Ohms]} \qquad (1)$$

From (1) it is immediately obvious that any changes in length or cross section of the conductor results in a change in its resistance. A crack in any given cross-section of an inductor reduces the effective travel path of electrons in that particular area, effectively creating a bottleneck. Any increase in the number of cracks in the course of the conductor sums up as increased total resistance of the conductor. These cracks can eventually breakoff completely and result in a total disconnect. Therefore, the worst-case scenario for an interconnect in a PCB is fail open as it is the case in Fig. 3. In some applications, even a slight increase of the resistance of interconnects can directly affect the expected performance of the system. For example, an interconnect which is part of an RF filter stage, can shift the filter response drastically.

Resistance is not the only electrical characteristic that might change due to external effects. Capacitance and inductance are other characteristics of an interconnect or conductor plane that will change based on the above-mentioned stress causes. Based on (2), change in capacitance is only significant with regard to conductive planes in a PCB, as the thin interconnecting traces often does not provide enough surface area to produce a meaningful capacitance.

$$C = \frac{\varepsilon A}{d} \text{ [Farads]} \qquad (2)$$

Where:

ε = complex permittivity

d = distance between planes of two planes of area A

Inductance is another important characteristic of a conductor. Layout designers try to lower parasitic inductance of conducting path by keeping the interconnects as short as possible and keeping the loop area at a minimum. When using vias to connect different layers of a PCB, inductance of vias becomes significant. It is a common practice to use multiple vias in parallel for ground return paths if the target power plane resides on a different layer (e.g. for decoupling capacitors close to an integrated circuit) and in this case a damaged via increases the total inductance. Increasing numbers of serial vias when routing a signal between multiple layers could result in more susceptibility to stress induced failures as a breakdown of any via causes the interconnect to fail open.

III. Failure Monitoring and Prediction Schemes

A. State of the art

IST (Interconnect Stress Testing) is one of the known methods for identifying failures in vias and multilayer interconnects. IST uses DC current induced thermal cycling for accelerated aging of a target PCB and its interconnects to provide knowledge and reliability estimation of the target PCB interconnects[5]. IST is an offline method operating on a test coupon which is in turn a sample material of the PCB of the product that will be deployed in the field. Test coupons are multilayer arrangement of vias and traces arranged in a small area with headers for connecting to a standard or custom made IST tester (Fig. 4).

TCT (Thermal Cycling Test) is another method of inducing stress and detection of changes of PCB

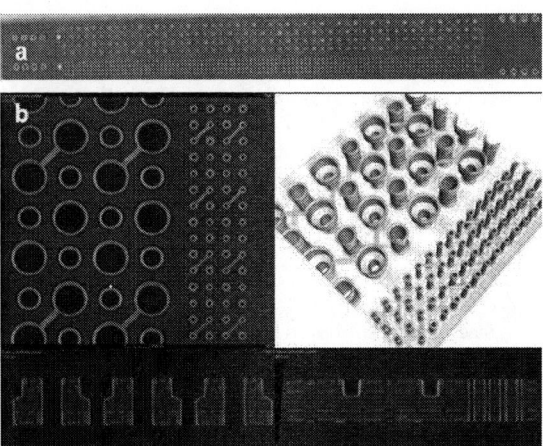

Fig. 4. (a) A Typicall coupon for IST and (b) 3D model and cross-section of the same coupon as illustraded in [5, Fig. 2] and [5, Fig. 3].

interconnects in a PCB. In TCT, the stress is applied in a thermal chamber (heating oven) instead of using forced direct current into the interconnects to increase temperature.

B. IST Performance

Test coupons used in IST testing should be from the same production lot, as part of the production panel. The placement of the test coupon in the panel can lead to variations in the IST test result as uniformity over the whole panel is rather difficult to maintain.

While IST can identify weaknesses both in design and fabrication processes and provide means for correcting the problems, as an offline method, it cannot limit the risk of poor interconnects escaping into the final products [6] delivered to the customer or detect the flaws during operational phases of the system [7].

Another issue with IST is that thermal stress and thermal cycling is induced through interconnects, i.e. transfer of heat is from the interconnect to the surrounding substrates, whereas in real-world environment the heat or temperature cycling could come from an external source affecting the entire PCB as a whole.

Furthermore, IST generally considers 10% increase in resistance of a interconnect after a fixed number of thermal cycling as a failure [8]. The exact amount of resistant increase while the product is in use might be more valuable to the end-user or maintenance operators.

C. TCT Performance

In contrast to IST, TCT approximates real-world conditions thanks to use of thermal chambers, yet it also is considered an offline, pre-release, test of product samples.

IV. ONLINE MONITORING AND WEAR OUT DETECTION

Both IST and ICT methods are offline methods for identifying interconnect problems before the final product is in the field. Today's safety-critical application areas call for a method that can monitor and predict the upcoming failures far before an unrecoverable situation can occur.

Our solution is based on in-circuit interconnect resistance monitoring within the PCB itself. High precision resistance measurement is available thanks to current sources and CSA

(Current Sense Amplifier) integrated circuits. In this approach, the resistance measurement is part of the final product to allow online monitoring of interconnect resistance at any given time.

The interconnect under test will effectively become a shunt resistor of a CSA IC. An adjustable constant current source is used to apply current into the interconnect as shunt resistor, allowing the CSA to measure the voltage drop across the interconnect and calculate the resistance of the interconnect with high precision.

Fig. 5 depict the proposed structure. A known current is applied to the interconnect under test (R1) and the voltage drop caused by the current moving through the resistor is measured by a CSA. A high precision CSA with digital interface (e.g. I2C) is used to allow logging of the measured resistance in a microcontroller.

The logged data overtime are processed for reasoning about the state of health of the interconnect under test. The CSA is configured in high-side mode and an optional load resistor (R2) is necessary if the CSA requires a minimum load that cannot be satisfied by the nominal resistance of the interconnect under test alone. A few centimeters of trace with tens of vias can already reach well above 1Ω, which is an acceptable shunt value for most of the CSA chips in the market.

In this arrangement, the interconnect is defined as a dummy trace on the board and serves no other purpose than being a state-of-health representative of the entire PCB. The real-estate of a PCB is precious but du to the availability of current sources and CSA in the form of integrated circuits in small packages we can implement such structure in any PCB. If the space constraint allows for it, this structure can be added to multiple places of the PCB, especially in the vicinity of components that have a high likelihood of overheating (e.g. DC-DC Converters, output drivers, etc.).

This method is capable of monitoring interconnects as part of the system (e.g. supply rails or signal traces). In this case, relays or other kind of switches can be used to connect and disconnect the interconnect to switch between current source and CSA line or intended signal line. In such an arrangement the resistance of the interconnect can be measured at system start and stored to a non-volatile memory. These data can be either automatically processed by firmware or reported to maintenance operator.

Using this structure, in conjunction with a few strategically placed interconnects can be representative of the overall health status of the entire PCB.

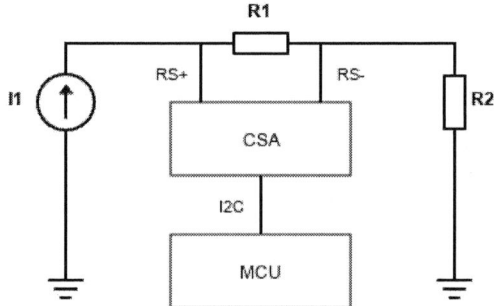

Fig. 5. Proposed measurement structure for a dummy trace.

V. IMPLEMENTATION, TESTING AND RESULTS

Fig. 6 shows the implemented structure on a two-layer FR4 (Tg 135°C) PCB and inclusion of via chains. We used LM134 [9] as a current source in zero temperature coefficient mode and a MAX9611[10] as CSA. The PCB was used in a basic test to assert reliability of the measurement structure. An MCU was used to continuously record resistance of the dummy traces during the test. The test started at room temperature and hot air was applied to the dummy traces in multiple steps until failing open (Fig. 7). The constant current source was operated in pulse mode, i.e. the current is only applied to the traces at the time of sampling for measurement.

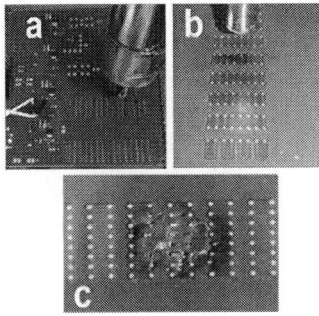

Fig. 6. First implemntation of the concept. (a) Test PCB with source and measurement ICs (b) A moment in the middle of basic destructive test and (c) the final result of the destructive test where the dummy trace failed open as expected.

After proof of concept, a 12-layer test coupon with a total of 25 discrete traces was developed (using reinforced FR4 with Tg of 185°C and Z expansion factor of 40 ppm/K) as seen in Fig. 8. 24 of these traces contain between 50 and 200 PTH vias in series, routed through various layers and the remaining trace connects all pins of a BGA socket in series. Fig. 8 shows a especial version of the measurement structure in order to allow measurements of larger test coupons. Test coupon and measurement structure were placed inside a climate chamber and stressed for a total duration of one week. The thermal cycling was set to 25°C to 125°C (maximum rated operational temperature of on-board ICs) once per hour

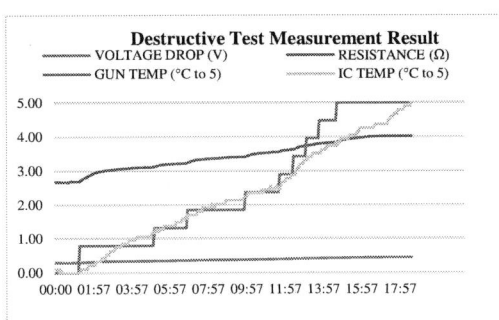

Fig. 7. Result of the basic destructive test. *X* Axis represetns the resistance in Ohms and *Y* axis represents the time elapsed sicne test start. The hot air temperature is normalized to 5 to fit in the chart. At the peak of the test the hot air temperature was set to 500°C. after 16 minutes, the dummy trace failed open and test was stopped (the rest of datapoints are the last measured resistance before failure). IC Temperature refers to the onboard temperature sensor of MAX9611 IC in the vicinity of the focused heat and registered peak temperature of 74°C. Resistance of dummy trace at room was 2.67Ω and moments before sudden failure reached to 4.16Ω, almost doubling the initial value.

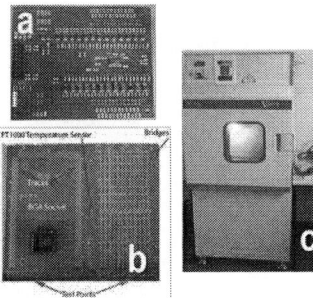

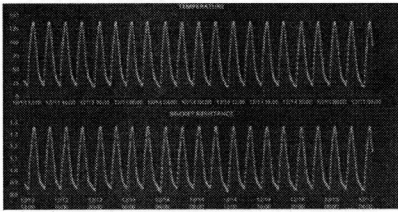

Fig. 8. (a) Measurement PCB (b) Special Coupon (30cm by 30cm) and (c) climate chamber.

with rapid cooldown. After one week of cycling, the test coupon did not show any signs of degradation, thanks to high Tg of 185°C and relatively low Z expansion factor. Fig. 9. Shows a snapshot of this measurement session.

Fig. 9. Snapshot of measurement data. Top: PCB Temperature captured by PT1000 sensor (22°C to 125°C, peak at 130°C). Botom: resistance of one of the dummy traces (0.85Ω to 1.35Ω). If the test trace is to wear out, higher highs and higher lows are anticipated for.

VI. CONCLUSION

Our proposed method of online monitoring of interconnects can be thought as a compliment to IST method, allowing continuous, online, monitoring of interconnect health in PCBs after deployment of the product in the field. Miniature size current sources and smart current sense amplifiers allow for in-circuit monitoring of live or dummy traces and vias and can help with identifying early signs of wear out in interconnects, preventing potential hazards attributed to interconnect failure in the operation field.

REFERENCES

[1] L. Ji, and Z. Yang, "Analysis on cracking blind vias of PCB for mobile phones," ICEPT-HDP, 2008.

[2] D. Slee, J. Stepan, W. Wei and J. Swart, "Introduction to printed circuit board failures," IEEE PSES, 2009.

[3] N. Park, J. Kim, C. Oh, C. Han, B. Song, and W. Hong, "Fatigue life prediction of plated throu holes (PTH) under thermal cycling," IEEE EMPC, 2009.

[4] Y. Lu, Z. Ming, M. Liang and C.Zhang, "Failure case analysis for abnormal delamination of printeed circuit board," QR2MSE, 2019.

[5] Y. Yang, "Reliabilities and failure analysis of printed circuit boards inteerconnect stress test," ICEPT, 2018.

[6] E. Arthur, C. Busa, W. Goldman and A. Grubbs, "Review of interconnect stress testing protocols and their effectiveness in screening microvias," IPC APEX EXPO, 2016.

[7] M. Yeh, and D. Jiang, "An interconnect stress testing (IST) and themral cycle test (TCT) approach on printed circuit board reliability," IMPACT, 2007.

[8] IPC-TM-650 Test Method Manual, Number 2.6.26.

[9] LM134, Adjustable Current Source, Texas Instruments.

[10] MAX9611, Current Sense Amplifier, Maxim Integrated.

© VDE VERLAG GMBH · Berlin · Offenbach

Experimental Investigation of Dielectric Loss Induced Noise in Charge Detection Systems for Cosmic Dust

Sebastian Kelz, Markus Grözing, Manfred Berroth
Institute of Electrical and Optical Communications Engineering
University of Stuttgart
Stuttgart, Germany
sebastian.kelz@int.uni-stuttgart.de

Abstract—**In this paper the influence of dielectric loss induced noise on a differential charge detection system for cosmic dust trajectory sensors is analyzed. For the sake of generality the distributed charge detector is replaced by compact capacitors with different dielectric materials. It is shown that for low-frequency high-impedance systems the selection of very low loss dielectric materials is of crucial importance to minimize the overall noise charge of the system.**

Index Terms—**dielectric loss, noise, charge sensitive amplifier, dielectric dissipation, flicker noise, cosmic dust**

I. INTRODUCTION

Charge detection is widely used in very diverse areas of application. While classical applications include x-ray detectors and particle detectors, purely capacitively coupled neural signal amplifiers can also be considered as charge amplifiers [1]. A once again very different application is the charge induction based trajectory detection of charged particles in space [2]. The two main noise sources which are typically considered when optimizing the noise performace of charge sensitive amplifers (CSAs) are the thermal and flicker noise of the input transistor and the thermal noise of the feedback resistor of the amplifier [3].

In applications with leaky (semiconductor based) detectors leakage currents of the detector and their shot noise likely dominate the overall noise power at low frequencies and relax constraints on the value of the feedback resistor.

In this paper the focus is on purely capacitive detectors, which show significantly lower leakage currents, dominated by the gate leakage of the amplifier input MOSFET and insulator leakage currents. These detectors enable the use of significantly higher feedback resistors in the Teraohm domain [4]. Therefore thermal noise of the leakage-path resistances and the feedback resistor becomes less important. Under these conditions dielectric loss appears as a new relevant and potentially dominant noise source [5] [6].

The goal of this paper is to define the conditions, under which dielectric loss induced noise (DLN) becomes relevant

Partially supported by Deutsche Forschungsgemeinschaft (DFG) grant BE 2256/31-1.

for low noise CMOS charge sensitive amplifiers for charge based cosmic dust particle sensors.

First the concept of DLN [7] and DLN in charge amplifiers [5] will be reviewed. The second part will show measurements of DLN for different insulators materials used in past trajectory detector prototypes.

II. NOISE IN LOSSY DIELECTRICS

It is well known that a resistor generates a thermal noise power which can be described by a parallel noise current source with the spectral density given by:

$$S_{I_R}(f) = \frac{4\,k\,T}{R} = 4\,k\,T\,G \tag{1}$$

with the Boltzmann constant k and the absolute temperature T. The actual loss mechanism of the resistor thereby is of no importance. Additional noise could arise [8], but is not considered here.

By repeating the simple thought experiment stated by Nyquist [9], it can be shown that dielectric loss actually generates the same thermal noise as any other electrical loss mechanism. In this case the experimental setup consists of a resistor and a lossy capacitor which are electrically connected by ideal wires. Both components are thermally isolated from each other and have the same temperature T. For this equilibrium condition the second law of thermodynamics states, that the temperatures of the two components have to stay the same. Therefore the noise power released by the resistor and being dissipated by the lossy capacitor has to be exactly the same as the noise power given off by the lossy capacitor and being absorbed by the resistor.

This condition has to hold at any frequency [9], as any type of lossless filter could be added between the two components without affecting the outcome of the experiment.

It follows, that any type of electrical loss generates the same noise as its equivalent loss resistance at any given frequency. A simple noise model of a lossy capacitor therefore consists of an ideal capacitor, a frequency dependent parallel equivalent loss resistance and the corresponding noise source of this

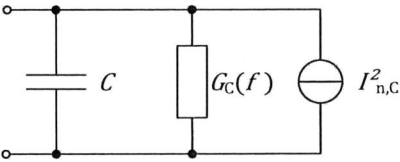

Fig. 1. Equivalent schematic of a lossy capacitor with the capacitance C. Note that G_C represents the dielectric loss of the capacitor and not a DC-leakage. Therefore $\lim_{f \to 0} G_C(f) = 0$.

resistance, as shown in Fig. 1 [7]. For a capacitor with the loss tangent $\tan \delta$ the equivalent loss conductance can be derived from [7] as:

$$G_C = \omega \, C \tan(\delta) \qquad (2)$$

For the equivalent current noise power spectral density (PSD) it follows:

$$S_{I_{n,C}} = 4 \, k \, T \, \omega \, C \tan(\delta) \qquad (3)$$

Assuming a constant value for $\tan \delta$ in the observed frequency range, the resulting current noise PSD increases proportional to the noise frequency. It is interesting to note that this behavior is quite the opposite to the one observed in high-k gate stacks, which show a $1/f$-like gate current noise PSD at low frequencies [10]. The $1/f$-noise in high-k gate stacks is more similar to excess noise in resistors, which occurs only if an external current is applied to the capacitor, while the DLN is purely thermal and does not require an additional power source.

Refering the current noise PSD of the capacitor to its open circuit terminal noise voltage, the result is [7]:

$$S_{U_C} = \frac{4 \, k \, T \tan \delta}{\omega \, C \, (1 + \tan^2 \delta)} \approx \frac{4 \, k \, T \tan \delta}{\omega \, C} \qquad (4)$$

This equation shows an $1/f$-like behavior, if $\tan \delta$ is assumed to be constant. While $\tan \delta$ of dielectric materials is not typically constant over the frequency, its slope is usually significantly flatter than $1/f$, e.g. [11]. It is therefore reasonable to assume an $1/f$-like slope for the open circuit noise voltage in a limited frequency range.

III. DIELECTRIC LOSS IN CHARGE DETECTION SYSTEMS

A classic capacitive-feedback based CSA is given in Fig. 2. The detector itself is represented by the current source i_{in} and the lossy detector capacitance C_{det} with a DC-leakage-resistance R_{leak}. The CSA consists of a feedback-coupled voltage sensitive amplifier with a feedback capacitance C_f, a feedback resistance R_f and the input capacitance C_{in}.
For the frequency range of interest it holds, that $R_f >> \frac{1}{\omega C_f}$, therefore the feedback can be considered as primarily capacitive.

Both the detector capacitance C_{det} and the feedback capacitance C_f contribute DLN-current to the input of the amplifier. The noise potentially caused by the input capacitance of the amplifier is not considered in this paper and is subject to

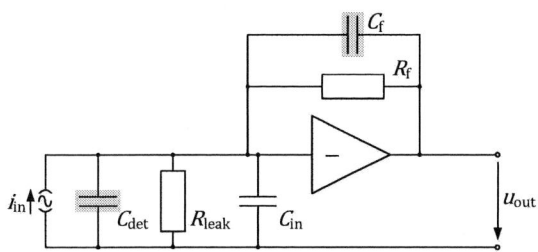

Fig. 2. Simplified schematic of a classic CSA. Capacitors with grey background (C_{det} and C_f) are considered to be lossy and can be represented by the equivalent circuit given in Fig. 1.

further investigation. For the two other capacitors the noise current can be assumed to be directly flowing into the input node of the amplifier (given a simplified voltage amplifier with zero output impedance). The equivalent input noise current PSD due to DLN is then given by:

$$S_{I,\text{DLN,input}} = 4 \, k \, T \, \omega \left(C_{\text{det}} \tan(\delta_{\text{det}}) + C_f \tan(\delta_{C_f}) \right) \quad (5)$$

For CSAs the noise of the system is usually expressed as an equivalent rms noise charge, which represents the output rms noise voltage of the CSA, divided by the charge gain of the CSA. For the sake of generality a spectral approach is taken here. The output voltage noise PSD is divided by the charge gain to yield the equivalent noise charge PSD (ENC-PSD). By filtering and integrating this value, the corresponding ENC can be calculated.

If an ideal voltage amplifier is assumed and R_f is neglected inside the considered frequency range, the equivalent noise charge PSD for DLN is given by:

$$S_{\text{ENC,DLN}} = \frac{4 \, k \, T \, (C_{\text{det}} \tan \delta_{\text{det}} + C_f \tan \delta_f)}{2 \, \pi \, f} \qquad (6)$$

The result again shows an $1/f$-slope. In a typical CSA design it holds that $C_{\text{det}} >> C_f$, indicating that C_{det} is more likely to dominate the overall DLN. On the other hand the dielectrics of both capacitors are typically made from different materials. Therefore C_f is not automatically irrelevant. To be able to distinguish DLN from regular thermal noise during measurements, we also consider the noise contribution of the feedback and leakage resistances with the corresponding ENC-PSD:

$$S_{\text{ENC,R}} = \frac{4 \, k \, T}{R_f \cdot (2\pi f)^2} + \frac{4 \, k \, T}{R_{\text{leak}} \cdot (2\pi f)^2} \qquad (7)$$

Therefore the leakage resistors generate an $1/f^2$-slope PSD, which can be easily distinguished from the $1/f$-slope of the DLN.

Considering flicker noise only, it is well known, that the optimal effective ENC is obtained for the matching condition $C_{\text{in}} = C_{\text{det}} + C_f$ [3]. This condition is not affected by DLN [5], as $S_{\text{ENC,DLN}}$ is not a function of the the input transistor parameters. Therefore the influence of DLN on the ENC of

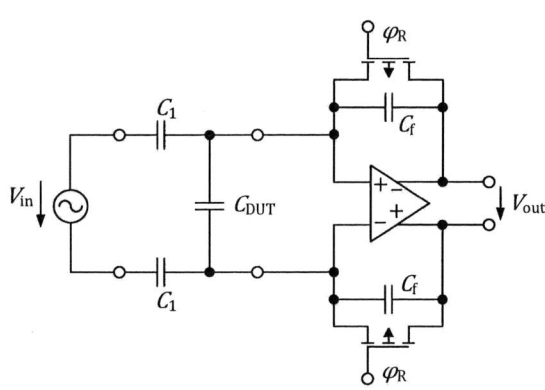

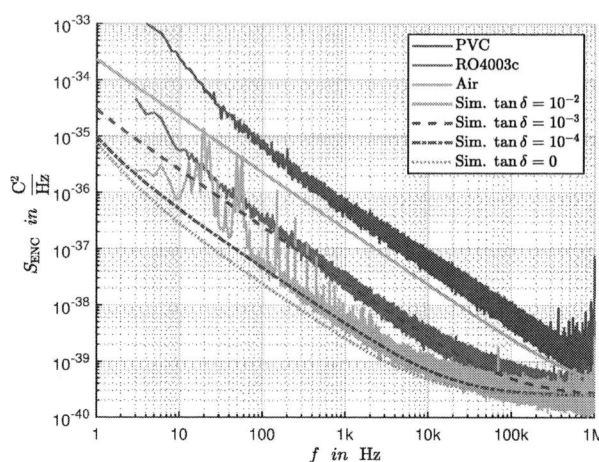

Fig. 3. Simplified measurement setup. The input charge signal is derived from the capacitively coupled differential signal generator V_{in}. The output voltage V_{out} is fed into a high input impedance active high-pass filter and then recorded using a digital storage oscilloscope. Prior to each measurement, the amplifier is reset by applying a reset pulse to ϕ_{R}. The value of the test capacitors is $C_1 = 1.1\,\text{pF}$. The feedback capacitors have a value of $C_{\text{f}} = 100\,\text{fF}$

Fig. 4. Measured ENC-PSD for the compared dielectric materials and simulation results. In each case $C_{\text{DUT}} + C_{\text{ext,para}} = 8.6\text{pF}$. The simulation results include both DLN and regular flicker-noise from the amplifier. The measurements of the air-capacitor are distorted at low frequencies due to mechanical oscillations.

the amplifier can only be reduced by selecting the appropriate dielectric materials with a low loss tangent.

IV. Measurements

A. Measurement Setup

Measurements of DLN are conducted using an amplifier as given in [4], but with an increased MOSFET input capacitance of $C_{\text{in,diff}} = 3.75\,\text{pF}$. The resulting measurement setup is schematically shown in Fig. 3. During noise measurements C_1 is left floating to reduce the influence of its lossy dielectric on the noise measurement results. The remaining contribution of C_1 to the input capacitance and all further external parasitic capacitances are summed up to one equivalent differential capacitance $C_{\text{ext,para}}$ when measuring the noise voltage.

Three different dielectrics, air, a Rogers RO4003c PCB material and a generic (hard) PVC material are compared for the DUT capacitor with $C_{\text{DUT}} \approx 8.0\,\text{pF}$.

The DUT capacitor setup consists of two parallel copper plates with an adjustable gap. The plates are suspended on two short traces on a RO4003c PCB, which connect the capacitor to the input of the CSA and C_1. RO4003c is choose for the PCB due to its low dielectric loss and high resistivity.

To reduce electrical interference and acoustic noise, the entire design is placed in a solid aluminum case. As the DUT capacitor relies on the PCB for its structural integrity, it is not possible to measure its capacitance without the parasitic capacitance of the PCB-traces and stray capacitances between the copper plates of the DUT and the grounded case of the setup.

Therefore the value of C_{DUT} cannot be measured and it is attempted to keep the total external capacitance, including parasitics, at a constant value of $C_{\text{ext,para}} + C_{\text{DUT}} = 8.6\,\text{pF}$.

While the contribution of the PCB-traces and C_1 to $C_{\text{ext,para}}$ is constant, very slight variations of the remaining parasitic capacitances will occur due to the different spacing of the DUT plates required for different dielectric materials.

The measurements for each dielectric material are conducted in three steps: Fist a layer of the dielectric material to be evaluated is placed between the plates of the DUT. The thickness of this layer is then adjusted until the measured value of capacitance is equal to $C_{\text{ext,para}} + C_{\text{DUT}} = 8.6\,\text{pF}$.

In the second step the input of the CSA is bonded to the DUT by wire bonding. The output of the CSA is connected to an active high-pass filter to further amplify the signal. The amplified noise signal is then measured by a digital storage oscilloscope.

In the final step both C_1 capacitances are connected to a function generator by wire bonding to measure the transfer function. The ENC-PSD is then calculated by dividing the output noise voltage through the previously measured charge gain transfer function.

In addition to measurements, simulation results are shown which are based on a schematic model of the CSA and a corresponding test-bench, implemented using lossy capacitors with the stated dielectric loss tangent. The feedback capacitors are considered to be ideal in all simulation runs, which is an acceptable approximation for our results, due to the small relative contribution of C_{f} to the overall capacitance.

B. Results

The measurement results are shown in Fig. 4. An approximately tenfold increase of the noise power in the 1/f-domain between the different test cases is apparent. For the comparison of PVC and RO4003c this roughly corresponds to the limited data available on the loss tangents of the two materials [12]

© VDE VERLAG GMBH · Berlin · Offenbach

[13], confirming the theoretical calculations, with RO4003c showing a $\tan\delta \approx 10^{-3}$.

For the test case of the air-capacitor the measurements show approximately twice the noise power simulated for an ideal capacitor. Part of this deviation is caused by the DLN of $C_{\text{ext,para}}$, which in parts contains RO4003c as its dielectric, but should not cause a dielectric loss above $\tan\delta = 7 \cdot 10^{-5}$. Further reasons are process corner related inaccuracies of the transistor flicker noise model or DLN of the amplifier input capacitance. Still, the results show a good accuracy of the transistor flicker noise model at high impedances.

Referring to the design of the charge detection sensor, the air capacitor setup shows a noise level comparable to a simulated capacitor with $\tan\delta = 10^{-4}$ and a significant contribution by $C_{\text{ext,para}}$. Therefore, for the presented setup, the dielectric loss of the charge detector should stay substantially below $\tan\delta = 10^{-4}$ to avoid a DLN-dominated design. The observed change in the slope of the noise PSDs for frequencies below approximately 10 Hz is due to resistive noise as given by eq. (7).

V. CONCLUSION

The measurement results clearly indicate the importance of DLN for the design of ultra low noise charge sensitive systems. Even if only a small fraction of the electric field of a charge detector passes through a high loss material like PVC-insulated cable, the noise performance of the system can be severely degraded. It should be noted that DLN can only be observed for very high DC-resistance systems, as otherwise thermal noise of the leakage resistance given by eq. (7) will dominate the low frequency noise.

As indicated by eq. (6), the ENC caused by DLN is not a function of the CSA, therefore mathematically it does not affect the CSA-ENC-optimization problem. However, the individual contributions of the detector and the CSA to the overall ENC should strongly affect the share of resources invested into optimizing the individual components of the system.

To minimize the DLN of a detector, insulators with a low dielectric constant and a very low dielectric loss, like vacuum or PTFE, should be employed. A further trade-off occurs between the amount of suspension material used inside the detector (causing DLN) and the rigidity of the electrodes, with mechanical oscillations becoming more severe in low rigidity designs.

REFERENCES

[1] S. Schroder, C. Cecchetto, S. Keil, M. Mahmud, E. Brose, O. Dogan, G. Bertotti, D. Wolanski, B. Tillack, J. Schneidewind, H. Gargouri, M. Arens, J. Bruns, S. Szyszka, S. Vassanelli, and R. Thewes, "CMOS-compatible purely capacitive interfaces for high-density in-vivo recording from neural tissue," in *2015 IEEE Biomedical Circuits and Systems Conference (BioCAS)*. Piscataway, NJ: IEEE, 2015, pp. 1–4.

[2] S. Auer, E. Grün, R. Srama, S. Kempf, and R. Auer, "The charge and velocity detector of the cosmic dust analyzer on Cassini," *Planetary and Space Science*, vol. 50, no. 7-8, pp. 773–779, 2002.

[3] P. O'Connor and G. de Geronimo, "Prospects for charge sensitive amplifiers in scaled CMOS," *Nuclear Instruments and Methods in Physics Research Section A: Accelerators, Spectrometers, Detectors and Associated Equipment*, vol. 480, no. 2-3, pp. 713–725, 2002.

[4] S. Kelz, T. Veigel, M. Grözing, and M. Berroth, "A fully differential charge-sensitive amplifier for dust-particle detectors," in *2018 14th Conference on Ph.D. Research in Microelectronics and Electronics (PRIME)*, July 2018, pp. 13–16.

[5] V. Radeka, "Field effect transistors for charge amplifiers," *IEEE Transactions on Nuclear Science*, vol. 20, no. 1, pp. 182–189, 1973.

[6] B. G. Lowe and R. A. Sareen, *Semiconductor x-ray detectors.* Boca Raton, FL: CRC Press, 2014.

[7] A. van der Ziel, "Flicker noise in electronic devices," in *Advances in electronics and electron physics*, ser. Advances in Electronics and Electron Physics, C. Marton and L. Marton, Eds. New York, N.Y: Academic Press, 1979, vol. 49, pp. 225–297.

[8] F. Hooge and A. Hoppenbrouwers, "1/f noise in continuous thin gold films," *Physica*, vol. 45, no. 3, pp. 386–392, 1969.

[9] H. Nyquist, "Thermal agitation of electric charge in conductors," *Physical review*, vol. 32, no. 1, p. 110, 1928.

[10] P. Magnone, F. Crupi, G. Giusi, C. Pace, E. Simoen, C. Claeys, L. Pantisano, Maji, V. R. Rao, and P. Srinivasan, "1/f noise in drain and gate current of MOSFETs with high-k gate stacks," *IEEE Transactions on Device and Materials Reliability*, vol. 9, no. 2, pp. 180–189, 2009.

[11] W. B. Westphal and A. Sils, "Dielectric constant and loss data," Massachusetts Inst. of Tech. Cambridge Lab for Insulation Research, Tech. Rep., 1972.

[12] *RO4000® series datasheet*, Rogers Corporation, 2018.

[13] L. A. Utracki, J. A. Jukes, "Dielectric studies of poly(vinyl chloride)," *Journal of Vinyl Technology*, vol. 6, no. 2, 1984.

Generalized comparison of the accessible emission limits of flash- and scanning LiDAR-systems

Roman Burkard
Department of Electronic Components and Circuits
University of Duisburg-Essen
Duisburg, Germany
roman.burkard@uni-due.de

Reinhard Viga
Department of Electronic Components and Circuits
University of Duisburg-Essen
Duisburg, Germany
reinhard.viga@uni-due.de

Jennifer Ruskowski
Department of CMOS Image Sensors
Fraunhofer Institute for Microelectronic Circuits and Systems
Duisburg, Germany
jennifer.ruskowski@ims.fraunhofer.de

Anton Grabmaier
Department of Electronic Components and Circuits
University of Duisburg-Essen
Duisburg, Germany
and
Fraunhofer Institute for Microelectronic Circuits and Systems
Duisburg, Germany
anton.grabmaier@uni-due.de

Abstract—In the field of autonomous driving and human-robot collaboration applications the demand for three-dimensional imaging systems, that are reliable, small and low-cost, is rising. A promising technology to satisfy these demands are scanning- or flash-based light detection and ranging (LiDAR)-systems, which differ mainly in the illumination of the field-of-view. A scanning LiDAR-system illuminates the field-of-view sequentially by deflecting a laser beam. In a flash LiDAR-system the laser beam is extended to illuminate the whole field-of-view with every emitted laser pulse. Both illumination principles are extensively treated in the recent literature separately and without the inclusion of the limits defined by the laser safety standard IEC 60825-1:2014. In this work a generalized model is derived from the standard. This model is able to determine the emission limits of the standard for both LiDAR-systems at the same time and it is used to compare the maximum output power and the intensities in the field-of-view for both LiDAR-systems.

Index Terms—light detection and ranging (LiDAR), scanning LiDAR, flash LiDAR, IEC 60825-1, eye-safety, accessible emission limit

I. INTRODUCTION

In the field of autonomous driving and human-robot collaboration applications the demand for systems, that are reliable, small and low-cost, is expected to rise in the next years [1]. A promising technology to satisfy these demands are light detection and ranging (LiDAR)-systems, which employ the direct time-of-flight measurement principle. This measurement principle is based on measuring the time between the emission and the reception of the reflection of a short and high-power laser pulse. The main advantage of LiDAR-systems is the ability to offer higher spatial and angular resolutions than other commonly used three-dimensional imaging techniques, such as stereo vision, RADAR and ultra-sonic. Furthermore, distant low reflective objects can be detected. [2]

Currently there are mainly two illumination principles for LiDAR-systems. These are the so-called flash- and scanning LiDAR-systems. A scanning LiDAR-system consists usually of a macro or micro mechanical mirror to deflect a laser beam. This creates a scanning trajectory, which is used to sequentially illuminate points or areas of a field-of-view. The recent advantages in the fabrication of large photodetector matrices offered the possibility to create flash LiDAR-systems. These illuminate the whole field-of-view with every emitted laser pulse and the photodetector matrix is used to create a three-dimensional image of the scene.

Both illumination principles utilize a high-power laser diode to illuminate the field-of-view. It is well known from the commonly used LiDAR-equation (1) that higher laser peak powers P_L yield a higher returned power P_R at the detector, where R is the distance to the target and Ω_L is the solid angle of the laser beam. The remaining terms of equation (1) account for the target cross section σ, which describes how the power is reflected by the illuminated target, the receiver geometry A_{Rec}, the losses of the optics η_{opt} and the atmospheric attenuation η_{atm}. [3]

$$P_R = P_L \cdot \frac{\sigma}{R^2 \Omega_L} \frac{A_{Rec}}{\pi R^2} \eta_{opt} \eta_{atm} \qquad (1)$$

The aforementioned improvement of the measurement range by increasing the laser output power is extensively treated in the recent literature mostly for both illumination principles separately. Furthermore, the inclusion of the limits, defined by the standard IEC 60825-1:2014, is typically omitted. The following presents a generalized calculation and comparison of the accessible emission limits (AEL) defined by the standard. To achieve this the standard IEC 60825-1:2014 is briefly

© VDE VERLAG GMBH · Berlin · Offenbach

summarized and a method is derived to estimate the AEL of a scanning LiDAR-system without the detailed evaluation of the utilized scanning trajectory. Using the limits defined by the standard and the derived method a comparison between the AELs of a flash and a scanning LiDAR-system is made at the end of this paper.

II. Standard IEC 60825-1:2014

The standard IEC 60825-1:2014 "Safety of laser products Part 1: Equipment classification and requirements" is used to classify laser systems that emit laser radiation in the wavelength range from 180 nm to 1 mm. The standard defines four different classes and only laser systems classified as class one laser products are considered eye-safe under normal operating conditions. The classification process of a laser product must include the full range of operation modes and it shall consider any reasonably foreseeable single-fault condition during operation. A laser product emitting a single wavelength, with a narrow spectral range, is assigned to a class when the accessible laser radiation exceeds the AELs of all lower classes but does not exceed that of the class assigned. [4]

To account for the different operation modes and determine the AEL there are three criteria to be evaluated. These criteria depend on the parameters of the output beam of the laser system and limit the energy of the laser pulse E_L. The energy E_L can be determined if the laser pulse duration τ_p, the temporal and spatial laser pulse shape, the pulse repetition frequency PRF and the peak power P_{Peak} are known. To evaluate the emitted laser energy the standard proposes a measurement geometry as depicted in figure 1. A circular measurement aperture with a diameter of 7 mm is placed 100 mm away from the laser source. Using the terms and definitions of the measurement geometry, the criteria for the determination of the AEL can be briefly summarized as follows. [4]

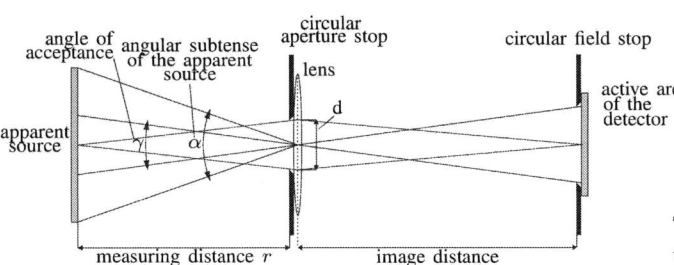

Fig. 1. Measurement geometry proposed by the standard IEC 60825-1:2014 [4]

1. Single pulse criterion

The single pulse criterion limits the energy of a single laser pulse. The AEL is calculated using equation (2) and the correction coefficients C_4 and C_6 which can be determined using equation (3) and equation (4) respectively. [4]

$$\text{AEL}_{\text{single}} = 7.7 \cdot 10^{-8} \cdot C_4 \cdot C_6 \tag{2}$$

$$C_4 = 10^{0.002(\lambda - 700)} \tag{3}$$

$$C_6 = \begin{cases} 1 & \text{for } \alpha \leq \alpha_{\min} \\ \alpha/\alpha_{\min} & \text{for } \alpha_{\min} \leq \alpha \leq \alpha_{\max} \\ \alpha_{\max}/\alpha_{\min} & \text{for } \alpha_{\max} < \alpha \end{cases} \tag{4}$$

2. Average power criterion

The average power criterion limits the energy of a pulse train with an emission duration T. This energy may not exceed the AEL_T of a single pulse with duration T. To determine the emission duration T the emission pattern of the laser system has to be considered. If every pulse of the pulse train hits the measurement aperture with a fixed pulse repetition frequency, it is considered to be a regular emission pattern and it is sufficient to average over the emission duration T. If the emission pattern is irregular, for example by deflecting the laser beam with a macro or micro mechanical mirror, not every laser pulse hits the measurement aperture and the most restrictive point may lie outside the measurement aperture. To account for this fact the standard requires that the measurement aperture is placed at the most restrictive point and the emission duration T is varied between the minimum emission duration T_i and maximum emission duration T_2. Both emission patterns can be described using equation (5) if the variation of the emission duration T and the number of pulses N that hit the measurement aperture during that emission duration is considered. The minimum emission duration T_i is given by the standard. In the wavelength range of 400 nm $\leq \lambda \leq$ 1050 nm the emission duration T_i is equal to 5 µs. The maximum emission duration is dependent on the angular extension of the laser source and can be determined using equation (6), where α denotes the angular subtense of the source and $\alpha_{\min}$ is the minimum angular subtense of the source, which is equal to 1.5 mrad. [4]

$$\text{AEL}_T = \begin{cases} \dfrac{7 \cdot 10^{-4} \cdot T^{0.75} \cdot C_4}{N} & \text{for } T < 10 \text{ s} \\ \dfrac{3.9 \cdot 10^{-4} \cdot T \cdot C_4}{N} & \text{for } T \geq 10 \text{ s} \end{cases} \tag{5}$$

$$T_2 = \begin{cases} 10 \text{ s} & \text{for } \alpha \leq \alpha_{\min} \\ 10 \cdot 10^{\left(\frac{\alpha - \alpha_{\min}}{98.5}\right)} \text{ s} & \text{for } \alpha_{\min} < \alpha < 100 \text{ mrad} \\ 100 \text{ s} & \text{for } \alpha > 100 \text{ mrad} \end{cases} \tag{6}$$

3. Reduced single pulse criterion

The reduced single pulse criterion is a further limitation of the single pulse criterion. A correction factor C_5 is introduced. This factor C_5 is always less than or equal to one and therefore the reduced single pulse criterion is at least as restrictive as the single pulse criterion. The determination of the $\text{AEL}_{\text{s.p.train}}$ can be described by equation (7). [4]

$$\text{AEL}_{\text{s.p.train}} = \text{AEL}_{\text{single}} \cdot C_5 \tag{7}$$

To determine the correction coefficient C_5 the number of effective pulses N, during the fixed time base T_2, has to be determined. The number of pulses is reduced, if multiple pulses occur during the minimum emission duration T_i. This reduction leads to the effective number of pulses. If the pulse repetition frequency PRF is less than or equal to 200 kHz and the angular extension of the source is less than α_{min}, the calculation of the correction coefficient C_5 reduces to equation (8), where the number of effective pulses can be determined using equation (9). [4]

$$C_5 = \begin{cases} 1 & \text{for } N \leq 600 \\ \max\left[0.4,\ 5 \cdot N^{-0.25}\right] & \text{else} \end{cases} \quad (8)$$

$$N = T_2 \cdot \text{PRF} \quad (9)$$

III. Generalized Modelling and Comparison

Using the equations for the calculation of the AELs from section II and the restrictions of the values, which are summarized in table I, a generalized mathematical model can be derived. Following this modelling a comparison between the AELs and the intensities in the plane of the field of view for a flash and a scanning LiDAR-system is made. [4]

TABLE I
Symbols and values used in the standard IEC 60825-1:2014. The value/range column states the values of the IEC 60825-1:2014 for which the following equations and AELs are applicable.

Symbol	Unit	Definition	Value/range
λ	m	wavelength	$700\text{ nm} \leq \lambda \leq 1050\text{ nm}$
τ_p	s	duration of the laser pulse	$10^{-11}\text{ s} \leq \tau_p \leq 5 \cdot 10^{-6}\text{ s}$
PRF	Hz	pulse repetition frequency of the laser	$\text{PRF} \leq 200\text{ kHz}$
T	s	time base	
T_i	s	time below which pulse groups are summed up	$T_i = 5\text{ µs}$
T_2	s	maximum emission duration that needs to be considered	cf. equation (6)
α	rad	angular extension of the source	cf. equation (4)
α_{min}	rad	minimum angular extension of the source	$\alpha_{min} = 1.5\text{ mrad}$
α_{max}	rad	maximum angular extension of the source	for τ_p less than 625 µs α_{max} equals 5 mrad
d	m	diameter of the measurement aperture	$d = 7\text{ mm}$

A. Generalized modelling

The AEL is determined by finding the most restrictive of the above summarized criteria. To generalize the modelling of this determination equation (10) is used as a starting point.

$$\text{AEL} = \min\left[\text{AEL}_{\text{single}}, \text{AEL}_T, \text{AEL}_{\text{s.p.train}}\right] \quad (10)$$

Since the correction coefficient C_5 is always less than or equal to one the term $\text{AEL}_{\text{single}}$ of equation (10) can be

neglected. Furthermore, a value of $\alpha = 1.5$ mrad is assumed for the angular subtense of the source, which leads to the most restrictive value of C_6. Therefore, a value of 10 s can be determined for the maximum emission duration T_2 and a value of one can be determined for the correction coefficient C_6. Since it is assumed that the pulse repetition frequency is less than or equal to 200 kHz the effective number of pulses equals the number of pulses emitted during the emission duration and can be determined using equation (9).

Using these restrictions and assumptions equation (10) can be further evaluated and four different regions of the most restrictive AEL, as a function of the pulse repetition frequency, can be distinguished. The resulting equation is given in (11) and is schematically shown in figure 2.

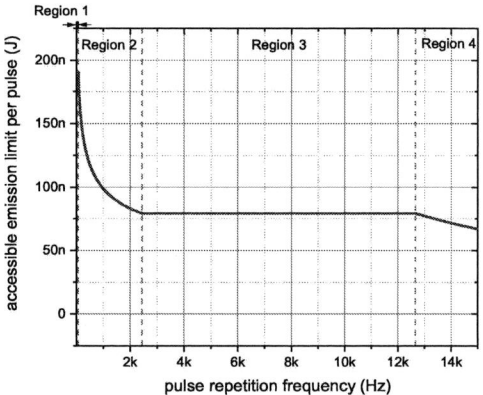

Fig. 2. Schematic representation of the AEL calculated using equation (11). The four distinguishable regions are marked.

$$\text{AEL} = \begin{cases} 7.7 \cdot 10^{-8} \cdot C_4 & \text{for PRF} \leq 60\text{ Hz} \\ 7.7 \cdot 10^{-8} \cdot C_4 \cdot 5 \cdot N^{-0.25} & \text{for } 60\text{ Hz} < \text{PRF} \leq 2441.4\text{ Hz} \\ 7.7 \cdot 10^{-8} \cdot C_4 \cdot 0.4 & \text{for } 2441.4\text{ Hz} < \text{PRF} \leq 12.6\text{ kHz} \\ \frac{3.9 \cdot 10^{-4} \cdot C_4}{\text{PRF}} & \text{for } 12662.34\text{ Hz} < \text{PRF} \end{cases} \quad (11)$$

B. Comparison

To compare the AELs of a flash- and a scanning LiDAR-system equation (11) is evaluated for pulse repetition frequencies from 1 Hz to 200 kHz. To compare scanning LiDAR-systems independent of the scanning trajectory, the AEL is evaluated for a different number of pulses N. The wavelength of the laser source used for the calculation is the typically used wavelength in LiDAR applications of 905 nm. The resulting AELs per pulse are depicted in figure 3 and figure 4. These figures show the pulse repetition frequency range from 10 Hz to 20 kHz and the pulse repetition frequency range from 20 kHz to 200 kHz respectively. The AELs of the flash LiDAR-system case can be described by the diagonal, which is drawn in the figures. This diagonal equals the case that every emitted laser pulse hits the measurement aperture. The AELs of a scanning LiDAR-system can be estimated from the figure if the number of pulses N is known. A more in-depth analysis to approximate N is given in [5]. The labelled values

© VDE VERLAG GMBH · Berlin · Offenbach

give the AELs for the cases that 10 %, 50 % or 100 % of the emitted pulses hit the measurement aperture. It is apparent that, especially in the higher pulse repetition frequency ranges, the AELs significantly differ.

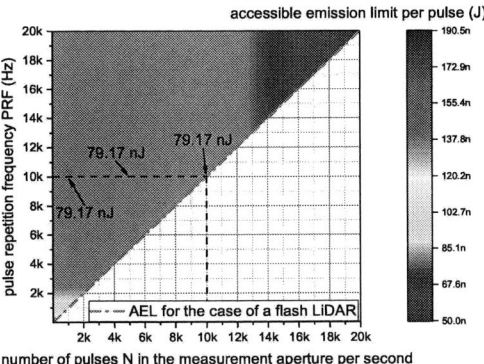

Fig. 3. Determined values of the AEL per pulse in the pulse repetition frequency range from 10 Hz to 20 kHz with varying number of pulses N that hit the measurement aperture per second.

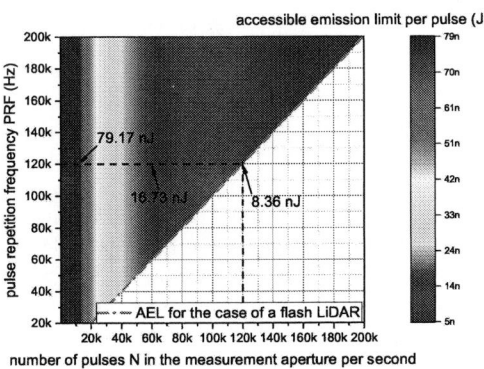

Fig. 4. Determined values of the AEL per pulse in the pulse repetition frequency range from 20 kHz to 200 kHz with varying number of pulses N that hit the measurement aperture per second.

A further comparison can be made if the intensity at the plane of the measurement aperture is considered as schematically depicted in figure 5. Using the measurement geometry of the standard and the determination rules for the AELs it can be readily determined that the maximum power is limited to $P_{\text{rest.}}$, which is proportional to the most restrictive AEL. If the illuminated area is larger than the area of the measurement aperture the emitted power may be increased as well, because this radiation will not hit the measurement aperture. This yields the term $P_{\text{flash}}/A_{\text{FOV}}$. This intensity is related to $P_{\text{rest.}}/A_{\text{meas.aperture}}$ and the relation can be described by equation (12). This equation yields the equality of both intensities for the case of an extended illumination. Using the same argument for the case of a smaller illuminated area A_{spot} than the area of the measurement aperture $A_{\text{meas.aperture}}$ the intensity will be much larger, which is usually favourable for LiDAR applications.

$$P_{\text{flash}} = P_{\text{rest.}} \cdot \frac{A_{\text{FOV}}}{A_{\text{meas.aperture}}} \qquad (12)$$

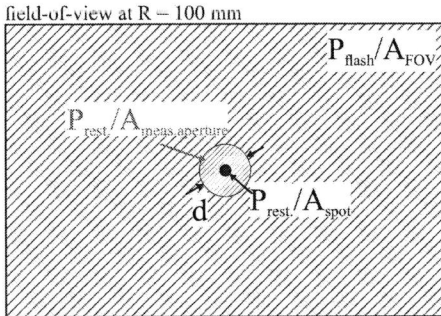

Fig. 5. Schematic representation of the measurement geometry proposed by the standard to evaluate and compare the intensities in the plane of the measurement aperture.

IV. CONCLUSION

This paper presented an approach to model the AELs of scanning LiDAR-systems without the detailed evaluation of the utilized scanning trajectories. Furthermore, a comparison was made between the case of a flash and a scanning LiDAR-system. This comparison complies with the determination rules of the IEC 60825-1:2014 standard. This comparison showed that especially in the higher pulse repetition frequency ranges the AELs of a flash LiDAR are significantly lower than those for a scanning LiDAR-system. A further comparison considering the intensities in the plane of the field-of-view was carried out. This comparison showed that a scanning LiDAR-system using a point illumination may emit less power than a flash LiDAR-system but the allowed intensities may be much larger in the case of a scanning LiDAR. A major drawback of the scanning LiDAR-systems is generally the coverage of larger field-of-views with a sufficient frame rate. This remark is subject to further studies because the comparison involves the detailed evaluation of the scanning trajectories and the availability of large and low-cost photodetector matrices.

REFERENCES

[1] P. Boulay and A. Debray, *LiDAR for automotive and industrial applications - challenges and market opportunities : From technologies to market 2019 report.* Lyon-Villeurbanne: Yole Développement, 2019.

[2] R. Kokozinski, O. M. Schrey, M. Beer, J. Ruskowski, W. Brockherde, B. J. Hosticka, and J. F. Haase, "SPAD-based flash LiDAR sensor with high ambient light rejection for automotive applications," in *Quantum Sensing and Nano Electronics and Photonics XV*, M. Razeghi, G. J. Brown, G. Leo, and J. S. Lewis, Eds. SPIE, 27.01.2018 - 01.02.2018, p. 85.

[3] P. McManamon, *LiDAR Technologies and Systems.* Bellingham, Washington, USA: SPIE Press, 2019.

[4] International Electrotechnical Commission, "IEC 60825-1:2014: Safety of laser products - Part 1: Equipment classification and requirements," Berlin, 2014-05.

[5] R. Burkard, R. Viga, J. Ruskowski, and A. Grabmaier, "Eye safety considerations and performance comparison of flash- and MEMS-based lidar systems," in *Optics, Photonics and Digital Technologies for Imaging Applications VI*, ser. Proceedings of SPIE. 5200-, P. Schelkens and T. Kozacki, Eds. Bellingham, Washington: SPIE, 2020, p. 39.

A Mixed-Precision Binary Neural Network Architecture for Touch Modality Classification

[1,2]Hamoud Younes, [1,2]Ali Ibrahim, [2,3]Mostafa Rizk, [1]Maurizio Valle

[1]COSMIC Laboratory, Department of Electrical, Electronic and Telecommunications Engineering and Naval Architecture
University of Genova, Genoa, Italy
[2]Department of Computer and Communication Engineering, Lebanese International University, Bekaa, Lebanon
[3]Department of Physics and Electronics, Faculty of Sciences, Lebanese University, Lebanon

Abstract—**Binary Neural Networks (BNN) have been proposed to address the computational complexity and memory requirements of Convolutional Neural Networks (CNN). However, in most of the applications, BNNs suffer from severe accuracy loss due to the 1-bit quantization. In this paper, a Mixed-Precision Binary Weight Network (MP-BWN) is proposed as a compromise between CNN and BNN. Compared to traditional binary networks, MP-BWN offers better performance with an acceptable increase in the network size. MP-BWN achieves up to 99% reduction in both the number of operations and the network size compared to similar state-of-the-art solutions. When validated on a touch modality classification problem, the MP-BWN surpassed similar existing solutions by achieving a classification accuracy of 77.8%.**

Index Terms—**Binary Neural Networks (BNN), Convolutional Neural Networks (CNN), Quantization, Touch Modality**

I. INTRODUCTION

Convolutional Neural Networks (CNN) are a promising solution in many application domains such as Internet of Things (IoT), image processing, tactile processing, etc. However, the computational complexity and memory requirements are the main challenge in the deployment of CNNs on resource-limited devices for energy-constrained applications. For instance, the VGG-16 network contains about 140 million 32-bit floating-point parameters and implements 1.6×10^{10} arithmetic operations [1]. There have been numerous efforts on the complexity reduction of CNNs such as network pruning [2], knowledge distillation [3], and weight quantization [4].

Quantization of CNNs may cause an information loss especially if it is applied to the extreme using 1-bit representation i.e. binarization. To address this issue, a variety of methods have been proposed in recent years [1]. These methods aim to: 1) minimize the quantization error, for instance by only quantizing the weights , 2) improve the network loss function to adapt to the binary values propagating through the network, and 3) reduce the gradient error by the adjustment of the back propagation training algorithm to adapt with binarization functions. Among these efforts, minimizing the quantization error is the most used technique since it leads to relevant memory saving and complexity reductions [1], [4]. In this paper, a new architecture based on mixed-precision representation is proposed as a compromise between the reliability of CNNs and the low complexity of Binary Neural Networks (BNN). The performance of the proposed architecture is evaluated on

a touch modality classification showing high performance in terms of accuracy and complexity.

The main contributions of this paper are summarized as follows:

- It proposes a mixed-precision architecture for a binarized network that offers a trade-off between classification accuracy and computational requirements.
- It presents the architecture, training, and tuning of the proposed network. Experimental results show that the proposed network overcomes similar state-of-the-art solutions [5], [6] in terms classification accuracy, model parameters, and number of operations.

The rest of the paper is organized as follows: Section II presents an overview of the existing algorithms used for touch modality classification. Section III presents a definition for BNNs and details the architecture of the proposed mixed-precision binary weight network. Section IV provides the touch modality classification problem, the data pre-processing techniques, and the architecture tuning and training. An assessment of the proposed architecture performance compared to similar solutions is also highlighted. Section V concludes the paper and illustrates on some observations.

II. RELATED WORK

Touch modality classification using machine and deep learning has been the focus of several works in the literature. In [7], a touch recognition system featuring a temporal decision tree classifier intended for online touch modality recognition has been proposed. The system provided an average recognition rate of 83% with respect to the 4 touch patterns of hit, beat, rub and push. Another set of touch modalities have been used in [8] when the LogitBoost algorithm has been adopted for classification. Experimental results showed that a touch was correctly classified in approximately 71% of the trials. The authors in [9] have proposed a Support Vector Machine (SVM) classifier with feature descriptors. The classifier is able to differentiate between nine modalities (e.g. scratch, slap, poke,etc.) with a recognition rate up to 96%. The same set of modalities has been successfully recognized via acoustic sensing and logistic model trees via a model proposed in [10].

Gastaldo *et. al* have presented two computationally intelligent techniques for touch modality classification in [5]. The techniques are based on the tensorial SVM and Regularized

© VDE VERLAG GMBH · Berlin · Offenbach

Least Square (RLS) algorithms, while the modalities under test are: brushing a paintbrush, sliding a finger, and rolling a washer. A classification accuracy above 80% has been achieved in all the experimental trials. In [6], the authors have investigated the use of Deep Convolutional Neural Network (DCNN) algorithms for the same classification problem described in [5]. The DCNN model used transfer learning to utilize the Inception Resnet neural network and achieved a higher classification accuracy compared to similar algorithms.

A common characteristic of the mentioned works is the computational complexity of the used learning algorithms. Such complexity is the main bottleneck for embedded hardware implementation especially for applications with memory and power constraints such as electronic-skin. Hence, our motivation is to propose a new architecture based on deep learning algorithms that meet these constraints.

III. Mixed-Precision Binary Weight Network

A. Definition

A binary neural network is a quantized convolutional neural network where the weights **w** and activations **a** are represented by 1-bit i.e. their values are constrained to $\{-1, +1\}$. This is achieved using a binarization function q given as:

$$\begin{cases} q_w(\mathbf{w}) = \alpha b_{\mathbf{w}} \\ q_a(\mathbf{a}) = \beta b_{\mathbf{a}} \end{cases} \quad (1)$$

where $b_{\mathbf{w}}$ and $b_{\mathbf{a}}$ are the binary weights and binary activations, with α and β are their corresponding scalars respectively. The *sign* binarization function is defined by:

$$sign(z) = \begin{cases} 1 & z >= 0 \\ -1 & otherwise \end{cases} \quad (2)$$

In forward propagation, the convolutionl process of a CNN is translated into a bit-wise XNOR operation with bit-count in BNN. The process can be written as:

$$y = \sigma(q_w(\mathbf{w}) \otimes q_a(\mathbf{a})) = \sigma(\alpha\beta(b_{\mathbf{w}} \odot b_{\mathbf{a}})) \quad (3)$$

where σ is a non-linear function, $\otimes$ is the floating-point convolution, and $\odot$ is the XNOR operation. In backward propagation, a BNN is trained through a function called Straight Through Estimator (STE) [11] using the gradient descent adopted in full-precision networks. The STE function is denoted by:

$$clip(x, -1, +1) = max(-1, min(1, x)) \quad (4)$$

This function has been introduced to avoid the undefined or zero-valued derivative of the binarization functions.

B. Proposed Architecture

To overcome the accuracy loss of BNNs, two main extensions have been presented in the literature. The first extension is called Binary Weight Network (BWN) where only the weights **w** are binarized [12]. The second extension is called Binary Activation Networks (BAN) where only the activations **a** are binarized [1]. To accommodate the accumulated precision loss due to binarization, a mixed-precision architecture for binarized networks as an extension for the BWN is proposed here as shown in Fig. 1. The configuration of each layer in the proposed architecture is as follows:

- The input convolution and the output fully connected (FC) layers weights are binarized while keeping the activations with 32-bit floating-point. All hidden layers are completely binarized. This design choice is due to the fact that the first convolution layer is responsible to extract as much features from the input as possible. Hence, no quantization is performed. Similarly, a full-precision FC output layer provides precised classification output.
- The number, size and type (convolution/FC) of all layers are configured to control network complexity and memory storage requirements without sacrificing significant accuracy loss.
- Batch normalization is used in all layers to compensate the unbalanced output and thus the mean and variance are kept within a reasonable range and the training process becomes much smoother.
- The input is padded with +1/-1 for the all the layers so that the output of the layer has the same size as the input i.e. 4×4.

IV. Validation and Results

A. Touch Modality Classification

The dataset collected in [5] is adopted in this work. The dataset includes 260 samples for each of the three touch modalities: brushing a paintbrush, sliding a finger, and rolling a washer. Each sample is recorded at 3KHz sampling frequency using a 4×4 tactile sensor array based on polyvinylidene fluoride (PVDF) sensing elements for a duration of 10 seconds. Thus, a sample is modeled as a tensor $\phi(4 \times 4 \times 30000)$. A set of modifications are applied on the dataset in order to increase the dataset size and reduce the input dimensionality:

- Each sample has been truncated from 10 to 3.5 seconds by eliminating all the readings outside the interval [3.5s,

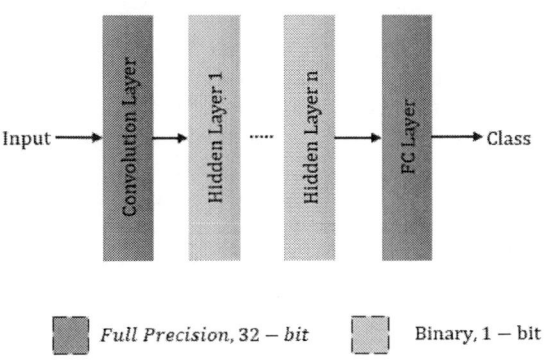

Fig. 1. Proposed Mixed-Precision BWN

TABLE I
QUANTIZERS FOR BNN TRAINING

	BNN	BWN	BAN
Input Quantizer	Approx-sign	None	Approx-sign
Kernel Quantizer	Approx-sign	Approx-sign	None
Kernel Constraint	Weight clip	Weight clip	None

kernel constraint represents the clip function defined in (4). The networks are trained on an Intel i-7 based PC equipped with NVIDIA GTX 1650 graphics card. Adam [16] with a learning rate of 0.01 is used as an optimizer with "categorical-crossentropy" loss function.

C. Results and Assessment

Fig. 3 shows the tuned MP-BWN to achieve the highest accuracy in classifying touch modalities. The architecture is characterized by:

- Starting from an input size $4 \times 4 \times 8$, the maximum number of pooling layers that could be used is two layers, then the output of the layer will be a single element. To keep as much features as possible, the pooling layer is not used at the first layer. The adopted pooling size is 2×2.
- The number of convolution layers is two and each layer has 7 filters with size 3×3.
- The number of FC layers is two. The first FC has 15 neurons while the output FC layer has 3 neurons corresponding to each class (rolling, brushing, and sliding).

Fig. 4 reports the average classification accuracy among 10 runs for the binarized networks under batch-size=100 and number of epochs = 500. The obtained results show that BNN, BWN, and BAN were not able to achieve an accuracy above 50%. However, the proposed mixed-precision methodology has significantly increased the accuracy reaching 71% for MP-BWN. Such results account for a compensation of the precision loss due to the binarization of the network. The compensation is at the expense of increasing the memory storage requirements from 2.3 KB to 4.8 KB while keeping the same number of trainable parameters as shown in Table II; however, such storage is even available in mainstream microcontrollers (e.g. STM32F0x2 with 16 KB memory storage). When compared to similar solutions for the same touch modality classification problem, the proposed MP-BWN is

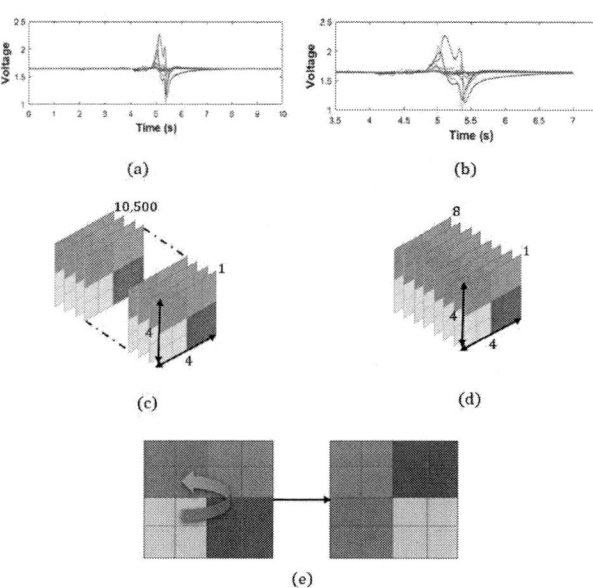

(a)   (b)

(c)   (d)

(e)

Fig. 2. Data Pre-processing: (a) Touch Modality, (b) Truncated Touch Modality, (c) Tensor Representation, (d) Sampled Tensor, (e) Data Augmentation

7s] where no touch is present (See Fig. 2 (a), (b)). Hence, the new sample size is reduced to $\phi(4 \times 4 \times 10500)$ as shown in Fig. 2 (c).

- Subsampling is applied on the truncated tensor to obtain a new tensor $T(4 \times 4 \times 8)$ (see Fig.2 (d)) where each $P = 1312$ readings are averaged into one reading according to the following equation:

$$\begin{cases} T(:,:,i) = (1/P) * \sum_{i=k}^{P+k} \phi(:,:,i) \\ k+ = P \end{cases} \quad (5)$$

starting with $k = 0$ and iterating $i = 1, 2..., 8$.

- The size of the dataset is doubled i.e from 780 to 1560 samples through data augmentation [13]. Specifically, each sample is rotated by 90 degrees as shown in Fig.2 (e).

B. Architecture Tuning

The traditional BNN, BWN, BAN, and the proposed mixed-precision networks have been modeled in Python using the Larq framework [14]. Each network model is trained using the modified dataset presented in the previous sub-section. The dataset has been divided into 5 folds where each time the network is trained using 4 folds and tested using the remaining one. The classification accuracy is determined as the average of the 5 folds. This process has been repeated 10 times to determine the best number/type of hidden layers, size of each layer, batch-size, number of epochs, and the number of filters required to achieve the highest possible accuracy. Table I shows the quantizer type for the different network configurations. The "Approx-Sign" quantizer has been used instead of STE as it has been proposed to enhance the performance of binarized neural networks [15]. The "weight_clip"

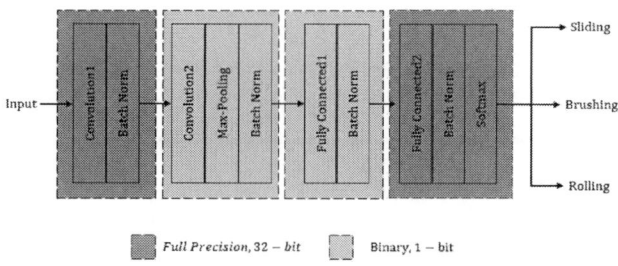

Fig. 3. Proposed Architecture Tuned for Touch Modality Classification

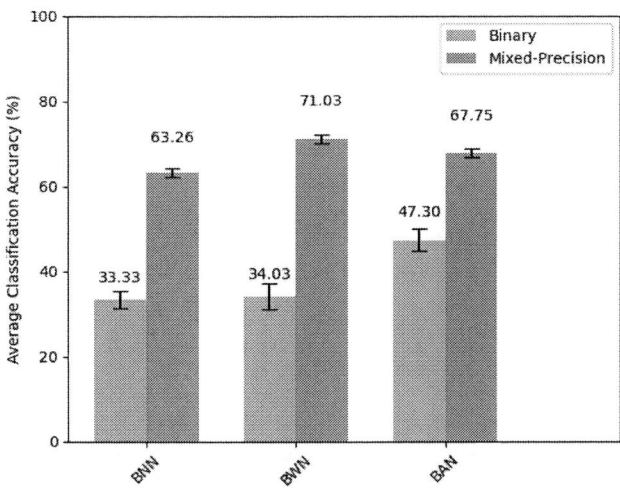

Fig. 4. Model Performance for Different Network Architecture

TABLE II
MODEL REQUIREMENTS

Representation	Binary			Mixed-Precision		
Model	BNN	BWN	BAN	BNN	BWN	BAN
Memory Storage (KB)	2	2.3	2.91	3.3	4.8	5.21
Model Parameters	1.42K					

superior in the aspects of accuracy, number of operations, and the trainable model parameters. In Table. III, MP-BWN offers up to 99.9% reduction in the number of operations compared to SVM and DCNN algorithms [6]. Similarly, the MP-BWN model size represents a 97.8% and 99.9% compression rate compared to that of SVM and DCNN respectively. The achieved computational reduction and compression are accompanied with the highest classification accuracy of 77.88% among 10 folds.

V. CONCLUSION

This paper introduces a mixed-precision architecture for binary weight network. The MP-BWN architecture consists of two convolution and two fully connected layers; the first and last layers use 32-bit floating-point representation while the hidden layers are binarized. MP-BWN provides a classification accuracy of 77.8% exceeding existing similar solutions. The MP-BWN performance is accompanied with a 99% reduction in the number of operations and model size compared to SVM

TABLE III
BEST MP-BWN PERFORMANCE IN COMPARISON WITH THE STATE OF THE ART

	MP-BWN	SVM [5]	DCNN [6]
Highest Classification Accuracy	77.88%	76.6%	76.9%
Number of Gops	3.2×10^{-5}	0.545	109×10^{-3}
Reduction in Gops	NA	99.99%	99.97%
Model Parameters	1.42K	67.2K	540M
Reduction in Model Parameters	NA	97.8%	99.9%

and DCNN algorithms. Such results pave the way towards the hardware implementation on resource-limited devices and energy-constrained applications. Similarly, the obtained model parameters shows that such network doesn't require massive computing platform for training and there is no need for extra memory (e.g. DRAM) to save the network's parameters.

ACKNOWLEDGEMENT

The authors acknowledge partial financial support from TACTIle feedback enriched virtual interaction through virtual realITY and beyond (TACTILITY) project: EU H2020, Topic ICT-25-2018-2020, RIA, Proposal ID 856718.

REFERENCES

[1] H. Qin, R. Gong, X. Liu, X. Bai, J. Song, and N. Sebe, "Binary neural networks: A survey," *Pattern Recognition*, vol. 105, p. 107281, Sept. 2020.

[2] Y. He, X. Zhang, and J. Sun, "Channel Pruning for Accelerating Very Deep Neural Networks," in *2017 IEEE International Conference on Computer Vision (ICCV)*, (Venice), pp. 1398–1406, IEEE, Oct. 2017.

[3] J. Yim, D. Joo, J. Bae, and J. Kim, "A Gift from Knowledge Distillation: Fast Optimization, Network Minimization and Transfer Learning," in *2017 IEEE Conference on Computer Vision and Pattern Recognition (CVPR)*, (Honolulu, HI), pp. 7130–7138, IEEE, July 2017.

[4] Y. Gong, L. Liu, M. Yang, and L. Bourdev, "Compressing Deep Convolutional Networks using Vector Quantization," *arXiv:1412.6115 [cs]*, Dec. 2014. arXiv: 1412.6115.

[5] P. Gastaldo, L. Pinna, L. Seminara, M. Valle, and R. Zunino, "Computational Intelligence Techniques for Tactile Sensing Systems," *Sensors*, vol. 14, pp. 10952–10976, June 2014.

[6] M. Alameh, A. Ibrahim, M. Valle, and G. Moser, "DCNN for Tactile Sensory Data Classification based on Transfer Learning," in *2019 15th Conference on Ph.D Research in Microelectronics and Electronics (PRIME)*, (Lausanne, Switzerland), pp. 237–240, IEEE, July 2019.

[7] Seong-yong Koo, Jong Gwan Lim, and Dong-soo Kwon, "Online touch behavior recognition of hard-cover robot using temporal decision tree classifier," in *RO-MAN 2008 - The 17th IEEE International Symposium on Robot and Human Interactive Communication*, (Munich, Germany), pp. 425–429, IEEE, Aug. 2008.

[8] D. Silvera Tawil, D. Rye, and M. Velonaki, "Touch modality interpretation for an EIT-based sensitive skin," in *2011 IEEE International Conference on Robotics and Automation*, (Shanghai, China), pp. 3770–3776, IEEE, May 2011.

[9] M. Kaboli, A. Long, and G. Cheng, "Humanoids learn touch modalities identification via multi-modal robotic skin and robust tactile descriptors," *Advanced Robotics*, vol. 29, pp. 1411–1425, Nov. 2015.

[10] Fernando Alonso-Martín, Juan Gamboa-Montero, José Castillo, Álvaro Castro-González, and Miguel Salichs, "Detecting and Classifying Human Touches in a Social Robot Through Acoustic Sensing and Machine Learning," *Sensors*, vol. 17, p. 1138, May 2017.

[11] Y. Bengio, N. Léonard, and A. Courville, "Estimating or Propagating Gradients Through Stochastic Neurons for Conditional Computation," *arXiv:1308.3432 [cs]*, Aug. 2013. arXiv: 1308.3432.

[12] M. Rastegari, V. Ordonez, J. Redmon, and A. Farhadi, "XNOR-Net: ImageNet Classification Using Binary Convolutional Neural Networks," in *Computer Vision – ECCV 2016* (B. Leibe, J. Matas, N. Sebe, and M. Welling, eds.), vol. 9908, pp. 525–542, Cham: Springer International Publishing, 2016. Series Title: Lecture Notes in Computer Science.

[13] C. Shorten and T. M. Khoshgoftaar, "A survey on Image Data Augmentation for Deep Learning," *J Big Data*, vol. 6, p. 60, Dec. 2019.

[14] L. Geiger and P. Team, "Larq: An Open-Source Library for Training Binarized Neural Networks," *JOSS*, vol. 5, p. 1746, Jan. 2020.

[15] Z. Liu, B. Wu, W. Luo, X. Yang, W. Liu, and K.-T. Cheng, "Bi-Real Net: Enhancing the Performance of 1-bit CNNs With Improved Representational Capability and Advanced Training Algorithm," *arXiv:1808.00278 [cs]*, Sept. 2018. arXiv: 1808.00278.

[16] D. P. Kingma and J. Ba, "Adam: A Method for Stochastic Optimization," *arXiv:1412.6980 [cs]*, Jan. 2017. arXiv: 1412.6980.

© VDE VERLAG GMBH · Berlin · Offenbach

A CMOS SPAD pixel with an integrated mixed-signal TDC

Sergio Moreno
Department of Electronic and Biomedical Engineering
University of Barcelona
Barcelona, Spain
sergiomoreno@ub.edu

Victor Moro
Department of Electronic and Biomedical Engineering
University of Barcelona
Barcelona, Spain
vmoro@ub.edu

Angel Dieguez
Department of Electronic and Biomedical Engineering
University of Barcelona
Barcelona, Spain
angel.dieguez@ub.edu

Abstract—During the last century, much of the scientific discoveries and advances in the biological field have been derived from the exploration of fluorescence phenomena. This has been possible, in part, thanks to the development of a wide variety of measurement techniques accompanied by high-performance sensing devices. Among these are CMOS Single Photon Avalanche Diodes (SPADs) image sensors, which offer fast response and low fabrication costs. In this work we have developed a pixel with dimensions of 40 μm x 40 μm, which together with a 10 μm SPAD, allows to obtain a 5% fill-factor. In addition, it has an integrated Time-to-Digital Converter (TDC) with a programmable bin width from 100 ps up to 1ns with a total measurement range of up to 72 ns. The pixel architecture will enable fluorescence lifetime measurements of organic and inorganic dyes. The photon arrival time will be used to generate a histogram.

Keywords—SPAD, CMOS camera, in-pixel TDC, analog counter

I. INTRODUCTION

The world of sensors, and in particular Single Photon Avalanche Diodes (SPADs), has experienced a pronounced improvement in performance over the last decades, mostly due to the high demand for these devices in the commercial sector and in the most popular experimental sciences, like the biological and medical fields.

SPAD technology saw a strong demand increase in the commercial sector, where image sensors are used for automotive applications related to Light Detection and Ranging (LIDAR) [1] or Advanced Driver Assistance Systems (ADAS) [2]. For example, it has led to the emergence of various unconventional imaging techniques to achieve compact 3D cameras or high dynamic range vision. Nevertheless, one of the most important applications is in the biomedical field, where they are used for imaging, including Time-of-Flight Positron Emission Tomography (TOF PET) [3], super-resolution microscopy [4] and Near-Infrared Optical Tomography (NIRI/NIROT) [5].

Related to the biomedical field and diagnosis, SPADs are used for time correlated single photon counting (TCSPC). This is the most accurate technique available for determining fluorescence decays. This technique has expanded in a large range of biomedical applications [6]. Moreover, it allows the detection of single photons and the measurement of their arrival times in respect to a reference signal. Thus, it is used from fluorescence lifetime imaging (FLIM) [7] to Point of Care (PoC) devices [8].CMOS SPAD technology enables massively parallelized counting and timing of single photons. The per-pixel time resolution offered by SPAD sensors have allow the development of new applications in endoscopy, aerial monitoring, hyperspectral scanning systems in microscopy and new modalities in fluorescence lifetime and Raman spectroscopies [9]. Several options for on-chip integration of SPAD sensors with digital processing circuits implemented are explored in the literature, including gated counters [10], per-pixel time-to-digital converters (TDCs) [11], time-gated memories [12], in-pixel centre-of-mass computation [13], column parallel flash TDCs [14], multi-event folded flash TDCs [15], off-chip FPGA TDCs [16] and on-chip histogramming [17], [18]. Perenzoni, reported in [19] a 160x120 pixel SPAD array in which he replaced the classical 1-bit memory with an analog counter (multi-bit). Each detected photon injected a controlled charge into the analog memory, being able to adjust up to 750 ps the temporal window of observation. In [18], Dieguez *et al.* introduced an in-pixel analog histogram with 9 bins that can count up to 13b each, with a resolution of 0.16mV/photon.

In this work, a proposal for an in-pixel TDC has been designed in 150 nm CMOS standard process. One of the problems with in-pixel TDC circuits is the skew of the TDC signals from pixel to pixel. This circuit alleviates this problem by generating all the required signals for the quenching, readout and TDC in-pixel. In this work we present a novel technique where we use digital TDC techniques that allow high resolution, up to 100 ps, and analog solutions to extend the range of the digital TDC. Moreover, in the digital TDC, both the rising and falling edges of the in pixel generated signals are used to classify photons, which significantly reduces the non-sensing pixel area, thus, increasing the Fill-Factor (FF) and Pixel-Pitch (PP) of the pixel.

II. PIXEL ARCHITECURE

The pixel consists of a digital readout, a TDC and an auxiliary control circuit for its activation. Figure 1 shows the block diagram of the SPAD pixel architecture. The Start signal activates both the digital readout of the SPAD and the start / stop TDC circuit.

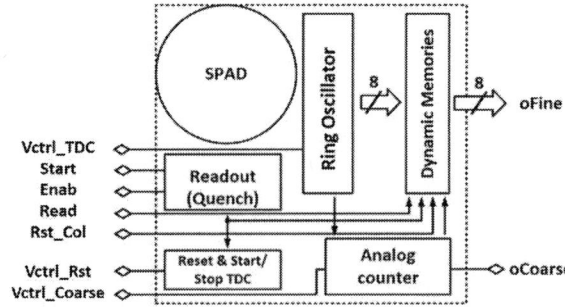

Figure 1. Block diagram of the proposed pixel based on analog count.

The TDC is composed by three main blocks: a ring oscillator, a set of dynamic memories and an analog counter. The ring oscillator (RO) works as the fine phase generator and allows the generation of 8 tunable time windows. The width of these time windows defines the resolution of the TDC. To extend the time range of the measure, an analog counter was added. The analog counter (AnC) increases its value for each cycle of the TDC-RO by performing a controlled charge injection on a capacitance.

III. SPAD READOUT

The digital readout of the SPAD is presented in Figure 2. It is based on an active quenching where the *Inh* and *Rst* signals controls the SPAD operation through the M_{P0} and M_{N0} transistors. The external trigger signal (*Start*) works as the *Inh* signal. Therefore, its rising edge will immediately activate a controlled pulse generator circuit that generates the in-pixel *Rst* signal [20]. Applying an external bias (*Vctrl_Rst*), the *Rst* pulse is managed from 230 ps to 2.5 ns. If a photon is detected, a logical 1 is set to the output (*PixOut*), as shown in Figure 3.

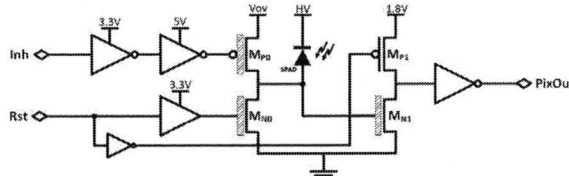

Figure 2. Schematic of the SPAD readout.

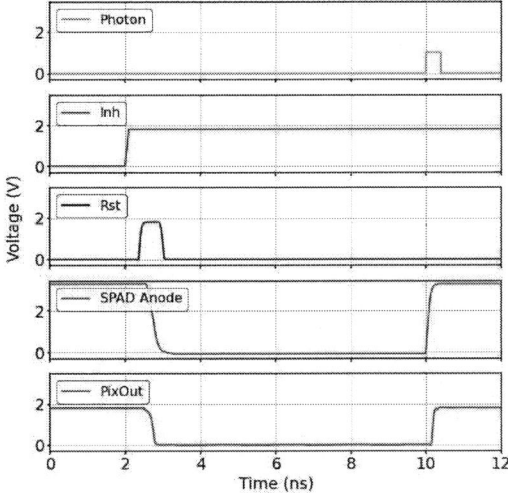

Figure 3. Time diagram of SPAD readout operation.

IV. TIME-TO-DIGITAL CONVERTER

Thus, by means of a simple structure of the delay cells, and by processing the data out of the sensing area, it is possible to classify the arrival of photons according to the moment in which they have produced an avalanche in the SPAD.

The TDC only operates during the exposure time, i.e., between the disabling of the *Rst* signal and the photon detection time. In Figure 4, the circuit that controls the start and stop of the TDC is shown.

The *Start/Stop_TDC* signal results from the combination of the previous internal signals *Inh*, *Rst* and *PixOut* (Figure 2).

If no photon has been detected during the measurement, the *Start/Stop_TDC* output signal will be disabled synchronously with the *Inh* signal status change. On the other hand, if a photon is detected, the state of *PixOut* signal will change. Therefore, the TDC is disabled to reduce the power consumption of the pixel.

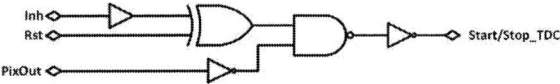

Figure 4. TDC start/stop circuit schematics.

A. Ring Oscillator

The first TDC block is the RO. Time resolution is given by the frequency of the RO, so it requires a very precise design capable of handling from picosecond to nanosecond resolutions.

The RO design consists in a delay line formed by 8 stages, as shown in Figure 5. The delay elements are based on a *Current Starved Voltage Control Oscillator* (CSVCO) architecture. The delay introduced by these stages define the bins, i.e., the elapsed time between two consecutive phases edge, that will be used to classify the photon arrival time.

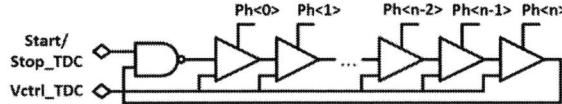

Figure 5. Schematic of the ring oscillator.

In order to obtain a pixel with a compact design, both output buffer transitions are involved. So, an equal bin width is reached with the rising and falling edges of the time windows. With this, it increases the number of bins by two per RO cycle, improving both FF and PP. Figure 6 shows the time diagram of the proposed RO operation.

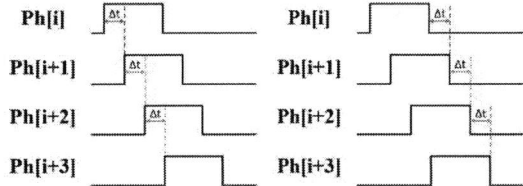

Figure 6. Principle of operation of the TDC ring oscillator. Each Ph[i] corresponds to the output of the RO delay cells. The bin width is defined by Δt.

Figure 7 shows the configuration of one of the RO cells. It consists of two consecutives current starved inverters. The gates of the transistors MN1 and MN3 are connected to a bias voltage, *Vctrl_TDC*. This changes the resistance of MN1 and MN3 modifying the amount of current flowing through their corresponding inverter. This adjusts the buffer switching times to achieve symmetrical response on rise/fall edges. The *Phase* signal is driven by a standard inverting buffer. Finally, the other output signal (*Next Stage*) will be connected to the input of the next delay element of the ring oscillator, *Prev Stage*.

Figure 8 shows the delay range of the first RO bin. It allows to observe in more detail the time differences between rising and falling delays. The size of the bin can be controlled between approximately 100 ps and 1 ns.

Table 1 shows the results obtained by performing a Monte Carlo simulation for a bias voltage of 1.8V, the bin width at the rising and falling cases and the sigma obtained.

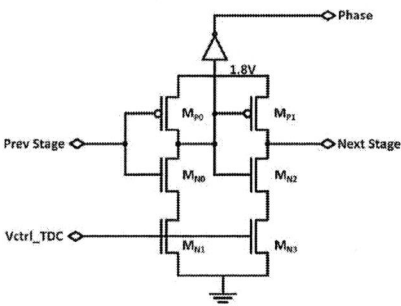

Figure 7. Schematic of the delay cell based on a CSVCO structure of TDC.

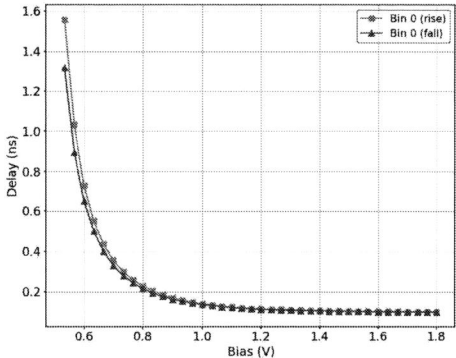

Figure 8. Variation of time delay per TDC stage as a function of bias voltage (*Vctrl TDC*). As all stages present a similar delay, only the first-time window is presented.

Bin 0	Nominal	σ
Rising	97.3 ps	8.51 ps
Falling	96.1 ps	8.51 ps

Table 1. Summary of the results obtained through Monte Carlo simulation for a bias voltage of 1.8V.

B. Dynamic Memories

The second main block of the TDC is the in-pixel memories, which registers the arrival time of the photon. In order to reduce the area as much as possible, a full-custom cell configuration has been developed. Figure 9 shows the schematic of a single in-pixel memory. The memory module consists of 8 individual integrated dynamic cells. The input of each memory cell is connected directly to the output of each delay buffer of the RO.

As in the rest of pixel blocks, when the local reset is performed, the critical nodes are pre-charged to a known voltage. In this case, the NMOS transistor MN1 will trigger the activation of the PMOS transistor MP2, which will pre-charge the *Store* node at 1.8 V. In turn, the NMOS transistor MN4 will discharge the output node, forcing the *oMem[i]* signal to 0 V.

The main inputs of the cell are *PixOut,* which is the SPAD readout output, and *Start/Stop_TDC,* which is obtained by the combination shown in Figure 4, and the corresponding ring oscillator phases, *Phase.* The memories will only store the value of the *PixOut* signal if the *Start/Stop_TDC* signal is a

logical 1, enabling write on the memory cell for the duration of the measurement until a photon is detected or the measurement is stopped.

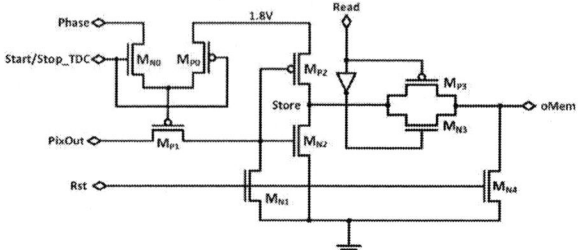

Figure 9. Schematic of the dynamic pixel memory cell.

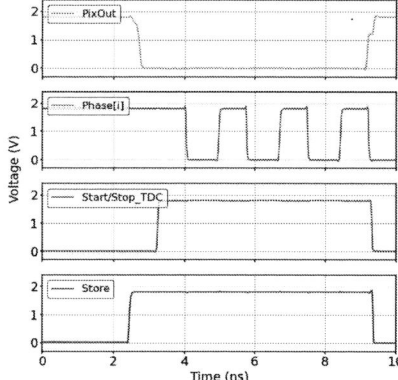

Figure 10. Temporal diagram of the control signals of the TDC memories.

If *Phase[i]* signal is in any state, i.e., high or low, and the photon has not been detected, a digital 1 is set in the *Store* node. On the other hand, if a photon is detected, a logical 1 is registered if *Phase[i]* is at high state and a logical 0 if *Phase[i]* is at low state. Finally, when the *Read* signal is activated, the *Store* value will be loaded into the output (*oMem[i]*). An example for the photon detection case can be seen in Figure 10. The photon arrival is produced when the *PixOut* signal goes high. The *Phase[i]* signal presents three transitions equivalent to the same RO cycles or AnC injections. Moreover, the *Phase[i]* is at high level, so the pre-charged value at *Store* is discharged. Thus, a logical 0 is read.

C. Analog Counter

In the same line as the RO and the memories designs, the integration of the AnC in-pixel allows to improve the area occupancy in front of a fully digital counter.

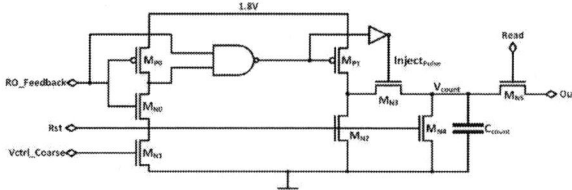

Figure 11. Analog counter schematic.

The AnC consists of a pulse generator circuit that injects a certain amount of charge into a CMIM capacitor with 190 fF capacitance (Figure 11). So, each time the RO completes a delay cycle, i.e., reaches the end of the chain, it triggers via the falling edge of the *RO_Feedback* signal (Figure 5) a charge injection into the C_{COUNT} capacitance. The amount of charge injected in the CMIM is managed by the bias voltage *Vctrl_Coarse*. Since the RO stops working when the SPAD detects a photon, the AnC stores an amount of charge, indicating how many cycles of the RO occurred during the measurement.

On the other hand, a characteristic of CMIMs is good linearity, and, in the case of a technology with a nominal voltage of 1.8 V, this linearity is preserved up to 1.2 V. Therefore, the design is focused on a working range from 0 V to 1.2 V. This range allows to perform up to 4 cycles of the digital TDC with controlled injections of ~300 mV, as proposed in Figure 12. At the end, up to a total of 72 bins are generated as a result of accumulating 4 injections of 16 bins.

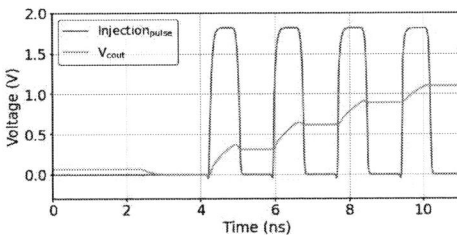

Figure 12. Simulation of the control signals and load injection of the AnC.

To ensure that the CMIM is started at 0 V, transistors MN2 and MN4 are activated by the internal reset signal (*Rst*). The *Read* signal enables the analog readout which is carried out by the NMOS pass transistor MN5. When the readout is disabled, MN5 isolates the output channel capacitor.

V. CONCLUSIONS

In summary, in this work we present a CMOS SPAD pixel of 40 μm x 40 μm with a FF of 5%. It includes a mixed-signal in-pixel TDC that allows the generation of temporary windows with a programmable width from 100ps to 1ns. So, a total measurement range of up to 72 ns is reached. The extension of the measurement range has been possible thanks to the combination of the phase generator, i.e., RO, with the AnC. On the other hand, Monte Carlo simulations show a promising standard deviation below 15 ps for narrower bins. In addition, the voltage drop measured in the memories does not affect the operation of the analog circuit for the extension of the desired range. By using counters on each bin, an on-chip histogram can be performed in the future. These results encourage us to design a CMOS camera based on this pixel with on-chip histogramming for LIDAR and FLIM applications.

ACKNOWLEDGMENT

This work has received funding from the European Union's Horizon 2020 research and innovation program under grant agreement No 737089 and from the Ministry of Science and Innovation, the Spanish State Research Agency and the European Regional Development Fund through project PID2019-105714RB-I00.

REFERENCES

[1] K. Kuzmenko *et al.*, "3D LIDAR imaging using Ge-on-Si single–photon avalanche diode detectors," *Opt. Express*, vol. 28, no. 2, p. 1330, 2020, doi: 10.1364/oe.383243.

[2] F. Villa *et al.*, "3D SP AD camera for Advanced Driver Assistance."

[3] F. Mattioli Della Rocca, T. Al Abbas, N. A. W. Dutton, and R. K. Henderson, "A high dynamic range SPAD pixel for time of flight imaging," *Proc. IEEE Sensors*, vol. 2017-Decem, pp. 1–3, 2017, doi: 10.1109/ICSENS.2017.8234049.

[4] N. Franch *et al.*, "Nano-Illumination Microscopy: a technique based on scanning with anarray of individually addressable nanoLEDs," *Opt. Express*, vol. 28, no. 13, pp. 19044–19057, 2020, doi: 10.1364/oe.391497.

[5] J. M. Pavia, M. Scandini, S. Lindner, M. Wolf, and E. Charbon, "A 1 × 400 Backside-Illuminated SPAD Sensor with 49.7 ps Resolution, 30 pJ/Sample TDCs Fabricated in 3D CMOS Technology for Near-Infrared Optical Tomography," *IEEE J. Solid-State Circuits*, vol. 50, no. 10, pp. 2406–2418, 2015, doi: 10.1109/JSSC.2015.2467170.

[6] W. Becker, *Advanced Time-Correlated Single Photon Counting Techniques*. Springer US, 2005.

[7] J. Canals, N. Franch, O. Alonso, A. Vilà, and A. Diéguez, "A point-of-care device for molecular diagnosis based on CMOS SPAD detectors with integrated microfluidics," *Sensors (Switzerland)*, vol. 19, no. 3, 2019, doi: 10.3390/s19030445.

[8] O. Alonso *et al.*, "An internet of things-based intensity and time-resolved fluorescence reader for point-of-care testing," *Biosens. Bioelectron.*, vol. 154, no. February, 2020, doi: 10.1016/j.bios.2020.112074.

[9] A. T. Erdogan *et al.*, "A CMOS SPAD Line Sensor with Per-Pixel Histogramming TDC for Time-Resolved Multispectral Imaging," *IEEE J. Solid-State Circuits*, vol. 54, no. 6, pp. 1705–1719, 2019, doi: 10.1109/JSSC.2019.2894355.

[10] L. Pancheri and D. Stoppa, "A SPAD-based pixel linear array for high-speed time-gated Fluorescence Lifetime Imaging," *ESSCIRC 2009 - Proc. 35th Eur. Solid-State Circuits Conf.*, pp. 428–431, 2009, doi: 10.1109/ESSCIRC.2009.5325948.

[11] I. Nissinen, J. Nissinen, J. Holma, and J. Kostamovaara, "A 4 × 128 SPAD array with a 78-ps 512-channel TDC for time-gated pulsed Raman spectroscopy," *Analog Integr. Circuits Signal Process.*, vol. 84, no. 3, pp. 353–362, 2015, doi: 10.1007/s10470-015-0592-1.

[12] Y. Maruyama, J. Blacksberg, and E. Charbon, "A 1024× 8, 700-ps time-gated spad line sensor for planetary surface exploration with laser raman spectroscopy and libs," *IEEE J. Solid-State Circuits*, vol. 49, no. 1, pp. 179–189, 2014, doi: 10.1109/JSSC.2013.2282091.

[13] N. Krstajić, J. Levitt, S. Poland, S. Ameer-Beg, and R. Henderson, "256 × 2 SPAD line sensor for time resolved fluorescence spectroscopy," *Opt. Express*, vol. 23, no. 5, p. 5653, 2015, doi: 10.1364/oe.23.005653.

[14] C. Niclass, M. Soga, H. Matsubara, and S. Kato, "A 100-m Range 10-Frame/s 340 x 96-Pixel Time-of-Flight Depth Sensor in 0.18-um CMOS," *IEEE J. Solid-State Circuits*, vol. 48, no. 2, pp. 559–572, 2013.

[15] T. Al Abbas, N. A. W. Dutton, O. Almer, N. Finlayson, F. M. Della Rocca, and R. Henderson, "A CMOS SPAD Sensor with a Multi-Event Folded Flash Time-to-Digital Converter for Ultra-Fast Optical Transient Capture," *IEEE Sens. J.*, vol. 18, no. 8, pp. 3163–3173, 2018, doi: 10.1109/JSEN.2018.2803087.

[16] N. Franch, O. Alonso, J. Canals, A. Vila, and A. Dieguez, "A low cost fluorescence lifetime measurement system based on SPAD detectors and FPGA processing," *J. Instrum.*, vol. 12, no. 2, 2017, doi: 10.1088/1748-0221/12/02/C02070.

[17] C. Niclass, M. Soga, H. Matsubara, M. Ogawa, and M. Kagami, "A 0.18-um CMOS SoC for a 100-m-Range 10-Frame/s 200 96-Pixel Time-of-Flight Depth Sensor," vol. 49, no. 1, pp. 1–16, 2014.

[18] A. Dieguez, J. Canals, N. Franch, J. Dieguez, O. Alonso, and A. Vila, "A Compact Analog Histogramming SPAD-Based CMOS Chip for Time-Resolved Fluorescence," *IEEE Trans. Biomed. Circuits Syst.*, vol. 13, no. 2, pp. 343–351, 2019, doi: 10.1109/TBCAS.2019.2892825.

[19] M. Perenzoni, N. Massari, D. Perenzoni, L. Gasparini, and D. Stoppa, "A 160 × 120 pixel analog-counting single-photon imager with time-gating and self-referenced column-parallel A/D conversion for fluorescence lifetime imaging," *IEEE J. Solid-State Circuits*, vol. 51, no. 1, pp. 155–167, 2016, doi: 10.1109/JSSC.2015.2482497.

[20] J. D. McKendry *et al.*, "Individually addressable AlInGaN micro-LED arrays with CMOS control and subnanosecond output pulses," *IEEE Photonics Technol. Lett.*, vol. 21, no. 12, pp. 811–813, 2009, doi: 10.1109/LPT.2009.2019114.

© VDE VERLAG GMBH · Berlin · Offenbach

Germanium–InGaZnO heterostructured thin-film phototransistor with high IR photoresponse

H. Ferhati, F. Djeffal and A. Bendjerad

Abstract— In this paper, the role of introducing Germanium (Ge)/IGZO heterostructure in enhancing the Infrared (IR) photodetection properties of thin-film phototransistor (Photo-TFT) is presented. Numerical models for the investigated device are developed using ATLAS device simulator. The influence of Ge photosensitive layer thickness on the sensor IR photoresponse is carried out. It is revealed that the optimized IR Photo-TFT based on p-Ge/IGZO heterojunction can offer improved IR responsivity of 4.1×10^2 A/W, and over 10^6 of sensitivity. These improvements are attributed to the role of the introduced p-Ge/IGZO heterostructure in promoting IR photodetection ability and improved separation and transfer mechanisms of photo-exited electron/hole pairs. The photosensor is then implemented in an optical inverter gate circuit in order to assess its switching capabilities. It is found that the proposed phototransistor shows an improved optical gain thus indicating its excellent performance. Therefore, providing high IR responsivity and low dark noise effects, the optimized Ge/IGZO IR Photo-TFT can be a potential alternative photosensor for designing optoelectronic systems with high-performance and ultralow power consumption.

Index Terms— IGZO; TFT;Optical inverter; Germanium;Infrared; Responsivity

I. INTRODUCTION

Phototransistors have received a great deal of attention during the last few years in various fields including Optical Wireless Communication Systems (OWCS), Internet of Things (IoTs), flame detection, environmental security and imaging [1-4]. Having benefited from the combination of photodetectors and transistor building block, phototransistors can offer high responsivity, less read-out circuits and maintain low dark noise effects [2-5]. In this context, several phototransistor structures were proposed based on various transistor platforms such as MOSFET, Organic transistors, Tunneling-FET and TFT [3-7]. The latter technology has recently received an enormous deal of attention due to its ability for providing low dark current, reduced power consumption and cost effectiveness properties [6-9]. Moreover, IGZO TFT demonstrated fascinating properties including UV photodetection ability, high field-effect carrier mobility, low leakage current, large-area applicability, low-cost, low-temperature processing and chemical stability [7-14]. Accordingly, various Photo-TFT structures based on different photoabsorbing materials were proposed according to the targeted spectral ranges. For instance, wide band gap materials such as ZnO and IGZO are used for UV sensing applications, whereas Peroviskite and selenium materials were used for Visible light detection [8-12]. Despite the ability of these photosensors for providing improved photoresponse characteristics, their photodetection is limited to the UV and Visible spectral bands due to the low absorbance efficiency of the photoabsorbing layers over NIR and IR ranges. Moreover, Perovskite/IGZO heterostructure exhibits high defect density at the interface and stability problems, making it challenging for reliable sensing applications [12]. Nanostructured devices based on nanowires, quantum dots and monolayer materials were also proposed to achieve broadband photoresponse characteristics [13-14]. However, the weak IR responsivity, noise effects, the high fabrication cost and processing complexity constitute the main problems preventing their large area production. Therefore, alternative inorganic, reliable and narrow band-gap absorbing materials that are able to provide high IR sensing properties are required to develop high-performance IGZO Photo-TFT devices. Interestingly, Germanium with an optical band-gap value of 0.66 eV can be effective for achieving a high IR photoresponse. Basically, this material shows high carrier mobility and high optical and electronic performances at the most commonly used communication band (1.3μm-1.55μm) making it promising for developing high-performance IR sensors. Photodetector and phototransistors based on Ge photosensitive layer are proposed, demonstrating a high IR photoresponse at λ=1.55μm [15-16]. However, noise effects and power consumption issues due to the high dark current can prevent its deployment for optoelectronic applications. Intuitively, combining the benefits of both Ge narrow band-gap material and IGZO TFT device can be efficient for developing high IR photoresponse sensors with low noise effects and energy dissipation. Accordingly, this work focuses on the design of a new IR Photo-TFT based on IGZO/Ge heterostructure. The impact of varying the Ge layer thickness on the performance of the IR phototransistor is carried out. Moreover, the optimized IGZO/p-Ge IR Photo-TFT is implemented in an optical inverter gate circuit to evaluate its optical switching characteristics. It is found that

H. Ferhati and A. Bendjerad are with the Advanced Electronics Laboratory (LEA), Department of Electronics, University of Batna 2, Algeria (ferhatihichem@gmail.com)

F. Djeffal is with the Advanced Electronics Laboratory (LEA), Department of Electronics, University of Batna 2, Algeria (faycaldzdz@hotmail.com).

© VDE VERLAG GMBH · Berlin · Offenbach

the proposed IR Photo-TFT demonstrates a high responsivity with superior optical gain, making it highly suitable for OWCS and optoelectronic applications.

II. DEVICE STRUCTURE AND NUMERICAL MODELS

The cornerstone of the proposed IR phototransistor resides on combining IGZO TFT with p-type Ge photosensitive film to form heterostructured design. Fig.1 shows a cross sectional view of the proposed Ge/IGZO Photo-TFT sensor. The investigated device consists of a three terminal transistor with n-doped IGZO channel and SiO_2 gate oxide with L and W refers to the channel length and width, respectively. In the proposed photosensor, t_{Ge} denotes the Ge film thickness, t_{IGZO} is the channel thickness and t_{SiO2} represents the gate oxide thickness. The adopted design parameters for the device numerical modeling are summarized in Table.1. The inserted Ge thin-layer acts as an absorber for the IR light owing to the high absorption coefficient of Germanium material in this spectral band. The photosensor is illuminated at normal incidence with a monochromatic light at the wavelength value of λ =1.55 μm.

To analyze the optoelectronic properties of the proposed Ge/IGZO Photo-TFT device, numerical models based on SILVACO commercial software are developed. In this context, ATLAS module is used to model the device transport mechanism by including the drift-diffusion model. The latter powerful and realistic tool has been extensively used to study the transport mechanism associated with several transistor designs, showing a great promise for reproducing the experimental results [17]. The continuity and Poisson equations given bellow are self-consistently solved using finite-element numerical method, which enables modeling carriers transport in the channel.

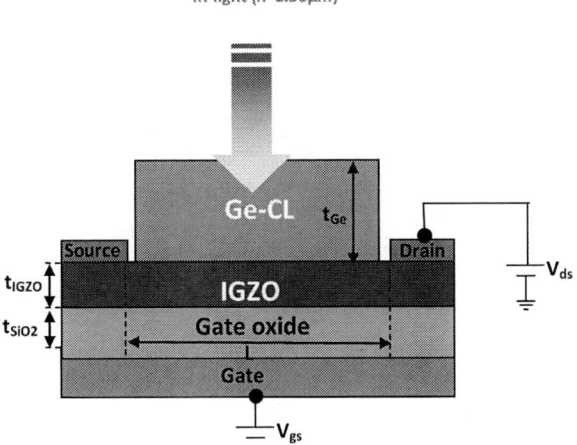

Fig. 1. Cross-sectional view of the analyzed IGZO NIR photo-TFT with Ge/IGZO heterostructure.

TABLE I
RECAPITULATION OF CONFIGURATION PARAMETERS FIXED DURING THE NUMERICAL INVESTIGATION.

Parameter	Symbol	Value
Channel length	L	100 μm
Channel width	W	1000μm
Oxide thickness	t_{SiO2}	80 nm
Germanium doping	N_a	10^{16} cm^{-3}
Germanium layer thickness	t_{Ge}	180nm
IGZO thickness	t_{IGZO}	20nm
Gate voltage	V_{gs}	(-5-20) V
Drain voltage	V_{ds}	1V

$$div(\varepsilon\nabla\psi) = -\rho \tag{1}$$

$$\text{with } \rho = q\left(p-n+n_t-p_t-p_d+N_d\right)$$

$$\frac{1}{q}div(\vec{j}_n) - R_n + G_n = 0 \tag{2}$$

$$-\frac{1}{q}div(\vec{j}_p) - R_p + G_p = 0 \tag{3}$$

where q denotes the electron charge, ε represents the absolute permittivity, ψ refers to the electrostatic potential, ρ is the net charge density determined by the n-channel doping N_d, the charged states n_t and p_t of conduction and valence band tail (CBT, VBT), the shallow donor-like densities (p_d) and free electron and hole densities (n and p). J_n and J_p denote respectively the current densities of electrons and holes, G_n, G_p, R_n and R_p are respectively the generation and recombination rates associated with electrons and holes.

The material models of the introduced films including Ge and IGZO are available in SILVACO software. Besides, the energy distribution of defect states in the band gap that depends on the degree of disorder should be taken into account to be very close to the realistic electric behavior of the investigated device [17-19]. To do so, ATLAS model provide the opportunity to include exponentially decayed band tail states and Gaussian distributions of mid-gap states by incorporating density of states (DOS) model. This model is composed of donor-like CBT, acceptor-like VBT and shallow donor-like deep level band induced by oxygen vacancies, which are modeled via Gaussian distribution. Besides, Electric Field-Dependent Mobility is introduced to consider the influence of the electric field on the carrier mobility. Besides, the optical absorption behavior and photogeneration mechanism is modeled by Finite-difference time-domain (FDTD) available in Luminous module [19].

After modeling the optical and electronic properties of the proposed Ge/IGZO Photo-TFT device, the photosensor is implemented in an optical inverter circuit to assess the device optical switching characteristics. For this purpose, SPICE module within Silvaco software is exploited. This circuit consisted of a drain load resistance and the investigated phototransistor. The applied gate voltage is fixed at 0 V and the load resistance is set to r_d=10^4 Ω. The optical power is varied and the corresponding output voltage is extracted to investigate the optical switching of the inverter circuit.

© VDE VERLAG GMBH · Berlin · Offenbach

III. RESULTS AND DISCUSSIONS

In order to elucidate the photodetection mechanism of the investigated Ge/IGZO Photo-TFT device, we show in Fig.2 I-V characteristics of the analyzed NIR photosensor with Ge photosensitive layer under dark and illumination conditions with $N_a=10^{16}$ cm^{-3}, $t_{Ge}=180$ nm, $V_{ds}=1$V, $\lambda=1.55$ μm and $P_i=1$ μW. It can be noticed that the proposed IR phototransistor exhibits a high photoresponse, where a superior photocurrent of 48 μA is recorded. Moreover, the device demonstrates a very low dark current due to the use of TFT based platform, thus offering a high ON to OFF ratio exceeding 10^6. This outstanding NIR photodetection characteristic is mainly attributed to the role of the inserted Ge/IGZO heterostructure in providing enhanced photogeneration and transfer mechanisms. In other words, electron/hole pairs are created in the Ge sensitive film and then transferred to the IGZO channel to create a current path. This leads to induce a threshold voltage shift, thus increasing the device derived current capability. On the other hand, it can be seen from this figure that by illuminating the device, the threshold voltage shift occurs, which enables enhancing the photocurrent and enlarge the operating voltage window. It is important to note that the conventional phototransistor without Ge photosensitive layer cannot provide an IR photoresponse owing to the wide band gap intrinsic property associated with amorphous IGZO, which is completely transparent over NIR and IR spectral ranges.

The sensor photodetection capabilities are mainly correlated with the Ge photoabsorbing layer thickness. In this sense, the variation of the device responsivity and sensitivity as a function of the Ge film thickness is depicted in Fig.3. To this extent, the device sensitivity and responsivity are estimated using the following formulas

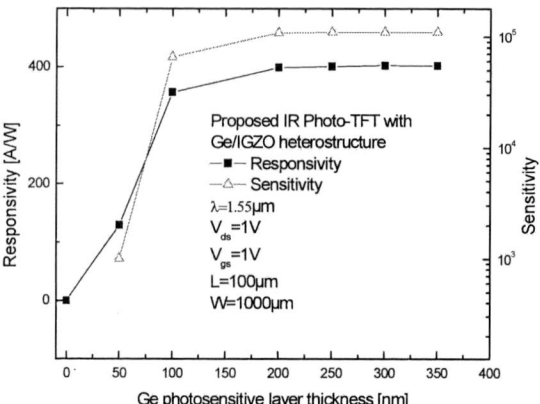

Fig.3. Device responsivity and sensitivity as a function of the Ge layer thickness with $V_{ds}=1$V, $V_{gs}=1$V, L=100μm, W=1000μm, $\lambda=1.55\mu$m, and $t_{IGZO}=20$nm.

$$S = \frac{I_{ds-ill} - I_{ds-dark}}{I_{ds-dark}} \times 100 \qquad (4)$$

$$R = \frac{I_{ds-ill}}{P_i} \ [A/W] \qquad (5)$$

where I_{ds-ill} and $I_{ds-dark}$ are drain currents under illumination and dark conditions and P_i denotes the optical power density.

It can be observed from Fig.3 that the phototransistor responsivity and sensitivity parameters increase with the Ge layer increase to reach their maximum values of 4×10^2 A/W and 10^5, respectively for $t_{Ge}=180$nm. This is mainly attributed to the relationship between the absorbance and the photosensitive layer, where the thickness increase leads to enlarge the quantity of the photo-induced carriers, thus allowing an enhanced NIR photoresponse. After this critical value of $t_{Ge}=180$ nm, the device FoM parameters are saturated due to the low minority carrier life-time of the Ge material.

Aiming at evaluating the optical switching characteristics of the proposed NIR Photo-TFT based on Ge/IGZO heterostructure, a circuit level performance analysis is performed by implementing the device in an optical inverter gate circuit as it is shown in Fig.4 (a). The associated output voltage as a function of the applied optical power density is extracted at $V_{gs}=0.5$ V. The obtained results are compared to the conventional Ge phototransistor design based on MOSFET platform [3] as it is shown in Fig.4 (b). It can be seen that the proposed TFT-based phototransitor demonstrates an improved dynamic range from high to low voltage states in comparison with that of the conventional counterpart based on MOSFET bulding block [3]. Moreover, the proposed photosensor based on TFT design enables achieving near 1V high state voltage due to the ultralow dark current promoted by this technology. It can be seen from this figure that the proposed phototransistor allows reaching an appropriate switching characteristics, where the optical power increase leads to the switch the response from OFF to ON states with high optical gain of 5.4. This outstanding results are attributed to the use of TFT platform. The latter allow achieving superior swing

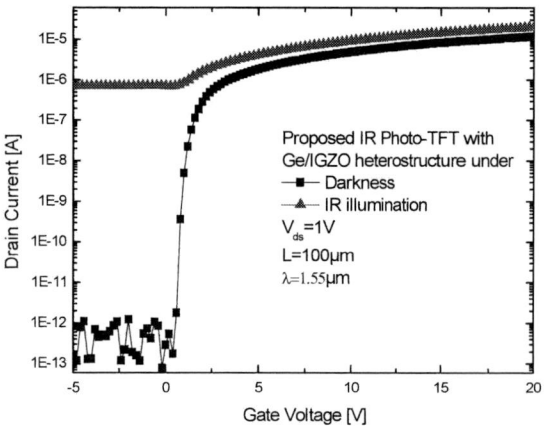

Fig. 2. I_{ds}-V_{gs} characteristics of the proposed Ge/IGZO IR Photo-TFT under dark and IR-light illumination with dissimilar optical power densities, $V_{ds}=1$V, L=100μm, W=1000μm, $\lambda=1.55\mu$m, $t_{Ge}=180$nm, $t_{IGZO}=20$nm and Pi=1μW.

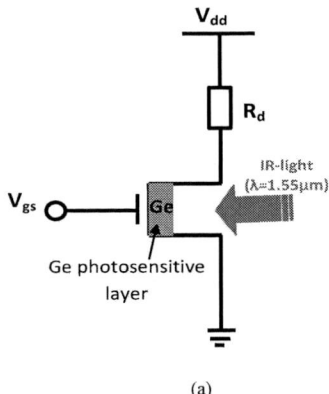

(a)

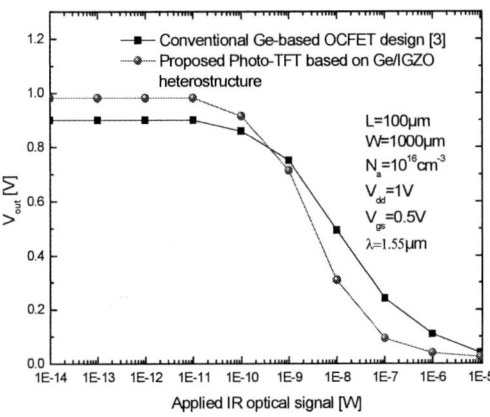

(b)

Fig. 4. (a) Optical inverter circuit design, (b) Output voltage as a function of the applied IR optical power for both optical inverter circuits based on the proposed Photo-TFT with Ge/IGZO heterojunction and the conventional Ge phototransistor with MOSFET platform.

properties. Therefore, the proposed IR phototransistor based on combining TFT platform with Ge/IGZO heterostructured design can be efficient for developing high-performance optical receivers for the next-generation OWCS.

IV. CONCLUSION

In this work, a new IR phototransistor based on Ge/IGZO heterostructure and TFT platform is proposed and numerically studied. The effect of the Ge photosensitive layer thickness on the device sensitivity and responsivity is carried out. The results show the IR photodetection ability of the proposed photosensor, where a high responsivity of 4×10^2 A/W is recorded over a wide operating voltage window. Moreover, the proposed device demonstrates a low dark noise effects due to the use of TFT platform. A circuit level performance analysis is performed by implementing the proposed Ge/IGZO Photo-TFT in an optical inverter to assess its optical switching capabilities. It was revealed that the proposed device shows an excellent optical gain with low power consumption. Therefore, the proposed IR phototransistor offers the possibility to bridge

gap between high photoresponse and ultra-low dark noise properties, making it highly attractive for OWCS.

REFERENCES

[1] C. Sun, et al, "Single-chip microprocessor that communicates directly using light," Nature, vol. 528, pp. 534-538, 2015.

[2] H. Ferhati and F. Djeffal, "Giant responsivity of a new InGaZnO ultraviolet thin-film phototransistor based on combined dual gate engineering and surface decorated Ag nanoparticles aspects," Sensors and Actuators A: Physical, vol.318, pp.112523, 2021.

[3] L. Colace, V. Sorianello and S. Rajamani, "Investigation of Static and Dynamic Characteristics of Optically Controlled Field Effect Transistors," Journal of lightwave technology, vol. 32, pp. 2233-2239, 2014.

[4] R. Pan, Q. Guo, J. Cao, G. Huang, Y. Wang, Y. Qin, Z. Tian, Z. An, Z. Did and Y. Mei, "Silicon Nanomembrane-based Near Infrared Phototransistor with Positive and Negative Photodetections," Nanoscale, vol.11, pp.16844-16851, 2019.

[5] H. Ferhati and F. Djeffal, "Boosting the optical performance and commutation speed of phototransistor using SiGe/Si/Ge tunneling structure," Mater. Res. Express, vol. 5, pp. 065902, 2018.

[6] Y. Lei, N. Li,W-K. E. Chan, B. S. Ong and F. Zhu, "Highly sensitive near infrared organic phototransistors based on conjugated polymer nanowire networks," Organic Electronics, vol.48, pp. 12-18, 2017.

[7] Y. J. Tak, D. J. Kim, W-G. Kim, J. H. Lee, S. J. Kim, J. H. Kim and H. J. Kim, "Boosting Visible Light Absorption of Metal-Oxide-Based Phototransistors via Heterogeneous In–Ga–Zn–O and CH3NH3PbI3 Films," ACS Applied Material Interfaces, vol. 10, pp.12854-12861, 2018.

[8] D. K. Hwang et al, "Ultrasensitive PbS quantum-dot-sensitized InGaZnO hybrid photo-inverter for near-infrared detection and imaging with high photogain," NPG Asia Materials, vol. 8, pp.233, 2016.

[9] T. H. Chang, C. J. Chiu, W. Y. Weng, S. J. Chang, T. Y. Tsai, and Z. D. Huang, "High responsivity of amorphous indium gallium zinc oxide phototransistor with Ta2O5 gate dielectric," Applied Physics Letters, vol. 101, pp. 261112, 2012.

[10] H-J. Na et al. "A visible light detector based on a heterojunction phototransistor with a highly stable inorganic CsPbIxBr3_x perovskite and In–Ga–Zn–O semiconductor double-layer," Materials Chemistry C, vol.7, pp.14223, 2019.

[11] H. Yu, X. Liu, L. Yan, T. Zou, H. Yang, C. Liu, S. Zhang and H. Zhou, "Enhanced UV-visible detection of InGaZnO phototransistors via CsPbBr3 quantum dots," Semiconductor Science and Technology, vol. 34, pp.25013, 2019.

[12] H. Yoo et al., "High Photosensitive Indium–Gallium–Zinc Oxide Thin-Film Phototransistor with a Selenium Capping Layer for Visible-Light Detection," ACS Appl. Mater. Interfaces, vol.12, pp.10673–10680, 2020.

[13] J. Yang, H. Kwak, Y. Lee, Y-S. Kang, M-H. Cho, J. H. Cho, Y-H. Kim, S-J. Jeong, S. Park, H-J. Lee and H. Kim, "MoS2–InGaZnO Heterojunction Phototransistors with Broad Spectral Responsivity," ACS Applied Material Interfaces, vol. 8, pp.8576–8582, 2016.

[14] X. Liu, X. Liu , J. Wang, C. Liao, X. Xiao , S. Guo, C. Jiang, Z. Fan, T. Wang, X. Chen , W. Lu, W. Hu, and L. Liao, "Transparent, High-Performance Thin-Film Transistors with an InGaZnO/Aligned-SnO 2-Nanowire Composite and their Application in Photodetectors, " Advanced Materials, vol. 26, pp. 7399–7404, 2014.

[15] X. Liu, et al, "High-Performance Ge Quantum Dot Decorated Graphene/Zinc-Oxide Heterostructure Infrared Photodetector," ACS Appl. Mater. Interfaces, vol.7, pp.2452–2458, 2015.

[16] H. Ferhati, F. Djeffal, "Planar junctionless phototransistor: A potential high-performance and low-cost device for optical-communications," Optics & Laser Technology,vol. 97, pp.29, 2017.

[17] Atlas User's manual, SILVACO TCAD, 2012.

[18] K. J. Saji, M. K. Jayaraj, K. Nomura, T. Kamiya and H. Hosono, "Optical and Carrier Transport Properties of Cosputtered Zn–In–Sn–O Films and Their Applications to TFTs," J. Electrochem. Soc., vol.155, pp.390–395, 2008.

[19] T. Kamiya, K. Nomura and H. Hosono, "Present status of amorphous In–Ga–Zn–O thin-film transistors," Sci. Technol. Adv. Mater., vol.11, pp.1–23, 2010.

© VDE VERLAG GMBH · Berlin · Offenbach

SMACD / PRIME 2021 | 19 – 22 July 2021, Online Event

Integrated Hysteretic Controlled Regulating Buck Converter with Capacitively Coupled Bootstrapping

Francarl Galea[1], Owen Casha, Ivan Grech, Edward Gatt and Joseph Micallef

Department of Microelectronics and Nanoelectronics

University of Malta

[1]Email: francarl.galea@um.edu.mt

Abstract—**This paper presents the circuit design, layout implementation and simulations of an on-chip hysteretic regulating buck converter. All the high voltage transistors of the buck converter were implemented using 45 V NMOS thin gate oxide layer transistors and are operated by means of a capacitively coupled bootstrap circuit. The circuit was implemented using the XFAB CMOS 0.35 μm high voltage technology and all the circuit blocks were designed using analog electronic techniques, with the transistors operating in the sub-threshold region, in order to minimize power consumption. Simulations show that the control circuit of the hysteretic regulator consumes around 0.5 μW, and the proposed buck converter operates at a peak efficiency of 82%. The input voltage range of the regulator is from 1.5 V to 45 V and has a power range from 10 μW to 200 mW. The output regulated voltage is tunable via a feedback resistor and can be varied from 0.6 V to 40 V, with a dropout voltage of 1 V.**

Index Terms—**Power Conditioning Circuits, Energy Harvesters, Sub-threshold analog operation, Buck Converter, Hysteretic Regulator, Capacitively Coupled Bootstrap.**

I. INTRODUCTION

Stepping down an input DC voltage is typically achieved using either linear regulators or switch-mode converters. The former is very inefficient especially at high voltage conversion ratios [1]. Switch-mode DC-DC converters are capable of converting DC voltages by temporarily storing the energy and then releasing it at a different voltage level. Improved DC-DC converters use synchronous switching, where the freewheeling diode is replaced by a transistor in order to obtain a higher efficiency as long as the converter remains operating in the continuous conduction mode (CCM). In discontinuous conduction mode (DCM) the efficiency is worse because the inductor current goes negative unless a zero current detector circuit is implemented [2]. Most DC-DC converters are close loop controlled where a constant output voltage can be obtained from a wider range of input voltage and output load current conditions [3], [4].

This paper proposes an on-chip hysteretic regulating buck converter which complements the design of an AC/DC-to-DC converter with Maximum Power Point Tracking (MPPT) function proposed in [5], [6]. By combining these two circuits, a power conditioning circuit for energy harvesters having MPPT, capacitive energy storage and a regulated output voltage is achieved. As shown in Fig. 1, the first stage of the power conditioning circuit consists of an AC/DC-to-DC converter controlled by a MPPT control. Prototypes of this integrated circuit were fabricated and its measured results are

characterization were presented in [7]This paper presents the design, layout considerations and simulations of the second stage which is a hysteretic regulating buck converter.

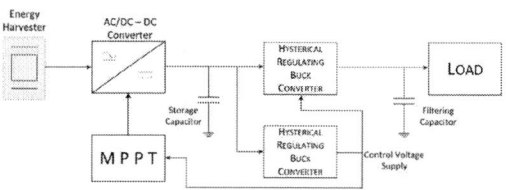

Fig. 1: Block diagram of the proposed power conditioning circuit.

The proposed control loop is designed on a converter operating in CCM so that its stability is maintained even if the converter shifts to DCM. A control loop tuned for a converter set to operate in DCM may become unstable if it shifts to CCM [8]. Most switch mode converters are operated with pulse width modulation at a fixed switching frequency. In some designs as in [9], even hysteretic controlled converters, use an oscillator, thus increasing the power consumption of the control circuitry. In this paper, the design uses a hysteretic control where only one comparator is capable of obtaining the required control, thus eliminating the need of a sawtooth generator or an oscillator, and so achieves a highly efficient operation. The design of the hysteretic control technique of a buck converter is mostly used to obtain a faster transient response, lower power dissipation in the control circuitry, a simpler design and simplification of the modelling of the steady state equations and dynamic analysis [10]. Till date no on-chip capacitively coupled bootstrapping circuits was reported in literature, however such technique has already been employed using discrete components as in [11].

The power conditioning circuit requires two identical buck converters. One is used to generate an output voltage as required by the load and another one is employed to generate the 2 V supply voltage for all the control circuitry. The architecture was implemented using the XFAB XH035 technology and is illustrated in Fig. 1. The power conditioning circuit control was designed to work with a 2 V supply voltage. This required the use of high voltage transistors having a thin gate oxide layer as their threshold voltage is low enough to be operated with such a low voltage. Since the chosen technology does not have any PMOS high voltage transistors with a thin gate oxide layer,

© VDE VERLAG GMBH · Berlin · Offenbach

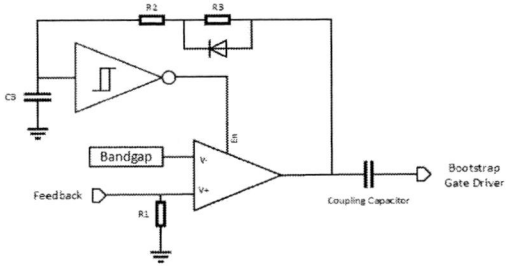

Fig. 2: Circuit diagram of the hysteretic control stage.

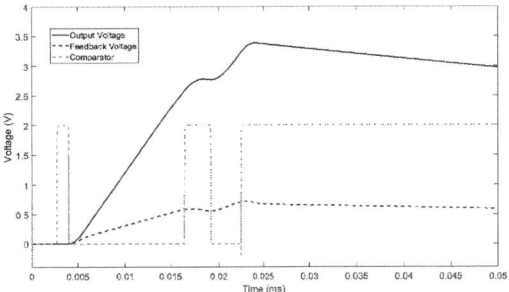

Fig. 3: Derivative feedback is used to lower the voltage overshoot during the startup phase.

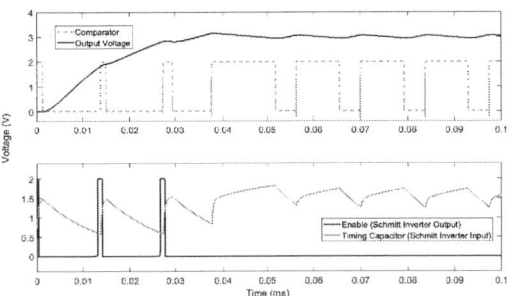

Fig. 4: The comparator is periodically disabled in order to switch off the bootstrap circuit and recharge capacitor C_{BOOT}.

the circuit was designed with only NMOS transistors while making use of a capacitively coupled bootstrap circuit in order to correctly switch on the high side NMOS transistors.

All the control circuitry was implemented using analogue electronics in order to obtain a high efficiency since microprocessors, ADCs, DACs and clock generators all consume a substantial amount of power which may sometimes be higher than the power available from energy harvesters. Additionally, all the transistors in the control circuit operate in the subthreshold region in order to limit the power dissipation in the transistors, thus further increasing the efficiency.

II. ARCHITECTURE OVERVIEW AND SIMULATIONS

The hysteretic controlled regulating buck converter consists of three main parts. The first part is the low voltage stage which provides the hysteretic control and operates with a supply voltage of 2 V. This stage is capacitively coupled to the second stage which is the bootstrap gate drive circuit. These two stages are capacitively coupled since the second stage is a high voltage stage and so the capacitor acts as an isolation between these stages. In fact, the voltage of the capacitor's plate on the high voltage side varies at a high rate during switching and can reach up to 45 V with respect to ground. The final stage is the buck converter which steps down any input voltage up to 45 V to the required output voltage. The hysteretic control stage is shown in Fig. 2. Control is achieved through the use of a comparator. The comparator design is very similar to the circuit proposed in [7]. The hysteresis element is not included in the comparator itself but hysteresis still happens in the L-C components of the buck converter which are externally connected. Once the output voltage reaches the reference voltage and the comparator switches off Q5, the energy stored in the inductor due to its forward current is transferred into capacitor C_O. As a result, C_O's voltage exceeds the reference voltage. The reference output voltage of the regulator can be set externally through the sizing of a feedback resistor. The comparator compares the output voltage which is fed back through the external feedback resistor with a reference voltage of 0.6 V generated via a bandgap circuit. Resistor R1's value is 50 kΩ. The feedback resistor's value in kΩ with respect to the reference output voltage V_O should be

$$R_{FB} = 83 \left(V_O - 0.6V \right) \qquad (1)$$

A capacitor can be connected in parallel with the feedback resistor in order to lower the overshoot of the output voltage during startup via derivative feedback as shown in Fig. 3. This switches off the comparator slightly before the output voltage reaches the reference voltage.

A problem with bootstrapping circuits is the discharging of the bootstrap capacitor. In order not to allow the bootstrap capacitor to discharge and in turn ensure that the high side NMOS transistor keeps operating in the triode region, the control circuit was designed such that it temporarily disables the comparator after long ON periods. The comparator's output goes high, making the bootstrap circuit switch off the Buck converter for an adequate time until the bootstrap capacitor charges again. This is achieved by the Schmitt inverter which upon the discharging of capacitor C_3, disables the comparator temporarily until C_3 is charged again as shown in Fig. 4. The Schmitt inverter design is similar to the circuit fabricated in [7]. The values of resistors R2 and R3 determine the duty cycle and the frequency of the enable pulses. During startup, when the ON period exceeds 12 µs, the Schmitt inverter temporarily disables the comparator and together with derivative feedback, lowers the startup voltage overshoot as seen in Fig. 4.

The bootstrap gate drive circuit is shown in Fig. 5. The coupling capacitor is driven by the comparator from the previous stage. V_{DD} is the supply voltage of 2 V whereas V_{IN} is the input voltage which can range from 1.5 V to 45 V. When the comparator pulls down its output from 2 V to 0 V, the gate driving stage charges the high side NMOS gate capacitance of transistor Q_5, through transistor Q_3, using the

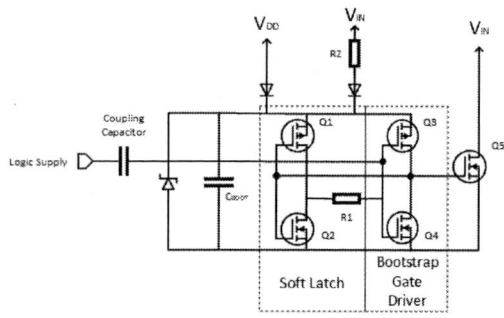

Fig. 5: Simplified bootstrap gate drive circuit diagram.

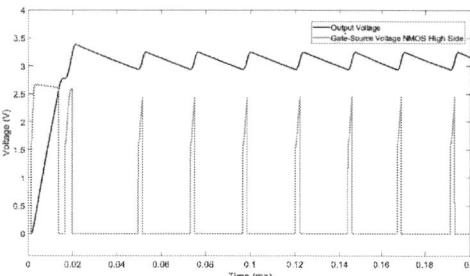

Fig. 6: V_{GS} of high side NMOS transistor generated by the bootstrap circuit.

capacitor's C_{BOOT} stored charge. As the source of transistor Q_5 increases towards V_{IN}, the gates of transistors Q_3 and Q_4 are pulled further down by the coupling capacitor, thus creating an indirect positive feedback latching mechanism. This ensures that the state is changed and maintained until Q_5 fully switches on to the triode region. Effectively the bootstrapping circuit is sensitive to the differentiation (pull up or pull down) of the coupling capacitor voltage. An additional soft latch is implemented in the circuit via transistors Q_1 and Q_2 and resistor R_1. This ensures that the state is latched until the comparator changes back its output to 2 V. A simulation of the bootstrap circuit is shown in Fig. 6 with the regulating buck converter maintaining a 3 V output voltage.

The buck converter implemented in this regulator is shown in Fig. 7. The output inductor and capacitor are not integrated and have to be externally connected. They can be sized specifically according to the voltage, current and ripple requirements. The feedback is to be taken from the voltage across capacitor C_O. Diode D1 is the freewheeling diode of the buck converter

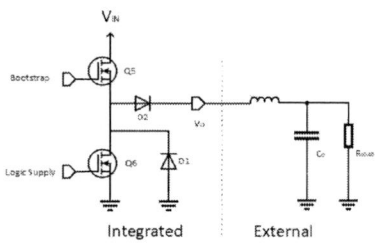

Fig. 7: Schematic diagram of the regulating buck converter.

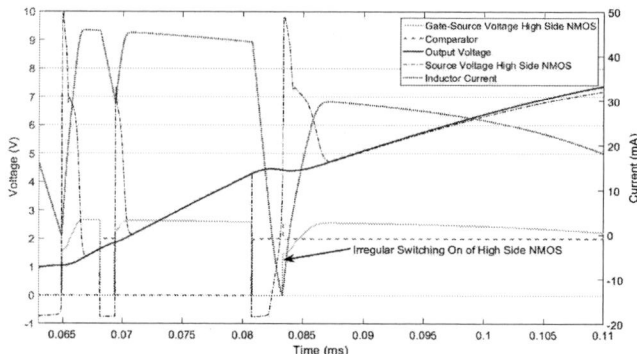

Fig. 8: A simulation showing the unreliable operation of the bootstrap circuit without diode D2, where the output voltage keeps increasing due to incorrect switching.

although most of its current is conducted by transistor Q_6. This achieves a high conversion efficiency. Diode D2 is essential for when the buck converter operates in DCM because as soon as the inductor current drops to zero, the source voltage of the high side NMOS transistor Q_5 increases to the voltage across capacitor C_O. As a result, the bootstrap circuit switches its state and turns on transistor Q_5 mainly because the bootstrap circuit senses a pull down from the coupling capacitor. This is shown in Fig. 8. By including diode D2 in the buck converter, it is ensured that the bootstrap circuit never switches on Q_5 as the inductor's current drops to zero. Diode D2 also prevents negative current from flowing from capacitor C_O to Q_6 during the OFF period when the converter is operating in DCM.

III. LAYOUT CONSIDERATIONS

Following the circuit design and simulations at various temperature, voltage and process conditions, the layout of the circuit was carried out. Particular attention was given to every component selected so that the absolute maximum voltage limit between the terminals of each component and between every terminal and substrate is not exceeded during the normal operation of the circuit. All the transistors, diodes and passive components in the bootstrap circuit and buck converter are isolated from the substrate. The hysteretic buck regulator layout is shown in Fig. 9, with all the respective components adequately labelled. Pads were included at the input and output of every block, so that each individual block could be monitored externally and maximize its testability. The dimensions of one hysteretic converter without pads is 530 μm by 284 μm, that is an area of 0.15 mm². The fabricated integrated circuit contains three copies of the same converter. The total dimensions of the three converters and pads is 1.6 mm by 1.8 mm, that is an area of 2.9 mm². The layout includes five high voltage pads and fourteen low voltage pads as shown in Figure 10.

Post-layout simulations were carried out in order to assess the effect of the pads on the total power consumption and the functionality of the circuit blocks. Fig. 11 shows the results of the simulation of the converter at an input voltage of 10 V, loaded with a 6 kΩ resistance and set to output

© VDE VERLAG GMBH · Berlin · Offenbach

310

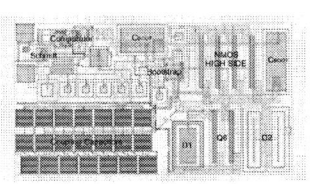

Fig. 9: Layout of the hysteretically controlled regulating buck converter.

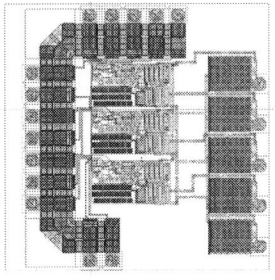

Fig. 10: Layout of the integrated circuit with the three regulators including all their respective pads.

a voltage of 3 V. The inductor and capacitor values in this simulation are 200 μH and 200 nF respectively. The inductor current of one ON cycle is shown. The ripple output voltage in this simulation is 5% and the conversion efficiency is 71%. Additional simulations were carried out and data was analysed in order to determine the operating efficiency of the hysteretic buck converter at various operating conditions. The efficiency of the converter varies depending on several parameters like the output current, the input voltage and the output voltage. The peak conversion efficiency of the buck converter is 82% at an input voltage of 25 V, an output voltage of 5 V and an output power of 2 mW whereas the control power consumption is 0.5 μW. The output voltage ripple is 2.3%.

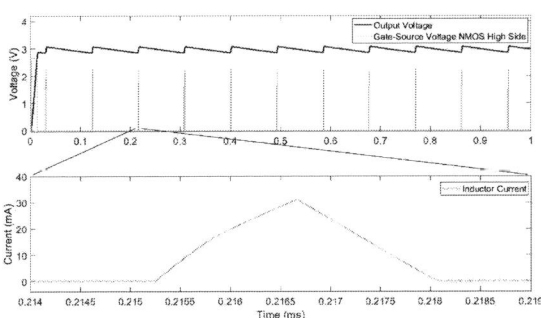

Fig. 11: Simulation of the buck converter with an input voltage of 10 V, an output voltage of 3 V and an output load resistance of 6 kΩ.

One of the three converters is capable to generate the 2 V supply voltage V_{DD} required for all the three converters to operate. Cold starting the converter supplying V_{DD} can be achieved externally by connecting an additional capacitor from the input voltage V_{IN} to the output of this converter (which in this case also has to be connected to V_{DD}) so that as soon as a voltage is applied on the input of the converters, a temporary voltage supply is generated on the comparator to startup the converters.

IV. CONCLUSIONS AND FUTURE WORK

This paper presents the design, implementation and simulation of a novel integrated hysteretic controlled regulating buck converter. The control is based on one comparator and

a Schmitt inverter and does not make use of any oscillators. Since most of the currently available fabrication technologies lack high voltage PMOS transistors with a thin gate oxide layer, the high side transistors can only be operated either by increasing the control supply voltage to 5 V and operate PMOS with medium gate oxide layer. This sacrifices power consumption as a result. As alternatively proposed in this work, one can make use of a bootstrapping circuit with an NMOS transistor. The bootstrapping circuit being proposed is completely integrated and is capacitively coupled between the low and high voltage sides. Since the feedback loop, the inductor and the output capacitor are connected externally, the control loop, the output voltage and the ripple voltage of the regulator can be easily tuned externally according to the required specifications.

Simulations show that the complete circuit works successfully and the conversion efficiency of the regulator reaches 82%. Simulation results have shown that the controller consumes a power of around 500 nW, which is considered negligible for microwatt and milliwatt range energy harvesters. This circuit is in the fabrication stage and experimental results obtained from this circuit will be published. Then it will be connected to the output of the AC/DC-to-DC converter with MPPT control, in order to generate a constant output voltage. Once the complete power conditioning circuit is connected, the performance of the whole circuit will be tested and evaluated with various types of energy harvesters.

REFERENCES

[1] R. Zinn, "High efficiency linear regulator," Aug. 1 2006, uS Patent 7,084,612.
[2] Yuan Gao, Shenglei Wang, Haiqi Li, Leicheng Chen, Shiquan Fan, and Li Geng, "A novel zero-current-detector for dcm operation in synchronous converter," in *2012 IEEE International Symposium on Industrial Electronics*, 2012, pp. 99–104.
[3] Y. Wang and D. Ma, "Ultra-fast on-chip load-current adaptive linear regulator for switch mode power supply load transient enhancement," in *2013 Twenty-Eighth Annual IEEE Applied Power Electronics Conference and Exposition (APEC)*, 2013, pp. 1366–1369.
[4] M. K. Kazimierczuk, *Pulse-width modulated DC-DC power converters*. John Wiley & Sons, 2015.
[5] F. Galea, O. Casha, I. Grech, E. Gatt, and J. Micallef, "A CMOS MPPT power conditioning circuit for energy harvesters," in *2017 24th IEEE International Conference on Electronics, Circuits and Systems (ICECS)*. IEEE, 2017, pp. 442–445.
[6] ——, "Ultra Low Frequency Low Power CMOS Oscillators for MPPT and Switch Mode Power Supplies," in *2018 14th Conference on Ph. D. Research in Microelectronics and Electronics (PRIME)*. IEEE, 2018, pp. 121–124.
[7] ——, "An Ultra Low Power CMOS MPPT Power Conditioning Circuit for Energy Harvesters," in *2020 Conference on IEEE International Symposium on Circuits and Systems (ISCAS)*. IEEE, 2020.
[8] L. Dinwoodie, "Reference Design: Isolated 50 Watt Flyback Converter Using the UCC3809 Primary Side Controller," *Unitrode Products from Texas Instruments*, 2003.
[9] K. Taniguchi, T. Sato, T. Nabeshima, and K. Nishijima, "Constant frequency hysteretic PWM controlled buck converter," in *2009 International Conference on Power Electronics and Drive Systems (PEDS)*, 2009, pp. 1194–1199.
[10] Yongxiao Liu, Jinbin Zhao, Keqing Qu, and Yang Fu, "Steady-state and dynamic analysis of hysteretic buck converter with indirect-feedforward and feedback control," in *IECON 2013 - 39th Annual Conference of the IEEE Industrial Electronics Society*, 2013, pp. 1296–1301.
[11] E. H. Wittenbreder Jr, "Capacitively coupled high side gate driver," Sep. 15 2009, uS Patent 7,589,571.

Single-Inductor Dual-Output Buck Converter with Charge Recycling

Kemal Ozanoglu and Gunhan Dundar
Department of Electronics and Communications Engineering,
Bogazici University, Istanbul, Turkey

Abstract—This paper presents a novel switching topology for SIDO buck converters aiming to achieve charge recycling during sink/source operation. Two switching sequence options have been demonstrated for low load current (DCM) and high load current (CCM) profiles. Simulation results show that the proposed architecture can achieve sink/source operation by re-using the charge stored in the output capacitors with continuous conduction through the inductor, demonstrating that this topology promises to be a convenient solution for SIDO/SIMO switching converters where sink/source operation and high power efficiency are essential.

Keywords — DC-DC converters, SIDO buck converter, power management integrated circuits.

I. INTRODUCTION

The increasing demand for portable applications has made switching converters one of the key blocks for most battery-operated systems due to their high efficiency voltage conversion performance [1]. In portable applications, form factor and cost are two of the most important system design criteria. Therefore, switching converter design topologies with smaller size and minimum number of extra components are preferred. To this end, single-inductor dual-output (SIDO) and single-inductor multiple-output (SIMO) buck converters have been proposed [2-4] which can supply more than one output voltage by using a single inductor, delivering high power efficiency at the same time, hence achieving longer battery life with small form factor. SIDO and SIMO buck topologies where the total load current can be supplied by a single inductor are utilized when it is required to save printed circuit board (PCB) area and cost by removing additional external components. Figure 1 shows a conventional SIDO buck converter switch topology. The buck switch S_{buck} charges the inductor and the load switches S_0 and S_1 distribute the inductor current to the outputs.

Design specifications of SIDO switching converters include supplying a wide current output range with high efficiency, low output voltage ripple, and sufficient line/load transient response. Similar to single output buck converters, negative current and dynamic voltage scaling (DVS) function capabilities [5] are also requested in SIDO/SIMO buck converters, necessitating the functionality to perform sink/source operation at its outputs.

However, saving from external components brings forth various drawbacks including: limited output current capability, need for complex control techniques, increased voltage ripple at the outputs, and single polarity output current operation:

Assuming a positive inductor current is flowing to one of the outputs (source operation), the switching converter will not be able to reduce the other output voltages (sink operation) if an overshoot happens or a negative DVS is requested. This is because the inductor element behaves like a current source and opposes any rapid changes e.g., changes in the polarity of the inductor current.

The straightforward solution to this limitation (charging/sourcing current to one of the outputs while discharging/sinking current from another output) is to use a pulldown switch for discharging: When an overshoot event occurs or a negative DVS event is requested, a pulldown switch discharges the output capacitor to the requested voltage level, but the major drawback of this approach is lost charge and reduced efficiency. In inductive switching converters, it is preferred to preserve charge by utilizing the inductor continuously, aiming to achieve high power efficiency. As a numeric example, assuming a SIDO buck converter with same voltage settings for both outputs, when one output is sourcing 1A and the other output is sinking 100mA, facilitating a re-route of the 100mA current back to the converter would improve the power efficiency by 10%!

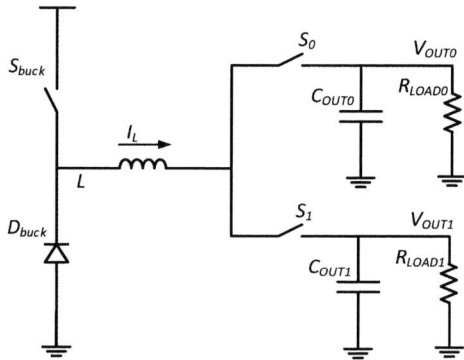

Fig. 1. Conventional SIDO buck converter.

This paper describes a novel technique aiming to achieve this functionality by using additional switches, making it possible to recycle a sinking current from one of the outputs while supplying a source current to the other output. Even though the technique describes a new topology through the example of a SIDO buck converter, the mentioned concepts can be applicable to other switching converters with multiple outputs.

© VDE VERLAG GMBH · Berlin · Offenbach

This paper is organized as follows: the next section describes the proposed SIDO topology and the proposed switching sequences followed by Section III about block level implementation details. Section IV gives simulation results, to be followed by conclusion.

II. PROPOSED TOPOLOGY

Figure 2 describes the proposed SIDO converter switch topology [6]. Switch S_B is the buck high side switch. The diode (located below S_B), also called the "active diode" is commonly implemented as a switch, allowing current flow when S_B is OFF. S_{Y0} and S_{Y1} are the conventional SIDO buck switches supplying current to outputs V_{OUT0} and V_{OUT1}. S_{X0}, S_{X1} and S_Z are the newly introduced switches aiming to achieve sink/source operation – allowing positive and negative currents at different outputs. During physical implementation, the switches S_{X0}, S_{X1}, and S_Z can be designed to be smaller in size (relative to the other pass device switches); the design trade-off here is silicon area and capacitive losses vs. power efficiency during sink (negative output current) operation.

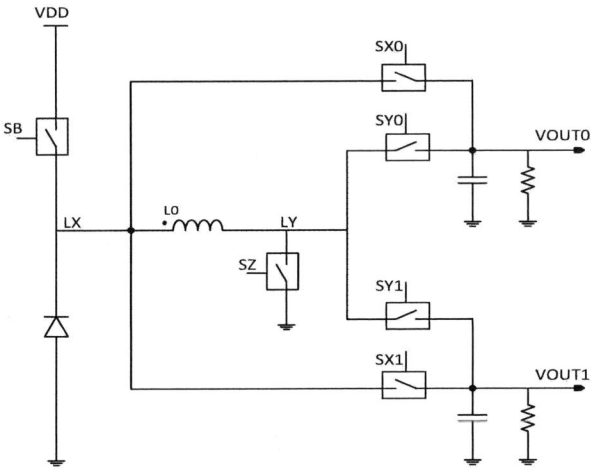

Fig. 2. Proposed SIDO converter switch topology.

During typical SIDO operation (source operation), switches S_{X0}, S_{X1} and S_Z will be OFF. For the case where one output (e.g., V_{OUT0}) is supplying a positive load current while the other output (e.g., V_{OUT1}) is requested to perform sink/negative current operation, the introduced switches will enable two different operation solutions as described in the following sections.

A. Negative SIDO Operation – Option 1 (DCM)

Figure 3 and Figure 4 give an example of source/sink SIDO operation. Figure 3 shows source/typical operation, where positive current is being sourced to V_{OUT0}. S_{Y0} is ON, S_{Y1} is OFF, S_B and the active diode is performing buck switching, the introduced switches S_{X0}, S_{X1} and S_Z are OFF.

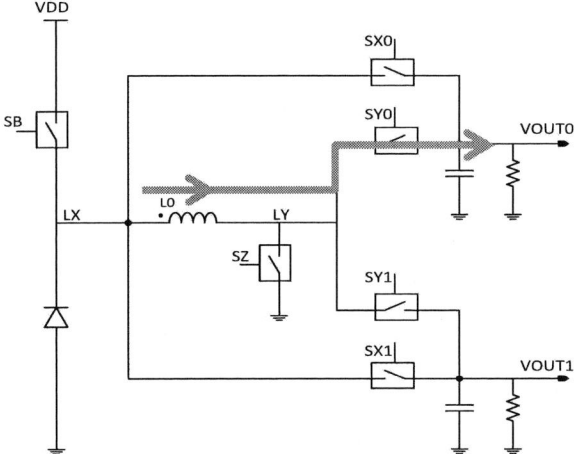

Fig. 3. SIDO current paths for source operation switch sequence.

Figure 4 describes sink/negative output current operation - *Option 1*, where the inductor current flows from V_{OUT1} to ground. S_{Y0} and S_{Y1} are OFF, S_B is OFF, S_{X0} is OFF, S_{X1} and S_Z are ON. S_{X1} and S_Z enable current sink operation from V_{OUT1} while maintaining a positive current flow at the inductor L_0. With this switch configuration, the lost charge on the discharged output capacitor is stored as current on the inductor (through its magnetic field). In the next phase, when SIDO is back to normal/source operation (as in Figure 3), the stored inductor current will continue to flow from S_{Y0} sourcing V_{OUT0}. Hence, subsequent sink/source operations can be achieved with continuous current flow on the inductor.

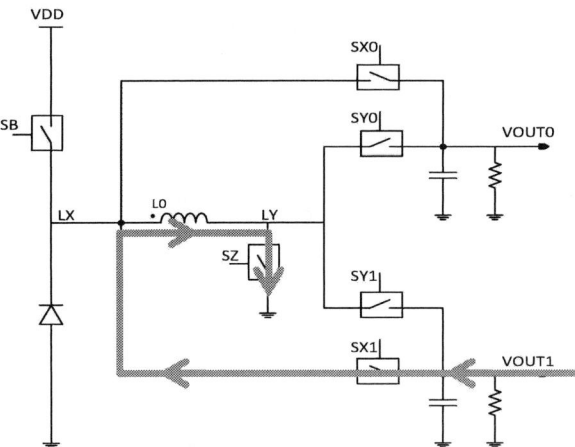

Fig. 4. SIDO current paths for sink operation switching sequence, *Option 1*.

This operation mode can be preferred at low load currents or for discontinuous conduction mode (DCM), as the connection of the inductor to ground by switch S_z guarantees that the inductor current will not decay to zero, hence no prior/stored inductor current is needed before the start of this switching sequence.

B. Negative SIDO Operation – Option 2 (CCM)

Figure 5 describes an alternative solution for sink operation, with higher power efficiency at high load currents. In the first phase, as in Figure 3, positive current is being sourced to V_{OUT0} (normal operation). S_{Y0} is ON, S_{Y1} is OFF, S_B is performing buck switching, S_{X0}, S_{X1} and S_Z are OFF.

In the second phase, shown in Figure 5, current is being drained from V_{OUT1} (sink operation) while at the same time sourcing a positive output current to V_{OUT0}. S_{X1} and S_{Y0} are ON, S_{X0}, S_{Y1}, S_B and S_Z are OFF. In this switch configuration S_{X1} and S_{Y0} enable current flow from V_{OUT1} to V_{OUT0} through L_0, maintaining a positive current flow at the inductor.

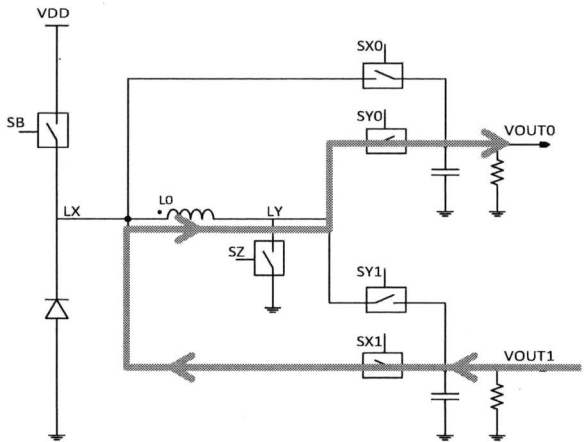

Fig. 5. SIDO current paths, sequence *Option 2*.

The advantage of this operation is that, the excess voltage in V_{OUT1} is used to supply load current to V_{OUT0} – not using any current from the input supply, but through recycling the stored energy in load capacitors. This advantage is expected to increase power efficiency in this operation mode and improve output voltage ripple (both outputs are being supplied at the same time – in normal SIDO operation only one output is supplied at a given phase, leading to higher output voltage ripple).

This operation mode can be preferred at high load currents or for continuous conduction mode (CCM), as the stored current on the inductor will continuously direct the discharge of capacitor at V_{OUT1} while charging the capacitor at V_{OUT0}. This mode might not be preferred when there is no/low load current at V_{OUT0}, since increasing V_{OUT0} could result in an over voltage (VOVER) condition. In such a condition *Option 1* described in Section II.A can be preferred over this solution.

III. BLOCK LEVEL IMPLEMENTATION

Figure 6 gives the block level implementation of the described topology. OTA_0 and OTA_1 generate error currents I_{ERR0} and I_{ERR1} which provide inputs to combiner and peak limit comparator. Using current mode control, peak limit (I_{PEAK}) will be I_{ERR0} and I_{ERR1} added together to be processed together with sensed inductor current I_{SENSE}. As OTA_0 and OTA_1 provide the

absolute value of error current as output, the error current will always be positive even if there is an undershoot or an overshoot. (without using the absolute value, an overshoot in one output could be cancelled with an undershoot in the other output, resulting in no response of the system).

V_{UNDER} and V_{OVER} comparators are utilized for both output voltages, monitoring if the output voltages are below/above the target (reference) output voltage. Sink/negative current operation decision will be given based on the outputs of V_{UNDER} and V_{OVER} comparators. The *Mode Select & Switch Control Logic* block sets the duty cycle of buck operation and also generates the switching sequences for sink/source operation.

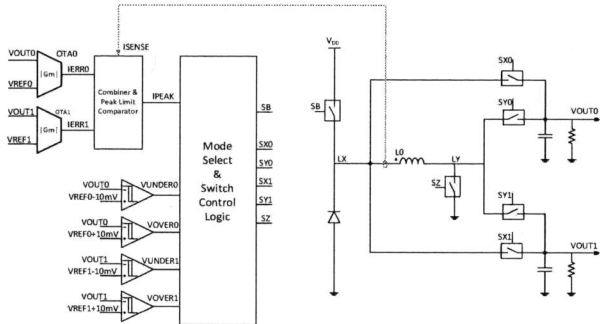

Fig. 6. Block level implementation.

System operation will further be explained through the simulation results provided in the next section.

IV. SIMULATION RESULTS

To validate the proposed topology, a SIDO buck converter model with the proposed switch control and operating as defined in Section III has been modelled with using Verilog-A and $100m\Omega$ switches.

In the following simulation results, switches S_{X0}, S_{X1}, S_{Y0} and S_{Y1} change logic positions synchronous with the clock. Though this is not a requirement of the proposed topology, this assumption helps building the macromodel systematically.

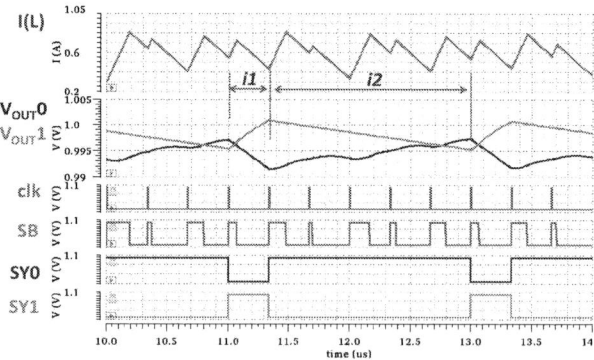

Fig. 7. Typical SIDO operation - both outputs having positive load.

Figure 7 gives typical SIDO operation where both outputs are loaded with positive load currents. Simulation conditions are as follows: V_{DD}=3.8V, V_{OUT0}=1V, V_{OUT1}=1V, L=1μH, C_{OUT0}=30μF, C_{OUT1}=30μF, f_{SW}=3MHz, I_{LOAD0}=500mA, and I_{LOAD1}=100mA.

As both output loads are positive, source operation will be performed (as in Figure 3). At simulation time 11μs, V_{OUT1} is lower, hence with the next clock S_{Y1} turns ON and the coil current supplies V_{OUT1} (interval $i1$ in the Figure). During interval $i2$, S_{Y0} is ON and the coil current supplies V_{OUT0} until 13μs.

Figure 8 gives negative SIDO operation with *Option 1* type switching (as in Figure 4). In this simulation I_{LOAD0}=50mA (low load current, the inductor current is expected to decay to zero as in DCM), and I_{LOAD1}= −100mA.

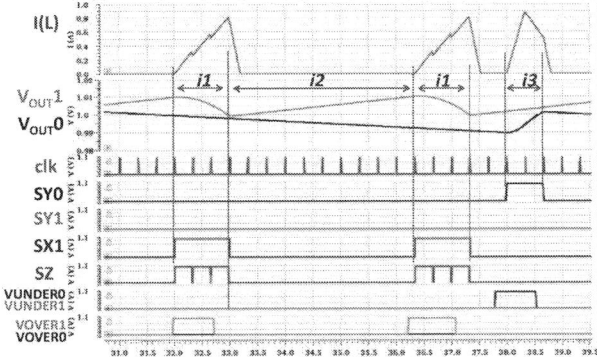

Fig. 8. Negative SIDO operation - *Option 1*.

During interval $i1$, at simulation time 32μs, V_{OVER1} becomes high, requesting sink/negative current operation for V_{OUT1}. S_{X1} and S_Z are ON, discharging V_{OUT1} as defined in Figure 4.

During interval $i2$, there are no V_{OVER}/V_{UNDER} signals, thus the system is not switching, staying in sleep mode. The build-up inductor current at the start of this interval can be discharged to input supply V_{DD} by an optional "recycling diode" connected at node L_Y.

During interval $i3$ (at simulation time 38μs), V_{UNDER0} is high, S_{Y0} turns ON and positive load current is supplied to V_{OUT0}.

Figure 9 gives negative SIDO operation with *Option 2* type switching (as in Figure 5). In this simulation I_{LOAD0}=1A (higher load current, inductor current will continuously stay above zero as in CCM), I_{LOAD1}= -250mA.

Throughout the simulation, V_{UNDER0} is continuously high as V_{OUT0} is loaded with 1A. During interval $i1$, V_{OVER1} becomes high, requesting negative operation for V_{OUT1}. S_{X1} turns ON, S_{Y0} continues to stay ON. Thus, V_{OUT1} is discharged with the coil current at the same time supplying load current to V_{OUT0} as described in Figure 5. During $i2$, V_{UNDER0} is high, load current is supplied to V_{OUT0}, S_{Y0} stays ON; buck switching (S_B switching) continues.

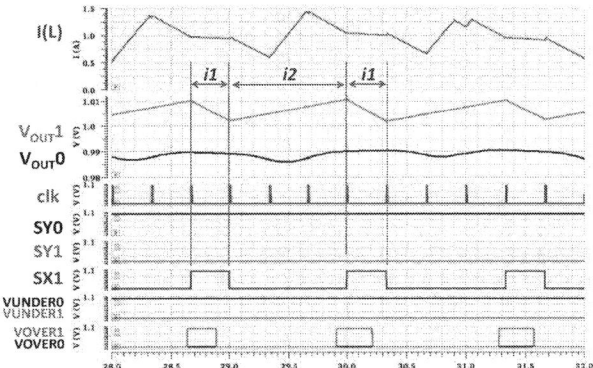

Fig. 9. Negative SIDO operation- *Option 2*.

V. Conclusion

A charge recycling switch topology for SIDO buck converters has been proposed. Two switching sequence options have been demonstrated for low load current (CCM) and high load current (DCM) profiles. Simulation results show that the proposed architecture achieves sink/source operation by re-using the charge stored in the output capacitors with continuous usage of the inductor, thus achieving high power efficiency and smooth transition between operation phases.

Even though the technique is described through the example of a SIDO buck converter, the technique can be applied to other Single-Inductor Multiple-Output switching converter topologies where sink/source operation is requested.

Acknowledgments

This work was supported by Dialog Semiconductor and Bogazici University.

References

[1] Pistoia, G., "Battery operated devices and systems: From portable electronics to industrial products," Elsevier, 2008.

[2] Kwon, D. and G. A. Rincón-Mora, "Single-inductor–multiple-output switching DC–DC converters", IEEE Transactions on Circuits and Systems II: Express Briefs, Vol. 56, No. 8, pp. 614-618, 2009.

[3] Belloni, M., E. Bonizzoni, and F. Maloberti. "On the design of single-inductor double-output dc-dc buck, boost and buck-boost converters." In 2008 15th IEEE International Conference on Electronics, Circuits and Systems, pp. 626-629. IEEE, 2008.

[4] Belloni, M., E. Bonizzoni, E. Kiseliovas, P. Malcovati, F. Maloberti, T. Peltola, and T. Teppo. "A 4-output single-inductor DC-DC buck converter with self-boosted switch drivers and 1.2 A total output current." In 2008 IEEE International Solid-State Circuits Conference-Digest of Technical Papers, pp. 444-626. IEEE, 2008.

[5] Lee, Yu-Huei, Shao-Chang Huang, Shih-Wei Wang, Wei-Chan Wu, Ping-Ching Huang, Hsin-Hsin Ho, Yuan-Tai Lai, and Ke-Horng Chen, "Power-tracking embedded buck–boost converter with fast dynamic voltage scaling for the SoC system", IEEE Transactions on Power Electronics 27, No. 3 (2010): 1271-1282.

[6] Ozanoglu, K., P. Cavallini, and B. Dundar, "SIDO Buck with Negative Current", U.S. Patent Application 16/687229, 20

© VDE VERLAG GMBH · Berlin · Offenbach

Design of an integrated Maximum Power Point Boost Converter for PV Submodules

Léon Weihs, Michael Hanhart, Leo Rolff, Ralf Wunderlich, Stefan Heinen

Chair of Integrated Analog Circuits and RF Systems Laboratory
RWTH Aachen University
Kopernikusstr. 16, D-52074 Aachen, Germany
E-Mail: mailbox@ias.rwth-aachen.de

Abstract—**The design constitutes an integrated boost controller with MPPT, targeting the optimization of power output of solar submodules, boosting system efficiency in partially shaded conditions compared to conventional PV panels. The negative impact on output power, caused by cell mismatch due to production and long-term degradation, can be minimized. The IC is designed for power throughputs up to 120 W with an input voltage range of 7 V - 24 V and output voltages up to 50 V. The converter operates at a switching frequency of 300 kHz in current mode control and is internally compensated. During low irradiance, the controller changes to pulse-frequency operation to further increase efficiency. The MPPT utilizes a delta modulator, avoiding the need for high resolution ADCs. The sense structure dynamically adjusts the MPP measurement resistance, further increasing tracking efficiency.**

Index Terms—**MPPT, Boost Converter, PV, Submodule, Integrated, DCDC, Current Mode Control, Perturb and Observe**

I. INTRODUCTION

The unceasing struggle for increased renewable energy contribution to the global supply of electricity is calling for increased efficiency at every link. While growth in wind power has stagnated in the past years, solar power, particularly generated through photovoltaic (PV) panels, is being seen as the strongest contributor in the years to come, with more than 1 TWp of added capacity projected in 2022 [1]. This is partly due to the fact that PV panels are becoming cheaper to manufacture, but also due to higher social acceptance [2]. A strong push has also been to integrate PV into structures already present throughout society, whose functions can be further expanded by the generating of electrical energy such as noise barriers. This brings rise to the challenge of optimising electric power extraction from large, far-spread PV panels.

By increasing the output voltage, transmission losses can be reduced over the large distance that the PV array covers. This is usually done by connecting multiple cells in series to form a submodule, which are again connected in series to form a panel, and then connecting multiple PV panels in series, summating their individual output voltages. This is known as a series or string configuration. However, string configurations can lead to drastic losses in efficiency due to PV panel, submodule, or even individual cell mismatches as well as sudden and unpredictable partial shading of the array [3], [4]. Within the string, the weakest link acts as a

throttle to the array's output. The use of bypass diodes is a simple approach, however being essentially binary, they cannot be used to regulate a submodule's output power, but only circumvent it.

By instead addressing individual submodules, each regulated to its maximum power point (MPP), the overall system efficiency is drastically improved [5], whilst also potentially increasing the panel's lifespan [6], as partial shading or mismatch no longer affects the entire array but can be constrained to that individual submodule. This can be achieved by using a microinverter, which acts as a maximum power point tracker (MPPT) for each submodule. By changing from DC to up-converted AC, the power extracted from the submodule can be carried away efficiently, as the boosted output voltage reduces transmission losses at the same power output. However, microinverters have remained large and costly due to the numerous large coils used to convert DC to AC, as well as their complex topology and regulating mechanisms, limiting their use case to panel and not submodular applications, as is the case with [7], as well as [8], where the DCDC converter is separated from the inverter design. Furthermore, the numerous instances required (one per single or multiple panels) compounds the costs associated with such systems.

The system presented in this paper aims to address both of these problems by improving the efficiency of PV panels under partially shaded, low irradiation conditions, whilst simultaneously boosting their output voltage. Where other designs, such as [9], resort to complex multiphase conversion, requiring multiple expensive inductors, the proposed system provides a low-cost, easy-to-integrate and extremely scalable solution, that can later be coupled with a single inverter, thereby reducing costs and conversion losses. It consists of a single integrated circuit (IC) containing a MPPT with a boost converter output stage, designed to track individual submodules and boost their output voltage.

II. SYSTEM DESIGN

The designed boost controller IC with integrated MPPT accommodates for a large range of input voltages, from 7–24 V, while allowing power throughputs ranging from near zero up to 120 W. This allows for a drop-in solution for

© VDE VERLAG GMBH · Berlin · Offenbach

SMACD / PRIME 2021 | 19 – 22 July 2021, Online Event

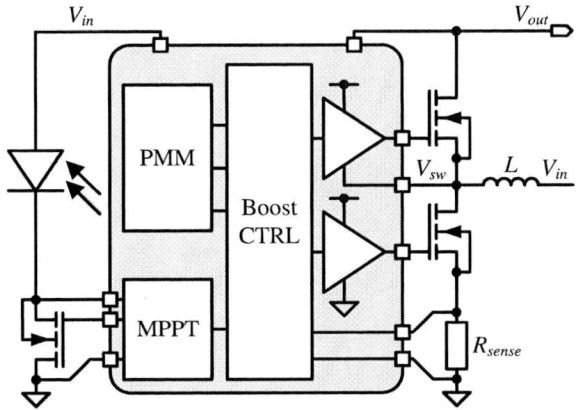

Fig. 1: Top-level system overview

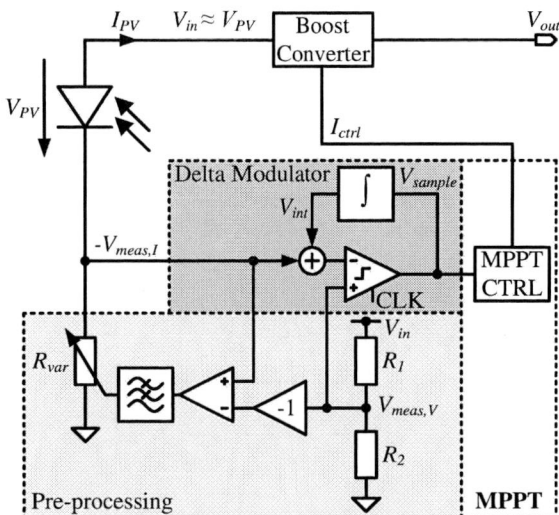

Fig. 2: MPPT system overview

a standard 30 V, 60 cell panel divided into three submodules. The output voltage can be boosted up to 50 V. The system is required to operate extremely reliably under harsh environmental conditions, calling for robust circuitry with minimised influence by temperature variations and mismatch. These technical requirements put high demands on the design of the control loop as well as the power management (PMM) to guarantee stable operation under all loads and environmental conditions. A top-level overview of the system can be seen in Fig. 1.

A. MPPT

The MPPT is an integrated incremental conductance MPPT based on a delta modulator with analog pre-processing as presented in [10] and shown in Fig. 2, providing a fast conversion time and high tracking efficiency over the required wide range of input power. The design is based on a perturb and observe algorithm. The input voltage is sensed using an internal resistive voltage divider, while input current is measured indirectly using a low side transistor. By adjusting the resistivity of said MOSFET, the voltage drop is regulated to the inverted, scaled-down, input voltage as defined in Eq. 1.

$$V_{meas,I} = R_{var} \cdot I_{PV} \stackrel{!}{=} V_{PV} \frac{-R_2}{R_1 + R_2} = -V_{meas,V} \quad (1)$$

This constant voltage sensing mechanism offers increased resolution for small input currents, while minimising the resistive losses for high PV current scenarios. Eq. 2 describes the condition for MPP operation. The equivalent expression is available to the MPPT as V_{sample} in Fig. 2. Depending on the actual relation between the two sides of Eq. 2 ($\leq$ / $\geq$), the MPPT algorithm adjusts the direction of desired input voltage change and signals this to the boost converter, thereby regulating the solar cell to its MPP voltage.

$$\frac{\partial I_{PV}}{\partial V_{PV}} = -\frac{I_{PV}}{V_{PV}} \quad (2)$$

B. Boost Converter

The boost converter consists of a low side driver, a high side driver, a control circuit as well as a resistive current measurement. The boost converter differs from typical implementations in that it regulates the input voltage, and not the output voltage. This is necessary in order to achieve the primary goal of operating the PV submodule in its MPP. Once set, the input voltage is boosted as much as possible, whilst monitoring the output voltage, not to exceed 50 V. The converter uses current mode control. The outer voltage loop sets the input voltage to Eq. 3, where I_{ctrl} is set by the MPPT.

$$V_{in} = V_{bg} \cdot \left(1 + \frac{R_1}{R_2}\right) + I_{ctrl} \cdot R_1 \quad (3)$$

The inner current loop monitors the current through a low side resistive current measurement using R_{sense}. This allows for reduced losses compared to a resistor in series to the inductor. However, it also means that no current information is available during the high side phase, creating the need for a ramp emulation circuit to properly detect the inductor current zero-crossing for accurate high side turn-off. A slope compensation is imperative to ensure stable operation without sub-harmonic oscillations. The converter's control mechanism can be separated into two scenarios: continuous conduction mode (CCM) and discontinuous conduction mode (DCM). In CCM, the boost controller operates at a fixed frequency of 300 kHz using peak current mode control. In DCM, the converter changes to pulse-frequency modulation. By lowering the frequency, a higher efficiency is thus obtained during light-load scenarios. The change to DCM occurs at a threshold value of the inductor ripple current and is a function of the output voltage.

C. Power Management

The complex control mechanisms of both the MPPT and the boost converter on the same IC create high demands on clean,

© VDE VERLAG GMBH · Berlin · Offenbach

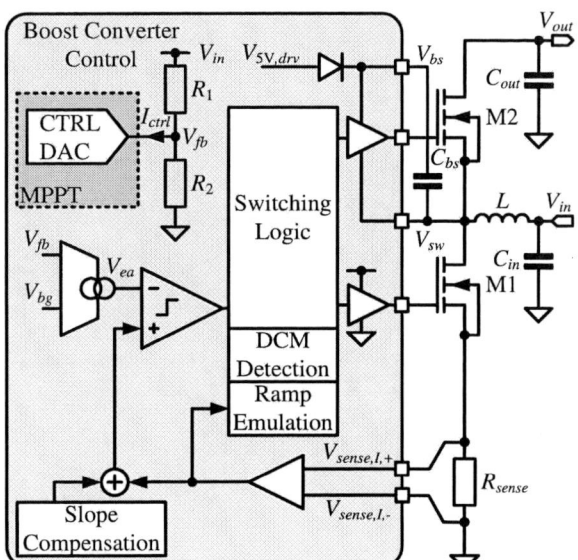

Fig. 3: Boost Converter Control system overview.

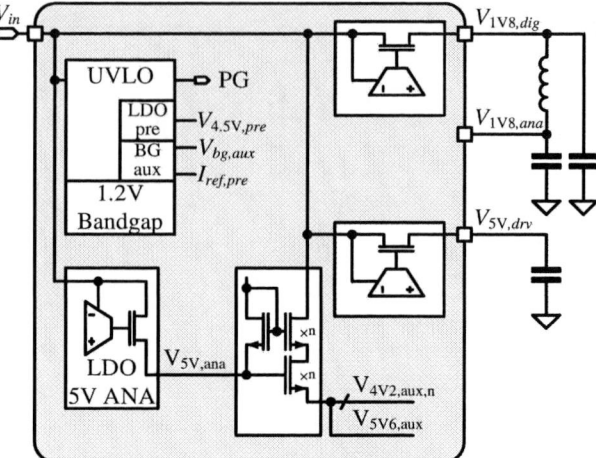

Fig. 4: Power Management system overview.

noise-free supply rails. Power supply rejection is a key factor for the sensitive analog control circuits, such as the current DAC used to set the input voltage to the MPPT. The multiple voltage levels required further pose a challenge on the system design: ensuring that cross-talk between power domains as well ground noise be minimised. The power management therefore provides two separate 1.8 V supply lines. The first is generated from the input voltage by a low-dropout regulator (LDO) and is used singularly for digital circuitry, which carries a lot of switching noise due to the fast 1.2 MHz switching speed. This voltage is supported by an external capacitor in close proximity to the chip. A second pin, supported by its own capacitor, brings this voltage back onto the chip as a clean and stable voltage for the sensitive analog circuitry used throughout the system. This separation by the bond and track inductances acts as a filter between the two voltage domains, while also allowing for a flexible choice in external filter and support components. The second power domain is provided as an extremely well-regulated 5 V analog supply, while a second 5 V supply is generated uniquely for the driver stage. The choice of two LDOs is necessary here, as the driver stage is expected to carry a lot of noise due to the fast, high current switching requirement. Furthermore, bias voltages of critical circuit blocks are generated locally, where needed, to further enhance performance and provide better isolation between different circuitry. These are provided with power through an auxiliary LDO, which, even though not very accurate in absolute terms, provides output voltages that are distinct, thereby decreasing the effect of cross-talk between circuitry as well as noise seen on the input voltage. All essential internal voltages are monitored by an under-voltage lock-out circuit. Presented in [11], it allows the system to undergo reliable start-up sequences while monitoring for consistent operation. It achieves this by having its own bandgap reference

voltage, with a strong focus on reliability and robustness. By monitoring the input and internal supply voltages, it is able to switch on or off the other power domains, as well as provide a backup reference voltage in case the main reference voltage fails, while also providing a reference voltage and current for the other power management circuits to achieve stable startup and shutdown, especially in low input voltage scenarios.

III. SIMULATION RESULTS

Where other approaches focus on a very low input voltage ripple in order to remain in the MPP, this design relaxes the requirements regarding input voltage ripple and instead lowers the switching frequency in DCM, which is much more favourable for overall efficiency, as illustrated in Fig. 5. This is further illustrated across a range of input currents in Fig. 7.

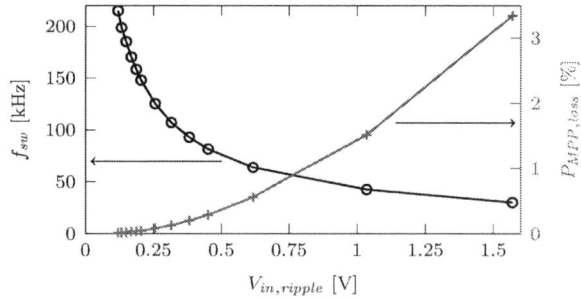

Fig. 5: Losses due to MPP deviation at $I_{in} = 8\% \cdot I_{sc}$

Simulated results in Fig. 6 show the fast transient response of the MPPT, as well as reliable tracking capability across sudden, major insolation changes in CCM. The tracking efficiency of the developed MPPT measures more than 99 % for input power larger 3 % of rated input power in CCM. By combining the electrical characteristics of a PV-panel with the boost converter, the overall system efficiency achieved lies at over 98 %.

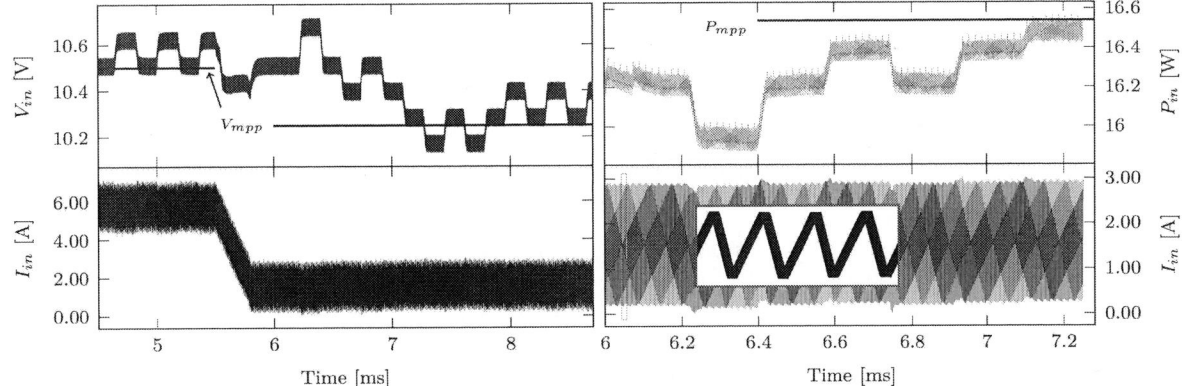

Fig. 6: Change in MPP regulation after illumination drop

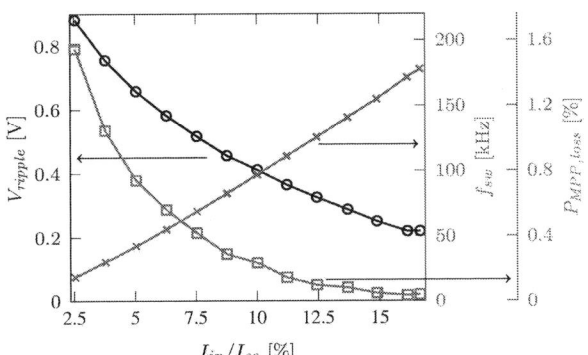

Fig. 7: Losses from MPP deviation at different input currents

TABLE I: Comparison to other approaches

Product	This Work	MAX20801	SPV1020	C2000
Topology	Boost	Buck	Boost	Boost + LLC
Power	120 W	120 W	320 W	500 W
Application	Submodule	Submodule	Submodule	Panel
Cost/Cmpl.	Low	Low	High	High
Multiphase	No	No	Yes	Yes
V_{in}	7–24 V	6.5–15.5 V	6.5–40 V	200–300 V
V_{out}	7–50 V	0.3–11 V	6.5–40 V	400 V

IV. CONCLUSION

A novel, simple and cost-effective solution for maximising submodular PV power extraction was presented. The proposed boost converter with integrated MPPT greatly increases the overall system efficiency of PV panels, especially under partially shaded conditions. Simulations in BCD 180 nm illustrate that this approach results in a very high efficiency for low to medium input irradiance. Compared to available designs, listed in Table I, the work presented is unique in both its function and cost. The vast majority of publications focus on discrete power optimizers, which are far more complex and occupy vastly more area. The only product matching this work's simplicity and cost is a newly released IC by Maxim Integrated [12], which, due to its buck converter topology, fails to meet efficiency targets in ensuing transmission losses.

ACKNOWLEDGEMENT

The authors acknowledge support for this work through the project Rolling Solar within the framework of the Interreg Euregio Meuse-Rhine V-A with financial support of the European Regional Development Fund.

REFERENCES

[1] S.P.Europe,"Global market outlook for solarpower", 2019. Available:https://www.solarpowereurope.org

[2] Agentur für Erneuerbare Energien, "Zustimmung für den Ausbau der Erneuerbaren Energien bleibt hoch", Jan 2021.
Available: https://unendlich-viel-energie.de/themen/akzeptanz-erneuerbarer/akzeptanz-umfrage/zustimmung-fuer-den-ausbau-der-erneuerbaren-energien-bleibt-hoch

[3] A. Shekharet al., "Harvesting roadway solar energy—performance of the installed infrastructure integrated pv bike path",IEEE Journal of Photovoltaics, vol. 8, no. 4, pp. 1066–1073, Jul. 2018.

[4] G. Zurbriggen and M. Ordonez, "Pv energy harvesting under extremely fast changing irradiance:State-plane direct mppt",IEEE Transactions on Industrial Electronics, vol. 66, no. 3, pp. 1852–1861, Mar. 2019.

[5] J. Galtieri and P. T. Krein, "Energy improvements from subpanel DC-DC converters in PV arrays with distributed mismatch," 2016 IEEE 43rd Photovoltaic Specialists Conference (PVSC), Portland, OR, USA, 2016, pp. 3213-3218, doi: 10.1109/PVSC.2016.7750259.

[6] P. Manganiello, M. Balato and M. Vitelli, "A Survey on Mismatching and Aging of PV Modules: The Closed Loop," in IEEE Transactions on Industrial Electronics, vol. 62, no. 11, pp. 7276-7286, Nov. 2015, doi: 10.1109/TIE.2015.2418731.

[7] Enphase Microinverters, IQ7x, 2019. Available: https://enphase.com/sites/default/files/downloads/support/IQ7X-DS-EN-US.pdf

[8] Texas Instruments, C2000, "C2000™ Solar DC/DC Converter with Maximum Power Point Tracking (MPPT)", 2014. Available: https://www.ti.com/tool/TIDM-SOLAR-DCDC

[9] STMicroelectronics, spv1020, "Interleaved DC-DC boost converter with built-in MPPT algorithm", 2012.
Available: https://www.st.com/content/st_com/en/products/power-management/photovoltaic-ics/mppt-dc-dc-converters/spv1020.html

[10] L. Rolff, M. Hanhart, R. Wunderlich and S. Heinen, "An Integrated Incremental Conductance MPPT based on a Delta Modulator with Analog Preprocessing," ANALOG 2020; 17th ITG/GMM-Symposium, Online, 2020, pp. 1-5.

[11] J. Grobe et al., "Design of a Flexible Bandgap Based High Voltage UVLO with Pre-Regulator," 2020 27th IEEE International Conference on Electronics, Circuits and Systems (ICECS), Glasgow, UK, 2020, pp. 1-4, doi: 10.1109/ICECS49266.2020.9294922.

[12] Maxim Integrated, MAX20801, 2020.
Available: https://www.maximintegrated.com/en/products/power/switching-regulators/MAX20801.html

SMACD / PRIME 2021 | 19 – 22 July 2021, Online Event

Design of a High PSRR Multistage LDO with On-Chip Output Capacitor

Jonas Zoche, Michael Hanhart, Jan Grobe, Léon Weihs, Leo Rolff, Ralf Wunderlich and Stefan Heinen
Chair of Integrated Analog Circuits and RF Systems
RWTH Aachen University
Kopernikusstrasse 16, D-52074 Aachen, Germany
Email: mailbox@ias.rwth-aachen.de

Abstract—**This paper proposes a high PSRR LDO design, implemented in a 0.18 μm BCD technology using, only on-chip capacitors. Multiple stages in series with PMOS pass transistors lead to high PSRR, small drop-out voltage, and low circuit complexity in comparison to designs relying on ripple feed-forward. Without external capacitors, the presented auxiliary LDO does not increase the pin count while the multistage approach leads to low circuit complexity. Overall, 54 dB PSRR are achieved across corners up to 10 GHz with a quiescent current of 15.5 μA and 0.192 μm² chip area.**

I. INTRODUCTION

BCD technologies offer a variety of low and high voltage transistors to provide both a high integration density and the ability for high voltage power management. Most analog building blocks consist of LV devices to benefit from the much better analog performance and matching as well as less area consumption. To obtain a supply voltage suited for the LV devices from the high voltage domain, a voltage regulator is required. By integrating this voltage regulator alongside the main functions in an IC, overall component count and PCB area can be reduced.

In this paper, the design of such an auxiliary LDO is shown, which provides the 5 V power supply for the analog core of a maximum power point tracking boost converter. With 6 V to 24 V input voltage, the drop-out voltage must be < 1 V. The switching of the dc-dc converter leads to a high noise level on the input voltage, requiring a high PSRR. Both static and transient load response are less important because the power consumption of the analog components exhibits no large load transients. On-chip capacitors are preferred because of the small maximum output current (500 μA) and the additional area consumption of bonding pads.

High PSRR results from a large ratio of input-to-output to output impedance for a large range of frequencies. A fast regulation loop leads to a decreased output impedance for low and mid frequencies while a large output capacitance provides a low impedance for mid to high frequencies. At highest frequencies, the PSRR is only determined by the parasitic capacitors of the pass transistor and the output capacitor, which

The authors acknowledge the financial support by the German Federal Ministry for Economic Affairs and Energy under 03ET1417B (Project EnEff: Stadt EnQM).

form a capacitive voltage divider. For a large input-to-output impedance, several approaches exist, resulting from two main ideas: either by stacking pass transistors or by feed-forwarding the input voltage ripple into the pass transistor.

In [1], a feed-forward is implemented by modulating V_{gs} of a PMOS transistor with the AC component of the input voltage. The corresponding current provides a compensation at the pass transistor gate, leading to a PSRR boost in the mid to high frequency range. For frequencies > 10 MHz, the feed-forward seems to be limited by parasitic capacitors, deteriorating the PSRR. [2] shows an improved design with increased output node feedback and thereby decreased output impedance in the mid-frequency range. The authors in [3] propose a different feed-forward scheme, which adds the amplified input ripple to the error amplifier (EA) output. All mentioned designs require precise matching of the feed-forward to the amplifier and pass transistor characteristics, resulting in complex circuit design. Furthermore, the feed-forwarding does not affect the high frequency PSRR significantly, making a large – and thereby off-chip – output capacitor mandatory.

Stacking pass transistors is shown in [4]. Because of the much smaller coupling capacitance between input and output, an on-chip capacitor can be used. The authors propose two pre-regulators in series based on NMOS source followers, followed by a PMOS output stage. The NMOS source followers require an auxiliary supply voltage above the input voltage, provided by a charge pump. Although the implementation of the three stages is quite simple, the charge pump complicates the overall circuit design and increases the current consumption. In [5], a two-stage PMOS design with ripple feed-forward is proposed, which shows good noise suppression without an external capacitor but again introduces complexity by the feed-forward path.

II. CIRCUIT IMPLEMENTATION

To achieve a very high PSRR without an external capacitor, a three-stage design is implemented, as depicted in Fig. 1. Although an NMOS source follower provides a very high PSRR, the required voltage headroom is too large to obtain a drop-out voltage < 1 V. Because a charge pump introduces additional circuit complexity and noise, PMOS stages are going to be used.

© VDE VERLAG GMBH · Berlin · Offenbach

SMACD / PRIME 2021 | 19 – 22 July 2021, Online Event

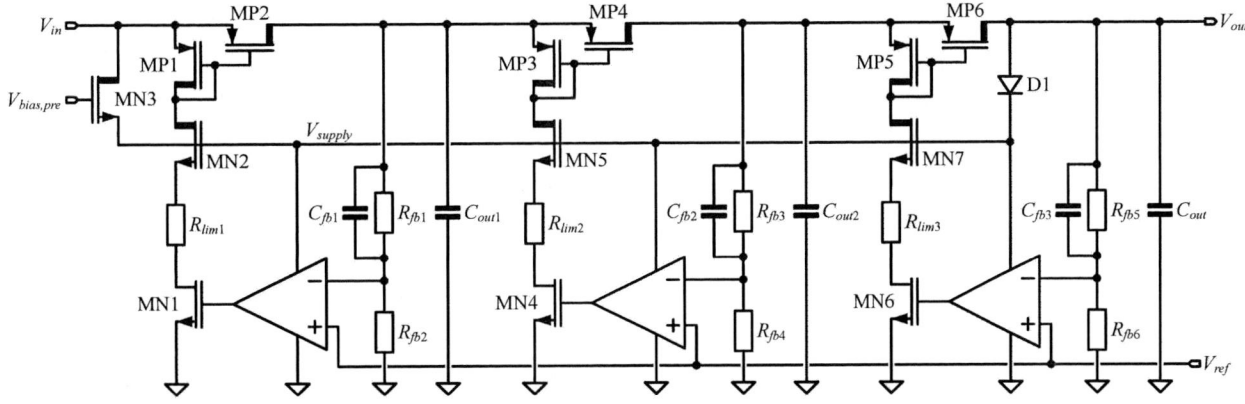

Fig. 1. Overall LDO circuit

The individual regulation loops of each stage can be analyzed and designed separately because of the block-wise concept. This simplifies the overall design compared to feed-forward approaches, where a high PSRR at high frequencies is achieved via an additional compensation network, leading to an increased small signal complexity. The transient performance for output current changes is determined mainly by the last stage, which further simplifies the design for the first two stages.

A. Output Stage

Fig. 2 shows a single stage of the design: the transistors MP1 and MP2 form a current mirror, multiplying the output current of the EA to the output. In the first stage, the width is scaled by $1:20$ and $1:10$ in the second and third stage. This design exhibits a high intrinsic PSRR: MN2 provides a high impedance current output, while MP1 exhibits a low impedance with respect to the input voltage. Therefore, the gate source voltage of MP2 is fixed when the input voltage changes. Only the finite output impedance of MP2 and MP1 limits the PSRR in the mid-frequency range. The EA determines the transient load regulation because MP2 provides a high impedance output.

A source follower, formed by MN2, separates the output stage from the LV domain, which contains the EA, R_{lim}, and MN1. The maximum output current is limited by R_{lim}

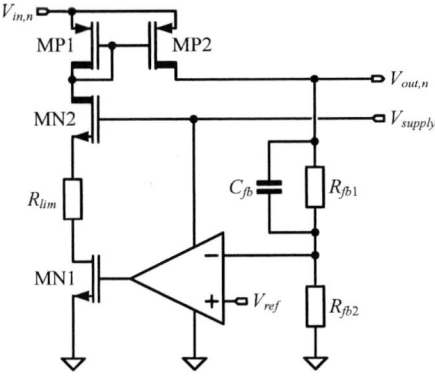

Fig. 2. The output stage of each sub-LDO

with respect to V_{DD}, to smooth the startup behavior and provide a basic over-current protection. With R_{fb1} and R_{fb2}, the output voltage is fed into the EA, closing the regulation loop. C_{fb} increases the phase margin, because the current mirror architecture induces a phase shift from the EA output to the output capacitor voltage. By C_{fb}, the bandwidth of the compensated design is improved severely.

The three stages of the LDO are designed similarly, differing only in minor device parameters. For the first two stages, the output ripple is reduced by 8 pF output capacitance each. The voltage feed-through via $C_{ds,P2}$ is inversely proportional to the capacitance; but taking the area consumption into account (see Sec. II-C), this value is a good compromise. Due to the lower voltage at the output of the last stage, capacitors with a much higher capacitance per area can be used. Therefore, the output capacitance with 40 pF is significantly larger than for the previous stages because it dominates the ripple at load transients. Fig. 1 shows the complete LDO with all three stages.

B. EA Design

The EA of each stage is implemented as a Miller OPA with a PMOS input stage, as depicted in Fig. 3. To reduce the offset voltage caused by mismatch of MP1 and MP2, a resistive source degeneration is added with R_{S+} and R_{S-}. Overall, $\sigma_{V_{out}}$ measures 17 mV. MP2 and MN2 form a first gain stage, followed by a second stage consisting of MN3 and MP5. A Miller compensation (R_Z and C_M) is added to increase the phase margin. A high amplification in the regulation loop is required to accommodate for the high output impedance; MN1 and MP1 (in Fig. 2) provide therefore a third gain stage to increase the amplification.

Noise on the EA supply is rejected to some extend by the top current mirror, but is limited at higher frequencies because of the finite C_{ds} of MP3, MP4, and MP5. Together with C_{gs}, a capacitive voltage divider is formed, which couples noise on the supply line into V_{gs}, getting amplified by g_{m5} over $g_{ds5} \parallel g_{ds3}$. During startup, the EAs are supplied by an NMOS source follower MN3 with an auxiliary supply voltage as gate bias potential (see Fig. 1).

© VDE VERLAG GMBH · Berlin · Offenbach

C. Layout

Fig. 4 shows the layout of the LDO: the output capacitors consume the most area. With 6.5 V and 6 V, the first two output voltages are higher than the maximum allowed voltage for MIM and MOS capacitors. Therefore, only MOM capacitors can be used for C_{out1} and C_{out2}, leading to the same area consumption as the much larger capacitance C_{out}, which is implemented via MOS capacitors. The feedback resistors contribute the second largest part to the overall area consumption, while the pass transistors are in comparison small. All in all, the LDO layout consumes an area of $600\,\mu\text{m} \times 320\,\mu\text{m}$.

III. SIMULATION RESULTS

A. Load Transient

Fig. 5 depicts the LDO response to a load current step with $50\,\mu\text{A}$ amplitude: the last stage regulates the output voltage without overshoot within a $110\,\text{mV}$ window and settles to $< 10\,\%$ deviation after $1.3\,\mu\text{s}$. The amplitude of the ripple is sufficient, because load current steps do not occur during operation. Both V_{out1} and V_{out2} exhibit a decent regulation as well; the minor ringing is filtered by the last stage. Overall, the LDO satisfies the requirements for load transient responses.

B. PSRR

To characterize the PSRR, both large signal (transient) and small signal (AC) simulations are carried out. Fig. 6 shows a transient simulation with a V_{in} step ($\Delta V = 100\,\text{mV}$, $t_r = t_f = 1\,\text{ns}$), which is filtered by each stage by one order

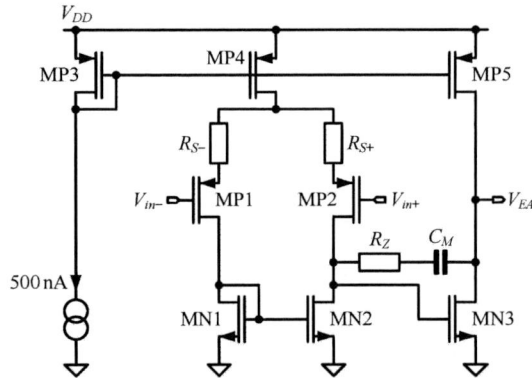

Fig. 3. Design of the individual EAs

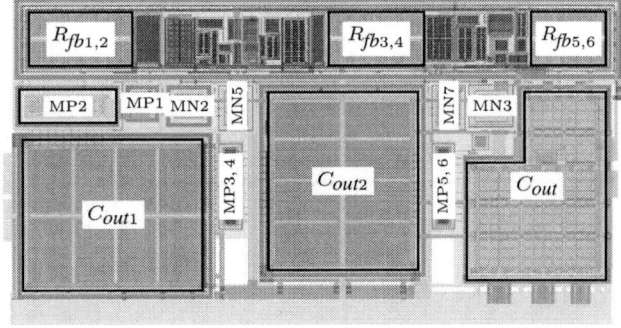

Fig. 4. Layout view of the LDO

of magnitude. This simulation demonstrates the performance of the multi-stage approach: the cascading leads to a reduction by a factor of 1000 (60 dB).

The AC simulation shows the positive effect of cascading as well, as shown in Fig. 7. Over process and temperature variations, the first stage exhibits a PSRR of $12\,\text{dB} @ 16\,\text{MHz}$ in the worst corner, which is increased to at least $29\,\text{dB}$ by the

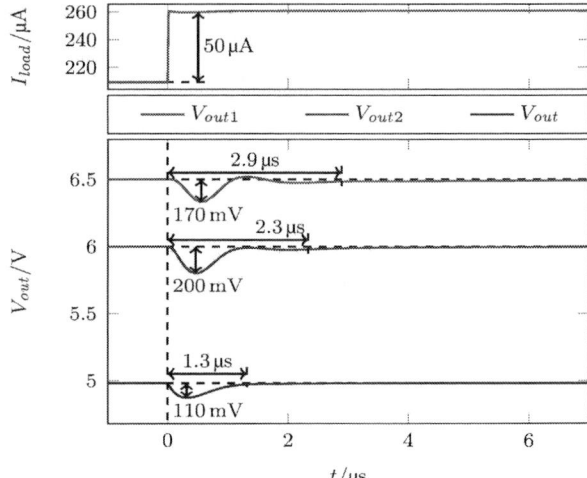

Fig. 5. Transient response of the three stages for a $50\,\mu\text{A}$ I_{load} step

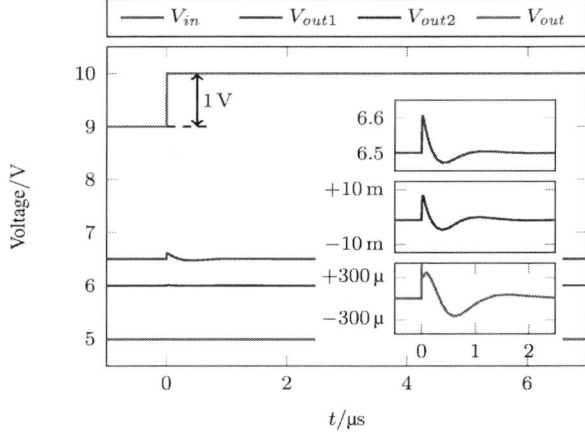

Fig. 6. Transient response of the three stages for a $1\,\text{V}$ V_{in} step, $t_r = 1\,\text{ns}$

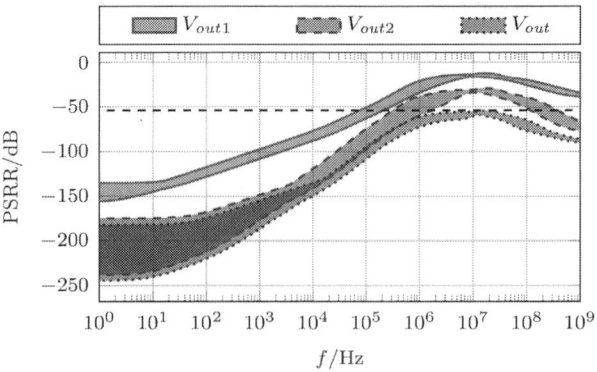

Fig. 7. PSRR across temperature and process corners, $I_{load} = 200\,\mu\text{A}$

© VDE VERLAG GMBH · Berlin · Offenbach

SMACD / PRIME 2021 | 19 – 22 July 2021, Online Event

TABLE I
PERFORMANCE OF THE PROPOSED LDO IN COMPARISON WITH OTHER DESIGNS

	This work	TCAS 2014 [4]	AICSP 2011 [5]	ASSCC 2013 [1]	SOCC 2015 [2]	JSSCC 2010 [3]
Technology	$0.18\,\mu m$	$0.35\,\mu m$	$0.18\,\mu m$	$0.18\,\mu m$	$0.18\,\mu m$	$0.13\,\mu m$
V_{in}	$5.4...24\,V$	$1.8\,V$	$> 2.1\,V$	$1.3...1.8\,V$	$1.3...1.8\,V$	$1.15...2\,V$
V_{out}	$5\,V$	$1.2\,V$	$1.8\,V$	$1.2\,V$	$1.2\,V$	$1\,V$
V_{DO}	$360\,mV$	$600\,mV$	$300\,mV$	$100\,mV$	$100\,mV$	$150\,mV$
C_{out}	$40\,pF$ (on-chip)	$0...100\,pF$ (on-chip)	$100\,pF$ (on-chip)	$4.7\,\mu F$	$4.7\,\mu F$	$2 \cdot 2\,\mu F$
Max. I_{load}	$500\,\mu A$	$12\,mA$	$4\,mA$	$25\,mA$	$25\,mA$	$25\,mA$
Line Reg.	$170\,nV/V@0.5I_{max}$	$280\,\mu V/V$	$240\,\mu V/V$	$3\,mV/V$	$500\,\mu V/V@I_{max}$	$30\,\mu V/V@5\,mA$
Load Reg.	$340\,\mu V/mA$	$680\,\mu V/mA$	$800\,\mu V/mA$	$50\,\mu V/mA$	$140\,\mu V/mA$	$48\,\mu V/mA$
PSRR	$54\,dB@18\,MHz$	$38\,dB@3\,MHz$	$30\,dB@5\,MHz$	$25\,dB@40\,MHz$	$55\,dB@30\,MHz$	$56\,dB@10\,MHz$
Chip Area	$0.192\,mm^2$	$0.084\,mm^2$	$0.104\,mm^2$	$0.041\,mm^2$	$0.021\,\mu m^2$	$0.049\,mm^2$
I_{qsc}	$15.5\,\mu A + 0.25 \cdot I_{load}$	$44\,\mu A$	$28\,\mu A$ (w/ BG)	$15\,\mu A$	$10\,\mu A$	$50\,\mu A$

second stage. Due to the larger output capacitor, the last stage contributes a higher noise suppression for high frequencies, which results in minimum $54\,dB$ PSRR at the output. At low frequencies, the output voltage is virtually independent on the input with a PSRR of $> 180\,dB$.

C. Stability Analysis

Fig. 8 depicts the loop gain and phase of the last stage. A phase margin of $77°$ and a transit frequency of $2.5\,MHz$ can be obtained from the bode plot with $200\,\mu A$ load current. Under no load condition, the bandwidth decreases to $660\,kHz$ with a phase margin of $67°$. Monte-Carlo simulations show a minor dependency for the phase margin on process parameters and temperature, whereas the transit frequency shows a strong correlation with the temperature (nominal $3.9\,MHz$ at $-40\,°C$ and $1.3\,MHz$ at $150\,°C$ with $200\,\mu A$ load).

IV. CONCLUSION

TABLE I compares the achieved key performance values with the previously mentioned designs. The three feed-forwarding designs [1], [2], and [3] exhibit a comparable PSRR with a very low drop-out voltage of $\leq 150\,mV$ and low quiescent current. But due to the required external capacitors,

their use as auxiliary supplies induces additional chip and PCB area. In [4], pass transistors are stacked as in this work, but due to the use of NMOS transistors for the first two stages, the drop-out voltage is higher and the charge pump increases power consumption as well as the design complexity. The design in [5] with two PMOS stages exhibits a smaller drop-out voltage with a comparable PSRR by using a multistage feed-forward architecture, but for both designs the resulting PSRR is not as high as in [2], [3] and this work. The proposed design consumes the most chip area, but the much higher input and output voltages induce additional area consumption for insulation rings, as can be seen in Fig. 4.

A high-PSRR, low quiescent current, low drop-out and external-capacitor-free linear voltage regulator has been presented in this paper. All requirements resulting from the boost converter characteristics are met and the LDO is thereby suited as its auxiliary supply. Due to the block-wise circuit, the implementation of this design is straightforward and can be easily adopted for other use cases and manufacturing technologies. Despite the simple design, the overall circuit performance is comparable to other high-PSRR implementations, providing the option for multiple auxiliary supplies in high-performance power management circuits.

REFERENCES

[1] J. Guo and K. N. Leung, "A 25mA CMOS LDO with −85dB PSRR at 2.5MHz," in *2013 IEEE Asian Solid-State Circuits Conference (A-SSCC)*, 2013, pp. 381–384.

[2] L. Chen, Q. Cheng, J. Guo, and M. Chen, "High-PSR CMOS LDO with Embedded Ripple Feed-Forward and Energy-Efficient Bandwidth Extension," in *2015 28th IEEE International System-on-Chip Conference (SOCC)*, 2015, pp. 384–389.

[3] M. El-Nozahi, A. Amer, J. Torres, K. Entesari, and E. Sanchez-Sinencio, "High PSR Low Drop-Out Regulator With Feed-Forward Ripple Cancellation Technique," *IEEE Journal of Solid-State Circuits*, vol. 45, no. 3, pp. 565–577, 2010.

[4] C. Zhan and W. Ki, "Analysis and Design of Output-Capacitor-Free Low-Dropout Regulators With Low Quiescent Current and High Power Supply Rejection," *IEEE Transactions on Circuits and Systems I: Regular Papers*, vol. 61, no. 2, pp. 625–636, 2014.

[5] V. Majidzadeh, K. M. Silay, A. Schmid, C. Dehollain, and Y. Leblebici, "A fully on-chip LDO voltage regulator with 37 dB PSRR at 1 MHz for remotely powered biomedical implants," *Analog Integrated Circuits and Signal Processing*, vol. 67, no. 2, pp. 157–168, May 2011.

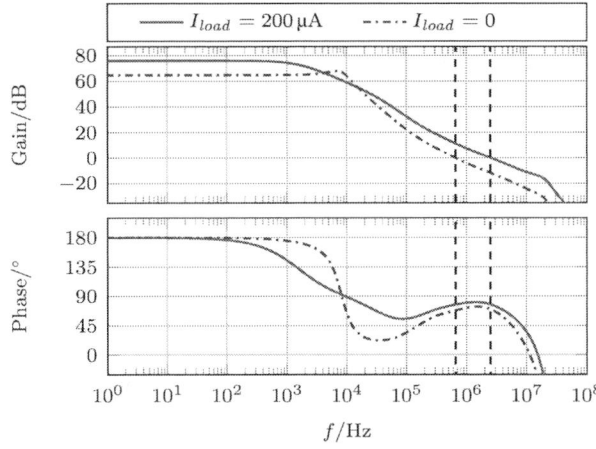

Fig. 8. Loop gain and phase of the last stage, $I_{load} = 200\,\mu A$

© VDE VERLAG GMBH · Berlin · Offenbach

SMACD / PRIME 2021 | 19 – 22 July 2021, Online Event

Skew and Jitter Performance in CMOS Clock Phase Splitter Circuits

Lorenzo Scaletti, Angelo Parisi and Luca Bertulessi

Dipartimento di Elettronica, Informazione e Bioingegneria
Politecnico di Milano
P.zza Leonardo da Vinci 32, Milano, Italy

Abstract—**This paper compares two different clock phase splitter architectures: the first is based on the standard XOR gate splitter topology, the other exploits the concept of phase interpolation. The comparison, supported by schematic simulations, shows that the phase interpolating splitter outperforms the XOR gate based topology with no penalty in power consumption. In particular, the output jitter is reduced by 15% while the skew between the output phases is reduced by about 80%. The phase interpolating splitter, implemented in 28nm bulk CMOS, achieves jitter lower than 35fs with power consumption lower than 5mW at 1GHz input frequency in post-layout simulations. The skew between the generated output signals is always lower than 10ps.**

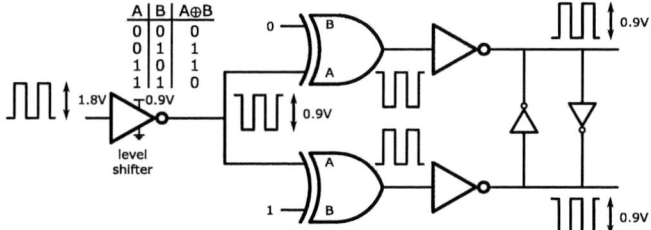

Fig. 1. XOR-based clock splitter

I. INTRODUCTION

Clock distribution architectures, in particular for high frequency and low jitter applications, such as in wide-band high-resolution ADCs which require very precise timing, are usually implemented in a differential fashion because of the increased noise and disturbance immunity with respect to single-ended architectures. Unfortunately, differential clocks are not always available, as many commercial low-cost frequency generators only come with a single-ended output. In this work, two alternative circuits performing the single-ended to differential conversion, called clock-phase splitters, are presented and their main performance metrics, i.e. output jitter, power consumption and skew between the differential outputs, are compared with the aid of transistor-level schematic simulations. Finally, the topology that achieves best performance is validated by post-layout simulations.

II. CLOCK PHASE SPLITTER ARCHITECTURES

In this section two alternative clock phase splitter architectures in 28nm bulk CMOS are presented. Both splitters are driven by an impedance matched clock buffer with 1.8V voltage swing. For this reason, the input stage is a level shifter from 1.8V to the 0.9V core voltage and it is implemented as a simple thick oxide CMOS inverter stage powered at 0.9V.

A. XOR based

The first topology considered is based on the logic properties of the XOR gate: the clock signal ('A' in Fig. 1) is either buffered or inverted when the control bit ('B') is set at logic level '0' or '1', respectively. This behavior can be exploited to split the input clock into two phases using two parallel paths: one generating the positive clock phase,

while the other generates the negative clock phase, each one with a XOR gate. In the non-inverting path, one input of the logic gate is always connected to ground, thus buffering the clock input; in the second path one terminal of the XOR gate is always connected to the power supply, thus inverting the clock input. The two XOR outputs are further passed through inverters with their outputs connected through a small cross-coupled inverter pair to ensure differential operation (Fig. 1). There are many possible architectures implementing a XOR gate in different logic styles [1]; the implementation used to test the splitter architecture is the one in Fig. 2, since it only uses four transistors and only one of the two inputs (B) is required with the complementary value ($\overline{B}$). This is necessary, since it is the splitter itself that generates the complementary clock phase. Even though this architecture may appear symmetrical at first sight, a closer look at how the XOR gates are actually implemented at transistor level reveals that on one path the clock signal passes through three inverters, whereas on the second path the middle inverter is substituted by a transmission gate that lets the signal through as is. Therefore, the symmetry is not preserved at transistor

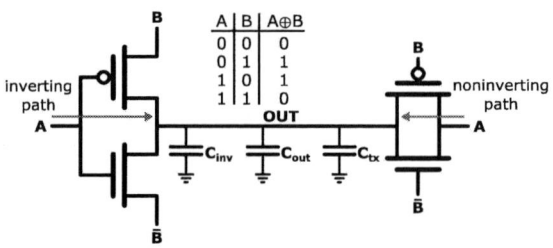

Fig. 2. XOR gate topology used in the comparison

© VDE VERLAG GMBH · Berlin · Offenbach

level, making the two paths poorly matched from the delay point of view. To size the transmission gate of the non-inverting path we have to consider that increasing its size reduces the equivalent resistance between its two terminals but, at the same time, the intrinsic capacitance of the gate increases (Fig. 2). We can expect the delay from node A to OUT to decrease, initially, due to the reduced resistance. At some point, though, the intrinsic capacitance C_{tx} will become dominant with respect to the gate capacitance of the following stage C_{out} and the delay will settle on the intrinsic τ_0, a constant value determined by technology parameters. This can be expressed mathematically using the following expression

$$\tau_{tx} = \tau_{tx,0} \left(1 + \frac{C_{out} + C_{inv}}{C_{tx}}\right) \tag{1}$$

In the inverting path, instead, the transmission gate is turned off, but acts as a load for the inverter (Fig. 2). Thus, the delay from node A to OUT will linearly increase as a function of the transmission gate capacitance with respect to the inverter intrinsic capacitance C_{inv}:

$$\tau_{inv} = \tau_{inv,0} \left(1 + \frac{C_{tx} + C_{out}}{C_{inv}}\right) \tag{2}$$

Fig. 3 shows the delay obtained from schematic simulations between node A and node OUT on both paths, as a function of the relative size of the transmission gate. The skew is minimized if the delay of the two paths is the same, i.e. the point in Fig. 3 where the two curves meet. Notice that the output signals are re-buffered by two inverters. The output skew is, therefore, low if the driving strength for rising and falling transitions is the same, which the case is if the pMOS and nMOS are correctly balanced. This means that we can expect a variability of skew against process variations, as the driving strength will not be perfectly balanced in every process corner. Moreover, from the jitter point of view there are two stages after the level shifter, each adding its own noise. In the inverting path the transmission gate acts like a resistance, which limits the current supplied to the load capacitance and adds noise. Since the output jitter of an inverter is inversely proportional to the current that charges the load capacitor [2], the transmission gate will degrade the jitter performance, due to the reduced slope on the load.

B. Phase-interpolation based

An alternative implementation of clock splitter was proposed in [3], even though sizing considerations and its jitter performance have not been discussed by the authors. This architecture is based on *phase interpolation*, which consists in averaging multiple phases generated by different means. Referring to the circuit schematic in Fig. 4, four signal paths can be identified after the level shifter: one passes through a single inverter (I_3), one through a single buffer (B_1), one through the series of an inverter (I_1) and a buffer (B_2), and the last one through the series of two inverters (I_1 and I_2). The two inverting paths are then connected together to interpolate their output and obtain the inverted output phase, and the

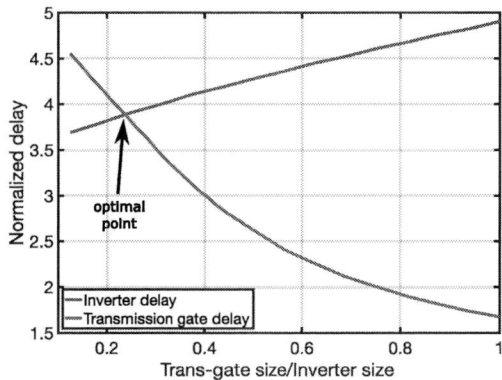

Fig. 3. Delay of the XOR gates in the two paths as a function of the transmission gate size, normalized to $\tau_{tx,0}$

same is done for the non-inverting paths. The two outputs are connected through cross-coupled inverters to ensure differential operation. The buffers used in the phase-interpolation-based splitter are similar to push-pull stages, hence their output would not be full-swing if used stand-alone. However, since it is connected to the output of an inverter, the latter restores the output rail-to-rail swing, effectively shutting off the buffers when the output reaches the supply or ground rails. Unlike the XOR-based phase splitter, this architecture is much more symmetrical even at transistor level since the clock signal, both in the inverting and non-inverting paths, passes through the parallel of one gate and the series of two gates. Therefore, the delay of the two paths will be similar, hence the skew between the two output phases will be reduced with respect to the alternative solution. The structure has to be sized such

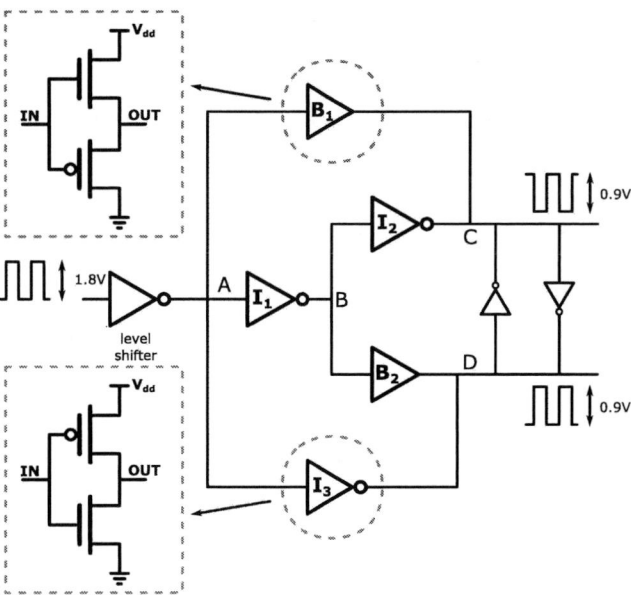

Fig. 4. Phase-interpolation based clock splitter

SMACD / PRIME 2021 | 19 – 22 July 2021, Online Event

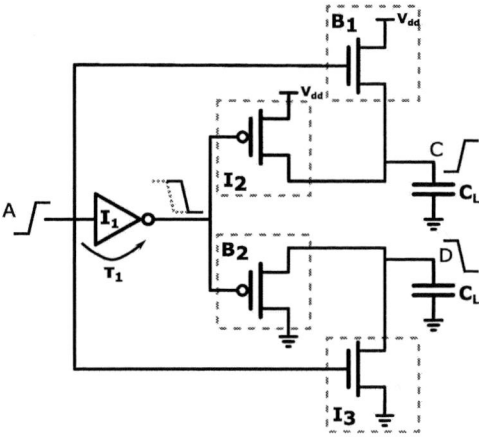

Fig. 5. Transistors turned on during an input logic low to high transition

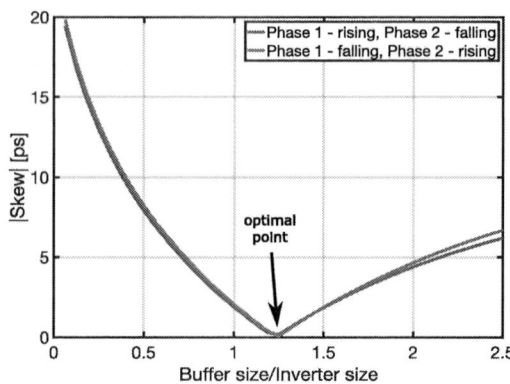

Fig. 6. Skew between the output signals as a function of the buffer sizing

that the delay from node A to the output nodes (i.e. C and D) is larger than the delay from node A to node B, which is easily obtained when the clock splitter drives large capacitive loads. After a short period of time during which the outputs are pulled up and down by the outer branches alone, the middle path will catch up and start pulling the output nodes in the same direction, carrying current with the complementary transistor. Considering, for example, the input toggling from logic low to high (Fig.5), node C starts being pulled up by the nMOS of buffer B_1, while node D is being pulled down by the nMOS of the I_3 inverter. After a delay τ_1 the slower paths turn on: the pMOS of I_2 starts pulling up node C while the pMOS of the buffer B_2 starts pulling down node D. From Fig. 5 it is clear that, after a delay τ_1, the output nodes are pulled in opposite directions by a pMOS in parallel to a nMOS on both sides. If the two building blocks are balanced, i.e. their driving strength is the same for upward and downward transitions, the current provided to the load will be the same on both sides, hence the delay will be equal. In order to minimize the skew between the output phases, it is necessary to have a pre-charge of the output nodes during τ_1 equal on both sides. The current provided by I_3 considering the transistor working in the velocity saturation region is equal to [4]

$$I_{i,n} = \mu_n C'_{ox} \left(\frac{W}{L}\right)_{i,n} \cdot V_{sat} \left(V_{DD} - V_{TH} - \frac{V_{sat}}{2}\right) \quad (3)$$

The current provided by B_1 is, similarly, given by

$$I_{b,n} = \mu_n C'_{ox} \left(\frac{W}{L}\right)_{b,n} V_{sat} \left(V_{DD} - V_{out} - V_{TH} - \frac{V_{sat}}{2}\right) \quad (4)$$

This current can be divided in two parts: a constant one, whose expression is the same as the inverter one, and a variable part, linearly dependent on V_{out}. With a rough approximation, we can assume that the current reduction during τ_1 is small compared to the total current. Therefore, the pre-charge is on both sides equal to

$$V(\tau_1) = I \cdot \frac{\tau_1}{C_L} \quad (5)$$

Thus, the pre-charge voltage is symmetrical if the current of the inverter is equal to the current of the buffer. Neglecting second order effects, this is obtained with

$$\left(\frac{W}{L}\right)_{i,n} = \left(\frac{W}{L}\right)_{b,n} \quad (6)$$

The same reasoning applies to the pMOS transistors when an input transition from logic high to logic low is considered. The relative sizing of the buffers with respect to the inverters can be refined by means of circuit simulations. Fig. 6 shows the output signals skew against the normalized buffer size, as obtained from schematic simulations. The minimum value of skew is achieved with buffers slightly larger than the inverters, due to the current reduction during the output voltage transition. In this architecture the output nodes are always either pulled by both types of transistors (after τ_1) or by the same type of transistor on both sides (during τ_1). This means that even in presence of process variations the strength of the gates driving the two sides will be similar, albeit different from nominal. Thus, the skew between the two output signals will not vary much against PVT variations. From the jitter point of view, on both paths the output is still ultimately pulled up or down by an inverter. The buffers help the inverters during the transitions, up to a threshold V_t from the supply or ground rails, where their transistors shut off. With respect to the XOR based topology, in this architecture there is only one stage after the level shifter, made by the parallel of a buffer and an inverter, since the phase splitting is accomplished by interpolation of the two. Since the number of stages is reduced, the output jitter will be reduced too, as every stage in a chain adds its own noise.

C. Comparison of the two architectures

In order to get a quantitative comparison between the two circuits, they were both tested in schematic-level simulations to assess their performance. The power consumption, output jitter and absolute value of the skew between the two output phases are compared with both splitters driven by the same impedance matched driver stage and with a capacitive load

© VDE VERLAG GMBH · Berlin · Offenbach

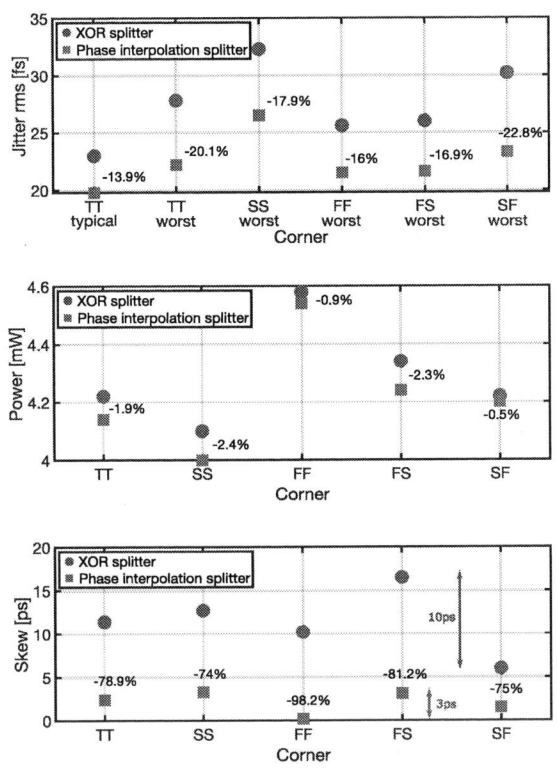

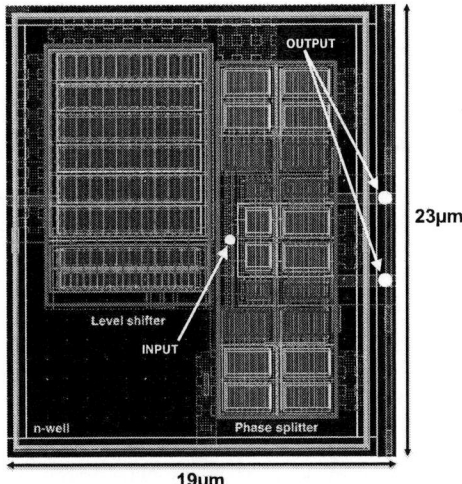

Fig. 8. Layout of the phase-interpolation clock splitter

Fig. 7. Performance comparison of the two topologies over different process corners (jitter, power and skew) and flicker noise corners (jitter)

of $C_L = 1\text{pF}$ per phase in both cases, which models the clock distribution tree the splitters are meant to drive. The performance metrics are compared in different process corners, to evaluate their stability with respect to process variations. The input frequency is 1GHz, hence jitter is calculated as the corresponding edge phase noise, obtained from periodic noise simulations, integrated from 1Hz to $f_{out}/2 = 500\text{MHz}$ [5]. As highlighted in Fig. 7, the phase-interpolation based structure achieves, as expected, a better jitter performance (15% reduction on average), with no penalty in power dissipation, thanks to the reduced number of stages with respect to the XOR-based topology. Moreover, the phase-interpolating architecture is definitely superior when the skew performance is compared. Thanks to the symmetry of the implementation the two output signals are much more balanced. This greatly reduces the skew both in the absolute value, which is on average 80% lower, and in its variability against process variations, which is reduced by a factor 3.

III. LAYOUT OF THE PHASE INTERPOLATION SPLITTER

Given the better performance obtained from schematic-level simulations, the phase interpolating splitter was implemented at layout level in 28nm bulk CMOS technology. Post-layout simulations allow to verify the behavior of the circuit in more realistic conditions which take into account the interconnection

parasitics. The complete layout of the splitter is presented in Fig. 8. It occupies about $440\mu\text{m}^2$, including both the level shifter and the splitter itself enclosed in a deep n-well to improve isolation from the rest of the chip. Post-layout simulations including both the input buffer and the splitter driving a 1pF per phase load showed a degradation of the output jitter due to the additional loads of the interconnections parasitics. Nevertheless, the output jitter is below 35fs even in the worst process corners, with a moderate increment in power consumption to, at most, 5mW. The skew between the two output signals remains below 10ps over all process corners.

IV. CONCLUSIONS

Two alternative clock phase splitter architectures have been analyzed and compared. Schematic-level simulations demonstrate that the phase-interpolation-based structure outperforms the XOR-based topology both from the jitter and the skew point of view. In particular, the output jitter 15% lower, while the skew between the output phases is 80% lower. A layout level implementation in 28nm bulk CMOS technology of the phase interpolating splitter shows, in post-layout simulations, a worst case output jitter below 35fs with maximum power consumption of 5mW while the skew between the output signals is lower than 10ps over all process corners.

REFERENCES

[1] S. S. Mishra, A. K. Agrawal, and R. Nagaria, "A Comparative Performance Analysis of Various CMOS Design Techniques for XOR and XNOR Circuits," *International Journal on Emerging Technologies*, vol. 1, no. 1, pp. 1–10, Jan. 2010.

[2] A. Abidi, "Phase Noise and Jitter in CMOS Ring Oscillators," *IEEE Journal of Solid-State Circuits*, vol. 41, no. 8, pp. 1803–1816, Aug. 2006.

[3] Y.-S. Park, S.-W. Lee *et al.*, "PVT-invariant single-to-differential data converter with minimum skew and duty-ratio distortion," in *2008 IEEE International Symposium on Circuits and Systems*, May 2008, pp. 1902–1905.

[4] J. M. Rabaey, *Digital integrated circuits: a design perspective*. Prentice Hall, 1996.

[5] N. Da Dalt and A. Sheikholeslami, *Understanding jitter and phase noise: a circuits and systems perspective*. Cambridge University Press, 2018.

Entropy Analysis of RO-based Physically Unclonable Functions

G. Diez-Senorans, M. Garcia-Bosque, C. Sánchez-Azqueta and S. Celma

Group of Electronic Design, University of Zaragoza

Zaragoza Spain

Email: {gds, mgbosque, csanaz, scelma}@unizar.es

Abstract—In this paper we estimate the probability distribution of a ring oscillator based PUF (RO-PUF) operating under different digitization schemes (topologies) of compensated measuring, and we quantify their performance and security properties by means of the entropy and entropy-related metrics: entropy per bit ratio, entropy per oscillator ratio and the product of both. We have studied the most common digitization schemes for this type of systems, and we propose a new construction designed to overcome some flaws found in these. Each topology has been tested with a large set of frequencies obtained from an array of three-stages oscillators implemented in FPGA.

Index Terms—Compensated measuring, Entropy, FPGA, Hardware security, Physically Unclonable Function, Ring oscillator

I. INTRODUCTION

The distributed nature of the Internet of Things (IoT) technology and its restrictions on power and area make the physical layer of these systems a major vulnerability. In this context, physically unclonable functions (PUFs) arise as a promising security primitive, capable of providing secure storage and identification of trusted instances.

Electronic physically unclonable functions (PUFs) are understood as pieces of hardware which provide a digital response when exposed to a suitable stimulus. Traditionally, PUFs have been classified depending on the size of this stimuli space as either weak if it is small (i.e. it can be exhausted in polynomial or less time) or strong (otherwise) [1].

As a consequence of this intrinsic difference, both types of PUFs behave in a dramatically different way, and thus they find application in very different fields: weak PUFs (also called Physical Obfuscated Keys) provide a secure key storage mechanism in which keys are re-generated from hardware-specific features of the device rather than stored in non-volatile memories [1]–[3], while strong PUFs can be used in identification/authentication protocols as well as key generation [1], [4].

In this paper, a new method to assess the security of PUFs based on a statistical entropy analysis is used to compare the security of three well-known topologies of RO-PUF: *1-out-of-2*, *N-1*, and *All-pairs*; as well as the new topology *k-modular* proposed in [5], which we have validated through experimental implementation in FPGA in this work. The experiments conducted on these systems and subsequent entropy analysis confirm that the new topology presents clear advantages with respect to the previous architectures.

It is remarkable that this study is of application to any type of PUF whose response bits are extracted from the outcome comparison of identically designed hardware (compensated measuring [4]). A main example of this scheme is the ring oscillator PUF (RO-PUF) [6].

II. BACKGROUND

The core of a RO-PUF is a bank of N ring oscillators designed to be identical, alongside with a system capable of measuring their natural frequencies. It is a common practice to apply compensated measuring techniques in order to improve the resistance of these systems against environmental variability. In the most simple case of compensated measuring one bit is extracted out of a pair of oscillators depending on which one runs faster [7], [8]. The entropy of this system is known to be upper bounded by $S_{max} = \log_2 N!$ [3]. However, the problem of how to select pairs of oscillators among all the $\frac{1}{2}N(N-1)$ possible choices (*topology* of the RO-PUF) in order to deliver a high entropy per bit ratio is hard because of the high correlation between bits in compensated measuring PUFs. This leaves two main design options, as reviewed in the literature: the *1-out-of-k* topology [3], [4], [9], and the *N-1* bits topology [10], [11]. The former takes one bit out of k different oscillators without repetition, by selecting the pair whose characteristic frequencies are further apart from each other. This scheme presents some advantages as it makes it possible to strengthen the system against environmental variations (particularly temperature changes), at the cost of a large decrease in performance since the extracted entropy per oscillator ratio reduces by a factor of k. On the other hand, the *N-1* bits strategy is an attempt to maintain the system within reasonable limits of environmental-robustness by performing comparisons between adjacent oscillators, while increasing the entropy outcome to approximately N bits.

III. EXPERIMENTAL SETUP

In this work, we investigate the probability distributions that emerge from different topologies $\mathcal{T}$ when applied to a set of ring oscillators. These are implemented in a FPGA model xc7z020 using three inverters and an AND gate as enable control 1. The measuring process is controlled by an Atmega micro controller which sets the enable signal to high and measures the time taken by each oscillator to complete 10^8 laps; for the set of 34×23 oscillators shown in Fig. 1 this

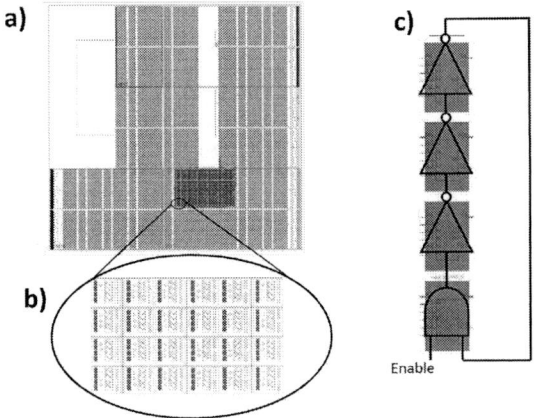

b)

c)

Enable

Fig. 1. a) Implementation of measuring system in FPGA. b) Zoom to matrix of ring oscillators. c) Scheme of an oscillator.

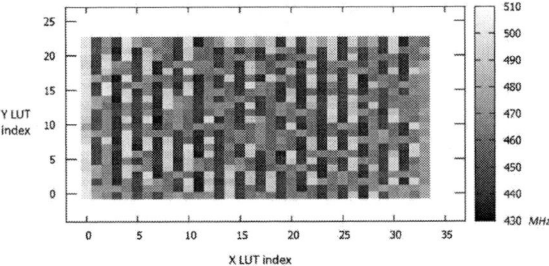

Fig. 2. Distribution of frequencies throughout the matrix of oscillators implemented in FPGA.

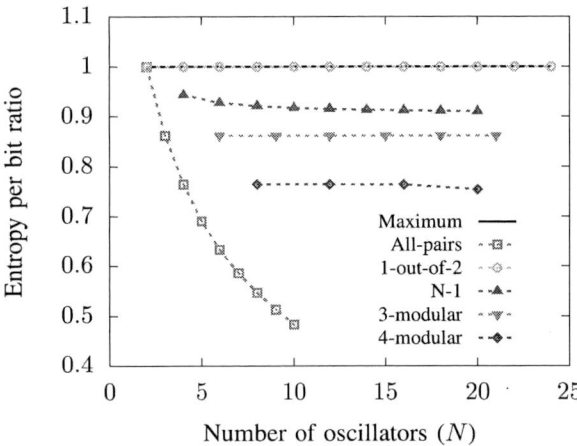

Fig. 3. Entropy per bit ratio against the number of oscillators.

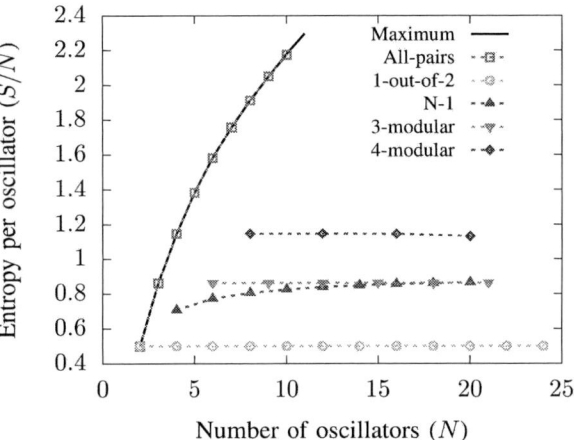

Fig. 4. Entropy per oscillator ratio against the number of oscillators.

leads to an average frequency of $\hat{f} = 463.93 \pm 0.01\,MHz$. The distribution of frequencies is shown in Fig. 2, which exposes some degree of correlation in the frequency spatial distribution.

PUF responses are generated in post processing by randomly picking N oscillators from the matrix and using them to obtain a binary word (PUF response) according to the instructions provided by the specific topologies that have been investigated: *1-out-of-k*, *N-1*, *All-pairs* (where all $N(N-1)/2$ possible comparisons are evaluated) and *k-modular* (an original topology proposed by the authors to overcome some of the flaws shown by the others schemes). We iterate this process to generate an estimation of the probability distribution underlying the topology, and this is repeated for different values of N to study how the distributions change depending on the number of oscillators.

From these distributions we extract a number of interesting metrics which have been used as figures of merit: the entropy per bit ratio S/bit in Fig. 3 might be used to characterize resistance against cryptanalysis, while the entropy per oscillator ratio S/N in Fig. 4 influences performance regarding area and power consumption. Additionally the total Shannon entropy S is presented in Fig. 5 for completeness. All these figures are presented along with the maximum

value (green line) for each metric. In the next section, we will show the probability distributions obtained for different topologies, proving these to be non-uniform, indicating a potential weakness for this kind of PUFs whose digitization system is based in pairs comparison.

IV. RESULTS

In our research we have explored four different topologies:

A. 1-out-of-2

In this topology, each bit is obtained comparing two adjacent oscillators without repetition, i.e. the i-th and $(i+1)$-th oscillators produce the $i/2$-th bit, so that $N/2$ bits are extracted from N oscillators. This kind of digitization and others within the *1-out-of-k* family (which are expected to behave in a similar way) are the most common topologies found in the literature. It produces a plain distribution (Fig. 6.a) because of the lack of repetition in oscillators comparison, i.e. a high entropy per bit ratio (~ 1, see red dotted line in Fig. 3) which

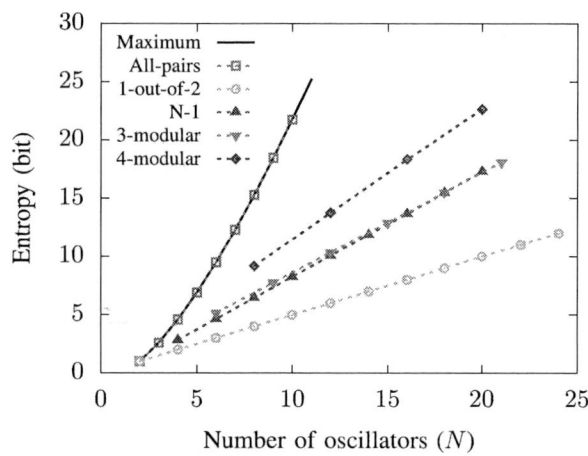

Fig. 5. Total entropy for each topology analyzed against the number of oscillators.

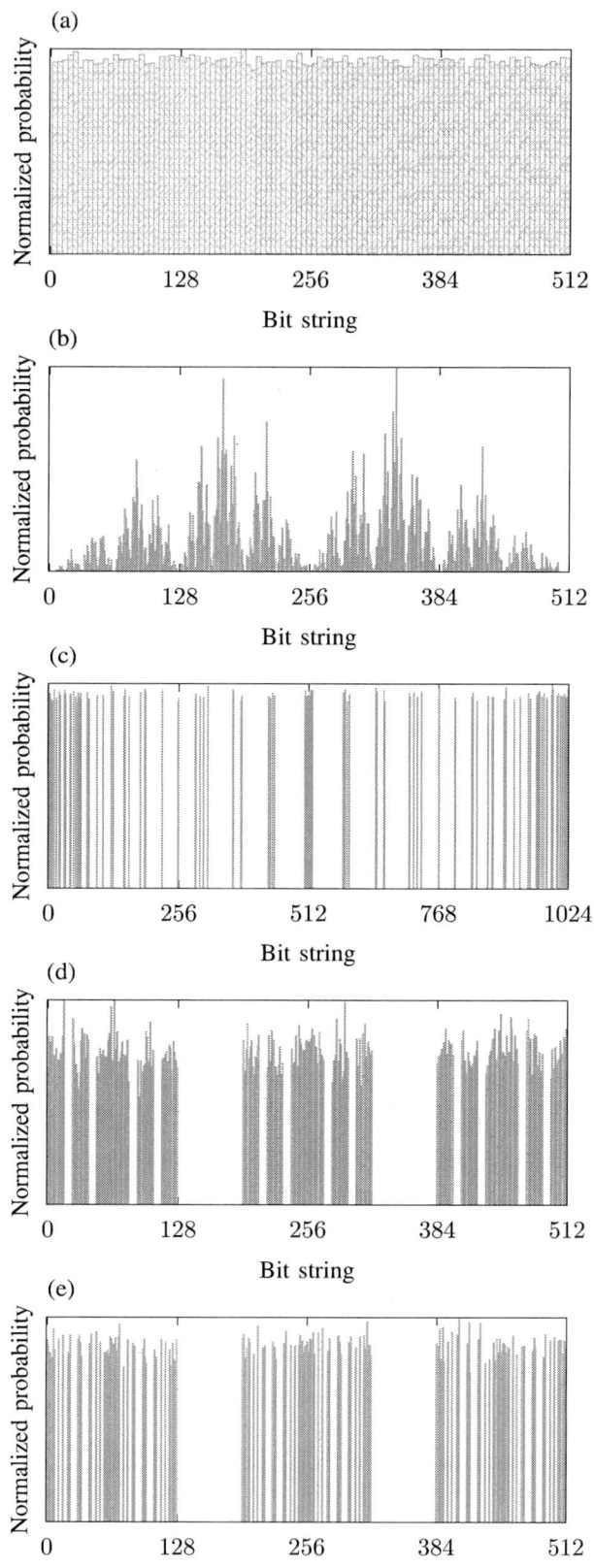

Fig. 6. Example of outcome distribution for topologies: (a) *1-out-of-2* , (b) *N-1*, (c) *All-pairs*, (d) *3-modular*, (e) *4-modular*.

however comes at the cost of a poor entropy per oscillator ratio of $\sim 1/2$, red dotted line in Fig. 4. These results suggest that this topology is a deeply conservative one, thus useful on systems without power or area restrictions, but might be of less use on mobile and wireless devices.

B. N-1 topology

In this topology $N-1$ bits are extracted from N oscillators comparing oscillators i-th and $(i+1)$-th in order to obtain the i-th bit, $1 \leq i \leq N-1$. This topology suffers a shortcoming that arises from the fact that the PUF is trying to accommodate $N!$ states on $N-1$ bits (which have room for 2^{N-1} different states, see Fig. 6.b), i.e. there exist some permutations of oscillators that leave the comparison pattern between oscillators pairs invariant, thus leading to an identical outcome for two different PUF realizations. If managed carefully, an adversary could take advantage from this collision probability, for example by presenting fraudulent keys cleverly chosen in a probability-descendant way: it would be expected that such an adversary would succeed in impersonating the legitimate PUF much faster than one trying to exhaust the space of keys in a random fashion (yet further research in this direction is necessary). The total entropy as a function of the number of oscillators was found to be linearly dependent (see black dotted line in Fig. 5), while entropy per bit (Fig. 3) and entropy per oscillator (Fig. 4) evolve to saturation.

C. All-pairs topology

In this case the digitization process is carried out by exhausting all possible comparisons in the matrix of oscillators, thus $N(N-1)/2$ pair bits are deployed. This way of extracting strings from the RO-PUF is infrequent in the literature, since it suffers from a deep correlation due to the transitivity of ordering. However, it is interesting to include it in this work for the sake of completeness.

Fig. 6.c shows the distribution probability of the *All-pairs* outcomes; this histogram leaves a large number of blank spaces throughout the possible decimal outcomes as expected, since the bit correlation prevents $2^{N(N-1)/2} - N!$ states from being visited; however, interestingly enough the system is uniformly distributed over the remaining $N!$ values. The escalation of the entropy with the number of oscillators is shown in Fig. 5 (pink dotted line)

Since each state consists of all possible pairs that can be made out of an N oscillator matrix, any other topology construction $\mathcal{T}_0$ will have a space of states which will be a subset of the *All-pairs* space, and thus the entropy of $\mathcal{T}_0$ will be minor or equal than that of *All-pairs*. This still stands for the construction that extracts the maximum possible entropy from the matrix, thus implying that *All-pairs* actually deploys the maximum possible entropy, $\log_2 N!$ (see pink dotted line overlapping green line in Fig. 5).

D. k-modular topology

The family of *k-modular* topologies, which have not been reported before to the best of our knowledge, represents an attempt to combine the benefits of avoiding repetition in oscillator comparison while keeping a high S/N ratio. In order to achieve that goal, the matrix of oscillators is divided in N/k groups of k oscillators, each of which is treated like an independent RO-PUF of $N = k$ oscillators, and is evaluated in an *All-pairs* fashion as to produce $k(k-1)/2$ bits. Thus, the total number of bits extracted from this topology is $N(k-1)/2$ bits. Since every group is unconnected to the rest and entropy turns out to be an additive magnitude, the total entropy deployed by this system is expected to be $S = N/k \times \log_2 k!$

Correspondingly, the entropy per bit will be $S/\text{bit} = 2\log_2 k!/k(k-1)$ and entropy per oscillator ratio is given by $S/N = \log_2 k!/k$.

The probability distributions for the specific cases of $k = 3$ and $k = 4$ are shown in Fig. 6.d and 6.e respectively; both of them show an improvement in the uniformity of the distribution with respect to the *N-1* topology shown in Fig. 6.b. As it was previously stated, the entropy is expected to be linearly dependent on N (see orange and blue dotted lines in Fig. 5), while entropy per bit ratio is constant on the number of oscillators and expected to equal $S/\text{bit} \approx 0.86$ for $k = 3$ and $S/\text{bit} \approx 0.76$ for $k = 4$ (Fig. 3), and $S/N \approx 0.86$ for $k = 3$ and $S/N \approx 1.15$ for $k = 4$ (Fig. 4).

Regarding the uniformity of the distribution, this topology promises to be more robust against a "clever search" attack as proposed against *N-1* topology. However, it is noticeable that the entropy extracted per bit is lower than that of either *1-out-of-2* or *N-1* topologies, which suggests the existence of different vulnerabilities other than the non-uniformity of probability distribution. Nevertheless the efficiency of the system in terms of entropy per oscillator seems to improve with respect to other topologies while keeping higher S/bit than *All-pairs* comparison, which suggests that this digitization proposal is a better alternative for resource-limited systems.

V. CONCLUSIONS

In this work we have analyzed the outcome probability distribution of compensated measurement PUFs, of which

the best known example is the RO-PUF, where the output frequencies of ring oscillators pairs are compared to generate an output binary string.

The PUF responses have been produced using different sets of N ring oscillators implemented in a xc7z020 FPGA as well as a topology, i.e. an algorithm specifying the exact way in which the comparison will be carried out. A large set of different PUF realizations is examined to ensure enough statistics about the output distribution.

The metrics used to evaluate each probability distribution were: the total entropy of the distribution, the entropy per bit and the entropy per oscillator ratios. From these figures of merit we can conclude that the proposed system *k-modular* performs better than others in terms of entropy per bit, while retaining a high entropy per oscillator ratio. Furthermore, it is noticeable that a system performing all possible comparisons (*All-pairs* topology) might be being underestimated in PUF design practice, since it reaches a good commitment between entropy per bit and oscillator ratios despite the large correlation between some bits.

REFERENCES

[1] C. Herder, M. Yu, F. Koushanfar and S. Devadas, "Physical Unclonable Functions and Applications: A Tutorial," in *P IEEE*, vol. 102, no. 8, pp. 1126-1141, Aug. 2014.

[2] M. García-Bosque, G. Díez-Señorans, C. Sánchez-Azqueta and S. Celma, "Overview of Physically Unclonable Functions: Properties and Applications" in ECCTD2020, 2020.

[3] G. E. Suh and S. Devadas, "Physical Unclonable Functions for Device Authentication and Secret Key Generation," in *2007 44th ACM IEEE D*, pp. 9-14, 2007.

[4] R. Maes, "Physically Unclonable Functions: Constructions, Properties and Applications", Ph.D. dissertation, Dept. of Electrical Engineering, K.U. Leuven, Belgium, 2012.

[5] G. Diez-Senorans, M. Garcia-Bosque, C. Sánchez-Azqueta and S. Celma, "A New Approach to Analysis the Security of Compensated Measuring PUFs" in ECCTD2020, 2020.

[6] T. McGrath, I. E. Bagci, Z. M. Wang, U. Roedig and R. J. Young, "A PUF Taxonomy", in *Appl. Phys. Rev.*, vol. 6, 2019.

[7] B. Gassend, "Physical random functions", M.S. thesis, Dept. of Electrical Engineering and Computer Science, MIT, Massachusetts, 2003.

[8] B. Gassend et al, "Silicon physical random functions" in *Proceedings of the 9th ACM conference on Computer and communications security*. ACM, pp. 148-160, 2002.

[9] C. D. Yin and G. Qu, LISA: Maximizing RO PUF's secret extraction, in *2010 IEEE HOST*, pp. 100-105, 2010.

[10] M. García-Bosque, G. Díez-Señorans, C. Sánchez-Azqueta and S. Celma, "Proposal and Analysis of a Novel Class of PUFs Based on Galois Ring Oscillators," in IEEE Access, vol. 8, pp. 157830-157839, 2020.

[11] A. Maiti and P. Schaumont, Improving the quality of a Physical Unclonable Function using configurable Ring Oscillators, *2009 INT CON FIELD PROG*, pp. 703-707, 2009.

On the Behavior of a Wide Set of Oscillators: PUFs or TRNGs?

M. Garcia-Bosque[1,2], A. Naya[1], G. Díez-Señorans[1], C. Sánchez-Azqueta[1] and S. Celma[1]

[1]Group of Electronic Design, University of Zaragoza, Zaragoza, Spain
[2]Centro Universitario de la Defensa, Zaragoza, Spain.
{mgbosque, abeln, gds, csanaz, scelma} @unizar.es

Abstract—In this paper, a generic structure that includes previously studied oscillators (such as ring oscillators) as well as many other new oscillators has been studied to evaluate their suitability as TRNGs or PUFs. The studied structure consists of an array of *n* combinational logic blocks in a loop where the output of each block is a function of the output of the previous block and the feedback signal. To perform this analysis, a novel implementation has been proposed where, using a single implementation, we can make each block perform any possible 2-input function using some external configuration inputs. By analyzing all possible configurations of size equal or smaller than 7, we have concluded that none of them behave as an ideal TRNG and that ring oscillators present a behavior closest to the ideal one. Regarding their suitability for being used as PUFs, none of the polynomials have shown an ideal behavior but some of them present a higher reproducibility than the classical ring-oscillator PUF. Furthermore, we have noticed that, by increasing the length, we can find configurations with better PUF properties.

Keywords—*Galois ring oscillators, hardware security, physically unclonable functions, ring oscillators, true random number generators*

I. INTRODUCTION

Nowadays, with the huge amount of confidential data that is being shared around the world, achieving the security of the communications has become a priority task. While, traditionally, most of the research works focused on designing encryption algorithms that were secure from a mathematical point of view, current efforts include other aspects such as generating secure keys, storing keys in a secure way or checking that the users are who they claim to be (user authentication).

In this context, two important cryptographic primitives are True Random Number Generators (TRNGs) and Physically Unclonable Functions (PUFs).

With regard to FPGA implementations, while many different structures have been proposed, most of the preferred solutions for both TRNGs and PUFs are based on ring oscillators (ROs). In the case of RO-TRNGs, they typically use the noise in frequency or phase (jitter) of ROs [1], [2]. In the case of RO-PUFs, they are often based in the small frequency differences between identical oscillators implemented in different locations [3], [4].

In [5], a new set of oscillators called Fibonacci Ring Oscillators (FIRO) and Galois Ring Oscillators (GARO) were proposed as fast TRNGs. These systems have been widely studied and several TRNGs based on them have been proposed [6]. However, this kind of structures are not completely understood, are not supported by a stochastic model and, therefore, there is not a way of guarantying a minimum entropy of these systems [7], [8]. Furthermore, some works have proven that the behavior of these systems can greatly depend on the location within the FPGA so that, in certain locations, these systems can present poor randomness results [9], [10]. Based on this fact, [11], [12] studied the possibility of using the variations with the location presented by the GAROs to construct a PUF. That work concluded that the bias of these systems varied with the location in a similar manner as the frequencies of a ring oscillator and, therefore, it was possible to use GAROs to construct a PUF in an analogous manner as an RO-PUF but comparing bias instead of frequencies. However, the uniqueness of the tested systems seemed to be smaller than the uniqueness presented by analogous RO-PUFs.

In this paper, an analysis of the bias of a much wider set of oscillators have been carried out to evaluate the suitability of these systems as both TRNGs and PUFs. The paper is organized as follows: Section II presents the generic structure of the oscillators that we have studied and a way to implement it in a configurable manner; Section III explains the experiment that we have carried out as well as the parameters calculated to evaluate the systems; Section IV presents the experimental results; finally, conclusions are drawn in Section V.

II. STUDIED GENERIC STRUCTURE

A. Background

In [5], with the aim of combining the true random properties of ROs and pseudo-random properties of Linear Feedback Shift Registers (LFSRs), Fibonacci Ring Oscillators (FIRO) and Galois Ring Oscillators (GARO) where proposed. Their structure was analogous as an LFSR (with a Fibonacci or Galois structure) but used inverters instead of registers.

In [11], it was shown experimentally that the bias of GAROs changed with their location in a reproducible way and, therefore, they could be used to construct a PUF. As a proof of concept, a 7-LUT PUF that compared the bias of neighboring GAROs was implemented achieving an average Intra-chip Hamming Distance (Intra-*HD*) of ~1% and an average Inter-chip Hamming Distance (Inter-*HD*) of ~39%.

In this work, a more generic structure has been studied to evaluate their suitability as both TRNGs and PUFs.

This work has been supported by MINECO-FEDER (TEC2017-85867-R), CUD (CUD-2020_04) and DGA fellowship to G. Díez-Señorans.

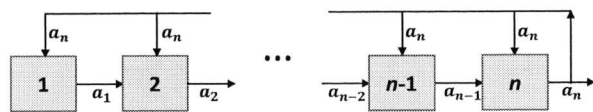

Fig. 1. Scheme the proposed generic structure.

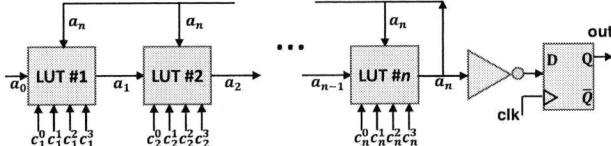

Fig. 2. Scheme of the proposed generic configurable structure.

B. Proposed generic configurable structure

The proposed generic structure consists of an array of logic blocks that perform a combinational operation where the output of each block, a_i, $i > 1$, can be any function of the feedback signal, a_n, and the output of the previous block, a_{i-1}. In case of the first block, its output, a_1, can only be the feedback signal a_n or its inverted signal, $\overline{a_n}$. The scheme is shown in Fig. 1. It can be trivially seen that this structure includes ROs (when all the blocks perform an inversion operation $a_i = \overline{a_{i-1}}$) and GAROs (when all the blocks perform an inversion or an XNOR operation) but also a huge number of additional oscillators.

The goal of this work is to analyze experimentally these oscillators to see if any of them can be used to construct a good TRNG or a good PUF.

Typically, when an oscillator is implemented in an FPGA, it has a fixed connectivity and can only perform a fixed function (a ring oscillator, a GARO with a certain feedback polynomial, etc.). However, as explained before, this work aims to analyze experimentally many of them. Therefore, creating a new implementation for each of them would require a huge amount of time and would be unfeasible. For this reason, a generic structure has been implemented in a configurable manner. Since each LUT can carry out any possible 6-input function, it is possible to use two of the inputs as the inputs of the logic block (a_n and a_{i-1} in Fig. 1) while using the other extra 4 inputs as configuration inputs $c_i = (c_i^0, c_i^1, c_i^2, c_i^3)$, which can be introduced externally, to determine the function $a_i = f_{c_i}(a_{i-1}, a_n)$ that the logic block is performing.

In case of LUT #1, it should be enough to use a single configuration input to determine if the LUT performs an inversion ($a_1 = \overline{a_n}$) or a delay operation ($a_1 = a_n$). However, to have a more symmetric structure, the first LUT also includes an a_0 signal that is introduced externally and four configuration inputs so that the first LUT can perform any logic operation $f_{c_1}(a_0, a_n)$. Nevertheless, during all our experiments, the external signal is always $a_0 = 0$ and the function f_{c_0} is always a delay or an inversion. Finally, an inverter followed by a flip-flop is used to sample the system. This inverter is used to avoid any possible frequency couplings. The scheme of the proposed configurable structure is shown in Fig. 2.

Regarding the implemented functions, there are 16 possible 2-input functions that can be configured with the configuration signals c_i. However, in practice, some functions are not interesting to be studied since they create fixed points or their only effect is to reduce the effective size of the system. For this reason, the only functions that have been considered are the XOR, XNOR, OR, NOR, AND, NAND, DEL ($f_{c_i}(a_{i-1}, a_n) = a_{i-1}$) and INV ($f_{c_i}(a_{i-1}, a_n) = \overline{a_{i-1}}$). In any case, the implemented structure can perform any operation.

It must be noticed that there are a couple of extra functions $f_{c_i}(a_{i-1}, a_n) = a_n$ and $f_{c_i}(a_{i-1}, a_n) = \overline{a_n}$ whose net effect is that the LUTs #1 to #i-1 do not have any influence in the output. Therefore, using these extra functions, it is possible to study any oscillator of size less than n.

III. Experiment Description

A. Experimental setup

To study the bias of these oscillators, a 7-LUT configurable structure has been implemented in 101 different locations in 20 different FPGAs (using Pynq Z2 boards). To carry out the experiments, a Python script has been used to send instructions to the FPGAs (choose the configuration, start each measurement, reset the systems, …) and to collect the data from the FPGAs. The communications between the computer and the FPGAs have been carried out through serial RS-232 standard. The current setup measures all 101 locations in parallel but, due to some limitations in our setup, we have only been capable of measuring 5 FPGAs at the same time.

To measure the bias of each system, the sampling frequency of the flip-flop is 100 kHz and, when the sampled value is 1, a counter is increased. After 100 000 samples (1 second), the final value of the counter is used as an estimation of the bias. These values for the sampling frequency as well as the total number of samples have been chosen for two reasons: first, there are the same as the ones used in [11] so, this way, it is easier to compare both works; second, according to [11], by choosing these values it is possible to estimate the bias with high precision without taking too much time to complete each measurement.

Since one of the properties that we want to measure is the reproducibility of the bias, each measurement is repeated 100 times. To sum up, for each configuration and each FPGA a matrix of integer numbers $A = \{A_i^j\}$ is generated where each element represents the final value of the counter at the ith measurement at the jth location.

B. Measured parameters

To evaluate whether each configuration can be used as a good PUF or as a good TRNG, several parameters have been calculated:

- **Randomness of the bias:** If a certain oscillator was an ideal TRNG, the measured values of the bias should follow a binomial distribution of $p = 0,5$ and $N = 100\ 000$. To measure how close the measured values are from a binomial distribution, we have calculated the Root Mean Square Error (RMSE) between the ideal

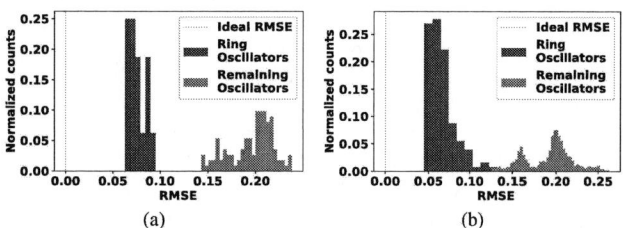

Fig. 3. Histograms of the obtained RMSE values for (a) 5-length oscillators and (b) n-length oscillators ($n \leq 7$).

binomial Cumulative Distribution Function (CDF) and the obtained CDF.

- **Reproducibility of the bias:** A PUF response is typically obtained by comparing the bias in two or more different locations so, in order to be reproducible, the differences between the column elements in A should be much smaller than the differences between the row elements in A. A possible way to quantify this reproducibility is to divide the average standard deviations of the rows and columns as done in [11]. However, it can be difficult to interpret how this parameter would exactly affect the average Intra-chip Hamming Distance (Intra-*HD*) of an actual PUF. For this reason, in this work, to measure the reproducibility, we have compared the bias of neighboring oscillators to obtain 100-bit responses and calculated their average Intra-chip Hamming Distances (Intra-*HD*s). This is in line with the analysis made in [13].

- **Uniqueness of the bias:** To check if a configuration can be used to construct a unique PUF, the average bias in a given location should be different in different FPGAs. In a similar way as explained before, this could be quantified comparing standard deviations but, again, we have chosen to generate 100-bit responses and calculate their average Inter-chip Hamming Distances (Inter-*HD*s).

IV. EXPERIMENTAL RESULTS

A. Preliminary test

With the chosen values of sampling frequency, number of samples and number of repetitions it takes 100 seconds to measure each configuration. With the initially chosen functions there is a total of $2 \times 8^6 = 524\,288$ configurations of length 7 so it is unfeasible to measure all of them. Therefore, to see which configurations are more interesting to be studied, a preliminary experiment has been made using only 5-length configurations in 5 different FPGAs. Furthermore, of all possible 5-length configurations ($2 \times 8^4 = 8192$) we have only measured those ones that do not have a logical fixed point (2048 in total).

From this initial test, some preliminary results have been obtained:

First, by looking at the obtained RMSE values, we have noticed that none of the configurations behave as an ideal TRNG (Fig. 3a). Furthermore, we have noticed that the ring oscillators (all configurations that only have an odd number of inverters and an even number of delays) have lower RMSE values than the

filename	Inter-HD	Intra-HD	RMSE
INVb XNOR INVa XOR DELa	30.267	1.357	0.975
INVb XNOR XNOR DELa DELa	30.000	1.374	0.559
DELa XOR XNOR DELa DELa	29.467	17.593	3.455
DELa XNOR INVa XNOR DELa	29.333	1.208	0.623
DELa XNOR DELa XOR DELa	29.233	1.051	0.596
DELa DELa INVa INVa NOR	29.067	14.612	2.701
INVb INVa INVa XOR XOR	28.800	1.817	0.574
INVb XOR XOR XOR XOR	28.533	1.471	0.596
INVb AND DELa NAND NAND	28.583	10.761	3.217
INVb XNOR DELa XNOR DELa	28.333	0.603	0.597
DELa INVa XNOR XNOR DELa	28.367	0.693	0.566
DELa NOR DELa OR XOR	29.267	19.999	3.934
DELa XOR DELa XNOR DELa	28.000	3.342	0.988
INVb INVa INVa XNOR XNOR	27.867	1.265	0.579
DELa DELa DELa DELa NOR	27.667	0.746	0.583
INVb INVa DELa XNOR XOR	27.600	1.998	0.573
INVb INVa XOR XNOR DELa	27.467	0.604	0.597
DELa DELa XOR XNOR DELa	27.467	1.032	0.615
INVb INVa DELa XOR XNOR	27.200	0.643	0.579
INVb XOR NAND INVa NAND	27.200	5.871	3.484

Fig. 4. Top 5-length configurations with the highest average Inter-*HD*s.

rest of the configurations, indicating that their CDFs are closer to the ideal binomial CDFs expected in case that the sampled bits were perfectly random. This does not necessarily mean that ring oscillators are always a better choice as TRNGs since their sampled bits could present higher autocorrelations. However, for slow sampling frequencies where the autocorrelations tend to disappear, this result indicates that ring oscillators would be better TRNGs.

The second thing that we have noticed is that the measured average Inter-*HD*s are all lower than the ideal value of 50%. In Fig. 4, we can see the top configurations ordered by their average Inter-*HD*. From these values it can be seen that most of the top elements have in common that they do not have an AND, NAND, OR or NOR gate. Furthermore, the few of them that have any of those functions and high average Inter-*HD*, also present a quite high average Intra-*HD*.

B. Final full experiment

Based on these preliminary results, we have carried out the full experiment with n-length configurations ($n \leq 7$) in 20 FPGAs but using only the XOR, XNOR, DEL and INV operations. Of all possible configurations, we have only measured those that do not present a fixed point (a total of 2730). From this experiment, several conclusions have been made.

First, by analyzing the RMSE values (Fig. 3b), the results are consistent with the results in the preliminary 5-length test (i.e., no configurations have a bias that follows a binomial distribution with $p = 0.5$ but the ring oscillators are the closest ones to this ideal binomial distribution). Therefore, none of these systems could work as an ideal TRNG.

Second, to study the reproducibility of possible PUFs, we have plotted the histograms of the average Intra-*HD*s of all configurations of each length in Fig. 5. From this figure, we can see that most of these oscillators tend to have a high reproducibility. Furthermore, we can see that there is a big influence of the length of the configuration in the measured Intra-*HD*s. This can be seen more clearly in Fig. 6a where we have plotted the mean value of the average Intra-*HD*s of all configurations of each length.

In a similar manner, by analyzing the obtained average Inter-*HD*s we have not found any configuration that achieves the ideal value of 50%. The highest obtained value has been 41% for the

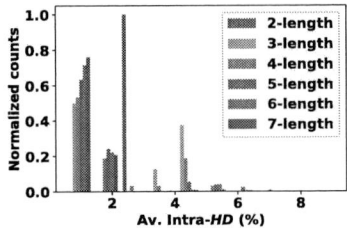

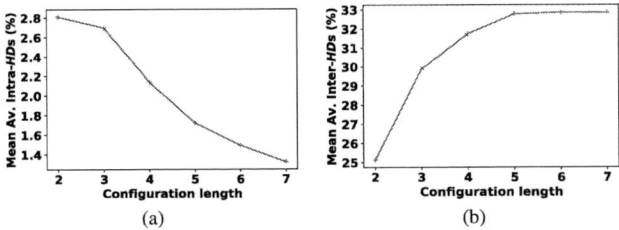

Fig. 5. Histograms of the average Intra-*HD*s for configurations of different lengths.

Fig. 6. Mean of the average (a) Intra-*HD*s and (b) Inter-*HD*s for different lengths.

configuration "DEL-DEL-DEL-DEL-XNOR-DEL-XOR". In this case, we have also noticed that bigger-length configurations tend to have higher average Inter-*HD*s. This can be seen in Fig. 6b where the mean value of the average Inter-*HD*s of the configurations of each length has been plotted. This tendency, however, seems to slow down for high lengths and there is not a big difference between the 6-length and 7-length configurations. However, even if the mean value did not change for further bigger lengths, since the number of possible configurations increases exponentially with their length, there could be bigger-length configurations with higher Inter-*HD*s (close to 50%).

It must be noticed that the oscillator with the highest average Inter-*HD* (41.2 %) also presents a very low average Intra-*HD*, 1.38 %. For comparison, by implementing a regular 7-LUT RO-PUF in the same FPGAs, we have obtained a better uniqueness (an average Inter-*HD* of 47.1 %) but a worse reproducibility (an average Intra-*HD* of 1.99 %).

V. Conclusions

In this work, we have proposed a generic structure that includes previously proposed oscillators (such as ROs and GAROs) as well as a new set of oscillators. Furthermore, we have proposed a way to implement this structure in a configurable manner so that, with the same implementation (i.e., the same bitstream file), it is possible to make the system work as any of the possible oscillators. Thanks to this configurable implementation, we have analyzed all configurations of length 7 or less to check their suitability as TRNGs or PUFs. From this analysis, several important conclusions have been extracted:

The first conclusion is that it is impossible to create an ideal TRNG based on sampling an oscillator of this kind (with 7 or less LUTs). Therefore, to generate perfect random sequences some kind of post-processing will always be needed. We believe that this result is very important since many previous works have proposed using ROs, GAROs or other similar oscillators as TRNGs. While we cannot rule out the fact that it might be

possible to build an ideal TRNG using one of these configurations in a particular FPGA or chip in a specific location with a certain routing, this could not be easily replicated in other implementations (such as ours). Secondly, in order to look for an oscillator to construct a good PUF, it seems advisable to try only XOR, XNOR, DEL and INV functions. Thirdly, with some 7-length configurations, it is possible to construct some PUFs with a quite high uniqueness (>40%) and very high reproducibility (some of them are better than a standard RO-PUF).

Finally, the reproducibility and uniqueness of these oscillators tend to increase when increasing the configuration length. Combining this fact with the fact that there is a huge number of oscillators with bigger lengths, it is likely that there are some configurations of bigger lengths that are suitable to construct much better PUFs.

REFERENCES

[1] B. Sunar, W. J. Martin and D. R. Stinson, "A Probably Secure True Random Number Generator with Built-In Tolerance to Active Attacks," *IEEE Transactions on Computers,* vol. 56, no. 1, pp. 109-119, 2007.

[2] C. Martínez-Gómez and I. Baturone, "Calibration of Ring Oscillator PUF and TRNG," in *2020 European Conference on Circuit Theory and Design (ECCTD),* Sofia, Bulgaria, 2020.

[3] M. Garcia-Bosque, G. Díez-Señorans, C. Sánchez-Azqueta and S. Celma, "Introduction to Physically Unclonable Fuctions: Properties and Applications," in *2020 European Conference on Circuit Theory and Design (ECCTD),* Sofia, Bulgaria, 2020.

[4] A. Maiti and P. Schaumont, "Improved Ring Oscillator PUF: An FPGA-friendly Secure Primitive," *Journal of Cryptology,* vol. 24, pp. 375-397, 2011.

[5] J. D. J. Golić, "New Methods for Digital Generation and Postprocessing of Random Data," *IEEE Transactions on Computers,* vol. 55, no. 10, pp. 1217-1229, 2006.

[6] M. Dichtl and J. D. J. Golić, "High-Speed True Random Number Generation with Logic Gates Only," in *2007 Cryptographic Hardware and Embedded Systems (CHES 2007),* Vienna, Austria, 2007.

[7] L. Matuszewski and M. Jessa, "An auxiliary source of randomness for combined TRNG based on ring oscillators," in *Proc. Poznaskie Warsztaty Telekomunikacyjne,* Poznanń, Poland, 2011.

[8] M. Dichtl, "Fibonacci Ring Oscillators as True Random Number Generators - A security Risk," *IACR report,* pp. 1-13, 2015.

[9] T. Addabbo, A. Fort, M. Mugnaini, V. Vignoli and M. Garcia-Bosque, "Digital Nonlinear Oscillators in PLDs: Pitfalls and Open Perspectives for a Novel Class of True Random Number Generators," in *2018 IEEE International Symposium on Ciruits and Systems (ISCAS),* 2018.

[10] T. Addabbo, A. Fort, R. Moretti, M. Mugnaini, V. Vignoli and M. Garcia-Bosque, "Lightweight True Random Bit Generators in PLDs: Figures of Merit and Performance Comparison," in *2019 IEEE International Symposium on Circuits and Systems (ISCAS),* Sapporo, Japan, 2019.

[11] M. Garcia-Bosque, G. Díez-Señorans, C. Sánchez-Azqueta and S. Celma, "Proposal and Analysis of a Novel Class of PUFs Based on Galois Ring Oscillators," *IEEE Access,* pp. 157830-157839, 2020.

[12] M. Garcia-Bosque, G. Díez-Señorans, C. Sánchez-Azqueta and S. Celma, "FPGA Implementation of a New PUF Based on Galois Ring Oscillators," in *12TH IEEE Latin American Symposium on Circuits and Systems (LASCAS),* Lima, Peru, 2021.

[13] A. Maiti, J. Casarona, L. Mchale and P. Schaumont, "A large scale characterization of RO-PUF," in *2010 IEEE International Symposium on Hardware Oriented Security and Trust (HOST),* Anaheim, USA, 2010.

A 55 MHz Integrated Crystal Oscillator with Chirp Injection Using a 28-nm Technology

Lantao Wang, Adrian Arnold, Jonas Meier, Markus Scholl, Ralf Wunderlich, and Stefan Heinen

Chair of Integrated Analog Circuits and RF Systems

RWTH Aachen University

Kopernikusstrasse 16, D-52074 Aachen, Germany

Email: mailbox@ias.rwth-aachen.de

Abstract—**This paper presents an integrated crystal oscillator to provide a 55 MHz reference frequency for applications such as phase-locked loops. The crystal oscillator uses a Pierce oscillator architecture and achieves a phase noise of -152 dBc/Hz at 1 kHz offset. A chirp frequency injection technique is also applied, achieving a start-up time of 300 us with 335 nJ energy consumption. The circuit is designed in a 28-nm technology, supplied by a 0.9 V voltage.**

Index Terms—**crystal oscillator, low-noise, 28-nm, chirp-injection**

I. INTRODUCTION

Crystal oscillators are widely used in communication circuits due to its capability to provide a stable, low-noise reference frequency. In particular, for the application such as a phase-locked loop (PLL), having a crystal oscillator as the reference clock is especially advantageous because it has the ability to fulfill the strict in-band phase noise requirement in a modern PLL system. With the rapid development of Internet of Things (IoT) and industry 4.0, new challenges and stricter design specifications are introduced to modern wireless communication systems. On the one hand, the start-up time of the crystal oscillator is a dominant contributor to the overall response time of a communication system. On the other hand, it is increasingly difficult to optimize the crystal oscillators with regard to the phase noise performance as the flicker noise induced by active devices increases with the technology node shrinking.

To meet the increasingly strict phase noise and short start-up time requirements in such applications in an advanced technology node, several efforts have been put into the optimization of crystal oscillator systems in recent years. [1] and [2] have proven that it is very effective to accelerate the oscillator start-up by injecting a signal with a frequency identical to the oscillation frequency, into the crystal. However, it is required to have a very precise match between the crystal's resonant frequency and the injected signal due to the crystal's high quality factor. In [2], the authors proposed to generate a chirp signal whose frequency continuously increases with time and in [1], a dithered signal between two frequencies are utilized, to address this challenge. [3] and [4], on the other hand, proposed methods with more sophisticated schemes by

frequency injection with multiple steps and Gaussian shaping the injected signal, respectively.

In this paper, an integrated crystal oscillator as the reference clock source for a phase locked loop (PLL) system in a 28-nm technology is proposed. The crystal oscillator is designed to provide a stable 55 MHz frequency with a phase noise of $-152\,\mathrm{dBc/Hz}$ at $1\,\mathrm{kHz}$ offset from the oscillation frequency. The chirp injection technique is selected for the purpose of the start-up acceleration due to its design simplicity, achieving a start-up time of less than $300\,\mu\mathrm{s}$.

II. FUNDAMENTALS OF CRYSTAL OSCILLATORS

In order to simulate the crystal oscillator, the crystal is modeled by a circuit consisting of a resistor R_s, a inductor L_m, two capacitors C_m and C_0, as shown in Fig. 1. The value of R_s is calculated with the equivalent series resistance ESR stated in the data sheet of the crystal by:

$$R_s = \frac{ESR}{(1 + \frac{C_0}{C_m})^2} \tag{1}$$

The motional capacitance C_m and the motional inductance L_m can be calculated based on the series resonant frequency

$$f_s = \frac{1}{2\pi\sqrt{L_m C_m}} \tag{2}$$

and the parallel resonant frequency

$$f_p = \frac{1}{2\pi\sqrt{L_m \frac{C_m(C_0+C_L)}{C_m+C_0+C_L}}} \tag{3}$$

where C_0 and C_L are the shunt capacitance and the load capacitance, respectively. The quality factor Q of the modeled crystal can hence be derived by:

$$Q = \frac{\sqrt{\frac{L_m}{C_m}}}{R_s} \tag{4}$$

To compensate the energy loss caused by the resistance in the crystal, an active circuit is necessary in the oscillator. In classic crystal oscillator design, the single-transistor based oscillator core, such as Pierce oscillator is preferred as it is able to provide a good phase noise performance. However, the lack of symmetry may degrade the performance in a system in

© VDE VERLAG GMBH · Berlin · Offenbach

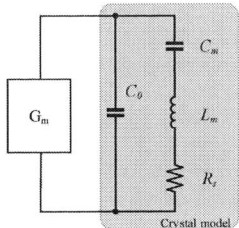

Fig. 1. General schematic of a crystal oscillator.

terms of spur suppression. In order to eliminate the asymmetry of this topology, a differential g_m stage is proposed in previous publications. But this architecture has the disadvantage that the oscillator could latch as no DC path is provided between the differential outputs. [6] introduced a structure with an active inductor to alleviate this problem at the cost of the design complexity and a tougher start-up condition.

III. PROPOSED CRYSTAL OSCILLATOR CIRCUIT

In Fig. 2, the schematic overview of the proposed crystal oscillator is shown. The crystal oscillator has a Pierce-oscillator architecture, using a single transistor M0 to provide g_m and compensate the losses in the crystal. M1 and M2 form a current mirror to bias the Pierce oscillator. The PMOS transistor M2 is designed to have a relatively large channel length to minimize the short-channel effect and also aid in the enhancement of power supply ripple rejection. Moreover, the width of M2 is also designed to be adequately large to ensure that the voltage drop across this transistor does not degrade the phase noise performance of the oscillator. The selection of PMOS type transistor, large width and length can also minimize the flick noise from the current source. In addition, a low-pass filter is inserted to mitigate the influence of the reference current noise. The capacitor in this filter is chosen to be a fringe capacitor of $12\,\mathrm{pF}$ in order to prevent any leakage current.

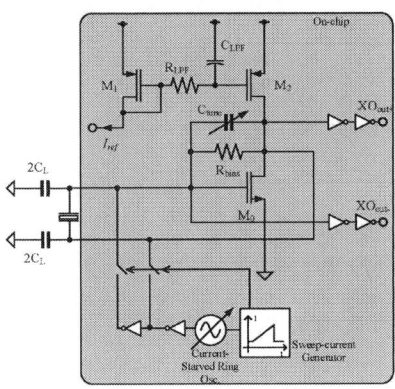

Fig. 2. Structure of the proposed crystal oscillator.

As the quality factor of the crystal is very high, it takes quite long time before the oscillation starts spontaneously. In order to accelerate the start-up, a chirp injection circuit consisting of a sweep current generator and a current-starved ring oscillator is developed in the crystal oscillator circuit.

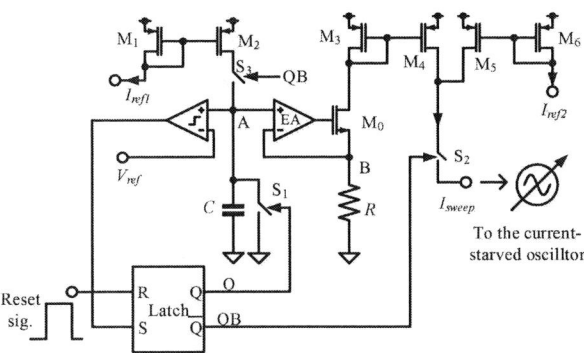

Fig. 3. Schematic of the sweep current generator.

The simplified schematic of the sweep current generator is depicted in Fig. 3. The circuits are biased by two reference currents I_{ref1} and I_{ref2} to M1 and M6, respectively. At the initial time t_0, the capacitor is discharged by switch S_1. The start-up sequence is controlled by a RS latch triggered by a $3\,\mu\mathrm{s}$ pulse fed into the reset terminal. When the RS latch detects the reset signal, it will set the output Q to the low level and QB to high level, thus, open the bypass switch S_1 and close S_2 and S_3. M2 mirrors the reference current I_{ref1} and charges the capacitor C, raising the voltage at node A. The voltage at node A, V_A can be hence expressed as:

$$V_A \approx \frac{I_{ref1}}{C}(t - t_0) \tag{5}$$

As the error amplifier, EA, and M0 form a negative feedback loop, the voltage at node B is equal to V_A. As a result, the current through the resistor R can be written as:

$$I_R = \frac{V_A}{R} = \frac{I_{ref1}}{RC}(t - t_0) \tag{6}$$

This current will then be duplicated by the current mirror constructed by M3 and M4 and added with I_{ref2}, yielding the current I_{sweep} that will be fed into the current-starved ring oscillator, whose schematic is illustrated in Fig. 4. I_{sweep} can hence be expressed as:

$$I_{sweep} = I_{ref2} + \frac{I_{ref1}}{RC}(t - t_0) \tag{7}$$

When the voltage at node A exceeds V_{ref}, the comparator generates a high voltage level, setting Q to high and QB to low. This will shut down the chirp injection circuit to save energy. Fig. 5 depicts the schematic of the comparator in Fig. 3. PMOS differential input stage is selected as the V_{ref} is designed as $200\,\mathrm{mV}$.

In addition, the simulation result shows that the oscillation frequency will vary ± 3 ppm due to process, voltage and

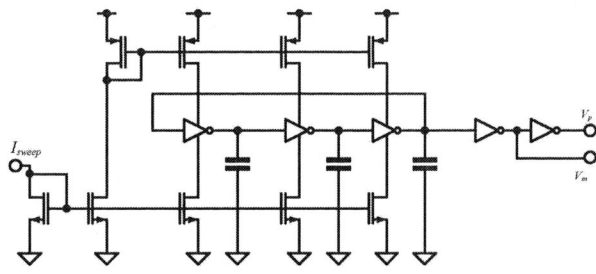

Fig. 4. Schematic of the current-starved ring oscillator.

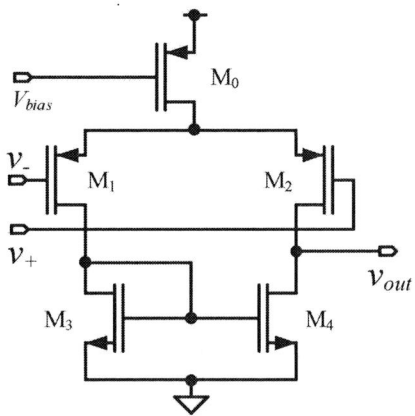

Fig. 5. Schematic of the comparator.

temperature (PVT) variation. Furthermore, the variation of the used crystal and the off-chip load capacitors C_L as well as the parasitics caused by soldering and wires will lead to a larger frequency variation. In order to tackle this problem, a switchable capacitor bank, C_{tune}, consisting of 128 unit cells, as shown in Fig. 6, is attached between the positive and negative terminals of the crystal.

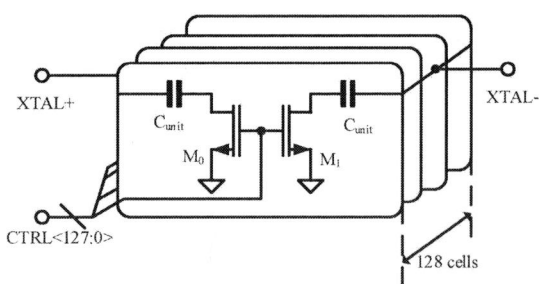

Fig. 6. Schematic of the capacitor array.

IV. SIMULATION RESULT

The proposed crystal oscillator has been developed in a TSMC 28 nm CMOS technology. The functionality of the

designed circuits are verified via simulations, with both global and local variations considered.

Fig. 7 shows the simulation result of the oscillation frequency of the chirp injection circuits. The frequency of the current-starved oscillation sweeps from 30 MHz to 115 MHz in less than 2 µs.

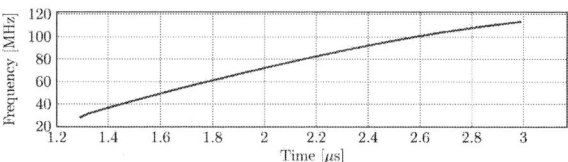

Fig. 7. Simulated sweep signal frequency.

Fig. 8 illustrates the tuning range of the crystal oscillator and its tuning linearity. In this design, 128 tuning steps are implemented using a thermometer code with the unitary switchable capacitor array, resulting in a linear tuning with ± 20 ppm tuning range to compensate the frequency variation.

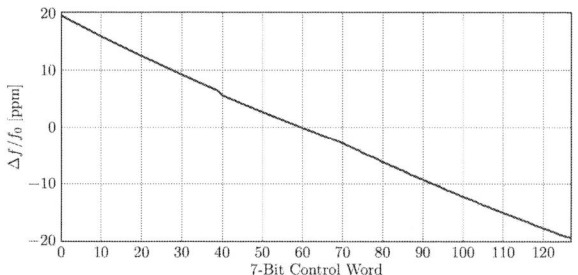

Fig. 8. Simulated tuning range of the crystal oscillator.

The simulated phase noise of the crystal oscillator is described in Fig. 9. The proposed crystal oscillator achieves a phase noise of -152 dBc/Hz at the 1 kHz offset from the 55 MHz oscillation frequency. The Monte-Carlo simulation result with 400 samples of the this performance parameter is depicted in Fig. 10 and the result shows the crystal oscillator has a good Monte-Carlo stability with a standard deviation of 2.49 dBc/Hz.

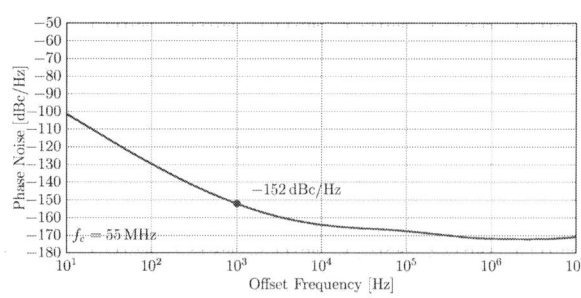

Fig. 9. Simulated phase noise of the crystal oscillator at TT corner.

TABLE I
PERFORMANCE COMPARISON WITH PREVIOUS PUBLICATIONS.

Parameter	[5]	[3]	[6]	[2]	[4]	This work
Technology	65 nm	65 nm	28 nm	180 nm	130 nm	28 nm
Oscillation frequency	39.25 MHz	54 MHz	48 MHz	39.25 MHz	32 MHz	55 MHz
Supply Voltage	3.3 V	1.0 V	1.0 V	1.5 V	1.2 V	0.9 V
PN@1 kHz offset	−139 dBc/Hz	−139.5 dBc/Hz	−114.3 dBc/Hz	−147 dBc/Hz	N.A.	−152 dBc/Hz
Start-up Time	3.9 ms	0.019 ms	N.A.	0.158 ms	0.037 ms	0.3 ms
Start-up Energy	N.A.	34.9 nJ	N.A.	349 nJ	31.7 nJ	335 nJ
Static Power	19 µW	198 µW	1.5 mW	181 µW	181.6 µW	449.1 µW

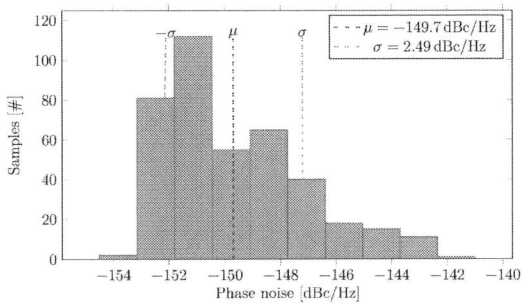

Fig. 10. Global and local mismatch of the oscillator phase noise at 1 kHz offset from 55 MHz nominal frequency.

Fig. 11 illustrates the start-up sequence of the proposed crystal oscillator. The start-up is triggered by a reset signal with width of 3 µs. The chirp signal is then generated and applied to the oscillator outputs to accelerate the oscillation start-up. After that, the reset signal will be pulled to low level and disable the chirp injection. The proposed circuit achieves a start-up time of 300 µs.

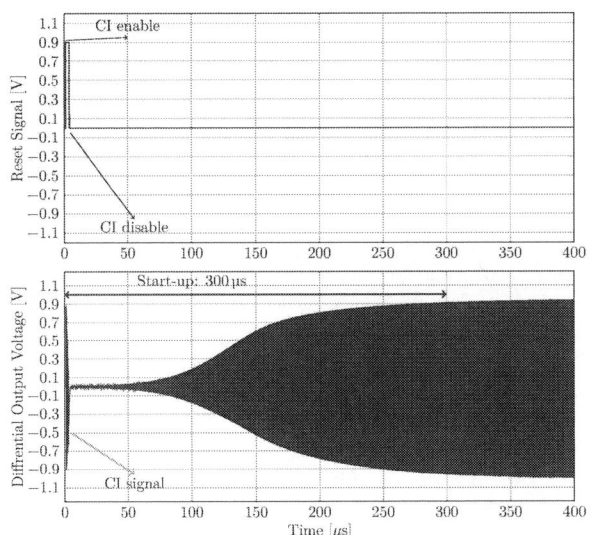

Fig. 11. Start-up behavior of the crystal oscillator.

Table I compares the prior works on crystal oscillators with similar specifications. The designs of the selected publications can provide a shorter start-up time with much more sophisticated start-up schemes and larger supply voltages. Nevertheless, the crystal oscillator proposed in this paper has a better performance in terms of phase noise.

V. CONCLUSION

In this paper, a 55 MHz crystal oscillator in a TSMC 28 nm CMOS technology is implemented. The crystal oscillator utilizes a Pierce oscillator core to ensure simple and robust start-up, achieving a phase noise of −152 dBc/Hz at 1 kHz offset from the nominal frequency. The proposed crystal oscillator shows good Monte-Carlo stability in terms of phase noise. In order to accelerate the start-up, a chirp injection technique is employed, resulting in 300 µs start-up time. In conclusion, the proposed crystal oscillator provides a simple and feasible solution to the wireless applications in an advanced technology node with the high demands on low phase-noise and short start-up time.

REFERENCES

[1] D. Griffith, J. Murdock and P. T. Rine, "5.9 A 24MHz crystal oscillator with robust fast start-up using dithered injection," 2016 IEEE International Solid-State Circuits Conference (ISSCC), San Francisco, CA, USA, 2016, pp. 104-105, doi: 10.1109/ISSCC.2016.7417928.

[2] S. Iguchi, H. Fuketa, T. Sakurai and M. Takamiya, "Variation-Tolerant Quick-Start-Up CMOS Crystal Oscillator With Chirp Injection and Negative Resistance Booster," in IEEE Journal of Solid-State Circuits, vol. 51, no. 2, pp. 496-508, Feb. 2016, doi: 10.1109/JSSC.2015.2499240.

[3] K. M. Megawer et al., "18.5 A 54MHz Crystal Oscillator With 30 Start-Up Time Reduction Using 2-Step Injection in 65nm CMOS," 2019 IEEE International Solid- State Circuits Conference - (ISSCC), San Francisco, CA, USA, 2019, pp. 302-304, doi: 10.1109/ISSCC.2019.8662403.

[4] M. Scholl et al., "A 32 MHz Crystal Oscillator with Fast Start-Up Using Dithered Injection and Negative Resistance Boost," ESSCIRC 2019 - IEEE 45th European Solid State Circuits Conference (ESSCIRC), Cracow, Poland, 2019, pp. 49-52, doi: 10.1109/ESSCIRC.2019.8902894.

[5] S. Iguchi, T. Sakurai and M. Takamiya, "A Low-Power CMOS Crystal Oscillator Using a Stacked-Amplifier Architecture," in IEEE Journal of Solid-State Circuits, vol. 52, no. 11, pp. 3006-3017, Nov. 2017, doi: 10.1109/JSSC.2017.2743174.

[6] Y. Rajavi, M. M. Ghahramani, A. Khalili, A. Kavousian, B. Kim and M. P. Flynn, "A 48-MHz Differential Crystal Oscillator With 168-fs Jitter in 28-nm CMOS," in IEEE Journal of Solid-State Circuits, vol. 52, no. 10, pp. 2735-2745, Oct. 2017, doi: 10.1109/JSSC.2017.2728781.

[7] E. A. Vittoz, M. G. R. Degrauwe and S. Bitz, "High-performance crystal oscillator circuits: theory and application," in IEEE Journal of Solid-State Circuits, vol. 23, no. 3, pp. 774-783, June 1988, doi: 10.1109/4.318.

© VDE VERLAG GMBH · Berlin · Offenbach

A low-power 26.56-GHz LC-based DCO for multi-gigabit communication systems

Pablo Jiménez-Fernández, Óscar Guerra, Rocío del Río
Instituto de Microelectrónica de Sevilla
(IMSE-CNM) CSIC, Universidad de Sevilla
{pabjimenez, guerra, rocio}@imse-cnm.csic.es

Alberto Rodríguez-Pérez, Enrique Prefasi
R&D department, KDPOF
{alberto.rodriguez, enrique.prefasi}@kdpof.com

Abstract—A voltage controlled oscillator (VCO) is one of the key building blocks in RF transceivers. By means of a Phase-Locked Loop (PLL) that controls the VCO, a clock signal at the desired frequency can be generated. Communications systems for multi-gigabit applications require high accuracy in the clock signal, so low phase noise of the VCO must be achieved. This paper presents the design of a 26.56-GHz digitally controlled VCO (DCO). The circuit is powered at 1.2 V and consumes 1.94 mW. Post-layout simulations based on a TSMC 65-nm CMOS RF process show a phase noise of -123.3 dBc/Hz at 10-MHz offset. By modelling the noise contributions of the digital PLL, a 242.84-fs rms jitter (integrated from 100 kHz to 100 MHz) has been estimated. The proposed DCO exhibits a FoM of -188.6 dBc/Hz at 1-MHz offset frequency.

Index Terms—CMOS integrated circuits, digital VCO, *LC* tank, phase noise, high-speed communications.

I. INTRODUCTION

The evolution of wired, wireless and cellular communications standards in recent years and the expected trend in terms of data rates is currently leading to an increased demand for data transmission systems with bandwidths in the tens of GHz range (in so-called multi-gigabit applications), such as serial data links for 1-G, 10-G and 100-G Ethernet, 60 GHz wireless receivers and 5G cellular receivers. Likewise, the progressive replacement of parallel buses between chips by very high-speed differential serial connections is becoming more and more imminent, as it allows simplifying the implementation of equipment by having a smaller number of tracks and, at the same time, obtaining good characteristics in terms of radiation, electromagnetic immunity or power consumption [1]. These serial interfaces require converter blocks between serial and parallel formats, i.e. serializers and deserializers (*SerDes*). Fig. 1 shows a representation of the typical blocks of a *SerDes* system.

Among these different elements, the clock signal control elements can be highlitghted. The PLL is located on the transmit side where the synthesizer generates the clock signal from a given reference signal. When it has a digital phase detector (based on a TDC, Time-to-Digital Converter) it is said to be a Digital Phase-Locked Loop (DPLL) [2], [3]. On the data reception side, the Clock Data Recovery (CDR) is responsible for determining the clock signal and synchronising the received data so that it can be correctly processed [1]. Both blocks share a common element: the oscillator. This element

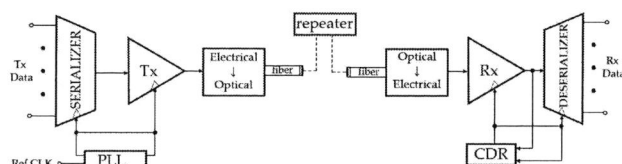

Fig. 1. Conceptual block diagram of a *SerDes* system.

provides at its output a periodic signal at a desired frequency ($v_{out}@f_0$ in Fig. 2), which depends on the setting value received at the input [4], [5]. Fig. 2 shows the block diagram of a Digital PLL. The TDC generates a digital representation of the error (e_d) caused by comparing the divided oscillator output (by the feedback divider block, FBD) and a reference frequency (F_{Ref}). This error is filtered by a digital loop filter (DLF) and is the setting value of the oscillator (V_{Ctrl}). When the TDC error is less than a fixed minimum value, the PLL is said to be phase-locked [2], [3].

This paper presents the design of an oscillator to be incorporated in a single-mode DPLL. The oscillator is designed to provide a periodic signal at a frequency f_o equal to 26.56 GHz. The tuning range is controllable by various fine/coarse adjustments [4], [5]. During the design, special attention has been paid to the phase noise of the oscillator, in order to minimize it to reduce its effect on the jitter noise of the DPLL [4]. Equation (1) shows the dependence of the phase noise on the quality factor (Q) of the oscillator [5],

$$\mathcal{L}(f_o, \Delta f) = 10 \log \left[\frac{kTFR_p}{V_{osc}^2 Q^2} \left(\frac{f_o}{\Delta f} \right)^2 \right] \quad (1)$$

where f_o stands for the carrier frequency, Δf for the offset frequency, k for the Boltzmann's constant, T for the temperature, F for the noise factor, R_p for a representation of *LC*

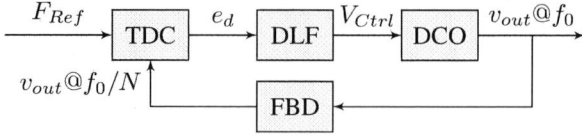

Fig. 2. Block diagram of a classical DPLL.

© VDE VERLAG GMBH · Berlin · Offenbach

tank losses, V_{osc} for the oscillation amplitude and Q for the quality factor of the oscillator. Thus, the larger the quality factor of the LC tank, the lower the phase noise of the DCO. Since the targeted application demands a considerably low phase noise, an LC tank is required. Although ring oscillators provide smaller area and higher programmability, they exhibit much lower quality factor than LC tank-based oscillators.

The proposed DCO has been implemented in TSMC 65-nm CMOS RF technology with 1.2 V of supply voltage. Post-layout simulations exhibit a phase noise of -103 dBc/Hz at 1-MHz offset frequency with a DC power consumption of 1.94mW, resulting in a FoM of -188.6 dBc/Hz at 1-MHz offset frequency.

This paper is organized as follows. The topological description of the oscillator and the implemented tuning methodology are described in section II. The circuit layout and the post-layout simulations are described in section III. Finally, section IV presents the conclusions of this work.

II. CIRCUIT DESIGN

The proposed LC tank-based DCO is shown in Fig. 3. The circuit can be divided into 3 functional blocks:

- The LC tank formed by the inductor and the equivalent capacitance implemented by the varactors.
- The NMOS cross-coupled pair formed by transistors M3 and M4 which implements the negative resistance, necessary to compensate the losses of the LC tank and sustain the oscillations [5].
- The current source consisting of transistors M1 and M2, which copies the reference current I_{Bias} of 50 μA and increases it by a factor 34 to bias the oscillator.

A symmetrical centered-tap inductor provided in the PDK library with a high quality factor ($Q = 35.76$) and a low self-inductance value ($L = 124.56$ pH) has been selected,

so that a wide capacitance range is available to compensate for the routing parasitic effects and provide a large tuning range of the DCO [4]. The design of varactors must be carefully considered, as the quantization noise caused by the discretisation of the steps in the frequency setting can significantly contribute to the total jitter noise of the DPLL [6]. Most of the DPLL noise will be come from TDC noise, oscillator phase noise and quantization noise. In this sense, a trade-off arises between the minimum frequency resolution that guarantees a minimum noise contribution and that, at the same time, allows the oscillator tuning to be performed in a given frequency range. Based on this, it is common practice in oscillator design to have a control element that provides a coarse frequency step that sets the frequency offset and a fine control element for more precise tuning around the former value. Using this approach, only the fine tuning control element will influence the quantization noise [6].

Since the oscillator is digitally controlled, the varactor control signal is a digital signal that can be either a logic zero (*AVSS*) or a logic one (*AVDD*). Fig. 4(a) shows an unitary varactor consisting of a pair of NMOS transistors with their bulk terminals grounded. Fig. 4(b) shows the capacitance curve between nodes X1 and X2 for a varactor with minimum dimensions (W/L = 120/60 nm) for a digital control voltage at high/low level. In practice, the NMOS transistors work in inversion region (highlighted in Fig. 4(b)), around 1.2 V, as this is the DC voltage at which the oscillator output nodes are set. In this region, a capacitance difference ΔC is achieved that defines the ultimate frequency resolution of the DCO.

Equation (2) shows the effect of ΔC on the frequency resolution Δf [6],

$$\Delta C = -\frac{1}{2} \frac{\Delta f}{L\pi^2 f_o^3} \tag{2}$$

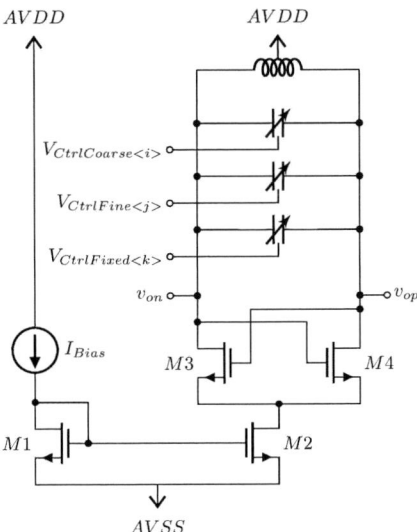

Fig. 3. Schematic of the proposed LC-based DCO.

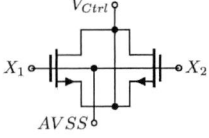

(a) Topology of n-MOSCap.

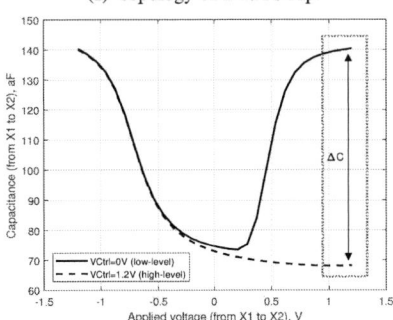

(b) Capacitance characteristic of an n-MOSCap with minimum size (W/L = 120/60 nm).

Fig. 4. Unitary varactor using a n-MOSCap and capacitance curve.

where f_o stands for the carrier frequency and L for the self-inductance value of the inductor. From this relationship, it has been calculated that to obtain a frequency resolution of the DCO of 11 MHz, it is necessary to implement a capacitance step around 240 aF with the fine-tuning varactor.

The frequency control element of the oscillator is obtained by constructing an array of unitary varactors that are selected following a thermometer code, obtaining the required capacitance value. The coarse-tuning varactor is designed with 4 control bits and the fine-tuning varactor with 6 control bits. In addition, a 5-bit varactor with larger capacitance resolution than the coarse varactor has been included with the intention of having a fixed control value that is not controlled by the DPLL and compensating for the loading capacitance of the subsequent buffer stage connected to the DCO outputs. The purpose of this buffer stage would be to increase the amplitude of the DCO output signal up to power rails, thus generating the clock signal.

Fig. 3 illustrates the 3 control elements. Each element is a representation of the array of unitary varactors shown in Fig. 4(a). The fine tuning is composed of 64 varactors, the coarse tuning by 16 varactors and the fixed capacitance is implemented with 32 varactors. The control signals in Fig. 3 indicate the number of varactors that are connected to *AVDD*, with the remainder fixed to *AVSS*. The control range will be from 0 (all unitary varactors to *AVSS*) to 2^{NBits} (all unitary varactors to *AVDD*), where $NBits$ is the number of bits of each control element.

III. Simulation Results

Fig. 5 shows the DCO layout, with the main parts highlighted. The inductor layout view is available in Cadence's PDK TSMC 65-nm CMOS RF with $6X1Z1U$ metallization (an ultra-thick, low-resistivity Mu-type metal to build the inductor).

The choice of the oscillator floorplan has been based on the inductor layout, since this element is the one that occupies most of the circuit's area (53.3% of a total equal to $230\mu\text{m} \times 174\mu\text{m}$). In this sense, and in order to avoid changes of direction of the inductor's paths, the metal 9 of the inductor's outputs has been used to limit the area where the rest of the components have been placed, as shown in Fig. 5. This is intended to eliminate corners in the signal path since these could cause undesired effects at very high frequencies. In addition, special attention has been paid to keep the symmetry and minimum distance of the components to reduce the parasitic effects of routing and improve the matching.

As design guidelines, the signal path of each component will be raised to higher metals whenever possible. This will decrease the effect of capacitive coupling to substrate thanks to a greater height than using lower metals [7]. The objective is to reduce the relative weight of parasitics on the capacitances implemented with varactors. With respect to the power supply or varactor control signals, this effect can be neglected as they are static signals, so lower metals have been used for them.

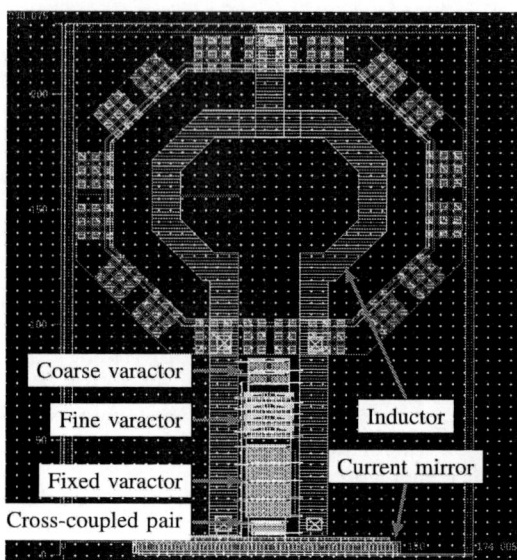

Fig. 5. Layout of the proposed *LC*-based DCO in TSMC 65-nm CMOS RF.

Fig. 6 shows the full tuning range of the oscillator for the nominal case for the different fine and coarse control values. The available tuning range for this configuration is 4.53 GHz (17.1% of f_o) and the obtained frequency resolution Δf in the coarse tuning varactor curve equal to 8 is 11 MHz. The tuning range has been designed with considerable overlap between curves to ensure that the whole frequency range is covered. Thus, in Fig. 6 the 3 configurations that allow tuning the DCO frequency to f_o (represented as the red dotted line in Fig. 6) can be seen.

Results of post-layout simulations in PVT corners are very promising (see bottom-left legend of Fig. 7 for PVT corners definition). Oscillations are guaranteed in all corners and in a range of frequencies that can be perfectly compensated with the designed varactors. These results are shown in Table I for a varactor configuration (*coarse–fine–fixed*) set to (8–49–25).

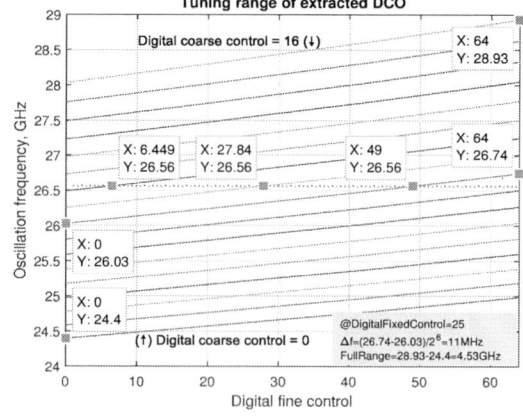

Fig. 6. Tuning range of extracted *LC*-based DCO in nominal case.

TABLE I
RESULTS OF DCO POST-LAYOUT SIMULATION

	Nominal	PVT Corners		
		min	max	std_dev
DC Power, mW	1.94	1.69	2.18	0.21
Frequency, GHz	26.56	26.44	26.68	0.07
Amplitude, mV	406.5	213.8	486.1	104.7
Phase noise @1MHz, dBc/Hz	-103.0	-105.1	-96.2	3.6
Phase noise @10MHz, dBc/Hz	-123.3	-125.8	-116.1	3.9
FoM* @1MHz, dBc/Hz	-188.6	-191.3	-181.3	3.8

$^{*}FOM = PhaseNoise - 10\log_{10}((f_o/\Delta f)^2 * 1mW/P_{diss})$

TABLE II
PERFORMANCE COMPARISON WITH OTHER OSCILLATORS

	This work	[8]	[9]	[10]
Technology	65nm	28nm	0.13μm	28nm
Implementation	DCO	DCO	VCO	VCO
Results	Sim	Sim	Lab	Lab
Frequency, GHz	24.4–28.8	23–30.8	26.5–29.7	24–30
Tuning Range	17%	29%	11%	22%
Phase noise*, dBc/Hz	-103	-102.5	-106.8	-104
Power Supply, V	1.2	0.9	1.3	1.2
Power, mW	1.94	7.1	14.4	6.6

*Phase noise measured at 1-MHz offset frequency.

Regarding power consumption, the nominal simulation shows a DC power dissipation of 1.94mW, with a standard deviation in PVT corners of 10.8%, resulting in a reasonable variability. Results also indicate a minimal variation of the oscillation frequency in PVT corners and, although the oscillation amplitude has a deviation of 25.8% with respect to the nominal value, this will be compensated by the buffer stage at the output of the DCO.

Fig. 7 shows the characteristic of phase noise versus frequency in PVT obtained by post-layout simulation. A two different concentrations in the -20 dB/dec zone can be seen. The upper group is for a supply voltage equal to 1.32 V and the lower group is for 1.08 V. This indicates that the phase noise of the oscillator will deteriorate in excess of the supply voltage. The phase noise in the nominal case is -103.0 dBc/Hz measured at an offset frequency of 1 MHz and -123.3 dBc/Hz for an offset frequency of 10 MHz, with a deviation due to PVT corners around 4%. Considering the phase noise measured at an offset frequency of 1 MHz and according to the DC power consumption, the proposed DCO gives a FoM of -188.6 dBc/Hz that is competitive with the state of the art.

Using the obtained phase noise and assuming a noise model of the TDC and the quantizer with the obtained frequency resolution of the DCO (11 MHz), the jitter noise of the DPLL has been estimated, resulting in a value of 242.84-fs rms (integrated from 100 kHz to 100 MHz), thus complying with the low-jitter requirements of the application.

Table II summarizes the performances of the proposed LC-based DCO compared with other reported oscillators.

IV. CONCLUSIONS

This paper presents the design of a low-power 26.56-GHz digitally controlled LC-based oscillator in TSMC 65-nm

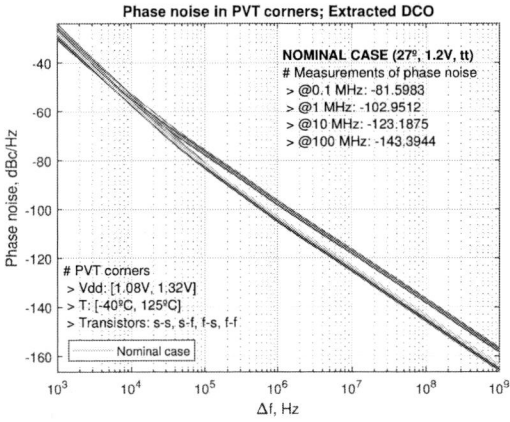

Fig. 7. Phase noise of extracted LC-based DCO in PVT corners.

CMOS RF technology. The frequency adjustment has been implemented using n-MOSCap arrays that allow different precision controls. Post-layout simulation shows that the tuning ranges from 24.4 to 28.8 GHz (17% of the carrier frequency) and a phase noise of -103.0 dBc/Hz at 1-MHz offset frequency. The low DC power consumption of 1.94 mW results in a figure of merit of -188.6 dBc/Hz at 1-MHz offset frequency. The area occupied by the proposed LC-based DCO is 230μm$\times$174μm.

ACKNOWLEDGMENT

This work was supported by the Office of Naval Research (USA) under grant N00014-19-1-2156.

REFERENCES

[1] A. Amirkhany, "Basics of clock and data recovery circuits - Exploring high-speed serial links," in IEEE Solid-State Circuits Magazine, pp. 25–38, Winter 2020.

[2] A. M. Fahim, "A compact, low-power low-jitter digital PLL," The 29th European Solid-State Circuits Conference, pp. 101–104, 2003.

[3] Y. Wu, M. Shahmohammadi, Y. Chen, P. Lu and R. B. Staszewski, "A 3.5–6.8-GHz wide-bandwidth DTC-assisted fractional-N all-digital PLL with a MASH $\Delta\Sigma$ -TDC for low in-band phase noise," in IEEE Journal of Solid-State Circuits, vol. 52, no. 7, pp. 1885–1903, July 2017.

[4] D. Leenaerts, J. van der Tang, and C. Vaucher, Circuit Design for RF Transceivers, 2nd ed., Springer, 2002.

[5] H. Darabi, Radio Frequency Integrated Circuits and Systems, 2nd ed., Cambridge University Press, 2002.

[6] D. Pfaff, R. Abbott, X. Wang, S. Moazzeni, R. Mason, and R. R. Smith, "A 14-GHz Bang-Bang digital PLL with sub-150-fs integrated jitter for wireline applications in 7-nm FinFET CMOS," in IEEE Journal of Solid-State Circuits, vol. 55, no. 3, pp. 580–591, March 2020.

[7] A. Goñi, J. del Pino, B. Gonzalez, and A. Hernandez, "An analytical model of electric substrate losses for planar spiral inductors on silicon," in IEEE Transactions on Electron Devices, vol. 54, no. 3, pp. 546–553, March 2007.

[8] Y. Li, J. Li, and Z. Hong, "A 23-30.8 GHz digital-controlled-oscillator in 28nm CMOS," 2018 14th IEEE International Conference on Solid-State and Integrated Circuit Technology (ICSICT), Qingdao, China, pp. 1–3, 2018.

[9] D. Shin, S. Raman, and K. Koh, "2.8 A mixed-mode injection frequency-locked loop for self-calibration of injection locking range and phase noise in 0.13μm CMOS," 2016 IEEE International Solid-State Circuits Conference (ISSCC), San Francisco, CA, USA, pp. 50–51, 2016.

[10] S. Ek et al., "A 28-nm FD-SOI 115-fs jitter PLL-based LO system for 24–30-GHz Sliding-IF 5G transceivers," in IEEE Journal of Solid-State Circuits, vol. 53, no. 7, pp. 1988–2000, July 2018.

© VDE VERLAG GMBH · Berlin · Offenbach

SMACD / PRIME 2021 | 19 – 22 July 2021, Online Event

A Wide-Tuning-Range 55 GHz CMOS VCO on 22 nm FD-SOI Technology

Zoltán Tibenszky, Corrado Carta, Frank Ellinger
Chair of Circuit Design and Network Theory
Technische Universität Dresden, 01069 Dresden, Germany
Email: zoltan.tibenszky@tu-dresden.de

Abstract—**This paper presents the design and characterization of a low-power 55 GHz oscillator using complementary transistors. It has the highest continuous tuning range in its frequency band reported to date. The tuning range is 27.4 % and 30.8 % for supply voltages 0.8 V and 1.4 V, respectively. Its core and buffers consume in average 3 mW and 8 mW power, respectively, from a supply voltage of 1.2 V. The peak DC-to-RF efficiency is about 6 % for supply voltages above 1 V. The circuit was manufactured on a 22 nm FD-SOI CMOS technology, and requires a total silicon area of 0.012 mm².**

I. INTRODUCTION

The deep-seated need for creating faster, smaller and more accurate machines has always been in heart of the technical evolution throughout its history. In the realm of communication and radar systems, this is manifested in pursuing ever higher data rates, and better spacial resolution. Both of which require higher available spectral bandwidth. Voltage controlled oscillators (VCOs) are building blocks present in the majority of communication and radar systems. They generate the high-frequency carrier signal which is modulated to transmit information or to detect objects. Their noise and frequency stability can be improved by coupling its output to a low-frequency, high-precision crystal reference with a phase-locked loop (PLL), but its output frequency can only be controlled within its tuning range.

Higher continuous bandwidths are allocated at the higher end of the used frequency spectrum once the bandwidths around the available frequencies are getting crowded and the technical development forecasts the possibility of high scale deployment in the new frequency band in a foreseeable time. Therefore to realize higher data rates and better spacial resolution, higher frequencies need to be used for these systems. Additionally, antennas become more compact with increasing frequency, making the full integration of systems on chip possible, which opens up new application possibilities. Such system on chips (SoCs) in the V-band have already been reported [1], [2]. To fully benefit from the wider bandwidth, the tuning range (TR) of the frequency generation has to cover the available bandwidth even in the presence of performance variations due to process, supply and temperature changes. This increases the importance of the tuning range further.

Frequency generation has one of the highest power consumption (P_{DC}) among the receiver circuit blocks [3], and often additional buffers are needed to adequately drive the

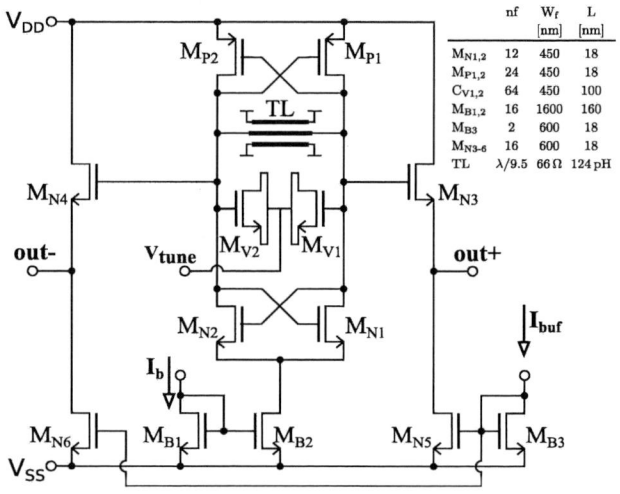

Fig. 1. Schematic of the presented circuit

mixers. For this reason both output power (P_{out}) and DC-to-RF efficiency (η) are important parameters of an oscillator beyond the phase noise ($\mathcal{L}$) and tuning range (TR). Our goal with the presented design was to investigate the achievable higher limit of a continuous tuning range at mm-wave frequencies, while aiming for a comparably good efficiency.

II. DESIGN

The transistor level schematic of the proposed design is shown in Fig. 1. The resonator core consists of a transmission line TL, providing the inductive component and varactors M_{V1-2} forming the capacitive part of the resonator. The losses of the resonator are compensated by the cross-coupled pairs M_{N1-2} and M_{P1-2}. The TL uses a minimal width thick copper trace instead of the top aluminium, to avoid any modeling inaccuracy due to the unplanarized chip top. A patterned ground shield [4] on the third metal layer to provide an underpass for the DC bias voltages and currents. The tail current source M_{B2} was added to reduce the sensitivity to the supply noise and to control the bias point. M_{B1} lowers the sensitivity of the tail current on the input bias. M_{B1} and M_{B2} are matched in size to avoid current noise multiplication from M_{B1} to the tail current. An increased voltage swing in the resonator reduces the phase noise, and improve the voltage efficiency. For this reason M_{B2}

© VDE VERLAG GMBH · Berlin · Offenbach

344

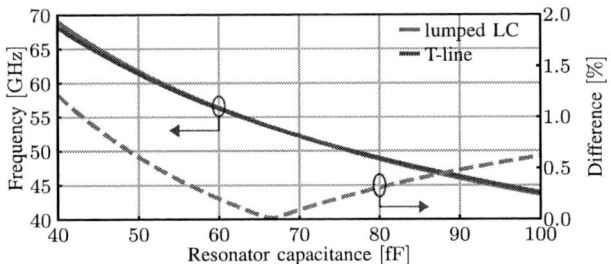

Fig. 2. Graphical representation of the oscillation frequency predictions according to (2) compared to the conventional LC formula as a function of resonator capacitance. The absolute value of the relative difference of the two curve is plotted with dashed line.

was sized for low voltage drop. Although the thermal noise from M_{B2} increases for a lower voltage drop, the benefits of the higher swing and higher voltage efficiency dominates and the phase noise was reduced [5]. The final size of M_{B2} was optimized through simulations. Though M_{B1-2} form a structure resembling a current mirror, no exact mirroring is required. The tail current will be smaller than I_b due to the smaller V_{DS} of M_{B2}. Therefore the core current are smaller than the I_b input values in the labels of the measurement figures. The bias inputs are connected to the core through $0\,\Omega$ lines [6].

Although the effective inductance of the TL is frequency dependent, the oscillation frequency (f_{osc}) can be calculated from the LC formula even for moderately wide frequency ranges. The equivalent inductance of a lossless transmission line is

$$L_{eff}(f) = \frac{Z_0}{2\pi f}\tan\left(\beta l\right) = \frac{Z_0}{2\pi f}\tan\left(2\pi f\frac{\sqrt{\epsilon_r}l}{c}\right), \quad (1)$$

where c stands for the speed of light in vacuum, Z_0, ϵ_r and l are the characteristic impedance, the effective permittivity and the physical length of the TL, respectively. The oscillation frequency (f_{osc}) can derived in implicit form as

$$f_{osc} = \frac{1}{2\pi\sqrt{LC}} = \frac{1}{2\pi(C_v+C_f)\,Z_0\tan\left(2\pi f_{osc}\frac{\sqrt{\epsilon_r}l}{c}\right)}, \quad (2)$$

where C_v and C_f denote the variable and fixed capacitances of the resonator. This formula has been evaluated numerically and is graphically compared in Fig. 2 with the conventional LC formula, where the effective inductance of the TL in the center frequency is used.

The varactors use the same RF-transistor types as the cross-coupled NMOS devices, and were optimized for a wide tuning range.

The output load is driven by the transistors M_{N3-4} in a common drain (CD) amplifier configuration. Their input are DC coupled to avoid losses related to the coupling capacitor and to bias the buffers in class-AB region. This increases power efficiency, while providing good linearity and isolation between the resonator and the output. The V_{GS} of M_{N3-4} and the V_{DS} of the biasing transistor need to be reduced for high output voltage swing and thus high output power, both of

which are related to their gate-drain overdrive voltage. M_{N3-6} were matched in size, because they share the same DC current as well. The same applies to the buffer biasing structure M_{N5-6} and M_{B3} as for the core tail current biasing. The buffers contribute with only 3 % to the overall capacitance of the resonator in the presented design. This reduces the oscillation frequency, and the tuning range by 1.5 %, and 0.75 %, respectively, while providing comparatively high output power.

A widely used figure of merit (FoM) for oscillators is defined as [7]–[12]:

$$FoM_T = \frac{\mathcal{L}(f_0, f_{\mathcal{L}})\cdot\frac{P_{DC}}{1\,\mathrm{mW}}}{\left(\frac{f_{osc}}{f_{\mathcal{L}}}\frac{TR}{10\,\%}\right)^2} \propto \frac{1}{(Q\cdot TR)^2}, \quad (3)$$

where $f_{\mathcal{L}}$ is the offset frequency. $\mathcal{L}$ is inversely proportional to the square of the quality factor (Q) [13], which is dominated by the varactor at the operating frequencies of the oscillator. The right hand side proportionality neglects only the effect of constants in the expression of $\mathcal{L}$. These constants are determined by the technology and circuit architecture, while the temperature, f_{osc} and $f_{\mathcal{L}}$ are assumed to be given as specification.

The operating frequency is approximately one fourth and one fifth of the transit frequency and the maximum oscillation frequency of the transistors, which have similar values. Due to the low gain of the active devices at these frequencies, the size of the active devices is increased. This in turn reduces the tuning range through the increased capacitance of the active devices. A high output power has similar effect on the tuning range. The input capacitance of a stronger output buffer would increase the fixed capacitances of the resonator, which will lead to a reduced tuning range.

A wide tuning range and a low phase noise are contradicting requirements at mm-wave frequencies due to the quality factor of the varactor, which dominates the quality factor of the resonator. A trade-off is present in the varactor geometry. Longer and wider channels offers higher tunable capacitance value for a similar parasitic capacitance, at the cost of increased channel and gate resistances, respectively. The increased series resistance reduces the quality factor of the resonator. This trade-off in the varactor geometry was analyzed in detail in [14]. Furthermore the weight of the parasitic capacitance of the transistors and the interconnections in the overall resonator capacitance increases with technology scaling. These make a higher tuning range without serious compromises more challenging to achieve.

III. Measurement Results

The proposed circuit has been fabricated in 22FDX®, a 22 nm fully depleted silicon on insulator (FD-SOI) CMOS technology from GLOBALFOUNDRIES [15]. The chip photograph is displayed in Fig. 3 (a) together with the core area, while Fig. 3 (b) shows the 3D view of the transmission line and the core. The proposed oscillator was characterized on a wafer prober for different supply voltages with a 67 GHz spectrum analyzer and a thermal power sensor. The measurement results

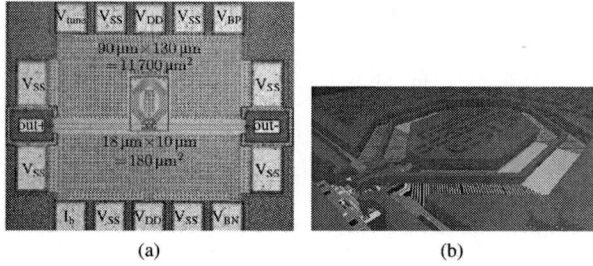

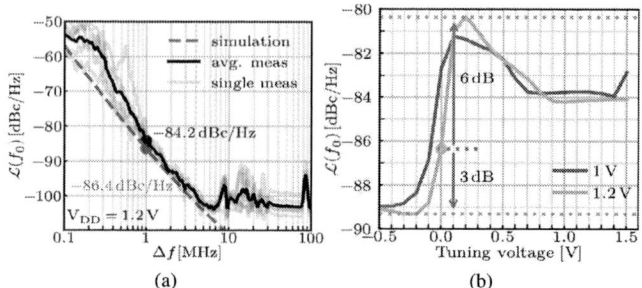

(a)   (b)

Fig. 3. (a) Micrograph of the circuit (b) 3D view of the circuit core

Fig. 5. (a) Measured $\mathcal{L}$ (b) simulated $\mathcal{L}$ at $\Delta f = 1\,\text{MHz}$ vs V_{tune}

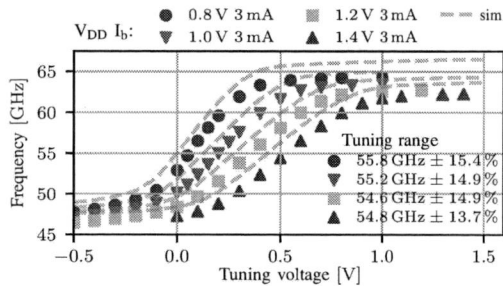

Fig. 4. Measured tuning characteristic

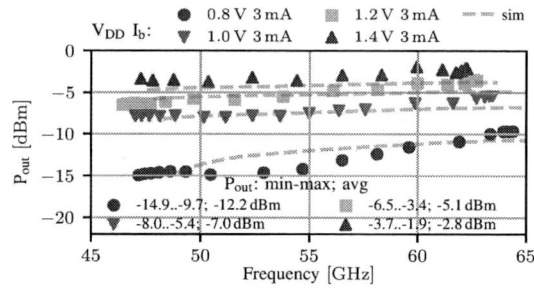

Fig. 6. Measured differential output power

were de-embedded to the pads with a two-tier calibration. The figures, which will be presented in this section, contain simulation results based on cross-capacitance (CC) post-layout extraction of the VCO core and an EM model of the resonator transmission line for comparison.

As shown in Fig. 4, the tuning range is relatively insensitive to changes in V_{DD}. A small reduction in the TR is present at higher supply voltages due to slight increase in the device capacitances at the increased DC voltages. The VCO pushing, which manifest itself in a horizontal shift in Fig. 4, is due to the supply dependent DC voltage at resonator terminals, where the varactor gates are connected to.

The measured phase noise curves shown in Fig. 5 (a) were obtained at $V_{tune} = 0\,\text{V}$, realized with a ground short at the probe. Measurements at different V_{tune} values were not reliable. Due to the long measurement time, the integrated excess noise from the DC source modulates the output frequency through the highly sensitive tuning terminal and corrupts the measurement results. An accurate phase noise measurements across the tuning range would require either an instrument with a different measurement principle or a PLL to stabilize the center frequency during the long measurement time. The simulated dependence of the phase noise on V_{tune} is shown in Fig. 5 (b). Offset frequencies above $5\,\text{MHz}$ were limited by the noise floor of the spectrum analyzer used for the phase noise measurement. Adding a preamplifier to the measurement setup to reduce the noise floor at higher offset frequencies has driven the IF chain of the spectrum analyzer into an overload condition. The resulting phase noise is sufficient for low power applications and it can be improved with higher signal power at the cost of proportionally higher P_{DC}.

The differential output power, presented in Fig. 6, is a strong function of the supply voltage. A higher supply voltage allows higher output swing and higher gate-source overdrive voltages and thus current to the load. Peak output powers of $-5.4\,\text{dBm}$, $-3.4\,\text{dBm}$ and $-1.9\,\text{dBm}$ were measured for $1.0\,\text{V}$, $1.2\,\text{V}$, and $1.4\,\text{V}$ supply voltages, respectively.

The average power consumption and the DC-to-RF efficiency of the oscillator including its buffers, are shown in Fig. 7 and 8, respectively, as functions of the oscillation frequency. Both the difference in the output power and the average power consumption is $1.8\,\text{dB}$ between $1\,\text{V}$ and $1.2\,\text{V}$ supply voltages at a tuning voltage of $0\,\text{V}$. This indicates that the oscillator is still in the current-limited regime at these supply voltages. At higher tuning voltages, Q of the varactors improves, which reduces the resonator losses and increases the voltage swing and thus P_{out} in the resonator. This moves the operation of the oscillator towards the voltage limited regime, which is verified by the positive slope in Fig. 6.

Table I shows a performance comparison with other CMOS publications in this frequency range sorted according to the DC-to-RF efficiency. P_{DC} of the presented circuit has been calculated based on the simulated ratio of the core and buffer current consumption, because the core and the buffers share their supply connection, and no direct core current measurement was possible. FoM_{T} is calculated from the core power consumption and does not take the power consumption of buffers into account, and therefore it differs from the similar system level FoM [9] using the total power consumption (P_{tot}).

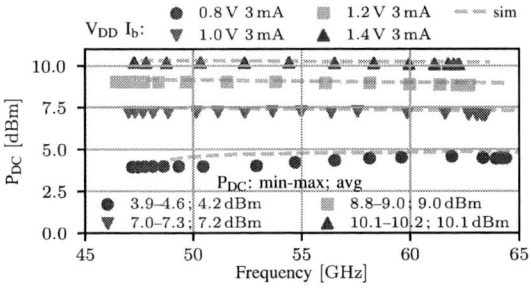

Fig. 7. Measured power consumption including the output buffer

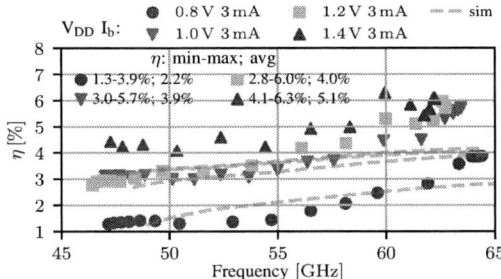

Fig. 8. Measured DC-to-RF efficiency including the output buffers

TABLE I
COMPARISON WITH THE STATE OF THE ART

	node	f_0	TR	P_{DC}[†]	V_{DD}	P_{out}	η	$\mathcal{L}$[‡]	FoM_T
	nm	GHz	%	mW	V	dBm	%	$\frac{dBc}{Hz}$	$\frac{dBc}{Hz}$
This[C]	22	54.6	29.8	3.0[§]	1.2	-3.4	6.0	-84.2	-183.7
[12][N,D]	22	61.5	34.0	7.5	0.7	—	—	-84.0	-181.7
[16][C]	22	58.5	26.7	2.7[§]	1.0	-0.2	14.7	-89.8	-189.4
[4][N]	22	64.3	25.5	20.0[§]	1.0	-0.4	—	-87.0	-178.3
[11][E,N,T]	45	60.5	19.0	40.0	1.0	-22.6	0.0	-101.7	-186.9
[17][E,N,D,T]	28	60.0	16.0	11.0	0.9	-23.0	0.0	-87.0	-176.2
[10][E,N,D,T]	32	52.0	11.7	54[§]	1.0	-6.0	0.5	-89.0	-167.4
[18][N,T]	28	64.0	11.1	3.1	0.9	—	—	-92.5	-184.5
[18][N]	28	67.0	10.9	3.1	0.9	—	—	-84.5	-176.8
[9][N,H]	90	64.0	8.8	3.2	0.6	-14.0	—	-95.0	-185.0
[8][N,H]	90	60.0	8.4	7.2	1.2	-2.5	3.6	-91.5	-177.0
[7][N]	45	69.4	5.8	4.5	1.0	-4.5	5.3	-94.4	-180.0

[N]: NMOS   [C]: CMOS   [H]: harmonic tank   [E]: harmonic extraction
[D]: digital tuning   [‡] $f_{\mathcal{L}} = 1$ MHz   [†] core only   [§] Estimated

$$FoM_T = \mathcal{L}(f_0, f_{\mathcal{L}}) \cdot \frac{P_{core}}{1\,\mathrm{mW}} \cdot \left(\frac{f_{\mathcal{L}}}{f_o}\right)^2 \cdot \left(\frac{TR}{10\,\%}\right)^2$$

IV. CONCLUSION

A low power CMOS oscillator with 55 GHz center frequency has been presented. Its wide tuning range is beneficial for multi-band operation or for radar applications, which require a continuous tuning range. The circuit has achieved the highest continuous tuning range and the second highest tuning range, while its core power consumption is comparable to the lowest power CMOS oscillators in the same frequency band.

ACKNOWLEDGEMENT

This work was funded by SMWK and SAB through project PROSECCO. The authors express their gratitude to Globalfoundries for the prototype fabrication.

REFERENCES

[1] T. Chi, F. Wang, S. Li, M. Y. Huang, J. S. Park, and H. Wang, "A 60GHz on-chip linear radiator with single-element 27.9dBm Psat and 33.1dBm peak EIRP using multifeed antenna for direct on-antenna power combining," *Dig. Tech. Pap. - IEEE Int. Solid-State Circuits Conf.*, vol. 60, pp. 296–297, 2017.

[2] A. Harutyunyan, "Analog frontend for ultra low power 60-ghz rfid tag for back-scattering communication." VDE, 2018, pp. 1–7.

[3] R. Wu *et al.*, "64-QAM 60-GHz CMOS transceivers for IEEE 802.11ad/ay," *IEEE Journal of Solid-State Circuits*, vol. 52, no. 11, pp. 2871–2891, nov 2017.

[4] Z. Tibenszky, D. Fritsche, C. Carta, and F. Ellinger, "An Efficient Wide Tuning Range -0.4 dBm 65 GHz NMOS VCO on 22 nm FD-SOI CMOS," in *2020 IEEE International Symposium on Radio-Frequency Integration Technology (RFIT)*, Sep. 2020, pp. 37–39.

[5] L. Fanori and P. Andreani, "Highly efficient class-C CMOS VCOs, including a comparison with class-B VCOs," *IEEE J. Solid-State Circuits*, vol. 48, no. 7, pp. 1730–1740, jul 2013.

[6] D. Fritsche, G. Tretter, C. Carta, and F. Ellinger, "Millimeter-wave low-noise amplifier design in 28-nm low-power digital CMOS," *IEEE Trans. Microw. Theory Tech.*, vol. 63, no. 6, pp. 1910–1922, Jun. 2015.

[7] Y. Wang, J. Xu, K. He, M. Wu, and R. Zhang, "A 67.4-71.2-GHz nMOS-only complementary VCO with buffer-reused feedback technique," *IEEE Microw. Wirel. Components Lett.*, vol. 29, no. 12, pp. 810–813, 2019.

[8] Y. C. Chiang and Y. H. Chang, "A 60 GHz CMOS VCO using a fourth-order resonator," *IEEE Microw. Wirel. Components Lett.*, vol. 25, no. 9, pp. 609–611, 2015.

[9] L. Li, P. Reynaert, and M. S. Steyaert, "A 60-GHz CMOS VCO using capacitance-splitting and gate-drain impedance-balancing techniques," *IEEE Trans. Microw. Theory Tech.*, vol. 59, no. 2, pp. 406–413, feb 2011.

[10] B. Sadhu, M. Ferriss, and A. Valdes-Garcia, "A 52 GHz frequency synthesizer featuring a 2nd harmonic extraction technique that preserves VCO performance," *IEEE J. Solid-State Circuits*, vol. 50, no. 5, pp. 1214–1223, 2015.

[11] J. Rimmelspacher, R. Weigel, A. Hagelauer, and V. Issakov, "A quad-core 60 GHz push-push 45 nm SOI CMOS VCO with-101.7 dBc/Hz phase noise at 1 MHz offset, 19 % continuous FTR and -187 dBc/Hz FoMT," *ESSCIRC 2018 - IEEE 44th Eur. Solid State Circuits Conf.*, pp. 138–141, 2018.

[12] C. Zhang and M. Otto, "A wide range 60 GHz VCO using back-gate controlled varactor in 22 nm FDSOI technology," *2017 IEEE SOI-3D-Subthreshold Microelectron. Unified Conf. S3S 2017*, vol. 2018-March, pp. 1–3, 2018.

[13] D. B. Leeson, "A simple model of feedback oscillator noise spectrum," *Proc. IEEE*, vol. 54, no. 2, pp. 329–330, 1966.

[14] Z. Tibenszky, C. Carta, and F. Ellinger, "Design of an efficient 6.5 dBm 55 GHz CMOS VCO with simultaneous phase noise and tuning range optimization," *IEEE Transactions on Microwave Theory and Techniques*, 2021, submitted.

[15] S. N. Ong *et al.*, "A 22nm FDSOI technology optimized for RF/mmWave applications," in *Dig. Pap. - IEEE Radio Freq. Integr. Circuits Symp.*, 2018, pp. 72–75.

[16] Z. Tibenszky, C. Carta, and F. Ellinger, "58 GHz CMOS VCO with 16% efficiency," *Electronics Letters*, vol. 56, no. 24, pp. 1301–1303, Nov. 2020.

[17] V. Issakov, F. Padovan, J. Rimmelspacher, R. Weigel, and A. Geisel-brechtinger, "A 52-to-61 GHz Push-Push VCO in 28 nm CMOS," *2018 48th Eur. Microw. Conf. EuMC 2018*, pp. 1009–1012, 2018.

[18] T. Forsberg, J. Wernehag, A. Nejdel, H. Sjöland, and M. Törmänen, "Two mm-wave VCOs in 28-nm UTBB FD-SOI CMOS," *IEEE Microw. Wirel. Components Lett.*, vol. 27, no. 5, pp. 509–511, 2017.

© VDE VERLAG GMBH · Berlin · Offenbach

SMACD / PRIME 2021 | 19 – 22 July 2021, Online Event

A Fully Integrated 28 GHz Class-J Doherty Power Amplifier in 130 nm BiCMOS

Simone Veni*, Michele Caruso†, David Seebacher†, Andrea Neviani*, Andrea Bevilacqua*

*DEI, University of Padova, Italy; †Infineon Technologies, Villach, Austria

simone.veni@studenti.unipd.it, {Michele.Caruso,David.Seebacher}@infineon.com, {neviani,bevilacqua}@dei.unipd.it

Abstract—A SiGe BiCMOS Class-J Doherty power amplifier operating at 28 GHz is presented. The class-J operation is chosen to maximize the efficiency of the system without degradation in terms of bandwidth. A dynamic bias circuit is used to progressively turn on the auxiliary amplifier and improve the efficiency at back-off. Supplied by a 2.1 V supply, the power amplifier features a saturation power as high as 25 dBm, a peak power added efficiency equal to 25.5 %, and a gain equal to 13.6 dB. The PAE at the 6 dB power back-off (PBO) is 17.6 %.

Index Terms—Power Amplifier, Doherty, Class-J, high output power, BiCMOS

I. Introduction

In the development of wireless communication systems, the design of high 1 dB compression point (P_{1dB}), high gain and high power added efficiency (PAE) power amplifiers (PAs) is a crucial target [1]. The efficiency of this building block usually sets the efficiency of the entire system because, typically, it has the highest power consumption. The class-J power amplifier has been proposed as a topology capable of achieving the same PAE of the conventional class-AB power amplifier but with several advantages [2], [3]. By engineering the first and second harmonic impedance terminations, class-J operation aims at reducing the overlap between the current and voltage waveforms of the active devices, thus maximizing the amplifier's efficiency [3]. The implementation of the circuit also becomes easier, due to the elimination of the need for a load impedance that acts as a short at higher harmonics, as required by other classes of operation. To further increase the efficiency at backoff, the class-J topology and the Doherty architecture can be combined, as reported in [4].

In this work, a class-J Doherty power amplifier operating at 28 GHz is proposed, presenting a topology that achieves a relatively high output power with improvements in terms of efficiency at power back-off. Post layout simulations on a design implemented in a 130 nm BiCMOS technology show a saturation power as high as 25 dBm, a peak power added efficiency equal to 25.5 %, and a power gain equal to 13.6 dB.

II. Class-J Doherty Power Amplier Topology

The Doherty architecture allows to improve the efficiency at power back-off by leveraging active load impedance modulation [3]. The block diagram of a conventional Doherty topology is shown in Fig. 1. To achieve active load modulation the Doherty topology makes use of an impedance inverter connected to the output of the main amplifier. When the

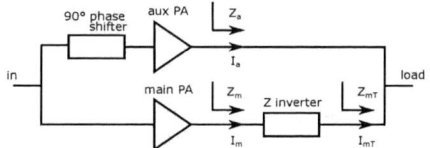

Fig. 1. Block diagram of a conventional Doherty PA.

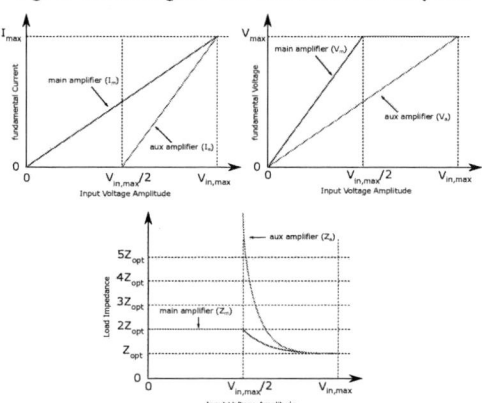

Fig. 2. Ideal behaviour of the Doherty amplifier.

auxiliary amplifier turns on and pushes some current I_a into the load, the load current that the main amplifier needs to supply, I_{mT}, decreases, which, equivalently, corresponds to an increase of the impedance Z_{mT}. Due to the presence of the impedance inverter, the actual impedance loading the active devices of the main amplifier, namely, Z_m in Fig. 1, decreases. This allows the main amplifier to output a larger power, while constantly operating at its maximum efficiency point, that is, close to saturation. The ideal behavior of the Doherty amplifier is shown in Fig. 2.

Assuming the auxiliary amplifier is turned on when the input signal is half of the one corresponding to the compression of the gain by 1 dB, $V_{in,max}$, the impedances that the main and auxiliary amplifier drive can be expressed as:

$$Z_m = \begin{cases} \frac{Z_T^2}{Z_L} & 0 < v_{in} < V_{in,max}/2 \\ \frac{Z_T^2}{Z_L\left(1+\frac{I_a}{I_{mT}}\right)} & V_{in,max}/2 < v_{in} < V_{in,max} \end{cases} \quad (1)$$

$$Z_a = \begin{cases} \infty & 0 < v_{in} < V_{in,max}/2 \\ Z_L\left(1 + \frac{I_{mT}}{I_a}\right) & V_{in,max}/2 < v_{in} < V_{in,max} \end{cases} \quad (2)$$

© VDE VERLAG GMBH · Berlin · Offenbach

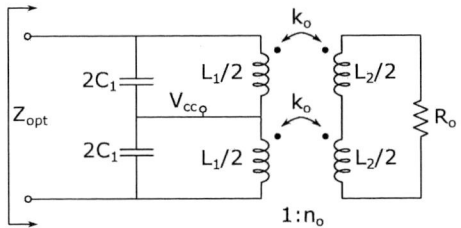

Fig. 3. Class-J output matching network.

where Z_T is the characteristic impedance of the impedance inverter and Z_L is the load of the system.

To achieve the class-J operation, the load impedance of the power amplifier (ideally) needs to be:

$$Z_{opt}(f_0) = R_{opt} + j R_{opt} \tag{3}$$

$$Z_{opt}(2f_0) = -j\frac{3\pi}{8} R_{opt} \tag{4}$$

An output matching network that allows to (approximately) meet the conditions for class-J operation and is amenable to integrated implementation was proposed in [2], [5]. It is shown in Fig. 3. The transformer's leakage inductance is used to implement the reactive part of the load impedance at the first harmonic. At the first harmonic, the capacitance C_1 resonates with the magnetizing inductance of the primary coil, L_1. The capacitance C_1 is the sum of the output parasitic capacitance of the power amplifier and an explicit output capacitance. Considering the differential mode equivalent half-circuit, The load impedance at the first harmonic is thus:

$$Z_{opt}(f_0) = \frac{R_o + j\omega L_2 \left(1 - k_o^2\right)}{2 \left(n_o k_o\right)^2} \tag{5}$$

At the second harmonic, the power amplifier is only loaded by the common mode equivalent circuit of the matching network, namely, the parallel combination of output capacitance and the common mode equivalent circuit of the primary coil, as the secondary coil is (ideally) not affected by the common mode operation. Considering the common mode equivalent half-circuit, the load impedance at the second harmonic is hence:

$$Z_{opt}(2f_0) = \frac{1}{j2\omega 2C_1} \,||\, j2\omega \frac{L_1}{2} \left(1 - k_{cm}\right) \tag{6}$$

where k_{cm} is the common mode magnetic coupling factor between the two halves ($L_1/2$) of the primary coil.

To implement the progressive turn-on of the auxiliary amplifier, as described in Fig. 2, a dynamic bias circuit is required. The amplitude of the input signal is sensed by a peak detector. When the input signal exceeds $V_{in,max}/2$, the auxiliary amplifier is progressively turned on.

III. CIRCUIT DESIGN

Both the main and auxiliary amplifiers are based on a differential cascode topology. A differential topology is chosen to increase the maximum output swing, and thus the output

TABLE I
LOAD IMPEDANCES AT THE MAXIMUM OUTPUT POWER, AND AT BACK-OFF.

	Back-Off	Peak
$Z_m(f_0)$	$2R_{opt} + jR_{opt}$	$R_{opt} + jR_{opt}$
$Z_m(2f_0)$	$-j\frac{3\pi}{8}R_{opt}$	$-j\frac{3\pi}{8}R_{opt}$
$Z_a(f_0)$	∞	$R_{opt} + jR_{opt}$
$Z_a(2f_0)$	∞	$-j\frac{3\pi}{8}R_{opt}$

power (for a given level of the load impedance). Moreover, a cascode structure results in higher reverse isolation, helping guaranteeing the stability of the circuit.

The optimal load resistance for both the main and the auxiliary amplifiers was determined by taking into consideration both the voltage rating of the used transistors and a target maximum output power for the overall power amplifier ($P_{sat} = 26 \, \text{dBm}$):

$$R_{opt} \approx 10 \, \Omega. \tag{7}$$

The output matching network is designed such that the the class-J condition is achieved at the maximum output power, as proposed in [4]. The reactive part of the load impedance at the first harmonic is determined by the magnetic coupling of the used transformers, and is not affected by the active load modulation of the Doherty architecture. The required class-J load impedances of the main and auxiliary amplfier at the maximum output power and at back-off (i.e., when the auxiliary amplifier is off) are reported in Table I.

A. Output Matching Network

The output matching network of the proposed Doherty amplifier is shown in Fig. 4. As discussed in Section II, a transformer based network is used to achieve the class-J operation. This network intrinsically scales the load impedance (R_o in Fig. 3) by a factor $1/(2n_o^2 k_o^2)$, as shown by (5). In the proposed design, this scaling factor should be equal to 0.4. However, to reduce the losses of the matching network, a unitary turn ratio $n_m = 1$ for the transformer was selected. This, combined with the achieved magnetic coupling, $k = 0.77$, does not yield the desired impedance transformation. The solution is to select a characteristic impedance equal to $Z_T = 34 \, \Omega$ for the impedance inverter. Hence, when the auxiliary amplifier is off, $Z_{mT} = 50 \, \Omega$ and, due to the impedance inverter, $Z_m = 23 \, \Omega$, which is then scaled by the transformer to the desired $2R_{opt} = 20 \, \Omega$ level. When the auxiliary amplifier is on, and the load modulation occurs, we have $Z_{mT} = 100 \, \Omega$, $Z_m = 11.5 \, \Omega$, and the load resistance of the main amplifier decreases to $R_{opt} = 10 \, \Omega$, as desired. To minimize the area consumption and to embed the parasitic capacitances, the impedance inverter was designed as a π-lumped-element transmission line. Its inductance value is $L_T = Z_T/\omega_0 = 197 \, \text{pH}$, while the capacitance value is $C_T = 1/(Z_T\omega_0) = 164 \, \text{fF}$.

The output matching network of the auxiliary amplifier is realized in a different way. Since it is supposed to scale down $Z_a = 100 \, \Omega$ to $R_{opt} = 10 \, \Omega$, a wide step-down transformation ratio is required. Such a large transformation ratio is achieved

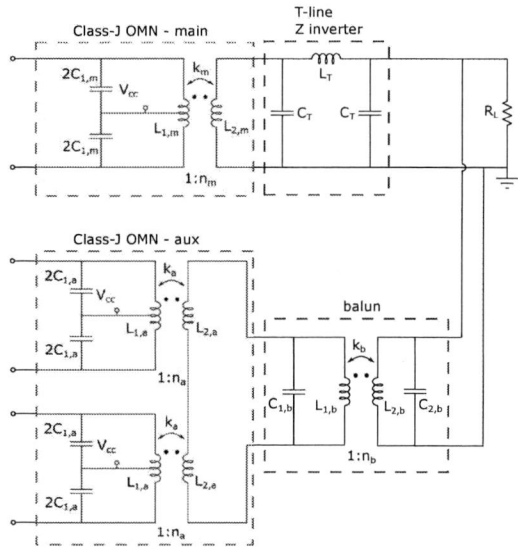

Fig. 4. Output matching network.

TABLE II
COMPONENT VALUES OF THE OUTPUT MATCHING NETWORK.

	k	n	L_1	L_2	C_1
main	0.77	1	$150pH$	$150pH$	$215fF$
aux	0.6	1.87	$126pH$	$444pH$	$254fF$

in two steps. First, a two-way series power combiner is used, as it inherently features a step-down impedance scaling equal to the number of combined amplifiers—in this case 2 [6]. Next, the remaining required step-down impedance scaling is implemented leveraging the transformer used to achieve the class-J operation. Due to the asymmetry of the series power combiner and the differential to single-ended conversion required to interface the output network of the auxiliary amplifier to the single-ended pad, a dedicated balun structure was deemed necessary. The balun is realized as a doubly-tuned transformer [7] with a stacked layout to maximize the magnetic coupling, its values are $C_{1,b} = C_{2,b} = 80$ fF, $L_{1,b} = L_{2,b} = 225$ pH and $k_b = 0.8$.

The design parameters of the output network of the proposed class-J Doherty amplifier are summarized in Table II.

B. Dynamic Bias Circuit

The schematic of the dynamic bias circuit is shown in Fig. 5. It is realized with a peak detector that outputs a voltage V_{PD} proportional to the signal at the input of the main power amplifier, and a comparator. The other input of the comparator, V_{ref}, sets the back-off point. If V_{PD} is much smaller than V_{ref}, the auxiliary amplifier is off. When V_{PD} is comparable to V_{ref}, the auxiliary amplifier turns on. As $V_{PD} - V_{ref}$ increases, so does the quiescent current of the auxiliary amplifier. V_{be_q} in Fig. 5 sets the bias voltage of the driver transistors of the auxiliary amplifier, while V_{c_q} sets the bias voltage of the cascode transistors.

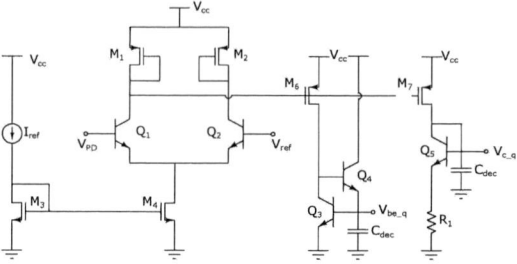

Fig. 5. Dynamic bias circuit for the auxiliary amplifier.

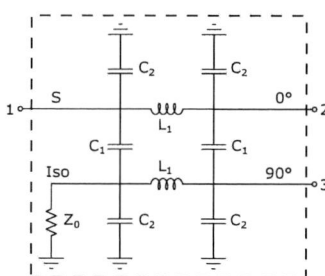

Fig. 6. Schematic of the input lumped-element quadrature hybrid coupler.

C. Input Matching Network

As shown in Fig. 1, at the input of the Doherty amplifier, the input signal has to be splitted between the main and auxiliary paths. Moreover, a 90° phase shift is to be introduced in the auxiliary path to compensate for the phase shift introduced by the impedance inverter in the main signal path. In the proposed design, the power division and phase shift are obtained by means of a lumped-element quadrature hybrid coupler, whose schematic is depicted in Fig. 6.

The combination of a high-pass and low-pass signal paths provides 90° phase difference between the outputs. The circuit components can be sized according to the following expressions:

$$
\begin{cases}
L_1 = \frac{Z_0}{\omega_0 \sqrt{2}} = 201pH \\
C_1 = \frac{1}{\omega_0 Z_0} = 114fF \\
C_2 = \frac{1}{\omega_0^2 L_1} - C_1 = 47fF
\end{cases}
\tag{8}
$$

where $Z_0 = 50\,\Omega$ is the characteristic impedance of the hybrid coupler.

To resonate the input capacitance of the main and auxiliary amplifiers and to match their input impedance to the characteristic impedance of the hybrid coupler, two input doubly-tuned transformer networks were used. These networks also implement single-ended to differential conversion, and provide bias to the driver transistors of the amplifiers via the center taps of the secondary coils.

IV. SIMULATION RESULTS

The proposed class-J Doherty power amplifier was designed in Infineon 130 nm SiGe BiCMOS technology.

Post layout small signal simulations are reported in Fig. 7. The input port is matched in a broadband fashion. The

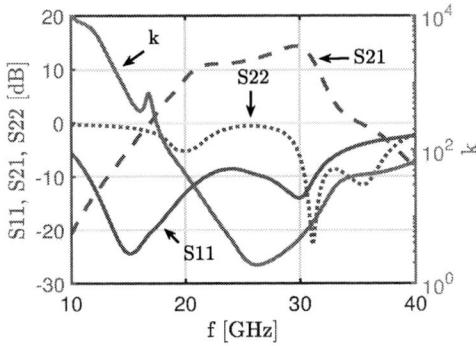

Fig. 7. Simulated S-parameters and stability factor.

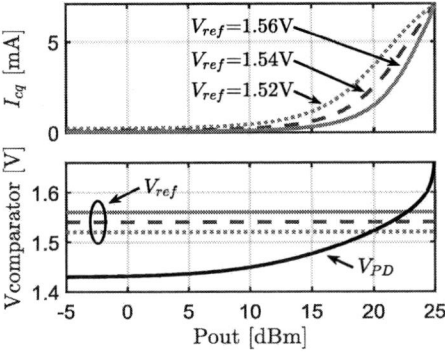

Fig. 8. Simulated behavior of the dynamic bias circuit.

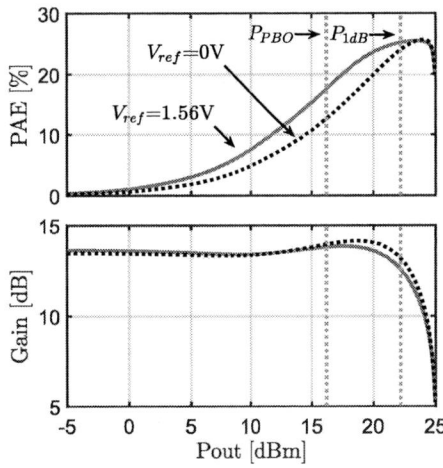

Fig. 9. Simulated large-signal behavior of the proposed class-J Doherty power amplifier.

V. Conclusion

A power amplifier that combines the advantages of class-J and Doherty operation been presented. Post layout simulations on a design implemented in a 130 nm SiGe BiCMOS technology show a remarkable performance: small-signal gain of 13.6 dB, and P_{1dB} equal to 22.2 dBm with a corresponding PAE of 25.2%. The PAE at 6 dB back-off is 17.6%. The circuit operates around 28 GHz with a 37% fractional bandwidth. The supply voltage is 2.1 V.

References

[1] V. Camarchia, R. Quaglia, A. Pacibello, D. P. Nguyen, H. Wang and A. Pham, "A Review of Technologies and Design Techniques of Millimeter-Wave Power Amplifiers," *IEEE Transactions on Microwave Theory and Techniques*, vol. 68, no. 7, pp. 2957–2983, 2020.

[2] P. Scaramuzza, C. Rubino, M. Tiebout, M. Caruso, M. Ortner, A. Neviani and A. Bevilacqua, "Class-AB and class-J 22 dBm SiGe HBT PAs for X-band radar systems," *ESSCIRC 2017 - 43rd IEEE European Solid State Circuits Conference*, pp. 187–190, 2017.

[3] S. C. Cripps, *RF Power Amplifier for Wireless Communication* (Artech House Microwave Library), Norwood, MA, USA: Artech House, 2006.

[4] N. Tuffy and L. Pattison, "A linearized, high efficiency 2.7 GHz wideband Doherty power amplifier with class-J based performance enhancement," *2015 European Microwave Conference (EuMC)*, pp. 215–218, 2015.

[5] P. Scaramuzza, C. Rubino, M. Caruso, M. Tiebout, A. Bevilacqua and A. Neviani, "Class-J SiGe X-Band Power Amplifier Using a Ladder Filter-Based AMPM Distortion Reduction Technique," *IEEE Transactions on Circuits and Systems I: Regular Papers*, vol. 65, no. 11, pp. 3780–3789, 2018.

[6] D. Manente, F. Padovan, D. Seebacher, M. Bassi and A. Bevilacqua, "A 28-GHz Stacked Power Amplifier with 20.7-dBm Output P1dB in 28-nm Bulk CMOS," *IEEE Solid-State Circuits Letters*, vol. 3, pp. 170–173, 2020.

[7] A. Mazzanti and A. Bevilacqua, "Second-Order Equivalent Circuits for the Design of Doubly-Tuned Transformer Matching Networks,"*IEEE Transactions on Circuits and Systems I: Regular Papers*, vol. 65, no. 12, pp. 4157–4168, 2018.

amplifier's gain is 13.6 dB. The 3 dB bandwidth of the PA is 21.1-31.5 GHz, corresponding to a fractional bandwidth of 37%. The simulated k-factor, reported in Fig. 7, shows that the amplifier is stable.

The simulated behavior of the dynamic bias circuit is illustrated in Fig. 8. As discussed, the voltage V_{PD} is the output of the peak detector, while the reference voltage V_{ref} (see Fig. 5) sets the power level at which the auxiliary amplifier turns on. I_{cq} is the quiescent current of (each branch of) the auxiliary power amplifier, which is proportional to the drain current of M_1(see Fig. 5). Figure 8 shows that, by increasing V_{ref}, the power level at which the auxiliary amplifier turns on increases. The nominal bias point is set for $V_{ref} = 1.56$ V.

The simulated large-signal behavior of the Doherty power amplifier is illustrated in Fig. 9 for operation at 28 GHz. The power gain and PAE are reported. Two conditions are compared: with the dynamic bias on ($V_{ref} = 1.56$ V), and with the auxiliary amplifier always on (corresponding to setting $V_{ref} = 0$ V). The saturated output power is 25 dBm. The output referred 1 dB compression point is $P_{1dB} = 22.2$ dBm, with a corresponding PAE of 25.2%. At 6 dB back-off from the P_{1dB}, the PAE is 17.6%. From Fig. 9, the advantage, in terms of improved PAE, of operating the circuit as a Doherty amplifier (that is, leveraging the dynamic bias) is evident.

© VDE VERLAG GMBH · Berlin · Offenbach

SMACD / PRIME 2021 | 19 – 22 July 2021, Online Event

A Scalable CPW Circuit Model in Advanced CMOS Technologies for mm-Wave frequencies

Carla Moran Guizan*[†], Peter Baumgartner*, Stefan Heinen[†]

*Intel, Munich, Germany

[†]RWTH Aachen University, Aachen, Germany

Email: carla.moran.guizan@intel.com

Abstract—This paper presents a physical circuit model for coplanar waveguide (CPW) transmission lines, scalable with line width and signal to ground spacing and suitable for circuit simulators such as SPICE. The circuit model components are fitted using EM simulation data, without the need to know materials and other stack up information that may be encrypted or hidden for the user. This is also helpful for complex layer stacks composed of many dielectric and conductive materials that are used in the latest CMOS technologies. The deviations of the transmission line main parameters, propagation constant and characteristic impedance, remain low for the mm-wave frequency band, from 20 GHz and up to 100 GHz. The scalability enables a fast circuit optimization.

Index Terms—transmission lines, CMOS technology, millimeter wave circuits, semiconductor device modeling

I. INTRODUCTION

In the last decade a number of communication technologies have been increasingly moved towards higher frequencies, such as Wi-Fi or mobile communications. This has led to advancements in CMOS processes that allow the implementation of transceivers in mm-wave band [1]. In these transceivers transmission lines are a common passive element. Therefore, their characterization is key for designs in these frequency bands, but difficult due to the complexity of present-day semiconductor fabrication process and materials [2].

Transmission lines are usually characterized in initial stages with electromagnetic (EM) simulators. These simulators have longer simulation times and use more resources than circuit simulators like SPICE. Additionally, when performing a circuit simulation of several components, the EM simulation results are included as S-parameters, which increases the circuit simulation computing time. There are transmission line models with good accuracy over large bandwidths [3]–[5], but some of the circuit components are dependent on frequency, which is not convenient for broadband circuit simulation.

The model we present is valid for a coplanar waveguide (CPW) transmission line in a 16 nm CMOS node. It is scalable with the signal width w_s and spacing s, as defined in Fig. 1. The circuit components values are independent of frequency. In contrast to models based on the structure and dimensions of the CPW [6], our model does not require knowledge of the layers and materials of the CMOS process. This is advantageous for advanced nodes composed of complicated layer sequences (dielectrics, conductors, liner and barrier materials) that are usually hidden from the user.

This paper describes the chosen circuit for the CPW and then continues explaining the method used to calculate a scalable function for each circuit component. Finally, we compare the CPW parameters resulting from the described model with training and validation EM simulated data, and analyze how the circuit section length influences the final deviations from the EM simulation.

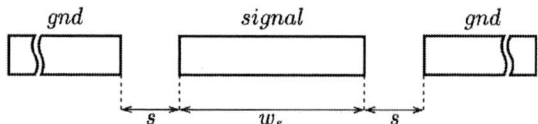

Fig. 1. Coplanar waveguide cross section showing signal width w_s and signal to ground spacing s.

II. CIRCUIT MODEL

The proposed circuit model of an infinitesimal section of the transmission line is shown in Fig. 2. It can be divided into a series part, representing the inductance and resistance of the conductors, and a parallel part, for the capacitance and substrate losses.

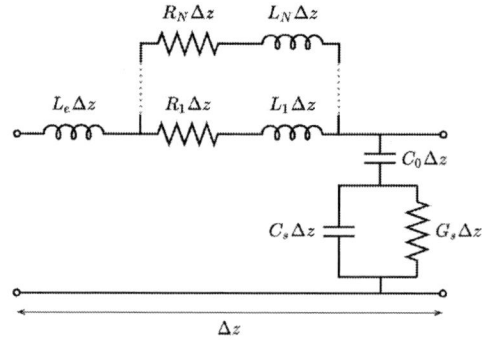

Fig. 2. Circuit model of each section.

The resistance and inductance of a CPW are dependent on frequency due to how the skin effect changes the current distribution in the conductor [4]. The inductance can be approximated as an internal inductance, affected by the skin effect, and external inductance, independent of the frequency.

© VDE VERLAG GMBH · Berlin · Offenbach

To mimic the internal inductance and the resistance, the circuit model uses an RL ladder of N branches [7]. It can be considered to represent the concentric shells of the conductor and it is used to reproduce the varying current distribution with frequency. The resistance of each branch R_i is the previous value multiplied by an x factor (1). The inductance L_i is the previous one divided by the same x factor (2). Our model uses five branches ($N = 5$); this is a compromise between circuit complexity and approximation accuracy. We use a scaling factor of $\sqrt{10}$, as suggested in [7].

$$R_i = R_{i-1} \cdot x = R_1 \cdot x^{i-1} \tag{1}$$

$$L_i = L_{i-1}/x = L_1/x^{i-1} \tag{2}$$

Therefore, the total impedance of the series subcircuit is the parallel combination of all the branches of the RL ladder plus the external inductance, as shown in (3).

$$Z_{series} = \left[\sum_{i=1}^{N} (R_i + j\omega L_i)^{-1} \right]^{-1} + j\omega L_e \tag{3}$$

where ω is the angular frequency, L_e the external inductance and Z_{series} the impedance of the series subcircuit.

The capacitance and conductance of a CPW are modeled by the three elements shown in Fig. 2. They are less dependent on frequency than the series subcircuit. The total impedance of the parallel subcircuit is:

$$Z_{parallel} = \frac{1}{G_s + j\omega C_s} + \frac{1}{j\omega C_0} \tag{4}$$

III. METHOD

The method to obtain the circuit model consist of three main steps. First, we perform EM simulations of a $200\,\mu m$ CPW for a set of w_s and s points using a 3D planar Method of Moments (MoM) solver. Then, we calculate the circuit model components from the RLCG parameters of the simulated lines normalized to the their length (see III-A). These RLCG values are the components of the classical lumped-element model and they are frequency dependent [8]. Finally, the component values are fitted to a function of w_s and s (see III-B).

A. Calculation of the circuit component values

The series part of the circuit model has to approximate the resistance and inductance of the simulated CPW. Therefore the real part of the RL ladder should be equal to the resistance of the simulated CPW, R_{EM}. Equation (5) is solved for L_1, with a fixed frequency point ω_0 and initial ladder resistance R_1. The ladder works better when the R_1 value is in the DC resistance range, e.g. $600\,\Omega$. The external inductance approximation is obtained by subtracting the inductance of the simulated data L_{EM} from the total inductance of the RL ladder at the selected frequency point. Fig. 3 shows the approximated resistance and inductance values using this subcircuit against the simulated data. The relative error of the approximation stays under $5\,\%$ for frequencies above $20\,GHz$.

$$R_{EM}(\omega_0) = \Re\left(\left[\sum_{i=1}^{N} \left(R_1 x^{i-1} + j\omega_0 \frac{L_1}{x^{i-1}} \right)^{-1} \right]^{-1} \right) \tag{5}$$

$$L_e = L_{EM}(\omega_0) - \Im\left(\left[\sum_{i=1}^{N} (R_i + j\omega_0 L_i)^{-1} \right]^{-1} \right) / \omega_0 \tag{6}$$

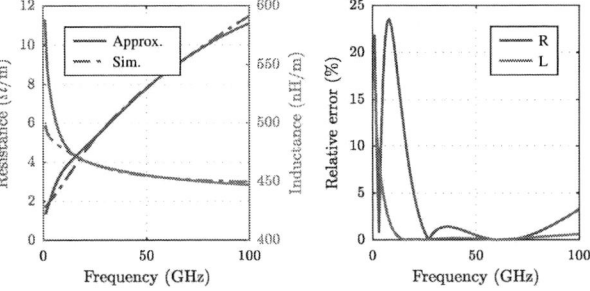

Fig. 3. Series subcircuit approximation and simulated data, for a CPW of $4\,\mu m$ width and $9\,\mu m$ spacing, and the relative error.

The real and imaginary parts of the parallel circuit impedance $Z_{parallel}$ are simplified to (7) and (8) at high frequencies and should equal the conductance G_{EM} and capacitance C_{EM} of the simulation data. Equation (9) is obtained by taking the G_{EM} at a low frequency point ω_L (in this case, $1\,GHz$) and solving the real part of $Z_{parallel}$. The relative error of the approximation is small, in particular for the capacitance, as shown in Fig. 4, because these components have a smaller dependency on the frequency than the resistance and inductance.

$$G_{EM}(\uparrow \omega) = \frac{C_0^2 G_s}{(C_0 + C_s)^2} \tag{7}$$

$$C_{EM}(\uparrow \omega) = \frac{C_0 C_s}{C_0 + C_s} \tag{8}$$

$$G_{EM}(\omega_L) = \frac{\omega_L^2 C_0^2 G_s}{G_s^2 + \omega_L^2 (C_0 + C_s)^2} \tag{9}$$

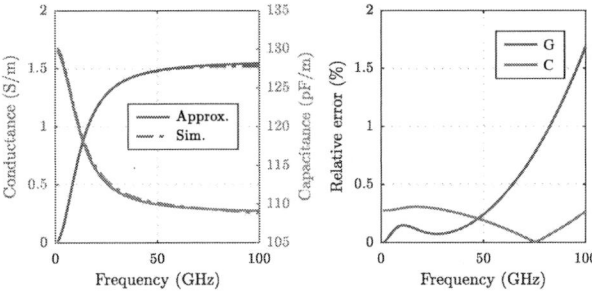

Fig. 4. Parallel subcircuit approximation and simulated data, for a CPW of $4\,\mu m$ width and $9\,\mu m$ spacing, and the relative error.

B. Fitting of the circuit components

Using equations (5) – (9), we calculated the values of L_1, L_e, C_0, G_s and C_s for a simulated set of width and space points. The following step is to model these components by a function that depends on width and space. The components of the circuit model seem to have a power dependency with spacing and vary linearly with the width, as it can be seen in Fig. 5 for the component C_0. The variation of the other components is similar, thus by scaling and offsetting them the same function can be applied for all the components. For this purpose we use a non-linear least squares algorithm, in particular the Levenberg-Marquardt, with the following fitting function:

$$f(s, w_s) = a_0 + a_1 w_s + \\ (b_0 + b_1 w_s)(c_0 + c_1 w_s + s)^{d_0 + d_1 w_s} \quad (10)$$

where $(a_0, a_1, ..., d_1)$ are the model parameters. Fig. 6 illustrates the obtained fit against the original C_0 values. It can be seen that there is no overfitting for the modeled range. Table I lists the parameters for this component. The coefficient of determination R^2 is higher than 0.999 for all the modeled elements.

TABLE I
MODEL PARAMETERS FOR C_0

Parameter	Value	Parameter	Value
a_0	-0.177	a_1	0.03
b_0	9.675	b_1	1.854
c_0	2.061	c_1	0.154
d_0	-1.759	d_1	-0.014
Scale	1.3×10^{-10}	Offset	1.1×10^{-10}

IV. RESULTS

We simulated CPW lines of different widths and spaces and used 24 for fitting the model and 16 for validation. Width values were between 4 μm and 10 μm, and the spacing range was 3 μm to 35 μm. The propagation constant γ and characteristic impedance Z_c are calculated from the equivalent RLCG using (11) and (12) [9]. The RLCG values are computed from the S-parameters for the simulation data and from the equivalent series and parallel impedances of the circuit model.

$$\gamma = \sqrt{(R + j\omega L)(G + j\omega C)} \quad (11)$$

$$Z_c = \sqrt{\frac{R + j\omega L}{G + j\omega C}} \quad (12)$$

The root-mean-square deviation (RMSD) of the model γ and Z_c with respect to the simulation data are shown in Fig. 7, where β is the phase constant and α the attenuation constant. The model performance at low frequencies is worse due to the bad approximation of the resistance for this band. The residuals of the phase constant increase with frequency. This is expected because $\beta \approx \omega\sqrt{LC}$. The RMSD of the validation set is close to the RMSD of the training one, demonstrating that the model generalizes to unseen data. The validation set deviations are lower because the model is less accurate in the extremes of the width and space ranges and the training set contains more extreme points compared to the validation set.

To make the model suitable for circuit simulation, the circuit of Fig. 2 must represent only a section of the whole transmission line and be concatenated until it reaches the desired length. As with the theory behind the lumped-element RLCG model, the section length should be ideally infinitesimally small to represent exactly the transmission line, but that would make an infinite number of sections and circuit components. Setting a definite length will add an error to that of the initial model. To illustrate this effect, Fig. 8 pictures the imaginary part of the characteristic impedance for the original simulation data and the reconstructed impedance from taking Δz long sections of the RLCG parameters, which were calculated from the same simulation data. As it can be observed, the longer the sections are, and thus the less circuit components there are, the more the impedance deviates from the original value.

The error due to the section length affects all the line parameters. Fig. 9 shows the relationship between γ and

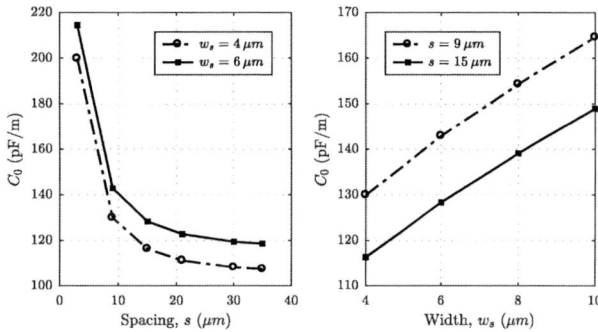

Fig. 5. Variation of C_0 with spacing (left) and width (right).

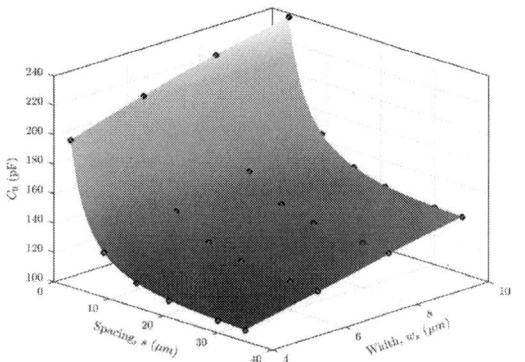

Fig. 6. Fit for C_0 (surface) and the data points used for it (markers).

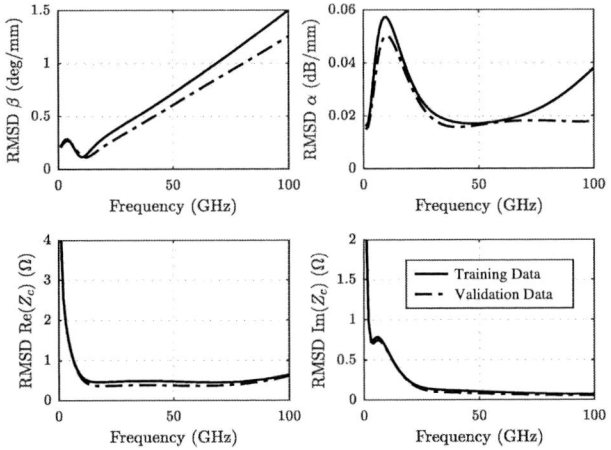

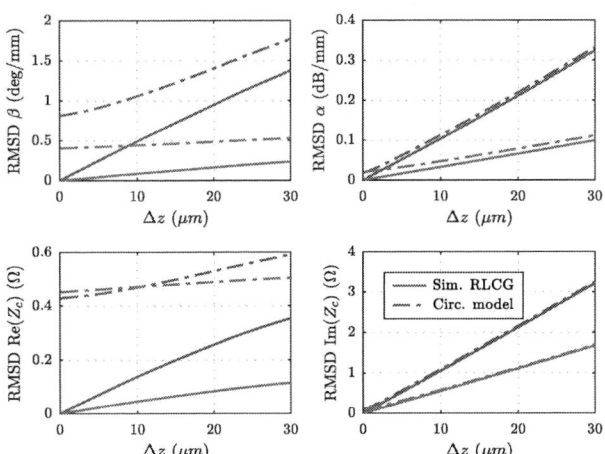

Fig. 7. RMSD for the propagation constant and characteristic impedance, using training data and validation data.

Fig. 9. RMSD for γ and Z_c against the section length calculated from the original RLCG values of the simulation results and from our circuit model, at 60 GHz (blue) and 30 GHz (red).

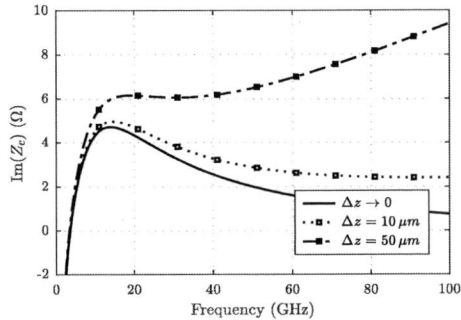

Fig. 8. Imaginary part of Z_c from a CPW simulation (noted as $\Delta z \to 0$) compared to its calculation from joining RLCG sections of 10 μm and 50 μm.

Z_c deviations and the circuit section length at two different frequency points, using directly the simulation RLCG and using the model components (training plus validation points). Two important conclusions can be extracted from the graph. First, the deviations at 60 GHz are higher than at 30 GHz, because the electrical length is longer at higher frequencies for the same physical length. Second, the deviations of the simulation-based parameters tend to zero as the length tends to zero and the model-based parameters tend to the model initial deviation. A shorter length implies higher accuracy but more sections needed for the same transmission line length, thus increasing the number of circuit components and simulation time. Nevertheless this time can still be in the seconds range even for thousands of passive components.

V. CONCLUSIONS

The proposed circuit model for CPW lines achieves low deviations from EM simulation data for mm-wave frequencies. It is scalable with the line width and spacing inside their specified range. Once the model is computed, it can be used in circuit simulations, instead of the more time and resource

consuming EM simulations. Additionally, the model does not require information about the layers and materials, because it is computed from simulated data.

REFERENCES

[1] T. Dinc, A. Chakrabarti, and H. Krishnaswamy, "A 60 GHz CMOS Full-Duplex Transceiver and Link with Polarization-Based Antenna and RF Cancellation," *IEEE Journal of Solid-State Circuits*, vol. 51, no. 5, pp. 1125–1140, 2016.

[2] T. S. Rappaport, J. N. Murdock, and F. Gutierrez, "State of the art in 60-GHz integrated circuits and systems for wireless communications," *Proceedings of the IEEE*, vol. 99, no. 8, pp. 1390–1436, 2011.

[3] W. Shu, H. Shichijo, and R. Henderson, "A Unified Equivalent-Circuit Model for Coplanar Waveguides with Silicon-Substrate Skin-Effect Modeling," *IEEE Transactions on Microwave Theory and Techniques*, vol. 64, no. 6, pp. 1727–1735, 2016.

[4] W. Heinrich, "Quasi-TEM description of MMIC coplanar lines including conductor-loss effects," *IEEE Transactions on Microwave Theory and Techniques*, vol. 41, no. 1, pp. 45–52, 1993.

[5] A. Bautista, A. L. Franc, and P. Ferrari, "Accurate Parametric Electrical Model for Slow-Wave CPW and Application to Circuits Design," *IEEE Transactions on Microwave Theory and Techniques*, vol. 63, no. 12, pp. 4225–4235, 2015.

[6] H. Wang, D. Zeng, D. Yang, L. Zhang, L. Zhang, Y. Wang, H. Qian, and Z. Yu, "A unified model for on-chip CPWs with various types of ground shields," *Digest of Papers - IEEE Radio Frequency Integrated Circuits Symposium*, pp. 2–5, 2011.

[7] B. K. Sen and R. L. Wheeler, "Skin effects models for transmission line structures using generic spice circuit simulators," in *IEEE 7th Topical Meeting on Electrical Performance of Electronic Packaging (Cat. No.98TH8370)*, 1998, pp. 128–131.

[8] D. M. Pozar, *Microwave Engineering, 4th Edition*, 2012, pp. 48–50.

[9] W. R. Eisenstadt and Y. Eo, "S-Parameter-Based IC Interconnect Transmission Line Characterization," *IEEE Transactions on Components, Hybrids, and Manufacturing Technology*, vol. 15, no. 4, pp. 483–490, 1992.

SMACD / PRIME 2021 | 19 – 22 July 2021, Online Event

A Sub-1µA Low-Power Low-Noise Amplifier with Tunable Gain and Bandwidth for EMG and EOG Biopotential Signals

Rafael Vieira[1], Ricardo Martins[1], Nuno Horta[1], Nuno Lourenço[1], Ricardo Póvoa[1,2]

[1]*Instituto de Telecomunicções*
Instituto Superior Técnico
Lisboa, Portugal
rafael.a.vieira@tecnico.ulisboa.pt

[2]*Escola Superior Náutica*
Infante D. Henrique
Paço de Arcos, Portugal
rpovoa@lx.it.pt

Abstract—**This paper presents the design of a low-power low-noise amplifier for biomedical and healthcare applications, focusing on electromyography and electrooculography. The signals operate in different broad bands, yet follow an impulse-shape transmission, being suitable to be applied and detected by the same receiver. The biopotential sensing amplifiers usually have a major impact in power and noise performance of an analog front end; hence, the development of a low-noise amplifier with low-power consumption is of great importance. In this paper, the state-of-the-art amplifiers for biomedical applications are overviewed, and the proposed solution is presented. The proposed design has tunable cutoff frequency (FC) and gain, being adjustable for each type of signal. The circuit is designed in UMC 130 nm CMOS technology, supplied by 1.2 V, and consumes less than 1 µA. Post-layout simulation results show that, at the high FC of 2 kHz, the gain is 34 dB, presenting an input-referred noise of 1.476 µVrms corresponding to a noise efficiency factor (NEF) of 1.27. Whereas at the low FC of 20.91 Hz, the gain is 52.35 dB, the input-referred noise is 0.202 µVrms, and the NEF is 1.70.**

Keywords— *Low-Power, Low-Noise, Biomedical, Healthcare, Biopotential Signals, Energy-Efficiency, Tunable, CMOS*

I. INTRODUCTION

Wearable and implantable devices to record biopotential signals have grown recently in the contexts of health monitoring, disease detection and brain stimulation. A device that enables analysis of multiple biopotential signals is more versatile, which reduces the cost, thus this paper focuses on the Electrooculography (EOG) and Electromyography (EMG) monitoring activities. These signals' amplitude and frequency characteristics are distinct from each other, as shown in Table I [1], yet they both follow an impulse-shape transmission and are sensed through electrodes attached to the skin. For everyday use and proper recording, the biomedical device must be portable, ergonomic and have a long operational lifetime in parallel with energy-efficiency. Ultra-low power consumption is a key feature in implanted devices to avoid excess heat flux tissue damage. Most front ends include three blocks: Low-Noise Amplifier (LNA), filtering, in most cases, low-pass type, and the programmable gain amplifier that outputs the signal to the Analog-to-Digital Converter (ADC), as shown in Fig. 1. This paper focus on the LNA, which is a significant contributor in terms of power consumption and noise contribution. Hence, it must provide voltage gain with maximum linearity while presenting low Input-Referred Noise (IRN) to ensure accurate detection. Since the amplifier is one of the most impactful blocks in terms of power performance, and given the importance of

This work is funded by FCT/MCTES through national funds and, when applicable co-funded EU funds under the project UIDB/EEA/50008/2020, including internal research projects HAICAS (X-0009-LX-20) and LAY(RF)² (X-0002-LX-20).

Table I. Characteristics of the focused biopotential signals.

	EOG	EMG
Amplitude (mV)	0.01-0.1	1-10
Frequency Range (Hz)	dc-10	20-2000

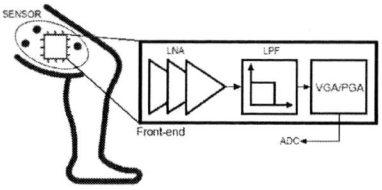

Fig. 1. Standard front end receiver block diagram.

energy efficiency, the NEF is a commonly used metric that quantifies the trade-off between power and IRN (1) [2]. The Power Efficiency Factor (PEF) introduced by [3], is a figure of merit to compare power efficiency (2), being notably helpful when different power supplies are used.

$$NEF = V_{rms,in}\sqrt{\frac{2 \cdot I_{tot}}{\pi \cdot U_T \cdot 4 \cdot k \cdot T \cdot BW}} \quad (1)$$

$$PEF = NEF^2 \cdot V_{DD} \quad (2)$$

Recent literature often refers fully differential amplifiers in closed-loop, the goal is to amplify neural signals while presenting great power-noise performance and removing the electrode potentials; showing higher linearity, better power supply rejection performance, and being less sensitive to process variations when compared to other topologies [4]. The most popular is the Capacitive-feedback structure, which enables the rejection of dc offset from the skin-electrode, without introducing noise [4-8]. These circuits present a band-pass response, where the mid-band gain is set by the ratio between the feedback capacitor and the input capacitor. Adding reset switches parallel to the feedback capacitor enables the amplifier to initialize the input and output nodes to common-mode voltage quickly, or reset when the output is saturated due to motion artifact or electrode fall-off [7-8]. Other techniques improved upon the capacitive-feedback, as the capacitive-coupled chopper in [9]. The chopper technique is commonly used to decrease noise impact as it can suppress the 1/f noise and offset [9-10]. However, it has some drawbacks, as they produce chopping ripple and have limited input impedance, requiring extra circuitry to mitigate it.

Regarding the LNA core, many authors choose current-reuse topologies to reduce power consumption [4-6, 11-12]. The current-reuse amplifier proposed in [4] is based on an inverter-based differential input stage for low noise and a class AB output stage for large output range and high *gm/I* efficiency. Since the class AB needs a driving circuit to stabilize the quiescent current at the output stage, a driving

© VDE VERLAG GMBH · Berlin · Offenbach

circuit is embedded in the inverter-based input stage to maximize the current efficiency. On the other hand, the authors of [5] and [11] based their designs on a folded cascode. The first has the advantage of allowing individual tuning of the channels noise level. The channel that receives the signal consumes more power to decrease the noise while the others consume less without regarding the noise at the cost of extra circuitry. The latter takes advantage of limited supply voltage by sharing the bias current and Mid-Rail Current Sink/Source between folded cascode input stages, being suitable for multi-channel biomedical acquisition systems. In [12], the authors expose a stacked current-reuse amplifier in a binary tree structure. This circuit exhibits several input differential pairs separated across stages, presenting upmost one stacked child input pair for each input. Since the output currents are independent, the output voltages are generated by summing the proper currents in the recombination output stage. Recently, the authors of [6] simplified this design, reducing the current consumption and circuit area. On the other hand, [13] shows a fully differential low-noise current mode Instrumentation Amplifier (IA). The IA is based on an inverter input and a transimpedance output and enables high Common-Mode Rejection Ratio (CMRR) as it is not dependent on capacitors or resistors matching.

This paper proposes a low power LNA with adaptive gain and frequency control for EOG and EMG recordings. The frequency tuning is implemented with a varactor-based circuit, where the control digit sets an FC of 2 kHz when EMG signals are desired and sets 21 Hz otherwise. The gain control is needed since the signals' amplitude range is a decade apart. Thus, tunable pseudo-resistors are used with a similar control procedure. Simulation results show that the circuit drains less than 1 µA from a 1.2 V supply, with an IRN of 1.476 V_{rms} and a NEF of 1.27 when in EMG mode, and 0.202 V_{rms} and a NEF of 1.70 when in EOG mode. The paper is organized as follows: Section II presents the circuit overview; in Section III the implementation of the circuit is described, and Section IV presents the post-layout simulation results. Finally in Section V, the conclusions are drawn.

II. PROPOSED LOW-POWER, LOW-NOISE AMPLIFIER

This paper proposes an LNA with embedded adaptive gain and FC tuning, shown in Fig. 2, extends the fully differential inverter-based current-mode IA presented in [13]. The Bandwidth (BW) is defined by C_H and C_L that implement CMOS external controlled varactors. The voltage control that sets the varactors' capacitance, also controls P_2's resistance, thereby tuning the gain and BW for each signal.

To discuss the implementation in detail, the LNA core is divided into six different blocks, current mirror (P_0), inverter (P_1, N_1), current source (N_2), common drain (N_3), common-mode feedback (N_4, N_5) and varactor-based circuit (C_H, C_L, P_H, P_L). The function of the current mirror is to bias with accurate current. Hence, P_0 should be well saturated. The inverter block is used as an input stage. This configuration, as an amplifying stage is biased in the sub-threshold region to obtain a high gm/I. The inverter, as input stage, reduces the IRN by doubling the transconductance under a given bias current. Furthermore, if the transconductance of both transistors is similar, the thermal noise is reduced, [4, 13]. The current source block sets the input stage bias current and stabilizes the common-mode voltage at the output of this stage. The common drain block represents the transimpedance amplifier presented at the circuit's output stage. This stage is

biased by the common-mode feedback block, which displays pseudo-resistors (N_5) and the transistors that bias the output stage (N_4). The pseudo-resistors sense and establish the output common-mode voltage along with the bias transistors. To enable adaptive tuning low FC is necessary and consequently high capacitance values. Thus, the varactor block contains CMOS capacitors, since they can achieve relatively high capacitance with low penalty in terms of chip area and does not introduce significant noise. The MOSFET varactors are implemented in a D=S=B structure, i.e., with transistor's drain, bulk, and source linked together, with an external control voltage. The capacitance depends on the bulk-gate voltage (V_{BG}) and equals C*S, where C is the capacitance per unit of area, and S is the transistor's channel area [14]. The varactors are designed to operate in the accumulation or inversion region. Only in these regions, the maximum capacitance per unit of area can be achieved [14].

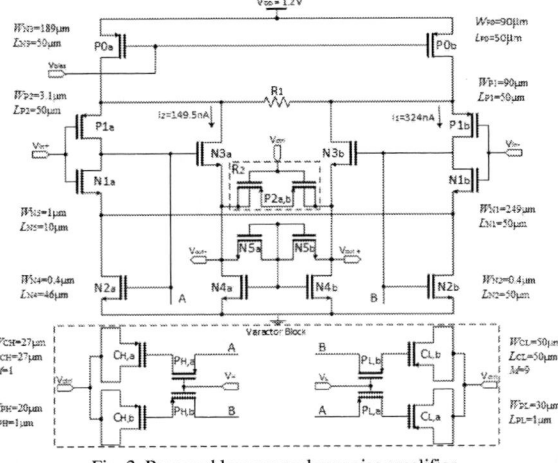

Fig. 2. Proposed low-power low-noise amplifier.

In order to simplify the small-signal analysis, and since the circuit is symmetrical, Bartlett's bisection theorem is used. The circuit is detached along the symmetry axis, the shared nodes are replaced with ground and considering P_2 as a resistor, both R_1 and R_2 are split in half, thereby having two identical networks. The small-signal circuit is attained from considering one network, as shown in Fig. 3. V_i and V_o are the differential input and output voltages, respectively, V_A is the small-signal equivalent circuit of P_0, which is considered as an ideal current source to simplify, while V_B is the first stage output voltage. The differential gain is given by (3), where the factors β, α, and α_1 are given by (4-6).

$$\frac{v_O}{v_i} = \frac{R_2\big(g_{m_p}(R_1 - 2g_{m_3}r_{o_3}r_{o_n})r_{o_p} - g_{m_n}r_{o_n}\beta\big)}{2(R_2 + 2r_{o_3} + g_{m_3}R_2r_{o_3})(r_{o_n} + r_{o_p}) + \alpha + \alpha_1} \tag{3}$$

$$\beta = R_1 + 2g_{m_3}r_{o_3}r_{o_p} + g_{m_3}R_1\left(r_{o_3} + g_{m_p}r_{o_3}r_{o_p}\right) \tag{4}$$

$$\alpha = R_1R_2\big(1 + g_{m_3}r_{o_3}\big)\left(1 + g_{m_p}r_{o_p}\right) \tag{5}$$

$$\alpha_1 = 2R_1(r_{o_n} + r_{o_p} + r_{o_3}(1 + g_{m_3}r_{o_n}(1 + g_{m_p}r_{o_p}))) \tag{6}$$

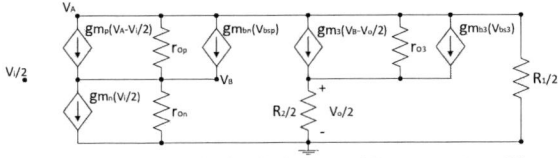

Fig. 3. Small-signal equivalent half circuit of the proposed amplifier.

III. DESIGN AND IMPLEMENTATION

For proof-of-concept, the circuit is implemented in UMC 130 nm CMOS technology. Special attention is needed to comply with low-power, aiming at portable solutions. Being biased at 1.2 V with a current budget below 1 μA and the common-mode output voltage at 500 mV. The dc biasing strategy is adapted to bias each branch with 475 nA, matching a current reference of 50 nA, which is achievable with the self-biased Widlar current source.

The biasing strategy influences the noise and linearity, hence having larger input transistors at the first stage reduces flicker noise, and by increasing the input voltage at the second stage the non-linear behavior inherent to the sub-threshold is reduced. Both R_1 and R_2 resistors are mainly linked to the noise and linearity, respectively, as well as gain and BW. Thus, by increasing R_1, the noise also increases. On the other hand, by increasing R_2, the linearity increases. As the gain is proportional to the R_2/R_1 ratio, in order to set a given gain, a trade-off between linearity and noise must be considered. As the EMG and EOG signals present different amplitudes, having the tunable pseudo-resistors P_2, enables tuning the gain for each signal. The pseudo-resistors operate in the sub-threshold region and may be tuned for high resistances in the order of giga-ohms (cutoff region) and for low resistances in the order of kilo-ohms (triode region), being controlled by the gate voltage [15]. Considering the threshold voltage dependency on the substrate/bulk potential, this configuration consists of a PMOS with bulk-drain connection, resulting in a finite equivalent resistance value [15]. R_1 is set to 50 kΩ, and the pseudo-resistors are sized to introduce 34 dB gain for a 0 V voltage control, *i.e.*, when the EMG signal is desired, and 52 dB gain for 1.2 V voltage control that targets EOG signals. The 34 dB gain is set so that the amplifier does not saturate when the EMG signal is applied. The tuning gain is done by applying external control voltage that tunes the FC to the gate of the pseudo-resistors.

Since a tunable BW is intended, the varactors are sized to tune the FC with a given control voltage. As the EOG presents a frequency range from dc to 10 Hz and the EMG signal from 20 Hz to 2 kHz, the LNA should enable tuning the FC at least at 20 Hz and 2 kHz. Yet, there is no varactor sizing that enables this tuning range. Therefore, two pairs of varactors are implemented, the first one cuts at 2 kHz (C_H) and the second at 21 Hz (C_L). Ideally, one pair would turn off, having no effect on the circuit, when the other is turned on, thus presenting its FC. At first sight, this could be implemented with a basic inverter connected between the control voltage and one pair of varactors. However, even if 0 V is applied, the varactor presents a given capacitance. Therefore, the equivalent capacitance *per* branch would be the sum of the varactors' capacitances, *i.e.*, would act as one capacitor. The implemented solution, shown in Fig. 2 varactor block, depicts a PMOS (P_H, P_L) after each varactor that turns the pairs of varactors on/off. Thus, the only influence in the BW would come from the capacitance of the branch that is on. However, the impedance is finite, so this implementation introduces additional poles and zeros to the system, which must be sized properly to have influence at frequencies outside of the BW. To enable $P_{H,L}$ turn on/off function, control voltages are added (V_H and V_L), thus when applied 0 V to P_H gate, turns on, while to the gate of P_L is applied 1.2 V, turning off. By implementing two more control voltages instead of only one and an inverter, the occupied area is reduced and, if applied to other application, introduces higher tuning frequency range.

IV. LAYOUT AND POST-LAYOUT SIMULATION RESULTS

The proposed LNA core amplifier layout comprises an area of 0.09811 mm², whereas the varactors cover 0.05564 mm² and the current source 0.00644 mm². All presented simulations are done post-layout. The ac response, shown in Fig. 5 a) and b) for high and low FC, respectively. The first one depicts a low-frequency gain of 34 dB and a 2 kHz BW. Whereas the other shows a low-frequency gain of 52.35 dB and a 21 Hz BW. In both cases the circuit is stable presenting a phase margin of 78.6° and 89° for the first and second case, respectively. The noise response for each mode is shown in Fig. 6 a) for the high FC and b) for the low FC, depicting an IRN of 1.476 μV$_{rms}$ and 0.202 μV$_{rms}$ integrated from 0.1 Hz to their respective FC. Corresponding to a NEF of 1.27 and a PEF of 1.94 for the high FC, on the other hand a NEF of 1.70 and PEF of 3.47 are obtained. A full transient analysis is carried out in open-loop. The discrete Fourier transform related to the LNA transient response is displayed in Fig. 7 a) and b) for the high and low FC cases, respectively. The first one, presents a Total Harmonic Distortion (THD) of 0.65 %, while the latter exposes a THD of 0.18 %. In terms of CMRR and PSRR, a Monte Carlo σ=3 simulation with 500 runs considering process and mismatch for the worst case in its BW, presents a mean value of 85.6 dB and 121.6 dB, respectively, for the high FC; 69.1 dB and 132.2 dB for the low FC, as shown in Figs. 8 a) and b). To test the circuit's robustness Monte Carlo σ=3 simulation with 500 runs is carried out considering process and mismatch variations and are presented in Fig. 10. In the high FC case Fig. 10a, the obtained gain results present a mean value of 34 dB and a standard deviation of 0.19 dB, regarding its BW it depicts a mean value of 2 kHz and a standard deviation of 53.4 Hz. Whereas, the low FC case, Fig. 10b, the gain and BW present a mean value of 52.2 dB and 21.7 Hz, matching a standard deviation of 2.25 dB and 5 Hz, respectively. The average power consumption and IRN were also tested with Monte Carlo simulation using the previous conditions, presenting for the high FC 1.194 μW and 1.477 μV$_{rms}$ mean value, with standard deviation of 63.92 nW and 20.57 nV$_{rms}$. Concerning de low FC, presents 1.197 μW and 0.208 μV$_{rms}$ mean value matching a standard deviation of 66.28 nW and 8.08 nV$_{rms}$. This is not presented since the variation is not significant.

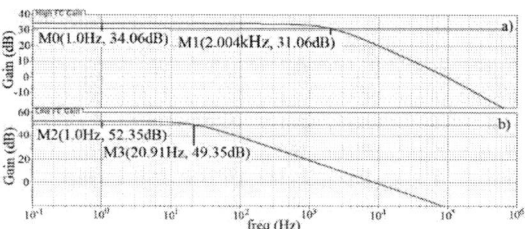

Fig. 5. Post-layout ac response: a) high FC; b) low FC.

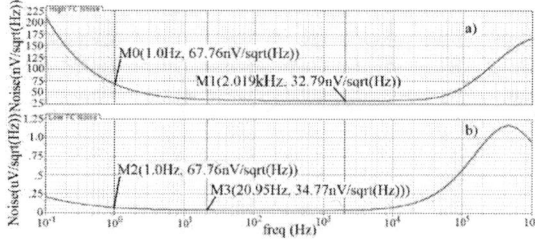

Fig. 6. Post-layout IRN: a) high FC; b) low FC.

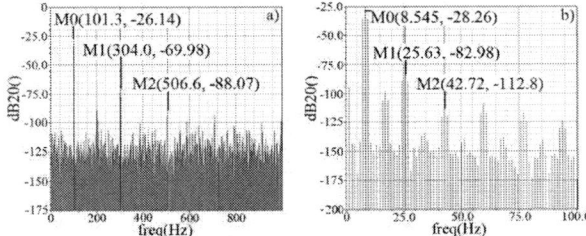

Fig. 7. Post-layout open-loop DFT: a) high FC; b) low FC.

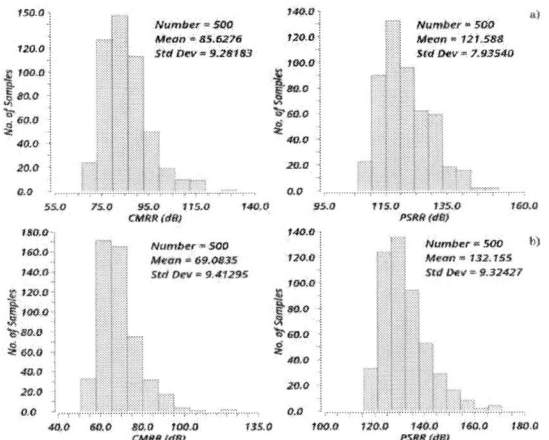

Fig. 8. Post-layout CMRR and PSRR: a) High FC; b) Low FC.

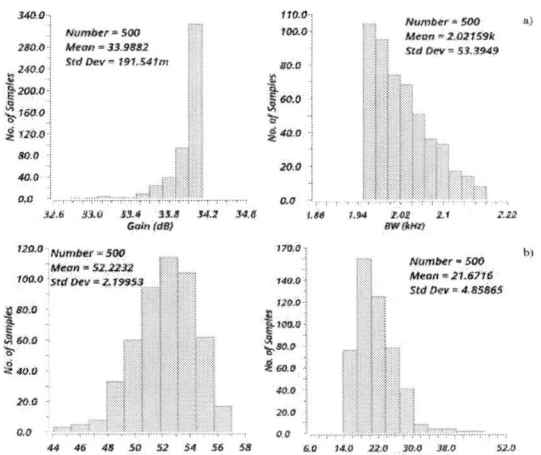

Fig. 9. Post-layout Monte Carlo $\sigma=3$ simulation: a) High FC; b) Low FC.

V. CONCLUSIONS

This paper proposed a low-power LNA to be used in biomedical and healthcare applications, prioritizing EMG and EOG signals. The LNA presents adaptive tuning gain and FC based on tunable pseudo-resistors and a varactor system, respectively. The gain is adjusted for each signal by the ratio between P_2's impedance and R_1. The varactor system, consists of two pairs to set each FC. Transistors were implemented after each varactor to turn them on/off. The circuit is supplied by a 1.2 V source draining less than 1 μA. From post-layout simulations, gains of 34 dB for the high FC at 2 kHz and 52.35 dB for the low FC at 21 Hz are obtained. Finally, an IRN of 1.476 μV$_{rms}$ corresponding to a NEF of 1.27 for the high FC case is obtained, while 0.202 μV$_{rms}$ matching a NEF of 1.70 is achieved otherwise. The results from the proposal is presented in Table II, along with other state-of-the-art LNAs.

Table II. Comparison of state-of-the-art LNAs with this work.

Work	[4]	[5]	[6]	[9]	[11]*	[13]	[16]**	This Work**	
Year	2018	2015	2018	2011	2014	2015	2020	2020	
Tech. (nm)	350	180	180	65	180	180	180	130	
Area (mm²)	0.18	1.6#	0.072	0.2	-	0.096	0.046	0.16	
Gain (dB)	39.8	33	35	40	30	40	39.22	34	52.35
BW [Hz]	0.2-200	0.7-182	9.3k	0.5-500	0.2-120	11k	3-5k	2004	20.91
NEF	2.3	1.7	1.94	3.3	1.2	1.88	2.4	1.27	1.70
PEF	10.58	4.05	6.77	10.89	1.44	6.36	8.06	1.94	3.47
Power (μW)	0.32	0.52/1.56	4.5	1.8	2.5	78.84	2.98	1.19	
Supply (V)	2	1.4	1.8	1	1	1.8	1.4	1.2	
CMRR (dB)	>65	>70	76	134	>60	100	>70	85.6	69.1
PSRR (dB)	>70	>70	80	120	>80	-	>77	121.6	132.2
THD (%)	1 @15 mV	1.5 @4.6 mV	0.07 @1 mV	-	0.3 @2 mV	-	1 @13.2 mV	0.65 @1 mV	0.18 @0.1 mV

*Simulation results ** Post-layout results #Includes extra circuitry

VI. REFERENCES

[1] N. Thakor, Biopotentials and Electrophysiology Measurement, CRC Press, 1999.

[2] M. S. J. Steyaert and W. M. C. Sansen, "A micropower low-noise monolithic instrumentation amplifier for medical purposes," IEEE J. Solid-State Circuits, vol. 22, no. 6, pp. 1163–1168, Dec 1987.

[3] R. Muller, S. Gambini, and J. M. Rabaey, "A 0.013 mm² 5 μw DC-coupled neural signal acquisition IC with 0.5 V supply," IEEE J. Solid-State Circuits, vol. 47, no. 1, pp. 232–243, Jan. 2012.

[4] J. Zhang, et al., "A low-noise, low-power amplifier with current-reused ota for ECG recordings," IEEE Trans. Biomed. Circuits Syst., vol. 12, no. 3, pp. 700-708, Jun. 2018.

[5] S. Song, et al., "A low-voltage chopper-stabilized amplifier for fetal ECG monitoring with 1.41 power efficiency factor," IEEE Trans. Biomed. Circuits Syst., vol. 9, no. 2, pp. 237-247, Apr. 2015.

[6] M. Rezaei, et al., "A low-power current-reuse analog front-end for high-density neural recording implants," IEEE Trans. Biomed. Circuits Syst., vol. 12, no. 2, pp. 271-280, Apr. 2018.

[7] X. Zhang, et al., "A 2.89 μw dry-electrode enabled clockless wireless ecg SoC for wearable applications," IEEE J. Solid-State Circuits, vol. 51, no. 10, pp. 2287-2298, Oct. 2016.

[8] R. R. Harrison, "The design of integrated circuits to observe brain activity," Proc. of the IEEE, vol. 96, no. 7, pp. 1203-1216, Jul. 2008.

[9] Q. Fan, et al., "A 1.8 μw 60 nv/√ hz capacitively-coupled chopper instrumentation amplifier in 65 nm CMOS for wireless sensor nodes," IEEE J. Solid-State Circuits, vol. 46, no. 7, pp. 1534 – 1543, Jul. 2011.

[10] F. M. Yaul and A. P. Chandrakasan, "A noise-efficient 36 nv/√ hz chopper amplifier using an inverter-based 0.2-v supply input stage," IEEE J. Solid-State Circuits, vol. 52, no. 11, pp. 3032-3042, Nov. 2017.

[11] S. Song, et al., "A multiple-channel frontend system with current reuse for fetal monitoring applications," in 2014 IEEE Int. Symposium on Circuits and Systems (ISCAS), Jun. 2014.

[12] H. Sepehrian and B. Gosselin, "A low-power current-reuse dual-band analog front-end for multi-channel neural signal recording," in 2014 36th Annu. Int. Conf. IEEE Eng. Med. Biol. Soc., Aug. 2014.

[13] D. Das, et al., "A novel low-noise fully differential CMOS instrumentation amplifier with 1.88 noise efficiency factor for biomedical and sensor applications," Elsevier's Microelectronics Journal, vol. 53, pp. 35-44, 2016.

[14] S. Li and T. Zhang, "Simulation and realization of mos varactors," Procedia Engineering, vol. 29, Dec. 2012.A.

[15] Tajalli, et al., "Implementing ultra-high-value floating tunable cmos resistors," Electronics Letters, vol. 44, no. 5, 2008.

[16] Naderi, Kebria, Erwin H. T. Shad, M. Molinas and A. Heidari. "A Power Efficient Low-noise and High Swing CMOS Amplifier for Neural Recording Applications." 2020 42nd Annu. Int. Conf. IEEE Eng. Med. Biol. Soc. (EMBC) (2020): 4298-4301.

© VDE VERLAG GMBH · Berlin · Offenbach

Transistor Downscaling toward Ultra-Low-Power, sub-100 μm^2 and sub-Hz Oscillators

Gian Luca Barbruni
Medtronic Chair in Neuroengineering
and Integrated Circuits Laboratory
École polytechnique fédérale de Lausanne
Switzerland
gianluca.barbruni@epfl.ch

Chiara Bielli, Danilo Demarchi
DET
Politecnico di Torino
Turin, Italy
danilo.demarchi@polito.it

Sandro Carrara
Integrated Circuits Laboratory
École polytechnique fédérale de Lausanne
Neuchâtel, Switzerland
sandro.carrara@epfl.ch

Abstract—This paper analyses and discusses the feasibility of implementing sub-Hz range oscillators in ultra-low-power and ultra-miniaturised implants. The final aim is the Body Dust application, in which multiple freestanding smart cubes are wirelessly powered and freely to move in the human blood for bio-sensing purpose. That system requires an overall size smaller than 100 μm^2 and ultra-low power consumption. Two different CMOS technologies have been compared, analyzing the effect of transistor down-scaling in sub-Hz range oscillators. Four CMOS-based oscillators have been tested in both 0.18 μm and 28 nm technologies. The best result is the Schmitt Trigger-based timer implemented in FD-SOI 28 nm and combined with a ten stages frequency divider that oscillates at 122.5 mHz with a reduced area of 80 μm^2. The study demonstrate that the effect of transistor downscaling opens the possibility to reach a sub-Hz oscillator with both ultra-low-power and ultra-low-area consumption, so suitable for Body Dust application.

Index Terms—Body Dust; CMOS design; Transistor Down-Scaling; sub-Hz oscillators; Ultra-Low Power CMOS systems; sub-100 μm^2 oscillators.

I. INTRODUCTION

Low frequency oscillators are very useful in different fields of application, such as biomedical [1], industrial [2] and environmental [3]. Our target application is the "Body Dust" [4], in which sub-Hz signals as required for the electrochemical biosensors, especially for drugs detection [5]. In this regard, each dust is thought as a drinkable and wirelessly powered CMOS cube for providing an in-body molecular detection combined with a wireless data transmission. Among the possible Wireless Power Transfer (WPT) methods toward ultra-miniaturised implants [6], Acoustic Power Transfer (APT) has been selected as the most appropriate for sub-100μm-sized implants operating at large penetration depth (i.e. higher than 2 cm). More in detail, the CMOS mote is ideated into different system-layers 3D integrated [7]: the power management unit to provide a stable supply starting from WPT via APT, the communication circuit for up-link data transmission [8], the front-end layer to manage the biosensor [9] and the multiplexer. The latter is required to switch among the different biosensors collecting information about different molecular concentrations. The multiplexer definitely needs a local clock, able to perform the switch in a sub-Hz range

of frequency. In this regard, electrochemical measurements may be performed in chronoamperometry [9], while for other molecules, in particular for exogenous compounds, cyclic voltammetry is definitely required to keep under control issues related to the system selectivity [10]. In particular, voltage scan in cyclic voltammetry is usually in the range 10-50 mV/s, while scanning on voltage ranges close to 1 V [10]. In other words, a mHz oscillator is needed for Body Dust application. System constrains regards the full size of the system which needs to be comparable with human cells (i.e. less than 100 μm^2) and the ultra-low-power consumption. These limitations are very challenging, especially combined together. Over the years, a wide range of solutions have been ideated to individually overcome these boundaries. On the one hand, power consumption could be minimized through gate-leakage currents [11]. On the other hand, thyristor-based oscillators [12], [13] which generate larger delays per stage, Dynamic Leakage Suppression (DLS) logic implementation with super-cut off operations [14] and delay elements [15] are among the most used methods to reduce the frequency oscillation. Size constrain may be satisfied using MEMS-based oscillators [16] or by exploiting the transistor down-scaling. The aim of this paper is to compare different existing oscillator-architectures, to verify the feasibility of implementing a sub-Hz oscillator with the Body Dust constrains. Four oscillator has been scaled from UMC 0.18 μm to FD-SOI 28 nm, testing its transient response to the optimal β (proportional to the ratio between width and length, W/L, of each transistor), capacitance and voltage supply. Results and limits of the implementations are compared using the two different technologies: UMC 0.18 μm and FD-SOI 28 nm. The first one is very cheap, commonly used in biomedical field and with many libraries already available. The second one is considered for investigating the advantages reachable by exploiting the lower nodes nowadays available.

II. UMC 0.18 μM CMOS TECHNOLOGY

In this section we are going to introduce the four architectures selected form literature and discuss their theoretical results derived from scaling the oscillators to UMC 0.18 μm while bringing only some modifications in the architectures.

© VDE VERLAG GMBH · Berlin · Offenbach

SMACD / PRIME 2021 | 19 – 22 July 2021, Online Event

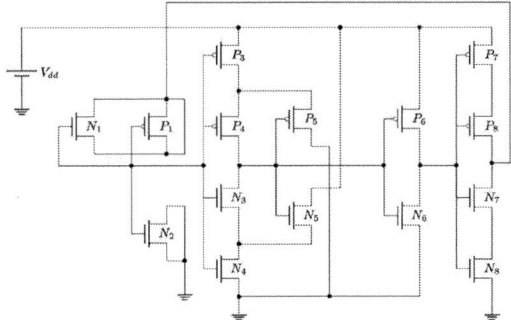

Fig. 1: Schematic of One-Hot Timer (re-drawn from [17])

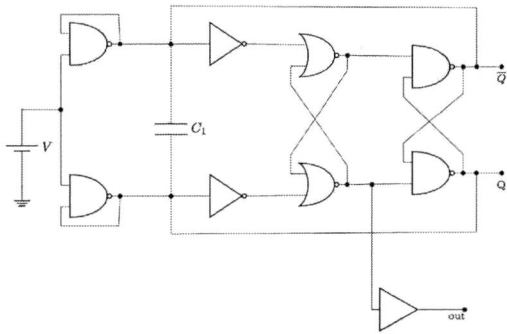

Fig. 2: Schematic of Relaxion Oscillator with DLS logic style (re-drawn from [14])

The first considered oscillator was originally implemented in 130 nm technology and was conceived for the periodic monitoring of intra-ocular pressure and, therefore, allowing measurements over time scale of hours [17]. The proposed architecture (Fig. 1) realizes a standard one-shot timer by using the gate leakage to reduce sensitivity to temperature and to generate a low frequency oscillation, as in [18], [19]. The simulation result in UMC 0.18 μm shows a total area of almost 13x8 μm^2 and a power consumption of 38.95 pW for an output frequency of 26 kHz, not yet suitable for sub-Hz applications in Body Dust. This result is justified by the negligible effect of gate-leakage currents which are instead more significant with the down-scaling in CMOS nodes [20].

The second tested architecture (Fig. 2) is the relaxion oscillator proposed in [14]. Its functioning is based on DLS, so the unpredictability of gate leakage is erased. This oscillator was already originally implemented in UMC 0.18 μm, using minimum size transistors. We verified that oscillator obtaining steady performances using a wide range of supply voltages (Vdd from 0.4 V to 1.8 V), confirming that the frequency, as well as the drain currents, are slightly dependent on Vdd, in opposite to the one-hot timer above considered. The DLS implementation allows to drain a very small on-current, which produces a lower frequency output while increasing the total number of transistors. In fact, considering a capacitance of 500 fF, we obtained an output frequency of 4 Hz with an overall area of 55x30 μm^2 and a ultra-low power consumption of 3.32 pW (for a Vdd of 0.4 V). Nevertheless, both the frequency and the area are not suitable for Body Dust application.

Then, the Voltage Control Oscillator (VCO) of Fig. 3 was considered [21]. We implemented this architecture and tested its reliability toward transient response to W/L ratio and Vdd changes. The obtained results show a frequency of less than 2 Hz, with a supply voltage of 1.8 V and an area improvement of 55x15 μm^2. The main limits of this VCO are the requirement of a big capacitor (C=500 fF) and the needs for high-length transistors to reduce the oscillation frequency, which determines an area that is definitely out of range with respect to the application. Moreover, VCO consumes 1.81 μW that is again too high for the Body Dust constrains.

Finally, with the aim of miniaturisation, we implemented a

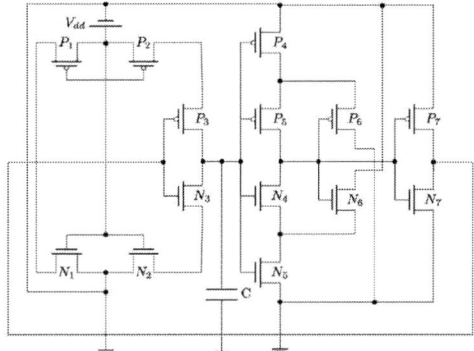

Fig. 3: Schematic of Voltage Controlled Oscillator (re-drawn from [21])

Schmitt Trigger-based Timer, originally designed using a 150 nm technology and an integrable capacitor, with total area of only 223 μm^2 and output frequency of 21.7 mHz [22]. We implemented the Timer in UMC 0.18 μm considering an on-chip capacitor instead, which causes the total area increment, as expected (Fig. 4).

The best result was obtained using a 1 pF capacitor and a voltage supply of 200 mV, which corresponds to an average frequency of less than 10 Hz, with 43x34 μm^2 of area occupied and 2.58 pW of power consumption. The results we reached with the 0.18 μm technology are remarkably low in frequency, but still unsatisfying for the target application. The frequency values obtained from the DLS relaxion oscillator, the VCO and Schmitt Trigger based Timer are very promising and can be further improved using a 10-stages frequency-divider, scarifying the total size instead. However, area con-strains represent a primary problem in UMC 0.18 μm. The simulations did not allow to find a trade-off among frequency, power and area. Therefore, we demonstrated that the present state-of-the-art in CMOS design of oscillators scaled in UMC 0.18 μm is unsuitable for the Body Dust application.

III. FDSOI-28 NM CMOS TECHNOLOGY

All the presented architectures were down-scaled using FD-SOI 28 nm CMOS technology. The same capacitance

© VDE VERLAG GMBH · Berlin · Offenbach

SMACD / PRIME 2021 | 19 – 22 July 2021, Online Event

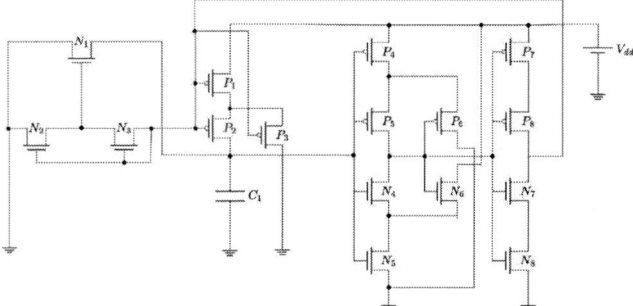

Fig. 4: Schematic of Schmitt Trigger based Oscillator (redrawn from [22])

values were used, while varying transistors sizes and voltage supplies. In this way we found the best trade-off among our target limitations for each configuration highlighting the advantages with respect to UMC 0.18 μm. Starting from the One-Hot Timer, the results shown a little improvement in term of frequency performance, with an output frequency equal to 522.48 Hz and power consumption of 0.42 pW. An outstanding area improvement of 2x3 μm^2 is observed, considering minimum-size transistors. The drawback is linked to the leakage currents, which depend on several different factors: transistor type (pMOS transistors conduct gate-currents almost 10 times smaller than nMOS transistors), transistor sizes (leakage currents are directly proportional to the gate area, W*L), applied bias, and relative position of the transistors ("structure dependence", [23], [24]). In the Relaxion Oscillator the parasitic currents compromise the right functioning of the DLS. As a result, that oscillator did not work anymore. Instead, the VCO architecture reached a frequency of 18.36 Hz, with a supply voltage of 280 mV, and an improved area of 23x10 μm^2, with total power consumption of only 2.51 pW. With this implementation, area and power fit the Body Dust application, while the frequency is not still in the right range. Finally, the Schmitt Trigger used as a Timer results in 83 mHz frequency oscillation by considering the same capacitance value of [22] (100 pF), but with a corresponding area of 13.59 mm^2. To further miniaturise the circuit, a smaller capacitor is used (i.e. down to 50 fF) obtaining a frequency of 125.49 Hz and consuming 9.45 pW in 8x7 μm^2, so suitable for Body Dust.

IV. RESULTS AND DISCUSSION

The final comparison between the two technologies is shown in Table I. The result shows the best trade-off of each oscillator in term of frequency, power and area consumption. Table I also highlights limits and advantages of each technology, with and without a ten stages frequency divider. From the analysis conducted so far, we confirmed that gate-leakage currents are almost negligible with a larger node, like UMC 0.18 μm. On the other hand, in the FD-SOI 28 nm, gate-leakage currents show more significant effects, but results are still dependent on many factors. The relaxion oscillator implemented with DLS logic style shows the expected behavior with

both technologies. In fact, in UMC 0.18 μm the super-cut-off operation performed by the DLS reduces power consumption to 3.32 pW, as well as the frequency (average value of 4 Hz). While with FD-SOI 28 nm the functioning of the oscillator is compromised by parasitic currents that are more significant in a lower technology. The VCO and the Schmitt Trigger-based Timer have similar performances in UMC 0.18 μm producing few Hz output and occupying out-range areas for our target. However, in FD-SOI 28 nm they present very small oscillating frequencies and quite contained sizes. As a rule of thumb, bigger capacitors lead to lower frequencies: implementing [22] in FD-SOI 28 nm we reached 83 mHz thanks to 100 pF capacitor and voltage supply of 0.4 V. Area is irremediably compromised by the on-chip capacitor, being larger than 13 mm^2. The VCO also shows the effect of transistors downscaling in area consumption: from 55x15 μm^2 to 23x10 μm^2 for UMC 0.18 μm and FD-SOI 28 nm respectively. Instead, using the Schmitt Trigger based Timer in FD-SOI 28 nm with 50 fF of capacitance we obtained an output frequency of 125.49 Hz with an area of only 8x7 μm^2 and 9.45 pW of power consumption. Nevertheless, the frequency remains in the order of few Hz in both technologies, being unsuitable for multiplexing in Body Dust. However, oscillation frequency is improved by using a 10 stages frequency divider. In UMC 0.18 μm the divider allows sub-Hz outputs, but has a total size of 11x70 μm^2. Using FD-SOI 28 nm the same divider size is around 1x3 μm^2. Therefore, the VCO reaches 17.93 mHz with almost the same area of 23x10 μm^2 and the Timer produces 122.5 mHz oscillation in 8x8 μm^2 only. Instead, temperature tests highlight swing of around 14000 ppm/°C in UMC 0.18 μm and around 25000 ppm/°C in FD-SOI 28 nm for the best cases, considering a temperature range from 0°C to 100°C. However, since the final application of our study is a dust mote that needs to be ingested and spread inside human body, the frequency dependence from temperature is not critical: the dust will work at almost constant temperature of 37°C. Moreover, none of the configurations tested is temperature-compensated. Improvement in this regard can be obtained considering so temperature compensated architectures, as in [25], [26]. Finally we recognized as the best configuration the Schmitt Trigger based Timer implemented in FD-SOI 28nm and combined with a ten stages frequency divider: both power and area consumption are suitable for the Body Dust multiplexer.

V. CONCLUSIONS

We studied the feasibility to develop an oscillator capable to generate a sub-Hz frequency output, while consuming low power and with an overall size smaller than 100 μm^2. Four oscillator-architectures have been implemented and compared in UMC 0.18 μm and FD-SOI 28 nm. Strengths and weaknesses of each circuit in both technologies have been discussed. The analysis conducted in UMC 0.18 μm highlights that the technology is not sufficient to address the typical needs for targeting Body Dust applications. Instead, our results demonstrate that the FD-SOI technology is capable to provide

© VDE VERLAG GMBH · Berlin · Offenbach

TABLE I: Oscillator's performances comparison in UMC 0.18 μm and FD-SOI 28 nm. *With 10 stages frequency divider.

Parameters	One-Hot Timer Osc.		DLS Relaxion Osc.		Voltage Control Oscillator		Schmitt Trigger Timer	
Technology [nm]	180	28	180	28	180	28	180	28
Voltage Supply [V]	0.45	0.45	0.4	N/A	1.8	0.28	0.4	0.4
Frequency [Hz]	$26.54 \cdot 10^3$	522.48	4	N/A	1.82	18.365	6.19	125.49
Power [pW]	38.95	0.42	3.32	N/A	$1.81 \cdot 10^3$	2.51	2.58	9.45
Area [μm^2]	13x8	2x3	55x30	N/A	55x15	23x10	43x34	8x7
Frequency* [mHz]	$25.92 \cdot 10^3$	510	3.91	N/A	1.77	17.93	6.04	122.5
Area* [μm^2]	13x67	3x3	55x44	N/A	55x29	23x10.2	43x52	8x8
f Variation with T [ppm/°C]	1600	31000	42000	N/A	0.0035	25000	14200	27800

a mHz oscillator suitable for both multiplexing and cyclic voltammetry on board, with an architecture that is contemporary contained in a small area and inside the specifications of ultra-low-power applications. The Schmitt Trigger based Timer implemented in FD-SOI 28 nm produces an output frequency of 122.5 mHz, consumes 9.45 pW and requires 8x8 μm^2 only. Future work include process and voltage variations analysis, corner and Monte-Carlo simulations and chip prototyping. On the other hand, circuit based on thyristors [22], [23] can be considered in the future. Alternatively, it is also possible to completely change the approach using an external transmitter for sending the signal to switch from one channel to another exploiting a wireless down-link communication contemporary with the wireless power transfer. At the very end, this paper demonstrated that a sub-Hz oscillator requiring less than 10 pW in power consumption is actually feasible in less than 10x10 μm^2 and, then, integrated in Body Dust motes.

REFERENCES

[1] X. Zhentao, W. Wang, L.W.M. Ning, Y. Liu and Q. Yu. *A supply voltage and temperature variation-tolerant relaxation oscillator for biomedical systems based on dynamic threshold and switched resistors*. IEEE Transactions on Very Large Scale Integration (VLSI) Systems 23.4 (2014), pp. 786–790.

[2] T. Srivyshnavi and A. Srinivasulu. *A current mode Schmitt trigger based on Current Differencing Transconductance Amplifier*. 3rd International Conference on Signal Processing, Communication and Networking (ICSCN). IEEE. (2015), pp. 1–4.

[3] M. Fojtik, et al. *A millimeter-scale energy-autonomous sensor system with stacked battery and solar cells*. IEEE Journal of Solid-State Circuits 48.3 (2013), pp. 801–813.

[4] S. Carrara. *Body Dust: Well Beyond Wearable and Implantable Sensors*. Sensors. IEEE Sensors Journal (2020).

[5] S. Carrara, A. Cavallini, V. Erokhin and G. De Micheli. *Multi-panel drugs detection in human serum for personalized therapy*. Biosensors and Bioelectronics 26.9 (2011), pp. 3914–3919.

[6] G.L. Barbruni, P. M. Ros, D. Demarchi, S. Carrara and D. Ghezzi. *Miniaturised Wireless Power Transfer Systems for Neurostimulation: A Review*. IEEE Transactions on Biomedical Circuits and Systems 14.6 (2020), pp. 1160-1178. doi=10.1109/TBCAS.2020.3038599

[7] S. Carrara and G. Pantelis. *Body Dust: Miniaturized Highly-integrated Low Power Sensing for Remotely Powered Drinkable CMOS Bioelectronics*. arXiv preprint arXiv:1805.05840 (2018).

[8] G.L. Barbruni, P. Motto Ros, S. Aiassa, D. Demarchi and S. Carrara. *Body Dust: Ultra-Low Power OOK Modulation Circuit for Wireless Data Transmission in Drinkable sub-100μm-sized Biochips*. arXiv preprint arXiv:1912.02670 (2019).

[9] J. Snoeijs, G. Pantelis and S. Carrara. *CMOS body dust—Towards drinkable diagnostics*. IEEE Biomedical Circuits and Systems Conference (BioCAS). IEEE. (2017), pp. 1–4.

[10] S. Carrara. *Bio/CMOS Interface and Co-Design*. Springer, 2013.

[11] H. Wang and P.P. Mercier. *A reference-free capacitive-discharging oscillator architecture consuming 44.4 pW/75.6 nW at 2.8 Hz/6.4 kHz*. IEEE Journal of Solid-State Circuits 51.6 (2016), pp. 1423–1435.

[12] A.D. Funke, P. Mayr, T. Maeke, J.S. McCaskill, A. Sharma, L. Straczek, J. Oehm, J. Tørresen and S. Aunet. *Ultra low-power, -area and -frequency CMOS thyristor based oscillator for autonomous microsystems*. Springer US Analog integrated circuits and signal processing, Vol.89 (2), p.347-356 (2016).

[13] P.P. Mendoza, S. Gayas, A.S. Lait, W.H. Krautschneider, K. Matthias. *A 1.9 nW Timer and Clock Generation Unit for Low Data-Rate Implantable Medical Devices*. IEEE 11th Latin American Symposium on Circuits Systems (LASCAS). IEEE (2020), pp 1-4 (2020).

[14] O. Aiello, P. Crovetti and M. Alioto. *Wake-up oscillators with pw power consumption in dynamic leakage suppression logic*. IEEE International Symposium on Circuits and Systems (ISCAS). IEEE. (2019), pp. 1–5.

[15] P.M. Nadeau, A. Paidimarri and A.P. Chandrakasan. *Ultra low-energy relaxation oscillator with 230 fJ/cycle efficiency*. IEEE Journal of Solid-State Circuits 51.4 (2016), pp. 789–799.

[16] J. Karim, A.Z. Alam and A.N. Nordin. *MEMS-based oscillators: a review*. IIUM Engineering Journal 15.1 (2014).

[17] Y.S. Lin, D. Sylvester and D. Blaauw. *A sub-pW timer using gate leakage for ultra low-power sub-Hz monitoring systems*. IEEE Custom Integrated Circuits Conference. IEEE. (2007), pp. 397–400.

[18] R. Woudsma and J.M. Noteboom. *The modular design of clock-generator circuits in a CMOS building-block system*. IEEE journal of solid-state circuits 20.3 (1985), pp. 770–774.

[19] P.K. Chan, L. Siek, C.S. Lee and K.L. Chan. *A programmable clock oscillator for integrated sensor applications*. Proceedings 1998 Hong Kong Electron Devices Meeting (Cat. No. 98TH8368). IEEE. (1998), pp. 144–147.

[20] H. Domenik, E. Reef, M. Malte, N. Wolfgang. *Leakage Models for High-Level Power Estimation*. IEEE Transactions on Computer-Aided Design of Integrated Circuits and Systems, pp 1627-1639 (2017).

[21] R.J. Baker. *CMOS Circuit Design Layout and Simulation 3rd Edition*. Wiley IEEE Press, 2010.

[22] H. Shahid and H. Alzaher. *A New pW CMOS Sub-Hertz Timer*. Arabian Journal for Science and Engineering 45.3 (2020), pp. 1379–1384.

[23] A. Zia, Z. Andleeb, M. Olivieri, A. Mastrandrea. *Geometry scaling impact on leakage currents in FinFET standard cells based on a logic-level leakage estimation technique*. Microelectronics, Electromagnetics and Telecommunications. Springer, Singapore,pp 283-294 (2018).

[24] V. Preeti, K.S. Ajay, S.P. Vinay, N. Arti, T. Anand. *Estimation of leakage power and delay in CMOS circuits using parametric variation*. Perspectives in Science, pp. 760-763 (2016).

[25] Y. Lee, et al. *A sub-nW multi-stage temperature compensated timer for ultra-low-power sensor nodes*. IEEE journal of solid-state circuits, pp. 48.10: 2511-2521 (2013).

[26] J. Lim, et al. *A 224 pW 260 ppm/° C gate-leakage-based timer for ultra-low power sensor nodes with second-order temperature dependency cancellation*. IEEE Symposium on VLSI Circuits, pp. 117-118. (2018).

Electronic solution to compensate the effects of the temperature and the humidity on the measurements of a capacitive sensor dedicated to an injection insulin pen

Sylvain JOLY
Innovation Group
Valtronic Technologies
Les Charbonnières, Switzerland
sjoly@valtronic.com

Albrecht LEPPLE-WIENHUES
Innovation Group
Valtronic Technologies
Les Charbonnières, Switzerland
alepple-wienhues@valtronic.com

Catherine DEHOLLAIN
RFIC Research Group
Ecole Polytechnique Fédérale de Lausanne
Lausanne, Switzerland
catherine.dehollain@epfl.ch

Abstract—This article presents the climatic effects on a capacitive sensor device which measures drug volume inside an injection pen. The variations of the temperature and the humidity have a direct impact on the value of the capacitance. Therefore, experiments have been performed by modifying the temperature from 32°C to 23 °C at a constant relative humidity of 30%, inducing a variation of absolute humidity. The errors due to temperature and humidity induce a capacitance variation corresponding to a volume error of ~130µL That represents a dose error of 13 International Units (IU) of insulin. Two main approaches are presented in this paper. Firstly, the effect of water absorption is decreased by encapsulating the analog Printed Circuit Board (PCB). Secondly, reference electrodes are added to the system to correct the climatic variation by differential measurement. It will be shown that these two combined techniques have a beneficial effect on the measurements with a reduction of the error by 70%. The remaining error corresponds to a volume error of 40µL, equivalent to 4IU of insulin.

Keywords—Capacitances, climatic effects, smart pen cap, diabetes

I. INTRODUCTION

Several diseases need treatments with daily injection, and sometime even multiple injections per day. This is difficult for the patient, who needs to pay attention to inject the right amount of drug at the right time. This paper will focus on diabetes cases because this is a frequent disease. In 2014, 422 million people had diabetes worldwide [1]. This is 108 million people more than in 1980, indicating an increasing incidence. In 2012, 1.5 million people died directly from diabetes and another 2.2 million people died from long-term complications caused by high blood glucose concentration. The cost of diabetes for the society is estimated in 2016 at 833 Billion $US/ year and that has tripled between 2003 to 2013. The World Health Organization estimates the cost of diabetes will increase at 1.7 trillion $US in 2030 [1]. So, it is important to create devices to limit the impacts of this disease and facilitate the treatment and life of patients.

Person who suffer of diabetes, have difficulties to regulate blood glucose due to metabolic problems caused by insulin which is the sugar's regulator hormone. Two main type of diabetes exist [2]:
- Diabetes type 1 is an autoimmune disease where the insulin-producing cells are destroyed. It can start during childhood; the pancreas does not produce any insulin anymore. These patients need insulin treatment for survival.
- Diabetes type 2 is typical for an older age and is correlated with overweight; it comprises a relative resistance to insulin.

To compensate this lack of insulin, patients need to inject themselves synthetic insulin. Thanks to injection pens, this task became easier, and people can live in a normal manner. They just need to take with them the injector pen [3].

But some problems still exist with these injector pens; the main problem is the difficulty to calculate the right dosage. Because this dosage depends on many parameters:
- Past parameters
  - When was the time of the last meal? And what was the type of food?
  - When was the last injection? How many doses were injected? What type of insulin was injected (fast or slow insulin)?
- Present parameter
  - What is the blood sugar level?
- Future parameter
  - Which activities will be planned after the injection (e.g. Staying at home or doing some sport)?

The patient can have a health problem (e.g. faint) if he performs a mistake when injecting the insulin. To help the patient with these problems, an electronic device has been developed to measure the volume of liquid inside the injector pen [4].

II. SMART PEN CAP DESCRIPTION

A. Objective

Objective of the smart pen cap is to follow drug volume inside the injector pen to provide treatment follow-up. The device must be as discrete as possible and easily usable by patients [3]. An example of a smart pen cap with an insulin pen is shown in Figure 1.

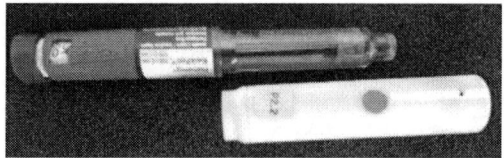

Figure 1: Smart pen (white part) with insulin pen

B. Technology

To measure the drug volume, the smart pen cap uses four electrodes with a capacitance-to-digital converter from Analog Device (AD7745). The electro-static field turns by 90° each 200ms to obtain an averaged electro-static field which reduces electric field asymmetry and inhomogeneity. This method to measure drugs has been presented in [5] [6] .

III. CHALLENGES

As described in the previous publications, it is possible to measure a dose of 1 IU insulin with the smart pen cap in a stable climatic environment; 1 IU is equivalent to 10μL of water as the concentration of insulin is very low. The following perturbations can impact the measurement:

- Climatic effect
- Pen/Cap mechanical precision
- Bubbles effects
- Internal pressures effects
- Ampoule movements

The variations of the temperature and the humidity have an impact on the value of the measured capacitance. That is due to the fact that these climatic variations change the relative permittivity of material and therefore modify the capacitance as it is shown in [7] in the case of the monitoring of the ice growth. This paper will focus its attention on the effects of the temperature and humidity on the capacitance.

IV. MATERIALS AND METHOD

A. Materials

The following devices have been used to perform climatic experiments:

- Climatic Chamber Vötsch® 4018
- Smart Pen Cap
- PCBs of the smart pen cap alone
- Full Injection pen (1.5mL)
- Bluetooth Low Energy (BLE) Receiver
- Computer

B. Methods

To characterize the climatic effect on the smart pen cap device, the following environment condition has been used.

1. 32°C – 30% of relative humidity
2. 23°C – 30% of relative humidity

The time duration of each step is equal to three hours to be sure that the climatic chamber is in stable climatic condition. Tests have been conducted with 5 different smart pen caps

1. PCBs of the smart pen cap alone
2. PCBs of the smart pen cap alone encapsulated
3. Smart pen cap
4. Smart pen cap with reference electrodes
5. Smart pen cap with reference electrodes and encapsulated PCB

V. RESULTS

A. PCBs of the smart pen cap alone

In Figure 2, the capacitive sensor (blue) and the PCB tracks capacitance (red) are plotted. These measurements show that the two capacitances have the same behavior.

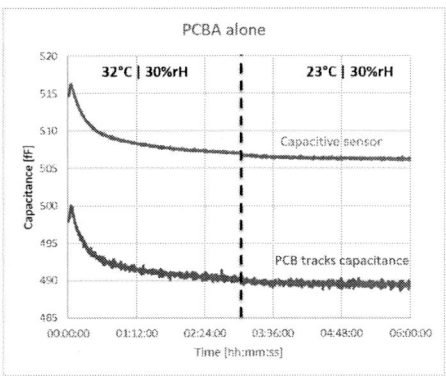

Figure 2 : PCBA alone at 23°C and 32°C

B. PCBs of the smart pen cap alone encapsulated

Figure 2 shows a long-term drift of the values of the two capacitances. This should be due to water absorption of the Smart Pen cap materials as FR-4 and plastic. Therefore, the PCB has been encapsulated with UV curable resin (Loctite 3211-LC) to reduce the long-term drift. Figure 3 proves that this encapsulation is very useful to limit this drift. This resin induces an increase of PCB tracks capacitance (red) and capacitive sensor (blue).

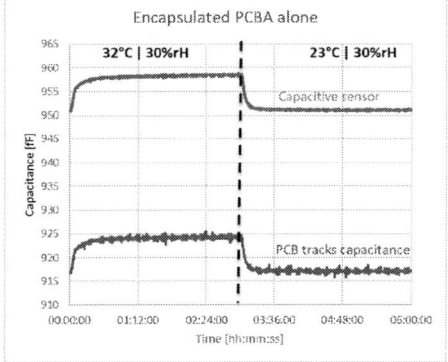

Figure 3: Encapsulated Analog PCBA at 23°C and 32°C

C. Smart Pen cap

Figure 4 represents the effect of the temperature and humidity variation on the value of the capacitive sensor and PCB tracks capacitance for a full assembled smart pen cap with a full injection pen inside. This full injection pen induces an offset on the capacitive sensor. The capacitance of the PCB tracks is equivalent to section V.A because it has no encapsulation.

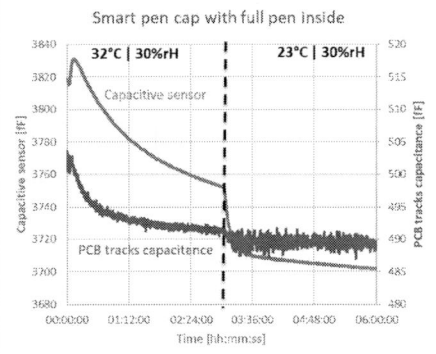

Figure 4: Smart pen cap with a full pen inside at 23°C and 32°C

D. Smart Pen cap with reference electrodes

Two copper plates have been soldered on the Analog PCB and are used as reference capacitor. This reference capacitor is measured through the AD7745 as a stand-alone capacitor and the compensation is done by using capacitive difference in post processing [8, 9]. This is a cheap solution and suitable for industrial design as presented in Figure 5.

Figure 5: Smart pen cap PCBA with reference electrodes

Then, the PCB has been assembled in a smart pen cap device. The reference capacitor (red) follows the capacitive sensor (blue) as plotted in Figure 6. The PCB is not encapsulated with UV resin.

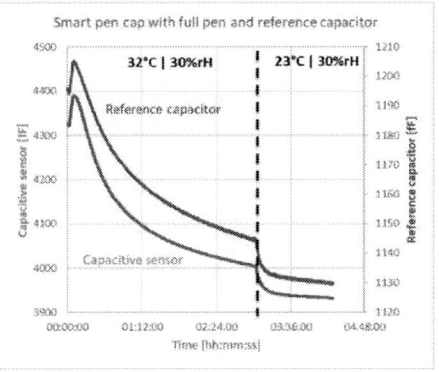

Figure 6: Pen cap with reference capacitor at 23°C and 32°C

E. Smart pen cap with reference electrodes and encapsulated PCB

The final test presented in Figure 7 used two reference electrodes and an encapsulated PCB, to have the best possible correlation between reference capacitor and capacitive sensor in order to do differential measurement.

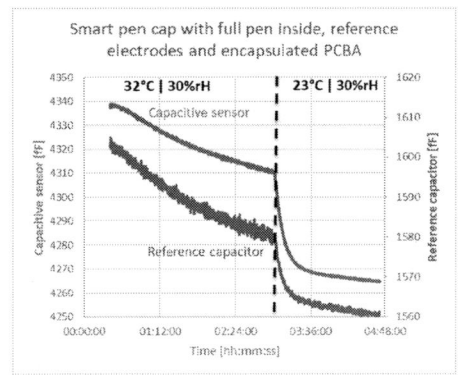

Figure 7 : Pen Cap with reference electrodes and encapsulated analog PCBA at 23°C and 32°C

VI. DISCUSSION

According to the climatic sensors in the chamber, the PCBs alone and smart pen cap devices can reach the stabilization after 30 minutes which corresponds to the stabilization time of the climatic chamber. The capacitance has long term drift which is not correlated to climatic variation as it is presented

in sections V.A and V.C. These long-term drifts could be due to water absorption of materials. Two main ways have been developed to solve this issue:

- Encapsulate the Analog PCB to reduce the water absorption effect
- Use reference electrodes to compensate measurement variation by a same nature variation.

In section V.B, the encapsulation seems to solve the long-term drift for the PCBs. A relation appears in Figure 8 between capacitive sensor and PCB tracks capacitor with a linear correlation factor R^2: 99.5%. This good correlation factor can be explained by the fact that PCB tracks for reference and measurement capacitance are quite similar. Comparing the measurement electrodes with the reference PCB tracks a poor correlation is shown in section V.C because these two capacitors do not measure anymore the same environment.

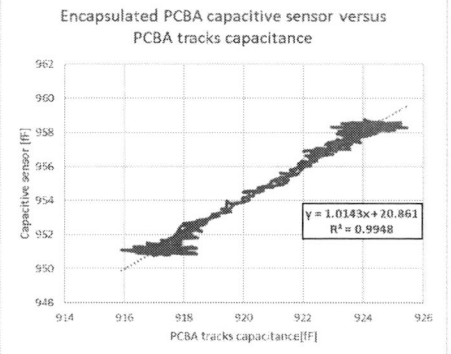

Figure 8: Encapsulated PCBA, capacitive sensor versus PCB tracks capacitance

To solve the difference between the capacitance of PCB tracks and the capacitive sensor as shown in Figure 4, reference electrodes have been placed in section V.D. For the same climatic change, the reference capacitor shows a variation of 70fF, versus 10fF for the PCB tracks capacitor. Therefore, the reference capacitor is more sensitive to temperature variation and it becomes possible to do a differential measurement to reduce the unwanted effects of temperature and humidity. A correlation appears in Figure 9 between reference and measurement electrodes with a R^2: 98.6%.

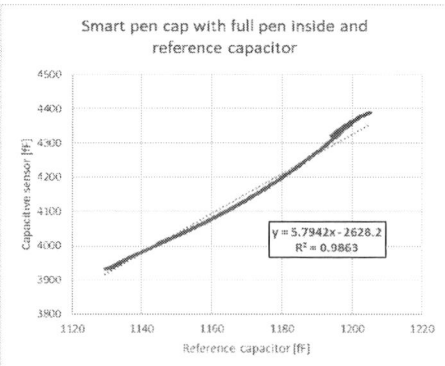

Figure 9: Capacitive sensor versus reference capacitor for a smart pen cap with a full pen inside

In the last section of results V.E, the reference electrodes and encapsulation are combined. A correlation appears in Figure

10 between capacitive sensor and reference capacitor with the PCB encapsulated. The correlation factor R^2 is 98.9% with a variation of 80fF for reference capacitor.

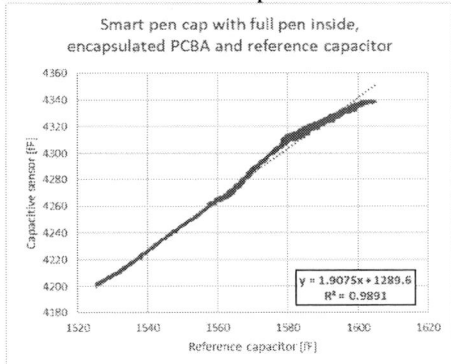

Figure 10 : Smart pen Cap with reference capacitor and encapsulated PCBA

VII. CONCLUSION

To correct the climatic variation due to the temperature and humidity, it has been shown that the unwanted climatic effects can be reduced when (i) adding two reference electrodes on the PCB to realize an additional capacitor, and (ii) encapsulating the PCB to cancel the effect of water absorption. By this way, it is possible to compensate the variation on the capacitive sensor by performing a differential measurement as it is shown in Figure 11 when the temperature drops from 32°C to 23 °C.

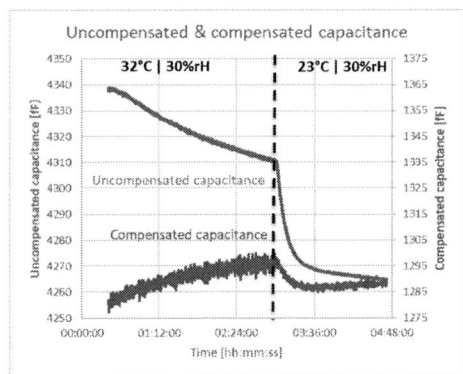

Figure 11: Uncompensated capacitance (blue) and compensated capacitance (red)

According to the results, the uncompensated variation due to climatic is about 72fF. By using the reference electrodes with differential measurement, the compensation of the measurement variation due to climatic is less than 22fF. This is a 70% improvement of the error. The capacitance value is converted in volume by calibrating the smart pen cap with five pens to obtain the conversion factor which is 0.55fF/μL This improvement allows to reduce the error from 130μL (13 IU of insulin) to 40μL (4 IU of insulin). This is a significant improvement and implementation of reference electrodes on a new prototype iteration will be studied.

ACKNOWLEDGMENT

This study has been financed by Valtronic Technologies and it has been supported by the CTI grant n°25386.1 PFLS-LS. The authors thank Dr. Matteo Simoncini, Mr. Alexis Pokorny, and Mrs. Adèle Richter for their advice and support.

REFERENCES

[1] World Health Organization, Global report on Diabetes, Geneva: World Health Organization, 2016.

[2] World Health Organization, Definition, diagnosis and classification of Diabetes mellitus and its complications, Geneva: World Health Organization, 1999.

[3] B. J. Anderson and M. J. Redondo, "What can we learn from patient-reported outcomes of insulin pen device," *Journal of diabetes science and technology,* vol. 5, no. 6, pp. 1563-1571, 2011.

[4] A. Lepple-Wienhues, "Device for attachment to a portable liquid injection device". Switzerland Patent US Patent US20170224922A1, European Patent EP 2 987 518 B1, August 2017.

[5] S. Joly, A. Lepple-Wienhues and C. Dehollain, "Capacitance measurement applied to the medical injection pen," in *Research in Microelectronics and Electronics (PRIME)*, Giardini Naxos, 2017.

[6] S. Joly, A. Lepple-Wienhues and C. Dehollain, "Modeling of a capacitive sensor dedicated to drug injection. In 2018," in *Ph. D. Research in Microelectronics and Electronics (PRIME)*, Prague, 2018.

[7] H. C. Cho, X. Zhi, B. Wang, C. H. Ahn and J. S. Go, "Development of a capacitive ice sensor to measure ice growth in a real time," in *Solid-State Sensors, Actuators and Microsystems*, Anchorage, 2015.

[8] H. K. Trieu, N. Kordas and W. Mokwa, "Fully CMOS compatible capacitive differential pressure sensors with on-chip programmabilities and temperature compensation," in *IEEE Sensors*, Orlando, 2002.

[9] I. Bord, P. Tardy and F. Menil, "Influence of the electrodes configuration on a differential capacitive rain sensor performances," *Sensors and Actuators,* vol. B, no. 114, pp. 640-645, 2005.

[10] Lilly USA, "Humalog insulin lispro Injection," 08 2017. [Online]. Available: https://www.humalog.com/index.aspx. [Accessed 02 2018].

[11] ISO 11608-1, "Needle-based injection systems for medical use - Requirements and test methods - Needle-based injection systems," ISO, 2014.

A scalable spike detection method for implantable high-density multielectrode array

Mattia Tambaro[1,*], Elia Arturo Vallicelli[2], Gerardo Saggese[3], Andrea La Gala[4],
Marta Maschietto[5], Alessandro Leparulo[5], Antonio Strollo[3], Marcello De Matteis[4], Andrea Baschirotto[4], Stefano Vassanelli[1,5]

*Mattia.Tambaro@PhD.UniPD.it

[1]Padova Neuroscience Center, University of Padua, Padua, Italy; [2]Institute for Nuclear Physics, University of Milano Bicocca, Milan, Italy;
[3]Department of Electrical Engineering and Information Technology, University of Naples Federico II, Naples, Italy;
[4]Department of Physics, University of Milano Bicocca, Milan, Italy; [5]Department of Biomedical Sciences, University of Padua, Padua, Italy.

Abstract—High-density CMOS-based Multielectrode Arrays (MEA) provide thousands of channels to record extracellular electrical activity of neuronal networks. Such a high channels count generates an amount of data that is difficult to manage with long-term fully implantable neural interfaces, where power and data transmission are provided wirelessly. To overcome this limitation, a low resources digital signal processor able to reduce the amount of transmitted data to a relevant subset represented by the spiking activity is essential. Unfortunately, the resources required to detect spikes linearly grow with the number of channels, limiting the total amount of MEA pixels in these devices. This work presents a method, here called Spike Detection-by-Difference (SDD), to drastically reduce this limit, for real-time spike detection with an impact on resources and consumption independent from the total channels count. It exploits the high resolution of MEAs to separate the highly localized extracellular action potential from the large-scale local field potential. The SDD is compared with the standard spike detection approaches as the Threshold Crossing (TC) and the Nonlinear Energy Operator (NEO). The detection accuracy is compared for different spiking amplitudes on a synthetic dataset generated from real recordings, showing a detection accuracy of 90% for spikes as low as 45 µV, with a noise in the frequency band from 300 Hz to 5 kHz of 10 µV$_{rms}$. Furthermore, the resource consumption shows a reduction of the 91,5% compared to the TC and of 94% compared to the NEO on a 32x32 pixels matrix. This reduction can be further accentuated increasing the matrix size or reducing the number of columns.

Keywords—Digital Signal Processing, Electrophysiology, Microelectrodes Array, Signal Detection, Low Power, Real-Time Systems

I. INTRODUCTION

In the recent years we assisted to a stunning increase in the brain-computer interface capability. Thanks to the use of CMOS technology in the field of electrophysiology, it is possible to actively sensing the voltage fluctuation of the neural activity with high resolution Multielectrode Arrays (MEAs) of thousands of pixels packed in only few mm^2 [1,2], with a spatial resolution that makes them able to discern the single axons of the neurons [3]. This technology gradually moved from the study of in-vitro culture to in-vivo intracortical implants, with MEA probes for acute or chronic implants wired to an external recording system for power delivery and power transmission [4,5]. Recently, this technology moved another step towards the long-term implantable devices. These devices open multiple scenarios of closed-loop implementations where the neural activity of a wide population of neurons can be monitored in real-time while an activity-dependent feedback can be provided through intracortical microstimulation, to create a new generation of prosthesis [6], sensory ability restoration [7], or disease mitigation [8,9]. On the other hand, the huge number of pixels causes also a dramatical increase in the computational throughput required for the elaboration and interpretation of the sampled neural activity [10]. This is a serious limitation

when it comes to implantable devices, powered wirelessly or by small batteries, and where the heating of the device can be seriously destructive for the brain cells [11]. An effective approach is to offload complex operations as spike sorting, clustering, and interpretation of the activity to an external device where power and performances constraints are less strict. Nevertheless, the transmission of the data generated by the entire MEA would be unsustainable for such low power devices, where the power from energy harvesting dedicated for the processing reaches only few µW [12,13]. To reduce the transfer to only a small subset of relevant information allows to drastically reduce the transmission bandwidth and the power required. In this view, a MEA-embedded detector able to reliably recognize the relevant signal events –basically the spiking activity– from thousands of channels, where resources available for detection are still significantly constrained, is a key aspect in the growth of the implantable neural interface in the next future. A reliable spike detector must reduce False Positives (FP, wrong detection) and increase the True Positives (TP, correct detections) for an optimal data transmission. Common approaches to spike detection range from the simple Threshold Crossing (TC) –where a detection occurs when the high pass filtered signal crosses a specified threshold–, to operator based on the energy of the filtered signal [14], up to matched filter –where the filtered signal is matched with spike templates, generated from extracellular Action Potentials (AP) classified by spike sorting [15]–. A figure of merit of different algorithms adapted for a MEA usage, evaluating the accuracy against the resource request, can be found in a previous work [16]. So, whatever the approach may be used, usually the first step in spike detection is a high-pass filtering –commonly above 300 Hz– of the signal, to work on the high frequency fluctuations and to avoid the predominant slow oscillation of the Local Field Potential (LFP). Unfortunately, even the filtering alone requires a growth in the memory proportional to the increase of channels count, due to the multiple samples to store for each channel to perform the computation. Energy operators, in addition, require more samples and usually involve a smoothing step on the output, that force to also store the previous outputs. Template matching expects to store several samples of the signal to continuously match different templates that must be also stored on chip. A MEA of thousands of channels relying on these methods for data transmission reduction, results in a request of memory incompatible with the few resources available on an implantable device. To overcome these limits, this work presents the Spike Detection-by-Difference (SDD), an approach that does not require to store states between the samples of same channels. This unties the request of memory from the pixels count, breaking the linear grow between channels and computational resources. The SDD assumes that APs are voltage fluctuations highly localized in space compared to the slow oscillations of the LFPs [17,18]. Thanks to the high spatial resolution of the MEA, the LFPs are sensed similarly on wider areas by adjacent pixels, with a negligible change in amplitude. For this, the SDD allows an embedded

© VDE VERLAG GMBH · Berlin · Offenbach

Digital Signal Processor (DSP) to discern an extracellular AP from a set of adjacent MEA pixels using only few consecutive samples, commonly digitized row by row in Time Division Multiplexing (TDM). Consequently, the DSP memory request will be constant and independent from the channels count, remaining related only to the distance of a pixel neighborhood in the stream.

II. METHODS

A. Spike Detection-by-Difference implementation

The SDD works on overlapping subgroups of a minimum of three neighboring channels. Since the LFP amplitude between adjacent MEA pixels slowly changes in space, its estimate is removed from each pixel as the mean of the samples of the neighbors, returning the amplitude of an eventual extracellular AP plus a LFP mismatch and the acquisition noise. The mismatch and the acquisition noise can be accounted comparing the result with a fixed threshold, that is also the minimum spike amplitude considered for detection. The threshold depends on the SNR of the acquisition system and sets a tradeoff between FP and False Negative (FN, missed detections). A lower threshold detects more FP and generates less FN, conversely, a higher threshold detects less FP but generate more FN. The SDD implementation can be found in (1), where x is the sample value of the pixel specified by the subscript at time t, $nbh(p)$ represents the neighborhood of the pixel p, and Th represents the minimum difference to detect a spike. The division operation can be replaced by multiplying both threshold and value of the pixel subject of detection as in (2) for a lighter implementation.

$$x_p(t) - \sum_{n \in nbh(p)} x_n(t)/|nbh| > Th \qquad (1)$$

$$\sum_{n \in nbh(p)} x_n(t) - x_p(t) * |nbh| > Th * |nbh| \qquad (2)$$

The SDD has the valuable advantage to require a fixed length for the acquisition queue, equal to the maximum distance between the pixels of the chosen neighborhood in the TDM stream. Samples can be in fact erased as soon as they are no longer required by the neighborhood of any of the following pixels, causing the total amount of samples to store being dependent only by the row length, regardless of the number of columns (figure 1). Increasing the neighborhood improves the estimation of the LFP value and enforces the detection against simultaneous spikes on adjacent channels or spikes sensed by multiple channels. Especially the latter is common in MEAs, due to the spatial resolution higher than a single neuron size; in this case the approach can be modified by selecting neighbors farther than the immediately adjacent pixels, or by averaging the spatial information also for the main detection channel, as done in other works [16,19]. In the SDD, each pixel requires to store only a detection state to prevent the trigger of multiple detections caused by the over-threshold spike samples following the first. This can be a small counter that is started whenever a value exceeds the threshold, or as a single bit commuting the state whenever the output changes from below threshold to above threshold and vice-versa, for a higher save of resources. Particularly, a counter can reduce multiple detections in case of a same noisy spike crossing the threshold multiple times and can be shared with the logic transmitting the samples following a detection to recreate a spike shape on an external device. Finally, for critical pixels as the ones on the MEA border, and therefore with a neighborhood that overflow to the opposite side of the matrix in the stream, detections can be ignored to avoid control logic providing a dynamic neighborhood.

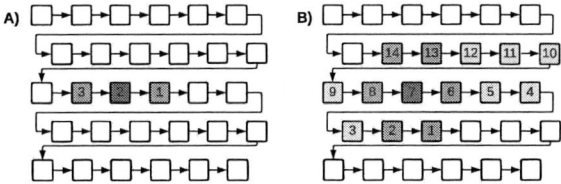

Fig. 1. Arrangements of pixel neighborhoods for the SDD on a MEA. The arrows indicate the pixels acquisition order in TDM. Active values in the stream are numbered from the last digitized value to the older that must be kept in memory. In blue is the pixel subject of spike detection, in green are the neighbors considered to estimate the LFP and in yellow are the pixels unused in the current detection but required for the following neighborhoods. A) An example of the smallest pixel subgroup required to apply the SDD, with a samples queue of 3. B) Example of a wider pixel subgroup, providing a better estimation of the LFP baseline but with a longer sample queue, equal to 2 times the row length plus 2.

B. Synthetic testbench dataset

A synthetic dataset provides the valuable advantage to allow testing the methods in different noise condition, where the ground truth of the spiking activity is well known. The testbench dataset is used to compare the SDD against the TC and the Nonlinear Energy Operator (NEO) and it is generated from the signal acquired during an intracortical recording from the rat's barrel cortex, acquired with a sampling frequency of 25 kHz using a passive linear probe of 32 channels spaced by 65 μm (Atlas Neuro E32+R-65-S1-L6 NT). This high spacing between the channels prevents the application of the SDD to be reliable due to the LFP fluctuations remarkably different on channels pairs crossing cortical layers. This difference would be highly mitigated by the lower spacing between pixels in a MEA. For this reason, data are reshaped to recreate a synthetic dataset representative of the expected behavior of a MEA recording, interpolating virtual channels with a smaller spacing. MEA recordings data acquired by the linear probe differ also by the number of channels, sampling rate, and signal quality. While the higher channel count does not influence the performance of the methods and can be ignored, the data must be down sampled, and the spikes amplitude modulated to test the methods capability in different conditions. Unfortunately, being the electrodes placed linearly on the shank of the probe, the greater neighborhood of a pixel that could be extract is limited to 2. This forces to test the SDD using the smallest required group of pixels as in figure 1-A, but it does not influence the TC and the NEO, that are applied to a single channel. To create the dataset, firstly, spikes template and recording noise are extracted. To achieve this, the signal is high pass filtered by a 2nd order Butterworth filter with -3dB cutoff at 300 Hz for two times, once in each direction, to obtain a linear phase response. The templates are then generated averaging the shape of hundred spikes detected from the filtered signal on a channel with 10 μV_{rms} noise, recording a firing neuron causing a depolarization up to 150 μV. The noise in the high frequency band is extracted from channels with a noise included between 9 and 11 μV_{rms}, that showed Multi Unit Activity (MUA) peaks below 50 $\mu V_{0\text{-peak}}$. MUA is considered a small voltage oscillation of few tens of μV that usually lasts few milliseconds, generated by a small population of neurons in the proximity of a channel [20]. Finally, the LFP signals are extracted by low pass filtering (2nd order Butterworth, -3dB at 300Hz, applied in both directions) three adjacent channels of the passive probe. To recreate a MEA acquisition, these LFPs are interpolated to emulate a recording from intermediate virtual pixels, spaced by half of the original electrodes spacing (recent MEAs provide spacing less than 8 times smaller than the probe used for the acquisition [1,2,4,5]).

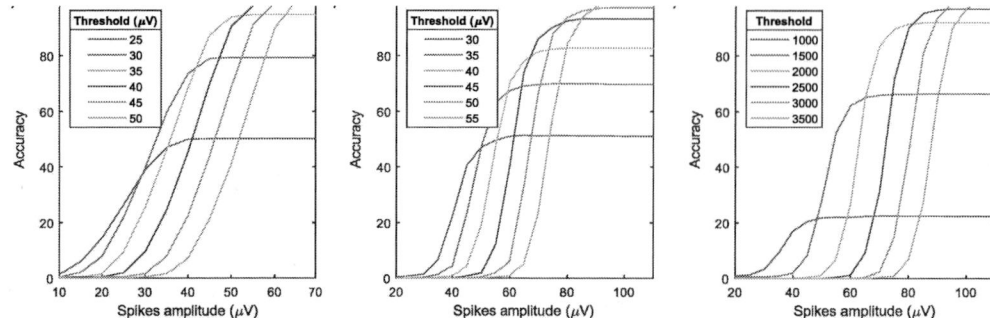

Fig. 2. Accuracy achieved on the testbench dataset by the SDD (A), by TC (B) and by the NEO (C). The accuracy is estimated with different threshold levels and spike amplitudes. The noise in the frequency between the 300 Hz to Nyquist is 11 µVRMS. Spike amplitude refers to the 0-peak value of the negative peak of the spike.

A synthetic signal comprehensive of every original component is recreated adding to the LFP the noise and the spike templates. The spikes are added at a fixed frequency of 25 Hz on the middle channel, rescaled to different amplitudes to test the accuracy at different SNR. Finally, the synthetic dataset is down sampled to 10 kHz, to reflect a realistic sampling rate of an implantable device of hundreds of channels. The generated testbench is 10 seconds long, and present 250 spikes.

III. RESULTS

A. Accuracy & comparison

The accuracy has been defined as the TP divided by the sum of the total amount of spikes and the FP [19], and it has been compared with the classical TC and with the NEO. The accuracy has been estimated varying both spikes amplitude to emulate different SNR conditions and thresholds level, to modulate the ratio between the FN and FP. For the TC implementation, the signal of the middle pixel is high pass filtered by a 2nd order Butterworth with a cutoff frequency of 300 Hz and the output is compared against a threshold. If the signal exceeds, an event is detected. For the NEO, the filtered signal s at time t is computed as in (3). The output, once again, is compared against a threshold Th.

$$s(t-1)^2 - s(t) * s(t-2) > Th \qquad (3)$$

A comparison between the performances of the SDD, TC, and NEO is shown in figure 2. The performances achieved with different thresholds are estimated for each detection method, to evaluate the tradeoff between TP and FP while varying the spike amplitude. The SDD shows better performances, reaching the highest accuracy compared with the other two. Particularly, an accuracy of 90% –a good tradeoff between data reduction (few FP to transmit) and precision (few FN)– is reached with a spike amplitude of ~45 µV. To achieve the same accuracy, both TC and NEO require a spike amplitude of ~75 µV. The greater accuracy of the SDD is largely due to the limited number of FP. As it can be observed in Figure 3, that shows a subset of the testbench signal and the detection of the three methods, the low threshold value needed to detect the spikes at their smaller amplitude (< ~55 µV), causes the detection of the MUA as events in both TC and NEO, increasing the FP. Being a less localized phenomena compared to the single AP, the SDD presented in this work reduces the MUA oscillations, since similar on adjacent channels, consequently reducing the FP and increasing the accuracy. Also, widespread artifacts and interferences are consistently reduced by this method, based on the same principles of being non-localized. It must be underlined that in case of wide spacing between pixels or in case of strong discrepancy on the LFP amplitude recorded by adjacent channel, the accuracy of the SDD can drastically

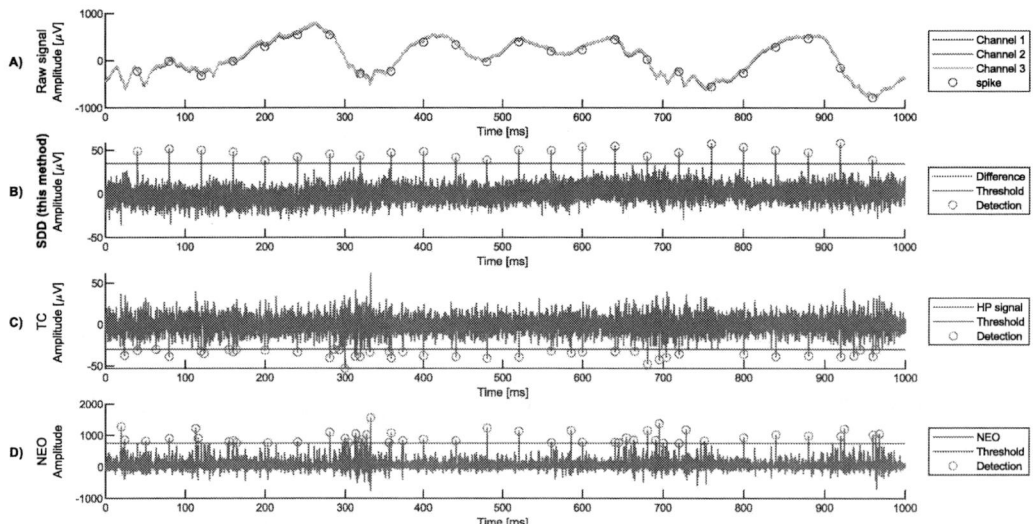

Fig. 3. Testbench subset of 2.5 seconds showing the detection performance for a spike template of 50 µV with 3 different detection methods. A) raw signal of 3 adjacent pixels and, circled, the position of the spikes. B) The output given by the SDD, compared against a threshold of 30 µV. C) The signal filtered by a high pass 2nd order Butterworth filter is compared against a threshold of -35 µV (TC). D) The ouput of the NEO is compared against a threshold of 750.

drop, forcing to increase the threshold and the FN as consequence; it is important then to have a reliable sensing pixel performance. As mentioned earlier, in case of a very high-density matrix, where a single pixel is smaller than a neuron soma, the signal can be averaged on pixel groups; this enforces the robustness against eventual defective pixels, other than reducing the overall noise and increasing the LFP estimation accuracy.

B. Resource usage & comparison

The resources required to implement the SDD are directly proportional to the number of samples acquired between one pixel and its neighbors and are independent from the total amount of pixels of the MEA. In addition, a state stored as 1-bit or counter per pixel must be accounted as discussed section II.A, depending basically on the implementation choice to transmit a chunk of samples to reconstruct the spike shapes for a detection or not. As comparison, the TC requires at least the last unfiltered and filtered samples of each pixel to implement an IIR filter of the lowest order; the NEO requires to store the samples for the filter and at least another filtered samples for the non-smoothed implementation. The smoothed NEO is neither compared, since it forces to store a consistent number of previous values, usually 9, to perform a smoothing operation on the output. To compare the resources consumption of the different detection methods, a square MEA of 1024 pixels, 32 by 32, with a sample width of 10 bits is used. The SDD, applied considering the entire neighborhood of a pixel, requires to simultaneously store 66 values (2 rows of 32 pixels plus 2) of 12 bits, plus a single bit for the detection state for each pixel, for a total of 1816 bits. As comparison, supposing a bit truncation to 8 bits for the filtered samples, the TC requires 1024 samples of 12 bits and 8 bits for a 1st order filter, plus a detection state single bit, for a total of 21504 bits. Finally, the NEO requires to store values as the previous for the filter plus another 8 bits filtered sample per pixel, for 29696 bits in total. The SDD, therefore, allows to save up to the 91.5% of resources compared to the TC and the 94% compared to the NEO for a MEA of 1024 pixel; increasing the number of pixels, or reducing the row size, will lead to a greater relative saving of resources.

IV. CONCLUSION & DISCUSSION

The SDD can be implemented to work at the same frequency of the sampling rate; pipelining the computation provides the output in 3 clock cycles with a neighborhood of 2 pixels (1 for the mean, 1 for the difference and 1 for the threshold comparison) with a delay in samples in the detection equal to half the samples stored. The same pipeline of operations can be used to elaborate each new sample. When a new spike is detected, a chunk of signal can be transmitted from the MEA. To stream the exact number of samples after a detection, the counter used to reject the following over-threshold samples can be shared. Moreover, the bit width of the transmitted data can be reduced subtracting the value of the first sample from the following. Assuming a high spiking rate of 100 spikes/second, the data reduction over a single channel achieved transmitting only the relevant information can reach the 68% (sampling of 10 kHz, 32 samples per spike), up to a 100% saving on a channel that does not record spiking activity. The greatest obstacles in the grown of pixel count and sensing area are currently the acquisition, transmission, processing, and storage of neural activity. The improvements in the productive process, biocompatible materials, and information processing as the object of this study, where the number of samples generated can be drastically reduced, are a plan road towards an evolution of the brain computer interfaces. The substantial decrease in resources dedicated to store the samples allows MEA to become larger and richer in pixels, while keeping the consumption, data transmission and heating low, for a new generation of low invasive, long-term, wireless implants.

REFERENCES

[1] Müller J, Ballini M, Livi P, et al. High-resolution CMOS MEA platform to study neurons at subcellular, cellular, and network levels. *Lab Chip.* 2015;15(13):2767-2780. doi:10.1039/c5lc00133a

[2] Tedjo W, Chen T. An Integrated Biosensor System With a High-Density Microelectrode Array for Real-Time Electrochemical Imaging. *IEEE Trans Biomed Circuits Syst.* 2020;14(1):20-35. doi:10.1109/TBCAS.2019.2953579

[3] Bakkum DJ, Frey U, Radivojevic M, et al. Tracking axonal action potential propagation on a high-density microelectrode array across hundreds of sites. *Nat Commun.* 2013;4:2181. doi:10.1038/ncomms3181

[4] Jun J, Steinmetz N, Siegle J, *et al.* Fully integrated silicon probes for high-density recording of neural activity. *Nature* 2017;**551**:232-236. doi:10.1038/nature24636.

[5] Scholvin J, Kinney JP, Bernstein JG, *et al.* Close-Packed Silicon Microelectrodes for Scalable Spatially Oversampled Neural Recording. *IEEE. Trans. Biomed. Eng. 2016*;63(1):120-130. doi:10.1109/TBME.2015.2406113.

[6] Bensmaia SJ, Miller LE. Restoring sensorimotor function through intracortical interfaces: progress and looming challenges. *Nat Rev Neurosci.* 2014;15(5):313-325. doi:10.1038/nrn3724

[7] Fernández E, Alfaro A, González-López P. Toward Long-Term Communication With the Brain in the Blind by Intracortical Stimulation: Challenges and Future Prospects. *Front Neurosci.* 2020;14:681. Published 2020 Aug 11. doi:10.3389/fnins.2020.00681

[8] Park YS, Cosgrove GR, Madsen JR, et al. Early Detection of Human Epileptic Seizures Based on Intracortical Microelectrode Array Signals. *IEEE Trans Biomed Eng.* 2020;67(3):817-831. doi:10.1109/TBME.2019.2921448

[9] Wan KR, Maszczyk T, See AAQ, Dauwels J, King NKK. A review on microelectrode recording selection of features for machine learning in deep brain stimulation surgery for Parkinson's disease. *Clin Neurophysiol.* 2019;130(1):145-154. doi:10.1016/j.clinph.2018.09.018

[10] Vallicelli EA, Reato M, Maschietto M, Vassanelli S, Guarrera D, Rocchi F, Collazuol G, Zeitler R, Baschirotto A, De Matteis M. Neural Spike Digital Detector on FPGA. *Electronics* 2018(7):392. Doi:10.3390/electronics7120392.

[11] Zarrintaj P, Saeb MR, Ramakrishna S, Mozafari M, Biomaterials selection for neuroprosthetics. *Curr Opin Biomed Eng.* 2018;6:99-109. doi:10.1016/j.cobme.2018.05.003.

[12] Dinis H, Mendes PM. A comprehensive review of powering methods used in state-of-the-art miniaturized implantable electronic devices. *Biosens Bioelectron.* 2021;172:112781. doi:10.1016/j.bios.2020.112781

[13] Das R, Moradi F, Heidari H. Biointegrated and Wirelessly Powered Implantable Brain Devices: A Review. *IEEE Trans Biomed Circuits Syst.* 2020;14(2):343-358. doi:10.1109/TBCAS.2020.2966920

[14] Mukhopadhyay S, Ray GC. A new interpretation of nonlinear energy operator and its efficacy in spike detection. *IEEE Trans Biomed Eng.* 1998;45(2):180-187. doi:10.1109/10.661266

[15] Carlson D, Carin L. Continuing progress of spike sorting in the era of big data, *Curr. Opin. Neurobiol.* 2019;55:90-96. doi:10.1016/j.conb.2019.02.007.

[16] Tambaro M, Vallicelli EA, Saggese G, Strollo A, Baschirotto A, Vassanelli S. Evaluation of In Vivo Spike Detection Algorithms for Implantable MTA Brain—Silicon Interfaces. *JLPEA.* 2020; 10(3):26. doi:10.3390/jlpea10030026

[17] Dubey A, Ray S. Spatial spread of local field potential is band-pass in the primary visual cortex. *J Neurophysiol.* 2016;116(4):1986-1999. doi:10.1152/jn.00443.2016.

[18] Kajikawa Y, Schroeder CE. How Local Is the Local Field Potential? *Neuron*, 2011;72(5):847-858. doi:10.1016/j.neuron.2011.09.029.

[19] Saggese G, Tambaro M, Vallicelli EA, Strollo AGM, Vassanelli S, Baschirotto A, De Matteis M. Comparison of Sneo-Based Neural Spike Detection Algorithms for Implantable Multi-Transistor Array Biosensors. *Electronics.* 2021; 10(4):410. doi:10.3390/electronics10040410

[20] Harrison BJ, Pantelis C. Multiunit Activity. *Encyclopedia of Psychopharmacology*; 2010. Stolerman IP, Ed.; Springer: Berlin/Heidelberg, Germany. doi:10.1007/978-3-540-68706-1_135

© VDE VERLAG GMBH · Berlin · Offenbach

SMACD / PRIME 2021 | 19 – 22 July 2021, Online Event

Current-reuse Low-Power Single-Ended to Differential LNA for Medical Ultrasound Imaging

1st Olivia Mirea
Department of Electronic Systems
NTNU[1]
Trondheim, Norway
olivia.mirea@ntnu.no

2nd Carsten Wulff
Department of Electronic Systems
NTNU[1]
Trondheim, Norway
carstenw@ntnu.no

3rd Trond Ytterdal
Department of Electronic Systems
NTNU[1]
Trondheim, Norway
trond.ytterdal@ntnu.no

Abstract—We present a low power, low noise current-reuse fully differential OTA (Operational Transconductance Amplifier) -designed in 180 nm CMOS technology- having an energy efficient architecture including a common-mode feedback which ensures the conversion from single-ended to differential with close to zero power - the power being used when the capacitors are reset, no power is consumed in active mode time, the amplifier being designed for an active time of 500 μs. The amplifier drives a capacitive load of 200 fF and achieves 55 μW, an input-referred noise of 19 nV/$\sqrt{Hz}$ and an input capacitance of 18 fF for a unity gain frequency (f_{ug}) of 280 MHz. Since this circuit can be designed for a large range of frequencies, we also investigate the way of optimizing a figure-of-merit (FOM).

Index Terms—Inverter-based amplifier, current-reuse OTA, low noise, low power, in-probe receivers, ultrasound probes, medical ultrasound imaging, 4D echocardiography, ultrasound heart imaging.

I. INTRODUCTION

It is a known fact that cardiovascular diseases are the lead diseases that cause death worldwide [1], [2]. To obtain a clear image of the heart muscle/chambers/valves/blood vessels, trans-esophageal echocardiography is used by medical doctors. To perform high volume rate of trans-thoracic acquisitions, an ultra-fast echocardiography acquisition (4D ultrasound image) is required [3]. Therefore, we aim to design in-probe receivers for 4D imaging.

The main blocks of the ultrasound front-end are: the transducer, transmitter, receiver, beamforming and signal processing. The receiver contains the LNA and ADC (see Fig. 1). LNA (Low Noise Amplifier) plays an important role in the overall function of the Ultrasound System (see Fig. 1): dynamic range of the ADC, signal swing, resolution of ultrasound image. Since the signal created by echoes is attenuated while travelling throughout the body, it is important for this signal to not be further attenuated at the input of the LNA.

So far, many LNA topologies were exploited: based on trans-impedance amplifier [4], [5], current-mirror OTA [6], inverter-based input stage amplifier [7], stacked inverter-based OTA [8], current-reuse amplifier [9] and inverter stacking current-reuse amplifier [10]. Nowadays, the challenge is to include an increasing amount of electronics inside of the probe while keeping power dissipation constant.

[1] Norwegian University of Science and Technology

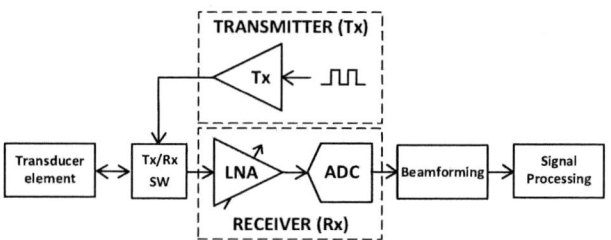

Fig. 1. Ultrasound front-end block diagram for single channel

Usually, for single-ended to differential-ended conversion, a high power (relatively comparable with OTA power) is used [5], [7]. Our goal is to avoid this issue, to design a low power and noise single-ended to differential amplifier with an energy efficient common-mode feedback circuit (CMFB). Based on current-reuse amplifier topologies, we propose an energy-efficient OTA (including the CMFB), our contribution being the CMFB circuit we are using. To quantify the impact of this topology, a figure-of-merit (see equation 1), containing the most important LNA's parameters, is defined:

- A_0 - the open loop gain of the OTA
- f_{ug} - the unity gain frequency of the OTA
- P - the power dissipation of the OTA
- V_{nin} - the input-referred noise of the OTA
- C_{in} - the input capacitance of the OTA

$$FOM = \frac{A_0 f_{ug}}{P V_{nin} C_{in}} \quad (1)$$

The structure of the paper is as follows: in Section II the LNA and proposed OTA architectures are described, in Section III the simulations underline what is previously discussed, including the way of optimizing the FOM defined (see equation 1) and, in Section IV, the conclusions are stated.

II. LNA AND OTA ARCHITECTURES

In Fig. 2 the block diagram of the LNA is illustrated. Even though it is single-ended input LNA, the circuit has to be symmetric. Thus, a capacitor C_1 is used between vinp and gnd.

© VDE VERLAG GMBH · Berlin · Offenbach

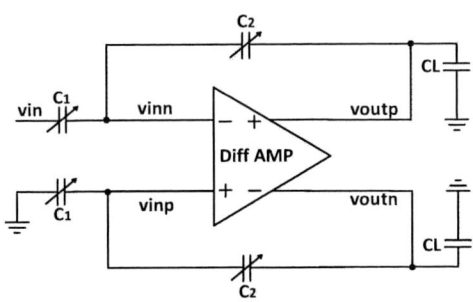

Fig. 2. Block diagram of the LNA

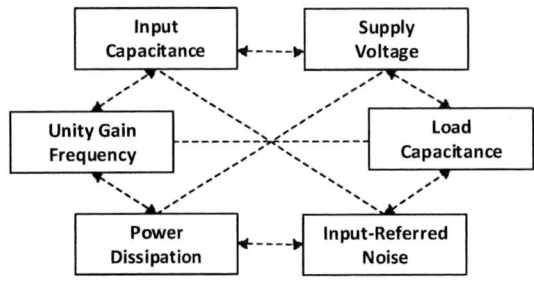

Fig. 3. Analog Design trade-off

To get a variable closed loop gain, variable capacitors C_1 and C_2 are used. To model the ADC's input impedance, we added the load capacitors C_L.

We are interested in finding out the best FOM this OTA can provide - being equivalent to finding the best trade-off between input capacitance, input-referred noise, supply voltage, unity gain frequency, load capacitance and supply voltage (see Fig. 3). In our design, we keep the C_L constant, thus we did not include it in equation 1.

Aiming for a low power dissipation, we propose an OTA structure (see Fig. 4) with almost zero power consumption for the active time of 500 μs.

Our CMFB (Fig. 4) combines the advantages of the two most common CMFB topologies: continuous-time CMFB made of 2 resistors between the gate and the output nodes [11] and switched-capacitor CMFB [12], [13]. This is made possible because in our application the active time is only 500 μs. The single-ended to differential conversion is done by the CMFB consisting of C_3, C_4, M7, M8, SW3 and SW4. The voltages of the C_3 and C_4 capacitors are updated only once/receive - in this way we avoid the switching - period (after reset, the voltage across the capacitor is set to zero).

The OTA uses inverter-based (through the M1p, M1n and M2p, M2n differential pair transistors) and current-reuse (using M3 and M4 nmos transistors) amplifier techniques [14] to decrease the OTA's input capacitance, by decreasing the transconductance of the input transistors.

By using current mirror topology (M4-M6 and M3-M5), the I_{Dbn} bias current is divided into 4 branches: the drain currents in the nmos differential transistors, M1n and M2n (I_{D1n} and I_{D2n}), M3 and M4 (I_{D3} and I_{D4}). The I_{D3} and I_{D4} currents are mirrored further through M5/M6, being used as active loads for the common-source amplifiers having M8 and M7 as input transistors. Therefore, we obtain the bias of the active load transistors, a branch current, without losing power. This is the advantage of using current-reuse technique.

The role of the switches SW1 and SW2 is to bias the input differential pair transistors during reset. In this way we minimize the power consumption since we do not require a dedicated bias circuit to set the DC input voltage of the differential amplifier.

Table I includes the device parameters based on g_m/I_D design method for a current gain, K, of the current mirrors M4-M6 and M3-M5 equal with 5. The rationale behind choosing K is described below. With a proper design, decreasing the ratio between the total input transconductance ($g_{m1} = g_{m1n} + g_{m1p}$) and the transconductance of M3/M4 transistor (g_{m1}/g_{m4}), the gate to drain capacitances of the input pair transistors (C_{GD1n}, C_{GD1p}, C_{GD2n}, C_{GD2p}) can be minimized, however there is always a trade-off. In this case, the input-referred noise is compromised (see Fig. 3). Therefore, the best compromise should be found. That is why we define the FOM including both parameters: the input capacitance and the input-referred noise. Thus, our concentration is focused on improving the FOM.

To get a noise figure close to 3dB, the noise of the amplifier should be equal with the one the source delivers to the input. Therefore, the product between the input capacitance and the noise given by the amplifier should be minimized.

The advantage of using the current-reuse method is that we can get the desired unity gain frequency by choosing the proper K, as shown in equation 3 derived for the unity gain frequency. Thus, by decreasing the transconductance of the input stage, g_{m1}, we obtain the desired unity gain frequency similar to using a regular amplifier having a higher input transconductance. An interesting aspect to be considered is how much the K can be increased.

The unity gain frequency of the OTA, f_{ug}, can be estimated by equation 3, where $C_{L,eff}$ is the effective load capacitance, including the load capacitance and parasitic capacitance of M5/M6 and M7/M8. Therefore, the required transconductance of the input stage is given by the equation 2. At first sight, by inspecting equation 2, the larger K is, the lower g_{m1} is needed, leading to a low input capacitance. Here, an important limitation of the technology used is encountered: by increasing K, while keeping the same $g_{m4}/I_{D4} = 8.8$ 1/V (see Table I), the gate to source and gate to drain parasitic capacitances of M5/M6/M7/M8 increase yielding the impedance in the nodes v1 and v2 decreases causing stability issues. Table I shows the dimensions of the transistors used for $K = 5$.

$$g_{m1} = \frac{2\pi C_{L,eff} f_{ug}}{K} \quad (2)$$

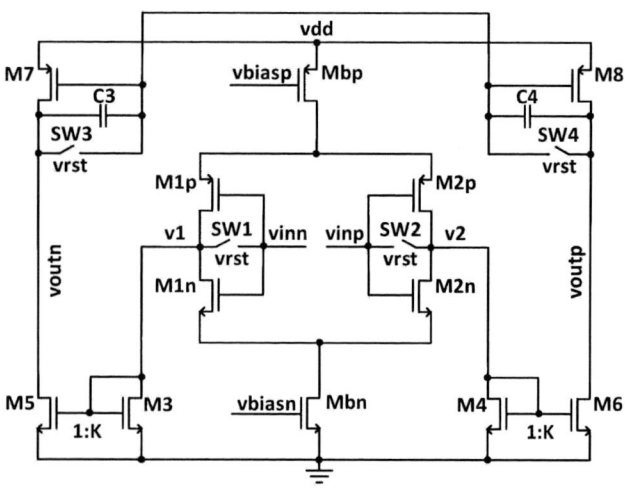

Fig. 4. Fully Differential Current-reuse OTA

TABLE I
DEVICE PARAMETERS BASED ON G_M/I_D METHOD

	Mbp	Mbn	M1p M2p	M1n M2n	M3 M4	M5 M6	M7 M8
W (μm)	3.52	0.92	6.8	1.22	0.22	1.1	3.85
L (μm)	0.52	1.52	0.24	0.24	0.43	0.43	0.92
g_m (μS)	87	23	86	26	28	140	87
g_m/I_d (1/V)	9.9	9.5	20	22	8.8	8.8	5.4

The gain of the OTA is given by equation 4. Even if the transconductance of M5, g_{m5}, is increased by a factor of K, the drain-source resistances of M5 and M7, r_{ds5}, r_{ds7}, are decreased by the same amount, resulting a constant open loop gain with respect to K.

$$f_{ug} = \frac{K g_{m1}}{2\pi C_{L,eff}} \qquad (3)$$

$$A_0 = g_{m1} r_{outf}, \text{ where } r_{outf} = K r_{out} = K(r_{ds5} || r_{ds7}) \qquad (4)$$

Assuming the noise coefficient (γ) is constant with respect to the bias current and equal for nmos and pmos transistors, the input-referred noise voltage squared per Hertz of the differential amplifier can be estimated by equation 5, where k is the Boltzman constant, T is the absolute temperature and g_{m1}, g_{m3} and g_{m7} are the transconductances of M1, M3, respectively M7 transistors. As can be seen, by keeping the same input pair transistor dimensions and increasing K, the overall V_{nin}^2 will decrease, the downside being a higher power dissipation (see Fig. 3).

$$V_{nin}^2 = 8kT \frac{\gamma}{g_{m1}} \left[1 + \frac{g_{m3}}{g_{m1}} \left(\frac{1}{K} + 1 \right) + \frac{1}{K^2} \frac{g_{m7}}{g_{m1}} \right] \qquad (5)$$

To minimize the input-referred noise, we may want to decrease more g_{m4}/I_{D4} (see Table I). As seen in Fig. 3, the trade-off is that the supply voltage will also increase leading to a higher power consumption. Therefore, we choose g_{m4}/I_{D4}

in such a way to ensure the right bias for the differential pair transistors during reset time (see equation 6).

$$V_{DS-sat,4} > V_{th,1n} - V_{th,4} + V_{DS-sat,1n} + V_{DS-sat,2n} \qquad (6)$$

The minimum supply voltage, $V_{DD,min}$, depends on the technology used and g_{m1p}/I_{D1p}, g_{m1n}/I_{D1n}, g_{m1bp}/I_{Dbp}, g_{m1bn}/I_{Dbn} ratios chosen, which give the drain-source saturation voltages of Mbp, Mbn, M1p and M1n transistors ($V_{DS-sat,1p}$, $V_{DS-sat,1n}$, $V_{DS-sat,bp}$, $V_{DS-sat,bn}$). The technology dependence can be seen in equation 7 through the input pair transistors threshold ($V_{th,1n}$ and $V_{th,1p}$). Therefore, to minimize the power consumption, a circuit must provide the supply voltage adjustment to address the process variation issue.

$$V_{DD,min} = V_{GS,1n} + V_{GS,1p} + V_{DS-sat,bp} + V_{DS-sat,bn},$$
$$\text{where } \begin{cases} V_{GS,1n} = V_{th,1n} + V_{DS-sat,1n} \\ V_{GS,1p} = V_{th,1p} + V_{DS-sat,1p} \end{cases} \qquad (7)$$

To have wide output swing, g_{m7}/I_{D7} is chosen in such a way that $V_{th,7} + V_{DS-sat,7} = V_{DD}/2$, where V_{DD} is the supply voltage.

In Fig. 5 we investigated the best FOM with respect to K, where the supply voltage is kept constant. By increasing K, the input-reffered noise decreases, but the power dissipation and f_{ug} increase. After simulating FOM with respect to K, we found out that $K = 5$ is the one which gives us the best FOM (see Fig. 5). As expected, for a larger K, the input-referred noise is smaller and the f_{ug} and power dissipation are larger.

Our proposed circuit is stable for up to at least K lower than 20 (where the phase margin is higher than 60°).

III. SIMULATION RESULTS AND DISCUSSION

The performance of our proposed OTA and previously published LNAs can be seen in Table II, where BW is the open loop bandwidth.

In [5] and our work a single-ended to differential LNA is presented and in [6] and [7] fully differential LNA is presented. An important result to be noted is the differential output voltage to common-mode output voltage swing ratio of 25, for a peak-to-peak input voltage equal with 160 mV

In [5], [7] and our current work, the inverter-based technique is used, while in [6] the current-mirror OTA topology is used. In addition, we use the current-reuse technique twice, succeeding to decrease the power consumption while increasing the open loop gain. With our design we pointed out how we can obtain the best FOM out of a current-reuse OTA with inverter-based input stage. There is still room for improving the FOM by decreasing the supply voltage by using floating gates [15], [16] to bias the input transistors during reset period. By using floating gates, we could halve the supply voltage currently used, yielding to a doubled FOM_c. There is a limitation in using the proposed LNA: when the lower limit of a transducer's bandwidth is lower than the noise corner. To avoid the flicker noise, the chopping technique (see [7]) is needed. In our case, the noise corner is lower than the lower limit of transducer's bandwidth.

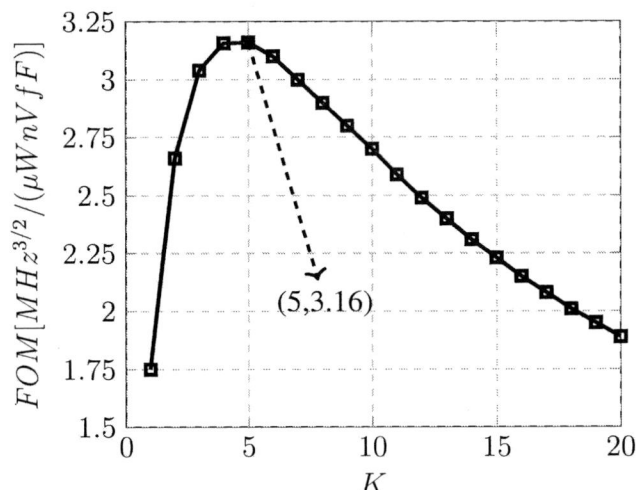

Fig. 5. FOM(K)

TABLE II
PERFORMANCE COMPARISON BETWEEN THIS AND OTHER WORKS

	This work*	[5]	[6]*	[7]
Process (nm)	180	65	180	180
Gain (dB)	46.5	26	62	57.8
Bandwidth (kHz)	1600	40000	130	0.67
Supply voltage (V)	1.35	0.5	0.8	0.8
Power (μW)	55	54	250	0.79
IRN $(nV/\sqrt{Hz})$	19@10 MHz	7@5MHz	17@1MHz	36
FOM$_c$ $\left(\frac{G(Hz)^{3/2}}{pWV}\right)$	0.32	2.1	0.038	0.018

*Simulation results.

The disadvantage of lowering the supply voltage [5] - [7] is a lower output swing, leading to a worse second harmonic distortion. To compare the designs, a figure-of-merit is defined, FOM$_c$:

$$FOM_c = \frac{A_0 BW}{PV_{nin}} \quad (8)$$

Other important aspects which contribute to our defined figure-of-merit, FOM, are load capacitance used and input capacitance of the differential amplifier. We intentionally removed C$_{in}$ from FOM because it is not mentioned in the literature. In [6], the estimated input capacitance is 0.4 pF which is approximately 22 times higher than the input capacitance of our proposed LNA.

IV. CONCLUSION

We presented a low power, low noise current-reuse fully differential OTA with inverter-based input stage having an energy efficient common-mode feedback which ensures the conversion from single-ended to differential with approximately zero power for a listening time of 500 μs. To broaden the applications this circuit can be used for, we also investigated what is the K for the best FOM. We obtained the best FOM, equal with 3.16 $MHz^{3/2}/(\mu WnVfF)$, for K equal with 5.

The simulation performances are: a power consumption of 55 μW, an input-referred noise of 19 $nV/\sqrt{Hz}$, an input capacitance of 18 fF and a unity gain frequency of 280 MHz, for a capacitive load of 200 fF.

REFERENCES

[1] Virani, Salim S., et al. "Heart disease and stroke statistics—2021 update: a report from the American Heart Association." Circulation (2021): CIR-0000000000000950.

[2] G. Zamzmi, L. -Y. Hsu, W. Li, V. Sachdev and S. Antani, "Harnessing Machine Intelligence in Automatic Echocardiogram Analysis: Current Status, Limitations, and Future Directions," in IEEE Reviews in Biomedical Engineering, vol. 14, pp. 181-203, 2021, doi: 10.1109/RBME.2020.2988295.

[3] C. Papadacci, V. Finel, O. Villemain, M. Tanter and M. Pernot, "4D Ultrafast Ultrasound Imaging of Naturally Occurring Shear Waves in the Human Heart," in IEEE Transactions on Medical Imaging, vol. 39, no. 12, pp. 4436-4444, Dec. 2020, doi: 10.1109/TMI.2020.3020147.

[4] S. Firouz, E. N. Aghdam and R. Jafarnejad, "A Low Power, Low Noise, Single-Ended to Differential TIA for Ultrasound Imaging Probes," in IEEE Transactions on Circuits and Systems II: Express Briefs, vol. 68, no. 2, pp. 607-611, Feb. 2021, doi: 10.1109/TCSII.2020.3018223.

[5] P. Wang and T. Ytterdal, "A 54- μW Inverter-Based Low-Noise Single-Ended to Differential VGA for Second Harmonic Ultrasound Probes in 65-nm CMOS," in IEEE Transactions on Circuits and Systems II: Express Briefs, vol. 63, no. 7, pp. 623-627, July 2016, doi: 10.1109/TCSII.2016.2530318.

[6] T. Lin, C. Wu and M. Tsai, "A 0.8-V 0.25-mW Current-Mirror OTA With 160-MHz GBW in 0.18-μm CMOS," in IEEE Transactions on Circuits and Systems II: Express Briefs, vol. 54, no. 2, pp. 131-135, Feb. 2007, doi: 10.1109/TCSII.2006.886465.

[7] F. M. Yaul and A. P. Chandrakasan, "A Noise-Efficient 36 nV/$\sqrt{}$ Hz Chopper Amplifier Using an Inverter-Based 0.2-V Supply Input Stage," in IEEE Journal of Solid-State Circuits, vol. 52, no. 11, pp. 3032-3042, Nov. 2017, doi: 10.1109/JSSC.2017.2746778.

[8] S. Mondal and D. A. Hall, "A 13.9-nA ECG Amplifier Achieving 0.86/0.99 NEF/PEF Using AC-Coupled OTA-Stacking," in IEEE Journal of Solid-State Circuits, vol. 55, no. 2, pp. 414-425, Feb. 2020, doi: 10.1109/JSSC.2019.2957193.

[9] J. Zhang, H. Zhang, Q. Sun and R. Zhang, "A Low-Noise, Low-Power Amplifier With Current-Reused OTA for ECG Recordings," in IEEE Transactions on Biomedical Circuits and Systems, vol. 12, no. 3, pp. 700-708, June 2018, doi: 10.1109/TBCAS.2018.2819207.

[10] L. Shen, N. Lu and N. Sun, "A 1-V 0.25- μW Inverter Stacking Amplifier With 1.07 Noise Efficiency Factor," in IEEE Journal of Solid-State Circuits, vol. 53, no. 3, pp. 896-905, March 2018, doi: 10.1109/JSSC.2017.2786724.

[11] Duque-Carrillo, J.F. Control of the common-mode component in CMOS continuous-time fully differential signal processing. Analog Integr Circ Sig Process 4, 131–140 (1993). https://doi.org/10.1007/BF01254864

[12] O. Choksi and L. R. Carley, "Analysis of switched-capacitor common-mode feedback circuit," in IEEE Transactions on Circuits and Systems II: Analog and Digital Signal Processing, vol. 50, no. 12, pp. 906-917, Dec. 2003, doi: 10.1109/TCSII.2003.820253.

[13] R. Castello and P. R. Gray, "A high-performance micropower switched-capacitor filter," in IEEE Journal of Solid-State Circuits, vol. 20, no. 6, pp. 1122-1132, Dec. 1985, doi: 10.1109/JSSC.1985.1052449.

[14] K. Naderi, E. H. T. Shad, M. Molinas, A. Heidari and T. Ytterdal, "A Very Low SEF Neural Amplifier by Utilizing a High Swing Current-Reuse Amplifier," 2020 XXXV Conference on Design of Circuits and Integrated Systems (DCIS), Segovia, Spain, 2020, pp. 1-4, doi: 10.1109/DCIS51330.2020.9268627.

[15] Y. Berg, T. S. Lande, O. Naess and H. Gundersen, "Ultra-low-voltage floating-gate transconductance amplifiers," in IEEE Transactions on Circuits and Systems II: Analog and Digital Signal Processing, vol. 48, no. 1, pp. 37-44, Jan 2001, doi: 10.1109/82.913185.

[16] T. Ytterdal and S. Aunet, "Compact low-voltage self-calibrating digital floating-gate CMOS logic circuits," 2002 IEEE International Symposium on Circuits and Systems. Proceedings (Cat. No.02CH37353), 2002, pp. V-V, doi: 10.1109/ISCAS.2002.1010723.

SMACD / PRIME 2021 | 19 – 22 July 2021, Online Event

Low Power High Linearity 14-23 GHz SiGe HBT Downconversion Mixer

Syed Sharfuddin Ahmed
Institute of Electron Devices and Circuits
Ulm University
89081 Ulm, Germany
syed.ahmed@uni-ulm.de

Hermann Schumacher
Institute of Electron Devices and Circuits
Ulm University
89081 Ulm, Germany
hermann.schumacher@uni-ulm.de

Abstract—This paper presents a low power, highly linear RF to IF down-conversion mixer operating at 14-23 GHz RF frequency in a $0.13\mu m$ SiGe:C BiCMOS technology. The mixer is designed as a modified Gilbert cell topology where the RF transconductance stage is replaced by a passive input matching network, thus reducing the supply voltage requirement and increasing linearity. The mixer consumes only 2.08 mW of power from a 1.5 V supply voltage while exhibiting a conversion gain of around -3 dB and a noise figure (NF) of <12 dB at the frequency of interest. The down-conversion mixer has input 1dB compression point (IP1dB) of -5 dBm and third-order intercept point (IIP3) of +2 dBm. The results show that the presented down-conversion mixer achieves comparable performance to state-of-the-art mixers at this frequency range consuming very low power.

Index Terms—K-band; mixer; down-conversion ; satellite communications; phased-array; low power ; high linearity

I. Introduction

In recent years, there has been increasing interest and demand for electronically steerable phased-array antenna systems in mobile satellite-enabled internet services (Satcom-on-the-move) due to its beamforming flexibility and capability to generate multiple beams to deliver multiple data streams. Such a system is often realized using high-performance microwave/millimeter-wave integrated circuits (MMICs) in the transmit/receive (T/R) modules. Due to the proximity of the antenna elements, especially in full-duplex systems, receive modules of such systems often require high linearity to handle substantial in-array cross-talk without degrading the SNR. Moreover, achieving EIRP and G/T specifications of satellite systems require hundreds of T/R modules, making their power consumption a critical challenge. Being one of the first blocks in such a receiver chain and frequently following a high gain LNA, a highly linear down-conversion mixer is often desired. Additionally, the mixer needs to meet the requirements of a low noise figure and sufficient conversion gain while having low power consumption.

In past years different techniques were proposed to increase the mixer linearity and reduce power consumption. Passive mixers, despite having better linearity, suffer from high conversion loss and need a higher local oscillator (LO) drive power, which negates the power consumption advantage of the mixer core [1]. A current mirror structure presented in [2] and a multiple-gate-transistor technique in [3] exhibits high linearity

but only for low-frequency operation. Low power and low LO drive level with a bias point in the weak inversion region is proposed for CMOS in [4], but it suffers from poor linearity.

In this paper, a down-conversion mixer working at the input frequency of 14 to 23 GHz is presented targeting highly linear operation while maintaining a relatively low noise figure (NF) and moderate conversion gain while consuming low dc power. The down-conversion mixer presented here uses the concept as seen in [5] [6]. A $0.13\mu m$ SiGe:C BiCMOS technology is used to design the mixer. This technology has a transit frequency f_t of 250 GHz, and maximum frequency of oscillation f_{max} of 300 GHz. The process offers two thick metal layers with relatively low sheet resistance and five thin metal layers. In the following section, the design steps of the mixer are explained. Later, post-layout simulation results are presented in section III.

II. Mixer Design

A. Circuit Design

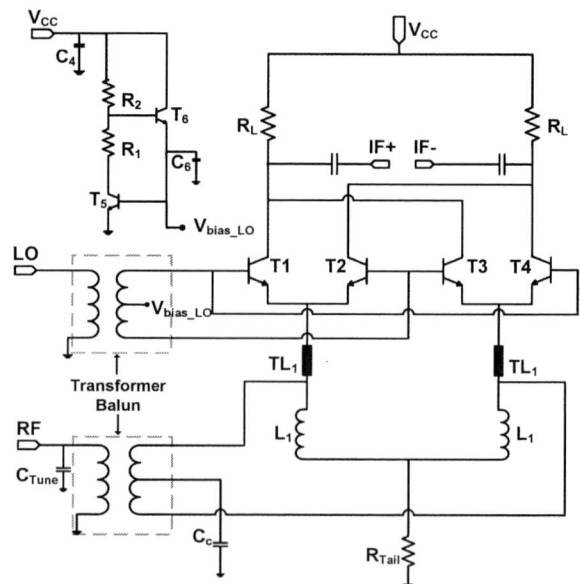

Fig. 1: Schematic of the Downconversion Mixer

© VDE VERLAG GMBH · Berlin · Offenbach

The circuit schematic of the down-conversion mixer is derived from the double-balanced Gilbert cell and presented in Fig.1.

In a conventional double-balanced Gilbert cell mixer topology three transistors are stacked to form the transconductance stage, mixing stage, and tail current source. The stacking of these transistors makes the supply voltage requirement high, increasing the power consumption while lowering the headroom thus limiting the linearity performance. Additionally, parasitic capacitance from the input stage narrows the operational bandwidth of the mixer at millimeter-wave frequencies. Thus, in our design, the transconductance stage is replaced with a passive matching circuit which increases linearity and reduces the required supply voltage resulting in reduced power consumption.

The on-chip baluns on the RF and LO ports perform single-ended to differential signal conversion. To achieve better input matching a tuning capacitor C_{Tune} is used at the RF input to tune the RF transformer balun. At the center tap of the secondary side of the RF balun, a large value capacitor C_C provides an ac ground resulting in better RF performance and enhance in the common-mode stability of the circuit. The inductors L_1 and transmission lines TL_1 along with the RF balun provide the input matching and perform the RF voltage to current conversion.

The multiplication of the RF and LO signals is performed by the LO switching quad, consisting of four HBT transistors T_1 to T_4. The optimal size of the transistors is determined through simulation. A small transistor has smaller junction capacitance which increases switching speed while a larger transistor has a lower base resistance that reduces the overall noise contribution. Also, biasing larger transistors for the same f_T requires a higher current than smaller transistors, resulting in higher conversion gain but increasing power consumption. Therefore, considering all the trade-offs size of the transistors T_1 to T_4 are set to emitter lengths of $l_e = 3 \times (0.12 \times 2\mu m^2)$.

The tail-current source of the mixer is removed to further reduce the supply voltage and instead, a small resistance R_{Tail} of 45Ω is used to provide further common-mode rejection. The resistance R_L of 510Ω is chosen as the load resistance performing the IF current to voltage conversion.

The mixer was biased with a current mirror with base current compensation technique and required a supply voltage V_{CC} of 1.5 V. Transistors T_5 and T_6 from the biasing network along with resistors R_1 and R_2.

B. Layout

Careful design and optimization of the passive input stage of the mixer consist of inductors, balun, and transmission line was done through ADS EM simulator Momentum. Using a passive transconductance stage enhances the linearity and decreases the power consumption at cost of the area. So the overall layout was performed to maximize mixer performance and reduce the total chip area.

The balun and the inductors L_1 are the most area-consuming component in the layout. The balun is designed as a trans-

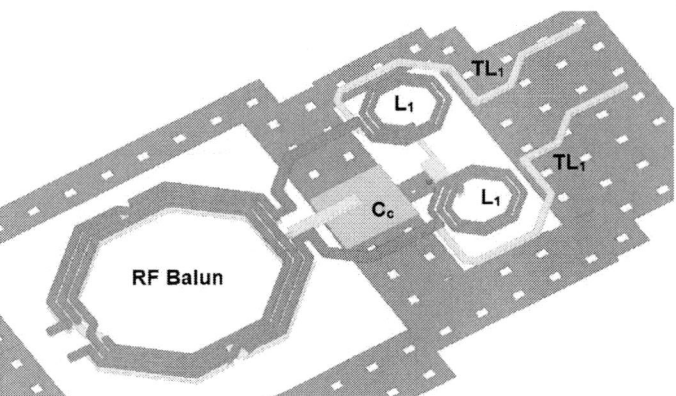

Fig. 2: 3D EM model of the transmission lines, inductors and the RF input balun.

former in an interleaved configuration, with 2:1 turns. The interleaving of metal traces provides the advantages of improved mutual coupling. To reduce the attenuation in the transmission path, other than the metal crossover sections, the transformers are designed by stacking the top two metal layers which have the lowest sheet resistance among the seven metal layers available in the technology. The balun exhibits maximum insertion loss of about -1.4 dB while having lower than 0.3 dB amplitude imbalance and less than 0.9^0 phase imbalance at the frequency of operation. The balun has an outer diameter of $148\mu m$.

The inductors L_1 are also designed with the topmost metal layer to have low resistance and a high-quality factor (Q) of 13.

Since the balun, inductors, and transmission lines make up the RF stage, all these components were EM simulated together in ADS Momentum and tuned accordingly to ensure accurate design. EM simulation results show that an optimized design is achieved by having inductor L_1 of 600 pH and TL_1 with a length of $98\mu m$. The symmetrical structure as shown in Fig.2 resulted in a more compact layout and better performance.

All the components of the mixer are carefully laid out to minimize the parasitics generated by the connection lines. The inner connection of quad transistors and the connection from the emitter of quad transistors to the matching network through TL1 lines are made symmetric, but the symmetry in the LO lines had to be compromised a little. In general, to reduce the interconnect parasitics, routing was performed as much as possible using the top two metal layers.

After the full layout, all the interconnects were fully EM simulated using ADS Momentum, and then the design was modified accordingly. To ensure good RF grounds for the supply and bias voltages, capacitances were distributed throughout the layout. The dc connections were routed using the two top metal layers to reduce the series resistance. The bottom metal layer Metal 1 was used for routing the ground.

III. Mixer Results

The initial design and simulation of the mixer were performed using Keysight ADS software. Design of the passive elements and full EM simulation of the layout was done using ADS Momentum. Cadence Virtuoso was used to finalize the layout and the Assura QRC tool was used for parasitic extraction. The circuit components were tuned and the layout was optimized after observing the effect of the layout and the pads. Finally, post-layout simulation was performed and the results are presented here. All the results include the RF balun insertion loss.

First, the required LO power was determined. The conversion gain of the mixer is a function of the LO power as shown in Fig 3a. With a low LO power, the switching quad will not fully turn on leading to less conversion efficiency, and thus less conversion gain. With increasing LO power the conversion gain of the mixer increases until the LO power is large enough to fully activate the switching quad. Increasing the LO power further will drive the switching devices into the saturation region. The optimum LO power for driving this mixer is found out to be -5 dBm while keeping the RF frequency (f_{RF}) fixed at 20 GHz and LO frequency at 20.3 GHz.

Then to determine the mixer bandwidth, the RF frequency was swept from 10 GHz to 30 GHz while keeping the LO frequency fixed at 19.9 GHz with -5 dBm power. The input power of the RF signal was kept well below the input compression point of the mixer at -30 dBm. Fig.3b shows the simulated result. For RF frequencies below the LO frequency, the conversion gain increases with increasing IF frequency and peaks at 2.5 GHz IF, then decreases with a further increase of IF frequency. Overall a flat conversion gain of -3 dB (± 0.8 dB) is achieved over a wide input frequency range from 14 GHz to 23 GHz.

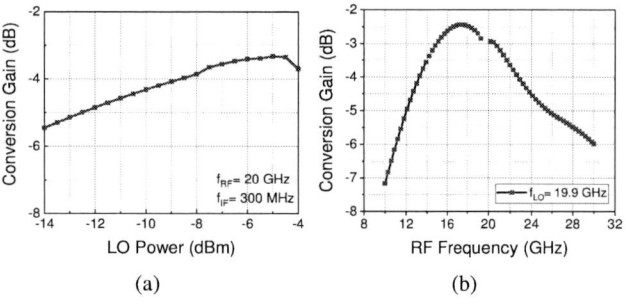

(a)

(b)

Fig. 3: Mixer (a) Conversion Gain vs LO power at fixed $f_{RF}=$ 20 GHz (b) Conversion Gain vs RF Frequency at fixed LO at 19.9 GHz with -5dBm of power

Fig. 4a and 4b correspondingly show the conversion gain and noise figure versus the RF frequency of the mixer where the IF frequency was fixed at 300 MHz and 2 GHz while the LO frequency was changed accordingly keeping LO power at -5 dBm. The mixer exhibits a minimum noise figure of 9.5 and overall has less than 12 dB of noise figure at the frequency range of interest.

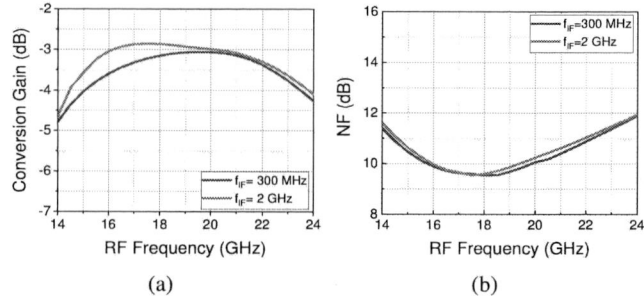

(a)

(b)

Fig. 4: Mixer (a) Conversion Gain vs RF Frequency (b) Noise Figure vs RF Frequency at fixed IF frequencies

The mixer port-to-port isolation was also checked through simulation. Ideally, such a double-balanced topology has a very good port-to-port isolation but a mismatch in the layout, parasitic capacitances lead to port-to-port leakage. To check the port-to-port isolation of the mixer, the IF frequency was kept fixed, and RF and LO frequencies were swept. As shown in Fig.5a the mixer shows LO-to-IF and RF-to-IF suppression of better than 35 dB. The mixer has high LO-to-RF isolation of more than 50 dB over the frequency of interest. One of the possible reasons for such high isolation is the removal of the active transconductance stage where LO to RF leakage occurs via the parasitic capacitance of the input transistors.

To determine the mixer linearity performance both input 1dB compression point and third-order intercept point are checked through harmonic balance simulation. To check the IP1dB of the mixer over the RF frequency range, RF input power was swept at different RF frequencies while keeping the IF frequency fixed at 300 MHz and 2 GHz. In both cases, IP1dB was found to be around -5.7 dBm over the input frequency range. The IIP3 was simulated with a two-tone harmonic balance simulation with a 400 kHz tone separation. The obtained IIP3 is around +2.2 dBm as shown in Fig.5b.

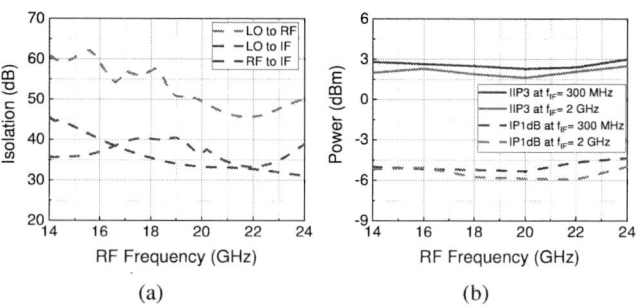

(a)

(b)

Fig. 5: Mixer (a) Port-to-Port Isolation vs RF Frequency (b)1-dB compression and third-order intercept point vs RF Frequency, at fixed IF frequencies

To evaluate the robustness of the mixer under PVT (Process, Voltage, Temperature) variation corner simulation was performed. The temperature was varied from -40^0C to 125^0C while the supply voltage variation was $\pm10\%$. The mixer

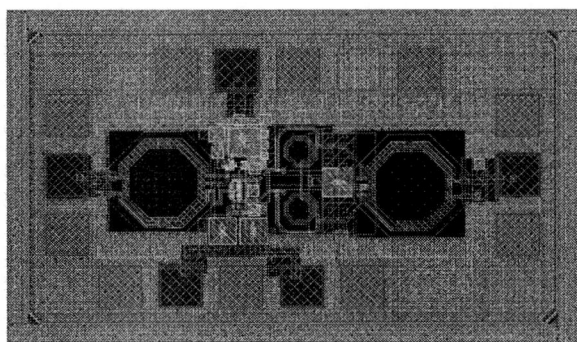

Fig. 6: Mixer Layout

conversion gain and 1 dB compression point are decreased by 0.8 dB and 1 dB respectively in the worst-case model.

The mixer consumes only 2.08 mW of power. In Fig. 6 layout of the mixer is shown. It has an area of $900 \times 480 \mu m^2$. The pads are also included in the area calculation. The comparisons between different published mixers are summarized in Table I. The presented mixer demonstrates a significant linearity improvement under low LO drive and low dc power while maintaining good CG and wide bandwidth.

IV. CONCLUSION

This paper demonstrated a low power, high linearity mixer operating at 14 to 23 GHz RF frequency. The modified Gilbert cell design methodology was explained briefly. Despite consuming a very low power of 2.08 mW, the mixer delivers gain and noise figure performance compared to other state-of-the-art active mixers. Its low power and high linearity make it very suitable for large scale phased array applications such as the SatCom. The mixer has been submitted for fabrication. Due to careful optimization during and post layout, the measurement results are expected to be very similar to the post-layout simulation results presented here.

REFERENCES

[1] Chen, Jung-Hau, et al. "A 15-50 GHz broadband resistive FET ring mixer using 0.18-μm CMOS technology." 2010 IEEE MTT-S International Microwave Symposium. IEEE, 2010.

[2] Shi, Long Xing, et al. "A 1.5-V current mirror double-balanced mixer with 10-dBm IIP3 and 9.5-dB conversion gain." IEEE transactions on circuits and systems II: express briefs 59.4 (2012): 204-208.

[3] Kim, Yeo Myung, Honggul Han, and Tae Wook Kim. "A 0.6-V+ 4 dBm IIP3 LC folded cascode CMOS LNA with g m linearization." IEEE Transactions on Circuits and Systems II: Express Briefs 60.3 (2013): 122-126.

[4] Zhang, Yaxin, et al. "12-mW 97-GHz low-power downconversion mixer with 0.7-V supply voltage." IEEE Microwave and Wireless Components Letters 29.4 (2019): 279-281.

[5] Ciocoveanu, Radu, et al. "A 1.8-mW low power, PVT-resilient, high linearity, modified Gilbert-cell down-conversion mixer in 28-nm CMOS." 2018 IEEE 18th Topical Meeting on Silicon Monolithic Integrated Circuits in RF Systems (SiRF). IEEE, 2018.

[6] Peng, Yao, et al. "A K-Band High-Gain and Low-Noise Folded CMOS Mixer Using Current-Reuse and Cross-Coupled Techniques." IEEE Access 7 (2019): 133218-133226.

[7] Testa, Paolo Valerio, et al. "A Low-Power Low-Voltage Down-Conversion Mixer for 5G Applications at 28 GHz in 22-nm FD-SOI CMOS Technology." 2020 IEEE Asia-Pacific Microwave Conference (APMC). IEEE, 2020.

[8] Y. -Chang, Yu-Teng, and Kun-You Lin. "A 28-GHz Bidirectional Active Gilbert-Cell Mixer in 90-nm CMOS." IEEE Microwave and Wireless Components Letters 31.5 (2021): 473-476.

[9] Zhu, Fang, et al. "A broadband low-power millimeter-wave CMOS downconversion mixer with improved linearity." IEEE Transactions on Circuits and Systems II: Express Briefs 61.3 (2014): 138-142.

[10] Mazor, N., et al. "Highly linear 60-GHz SiGe downconversion/upconversion mixers." IEEE Microwave and Wireless components letters 27.4 (2017): 401-403.

TABLE I :Performance Summary and Comparison

	RF Frequency (GHz)	CG (dB)	NF (dB)	IP1dB (dBm)	IIP3 (dBm)	PLO (mW)	Vcc (V)	Pdc (mW)	Technology
This Work	*14-23*	-3	11	-5	2	-5	1.5	2.08	130 nm SiGe BiCMOS
[8] 2021	25-31	-3.28	14	-13	-	1	1.2	6.4	90 nm CMOS
[7] 2020	25-31	12	12	-14.8	-	0	1.2	25	22 nm FD-SOI
[6] 2019	*23-25*	26	*7.7*	-17.8	-17.5	-3	1.5	16.8	130 nm SiGe BiCMOS
[4] 2019	97	5	13.6	-20	-	-5	1.2	12	130 nm SiGe BiCMOS
[10] 2017	57-66	-6	15	-	2	0	1.5	15	130 nm SiGe BiCMOS
[9] 2014	20-50	-1	16	-1	9.5	0	1.5	6	130 nm CMOS
[1] 2010	15-50	-15.5	15.5	5	-	10	-	0	180 nm CMOS

A Mixer-Embedded Low Noise Amplifier for Mixer-First Direct-Conversion Wake-Up Receivers

Christopher Nardi, Alexander Kronig, Ralf Wunderlich, and Stefan Heinen
Integrated Analog Circuits and RF Systems
RWTH Aachen University, Aachen, D-52062, Germany
Email: christopher.nardi@ias.rwth-aachen.de

Abstract—This paper presents the design of an integrated mixer-embedded low-noise amplifier (LNA) for mixer-first direct down-conversion wake-up receivers suited for IEEE 802.11ba. An efficient self-biased common gate-common source (CG-CS) coupling LNA with body coupling for improved noise figure (NF) was combined with three passive two-path mixers with capacitive load. This way, large decoupling capacitors can be avoided and antenna impedance matching without the need for external matching circuitry is achieved. Simulations show a high gain which can be varied between 52 dB and 16 dB, an NF of less than 9.4 dB and a current consumption of only 35 μA at 900 mV supply voltage. In its lowest gain setting, an IIP3 of −22 dBm is achieved. The structure is implemented in a 28 nm CMOS technology. Currently, only post-layout simulation data is available and a test die for collecting experiment data is being manufactured.

Index Terms—Mixer, LNA, Low Power, Wake-Up Receiver, IEEE 802.11ba, 28 nm CMOS

I. INTRODUCTION

The need for low power consumption is an ever present challenge in the design of modern wireless communication applications. Power consumption is directly connected to battery lifetime and, thus, crucial for mobile devices. A means to tackle this challenge without inducing additional latency, is the use of a wake-up receiver (WURx), a concept whose effectiveness has already been proven of value during the last decade and which is still evolving [1].

The project IEEE 802.11ba generalizes specifications for wake-up receivers used with the well established and widely adopted Wi-Fi standard [2]. The purpose of this is to make the wake-up approach a viable and well-defined option for legacy Wi-Fi devices without the need for additional hardware besides the wake-up receiver itself. The used modulation scheme is a combination of the simple on-off keying (OOK) commonly used in WURx and orthogonal frequency division multiplexing (OFDM), which is the default modulation in Wi-Fi devices. Requirements on the WURx frontend are given by the Wi-Fi requirements themselves. Hence, a minimum sensitivity of −82 dBm is necessary to not limit the main radio's communication range. Furthermore, the standard sets a maximum power budget of 1 mW to still provide a sufficient advantage over the main receiver's consumption [2]. A noise figure (NF) requirement can be determined by the aimed packet-error-rate (PER) via its corresponding signal-to-noise-ratio (SNR) at the end of the signal chain. Taking into account the target sensitivity, [2] and [3] calculated 14 dB.

In [2] and [3], mixer-first direct down-conversion WURx architectures were implemented. Given the aforementioned requirements, a mixer-first architecture seems to be well suited for 802.11ba WURx, shifting power-hungry RF gain and possible external high-Q filtering to baseband (BB) while the standard still provides a sufficiently relaxed NF requirement and enough power budget for a dedicated local oscillator (LO). Furthermore, the use of a mixer provides frequency selectivity as opposed to a tuned RF architecture where down-conversion is performed by a non-linear device like a diode [4].

The overall WURx frontend NF in a mixer-first configuration is dominated by the mixer or the combination of a mixer and the following low-noise amplifier (LNA) which, thus, has to provide low NF and high gain. In [3], a third-harmonic down-conversion approach is used which allows for a lower base frequency in the LO. However, they need three passive differential mixers and three baseband LNAs. By choosing the phases of the three LO signals properly, the components down-converted by the third harmonic add constructively in phase while the others cancel out each other. An additional transimpedance amplifier (TIA) is needed after the three baseband LNAs to perform the addition of the signals in the current domain and convert the result back into the voltage domain. An off-chip transformer is used for RF matching. Likewise, it provides additional passive voltage gain to still meet the sensitivity requirement.

Reference [2] uses a mixer-embedded interleaved dynamically biased baseband LNA to avoid large biasing circuitry. In a reset phase, a capacitor is charged by a current to then provide the bias voltage for the amplifier in an active phase. The interleaved LNA consists of two single amplifiers. The phases periodically change in such a way that one is always active when the other one is in reset mode. NF and gain are further improved by using a common gate-common source (CG-CS) coupling LNA architecture. Efficiency is enhanced by using it in a current reuse structure. Consequently, two double-balanced passive mixers, a pMOS one and an nMOS one, are used. RF matching is achieved by making use of the Wi-Fi main radio's matching network. Although this approach achieves an even better performance, additional circuit overhead in the form of an extra clock for the interleaved dynamically biased amplifier and a multiplexer to combine its two output signals is needed.

In this work, a new mixer-embedded baseband LNA is pre-

© VDE VERLAG GMBH · Berlin · Offenbach

SMACD / PRIME 2021 | 19 – 22 July 2021, Online Event

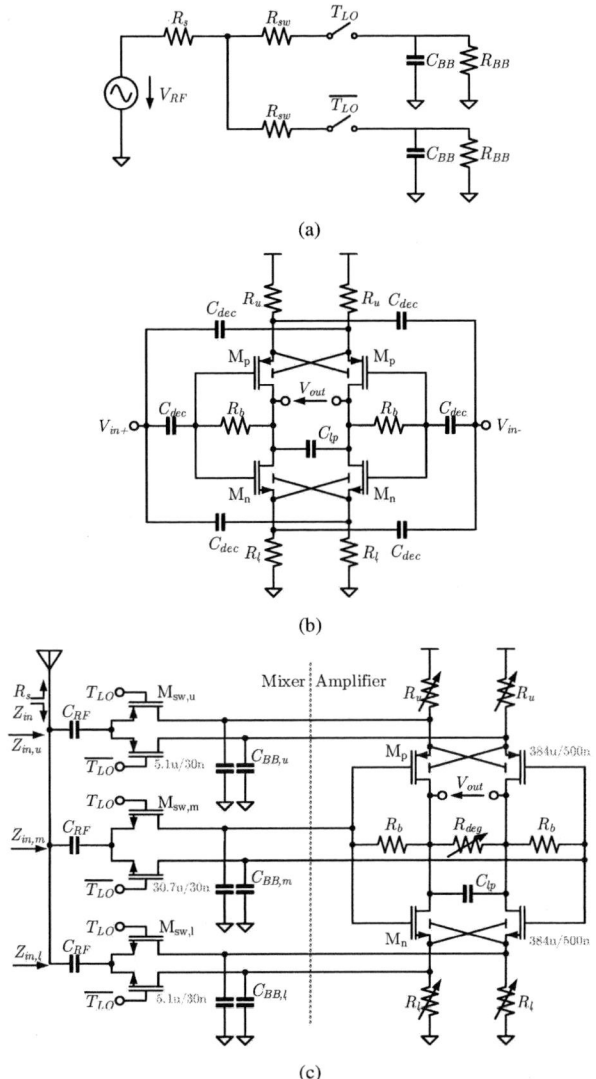

Fig. 1. (a) Two-path passive mixer with capacitive load, (b) Current reuse common gate-common source coupled LNA with shunt-feedback and body coupling, and (c) combined mixer-embedded LNA

sented. It combines the benefits of passive two-path mixers [5] and a CG-CS coupled current reuse LNA [6], achieving input impedance matching without any external matching circuitry, a higher voltage gain and a lower NF while consuming a similar amount of current as the corresponding structures in [2] and [3] without the need for a second amplifier or an extra clock signal for dynamic biasing. The present post-layout simulation results show an excellent suitability for potentially more advanced mixer-first wake-up receiver frontends that are suited for IEEE 802.11ba.

II. MIXER-EMBEDDED LOW NOISE AMPLIFIER

Fig. 1 shows the structures used for the mixer-embedded LNA. A single-balanced two-path passive mixer with capacitive load [7] is used for the mixing stage. A differential LO

signal is applied to it for down-converting the incoming RF signal and likewise up-converting the baseband impedance which can be described by a linear time invariant equivalent model [5]. The baseband impedance, which is in this case provided by the amplifying stage, and the switch resistance, which is given by the size of the mixer switches and their DC bias level, hence, determine the impedance seen at the RF side and can be used for antenna impedance matching.

For the amplifying stage, a current reuse CG-CS LNA with shunt-feedback and body-coupling as in [6] is used. It is shown in Fig. 1(b). The body-coupling further helps to avoid the body effect and boosts the amplifying transistors' g_m by their g_{mb} which increases gain and reduces NF. This LNA implementation is self-biased; the according DC voltage levels at the inputs of the CG stages are generated via the voltage drop at the upper and lower load resistors, R_u and R_l, respectively. Since the design is symmetrical, the DC voltage level at the amplifier's output sets at $V_{supply}/2$. It is fed back to the CS stages' inputs via large feedback resistors R_b to provide proper input biasing. However, the same down-converted RF signal has to be applied to all three inputs, thus, their different DC levels have to be separated from each other by decoupling capacitors C_{dec}. Especially at the intended low-frequency operating region in baseband, those have to become extremely large. Therefore, the structures of Fig. 1(a) and (b) are combined.

Three mixers are connected to the three differential inputs of the CG-CS LNA, that are indicated with subscripts u, m and l, as shown in Fig. 1(c). This way, the decoupling capacitors can be shifted to the RF side in front of the mixer when the switching transistors $M_{sw,u}$, $M_{sw,m}$ and $M_{sw,l}$ work at the corresponding DC levels of the LNA input ports. This significantly reduces the capacitors' size and now only half as much are needed, which also improves the LNA's performance at very low frequencies. The upper mixer is realized with pMOS switches since the DC level at the upper LNA input is close to V_{supply}. The middle and the lower mixers consist of nMOS switches. Although the middle one has to be dimensioned larger as it is biased at $V_{supply}/2$ which greatly increases its on-resistance by effectively lowering its maximum achievable gate-source voltage. The resulting RF impedance seen by the antenna then is the combination of the three impedances provided by the three mixers in parallel. When the mixers' capacitive loads together with switch and antenna resistances result in a time constant larger than the LO period, the impedance seen into one of the three mixers is given by [8]

$$Z_{in,x}(\omega_{RF}) = R_{sw,x} + (R_{sh,x}||R_{in,LNA,x}) || \left(\frac{1}{j\omega_{BB}C_{BB,x}} \right) \tag{1}$$

with the shunt resistance modeling the power loss due to backward mixing with LO harmonics (here only at the fundamental n=1)

$$R_{sh,x} = \frac{2\gamma(R_s + R_{sw,x})R_{in,LNA,x}}{2(R_s + R_{sw,x})(1-\gamma) + R_{in,LNA,x}(1-2\gamma)}, \tag{2}$$

© VDE VERLAG GMBH · Berlin · Offenbach

SMACD / PRIME 2021 | 19 – 22 July 2021, Online Event

$\gamma = 2/\pi^2$, R_s being the antenna or source resistance, $\omega_{RF} = \omega_{LO} + \omega_{BB}$ and the subscript x indicating u, m or l. The resulting impedance seen from the antenna side then is

$$Z_{in} = Z_{in,u}||Z_{in,m}||Z_{in,l}. \tag{3}$$

Thus, for RF matching, one mixer has to provide approximately three times the antenna impedance. In terms of power requirements of an LO buffer that has to drive the switches and noise contribution through switch resistance, this represents a good compromise.

In the next section the design of such a structure will be presented and the corresponding simulation results will be shown and discussed.

III. IMPLEMENTATION AND SIMULATION RESULTS

A. Implementation in 28 nm CMOS

The proposed mixer-embedded LNA was designed in a 28 nm CMOS technology with a supply voltage of 900 mV. It is supposed to be part of an 802.11ba-compliant WURx frontend laid out for the 2.4 GHz band. Therefore, the mixers are switched with an LO frequency of 2.4 GHz. In this implementation, each mixer path is driven by the same LO phase. However, other phase distributions could be applied. For example, driving each mixer path with a phase difference of 120° results in a third harmonic down-conversion scheme similar to the one used in [3]. However, since the LNA provides a different gain at its middle path compared to its upper and lower paths, there would be no proper suppression of the fundamental tone.

As per the standard, the signal bandwidth is 4 MHz. Consequently, ω_{BB} equals 2 MHz which represents the BB LNA's bandwidth since the mixer performs direct down-conversion. An antenna impedance R_s of 50 Ω is assumed. The three mixer paths were dimensioned in such a way that an input impedance matching of better than −10 dB within ±2 MHz of the LO frequency is achieved. Since the BB LNA inputs provide relatively high impedances, $R_{in,LNA,x}$ in (2) can be assumed to approach infinity leading to a simplified expression for $R_{sh,x}$. Furthermore, considering that the term $1/(\omega_{BB}C_{BB,x})$ is quite large as well, (1) can be written as

$$Z_{in,x}(\omega_{RF}) = \frac{1}{(1-2\gamma)}(R_{sw,x} + 2\gamma R_s), \tag{4}$$

which is solely dependent on the switch resistance and serves as a reasonable estimate for the impedance seen into one mixer path. Using (4) and (3), a switch resistance of 60-70 Ω is required for impedance matching. This relation can only be applied when the baseband capacitors $C_{BB,x}$ are large enough to provide a proper time constant for switching, as mentioned further above. Here, we can make use of the fact that the BB LNA has a significant input capacitance since its input MOSFETs have been dimensioned relatively large to diminish the impact of flicker noise [9]. Thus, $C_{BB,x}$ is supported and does not need to provide all the capacitance that is needed for proper mixer operation on its own. Simulations revealed that the effect of $C_{BB,m}$, however, is very low as the middle

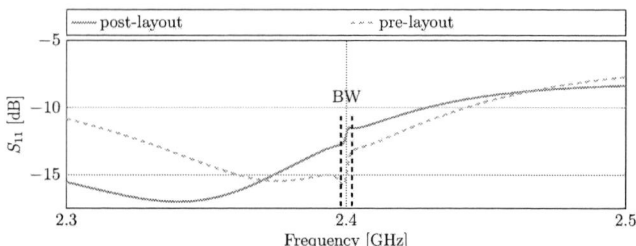

Fig. 2. Simulated post-layout and pre-layout S_{11} centered around $f_{LO} = 2.4$ GHz. The RF input matching center depends on the value of the LO frequency.

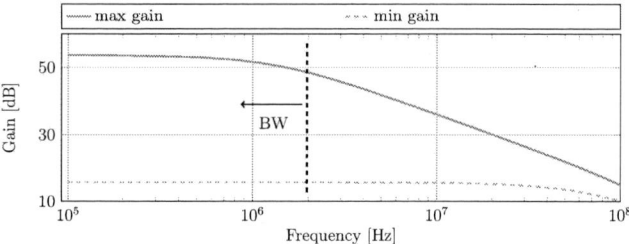

Fig. 3. Simulated post-layout voltage gain for maximum and minimum gain setting

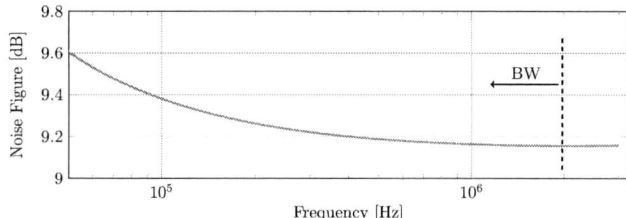

Fig. 4. Simulated post-layout noise figure

path of the mixer solely interfaces at the gates of the input MOSFETs. Hence, it was chosen to be smaller than $C_{BB,u}$ and $C_{BB,l}$ to save some die area. In fact, the RF blocking and the BB capacitors alongside with the BB LNA's input MOSFETs are the main contributors in terms of needed area. The mixer structure and the BB LNA each take up a space of approximately 0.01 mm² in layout.

B. Simulations Results

An ideal LO was assumed for simulating the structure as mixer and BB LNA are the main focus of this work. The extracted layout views were used, thus, all presented simulations are post-layout results.

The S_{11} can be seen in Fig. 2. Z_{in} is close to 50 Ω although it still has a small capacitive part mainly due to C_{RF}. The pre-layout result is also shown. It can be seen that due to parasitics in the layout, the capacitance seen at the input was increased and the frequency is shifted. This is sufficient, however, for the targeted S_{11} below −10 dB over the relevant bandwidth of 4 MHz without any additional matching circuitry. Simulations even show that a typical bondwire inductance of 1

© VDE VERLAG GMBH · Berlin · Offenbach

to 1.5 nH would further improve these values. It is possible to further narrow the matching bandwidth for better selectivity by increasing C_{BB}. This comes at the cost of an increased circuit area, however. Moreover, it should be kept in mind that due to the impedance transformation property of the passive two-path mixer, the RF properties of the structure shift with the value of the LO frequency which lies at 2.4 GHz here.

The overall voltage gain of the structure, i.e., from the RF input port to the output port of the BB LNA, is shown in Fig. 3. A high value of more than 52 dB in the operating bandwidth of 2 MHz is achieved due to the passive voltage gain provided by the impedance transformation performed by the mixer in combination with the high input impedance of the LNA. With such a high gain, to still provide a suitable linear range, an adjustable degeneration resistor R_{deg} is placed at the output of the structure. This way, the gain can be reduced down to 16 dB which is also shown in the plot. This results in an IIP3 of −22 dBm. It should be kept in mind that, since the used modulation is OOK-based, the linearity requirements on the frontend are rather low and can be achieved without additional effort. A small capacitor C_{lp} is placed at the LNA's output to support a BB filter that would follow in a complete system. Here, a value of 140 fF results in an additional gain reduction of 3 dB at 20 MHz, which is where the next Wi-Fi channel would be located. At the maximum gain setting, the gain difference between BB and 20 MHz is 21 dB, which is important for still achieving a high interferer rejection in a complete system as in [2].

Fig. 4 shows the simulated single-sideband NF. It is better than 9.4 dB between 150 kHz and 2 MHz. Due to the high gain of the structure, following stages will only have a limited impact on the overall NF, thus, providing excellent prerequisites for a high frontend sensitivity.

The structure consumes 35 μA solely through the BB LNA since the mixer is completely passive. However, the mixer determines the current consumption of the driving LO buffers. Several tests with differently dimensioned buffers that generate a square wave out of a sine wave revealed that a current consumption of around 190 μA has to be expected from a buffer that drives the mixers without significantly changing the performance.

The circuit performs well across PVT variations (500 Monte Carlo runs/3 sigma) and is functional in every corner with only some increase in current consumption in the worst case.

Table I compares the proposed structure with other works that implement a mixer-first architecture for a WURx frontend. At the time of the creation of this paper, a test die containing the mixer-embedded LNA is still being manufactured.

IV. CONCLUSION

An ultra-low power mixer-embedded LNA for mixer-first frontends was presented. A combination of three two-path passive mixers with a baseband CG-CS LNA significantly reduced the needed area for decoupling capacitors and improved achieved gain and noise performance. Simulations showed a maximum gain of more than 52 dB and a NF of less than 9.4 dB. Above the BB bandwidth of 2 MHz, the structure shows first order low-pass behavior, attenuating out-of-band interferers with respect to the passband supporting potentially following filter stages. RF impedance matching of better than −10 dB without the need for additional matching circuitry was achieved by making use of the impedance transformation property of passive mixers in each of the three parallel paths. The whole structure consumes only 35 μA. It shows an excellent suitability for the use in IEEE 802.11ba compliant wake-up receiver frontends. A layout has been designed and a prototype providing experiment data is currently being manufactured.

TABLE I
COMPARISON TO OTHER MIXER/LNAS IN MIXER-FIRST
IMPLEMENTATIONS

	This Work	[3]	[2]
Process	28 nm	40 nm	28 nm
Frequency	2.4 GHz	5.8 GHz	2.4 GHz
Supply Voltage	0.9 V	0.95 V	0.9 V
Current consumption*	35 μA**	47 μA	38 μA
Used structure	3 passive two-path mixers + Current reuse CG-CS coupled LNA with shunt-feedback and body coupling	3 passive two-path mixers + 3 IF-LNAs for 3rd-harmonic down-conversion + IF TIA	2 passive double-balanced mixers + Current reuse CG-CS coupled LNA with dynamic biasing
Gain*	16 - 52 dB**	N.A.	42 dB
Noise Figure*	9.4 dB**	14 dB	11 dB
Impedance matching	no additional matching circuitry	external transformer	matching network of Wi-Fi chip used

*Mixer and LNA structure only.
**Post-layout simulation results.

REFERENCES

[1] R. Piyare, A. L. Murphy, C. Kiraly, P. Tosato and D. Brunelli, "Ultra Low Power Wake-Up Radios: A Hardware and Networking Survey," in IEEE Communications Surveys & Tutorials, vol. 19, no. 4, pp. 2117-2157, 2017.

[2] R. Liu et al., "An 802.11ba-Based Wake-Up Radio Receiver With Wi-Fi Transceiver Integration," in IEEE Journal of Solid-State Circuits, vol. 55, no. 5, pp. 1151-1164, May 2020.

[3] J. Im, H. Kim and D. D. Wentzloff, "A 220-μW −83-dBm 5.8-GHz Third-Harmonic Passive Mixer-First LP-WUR for IEEE 802.11ba," in IEEE Trans. on Microwave Theory and Techniques, vol. 67, no. 7, pp. 2537-2545, July 2019.

[4] N. M. Pletcher, S. Gambini and J. Rabaey, "A 52 μ W Wake-Up Receiver With − 72 dBm Sensitivity Using an Uncertain-IF Architecture," in IEEE Journal of Solid-State Circuits, vol. 44, no. 1, pp. 269-280, Jan. 2009.

[5] C. Andrews and A. C. Molnar, "Implications of Passive Mixer Transparency for Impedance Matching and Noise Figure in Passive Mixer-First Receivers," in IEEE Transactions on Circuits and Systems I: Regular Papers, vol. 57, no. 12, pp. 3092-3103, Dec. 2010.

[6] S. B. T. Wang, A. M. Niknejad and R. W. Brodersen, "Design of a Sub-mW 960-MHz UWB CMOS LNA," in IEEE Journal of Solid-State Circuits, vol. 41, no. 11, pp. 2449-2456, Nov. 2006.

[7] Behzad Razavi. RF Microelectronics. Pearson education, Inc., 2012.

[8] C. Salazar, A. Cathelin, A. Kaiser and J. Rabaey, "A 2.4 GHz Interferer-Resilient Wake-Up Receiver Using A Dual-IF Multi-Stage N-Path Architecture," in IEEE Journal of Solid-State Circuits, vol. 51, no. 9, pp. 2091-2105, Sept. 2016.

[9] Behzad Razavi. Design of Analog CMOS Integrated Circuits. McGraw-Hill, 2001.

© VDE VERLAG GMBH · Berlin · Offenbach

SMACD / PRIME 2021 | 19 – 22 July 2021, Online Event

MAKE SOME NOISE: ENERGY-EFFICIENT 38 GBIT/S WIDE-RANGE FULLY-CONFIGURABLE LINEAR FEEDBACK SHIFT REGISTER

Christoph W. Wagner[□], Georg Gläser[△], Thomas Sasse[×], Gerald Kell[○], Giovanni Del Galdo[□][★]

[□]Technische Universität Ilmenau, Institute for Information Technology, Ilmenau, Germany
[△]IMMS Institut für Mikroelektronik- und Mechatronik-Systeme
gemeinnützige GmbH (IMMS GmbH), Ilmenau, Germany
[×]Technische Universität Ilmenau, Institute for Mathematics, Ilmenau, Germany
[○]Technische Hochschule Brandenburg, Fachbereich für Informatik und Medien, Brandenburg, Germany
[★]Fraunhofer IIS, Fraunhofer Institute for Integrated Circuits IIS, Ilmenau, Germany

E-Mail: christoph.wagner@tu-ilmenau.de, georg.glaeser@imms.de, gerald.kell@th-brandenburg.de, giovanni.delgaldo@iis.fraunhofer.de

ABSTRACT

Compressed Sensing (CS) and Radio Detection and Ranging (RADAR) Systems require stimulus signals with properties similar to true random signals, but deterministic and reproducible in hardware. Therefore, Pseudo-Random Noise (PRN) sequences render ideal for this purpose. Especially mm-Wave systems require very high symbol rates and hence operating frequencies. Being able to choose a PRN signal is key to achieving good system performance by means of high operating frequency and energy consumption. For operating near the extreme limits of the technology, we propose an energy-efficient fully-configurable Linear Feedback Shift Register (LFSR) architecture with synchronous reset for PRN generation based on Positive Emitter Coupled Logic (PECL). By choosing a shift-register based multi-data-rate (MDR) structure, we shift the logic paths to a low-frequency domain. Further, we construct the register from small elementary slices with two levels of Power-Shut-Off (PSO) functionality for reducing the power consumption from 12 % to 69 %. We prove our architecture in 130 nm SiGe BiCMOS technology, using transistor-level simulations and calibrated fab models. The register of length 24 is shown to operate correctly up to $f_0 = 38.5$ GHz, corresponding to $\approx$ $1/6$th of the process transition frequency. Our design draws 180 mW to 510 mW from a 2.50 V supply at a die area of 0.10 mm² (includes serializer).

Index Terms— LFSR, Hardware architecture, Multi-Data Rate, Logic-Shut-Down, Energy-Efficient, Integrated Circuit

1. INTRODUCTION

RADAR signals often are sparse, such that Compressed Sensing (CS) theory can be applied to reduce the sampling rate even below the limits known from the Shannon-Nyquist theorem [1]. Compressive signal acquisition systems rely on PRN sequences as mixing signals (see Fig. 1), which should behave like random stochastic signals albeit being

---

The PRIME research group is supported by the Carl-Zeiss Foundation. The HyLoC research group is supported by the Free State of Thuringia and the European Social Fund under the reference *2019 FGR 0100*.

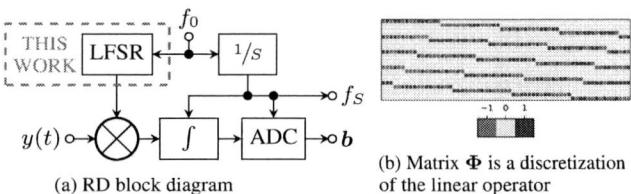

Fig. 1: Example application of this work: Random Demodulator (RD) [4] as analog hardware implementation of a linear operator

fully deterministic. An Ultra-Wideband (UWB) RADAR method, based on PRN sequences, is introduced in [2] as *M-Sequence RADAR*. A CS extension, compressing the RADAR signal in the analog domain and prior to digitization, is introduced in [3].

The core of this concept is the Random Demodulator (RD) [4] shown in Fig. 1a. The linear operator Φ (see Fig. 1b) is implemented in analog hardware to represent a set of pseudo-random linear combinations of a signal $y(t)$. According to CS theory, it is possible to retrieve the original signal, solely from observations of these combinations – if it can be assumed to be sparse in a known domain and if the matrix Φ satisfies certain conditions. The sequences are typically serialized from a variable storage or generated from a fast LFSRs with fixed (for speed) feedback coefficients. Having variable coefficients would allow for tuning the system performance by adaptively updating Φ during operation. Unfortunately this limits speed and the higher design complexity leads to and increased area and power footprint.

Especially in mm-Wave Radio Frequency (RF) bands, the PRN generator hardware realization must be able to run at the extreme limits of the technology. The maximum operating frequency is inherently limited by the used technology and logic family. Using PECL, said limits can be approached at the price of higher power consumption. Also, the fundamental issue of long feedback logic paths cannot be resolved from fast logic alone. With specialized feedback logic we tackle this problem by introducing the MDR concept for fully-configurable LFSRs. In every internal clock cycle multiple bits are produced and finally serialized to the OUTPUT bitstream at full rate.

Since multi-GHz systems consume large amounts of already tight power budgets, we propose a PSO logic scheme on two levels to improve our architecture's energy efficiency. By assembling a LFSR

© VDE VERLAG GMBH · Berlin · Offenbach

from 8 bit slices, we introduce PSO on the register level to dynamically shorten the register length to just enough. For constructing the feedback logic tree, we propose a special logic cell with complex function and an advanced switch-off scheme on gate level. An evaluation confirm a significant reduction of power consumption from our PSO measures.

2. LFSR FEEDBACK GENERATION

One class of PRN signals, that both exhibit desirable properties and that are well-implementable in hardware, are *shift register sequences* [5]. For this, a LFSR of length K implements a recursion, that describes the i-th OUTPUT bit $s[i]$ as

$$s_i = \sum_{j=0}^{K-1} t_j s_{i-K+j} \quad , \text{ where } \begin{aligned} \boldsymbol{s} &= (s_0, \ldots, s_{K-1}) \in \{0,1\}^K \\ \boldsymbol{t} &= (t_0, \ldots, t_{K-1}) \in \{0,1\}^K \end{aligned} \quad (1)$$

The sequence properties depend on the choice of $\boldsymbol{t}$ corresponding to the recursion's *characteristic polynomial* $p(x)$ of order k_p [5]. Given the finite initial state $\boldsymbol{s}$ and the chosen polynomial $p(x)$ or $\boldsymbol{t}$, the OUTPUT sequence is deterministic and periodic. The configuration vector signals START and TAPS determine $\boldsymbol{s}$ and $\boldsymbol{t}$ respectively.

Conventional configurable LFSR based generators (see Fig. 2a for $K = 4$) use a layer of AND gates to implement variable $\boldsymbol{t}$. Since the feedback logic must be fully connected, logic path timing is constrained by K. Extending to a double-data-rate (DDR) scheme, the feedback paths are allowed to be generated in half-speed, compared to the OUTPUT rate (see Fig. 2b). The OUTPUT bitstream is controlled by the CLK phase where a multiplexer acts as a simple 2-tap serializer.

This is possible, since the recursion of eq.1 can also be fully described through the linear operator/companion matrix $\boldsymbol{M}$, such that

$$\boldsymbol{\sigma}_i = \begin{bmatrix} s_i \\ \vdots \\ s_{i-K+1} \end{bmatrix} = \boldsymbol{M} \begin{bmatrix} s_{i-1} \\ \vdots \\ s_{i-K} \end{bmatrix}, \quad \boldsymbol{M} := \begin{bmatrix} t_3 & t_2 & t_1 & t_0 \\ 1 & 0 & 0 & 0 \\ 0 & 1 & 0 & 0 \\ 0 & 0 & 1 & 0 \end{bmatrix}. \quad (2)$$

In the DDR case, the register is advanced by two feedback bits per cycle, determined from the first two rows of $\boldsymbol{M}^2$. The natural extension of the DDR is the MDR case, where W feedback circuits yield W bits and OUTPUT is serialized from the W least significant bits. For the same OUTPUT rate as in the single-data-rate (SDR) case, the register runs at $1/w$ of the clock rate. The W feedback control vectors TAPSW...TAPS1 correspond to the first W rows of $\boldsymbol{M}_1^W \ldots \boldsymbol{M}_W^W$. This

resolves the timing problems associated with highest OUTPUT rates, while maintaining full LFSR configurability. However, requiring W times more combinatorial logic severely increases power consumption.

Implementing a MDR register in *Galois*-type does not yield similar speed improvements over the shown *Fibonacci*-type as one would expect from the SDR case. This not only results from more complex intra-register feedback structures (compared to the single XOR-gate in the SDR case), but also from an additionally required OUTPUT decoder.

2.1. PECL Gates with CMOS driven shut-off

The speed of digital circuitry is limited by the used logic gates. Choosing a well-suited circuit technology operating near the process node's transistor transition frequency f_t, is essential to maintain the speed advantages of the MDR approach. A logic type with speed approaching f_t is PECL, where logic is implemented using differential pairs of transistors [6]. Figure 3a shows a PECL buffer circuit.

A constant current I is steered predominantly through one of the two transistors Q1 or Q2, depending on the sign of the differential input signal $A = Ap - An$. The voltage drop over the corresponding collector resistor then results in a change of the differential output voltage $Q = Qp - Qn$. Since the transistors never fully switch nor saturate, this change can happen very fast, resulting in very high performance. If A is close to 0 V both Q1 and Q2 conduct, causing Q to also approach 0 V, effectively propagating an undesired $\mathbb{X}$ state. Additional transistors or differential pairs in the current path allow for the construction of complex logic functions and, due to differential signaling, logic inversion comes at no-cost. The constant current I is drawn from a shared current mirror, which is resembled by the *control voltage* signal CV, provided from a common biasing circuit.

One major concern of PECL designs is its high, albeit static, power consumption of each gate. In contrast to Complementary Metal Oxide Semiconductor (CMOS), logic transitions do no impact dynamic gate power consumption. Lowering the current I does, but is limited to gates outside the critical path, as it also reduces gate speed. Another technique, known as PSO, may temporarily zero a gate's power consumption by shutting it off entirely if the output is not required. Since NMOS transistors are employed as I current sources, we implement PSO by connecting the NMOS' gate either to the common CV net or ground with a specialized *break-before-make* transmission gate, which in turn is controlled from slow low-power CMOS enable signals. These must be routed alongside the differential

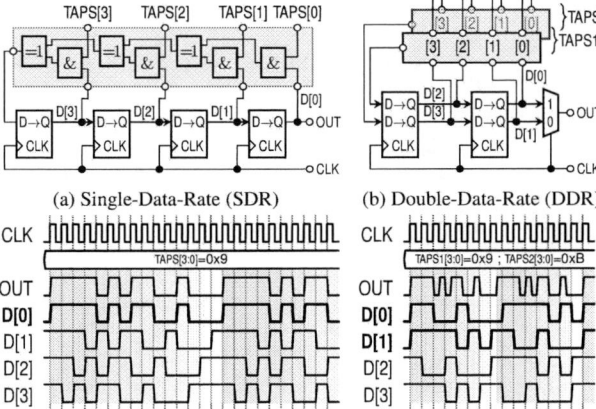

(a) Single-Data-Rate (SDR)   (b) Double-Data-Rate (DDR)

Fig. 2: Two LFSR architectures generating identical OUT bitstreams for $p(x) = x^4 + x^3 + x + 1$ (color indicates sequence period). The DDR LFSR produces two OUTPUT bits per CLK cycle.

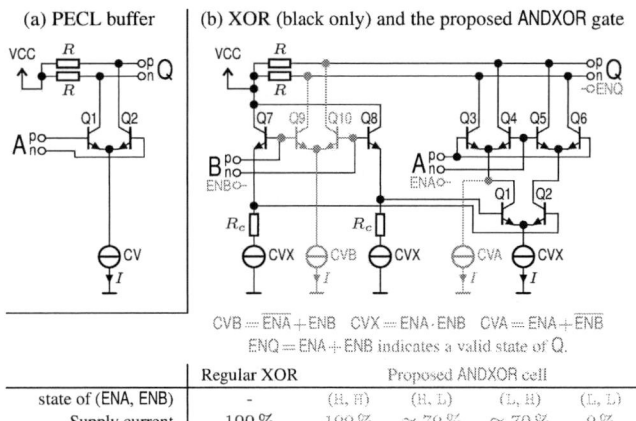

(a) PECL buffer   (b) XOR (black only) and the proposed ANDXOR gate

$CVB = \overline{ENA} + ENB \quad CVX = ENA \cdot ENB \quad CVA = ENA + \overline{ENB}$
$ENQ = ENA + ENB$ indicates a valid state of Q.

state of (ENA, ENB)	Regular XOR	Proposed ANDXOR cell			
	-	(H, H)	(H, L)	(L, H)	(L, L)
Supply current	100 %	100 %	≈ 70 %	≈ 70 %	0 %

Fig. 3: Implementation of standard PECL XOR gate (black) and its PSO extension for power-efficient LFSR feedback generation (black and red), implementing the logic function $Q = (ENA \cdot A) \oplus (ENB \cdot B)$.

signal pairs as validity indicators. An isolation cell at the end of each PSO logic path confines X states, that emerge from shut-off gates.

Figure 3b shows a XOR cell in PECL (black color), to which extra circuitry is added (red) to form our proposed ANDXOR cell. If either of the enable signals ENA or ENB is high, the corresponding input is propagated through the corresponding buffer pairs Q3/Q4 or Q9/Q10 and their switched current sources CVA or CVB. The CVX source, and therefore A⊕B functionality, is only active when both inputs are enabled at the same time. If no input is enabled, the gate is fully shut-off, resulting in Q=X and ENQ being low (indicating the invalid state of Q). The collector stages Q7/Q8 are shut-off during the direct mapping states (A/B↦Q), also reducing power consumption. One ANDXOR cell contains 56 MOS and 10 bipolar transistors in an area of only $15 \times 32\mu m$, including current sources and transmission gates.

3. PROPOSED LFSR GENERATOR ARCHITECTURE

After establishing the principles of the MDR and PSO design techniques, we propose our LFSR architecture in Fig. 4. We found a MDR factor of $W=4$ to be a good trade-off between speed (approaching f_t), power and area (we need W feedback trees). Increasing W further quickly increases total area and creates issues from excess wiring parasitics, which consume most of the gained timing improvements.

From eq. 1, and the realization of t to the characteristic polynomial, it is evident that already a small register is sufficient for sequences of small k_p. However, since silicon circuits are not (yet) dynamically adaptable, the LFSR must accommodate all circuitry required for a given maximum K in order to support it. This leaves most of the register unused in the small-k_p case, wasting power. Hence, we implement the actual LFSR from elementary slices of size $K=8$. In addition to gate-level PSO, we introduce a second PSO level for powering down whole register slices when they are not required to represent $p(x)$. Each slice also features FEEDBACK vector carry-in for carry-chaining the individual slice FEEDBACK output. We found that splitting the register in two $K=4$ sub-slices, arranged symmetrically around the feedback network that computes the required rows of M^4, allowing for a compact layout and reduced wire delays.

A tree of ANDXOR cells perform the required masking and adding operations of eq. 1 in the scope of a sub-slice. Note that the shift register sequences reside in the finite *Galois Field* GF(2), all additions must be performed modulo 2, rendering simple XOR gates sufficient for the task [5]. The input-enables of the first layer of PSO-enabled ANDXOR already implement the masking required by t of eq.1 and render a set of distinct AND gates obsolete. The output-enables of each cell are propagated to the successive layers of the tree, allowing for full shut-down of whole sub-trees within the feedback network, leading to large power savings at the gate level. Also, since the propagation paths A↦Q and B↦Q have different lengths (see Fig. 4b), proper swapping of A and B improves critical path timing.

To configure the register for a given $p(x)$, we select the active register size as $k' = 8 \cdot \lceil k_p/8 \rceil$, defining how many of the lower slices are shut down. Since all outputs of a shut-off slice have invalid X state and CMOS logic is not affected by PSO, all bits of a TAPS• vector corresponding to shut-off slices must be low to ensure that no feedback output is wrongly recognized as valid. Eventually, bypassing the data path directly prior to the first shut-off slice guarantees data path integrity by protecting from X states emitted from shut-off slices.

The difference of K' and k_p gives rise to an extra degree of freedom for power optimization. It is known from Galois Field theory, that all polynomials $p'(x) = p(x) \cdot g(x)$ produce identical sequences s_i [5]. Therefore, we select the $p'(x)$ exhibiting the lowest overall register power consumption, considering the PSO model.

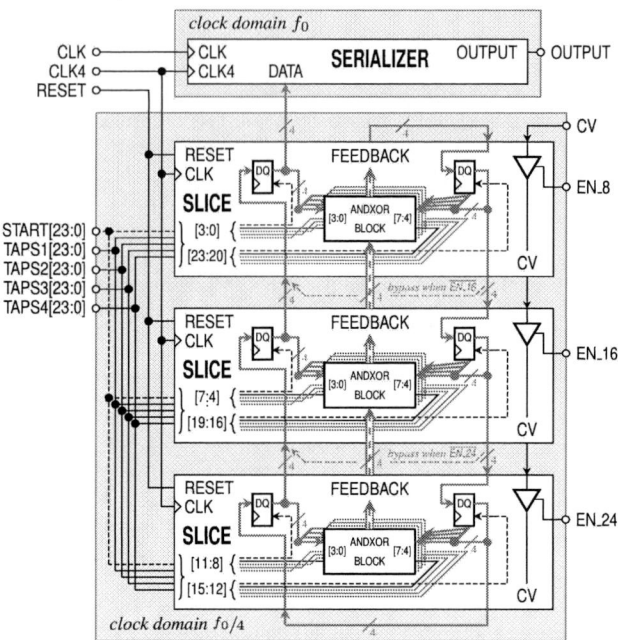

(a) Block diagram for a three SLICE design. Main data path is indicated in red, propagating through the feedback chain and registers, ending at the serializer.

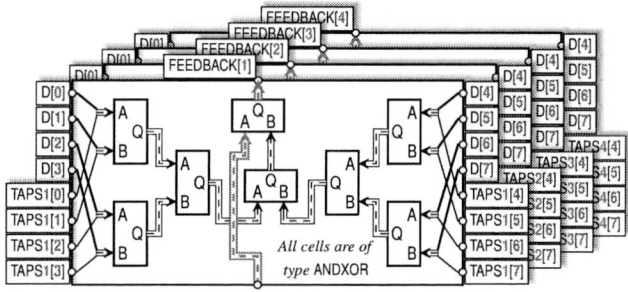

(b) The ANDXOR BLOCK block sums a masked 8 bit input vector D in GF(2).

Fig. 4: The proposed architecture, applied for a24 bit LFSR. Wire styles indicate CMOS (thin), differential PECL (thick) and the combination of both as PECL with dedicated enable (road-like).

4. EXAMPLE IMPLEMENTATION

We prove the architecture by implementation in a 130 nm SiGe process using the *Common-ECL* standard cell library of the research project *EuRISCOSi* [7]. Gate-level synthesis, placement and layout (see Fig. 6e) was performed by hand to allow for handling of RF design considerations. Due to lacking design automation tool support, timing verification was done using a hardware description language model, annotated by timing parameters derived from transistor-level logic gate simulations. The active area required for each register slice is $0.029\,\text{mm}^2$. The serializer requires additional $0.011\,\text{mm}^2$.

Power consumption is in the range of 180 mW to 510 mW for the shown $K=24$ design, depending on the chosen $p(x)$. To examine our power saving measures, we survey current consumption (after applying all described optimization measures) over all possible polynomials $p(x)$ of the orders $k_p = 4 \ldots 24$. We determine one histogram for every order k_p and compute the Cumulative Distribution Functions

(CDFs) shown in Fig. 5. The curves tell the probability (y-axis) of observing a given current consumption (x-axis), when randomly drawing a polynomial of given order k_p. For all curve points, the y-indicated proportion of all k_p polynomials exhibit lower current consumption than the x-indicated value. The current consumption of the register without PSO functions is 232 mA. The highlighted curve in Fig. 5 states a current consumption from 108 mA to 132 mA for polynomials of order $k_p = 12$. Half of those polynomials cause the register to consume less than 126 mA, indicating power savings of 46 % on average from applying the PSO scheme. Over all polynomials, the power savings range from 12 % to 69 %. Comparing all CDFs indicates that gate-level PSO attributes for a $\approx 25\%$ saving. Static consumption of one slice is ≈ 36 mA and ≈ 29 mA for the serializer.

The transistor-level simulation results shown in Fig. 6a confirm correct operation at 38.5 GHz for a Maximum Length Binary Sequence (MLBS) of order $k_p = 12$ at a supply voltage of 2.50 V and at 70 °C, using calibrated models and including noise. At a slightly reduced maximum clock rate, correct operation extends for a wide range of corners between 2.35 V to 2.65 V and 0 °C to 120 °C. The eye diagram of Fig. 6b indicates good integrity of the OUTPUT. Histogram (Fig. 6c) and spectrum (Fig. 6d) match the expected MLBS properties, indicating a correct bitstream. Since the register state is kept in the slow clock domain, OUTPUT fails gracefully when the maximum speed is exceeded, by exhibiting an elevated bit error rate.

In [8], a LFSR with fixed configuration is shown, producing a MLBS of order $k_p = 15$ at a rate up to 12 GHz, which is $\approx f_t/6$ of the used process. To the best of the authors knowledge, no designs with similar functionality were found to compare against.

5. CONCLUSION

Configurable LFSR generators are an important building blocks for advanced RADAR and CS systems. We have shown a circuit operating at the technology limits, while preserving full configurability along with low-power optimization. Due to the serializer based approach, essential register operation is moved to a slower clock domain, improving the overall stability. Timing issues may first arise from bit errors in the serializer. This also allows the less sensitive internal state to be synchronized to the external measurement equipment. Further improvements comprise the extension to multi-valued sequences and parallel generation of interleaved sequences.

6. REFERENCES

[1] J. H. G. Ender, "On compressive sensing applied to radar," *Signal Process.*, vol. 90, no. 5, p. 1402–1414, May 2010. [Online]. Available: https://doi.org/10.1016/j.sigpro.2009.11.009

[2] J. Sachs, P. Peyerl, and M. Rossberg, "A new UWB-principle for sensor-array application," in *IMTC/99. Proceedings of the 16th IEEE Instrumentation and Measurement Technology Conference (Cat. No.99CH36309)*, vol. 3, May 1999, pp. 1390–1395 vol.3.

[3] C. W. Wagner, S. Semper, F. Römer, A. Schönfeld, and G. Del Galdo, "Hardware architecture for ultra-wideband channel impulse response measurements using compressed sensing," in *2020 28th European Signal Processing Conference (EUSIPCO)*, 2021, pp. 1663–1667.

[4] J. N. Laska, S. Kirolos, M. F. Duarte, T. S. Ragheb, R. G. Baraniuk, and Y. Massoud, "Theory and implementation of an analog-to-information converter using random demodulation," in *2007 IEEE International Symposium on Circuits and Systems*, May 2007, pp. 1959–1962.

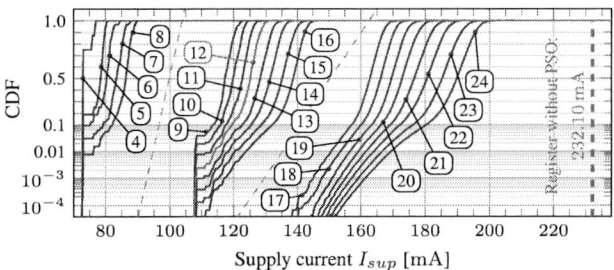

Fig. 5: Current consumption of the proposed architecture over k_p up to 24, including serializer. The text example of $k_p = 12$ is highlighted.

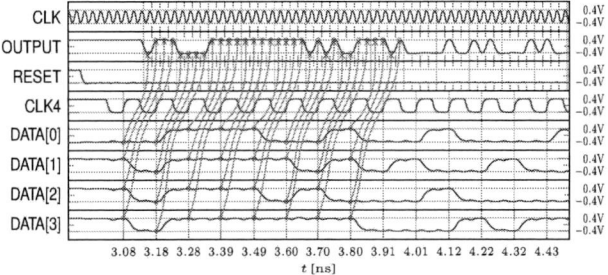

(a) Simulated OUTPUT signal for $t = $ 0x1891, $s = $ 0xFF0E and $W = 4$ using two SLICE blocks ($K = 16$), producing the MLBS $p(x) = x^{12}+x^{11}+x^7+x^4+1$

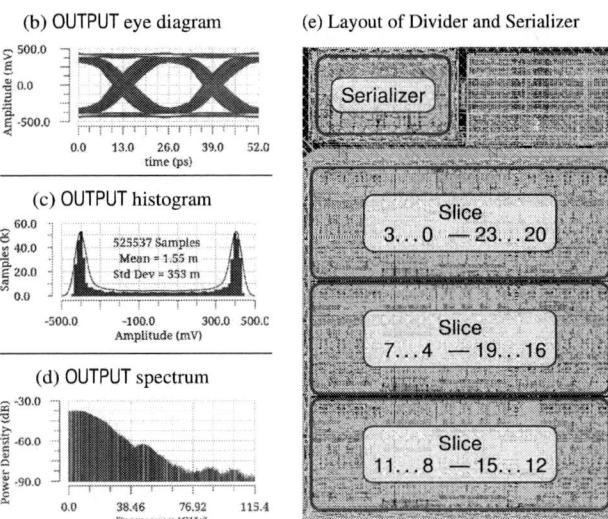

Fig. 6: Simulation results for implementation in a 130 nm SiGe BiCMOS process with $f_t = 250$ GHz, showing operation up to $f_t/6$

[5] S. W. Golomb, *Shift Register Sequences*. Laguna Hills, CA, USA: Aegean Park Press, 1981.

[6] A. Alvarez, *Introduction to BiCMOS*. Boston, MA: Springer US, 07 2011, pp. 1–20.

[7] Technische Hochschule Brandenburg (THB), Germany, "Schlussbericht zum Förderprojekt EuRISCOSi," Project Final Report 13FH069PX3, doi 10.2314/GBV:1024449688, 2018.

[8] M. Pečovský, M. Kmec, M. Sokol, P. Galajda, and S. Slovák, "New hardware components for m-sequence uwb channel sounder," in *2019 PhotonIcs Electromagnetics Research Symposium - Spring (PIERS-Spring)*, 2019, pp. 3696–3703.

SMACD / PRIME 2021 | 19 – 22 July 2021, Online Event

EVERY CLOCK COUNTS – 41 GHZ WIDE-RANGE INTEGER-N CLOCK DIVIDER

Christoph W. Wagner[□], Georg Gläser[△], Gerald Kell[×], Giovanni Del Galdo[○□]

[□]Technische Universität Ilmenau, Institute for Information Technology, Ilmenau, Germany
[△]IMMS Institut für Mikroelektronik- und Mechatronik-Systeme
gemeinnützige GmbH (IMMS GmbH), Ilmenau, Germany
[×]Technische Hochschule Brandenburg, Fachbereich für Informatik und Medien, Brandenburg, Germany
[○]Fraunhofer IIS, Fraunhofer Institute for Integrated Circuits IIS, Ilmenau, Germany

E-Mail: christoph.wagner@tu-ilmenau.de, georg.glaeser@imms.de, gerald.kell@th-brandenburg.de, giovanni.delgaldo@iis.fraunhofer.de

ABSTRACT

Current clock divider architectures suffer from either inflexible divider ranges or slow performance due to long logic paths. When implementing Compressed Sensing (CS) signal acquisition for systems operating at mm-wave Radio Frequency (RF), both flexibility and operation at the limits of the technology node are required.

We propose a configurable integer-N clock divider architecture with synchronous reset that satisfies this need. With a wide divider range of $S = 8 \ldots 1048583$, an output duty cycle of 50% is guaranteed for even, and approached for odd divider factors. The architecture is suited for operation with very high input clock frequencies, approaching the transit frequency of the technology.

The key to construct our design is a serializer based approach, that enables the control logic to operate on two levels of lower frequencies. A symbol generator provides the output symbol stream. Internal clocks are derived directly from internal state vectors. In this way, the 20 bit divider range is achieved with only 22 Flip Flops (FFs) (excluding the serializer) and no combinatory logic in the fastest clock domain.

We demonstrate our architecture in 130 nm SiGe BiCMOS technology using Positive Emitter Coupled Logic (PECL). We show that transistor-level simulation using calibrated fab models confirms successful operation up to $f_0 = 41.70$ GHz which corresponds to $\approx 1/6^{\text{th}}$ of the process transition frequency. At a die area of $0.06\,\text{mm}^2$ (including serializer), our design draws $\approx 225\,\text{mW}$ from a 2.50 V supply.

Index Terms— Clock Divider, Integer-N, Wide-Range, Hardware architecture, PECL, BiCMOS, Integrated Circuit, Signal Acquisition

1. INTRODUCTION

Clock Dividers are important building blocks, not just in RF, but also in many modern signal processing systems. In the field of Ultra-Wideband (UWB) Radio Detection and Ranging (RADAR), where system clock rates and bandwidths of several (tens of) GHz are common, clock dividers allow for employing Nyquist subsampling to reduce hardware complexity through sampling rate reduction [1].

The PRIME research group is supported by the Carl-Zeiss Foundation. The HyLoC research group is supported by the Free State of Thuringia and the European Social Fund under the reference *2019 FGR 0100*.

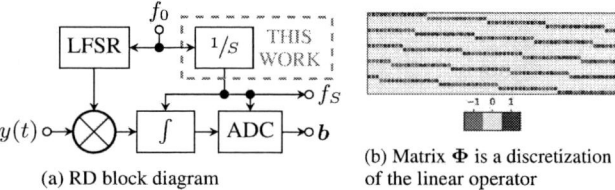

(a) RD block diagram

(b) Matrix **Φ** is a discretization of the linear operator

Fig. 1: Example application of this work: Random Demodulator (RD) [5] as analog hardware implementation of a linear operator

Since RADAR signals usually are sparse, compressed sensing theory can be applied to reduce the sampling rate even below the limits known from the Shannon-Nyquist theorem [2]. An UWB RADAR method based on Pseudo-Noise (PN) sequences is introduced in [3] as *M-Sequence RADAR*. A CS extension, compressing the radar signal in the analog domain and prior to digitization, is introduced in [4].

One core component in this concept is the Random Demodulator (RD) [5], shown in Fig. 1a. It implements a linear operator **Φ** (see Fig. 1b) in analog hardware to represent a set of pseudo-random linear combinations of the signal-of-interest. According to CS theory, it is possible to retrieve the original signal, which is assumed to be sparse in a known domain, solely from observations of these combinations. However, for this concept to function, the matrix **Φ** is required to satisfy some conditions known from CS theory. Therefore, the configuration of clock divider and Linear Feedback Shift Register (LFSR) in Fig. 1a must be chosen carefully. As the divider factor S controls the width of the apparent "stairs" in **Φ**, it has great impact on how the set of individual integration domains in **Φ** are distributed over the domain of signal $y(t)$. System performance can be tuned adaptively to measurement conditions by when S can be reconfigured.

The speed of amplifying circuits (including logic gates and modules) is limited by the transition frequency f_t, describing the frequency at which the transistor current gain drops to 1. In fact, reliable operation can only be attained up to a small fraction of $f_t = 3 \ldots 10 f_{\text{max}}$. One logic type with performance known to approach f_t is PECL. Here, logic gates are implemented using differential pairs of bipolar transistors [6]. Depending on the sign of the differential input signal, a constant current is steered predominantly through one of the two transistors, effecting change to the differential output. Hence, the transistors never fully switch, resulting in very high performance. Additional transistors or differential pairs in the current path allow for the construction of complex logic functions and, due to differential

© VDE VERLAG GMBH · Berlin · Offenbach

signaling, logic inversion comes at no-cost. However, one major drawback is high static power consumption as a result of the constant currents flowing through each gate.

To maintain high f_{max} from transistor performance and logic topology, also the higher-level architecture must be considered. For similar technology and gate design, architectures with shorter logic paths generally exhibit better performance. The design of clock dividers is usually done using simple counters or periodic LFSR [7]. Asynchronous counters are normally not used since their outputs may glitch. This includes simple Divide-by-N and Johnson counters that can be configured with a limit to a clock division ratio. Still, these counters need additional state transition logic and also exhibit long logic paths, making them a poor choice for use in a close-to-f_t regime.

2. PROPOSED CLOCK DIVIDER ARCHITECTURE

Optimizing logic circuits for speed generally offers few knobs to turn: Reducing logic path delay through flattening logic hierarchy improves speed by shortening critical logic paths. However, some paths, like carry chains in counters, cannot be reduced below a certain minimum. Multi-cycle logic aims to alleviate this problem by restructuring a design such that critical paths are permitted to complete in multiple clock cycles. All speed optimizations generally come at the expense of higher area usage and larger power consumption.

We combine these concepts in our proposed architecture by serializing symbols from a parallelized logic. The OUTPUT signal can be seen as a bit stream of $n_H = \lceil S/2 \rceil$ HIGH (H) and $n_L = \lfloor S/2 \rfloor$ LOW (L) bits with a rate determined by CLK, that divides the input clock CLK ($\mapsto f_0$) by an integer factor S. By introducing a second clock CLK4 ($\mapsto f_0/4$), we may now form a stream of four-bit-wide DATA symbols at the lower rate CLK4, that finally can be serialized to the correct OUTPUT. In this way, the timing constraints are greatly relaxed (compared to operation in the f_0 clock domain) in exchange for solving the problem of finding the right symbol sequence for a given S.

Defining $S \geq 8$ ensures that no DATA symbol can contain more than one logic transition (H$\leftrightharpoons$L)), which greatly simplifies design of the symbol generator logic. Figure 2 shows the OUTPUT waveform and the corresponding DATA symbol streams for $S = 8 \ldots 15$. We refer to this set as *fundamental streams*, since the stream $S+8$ can be constructed from the stream S by padding every sequence of consecutive H and L bits with one *wait symbol*, containing no logic transition. Further examining the streams reveal that only eight different DATA symbols occur, which can be mapped to a 3 bit SYMBOL vector (see Fig. 5). This also allows for introducing the symbol notation L0...L3 and H0...H3, where the letter indicates the state of the first bit and the number n tells the distance to the logic transition after the start bit.

S	DIV[2]	DIV[1]	DIV[0]	SYMBOL sequence, **STATE** and OUTPUT indicated
8	0	0	0	T:3H T:3L
9	0	0	1	W:3H T:0H T:0L T:1H T:1L T:2H T:2L T:3H T:3L
10	0	1	0	W:3H T:0H T:1L T:3L
11	0	1	1	W:3H T:1H T:2L W:3H T:0H T:1L T:3H W:3L T:0L T:2H T:3L
12	1	0	0	W:3H T:1H T:3L
13	1	0	1	W:3H T:2H W:3L T:0L T:3H W:3L T:1L W:3H T:0H T:2L W:3H T:1H T:3L
14	1	1	0	W:3H T:3H W:3L T:1L W:3H T:3L
15	1	1	1	W:3H T:3H W:3L T:2L W:3H T:2H W:3L T:1L W:3H T:1H W:3L T:0L W:3H T:0H W:3L T:3L

Fig. 2: Symbol and state sequence for different divider factors S. Background color in SYMBOL indicates a starting H bit.

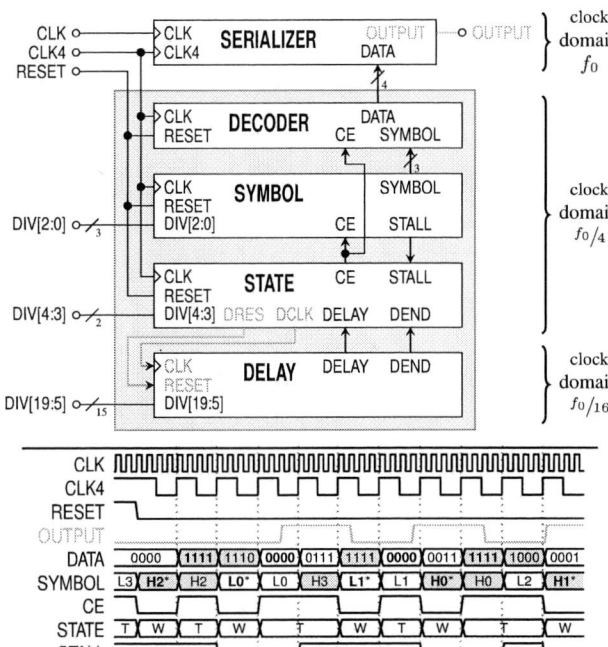

Fig. 3: Block diagram of the divider (gray) and the OUTPUT serializer with signal timings for $S = 13$. Color in SYMBOL and DATA symbols indicate a starting H bit. The serializer delays OUTPUT by 6 CLK cycles.

The proposed architecture (see Fig. 3) implements this stream coding approach in four stages distributed over three clock domains:

(1) A 4:1 serializer, producing OUTPUT from the DATA stream and the clocks CLK ($\mapsto f_0$) and CLK4 ($\mapsto f_0/4$),

(2) A symbol generator block (**SYMBOL** in Fig. 3), synthesizing the correct symbol stream in the second clock domain CLK4,

(3) A Finite State Machine (FSM) (**STATE** block in Fig. 3), controlling the injection of wait symbols to the SYMBOL stream by driving CE=H, while also generating the Extended Delay Unit (EDU) control signals DRES and DCLK ($\mapsto f_0/16$). Finally,

(4) the EDU (**DELAY** block in Fig. 3) requests the FSM to add extra wait symbols for a configurable amount of DCLK clock cycles.

The DECODER block performs the mapping DATA$\mapsto$SYMBOL according the Fig. 5, if no wait symbol shall be inserted (CE=H). During a wait symbol (CE=L), all bits in the DATA symbol will match the starting bit of SYMBOL (H3 or L3) and the symbol generator halts. Therefore, SYMBOL is only advanced when the state FSM does not request the insertion of wait states (CE=H). The next SYMBOL is then determined by the following set of rules:

(1) The next SYMBOL's start bit is always toggled.

(2) The bit distance $\hat{n}$ to the next logic transition (H$\leftrightharpoons$L) in the OUTPUT stream is determined by adding n_H or n_L (depending by the current SYMBOL start bit) to the n of the current SYMBOL.

(3) If $\hat{n} \geq 4$ the next logic transition is more than one symbol apart and the symbol generator signals the state FSM to insert an extra wait state by setting STALL=H.

(4) The remainder of $\hat{n}$ determines the next SYMBOL's logic transition distance as $n_+ = \hat{n} \bmod 4$.

Consult sec. 2.2 for more details on the workings of the state FSM. The signal timings in Fig. 3 illustrate the presented streaming concept.

2.1. Symbol Generation Logic

The SYMBOL advancement scheme introduced before requires to perform one addition whenever no wait symbol should by generated, to implement $n_+ = (n+S) \bmod 4$. A glitch-free clock gate therefore ensures that the internal state is only updated when CE=H. Also, for odd divider factors, n_H must be one bit longer than n_L. This is achieved by setting the adder chain input carry bit only when both DIV[0] and the SYMBOL start bit are H. Figure 4 shows the two-bit full-adder used to perform the addition and the described side logic. The control word bits DIV[2:1] set the addition argument. In this way, all three control bits select the fundamental SYMBOL stream, as shown in Fig. 2.

Conveniently, the carry-out bit of the adder chain directly tells when $\hat{n} \geq 4$, providing the STALL signal. In response to this condition the state FSM is supposed to add an extra wait state.

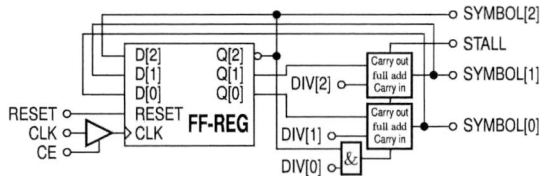

Fig. 4: Gate-Level Schematic of Symbol Block

SYMBOL	DATA	STATE	DRES	DCLK
H3 111	1111	T 1000	1	1
H2 110	1110	Z 0010	1	1
H1 101	1100	Y 0001	1	0
H0 100	1000	X 0011	1	0
L3 011	0000	W 0000	1	1
L2 010	0001	4 0110	0	1
L1 001	0011	3 0100	0	1
L0 000	0111	2 0111	0	0
		1 0101	0	0

Fig. 5: Vector mappings: SYMBOL ↦ DATA and STATE ↦ DRES , DCLK. The latter drive the $f_0/16$ clock domain directly from the STATE vector.

2.2. Wait State- and Clock Domain Generation State Machine

The state FSM controls the insertion of wait states to the SYMBOL stream, by controlling the CE signal. Since the starting bit of SYMBOL flips in every CE=H cycle, the corresponding state is named *Toggle*, or T state. The FSM further consists of eight more wait states (Z . . . W, 4 . . . 1, see Fig. 5) and implements the following functions:

(1) Insert up to three recurring wait states after every T state by consuming the control word bits DIV[4:3].
(2) Insert an additional wait state if requested by STALL=H.
(3) Provide the third clock domain control signals from its internal state vector to operate the EDU at a clock rate of $f_0/16$.

The wait state set Z...W implements the total number of to-be-inserted wait states, defined as $\delta = $ DIV[4:3] + STALL. If DELAY=H, the EDU indicates that an extended delay should be provided by the FSM. Then, the FSM de-asserts DRES by switching to cycle through a second set of wait states 4 . . . 1. Now the EDU is actively clocked by DCLK until DEND=H during the final 1 state of the cycle. DCLK and DRES can be directly mapped from the STATE vector bits with no extra logic.

To avoid race conditions, the state machine in Fig. 7 and the STATE vector mapping in Fig. 5 is designed to ensure that the rising edge of DCLK never coincides DRES de-asserting. A minimum DRES duration of two CLK4 cycles is ensured by the extra edge Z↦3. Figures 6 and 7 illustrate the FSM state progression and state diagram of the FSM.

δ	DIV[19:5]	DIV[4]	DIV[3]	STALL	STATE sequence, DRES and DCLK indicated	DELAY	wait states
0	0	0	0	0	T	L	0
1	0	0	0	1	W T	L	1
2	0	0	1	0	X W T	L	2
3	0	1	0	1	Y X W T	L	3
4	0	1	1	1	Z Y X W T	L	4
0	1	0	0	0	Z 3 2 1 T	H	4
1	1	0	0	1	W 4 3 2 1 T	H	5
2	1	0	1	0	X W 4 3 2 1 T	H	6
3	1	1	0	1	Y X W 4 3 2 1 T	H	7
4	1	1	1	1	Z Y X W 4 3 2 1 T	H	8
0	2	0	0	0	Z 3 2 1 4 3 2 1 T	H	8
...	...				...		...
4	2	1	1	1	Z Y X W 4 3 2 1 4 3 2 1 T	H	12
...	...				...	H	...

Fig. 6: State progression sequences. Colors illustrate the link between the STATE vector and the EDU control signals DCLK and DRES.

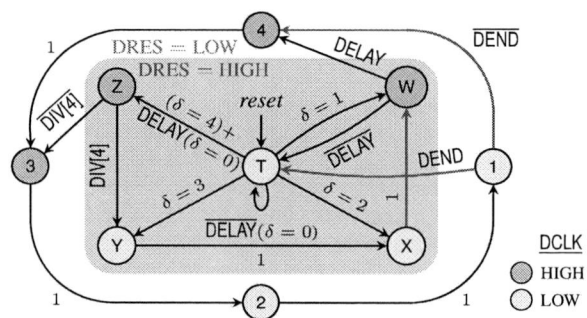

Fig. 7: Full state diagram for the STATE FSM block. Sensitive edges of the EDU register indicated in red.

2.3. Extended Long-Range Delay Stage

The purpose of the EDU is to provide one DEND pulse, after a configurable amount of DCLK cycles since DRES last de-asserted. As the state FSM guarantees long setup times for CLK and DRES, slower speed grades are used for register and gate cells in the final $f_0/16$ clock domain, leading to large power savings. The delay itself is solved with a preloadable LFSR (see Fig. 8). The characteristic polynomial is chosen as $x^{15} + x + 1$, which generates a maximum length sequence of period $2^{15} - 1$. The register feedback logic is minimal (one XOR gate), as the polynomial consists of only three terms.

The DEND signal is generated from checking the LFSR state for the all-H state, indicating the end of the delay period. The desired delay is now configured by choosing the preset state of the register to a value that produces the all-H register state after the desired amount

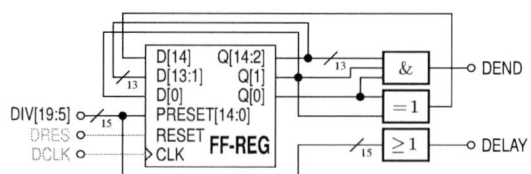

Fig. 8: Gate-Level Schematic of the Extended Delay Unit (EDU)

of DCLK cycles. Since the all-L state only progresses to itself with each DCLK cycle, it must be avoided. This is accomplished through the DELAY signal, which causes the FSM to bypass the EDU altogether.

Since only one FF must be added to double the EDU range, this concept easily extends to arbitrarily large division ranges with no impact on speed and only a very minute power increase per doubling.

3. EVALUATION OF IMPLEMENTATION EXAMPLE

We demonstrate the architecture by implementation in a 130 nm SiGe process using the *Common-ECL* standard cell library from the research project *EuRISCOSi* [8]. Gate-level synthesis, placement and layout (see Fig. 10c) was performed by hand to allow for handling of RF design considerations. Due to lacking design automation tool support, timing verification was done using a hardware description language model. Timing parameters, derived from transistor-level simulations of the logic cells, were annotated to the model. The design has an active area (excluding pads) of $165 \times 369 \, \mu m^2$ and draws $\approx 90 \, mA$ at a 2.50 V supply (power consumption is $\approx 225 \, mW$).

The transistor-level simulation results shown in Fig. 10 confirm correct operation at 41.67 GHz for various choices of S at a supply voltage of 2.50 V and at 70 °C, using calibrated models and including noise. At a slightly reduced maximum clock rate, correct operation extends for a wide range of corners between 2.35 V to 2.65 V and 0 °C to 120 °C. Due to the serializing approach and separated serializer layout, the signal integrity of the generated OUTPUT eye is excellent, independently of S (see Fig. 10b).

4. CONCLUSION

Our proposed clock divider design is the first to combine a wide configuration range with extreme high speed operation, close to technology node limits. Operation at one sixth the transition frequency of a SiGe technology was demonstrated. The hierarchical design approach makes the architecture highly portable and is also a promising choice for the implementation of high-performance CMOS clock dividers in low-cost technology nodes using design automation. In the field of RADAR signal acquisition, the architecture enables a new range of new research activity by allowing for the implementation of flexible CS-based systems operating up into the millimeter wave RF bands.

5. REFERENCES

[1] J. Sachs, *Ultra-Wideband Radar*. John Wiley & Sons, Ltd, 2012, ch. 4, pp. 363–584. [Online]. Available: https://onlinelibrary.wiley.com/doi/abs/10.1002/9783527651818.ch4

[2] J. H. G. Ender, "On compressive sensing applied to radar," *Signal Process.*, vol. 90, no. 5, p. 1402–1414, May 2010. [Online]. Available: https://doi.org/10.1016/j.sigpro.2009.11.009

[3] J. Sachs, P. Peyerl, and M. Rossberg, "A new UWB-principle for sensor-array application," in *IMTC/99. Proceedings of the 16th IEEE Instrumentation and Measurement Technology Conference (Cat. No.99CH36309)*, vol. 3, May 1999, pp. 1390–1395 vol.3.

[4] C. W. Wagner, S. Semper, F. Römer, A. Schönfeld, and G. Del Galdo, "Hardware architecture for ultra-wideband channel impulse response measurements using compressed sensing," in *2020 28th European Signal Processing Conference (EUSIPCO)*, 2021, pp. 1663–1667.

[5] J. N. Laska, S. Kirolos, M. F. Duarte, T. S. Ragheb, R. G. Baraniuk, and Y. Massoud, "Theory and implementation of an analog-to-information converter using random demodulation," in

	[9]	[10]	[11]	**This work**
Architecture	State-Lookahead	—	2/3 FF stages	Mixed
Range of S	2 ... 255	8 ... 512	64 ... 127	8 ... 1048583
Technology	CMOS	SiGe	SiGe BiCMOS	SiGe BiCMOS
Feature size	150 nm	—	130 nm	130 nm
Logic type	CMOS	CML	ECL	PECL
f_{max}	2.00 GHz	9.00 GHz	44 GHz	41.67 GHz
Power	15.70 mW	460 mW*	92 mW	225 mW
Active area	0.11 mm^2	—	0.09 mm^2	0.06 mm^2

* comprises complete chip, including input buffers and output drivers.

Fig. 9: Comparison of our design to other work, focussing on low power [9], wide range [10] and high operating frequency [11]

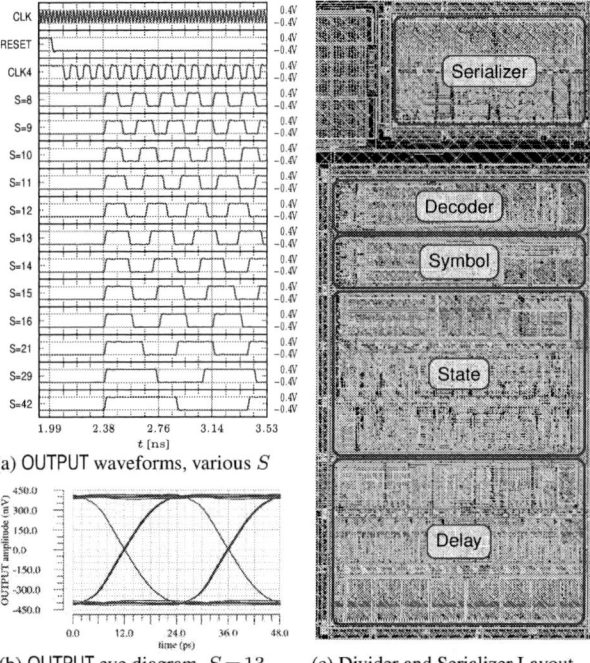

(a) OUTPUT waveforms, various S

(b) OUTPUT eye diagram, $S = 13$

(c) Divider and Serializer Layout

Fig. 10: Simulation results for an implementation in 130 nm SiGe BiCMOS technology with $f_t = 250$ GHz, for CLK $= 41.67$ GHz ($\approx f_t/6$)

2007 IEEE International Symposium on Circuits and Systems, May 2007, pp. 1959–1962.

[6] A. Alvarez, *Introduction to BiCMOS*. Boston, MA: Springer US, 07 2011, pp. 1–20.

[7] K. Mishra, *Advanced Chip Design: Practical Examples in Verilog*. CreateSpace Independent Publishing Platform, 2013.

[8] Technische Hochschule Brandenburg (THB), Germany, "Schlussbericht zum Förderprojekt EuRISCOSi," Project Final Report 13FH069PX3, doi 10.2314/GBV:1024449688, 2018.

[9] S. Abdel-hafeez and A. Gordon-Ross, "A gigahertz digital cmos divide-by-n frequency divider based on a state look-ahead structure," *Circuits Systems and Signal Processing*, vol. 30, pp. 1549–1572, 12 2011.

[10] *9 GHz Divide-by-8 to 511 Programmable Integer Divider*, Microsemi, 2014, sMD-00027 Rev F.

[11] A. Ergintav, Y. Sun, C. Scheytt, and Y. Gürbüz, "49 ghz 6-bit programmable divider in sige bicmos," in *2013 IEEE 13th Topical Meeting on Silicon Monolithic Integrated Circuits in RF Systems*, 2013, pp. 117–119.

Modeling Ni/β-Ga_2O_3 SBD interface properties

Madani Labed
Laboratory of Semiconducting and
Mettalic Materials
University of Biskra,07000
Algeria
madani.labed@univ-biskra.dz

Nouredine Sengouga
Laboratory of Semiconducting and
Mettalic Materials
University of Biskra,07000
Algeria
n.sengouga@univ-biskra.dz

Afak Meftah
Laboratory of Semiconducting and
Mettalic Materials
University of Biskra,07000
Algeria
af.meftah@univ-biskra.dz

Jun Hui Park
Department of Intelligent
Mechatronics Engineering, and
Convergence Engineering for
Intelligent Drone
Sejong university
Seoul 05006, Republic of Korea
julia980406@gmail.com

Sinsu Kyoung
Research and Development,
Powercubesemi Inc
Sujeong-gu, Seongnam-si,
Gyeonggi-do 13449
Republic of Korea
sskyoung@powercubesemi.com

Hojoong Kim
Department of Intelligent
Mechatronics Engineering, and
Convergence Engineering for
Intelligent Drone
Sejong university
Seoul 05006, Republic of Korea
hojoongkim@sejong.ac.kr

You Seung Rim
Department of Intelligent
Mechatronics Engineering, and
Convergence Engineering for
Intelligent Drone
Sejong university
Seoul 05006, Republic of Korea
youseung@sejong.ac.kr

Abstract— In this work, Ni/β-Ga₂O₃ Schottky diode deposited by electron-beam evaporation was studied. A detailed numerical simulation is carried out to reproduce the current-voltage measurement of Ni/β- Ga₂O₃ Schottky diode and extract the Ni/β- Ga₂O₃ interface properties. For more agreement between simulation and measurement the effect of Ni workfunction, Si-doped β-Ga₂O₃ surface electron affinity and traps concentration were studied.

Keywords— β-Ga_2O_3, SBD, Interface, Traps, Silvaco-Atlas

I. INTRODUCTION

Gallium oxide (Ga₂O₃) is a new oxide semiconductor material with a long rich history [1]. Pioneer studies were performed in the sixties but since then, it was almost forgotten for about three decades. However, in the last two decades, its ultra-wide bandgap (UWBG) of ~4.8 eV, high breakdown electric field of ~8 MV/cm, and high saturation velocity of $1 \times 10^7 cm/s$ brought the Ga₂O₃ to the fore again[2]–[4]. In addition, Ga₂O₃ has six polymorphs with the β-Ga₂O₃ being the most stable at high temperature. Furthermore, it can be grown directly from the melt, is of low cost, and allows large scale production compared to GaN, InGaN, and SiC [1], [5]. However, this material has a serious drawback in developing p-type [6], [7], thus hindering its applications in bipolar devices (p-n junction, BJT) [5]. Currently, β-Ga₂O₃ is therefore mainly used in unipolar devices (SBD [5], [9], MOSFET [10], Thin-Film Transistor (TFT) [11], and field emission [12]). For SBDs (Schottky Barrier Diodes), high work function metals are required for Schottky contact with β-Ga₂O₃ [13] such as Nickel [14] and Platinum [15].

In this work, Ni/β-Ga₂O₃ vertical Schottky diode was modeled. Ni workfunction, Si-doped β-Ga₂O₃ surface electron affinity and traps

effect on the electrical characteristics of Ni/β-Ga₂O₃ SBD were studied to reproduce the current-voltage measurement.

II. DEVICE STRUCTURE AND SIMULATION

The β-Ga₂O₃ Schottky barrier diode (SBD) structure, based on an experimental structure[16], is investigated. It consists of a 300 nm thick Nickel (its work function is Φ_m=5.25 eV), a Si-doped β-Ga₂O₃ layer deposited on a Sn-doped β-Ga₂O₃ substrate by halide vapor phase epitaxy (HVPE) which is used as a drift layer in this SBD. This layer is used due to its high purity and provides a low resistance, a low on-resistance and a high breakdown voltage. Then, 300 nm Nickel layer deposited on the top of the drift layer by electron beam evaporation, followed by annealing at 400 ℃[16]. The electrical current density-voltage (J-V) characteristics were measured by a semiconductor analyser and a source mater (SCS-4200A and 2410 Source meter, Keithley). A schematic representation of this SBD structure is shown in **Fig. 1**. The thicknesses of Si-doped β-Ga₂O₃ and Sn-doped β-Ga₂O₃ are 10 and 650 μm while their doping are 1×10^{18} and $3 \times 10^{16} cm^{-3}$, respectively.

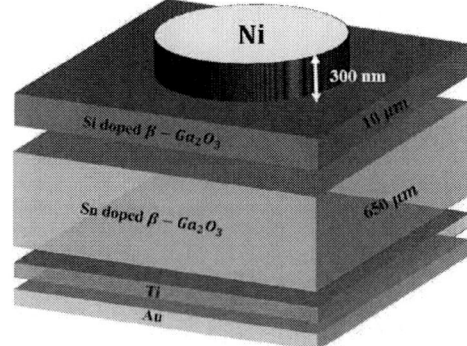

Fig. 1. A schematic representation of β-Ga₂O₃ SBD.

Identify applicable funding agency here. If none, delete this text box.

© VDE VERLAG GMBH · Berlin · Offenbach

In this work, the simulation is carried out using SILVACO TCAD. thermionic emission, Shockley-Read-Hall, Auger recombination, Klassen's and image force lowering models are considered in the simulation.

III. RESULTS AND DISCUSSION

The main work underlying this study was an investigation on the Ni/β-Ga_2O_3 interface modeling based on the interface traps concentration study, surface electron affinity and Ni workfunction. As presented in **Fig. 2,** a huge disagreement between simulation and measurement when traps presented in **Table I** are considered. This deviation related to the Nickel workfunction, interface traps concentrations and β-Ga_2O_3 surface electron affinity deviation from the known values as we will show in the next steps.

TABLE I. Properties of each layer of the studied SBD[3]–[5].

Traps	Trap level $(E_c - E)$ (eV)	Trap concentration (cm^{-3})	Capture cross section σ_n (cm^2)	σ_n/σ_p
Sn doped β-Ga₂O₃	0.55	3×10^{13}	5.4×10^{-11}	100
	0.74	2×10^{16}	0.5×10^{-12}	100
	1.04	4×10^{16}	2×10^{-14}	10
Si doped β-Ga₂O₃ thin layer	0.60	3.6×10^{13}	5.4×10^{-11}	100
	0.75	4.6×10^{13}	0.5×10^{-12}	100
	0.72	4.6×10^{13}	0.5×10^{-12}	100
	1.05	1.1×10^{14}	2×10^{-14}	10

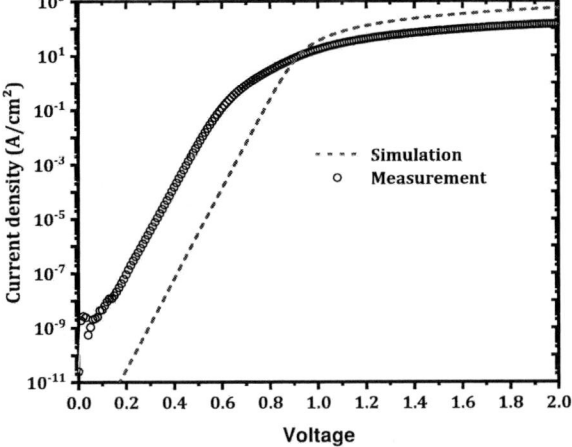

Fig. 2. Comparsison between simulation and measurement.

The effect of Ni workfunction is studied as presented in **Figure 3**. When workfunction increased from 4.8 to 5.5 eV a decrease in the SBD thermionic current. This decrease related to the Schottky barrier height (ϕ_B) increasing as the following equation shows[17]–[19]:

$$\phi_B = \phi_{Ni} - \chi_{Ga_2O_3} \tag{1}$$

Where ϕ_{Ni} and $\chi_{Ga_2O_3}$ are Nickel workfunction and β-Ga_2O_3 electron affinity respectively. An agreement between simulation and measurement results is observed For low voltage for ϕ_{Ni}=5 eV as presented in **Fig. 3**. But for high voltage the agreement is not achieved. Hence other possible causes (Ni/β-Ga_2O_3 interface traps concentrations and β-Ga_2O_3 surface electron affinity) are investigated.

The concentration of E_2 ($E_c - 0.75$), E_2^* ($E_c - 0.72$) and E_3 ($E_c - 1.05$) traps in Ni/β-Ga_2O_3 interface effect on the thermionic current is studied. The choice of these traps is due to the fact that this traps was a dominating traps in in most samples[20]. The effect of this traps is shown in **Fig. 4 (a),(b)** and **(c)** for the traps E_2, E_2^* and E_3 respectively. First with increase the traps concentrations the thermionic current decreased. The best obtianed result for concentrtions E_2, E_2^* and E_3 are $8 \times 10^{15} cm^{-3}$ for each trap. But a disagreement between the simulation and measurement is still existes.

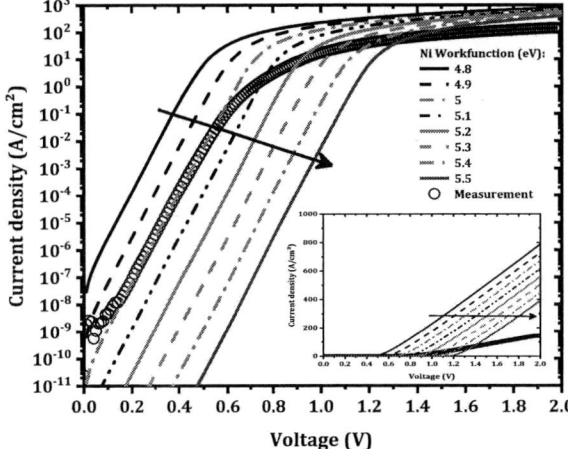

Fig. 3. Simulated J-V characteristics for different Ni work function, compared to measurement.

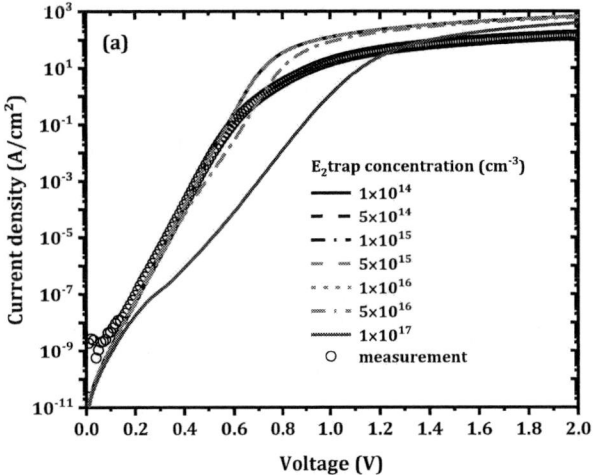

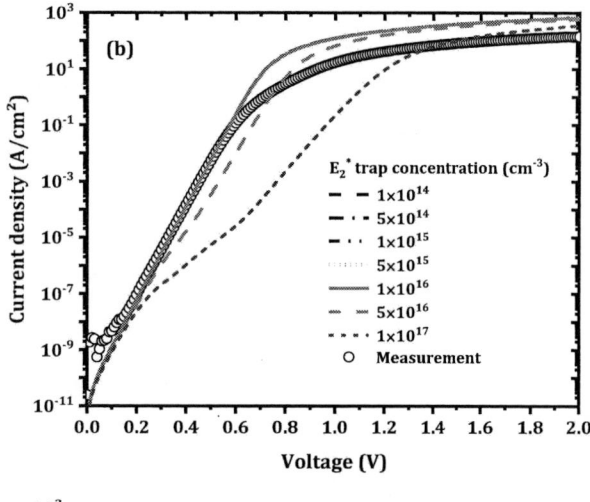

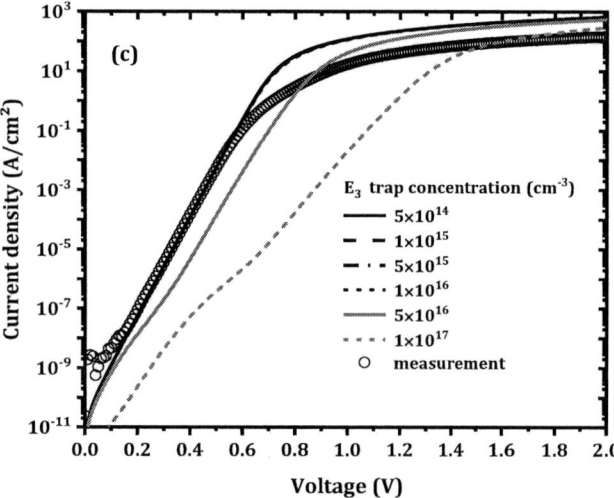

Fig. 4. Effects of trap density on SBD **(a)** E_2 $(E_c - 0.75)$, **(b)** E_2^* $(E_c - 0.72)$ and **(c)** E_3 $(E_c - 1.05)$

Now, the surface electron affinity of β-Ga_2O_3 is studied. When the electron affinity decreased, the most effected region is the serie resistance region as presented in **Fig. 5**. This result is interpreted by the increase in the serie resistance with decrease the surface electron affinity[3].

For $\chi_{S,Ga_2O_3} = 3.89$ eV a good agrement between simulation and measurement results is obtained as **Fig. 6** schows. The deviation between the simulation and measurement for low voltage related to the low current density measurement which is outside the measurement domain of the analyzer (SCS-4200A, Keithley), or probably results from the effect of the metal-semiconductor contact[3]. The extracted SBD paramaters for simulation and measurement presented in **Table II**.

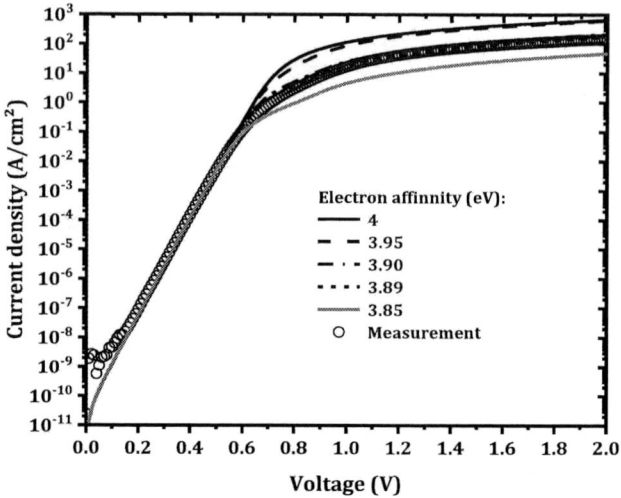

Fig. 5. Simulated J-V characteristics for different surface electron affinity, compared to measurement.

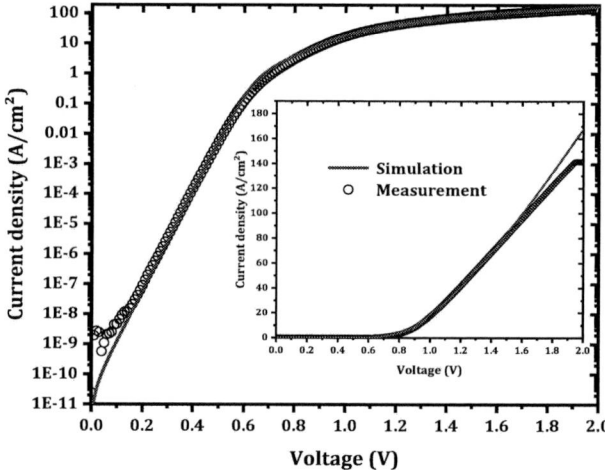

Fig. 6. Comparison between simulation and measurement.

TABLE II Outputs paramaters comparison for simulation and measurement.

	n	ϕ_B(eV)	R_s (mΩ.cm²)	R_{on} (mΩ.cm²)
Measurement	1.07	1.30	67.8	7.19
Simulation	1.06	1.32	60.3	7.47

IV. CONCLUSIONS

In summary, the J-V characteristics of an Ni/β-Ga_2O_3 Schottky barrier diode was simulated by SILVACO-Atlas and compared with measurement for interface properties extraction. The effect of Ni workfunction and Si-doped β-Ga₂O₃ surface electron affinity and traps concentration were studied for further agreement with measurement.

Acknowledgment

This work was supported by the National Research Foundation of Korea (NRF) grant funded by the Korea government (MSIT) (No. 2020R1A2C1013693) and Korea Institute for Advancement of Technology (KIAT) grant funded by the Korea Government (MOTIE)(P0012451, The Competency Development Program for Industry Specialist).

References:

[1] H. Xue, Q. He, G. Jian, S. Long, T. Pang, and M. Liu, "An Overview of the Ultrawide Bandgap Ga2O3 Semiconductor-Based Schottky Barrier Diode for Power Electronics Application," *Nanoscale Res. Lett.*, vol. 13, no. 1, p. 290, Dec. 2018, doi: 10.1186/s11671-018-2712-1.

[2] M. Labed *et al.*, "Leakage Current Modelling and Optimization of β-Ga 2 O 3 Schottky Barrier Diode with Ni Contact under High Reverse Voltage," *ECS J. Solid State Sci. Technol.*, vol. 9, no. 12, p. 125001, 2020, doi: 10.1149/2162-8777/abc834.

[3] M. Labed *et al.*, "Modeling a Ni/β-Ga2O3 Schottky barrier diode deposited by confined magnetic-field-based sputtering," *J. Phys. D. Appl. Phys.*, vol. 54, no. 11, p. 115102, 2021, doi: 10.1088/1361-6463/abce2c.

[4] M. Labed *et al.*, "Modeling and analyzing temperature-dependent parameters of Ni/β-Ga2O3 Schottky barrier diode deposited by confined magnetic field-based sputtering," *Semicond. Sci. Technol.*, vol. 36, no. 3, p. 35020, 2021, doi: 10.1088/1361-6641/abe059.

[5] Z. Galazka, "β -Ga 2 O 3 for wide-bandgap electronics and optoelectronics," *Semicond. Sci. Technol.*, vol. 33, no. 11, p. 113001, Nov. 2018, doi: 10.1088/1361-6641/aadf78.

[6] A. Kyrtsos, M. Matsubara, and E. Bellotti, "On the feasibility of p-type Ga 2 O 3," *Appl. Phys. Lett.*, vol. 112, no. 3, p. 032108, Jan. 2018, doi: 10.1063/1.5009423.

[7] E. Chikoidze *et al.*, "Enhancing the intrinsic p-type conductivity of the ultra-wide bandgap Ga 2 O 3 semiconductor," *J. Mater. Chem. C*, vol. 7, no. 33, pp. 10231–10239, 2019, doi: 10.1039/C9TC02910A.

[8] A. Y. Polyakov *et al.*, "Defects at the surface of β-Ga 2 O 3 produced by Ar plasma exposure," *APL Mater.*, vol. 7, no. 6, p. 061102, Jun. 2019, doi: 10.1063/1.5109025.

[9] G. Jian *et al.*, "Characterization of the inhomogeneous barrier distribution in a Pt/(100) β -Ga 2 O 3 Schottky diode via its temperature-dependent electrical properties," *AIP Adv.*, vol. 8, no. 1, p. 015316, Jan. 2018, doi: 10.1063/1.5007197.

[10] S. J. Pearton, F. Ren, M. Tadjer, and J. Kim, "Perspective: Ga 2 O 3 for ultra-high power rectifiers and MOSFETS," *J. Appl. Phys.*, vol. 124, no. 22, p. 220901, Dec. 2018, doi: 10.1063/1.5062841.

[11] S. R. Thomas *et al.*, "High electron mobility thin-film transistors based on Ga 2 O 3 grown by atmospheric ultrasonic spray pyrolysis at low temperatures," *Appl. Phys. Lett.*, vol. 105, no. 9, p. 092105, Sep. 2014, doi: 10.1063/1.4894643.

[12] A. Grillo *et al.*, "High field-emission current density from β-Ga 2 O 3 nanopillars," *Appl. Phys. Lett.*, vol. 114, no. 19, p. 193101, May 2019, doi: 10.1063/1.5096596.

[13] E. Farzana, Z. Zhang, P. K. Paul, A. R. Arehart, and S. A. Ringel, "Influence of metal choice on (010) β-Ga2O3 Schottky diode properties," *Appl. Phys. Lett.*, vol. 110, no. 20, p. 202102, May 2017, doi: 10.1063/1.4983610.

[14] Y. Gao *et al.*, "High-Voltage β-Ga2O3 Schottky Diode with Argon-Implanted Edge Termination," *Nanoscale Res. Lett.*, vol. 14, no. 1, p. 8, Dec. 2019, doi: 10.1186/s11671-018-2849-y.

[15] Q. He *et al.*, "Schottky barrier diode based on β -Ga 2 O 3 (100) single crystal substrate and its temperature-dependent electrical characteristics," *Appl. Phys. Lett.*, vol. 110, no. 9, p. 093503, Feb. 2017, doi: 10.1063/1.4977766.

[16] H. Kim, S. Kyoung, T. Kang, J. Y. Kwon, K. H. Kim, and Y. S. Rim, "Effective surface diffusion of nickel on single crystal β-Ga2O3 for Schottky barrier modulation and high thermal stability," *J. Mater. Chem. C*, vol. 7, no. 35, pp. 10953–10960, 2019, doi: 10.1039/c9tc02922b.

[17] S. M. Sze and K. K. Ng, *Physics of semiconductor devices*, 3rd ed. John wiley & sons, 2006.

[18] R. T. (董梓則) Tung, "The physics and chemistry of the Schottky barrier height," *Appl. Phys. Rev.*, vol. 1, no. 1, p. 11304, Jan. 2014, doi: 10.1063/1.4858400.

[19] Donald A. Neamen, *Semiconductor physics and devices: Basic principles*, 4th ed., vol. 53, no. 9. 2012.

[20] Z. Galazka, "β-Ga2O3 for wide-bandgap electronics and optoelectronics," *Semiconductor Science and Technology*, vol. 33, no. 11. p. 113001, Nov. 01, 2018, doi: 10.1088/1361-6641/aadf78.

Performance assessment of a new low-cost RF sputtered Schottky diode based on a-Si/Ti structure

H. Ferhati, F. Djeffal, A. Bendjerad and A. Benhaya

Abstract— In this paper, a new efficient and low-cost Schottky Diode (SD) based on a-Si/Ti structure was elaborated using RF magnetron sputtering technique. An exhaustive investigation of structural and electrical properties was performed, where the sputtered device was characterized using X-ray diffraction (XRD) and Keithley (4200-SCS) to measure the current-voltage characteristics. Moreover, a comprehensive study regarding the impact of the Ti layers on the device characteristics is carried out. It was demonstrated that implementing Ti intermediate layers could induce depletion regions at the interfaces, leading to significantly enlarged voltage barrier height. Furthermore, the elaborated SD exhibits a rectification behavior providing an appropriate current with a favorable ideality factor. This is mainly due to the reduced series resistance of the multilayer structure as confirmed by electrical analysis. Therefore, the proposed SD structure based on Ti intermediate layers provides improved performance and can open a new route for the fabrication of promising alternative devices for microelectronic and sensing applications.

Index Terms— RF sputtering; a-Si/Ti; Schottky Diode; intermediate layers

I. INTRODUCTION

Silicon-based microelectronic technology has attracted an enormous research interest due to its capability for offering high performance and an appropriate fabrication cost in comparison to that based on composite and alloy-based materials (GaAs, InP, AlGaAs, ...) [1-4]. Schottky Diodes (SDs), which are the key structures of microelectronic devices (optoelectronics, power electronics, communication systems, ..), have received an important research interest [3-8]. In this context, the development of high performance microelectronic devices requires new approaches and design strategies to enhance the Schottky diode electrical properties for high performance and low cost microelectronic devices [6-9]. SDs have been well investigated in numerous microelectronic domains, but conventional silicon diodes have now reached technological saturation [1-3]. This is because of the intrinsic drawbacks of the Si material for high commutation speed and large applied voltages. These limitations are mainly due to the low saturation velocity at superior electric field intensities and narrow band gap of the silicon material [1-3]. Therefore, it is important to develop new structures to potentially improve the SD performances. In this framework, amorphous-Si/Metal-based multilayer structures are proposed as new alternative materials to overcome the limitations imposed by Si-SD, although the Silicon-based technology has attained better degree of integration circuits as compared to SD-based on wide band gap materials such as SiC, GaN and ZnO [1-2]. Moreover, the a-Si-based devices not only offer closer electrical properties in comparison to that of c-Si-based counterpart, but also provide exciting opportunities for developing low-cost and thinner microelectronic devices, which is advantageous for flexible microelectronic industry. These advantages can open new pathways to elaborate a-Si-based SDs as alternative candidates to the conventional SD devices for sensing and microelectronic applications [3-10]. Although these attractive performances, more enhancement becomes crucial in order to boost up the device electrical characteristics such as derived current and ideality factor. Motivated by this concept, the main objective of the present work is to develop a new highly efficient and low-cost SD based on RF sputtered a-Si/Ti multilayer structure for Si-based platform applications. To do so, the a-Si/Ti multilayer-based SD was deposited using RF magnetron sputtering technique. Electrical and structural characterizations of the fabricated a-Si/Ti-SD were performed. It is demonstrated that the elaborated SD exhibits high performances in terms of structural and electrical properties. Consequently, the proposed SD structure provides a new pathway for developing thin and efficient SD devices for high performance Si-based technology.

II. DESIGN AND EXPERIMENTS

The elaboration of high-performance a-Si-based SDs is difficult. This is mainly due to the intrinsic limited electrons transport and low mobility of this material. To overcome this electrical limitation, in the proposed design we study a new thin film SD based on non-hydrogenated a-Si with

H. Ferhati, F. Djeffal and A. Benhaya are with the Advanced Electronics Laboratory (LEA), Department of Electronics, University of Batna 2, Algeria (faycaldzdz@hotmail.com)

A. Bendjerad is with the Advanced Electronics Laboratory (LEA), Department of Electronics, University of Batna 2, Algeria (a-bendjerad @hotmail.com).

intermediate Ti ultra-thin layers. Fig.1 shows a cross-sectional view of the proposed SD device including three Ti intermediate layers, where the a-Si/Ti multilayer structure is sputtered on a glass substrate. From this Figure, t_{a-Si} and t_{Ti} represent the thickness values of the amorphous silicon and titanium layers, respectively. The thickness of the SD is $t_{a-Si/Ti}$.

The fabrication of the SD device is based on the following manufacturing processes: 1. glass substrate was cleaned up using an ultrasonic bath and commercial detergent and then dried under a nitrogen jet procedure; 2. successive deposition of deferent layers (a-Si and Ti) on the glass substrate was performed using RF magnetron sputtering technique (MOORFIELD MiniLab 060). The latter tool is an efficient experimental facility, which able to realize defect free interfaces with high structural properties. P-type Si and Ti targets with high purity of 99.99% were used for depositing different layers. The sputtering process was carried out in a pure Ar atmosphere. Moreover, the targets to substrate distances were kept at 6.5 cm and 5.1 cm for the Titanium and Si materials, respectively. The top and bottom gold contacts were evaporated via E-beam evaporation technique. The sputtering elaboration parameters used in our experimental study are summarized in Table.1. The ellipsometry technique is used to calculate the thickness of the sputtered structure.

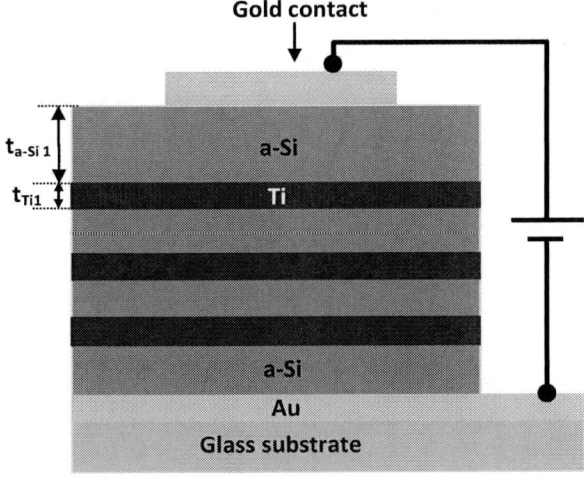

Fig. 1. Cross sectional view of the investigated Schottky Diode based on a-Si/ Ti multilayer structure.

TABLE I. DEPOSITION PARAMETERS OF THE SPUTTERED A-SI/TI MULTILAYER STRUCTURE

Parameter	a-Si	Ti
Target	99.99% p-Si	99.99% Ti
Target to substrate distance (cm)	5.1	6.5
Gas composition (Ar: O₂)	/	/
Substrate temperature (K)	300	300
power of *RF* source (W)	250	250
Working pressure (Pa)	1.5	1.5
deposition rate (nm/s)	0.4	0.7

III. RESULTS AND DISCUSSIONS

In order to evaluate the electrical performances of the elaborated device, structural and electrical characterizations were carried out using X-ray diffraction measurements (ARL Equinox 3000) for 2θ diffraction angle scans of [25°-80°] and the semiconductor characterization system (Keithley 4200-SCS) to measure the current-voltage characteristics of the fabricated device. In this context, Fig.2 shows XRD measurements of the prepared device based on a-Si/Ti multilayered structure. It is revealed from the XRD pattern, the amorphous state of the silicon sputtered layers was recorded. This crystalline characteristic can be explained by the ultralow thickness of the deposited silicon sub-layers, which can prevent reaching the crystallization phase of the Si material. On the other hand, this pattern demonstrates the presence of crystallized Ti metal, manifesting (101)-oriented film at the diffraction angle of 39.5°.

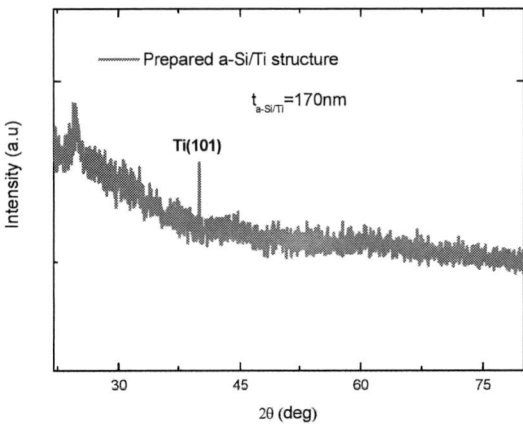

Fig. 2. X-ray diffraction patterns of the elaborated a-Si/Ti structure with $t_{a-Si/Ti}$=170nm.

To show the impact of the introduced Ti ultra-thin layers on the SD performance, it is important to study and compare the I-V characteristics of the elaborated device based on multilayer structure. In this perspective, Fig.3 compares the I-V curves related to a-Si based SD reported in [3] and the proposed device design including multi-layer structure. From this figure, it is shown that the elaborated diode device exhibits a clear rectification behavior with a high rectifying current ratio surpassing 10^5. Furthermore, the elaborated a-Si/Ti multilayer device exhibits an enhanced derived current ability over that of the conventional amorphous Si SD device. This is mainly due to the improved conductivity of the device, where introducing Ti metal layer induces significant enhancement in the device electrical behavior, which can lead to reduce the series resistance of the SD device. Moreover, the proposed design provides an appropriate leakage, where a very low OFF- current around 10^{-12}A is achieved. This improvement can be attributed to the formation of depletion regions at the a-

Si/Ti interfaces imposed by the difference work-function values, enabling the Schottky barrier height increasing. The latter leads to reduce the leakage current value.

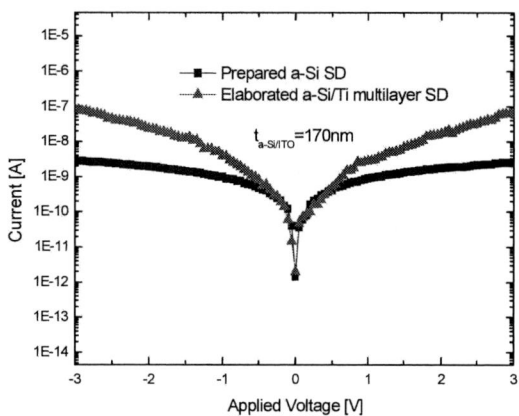

Fig. 3. Measured I-V characteristics of the conventional device with a-Si active layer and the prepared SD based on embedded a-Si and Ti sub-layers.

To get a profound insight concerning the electrical properties of the proposed Schottky diode based on a-Si/Ti multilayer structure, an electrical characterization should be performed to extract the min electrical parameters including Schottky barrier height (*SBH*) and the ideality factor (*n*) of the device. For this propose, the latter parameters can be extracted from the measured current-voltage curves plotted in Fig.2, which is governed by the thermionic emission model expressed by [11-13]

$$I = I_s \exp\left(\frac{q(V - R_s I)}{n\,KT} - 1\right) \qquad (1)$$

where R_s is the series resistance, q represents the electron charge, I_s refers to the saturation current, K denotes the Boltzmann constant, and T is the absolute temperature.

Afterwards, the evolution of the ln(I) = f(V) function is used to extract the series resistance and the ideality factor of the prepared diode based on a-Si/Ti multilayered structure. This function is extracted after some mathematical manipulations and can be described as follows

$$F(V) = \ln\left[I/1 - \left(\exp\left(\frac{-qV}{KT}\right)\right)\right] \qquad (2)$$

It is worth mentioning that the above-detailed characterization technique is extensively exploited to estimate the diode electrical parameters with a high accuracy [11-13].

Fig.4 depicts the logarithmic current function provided by Eq.2 for both investigated a-Si-based SDs with and without Ti intermediate sub-layers. Two straight lines zones can be clearly achieved for both devices, where the slope of the first one associated with low voltages allows the determination of

the ideality factor, while its intersection with the current axis represents the diode saturation current. Besides, the slope of the second line associated with high applied voltages yields the series resistance. The extracted saturation current value is then used to find out the barrier height by using the thermionic emission model [11]. The results show that the proposed SD device with a-Si/Ti exhibits better electrical parameters as compared to the conventional structure. In this context, a lower series resistance of 8×10^3 Ω, high SBH of 0.98eV and low saturation current of 0.1pA were achieved. While, the conventional a-Si SD exhibits a high series resistance of 2×10^4 Ω, lower SBH of 0.95eV. The improved electrical parameters associated with the proposed device are attributed to the effect of the inserted Ti metallic layer in enhancing the structure resistive behavior. On the other hand, the proposed device shows higher ideality factor of 6.5 as compared to the conventional structure (n=5.3). This behavior is related to the device structural properties, where the introduction of several Ti sub-layers can affect the layers uniformity and can increase the defect density within the structure. Therefore, the proposed SD based on a-Si/Ti structure offers new opportunities for the realization of high-performance SDs, which are highly attractive for microelectronic and optoelectronic applications

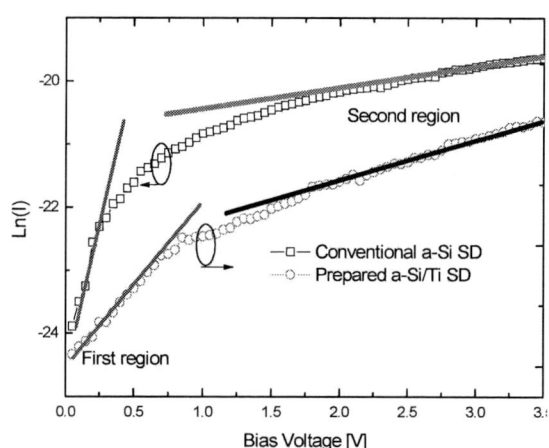

Fig. 4. Variations of ln(I) as a function of the bias voltage of the elaborated SD structures with and without Ti inter-layers extracted from the experimental data.

IV. CONCLUSION

In this paper, we have elaborated a new SD based on a-Si/Ti multilayer structure. The proposed design was deposited using RF magnetron sputtering technique. The device structural and electrical properties were analyzed using XRD and I-V measurements. It was found that the use of Ti intermediate layers induces depletion regions at the interfaces between both Ti and a-Si materials, which can leads to an increasing in the

Schottky barrier and thereby improving the current leakage behavior. Moreover, the proposed multilayer SD device provided a suitable rectifying behavior. This result can be attributed to the improved conductivity of the a-Si material by introducing Ti intermediate layers. Therefore, the presented experimental study not only provides a new approach to overcome the major limitations of a-Si based devices but also opens up the way for improving conductivity behavior of the device and reducing the self-heating losses. It is to note that our investigation can be extended by studying the impact of the number of inserted intermediate metallic layer on the device behavior. However, new samples and measurements should be developed using adequate elaboration process and characterization techniques. The obtained results make the developed multilayer SD promising alternative structure for microelectronic and sensing applications.

REFERENCES

[1] S. Manipatruni, M. Lipson, and I. A. Young, "Device Scaling Considerations for Nanophotonic CMOS Global Interconnects," IEEE J. Sel. Top. Quantum Electron., vol. 19, pp. 1077-1086 , 2013.

[2] D. A. B. Miller, "Optical interconnects to silicon," IEEE J. Sel. Top. Quantum Electron., vol. 6, pp. 1312-1317, 2000.

[3] H. Ferhati, F. Djeffal, A. Saidi, A. Benhaya, A. Bendjerad, "Effects of annealing process on the structural and photodetection properties of new thin-film solar-blind UV sensor based on Si-photonics technology," Mater. Sci. Semicond. Process., Vol.121, pp. 105331, 2021.

[4] N. Naderi, M. Moghaddam, "Ultra-sensitive UV sensors based on porous silicon carbide thin films on silicon substrate," Ceramics International., Vol. 46, pp. 13821-138260, 2020.

[5] C. Li, W. Huang, L. Gao, H. Wang, L. Hu, T. Chen and H. Zhang, "Recent advances in solution-processed photodetectors based on inorganic and hybrid photo-active materials," Nanoscale, vol.12, pp. 2201-2227, 2020.

[6] K. Benyahia, F. Djeffal, H. Ferhati, A. Bendjerad, A. Benhaya, A. Saidi," Self-powered photodetector with improved and broadband multispectral photoresponsivity based on ZnO-ZnS composite," J. Alloy. Compd., vol. 859, pp.158242, 2021.

[7] J. Chen and J. Lv, Q. "Wang, Electronic properties of Al/MoO3/p-InP enhanced Schottky barrier contacts," Thin Solid Films, vol.616, pp.145-150, 2016.

[8] T. Teraji, Y. Koide and T. Ito ", Schottky barrier height and thermal stability of p-diamond (100) Schottky interfaces," Thin Solid Films, vol.557, pp. 241-248, 2014.

[9] O. Demircioglua, et al, "Temperature dependent current voltage and capacitance voltage characteristics of chromium Schottky contacts formed by electrodeposition technique on n type Si," Journal of Alloys Compounds, vol.509, pp.6433-6439, 2011.

[10] E. Márquez, et al "The influence of Ar pressure on the structure and optical properties of nonhydrogenated a-Si thin films grown by rf magnetron sputtering onto room temperature glass substrates", J. Non-Cryst Solids, vol. 517, pp. 32-43, 2019.

[11] A. Benhaya, F. Djeffal, K. Kacha, H. Ferhati and A. Bendjerad, "Role of ITO ultra-thin layer in improving electrical performance and thermal reliability of Au/ITO/Si/Au structure: An experimental investigation," Superlattices and Microstructures, vol.120, pp. 419-426, 2018.

[12] R. Kumar and S. Chand, "Fabrication and electrical characterization of nickel/p-Si Schottky diode at low temperature," Solid State Sciences, vol.58, pp. 115-121, 2016.

[13] A. H. Kacha, B. Akkal, Z. Benamara, M. Amrani, A. Rabhi, G. Monier, C. Robert-Goumet, L. Bideux, B. Gruzza, "Effects of the GaN layers and the annealing on the electrical properties in the Schottky diodes based on nitrated GaAs," Superlattices and Microstructures, vol. 83, pp. 827-833, 2015.

Digitally Programmable Potentiometer Multistage Architecture with Switch Independent Linearity

Giorgiana-Catalina Ilie (Chiranu)
Power Solutions Group – ON Semiconductor
Univerisity "Politehnica" of Bucharest
Bucharest, Romania
GiorgianaCatalina.Ilie@onsemi.com

Cristian Tudoran
Power Solutions Group – ON Semiconductor
Bucharest, Romania
Cristian.Tudoran@onsemi.com

Otilia Neagoe
formerly of Power Solutions Group – ON Semiconductor
Bucharest, Romania
Otilia.Neagoe@onsemi.com

Gheorghe Pristavu
Dept. of Electrical Devices and Circuits
University "Politehnica" of Bucharest
Bucharest, Romania
gheorghe.pristavu@upb.ro

Gheorghe Brezeanu
Dept. of Electrical Devices and Circuits
University "Politehnica" of Bucharest
Bucharest, Romania
Gheorghe.Brezeanu@dce.pub.ro

Abstract—**This paper describes a multistage architecture for high resolution digitally programmable potentiometers. Its linearity characteristics dependence on switch's on-resistance is highly diminished. The proposed topology was implemented in a 0.18μm CMOS process and used in fabricating an 8-bit resolution digital potentiometer with I2C interface. The maximum measured values for its non-linearity errors are 0.25LSB for INL and 0.1LSB for DNL. A theoretical model for determining the non-linearity errors was developed for this architecture. It is proven that the linearity characteristics of the potentiometer are not affected by switches on-resistances. Moreover, the experimental non-linearity error values are attributed to deviations in unit resistances caused by process variations.**

Keywords— *digitally programmable potentiometer; high resolution; multistage architecture; INL; DNL.*

I. INTRODUCTION

The expanding popularity of digital trimming functions within analog and mixed signal systems resulted in a demand for increasing the resolution of digitally programmable potentiometers (DPPs) to more than 7 bits. Using the classical approach (single stage architecture) for achieving high resolution potentiometers led to an increased number of components (resistors, switches and decoding circuits) leading to practical limitations and excessive area consumption [1], [2], [3].

The solution to obtaining high resolution together with reasonable production costs, multistage potentiometer architectures have been developed. They rely on dramatically reducing the number of required switches by employing a structure with carefully tuned resistor groupings. Thus, a high precision can be obtained while keeping the occupied area at manageable levels [1], [2], [3]. The drawbacks stem from the non-ideality of these switches, particularly their on-resistance fluctuations and consist of increased non-linearity errors (INL and DNL) [1], [3].

This paper presents a multistage architecture which compensates for variations in switch on-resistances, leading to significantly decreased INL and DNL errors. It is based on redistributing resistor values and switch positions in the potentiometer structure in order to maintain stable bias conditions. The proposed architecture was used in fabricating

an 8-bit resolution digitally programmable potentiometer with I2C interface in a 0.18μm CMOS process. All its features are confirmed by analytical, simulated and experimental results.

II. PROPOSED POTENTIOMETER ARCHITECTURE

The starting point of the multistage potentiometer was an architecture recently described in [1], [3], and depicted in Fig. 1A. The proposed solution is given in Fig. 1B. Both configurations are designed for implementation in a 0.18μm CMOS process. All switches are identical and use the complementary CMOS transfer gate configuration, while all resistors are made of polysilicon. A total resistance of 50kΩ and 8-bit resolution (256 steps) are sought.

Compared to circuit A, the proposed architecture (B) has only three stages: one wiper stage comprising 15 incremental resistances (R0=196Ω) and 16 switches (SW0-SW15, Fig. 1B), and two shunt stages. Each shunt stage includes 15 resistances equal to 16×R0, 15 switches (SWL0 to SWL14 or SWH0 to SWH14, Fig. 1B) and one compensation block (red dashed box Fig. 1B). The latter contains a dummy resistor and a permanently-on switch. The block's purpose is to mirror the resistive behavior of similar switches across the shunt stage [2].

For steps 0-15, all the switches from the low shunt stage are closed, while those from the high shunt stage are open. Above 15, SWL0 opens and SWH14 closes. This process continues for switches SWL1-14 and SWH13-0, respectively, every further 16 steps. Thus, the total resistance measured between H and L (Fig. 1B) is always equal to 255×R0. For each step, there is a single closed switch in the wiper stage.

In comparison to the topology in Fig. 1A, the proposed architecture integrated the two bulk resistances of 128×R0 into the shunt stages. It is ensured that, regardless of step value, all shunt-stage switches are biased at constant levels, which greatly reduces their on-resistance fluctuations and, ultimately INL and DNL errors. This effect is further enhanced by the compensation blocks which maintain the equivalent resistance of the potentiometer at a constant value independent of the step level. In order to achieve this linearity improvement, the number of switches had to be increased to 48 (the topology in Fig. 1A required 32).

© VDE VERLAG GMBH · Berlin · Offenbach

SMACD / PRIME 2021 | 19 – 22 July 2021, Online Event

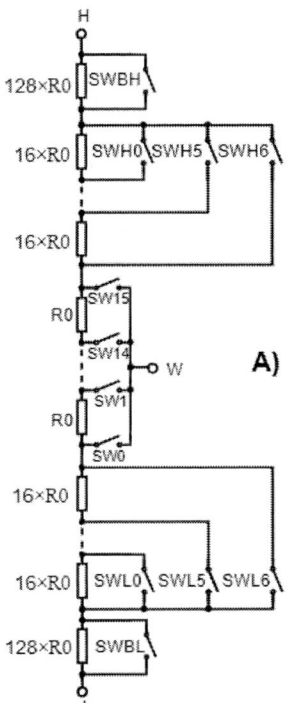
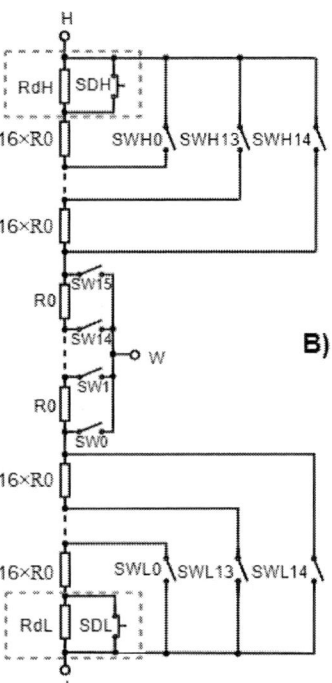

Fig. 1. Potentiometer multistage architectures: A) prior art [1, 3] and B) proposed topology.

The impact of the compensation blocks was proven by simulations. Two versions were studied for the configuration in Fig. 1B: one, implemented as-is in Fig. 1B (denoted topology B) and the other, where the compensation blocks were not included (referred to as topology B$_\varepsilon$).

Simulations for determining the non-linearity errors were performed on all three topologies (A, B and B$_\varepsilon$) considering the voltage applied on the potentiometer (V$_{HL}$) equal to the supply voltage (V$_{CC}$=V$_{HL}$=5.5V). Two versions for each topology were simulated, one where switches were sized to meet the maximum on-resistance of 8Ω (SW$_x$ in Fig. 2) and the other where switches were smaller by a factor of 20 (SW$_{x/20}$ in Fig. 2) corresponding to a proportional increase in the on-resistance. Comparative analysis was performed between all potentiometer versions in order to emphasize the effect of switch on-resistance on linearity.

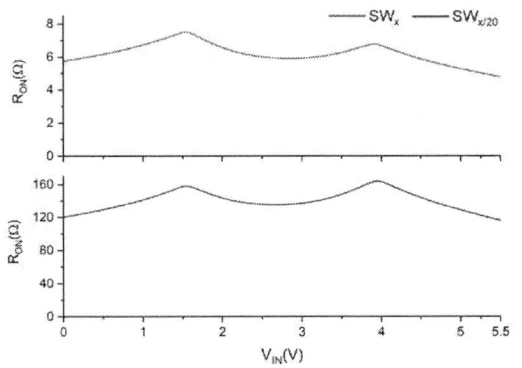

Fig. 2. SW$_x$ and SW$_{x/20}$ on-resistances as a function of the applied voltage.

Fig. 3 illustrates the simulated results for INL and DNL errors as a function of potentiometer step for the first configuration (Fig. 1A). Maximum absolute values were INL = 0.015LSB, DNL = 0.02LSB (SW$_x$ variant) and INL = 0.3LSB, DNL = 0.4LSB (SW$_{x/20}$ variant). Thus, it is proven that the linearity of topology A is strongly affected by switch size.

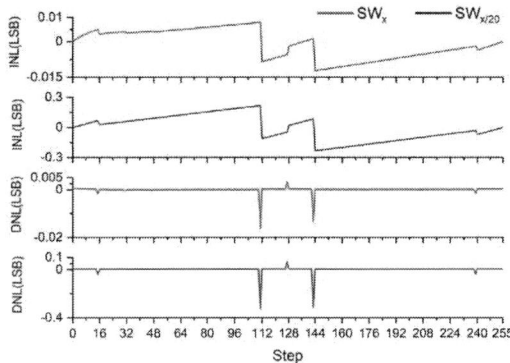

Fig. 3. Simulated INL and DNL errors for topology A.

The simulations for INL and DNL errors corresponding to configuration B$_\varepsilon$ are presented in Fig. 4.

Now, the maximum absolute errors are much smaller compared to those offered by the architecture in Fig. 1A, under 0.04LSB for both INL and DNL (Fig. 4). In this case, the constant bias of the switches, which eliminates on-resistance fluctuations contribute to the improved errors values. However, the effective switch on-resistance levels still play an important part in determining potentiometer linearity,

© VDE VERLAG GMBH · Berlin · Offenbach

albeit less so than for topology A. A difference of approximately one order of magnitude is noted between $SW_{x/20}$ and SW_x B_ε variants, favoring the largest switches.

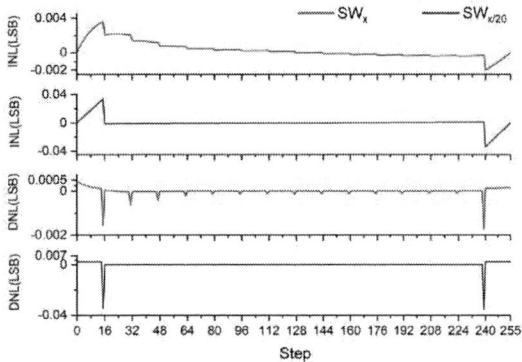

Fig. 4. Simulated INL and DNL errors for topology B_ε.

Results of non-linearity error simulations, performed on topology B, are illustrated in Fig. 5.

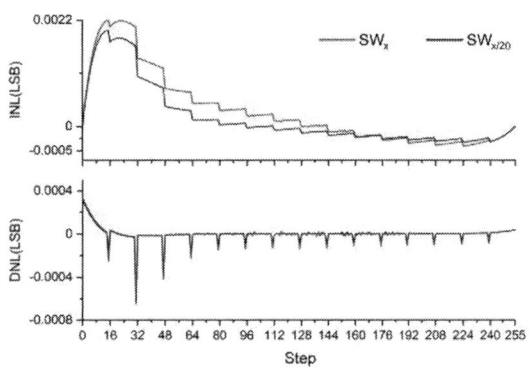

Fig. 5. Simulated INL and DNL errors for topology B.

The introduction of the compensation blocks into the shunt stages, leads to a further improvement in potentiometer linearity, with INL and DNL errors below 0.0022LSB (Fig. 5). Additionally, this time, linearity is comparable for both $SW_{x/20}$ and SW_x versions (Fig. 5). Thus, for the proposed architecture (Fig. 1B) switch size is not a decisive contributing factor in potentiometer accuracy. Larger switches are still preferable in order to obtain smaller zero and full-scale errors.

III. EXPERIMENTAL RESULTS AND DISCUSSION

The proposed multistage architecture (Fig. 1B) was implemented in a 0.18µm CMOS process as a digital potentiometer with I2C interface. The layout of the DPP, shown in Fig. 6, occupies 0.49mm². Within the yellow frame the architecture B area is emphasized. Switch sizes correspond to the SW_x simulated version.

Linearity parameters were measured for the same supply voltage of 5.5V used in the simulations. Experimental results are presented in Fig. 7 (INL error) and Fig. 8 (DNL error) respectively. Maximum absolute measured value is around 0.25LSB for INL and 0.1LSB for DNL, much lower than the common datasheet limit of ±1LSB (INL) and ±0.5LSB (DNL).

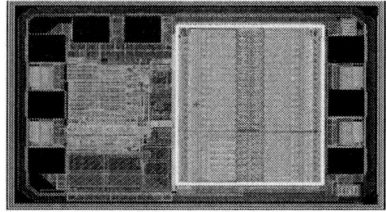

Fig. 6. Layout of the DPP implemented using topology B.

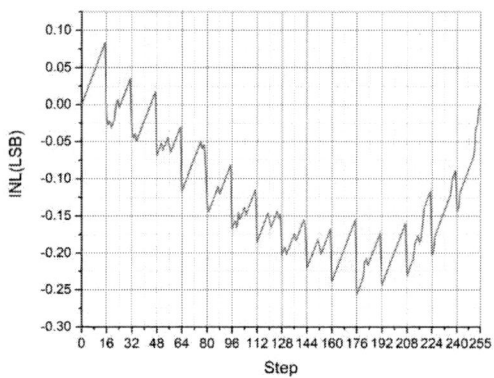

Fig. 7. Measured INL for topology B.

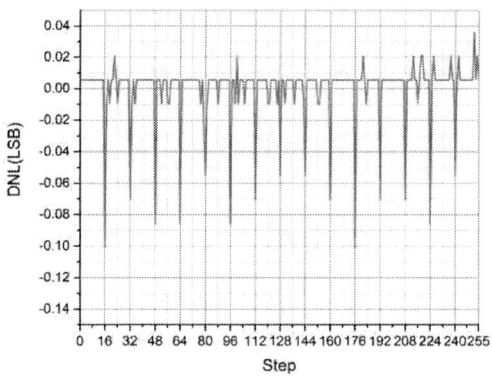

Fig. 8. Measured DNL for topology B.

Compared to the simulated values (Fig. 5) the measured errors are higher. In order to explain these discrepancies, theoretical formulas for determining INL and DNL as a function of DPP step (k) are used [4, 5]:

$$INL(k) = \frac{V_W(k) - V_W(0)}{V_{LSB}} - k, k = 0 \ to \ 255 \qquad (1)$$

$$DNL(k) = \frac{V_W(k) - V_W(k-1)}{V_{LSB}} - 1, k = 1 \ to \ 255 \qquad (2)$$

where $V_W(k)$ is the wiper voltage at step k. Its expression can be obtained by evaluating the equivalent resistances between the W-L and H-L terminals, respectively (see Fig. 1):

$$V_W(k) = \frac{R_{WL}(k) + R_{SL}}{R_{HL}(k) + R_{SL} + R_{SH}} V_{HL}, k = 0 \; to \; 255 \qquad (3)$$

R_{WL} represents the lumped polysilicon resistances under the wiper, while R_{HL} is the sum of all polysilicon resistances in the potentiometer. R_{SL} and R_{SH} are the low and high shunt switch on-resistances. Because of the compensation block they have quasi-constant values regardless of step value (k).

V_{LSB} is the potentiometer voltage step equal to:

$$V_{LSB} = \frac{V_W(255) - V_W(0)}{255} \qquad (4)$$

Using (3) in (1) and (2) will yield null INL and DNL errors for every k value, regardless of R_{SL} and R_{SH}. This is because $R_{WL}(k) = k \times R0$, while $R_{HL}(k) = 255 \times R0$.

Thus, it is also theoretically proven that the linearity of the proposed DPP is unaffected by the on-resistance of the shunt switches.

Considering the above, the experimental values of the INL and DNL errors (Figs. 7-8) can be explained only by deviations in polysilicon incremental resistance (R0). These deviations can be caused by technological variations and they are reflected in the potentiometer linearity performances.

A theoretical model is proposed for determining R_{WL} and R_{HL} variations with k in the context of a fluctuating R0. Thus, α_{SL}, α_W and α_{SH} are defined as the deviations corresponding to incremental resistance in shunt-low, wiper and shunt-high stages, respectively (see Fig. 1B). R_{WL} and R_{HL} can be expressed as:

$$R_{WL}(k) = 16j(R0 + \alpha_{SL}) + i(R0 + \alpha_W) \qquad (5)$$

$$R_{HL}(k) = 16j(R0 + \alpha_{SL}) + 15(R0 + \alpha_W) + \\ 16(15 - j)(R0 + \alpha_{SH}) \qquad (6)$$

In these equations, k is expressed by:

$$k = j2^4 + i2^0 \qquad (7)$$

where, $i, j \in \{0, 1, \ldots, 15\}$.

Fig. 9 presents calculated values for INL and DNL errors for $\alpha_{SL} = -0.3\Omega$, $\alpha_W = 1.2\Omega$ and $\alpha_{SH} = 0.3\Omega$ (chosen using a best-fit approach to the measurement results), alongside their simulated and measured conterparts.

The simulations depicted in Fig. 9 also considered the R0 deviations. Strong agreement is evinced between theoretical curves, measured data and simulations (Fig. 9).

As such, it was demonstrated that resistor mismatch is the main cause of linearity error in the proposed architecture. In order to aleviate this issue, future designs can consider layout optimization techniques [6], [7].

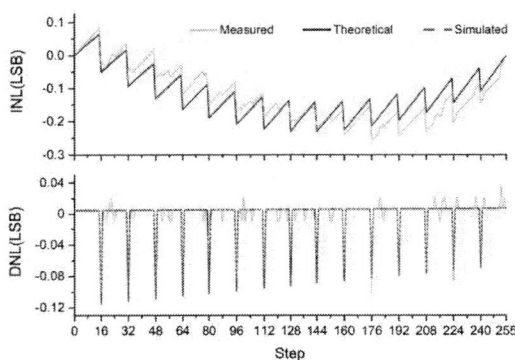

Fig. 9. Measured vs. theoretical vs. simulated INL and DNL errors for topology B when resistance deviations are considered.

IV. CONCLUSIONS

In this paper, a multistage architecture for digital programmable potentiometers was presented. It focuses on maintaining a constant bias level on its component switches, ensuring stable on-resistance values and minimizing INL and DNL errors. Additionally, it contains a compensation block which greatly reduces the dependence of the topology's linearity on switch size, as validated in simulations.

An 8-bit DPP with I2C interface, implemented using the proposed topology, was fabricated in a 0.18μm CMOS technology. The maximum values of non-linearity errors were measured at 0.25LSB for INL and 0.1LSB for DNL. Following a theoretical modeling and analysis, these values were mainly attributed to resistance mismatches caused by systematic, random and gradient errors in the technological process.

REFERENCES

[1] G.-C, Ilie (Chiranu), C. Tudoran, O. Neagoe, G. Brezeanu, "Performance Analysis for High Resolution Digitally Programmable Potentiometers" in Proceedings of IEEE International Semiconductor Conference (CAS), Sinaia, Romania, 9-11 Oct. 2019.

[2] R. Iacob, O. Neagoe, A. Manolescu, "Multistage architectures for high resolution digital potentiometers" in Proceedings of International Symposium on Signals, Circuits and Systems (ISSCS), Iasi, Romania, 9-10 July 2009.

[3] G.-C, Ilie (Chiranu), C. Tudoran, O. Neagoe, G. Brezeanu, "Low cost approaches for High Resolution Digitally Programmable Potentiometers". Romanian J. Inf. Sci. Technol., vol. 23, pp. 157-175, 2020.

[4] F. Maloberti, "Data Converters", Springer, Boston, MA, 2007.

[5] Texas Instruments Datasheet, "TPL0202 256-Taps Dual Channel Digital Potentiometer With SPI and Non-Volatile Memory", accesed online at https://www.ti.com/lit/ds/symlink/tpl0202-10.pdf?ts=1616157443935&ref_url=https%253A%252F%252Fwww.google.com%252F.

[6] N. Liu, C. Lash, J. Todsen, D. Chen, "Practical linear and Quadratic Errors Suppression Techniques in string DACs" in Proceedings of IEEE International Midwest Symposium on Circuits and Systems (MWSCAS), Boston, MA, USA, 6-9Aug. 2017.

[7] N. Liu, C. Lash, J. Todsen, D. Chen, "An 8-bit Low-Cost String DAC With Gradient Errors Suppression to achieve 16-bit Linearity", IEEE Trans Circuits Syst I Regul Pap, vol. 67, pp. 2157-2168, 2020.

Reliability Investigation of 0.18μm CMOS for Oilfield Applications

Yen Tran
Etudes et Production Schlumberger
Clamart, France
Univ. Bordeaux, Bordeaux INP, UMR
CNRS 5218, IMS Laboratory,
Talence, France
Ttran50@slb.com

Toshihiro Nomura
Etudes et Production Schlumberger
Clamart, France
TNomura@slb.com

Mohamed Salim Cherchali
Etudes et Production Schlumberger
Clamart, France
MCherchali@slb.com

Claire Tassin
Etudes et Production Schlumberger
Clamart, France
CTassin@slb.com

Yann Deval
Univ. Bordeaux, Bordeaux INP, UMR
CNRS 5218, IMS Laboratory,
Talence, France
yann.deval@ims-bordeaux.fr

Cristell Maneux
Univ. Bordeaux, Bordeaux INP, UMR
CNRS 5218, IMS Laboratory,
Talence, France
cristell.maneux@u-bordeaux.fr

Abstract—We investigated the degradation due to bias temperature instability (BTI) and hot carrier injection (HCI) for 0.18μm CMOS (Complementary Metal-Oxide-Semiconductor) under extreme temperature operations (150°C and 210°C). The transistors have been applied dedicated DC bias and temperature conditions to investigate each intrinsic wear-out mechanism in specific severe environment for oilfield applications. The aging tests have been monitored for up to 1,000 hours. These results are preliminarily used to develop equations reflecting aging laws to be included in commercial software tool for further investigation at logic circuit level.

Keywords—Reliability, oilfield, 0.18μm CMOS, extreme temperature, DC characterization, HCI, BTI

I. INTRODUCTION

CMOS transistor reliability is becoming more and more critical in IC (Integrated Circuit) technology domain due to high-density and mixed-signal integration in the context of extreme temperature application. Although the device failure mechanisms have been intensively investigated for many years, the range of investigations in term of operating conditions is usually limited to conventional application domains involving mass markets. Therefore, the application domain involving specific extreme temperatures are seldom studied even though the reliability of the associate electronic circuits turned out to be crucial for the application. For oilfield applications, downhole tools require reliable and stable performance of electronic devices during several years at high temperature. Therefore, the focus of this work is to develop compact model including BTI and HCI ageing laws for 0.18μm CMOS technology to optimize circuit design for a targeted lifetime under extreme temperature. For that purpose, we have conducted an intensive aging test campaign for both nMOS and pMOS featuring two different gate lengths. Two most important intrinsic wear-out mechanisms namely BTI and HCI were characterized and modeled under operating voltage biases and elevated stress temperatures.

It is well accepted that both BTI and HCI induce threshold voltage shift and this shift is usually considered as the best monitor of these degradation [1-3,10]. Therefore, in this study, the threshold voltage shift was investigated through its extraction and fitted to a power-law model based on experimental results. After that, this model was implemented into the simulator using tool-specific application

programming interfaces (APIs) provided by electronic design automation (EDA) commercial software. Finally, simulation results were compared to measurements to validate the accuracy of these BTI and HCI models for extreme temperatures for the specific technology under investigation.

BTI and HCI characterization and modeling has been extensively studied so far; however, most of these research works were conducted at elevated voltages beyond the safe operating conditions to activate the degradation within a short time. However, this approach presents the risk of over-accelerating other different wear-out mechanisms that are not supposed to be experienced in normal operating voltages. Therefore, our experiment is a long-term test in which we characterized BTI and HCI degradation at the operating voltages with the stress time up to 1,000 hours. These experiment results are expected to reflect the real wear-out in practical application.

The second highlight of this study is that BTI and HCI tests were performed under severe temperatures (150°C and 210°C) to simulate the oilfield application in extreme thermal conditions. Note that the stress temperatures of already published works on BTI and HCI degradation are commonly achieved between the range of 125°C to 175°C [1-7].

II. THEORITICAL BACKGROUND

BTI and HCI are the two major reliability issues in advanced CMOS technologies. The main impacts of HCI and BTI on DC characteristics are threshold voltage shift and a decrease in channel carrier mobility μ, which finally lead to a drop of the drain current.

BTI is a physical phenomenon related to the generation and activation of the interface traps and oxide trapped charges inside the dielectric layer [10]. The mechanism is accelerated by elevated temperature (between the range of 100°C to 250°C) and high vertical electric field [4]. BTI is observed in both nMOS and pMOS. Negative Bias Temperature Instability (NBTI) occurs in pMOS with negative gate voltage and Positive Bias Temperature Instability (PBTI) occurs in nMOS with positive gate voltage at elevated temperature.

Hot carriers (electrons in nMOS or holes in pMOS) are channel carriers that achieve large kinetic energies from being accelerated by high electric field. Under HCI bias conditions, the lateral electric field is responsible for carrier heating while vertical electric field supports the injection of hot carriers into the gate oxide causing the damage at the interface Si/SiO$_2$ [1,8]. Besides, the downscaling of gate length (L) can increase the lateral electric field and lead to the serious HCI degradation [9]. The diagram in Figure 1 describes both types of degradation, their accelerating factors and impacts on the transistor characteristics.

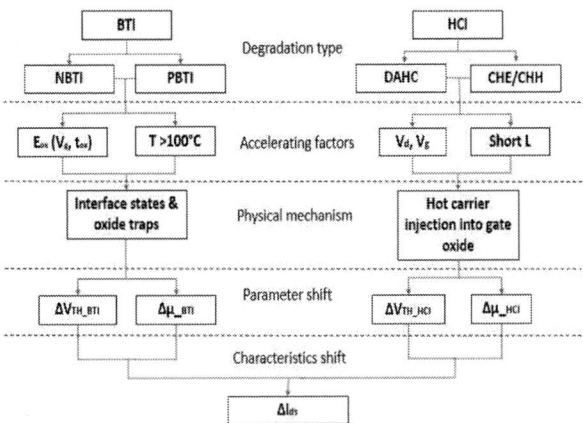

Figure 1: BTI and HCI mechanisms

III. AGING TEST DESCRIPTION

A. Device Under Test

The technology under test is a CMOS 0.18 um qualified for high temperatures. The devices under test are 1.8 V isolated n-type and p-type MOSFETs. Two sizes of each transistor type have been investigated: W/L=20um/0.18um and 20um/0.5um. The gate oxide thickness (typical value) is t_{ox} = 4.1nm for nMOS and t_{ox} = 3.9nm for pMOS. At least 10 devices were characterized at wafer level to evaluate the technology dispersion. The average dispersion of 5% indicates the maturity of the technology. All dies were packaged into ceramic DIL-40 allowing performing the aging tests at elevated temperature.

B. Stress Bias and Temperature Conditions

The characterization includes periodical measurements of the DC characteristics. In this aging test campaign, 2 to 3 devices were used for each stress bias conditions. The threshold voltage shift ΔV_{th} has been chosen to capture the BTI and HCI degradation, and the shifts of the threshold voltage were extracted from I$_d$-V$_g$ curves of MOSFETs biased in saturation region based on procedure described in [11]. BTI recovery modeling is not considered in this paper and will be considered in future work. The aging test conditions are listed in Table 1. The body is connected to the source and the substrate is grounded in all cases to reflect the real practical applications. The aging tests were performed at three different temperatures: 25°C, 150°C and 210°C.

For BTI characterization, NBTI_1 and NBTI_2 correspond to two different |V$_{gs}$| values (1.8V and 2.0V, respectively) and have been considered to activate uniform BTI degradation in which |V$_{ds}$| is kept quite small (0.05V).

More aggressive |V$_{ds}$| would have triggered unwanted HCI causing the overestimation of the BTI degradation.

Table 1: *Bias and temperature stress condition*

Type	|Vds|(V)	|Vgs|(V)	Bias Conditions	T°C
nMOS	2.0	1.0	[1]DAHC	
	2.0	2.0	[2]CHE	
	0.05	2.0	[3]PBTI	25°C
pMOS	0.05	1.8	[4]NBTI_1	150°C
	0.05	2.0	NBTI_2	210°C
	2.0	2.0	[5]CHH or NBTI_3	

[1] Drain Avalanche Hot Carrier
[2] Channel Hot Electron
[3] Positive Bias Temperature Instability
[4] Negative Bias Temperature Instability
[5] Channel Hot Hole

For HCI characterization, drain avalanche hot carrier (DAHC) in nMOS has been performed with V$_{gs}$=0.5V$_{ds}$ corresponding to the maximum body current. This stress condition is commonly used in standard hot carrier investigation for low-voltage MOSFETs, where the generated hot carriers are proportional to the body current and therefore, HCI degradation is maximized when body current achieves its peak value.

Channel hot electron (CHE) in nMOS or channel hot hole (CHH) in pMOS has been performed with maximum drain voltage and maximum gate voltage |V$_{gs}$|=|V$_{ds}$|=2.0V, which corresponds to maximum device biases recommended in datasheet.

Because the CHH bias condition combined with the presence of high temperature can activate NBTI, the CHH bias condition is also considered as NBTI_3 to investigate the contribution of high |V$_{ds}$| to NBTI degradation.

IV. EXPERIMENTAL RESULTS AND DISCUSSION

A. Bias Temperature Instability

Three configurations NBTI_1, NBTI_2 and NBTI_3 bias conditions were applied at 150°C and 210°C to characterize NBTI degradation. According to our results, NBTI in pMOS is considered as the most serious reliability issue of 0.18um

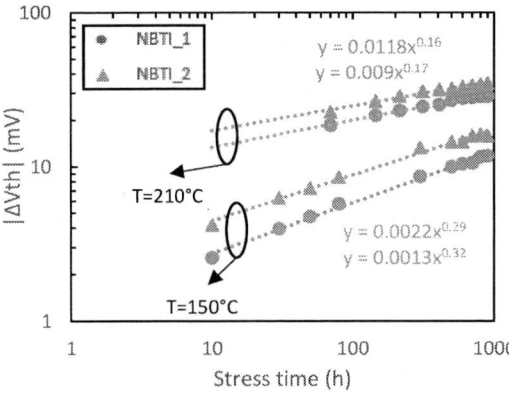

Figure 2: *Threshold voltage shift, ΔVth, under NBTI_1 and NBTI_2 bias condition at 150°C and 210°C of 0.18um pMOS. The dotted lines are fitted to the results.*

CMOS. The threshold voltage shifts due to NBTI at 210°C can achieve up to 9% after 1,000 hours applying stress.

The threshold voltage shifts due to NBTI_1 and NBTI_2 bias voltages have been observed following a power-law as function of stress time (Figure 2):

$$\Delta V_{th} = A \times t^n \qquad (1)$$

where A is the prefactor and n is the time exponent parameter. At 150°C, the value of n is between the range of 0.29 to 0.32. However, this value decreases significantly down to 0.16-0.17 under 210°C with the similar stress voltages applied.

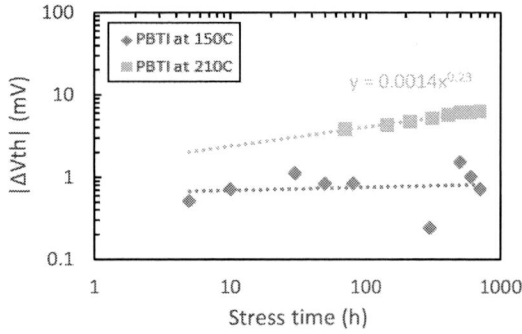

Figure 3: Threshold voltage shift, ΔVth, under PBTI bias condition at 150°C and 210°C of 0.18um nMOS. The dotted lines are fitted to the results.

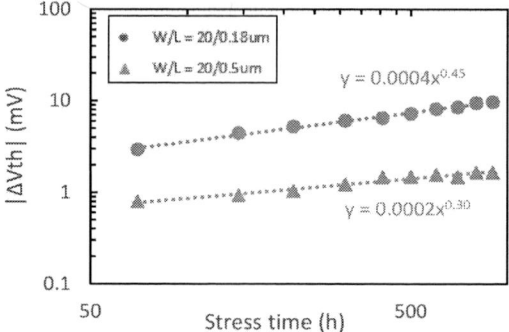

Figure 5: Threshold voltage shift, ΔVth, under CHE bias condition at 210°C of 0.18um and 0.5um nMOS. The dotted lines are fitted to the results.

The NBTI tests conducted on 0.18um and 0.5um (not shown here) revealed the independence of BTI degradation on gate length, i.e., the similar threshold voltage shifts have been obtained when applying the same stress conditions.

PBTI degradation in nMOS is usually neglected due to its limited impact. Our results after 1,000 hours stress show that PBTI degradation under 150°C show limited ΔV_{th} below 0.5%. However, PBTI becomes more pronounced at 210°C in which ΔV_{th} shift is around 2.5% after the same stress period and bias conditions as the PBTI test at 150°C. PBTI degradation at 210°C also follows power-law with stress time which is demonstrated in Figure 3.

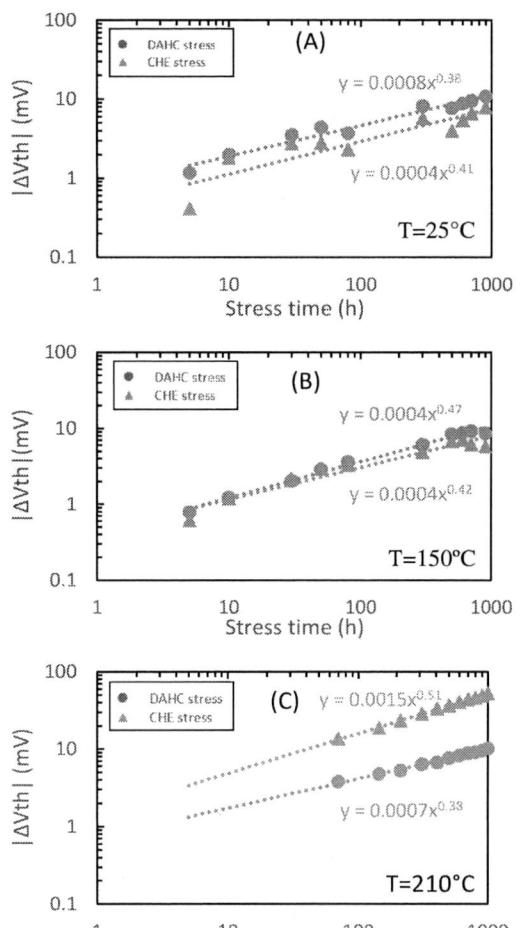

Figure 4: The threshold voltage shift ΔVth under DAHC and CHE bias conditions at 25°C (A), 150°C (B) and 210°C (C) of 0.18um nMOS. The dotted lines are fitted to the results.

B. Hot Carrier Injection

The body current is often considered as a criterion of HCI degradation. Therefore, the worst-case scenario is usually realized at DAHC bias conditions in which the maximum body current is obtained. Nevertheless, with technology scaling ($L \leq 0.1\mu m$), the worst-case situation occurs at CHE/CHH bias condition ($|V_{gs}| \approx |V_{ds}|$) for both p- and n-channel [9].

We performed both DAHC and CHE bias condition at 25°C, 150°C and 210°C. Like BTI, the threshold voltage shifts due to HCI also obey the power law with stress time. As expected, DAHC causes the worst-case degradation at 25°C. However, CHE damage exceeds greatly DAHC damage at 210°C which is unforeseen. Figure 4 presents the competition between DAHC and CHE degradation in which DAHC exhibit poor relationship with temperature while CHE degradation becomes more and more serious at high temperature.

It is accepted that HCI degradation (both CHE and DAHC) is weakly dependent on the stress temperature [12-13]. However, HCI degradation at extreme temperature (210°C) exhibits a complex behavior, which is the combination of PBTI and intrinsic CHE damage. Nevertheless, PBTI cannot bear full responsibility for the dominant CHE degradation at 210°C. According to our results, the magnitude of the V_{th}

shifts at 210ºC due to PBTI (Figure 2) is pretty small compared with the one caused by CHE (Figure 4C). Therefore, we expected that CHE mechanism is strongly dependent on temperature. So far, the relationship of CHE degradation and elevated temperature has not been intensively studied.

Our HCI characterization on different transistor sizes (W/L=20/0.18um and 20/0.5um) presents an important dependence of the degradation on gate length. As shown in Figure 5, the smaller transistor exhibits larger threshold voltage shift which is in accordance with literature [9].

HCI in pMOS can be negligible since threshold voltage shift due CHH degradation is below 1% after 1,000 hours applying stress at 25°C and even at 210°C.

From these experimental results, the dependence of threshold voltage shifts due to BTI and HCI degradation on temperature, bias voltage, gate length and stress time has been modeled (not describe here for the sake of paper length) to implement into circuit simulator.

V. INTEGRATION OF AGING MODELS INTO CIRCUIT SIMULATOR

A compact model including aging laws has been proposed based on the aging test results to simulate the degradation behavior observed from experiments. These aging laws were adapted to the underlying BSIM3 parameter, V_{th0}, with the fitted parameters extracted from experimental results. The degraded V_{th0} in BSIM3 is defined as:

$$V_{th0}(aged) = V_{th0} - \Delta V_{th0} \qquad (2)$$

Where V_{th0} is the threshold voltage at zero-volt substrate bias and $V_{th0} < 0$ for pMOS.

In this paper, Cadence Unified Reliability Interface (URI) was used to implement NBTI model for the reliability simulation.

The compact model provided by the foundry just ensures DC characteristics accuracy lower than T=175°C. Beyond this critical temperature, there is a large difference between simulation and measurement (~14%). Therefore, before performing aging simulation, some BSIM3 parameters (W, L, V_{th0}) have been slightly modified to obtain the good agreement between measurement and simulation of fresh device.

Reliability simulations of a single pMOS at NBTI_2 were performed to validate the proposed NBTI model. We obtained very good accuracy compared to experimental results (Figure 6) even at extreme temperature close to 210°C.

VI. CONCLUSION

For aging tests achieved at temperature lower than conventional aging test (typically around 150°C), our results confirm the ones reported in the literature. For aging tests performed under extreme temperatures (typically 210°C), the results show significant deviations compared with conventional values of aging model parameter. In this study, we have demonstrated the aging models for BTI and HCI degradation and implemented it into a commercial software

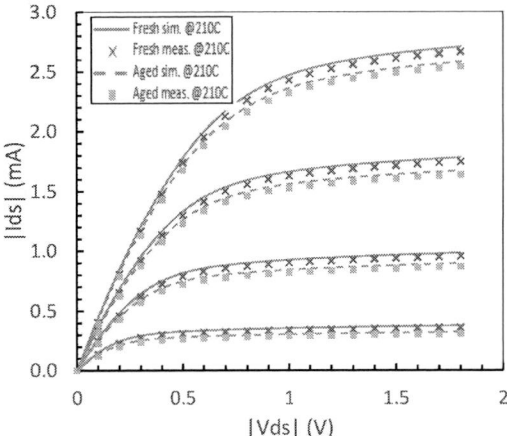

Figure 6: *Comparison between aged simulation and aged measurement of 0.18um pMOS under NBTI_2 bias condition at 210°C after 900h stress. The values of |Vgs| are 0.9V, 1.2V, 1.5V and 1.8V, respectively.*

tool. The simulated DC characteristics were compared with measurements at 210°C. A good agreement between simulation and experimental results implies that the extracted parameter value combined with BSIM3 model ones can provide a good description of BTI and HCI degradation. The future research will further focus on circuit level simulation and comparison with actual logic circuit.

REFERENCES

[1] G. Chen et al. "Dynamic NBTI of PMOS transistors". Proc. IEEE Reliab. Phys. Symp 2003, pp 196–202.

[2] D. Schroder et al., "NBTI: Road to cross in deep submicron silicon semiconductor manufacturing". J. Appl. Phys., 94(1)1, 1-18, 2003.

[3] X. Li et al., "Compact modeling of MOSFET Wearout Mechanisms for Circuit-Reliability Simulation". IEEE Transactions on device and materials reliability, Vol. 8, No.1, March 2008

[4] V. Huard, et al., "NBTI degradation from physical mechanisms to modeling". Microelectronics Reliability, 46: 2006, pp 1–23

[5] G. Chen et al. "Dynamic NBTI of PMOS transistors". Proc. IEEE Reliab. Phys. Symp 2003, pp 196–202.

[6] Schroder et al., "NBTI: What do we understand". Microelectron. Reliab. 2007,47,841-852.

[7] J. Franco, B. Kaczer, G. Groeseneken, "Reliability of High Mobility SiGe Channel MOSFETs for Future CMOS Applications". Springer Series in Advanced Microelectronics.

[8] V. Huard et al., "Design-in-Reliability Approach for Negative Bias Temperature Instability and Hot Carrier Degradation in Advanced Nodes". Transactions on Device & Materials Reli.,7(4), 558-570, 2007.

[9] Grasser, T. (2015). "Hot Carrier Degradation in Semiconductor Devices" (1st ed. 2015.. ed.).

[10] Alvin W. Strong et al., ed. (2009), "Reliability Wearout Mechanisms in Advanced CMOS Technologies". Wiley-IEEE Press.

[11] Ortiz-Conde et al., "A review of recent MOSFET threshold voltage extraction methods". Microelec. and Reli., 42(4), 583-596, 2002.

[12] E. Takeda, N. Suzuki, "An empirical model for device degradation due to Hot-Carrier injection". IEEE Electron Dev. Lett. 4(4), 111–113 (1983)

[13] C. Hu et al., "Hot-electron-induced MOSFET degradation–model, monitor, and improvement". IEEE Trans. Electron Dev. 32(2), 375–385 (1985)

© VDE VERLAG GMBH · Berlin · Offenbach

A Cryogenic High-Voltage Amplifier for Ion Traps

Michael Sieberer
Infineon Technologies Austria AG
Villach, Austria
michael.sieberer@infineon.com

Christoph Sandner
Infineon Technologies Austria AG
Villach, Austria
christoph.sandner@infineon.com

Peter Hadley
Institute of Solid State Physics
Graz University of Technology
p.hadley@tugraz.at

Abstract—Ion traps for quantum computers need high voltages to confine ions in an electrostatic potential. This work presents a high voltage current-feedback instrumentation amplifier, implemented in 130 nm-CMOS, that is fully functional at temperatures below 20 K and dissipates only 1.5 mW when powered with ± 12 V. Since drain-extended MOS transistors cannot be used at cryogenic temperatures, high-voltage compliance is achieved by stacking low-voltage transistors. The high-frequency noise is below $8\,\text{nV}/\sqrt{\text{Hz}}$ to avoid ion heating. Flicker noise and offset is reduced by chopping of the input gain stages.

Index Terms—cryogenic electronics, chopper amplifier, high voltage, ion traps, quantum computing

I. Introduction

Trapped ions are a promising option for practical quantum computers, featuring long-lived quantum states and general purpose gates [1]. However, like many other quantum computing architectures, the current systems are limited to a few tens of qubits. Enlarging the total number of qubits is necessary to implement useful noisy intermediate-scale quantum (NISQ) algorithms or to enable quantum error correction.

Paul ion traps, like the one sketched in Fig. 1, generate a trapping potential Φ_t by combining a radio-frequency (RF) field with a low-frequency (DC) field [1]–[3]. In our surface ion traps, the RF field mostly provides vertical (z-axis) confinement while the DC fields controls lateral (x-y-plane) confinement. Following this principle, the ion can be moved laterally by altering the voltages that generate the DC field. This ion movement is thought to be a key enabler for scaling up trapped ion quantum computers.

Currently, the DC voltages are generated far from the experiment, using off-the-shelf digital-to-analog converters (DACs). Since many different DC voltages are needed to scale up the number of qubits, wiring will eventually become a major obstacle. This issue can be solved by moving the DACs closer to the ion trap. However, moving the DACs closer to the experiment involves cooling the DACs to cryogenic temperatures, because most ion traps are operated well below 20 K.

Apart from being functional at cryogenic temperatures, the system must exhibit very low noise at the trap frequency

This project has received funding from the European Union's Horizon 2020 research and innovation programme under Grant Agreement No. 801285 (PIEDMONS).

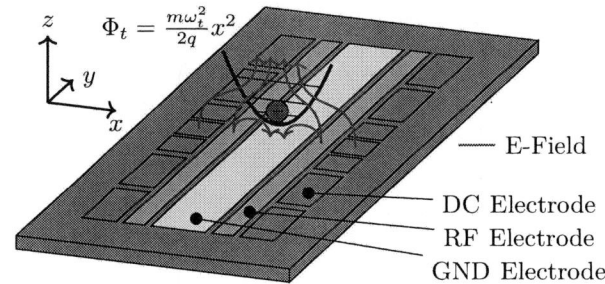

Fig. 1: Operating Principle of an Ion Trap

ω_t to avoid motional heating. The heating rate due to voltage noise S_V is given by [2], [3]

$$\Gamma_\text{h} = \frac{q^2}{4m\hbar} \cdot \frac{S_V(\omega_t)}{\omega_t \delta_c^2}, \tag{1}$$

where m and q is the ion's mass and charge, $\omega_t \approx 2\pi{\cdot}1\,\text{MHz}$ is the trap frequency and δ_c is the characteristic electrode-ion distance. For the targeted ion traps, the characteristic distance is on the order of $\delta_c \approx 2\,\text{mm}$ [2].

Integrating DACs into an ion trap has been presented by [3], however the implementation shows large flicker noise and power dissipation. In the following sections, after giving a brief introduction into cryogenic electronics, we present a low-power chopper amplifier as a core building block for a DAC that remedies both drawbacks.

II. Cryogenic Electronics

To identify relevant parameters, we measured several different MOS-transistors of a 130 nm-process down to a temperature of about 16 K, which was the lower limit of our measurement setup [4]. We then used the measurement results to adapt the existing BSIM4-models. Simulations using these models allow us to identify headroom and stability issues early.

In cryogenic operation, two noteworthy effects are a significant increase of the threshold voltage V_th and a strong increase of the transconductance efficiency g_m/I_D.

A. Threshold Voltage at Cryogenic Temperatures

With decreasing temperature, as the intrinsic carrier concentration n_i decreases exponentially, the bulk Fermi-energy E_f moves closer to the conduction- (valence-) band

© VDE VERLAG GMBH · Berlin · Offenbach

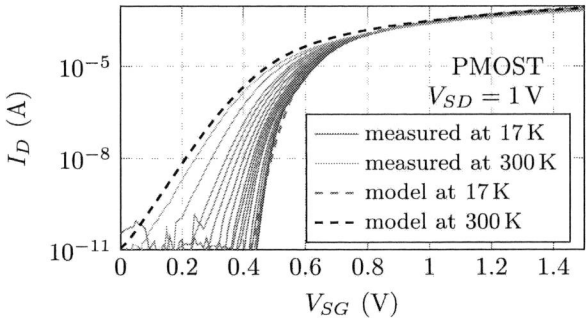

Fig. 2: I_D in saturation, measured down to 17 K

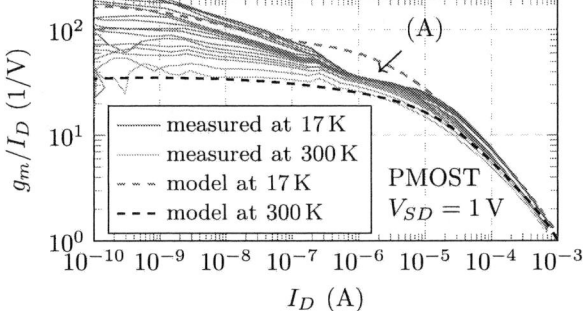

Fig. 3: Measured g_m/I_D over I_D

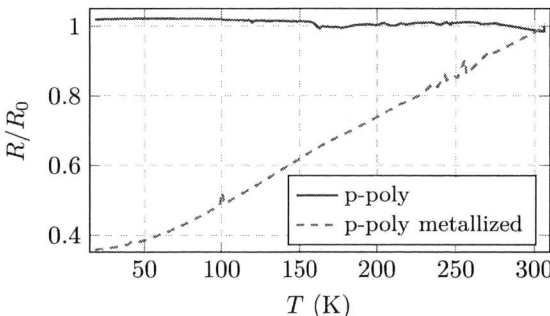

Fig. 4: Measured relative change of an integrated p-doped poly-silicon resistor.

edge in n- (p-) doped body regions. Consequently, it takes a larger gate-potential to invert the channel region, effectively increasing the threshold-voltage V_{th} [5]. In practice, this means V_{th} increases by about 200 mV at 17 K compared to room temperature in our process. The change of threshold voltage is clearly visible in Fig. 2.

A second effect that follows from the exponential decrease in n_i is a huge degradation of the bulk conductance, which increases the likelihood of latch-up [6].

B. Increase of Transconductance Efficiency

Traditionally, the maximum transconductance efficiency of any MOST is given by

$$\left(\frac{g_m}{I_D}\right)_{\text{max}} = \left(\frac{\partial \log I_D}{\partial V_{GS}}\right)_{\text{max}} = \frac{1}{n \cdot V_T}, \qquad (2)$$

where $n \approx 1.2$ is the subthreshold slope factor and $V_T = kT/e$ is the thermal voltage.

Cooling down the transistor to tens of Kelvins increases g_m/I_D noticeably. Fig. 3 shows the measured behavior of a PMOST, the maximum achievable g_m/I_D exceeds 200 /V. This maximum g_m/I_D is still less than would be expected from eq. (2), a suggested reason for this behavior is a band tail at the edge of the conduction and valence bands, limiting eq. (2) to about $e/4\,\text{meV} = 250\,/\text{V}$ [7].

A second effect that the measurements of Fig. 3 display (A) is a sudden degradation of g_m/I_D right at the onset of weak inversion. We attribute this behavior to current flowing along the edges of the transistor in so-called

parasitic corner devices [8]. Matching in this region is poor, especially if the width of the transistors differs. Since BSIM4 does not provide a parameter to model this behavior, it will not be visible in our simulations.

C. Resistors and Capacitors

Resistors and capacitors are, for example, useful for implementing continuous time filters. As the corner frequency $\omega_c = 1/RC$ depends on both R and C, a temperature model for both devices is crucial. The measurements in Fig. 4 confirm that the metallized (salicided) p-poly-silicon resistors show a strong temperature dependence, whereas the resistivity of un-salicided poly-silicon resistors remains virtually unchanged over the whole measured temperature range. This is due to the fact that the mobility in poly-silicon is limited by scattering on grain boundaries rather than phonon scattering [9]. We also confirmed that the change in MOS- and MIM- (metal-insulator-metal) capacitances over temperature is negligible.

Following these observations, we decided to use source degeneration for current mirrors and critical transconductors, as this will effectively stabilize the transconductance over the full temperature range.

III. High Voltage Output Stage

Our cryogenic measurements indicate that the drain-extended high voltage MOS transistors available in our process become unreliable at cryogenic temperatures. A way to overcome this limitation is to stack transistors such that each drops a fraction of the total voltage. An output stage that can supply ± 10 V is depicted in Fig. 5. The bulk of all transistors is shorted to its source; the NMOST are placed in triple-well structures.

The following explanation focuses on the drawn negative-voltage side only; the whole structure is symmetrical around the center transistors and the discussion applies to the positive-voltage side with complementary transistors as well. At the structure's core is a floating voltage follower consisting of M_{10}, M_{20} and their corresponding NMOST, that copies the voltage of the high-impedance node V_{int} to the low-impedance output V_{out}. The NMOST

SMACD / PRIME 2021 | 19 – 22 July 2021, Online Event

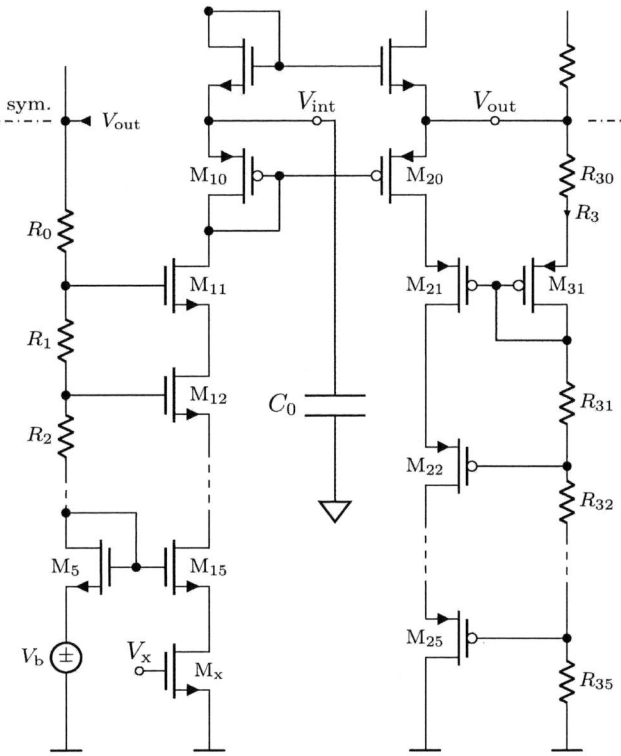

Fig. 5: Stacked-Transistor Output Stage

M_{11}–M_{15} protect the input transistor M_x from a drain-to-gate breakthrough, the PMOST M_{21}–M_{25} protect M_{20}. The resistors evenly distribute the output voltage on all protection transistors, for example R_{30} – R_{35} evenly distribute V_{out} over M_{20} – M_{25}. When going close to the rails, M_x and M_{10} need to remain in saturation to ensure a satisfactory loop gain. For M_x, this is ensured by the voltage source V_b and M_{15} limiting V_{DS} of M_x. A voltage follower provides V_b. M_{31} and M_{21} ensure that M_{20} remains in saturation for the targeted output range.

To make the structure stable, a large capacitor C_0 is connected to V_{int}. This capacitor will also be the final filtering capacitor (see Fig. 6) for the overall amplifier.

IV. Amplifier Design

The overall implemented structure, outlined in Fig. 6, is a chopped current-feedback instrumentation amplifier (CFIA) [10, fig. 10.40]. It uses 4 gain stages G_{in} – G_1 to achieve both a high DC gain and a third-order filter function from the input to the output. G_1 and A_0 are implemented using the presented stacked-transistor design, the other blocks reside in the low-voltage domain. The closed loop transfer function is given by [10, chap. 6.4.2]

$$H_{cl} = \frac{A}{1 + A\beta} = \frac{1}{\beta} \cdot \frac{K_1 K_2 p_1'}{s^3 + p_1' s^2 + K_1 p_1' s + K_1 K_2 p_1'} \quad (3)$$

$$\beta = \frac{R_2}{R_1 + R_2} \cdot \frac{g_{m,\text{fb}}}{g_{m,\text{in}}}, \quad (4)$$

where $p_1' = g_{m1}/C_0$, $K_1 = C_{m3} g_{m2}/C_{m2} C_{m1}$ and $K_2 = \beta g_{m,\text{in}}/C_{m3}$.

Chopping of G_{in} and G_{fb} moves the input-transistors' low-frequency (flicker) noise out of band. The now up-modulated low-frequency noise and amplifier offset will be visible at the output around the chopping frequency as ripple. To limit the ripple amplitude the chopper frequency is much higher than the amplifier bandwidth and a ripple-reduction filter [10, chap. 10.9] actively suppresses this ripple.

The large MIM capacitors C_0 and C_{m1} are needed to filter the high-frequency noise.

V. Simulation Results

Fig. 7 shows the transient response to a voltage ramp on V_{in}. The slew rate is limited by g_{m2} and C_{m1} to about $250\,\text{kV/s}$. Applying a higher slew rate causes strong nonlinear effects, thus the input will be band-limited in the application. The output stage draws about $50\,\mu\text{A}$ and $55\,\mu\text{A}$ from the $\pm 12\,\text{V}$ supplies respectively. The low voltage parts of the circuit consume another $170\,\mu\text{A}$ at $1.5\,\text{V}$ for a total power consumption of about $1.5\,\text{mW}$. Monte-carlo simulation indicates a statistical (1σ) gain error of $0.1\,\%$, caused by mismatch between G_{in} and G_{fb}.

Fig. 8 shows the results of noise simulations. As at the time of writing this paper, the cryogenic models were not ready for periodic steady-state (pss) simulations, only the high-frequency noise component was considered for $T = 20\,\text{K}$. A load capacitance of $C_L = 100\,\text{pF}$ was assumed for all but the fourth simulation. The voltage noise at room temperature and $\omega_t = 2\pi \cdot 1\,\text{MHz}$ is below $\sqrt{S_V} \approx 8\,\text{nV}/\sqrt{\text{Hz}}$, corresponding to a motional heating rate of less than $\Gamma_h \approx 2500\,/\text{s}$. Without adapting the output capacitor to the lower output resistance, this figure reduces to $\Gamma_h \approx 500\,/\text{s}$ at $20\,\text{K}$. Although S_V is at least reduced by a factor of 10 compared to [3], the expected motional heating rate is similar because of the lower ω_t and δ_c considered in this work.

The voltage-dependent $1/f$-noise is due to R_1 and R_2 having significant flicker noise [9].

VI. Conclusion

Based on cryo-adapted device models, this work shows that it is possible to implement a high-voltage circuit that works down to deep cryogenic temperatures where drain-extended MOS devices become unreliable. It then presents the simulation results of a low-power chopping amplifier that has an adequate noise performance for ion traps. These properties allow the circuit to generate high voltages close to the ion trap, which is a prerequisite for large-scale trapped-ion quantum computing.

Acknowledgment

We would like to thank Andreas Pawlak for adapting the BSIM4-models and making them fit for cryogenic simulation. We would also like to thank the PIEDMONS team for their input on ion traps.

© VDE VERLAG GMBH · Berlin · Offenbach

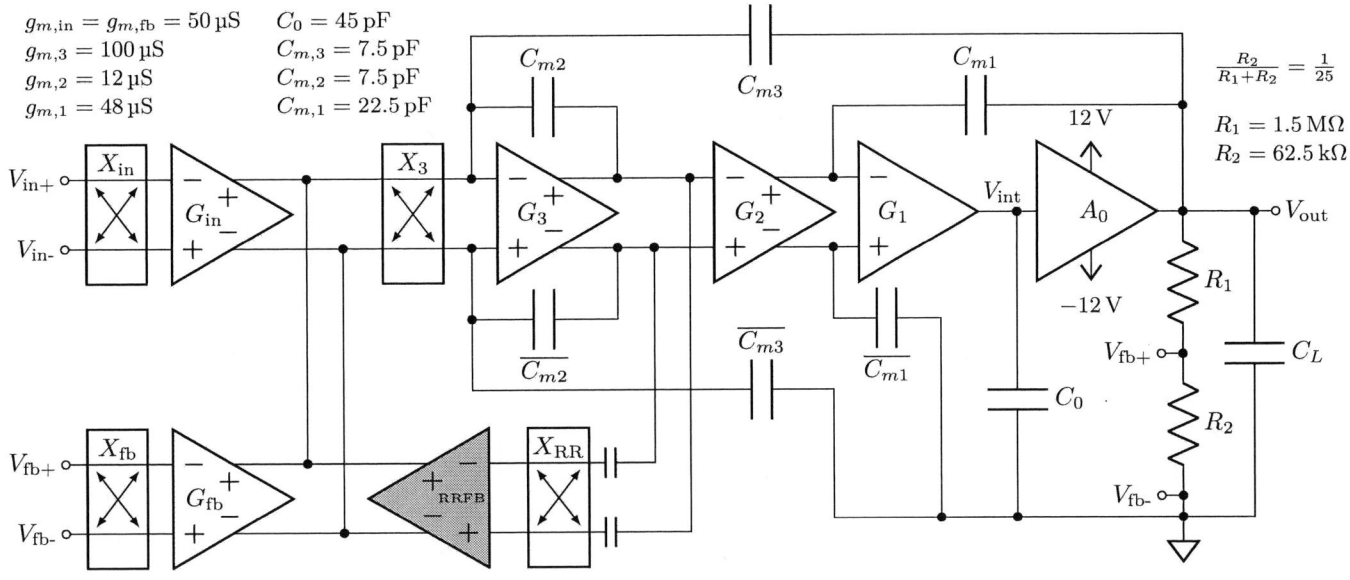

Fig. 6: Chopper Concept

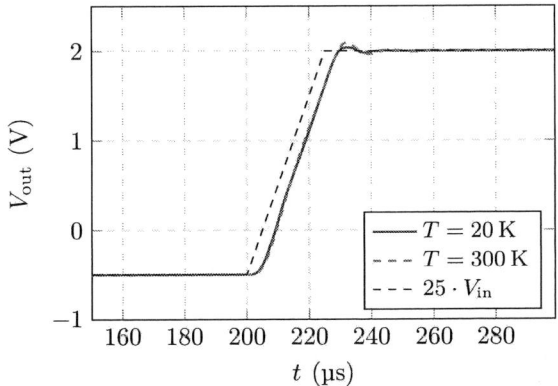

Fig. 7: Transient Response to a Ramp

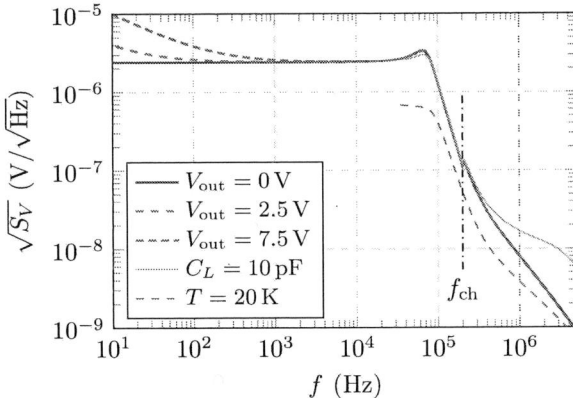

Fig. 8: Simulated Output Noise

If not stated, $T = 300\,\text{K}$ and $C_L = 100\,\text{pF}$.
Low-frequency noise is voltage dependent.

References

[1] C. D. Bruzewicz, J. Chiaverini, R. McConnell, and J. M. Sage, Trapped-ion quantum computing: Progress and challenges, Apr. 8, 2019. [Online]. Available: http://arxiv.org/pdf/1904.04178v1.

[2] P. C. Holz, S. Auchter, G. Stocker, et al., "2d linear trap array for quantum information processing," Advanced Quantum Technologies, vol. 3, no. 11, p. 2 000 031, 2020, issn: 2511-9044. doi: 10.1002/qute.202000031.

[3] J. Stuart, R. Panock, C. D. Bruzewicz, et al., "Chip-integrated voltage sources for control of trapped ions," Physical Review Applied, vol. 11, no. 2, 2019. doi: 10.1103/PhysRevApplied.11.024010.

[4] P. Stampfer, "Characterization and modeling of semiconductor devices at cryogenic temperatures," Master, TU Graz, Graz, 2020.

[5] A. Beckers, F. Jazaeri, and C. Enz, Cryogenic mosfet threshold voltage model, Apr. 22, 2019. [Online]. Available: http://arxiv.org/pdf/1904.09911v1.

[6] E. Schriek, F. Sebastiano, and E. Charbon, "A cryo-cmos digital cell library for quantum computing applications," IEEE Solid-State Circuits Letters, vol. 3, pp. 310–313, 2020. doi: 10.1109/LSSC.2020.3017705.

[7] A. Beckers, F. Jazaeri, and C. Enz, "Theoretical limit of low temperature subthreshold swing in field-effect transistors," IEEE Electron Device Letters, p. 1, 2020, issn: 0741-3106. doi: 10.1109/LED.2019.2963379.

[8] Y. Joly, L. Lopez, J.-M. Portal, et al., "Impact of hump effect on mosfet mismatch in the sub-threshold area for low power analog applications," in 2010 10th ICSICT, IEEE, 112010, pp. 1817–1819, isbn: 978-1-4244-5797-7. doi: 10.1109/ICSICT.2010.5667684.

[9] R. Brederlow, W. Weber, C. Dahl, D. Schmitt-Landsiedel, and R. Thewes, "Low-frequency noise of integrated polysilicon resistors," IEEE Transactions on Electron Devices, vol. 48, no. 6, pp. 1180–1187, 2001, issn: 0018-9383. doi: 10.1109/16.925245.

[10] J. H. Huijsing, Operational amplifiers: Theory and design / Johan Huijsing. Switzerland: Springer, 2016, isbn: 978-3-319-28126-1.

Cryogenic RF Transimpedance Amplifier in 22 nm SOI-CMOS for Control of a Qubit

Ricardo Heinen[1,2], Dennis Nielinger[1], Christian Grewing[1], Ralf Wunderlich[2] and Stefan Heinen[2], *Fellow*, IEEE

[1]*Central Institute of Engineering, Electronics and Analytics, Electronic Systems*
Forschungszentrum Jülich GmbH, Jülich, Germany
[2]*Chair of Integrated Analog Circuits and RF Systems, RWTH Aachen University, Aachen, Germany*
ricardo.heinen@rwth-aachen.de

Abstract—The design and simulation of a cryogenic RF transimpedance amplifier (TIA) for signals around 20 GHz with millivolt amplitudes is presented in this paper. The TIA is part of an IC with a modulator system, where its primary application is to drive a semiconductor based spin qubit at temperatures around 100 mK. A 22 nm FDSOI CMOS technology is used for the circuit development. In post-layout simulations the TIA showed transimpedance magnitudes over 40 dBΩ and bandwidths larger than 1 GHz while dissipating less than 170 µW. Tuning options for the operating point and the frequency behavior where implemented to ensure functionality at cryogenic temperatures. The TIA performance allows to flip chip bond the modulator system in close vicinity to the qubit inside a dilution refrigerator. This setup can verify if the concept is feasible to upscale qubit numbers.

Index Terms—TIA, RF, cryogenic, SOI, low power, quantum computing, qubit

I. INTRODUCTION

For specific fields, e. g. modeling of chemical reactions and materials, sampling of probability distributions or optimization tasks, using the quantum mechanic properties of matter rather than classical binary computing, can solve the related computational problems exponentially faster [1]. From the idea of using quantum mechanic properties for computation, concepts for quantum computers based on these properties emerged. The unit of information in quantum computers is the qubit and up to now 53 of these are operated together [2], [3]. To solve practical problems efficiently quantum computers consisting of millions of qubits are required [1]. That is why research has been carried out for more than 20 years to implement these qubits in a scalable way and thus lead to universal quantum computers [4].

One of the most promising implementations of qubits are semiconductor spin qubits due to their possible scalability [5]. This technology relies on the trapping of single electrons by isolated donor atoms or by gate defined potentials [6]. The electron wave function in semiconductors extends over relatively large distances [7], [8]. This leads to electron-mediated or indirect nuclear spin coupling making the spin of the trapped electron sensitive to externally applied electric fields [7], [8]. That is why the trapped electron can be used as a qubit representation for quantum computation [7], [8].

The RF control signals to operate the qubits can be generated in close vicinity by a CMOS IC. This avoids large setups

of standard laboratory equipment at room temperature, thus taking advantage of the scalable semiconductor qubit. The near by placement of control and qubit IC can be realized by flip chip bonding both ICs on a common interposer. For the reliable operation of the qubit the temperature of the setup is around 100 mK. The dissipated power of the developed control IC must be below the cooling limits of the dilution refrigerator which is less than 1 mW.

The presented work is part of an integrated modulator system (SQuBiC 3) to mix a baseband signal with a carrier signal. Additionally, the phase of the resulting signal is switched by the modulator to operate the qubit. The analog part of the system is shown in Fig. 1, where the interface to the qubit chip is modeled as a capacitor. The image reject mixer is depicted here simplified. Since the mixer is not able to drive the high capacitive load, an RF amplifier to drive the qubit was developed. The core part of the amplifier is a single ended inverter based TIA with resistive feedback. The differential output signal of the mixer is coupled with a balanced to unbalanced transformer (balun) into the feedback path of the TIA. To the best of the authors' knowledge this is a new way to include the TIA feedback into the balun circuitry.

The overview of the system which this work is part of and the resulting requirements are explained in section II. In section III the design of the circuit is described. Following in section IV, the results from simulations to characterize the circuit and study the system performance.

II. SYSTEM OVERVIEW AND REQUIREMENTS

At the input of the mixer in Fig. 1 the RF signal v_{center} with a frequency of 20 GHz and the baseband signal v_{BB} of 10 MHz frequency are applied differentially. This results at the output of this block in frequencies of 20.01 GHz or 19.99 GHz depending on the configuration.

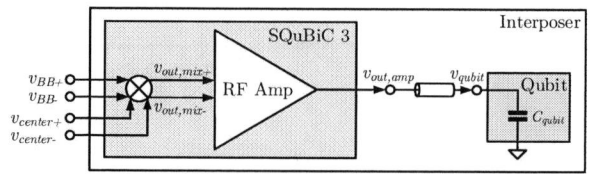

Fig. 1. Analog block diagramm of SQuBiC 3 system with qubit load.

The output of the amplifier connects single ended to the qubit chip through a transmission line on the interposer which has an impedance of $100\,\Omega$. The input of the qubit chip can be approximated by a capacitive load of $80\,\mathrm{fF}$. A voltage amplitude v_{qubit} of 3 to 10 mV is required at the qubit input. The qubit will be operated at $100\,\mathrm{mK}$, but it is also necessary to test the control IC at room temperature to compare the performance and study the effects of cryogenic temperatures. To allow the cryogenic operation the amplifier should not dissipate more than $200\,\mu\mathrm{W}$.

A change in phase of $90°$ controls the state of the qubit. The output of the mixer will be switched off during the phase changes by an external pulse signal requiring the amplifier to have a bandwidth BW greater than $200\,\mathrm{MHz}$ to avoid delays. The suppression of the opposite sideband and of the center frequency are specified to be more than $40\,\mathrm{dBc}$.

Apart from the amplifier used in the system of Fig. 1, a standalone test amplifier is placed on the control IC. In this way the amplifier can be characterized with dedicated measurements allowing to verify the functionality and performance of the system. Therefore, the amplifier must be able to drive a $50\,\Omega$ load. The test amplifier should be turned off during the operation of the system to meet the power budget of the control IC.

The most relevant requirements for the design of the amplifier are summarized in Table I. Despite the low output amplitude of less than $10\,\mathrm{mV}$, the low power consumption at $20\,\mathrm{GHz}$ is critical.

III. IMPLEMENTATION

At frequencies around $20\,\mathrm{GHz}$ the transistors gate capacitances result in low input impedance magnitudes $|Z_{in,amp}|$ of a voltage amplifier stage compared to the output impedance magnitude of the mixer $|Z_{out,mix}|$. This leads to low voltage amplitudes at the input of the amplifier. Therefore, the amplifier is implemented as TIA, optimized such that the mixer provides the largest possible current to the input $i_{in,amp}$. The TIA converts $i_{in,amp}$ to a defined voltage amplitude at its output $v_{out,amp}$ while driving the load. The current to voltage conversion factor, also called transimpedance, is defined by

$$Z_T = \frac{v_{out,amp}}{i_{in,amp}}. \qquad (1)$$

The TIA developed in this work is based on the circuit topology of Fig. 2. A TIA composed of multiple stages of this circuit was presented in [9] for spin qubit readout at $2\,\mathrm{K}$. The inverter is self-biased through the feedback resistor R_{fb} at the middle-point of its DC-characteristic [10]. At this operating point both transistors, M1 and M2, are in saturation and the gate-source voltages are above their respective threshold voltage values. Therefore, the inverter acts as a linear common source amplifier with an effective total transconductance $g_{m,t}$ being the sum of NMOS $g_{m,n}$ and PMOS $g_{m.p}$ transconductance, as described by

$$g_{m,t} = g_{m,n} + g_{m,p} \qquad [10]. \qquad (2)$$

TABLE I
REQUIREMENTS FOR AMPLIFIER

Parameter	Requirement
f	20 GHz
BW_{min}	200 MHz
v_{qubit}	3-10 mV
$|Z_{qubit}|$	~100 Ω
Z_{test}	50 Ω
$P_{DC,max}$	200 μW
T	0.1-300 K

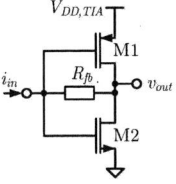

Fig. 2. TIA consisting of inverter with resistive feedback.

The total output resistance of the TIA is given by

$$r_{DS,t} = (r_{DS,n} \parallel r_{DS,p}), \qquad (3)$$

where $r_{DS,n}$ is the NMOS and $r_{DS,p}$ the PMOS drain-source resistance. With the defined quantities the open loop voltage gain for low frequencies results in

$$A_{v0,OL} = -g_{m,t} (r_{DS,t} \parallel R_L), \qquad (4)$$

where R_L is the load resistance connected to the output [11]. With the open loop gain the transimpedance for low frequencies can be approximated by

$$Z_{T,0} \approx \frac{-R_{fb} |A_{v0,OL}|}{1 + |A_{v0,OL}|} \qquad [10]. \qquad (5)$$

For the application in the gigahertz range a two pole behavior has to be considered. Nonetheless, the advantage of this topology can already be seen for the low frequency approximation in (5). Z_T is directly proportional to R_{fb} allowing large transimpedance values compared e.g. to a common gate circuit with a low ohmic load. It has to be considered that the first pole is formed by R_{fb} together with the gate capacitances, thus R_{fb} is limited by bandwidth requirements [9]. In simple common source stages the power dissipation requirement limits the transconductance achieved by the technology to $g_m < 8\,\mathrm{mS}$, effectively damping the voltage. In contrast to this, from (2) results in first schematic simulations $g_{m,t} \approx 12\,\mathrm{mS}$ leading to $|A_{v0,OL}| > 1$.

Since the topology of Fig. 2 showed a promising performance in preliminary schematic simulations, it was further extended to the complete design showed in Fig. 3. The differential output signal of the mixer is converted into a single ended signal through an integrated balun. Digitally configurable capacitances are set to resonate with the inductance of the primary side of the balun at the operation frequency. This resonant behavior enhances the output signal of the mixer, resulting in $i_{in,amp} \approx 20\,\mu\mathrm{A}$. The secondary side

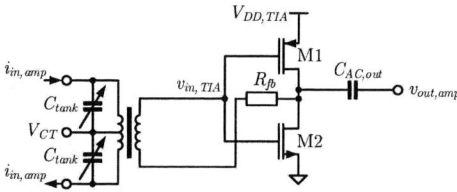

Fig. 3. TIA with capacitively tuned balun at input and feedback inductance.

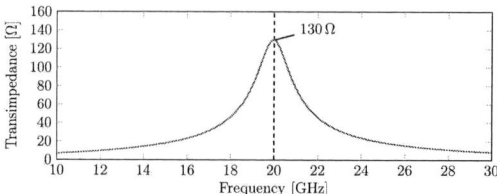

Fig. 4. Transimpedance over frequency for post-layout simulation.

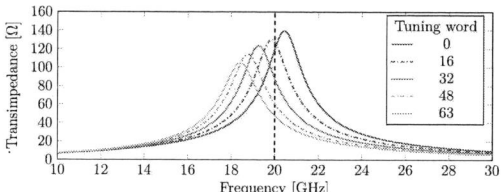

Fig. 5. Tuning behavior for post-layout simulation.

of the balun is connected in the feedback path to avoid an AC-coupling. Furthermore, R_{fb} together with the inductance partially compensate the first input pole so that the bandwidth is increased [10]. An output coupling capacitance $C_{AC,out}$ is required to not affect the operating point by connecting a load. This final implementation was layouted and verified in simulations.

IV. SIMULATION RESULTS

A. TIA Characterization

The TIA was characterized in simulations including the parasitic effects extracted from the layout. The most important figure of merit for this circuit is the transimpedance shown in Fig. 4 for the LC-tank configured to resonate at 20 GHz. The peak value is around $130\,\Omega$ or $42\,\mathrm{dB}\,\Omega$. With the expected $i_{in,amp} \approx 20\,\mu\mathrm{A}$ results $v_{out,amp} \approx 5\,\mathrm{mV}$. This indicates that the amplifier itself fulfills the required voltage amplitude for the qubit.

The frequency tuning range is displayed in Fig. 5, where the transimpedance characteristic is shown for the extreme and some intermediate tuning words. The bandpass corresponding to each tuning word has different peak values for the transimpedance from $105\,\Omega$ to $140\,\Omega$ and 3 dB bandwidths ranging from 1.67 GHz for the lowest tuning word to 1.49 GHz for the largest one. The whole tuning range of the test amplifier setup is shifted by 750 MHz to the left due to the pad capacitance at the input. This is not the case for the amplifier connected to the mixer and the frequencies from 18 GHz to 21 GHz are still covered. The effective tuning resolution is approximately 33 MHz. This resolution is suitable for fine tuning at cryogenic temperatures, where the quality factor of the LC-tank is likely to increase lowering the bandwidth [9].

The supply $V_{DD,TIA}$ can be varied from 600 mV to 900 mV improving the performance, but this also results in a larger power dissipation illustrated in Fig. 6. For the application as qubit driver $V_{DD,TIA}$ is set to 600 mV dissipating a power of $165\,\mu\mathrm{W}$. If the power requirement is relaxed, $V_{DD,TIA}$ can be increased up to 900 mV causing the TIA to have a transimpedance of around $230\,\Omega$ at 20 GHz but consuming 1.23 mW.

With lower temperatures the threshold voltages of the transistors increase approximately linear reducing the quiescent current [13]. Therefore, the TIA consumes less power but also has a lower transimpedance, which can be seen in Fig. 7 and Fig. 8. To counteract this the SOI technology provides back-gate contacts for each transistor to tune their threshold voltages, these are denoted by $V_{BG,N}$ and $V_{BG,P}$ for NMOS and PMOS respectively [13]. These voltages can be adjusted to keep the transimpedance or the power consumption over temperature constant.

Displayed in Table II are the back-gate voltages to keep a constant transimpedance over the temperature range. For lower temperatures the devices require less current to produce the same tranismpedance so that the power consumption lowers. The device models provide reliable simulation results down to $-50\,°\mathrm{C}$. The back gates need to be shifted for a constant transimpedance by maximum 500 mV at $-50\,°\mathrm{C}$. This value together with the linear increase in the threshold voltages indicates that the back gates will have to be shifted by less than 2 V for temperatures around 100 mK which is in the specified range for the technology. The back gate voltages for constant

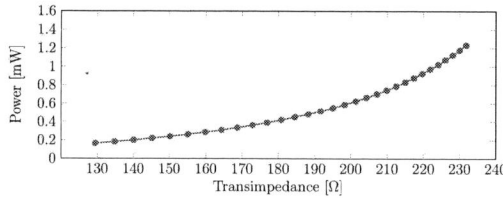

Fig. 6. Power dissipation over transimpedance at 20 GHz for post-layout simulation.

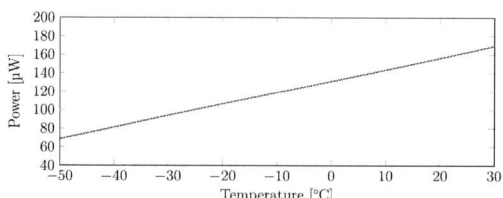

Fig. 7. Power dissipation over temperature for fixed back-gate and supply voltages.

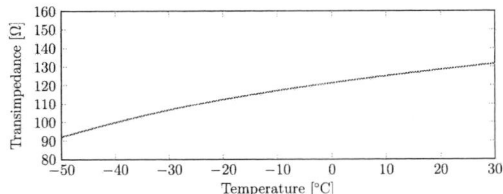

Fig. 8. Transimpedance over temperature for fixed back-gate and supply voltages.

TABLE II
BACK-GATE VOLTAGES FOR CONSTANT TRANSIMPEDANCE

T [°C]	P_{DC} [µW]	Z_T [Ω]	$V_{BG,N}$ [mV]	$V_{BG,P}$ [mV]
30	169	130	0	0
10	152	130	0	-100
-10	136	130	100	-100
-30	124	131	100	-300
-50	108	130	100	-500

TABLE III
BACK GATE VOLTAGES FOR CONSTANT POWER CONSUMPTION

T [°C]	P_{DC} [µW]	Z_T [Ω]	$V_{BG,N}$ [mV]	$V_{BG,P}$ [mV]
30	169	130	0	0
10	170	137	100	-200
-10	168	144	100	-500
-30	169	152	300	-600
-50	166	160	600	-600

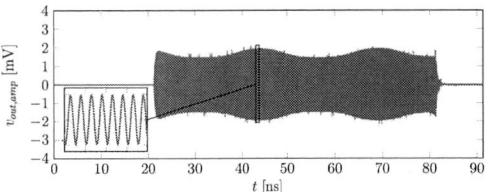

Fig. 9. Transient output of system for post-layout simulation.

power consumption are given in Table III. The g_m and general RF-performance of the devices increase for lower temperatures [12], [14]. Thus, the overall transimpedance of the amplifier increases for the same power consumption.

The noise figure at room temperature is approximately 9 dB around the center frequency which is comparable to state of the art TIA circuits [10]. At −50 °C this noise figure is below 8 dB, mainly due to the lower power consumption and thermal noise but also influenced by the improved noise performance of the devices. Therefore, it is important to show in measurements at which temperature effects like shot noise begin to dominate and saturate the improvement which is critical for the qubit operation at 100 mK.

B. System Performance

In Fig. 9 the transient output voltage amplitude of the system $v_{out,amp}$ is depicted for an exemplary waveform. An external pulse switches $v_{out,mix}$ on and off. The envelope of the output signal oscillates with approximately 20 MHz. This oscillation is already present at the input of the TIA and originates from nonlinearities in the mixer.

The achieved $v_{out,amp}$ in Fig. 9 is around 2.1 mV equivalent to 36 % of the 5.9 mV achieved in schematic simulations. Simulating the mixer schematic with the TIA including its parasitic effects from layout results in $v_{out,amp} \approx 4.6$ mV, so that the parasitic effects of the mixer are responsible for the remaining 2.5 mV in output amplitude reduction. Since the TIA consumes at −50 °C half of the specified power, the remaining power can be used to apply larger input signals. It was verified that using this approach the desired output voltage could be achieved.

V. CONCLUSION

It was shown in post-layout simulations that a TIA consisting of an inverter with resistive feedback and a tuned input balun fulfills the requirements to drive a semiconductor based qubit. The TIA has a transimpedance of more than 40 dBΩ

while consuming 165 µW at room temperature and less than 110 µW at −50 °C. The power consumptions are appropriate to place the circuit in a dilution refrigerator and drive a qubit at temperatures around 100 mK. The tuning options for low temperatures present the desired behavior down to −50 °C and offer sufficient margin for a further temperature decrease based on the expected cryogenic behavior from the literature. However, the final functionality and performance at cryogenic temperatures has still to be shown in measurements.

REFERENCES

[1] M. Mohseni, P. Read and H. Neven, "Commercialize early quantum technologies," Nature, vol. 543, pp. 171-174, 2017.

[2] R. Feynman, "Simulating Physics with Computers," International Journal of Theoretical Physics, vol. 21, pp. 467-488, 1982.

[3] F. Arute et al., "Quantum supremacy using a programmable superconducting processor," Nature, vol. 574, pp. 505-510, 2019.

[4] C. Degenhardt et al., "CMOS based scalable cryogenic Control Electronics for Qubits," IEEE International Conference on Rebooting Computing, 2017.

[5] L. Vandersypen et al., "Interfacing spin qubits in quantum dots and donors–hot, dense, and coherent," Design, npj Quantum Information, vol. 3, 2017.

[6] C. Almudever et al., "The Engineering Challenges in Quantum Computing," Design, Automation and Test in Europe Conference and Exhibition, pp. 836-845, 2017.

[7] B. Kane, "A silicon-based nuclear spin quantum computer," Nature, vol. 574, pp. 133-137, 1998.

[8] F. Zwanenburg et al., "Silicon Quantum Electronics," Reviews of Modern Physics, vol. 85, 2013.

[9] M. Gong, "Design Considerations for Spin Readout Amplifiers in Monolithically Integrated Semiconductor Quantum Processors," IEEE Radio Frequency Integrated Circuits Symposium, pp. 111-114, 2019.

[10] S. Voinigescu, "High-Frequency Integrated Circuits," Cambridge University Press, 2013.

[11] B. Razavi, "Design of Analog CMOS Integrated Circuits," McGraw-Hill, 2001.

[12] A. Beckers, F. Jazaert and C. Enz, "Characterization and Modeling of 28-nm Bulk CMOS Technology Down to 4.2 K," IEEE Journal of the Electron Devices Society, vol. 6, pp. 1007-1018, 2018.

[13] R. Carter et al., "22nm FDSOI Technology for Emerging Mobile, Internet-of-Things, and RF Applications," IEEE International Electron Devices Meeting, pp. 27-30, 2016.

[14] S. Hong et al., "Low-Temperature Performance of Nanoscale MOSFET for Deep-Space RF Applications," IEEE Electron Device Letters, vol. 29, pp. 775-777, 2008.

SMACD / PRIME 2021 | 19 – 22 July 2021, Online Event

A First Order-Curvature Compensation 5ppm/°C Low-Voltage & High PSR 65nm-CMOS Bandgap Reference with one-point 4-bits Trimming Resistor

1st Edoardo Barteselli
Physics Department
"G.Occhialini"
University of Milano - Bicocca
Milan, Italy
e.barteselli@campus.unimib.it

2nd Luca Sant
Infineon Technologies
Villach, Austria
Luca.Sant@infineon.com

3rd Richard Gaggl
Infineon Technologies
Villach, Austria
Richard.Gaggl@infineon.com

4th Andrea Baschirotto
Physics Department
"G.Occhialini"
University of Milano - Bicocca
Milan, Italy
andrea.baschirotto@unimib.it

Abstract—**Natural bandgaps produce 1.25V reference voltage that does not operate in low-voltage applications. This paper presents a current-mode bandgap reference circuit operating down to 1V supply (nominal = 1.2V) with low temperature coefficient, low power consumption, high Power-Supply-Rejection (PSR) and high performance robustness for industrial production in the consumer microphones field. The voltage reference is generated by the sum of two currents over a matched resistor: one current is proportional to V_{EB}, the other one is proportional to V_T. In 65nm node, a 600mV bandgap (BG) reference voltage consumes 5.2μW (4.3μA) at 1.2V supply. The simulated PSR is -91dB, -43dB and -29dB at DC, 1kHz and 10kHz respectively. Montecarlo simulations show a variation of 1% at 3σ after 4-bits trimming over a temperature range between -40°C and 100°C without any high-order curvature compensation, performing 5ppm/°C temperature coefficient.**

Keywords— Bandgap, Low-Voltage, Voltage Reference, 65nm CMOS, Trimming, Low-Current, Current-mode

I. INTRODUCTION

Bandgap (BG) voltage references are widely used in integrated circuits since they provide a constant voltage regardless of temperature, process and power supply variations. Nowadays, the electronic field is pushing towards the reduction of power supply in electronic devices, maintaining the same performance or improving them, as robustness and current consumption. In this scenario, the target of this paper is developing a current-mode bandgap (also defined as Reverse Bandgap) with low temperature coefficient, low power consumption, high PSR [1] guaranteeing at the same time production robustness performance in the consumer microphones filed. The bandgap works with a VDD value down to 1V. Therefore, a low voltage structure [2] is used to provide a 600mV reference voltage (V_{REF}) with a precision of 6mV (1%) at 3σ after 4-bits trimming. Under nominal conditions (i.e. 27°C & 1.2V supply) the optimized design consumes 5.2μW (i.e. 4.3μA). The proposed BG performs high DC-PSR of -91dB and low temperature coefficient of 5ppm/°C from -40°C to 100°C without high-order curvature compensation, saving circuit complexity, area and power. The proposed design is challenging since the above performance are guaranteed in industrial applications, since within full PVT variation.

This paper is organized as follows. Section II presents the implementation of the proposed bandgap reference and Section III shows the obtained results and compares them with the State of the Art. Finally, the conclusions are written in section IV.

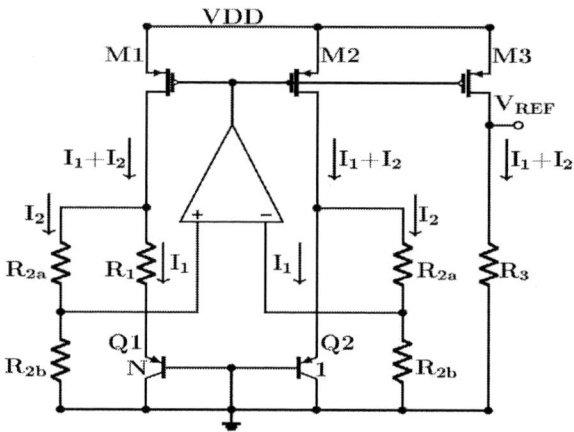

Fig. 1. Low Voltage Bandgap Schematic.

II. LOW VOLTAGE BANDGAP DESIGN

Fig. 1 shows the low-voltage BG voltage reference structure that has been aggressively designed to guarantee at the same time high performance level with industrial production robustness in the consumer microphones field. Nonetheless tested temperature range is [-20, 70]°C, in this design wider temperature range [-40, 100]°C is considered to have a safer margin. Moreover, power consumption reduction and, then, current consumption are crucial for the developed BG.

The structure can be divided in two main parts: the BG branches (including Q2 of single PNP device and Q1 with N PNP devices) and the operational amplifier.

A. Bandgap branches

In the BG branches, the minimum current is defined by the PNP bipolar transistors (BJT) made by parasitic vertical bipolar junction structure. In the adopted technology, performance robustness imposes minimum 85nA-per-unit-PNP-device current to avoid recombination region with consequent β factor reduction. The current in each BG branch is then defined by such minimum current per unit-PNP multiplied by the Q1 PNP-devices, i.e. N.

Key design parameter is then the device number ratio (N) between Q1 and Q2, which enters in the V_T (= k·T/q) product (i.e. PTAT component is V_T·ln(N)). For BJT matching reasons, the best N is a number that follows the equation N = n² − 1, with *n* odd to design the layout with common centroid structures. Moreover, the current in each BG branch

© VDE VERLAG GMBH · Berlin · Offenbach

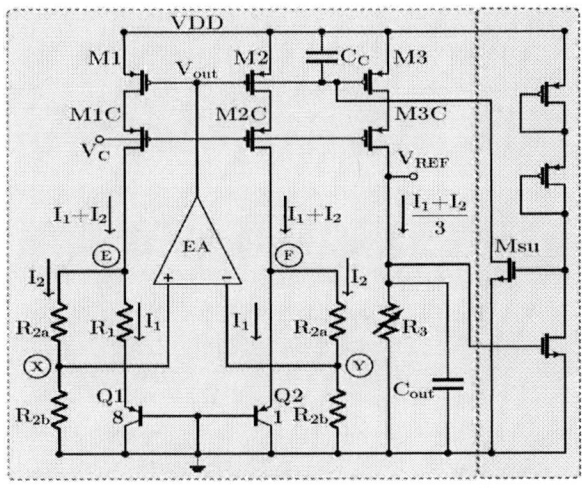

Fig. 2. Improved Low Voltage topology.

would be N times 85nA. Therefore, lower N value would reduce the power consumption.

On the other hand, higher N value, increasing PTAT component, would improve robustness, reducing the importance of circuit non-idealities (like opamp offset, and component mismatch effects). However due to the *ln* (natural logarithm) operation, significant advantage would require an excessive increase in the N value (and then in area and power consumption). As a trade-off between power consumption and performance robustness N = 8 (Fig. 2) is adopted, which offers also an excellent common-centroid layout solution since $8 = (3^2-1)$. The current flowing in each PNP is then $I_1 = 8 \times 85nA = 680nA$. Upon this choice, resistance values R_1 and R_2 (= $R_{2a} + R_{2b}$) can be designed as follows:

$$R_1 = \frac{\Delta V_{EB}}{I_1} = V_T \frac{\ln(N)}{I_1} \approx 81.5k\Omega. \qquad (1)$$

The value of *m*, defined as $(\partial V_{EB}/\partial T)/(\partial V_T/\partial T)$ is ~20. So, equating *m* to $(R_2/R_1) \cdot \ln(N)$, the value of R_2 is:

$$R_2 = m \frac{R_1}{\ln(N)} \approx 790k\Omega. \qquad (2)$$

The current through R_2 results to be $I_2 = 860nA$. In conclusion, the total current in each BG branches (flowing through PMOS current mirrors M1–M1C, and M2–M2C) is 1.54µA.

The above current is mirrored to the output branch (M3-M3C) with a 3x reduction factor. For a 600mV output voltage V_{REF}, R_3 is designed as:

$$V_{REF} = \frac{1}{3} \cdot \left\{ \frac{R_3}{R_2} \left[V_{EB} + \frac{R_2}{R_1} \ln(N) \cdot V_T \right] \right\} \qquad (3)$$

The resistor R_3 is 4-bit trimmable such that the V_{REF} value, dependent on process and mismatch, can be subsequently adjusted.

The cascode current mirror improves the current matching between the BG branches and the output branch. Cascode current mirror use, which improves the gain and the current matching, is enabled by operating all the transistors in the subthreshold region, which results in $V_{GS} < V_{TH}$. The voltage divider made by R_{2a} and R_{2b} introduces a voltage shift at the input of the Error Amplifier (EA). This allows proper biasing of the differential PMOS input pair, nonetheless the low VDD value. Same value of $R_{2a} = R_{2b} = R_2/2$ is adopted also for R_{2a}–R_{2b} matching purpose.

B. Error Amplifier

Theoretical analysis and simulations are exploited to optimize error amplifier specifications for the target BG performance. EA gain has to be optimized in consideration of the ΔV_{REF} (defined as $V_{REF,max} - V_{REF,min}$ in [-40, 100]°C temperature range), and of the absolute V_{REF} variation to be trimmed. Fig. 3a shows ΔV_{REF} as a function of EA gain. It shows that EA gain larger than 70dB is needed to keep ΔV_{REF} lower than 1mV. In addition, Fig. 3b shows V_{REF} value as a function of EA gain. Assuming a target value after trimming of $V_{REF} = 600mV$, and assuming a maximum trimming range of about 30mV (for 4-bits control to limit trimming complexity), the EA gain has to be larger than 50dB. Based on these analysis, EA gain larger than 70dB has to be guaranteed. This high gain permits to have 5ppm/°C without high-order curvature compensation. To avoid values that exceed this error, a gain of 80dB has been chosen for the error amplifier. This means a ΔV_{REF} of 615.8µV and a V_{REF} of 602mV at 27°C.

DC-gain is given by $(g_m \cdot r_{ds})^2 \approx 79dB$, satisfying accuracy requirement. Offset specification is guaranteed by exploiting the symmetrical structure of the EA. To reduce systematic offset due to input pair V_{DS} variation, the symmetrical opamp in Fig. 4 has been adopted [10]. Furthermore, large input devices size [9] with W=5.6µm, L=4.8µm and a multiplicity of 36 have been used. Moreover, the mismatch between the two resistors R_2 is reduced by realizing each resistor as an array of 10kΩ resistors with non-minimum W.

Performance robustness is improved by biasing input PMOS with $V_{SB}=0$, to avoid body effect. The minimum VDD for this structure is then:

$$VDD_{minBG} = V_X + |V_{GS,sth}| + 2V_{DS,sat} \qquad (4)$$

in which V_X is the voltage at X (= Y) node and $V_{GS,sth}$ is the subthreshold region V_{GS}. Since $R_{2a} = R_{2b}$, $V_X = V_{BE}/2$, in nominal conditions the VDD_{min} is about 950mV. Safe margin to reach 1V avoids improper operation in PVT and Montecarlo simulations.

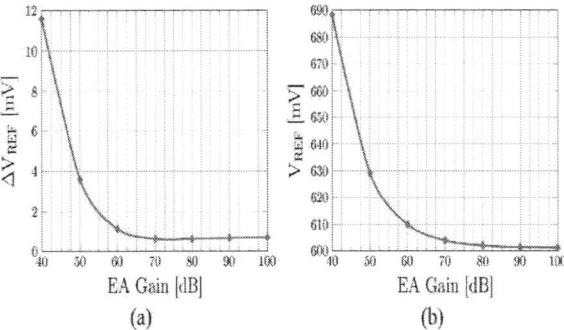

Fig. 3. (a) V_{ref} error versus EA gain. (b) V_{ref} value at 27°C versus EA gain.

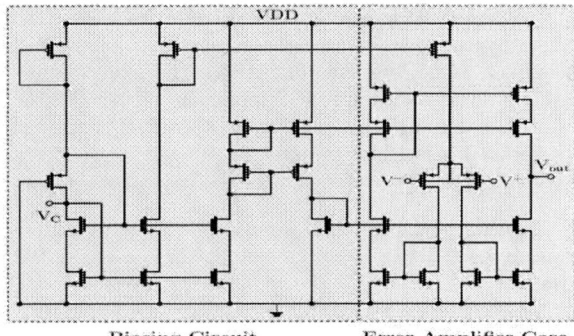

Fig. 4. Schematic of the Error Amplifier.

C. PSR optimization

Fig. 6 represents the PSRs of the bandgap with Coupling Capacitor C_C connected from V_{out} to VDD and from V_{out} to GND (ground). The coupling capacitor, shown in Fig. 2, is connected between the output of the EA and the supply voltage VDD [1], instead of being connected to ground. This allows to increase the coupling between the VDD and the output of the amplifier. By means of C_C, the zero z_1 increases of a quantity $(1+g_m R_0)$, while the poles p_1 and p_2 maintain the same values. Equations are written in (5) [1]:

$$z_1 = \frac{1 + g_m R_0}{r_{ea} C_C},$$

$$p_1 = \frac{1}{R_{0||3} C_{out}}, \quad p_2 = \frac{g_m g_{m,ea}(R_X - R_Y)}{C_c} \quad (5)$$

where g_m is the trans-conductance of M3, R_0 the output resistance of the bandgap (at V_{REF}), $g_{m,ea}$ the trans-conductance of the EA, r_{ea} the output resistance of the EA and R_X and R_Y the resistances at the nodes X and Y respectively. By increasing the position of the zero, the PSR at higher frequencies is improved. The PSR is reduced by almost 20dB between 1kHz and 30kHz. In DC the PSR is given by the following equation:

$$PSR_{DC} = \frac{R_3(1 + g_m R_0)}{R_3 + R_0} \cdot \frac{1}{1 + g_{m,ea} r_{ea} g_m(R_x - R_y)} \quad .(6)$$

In Fig. 6 the Post–Layout simulation (solid line) of the PSR with C_C connected from V_{out} to VDD is also represented. As can be seen z_1 is shifted to the left. This is given by parasitic capacitor coupling from V_{out} to ground, as resulting in the compact layout of 160μm (horizontally) x 209μm (vertically) (0.033mm²) in Fig. 5.

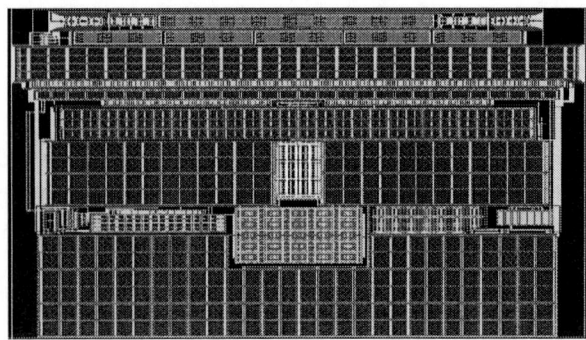

Fig. 5. Bandgap Layout.

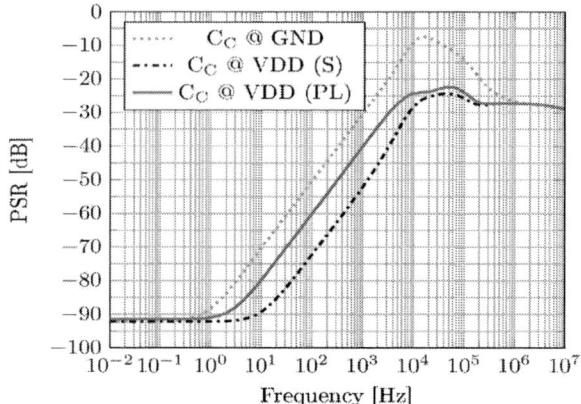

Fig. 6. PSR with C_C at VDD (dash-dotted, solid) and GND (dotted). (S) means Schematic and (PL) Post-Layout.

D. 4-bits trimming resistor

To adjust V_{REF} value, which spreads due to mismatch and process, a 4-bits trimming resistor, composed of 15 resistors and 15 switches, has been added. Each resistor is equal to 10kΩ, which means a V_{REF} step control of 5mV. Theoretically, this variation implies a maximum error of about 2.5mV compared to the nominal V_{REF} but due to the intrinsic resistance of the switches and the non-ideal resistors it can be higher during Montecarlo simulations.

To simplify the trimming resistor array, the worst case was set to 600mV. After this, the nominal bit was chosen. This solution allows to have all the resistors in series saving occupied area.

III. SIMULATION RESULTS

In this section the simulation results of V_{REF} vs. temperature Pre- and Post-Layout are presented. Then the Montecarlo simulations of the value of V_{REF} before and after trimming.

A. DC Simulation

Fig. 7 shows V_{REF} in the temperature range [-40, 100] °C. There is a variation of 5ppm/°C over the temperature range with a minimum value of 600.49mV and a maximum of 600.96mV. Hence, the total variation over the range is: $\Delta V = 0.47$mV. By substituting in (3) the following values: $R_1 = 81.5$kΩ, $R_2 = 790$kΩ, $R_3 = 1170$kΩ and $V_{BE} = 690.9$mV, the resulting V_{REF} is the one expected, as in Fig. 7.

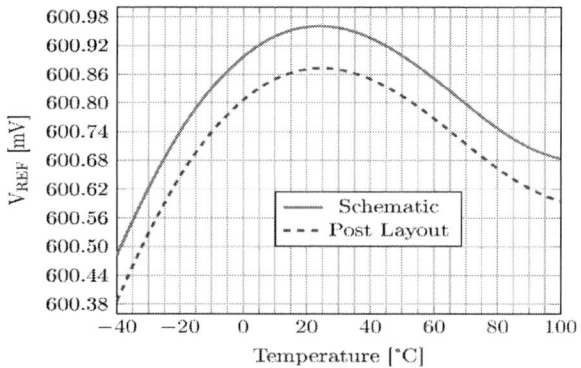

Fig. 7. V_{REF} curvature in nominal conditions.

TABLE I. SIMULATED PERFORMANCE COMPARED WITH OTHER WORKS

	[1] 2016	[3] 2016	[4] 2018	[5] 2014	[6] 2020	[7] 2016	[8] 2013	This work 2020
CMOS process [nm]	180	65	65	65	65	65	65	65
VDD [V]	1.1~2.2	1.1~1.3	1.2	N.A.	0.5	0.3	0.6~1.2	1.0~1.4
V$_{REF}$ [mV]	800	466	730	~441.5	495	168	435	600
Temp. Range [°C]	-40~125	-55~125	-20~100	-45~125	-40~120	-20~100	-40~125	-40~100
Power Cons. [μW]	19.8	N.A.	N.A.	104	0.036	0.07	0.22	5.2
TC [ppm/°C]	9	30.9	9.8	10.65	42	142	30	5
PSR DC [dB]	-108	-61	-79	-20.21	-50	N.A.	-38	-91
PSR 10kHz [dB]	-68	N.A.	~-30	-20.21	N.A.	N.A.	-27.5	-28.8

B. Montecarlo simulation

The Montecarlo simulation is useful to predict the behavior of the circuit under worst cases, including the transistor mismatches. Before trimming, the V$_{REF}$ value varies from 575.2mV to 630.3mV with σ of 7.77mV. After trimming, instead, the interval is reduced: V$_{REF}$ varies from 596.3mV to 602.6mV with σ of 1.5mV (considering a 2000 points simulation). This means a variation at 3σ of ~1% instead of 4.5% without trimming. A Verilog-A code able to simulate an A/D converter, has been used to simulate the trimming. Fig. 8 shows the histogram collecting the simulations at 27°C, before and after trimming. Moreover, Montecarlo simulations reveal an error amplifier input referred offset of 471.2μV at 1σ.

Table I compares the proposed BG performance with SoA. The aggressive 5ppm/°C outperforms SoA BGs with comparable power consumption.

TABLE II. SIMULATED PERFORMANCE OF THIS WORK

Parameter	Value
CMOS process	65nm
V$_{REF}$	600mV
3σ$_{VREF}$	6mV (1%)
Supply Voltage Range	[1,1.4]V
Total Current	4.3μA
Power Consumption	5.2μW
DC Gain	79.21dB
Phase Margin	69.64°
EA Input Referred Offset	471.2μV (at 1σ)
Temperature Range	[-40, 100]°C
Temp. Coefficient	5ppm/°C
VDD Dependence	0.91mV/V
PSR DC / 1kHz / 10kHz	-91.0dB / -43.0dB / 28.8dB

IV. CONCLUSIONS

A low voltage and low current 65nm CMOS current mode bandgap with frequency compensation and low temperature coefficient has been presented. The proposed bandgap merges high performance with low power consumption. The circuit generates a voltage reference of 600mV with a power supply of 1.2V and without high order curvature compensation. Montecarlo simulations reveal an average current consumption of 4.3μA at 27°C. The bandgap exhibits a temperature coefficient of 5ppm/°C and a supply voltage dependence of 0.91mV/V.

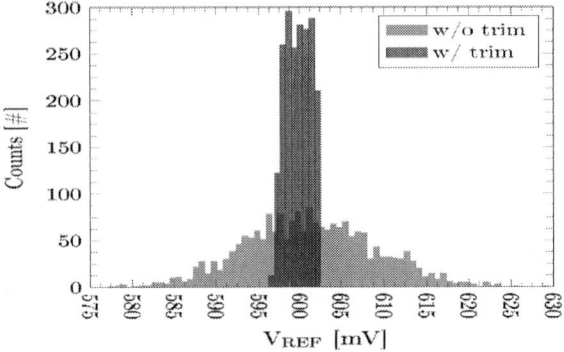

Fig. 8. Combined histogram of Montecarlo V$_{REF}$ with and without trimming at 27°C.

REFERENCES

[1] L. Wang et al., "Design of high-PSRR current-mode bandgap reference with improved frequency compensation", 2016 IEEE International Conference on Electron Devices and Solid-State Circuits (EDSSC), Hong Kong, 2016, pp. 410-413.

[2] K. N. Leung and P. K. T. Mok, "A sub-1-V 15-ppm//spl deg/C CMOS bandgap voltage reference without requiring low threshold voltage device", in IEEE Journal of Solid-State Circuits, vol. 37, no. 4, pp. 526-530, April 2002.

[3] Z. Jun-an et al., "A bandgap reference in 65nm CMOS", 2016 IEEE International Nanoelectronics Conference (INEC), Chengdu, 2016, pp. 1-2.

[4] K. Peng and Y. Xu, "Design of Low-Power Bandgap Voltage Reference for IoT RFID Communication," 2018 IEEE 3rd International Conference on Integrated Circuits and Microsystems (ICICM), Shanghai, 2018, pp. 345-348, doi: 10.1109/ICAM.2018.8596364.

[5] S. Yang, P. Mak and R. P. Martins, "A 104μW EMI-resisting bandgap voltage reference achieving −20dB PSRR, and 5% DC shift under a 4dBm EMI level," 2014 IEEE Asia Pacific Conference on Circuits and Systems (APCCAS), Ishigaki, 2014, pp. 57-60, doi: 10.1109/APCCAS.2014.7032718.

[6] C. U, W. Zeng, M. Law, C. Lam and R. P. Martins, "A 0.5-V Supply, 36 nW Bandgap Reference With 42 ppm/°C Average Temperature Coefficient Within -40 °C to 120 °C," in IEEE Transactions on Circuits and Systems I: Regular Papers, doi: 10.1109/TCSI.2020.3010998.

[7] T. Lu, M. Ker and H. Zan, "A 70nW, 0.3V temperature compensation voltage reference consisting of subthreshold MOSFETs in 65nm CMOS technology", 2016 International Symposium on VLSI Design, Automation and Test (VLSI-DAT), Hsinchu, 2016, pp. 1-4.

[8] X. Li et al., "A novel voltage reference with an improved folded cascode current mirror OpAmp dedicated for energy harvesting application", 2013 International SoC Design Conference (ISOCC), Busan, 2013, pp. 318-321.

[9] M. J. M. Pelgrom, A. C. J. Duinmaijer and A. P. G. Welbers, "Matching properties of MOS transistors", in IEEE Journal of Solid-State Circuits, vol. 24, no. 5, pp. 1433-1439, Oct. 1989.

[10] D. M. Colombo and G. I. Wirth, "Impact of different op-amps in CMOS bandgap references implemented in 0.18μM technology".

Resource Efficient Sub-V_T Level Shifter Circuit Design Using a Hybrid Topology in 28 nm

Saikat Chatterjee, Ulrich Rückert
Cognitronics and Sensor Systems Group
CITEC, Bielefeld University
33619, Germany
Email: schatterjee@cit-ec.uni-bielefeld.de

Abstract—This paper presents a resource efficient level shifter circuit, which is capable of converting input voltages below subthreshold to above threshold voltages, making it suitable for ultra low power applications such as wireless sensor networks, biomedical implants, environmental sensors, to name a few. The proposed circuit topology has two stages. The first stage comprises of a Wilson current mirror, whereas the second stage has a cross coupled PMOS circuit. The two staged topology helps to overcome the challenges of deep nano process, when operated with ultra low input supply voltage. The circuit presented here, is implemented in 28 nm FDSOI technology from ST Microelectronics. The proposed level shifter is capable of converting an input voltage as low as 150 mV to 1 V. The static power consumption is measured to be 100 pW, when the circuit is operated with the minimum possible input supply and operational frequency of 500 kHz.

Keywords—ultra low-power, subthreshold design, level shifter, 28 nm FDSOI technology

I. INTRODUCTION

In the recent era of nanoelectronics, power consumption has been the most concerning factor. Subthreshold operation is considered to be one solution [1] to this, as the dynamic power consumption varies quadratically with the supply voltage while the static power consumption varies exponentially with the same. However, the benefit of subthreshold operation comes at the cost of reduced performance and robustness. Multi-voltage subthreshold system-of-chip (SoC) helps to trade off the system performance with the power consumption, where some parts of the system are operated at above-threshold level while the rest of it is maintained at below threshold. Also, an above threshold supply voltage is always necessary to connect to the digital I/O pad cells even if the entire SoC will be operated at the subthreshold domain. Therefore, a resource efficient and robust on-chip level shifter is required to interface subthreshold circuit parts with the above threshold modules.

As the era of nanotechnology approaches, there grows a need of scaling down the sub-microsystems into sub-nanosystems. Several researches related to level shifter operation have been done in last few years. Few topologies are proposed in 65 nm technologies. Below 65 nm, changes in current in the saturation region is more. As a result, the transistor has less control over the current; making the circuit design more difficult. As the feasible gain per stage is significantly reduced, circuits need to be more sophisticated. This, of course, accounts for larger area and higher power consumption.

Fully depleted silicon-on-insulator (FDSOI) technology from ST Microelectronics is chosen for our work so as to overcome the challenges related to deep nano process. The added advantage of FDSOI technology is that it offers power efficient solution without any sacrifice in the speed. The topology is simulated using 28 nm process. We intend to have a solution which can work with below 200 mV input voltages.

The following Section II depicts the state of art. Section III introduces the proposed level shifter with the parameter values. The simulation results will be discussed in Section IV along with the conclusion in Section V.

II. STATE OF ART

The conventional level shifter circuit consists of two PMOS and two NMOS transistors as shown in figure 1. The drive strength of PMOS is overcome by increasing the width of the corresponding NMOS transistors.

NMOS-to-PMOS ratio grows exponentially in the sub-threshold region due to the drive strength of the pull-up path lying in above-threshold region. The pull-down path transistors allow only a poor subthreshold ON current to flow [2]. Few solutions has been proposed so that the drawback related to the transistor dimension can be overcome.

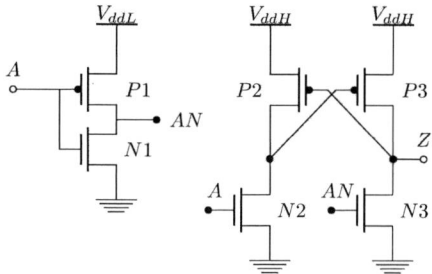

Fig. 1. Half-latch based conventional level shifter schematic

One of the effective solutions has been the replacement of the PMOS latch by a current mirror load [3]. This circuit at on-state produces an on-current, increasing the static power dissipation. However, the circuit can be implemented only in SOI CMOS, because each transistor requires individual body connections. Another approach [4] is to limit the current drive capability of the PMOS half-latch. A self-controlled supply feedback loop based design was proposed by [5], which can convert a signal as low as 80 mV to 1.2 V. The drive strength of the pull-up network is adjusted using diode-connected and off-biased PMOS transistors. An optimal tradeoff between speed and power consumption is found out using multithreshold CMOS (MTCMOS) design. However, due to the off-biased PMOS transistors, propagation delay does not scale well when the supply voltage is raised above the threshold voltage.

To combat the leakage current, a Wilson current mirror (WCM) based design in 90 nm bulk CMOS technology [6] was proposed by our group, which consists of five transistors (three PMOS and two NMOS). This concept was further modified in [7][8] using mixed-V_T devices (1.8 V normal, 3.3 V normal, 3.3 V native). Luo et al. [9] modified this concept further into a modified Wilson current mirror hybrid buffer (MWCMHB) based level shifter; which is bidirectional and consists of three parts; a modified Wilson current mirror block, a delay path and a complementary OR-gate .

In [10], a virtual platform was proposed by the authors in III-V nanowire TFET technology for the design of level shifter circuits with sub-0.4 V operation.

III. PROPOSED DESIGN

In this work, we intended to design a level shifter which can operate with 100 mV input. In higher technology nodes there are existing solutions which are implemented in CMOS process. However, we wanted to explore the low leakage benefits of FDSOI technology. Here the primary challenge was to build a strong pull-down network. Although, we preferred not to have an area overhead. Therefore our next challenge was to limit the number of transistors in the circuit.

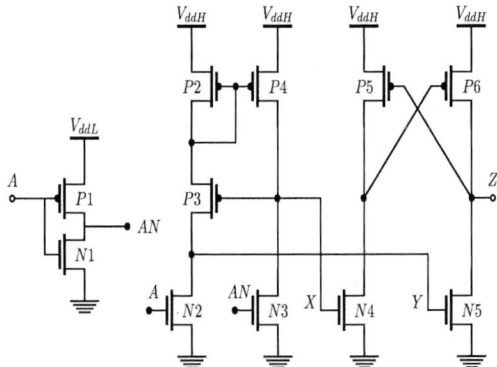

Fig. 2. WCM based level shifter schematic using LVT cells

The benefit of using the Wilson current mirror is to reduce the leakage current. Therefore, it is used as the first stage of our design. This stage helps to raise the low input voltage to a voltage close to the threshold voltage of NMOS transistors. The cross-coupled structure in the second stage of the design helps to maintain the full swing of the output. The operating speed of the circuit can be varied by modifying the W/L ratios of the output transistors. When input A is high, N2 is turned on. Therefore, the current will flow through P2, P4 and N2 and it will be mirrored in P4. As N3 is off, the node X will be charged till P3 is turned off.

TABLE I. TRANSISTOR SIZE

Transistor	Width [μm]	Length [μm]
P1	1.080	0.048
N1	0.110	0.048
P2	0.080	0.048
P3	0.080	0.048
P4	0.080	0.048
N2	0.150	0.048
N3	0.150	0.048
P5	0.080	0.048
P6	0.080	0.048
N4	0.500	0.048
N5	0.080	0.048

When A is low and AN is high, N2 will be off and N3 will be turned on. There will be no current flowing through P2, P3 and N2, forcing X to discharge. This will help charging the point Y. As a result, there will be differential inputs on nodes X and Y. Due to the off-biased transistors, the amplitude of these signals will be at near or above threshold of the transistors. The cross-coupled stage further helps to raise the output from near or above threshold to VDDH. The drive strength of P5 and P6 are chosen such that the nodes X and Y can exceed them easily. Table I contains the dimension of the transistors, which gives the desired performance at room temperature.

FDSOI technology from STMicroelectronics offers more speed compared to bulk technology. It offers a conventional variant;- Regular-V_T (RVT) transistors and a flip-Well variant, known as Low-V_T (LVT). LVT transistors can be forward body-biased to reduce active power or enhance speed where RVT transistors help to reduce leakage current. For our work, LVT cells have been chosen. As, the operating speed of the system is also a concern.

IV. SIMULATION RESULTS

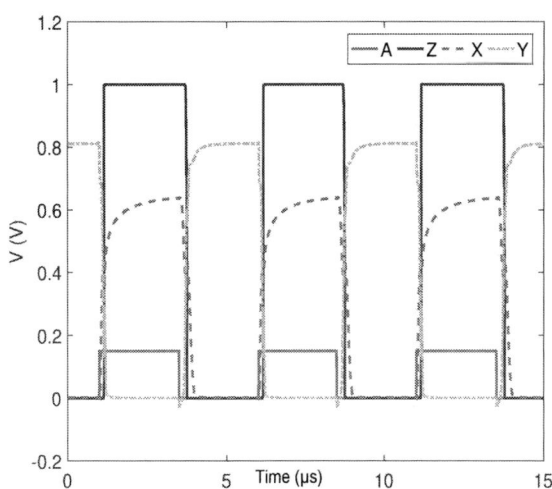

Fig. 3. Transient behaviour of the proposed level shifter using LVT cells

The basic operation of the level shifter is shown in Figure 3. The plot describes the behaviour at input node A, output node Z and the intermediate nodes X and Y. A load of 4 fF is applied at the output node. The input rise and fall times of the source connected to node A are chosen as 10 ns at 200 kHz input signal frequency. The simulation time for transient response is chosen as 15 µs.

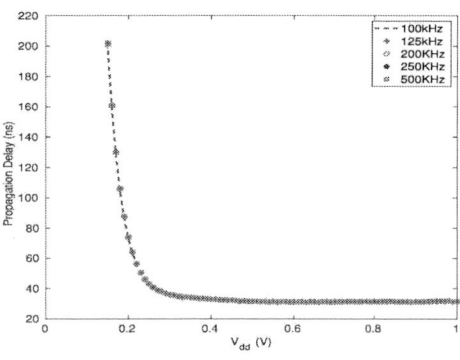

(a) Propagation delay variation

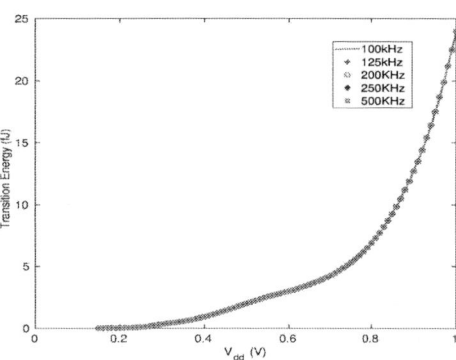

(b) Variation of total energy per transition

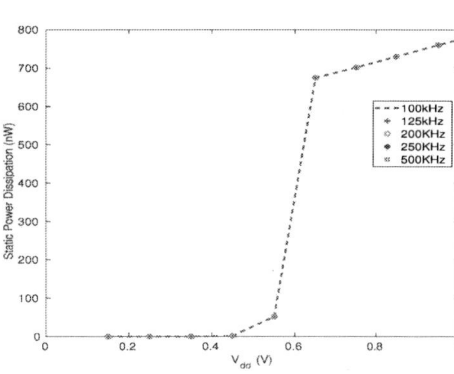

(c) Static power dissipation variation

Fig. 4. Comparison of the performance in terms of propagation delay, energy per transition and static power dissipation across different frequency of operations

The simulation results focus around the following design parameters: Propagation delay, energy transition per cycle and static power dissipation. The input supply voltage (VDDL) is scaled so as to observe the performance variation in context

with the aforementioned parameters. The temperature of the simulation environment is maintained at 27 °C.

Figure 4(a) shows the propagation delay as a function of VDDL. Here, it is clearly visible that between 150 mV and 300 mV, there is an exponential drop in propagation delay. This is primarily because of the two stages used in the circuit. This also confirms that the circuit is capable of working at higher operational frequencies as the VDDL increases. P1 and N1 of the inverter circuit also contributes significantly to the propagation delay. The energy per transition increases with VDDL as shown in figure 4(b). This happens primarily because of the increase in the static current in the pull down transistors lying in the above threshold domain i.e. N2,N3,N4 and N5, also accounting for the increase of static power dissipation as visible in figure 4(c).

Figure 4 indicates that the circuit performance does not vary significantly with the operating frequency, the plots corresponding to different operating frequencies being not distinguishable from each othe. At 150 mV with an input frequency of 250 kHz; propagation delay, energy per transition, and static power dissipation are measured as 200 ns, 29 aJ and 107 pW, respectively.

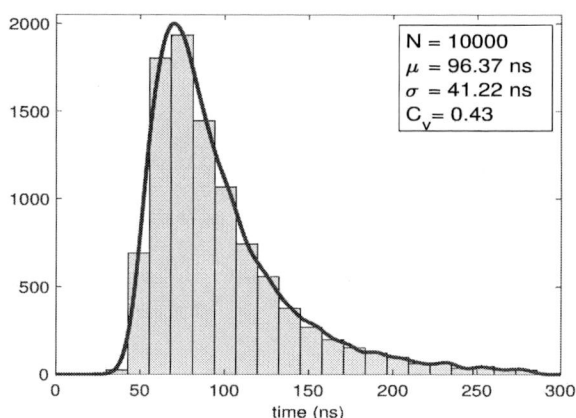

(a) Distribution of the propagation delay

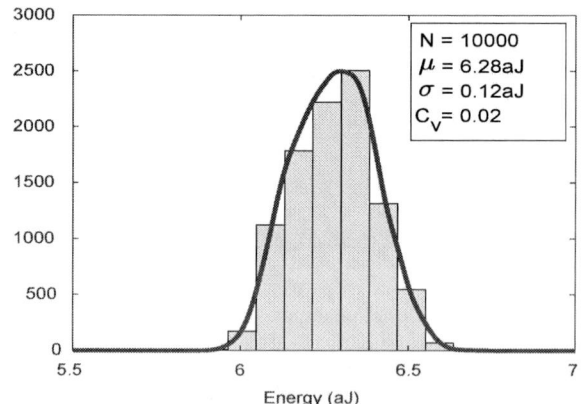

(b) Distribution of the energy per transition

Fig. 5. Monte Carlo simulation representation of the proposed level shifter

TABLE II. COMPARISON OF LS DESIGNS

	Technology	Min. VDDL (V)	VDDH (V)	Delay	Energy/op	Static Power (nW)
This Work	28 nm	0.150	1	200 ns @ 0.15 V	29 aJ @ 0.15V, 250 kHz	0.107
[9]	65 nm	0.200	1.2	300 ns @ 0.3 V	1 pJ @ 0.3V, 10 kHz	N.A.[1]
[11]	65 nm	0.140	1.2	25 ns @ 0.3 V	30.7 fJ @ 0.3V, 1 MHz	2.5
[5]	65 nm	0.080	1.2	36.1 ns @ 0.2 V	85 fJ @ 0.2V, 1 MHz	2.12
[12]	65 nm	0.120	1.2	66 ns @ 0.2 V	28 fJ @ 0.3V, 72 MHz	0.64
[13]	90 nm	0.100	1.0	16.6 ns @ 0.2 V	77 fJ @ 0.2V, 1 MHz	8.7
[6]	90 nm	0.200	1.0	18.4 ns @ 0.2 V	93.9 fJ @ 0.2V, 1 MHz	6.6
MDCVSHS [4]	90 nm	0.180	1.0	32 ns @ 0.35 V	17 fJ @ 0.35V, 1 MHz	2.5
[2]	130 nm	0.100	1.2	50 ns @ 0.2 V	25 pJ @ 0.2V, 50 kHz	5
[14]	130 nm	0.100	2.5	125 ns @ 0.3 V	1.7 pJ @ 0.3V, 8 MHz	13600
[7]	180 nm	0.210	3.3	162 ns @ 0.3 V	39 fJ @ 0.3V, 100 kHz	0.160
DSLS2 [3]	250 nm	0.350	1.2	125 ns @ 0.35 V	21.4 pJ @ 0.35V, 100 kHz	N.A.[1]
DSLS2b [3]	250 nm	0.350	1.2	110 ns @ 0.35 V	21.8 pJ @ 0.35V, 100 kHz	N.A.[1]
SSLSb [3]	250 nm	0.350	1.2	161 ns @ 0.35 V	40.5 pJ @ 0.35V, 100 kHz	N.A.[1]

[1] N.A. = Data not available

A 10000-point Monte Carlo simulation was performed for a supply voltage of 150 mV with local variations, so as to understand the effects of process variations on the level shifter characteristics. The yield generated was 100%. Figure 5(a) shows the distribution of the propagation delay. As can be seen, the mean delay (μ) obtained is 96.37 ns and the standard deviation (σ) is 41.22 ns.

Figure 5(b) illustrates the Monte Carlo simulation of energy consumption per transition. Results show that the mean value (μ) of the distribution is 6.28 aJ with the standard deviation (σ) being 0.12 aJ. Therefore, the coefficient of variance (μ/σ) is really low. Therefore, it can be concluded that the sensitivity of our design towards process variation is really low in terms of energy consumption.

Table II shows the comparison of several state of the art designs with the proposed design in terms of different performance characteristics. Data related to other designs have been directly taken from original papers. Technology variation from higher to lower transistor channel length contribute significantly to leakage current, propagation delay etc. Therefore, the performance variation is not linear at all. Also, different types of technologies such as FDSOI, bulk etc., have different advantages on the circuit performance. The emphasis was taken on the value of VDDL while choosing different topologies. Our design achieves an overall improvement in terms total energy consumption and static power dissipation. The propagation delay is a bit high. But, owing to the low frequency of operation, the delay does not come into play in the circuit performance.

V. CONCLUSION

Our work represents a novel circuit topology in 28 nm FDSOI technology which is capable of converting an input as low as 150 mV to 1 V. While designing the circuit, the aim was to limit the number of transistors as well as increasing the operating range. Both the current mirror and cross coupled topology are exploited here so as to get the best of both worlds. According to the simulation results, the power consumption is around 100 pW when an input supply with operating frequency of 250 kHz is applied. From the measurement results we can conclude that the level shifter circuit has a wide range (150 mV to 1 V) of voltage conversion. The frequency of the input supply was varied from 100 kHz to 500 kHz while

simulating the circuit. It is obsserved that the performance did not degenerate across the frequency range.

REFERENCES

[1] D. Blaauw and B. Zhai, "Energy efficient design for subthreshold supply voltage operation," in *IEEE International Symposium on Circuits and Systems*, 2006.

[2] T.-H. Chen, J. Chen, and L. T. Clark, "Subthreshold to above threshold level shifter design," in *J. Low Power Electron.*, vol. 2, no. 2, pp. 251–258, Aug. 2006.

[3] A. Chavan and E. MacDonald, "Ultra low voltage level shifters to interface sub and super threshold reconfigurable logic cells," in *IEEE Aerospace Conference*, 2008.

[4] A. Hasanbegovic and S. Aunet, "Low-power subthreshold to above threshold level shifter in 90 nm process," in *NORCHIP*, 2009.

[5] L. Wen, H. Wen and X. Zeng, "Sub-threshold level converter with internal supply feedback for multi-voltage applications," in *IET Circuits, Devices & Systems*, 2016.

[6] S. Lütkemeier and U. Rückert, "A subthreshold to above-threshold level shifter comprising a Wilson current mirror," in *IEEE Transactions On Circuits and Systems-II: Express Briefs, VOL. 57, NO.9*, 2010.

[7] J. Zhou, C. Wang, X. Liu, X. Zhang, and M. Je, "A fast and energy-efficient level shifter with wide shifting range from sub-threshold up to I/O voltage," in *Solid-State Circuits Conference (A-SSCC), 2013 IEEE Asian, Singapore, 2013, pp. 137-140*.

[8] J. Zhou, C. Wang, X. Liu, and M. Je, "Fast and energy-efficient low-voltage level shifters," in *Microelectronics Journal, Volume 46, Issue 1, January 2015, Pages 75-80*.

[9] S. Luo, C. Huang, and Y. Chu,"A wide-range level shifter using a modified wilson current mirror hybrid buffer," in *IEEE Transactions on Circuits and Systems I: Regular Papers, vol. 61, no. 6, pp. 1656-1665, June 2014*

[10] F. Settino, M. Lanuzza, S. Strangio, F. Crupi, P. Palestri and D. Esseni, "A virtual III-V tunnel FET technology platform for ultra-low voltage comparators and level shifters," in *13th Conference on Ph.D. Research in Microelectronics and Electronics (PRIME), Giardini Naxos, 2017, pp. 145-148*.

[11] W. Zhao, A.-B. Alvarez and Y. Ha, "A 65-nm 25.1-ns 30.7-fJ robust subthreshold level shifter with wide conversion range," in *IEEE Transactions on Circuits and Systems II: Express Briefs 62.7: 671-675, 2015*.

[12] B. Mohammadi and J. Rodrigues, "A 65 nm single stage 28 fJ/cycle 0.12 to 1.2 V level-shifter," in *IEEE International Symposium on Circuits and Systems 2014*.

[13] M. Lanuzza, P. Corsonello and S. Perri, "Fast and wide range voltage conversion in multisupply voltage designs," in *IEEE Transactions on Very Large Scale Integration (VLSI) Systems 23.2: 388-391, 2015*.

[14] I.J. Chang, J.J. Kim, K. Kim and K. Roy, "Robust level converter for sub-threshold/super-threshold operation:100 mV to 2.5 V," in *IEEE Transactions on Very Large Scale Integration (VLSI) Systems: 1429-1437, 2011*.

© VDE VERLAG GMBH · Berlin · Offenbach

VDE VERLAG GMBH
Bismarckstr. 33
P.O.B. 12 01 43
10625 Berlin, Germany

ISBN 978-1-7138-3572-1